Handbook of
NANOSCIENCE, ENGINEERING, and TECHNOLOGY

The Electrical Engineering Handbook Series

Series Editor
Richard C. Dorf
University of California, Davis

Titles Included in the Series

The Avionics Handbook, Cary R. Spitzer
The Biomedical Engineering Handbook, 2nd Edition, Joseph D. Bronzino
The Communications Handbook, 2nd Edition, Jerry Gibson
The Control Handbook, William S. Levine
The Digital Signal Processing Handbook, Vijay K. Madisetti & Douglas Williams
The Electrical Engineering Handbook, 2nd Edition, Richard C. Dorf
The Electric Power Engineering Handbook, Leo L. Grigsby
The Electronics Handbook, Jerry C. Whitaker
The Engineering Handbook, Richard C. Dorf
The Handbook of Formulas and Tables for Signal Processing, Alexander D. Poularikas
The Industrial Electronics Handbook, J. David Irwin
The Measurement, Instrumentation, and Sensors Handbook, John G. Webster
The Mechanical Systems Design Handbook, Osita D.I. Nwokah and Yidirim Hurmuzlu
The RF and Microwave Handbook, Mike Golio
The Mobile Communications Handbook, 2nd Edition, Jerry D. Gibson
The Ocean Engineering Handbook, Ferial El-Hawary
The Technology Management Handbook, Richard C. Dorf
The Transforms and Applications Handbook, 2nd Edition, Alexander D. Poularikas
The VLSI Handbook, Wai-Kai Chen
The Mechatronics Handbook, Robert H. Bishop
The Computer Engineering Handbook, Vojin G. Oklobdzija
The Handbook of Nanoscience, Engineering, and Technology, William A. Goddard, III,
   Donald W. Brenner, Sergey E. Lyshevski, and Gerald J. Iafrate

Forthcoming Titles

The Circuits and Filters Handbook, 2nd Edition, Wai-Kai Chen
The Handbook of Ad hoc Wireless Networks, Mohammad Ilyas
The Handbook of Optical Communication Networks, Mohammad Ilyas and
   Hussein T. Mouftah
The Engineering Handbook, Second Edition, Richard C. Dorf

Handbook of
NANOSCIENCE, ENGINEERING, and TECHNOLOGY

Edited by

William A. Goddard, III
California Institute of Technology
The Beckman Institute
Pasadena, California

Donald W. Brenner
North Carolina State University
Raleigh, North Carolina

Sergey Edward Lyshevski
Rochester Institute of Technology
Rochester, New York

Gerald J. Iafrate
North Carolina State University
Raleigh, North Carolina

CRC PRESS

Boca Raton   London   New York   Washington, D.C.

The **front cover** depicts a model of a gramicidin ionic channel showing the atoms forming the protein, and the conduction pore defined by a representative potential isosurface. The **back cover** (left) shows a 3D simulation of a nano-arch termination/zipping of a graphite crystal edge whose structure may serve as an element for a future nanodevice, and as a template for nanotube growth. The back cover (right) shows five figures explained within the text.

Cover design by Benjamin Grosser, Imaging Technology Group, Beckman Institute for Advanced Science and Technology, University of Illinois at Urbana-Champaign. Ionic channel image (front) by Grosser and Janet Sinn-Hanlon; data by Munoj Gupta and Karl Hess. Graphite nano-arch simulation image (back left) by Grosser and Slava V. Rotkin; data by Rotkin. Small figure images by (from top to bottom): 1) T. van der Straaten; 2) Rotkin and Grosser; 3) Rotkin and Grosser; 4) B. Tuttle, Rotkin and Grosser; 5) Rotkin and M. Dequesnes. Background image by Glenn Fried.

Library of Congress Cataloging-in-Publication Data

Handbook of nanoscience, engineering, and technology / edited by William A. Goddard, III ... [et al.].
    p. cm. — (Electrical engineering handbook series)
    Includes bibliographical references and index.
    ISBN 0-8493-1200-0 (alk. paper)
    1. Molecular electronics. 2. Nanotechnology. I. Goddard, William A., 1937– II. Series.

TK7874.8 .H35 2002
620′.5—dc21
    2002073329

Visit the CRC Press Web site at www.crcpress.com

© 2003 by CRC Press LLC

No claim to original U.S. Government works
International Standard Book Number 0-8493-1200-0
Library of Congress Card Number 2002073329
Printed in the United States of America 1 2 3 4 5 6 7 8 9 0
Printed on acid-free paper

Dedication

For my wife Karen, for her dedication and love, and for Sophie and Maxwell.

Donald W. Brenner

For my dearest wife Marina, and for my children Lydia and Alexander.

Sergey E. Lyshevski

To my wife, Kathy, and my family for their loving support and patience.

Gerald J. Iafrate

Preface

In the now-famous talk given in 1959, "There's Plenty of Room at the Bottom," Nobel Prize laureate Richard Feynman outlined the promise of nanotechnology. It took over two decades, but the development of the scanning tunneling microscope by IBM researchers in the early 1980s gave scientists and engineers the ability not only to image atoms but also to manipulate atoms and clusters with a precision equal to that of a chemical bond. Also in the 1980s, Eric Drexler wrote several books that went beyond Feynman's vision to outline a fantastic technology that includes pumps, gears, and molecular assemblers consisting of only hundreds to thousands of atoms that, if built, promised to revolutionize almost every aspect of human endeavor. While Drexler's vision continues to stir controversy and skepticism in the science community, it has served to inspire a curious young generation to pursue what is perceived as the next frontier of technological innovation. Fueled by breakthroughs such as in the production and characterization of fullerene nanotubes, self-assembled monolayers, and quantum dots — together with advances in theory and modeling and concerted funding from the National Nanotechnology Initiative in the U.S. and similar programs in other countries — the promise of nanotechnology is beginning to come true. Will nanotechnology revolutionize the human condition? Only time will tell. Clearly, though, this is an exciting era in which to be involved in science and engineering at the nanometer scale.

Research at the nanometer scale and the new technologies being developed from this research are evolving much too rapidly for a book like this to provide a complete picture of the field. Many journals such as *Nature*, *Science*, and *Physical Review Letters* report critical breakthroughs in nanometer-scale science and technology almost weekly. Instead, the intent of this handbook is to provide a wide-angle snapshot of the state of the field today, including basic concepts, current challenges, and advanced research results, as well as a glimpse of the many breakthroughs that will assuredly come in the next decade and beyond. Specifically, visionary research and developments in nanoscale and molecular electronics, biotechnology, carbon nanotubes, and nanocomputers are reported. This handbook is intended for a wide audience, with chapters that can be understood by laymen and educate and challenge seasoned researchers. A major goal of this handbook is to further develop and promote nanotechnology by expanding its horizon to new and exciting areas and fields in engineering, science, medicine, and technology.

Acknowledgments

Dr. Brenner would like to thank his current and former colleagues for their intellectual stimulation and personal support. Especially thanked are Dr. Brett Dunlap, Professor Barbara Garrison, Professor Judith Harrison, Professor John Mintmire, Professor Rod Ruoff, Dr. Peter Schmidt, Professor Olga Shenderova, Professor Susan Sinnott, Dr. Deepak Srivastava, and Dr. Carter White. Professor Brenner also wishes to thank the Office of Naval Research, the National Science Foundation, the NASA Ames and NASA Langley Research Centers, the Army Research Office, and the Department of Energy for supporting his research group over the last 8 years.

Donald W. Brenner

This handbook is the product of the collaborative efforts of all contributors. Correspondingly, I would like to acknowledge the authors' willingness, commitment, and support of this timely project. The support and assistance I have received from the outstanding CRC team, lead by Nora Konopka, Helena Redshaw, and Gail Renard, are truly appreciated and deeply treasured. In advance, I would like also to thank the readers who will provide feedback on this handbook.

Sergey Edward Lyshevski

I would like to acknowledge the career support and encouragement from my colleagues, the Department of Defense, the University of Notre Dame, and North Carolina State University.

Gerald J. Iafrate

About the Editors

William A. Goddard, III, obtained his Ph.D. in Engineering Science (minor in Physics) from the California Institute of Technology, Pasadena, in October 1964, after which he joined the faculty of the Chemistry Department at Caltech and became a professor of theoretical chemistry in 1975.

In November 1984, Goddard was honored as the first holder of the Charles and Mary Ferkel Chair in Chemistry and Applied Physics. He received the Badger Teaching Prize from the Chemistry and Chemical Engineering Division for Fall 1995.

Goddard is a member of the National Academy of Sciences (U.S.) and the International Academy of Quantum Molecular Science. He was a National Science Foundation (NSF) Predoctoral Fellow (1960–1964) and an Alfred P. Sloan Foundation Fellow (1967–69). In 1978 he received the Buck–Whitney Medal (for major contributions to theoretical chemistry in North America). In 1988 he received the American Chemical Society Award for Computers in Chemistry. In 1999 he received the Feynman Prize for Nanotechnology Theory (shared with Tahir Cagin and Yue Qi). In 2000 he received a NASA Space Sciences Award (shared with N. Vaidehi, A. Jain, and G. Rodriquez).

He is a fellow of the American Physical Society and of the American Association for the Advancement of Science. He is also a member of the American Chemical Society, the California Society, the California Catalysis Society (president for 1997–1998), the Materials Research Society, and the American Vacuum Society. He is a member of Tau Beta Pi and Sigma Xi.

His activities include serving as a member of the board of trustees of the Gordon Research Conferences (1988–1994), the Computer Science and Telecommunications Board of the National Research Council (1990–1993), and the Board on Chemical Science and Technology (1980s), and a member and chairman of the board of advisors for the Chemistry Division of the NSF (1980s). In addition, Goddard serves or has served on the editorial boards of several journals (*Journal of the American Chemical Society, Journal of Physical Chemistry, Chemical Physics, Catalysis Letters, Langmuir,* and *Computational Materials Science*).

Goddard is director of the Materials and Process Simulation Center (MSC) of the Beckman Institute at Caltech. He was the principal investigator of an NSF Grand Challenge Application Group (1992–1997) for developing advanced methods for quantum mechanics and molecular dynamics simulations optimized for massively parallel computers. He was also the principal investigator for the NSF Materials Research Group at Caltech (1985–1991).

Goddard is a co-founder (1984) of Molecular Simulations Inc., which develops and markets state-of-the-art computer software for molecular dynamics simulations and interactive graphics for applications to chemistry, biological, and materials sciences. He is also a co-founder (1991) of Schrödinger, Inc., which develops and markets state-of-the-art computer software using quantum mechanical methods for applications to chemical, biological, and materials sciences. In 1998 he co-founded Materials Research Source LLC, dedicated to development of new processing techniques for materials with an emphasis on nanoscale processing of semiconductors. In 2000 he co-founded BionomiX Inc., dedicated to predicting the structures and functions of all molecules for all known gene sequences.

Goddard's research activities focus on the use of quantum mechanics and of molecular dynamics to study reaction mechanisms in catalysis (homogeneous and heterogeneous); the chemical and electronic properties of surfaces (semiconductors, metals, ceramics, and polymers); biochemical processes; the structural, mechanical, and thermodynamic properties of materials (semiconductors, metals, ceramics, and polymers); mesoscale dynamics; and materials processing. He has published over 440 scientific articles.

Donald W. Brenner is currently an associate professor in the Department of Materials Science and Engineering at North Carolina State University. He earned his B.S. from the State University of New York College at Fredonia in 1982 and his Ph.D. from Pennsylvania State University in 1987, both in chemistry. He joined the Theoretical Chemistry Section at the U.S. Naval Research Laboratory as a staff scientist in 1987 and the North Carolina State University faculty in 1994. His research interests focus on using atomic and mesoscale simulation and theory to understand technologically important processes and materials. Recent research areas include first-principles predictions of the mechanical properties of polycrystalline ceramics; crack dynamics; dynamics of nanotribology, tribochemistry, and nanoindentation; simulation of the vapor deposition and surface reactivity of covalent materials; fullerene-based materials and devices; self-assembled monolayers; simulations of shock and detonation chemistry; and potential function development. He is also involved in the development of new cost-effective virtual reality technologies for engineering education.

Brenner's awards include the Alcoa Foundation Engineering Research Achievement Award (2000), the Veridian Medal Paper (co-author, 1999), an Outstanding Teacher Award from the North Carolina State College of Engineering (1999), an NSF Faculty Early Career Development Award (1995), the Naval Research Laboratory Chemistry Division Young Investigator Award (1991), the Naval Research Laboratory Chemistry Division Berman Award for Technical Publication (1990), and the Xerox Award from Penn State for the best materials-related Ph.D. thesis (1987). He was the scientific co-chair for the Eighth (2000) and Ninth (2001) Foresight Conferences on Molecular Nanotechnology; and he is a member of the editorial board for the journal *Molecular Simulation*, the Scientific Advisory Boards of Nanotechnology Partners and of L.P. and Apex Nanotechnologies, and the North Carolina State University Academy of Outstanding Teachers.

Sergey Edward Lyshevski earned his M.S. (1980) and Ph.D. (1987) degrees from Kiev Polytechnic Institute, both in electrical engineering. From 1980 to 1993 Dr. Lyshevski held faculty positions at the Department of Electrical Engineering at Kiev Polytechnic Institute and the Academy of Sciences of Ukraine. From 1989 to 1993 he was head of the Microelectronic and Electromechanical Systems Division at the Academy of Sciences of Ukraine. From 1993 to 2002, he was with Purdue University/Indianapolis. In 2002, Dr. Lyshevski joined Rochester Institute of Technology, where he is a professor of electrical engineering.

Lyshevski serves as the senior faculty fellow at the U.S. Surface and Undersea Naval Warfare Centers. He is the author of eight books including *Nano- and Micro-Electromechanical Systems: Fundamentals of Micro- and Nano- Engineering* (for which he also acts as CRC series editor; CRC Press, 2000); *MEMS and NEMS: Systems, Devices, and Structures* (CRC Press, 2002); and author or co-author of more than 250 journal articles, handbook chapters, and regular conference papers. His current teaching and research activities are in the areas of MEMS and NEMS (CAD, design, high-fidelity modeling, data-intensive analysis, heterogeneous simulation, fabrication), intelligent large-scale microsystems, learning configurations, novel architectures, self-organization, micro- and nanoscale devices (actuators, sensors, logics, switches, memories, etc.), nanocomputers and their components, reconfigurable (adaptive) defect-tolerant computer architectures, and systems informatics. Dr. Lyshevski has been active in the design, application, verification, and implementation of advanced aerospace, automotive, electromechanical, and naval systems.

Lyshevski has made 29 invited presentations (nationally and internationally) and has taught undergraduate and graduate courses in NEMS, MEMS, microsystems, computer architecture, motion devices, integrated circuits, and signals and systems.

Gerald J. Iafrate joined the faculty of North Carolina State University in August 2001. Previously, he was a professor at the University of Notre Dame; he also served as Associate Dean for Research in the College of Engineering, and as director of the newly established University Center of Excellence in Nanoscience and Technology. He has extensive experience in managing large interdisciplinary research programs. From 1989 to 1997, Dr. Iafrate served as the Director of the U.S. Army Research Office (ARO). As director, he was the Army's key executive for the conduct of extramural research in the physical and engineering sciences in response to DoD-wide objectives. Prior to becoming Director of ARO, Dr. Iafrate was the Director of Electronic Devices Research at the U.S. Army Electronics Technology and Devices Laboratory (ETDL). Working with the National Science Foundation, he played a key leadership role in establishing the first-of-its-kind Army–NSF–University consortium.

He is currently a professor of electrical and computer engineering at North Carolina State University, Raleigh, where his research interests include quantum transport in nanostructures such as resonant tunneling diodes and quantum dots. He is also conducting studies in the area of quantum dissipation, with emphasis on ratchet-like transport phenomena and nonequilibrium processes in nanosystems. Dr. Iafrate is a fellow of the IEEE, APS, and AAAS.

Contributors

S. Adiga
North Carolina State University
Department of Materials Science
and Engineering
Raleigh, NC

Damian G. Allis
Syracuse University
Department of Chemistry
Syracuse, NY

Narayan R. Aluru
University of Illinois
Beckman Institute for Advanced
Science and Technology
Urbana, IL

D.A. Areshkin
North Carolina State University
Department of Materials Science
and Engineering
Raleigh, NC

Rashid Bashir
Purdue University
School of Electrical and
Computer Engineering
Department of Biomedical
Engineering
West Lafayette, IN

Donald W. Brenner
North Carolina State University
Department of Materials Science
and Engineering
Raleigh, NC

Kwong–Kit Choi
U.S. Army Research Laboratory
Adelphi, MD

**Almadena Y.
Chtchelkanova**
Strategic Analysis, Inc.
Arlington, VA

Supriyo Datta
Purdue University
School of Electrical and
Computer Engineering
West Lafayette, IN

James C. Ellenbogen
The MITRE Corporation
Nanosystems Group
McLean, VA

R. Esfand
Central Michigan University
Dendritic Nanotechnologies Ltd.
Mt. Pleasant, MI

Michael Falvo
University of North Carolina
Department of Physics and
Astronomy
Chapel Hill, NC

Richard P. Feynman
California Institute of Technology
Pasadena, CA

J.A. Harrison
U.S. Naval Academy
Chemistry Department
Annapolis, MD

Scott A. Henderson
Starpharma Limited
Melbourne, Victoria, Australia

Karl Hess
University of Illinois
Beckman Institute for Advanced
Science and Technology
Urbana, IL

G. Holan
Starpharma Limited
Melbourne, Victoria, Australia

Michael Pycraft Hughes
University of Surrey
School of Engineering
Guildford, Surrey, England

Dustin K. James
Rice University
Department of Chemistry
Houston, TX

Jean-Pierre Leburton
University of Illinois
Beckman Institute for Advanced
Science and Technology
Urbana, IL

Wing Kam Liu
Northwestern University
Department of Mechanical
  Engineering
Evanston, IL

J. Christopher Love
Harvard University
Cambridge, MA

Sergey Edward Lyshevski
Rochester Institute of
  Technology
Department of Electrical
  Engineering
Rochester, NY

Karen Mardel
Starpharma Limited
Melbourne, Victoria, Australia

William McMahon
University of Illinois
Beckman Institute for Advanced
  Science and Technology
Urbana, IL

Meyya Meyyappan
NASA Ames Research Center
Moffett Field, CA

Vladimiro Mujica
Northwestern University
Department of Chemistry
Evanston, IL

Radik R. Mulyukov
Russian Academy of Science
Institute for Metals
Superplasticity Problems
Ufa, Russia

Airat A. Nazarov
Russian Academy of Science
Institute for Metals
Superplasticity Problems
Ufa, Russia

Gregory N. Parsons
North Carolina State University
Department of Chemical
  Engineering
Raleigh, NC

Magnus Paulsson
Purdue University
School of Electrical and
  Computer Engineering
West Lafayette, IN

Wolfgang Porod
University of Notre Dame
Department of Electrical
  Engineering
Notre Dame, IN

Dennis W. Prather
University of Delaware
Department of Electrical and
  Computer Engineering
Newark, DE

Dong Qian
Northwestern University
Department of Mechanical
  Engineering
Evanston, IL

Mark A. Ratner
Northwestern University
Department of Chemistry
Evanston, IL

Umberto Ravaioli
University of Illinois
Beckman Institute for Advanced
  Science and Technology
Urbana, IL

Slava V. Rotkin
University of Illinois
Beckman Institute for Advanced
  Science and Technology
Urbana, IL

Rodney S. Ruoff
Northwestern University
Department of Mechanical
  Engineering
Evanston, IL

J.D. Schall
North Carolina State University
Department of Materials Science
  and Engineering
Raleigh, NC

Ahmed S. Sharkawy
University of Delaware
Department of Electrical and
  Computer Engineering
Newark, DE

O.A. Shenderova
North Carolina State University
Department of Materials Science
  and Engineering
Raleigh, NC

Shouyuan Shi
University of Delaware
Department of Electrical and
  Computer Engineering
Newark, DE

James T. Spencer
Syracuse University
Department of Chemistry
Syracuse, NY

Deepak Srivastava
NASA Ames Research Center
Moffett Field, CA

Martin Staedele
Infineon Technologies AG
Corporate Research ND
Munich, Germany

S.J. Stuart
Clemson University
Department of Chemistry
Clemson, SC

Richard Superfine
University of North Carolina
Department of Physics and
    Astronomy
Chapel Hill, NC

Russell M. Taylor, II
University of North Carolina
Department of Computer
    Science, Physics, and
    Astronomy
Chapel Hill, NC

Donald A. Tomalia
Central Michigan University
Dendritic Nanotechnologies Ltd.
Mt. Pleasant, MI

James M. Tour
Rice University
Center for Nanoscale Science
    and Technology
Houston, TX

Daryl Treger
Strategic Analysis, Inc.
Arlington, VA

Blair R. Tuttle
Pennsylvania State University
Behrend College
School of Science
Erie, PA

Trudy van der Straaten
University of Illinois
Beckman Institute for Advanced
    Science and Technology
Urbana, IL

Gregory J. Wagner
Northwestern University
Department of Mechanical
    Engineering
Evanston, IL

Sean Washburn
University of North Carolina
Department of Physics and
    Astronomy
Chapel Hill, NC

Stuart A. Wolf
DARPA/DSO, NRL
Arlington, VA

Boris I. Yakobson
Rice University
Center for Nanoscale Science
    and Technology
Houston, TX

Min–Feng Yu
University of Illinois
Department of Mechanical and
    Industrial Engineering
Urbana, IL

Ferdows Zahid
Purdue University
School of Electrical and
    Computer Engineering
West Lafayette, IN

Contents

Section 3 Molecular Electronics: Fundamental Processes

Section 4　Manipulation and Assembly

Section 5 Functional Structures and Mechanics

Section 1

The Promise of Nanotechnology and Nanoscience

1

There's Plenty of Room at the Bottom: An Invitation to Enter a New Field of Physics

Richard P. Feynman
California Institute of Technology

CONTENTS

This transcript of the classic talk that Richard Feynman gave on December 29, 1959, at the annual meeting of the American Physical Society at the California Institute of Technology (Caltech) was first published in the February 1960 issue (Volume XXIII, No. 5, pp. 22–36) of Caltech's *Engineering and Science*, which owns the copyright. It has been made available on the web at http://www.zyvex.com/nanotech/feynman.html with their kind permission.

For an account of the talk and how people reacted to it, see Chapter 4 of *Nano!* by Ed Regis. An excellent technical introduction to nanotechnology is *Nanosystems: Molecular Machinery, Manufacturing, and Computation* by K. Eric Drexler.

1.1 Transcript

I imagine experimental physicists must often look with envy at men like Kamerlingh Onnes, who discovered a field like low temperature, which seems to be bottomless and in which one can go down and down. Such a man is then a leader and has some temporary monopoly in a scientific adventure. Percy Bridgman, in designing a way to obtain higher pressures, opened up another new field and was able to move into it and to lead us all along. The development of ever-higher vacuum was a continuing development of the same kind.

I would like to describe a field in which little has been done but in which an enormous amount can be done in principle. This field is not quite the same as the others in that it will not tell us much of fundamental physics (in the sense of "what are the strange particles?"); but it is more like solid-state physics in the sense that it might tell us much of great interest about the strange phenomena that occur in complex situations. Furthermore, a point that is most important is that it would have an enormous number of technical applications.

What I want to talk about is the problem of manipulating and controlling things on a small scale.

As soon as I mention this, people tell me about miniaturization, and how far it has progressed today. They tell me about electric motors that are the size of the nail on your small finger. And there is a device on the market, they tell me, by which you can write the Lord's Prayer on the head of a pin. But that's nothing; that's the most primitive, halting step in the direction I intend to discuss. It is a staggeringly small world that is below. In the year 2000, when they look back at this age, they will wonder why it was not until the year 1960 that anybody began seriously to move in this direction.

Why cannot we write the entire 24 volumes of the *Encyclopaedia Britannica* on the head of a pin?

Let's see what would be involved. The head of a pin is a sixteenth of an inch across. If you magnify it by 25,000 diameters, the area of the head of the pin is then equal to the area of all the pages of the *Encyclopaedia Britannica*. Therefore, all it is necessary to do is to reduce in size all the writing in the encyclopedia by 25,000 times. Is that possible? The resolving power of the eye is about 1/120 of an inch — that is roughly the diameter of one of the little dots on the fine half-tone reproductions in the encyclopedia. This, when you demagnify it by 25,000 times, is still 80 angstroms in diameter — 32 atoms across, in an ordinary metal. In other words, one of those dots still would contain in its area 1000 atoms. So, each dot can easily be adjusted in size as required by the photoengraving, and there is no question that there is enough room on the head of a pin to put all of the *Encyclopaedia Britannica*. Furthermore, it can be read if it is so written. Let's imagine that it is written in raised letters of metal; that is, where the black is in the encyclopedia, we have raised letters of metal that are actually 1/25,000 of their ordinary size. How would we read it?

If we had something written in such a way, we could read it using techniques in common use today. (They will undoubtedly find a better way when we do actually have it written, but to make my point conservatively I shall just take techniques we know today.) We would press the metal into a plastic material and make a mold of it, then peel the plastic off very carefully, evaporate silica into the plastic to get a very thin film, then shadow it by evaporating gold at an angle against the silica so that all the little letters will appear clearly, dissolve the plastic away from the silica film, and then look through it with an electron microscope!

There is no question that if the thing were reduced by 25,000 times in the form of raised letters on the pin, it would be easy for us to read it today. Furthermore, there is no question that we would find it easy to make copies of the master; we would just need to press the same metal plate again into plastic and we would have another copy.

How Do We Write Small?

The next question is, how do we write it? We have no standard technique to do this now. But let me argue that it is not as difficult as it first appears to be. We can reverse the lenses of the electron microscope in order to demagnify as well as magnify. A source of ions, sent through the microscope lenses in reverse, could be focused to a very small spot. We could write with that spot like we write in a TV cathode ray oscilloscope, by going across in lines and having an adjustment that determines the amount of material which is going to be deposited as we scan in lines.

This method might be very slow because of space charge limitations. There will be more rapid methods. We could first make, perhaps by some photo process, a screen that has holes in it in the form of the letters. Then we would strike an arc behind the holes and draw metallic ions through the holes; then we could again use our system of lenses and make a small image in the form of ions, which would deposit the metal on the pin.

A simpler way might be this (though I am not sure it would work): we take light and, through an optical microscope running backwards, we focus it onto a very small photoelectric screen. Then electrons come away from the screen where the light is shining. These electrons are focused down in size by the electron microscope lenses to impinge directly upon the surface of the metal. Will such a beam etch away the metal if it is run long enough? I don't know. If it doesn't work for a metal surface, it must be possible to find some surface with which to coat the original pin so that, where the electrons bombard, a change is made which we could recognize later.

There is no intensity problem in these devices — not what you are used to in magnification, where you have to take a few electrons and spread them over a bigger and bigger screen; it is just the opposite. The light which we get from a page is concentrated onto a very small area so it is very intense. The few electrons which come from the photoelectric screen are demagnified down to a very tiny area so that, again, they are very intense. I don't know why this hasn't been done yet!

That's the *Encyclopedia Britannica* on the head of a pin, but let's consider all the books in the world. The Library of Congress has approximately 9 million volumes; the British Museum Library has 5 million volumes; there are also 5 million volumes in the National Library in France. Undoubtedly there are duplications, so let us say that there are some 24 million volumes of interest in the world.

What would happen if I print all this down at the scale we have been discussing? How much space would it take? It would take, of course, the area of about a million pinheads because, instead of there being just the 24 volumes of the encyclopedia, there are 24 million volumes. The million pinheads can be put in a square of a thousand pins on a side, or an area of about 3 square yards. That is to say, the silica replica with the paper-thin backing of plastic, with which we have made the copies, with all this information, is on an area approximately the size of 35 pages of the encyclopedia. That is about half as many pages as there are in this magazine. All of the information which all of mankind has ever recorded in books can be carried around in a pamphlet in your hand — and not written in code, but a simple reproduction of the original pictures, engravings, and everything else on a small scale without loss of resolution.

What would our librarian at Caltech say, as she runs all over from one building to another, if I tell her that, 10 years from now, all of the information that she is struggling to keep track of — 120,000 volumes, stacked from the floor to the ceiling, drawers full of cards, storage rooms full of the older books — can be kept on just one library card! When the University of Brazil, for example, finds that their library is burned, we can send them a copy of every book in our library by striking off a copy from the master plate in a few hours and mailing it in an envelope no bigger or heavier than any other ordinary airmail letter. Now, the name of this talk is "There Is *Plenty* of Room at the Bottom" — not just "There Is Room at the Bottom." What I have demonstrated is that there is room — that you can decrease the size of things in a practical way. I now want to show that there is *plenty* of room. I will not now discuss how we are going to do it, but only what is possible in principle — in other words, what is possible according to the laws of physics. I am not inventing antigravity, which is possible someday only if the laws are not what we think. I am telling you what could be done if the laws are what we think; we are not doing it simply because we haven't yet gotten around to it.

Information on a Small Scale

Suppose that, instead of trying to reproduce the pictures and all the information directly in its present form, we write only the information content in a code of dots and dashes, or something like that, to represent the various letters. Each letter represents six or seven "bits" of information; that is, you need only about six or seven dots or dashes for each letter. Now, instead of writing everything, as I did before, on the surface of the head of a pin, I am going to use the interior of the material as well.

Let us represent a dot by a small spot of one metal, the next dash by an adjacent spot of another metal, and so on. Suppose, to be conservative, that a bit of information is going to require a little cube of atoms $5 \times 5 \times 5$ — that is 125 atoms. Perhaps we need a hundred and some odd atoms to make sure that the information is not lost through diffusion or through some other process.

I have estimated how many letters there are in the encyclopedia, and I have assumed that each of my 24 million books is as big as an encyclopedia volume, and have calculated, then, how many bits of information there are (10^{15}). For each bit I allow 100 atoms. And it turns out that all of the information that man has carefully accumulated in all the books in the world can be written in this form in a cube of material 1/200 of an inch wide — which is the barest piece of dust that can be made out by the human eye. So there is plenty of room at the bottom! Don't tell me about microfilm! This fact — that enormous amounts of information can be carried in an exceedingly small space — is, of course, well known to the biologists and resolves the mystery that existed before we understood all this clearly — of how it could be that, in the tiniest cell, all of the information for the organization of a complex creature such as ourselves can be stored. All this information — whether we have brown eyes, or whether we think at all, or that in the embryo the jawbone should first develop with a little hole in the side so that later a nerve can grow through it — all this information is contained in a very tiny fraction of the cell in the form of long-chain DNA molecules in which approximately 50 atoms are used for one bit of information about the cell.

Better Electron Microscopes

If I have written in a code with $5 \times 5 \times 5$ atoms to a bit, the question is, how could I read it today? The electron microscope is not quite good enough — with the greatest care and effort, it can only resolve about 10 angstroms. I would like to try and impress upon you, while I am talking about all of these things on a small scale, the importance of improving the electron microscope by a hundred times. It is not impossible; it is not against the laws of diffraction of the electron. The wavelength of the electron in such a microscope is only 1/20 of an angstrom. So it should be possible to see the individual atoms. What good would it be to see individual atoms distinctly? We have friends in other fields — in biology, for instance. We physicists often look at them and say, "You know the reason you fellows are making so little progress?" (Actually I don't know any field where they are making more rapid progress than they are in biology today.) "You should use more mathematics, like we do." They could answer us — but they're polite, so I'll answer for them: "What *you* should do in order for us to make more rapid progress is to make the electron microscope 100 times better."

What are the most central and fundamental problems of biology today? They are questions like, what is the sequence of bases in the DNA? What happens when you have a mutation? How is the base order in the DNA connected to the order of amino acids in the protein? What is the structure of the RNA; is it single-chain or double-chain, and how is it related in its order of bases to the DNA? What is the organization of the microsomes? How are proteins synthesized? Where does the RNA go? How does it sit? Where do the proteins sit? Where do the amino acids go in? In photosynthesis, where is the chlorophyll; how is it arranged; where are the carotenoids involved in this thing? What is the system of the conversion of light into chemical energy?

It is very easy to answer many of these fundamental biological questions; you just look at the thing! You will see the order of bases in the chain; you will see the structure of the microsome. Unfortunately, the present microscope sees at a scale which is just a bit too crude. Make the microscope one hundred times more powerful, and many problems of biology would be made very much easier. I exaggerate, of course, but the biologists would surely be very thankful to you — and they would prefer that to the criticism that they should use more mathematics.

The theory of chemical processes today is based on theoretical physics. In this sense, physics supplies the foundation of chemistry. But chemistry also has analysis. If you have a strange substance and you want to know what it is, you go through a long and complicated process of chemical analysis. You can analyze almost anything today, so I am a little late with my idea. But if the physicists wanted to, they could also dig under the chemists in the problem of chemical analysis. It would be very easy to make an analysis of any complicated chemical substance; all one would have to do would be to look at it and see where the atoms are. The only trouble is that the electron microscope is 100 times too poor. (Later, I would like to ask the question: can the physicists do something about the third problem of chemistry — namely, synthesis? Is there a physical way to synthesize any chemical substance?)

The reason the electron microscope is so poor is that the f-value of the lenses is only 1 part to 1000; you don't have a big enough numerical aperture. And I know that there are theorems which prove that it is impossible, with axially symmetrical stationary field lenses, to produce an f-value any bigger than so and so; and therefore the resolving power at the present time is at its theoretical maximum. But in every theorem there are assumptions. Why must the field be symmetrical? I put this out as a challenge: is there no way to make the electron microscope more powerful?

The Marvelous Biological System

The biological example of writing information on a small scale has inspired me to think of something that should be possible. Biology is not simply writing information; it is doing something about it. A biological system can be exceedingly small. Many of the cells are very tiny, but they are very active; they manufacture various substances; they walk around; they wiggle; and they do all kinds of marvelous things — all on a very small scale. Also, they store information. Consider the possibility that we too can make a thing very small which does what we want — that we can manufacture an object that maneuvers at that level!

There may even be an economic point to this business of making things very small. Let me remind you of some of the problems of computing machines. In computers we have to store an enormous amount of information. The kind of writing that I was mentioning before, in which I had everything down as a distribution of metal, is permanent. Much more interesting to a computer is a way of writing, erasing, and writing something else. (This is usually because we don't want to waste the material on which we have just written. Yet if we could write it in a very small space, it wouldn't make any difference; it could just be thrown away after it was read. It doesn't cost very much for the material).

Miniaturizing the Computer

I don't know how to do this on a small scale in a practical way, but I do know that computing machines are very large; they fill rooms. Why can't we make them very small, make them of little wires, little elements — and by little, I mean *little*. For instance, the wires should be 10 or 100 atoms in diameter, and the circuits should be a few thousand angstroms across. Everybody who has analyzed the logical theory of computers has come to the conclusion that the possibilities of computers are very interesting — if they could be made to be more complicated by several orders of magnitude. If they had millions of times as many elements, they could make judgments. They would have time to calculate what is the best way to make the calculation that they are about to make. They could select the method of analysis which, from their experience, is better than the one that we would give to them. And in many other ways, they would have new qualitative features.

If I look at your face I immediately recognize that I have seen it before. (Actually, my friends will say I have chosen an unfortunate example here for the subject of this illustration. At least I recognize that it is a man and not an apple.) Yet there is no machine which, with that speed, can take a picture of a face and say even that it is a man; and much less that it is the same man that you showed it before — unless it is exactly the same picture. If the face is changed; if I am closer to the face; if I am further from the face; if the light changes — I recognize it anyway. Now, this little computer I carry in my head is easily able to do that. The computers that we build are not able to do that. The number of elements in this bone box of mine are enormously greater than the number of elements in our "wonderful" computers. But our mechanical computers are too big; the elements in this box are microscopic. I want to make some that are submicroscopic.

If we wanted to make a computer that had all these marvelous extra qualitative abilities, we would have to make it, perhaps, the size of the Pentagon. This has several disadvantages. First, it requires too much material; there may not be enough germanium in the world for all the transistors which would have to be put into this enormous thing. There is also the problem of heat generation and power consumption; TVA would be needed to run the computer. But an even more practical difficulty is that the computer would be limited to a certain speed. Because of its large size, there is finite time required to get the information from one place to another. The information cannot go any faster than the speed of light — so, ultimately, when our computers get faster and faster and more and more elaborate, we will have to make them smaller and smaller. But there is plenty of room to make them smaller. There is nothing that I can see in the physical laws that says the computer elements cannot be made enormously smaller than they are now. In fact, there may be certain advantages.

Miniaturization by Evaporation

How can we make such a device? What kind of manufacturing processes would we use? One possibility we might consider, since we have talked about writing by putting atoms down in a certain arrangement, would be to evaporate the material, then evaporate the insulator next to it. Then, for the next layer, evaporate another position of a wire, another insulator, and so on. So, you simply evaporate until you have a block of stuff which has the elements — coils and condensers, transistors and so on — of exceedingly fine dimensions.

But I would like to discuss, just for amusement, that there are other possibilities. Why can't we manufacture these small computers somewhat like we manufacture the big ones? Why can't we drill holes, cut things, solder things, stamp things out, mold different shapes all at an infinitesimal level? What are the limitations as to how small a thing has to be before you can no longer mold it? How many times when you are working on something frustratingly tiny, like your wife's wristwatch, have you said to yourself, "If I could only train an ant to do this!" What I would like to suggest is the possibility of training an ant to train a mite to do this. What are the possibilities of small but movable machines? They may or may not be useful, but they surely would be fun to make.

Consider any machine — for example, an automobile — and ask about the problems of making an infinitesimal machine like it. Suppose, in the particular design of the automobile, we need a certain precision of the parts; we need an accuracy, let's suppose, of 4/10,000 of an inch. If things are more inaccurate than that in the shape of the cylinder and so on, it isn't going to work very well. If I make the thing too small, I have to worry about the size of the atoms; I can't make a circle of "balls" so to speak, if the circle is too small. So if I make the error — corresponding to 4/10,000 of an inch — correspond to an error of 10 atoms, it turns out that I can reduce the dimensions of an automobile 4000 times, approximately, so that it is 1 mm across. Obviously, if you redesign the car so that it would work with a much larger tolerance, which is not at all impossible, then you could make a much smaller device.

It is interesting to consider what the problems are in such small machines. Firstly, with parts stressed to the same degree, the forces go as the area you are reducing, so that things like weight and inertia are of relatively no importance. The strength of material, in other words, is very much greater in proportion. The stresses and expansion of the flywheel from centrifugal force, for example, would be the same proportion only if the rotational speed is increased in the same proportion as we decrease the size. On the other hand, the metals that we use have a grain structure, and this would be very annoying at small scale because the material is not homogeneous. Plastics and glass and things of this amorphous nature are very much more homogeneous, and so we would have to make our machines out of such materials.

There are problems associated with the electrical part of the system — with the copper wires and the magnetic parts. The magnetic properties on a very small scale are not the same as on a large scale; there is the "domain" problem involved. A big magnet made of millions of domains can only be made on a small scale with one domain. The electrical equipment won't simply be scaled down; it has to be redesigned. But I can see no reason why it can't be redesigned to work again.

Problems of Lubrication

Lubrication involves some interesting points. The effective viscosity of oil would be higher and higher in proportion as we went down (and if we increase the speed as much as we can). If we don't increase the speed so much, and change from oil to kerosene or some other fluid, the problem is not so bad. But actually we may not have to lubricate at all! We have a lot of extra force. Let the bearings run dry; they won't run hot because the heat escapes away from such a small device very, very rapidly.

This rapid heat loss would prevent the gasoline from exploding, so an internal combustion engine is impossible. Other chemical reactions, liberating energy when cold, can be used. Probably an external supply of electrical power would be most convenient for such small machines.

What would be the utility of such machines? Who knows? Of course, a small automobile would only be useful for the mites to drive around in, and I suppose our Christian interests don't go that far. However, we did note the possibility of the manufacture of small elements for computers in completely automatic factories, containing lathes and other machine tools at the very small level. The small lathe would not have to be exactly like our big lathe. I leave to your imagination the improvement of the design to take full advantage of the properties of things on a small scale, and in such a way that the fully automatic aspect would be easiest to manage.

A friend of mine (Albert R. Hibbs) suggests a very interesting possibility for relatively small machines. He says that although it is a very wild idea, it would be interesting in surgery if you could swallow the surgeon. You put the mechanical surgeon inside the blood vessel and it goes into the heart and "looks" around. (Of course the information has to be fed out.) It finds out which valve is the faulty one and takes a little knife and slices it out. Other small machines might be permanently incorporated in the body to assist some inadequately functioning organ.

Now comes the interesting question: how do we make such a tiny mechanism? I leave that to you. However, let me suggest one weird possibility. You know, in the atomic energy plants they have materials and machines that they can't handle directly because they have become radioactive. To unscrew nuts and put on bolts and so on, they have a set of master and slave hands, so that by operating a set of levers here, you control the "hands" there, and can turn them this way and that so you can handle things quite nicely.

Most of these devices are actually made rather simply, in that there is a particular cable, like a marionette string, that goes directly from the controls to the "hands." But, of course, things also have been made using servo motors, so that the connection between the one thing and the other is electrical rather than mechanical. When you turn the levers, they turn a servo motor, and it changes the electrical currents in the wires, which repositions a motor at the other end.

Now, I want to build much the same device — a master–slave system which operates electrically. But I want the slaves to be made especially carefully by modern large-scale machinists so that they are 1/4 the scale of the "hands" that you ordinarily maneuver. So you have a scheme by which you can do things at 1/4 scale anyway — the little servo motors with little hands play with little nuts and bolts; they drill little holes; they are four times smaller. Aha! So I manufacture a 1/4-size lathe; I manufacture 1/4-size tools; and I make, at the 1/4 scale, still another set of hands again relatively 1/4 size! This is 1/16 size, from my point of view. And after I finish doing this I wire directly from my large-scale system, through transformers perhaps, to the 1/16-size servo motors. Thus I can now manipulate the 1/16 size hands.

Well, you get the principle from there on. It is rather a difficult program, but it is a possibility. You might say that one can go much farther in one step than from one to four. Of course, this all has to be designed very carefully, and it is not necessary simply to make it like hands. If you thought of it very carefully, you could probably arrive at a much better system for doing such things.

If you work through a pantograph, even today, you can get much more than a factor of four in even one step. But you can't work directly through a pantograph which makes a smaller pantograph which then makes a smaller pantograph — because of the looseness of the holes and the irregularities of construction. The end of the pantograph wiggles with a relatively greater irregularity than the irregularity with which you move your hands. In going down this scale, I would find the end of the pantograph on the end of the pantograph on the end of the pantograph shaking so badly that it wasn't doing anything sensible at all.

At each stage, it is necessary to improve the precision of the apparatus. If, for instance, having made a small lathe with a pantograph, we find its lead screw irregular — more irregular than the large-scale one —we could lap the lead screw against breakable nuts that you can reverse in the usual way back and forth until this lead screw is, at its scale, as accurate as our original lead screws, at our scale.

We can make flats by rubbing unflat surfaces in triplicates together — in three pairs — and the flats then become flatter than the thing you started with. Thus, it is not impossible to improve precision on a small scale by the correct operations. So, when we build this stuff, it is necessary at each step to improve the accuracy of the equipment by working for a while down there, making accurate lead screws, Johansen blocks, and all the other materials which we use in accurate machine work at the higher level. We have to stop at each level and manufacture all the stuff to go to the next level — a very long and very difficult program. Perhaps you can figure a better way than that to get down to small scale more rapidly.

Yet, after all this, you have just got one little baby lathe 4000 times smaller than usual. But we were thinking of making an enormous computer, which we were going to build by drilling holes on this lathe to make little washers for the computer. How many washers can you manufacture on this one lathe?

A Hundred Tiny Hands

When I make my first set of slave "hands" at 1/4 scale, I am going to make ten sets. I make ten sets of "hands," and I wire them to my original levers so they each do exactly the same thing at the same time in parallel. Now, when I am making my new devices 1/4 again as small, I let each one manufacture ten copies, so that I would have a hundred "hands" at the 1/16 size.

Where am I going to put the million lathes that I am going to have? Why, there is nothing to it; the volume is much less than that of even one full-scale lathe. For instance, if I made a billion little lathes, each 1/4000 of the scale of a regular lathe, there are plenty of materials and space available because in the billion little ones there is less than 2% of the materials in one big lathe.

It doesn't cost anything for materials, you see. So I want to build a billion tiny factories, models of each other, which are manufacturing simultaneously, drilling holes, stamping parts, and so on.

As we go down in size, there are a number of interesting problems that arise. All things do not simply scale down in proportion. There is the problem that materials stick together by the molecular (Van der Waals) attractions. It would be like this: after you have made a part and you unscrew the nut from a bolt, it isn't going to fall down because the gravity isn't appreciable; it would even be hard to get it off the bolt. It would be like those old movies of a man with his hands full of molasses, trying to get rid of a glass of water. There will be several problems of this nature that we will have to be ready to design for.

Rearranging the Atoms

But I am not afraid to consider the final question as to whether, ultimately — in the great future — we can arrange the atoms the way we want; the very atoms, all the way down! What would happen if we could arrange the atoms one by one the way we want them (within reason, of course; you can't put them so that they are chemically unstable, for example).

Up to now, we have been content to dig in the ground to find minerals. We heat them and we do things on a large scale with them, and we hope to get a pure substance with just so much impurity, and so on. But we must always accept some atomic arrangement that nature gives us. We haven't got anything, say, with a "checkerboard" arrangement, with the impurity atoms exactly arranged 1000 angstroms apart, or in some other particular pattern. What could we do with layered structures with just the right layers? What would the properties of materials be if we could really arrange the atoms the way we want them? They would be very interesting to investigate theoretically. I can't see exactly what would happen, but I can hardly doubt that when we have some control of the arrangement of things on a small scale we will get an enormously greater range of possible properties that substances can have, and of different things that we can do.

Consider, for example, a piece of material in which we make little coils and condensers (or their solid state analogs) 1,000 or 10,000 angstroms in a circuit, one right next to the other, over a large area, with little antennas sticking out at the other end — a whole series of circuits. Is it possible, for example, to emit light from a whole set of antennas, like we emit radio waves from an organized set of antennas to beam the radio programs to Europe? The same thing would be to beam the light out in a definite direction with very high intensity. (Perhaps such a beam is not very useful technically or economically.)

I have thought about some of the problems of building electric circuits on a small scale, and the problem of resistance is serious. If you build a corresponding circuit on a small scale, its natural frequency goes up, since the wavelength goes down as the scale; but the skin depth only decreases with the square root of the scale ratio, and so resistive problems are of increasing difficulty. Possibly we can beat resistance through the use of superconductivity if the frequency is not too high, or by other tricks.

Atoms in a Small World

When we get to the very, very small world — say circuits of seven atoms — we have a lot of new things that would happen that represent completely new opportunities for design. Atoms on a small scale behave like nothing on a large scale, for they satisfy the laws of quantum mechanics. So, as we go down and fiddle around with the atoms down there, we are working with different laws, and we can expect to do different things. We can manufacture in different ways. We can use not just circuits but some system involving the quantized energy levels, or the interactions of quantized spins, etc.

Another thing we will notice is that, if we go down far enough, all of our devices can be mass produced so that they are absolutely perfect copies of one another. We cannot build two large machines so that the dimensions are exactly the same. But if your machine is only 100 atoms high, you only have to get it correct to 1/2% to make sure the other machine is exactly the same size — namely, 100 atoms high!

At the atomic level, we have new kinds of forces and new kinds of possibilities, new kinds of effects. The problems of manufacture and reproduction of materials will be quite different. I am, as I said, inspired by the biological phenomena in which chemical forces are used in repetitious fashion to produce all kinds of weird effects (one of which is the author).

The principles of physics, as far as I can see, do not speak against the possibility of maneuvering things atom by atom. It is not an attempt to violate any laws; it is something, in principle, that can be done; but in practice, it has not been done because we are too big.

Ultimately, we can do chemical synthesis. A chemist comes to us and says, "Look, I want a molecule that has the atoms arranged thus and so; make me that molecule." The chemist does a mysterious thing when he wants to make a molecule. He sees that it has that ring, so he mixes this and that, and he shakes it, and he fiddles around. And, at the end of a difficult process, he usually does succeed in synthesizing what he wants. By the time I get my devices working, so that we can do it by physics, he will have figured out how to synthesize absolutely anything, so that this will really be useless. But it is interesting that it would be, in principle, possible (I think) for a physicist to synthesize any chemical substance that the chemist writes down. Give the orders and the physicist synthesizes it. How? Put the atoms down where the chemist says, and so you make the substance. The problems of chemistry and biology can be greatly helped if our ability to see what we are doing, and to do things on an atomic level, is ultimately developed — a development which I think cannot be avoided.

Now, you might say, "Who should do this and why should they do it?" Well, I pointed out a few of the economic applications, but I know that the reason that you would do it might be just for fun. But have some fun! Let's have a competition between laboratories. Let one laboratory make a tiny motor which it sends to another lab which sends it back with a thing that fits inside the shaft of the first motor.

High School Competition

Just for the fun of it, and in order to get kids interested in this field, I would propose that someone who has some contact with the high schools think of making some kind of high school competition. After all, we haven't even started in this field, and even the kids can write smaller than has ever been written before. They could have competition in high schools. The Los Angeles high school could send a pin to the Venice high school on which it says, "How's this?" They get the pin back, and in the dot of the "i" it says, "Not so hot."

Perhaps this doesn't excite you to do it, and only economics will do so. Then I want to do something, but I can't do it at the present moment because I haven't prepared the ground. It is my intention to offer a prize of $1000 to the first guy who can take the information on the page of a book and put it on an area 1/25,000 smaller in linear scale in such manner that it can be read by an electron microscope.

And I want to offer another prize — if I can figure out how to phrase it so that I don't get into a mess of arguments about definitions — of another $1000 to the first guy who makes an operating electric motor — a rotating electric motor which can be controlled from the outside and, not counting the lead-in wires, is only a 1/64-inch cube.

I do not expect that such prizes will have to wait very long for claimants.

2

Room at the Bottom, Plenty of Tyranny at the Top

CONTENTS

Karl Hess
University of Illinois

2.1 Rising to the Feynman Challenge

Richard Feynman is generally regarded as one of the fathers of nanotechnology. In giving his landmark presentation to the American Physical Society on December 29, 1959, at Caltech, his title line was, "There's Plenty of Room at the Bottom." At that time, Feynman extended an invitation for "manipulating and controlling things on a small scale, thereby entering a new field of physics which was bottomless, like low-temperature physics." He started with the question, can we "write the Lord's prayer on the head of a pin," and immediately extended the goal to the entire 24 volumes of the *Encyclopaedia Britannica*. By following the Gedanken Experiment, Feynman showed that there is no physical law against the realization of such goals: if you magnify the head of a pin by 25,000 diameters, its surface area is then equal to that of all the pages in the *Encyclopaedia Britannica*.

Feynman's dreams of writing small have all been fulfilled and even exceeded in the past decades. Since the advent of scanning tunneling microscopy, as introduced by Binnig and Rohrer, it has been repeatedly demonstrated that single atoms can not only be conveniently represented for the human eye but manipulated as well. Thus, it is conceivable to store all the books in the world (which Feynman estimates to contain 10^{15} bits of information) on the area of a credit card! The encyclopedia, having around 10^9 bits of information, can be written on about 1/100 the surface area of the head of a pin.

One need not look to atomic writing to achieve astonishing results: current microchips contain close to 100 million transistors. A small number of such chips could not only store large amounts of information (such as the *Encyclopaedia Britannica*); they can process it with GHz speed as well. To find a particular word takes just a few nanoseconds. Typical disk hard drives can store much more than the semiconductor chips, with a trade-off for retrieval speeds. Feynman's vision for storing and retrieving information on a small scale was very close to these numbers. He did not ask himself what the practical difficulties were in achieving these goals, but rather asked only what the principal limitations were. Even he could not possibly foresee the ultimate consequences of writing small and reading fast: the creation of the Internet. Sifting through large databases is, of course, what is done during Internet browsing. It is not only the microrepresentation of information that has led to the revolution we are witnessing but also the ability to browse through this information at very high speeds.

Can one improve current chip technology beyond the achievements listed above? Certainly! Further improvements are still expected just by scaling down known silicon technology. Beyond this, if it were possible to change the technology completely and create transistors the size of molecules, then one could fit hundreds of billions of transistors on a chip. Changing technology so dramatically is not easy and less likely to happen. A molecular transistor that is as robust and efficient as the existing ones is beyond current implementation capabilities; we do not know how to achieve such densities without running into problems of excessive heat generation and other problems related to highly integrated systems. However, Feynman would not be satisfied that we have exhausted our options. He still points to the room that opens if the third dimension is used. Current silicon technology is in its essence (with respect to the transistors) a planar technology. Why not use volumes, says Feynman, and put all books of the world in the space of a small dust particle? He may be right, but before assessing the chances of this happening, I would like to take you on a tour to review some of the possibilities and limitations of current planar silicon technology.

2.2 Tyranny at the Top

Yes, we do have plenty of room at the bottom. However, just a few years after Feynman's vision was published, J. Morton from Bell Laboratories noticed what he called the *tyranny of large systems*. This tyranny arises from the fact that scaling is, in general, not part of the laws of nature. For example, we know that one cannot hold larger and larger weights with a rope by making the rope thicker and thicker. At some point the weight of the rope itself comes into play, and things may get out of hand. As a corollary, why should one, without such difficulty, be able to make transistors smaller and smaller and, at the same time, integrate more of them on a chip? This is a crucial point that deserves some elaboration.

It is often said that all we need is to invent a new type of transistor that scales to atomic size. The question then arises: did the transistor, as invented in 1947, scale to the current microsize? The answer is no! The point-contact transistor, as it was invented by Bardeen and Brattain, was much smaller than a vacuum tube. However, its design was not suitable for aggressive scaling. The field-effect transistor, based on planar silicon technology and the hetero-junction interface of silicon and silicon dioxide with a metal on top (MOS technology), did much better in this respect. Nevertheless, it took the introduction of many new concepts (beginning with that of an inversion layer) to scale transistors to the current size. This scalability alone would still not have been sufficient to build large integrated systems on a chip. Each transistor develops heat when operated, and a large number of them may be better used as a soldering iron than for computing. The saving idea was to use both electron and hole-inversion layers to form the CMOS technology. The transistors of this technology create heat essentially only during switching operation, and heat generation during steady state is very small. A large system also requires interconnection of all transistors using metallic "wires." This becomes increasingly problematic when large numbers of transistors are involved, and many predictions have been made that it could not be done beyond a certain critical density of transistors. It turned out that there never was such a critical density for interconnection, and we will discuss the very interesting reason for this below. Remember that Feynman never talked about the tyranny at the top. He only was interested in fundamental limitations. The exponential growth of silicon technology with respect to the numbers of transistors on a chip seems to prove Feynman right, at least up to now. How can this be if the original transistors were not scalable? How could one always find a modification that permitted further scaling?

One of the reasons for continued miniaturization of silicon technology is that its basic idea is very flexible: use solids instead of vacuum tubes. The high density of solids permits us to create very small structures without hitting the atomic limit. Gas molecules or electrons in tubes have a much lower density than electrons or atoms in solids typically have. One has about 10^{18} atoms in a cm^3 of gas but 10^{23} in a cm^3 of a solid. Can one therefore go to sizes that would contain only a few hundred atoms with current silicon technology? I believe not. The reason is that current technology is based on the doping of silicon with donors and acceptors to create electron- and hole-inversion layers. The doping densities are much lower than the densities of atoms in a solid, usually below 10^{20} per cm^3. Therefore, to go to the ultimate limits of atomic size, a new type of transistor, without doping, is needed. We will discuss such possibilities below. But even if we have such transistors, can they be interconnected? Interestingly enough, interconnection problems have always been overcome in the past. The reason was that use of the third dimension has been made for interconnects. Chip designers have used the third dimension — not to overcome the limitations that two dimensions place on the number of transistors, but to overcome the limitations that two dimensions present for interconnecting the transistors. There is an increasing number of stacks of metal interconnect layers on chips — 2, 5, 8. How many can we have? (One can also still improve the conductivity of the metals in use by using, for example, copper technology.)

Pattern generation is, of course, key for producing the chips of silicon technology and represents another example of the tyranny of large systems. Chips are produced by using lithographic techniques. Masks that contain the desired pattern are placed above the chip material, which is coated with photosensitive layers that are exposed to light to engrave the pattern. As the feature sizes become smaller and smaller, the wavelength of the light needs to be reduced. The current work is performed in the extreme ultraviolet, and future scaling must overcome considerable obstacles. Why can one not use the atomic resolution of scanning tunneling microscopes? The reason is, of course, that the scanning process takes time; and this would make efficient chip production extremely difficult. One does need a process that works "in parallel" like photography. In principle there are many possibilities to achieve this, ranging from the use of X-rays to electron and ion beams and even self-organization of patterns in materials as known in chemistry and biology. One cannot see principal limitations here that would impede further scaling. However, efficiency and expense of production do represent considerable tyranny and make it difficult to predict what course the future will take. If use is made of the third dimension, however, optical lithography will go a long way.

Feynman suggested that there will be plenty of room at the bottom only when the third dimension is used. Can we also use it to improve the packing density of transistors? This is not going to be so easy. The current technology is based on a silicon surface that contains patterns of doping atoms and is topped by silicon dioxide. To use the third dimension, a generalization of the technology is needed. One would need another layer of silicon on top of the silicon dioxide, and so forth. Actually, such technology does already exist: silicon-on-insulator (SOI) technology. Interestingly enough, some devices that are currently heralded by major chip producers as devices of the future are SOI transistors. These may be scalable further than current devices and may open the horizon to the use of the third dimension. Will they open the way to unlimited growth of chip capacity? Well, there is still heat generation and other tyrannies that may prevent the basically unlimited possibilities that Feynman predicted. However, billions of dollars of business income have overcome most practical limitations (the tyranny) and may still do so for a long time to come. Asked how he accumulated his wealth, Arnold Beckman responded: "We built a pH-meter and sold it for three hundred dollars. Using this income, we built two and sold them for $600 … and then 4, 8, … ." This is, of course, the well-known story of the fast growth of a geometric series as known since ages for the rice corns on the chess board. Moore's law for the growth of silicon technology is probably just another such example and therefore a law of business rather than of science and engineering. No doubt, it is the business income that will determine the limitations of scaling to a large extent. But then, there are also new ideas.

2.3 New Forms of Switching and Storage

Many new types of transistors or switching devices have been investigated and even mass fabricated in the past decades. Discussions have focused on GaAs and III-V compound materials because of their special properties with respect to electron speed and the possibility of creating lattice-matched interfaces and layered patterns of atomic thickness. Silicon and silicon-dioxide have very different lattice constants (spacing between their atoms). It is therefore difficult to imagine that the interface between them can be electronically perfect. GaAs and AlAs on the other side have almost equal lattice spacing, and two crystals can be perfectly placed on top of each other. The formation of superlattices of such layers of semiconductors has, in fact, been one of the bigger achievements of recent semiconductor technology and was made possible by new techniques of crystal growth (molecular beam epitaxy, metal organic chemical vapor deposition, and the like). Quantum wells, wires, and dots have been the subject of extremely interesting research and have enriched quantum physics for example, by the discovery of the Quantum Hall Effect and the Fractional Quantum Hall effect. Use of such layers has also brought significant progress to semiconductor electronics. The concept of modulation doping (selective doping of layers, particularly involving pseudomorphic InGaAs) has led to modulation-doped transistors that hold the current speed records and are used for microwave applications. The removal of the doping to neighboring layers has permitted the creation of the highest possible electron mobilities and velocities. The effect of resonant tunneling has also been shown to lead to ultrafast devices and applications that reach to infrared frequencies, encompassing in this way both optics and electronics applications. When it comes to large-scale integration, however, the tyranny from the top has favored silicon technology. Silicon dioxide, as an insulator, is superior to all possible III-V compound materials; and its interface with silicon can be made electronically perfect enough, at least when treated with hydrogen or deuterium.

When it comes to optical applications, however, silicon is inefficient because it is an indirect semiconductor and therefore cannot emit light efficiently. Light generation may be possible by using silicon. However, this is limited by the laws of physics and materials science. It is my guess that silicon will have only limited applications for optics, much as III-V compounds have for large-scale integrated electronics. III-V compounds and quantum well layers have been successfully used to create efficient light-emitting devices including light-emitting and semiconductor laser diodes. These are ubiquitous in every household, e.g., in CD players and in the back-lights of cars. New forms of laser diodes, such as the so-called vertical cavity surface emitting laser diodes (VCSELs), are even suitable to relatively large integration. One can put thousands and even millions of them on a chip. Optical pattern generation has made great advances by use of selective superlattice intermixing (compositionally disordered III-V compounds and superlattices have a different index of refraction) and by other methods. This is an area in great flux and with many possibilities for miniaturization. Layered semiconductors and quantum well structures have also led to new forms of lasers such as the quantum cascade laser. Feynman mentioned in his paper the use of layered materials. What would he predict for the limits of optical integration and the use of quantum effects due to size quantization in optoelectronics?

A number of ideas are in discussion for new forms of ultrasmall electronic switching and storage devices. Using the simple fact that it takes a finite energy to bring a single electron from one capacitor plate to the other (and using tunneling for doing so), single-electron transistors have been proposed and built. The energy for this single-electron switching process is inversely proportional to the area of the capacitor. To achieve energies that are larger than the thermal energy at room temperature (necessary for robust operation), extremely small capacitors are needed. The required feature sizes are of the order of one nanometer. There are also staggering requirements for material purity and perfection since singly charged defects will perturb operation. Nevertheless, Feynman may have liked this device because the limitations for its use are not due to physical principles. It also has been shown that memory cells storing only a few electrons do have some very attractive features. For example, if many electrons are stored in a larger volume, a single material defect can lead to unwanted discharge of the whole volume. If, on the other hand, all these electrons are stored in a larger number of quantum dots (each carrying few electrons), a single defect can discharge only a single dot, and the remainder of the stored charge stays intact.

Two electrons stored on a square-shaped "quantum dot" have been proposed as a switching element by researchers at Notre Dame. The electrons start residing in a pair of opposite corners of the square and are switched to the other opposite corner. This switching can be effected by the electrons residing in a neighboring rectangular dot. Domino-type effects can thus be achieved. It has been shown that architectures of cellular neural networks (CNNs) can be created that way as discussed briefly below.

A new field referred to as *spintronics* is developing around the spin properties of particles. Spin properties have not been explored in conventional electronics and enter only indirectly, through the Pauli principle, into the equations for transistors. Of particular interest in this new area are particle pairs that exhibit quantum entanglement. Consider a pair of particles in a singlet spin-state sent out to detectors or spin analyzers in opposite directions. Such a pair has the following remarkable properties: measurements of the spin on each side separately give random values of the spin (up/down). However, the spin of one side is always correlated to the spin on the other side. If one is up, the other is down. If the spin analyzers are rotated relative to each other, then the result for the spin pair correlation shows rotational symmetry. A theorem of Bell proclaims such results incompatible with Einstein's relativity and suggests the necessity of instantaneous influences at a distance. Such influences do not exist in classical information theory and are therefore considered a quantum addition to classical information. This quantum addition provides part of the novelty that is claimed for possible future quantum computers. Spintronics and entanglement are therefore thought to open new horizons for computing.

Still other new device types use the wave-like nature of electrons and the possibility to guide these waves by externally controllable potential profiles. All of these devices are sensitive to temperature and defects, and it is not clear whether they will be practical. However, new forms of architectures may open new possibilities that circumvent the difficulties.

2.4   New Architectures

Transistors of the current technology have been developed and adjusted to accommodate the tyranny from the top, in particular the demands set forth by the von Neuman architecture of conventional computers. It is therefore not surprising that new devices are always looked at with suspicion by design engineers and are always found wanting with respect to some tyrannical requirement. Many regard it extremely unlikely that a completely new device will be used for silicon chip technology. Therefore, architectures that deviate from von Neuman's principles have received increasing attention. These architectures invariably involve some form of parallelism. Switching and storage is not localized to a single transistor or small circuit. The devices are connected to each other, and their collective interactions are the basis for computation. It has been shown that such collective interactions can perform some tasks in ways much superior to von Neuman's sequential processing.

One example for such new principles is the cellular neural network (CNN) type of architectures. Each cell is connected by a certain coupling constant to its nearest neighbors, and after interaction with each other, a large number of cells settle on a solution that hopefully is the desired solution of a problem that cannot easily be done with conventional sequential computation. This is, of course, very similar to the advantages of parallel computation (computation by use of more than one processor) with the difference that it is not processors that interact and compute in parallel but the constituent devices themselves. CNNs have advantageously been used for image processing and other specialized applications and can be implemented in silicon technology. It appears that CNNs formed by using new devices, such as the coupled square quantum dots discussed above, could (at least in principle) be embedded into a conventional chip environment to perform a certain desired task; and new devices could be used that way in connection with conventional technology. There are at least three big obstacles that need to be overcome if this goal should be achieved. The biggest problem is posed by the desire to operate at room temperature. As discussed above, this frequently is equivalent to the requirement that the single elements of the CNN need to be extremely small, on the order of one nanometer. This presents the second problem — to create such feature sizes by a lithographic process. Third, each element of the CNN needs to be virtually perfect and free of defects that would impede its operation. Can one create such a CNN by the organizing and self-organizing principles of chemistry on semiconductor surfaces? As Dirac once said (in connection with difficult problems), "one must try." Of course, it will be tried only if an important problem exists that defies conventional solution. An example would be the cryptographically important problem of factorizing large numbers. It has been shown that this problem may find a solution through quantum computation.

The idea of quantum computation has, up to now, mainly received the attention of theoreticians who have shown the superior power of certain algorithms that are based on a few quantum principles. One such principle is the unitarity of certain operators in quantum mechanics that forms a solid basis for the possibility of quantum computing. Beyond this, it is claimed that the number of elements of the set of parameters that constitutes quantum information is much larger than the comparable set used in all of classical information. This means there are additional quantum bits (qubits) of information that are not covered by the known classical bits. In simpler words, there are instantaneous action at a distance and connected phenomena, such as quantum teleportation, that have not been used classically but can be used in future quantum information processing and computation. These claims are invariably based on the theorem of Bell and are therefore subject to some criticism. It is well known that the Bell theorem has certain loopholes that can be closed only if certain time dependencies of the involved parameters are excluded. This means that even if the Bell theorem were general otherwise, it does not cover the full classical parameter space. How can one then draw conclusions about the number of elements in parameter sets for classical and quantum information? In addition, recent work has shown that the Bell theorem excludes practically all time-related parameters — not only those discussed in the well-known loopholes. What I want to say here is that the very advanced topic of quantum information complexity will need further discussion even about its foundations. Beyond this, obstacles exist for implementation of qubits due to the tyranny from the top. It is necessary to have a reasonably large number of qubits in order to implement the quantum computing algorithms and make them applicable to large problems. All of these qubits need to be connected in a quantum mechanical coherent way. Up to now, this coherence has always necessitated the use of extremely low temperatures, at least when electronics (as opposed to optics) is the basis for implementation. With all these difficulties, however, it is clear that there are great opportunities for solving problems of new magnitude by harnessing the quantum world.

2.5   How Does Nature Do It?

Feynman noticed that nature has already made use of nanostructures in biological systems with greatest success. Why do we not copy nature? Take, for example, biological ion channels. These are tiny pores formed by protein structures. Their opening can be as small as a few one-tenths of a nanometer. Ion currents are controlled by these pores that have opening and closing gates much as transistors have. The on/off current ratio of ion channels is practically infinite, which is a very desirable property for large systems. Remember that we do not want energy dissipation when the system is off. Transistors do not come close to an infinite on/off ratio, which represents a big design problem. How do the ion channels do it? The various gating mechanisms are not exactly understood, but they probably involve changes in the aperture of the pore by electrochemical mechanisms. Ion channels do not only switch currents perfectly. They also can choose the type of ions they let through and the type they do not. Channels perform in this way a multitude of functions. They regulate our heart rate, kill bacteria and cancer cells, and discharge and recharge biological neural networks, thus forming elements of logic and computation. The multitude of functions may be a great cure for some of the tyranny from the top as Jack Morton has pointed out in his essay "From Physics to Function." No doubt, we can learn in this respect by copying nature. Of course, proteins are not entirely ideal materials when it comes to building a computer within the limits of a preconceived technology. However, nature does have an inexpensive way of pattern formation and replication — a self-organizing way. This again may be something to copy. If we cannot produce chip patterns down to nanometer size by inexpensive photographic means, why not produce them by methods of self-organization? Can one make ion channels out of materials other than proteins that compare more closely to the solid-state materials of chip technology? Perhaps carbon nanotubes can be used. Material science has certainly shown great inventiveness in the past decades.

Nature also has no problems in using all three dimensions of space for applying its nanostructures. Self-organization is not limited to a plane as photography is. Feynman's ultimate frontier of using three dimensions for information storage is automatically included in some biological systems such as, for example, neural networks. The large capacity and intricate capability of the human brain derives, of course, from this fact.

The multitude of nanostructure functionalities in nature is made possible because nature is not limited by disciplinary boundaries. It uses everything, whether physics or chemistry, mechanics or electronics

— and yes, nature also uses optics, e.g., to harvest energy from the sun. I have not covered nanometer-size mechanical functionality because I have no research record in this area. However, great advances are currently made in the area of nanoelectromechanical systems (NEMS). It is no problem any more to pick up and drop atoms, or even to rotate molecules. Feynman's challenge has been far surpassed in the mechanical area, and even his wildest dreams have long since become reality. Medical applications, such as the insertion of small machinery to repair arteries, are commonplace. As we understand nature better, we will not only be able to find new medical applications but may even improve nature by use of special smart materials for our bodies. Optics, electronics, and mechanics, physics, chemistry, and biology need to merge to form generations of nanostructure technologies for a multitude of applications.

However, an area exists in which man-made chips excel and are superior to natural systems (if man-made is not counted as natural). This area relates to processing speed. The mere speed of a number-crunching machine is unthinkable for the workings of a biological neural network. To be sure, nature has developed fast processing; visual evaluations of dangerous situations and recognition of vital patterns are performed with lightening speed by some parallel processing of biological neural networks. However, when it comes to the raw speed of converting numbers, which can also be used for alphabetical ordering and for a multitude of algorithms, man-made chips are unequaled. Algorithmic speed and variability is a very desirable property, as we know from browsing the Internet, and represents a great achievement in chip technology.

Can we have both —the algorithmic speed and variability of semiconductor-based processors and, at the same time, three-dimensional implementations and the multitude of functionality as nature features it in her nanostructure designs? I would not dare to guess an answer to this question. The difficulties are staggering! Processing speed seems invariably connected to heat generation. Cooling becomes increasingly difficult when three-dimensional systems are involved and the heat generation intensifies.

But then, there are always new ideas, new materials, new devices, new architectures, and altogether new horizons. Feynman's question as to whether one can put the *Encyclopaedia Britannica* on the head of a pin has been answered in the affirmative. We have proceeded to the ability to sift through the material and process the material of the encyclopedia with lightning speed. We now address the question of whether we can process the information of three-dimensional images within the shortest of times, whether we can store all the knowledge of the world in the smallest of volumes and browse through gigabits of it in a second. We also proceed to the question of whether mechanical and optical functionality can be achieved on such a small scale and with the highest speed. Nature has shown that the smallest spatial scales are possible. We have to search for the greatest variety in functionality and for the highest possible speed in our quest to proceed in science from what is possible in principle to a function that is desirable for humanity.

Section 2

Molecular and Nano-Electronics: Concepts, Challenges, and Designs

3

Engineering Challenges in Molecular Electronics

CONTENTS

Gregory N. Parsons

North Carolina State University

Abstract

This article discusses molecular electronics in the context of silicon materials and device engineering. Historic trends in silicon devices — and engineering scaling rules that dictate these trends — give insight into how silicon has become so dominant an electronic material and how difficult it will be to challenge silicon devices with a distinctly different *leapfrog* technology. Molecular electronics presents some intriguing opportunities, and it is likely that these opportunities will be achieved through hybrid silicon/molecular devices that incorporate beneficial aspects of both material systems.

3.1   Introduction

Manufacturing practices for complementary metal oxide semiconductor (CMOS) devices are arguably the most demanding, well developed, and lucrative in history. Even so, it is well recognized that historic trends in device scaling that have continued since the 1960s are going to face serious challenges in the next several years. Current trends in Moore's Law scaling are elucidated in detail in the Semiconductor

Industry Association's *The International Technology Roadmap for Semiconductors.*[1] The 2001 roadmap highlights significant fundamental barriers in patterning, front-end processes, device structure and design, test equipment, interconnect technology, integration, assembly and packaging, etc.; and there are significant industry and academia research efforts focused on these challenges. There is also significant growing interest in potential leapfrog technologies, including quantum-based structures and molecular electronics, as possible means to redefine electronic device and system operation. The attention (and research funds) applied to potential revolutionary technologies is small compared with industrial efforts on silicon. This is primarily because of the tremendous manufacturing infrastructure built for silicon technology and the fact that there is still significant room for device performance improvements in silicon — even though many of the challenges described in the roadmap still have "no known solution." Through continued research in leapfrog approaches, new materials and techniques are being developed that could significantly impact electronic device manufacturing. However, such transitions are not likely to be realized in manufacturing without improved insight into the engineering of current high-performance electronic devices.

Silicon devices are highly organized inorganic structures designed for electronic charge and energy transduction.[2-4] Organic molecules are also highly organized structures that have well-defined electronic states and distinct (although not yet well-defined) electronic interactions within and among themselves. The potential for extremely high device density and simplified device fabrication has attracted attention to the possibility of using individual molecules for advanced electronic devices (see recent articles by Ratner;[5] Kwok and Ellenbogen;[6] and Wada[7]). A goal of molecular electronics is to use fundamental molecular-scale electronic behavior to achieve electronic systems (with functional logic and/or memory) composed of individual molecular devices. As the field of molecular electronics progresses, it is important to recognize that current silicon circuits are likely the most highly engineered systems in history, and insight into the engineering driving forces in silicon technology is critical if one wishes to build devices more advanced than silicon. The purpose of this article, therefore, is to give a brief overview of current semiconductor device operation, including discussion of the strengths and weaknesses of current devices and, within the context of current silicon device engineering, to present and discuss possible routes for molecular electronics to make an impact on advanced electronics engineering and technology.

3.2 Silicon-Based Electrical Devices and Logic Circuits

3.2.1 Two-Terminal Diode and Negative Differential Resistance Devices

The most simple silicon-based solid-state electronic device is the p/n junction diode, where the current through the two terminals is small in the reverse direction and depends exponentially on the applied voltage in the forward direction. Such devices have wide-ranging applications as rectifiers and can be used to fabricate memory and simple logic gates.[8,9] A variation on the p/n diode is a resonant tunneling diode (RTD) where well-defined quantum states give rise to negative differential resistance (NDR). A schematic current vs. voltage trace for an NDR device is shown in Figure 3.1. Such devices can be made with inorganic semiconductor materials and have been integrated with silicon transistors[10-13] for logic devices with multiple output states to enhance computation complexity.

An example circuit for an NDR device with a load resistor is shown in Figure 3.1. This circuit can act as a switch, where V_{out} is determined by the relative voltage drop across the resistor and the diode. The resistance of the diode is switched from high to low by applying a short voltage pulse in excess of V_{dd} across the series resistor and diode, and the smaller resistance results in a small V_{out}. These switching circuits may be useful for molecular logic gates using RTD molecules, but several important issues need to be considered for applications involving two-terminal logic. One concern is the size of the output impedance. If the outlet voltage node is connected to a resistance that is too small (i.e., similar in magnitude to the RTD impedance), then the outlet voltage (and voltage across the RTD) will shift from the expected value; and this error will propagate through the circuit network. Another concern is that full logic gates fabricated with RTDs require an additional clock signal, derived from a controlled oscillator

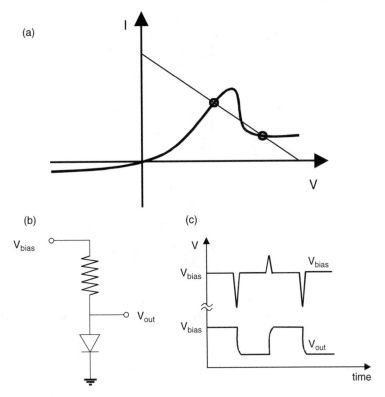

FIGURE 3.1 Schematic of one possible NDR device. (a) Schematic current vs. voltage curve for a generic resonant tunneling diode (RTD) showing negative differential resistance (NDR). The straight line is the resistance load line, and the two points correspond to the two stable operating points of the circuit. (b) A simple circuit showing an RTD loaded with a resistor. (c) Switching behavior of resistor/RTD circuit. A decrease in V_{bias} leads to switching of the RTD device from high to low impedance, resulting in a change in V_{out} from high to low state.

circuit. Such oscillators are readily fabricated using switching devices with gain, but to date they have not been demonstrated with molecular devices. Possibly the most serious concern is the issue of power dissipation. During operation, the current flows continuously through the RTD device, producing significant amounts of thermal energy that must be dissipated. As discussed below in detail, power dissipation in integrated circuits is a long-standing problem in silicon technology, and methodologies to limit power in molecular circuits will be a critical concern for advanced high-density devices.

3.2.2 Three-Terminal Bipolar, MOS, and CMOS Devices

The earliest solid-state electronic switches were bipolar transistors, which in their most simple form consisted of two back-to-back p/n junctions. The devices were essentially solid-state analogs of vacuum tube devices, where a current on a base (or grid) electrode modulated the current between the emitter and collector contacts. Because a small change in the base voltage, for example, could enable a large change in the collector current, the transistor enabled signal amplification (similar to a vacuum tube device) and, therefore, current or voltage gain. In the 1970s, to reduce manufacturing costs and increase integration capability, industry moved away from bipolar and toward metal-oxide-semiconductor field effect transistor (MOSFET) structures, shown schematically in Figure 3.2. For MOSFET device operation, voltage applied to the gate electrode produces an electric field in the semiconductor, attracting charge to the silicon/dielectric interface. A separate voltage applied between the source and drain then enables current to flow to the drain in a direction perpendicular to the applied gate field. Device geometry is determined by the need for the field in the channel to be determined primarily by the gate voltage and

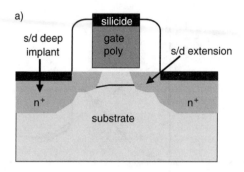

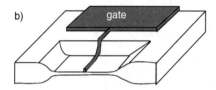

FIGURE 3.2   (a) Cross section of a conventional MOS transistor. (b) A three-dimensional representation of a MOS transistor layout. Two transistors, one NMOS and one PMOS, can be combined to form a complementary MOS (CMOS) device.

not by the voltage between the source and drain. In this structure, current flow to or from the gate electrode is limited by leakage through the gate dielectric. MOS devices can be either NMOS or PMOS, depending on the channel doping type (p- or n-type, respectively) and the charge type (electrons or holes, respectively) flowing in the inversion layer channel. Pairing of individual NMOS and PMOS transistors results in a complementary MOS (CMOS) circuit.

3.2.3 Basic Three-Terminal Logic Circuits

A basic building block of MOS logic circuits is the signal inverter, shown schematically in Figure 3.3. Logic elements, including, for example, NOR and NAND gates, can be constructed using inverters with multiple inputs in parallel or in series. Early MOS circuits utilized single-transistor elements to perform the inversion function utilizing a load resistor as shown in Figure 3.3a. In this case, when the NMOS is off (V_{in} is less than the device threshold voltage V_{th}), the supply voltage (V_{dd}) is measured at the outlet. When V_{in} is increased above V_{th}, the NMOS turns on and V_{dd} is now dropped across the load resistor; V_{out} is now in common with ground, and the signal at V_{out} is inverted relative to V_{in}. The same behavior is observed in enhancement/depletion mode circuits (Figure 3.3b) where the load resistor is replaced with another NMOS device. During operation of these NMOS circuits, current is maintained between V_{dd} and ground in either the high- or low-output state. CMOS circuits, on the other hand, involve combinations of NMOS and PMOS devices and result in significantly reduced power consumption as compared with NMOS-only circuits. This can be seen by examining a CMOS inverter structure as shown in Figure 3.3c. A positive input voltage turns on the NMOS device, allowing charge to flow from the output capacitance load to ground and producing a low V_{out}. A low-input voltage likewise enables the PMOS to turn on, and the output to go to the level of the supply voltage, V_{dd}. During switching, current is required to charge and discharge the channel capacitances, but current stops flowing when the channel and output capacitances are fully charged or discharged (i.e., when V_{out} reaches 0 or V_{dd}). In this way, during its static state, one of the two transistors is always off, blocking current from V_{dd} to ground. This means that the majority of the power consumed in an array of these devices is determined by the rate of switching and not by the number of inverters in the high- or low-output state within the array. This is a tremendously important outcome of the transition in silicon technology from NMOS to CMOS: *the*

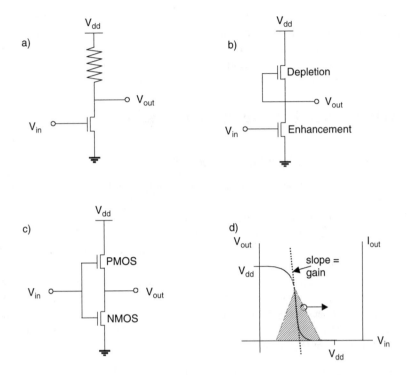

FIGURE 3.3 Inverter circuit approaches for NMOS and CMOS devices. (a) A simple inverter formed using an NMOS device and a load resistor, commonly used in the 1970s. (b) An inverter formed using an NMOS device and a depletion-mode load transistor. (c) Complementary MOS (CMOS) inverter using an NMOS and a PMOS transistor, commonly used since the 1980s. (d) A schematic voltage inverter trace for a CMOS inverter. The slope of the V_{out} vs. V_{in} gives the inverter gain. Also shown is the net current through the device as a function of V_{in}. Note that for the CMOS structure, current flows through the inverter only during the switching cycle, and no current flows during steady-state operation.

power produced per unit area of a CMOS chip can be maintained nearly constant as the number density of individual devices on the chip increases. The implications for this in terms of realistic engineered molecular electronic systems will be discussed in more detail below.

3.2.4   The Importance of Gain

Gain in an electronic circuit is generally defined as the ratio of output voltage change to input voltage change (i.e., voltage gain) or ratio of output current change to input current change (current gain). For a simple inverter circuit, therefore, the voltage gain is the slope of the V_{out} vs. V_{in} curve, and the maximum gain corresponds to the value of the maximum slope. A voltage gain in excess of one indicates that if a small-amplitude voltage oscillation is placed on the input (with an appropriate dc bias), a larger oscillating voltage (of opposite phase) will be produced at the output. Typical silicon devices produce gain values of several hundred. Power and voltage gain are the fundamental principles behind amplifier circuits used in common electronic systems, such as radios and telephones.

Gain is also critically important in any electronic device (such as a microprocessor) where voltage or current signals propagate through a circuit. Without gain, the total output power of any circuit element will necessarily be less than the input; the signal would be attenuated as it moved through the circuit, and eventually *high* and *low* states could not be differentiated. Because circuit elements in silicon technology produce gain, the output signal from any element is "boosted" back up to its original input value; and the signal can progress without attenuation.

In any system, if the power at the output is to be greater than the power at an input reference, then an additional input signal will be required. Therefore, at least three contact terminals are required to achieve gain in any electronic element: high voltage (or current) input, small voltage or current input, and large signal output. The goal of demonstrating molecular structures with three terminals that produce gain continues to be an important challenge.

3.3   CMOS Device Parameters and Scaling

3.3.1   Mobility and Subthreshold Slope

The speed of a circuit is determined by how fast a circuit output node is charged up to its final state. This is determined by the transistor drive current, which is related to the device dimension and the effective charge mobility, μ_{eff} (or transconductance), where mobility is defined as charge velocity per unit field. In a transistor operating under sufficiently low voltages, the velocity will increase in proportion to the lateral field (V_{sd}/L_{ch}), where V_{sd} is the source/drain voltage and L_{ch} is the effective length of the channel. At higher voltages, the velocity will saturate, and the lateral field becomes nonuniform. This *saturation velocity* is generally avoided in CMOS circuits but becomes more problematic for very short device lengths. Saturation velocity for electrons in silicon at room temperature is near 10^7cm/s, and it is slightly smaller for holes. For a transistor operating with low voltages (i.e., in the linear regime), the mobility is a function of the gate and source/drain voltages (V_{gs} and V_{sd}), the channel length and width (L_{ch} and W_{ch}) and the gate capacitance per unit area (C_{ox}; Farads/cm^2):

$$\mu_{eff} = \frac{I_{sd}L_{ch}}{C_{ox}W_{ch}(V_{gs} - V_{th})V_{sd}} \tag{3.1}$$

The mobility parameter is independent of device geometry and is related to the current through the device. Because the current determines the rate at which logic signals can move through the circuit, the effective mobility is an important figure of merit for any electronic device.

Another important consideration in device performance is the subthreshold slope, defined as the inverse slope of the log (I_{sd}) vs. V_g curve for voltages below V_{th}. A typical current vs. voltage curve for an NMOS device is shown in Figure 3.4. In the subthreshold region, the current flow is exponentially dependent on voltage:

$$I_{sd} \propto \exp(qV/nkT) \tag{3.2}$$

where n is a number typically greater than 1. At room temperature, the ideal case (i.e., minimal charge scattering and interface charge trapping) results in n = 1 and an inverse slope $(2.3 \cdot kT)/q \approx 60$mV/decade. The subthreshold slope of a MOS device is a measure of the rate at which charge diffuses from the channel region when the device turns off. Because the rate of charge diffusion does not change with device dimension, the subthreshold slope will not change appreciably as transistor size decreases. The nonscaling of subthreshold slope has significant implications for device scaling limitations. Specifically, it puts a limit on how much V_{th} can be reduced because the current at zero volts must be maintained low to control off-state current, I_{off}. This in turn puts a limit on how much the gate voltage can be decreased, which means that the ideal desired constant field scaling laws, described below, cannot be precisely followed.

Neither the mobility nor subthreshold slope is significantly affected by reduction in transistor size. Under ideal constant field scaling, the channel width, length, and thickness all decrease by the same factor (κ). So how is it that smaller transistors with less current are able to produce faster circuits? The speed of a transistor circuit is determined by the rate at which a logic element (such as an inverter) can change from one state to another. This switching requires the charging or discharging of a capacitor at the output node (i.e., another device in the circuit). Because capacitance $C = (W_{ch} \cdot L_{ch} \cdot \varepsilon\varepsilon_o)/t_{ox}$, the total gate capacitance will decrease by the scaling factor κ. The charge required to charge a capacitor is Q = CV, which

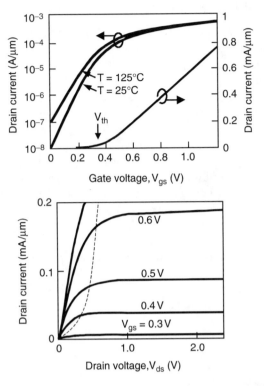

FIGURE 3.4 Operating characteristics for a typical MOS device. (a) Current measured at the drain as a function of voltage applied to the gate. The same data is plotted on linear and logarithmic scales. An extrapolation of the linear data is an estimate of the threshold voltage V_{th}. The slope of the data on the logarithmic scale for $V < V_{th}$ gives the subthreshold slope. Note that the device remains operational at temperatures >100°C, with an increase in inverse subthreshold slope and an increase in off-current (I_{off}) as temperature is increased. (b) Current measured at the drain as a function of source/drain voltage for various values of gate voltage for the same device as in (a). The dashed line indicates the transition to current saturation.

means that a reduction in C and a reduction in V by the factor κ results in a decrease in the total charge required by a factor of κ^2. Therefore, the decrease in current by κ still results in an increase in charging rate by the scaling factor κ, enabling the circuit speed to increase by a similar factor.

Another way to think about this is to directly calculate the time that it takes to charge the circuit output node. The charging rate of a capacitor is:

$$\frac{\partial V}{\partial t} = \frac{I}{C} \tag{3.3}$$

where C is the total capacitance in Farads. Because I and C both decrease with size, the charging rate is also independent of size. However, because the smaller device will operate at a smaller voltage, then the time that it takes to charge will decrease (and the circuit will become faster) as size decreases. An expression for charging time is obtained by integrating Equation 3.3, where $V = V_{sd}$, $I = I_{sd}$, and $C = (C_{ox} \cdot L_{ch} \cdot W_{ch})$; and substituting in for current from the mobility expression (Equation 3.1):

$$t = \frac{L_{ch}^2}{\mu_{eff}(V_{gs} - V_{th})} \tag{3.4}$$

This shows that as V_{gs} and L_{ch} decrease by κ, the charging time will also decrease by the same factor, leading to an increase in circuit speed.

3.3.2 Constant Field Scaling and Power Dissipation

As discussed in the SIA Roadmap, significant attention is currently paid to issues in front-end silicon processing. One of the most demanding challenges is new dielectric materials to replace silicon dioxide to achieve high gate capacitance with low gate leakage.[14] The need for higher capacitance with low leakage is driven primarily by the need for improved device speed while maintaining low power operation. Power has been an overriding challenge in MOS technologies since the early days of electronics, long before the relatively recent interest in portable systems. For example, reduced power consumption was one of the important problems that drove replacement of vacuum tubes with solid-state electronics in the 1950s and 1960s. Low power is required primarily to control heat dissipation, since device heating can significantly reduce device performance (especially if the device gets hot enough to melt). For current technology devices, dissipation of several watts of heat from a logic chip can be achieved using air cooling; and inexpensive polymer-based packaging approaches can be used. The increase in portable electronics has further increased the focus on problems associated with low power operation.

There are several possible approaches to consider when scaling electronic devices, including constant field and constant voltage scaling. Because of the critical need for low heat generation and power dissipation, current trends in shrinking CMOS transistor design are based on the rules of constant field scaling. Constant field scaling is an idealized set of scaling rules devised to enable device dimensions to decrease while output power density remains fixed. As discussed briefly above, constant field scaling cannot be precisely achieved, primarily because of nonscaling of the voltage threshold. Therefore, modifications in the ideal constant field scaling rules are made as needed to optimize device performance. Even so, the trends of constant field scaling give important insight into engineering challenges facing any advanced electronic device (including molecular circuits); therefore, the rules of constant field scaling are discussed here.

Heat generation in a circuit is related to the product of current and voltage. In a circuit operating at frequency f, the current needed to charge a capacitor C_T in half a cycle time is $(f \cdot C_T \cdot V_{dd})/2$, where V_{dd} is the applied voltage. Therefore, the power dissipated in one full cycle of a CMOS switching event (the *dynamic power*) is proportional to the total capacitance that must be charged or discharged during the switching cycle, C_T (i.e., the capacitance of the output node of the logic gate), the power supply voltage squared, V_{dd}^2, and the operation frequency, f:

$$P_{switch} = f C_T V_{dd}^2 \qquad (3.5)$$

The capacitance of the output node is typically an input node of another logic gate (i.e., the gate of another CMOS transistor). If a chip has $\sim 10^8$ transistors/cm^2, a gate length of ~ 130 nm, and width/length ratio of 3:1, then the total gate area is $\sim 5\%$ of the chip area. Therefore, the total power consumed by a chip is related to the capacitance density of an individual transistor gate ($C/A = \varepsilon \varepsilon_o / t_{ox}$).

$$\frac{P_{switch}}{A_{gate}} = \frac{f \varepsilon \varepsilon_O V_{dd}^2}{t_{ox}}$$

$$\frac{P_{chip}}{A_{chip}} \approx 0.05 \left(\frac{P_{switch}}{A_{gate}} \right) \qquad (3.6)$$

where t_{ox} is the thickness and ε is the dielectric constant of the gate insulator. This is a highly simplified analysis (a more complete discussion is given in Reference 15). It does not include the power loss associated with charging and discharging the interconnect capacitances, and it does not include the fact that in CMOS circuits, both NMOS and PMOS transistors are partially on for a short time during the switching transition, resulting in a small current flow directly to ground, contributing to additional power loss. It also assumes that there is no leakage through the gate dielectric and that current flow in the off state is negligible. Gate leakage becomes a serious concern as dielectric thickness decreases and tunneling increases, and off-state leakage becomes more serious as the threshold voltage decreases. These two processes result in an additional *standby power* term ($P_{standby} = I_{off} \cdot V_{dd}$) that must be added to the power loss analysis.

Even with these simplifications, the above equation can be used to give a rough estimate of power consumption by a CMOS chip. For example, for a 1GHz chip with V_{dd} of ~1.5V and gate dielectric thickness of 2 nm, the above equation results in ~150 W/cm², which is within a factor 3 of the ~50W/cm² dissipated for a 1GHz chip.[16] The calculation assumes that all the devices are switching on each cycle. Usually, only a fraction of the devices will switch per cycle, and the chip will operate with maximum output for only short periods, so the total power output will be less than the value calculated. A more complete calculation must also consider additional capacitances related to fan-out and loading of interconnect transmission lines, which will lead to power dissipation in addition to that calculated above. It is clear that heat dissipation problems in high-density devices are significant, and most high-performance processors require forced convection cooling to maintain operating temperatures within the maximum operation range of 70–80°C.[16]

The above relation for power consumption (Equation 3.5) indicates that in order to maintain power density, increasing frequency requires a scaled reduction in supply voltage. The drive to reduce V_{dd} in turn leads to significant challenges in channel and contact engineering. For example, the oxide thickness must decrease to maintain sufficient charge density in the channel region, but it must not allow significant gate leakage (hence the drive toward high dielectric constant insulators). Probably the most challenging problem in V_{dd} reduction is engineering of the device threshold voltage, V_{th}, which is the voltage at which the carrier velocity approaches saturation and the device turns on. If V_{th} is too small, then there is significant off-state leakage; and if it is too close to V_{dd}, then circuit delay becomes more problematic.

The primary strength of silicon device engineering over the past 20 years has been its ability to meet the challenges of power dissipation, enabling significant increases in device density and speed while controlling the temperature increases associated with packing more devices into a smaller area. To realize viable molecular electronic devices and systems, technologies for low-power device operation and techniques that enable power-conscious scaling methodologies must be developed. Discussion of power dissipation in molecular devices — and estimations of power dissipation in molecular circuits in comparison with silicon — have not been widely discussed, but are presented in detail below.

The steady-state operating temperature at the surface of a chip can be roughly estimated from Fourier's law of heat conduction:

$$Q = U \, \Delta T \tag{3.7}$$

where Q is the power dissipation per unit area, U is the overall heat transfer coefficient, and ΔT is the expected temperature rise in the system. The heat generated per unit area of the chip is usually transferred by conduction to a larger area where it is dissipated by convection. If a chip is generating a net ~100mW/cm² and cooling is achieved by natural convection (i.e., no fan), then U~20W/(m²K),[17] and a temperature rise of 50°C can be expected at the chip surface. (Many high-performance laptops are now issued with warnings regarding possible burns from contacting the hot casing surface.) A fan will increase U to 50W/(m²K) or higher. Because constant field scaling cannot be precisely achieved in CMOS, the power dissipation in silicon chips is expected to increase as speed increases; and there is significant effort under way to address challenges specific to heat generation and dissipation in silicon device engineering. Organic materials will be much more sensitive to temperature than current inorganic electronics, so *the ability to control power consumption and heat generation will be one of the overriding challenges that must be addressed to achieve viable high-density and high-speed molecular electronic systems.*

3.3.3 Interconnects and Parasitics

For a given supply voltage, the signal delay in a CMOS circuit is determined by the charge mobility in the channel and the capacitance of the switch. Several other *parasitic* resistance and capacitance elements can act to impede signal transfer in the circuit. The delay time of a signal moving through a circuit (τ) is given by the product of the circuit resistance and capacitance: $\tau = RC$, and parasitic elements add resistance and capacitance on top of the intrinsic R and C in the circuit. Several parasitic resistances and capacitances exist within the MOS structure itself, including contact resistance, source/drain and "spreading" resistance, gate/source overlap capacitance, and several others.

Also important are the resistance and capacitance associated with the lines that connect one circuit to another. These interconnects can be local (between devices located close to each other on the chip) or global (between elements across the length of the chip). As devices shrink, there are significant challenges in interconnect scaling. Local interconnects generally scale by decreasing wire thickness and decreasing (by the same factor) the distance between the wires. When the wires are close enough that the capacitance between neighboring wires is important (as it is in most devices), the total capacitance per unit length C_L does not depend on the scaling factor. Because the resistance per unit length R_L increases as the wire diameter squared and the wire length decreases by L, then the delay time $\tau \sim R_L C_L L^2$ is not changed by device scaling. Moreover, using typical materials (Cu and lower-k dielectrics) in current device generations, local interconnect RC delay times do not significantly affect device speed. In this way, local interconnect signal transfer rates benefit from the decreasing length of the local interconnect lines. Global interconnects, on the other hand, generally increase in length as devices shrink due to increasing chip sizes and larger numbers of circuits per chip. This leads to significant signal delay issues across the chip. Solving this problem requires advanced circuit designs to minimize long interconnects and to reduce R_L and C_L by advanced materials such as high-conductivity metals and low dielectric constant insulators. This also implies that if sufficient function could be built into very small chips (much less than a few centimeters), using molecular components for example, issues of signal delay in global interconnects could become less of a critical issue in chip operation. However, sufficient current is needed in any network structure to charge the interconnecting transfer line. If designs for molecular devices focus on low current operation, then there may not be sufficient currents to charge the interconnect in the cycle times needed for ultrafast operation.

In addition to circuit performance, another concern for parasitic resistance and capacitance is in structures developed for advanced device testing. This will be particularly important as new test structures are developed that can characterize small numbers of molecules. As device elements decrease in size, intrinsic capacitances will increase. If the test structure contains any small parasitic capacitance in series or large parasitic capacitance in parallel with the device under test, the parasitic can dominate the signal measured. Moreover, in addition to difficulties associated with small current measurements, there are some significant problems associated with measuring large-capacitance devices (including ultrathin dielectric films), where substantial signal coupling between the device and the lead wires, for example, can give rise to spurious parasitic-related results.

3.3.4 Reliability

A hallmark attribute of solid-state device technology recognized in the 1940s was that of reliability. Personal computer crashes may be common (mostly due to software problems), but seldom does a PC processor chip fail before it is upgraded. Mainframe systems, widely used in finance, business, and government applications, have reliability requirements that are much more demanding than PCs; and current silicon technology is engineered to meet those demands. One of the most important modes of failure in CMOS devices is gate dielectric breakdown.[14] Detailed mechanisms associated with dielectric breakdown are still debated and heavily studied, but most researchers agree that charge transport through the oxide, which occurs in very small amounts during operation, helps create defects which eventually create a shorting path (breakdown) across the oxide. Defect generation is also enhanced by other factors, such as high operation temperature, which links reliability to the problem of power dissipation. Working with these restrictions, silicon devices are engineered to minimize oxide defect generation; and systems with reliable operation times exceeding 10 years can routinely be manufactured.

3.3.5 Alternate Device Structures for CMOS

It is widely recognized that CMOS device fabrication in the sub-50 nm regime will put significant pressure on current device designs and fabrication approaches. Several designs for advanced structures have been proposed, and some have promising capabilities and potential to be manufacturable.[18–22] Some of these structures include dual-gate designs, where gate electrodes on top and bottom of the

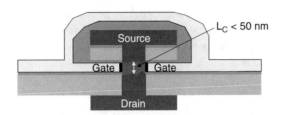

FIGURE 3.5 Schematic diagram of an example vertical field-effect transistor. For this device, the current between the source and drain flows in the vertical direction. The channel material is formed by epitaxial growth, where the channel length (L_c) is controlled by the film thickness rather than by lithography.

channel (or surrounding the channel) can increase the current flow by a factor of 2 and reduce the charging time by a similar factor, as compared with the typical single-gate structure. Such devices can be partially depleted or fully depleted, depending on the thickness of the channel layer, applied field, and dielectric thickness used. Many of these devices rely on silicon-on-insulator (SOI) technology, where very thin crystalline silicon layers are formed or transferred onto electrically insulating amorphous dielectric layers. This electrical isolation further reduces capacitance losses in the device, improving device speed.

Another class of devices gaining interest is vertical structures. A schematic of a vertical device is shown in Figure 3.5. In these devices, the source and drain are on top and bottom of the channel region; and the thickness of the channel region is determined relatively easily by controlling the thickness of a deposited layer rather than by lithography. Newly developed thin film deposition approaches, such as atomic layer deposition, capable of highly conformal coverage of high dielectric constant insulators, make these devices more feasible for manufacturing with channel length well below 50 nm.[19]

3.4 Memory Devices

3.4.1 DRAM, SRAM, and Flash

The relative simplicity of memory devices, where in principle only two contact terminals are needed to produce a memory cell, makes memory an attractive possible application for molecular electronic devices. Current computer random access memory (RAM) is composed of dynamic RAM (DRAM), static RAM (SRAM), and flash memory devices. SRAM, involving up to six transistors configured as cross-coupled inverters, can be accessed very quickly; but it is expensive because it takes up significant space on the chip. SRAM is typically small (~1MB) and is used primarily in processors as cache memory. DRAM uses a storage capacitor and one or two transistors, making it more compact than SRAM and less costly to produce. DRAM requires repeated refreshing, so cycle times for data access are typically slower than with SRAM. SRAM and DRAM both operate at typical supply voltage, and the issue of power consumption and heat dissipation with increased memory density is important, following a trend similar to that for processors given in Equations 3.5 and 3.6. Overall power consumption per cm^2 for DRAM is smaller than that for processors. Flash memory requires higher voltage, and write times are slower than for DRAM; but it has the advantage of being nonvolatile. Power consumption is important for flash since it is widely used in portable devices.

3.4.2 Passive and Active Matrix Addressing

It is important to note that even though DRAM operates by storing charge in a two-terminal capacitor, it utilizes a three-terminal transistor connected to each capacitor to address each memory cell. This *active matrix* approach can be contrasted with a *passive matrix* design, where each storage capacitor is addressed by a two-terminal diode. The passive addressing approach is much more simple to fabricate, but it suffers from two critical issues: cross-talk and power consumption. Cross-talk is associated with fringing fields, where the voltage applied across a cell results in a small field across neighboring cells; and this becomes more dominant at higher cell density. The fringing field is a problem in diode-addressed arrays because of the slope of the diode current vs. voltage (IV) curve and the statistical control of the diode turn-on voltage. Because the diode IV trace has a finite slope, a small voltage drop resulting from a fringing field will give rise to a small current that can charge or discharge neighboring cells. The diode-addressing scheme also results in a small voltage drop across all the cells in the row and column addressed. Across thousands of cells, this small current can result in significant chip heating and power dissipation problems.

An active addressing scheme helps solve problems of cross-talk and power dissipation by minimizing the current flow in and out of cells not addressed. The steep logarithmic threshold of a transistor (Equation 3.2) minimizes fringing field and leakage problems, allowing reliable operation at significantly higher densities. The importance of active addressing schemes is not limited to MOS memory systems. The importance of active matrix addressing for flat panel displays has been known for some time,[23] and active addressing has proven to be critical to achieve liquid crystal displays with resolution suitable for most applications. Well-controlled manufacturing has reduced costs associated with active addressing, and many low-cost products now utilize displays with active matrix addressing.

Molecular approaches for matrix-array memory devices are currently under study by several groups. Challenges addressed by silicon-based memory, including the value and statistical control of the threshold slope, will need to be addressed in these molecular systems. It is likely that, at the ultrahigh densities proposed for molecular cross-bar array systems, some form of active matrix addressing will be needed to control device heating. This points to the importance of three-terminal switching devices for molecular systems.

3.5   Opportunities and Challenges for Molecular Circuits

The above discussion included an overview of current CMOS technology and current directions in CMOS scaling. The primary challenges in CMOS for the next several generations include lithography and patterning, tolerance control, scaling of threshold and power supply voltages, controlling high-density dopant concentration and concentration profiles, improving contact resistance and capacitances, increasing gate capacitance while reducing gate dielectric tunneling leakage, and maintaining device performance (i.e., mobility and subthreshold slope). These issues can be summarized into (at least) six distinct engineering challenges for any advanced electronic system:

- Material patterning and tolerance control
- Reliability
- Interconnects and parasitics
- Charge transport (including device speed and the importance of contacts and interfaces)
- Power and heat dissipation
- Circuit and system design and integration (including use of gain)

These challenges are not unique to CMOS or silicon technology, but they will be significant in any approach for high-density, high-speed electronic device technology (including molecular electronics). For molecular systems, these challenges are in addition to the overriding fundamental material challenges associated with design and synthesis, charge transport mechanisms, control of electrostatic and contact potentials, etc.

It is possible that alternate approaches could be developed to circumvent some of these challenges. For example, high-density and highly parallel molecular computing architectures could be developed such that the speed of an individual molecular device may not need to follow the size/speed scaling rules. It is important to understand, however, that the engineering challenges presented above must be addressed together. An increase in parallelism may enable lower device speeds, but it puts additional demands on interconnect speed and density, with additional problems in parasitics, heat and power dissipation, contacts, etc.

Lithography-based approaches to form CMOS device features less than 20 nm have been demonstrated[24,25] (but not perfected), and most of the other engineering challenges associated with production of 10–20 nm CMOS devices have yet to be solved. It is clear that Moore's law cannot continue to atomic-scale silicon transistors. This raises some natural questions:

1. At what size scale will alternate material technologies (such as quantum or molecular electronics) have a viable place in engineered electronic systems?
2. What fundamental material challenges should be addressed now to enable required engineering challenges to be met?

The theoretical and practical limits of silicon device speed and size have been addressed in several articles,[2–4,26–29] and results of these analyses will not be reviewed here. In this section, several prospective molecular computing architectures will be discussed in terms of the six engineering challenges described above. Then a prospective hybrid silicon/molecular electronics approach for engineering and implementing molecular electronic materials and systems will be presented and discussed.

3.5.1 Material Patterning and Tolerances

As devices shrink and numbers of transistor devices in a circuit increase, and variations occur in line width, film thickness, feature alignment, and overlay accuracy across a chip, significant uncertainty in device performance within a circuit may arise. This uncertainty must be anticipated and accounted for in circuit and system design, and significant effort focuses on statistical analysis and control of material and pattern tolerances in CMOS engineering. Problems in size and performance variations are expected to be more significant in CMOS as devices continue to shrink. Even so, alignment accuracy and tolerance currently achieved in silicon processing are astounding (accuracy of ~50 nm across 200 mm wafers, done routinely for thousands of wafers). The attraction of self-assembly approaches is in part related to prospects for improved alignment and arrangement of nanometer-scale objects across large areas. Nanometer scale arrangement has been demonstrated using self-assembly approaches, but reliable self-assembly at the scale approaching that routinely achieved in silicon manufacturing is still a significant challenge. Hopefully, future self-assembly approaches will offer capabilities for feature sizes below what lithography can produce at the time. What lithography will be able to achieve in the future is, of course, unknown.

3.5.2 Reliability

As discussed above, reliable operation is another hallmark of CMOS devices; and systems with reliable operation times exceeding 10 years can be routinely manufactured. Achieving this level of reliability in molecular systems is recognized as a critical issue, but it is not yet widely discussed. This is because molecular technology is not yet at the stage where details of various approaches can be compared in terms of reliability, and fundamental mechanisms in failure of molecular systems are not yet discernible. Even so, some general observations can be made. Defect creation energies in silicon-based inorganic materials are fairly well defined. Creating a positive charged state within the silicon band gap, for example, will require energies in excess of silicon's electron affinity (> 4.1 eV). Ionization energies for many organic electronic materials are near 4–5 eV, close to that for silicon; but deformation energies are expected to be smaller in the organics, possibly leading to higher energy defect structures where less excess energy is needed to create active electronic defects. Therefore, reliability issues are expected to be more problematic in molecular systems as compared with silicon. Other factors such as melting temperatures and heat capacity also favor inorganic materials for reliable and stable operation. *Defect tolerant* designs are being considered that could overcome some of the problems of reliably interconnecting large numbers of molecular scale elements.[30] However, it is not clear how, or if, such an approach would manage a system with a defect density and distribution that changes relatively rapidly over time.

3.5.3　Interconnects, Contacts, and the Importance of Interfaces

Interconnection of molecular devices is also recognized as a critical problem. Fabrication and manipulation of molecular scale wires is an obvious concern. The size and precision for manipulating the wires must reach the same scale as the molecules. Otherwise, the largest device density would be determined by the density of the interconnect wire packing (which may not be better than future silicon devices) — not by the size of the molecules. Multilevel metallization technology is extremely well developed and crucial for fully integrated silicon systems; but as yet, no methodologies for multilevel interconnect have been demonstrated. Approaches such as metal nanocluster-modified viral particles[6] under study at the Naval Research Lab are being developed, in part, to address this issue. Also, as discussed above in relation to silicon technology, approaches will be needed to isolate molecular interconnections to avoid cross-talk and RC signal decay during transmission across and among chips. These problems with interconnect technology have the potential to severely limit realistic implementation of molecular-scale circuits.

The need to connect wires to individual molecules presents another set of problems. Several approaches to engineer linker elements within the molecular structure have been successful to achieve high-quality molecular monolayers on metals and other surfaces, and charge transport at molecule/metal surfaces is well established. However, the precise electronic structure of the molecule/metal contact and its role in the observed charge transport are not well known. As more complex material designs are developed to achieve molecular-scale arrays, other materials will likely be needed for molecular connections; and a more fundamental understanding of molecule/solid interfaces will be crucial. One specific concern is molecular conformational effects at contacts. Molecules can undergo a change in shape upon contact with a surface, resulting in a change in the atomic orbital configurations and change in the charge transfer characteristics at the interface. The relations among adsorption mechanisms, interface bond structure, configuration changes, and interface charge transfer need to be more clearly understood.

3.5.4　Power Dissipation and Gain

3.5.4.1　Two-Terminal Devices

The simplest device structures to utilize small numbers of molecules and to take advantage of self-assembly and "bottom-up" device construction involve linear molecules with contacts made to the ends. These can operate by quantum transport (i.e., conduction determined by tunneling into well-defined energy levels) or by coulomb blockade (i.e., conduction is achieved when potential is sufficient to overcome the energy of charge correlation). Molecular "shuttle" switches[31,32] are also interesting two-terminal structures. Two-terminal molecules can be considered for molecular memory[33–35] and for computation using, for example, massively parallel crossbar arrays[30] or nanocells.[36] All of the engineering challenges described above will need to be addressed for these structures to become practical, but the challenges of gain and power dissipation are particularly demanding.

The lack of gain is a primary problem for two-terminal devices, and signal propagation and fan-out must be supported by integration of other devices with gain capability. This is true for computation devices and for memory devices, where devices with gain will be needed to address and drive a memory array. Two-terminal logic devices will also suffer from problems of heat dissipation. The power dissipation described above in Equations 3.5 and 3.6 corresponds only to power lost in capacitive charging and discharging (dynamic power dissipation) and assumes that current does not flow under steady-state operation. The relation indicates that the dynamic power consumption will scale with the capacitance (i.e., the number of charges required to change the logic state of the device), which in principle could be small for molecular devices. However, quantum transport devices can have appreciable current flowing at steady state, adding another *standby power* term:

$$P_{standby} = I_{off} \cdot V_{dd} \qquad (3.8)$$

where I_{off} is the integrated current flow through the chip per unit area during steady-state operation. As discussed in detail above, complementary MOS structures are widely used now primarily because they can be engineered to enable negligible standby current (I_{off}). Standby power will be a serious concern if molecular devices become viable, and approaches that enable complementary action will be very attractive.

The relations above show that operating voltage is another important concern. Most molecular devices demonstrated to date use fairly high voltages to produce a switching event, and the parameters that influence this threshold voltage are not well understood. Threshold parameters in molecular devices must be better understood and controlled in order to manage power dissipation. Following the heat transfer analysis in Equation 3.7, generation of only ~200 mW/cm^2 in a molecular system cooled by natural convection would likely be sufficient to melt the device!

3.5.4.2 Multiterminal Structures

There are several examples of proposed molecular-scale electronic devices that make use of three or more terminals to achieve logic or memory operation. Approaches include the single electron transistor (SET),[11] the nanocell,[36] quantum cellular automata (QCA),[37] crossed or gated nanowire devices,[38–41] and field-effect devices with molecular channel regions.[42,43] Of these, the SET, nanowire devices, and field-effect devices are, in principle, capable of producing gain.

The nanocell and QCA structures are composed of sets of individual elements designed and organized to perform as logic gates and do not have specific provisions for gain built into the structures. In their simplest operating forms, therefore, additional gain elements would need to be introduced between logic elements to maintain signal intensity through the circuit. Because the QCA acts by switching position of charge on quantum elements rather than by long-range charge motion, the QCA approach is considered attractive for low power consumption. However, power will still be consumed in the switching events, with a value determined by the dynamic power equation (Equation 3.6). This dynamic power can be made small by reducing the QCA unit size (i.e., decreasing capacitance). However, reducing current will also substantially affect the ability to drive the interconnects, leading to delays in signal input and output. Also, the switching voltage must be significantly larger than the thermal voltage at the operating temperature, which puts a lower limit on V_{dd} and dynamic power consumption for these devices. The same argument for dynamic power loss and interconnect charging will apply to the switching processes in the nanocell device. The nanocell will also have the problem of standby power loss as long as NDR molecules have significant off-state leakage (i.e., low output impedance and poor device isolation). Output currents in the high impedance state are typically 1 pA, with as much as 1000 pA in the low impedance state. If a nanocell 1 μm × 1 μm contains 10^4 molecules,[36] all in the off state (I_{off} = 1pA) with V_{dd} = 2V, then the power dissipation is expected to be ~1 W/cm^2 (presuming 50% of the chip area is covered by nanocells). This power would heat the chip and likely impair operation substantially (from Equation 3.7, ΔT would be much greater than 100°C under natural convection cooling). Stacking devices in three-dimensional structures could achieve higher densities, but it would make the heat dissipation problem substantially worse. Cooling of the center of a three-dimensional organic solid requires conductive heat transfer, which is likely significantly slower than cooling by convection from a two-dimensional surface. To address these problems, molecular electronic materials are needed with smaller operating voltage, improved off-state leakage, better on/off ratios, and sharper switching characteristics to enable structures with lower load resistances. It is important to note one aspect of the nanocell design: it may eventually enable incorporation of pairs of RTD elements in "Goto pairs," which under a limited range of conditions can show elements of current or voltage gain.[36,44]

Some devices, including the single-electron transistor (SET),[45,46] crossed and gated nanowire devices,[38–41] and field-effect devices with molecular channel regions[42,43] show promise for gain at the molecular scale; but none has yet shown true molecular-scale room temperature operation. The nanowire approach is attractive because of the possible capability for complementary operation (i.e., possibly eliminating standby power consumption). Field-effect devices with molecular channel regions show intriguing results,[42,43] but it is not clear how a small voltage applied at a large distance (30 nm) can affect transport across a small molecule (2 nm) with a larger applied perpendicular field. Even so, approaches such as these, and others with the prospect of gain, continue to be critical for advanced development of fully engineered molecular-scale electronic devices and systems.

3.5.5 Thin-Film Electronics

Thin-film electronics, based on amorphous silicon materials, is well established in device manufacturing. Charge transport rates in organic materials can challenge those in amorphous silicon; and organic materials may eventually offer advantages in simplicity of processing (including solution-based processing, for example), enabling very low-cost large-area electronic systems. The materials, fabrication, and systems engineering challenges for thin-film electronics are significantly different from those of potential single-molecule structures. Also, thin-film devices do not address the ultimate goal of high-density, high-speed electronic systems with molecular-scale individual elements. Therefore, thin-film electronics is typically treated as a separate topic altogether. However, some thin-film electronic applications could make use of additional functionality, including local computation or memory integrated within the large-area system. Such a system may require two distinctly different materials (for switching and memory, for example), and there may be some advantages of utilizing silicon and molecules together in hybrid thin-film inorganic/molecular/organic systems.

3.5.6 Hybrid Silicon/Molecular Electronics

As presented above, silicon CMOS technology is attractive, in large part, because of its ability to scale to very high densities and high speed while maintaining low power generation. This is achieved by use of: (1) complementary device integration (to achieve minimal standby power) and (2) devices with capability of current and voltage gain (to maintain the integrity of a signal as it moves within the circuit). Molecular elements are attractive for their size and possible low-cost chemical routes to assembly. However, the discussion above highlights many of the challenges that must be faced before realistic all-molecular electronic systems can be achieved.

A likely route to future all-molecular circuits is through engineered hybrid silicon/molecular systems. Realizing such hybrid systems presents additional challenges that are generally not addressed in studies of all-molecular designs, such as semiconductor/molecule chemical and electrical coupling, semiconductor/organic processing integration, and novel device, circuit, and computational designs. However, these additional challenges of hybrid devices are likely more surmountable in the near term than those of all-molecule devices and could give rise to new structures that take advantage of the benefits of molecules and silicon technology. For example, molecular RTDs assembled onto silicon CMOS structures could enable devices with multiple logic states, so more complex computation could be performed within the achievable design rules of silicon devices. Also, molecular memory devices integrated with CMOS transistors could enable ultrahigh density and ultrafast memories close-coupled with silicon to challenge SRAM devices in cost and performance for cache applications in advanced computing. Coupling molecules with silicon could also impact the problem of molecular characterization. As discussed above, parasitic effects in device characterization are a serious concern; and approaches to intimately couple organic electronic elements with silicon devices would result in structures with well-characterized parasitics, leading to reliable performance analysis of individual and small ensembles of molecules — critically important for the advance of any molecular-based electronic technology.

3.6 Summary and Conclusions

Present-day silicon technology is a result of many years of tremendously successful materials, device, and systems engineering. Proposed future molecular-based devices could substantially advance computing technology, but the engineering of proposed molecular electronic systems will be no less challenging than what silicon has overcome to date. It is important to understand the reasons why silicon is so successful and to realize that silicon has and will continue to overcome many substantial "show-stoppers" to successful production. Most of the engineering issues described above can be reduced to challenges in materials and materials integration. For example: can molecular switches with sufficiently low operating voltage and operating current be realized to minimize heat dissipation problems? Or: can charge coupling and transport through interfaces be understood well enough to design improved electrical contacts to molecules to control parasitic losses? It is likely that an integrated approach, involving close-coupled studies of fundamental materials and engineered systems including hybrid molecular/silicon devices, will give rise to viable and useful molecular electronic elements with substantially improved accessibility, cost, and capability over current electronic systems.

Acknowledgments

The author acknowledges helpful discussions with several people, including Veena Misra, Paul Franzon, Chris Gorman, Bruce Gnade, John Hauser, David Nackashi, and Carl Osburn. He also thanks Bruce Gnade, Carl Osburn, and Dong Niu for critical reading of the manuscript.

References

1. Semiconductor Industry Association, *The International Technology Roadmap for Semiconductors* (Austin, TX, 2001) (http://public.itrs.net).
2. R.W. Keyes, Fundamental limits of silicon technology, *Proc. IEEE* **89,** 227–239 (2001).
3. J.D. Plummer and P.B. Griffin, Material and process limits in silicon VLSI technology, *Proc. IEEE* **89,** 240–258 (2001).
4. R.L. Harriott, Limits of lithography, *Proc. IEEE* **89,** 366–374 (2001).
5. M.A. Ratner, Introducing molecular electronics, *Mater. Today* **5,** 20–27 (2002).
6. K.S. Kwok and J.C. Ellenbogen, Moletronics: future electronics, *Mater. Today* **5,** 28–37 (2002).
7. Y. Wada, Prospects for single molecule information processing devices, *Proc. IEEE* **89,** 1147–1173 (2001).
8. J.F. Wakerly, *Digital Design Principles and Practices* (Prentice Hall, Upper Saddle River, NJ, 2000).
9. P. Horowitz and W. Hill, *The Art of Electronics* (Cambridge University Press, New York, 1980).
10. L.J. Micheel, A.H. Taddiken, and A.C. Seabaugh, Multiple-valued logic computation circuits using micro- and nanoelectronic devices, *Proc. 23rd Intl. Symp. Multiple-Valued Logic,* 164–169 (1993).
11. K. Uchida, J. Koga, A. Ohata, and A. Toriumi, Silicon single-electron tunneling device interfaced with a CMOS inverter, *Nanotechnology* **10,** 198–200 (1999).
12. R.H. Mathews, J.P. Sage, T.C.L.G. Sollner, S.D. Calawa, C.-L. Chen, L.J. Mahoney, P.A. Maki, and K.M. Molvar, A new RTD-FET logic family, *Proc. IEEE* **87,** 596–605 (1999).
13. D. Goldhaber–Gordon, M.S. Montemerlo, J.C. Love, G.J. Opiteck, and J.C. Ellenbogen, Overview of nanoelectronic devices, *Proc. IEEE* **85,** 521–540 (1997).
14. D.A. Buchanan, Scaling the gate dielectric: materials, integration, and reliability, *IBM J. Res. Dev.* **43,** 245 (1999).
15. Y. Taur and T.H. Ning, *Fundamentals of Modern VLSI Devices* (Cambridge University Press, Cambridge, U.K., 1998).
16. Intel Pentium III Datasheet (ftp://download.intel.com/design/PentiumIII/datashts/24526408.pdf).
17. C.O. Bennett and J.E. Myers, *Momentum, Heat, and Mass Transfer* (McGraw-Hill, New York, 1982).
18. H. Takato, K. Sunouchi, N. Okabe, A. Nitayama, K. Hieda, F. Horiguchi, and F. Masuoka, High performance CMOS surrounding gate transistor (SGT) for ultra high density LSIs, *IEEE IEDM Tech. Dig.,* 222–226 (1988).
19. J.M. Hergenrother, G.D. Wilk, T. Nigam, F.P. Klemens, D. Monroe, P.J. Silverman, T.W. Sorsch, B. Busch, M.L. Green, M.R. Baker, T. Boone, M.K. Bude, N.A. Ciampa, E.J. Ferry, A.T. Fiory, S.J. Hillenius, D.C. Jacobson, R.W. Johnson, P. Kalavade, R.C. Keller, C.A. King, A. Kornblit, H.W. Krautter, J.T.-C. Lee, W.M. Mansfield, J.F. Miner, M.D. Morris, O.-H. Oh, J.M. Rosamilia, B.T. Sapjeta, K. Short, K. Steiner, D.A. Muller, P.M. Voyles, J.L. Grazul, E.J. Shero, M.E. Givens, C. Pomarede, M. Mazanec, and C. Werkhoven, 50nm vertical replacement gate (VRG) nMOSFETs with ALD HfO$_2$ and Al$_2$O$_3$ gate dielectrics, *IEEE IEDM Tech. Dig.,* 3.1.1–3.1.4 (2001).
20. D. Hisamoto, W.-C. Lee, J. Kedzierski, H. Takeuchi, K. Asano, C. Kuo, E. Anderson, T.-J. King, J. Bokor, and C. Hu, FinFET — a self-aligned double-gate MOSFET scalable to 20 nm, *IEEE Trans. Electron Devices* **47,** 2320–25 (2000).

21. C.M. Osburn, I. Kim, S.K. Han, I. De, K.F. Yee, J.R. Hauser, D.-L. Kwong, T.P. Ma, and M.C. Öztürk, Vertically-scaled MOSFET gate stacks and junctions: how far are we likely to go?, *IBM J. Res. Dev.,* *(accepted)* (2002).

22. T. Schulz, W. Rösner, L. Risch, A. Korbel, and U. Langmann, Short-channel vertical sidewall MOSFETs, *IEEE Trans. Electron Devices* **48** (2001).

23. P.M. Alt and P. Pleshko, Scanning limitations of liquid crystal displays, *IEEE Trans. Electron Devices* **ED-21,** 146–155 (1974).

24. R. Chau, J. Kavalieros, B. Doyle, A. Murthy, N. Paulsen, D. Lionberger, D. Barlage, R. Arghavani, B. Roberds, and M. Doczy, A 50nm depleted-substrate CMOS transistor (DST), *IEEE IEDM Tech. Dig.* **2001,** 29.1.1–29.1.4 (2001).

25. M. Fritze, B. Tyrrell, D.K. Astolfi, D. Yost, P. Davis, B. Wheeler, R. Mallen, J. Jarmolowicz, S.G. Cann, H.-Y. Liu, M. Ma, D.Y. Chan, P.D. Rhyins, C. Carney, J.E. Ferri, and B.A. Blachowicz, 100-nm node lithography with KrF? *Proc. SPIE* **4346,** 191–204 (2001).

26. D.J. Frank, S.E. Laux, and M.V. Fischetti, Monte Carlo simulation of 30nm dual-gate MOSFET: how short can Si go?, *IEEE IEDM Tech. Dig.,* 21.1.1–21.1.3 (1992).

27. J.R. Hauser and W.T. Lynch, Critical front end materials and processes for 50 nm and beyond IC devices, *(unpublished).*

28. T.H. Ning, Silicon technology directions in the new millennium, *IEEE 38 Intl. Reliability Phys. Symp.* (2000).

29. J.D. Meindl, Q. Chen, and J.A. Davis, Limits on silicon nanoelectronics for terascale integration, *Science* **293,** 2044–2049 (2001).

30. J.R. Heath, P.J. Kuekes, G.S. Snider, and S. Williams, A defect-tolerant computer architecture: opportunities for nanotechnology, *Science* **280,** 1716–1721 (1998).

31. P.L. Anelli et al., Molecular meccano I. [2] rotaxanes and a [2]catenane made to order, *J. Am. Chem. Soc.* **114,** 193–218 (1992).

32. D.B. Amabilino and J.F. Stoddard, Interlocked and intertwined structures and superstructures, *Chem. Revs.* **95,** 2725–2828 (1995).

33. D. Gryko, J. Li, J.R. Diers, K.M. Roth, D.F. Bocian, W.G. Kuhr, and J.S. Lindsey, Studies related to the design and synthesis of a molecular octal counter, *J. Mater. Chem.* **11,** 1162 (2001).

34. K.M. Roth, N. Dontha, R.B. Dabke, D.T. Gryko, C. Clausen, J.S. Lindsey, D.F. Bocian, and W.G. Kuhr, Molecular approach toward information storage based on the redox properties of porphyrins in self-assembled monolayers, *J. Vacuum Sci. Tech. B* **18,** 2359–2364 (2000).

35. M.A. Reed, J. Chen, A.M. Rawlett, D.W. Price, and J.M. Tour, Molecular random access memory cell, *Appl. Phys. Lett.* **78,** 3735–3737 (2001).

36. J.M. Tour, W.L.V. Zandt, C.P. Husband, S.M. Husband, L.S. Wilson, P.D. Franzon, and D.P. Nackashi, Nanocell logic gates for molecular computing, *(submitted)* (2002).

37. G. Toth and C.S. Lent, Quasiadiabatic switching for metal-island quantum-doc cellular automata, *J. Appl. Phys.* **85,** 2977–2181 (1999).

38. X. Liu, C. Lee, C. Zhou, and J. Han, Carbon nanotube field-effect inverters, *Appl. Phys. Lett.* **79,** 3329–3331 (2001).

39. Y. Cui and C.M. Lieber, Functional nanoscale electronic devices assembled using silicon nanowire building blocks, *Science* **291,** 851–853 (2001).

40. Y. Huang, X. Duan, Y. Cui, L.J. Lauhon, K.-H. Kim, and C.M. Lieber, Logic gates and computation from assembled nanowire building blocks, *Science* **294,** 1313–1317 (2001).

41. A. Bachtold, P. Hadley, T. Nakanishi, and C. Dekker, Logic circuits with nanotube transistors, *Science* **294,** 1317–1320 (2001).

42. J.H. Schön, H. Meng, and Z. Bao, Self-assembled monolayer organic field-effect transistors, *Nature* **413,** 713–716 (2001).

43. J.H. Schön, H. Meng, and Z. Bao, Field effect modulation of the conductance of single molecules, *Science* **294,** 2138–2140 (2001).

44. J.C. Ellenbogen and J.C. Love, Architectures for molecular electronic computers: 1. logic structures and an adder designed from molecular electronic diodes, *Proc. IEEE* **88,** 386–426 (2000).
45. M.A. Kastner, The single electron transistor, *Rev. Mod. Phys.* **64,** 849–858 (1992).
46. H. Ahmed and K. Nakazoto, Single-electronic devices, *Microelectron. Eng.* **32,** 297–315 (1996).

4

Molecular Electronic Computing Architectures

CONTENTS

James M. Tour
Rice University

Dustin K. James
Rice University

4.1 Present Microelectronic Technology

Technology development and industrial competition have been driving the semiconductor industry to produce smaller, faster, and more powerful logic devices. That the number of transistors per integrated circuit will double every 18–24 months due to advancements in technology is commonly referred to as *Moore's Law*, after Intel founder Gordon Moore, who made the prediction in a 1965 paper with the prophetic title "Cramming More Components onto Integrated Circuits."[1] At the time he thought that his prediction would hold until at least 1975; however, the exponentially increasing rate of circuit densification has continued into the present (Graph 4.1). In 2000, Intel introduced the Pentium 4 containing 42 million transistors, an amazing engineering achievement. The increases in packing density of the circuitry are achieved by shrinking the line widths of the metal interconnects, by decreasing the size of other features, and by producing thinner layers in the multilevel device structures. These changes are only brought about by the development of new fabrication techniques and materials of construction. As an example, commercial metal interconnect line widths have decreased to 0.13 μm. The resistivity of Al at 0.13 μm line width, combined with its tendency for electromigration (among other problems), necessitated the substitution of Cu for Al as the preferred interconnect metal in order to achieve the 0.13 μm line width goal. Cu brings along its own troubles, including its softness, a tendency to migrate into silicon dioxide (thus requiring a barrier coating of Ti/TiN), and an inability to deposit Cu layers via the

0-8493-1200-0/03/$0.00+$1.50
© 2003 by CRC Press LLC

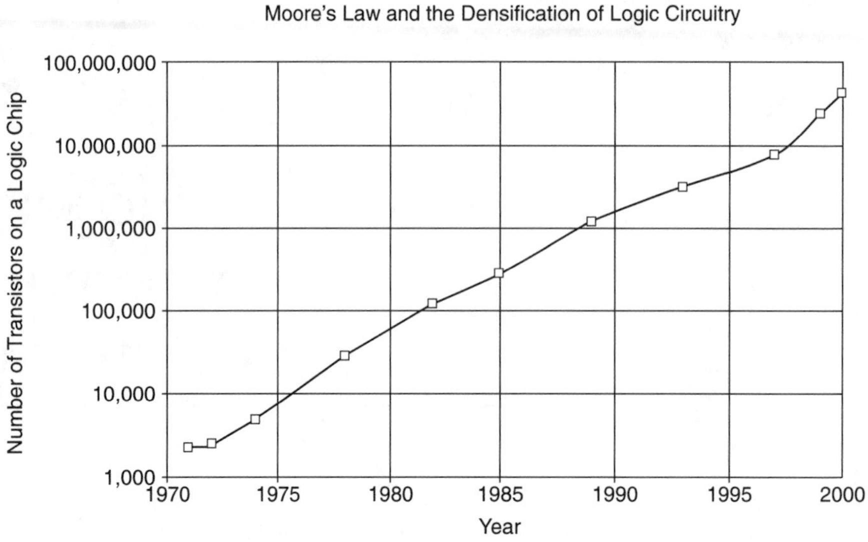

Moore's Law and the Densification of Logic Circuitry

GRAPH 4.1   The number of transistors on a logic chip has increased exponentially since 1972. (Courtesy of Intel Data.)

vapor phase. New tools for depositing copper using electroless electroplating and new technologies for removing the metal overcoats — because copper does not etch well — had to be developed to meet these and other challenges. To integrate Cu in the fabrication line, innovations had to be made all the way from the front end to the back end of the process. These changes did not come without cost, time, and Herculean efforts.

4.2   Fundamental Physical Limitations of Present Technology

This top-down method of producing faster and more powerful computer circuitry by shrinking features cannot continue because there are fundamental physical limitations, related to the material of construction of the solid-state-based devices, that cannot be overcome by engineering. For instance, charge leakage becomes a problem when the insulating silicon oxide layers are thinned to about three silicon atoms deep, which will be reached commercially by 2003–2004. Moreover, silicon loses its original band structure when it is restricted to very small sizes. The lithography techniques used to create the circuitry on the wafers has also neared its technological limits, although derivative technologies such as e-beam lithography, extreme ultraviolet lithography (EUV),[2] and x-ray lithography are being developed for commercial applications. A tool capable of x-ray lithography in the sub-100 nm range has been patented.[3]

Financial roadblocks to continued increases in circuit density exist. Intel's Fab 22, which opened in Chandler, Arizona, in October 2001, cost $2 billion to construct and equip; and it is slated to produce logic chips using copper-based 0.13 μm technology on 200 mm wafers. The cost of building a Fab is projected to rise to $15–30 billion by 2010[4] and could be as much as $200 billion by 2015.[5] The staggering increase in cost is due to the extremely sophisticated tools that will be needed to form the increasingly small features of the devices. It is possible that manufacturers may be able to take advantage of infrastructure already in place in order to reduce the projected cost of the introduction of the new technologies, but much is uncertain because the methods for achieving further increases in circuit density are unknown or unproven.

As devices increase in complexity, defect and contamination control become even more important as defect tolerance is very low — nearly every device must work perfectly. For instance, cationic metallic impurities in the wet chemicals such as sulfuric acid used in the fabrication process are measured in the part per billion (ppb) range. With decreases in line width and feature size, the presence of a few ppb of metal contamination could lead to low chip yields. Therefore, the industry has been driving suppliers to produce chemicals with part per trillion (ppt) contamination levels, raising the cost of the chemicals used.

Depending on the complexity of the device, the number of individual processing steps used to make them can be in the thousands.[6] It can take 30–40 days for a single wafer to make it through the manufacturing process. Many of these steps are cleaning steps, requiring some fabs to use thousands of gallons of ultra-pure water per minute.[7] The reclaim of waste water is gaining importance in semiconductor fab operations.[8] The huge consumption of water and its subsequent disposal can lead to problems where aquifers are low and waste emission standards require expensive treatment technology.

A new technology that addressed only one of the potential problems we have discussed would be of interest to the semiconductor industry. A new technology would be revolutionary if it produced faster and smaller logic and memory chips, reduced complexity, saved days to weeks of manufacturing time, and reduced the consumption of natural resources.

4.3 Molecular Electronics

How do we overcome the limitations of the present solid-state electronic technology? Molecular electronics is a fairly new and fascinating area of research that is firing the imagination of scientists as few research topics have.[9] For instance, *Science* magazine labeled the hook-up of molecules into functional circuits as the breakthrough of the year for 2001.[10] Molecular electronics involves the search for single molecules or small groups of molecules that can be used as the fundamental units for computing, i.e., wires, switches, memory, and gain elements.[11] The goal is to use these molecules, designed from the bottom up to have specific properties and behaviors, instead of present solid-state electronic devices that are constructed using lithographic technologies from the top down. The top-down approach is currently used in the silicon industry, wherein small features such as transistors are etched into silicon using resists and light; the ever-increasing demand for densification is stressing the industry. The bottom-up approach, on the other hand, implies the construction of functionality into small features, such as molecules, with the opportunity to have the molecules further self-assemble into the higher ordered structural units such as transistors. Bottom-up methodologies are quite natural in that all systems in nature are constructed bottom-up. For example, molecules with specific features assemble to form higher order structures such as lipid bilayers. Further self-assembly, albeit incomprehensibly complex, causes assembly into cells and further into high life forms. Hence, utilization of a diversity of self-assembly processes could lead to enormous advances in future manufacturing processes once scientists learn to further control specific molecular-level interactions.

Ultimately, given advancements in our knowledge, it is thought by the proponents of molecular electronics that its purposeful bottom-up design will be more efficient than the top-down method, and that the incredible structure diversity available to the chemist will lead to more effective molecules that approach optional functionality for each application. A single mole of molecular switches, weighing about 450 g and synthesized in small reactors (a 22-L flask might suffice for most steps of the synthesis), contains 6×10^{23} molecules — more than the combined number of transistors ever made in the history of the world. While we do not expect to be able to build a circuit in which each single molecule is addressable and is connected to a power supply (at least not in the first few generations), the extremely large numbers of switches available in a small mass illustrate one reason molecular electronics can be a powerful tool for future computing development.

The term *molecular electronics* can cover a broad range of topics. Petty, Bryce, and Bloor recently explored molecular electronics.[12] Using their terminology, we will focus on molecular-scale electronics instead of molecular materials for electronics. Molecular materials for electronics deal with films or crystals (i.e., thin-film transistors or light-emitting diodes) that contain many trillions of molecules per functional unit, the properties of which are measured on the macroscopic scale, while molecular-scale electronics deals with one to a few thousand molecules per device.

4.4 Computer Architectures Based on Molecular Electronics

In this section we will initially discuss three general architectural approaches that researchers are considering to build computers based on molecular-scale electronics and the advances made in these three areas in the years 1998–2001. In addition, we will touch upon progress made in measuring the electrical characteristics of molecular switches and in designing logic devices using molecular electronics components.

The first approach to molecular computing, based on quantum cellular automata (QCA), was briefly discussed in our prior review.[11] This method relies on electrostatic field repulsions to transport information throughout the circuitry. One major benefit of the QCA approach is that heat dissipation is less of an issue because only one to fractions of an electron are used rather than the 16,000 to 18,000 electrons needed for each bit of information in classical solid-state devices.

The second approach is based on the massively parallel solid-state Teramac computer developed at Hewlett-Packard (HP)[4] and involves building a similarly massively parallel computing device using molecular electronics-based crossbar technologies that are proposed to be very defect tolerant.[13] When applied to molecular systems, this approach is proposed to use single-walled carbon nanotubes (SWNT)[14–18] or synthetic nanowires[14,19–22] for crossbars. As we will see, logic functions are performed either by sets of crossed and specially doped nanowires or by molecular switches placed at each crossbar junction.

The third approach uses molecular-scale switches as part of a nanocell, a new concept that is a hybrid between present silicon-based technology and technology based purely on molecular switches and molecular wires (in reality, the other two approaches will also be hybrid systems in their first few generations).[23] The nanocell relies on the use of arrays of molecular switches to perform logic functions but does not require that each switching molecule be individually addressed or powered. Furthermore, it utilizes the principles of chemical self-assembly in construction of the logic circuitry, thereby reducing complexity. However, programming issues increase dramatically in the nanocell approach.

While solution-phase-based computing, including DNA computing,[24] can be classified as molecular-scale electronics, it is a slow process due to the necessity of lining up many bonds, and it is wedded to the solution phase. It may prove to be good for diagnostic testing, but we do not see it as a commercially viable molecular electronics platform; therefore, we will not cover it in this review.

Quantum computing is a fascinating area of theoretical and laboratory study,[25–28] with several articles in the popular press concerning the technology.[29,30] However, because quantum computing is based on interacting quantum objects called *qubits*, and not molecular electronics, it will not be covered in this review. Other interesting approaches to computing such as "spintronics"[31] and the use of light to activate switching[32] will also be excluded from this review.

4.4.1 Quantum Cellular Automata (QCA)

Quantum dots have been called *artificial atoms* or *boxes for electrons*[33] because they have discrete charge states and energy-level structures that are similar to atomic systems and can contain from a few thousand to one electron. They are typically small electrically conducting regions, 1 μm or less in size, with a variety of geometries and dimensions. Because of the small volume, the electron energies are quantized. No shell structure exists; instead, the generic energy spectrum has universal statistical properties associated with quantum chaos.[34] Several groups have studied the production of quantum dots.[35] For example, Leifeld and coworkers studied the growth of Ge quantum dots on silicon surfaces that had been precovered with 0.05–0.11 monolayer of carbon,[36] i.e., carbon atoms replaced about five to ten of every 100 silicon atoms at the surface of the wafer. It was found that the Ge dots grew directly over the areas of the silicon surface where the carbon atoms had been inserted.

Heath discovered that hexane solutions of Ag nanoparticles, passivated with octanethiol, formed spontaneous patterns on the surface of water when the hexane was evaporated;[37] and he has prepared superlattices of quantum dots.[38,39] Lieber has investigated the energy gaps in "metallic" single-walled carbon nanotubes[16] and has used an atomic-force microscope to mechanically bend SWNT in order to create quantum dots less than 100 nm in length.[18] He found that most metallic SWNT are not true metals and that, by bending the SWNT, a defect was produced that had a resistance of 10 to 100 kΩ. Placing two defects less than 100 nm apart produced the quantum dots.

One proposed molecular computing structural paradigm that utilizes quantum dots is termed a *quantum cellular automata* (QCA) wherein four quantum dots in a square array are placed in a cell such that electrons are able to tunnel between the dots but are unable to leave the cell.[40] As shown in Figure 4.1, when two excess electrons are placed in the cell, Coulomb repulsion will force the electrons to occupy dots on opposite corners. The two ground-state polarizations are energetically equivalent and can be labeled logic "0" or "1." Flipping the logic state of one cell, for instance by applying a negative potential to a lead near the quantum dot occupied by an electron, will result in the next-door cell flipping ground states in order to reduce Coulomb repulsion. In this way, a line of QCA cells can be used to do computations. A simple example is shown in Figure 4.2, the structure of which could be called a *binary wire*, where a "1" input gives a "1" output. All of the electrons occupy positions as far away from their neighbors as possible, and they are all in a ground-state polarization. Flipping the ground state of the cell on the left end will result in a domino effect, where each neighboring cell flips ground states until the end of the wire is reached. An inverter built from QCA cells is shown in Figure 4.3 — the output is "0" when the input is "1." A QCA topology that can produce *AND* and *OR* gates is called a *majority gate*[41] and is shown in Figure 4.4, where the three input cells "vote" on the polarization of the central cell. The polarization of the central cell is then propagated as the output. One of the inputs can be

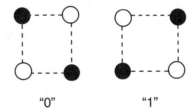

"0"   "1"

FIGURE 4.1   The two possible ground-state polarizations, denoted "0" and "1," of a four-dot QCA cell. Note that the electrons are forced to opposite corners of the cells by Coulomb repulsion.

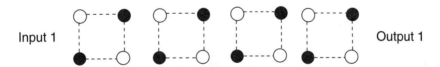

Input 1   Output 1

FIGURE 4.2   Simple QCA cell logic line where a logic input of 1 gives an logic output of 1.

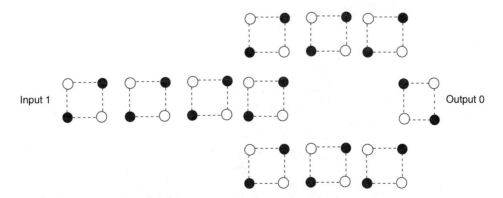

Input 1   Output 0

FIGURE 4.3   An inverter built using QCA cells such that a logic input of 1 yields a logic out of 0.

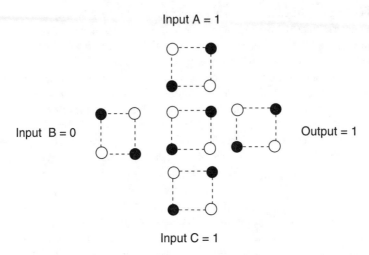

FIGURE 4.4   A QCA majority cell in which the three input cells A, B, and C determine the ground state of the center cell, which then determines the logic of the output. A logic input of 0 gives a logic output of 1.

designated a programming input and determines whether the majority gate produces an AND or an OR. If the programming gate is a logic 0, then the result shown in Figure 4.4 is OR while a programming gate equal to logic 1 would produce a result of AND.

A QCA fan-out structure is shown in Figure 4.5. Note that when the ground state of the input cell is flipped, the energy put into the system may not be enough to flip all the cells of both branches of the structure, producing long-lived metastable states and erroneous calculations. Switching the cells using a quasi-adiabatic approach prevents the production of these metastable states.[42]

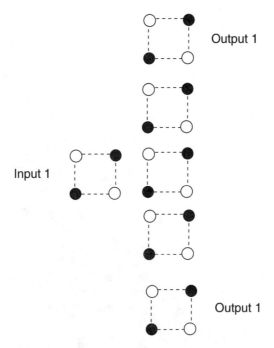

FIGURE 4.5   A fan-out constructed of QCA cells. A logic input of 1 produces a logic output of 1 at both ends of the structure.

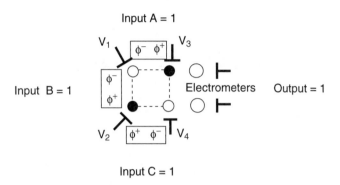

FIGURE 4.6 A QCA majority cell as set up experimentally in a nonmolecular system.

Amlani and co-workers have demonstrated experimental switching of 6-dot QCA cells.[43–45] The polarization switching was accomplished by applying biases to the gates of the input double-dot of a cell fabricated on an oxidized Si surface using standard Al tunnel junction technology, with Al islands and leads patterned by e-beam lithography, followed by a shadow evaporation process and an *in situ* oxidation step. The switching was experimentally verified in a dilution refrigerator using the electrometers capacitively coupled to the output double-dot.

A functioning majority gate was also demonstrated by Amlani and co-workers,[46] with logic AND and OR operations verified using electrometer outputs after applying inputs to the gates of the cell. The experimental setup for the majority gate is shown in Figure 4.6, where the three input tiles A, B, and C were supplanted by leads with biases that were equivalent to the polarization states of the input cells. The negative or positive bias on a gate mimicked the presence or absence of an electron in the input dots of the tiles A, B, and C that were replaced. The truth table for all possible input combinations and majority gate output is shown in Figure 4.7. The experimental results are shown in Figure 4.8. A QCA binary wire has been experimentally demonstrated by Orlov and co-workers,[47] and Amlani and co-workers have demonstrated a leadless QCA cell.[48] Bernstein and co-workers have demonstrated a latch in clocked QCA devices.[49]

While the use of quantum dots in the demonstration of QCA is a good first step in reduction to practice, the ultimate goal is to use individual molecules to hold the electrons and pass electrostatic potentials down QCA wires. We have synthesized molecules that have been shown by *ab initio* computational methods to have the capability of transferring information from one molecule to another through electrostatic potential.[50] Synthesized molecules included three-terminal molecular junctions, switches, and molecular logic gates.

The QCA method faces several problems that need to be resolved before QCA-based molecular computing can become reality. While relatively large quantum-dot arrays can be fabricated using existing methods, a major problem is that placement of molecules in precisely aligned arrays at the nanoscopic level is very difficult to achieve with accuracy and precision. Another problem is that degradation of only one molecule in the array can cause failure of the entire circuit. There has also

A	B	C	Output
0	0	0	0
0	0	1	0
0	1	1	1
0	1	0	0
1	1	0	1
1	1	1	1
1	0	1	1
1	0	0	0

FIGURE 4.7 The logic table for the QCA majority cell.

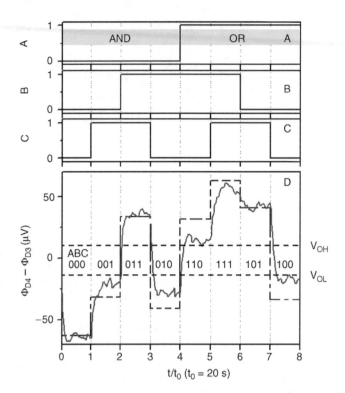

FIGURE 4.8 Demonstration of majority gate operation, where A to C are inputs in Gray code. The first four and last four inputs illustrate AND and OR operations, respectively. (D) Output characteristic of majority gate where t_0 = 20 s is the input switching period. The dashed stair-step-like line shows the theory for 70 mK; the solid line represents the measured data. Output high (V_{OH}) and output low (V_{OL}) are marked by dashed horizontal lines. (Reprinted from Amlani, I., Orlov, A.O., Toth, G., Bernstein, G.H., Lent, C.S., and Snider, G.L. *Science*, 284, 289, 1999. ©1999 American Association for the Advancement of Science. With permission.)

been some debate about the unidirectionality (or lack thereof) of QCA designs.[47,51–52] Hence, even small examples of 2-dots have yet to be demonstrated using molecules, but hopes remain high and researchers are continuing their efforts.

4.4.2 Crossbar Arrays

Heath, Kuekes, Snider, and Williams recently reported on a massively parallel experimental computer that contained 220,000 hardware defects yet operated 100 times faster than a high-end single processor workstation for some configurations.[4] The solid-state-based (not molecular electronic) Teramac computer built at HP relied on its fat-tree architecture for its logical configuration. The minimum communication bandwidth needed to be included in the fat-tree architecture was determined by utilizing Rent's rule, which states that the number of wires coming out of a region of a circuit should scale with the power of the number of devices (n) in that region, ranging from $n^{1/2}$ in two dimensions to $n^{2/3}$ in three dimensions. The HP workers built in excess bandwidth, putting in many more wires than needed. The reason for the large number of wires can be understood by considering the simple but illustrative city map depicted in Figure 4.9. To get from point A to point B, one can take local streets, main thoroughfares, freeways, interstate highways, or any combination thereof. If there is a house fire at point C, and the local streets are blocked, then by using the map it is easy to see how to go around that area to get to point B. In the Teramac computer, *street blockages* are stored in a defect database; when one device needs to communicate with another device, it uses the database and the map to determine how to get there. The Teramac design can therefore tolerate a large number of defects.

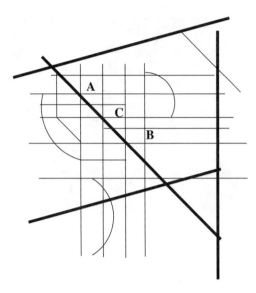

FIGURE 4.9 A simple illustration of the defect tolerance of the Teramac computer. In a typical city, many routes are available to get from point A to point B. One who dislikes traffic might take only city streets (thin lines) while others who want to arrive faster may take a combination of city streets and highways (thick lines). If there were a house fire at point C, a traveler intent on driving only on city streets could look at the map and determine many alternate routes from A to B.

In the Teramac computer (or a molecular computer based on the Teramac design), the wires that make up the address lines controlling the settings of the configuration switches and the data lines that link the logic devices are the most important and plentiful part of the computer. It is logical that a large amount of research has been done to develop nanowires (NW) that could be used in the massively parallel molecular computer. Recall that nanoscale wires are needed if we are to take advantage of the smallness in size of molecules.

Lieber has reviewed the work done in his laboratory to synthesize and determine the properties of NW and nanotubes.[14] Lieber used Au or Fe catalyst nanoclusters to serve as the nuclei for NW of Si and GeAs with 10 nm diameters and lengths of hundreds of nm. By choosing specific conditions, Lieber was able to control both the length and the diameter of the single crystal semiconductor NW.[20] Silicon NW doped with B or P were used as building blocks by Lieber to assemble semiconductor nanodevices.[21] Active bipolar transistors were fabricated by crossing *n*-doped NW with *p*-type wire base. The doped wires were also used to assemble complementary inverter-like structures.

Heath reported the synthesis of silicon NW by chemical vapor deposition using SiH_4 as the Si source and Au or Zn nanoparticles as the catalytic seeds at 440°C.[22,53] The wires produced varied in diameter from 14 to 35 nm and were grown on the surface of silicon wafers. After growth, isolated NW were mechanically transferred to wafers; and Al contact electrodes were put down by standard e-beam lithography and e-beam evaporation such that each end of a wire was connected to a metallic contact. In some cases a gate electrode was positioned at the middle of the wire (Figure 4.10). Tapping AFM indicated the wire in this case was 15 nm in diameter.

Heath found that annealing the Zi-Si wires at 550°C produced increased conductance attributed to better electrode/nanowire contacts (Figure 4.11). Annealing Au-Si wires at 750°C for 30 min increased current about 10^4, as shown in Figure 4.12 — an effect attributed to doping of the Si with Au and lower contact resistance between the wire and Ti/Au electrodes.

Much research has been done to determine the value of SWNT as NW in molecular computers. One problem with SWNT is their lack of solubility in common organic solvents. In their synthesized state, individual SWNT form ropes[54] from which it is difficult to isolate individual tubes. In our laboratory some solubility of the tubes was seen in 1,2-dichlorobenzene.[55] An obvious route to better solubilization

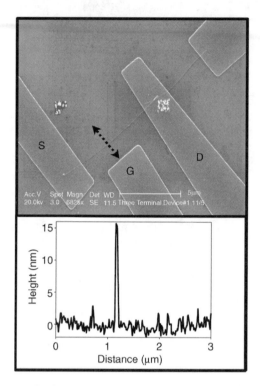

FIGURE 4.10 (Top) SEM image of a three-terminal device, with the source (S), gate (G), and drain (D) labeled. (Bottom) Tapping mode AFM trace of a portion of the silicon nanowire (indicated with the dashed arrow in the SEM image), revealing the diameter of the wire to be about 15 nm. (Reprinted from Chung, S.-W., Yu, J.-Y, and Heath, J.R., *Appl. Phys. Lett.*, 76, 2068, 2000. ©2000 American Institute of Physics. With permission.)

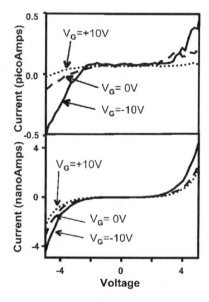

FIGURE 4.11 Three-terminal transport measurements of an as-prepared 15 nm Si nanowire device contacted with Al electrodes (top) and the same device after annealing at 550°C (bottom). In both cases, the gating effect indicates *p*-type doping. (Reprinted from Chung, S.-W., Yu, J.-Y., and Heath, J.R., *Appl. Phys. Lett.*, 76, 2068, 2000. ©2000 American Institute of Physics. With permission.)

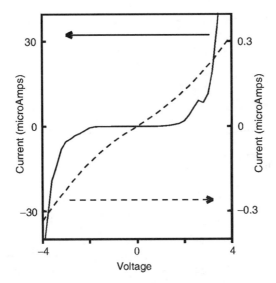

FIGURE 4.12 I(V) characteristics of Au-nucleated Si nanowires contacted with Ti/Au electrodes, before (solid line, current axis on left) and after (dashed line, current axis on right) thermal treatment (750°C, 1 h). After annealing the wire exhibits metallic-like conductance, indicating that the wire has been heavily doped. (Reprinted from Chung, S.-W., Yu, J.-Y, and Heath, J.R., *Appl. Phys. Lett.*, 76, 2068, 2000. ©2000 American Institute of Physics. With permission.)

is to functionalize SWNT by attachment of soluble groups through covalent bonding. Margrave and Smalley found that fluorinated SWNT were soluble in alcohols,[56] while Haddon and Smalley were able to dissolve SWNT by ionic functionalization of the carboxylic acid groups present in purified tubes.[57]

We have found that SWNT can be functionalized by electrochemical reduction of aryl diazonium salts in their presence.[58] Using this method, about one in 20 carbon atoms of the nanotube framework are reacted. We have also found that the SWNT can be functionalized by direct treatment with aryl diazonium tetrafluoroborate salts in solution or by *in situ* generation of the diazonium moiety using an alkyl nitrite reagent.[59] These functional groups give us handles with which we can direct further, more selective derivatization.

Unfortunately, fluorination and other sidewall functionalization methods can perturb the electronic nature of the SWNT. An approach by Smalley[54,60] and Stoddart and Heath[17] to increasing the solubility without disturbing the electronic nature of the SWNT was to wrap polymers around the SWNT to break up and solubilize the ropes but leave individual tube's electronic properties unaffected. Stoddart and Heath found that the SWNT ropes were not separated into individually wrapped tubes; the entire rope was wrapped. Smalley found that individual tubes were wrapped with polymer; the wrapped tubes did not exhibit the roping behavior. While Smalley was able to demonstrate removal of the polymer from the tubes, it is not clear how easily the SWNT can be manipulated and subsequently used in electronic circuits. In any case, the placement of SWNT into controlled configurations has been by a top-down methodology for the most part. Significant advances will be needed to take advantage of controlled placement at dimensions that exploit a molecule's small size.

Lieber proposed a SWNT-based nonvolatile random access memory device comprising a series of crossed nanotubes, wherein one parallel layer of nanotubes is placed on a substrate and another layer of parallel nanotubes, perpendicular to the first set, is suspended above the lower nanotubes by placing them on a periodic array of supports.[15] The elasticity of the suspended nanotubes provides one energy minima, wherein the contact resistance between the two layers is zero and the switches (the contacts between the two sets of perpendicular NW) are OFF. When the tubes are transiently charged to produce attractive electrostatic forces, the suspended tubes flex to meet the tubes directly below them; and a contact is made, representing the ON state. The ON/OFF state could be read by measuring the resistance

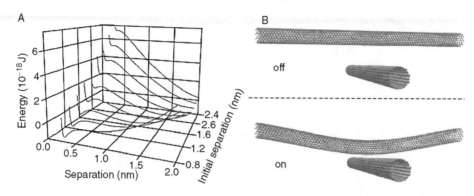

FIGURE 4.13 Bistable nanotubes device potential. (A) Plots of energy, $E_t = E_{vdW} + E_{elas}$, for a single 20 nm device as a function of separation at the cross point. The series of curves corresponds to initial separations of 0.8, 1.0, 1.2, 1.4, 1.6, 1.8, 2.0, 2.2, and 2.4 nm, with two well-defined minima observed for initial separations of 1.0 to 2.0 nm. These minima correspond to the crossing nanotubes being separated and in cdW contact. (B) Calculated structures of the 20 nm (10, 10) SWNT device element in the OFF (top) and ON (bottom) states. The initial separation for this calculation was 2.0 nm; the silicon support structures (elastic modulus of 168 Gpa) are not shown for clarity. (Reprinted from Rueckes, T., Kim, K., Joselevich, E., Tseng, G.Y., Cheung, C.–L., and Lieber, C.M., *Science*, 289, 94, 2000. © 2000 American Association for the Advancement of Science. With permission.)

at each junction and could be switched by applying voltage pulses at the correct electrodes. This theory was tested by mechanically placing two sets of nanotube bundles in a crossed mode and measuring the I(V) characteristics when the switch was OFF or ON (Figure 4.13). Although they used nanotube bundles with random distributions of metallic and semiconductor properties, the difference in resistance between the two modes was a factor of 10, enough to provide support for their theory.

In another study, Lieber used scanning tunneling microscopy (STM) to determine the atomic structure and electronic properties of intramolecular junctions in SWNT samples.[16] Metal–semiconductor junctions were found to exhibit an electronically sharp interface without localized junction states while metal–metal junctions had a more diffuse interface and low-energy states.

One problem with using SWNT or NW as wires is how to guide them in formation of the device structures — i.e., how to put them where you want them to go. Lieber has studied the directed assembly of NW using fluid flow devices in conjunction with surface patterning techniques and found that it was possible to deposit layers of NW with different flow directions for sequential steps.[19] For surface patterning, Lieber used NH_2-terminated surface strips to attract the NW; in between the NH_2- terminated strips were either methyl-terminated regions or bare regions, to which the NW had less attraction. Flow control was achieved by placing a poly(dimethylsiloxane) (PDMS) mold, in which channel structures had been cut into the mating surface, on top of the flat substrate. Suspensions of the NW (GaP, InP, or Si) were then passed through the channels. The linear flow rate was about 6.40 mm/s. In some cases the regularity extended over mm-length scales, as determined by scanning electron microscopy (SEM). Figure 4.14 shows typical SEM images of their layer-by-layer construction of crossed NW arrays.

While Lieber has shown that it is possible to use the crossed NW as switches, Stoddart and Heath have synthesized molecular devices that would bridge the gap between the crossed NW and act as switches in memory and logic devices.[61] The UCLA researchers have synthesized catenanes (Figure 4.15 is an example) and rotaxanes (Figure 4.16 is an example) that can be switched OFF and ON using redox chemistry. For instance, Langmuir–Blodgett films were formed from the catenane in Figure 4.15, and the monolayers were deposited on polysilicon NW etched onto a silicon wafer photolithographically. A second set of perpendicular titanium NW was deposited through a shadow mask, and the I(V) curve was determined. The data, when compared to controls, indicated that the molecules were acting as solid-state molecular switches. As yet, however, there have been no demonstrations of combining the Stoddart switches with NW.

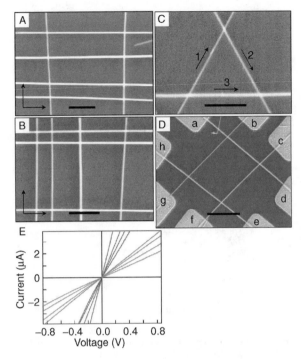

FIGURE 4.14 Layer-by-layer assembly and transport measurements of crossed NW arrays. (A and B) Typical SEM images of crossed arrays of InP NW obtained in a two-step assembly process with orthogonal flow directions for the sequential steps. Flow directions are highlighted by arrows in the images. (C) An equilateral triangle of GaP NW obtained in a three-step assembly process, with 60° angles between flow directions, which are indicated by numbered arrows. The scale bars correspond to 500 nm in (A), (B), and (C). (D) SEM image of a typical 2-by-2 cross array made by sequential assembly of n-type InP NW with orthogonal flows. Ni/In/Au contact electrodes, which were deposited by thermal evaporation, were patterned by e-beam lithography. The NW were briefly (3 to 5 s) etched in 6% HF solution to remove the amorphous oxide outer layer before electrode deposition. The scale bar corresponds to 2 μm. (E) Representative I(V) curves from two terminal measurements on a 2-by-2 crossed array. The solid lines represent the I(V) of four individual NW (ad, by, cf, eh), and the dashed lines represent I(V) across the four n–n crossed junctions (ab, cd, ef, gh). (Reprinted from Huang, Y., Duan, X., Wei, Q., and Lieber, C.M., *Science*, 291, 630, 2001. © 2001 American Association for the Advancement of Science. With permission.)

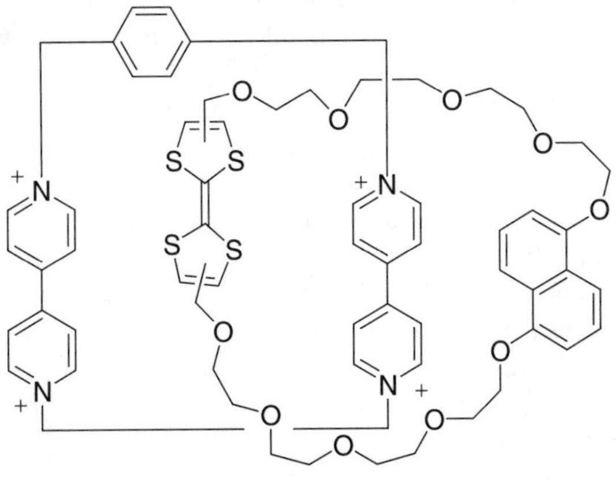

FIGURE 4.15 A catenane. Note that the two ring structures are intertwined.

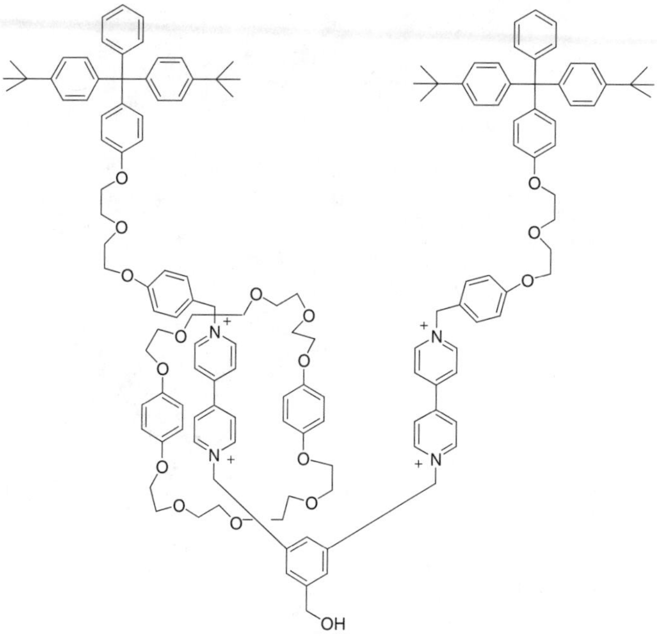

FIGURE 4.16   A [2] rotaxane. The two large end groups do not allow the ring structure to slip off either end.

Carbon nanotubes are known to exhibit either metallic or semiconductor properties. Avouris and coworkers at IBM have developed a method of engineering both multiwalled nanotubes (MWNT) and SWNT using electrical breakdown methods.[62] Shells in MWNT can vary between metallic or semiconductor character. Using electrical current in air to rapidly oxidize the outer shell of MWNT, each shell can be removed in turn because the outer shell is in contact with the electrodes and the inner shells carry little or no current. Shells are removed until arrival at a shell with the desired properties.

With ropes of SWNT, Avouris used an electrostatically coupled gate electrode to deplete the semiconductor SWNT of their carriers. Once depleted, the metallic SWNT can be oxidized while leaving the semiconductor SWNT untouched. The resulting SWNT, enriched in semiconductors, can be used to form nanotubes-based field-effect transistors (FETs) (Figure 4.17).

The defect-tolerant approach to molecular computing using crossbar technology faces several hurdles before it can be implemented. As we have discussed, many very small wires are used in order to obtain the defect tolerance. How is each of these wires going to be accessed by the outside world? Multiplexing, the combination of two or more information channels into a common transmission medium, will have to be a major component of the solution to this dilemma. The directed assembly of the NW and attachment to the multiplexers will be quite complicated. Another hurdle is signal strength degradation as it travels along the NW. Gain is typically introduced into circuits by the use of transistors. However, placing a transistor at each NW junction is an untenable solution. Likewise, in the absence of a transistor at each cross point in the crossbar array, molecules with very large ON:OFF ratios will be needed. For instance, if a switch with a 10:1 ON:OFF ratio were used, then ten switches in the OFF state would appear as an ON switch. Hence, isolation of the signal via a transistor is essential; but presently the only solution for the transistor's introduction would be for a large solid-state gate below each cross point, again defeating the purpose for the small molecules.

Additionally, if SWNT are to be used as the crossbars, connection of molecular switches via covalent bonds introduces sp³ linkages at each junction, disturbing the electronic nature of the SWNT and possibly obviating the very reason to use the SWNT in the first place. Noncovalent bonding will not provide the conductance necessary for the circuit to operate. Therefore, continued work is being done to devise and construct crossbar architectures that address these challenges.

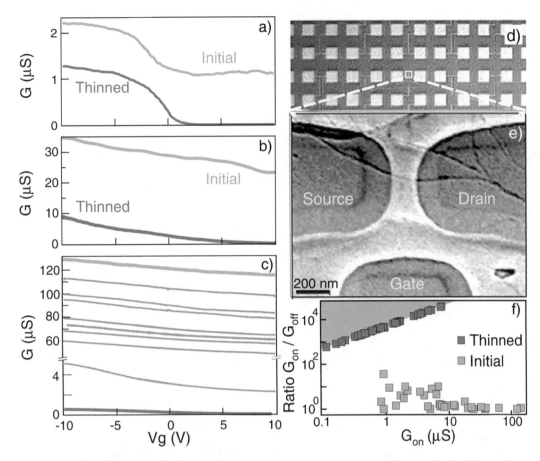

FIGURE 4.17 (a and b) Stressing a mixture of s- and m-SWNT while simultaneously gating the bundle to deplete the semiconductors of carriers resulted in the selective breakdown of the m-SWNT. The G(Vg) curve rigidly shifted downward as the m-SWNT were destroyed. The remaining current modulation is wholly due to the remaining s-SWNTs. (c) In very thick ropes, some s-SWNT must also be sacrificed to remove the innermost m-SWNT. By combining this technique with standard lithography, arrays of three-terminal, nanotubes-based FETs were created (d and e) out of disordered bundles containing both m- and s-SWNT. Although these bundles initially show little or no switching because of their metallic constituents, final devices with good FET characteristics were reliably achieved (f). (Reprinted from Collins, P.G., Arnold, M.S., and Avouris, P., *Science*, 292, 706, 2001. © 2001 American Association for the Advancement of Science. With permission.)

4.4.3 The Nanocell Approach to a Molecular Computer: Synthesis

We have been involved in the synthesis and testing of molecules for molecular electronics applications for some time.[11] One of the synthesized molecules, the nitro aniline oligo(phenylene ethynylene) derivative (Figure 4.18), exhibited large ON:OFF ratios and negative differential resistance (NDR) when placed in a nanopore testing device (Figure 4.19).[63] The peak-to-valley ratio (PVR) was 1030:1 at 60 K.

The same nanopore testing device was used to study the ability of the molecules to hold their ON states for extended periods of time. The performance of molecules 1–4 in Figure 4.20 as molecular memory devices was tested, and in this study only the two nitro-containing molecules 1 and 2 were found to exhibit storage characteristics. The write, read, and erase cycles are shown in Figure 4.21. The I(V) characteristics of the Au-(1)-Au device are shown in Figure 4.22. The characteristics are repeatable to high accuracy with no degradation of the device noted even after 1 billion cycles over a one-year period.

The I(V) characteristics of the Au-(2)-Au were also measured (Figure 4.23, A and B). The measure logic diagram of the molecular random access memory is shown in Figure 4.24.

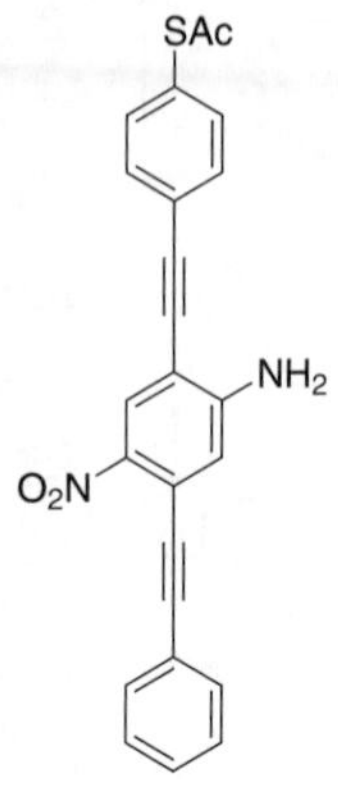

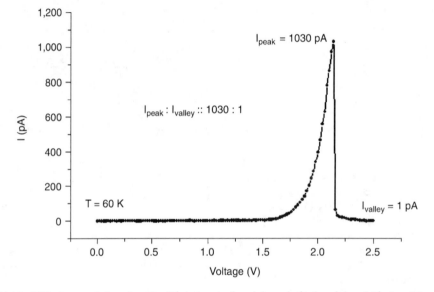

FIGURE 4.18   The protected form of the molecule tested in Reed and Tour's nanopore device.

FIGURE 4.19   I(V) characteristics of an Au-(2′-amino-4-ethynylphyenyl-4′-ethynylphenyl-5′-nitro-1-benzenethi-olate)-Au device at 60 K. The peak current density is ~50 A/cm^2, the NDR is ~– 400 µohm•cm^2, and the PVR is 1030:1.

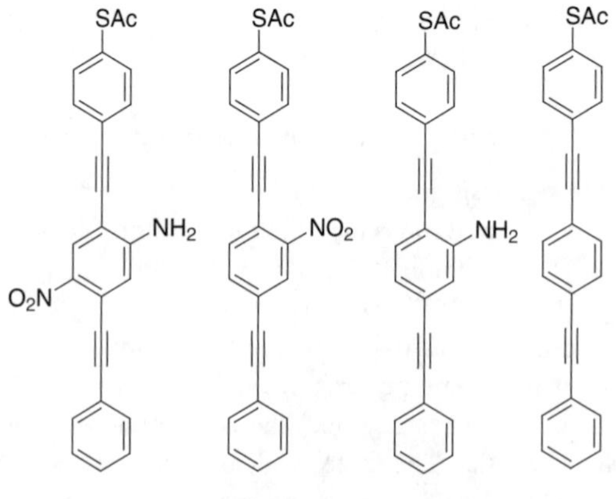

FIGURE 4.20   Molecules 1–4 were tested in the nanopore device for storage of high- or low-conductivity states. Only the two nitro-containing molecules 1 and 2 showed activity.

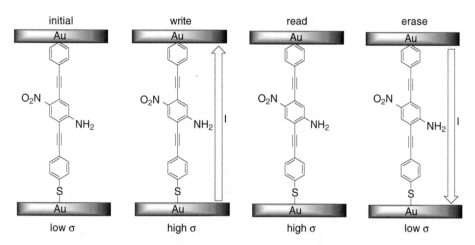

FIGURE 4.21 The memory device operates by the storage of a high- or low-conductivity state. An initially low-conductivity state (low σ) is changed into a high-conductivity state (high σ) upon application of a voltage. The direction of current that flows during the write and erase pulses is diagrammed by the arrows. The high σ state persists as a stored bit. (Reprinted from Reed, M.A., Chen, J., Rawlett, A.M., Price, D.W., and Tour, J.M. *Appl. Phys. Lett.*, 78, 3735, 2001. © 2001 American Institute of Physics. With permission.)

Seminario has developed a theoretical treatment of the electron transport through single molecules attached to metal surfaces[64] and has subsequently done an analysis of the electrical behavior of the four molecules in Figure 4.20 using quantum density functional theory (DFT) techniques at the B3PW91/6–31G* and B3PW91/LAML2DZ levels of theory.[65] The lowest unoccupied molecular orbit (LUMO) of nitro-amino functionalized molecule 1 was the closest orbital to the Fermi level of the Au. The LUMO of neutral 1 was found to be localized (nonconducting). The LUMO became delocalized (conducting) in the −1 charged state. Thus, ejection of an electron from the Au into the molecule to form a radical anion leads to conduction through the molecule. A slight torsional twist of the molecule allowed the orbitals to line up for conductance and facilitated the switching.

Many new molecules have recently been synthesized in our laboratories, and some have been tested in molecular electronics applications.[66–69] Since the discovery of the NDR behavior of the nitro aniline derivative, we have concentrated on the synthesis of oligo(phenylene ethynylene) derivatives. Scheme 4.1 shows the synthesis of a dinitro derivative. Quinones, found in nature as electron acceptors, can be easily reduced and oxidized, thus making them good candidates for study as molecular switches. The synthesis of one such candidate is shown in Scheme 4.2.

The acetyl thiol group is called a protected *alligator clip*. During the formation of a self-assembled monolayer (SAM) on a gold surface, for instance, the thiol group is deprotected *in situ*, and the thiol forms a strong bond (~2 eV, 45 kcal/mole) with the gold.

Seminario and Tour have done a theoretical analysis of the metal–molecule contact[70] using the B3PW91/LANL2DZ level of theory as implemented in Gaussian-98 in conjunction with the Green function approach that considers the "infinite" nature of the contacts. They found that Pd was the best metal contact, followed by Ni and Pt; Cu was intermediate, while the worst metals were Au and Ag. The best alligator clip was the thiol clip, but they found it was not much better than the isonitrile clip.

We have investigated other alligator clips such as pyridine end groups,[68] diazonium salts,[67] isonitrile, Se, Te, and carboxylic acid end groups.[66] Synthesis of an oligo(phenylene ethynylene) molecule with an isonitrile end group is shown in Scheme 4.3.

We have previously discussed the use of diazonium salts in the functionalization of SWNT. With modifications of this process, it might be possible to build the massively parallel computer architecture

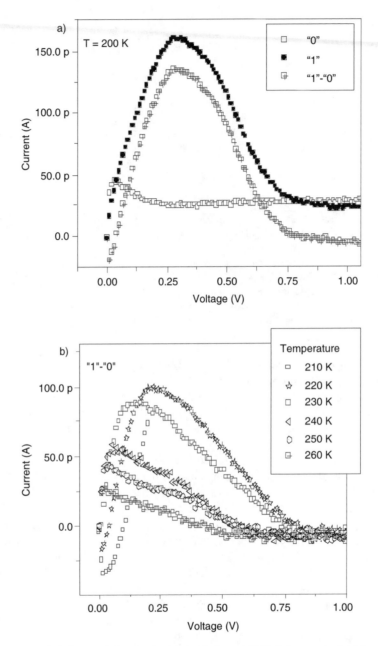

FIGURE 4.22   (a) The I(V) characteristics of a Au-(1)-Au device at 200 K. 0 denotes the initial state, 1 the stored written state, and 1–0 the difference of the two states. Positive bias corresponds to hole injection from the chemisorbed thiol-Au contact. (b) Difference curves (1–0) as a function of temperature. (Reprinted from Reed, M.A., Chen, J., Rawlett, A.M., Price, D.W., and Tour, J.M. *Appl. Phys. Lett.*, 78, 3735, 2001. © 2001 American Institute of Physics. With permission.)

using SWNT as the crosswires and oligo(phenylene ethynylene) molecules as the switches at the junctions of the crosswires, instead of the cantenane and rotaxane switches under research at UCLA (see Figure 4.25). However, the challenges of the crossbar method would remain as described above. The synthesis of one diazonium switch is shown in Scheme 4.4. The short synthesis of an oligo(phenylene ethynylene) derivative with a pyridine alligator clip is shown in Scheme 4.5.

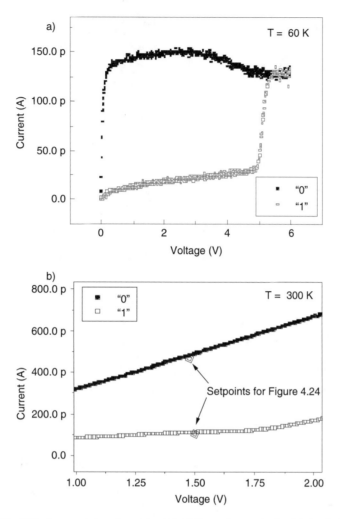

FIGURE 4.23 (a) The I(V) characteristics of stored and initial/erased states in Au-(2)-Au device at 60 K and (b) ambient temperatures (300 K). The setpoints indicated are the operating point for the circuit of Figure 4.24. (Reprinted from Reed, M.A., Chen, J., Rawlett, A.M., Price, D.W., and Tour, J.M. *Appl. Phys. Lett.*, 78, 3735, 2001. ©2001 American Institute of Physics. With permission.)

4.4.4 The Nanocell Approach to a Molecular Computer: The Functional Block

In our conceptual approach to a molecular computer based on the nanocell, a small 1 μm² feature is etched into the surface of a silicon wafer. Using standard lithography techniques, 10 to 20 Au electrodes are formed around the edges of the nanocell. The Au leads are exposed only as they protrude into the nanocell's core; all other gold surfaces are nitride-coated. The silicon surface at the center of the nanocell (the molehole — the location of "moleware" assembly) is functionalized with $HS(CH_2)_3SiO_x$. A two-dimensional array of Au nanoparticles, about 30–60 nm in diameter, is deposited onto the thiol groups in the molehole. The Au leads (initially protected by alkane thiols) are then deprotected using UV/O_3; and the molecular switches are deposited from solution into the molehole, where they insert themselves between the Au nanoparticles and link the Au nanoparticles around the perimeter with the Au electrodes. The assembly of nanoparticles combined with molecular switches in the molehole will form hundreds to thousands of complete circuits from one electrode to another; see Figure 4.26 for a simple illustration. By applying voltage pulses to selected nanocell electrodes, we expect to be able to turn interior switches

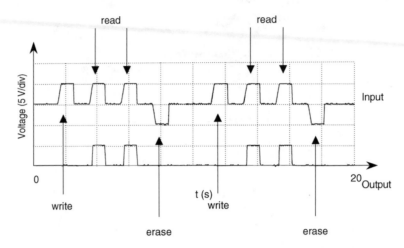

FIGURE 4.24   Measured logic diagram of the molecular random access memory. (Reprinted from Reed, M.A., Chen, J., Rawlett, A.M., Price, D.W., and Tour, J.M. *Appl. Phys. Lett.*, 78, 3735, 2001. © 2001 American Institute of Physics. With permission.)

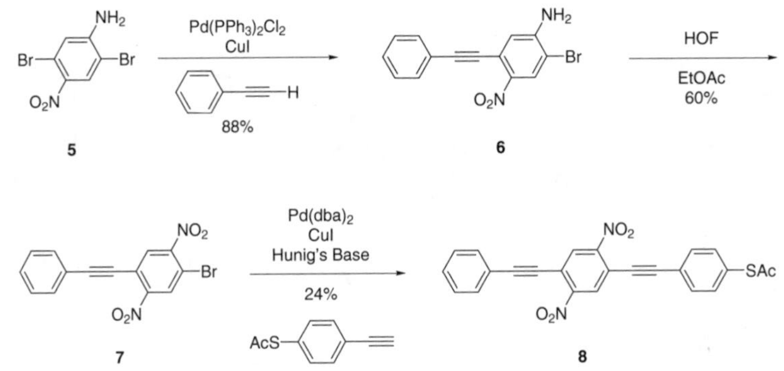

SCHEME 4.1   The synthesis of a dinitro-containing derivative. (Reprinted from Dirk, S.M., Price, D.W. Jr., Chanteau, S., Kosynkin, D.V., and Tour, J.M., *Tetrahedron*, 57, 5109, 2001. © ©2001 Elsevier Science. With permission.)

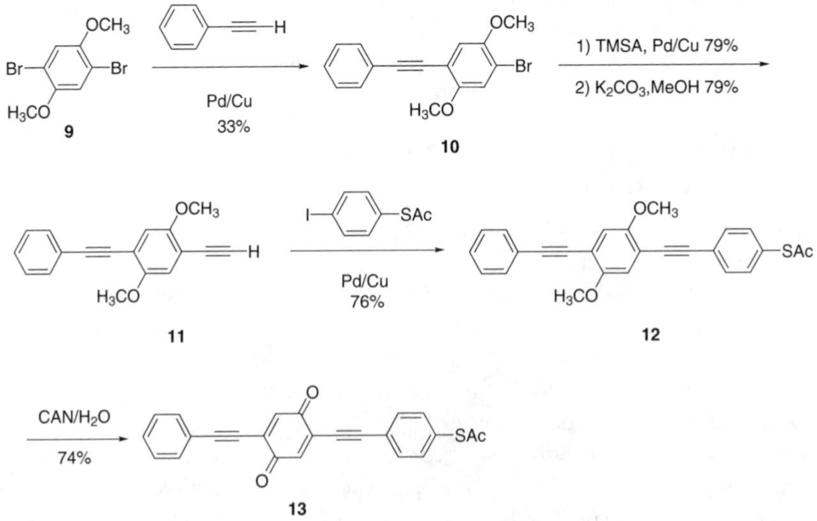

SCHEME 4.2   The synthesis of a quinone molecular electronics candidate. (Reprinted from Dirk, S.M., Price, D.W. Jr., Chanteau, S., Kosynkin, D.V., and Tour, J.M., *Tetrahedron*, 57, 5109, 2001. © ©2001 Elsevier Science. With permission.)

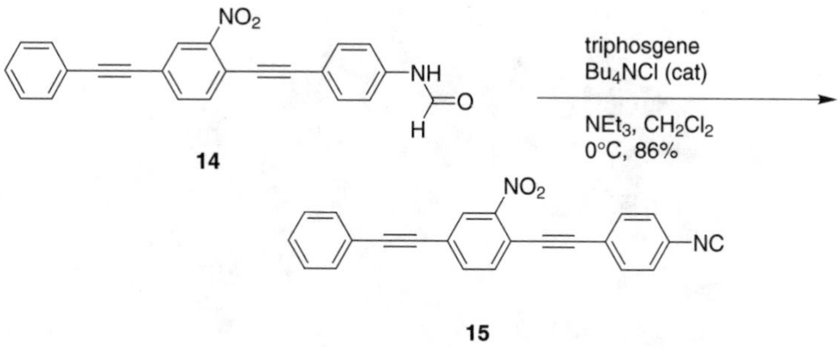

SCHEME 4.3 The formation of an isonitrile alligator clip from a formamide precursor.

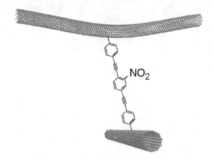

FIGURE 4.25 Reaction of a bis-diazonium-derived nitro phenylene ethynylene molecule with two SWNT could lead to functional switches at cross junctions of SWNT arrays.

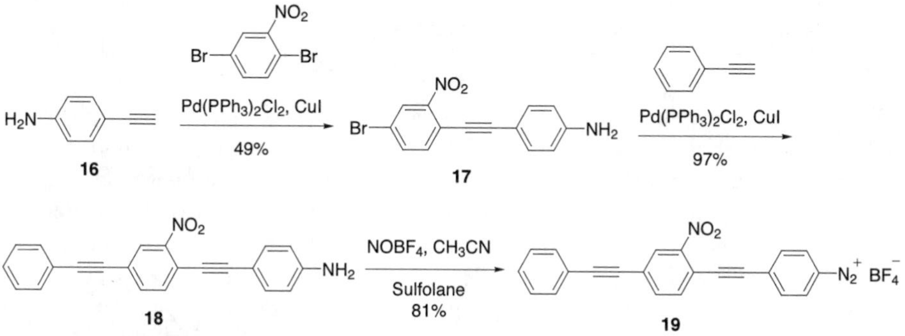

SCHEME 4.4 The synthesis of a diazonium containing molecular electronics candidate. (Reprinted from Dirk, S.M., Price, D.W. Jr., Chanteau, S., Kosynkin, D.V., and Tour, J.M., *Tetrahedron*, 57, 5109, 2001. © 2001 Elsevier Science. With permission.)

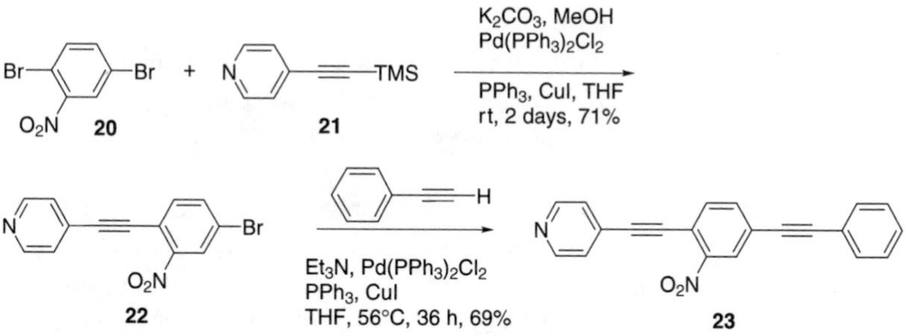

SCHEME 4.5 The synthesis of a derivative with a pyridine alligator clip. (Reprinted from Chanteau, S. and Tour, J.M., Synthesis of potential molecular electronic devices containing pyridine units, *Tet. Lett.*, 42, 3057, 2001. ©2001 Elsevier Science. With permission.)

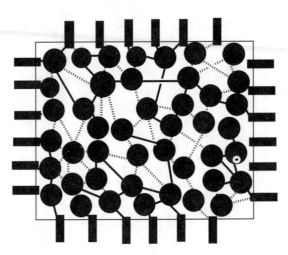

FIGURE 4.26 The proposed nanocell, with electrodes (black rectangles) protruding into the square molehole. Our simulations involve fewer electrodes. The metallic nanoparticles, shown here as black circles with very similar sizes, are deposited into the molehole along with organic molecular switches, not all of which are necessarily the same length or contain the same functionality. The molecular switches, with alligator clips on both ends, bridge the nanoparticles. Switches in the ON state are shown as solid lines while switches in the OFF state are shown as dashed lines. Because there would be no control of the nanoparticle or switch deposition, the actual circuits would be unknown. However, thousands to millions of potential circuits would be formed, depending on the number of electrodes, the size of the molehole, the size of the nanoparticles, and the concentration and identity of the molecular switches. The nanocell would be queried by a programming module after assembly in order to set the particular logic gate or function desired in each assembly. Voltage pulses from the electrodes would be used to turn switches ON and OFF until the desired logic gate or function was achieved.

ON or OFF, especially with the high ON:OFF ratios we have achieved with the oligo(phenylene ethynylene)s. In this way we hope to train the nanocell to perform standard logic operations such as AND, NAND, and OR. The idea is that we construct the nanocell first, with no control over the location of the nanoparticles or the bridging switches, and train it to perform certain tasks afterwards. Training a nanocell in a reasonable amount of time will be critical. Eventually, trained nanocells will be used to teach other nanocells. Nanocells will be tiled together on traditional silicon wafers to produce the desired circuitry. We expect to be able to make future nanocells 0.1 μm^2 or smaller if the input/output leads are limited in number, i.e., one on each side of a square.

While we are still in the research and development phase of the construction of an actual nanocell, we have begun a program to simulate the nanocell using standard electrical engineering circuit simulation programs such as SPICE and HSPICE, coupled with genetic algorithm techniques in three stages:[23]

1. With complete omnipotent programming, wherein we know everything about the interior of the constructed nanocell such as the location of the nanoparticles, how many switches bridge each nanoparticle pair, and the state of the conductance of the switches, and that we have control over turning specific switches ON or OFF to achieve the desired outcome without using voltage pulses from the outside electrodes;

2. With omniscient programming, where we know what the interior of the nanocell looks like and know the conductance state of the switches, but we have to use voltage pulses from the surrounding electrodes to turn switches ON and OFF in order to achieve the desired outcome; and

3. With mortal programming, where we know nothing about the interior of the nanocell and have to guess where to apply the voltage pulses. We are just beginning to simulate mortal programming; however, it is the most critical type since we will be restricted to this method in the actual physical testing of the nanocell.

Our preliminary results with omnipotent programming show that we can simulate simple logic functions such as AND, OR, and half-adders.

The nanocell approach has weaknesses and unanswered questions just as do the other approaches. Programming the nanocell is going to be our most difficult task. While we have shown that in certain circumstances our molecular switches can hold their states for extended periods of time, we do not know if that will be true for the nanocell circuits. Will we be able to apply voltage pulses from the edges that will bring about changes in conductance of switches on the interior of the nanocell, through extended distances of molecular arrays? Deposition of the SAMs and packaging the completed nanocells will be monumental development tasks. However, even with these challenges, the prospects for a rapid assembly of molecular systems with few restrictions to fabrication make the nanocell approach enormously promising.

4.5 Characterization of Switches and Complex Molecular Devices

Now that we have outlined the major classes of molecular computing architectures that are under consideration, we will touch upon some of the basic component tests that have been done. The testing of molecular electronics components has been recently reviewed.[11,71] Seminario and Tour developed a density functional theory calculation for determination of the I(V) characteristics of molecules, the calculations from which corroborated well with laboratory results.[72]

Stoddart and Heath have formed solid-state, electronically addressable switching devices using bistable [2] catenane-based molecules sandwiched between an n-type polycrystalline Si bottom electrode and a metallic top electrode.[73] A mechanochemical mechanism, consistent with the temperature-dependent measurements of the device, was invoked for the action of the switch. Solid-state devices based on [2] or [3] rotaxanes were also constructed and analyzed by Stoddart and Heath.[74,75]

In collaboration with Bard, we have shown that it is possible to use tuning-fork-based scanning probe microscope (SPM) techniques to make stable electrical and mechanical contact to SAMs.[76] This is a promising technique for quick screening of molecular electronics candidates. Frisbie has used an Au-coated atomic-force microscope (AFM) tip to form metal–molecule–metal junctions with Au-supported SAMs. He has measured the I(V) characteristics of the junctions, which are approximately 15 nm,[2] containing about 75 molecules.[77] The I(V) behavior was probed as a function of the SAM thickness and the load applied to the microcontact. This may also prove to be a good method for quick screening of molecular electronics candidates.

In collaboration with Allara and Weiss, we have examined conductance switching in molecules 1, 2, and 4 (from Figure 4.20) by scanning tunneling microscopy (STM).[78] Molecules 1 and 2 have shown NDR effects under certain conditions, while molecule 4 did not.[63] SAMs made using dodecanethiol are known to be well packed and to have a variety of characteristic defect sites such as substrate step edges, film-domain boundaries, and substrate vacancy islands where other molecules can be inserted. When 1, 2, and 4 were separately inserted into the dodecanethiol SAMs, they protruded from the surrounding molecules due to their height differences. All three molecules had at least two states that differed in height by about 3 Å when observed by STM over time. Because topographic STM images represent a combination of the electronic and topographic structure of the surface, the height changes observed in the STM images could be due to a change in the physical height of the molecules, a change in the conductance of the molecules, or both. The more conductive state was referred to as ON, and the less conductive state was referred to as OFF. SAM formation conditions can be varied to produce SAMs with lower packing density. It was found that all three molecules switched ON and OFF more often in less ordered SAMs than in more tightly packed SAMs. Because a tightly packed SAM would be assumed to hinder conformational changes such as rotational twists, it was concluded that conformational changes controlled the conductance switching of all three molecules.

McCreery has used diazonium chemistry to form tightly packed monolayers on pyrolyzed photoresist film (PPF), a form of disordered graphitic material similar to glassy carbon.[79] Electrochemical reduction of stilbene diazonium salt in acetonitrile solvent in the presence of PPF forms a strong C–C bond between the stilbene molecule and carbons contained in the PPF. The I(V) characteristics of the stilbene junction was measured using Hg-drop electrode methods.

Lieber and coworkers constructed logic gates using crossed NW, which demonstrated substantial gain and were used to implement basic computations.[80] Avouris used SWNT that had been treated to prepare both p- and n-type nanotubes transistors to build voltage inverters, the first demonstration of nanotube-based logic gates.[81] They used spatially resolved doping to build the logic function on a single bundle of SWNT. Dekker and coworkers also built logic circuits with doped SWNT.[82] The SWNT were deposited from a dichloroethane suspension, and those tubes having a diameter of about 1 nm and situated atop preformed Al gate wires were selected by AFM. Schön and coworkers demonstrated gain for electron transport perpendicular to a SAM by using a third gate electrode.[83] The field-effect transistors based on SAMs demonstrate five orders of magnitude of conductance modulation and gain as high as six. In addition, using two-component SAMs, composed of both insulating and conducting molecules, three orders of magnitude changes in conductance can be achieved.[84]

4.6 Conclusion

It is clear that giant leaps remain to be made before computing devices based on molecular electronics are commercialized. The QCA area of research, which has seen demonstrations of logic gates and devices earlier than other approaches, probably has the highest hurdle due to the need to develop nanoscopic quantum dot manipulation and placement. Molecular-scale quantum dots are in active phases of research but have not been demonstrated. The crossbar-array approach faces similar hurdles since the advances to date have only been achieved by mechanical manipulation of individual NWs, still very much a research-based phenomenon and nowhere near the scale needed for commercialization. Pieces of the puzzle, such as flow control placement of small arrays, are attractive approaches but need continued development. To this point, self-assembly of the crossbar arrays, which would simplify the process considerably, has not been a tool in development. The realization of mortal programming and development of the overall nanocell assembly process are major obstacles facing those working in the commercialization of the nanocell approach to molecular electronics. As anyone knows who has had a computer program crash for no apparent reason, programming is a task in which one must take into account every conceivable perturbation while at the same time not knowing what every possible perturbation is — a difficult task, to say the least. Many cycles of testing and feedback analysis will need to occur with a working nanocell before we know that the programming of the nanocell is successful.

Molecular electronics as a field of research is rapidly expanding with almost weekly announcements of new discoveries and breakthroughs. Those practicing in the field have pointed to Moore's Law and inherent physical limitations of the present top-down process as reasons to make these discoveries and breakthroughs. They are aiming at a moving target, as evidenced by Intel's recent announcements of the terahertz transistor and an enhanced 0.13 μm process.[85–87] One cannot expect that companies with "iron in the ground" will stand still and let new technologies put them out of business. While some may be kept off the playing field by this realization, for others it only makes the area more exciting. Even as we outlined computing architectures here, the first insertion points for molecular electronics will likely not be for computation. Simpler structures such as memory arrays will probably be the initial areas for commercial molecular electronics devices. Once simpler structures are refined, more precise methods for computing architecture will be realized. Finally, by the time this review is published, we expect that our knowledge will have greatly expanded, and our expectations as to where the technology is headed will have undergone some shifts compared with where we were as we were writing these words. Hence, the field is in a state of rapid evolution, which makes it all the more exciting.

Acknowledgments

The authors thank DARPA administered by the Office of Naval Research (ONR); the Army Research Office (ARO); the U.S. Department of Commerce, National Institute of Standards and Testing (NIST); National Aeronautics and Space Administration (NASA); Rice University; and the Molecular Electronics Corporation for financial support of the research done in our group. We also thank our many colleagues for their hard work and dedication. Dustin K. James thanks David Nackashi for providing some references on semiconductor manufacturing. Dr. I. Chester of FAR Laboratories provided the trimethylsilylacetylene used in the synthesis shown in Scheme 4.2.

References

1. Moore, G.E., Cramming more components onto integrated circuits, *Electronics*, 38, 1965.
2. Hand, A., EUV lithography makes serious progress, *Semiconductor Intl.*, 24(6),15, 2001.
3. Selzer, R.A. et al., Method of improving X-ray lithography in the sub-100 nm range to create high-quality semiconductor devices, U.S. patent 6,295,332, 25 September 2001.
4. Heath, J.R., Kuekes, P.J. Snider, G.R., and Williams, R.S., A defect-tolerant computer architecture: opportunities for nanotechnology, *Science*, 280, 1716, 1998.
5. Reed, M.A. and Tour, J.M., Computing with molecules, *Sci. Am.*, 292, 86, 2000.
6. Whitney, D.E., Why mechanical design cannot be like VLSI design, *Res. Eng. Des.*, 8, 125, 1996.
7. Hand, A., Wafer cleaning confronts increasing demands, *Semiconductor Intl.*, 24 (August), 62, 2001.
8. Golshan, M. and Schmitt, S., Semiconductors: water reuse and reclaim operations at Hyundai Semiconductor America, *Ultrapure Water*, 18 (July/August), 34, 2001.
9. Overton, R., Molecular electronics will change everything, *Wired*, 8(7), 242, 2000.
10. Service, R.F., Molecules get wired, *Science*, 294, 2442, 2001.
11. Tour, J.M., Molecular electronics, synthesis and testing of components, *Acc. Chem. Res.*, 33, 791, 2000.
12. Petty, M.C., Bryce, M.R., and Bloor, D., *Introduction to Molecular Electronics*, Oxford University Press, New York, 1995.
13. Heath, J.R., Wires, switches, and wiring: a route toward a chemically assembled electronic nano-computer, *Pure Appl. Chem.*, 72, 11, 2000.
14. Hu, J., Odom, T.W., and Lieber, C.M., Chemistry and physics in one dimension: synthesis and properties of nanowires and nanotubes, *Acc. Chem. Res.*, 32, 435, 1999.
15. Rueckes, T., Kim, K., Joselevich, E., Tseng, G.Y., Cheung, C.–L., and Lieber, C.M., Carbon nano-tubes-based nonvolatile random access memory for molecular computing, *Science*, 289, 94, 2000.
16. Ouyang, M., Huang, J.–L., Cheung, C.-L., and Lieber, C.M., Atomically resolved single-walled carbon nanotubes intramolecular junctions, *Science*, 291, 97, 2001.
17. Star, A. et al., Preparation and properties of polymer-wrapped single-walled carbon nanotubes, *Angew. Chem. Intl. Ed.*, 40, 1721, 2001.
18. Bozovic, D. et al., Electronic properties of mechanically induced kinks in single-walled carbon nanotubes, *App. Phys. Lett.*, 78, 3693, 2001.
19. Huang, Y., Duan, X., Wei, Q., and Lieber, C.M., Directed assembly of one-dimensional nanostructures into functional networks, *Science*, 291, 630, 2001.
20. Gudiksen, M.S., Wang, J., and Lieber, C.M., Synthetic control of the diameter and length of single crystal semiconductor nanowires, *J. Phys. Chem. B*, 105, 4062, 2001.
21. Cui, Y. and Lieber, C.M., Functional nanoscale electronic devices assembled using silicon nanowire building blocks, *Science*, 291, 851, 2001.
22. Chung, S.-W., Yu, J.-Y, and Heath, J.R., Silicon nanowire devices, *App. Phys. Lett.*, 76, 2068, 2000.
23. Tour, J.M., Van Zandt, W.L., Husband, C.P., Husband, S.M., Libby, E.C., Ruths, D.A., Young, K.K., Franzon, P., and Nackashi, D., A method to compute with molecules: simulating the nanocell, submitted for publication, 2002.

24. Adleman, L.M., Computing with DNA, *Sci. Am.*, 279, 54, 1998.

25. Preskill, J., Reliable quantum computing, *Proc.R. Soc. Lond. A*, 454, 385, 1998.

26. Preskill, J., Quantum computing: pro and con, *Proc.R. Soc. Lond. A*, 454, 469, 1998.

27. Platzman, P.M. and Dykman, M.I., Quantum computing with electrons floating on liquid helium, *Science*, 284, 1967, 1999.

28. Kane, B., A silicon-based nuclear spin quantum computer, *Nature*, 393, 133, 1998.

29. Anderson, M.K., Dawn of the QCAD age, *Wired*, 9(9), 157, 2001.

30. Anderson, M.K., Liquid logic, *Wired*, 9(9), 152, 2001.

31. Wolf, S.A. et al., Spintronics: a spin-based electronics vision for the future, *Science*, 294, 1488, 2001.

32. Raymo, F.M. and Giordani, S., Digital communications through intermolecular fluorescence modulation, *Org. Lett.*, 3, 1833, 2001.

33. McEuen, P.L., Artificial atoms: new boxes for electrons, *Science*, 278, 1729, 1997.

34. Stewart, D.R. et al., Correlations between ground state and excited state spectra of a quantum dot, *Science*, 278, 1784, 1997.

35. Rajeshwar, K., de Tacconi, N.R., and Chenthamarakshan, C.R., Semiconductor-based composite materials: preparation, properties, and performance, *Chem. Mater.*, 13, 2765, 2001.

36. Leifeld, O. et al., Self-organized growth of Ge quantum dots on Si(001) substrates induced by sub-monolayer C coverages, *Nanotechnology*, 19, 122, 1999.

37. Sear, R.P. et al., Spontaneous patterning of quantum dots at the air–water interface, *Phys. Rev. E*, 59, 6255, 1999.

38. Markovich, G. et al., Architectonic quantum dot solids, *Acc. Chem. Res.*, 32, 415, 1999,

39. Weitz, I.S. et al., Josephson coupled quantum dot artificial solids, *J. Phys. Chem. B*, 104, 4288, 2000.

40. Snider, G.L. et al., Quantum-dot cellular automata: review and recent experiments (invited), *J. Appl. Phys.*, 85, 4283, 1999.

41. Snider, G.L. et al., Quantum-dot cellular automata: line and majority logic gate, *Jpn.J. Appl. Phys. Part I*, 38, 7227, 1999.

42. Toth, G. and Lent, C.S., Quasiadiabatic switching for metal-island quantum-dot cellular automata, *J. Appl. Phys.*, 85, 2977, 1999.

43. Amlani, I. et al., Demonstration of a six-dot quantum cellular automata system, *Appl. Phys. Lett.*, 72, 2179, 1998.

44. Amlani, I. et al., Experimental demonstration of electron switching in a quantum-dot cellular automata (QCA) cell, *Superlattices Microstruct.*, 25, 273, 1999.

45. Bernstein, G.H. et al., Observation of switching in a quantum-dot cellular automata cell, *Nanotechnology*, 10, 166, 1999.

46. Amlani, I., Orlov, A.O., Toth, G., Bernstein, G.H., Lent, C.S., and Snider, G.L., Digital logic gate using quantum-dot cellular automata, *Science*, 284, 289, 1999.

47. Orlov, A.O. et al., Experimental demonstration of a binary wire for quantum-dot cellular automata, *Appl. Phys. Lett.*, 74, 2875, 1999.

48. Amlani, I. et al., Experimental demonstration of a leadless quantum-dot cellular automata cell, *Appl. Phys. Lett.*, 77, 738, 2000.

49. Orlov, A.O. et al., Experimental demonstration of a latch in clocked quantum-dot cellular automata, *Appl. Phys. Lett.*, 78, 1625, 2001.

50. Tour, J.M., Kozaki, M., and Seminario, J.M., Molecular scale electronics: a synthetic/computational approach to digital computing, *J. Am. Chem. Soc.*, 120, 8486, 1998.

51. Lent, C.S., Molecular electronics: bypassing the transistor paradigm, *Science*, 288, 1597, 2000.

52. Bandyopadhyay, S., Debate response: what can replace the transistor paradigm?, *Science*, 288, 29, June, 2000.

53. Yu, J.-Y., Chung, S.-W., and Heath, J.R., Silicon nanowires: preparation, devices fabrication, and transport properties, *J. Phys. Chem.B.*, 104, 11864, 2000.

54. Ausman, K.D. et al., Roping and wrapping carbon nanotubes, *Proc. XV Intl. Winterschool Electron. Prop. Novel Mater.*, Euroconference Kirchberg, Tirol, Austria, 2000.

55. Bahr, J.L. et al., Dissolution of small diameter single-wall carbon nanotubes in organic solvents? *Chem. Commun.*, 2001, 193, 2001.

56. Mickelson, E.T. et al., Solvation of fluorinated single-wall carbon nanotubes in alcohol solvents, *J. Phys. Chem. B.*, 103, 4318, 1999.

57. Chen, J. et al., Dissolution of full-length single-walled carbon nanotubes, *J. Phys. Chem. B.*, 105, 2525, 2001.

58. Bahr, J.L. et al., Functionalization of carbon nanotubes by electrochemical reduction of aryl diazonium salts: a bucky paper electrode, *J. Am. Chem. Soc.*, 123, 6536, 2001.

59. Bahr, J.L. and Tour, J.M., Highly functionalized carbon nanotubes using *in situ* generated diazonium compounds, *Chem. Mater.*, 13, 3823, 2001,

60. O'Connell, M.J. et al., Reversible water-solubilization of single-walled carbon nanotubes by polymer wrapping, *Chem. Phys. Lett.*, 342, 265, 2001.

61. Pease, A.R. et al., Switching devices based on interlocked molecules, *Acc. Chem. Res.*, 34, 433, 2001.

62. Collins, P.G., Arnold, M.S., and Avouris, P., Engineering carbon nanotubes and nanotubes circuits using electrical breakdown, *Science*, 292, 706, 2001.

63. Chen, J., Reed, M.A., Rawlett, A.M., and Tour, J.M., Large on-off ratios and negative differential resistance in a molecular electronic device, *Science*, 286, 1550, 1999.

64. Derosa, P.A. and Seminario, J.M., Electron transport through single molecules: scattering treatment using density functional and green function theories, *J. Phys. Chem. B.*, 105, 471, 2001.

65. Seminario, J.M., Zacarias, A.G., and Derosa, P.A., Theoretical analysis of complementary molecular memory devices, *J. Phys. Chem. A.*, 105, 791, 2001.

66. Tour, J.M. et al., Synthesis and testing of potential molecular wires and devices, *Chem. Eur. J.*, 7, 5118, 2001.

67. Kosynkin, D.V. and Tour, J.M., Phenylene ethynylene diazonium salts as potential self-assembling molecular devices, *Org. Lett.*, 3, 993, 2001.

68. Chanteau, S. and Tour, J.M., Synthesis of potential molecular electronic devices containing pyridine units, *Tet. Lett.*, 42, 3057, 2001.

69. Dirk, S.M., Price, D.W. Jr., Chanteau, S., Kosynkin, D.V., and Tour, J.M., Accoutrements of a molecular computer: switches, memory components, and alligator clips, *Tetrahedron*, 57, 5109, 2001.

70. Seminario, J.M., De La Cruz, C.E., and Derosa, P.A., A theoretical analysis of metal–molecule contacts, *J. Am. Chem. Soc.*, 123, 5616, 2001.

71. Ward, M.D., Chemistry and molecular electronics: new molecules as wires, switches, and logic gates, *J. Chem. Ed.*, 78, 321, 2001.

72. Seminario, J.M., Zacarias, A.G., and Tour, J.M., Molecular current–voltage characteristics, *J. Phys. Chem.*, 103, 7883, 1999.

73. Collier, C.P. et al., A [2]catenane-based solid-state electronically reconfigurable switch, *Science*, 289, 1172, 2000.

74. Wong, E.W. et al., Fabrication and transport properties of single-molecule thick electrochemical junctions, *J. Am. Chem. Soc.*, 122, 5831, 2000.

75. Collier, C.P., Molecular-based electronically switchable tunnel junction devices, *J. Am. Chem. Soc.*, 123, 12632, 2001.

76. Fan, R.-F.F. et al., Determination of the molecular electrical properties of self-assembled monolayers of compounds of interest in molecular electronics, *J. Am. Chem. Soc.*, 123, 2424, 2001.

77. Wold, D.J. and Frisbie, C.D., Fabrication and characterization of metal–molecule–metal junctions by conducting probe atomic force microscopy, *J. Am. Chem. Soc.*, 123, 5549, 2001.

78. Donahauser, Z.J. et al., Conductance switching in single molecules through conformational changes, *Science*, 292, 2303, 2001.

79. Ranganathan, S., Steidel, I., Anariba, F., and McCreery, R.L., Covalently bonded organic monolayers on a carbon substrate: a new paradigm for molecular electronics, *Nano Lett.*, 1, 491, 2001.

80. Huang, Y. et al., Logic gates and computation from assembled nanowire building blocks, *Science*, 294, 1313, 2001.
81. Derycke, V., Martel, R., Appenzeller, J., and Avouris, P., Carbon nanotubes inter- and intramolecular logic gates, *Nano Lett.*, 1, 453, 2001.
82. Bachtold, A., Hadley, P., Nakanishi, T., and Dekker, C., Logic circuits with carbon nanotubes transistors, *Science*, 294, 1317, 2001.
83. Schön, J.H., Meng, H., and Bao, Z., Self-assembled monolayer organic field-effect transistors, *Nature*, 413, 713, 2001.
84. Schön, J.H., Meng, H., and Bao, Z., Field-effect modulation of the conductance of single molecules, *Science*, 294, 2138, 2001.
85. Chau, R. et al., A 50 nm Depleted-Substrate CMOS Transistor (DST), International Electron Devices Meeting, Washington, D.C., December 2001.
86. Barlage, D. et al., High-Frequency Response of 100 nm Integrated CMOS Transistors with High-K Gate Dielectrics, International Electron Devices Meeting, Washington, D.C., December 2001.
87. Thompson, S. et al., An Enhanced 130 nm Generation Logic Technology Featuring 60 nm Transistors Optimized for High Performance and Low Power at 0.7–1.4 V, International Electron Devices Meeting, Washington, D.C., December 2001.

5

Nanoelectronic Circuit Architectures

CONTENTS

Wolfgang Porod
University of Notre Dame

Abstract

In this chapter we discuss proposals for nanoelectronic circuit architectures, and we focus on those approaches which are different from the usual wired interconnection schemes employed in conventional silicon-based microelectronics technology. In particular, we discuss the *Quantum-Dot Cellular Automata* (QCA) concept, which uses physical interactions between closely spaced nanostructures to provide local connectivity. We also highlight an approach to molecular electronics that is based on the (imperfect) chemical self-assembly of atomic-scale switches in crossbar arrays and software solutions to provide defect-tolerant reconfiguration of the structure.

5.1   Introduction

The integrated circuit (IC), manufactured by optical lithography, has driven the computer revolution for three decades. Silicon-based technology allows the fabrication of electronic devices with high reliability and circuits with near-perfect precision. In fact, the main challenges facing conventional IC technology are not so much in making the devices but in interconnecting them and managing power dissipation.

IC miniaturization has provided the tools for imaging, manipulating, and modeling on the nanometer scale. These new capabilities have led to the discovery of new physical phenomena, which have been the basis for new device proposals.[1,2] Opportunities for nanodevices include low power, high packing densities, and speed. While there has been significant attention paid to the physics and chemistry of nanometer-scale device structures, there has been less appreciation for the need for new interconnection strategies for these new kinds of devices. In fact, the key problem is not so much how to make individual devices but how to interconnect them in appropriate circuit architectures.

Nanotechnology holds the promise to put a trillion molecular-scale devices in a square centimeter.[3] How does one assemble a trillion devices per square centimeter? Moreover, this needs to be done quickly, inexpensively, and sufficiently reliably. What does one do with a trillion devices? If we assume that one can make them (and they actually work), how can this massive amount of devices be harnessed for useful computation? These questions highlight the need for innovative nanoelectronic circuit architectures.

Recently there has been significant progress in addressing the above issues. To wit, *nanocircuits* have been featured as the "Breakthrough of the Year 2001" in *Science* magazine.[4] Recent accomplishments include the fabrication of molecular circuits that are capable of performing logic operations.

The focus of this chapter is on the architectural aspects of nanometer-scale device structures. Device and fabrication issues will be referred to the literature. As a note of caution, this chapter attempts to survey an area that is under rapid development. Some of the architecture ideas described here have not yet been realized due to inherent fabrication difficulties. We attempt to highlight ideas that are at the forefront of the development of circuit architectures for nanoelectronic devices.

5.2 Quantum-Dot Cellular Automata (QCA)

As device sizes shrink and packing densities increase, device–device interactions are expected to become ever more prominent.[5,6] While such parasitic coupling represents a problem for conventional circuitry, and efforts are being made to avoid it, such interactions may also represent an opportunity for alternate designs that utilize device–device coupling. Such a scheme appears to be particularly well suited for closely spaced quantum-dot structures, and the general notion of single-electrons switching on interacting quantum dots was first formulated by Ferry and Porod.[7]

Based upon the emerging technology of quantum-dot fabrication, the Notre Dame NanoDevices group has proposed a scheme for computing with cells of coupled quantum dots,[8] which has been termed *Quantum-Dot Cellular Automata* (QCA). To our knowledge, this is the first concrete proposal to utilize quantum dots for computing. There had been earlier suggestions that device–device coupling might be employed in a cellular-automaton-like scheme, but without accompanying concrete proposals for a specific implementation.[7,9–11]

The QCA cellular architecture is similar to other cellular arrays, such as *Cellular Neural/Nonlinear Networks* (CNN)[12,13] in that they repeatedly employ the same basic cell with its associated near-neighbor interconnection pattern. The difference is that CNN cells have been realized by conventional CMOS circuitry, and the interconnects are provided by wires between cells in a local neighborhood.[14] For QCA, on the other hand, the coupling between cells is given by their direct physical interactions (and not by wires), which naturally takes advantage of the fringing fields between closely spaced nanostructures. The physical mechanisms available for interactions in such field-coupled architectures are electric (Coulomb) or magnetic interactions, in conjunction with quantum-mechanical tunneling.

5.2.1 Quantum-Dot Cell

The original QCA proposal is based on a quantum-dot cell which contains five dots[15] as schematically shown in Figure 5.1(a). The quantum dots are represented by the open circles, which indicate the confining electronic potential. In the ideal case, each cell is occupied by two electrons, which are schematically shown as the solid dots. The electrons are allowed to tunnel between the individual quantum dots inside an individual cell, but they are not allowed to leave the cell, which may be controlled during fabrication by the physical dot–dot and cell–cell distances.

This quantum-dot cell represents an interesting dynamical system. The two electrons experience their mutual Coulombic repulsion, yet they are constrained to occupy the quantum dots inside the cell. If left alone, they will seek, by hopping between the dots, the configuration that corresponds to the physical ground state of the cell. It is clear that the two electrons will tend to occupy different dots on opposing corners of the cell because of the Coulomb energy cost associated with having them on the same dot or

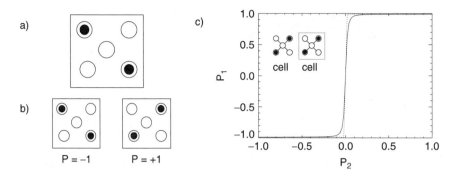

FIGURE 5.1 (a) Schematic diagram of a QCA cell consisting of 5 quantum dots and occupied by 2 electrons. (b) The two basic electronic arrangements in the cell, which can be used to represent binary information, P = +1 and P = −1. (c) Cell–cell response, which indicates that cell 1 abruptly switches when "driven" by only a small charge asymmetry in cell 2.

bringing them together in closer proximity. It is easy to see that the ground state of the system will be an equal superposition of the two basic configurations, as shown in Figure 5.1(b).

We may associate a *polarization* (P = +1 or P = −1) with either basic configuration of the two electrons in each cell. Note that this polarization is not a dipole moment but a measure for the alignment of the charge along the two cell diagonals. These two configurations may be interpreted as binary information, thus encoding bit values in the electronic arrangement inside a single cell. Any polarization between these two extreme values is possible, corresponding to configurations where the electrons are more evenly "smeared out" over all dots. The ground state of an isolated cell is a superposition with equal weight of the two basic configurations and therefore has a net polarization of zero.

As described in the literature,[16–18] this cell has been studied by solving the Schrödinger equation using a quantum-mechanical model Hamiltonian. Without going into the details, the basic ingredients of the theory are (1) the quantized energy levels in each dot, (2) the coupling between the dots by tunneling, (3) the Coulombic charge cost for a doubly occupied dot, and (4) the Coulomb interaction between electrons in the same cell and also with those in neighboring cells. Numerical solutions of the Schrödinger equation confirm the intuitive understanding that the ground state is a superposition of the P = +1 and P = −1 states. In addition to the ground state, the Hamiltonian model yields excited states and cell dynamics.

The properties of an isolated cell were discussed above. Figure 5.1(c) shows how one cell is influenced by the state of its neighbor. As schematically depicted in the inset, the polarization of cell 2 is presumed to be fixed at a given value P_2, corresponding to a specific arrangement of charges in cell 2; and this charge distribution exerts its influence on cell 1, thus determining its polarization P_1. Quantum-mechanical simulations of these two cells yield the polarization response function shown in the figure. The important finding here is the strongly nonlinear nature of this cell–cell coupling. As can be seen, cell 1 is almost completely polarized even though cell 2 might only be partially polarized, and a small asymmetry of charge in cell 2 is sufficient to break the degeneracy of the two basic states in cell 1 by energetically favoring one configuration over the other.

5.2.2 QCA Logic

This bistable saturation is the basis for the application of such quantum-dot cells for computing structures. The above conclusions regarding cell behavior and cell–cell coupling are not specific to the five-dot cell discussed so far, but they generalize to other cell configurations. Similar behavior is found for alternate cell designs, such as cells with only four dots in the corners (no central dot), or even cells with only two dots (molecular dipole). Based upon this bistable behavior of the cell–cell coupling, the cell polarization can be used to encode binary information. It has been shown that arrays of such physically interacting cells may be used to realize any Boolean logic functions.[16–18]

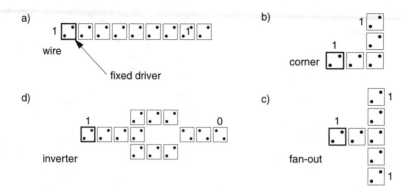

FIGURE 5.2 Examples of simple QCA structures showing (a) a binary wire, (b) signal propagation around a corner, (c) wire splitting and fan-out, and (d) an inverter.

Figure 5.2 shows examples of some basic QCA arrays. In each case, the polarization of the cell at the edge of the array is kept fixed; this so-called driver cell represents the input polarization, which determines the state of the whole array. Each figure shows the cell polarizations that correspond to the physical ground-state configuration of the whole array. Figure 5.2(a) shows that a line of cells allows the propagation of information, thus realizing a binary wire. Note that only information but no electric current flows down the line. Information can also flow around corners, as shown in Figure 5.2(b), and fan-out is possible as shown in Figure 5.2(c). A specific arrangement of cells, such as the one shown in Figure 5.2(d), may be used to realize an inverter. In each case, electronic motion is confined to within a given cell but not between different cells. Only information, but not charge, is allowed to propagate over the whole array. This absence of current flow is the basic reason for the low power dissipation in QCA structures.

The basic logic function that is *native* to the QCA system is majority logic.[16–18] Figure 5.3 shows a majority logic gate, which simply consists of an intersection of lines, and the *device cell* is only the one in the center. If we view three of the neighbors as inputs (kept fixed), then the polarization of the output cell is the one which *computes* the majority votes of the inputs. The figure also shows the majority logic truth table, which was computed (using the quantum-mechanical model) as the physical ground-state polarizations for a given combination of inputs. Using conventional circuitry, the design of a majority logic gate would be significantly more complicated. The new physics of quantum mechanics gives rise to new functionality, which allows a rather compact realization of majority logic. Note that conventional AND and OR gates are hidden in the majority logic gate. Inspection of the majority logic truth table reveals that if input A is kept fixed at 0, the remaining two inputs B and C realize an AND gate. Conversely, if A is held at 1, inputs B and C realize a binary OR gate. In other words, majority logic gates may be

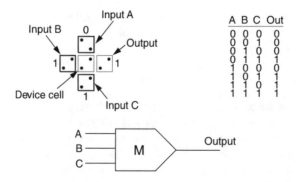

A	B	C	Out
0	0	0	0
0	0	1	0
0	1	0	0
0	1	1	1
1	0	0	0
1	0	1	1
1	1	0	1
1	1	1	1

FIGURE 5.3 Majority logic gate, which basically consists of an intersection of lines. Also shown are the computed majority logic truth table and the logic symbol.

viewed as programmable AND and OR gates. This opens up the interesting possibility that the functionality of the gate may be determined by the state of the computation itself.

One may conceive of larger arrays representing more complex logic functions. The largest structure simulated so far (containing some 200 cells) is a single-bit full adder, which may be designed by taking advantage of the QCA majority logic gate as a primitive.[16]

5.2.3 Computing with QCA

In a QCA array, cells interact with their neighbors; and neither power nor signal wires are brought to each cell. In contrast to conventional circuits, one does not have external control over each and every interior cell. Therefore, a new way is needed of using such QCA arrays for computing. The main concept is that the information in a QCA array is contained in the physical ground state of the system. The two key features that characterize this new computing paradigm are *computing with the ground state* and *edge-driven computation*, which will be discussed in further detail below. Figure 5.4 schematically illustrates the main idea.

5.2.3.1 Computing with the Ground State

Consider a QCA array before the start of a computation. The array, left to itself, will have assumed its physical ground state. Presenting the input data, i.e., setting the polarization of the input cells, will deliver energy to the system, thus promoting the array to an excited state. The computation consists in the array reaching the new ground-state configuration, compatible with the boundary conditions given by the fixed input cells. Note that the information is contained in the ground state itself and not in how the ground state is reached. This relegates the question of the dynamics of the computation to one of secondary importance although it is of significance, of course, for actual implementations. In the following, we will discuss two extreme cases for these dynamics — one where the system is completely left to itself, and another where exquisite external control is exercised.

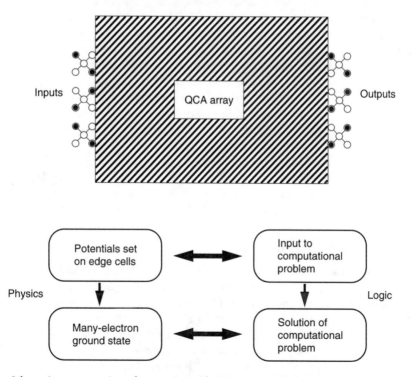

FIGURE 5.4 Schematic representation of computing with QCA arrays. The key concepts are *computing with the ground state* and *edge-driven computation*. The physical evolution of the QCA structure is designed to mimic the logical solution path from input to output.

- *Let physics do the computing*: the natural tendency of a system to assume the ground state may be used to drive the computation process. Dissipative processes due to the unavoidable coupling to the environment will relax the system from the initial excited state to the new ground state. The actual dynamics will be tremendously complicated because all the details of the system–environment coupling are unknown and uncontrollable. However, we do not have to concern ourselves with the detailed path in which the ground state is reached, as long as the ground state is reached. The attractive feature of this relaxation computation is that no external control is needed. However, there also are drawbacks in that the system may get "stuck" in metastable states and that there is no fixed time in which the computation is completed.

- *Adiabatic computing*: due to the above difficulties associated with metastable states, Lent and co-workers have developed a clocked adiabatic scheme for computing with QCAs. The system is always kept in its instantaneous ground state, which is adiabatically transformed during the computation from the initial state to the desired final state. This is accomplished by lowering or raising potential barriers within the cells in concert with clock signals. The modulation of the potential barriers allows or inhibits changes of the cell polarization. The presence of clocks makes synchronized operation possible, and pipelined architectures have been proposed.[16] As an alternative to wired clocking schemes, optical pumping has been investigated as a means of providing power for signal restoration.[19]

5.2.3.2 Edge-Driven Computation

Edge-driven computation means that only the periphery of a QCA array can be contacted, which is used to write the input data and to read the output of the computation. No internal cells may be contacted directly. This implies that no signals or power can be delivered from the outside to the interior of an array. All interior cells only interact within their local neighborhood. The absence of signal and power lines to each and every interior cell has obvious benefits for the interconnect problem and the heat dissipation. The lack of direct contact to the interior cells also has profound consequences for the way such arrays can be used for computation. Because no power can flow from the outside, interior cells cannot be maintained in a far-from-equilibrium state. Because no external signals are brought to the inside, internal cells cannot be influenced directly. These are the reasons why the ground state of the whole array is used to represent the information, as opposed to the states of each individual cell. In fact, edge-driven computation *necessitates* computing with the ground state. Conventional circuits, on the other hand, maintain devices in a far-from-equilibrium state. This has the advantage of noise immunity, but the price to be paid comes in the form of the wires needed to deliver the power (contributing to the wiring bottleneck) and the power dissipated during switching (contributing to the heat dissipation problem).

A formal link has been established between the QCA and CNN paradigms, which share the common feature of near-neighbor coupling. While CNN arrays obey completely classical dynamics, QCAs are mixed classical/quantum-mechanical systems. We refer to the literature for the details on such *Quantum Cellular Neural Network* (Q-CNN) systems.[20,21]

5.2.4 QCA Implementations

The first QCA cell was demonstrated using Coulomb-coupled metallic islands.[22] These experiments showed that the position of one single electron can be used to control the position (switching) of a neighboring single electron. This demonstrated the proof-of-principle of the QCA paradigm — that information can be encoded in the arrangements of electronic charge configurations. In these experiments, aluminum Coulomb-blockade islands represented the dots, and aluminum tunnel junctions provided the coupling. Electron-beam lithography and shadow evaporation were used for the fabrication. In a similar fashion, a binary QCA was realized.[23] The binary wire consisted of capacitively coupled double-dot cells charged with single electrons. The polarization switch caused by an applied input signal in one cell led to the change in polarization of the adjacent cell and so on down the line, as in falling dominos. Wire polarization was measured using single islands as electrometers. In addition, a functioning logic gate was also realized,[24] where digital data was encoded in the positions of only two electrons. The logic gate consisted of a cell, composed of four dots connected in a ring by tunnel junctions, and two single-dot electrometers. The device operated by applying voltage inputs to the gates of the cell. Logic AND and OR operations have been verified using the electrometer outputs. A drawback of these QCA realizations using metallic Coulomb-blockade islands are operation at cryogenic temperatures. Recent experimental progress in this area is summarized elsewhere.[25]

Molecular-scale QCA implementations hold the promise of room-temperature operation. The small size of molecules means that Coulomb energies are much larger than for metallic dots, so operation at higher temperatures is possible. QCA molecules must have several redox centers, which act as quantum dots and which are arranged in the proper geometry. Furthermore, these redox centers must be able to respond to the local field created by another nearby QCA molecule. Several classes of molecules have been identified as candidates for possible molecular QCA operation.[26] It is emphasized that QCA implementations for molecular electronics represent an alternate viewpoint to the conventional approaches taken in the field of molecular electronics, which commonly use molecules as wires or switches. QCA molecules are not used to conduct electronic charge, but they represent structured *charge containers* that communicate with neighboring molecules through Coulombic interactions generated by particular charge arrangements inside the molecule.

Magnetic implementations appear to be another promising possibility for room-temperature operation. Recent work demonstrated that QCA-like arrays of interacting submicrometer magnetic dots can be used to perform logic operations and to propagate information at room temperature.[27,28] The logic states are represented by the magnetization directions of single-domain magnetic dots, and the dots couple to their nearest neighbors through magnetostatic interactions.

5.3 Single-Electron Circuits

The physics of single-electron tunneling is well understood,[29,30] and several possible applications have been explored.[31] Single-electron transistors (SETs) can, in principle, be used in circuits similar to conventional silicon field-effect transistors (MOSFETs),[32,33] including complementary CMOS-type circuits.[34] In these applications the state of each node in the circuit is characterized by a voltage, and one device communicates with other devices through the flow of current. The peculiar nonlinear nature of the SET I-V characteristic has led to proposals of SETs in synaptic neural-network circuits and for implementations of cellular neural/nonlinear networks.[35,36]

There are, however, practical problems in using single-electron transistors as logic devices in conventional circuit architectures. One of the main problems is related to the presence of stray charges in the surrounding circuitry, which change the SET characteristics in an uncontrollable way. Because the SET is sensitive to the charge of one electron, only a small fluctuation in the background potential is sufficient to change the Coulomb-blockade condition. Also, standard logic devices rely on the high gain of conventional metal–oxide–semiconductor (MOS) transistors, which allow circuit design with fan-out. In contrast, the gain of SETs is rather small, which limits its usefulness in MOSFET-like circuit architectures.

An interesting SET logic family has been proposed that does not rely on high gain. This architecture is based on the binary decision diagram (BDD).[37,38] The BDD consists of an array of current pathways connected by Coulomb-blockade switching nodes. These nodes do not need high gain but only distinct ON-OFF switching characteristics that can be realized by switching between a blockaded state and a completely pinched-off state (which minimizes the influence of stray potentials). The functioning of such a CB BDD structure was demonstrated in experiment, which included the demonstration of the AND logic function.[39]

In contrast to the above SET-based approaches, the quantized electronic charge can be used to directly encode digital information.[40,41] Korotkov and Likharev proposed the SET parametron,[41] which is a wireless single-electron logic family. As schematically shown in Figure 5.5, the basic building block of this logic family consists of three conducting islands, where the middle island is slightly shifted off

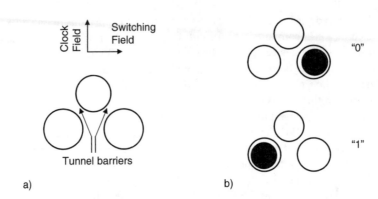

FIGURE 5.5 Schematic diagram of the SET parametron. (a) The basic cell consisting of 3 quantum dots and occupied by a single electron. (b) The two basic electronic arrangements in the cell, which can be used to represent binary information. (Adapted from A.N. Korotkov and K.K. Likharev, Single-electron-parametron-based logic devices, *J. Appl. Phys.*, 84(11), 6114–6126, 1998.)

the line passing through the centers of the edge island. Electrons are allowed to tunnel through small gaps between the middle and edge islands but not directly between the edge islands (due to their larger spatial separation).

Let us assume that each cell is occupied by one additional electron and that a *clock* electric field is applied that initially pushes this electron onto the middle island (the direction of this clock field is perpendicular to the line connecting the edge islands). Now that the electron is located on the central island, the clock field is reduced, and the electron eventually changes direction. At some point in time during this cycle, it will be energetically favorable for the electron to tunnel off the middle island and onto one of the edge islands. If both islands are identical, the choice of island will be random. However, this symmetry can be broken by a small switching field that is applied perpendicular to the clock field and along the line of the edge cells. This control over the left–right final position of the electron can be interpreted as one bit of binary information; the electron on the right island might mean logical "1" and the left island logical "0." (This encoding of logic information by the arrangement of electronic charge is similar to the QCA idea.) An interesting consequence of the asymmetric charge configuration in a switched cell is that the resulting electric dipole field can be used to switch neighboring cells during their decision-making moment.

Based on this near-neighbor interacting single-electron-parametron cell, a family of logic devices has been proposed.[41] A line of cells acts as an inverter chain and can be thought of as a shift register. Lines of cells can also split into two, thus providing fan-out capability. In addition, SET parametron logic gates have been proposed, which include NAND and OR gates.

Another interesting possibility for single-electron circuits is the proposal by Kiehl and Oshima to encode information by the phase of bistable phase-locked elements. We refer to the literature for the details.[42,43]

5.4 Molecular Circuits

Chemical self-assembly processes look promising because they, in principle, allow vast amounts of devices to be fabricated very cheaply. But there are key problems: (1) the need to create complex circuits for computers appears to be ill suited for chemical self-assembly, which yields mostly regular (periodic) structures; and (2) the need to deal with very large numbers of components and to arrange them into useful structures is a hard problem (NP-hard problem).

One approach to molecular electronics is to build circuits in analogy to conventional silicon-based electronics. The idea is to find molecular analogues of electronic devices (such as wires, diodes, transistors) and to then assemble these into molecular circuits. This approach is reviewed and described in work by Ellenbogen and Love.[44] An electronically programmable memory device based on self-assembled molecular monolayers was recently reported by the groups of Reed and Tour.[45–47]

The discovery of carbon nanotubes[48] provided a new building block for the construction of molecular-scale circuits. Dekker's group demonstrated a carbon nanotube single-electron transistor operating at room temperature.[49] In subsequent work, the same group constructed logic circuits with field-effect transistors based on single-wall carbon nanotubes, which exhibited power gain (> 10) and large on-off ratios (>10^5). A local-gate layout allowed for integration of multiple devices on a single chip; and one-, two-, and three-transistor circuits were demonstrated that exhibited digital logic operations, such as an inverter, a logic NOR, and a static random-access memory cell.[50] In related work, Lieber's group has demonstrated logic circuits based on nanotube and semiconductor nanowires,[51–53] and Avouris' group built an inverter from chemically doped nanotubes on a silicon substrate.[54]

Another idea of a switch (and related circuitry) at the molecular level is the (mechanical) concept of an atom relay, which was proposed by Wada and co-workers.[55,56] The atom relay is a switching device based upon the controlled motion of a single atom. The basic configuration of an atom relay consists of a conducting atom wire, a switching atom, and a switching gate. The operation principle of the atom relay is that the switching atom is displaced from the atom wire due to an applied electric field on the switching gate ("off" state of the atom relay). Memory cell and logic gates (such as NAND and NOR functions) based on the atom relay configuration have been proposed, and their operation was examined through simulation.

The above circuit approaches are patterned after conventional microelectronic circuit architectures, and they require the same level of device reliability and near-perfect fabrication yield. This is an area of concern because it is far from obvious that future molecular-electronics fabrication technologies will be able to rival the successes of the silicon-based microelectronics industry. There are several attempts to address these issues, and we will discuss the approach taken by the Hewlett-Packard and University of California research team.[57,58] This approach uses both chemistry (for the massively parallel construction of molecular components, albeit with unavoidable imperfections) and computer science (for a defect-tolerant reconfigurable architecture that allows one to download complex electronic designs).

This reconfigurable architecture is based on an experimental computer which was developed at Hewlett-Packard Laboratories in the middle 1990s.[59] Named *Teramac*, it was first constructed using conventional silicon integrated-circuit technology in an attempt to develop a fault-tolerant architecture based on faulty components. Teramac was named for *Tera*, 10^{12} operations per second (e.g., 10^6 logic elements operating at 10^6 Hz), and *mac* for "multiple architecture computer." It basically is a large custom-configurable, highly parallel computer that consists of field-programmable gate arrays (FPGAs), which can be programmed to reroute interconnections to avoid faulty components.

The HP design is based on a crossbar (Manhattan) architecture, in which two sets of overlapping nanowires are oriented perpendicularly to each other. Each wire crossing becomes the location of a molecular switch, which is sandwiched between the top and bottom wires. Erbium disilicide wires are used, which are 2 nm in diameter and 9 nm apart. The switches are realized by rotaxane molecules, which preferentially attach to the wires and are electrically addressable, i.e., electrical pulses lead to solid-state processes (analogous to electrochemical reduction or oxidation). These set or reset the switches by altering the resistance of the molecule. Attractive features of the crossbar architecture include its regular structure, which naturally lends itself to chemical self-assembly of a large number of identical units and to a defect-tolerant strategy by having large numbers of potential "replacement parts" available.

Crossbars are natural for memory applications but can also be used for logic operations. They can be configured to compute Boolean logic expressions, such as wired ANDs followed by wired ORs. A 6×6 diode crossbar can perform the function of a 2-bit adder.[60]

The ability of a reconfigurable architecture to create a functional system in the presence of defective components represents a radical departure from today's microelectronics technology.[60] Future nanoscale information processing systems may not have a central processing unit, but they may instead contain extremely large configuration memories for specific tasks which are controlled by a tutor that locates and avoids the defects in the system. The traditional paradigm for computation is to design the computer, build it perfectly, compile the program, and then run the algorithm. On the other hand, the Teramac paradigm is to build the computer (imperfectly), find the defects, configure the resources with software, compile the program, and then run it. This new paradigm moves tasks that are difficult to do in hardware into software tasks.

5.5 Summary

If we are to continue to build complex systems of ever-smaller components, we must find new technologies, in conjunction with appropriate circuit architectures, that will allow massively parallel construction of electronic circuits at the atomic scale. In this chapter we discussed several proposals for nanoelectronic circuit architectures, and we focused on those approaches that are different from the usual wired interconnection schemes employed in conventional silicon-based microelectronics technology. In particular, we discussed the *Quantum-Dot Cellular Automata* (QCA) concept, which uses physical interactions between closely spaced nanostructures to provide local connectivity.[61,62] We also highlighted an approach to molecular electronics that is based on the (imperfect) chemical self-assembly of atomic-scale switches in crossbar arrays and software solutions to provide defect-tolerant reconfiguration of the structure.

Acknowledgments

I would like to acknowledge many years of fruitful collaborations with my colleagues in Notre Dame's Center for Nano Science and Technology. Special thanks go to Professor Arpad Csurgay for many discussions that have strongly shaped my views of the field. This work was supported in part by grants from the Office of Naval Research (through a MURI project) and the W.M. Keck Foundation.

References

1. D. Goldhaber–Gordon, M.S. Montemerlo, J.C. Love, G.J. Opiteck, and J.C. Ellenbogen, Overview of nanoelectronic devices, *Proc. IEEE*, 85(4), 541–557 (1997).
2. T. Ando, Y. Arakawa, K. Furuya, S. Komiyama, and H. Nakashima (Eds.), *Mesoscopic Physics and Electronics*, NanoScience and Technology Series, Springer Verlag, Heidelberg (1998).
3. G.Y. Tseng and J.C. Ellenbogen, Towards nanocomputers, *Science*, 294, 1293–1294 (2001).
4. Editorial: Breakthrough of the year 2001, molecules get wired, *Science*, 294, 2429–2443 (2001).
5. J.R. Barker and D.K. Ferry, Physics, synergetics, and prospects for self-organization in submicron semiconductor device structures, in *Proc. 1979 Intl. Conf. Cybernetics Soc.*, IEEE Press, New York (1979), p. 762.
6. J.R. Barker and D.K. Ferry, On the physics and modeling of small semiconductor devices – II, *Solid-State Electron.*, 23, 531–544 (1980).
7. D.K. Ferry and W. Porod, Interconnections and architecture for ensembles of microstructures, *Superlattices Microstruct.*, 2, 41 (1986).
8. C.S. Lent, P.D. Tougaw, W. Porod, and G.H. Bernstein, Quantum cellular automata, *Nanotechnology*, 4, 49–57 (1993).
9. R.O. Grondin, W. Porod, C.M. Loeffler, and D.K. Ferry, Cooperative effects in interconnected device arrays, in F.L. Carter (Ed.), *Molecular Electronic Devices II*, Marcel Dekker, (1987), pp. 605–622.
10. V. Roychowdhuri, D.B. Janes, and S. Bandyopadhyay, Nanoelectronic architecture for Boolean logic, *Proc. IEEE*, 85(4), 574–588 (1997).
11. P. Bakshi, D. Broido, and K. Kempa, Spontaneous polarization of electrons in quantum dashes, *J. Appl. Phys.*, 70, 5150 (1991).
12. L.O. Chua and L. Yang, Cellular neural networks: theory, and CNN applications, *IEEE Trans. Circuits Systems*, CAS-35, 1257–1290 (1988).
13. L.O. Chua (Ed.), Special issue on nonlinear waves, patterns and spatio-temporal chaos in dynamic arrays, *IEEE Trans. Circuits Systems I. Fundamental Theory Appl.*, 42, 10 (1995).
14. J. A. Nossek and T. Roska, Special issue on cellular neural networks, *IEEE Trans. Circuits Systems I. Fundamental Theory Appl.*, 40, 3 (1993).
15. C.S. Lent, P.D. Tougaw, and W. Porod, Bistable saturation in coupled quantum dots for quantum cellular automata, *Appl. Phys. Lett.*, 62, 714–716 (1993).

16. C.S. Lent and P.D. Tougaw, A device architecture for computing with quantum dots, *Proc. IEEE*, 85(4), 541–557 (1997).

17. W. Porod, Quantum-dot cellular automata devices and architectures, *Intl. J. High Speed Electron. Syst.*, 9(1), 37–63 (1998).

18. W. Porod, C.S. Lent, G.H. Bernstein, A.O. Orlov, I. Amlani, G.L. Snider, and J.L. Merz, Quantum-dot cellular automata: computing with coupled quantum dots, invited paper in the Special Issue on Single Electronics, *Intl. J. Electron.*, 86(5), 549–590 (1999).

19. G. Csaba, A.I. Csurgay, and W. Porod, Computing architecture composed of next-neighbor-coupled optically-pumped nanodevices, *Intl. J. Circuit Theory Appl.*, 29, 73–91 (2001).

20. G. Toth, C.S. Lent, P.D. Tougaw, Y. Brazhnik, W. Weng, W. Porod, R.-W. Liu and Y.-F. Huang, Quantum cellular neural networks, *Superlattices Microstruct.*, 20, 473–477 (1996).

21. W. Porod, C.S. Lent, G. Toth, H. Luo, A. Csurgay, Y.-F. Huang, and R.-W. Liu, (Invited), Quantum-dot cellular nonlinear networks: computing with locally-connected quantum dot arrays, *Proc. 1997 IEEE Intl. Symp. Circuits Syst.: Circuits Systems Inform. Age* (1997), pp. 745–748.

22. A.O. Orlov, I. Amlani, G.H. Bernstein, C.S. Lent, and G.L. Snider, Realization of a functional cell for quantum-dot cellular automata, *Science*, 277, 928–930 (1997).

23. O. Orlov, I. Amlani, G. Toth, C.S. Lent, G.H. Bernstein, and G.L. Snider, Experimental demonstration of a binary wire for quantum-dot cellular automata, *Appl. Phys. Lett.*, 74(19), 2875–2877 (1999).

24. I. Amlani, A.O. Orlov, G. Toth, G.H. Bernstein, C.S. Lent, and G.L. Snider, Digital logic gate using quantum-dot cellular automata, *Science*, 284, 289–291, 1999.

25. G.L. Snider, A.O. Orlov, I. Amlani, X. Zuo, G.H. Bernstein, C.S. Lent, J. L. Merz, and W. Porod, Quantum-dot cellular automata: Review and recent experiments (invited), *J. Appl. Phys.*, 85(8), 4283–4285 (1999).

26. M. Lieberman, S. Chellamma, B. Varughese, Y. Wang, C.S. Lent, G.H. Bernstein, G.L. Snider, and F.C. Peiris, Quantum-dot cellular automata at a molecular scale, *Molecular Electronics II*, Ari Aviram, Mark Ratner, and Vladimiro Mujia (Eds.), *Ann. N.Y. Acad. Sci.*, 960, 225–239 (2002).

27. R.P. Cowburn and M.E. Welland, Room temperature magnetic quantum cellular automata, *Science*, 287, 1466–1468 (2000).

28. G. Csaba and W. Porod, Computing architectures for magnetic dot arrays, presented at the First International Conference and School on Spintronics and Quantum Information Technology, Maui, Hawaii, May 2001.

29. D.V. Averin and K.K. Likharev, in B.L. Altshuler et al. (Eds.), *Mesoscopic Phenomena in Solids*, Elsevier, Amsterdam, (1991), p. 173.

30. H. Grabert and M.H. Devoret (Eds.), *Single Charge Tunneling*, Plenum, New York (1992).

31. D.V. Averin and K.K. Likharev, Possible applications of single charge tunneling, in H. Grabert and M.H. Devoret (Eds.), *Single Charge Tunneling*, Plenum, New York (1992), pp. 311–332.

32. K.K. Likharev, Single-electron transistors: electrostatic analogs of the DC SQUIDs, *IEEE Trans. Magn.*, 2, 23, 1142–1145 (1987).

33. R.H. Chen, A.N. Korotkov, and K.K. Likharev, Single-electron transistor logic, *Appl. Phys. Lett.*, 68, 1954–1956 (1996).

34. J.R. Tucker, Complementary digital logic based on the Coulomb blockade, *J. Appl. Phys.* 72, 4399–4413 (1992).

35. C. Gerousis, S.M. Goodnick, and W. Porod, Toward nanoelectronic cellular neural networks, *Intl. J. Circuit Theory Appl.*, 28, 523–535 (2000).

36. X. Wang and W. Porod, Single-electron transistor analytic I-V model for SPICE simulations, *Superlattices Microstruct.*, 28(5/6), 345–349 (2000).

37. N. Asahi, M. Akazawa, and Y. Amemiya, Binary-decision-diagram device, *IEEE Trans. Electron Devices*, 42, 1999–2003 (1995).

38. N. Asahi, M. Akazawa, and Y. Amemiya, Single-electron logic device based on the binary decision diagram, *IEEE Trans. Electron Devices*, 44, 1109–1116 (1997).

39. K. Tsukagoshi, B.W. Alphenaar, and K. Nakazato, Operation of logic function in a Coulomb blockade device, *Appl. Phys. Lett.*, 73, 2515–2517 (1998).

40. M.G. Ancona, Design of computationally useful single-electron digital circuits, *J. Appl. Phys.* 79, 526–539 (1996).

41. A.N. Korotkov and K.K. Likharev, Single-electron-parametron-based logic devices, *J. Appl. Phys.*, 84(11), 6114–6126 (1998).

42. R.A. Kiehl and T. Oshima, Bistable locking of single-electron tunneling elements for digital circuitry, *Appl. Phys. Lett.*, 67, 2494–2496 (1995).

43. T. Oshima and R.A. Kiehl, Operation of bistable phase-locked single-electron tunneling logic elements, *J. Appl. Phys.*, 80, 912 (1996).

44. J.C. Ellenbogen and J.C. Love, Architectures for Molecular Electronic Computers: 1. Logic Structures and an Adder Built from Molecular Electronic Diodes; and J. C. Ellenbogen, Architectures for Molecular Electronic Computers: 2. Logic Structures Using Molecular Electronic FETs, MITRE Corporation Reports (1999), available at http://www.mitre.org/technology/nanotech/.

45. J. Chen, M.A. Reed, A.M. Rawlett, and J.M. Tour, Observation of a large on-off ratio and negative differential resistance in an electronic molecular switch, *Science*, 286, 1550–1552 (1999).

46. J. Chen, W. Wang, M.A. Reed, A.M. Rawlett, D.W. Price, and J.M. Tour, Room-temperature negative differential resistance in nanoscale molecular junctions, *Appl. Phys. Lett.*, 77, 1224–1226, (2000).

47. M.A. Reed, J. Chen, A.M. Rawlett, D.W. Price, and J.M. Tour, Molecular random access memory cell, *Appl. Phys. Lett.*, 78, 3735–3737 (2001).

48. S. Iijima, Helical microtubules of graphitic carbon, *Nature*, 354, 56 (1991).

49. H.W.Ch. Postma, T. Teepen, Z. Yao, M. Grifoni, and C. Dekker, Carbon nanotube single-electron transistors at room temperature, *Science*, 293, 76–79 (2001).

50. A. Bachtold, P. Hadley, T. Nakanishi, and C. Dekker, Logic circuits with carbon nanotube transistors, *Science*, 294, 1317–1320 (2001).

51. T. Rueckes, K. Kim, E. Joselevich, G. Tseng, C.-L. Cheung, and C. Lieber, Carbon nanotube-based nonvolatile random access memory for molecular computing, *Science*, 289, 94–97 (2000).

52. Y. Cui and C. Lieber, Functional nanoscale electronic devices assembled using silicon nanowire building blocks, *Science*, 291, 851–853 (2001).

53. Y. Huang, X. Duan, Y. Cui, L.J. Lauhon, K-H. Kim, and C.M. Lieber, Logic gates and computation from assembled nanowire building blocks, *Science*, 294, 1313–1317 (2001).

54. V. Derycke, R. Martel, J. Appenzeller, and Ph. Avouris, Carbon nanotube inter- and intramolecular logic gates, *Nano Lett.*, (Aug 2001).

55. Y. Wada, T. Uda, M. Lutwyche, S. Kondo, and S. Heike, A proposal of nano-scale devices based on atom/molecule switching, *J. Appl. Phys.*, 74, 7321–7328 (1993).

56. Y. Wada, Atom electronics, *Microelectron. Eng.*, 30, 375–382 (1996).

57. J.R. Heath, P.J. Kuekes, G.S. Snider, and R.S. Williams, A defect tolerant computer architecture: opportunities for nanotechnology, *Science*, 280, 1716–1721 (1998).

58. C.P. Collier, E.W. Wong, M. Belohradsky, F.M. Raymo, J.F. Stoddart, P.J. Kuekes, R.S. Williams, and J.R. Heath, Electronically configurable molecular-based logic gates, *Science*, 285 (1999).

59. R. Amerson, R.J. Carter, W.B. Culbertson, P. Kuekes, G. Snider, Teramac — configurable custom computing, *Proc. 1995 IEEE Symp. FPGAs Custom Comp. Mach.*, (1995), pp. 32–38.

60. P. Kuekes and R.S. Williams, Molecular electronics, in *Proc. Eur. Conf. Circuit Theory Design*, ECCTD (1999).

61. A.I. Csurgay, W. Porod, and C.S. Lent, Signal processing with near-neighbor-coupled time-varying quantum-dot arrays, *IEEE Trans. Circuits Syst. I*, 47(8), 1212–1223 (2000).

62. A.I. Csurgay and W. Porod, Equivalent circuit representation of arrays composed of Coulomb-coupled nanoscale devices: modeling, simulation and realizability, *Intl. J. Circuit Theory Appl.*, 29, 3–35 (2001).

6

Nanocomputer Architectronics and Nanotechnology

CONTENTS

Sergey Edward Lyshevski
Rochester Institute of Technology

Abstract

Significant progress has been made over the last few years in various applications of nanotechnology. One of the most promising directions that will lead to benchmarking progress and provide far-reaching benefits is devising and designing nanocomputers using the recent pioneering developments. In this chapter we examine generic nanocomputer architectures, which include the following major components: the arithmetic-logic unit, the memory unit, the input/output unit, and the control unit. Mathematical models were examined based on the behavior description using the finite-state machines. It is illustrated that this model can be applied to perform analysis, design, simulation, and optimization of nanocomputers. Innovative methods in design of nanocomputers and their components are documented. The basic motivation of this chapter is to further develop and apply the fundamental theory of nanocomputers, further expand the basic research toward the sound nanocomputer theory and practice, and report on the application of nanotechnology to fabricate nanocomputers and their components. In fact, to increase the computer performance, novel logic and memory nanoscale integrated circuits can be fabricated and implemented. These advancements and progress are ensured using novel materials, fabrication processes, techniques, and technologies. Fundamental and applied results researched in this chapter further expand the horizon of nanocomputer theory and practice.

6.1   Introduction

Computer engineering and science emerged as fundamental and exciting disciplines. Tremendous progress has been made within the last 50 years, e.g., from invention of the transistor to building computers with 2 cm² (quarter-size) processors. These processors include hundreds of millions of transistors. Excellent books[1–10] report the basic theory of computers, but further revolutionary developments are needed to satisfy Moore's first law. This can be accomplished by devising and applying novel theoretical fundamentals and utilizing superior performance of nanoscale integrated circuits (nanoICs). Due to the recent basic theoretical developments and nanotechnology maturity, the time has come to further expand the theoretical, applied, and experimental horizons.

Nanocomputer theory and practice are both revolutionary and evolutionary pioneering advances compared with the existing theory and semiconductor integrated circuit (IC) technology. The current ICs are very large-scale integration circuits (VLSI). Though 90 nm fabrication technologies have been developed and implemented by the leading computer manufacturers (Dell, IBM, Intel, Hewlett-Packard, Motorola, Sun Microsystems, Texas Instruments, etc.), and billions of transistors can be placed on a single multilayered die, the VLSI technology approaches its physical limits. Alternative affordable, high-yield, and robust technologies are sought, and nanotechnology promises further far-reaching revolutionary progress.

It is envisioned that nanotechnology will lead to three-dimensional nanocomputers with novel computer architectures to attain the superior overall performance level. Compared with the existing most advanced computers, in nanocomputers the execution time, switching frequency, and size will be decreased by the order of millions, while the memory capacity will be increased by the order of millions. However, significant challenges must be overcome. Many difficult problems must be addressed, researched, and solved such as novel nanocomputer architectures, advanced organizations and topologies, high-fidelity modeling, data-intensive analysis, heterogeneous simulations, optimization, control, adaptation, reconfiguration, self-organization, robustness, utilization, and other problems. Many of these problems have not even been addressed yet. Due to tremendous challenges, much effort must be focused to solve these problems.

This chapter formulates and solves some long-standing fundamental and applied problems in design, analysis, and optimization of nanocomputers. The fundamentals of *nanocomputer architectronics* are reported; and the basic organizations and topologies are examined, progressing from the general system-level consideration to the nanocomputer subsystem/unit/device-level study. Specifically, nanoICs are examined using nanoscale field-effect transistors (NFET). It is evident that hundreds of books will soon be written on nanocomputers; correspondingly, the author definitely cannot cover all aspects or attempt to emphasize, formulate, and solve all challenging problems. Furthermore, a step-by-step approach will be the major objective rather than to formulate and attempt to solve the abstract problems with minimal chance to succeed, validate, and implement the results. Therefore, it is my hope that this chapter will stimulate research and development focusing on well-defined current and future nanocomputer perspectives emphasizing the near- and long-term prospect and vision.

6.2   Brief History of Computers: Retrospects and Prospects

Having mentioned a wide spectrum of challenges and problems, it is likely that the most complex issues are devising and designing nanocomputer architectures — organizations and topologies that the author names as *nanocomputer architectronics*. Before addressing its theory and application, let us turn our attention to the past, and then focus on the prospects, remarkable opportunities, and astonishing futures.

The history of computers is traced back to thousands of years ago. To enter the data and perform calculations, people used a wooden rack holding two horizontal wires with beads strung on them. This mechanical "tool," called an abacus, was used for counting, keeping track, and recording the facts even before numbers were invented. The early abacus, known as a counting board, was a piece of wood, stone, or metal with carved grooves or painted lines between which beads, pebbles or wood, bone, stone, or metal disks could be moved (Figure 6.1.a). When these beads were moved around, according to the *programming rules* memorized by the user, arithmetic and recording problems could be solved. The oldest counting board found, called the *Salamis tablet*, was used by the Babylonians around 300 B.C. This board was discovered in 1899 on the island of Salamis. As shown in Figure 6.1.a, the Salamis tablet abacus is a slab of marble marked with two sets of eleven vertical lines (10 columns), a blank space between them, a horizontal line crossing each set of lines, and Greek symbols along the top and bottom.

Another important invention around the same time was the astrolabe for navigation.

In 1623, Wilhelm Schickard (Germany) built his "calculating clock," which was a six-digit machine that could add, subtract, and indicate overflow by ringing a bell. Blaise Pascal (France) is usually credited

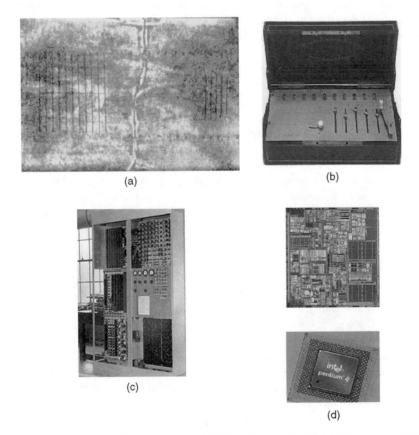

(a)

(b)

(c)

(d)

FIGURE 6.1   Evolution: from (a) abacus (around 300 B.C.) to (b) the Thomas *Arithmometer* (1820), and from (c) the Electronic Numerical Integrator and Computer processor (1946) to (d) 1.5 × 1.5 cm 478-pin Intel® Pentium® 4 processor with 42 million transistors (2002), http://www.intel.com/.

with building the first digital computer. His machine was made in 1642 to help his father, a tax collector. It was able to add numbers entered with dials. Pascal also built a *Pascaline* machine in 1644. These five- and eight-digit machines devised by Pascal used a different concept compared with the Schickard's calculating clock. In particular, rising and falling weights instead of a direct gear drive were used. The Pascaline machine could be extended to more digits, but it could not subtract. Pascal sold more than 10 machines, and several of them still exist. In 1674, Gottfried Wilhelm von Leibniz (Germany) introduced a "Stepped Reckoner," using a movable carriage to perform multiplications. Charles Xavier Thomas (France) applied Leibniz's ideas and in 1820 initiated fabrication of mechanical calculators (Figure 6.1.b).

In 1822, Charles Babbage (England) built a six-digit calculator that performed mathematical operations using gears. For many years, from 1834 to 1871, Babbage envisioned and carried out the "Analytical Engine" project. The design integrated the stored-program (memory) concept, envisioning that the memory would hold more than 100 numbers. The machine proposed had read-only memory in the form of punch cards. This device can be viewed as the first programmable calculator. Babbage used several punch cards for both programs and data. These cards were chained, and the motion of each chain could be reversed. Thus, the machine was able to perform the conditional jumps. The microcoding features were also integrated (the "instructions" depended on the positioning of metal studs in a slotted barrel, called the *control barrel*). These machines were implemented using mechanical analog devices. Babbage only partially implemented his ideas because these innovative initiatives were far ahead of the capabilities of the existing technology, but the idea and goal were set.

In 1926, Vannevar Bush (MIT) devised the "product integraph," a semiautomatic machine for solving problems in determining the characteristics of complex electrical systems. International Business Machines introduced the IBM 601 in 1935 (IBM made more than 1500 of these machines). This was a punch-card machine with an arithmetic unit based on relays and capable of doing a multiplication in 1 sec. In 1937, George Stibitz (Bell Telephone Laboratories) constructed a one-bit binary adder using relays. Alan Turing (England) in 1937 published a paper reporting "computable numbers." This paper solved mathematical problems and proposed a mathematical computer model known as a *Turing machine*.

The idea of an electronic computer is traced back to the late 1920s. However, the major breakthroughs appeared in the 1930s. In 1938, Claude Shannon published in the *AIEE Transactions* his article that outlined the application of electronics (relays). He proposed an "electric adder to the base of two." George Stibitz (Bell Telephone Laboratories) built and tested the proposed adding device in 1940. John V. Atanasoff (Iowa State University) completed a prototype of the 16-bit adder using vacuum tubes in 1939. In the same year, Zuse and Schreyer (Germany) examined the application of relay logic. In 1940 Schreyer completed a prototype of the 10-bit adder using vacuum tubes, and he built memory using neon lamps. Zuse demonstrated the first operational programmable calculator in 1940. This calculator was demonstrated for floating-point numbers having a seven-bit exponent, 14-bit mantissa, sign bit, 64 words of memory with 1400 relays, and arithmetic and control units with 1200 relays.

Howard Aiken (Harvard University) proposed an immense calculating machine that could solve some problems of relativistic physics. Funded by Thomas J. Watson (IBM president), Aiken built an "Automatic Sequence Controlled Calculator Mark I." This project was finished in 1944, and "Mark I" was used to calculate ballistics problems for the U.S. Navy. This electromechanical machine was 15 m long, 5 tons in weight, and had 750,000 parts (72 accumulators with arithmetic units and mechanical registers with a capacity of 23 digits plus sign). The arithmetic was fixed-point, with a plugboard setting determining the number of decimal places. The input–output unit included card readers, card punch, paper tape readers, and typewriters. There were 60 sets of rotary switches, each of which could be used as a constant register (mechanical read-only memory). The program was read from a paper tape, and data could be read from the other tapes, card readers, or constant registers.

In 1943, the U.S. government contracted John W. Mauchly and J. Presper Eckert (University of Pennsylvania) to build the Electronic Numerical Integrator and Computer, which likely was the first true electronic digital computer. The machine was built in February 1946 (Figure 6.1.c). This computer performed 5000 additions or 400 multiplications per second, showing the enormous capabilities of electronic computers at that time. This computing performance does not fascinate us these days, but the

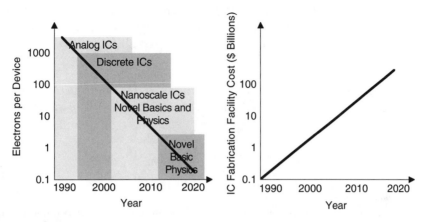

FIGURE 6.2 Moore's laws.

facts that the computer weight was 30 tons, it consumed 150 kW power, and had 18,000 vacuum tubes are likely interesting information. John von Neumann and his team built an Electronic Discrete Variable Automatic Computer in 1945 using "von Neumann computer architecture."

It is virtually impossible to cover the developments in discoveries (from semiconductor devices to implementation of complex logics using VLSI in advanced processors), to chronologically and thoroughly report meaningful fundamental discoveries (quantum computing, neurocomputing, etc.), and to cover the history of software development. First-, second-, third-, and fourth-generations of computers emerged, and tremendous progress has been achieved. The Intel Pentium 4 (2.4 GHz) processor, illustrated in Figure 6.1.d, was built using advanced Intel NetBurst™ microarchitecture. This processor ensures high-performance processing and is fabricated with 0.13 micron technology. The processor is integrated with high-performance memory systems, e.g., 8KB L1 data cache, 12K μops L1 execution trace cache, 256 KB L2 advanced transfer cache, and 512 KB advance transfer cache.

The fifth generation of computers will be built using emerging nanoICs. Currently, 70 nm and 50 nm technologies are emerging to fabricate high-yield high-performance ICs with billions of transistors on a single 1 cm^2 die. Further progress is needed, and novel developments are emerging. This chapter studies the application of nanotechnology to design nanocomputers with nanoICs. Though there are tremendous challenges, they will be overcome in the next 10 or 20 years. Synthesis, integration, and implementation of new affordable high-yield nanoICs are critical to meet Moore's first law. Figure 6.2 illustrates the first and second Moore laws. Despite the fact that some data and foreseen trends can be viewed as controversial and subject to adjustments, the major trends and tendencies are obvious and most likely cannot be seriously argued and disputed.

6.3 Nanocomputer Architecture and Nanocomputer Architectronics

The critical problems in the design of nanocomputers are focused on devising, designing, analyzing, optimizing, and fully utilizing hardware and software. Numbers in digital computers are represented as a string of zeros and ones, and hardware can perform Boolean operations. Arithmetic operations are performed on a hierarchical basis that is built upon simple operations. The methods to compute and the algorithms used are different. Therefore, speed, robustness, accuracy, and other performance characteristics vary. There is a direct relationship between the fabrication technology to make digital ICs or nanoICs and computational performance.

The information in digital computers is represented as a string of bits (zeros and ones). The number of bits depends on the length of the computer word (quantity of bits on which hardware can operate). The correspondence between bits and a number can be established. The properties in a particular number representation system are satisfied and correspond to the operations performed by hardware over the string of bits. This relationship is defined by the rule that associates one numerical value denoted as X with the corresponding bit string denoted as x, $x=\{x_0, x_1, ..., x_{n-2}, x_{n-1}\}$, $x_i \in 0, 1$. The associated word (the string of bits) is n bits long.

If for every value X there exists one, and only one, corresponding bit string x, the number system is nonredundant. If there is more than one bit string x that represents the same value X, the number system is redundant. A *weighted* number system is used, and a numerical value is associated with the bit string x as:

$$x = \sum_{i=0}^{n-1} x_i w_i, \; w_0 = 1, ..., w_i = (w_i - 1)(r_i - 1),$$

where r_i is the *radix* integer.

Nanocomputers can be classified using different classification principles. For example, making use of the multiplicity of instruction and data streams, the following classifications can be applied:

1. Single instruction stream / single data stream — conventional word-sequential architecture including pipelined nanocomputers with parallel arithmetic logic unit (ALU)
2. Single instruction stream / multiple data stream — multiple ALU architectures, e.g., parallel-array processor (ALU can be either bit-serial or bit-parallel)
3. Multiple instruction stream / single data stream
4. Multiple instruction stream / multiple data stream — the multiprocessor system with multiple control unit

The author does not intend to classify nanocomputers because there is no evidence regarding the abilities of nanotechnology to fabricate affordable and robust nanocomputers yet. Therefore, different nanocomputer architectures must be devised, analyzed, and tested. The tremendous challenges emphasized in this chapter illustrate that novel architectures, organizations, and topologies can be synthesized; and then the classification problems can be addressed and solved.

The nanocomputer architecture integrates the functional, interconnected, and controlled hardware units and systems that perform propagation (flow), storage, execution, and processing of information (data). Nanocomputers accept digital or analog input information, process and manipulate it according to a list of internally stored machine instructions, store the information, and produce the resulting output. The list of instructions is called a *program*, and internal storage is called *memory*.

A program is a set of instructions that one writes to order a computer what to do. Keeping in mind that the computer consists of on and off logic switches, one can assign: first switch off, second switch off, third switch off, fourth switch on, fifth switch on, sixth switch on, seventh switch on, and eighth switch on to receive an eight-bit signal as given by 00011111. A program commands millions of switches and is written in the machine circuitry-level complex language. Programming has become easier with the use of high-level programming languages. A high-level programming language allows one to use a vocabulary of terms, e.g., read, write or do instead of creating the sequences of on-off switching that perform these tasks. All high-level languages have their own syntax (rules), provide a specific vocabulary, and give an explicitly defined set of rules for using this vocabulary. A compiler is used to translate (interpret) the high-level language statements into machine code. The compiler issues error messages each time the programmer uses the programming language incorrectly. This allows one to correct the error and perform another translation by compiling the program again. The programming logic is an important issue because it involves executing various statements and procedures in the correct order to produce the desired results. One must use the syntax correctly and execute a logically constructed, workable programs. Two common approaches used to write computer programs are procedural and object-oriented programming. Through procedural programming, one defines and executes computer memory locations (variables) to hold values and writes sequential steps to manipulate these values. Object-oriented programming is the extension of the procedural programming because it creates objects (program components) and applications that use these objects. Objects are made up of states, and states describe the characteristics of an object. It will likely be necessary to develop advanced software environments that ideally are architecturally neutral, e.g., software that can be used or will be functional on any platform. We do not consider quantum computing in this chapter because it is still a quite controversial issue. There is no doubt that if quantum computing could be attained, other software would be necessary. This chapter concentrates on the *nanocomputer architectronics*. Therefore, the software issues are briefly discussed to demonstrate the associations between hardware and software.

Nanocomputer architecture integrates the following major systems: input–output, memory, arithmetic and logic, and control units. The input unit accepts information from electronic devices or other computers through the cards (electromechanical devices, such as keyboards, can be also interfaced). The information received can be stored in memory and then manipulated and processed by the arithmetic and logic unit (ALU). The results are output using the output unit. Information flow, propagation, manipulation, processing, and storage are coordinated by the control unit. The arithmetic and logic unit, integrated with the control unit, is called the processor or central processing unit (CPU). Input and output systems are called the input–output (I/O) unit. The memory unit, which integrates memory systems, stores programs and data. There are two main classes of memory called *primary* (main) and *secondary* memory. In nanocomputers, the primary memory is implemented using nanoICs that can consist of billions of nanoscale storage cells (each cell can store one bit of information). These cells are accessed in groups of fixed size called *words*. The main memory is organized such that the contents of one word can be stored or retrieved in one basic operation called a *memory cycle*. To provide consistent direct access to any word in the main memory in the shortest time, a distinct address number is associated with each word location. A word is accessed by specifying its address and issuing a control command that starts the storage or retrieval process. The number of bits in a word is called the *word length*. Word lengths vary, for example, from 16 to 64 bits. Personal computers and workstations usually have a few million words in main memory, while nanocomputers can have hundreds of millions of words, with the time required to access a word for reading or writing within psec range. Although the main memory is essential, it tends to be expensive and volatile. Therefore, nanoICs can be effectively used to implement the additional memory systems to store programs and data, forming secondary memory.

The execution of most operations is performed by the ALU. In the ALU, the logic nanogates and nanoregisters are used to perform the basic operations (addition, subtraction, multiplication, and division) of numeric operands and the comparison, shifting, and alignment operations of general forms of numeric and nonnumeric data. The processors contain a number of high-speed registers that are used for temporary storage of operands. The register, as a storage device for words, is a key sequential component; and registers are connected. Each register contains one word of data, and its access time is at least 10 times faster than main memory access time. A register-level system consists of a set of registers connected by combinational data processing and data processing nanoICs. Each operation is implemented as given by the following statement:

$$\text{cond: } X := f(x_1, x_2, \ldots, x_{i-1}, x_i),$$

i.e., when the condition cond holds, compute the combinational function of f on $x_1, x_2, \ldots, x_{i-1}, x_i$ and assign the resulting value to X. Here, cond is the control condition prefix which denotes the condition that must be satisfied; $X, x_1, x_2, \ldots, x_{i-1}, x_i$ are the data words or the registers that store them; f is the function to be performed within a single clock cycle.

Suppose that two numbers located in main memory should be multiplied, and the result must be stored back into the memory. Using instructions determined by the control unit, the operands are first fetched from the memory into the processor. They are then multiplied in the ALU, and the result is stored back in memory. Various nanoICs can be used to execute data processing instructions. The complexity of ALU is determined by the arithmetic instruction implementation. For example, ALUs that perform fixed-point addition and subtraction, and word-based logics can be implemented as combinational nanoICs. The floating-point arithmetic requires complex implementation, and arithmetic coprocessors to perform complex numerical functions are needed. The floating-point numerical value of a number X is (X_m, X_e), where X_m is the mantissa and X_e is the fixed-point number. Using the base b (usually, $b=2$), we have $X = X_m \times b^{X_e}$. Therefore, the general basic operations are quite complex, and some problems (biasing, overflow, underflow, etc.) must be resolved.

6.3.1. Basic Operation

To perform computing, specific programs consisting of a set of machine instructions are stored in main memory. Individual instructions are fetched from the memory into the processor for execution. Data used as operands are also stored in the memory. A typical instruction — move memory_location1, R1 — loads a copy of the operand at memory_location1 into the processor register R1. The instruction requires a few basic steps to be performed. First, the instruction must be transferred from the memory into the processor, where it is decoded. Then the operand at memory_location1 must be fetched into the processor. Finally, the operand is placed into register R1. After operands are loaded into the processor registers, instructions such as add R1, R2, R3 can be used to add the contents of registers R1 and R2 and then place the result into register R3.

The connection between the main memory and the processor that guarantees the transfer of instructions and operands is called the *bus*. A bus consists of a set of address, data, and control lines. The bus permits transfer of program and data files from their long-term location (virtual memory) to the main memory. Communication with other computers is ensured by transferring the data through the bus. Normal execution of programs may be preempted if some I/O device requires urgent control. To perform this, specific programs are executed, and the device sends an interrupt signal to the processor. The processor temporarily suspends the program that is being executed and runs the special interrupt service routine instead. After providing the required interrupt service, the processor switches back to the interrupted program. During program loading and execution, the data should be transferred between the main memory and secondary memory. This is performed using the direct memory access.

6.3.2 Performance Fundamentals

In general, reversible and irreversible nanocomputers can be designed. Today all existing computers are irreversible. The system is reversible if it is deterministic in the reverse and forward time directions. The reversibility implies that no physical state can be reached by more than one path (otherwise, reverse evolution would be nondeterministic). Current computers constantly irreversibly erase temporary results, and thus the entropy changes. The average instruction execution speed (in millions of instructions executed per second I_{PS}) and cycles per instruction are related to the time required to execute instructions as given by $T_{inst} = 1/f_{clock}$, where the clock frequency f_{clock} depends mainly on the ICs or nanoICs used and the fabrication technologies applied (for example, f_c is 2.4 GHz for the existing Intel Pentium processors). The quantum mechanics implies an upper limit on the frequency at which the system can switch from one state to another. This limit is found as the difference between the total energy E of the system and ground state energy E_0, i.e.,

$$f_l \leq \frac{4}{h}(E - E_0),$$

where h is the Planck constant, $h = 6.626 \times 10^{-34}$ J-sec or J/Hz. An isolated nanodevice, consisting of a single electron at a potential of 1V above its ground state, contains 1 eV of energy (1eV $= 1.602 \times 10^{-19}$ J) and therefore, cannot change its state faster than:

$$f_l \leq \frac{4}{h}(E - E_0) = \frac{4}{6.626 \times 10^{-34}} 1.602 \times 10^{-19} \approx 0.97 \times 10^{15} \text{ Hz},$$

i.e., the switching frequency is less than 1×10^{15} Hz. Correspondingly, the switching frequency of nanoICs can be significantly increased compared with the currently used CMOS ICs.

In asymptotically reversible nanocomputers, the generated entropy is $S = b/t$, where b is the entropy coefficient (b varies from 1×10^7 to 1×10^6 bits/GHz for ICs, and from 1 to 10 bits/GHz for quantum FETs) and t is the length of time over which the operation is performed. Correspondingly, the minimum entropy and processing (operation) rate for quantum devices are $S = 1$ bit/operation and $r_e = 1 \times 10^{26}$ operation/sec-cm^2, while CMOS technology allows one to achieve $S = 1 \times 10^6$ bits/operation and $r_e = 3.5 \times 10^{16}$ operation/sec-cm^2.

Using the number of instructions executed (N), the number of cycles per instruction (C_{PI}) and the clock frequency (f_{clock}), the program execution time is found to be

$$T_{ex} = \frac{N \times C_{PI}}{f_{fclock}}.$$

In general, the hardware defines the clock frequency f_{clock}, the software influences the number of instructions executed N, while the nanocomputer architecture defines the number of cycles per instruction C_{PI}.

One of the major performance characteristics for computer systems is the time it takes to execute a program. Suppose N_{inst} is the number of machine instructions to be executed. A program is written in high-level language, translated by compiler into machine language, and stored. An operating system software routine loads the machine language program into the main memory for execution. Assume that each machine language instruction requires N_{step} basic steps for execution. If basic steps are executed at the constant rate of R_T [steps/sec], then the time to execute the program is

$$T_{ex} = \frac{N_{inst} \times N_{step}}{R_T}.$$

The main goal is to minimize T_{ex}. Optimal memory and processor design allows one to achieve this goal. The access to operands in processor registers is significantly faster than access to main memory. Suppose that instructions and data are loaded into the processor. Then they are stored in a small and fast cache memory (high-speed memory for temporary storage of copies of the sections of program and data from the main memory that are active during program execution) on the processor. Hence, instructions and data in cache are accessed repeatedly and correspondingly. The program execution will be much faster. The cache can hold small parts of the executing program. When the cache is full, its contents are replaced by new instructions and data as they are fetched from the main memory. A variety of cache replacement algorithms are used. The objective of these algorithms is to maximize the probability that the instructions and data needed for program execution can be found in the cache. This probability is known as the *cache hit ratio*. High hit ratio means that a large percentage of the instructions and data is found in the cache, and the requirement to access the slower main memory is reduced. This leads to a decrease in the memory access basic step time components of N_{step}, and this results in a smaller T_{ex}. The application of different memory systems results in a memory hierarchy concept. The nanocomputer memory hierarchy is shown in Figure 6.3. As was emphasized, to attain efficiency and high performance, the main memory should not store all programs and data. Specifically, caches are used. Furthermore, virtual memory, which has the largest capacity but the slowest access time, is used. Segments of a program are transferred from the virtual memory to the main memory for execution. As other segments are needed, they may replace the segments existing in main memory when main memory is full. The sequential controlled movement of large programs and data between the cache, main, and virtual memories, as programs execute, is managed by a combination of operating system software and control hardware. This is called *memory management*.

Using the memory hierarchy illustrated in Figure 6.3, it is evident that the CPU can communicate directly with only $M1$, and $M1$ communicates with $M2$, and so on. Therefore, for the CPU to access the information stored in the memory M_j, the sequence of j data transfer required is given given by:

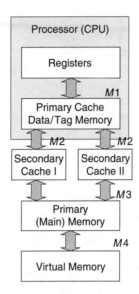

FIGURE 6.3 Memory hierarchy in nanocomputer with cache (primary and secondary), primary (main) memory, and virtual memory.

$$M_{j-1} := M_j, \; M_{j-2} := M_{j-1}, \; \dots \; M_1 := M_2, \; \text{CPU} := M_1.$$

However, the memory bypass can be implemented and effectively used.

6.3.3 Memory Systems

A memory unit that integrates different memory systems stores the information (data). The processor accesses (reads or loads) the data from the memory systems, performs computations, and stores (writes) the data back to memory. The memory system is a collection of storage locations. Each storage location (memory word) has a numerical address. A collection of storage locations forms an address space. Figure 6.4 documents the data flow and its control, representing how a processor is connected to a memory system via address, control, and data interfaces. High-performance memory systems should be able to serve multiple requests simultaneously, particularly for vector nanoprocessors.

When a processor attempts to load or read the data from the memory location, the request is issued, and the processor stalls while the request returns. While nanocomputers can operate with overlapping memory requests, the data cannot be optimally manipulated if there are long memory delays. Therefore, a key performance parameter in the design of nanocomputers is the effective speed of their memory. The following limitations are imposed on any memory systems: the memory cannot be infinitely large, cannot contain an arbitrarily large amount of information, and cannot operate infinitely fast. Hence, the major characteristics are speed and capacity.

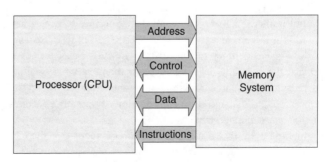

FIGURE 6.4 Memory–processor interface.

The memory system performance is characterized by the latency (τ_1) and bandwidth (B_w). The memory latency is the delay as the processor first requests a word from memory until that word arrives and is available for use by the processor. The bandwidth is the rate at which information can be transferred from the memory system. Taking note of the number of requests that the memory can service concurrently, $N_{request}$, we have:

$$B_w = \frac{N_{request}}{\tau_l}.$$

Using nanoICs, it becomes feasible to design and build superior memory systems with desired capacity, low latency, and high bandwidth approaching the physical limits. Furthermore, it will be possible to match the memory and processor performance requirements.

Memory hierarchies provide decreased average latency and reduced bandwidth requirements, whereas parallel memories provide higher bandwidths. As was emphasized, nanocomputer architectures, based upon small and fast memory located in front of large but relatively slow memory, can satisfy the requirements on speed and memory capacity. This results in application of registers in the CPU, and most commonly accessed variables should be allocated at registers. A variety of techniques, employing either hardware, software, or a combination of hardware and software, are employed to ensure that most references to memory are fed by the faster memory. The locality principle is based on the fact that some memory locations are referenced more often than others. The implementation of spatial locality, due to the sequential access, provides the property that an access to a given memory location increases the probability that neighboring locations will soon be accessed. Making use of the frequency of program looping behavior, temporal locality ensures access to a given memory location, increasing the probability that the same location will be accessed again soon. It is evident that if a variable is not referenced for a while, it is unlikely that this variable will be needed soon.

Let us return our attention to the issue of the memory hierarchy (Figure 6.3). At the top of the hierarchy are the superior speed CPU registers. The next level in the hierarchy is a high-speed cache memory. The cache can be divided into multiple levels, and nanocomputers will likely have multiple cache levels with cache fabricated on the CPU nanochip. Below the cache memory is the slower but larger main memory, and then the large virtual memory, which is slower than the main memory. Three performance characteristics (access time, bandwidth, and capacity) and many factors (affordability, robustness, etc.) support the application of multiple levels of cache memory and the memory hierarchy in general. The time needed to access the primary cache should match with the clock frequency of the CPU, and the corresponding nanoICs must be used. We place a smaller first-level (primary) cache above a larger second-level (secondary) cache. The primary cache is accessed quickly, and the secondary cache holds more data close to the CPU. The nanocomputer architecture depends on the technologies available. For example, primary cache can be fabricated on the CPU chip, while the secondary caches can be an on-chip or out-of-chip solution.

Size, speed, latency, bandwidth, power consumption, robustness, affordability, and other performance characteristics are examined to guarantee the desired overall nanocomputer performance based upon the specifications imposed. The performance parameter, which can be used to quantitatively examine different memory systems, is the effective latency τ_{ef}. We have

$$\tau_{ef} = \tau_{hit} R_{hit} + \tau_{miss} (1 - R_{hit}),$$

where τ_{hit} and τ_{miss} are the hit and miss latencies; R_{hit} is the hit ratio, $R_{hit} < 1$.

If the needed word is found in a level of the hierarchy, it is called a *hit*. Correspondingly, if a request must be sent to the next lower level, the request is said to be a *miss*. The miss ratio is given as $R_{miss} = (1 - R_{hit})$. R_{hit} and R_{miss} are strongly influenced by the program executed and by the high-/low-level memory capacity ratio. The access efficiency E_{ef} of multiple-level memory ($i-1$ and i) is found using the access time and hit and miss ratios. In particular,

$$E_{ef} = \cfrac{1}{\cfrac{t_{\text{access time } i-1}}{t_{\text{access time } i}} R_{miss} + R_{hit}}.$$

The hardware can dynamically allocate parts of the cache memory for addresses likely to be accessed soon. The cache contains only redundant copies of the address space. The cache memory can be associative or content-addressable. In an associative memory, the address of a memory location is stored along with its content. Rather than reading data directly from a memory location, the cache is given an address and responds by providing data which might or might not be the data requested. When a cache miss occurs, the memory access is then performed from main memory, and the cache is updated to include the new data. The cache should hold the most active portions of the memory, and the hardware dynamically selects portions of main memory to store in the cache. When the cache is full, some data must be transferred to the main memory or deleted. Therefore, a strategy for cache memory management is needed. Cache management strategies are based on the locality principle. In particular, spatial (selection of what is brought into the cache) and temporal (selection of what must be removed) localities are embedded. When a cache miss occurs, hardware copies a contiguous block of memory into the cache, which includes the word requested. This fixed-size memory block can be small (bit or word) or hundreds of bytes. Caches can require all fixed-size memory blocks to be aligned. When a fixed-size memory block is brought into the cache, it is likely that another fixed-size memory block must be removed. The selection of the removed fixed-size memory block is based on an effort to capture temporal locality. In general, this is difficult to achieve. Correspondingly, viable methods are used to predict future memory accesses. A least-recently-used concept can be the preferred choice for nanocomputers.

The cache can integrate the data memory and the tag memory. The address of each cache line contained in the data memory is stored in the tag memory (the state can also track which cache lines is modified). Each line contained in the data memory is allocated by a corresponding entry in the tag memory to indicate the full address of the cache line. The requirement that the cache memory be associative (content-addressable) complicates the design because addressing data by content is more complex than by its address (all tags must be compared concurrently). The cache can be simplified by embedding a mapping of memory locations to cache cells. This mapping limits the number of possible cells in which a particular line may reside. Each memory location can be mapped to a single location in the cache through direct mapping. Because there is no choice of where the line resides and which line must be replaced, poor utilization results. In contrast, a two-way set-associative cache maps each memory location into either of two locations in the cache. Hence, this mapping can be viewed as two identical directly mapped caches. In fact, both caches must be searched at each memory access, and the appropriate data selected and multiplexed on a tag match — hit and on a miss. Then a choice must be made between two possible cache lines as to which is to be replaced. A single least-recently-used bit can be saved for each such pair of lines to remember which line has been accessed more recently. This bit must be toggled to the current state each time. To this end, an M-way associative cache maps each memory location into M memory locations in the cache. Therefore, this cache map can be constructed from M identical direct-mapped caches. The problem of maintaining the least-recently-used ordering of M cache lines is primarily due to the fact that there are $M!$ possible orderings. In fact, it takes at least $\log_2 M!$ bits to store the ordering. It can be envisioned that two-, three- or four-way associative cache will be implemented in nanocomputers.

Let us focus our attention on the write operation. If the main memory copy is updated with each write operation, then a write-through or store-through technique is used. If the main memory copy is not updated with each write operation, a write-back, copy-back, or deferred-write algorithm is enforced. In general, the cache coherence or consistency problem must be examined due to implementation of different bypass techniques.

6.3.4 Parallel Memories

Main memories can comprise a series of memory nanochips or nanoICs on a single nanochip. These nanoICs form a *nanobank*. Multiple memory *nanobanks* can be integrated together to form a parallel main memory system. Because each nanobank can service a request, a parallel main memory system with N_{mb} nanobanks can service N_{mb} requests simultaneously, increasing the bandwidth of the memory system by N_{mb} times the bandwidth of a single nanobank. The number of nanobanks is a power of two, that is, $N_{mb} = 2^p$. An n-bit memory word address is broken into two parts: a p-bit nanobank number and an m-bit address of a word within a nanobank. The p bits used to select a nanobank number could be any p bits of the n-bit word address. Let us use the low-order p address bits to select the nanobank number and the higher order $m = n - p$ bits of the word address to access a word in the selected nanobank.

Multiple memory nanobanks can be connected using *simple paralleling* and *complex paralleling*. Figure 6.5 shows the structure of a simple parallel memory system where m address bits are simultaneously supplied to all memory nanobanks. All nanobanks are connected to the same read/write control line. For a read operation, the nanobanks perform the read operation and deposit the data in the latches. Data can then be read from the latches one by one by setting the switch appropriately. The nanobanks can be accessed again to carry out another read or write operation. For a write operation, the latches are loaded one by one. When all latches have been written, their contents can be written into the memory nanobanks by supplying m bits of address. In a simple parallel memory, all nanobanks are cycled at the same time. Each *nanobank* starts and completes its individual operations at the same time as every other nanobank, and a new memory cycle starts for all nanobanks once the previous cycle is complete.

A complex parallel memory system is documented in Figure 6.5. Each nanobank is set to operate on its own, independent of the operation of the other nanobanks. For example, ith nanobank performs a read operation on a particular memory address, while $(i+1)$th nanobank performs a write operation on a different and unrelated memory address. Complex paralleling is achieved using the address latch and a read/write command line for each nanobank. The *memory controller* handles the operation of the complex parallel memory. The processing unit submits the memory request to the memory controller, which determines which nanobank needs to be accessed. The controller then determines if the nanobank is busy by monitoring a busy line for each nanobank and holds the request if the nanobank is busy, submitting it when the nanobank becomes available to accept the request. When the nanobank responds to a read request, the switch is set by the controller to accept the request from the nanobank and forward it to the processing unit. It can be foreseen that complex parallel main memory systems will be implemented as in vector nanoprocessors. If consecutive elements of a vector are present in a different memory nanobank, then the memory system can sustain a bandwidth of one element per clock cycle. Memory

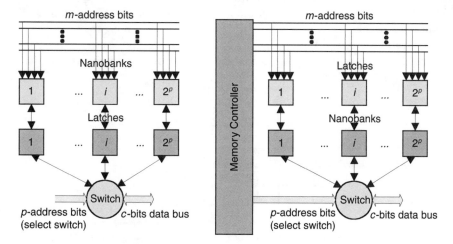

FIGURE 6.5 Simple and complex parallel main memory systems.

systems in nanocomputers can have hundreds of nanobanks with multiple memory controllers that allow multiple independent memory requests at every clock cycle.

6.3.5 Pipelining

Pipelining is a technique to increase the processor throughput with limited hardware in order to implement complex *datapath* (data processing) units (multipliers, floating-point adders, etc.). In general, a pipeline processor integrates a sequence of i data-processing nanoICs (nanostages) that cooperatively perform a single operation on a stream of data operands passing through them. Design of pipelining nanoICs involves deriving multistage, balanced, sequential algorithms to perform the given function. Fast buffer registers are placed between the nanostages to ensure the transfer of data between nanostages without interfering with one another. These buffers should be clocked at the maximum rate that guarantees reliable data transfer between nanostages.

As illustrated in Figure 6.6, the nanocomputers must be designed to guarantee the robust execution of overlapped instructions using pipelining. Four basic steps (fetch F_i, decode D_i, operate O_i, and write W_i) with specific hardware units are needed to achieve this. As a result, the execution of instructions can be overlapped. When the execution of some instruction I_i depends on the results of a previous instruction I_{i-1} that is not yet completed, instruction I_i must be delayed. The pipeline is said to be stalled, waiting for the execution of instruction I_{i-1} to be completed. While it is not possible to eliminate such situations, it is important to minimize the probability of occurrence. This is a key consideration in the design of the instruction set for nanocomputers and the design of the compilers that translate high-level language programs into machine language.

6.3.6 Multiprocessors

Multiple functional units can be applied to attain the nanocomputer operation when more than one instruction is in the operating stage. The parallel execution capability, when added to pipelining of the individual instructions, means that more than one instruction can be executed per basic step. Thus, the execution rate can be increased. This enhanced processor performance is called *superscalar processing*. The rate R_T of performing basic steps in the processor depends on the processor clock rate. This rate is on the order of billions of steps per second in current high-performance nanoICs. It was illustrated that physical limits prevent speeding up single processors indefinitely. Nanocomputers with multiprocessors will speed up the execution of large programs by executing subtasks in parallel. The main difficulty in achieving this is decomposition of given tasks into parallel subtasks and ordering these subtasks to the individual processors in such a way that communication among the subtasks will be performed efficiently and robustly. Figure 6.7 documents a block diagram of a multiprocessor system with the interconnection network needed for data sharing among the processors P_i. Parallel paths are needed in this network in order for parallel activity to proceed in the processors as they access the global memory space represented by the multiple memory units M_i. The basic parallel organization is represented in Figure 6.8.

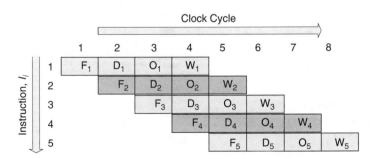

FIGURE 6.6 Pipelining of instruction execution.

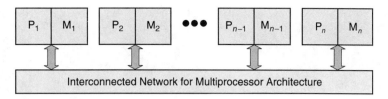

FIGURE 6.7   Multiprocessor architecture.

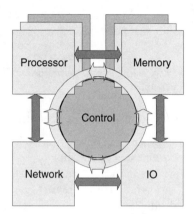

FIGURE 6.8   Parallel nanocomputer organization to maximize the computer concurrency.

6.4   Nanocomputer Architectronics and Neuroscience

The theory of computing, computer architecture, and networking is the study of efficient robust processing and communication, modeling, analysis, optimization, adaptive networks, architecture, organization synthesis, and other problems of hardware and software design. These studies have emerged as a synergetic fundamental discipline (computer engineering and science), and many problems have not been solved yet. Correspondingly, many questions are still unanswered.

Nanocomputer architectronics is the theory of nanocomputers devised and designed using fundamental theory and applying nanoICs. Our goal is to develop the nanocomputer architectronics basics — i.e., fundamental methods in synthesis, design, analysis, neuroanalogies, neuromimicking, neuroprototyping, computer-aided design, etc. Nanocomputer architectronics, which is a computer science and engineering frontier, will allow one to solve a wide spectrum of fundamental and applied problems. Applying the nanocomputer architectronics paradigm, this section spans the theory of nanocomputers, related areas, and their applications — e.g., information processing, communication paradigms, computational complexity theory, combinatorial optimization, architecture synthesis, and optimal organization.

In the development of nanocomputer architectronics paradigms, one can address and study fundamental problems in nanocomputer architecture synthesis, design, and optimization applying three-dimensional organization, application of nanoICs, multithreading, error recovery, massively parallel computing organization, shared memory parallelism, message passing parallelism, and more. The nanocomputer fundamentals, operation, and functionality can be devised through neuroscience. The key to understanding adaptation, control, architecture, hierarchy, organization, memory, intelligence, diagnostics, self-organization, computing, and other system-level basics lies in the study of phenomena and effects in the central nervous system, its components (brain and spinal cord), and the fundamental building blocks (neurons).

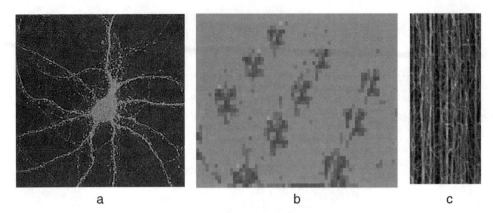

a   b   c

FIGURE 6.9   (a) *Vertebrate neuron (*soma, axon with synaptic terminals, dendrites, and synapses), (b) three-dimensional nanoICs, and (c) aligned carbon nanotubes.

6.4.1   Communication and Information Processing among Nerve Cells

Neuronal cells have a large number of synapses. A typical nerve cell in the human brain has thousands of synapses which establish communication and information processing. The communication and processing are not fixed; they constantly change and adapt. Neurons function in the hierarchically distributed, robust, adaptive network manner. During information transfer, some synapses on a cell are selectively triggered to release specific neurotransmitters, while other synapses remain passive. Neurons consist of a cell body with a nucleus (soma), axon (which transmits information from the soma), and dendrites (which transmit information to the soma) (Figure 6.9). It will be illustrated in the next section that it becomes possible to implement three-dimensional nanoIC structures using biomematic analogies as illustrated in Figure 6.9. For example, the complex inorganic dendrite-like trees can be implemented using the carbon nanotube-based technology (the Y-junction-branching carbon-nanotube networks ensure robust ballistic switching and amplification behavior desired in nanoICs).

Nerve cells (neurons) collect and transmit information about the internal state of the organism and the external environment, process and evaluate this information, and perform learning, diagnostics, adaptation, evolutionary decision making, coordination, and control. For example, sensory neurons transmit information received from sensory receptors to the interneurons of the central nervous system, while motor neurons transmit the information from the interneurons (located within the central nervous system) to motor (effector) cells.

The information between transmitting and receiving neurons is transferred along synapses, — synaptic terminals on the transmitting axon, synaptic vesicles containing neurotransmitter molecules or ions, synaptic gap dendrites of the receiving neuron, postsynaptic membrane channels, or neuron cytoplasm. Thus, synaptic terminals (axonal endings of transmitting neurons) transfer information to the dendrites of the receiving neuron through the synaptic gap (analogy of wireless communication). Neurons receive and transmit information by means of electrochemical phenomena, and synapses connect neurons creating a networked information bus. Information processing is also based on electrochemical phenomena.

A signal propagates along a neuronal axon and synaptic terminals as ionic current. The ionic gradient across the axonal plasma membrane is a function of so-called membrane permeability, and the permeability coefficients for various cations (K^+, Na^+, and Ca^{2+}) and anions (Cl^-) are known.

The Lorenz equation,

$$\nabla \cdot \mathbf{A} = -\frac{\partial V}{\partial t},$$

allows one to explicitly express the magnetic vector potential A as a function of the scalar electric potential V. The vector potential wave equation is[11]

$$-\nabla^2\mathbf{A} + \mu\sigma\frac{\partial\mathbf{A}}{\partial t} + \mu\varepsilon\frac{\partial^2\mathbf{A}}{\partial t^2} = -\mu\sigma\nabla V.$$

The electric potential may vary in the range of tens of mV. However, taking into the account that the membrane thickness is 50 Å, the electric field can be up to 20 million V/m. The K^+ and Na^+ channels (gates) open and close, varying the potential. The open time varies from ~0.5 to ~1 msec, and the Na^+ ions diffuse into the cell at the rate of ~ 6000 ions/msec. The axon can transmit impulses with a maximum frequency of ~ 200 Hz, and the propagation speed is from 10 to 100 m/sec. The author hypothesizes that because the nerve impulses have almost the same amplitude, the data is transmitted using the pulse–width–modulation mechanism.

The potential at the presynaptic membrane triggers the release of neurotransmitters (synaptic vesicles contain neurotransmitter molecules or ions) into the synaptic gap. For example, the well-studied acetyl-choline (ACh) ~400 Å synaptic vesicle contains ~100,000 ACh molecules. The released ACh molecules trigger the opening of the Ca^{2+} channels on the postsynaptic membrane of dendrite of the receiving neuron. The nanotransmitters cross the fluid-filled synaptic cleft (20–30 nm) and bind to the corresponding receptors on the postsynaptic membrane, changing the membrane potential. It must be emphasized that there are excitatory and inhibitory synapses with selective neurotransmitters triggered by distinct receptors. Anion channels are inhibitory, while cation channels are excitatory. The neuron response depends on the specific receptor and neurotransmitter identity. Hence, the precise selectivity and control are established. The receiving neurons transmit the electric current (signal) to the soma. A neuron can have thousands of postsynaptic potentials to simultaneously process. Signals (information) propagate from one neuron to another due to electrochemical information processing, and neurons form hierarchically interconnected adaptive reconfigurable networks.

In addition to information transfer, neurons of the central nervous system perform information processing (computing). Each brain neurotransmitter works within a widely spread but specific brain region, has a different effect and function, and encompasses a distinctive physiological role. There are more than 60 human neurotransmitters that have been identified within the following four classes: (1) cholines (ACh is the most important one); (2) biogenic amines (serotonin, histamine, and catecholamines); (3) amino acids (glutamate and aspartate are the excitatory transmitters, while gamma-aminobutyric acid/GABA, glycine, and taurine are inhibitory neurotransmitters); (4) neuropeptides (more than 50 neuropeptides are involved in modulation, transmission, and processing).

It is possible to map the complex electrochemical information transfer, performed by the sensory, motor, and interneurons, by two stream-oriented I/O operators. In particular, we have *receive* and *send*. In fact, these neurons maintain communication using *receive* and *send* messages, and the execution of these messages is due to electrochemical processes. The data is transferred between the brain (processor) and sensor/receptor (I/O device) using a communication bus.

The processing brain neurons perform computing. Though the specific mechanism of the data processing is under research, it seems that neurons process the data in a hybrid form. Two triggering states in neurons can be mapped as *off* (closed) and *on* (open), i.e., 0 and 1. Therefore, the strings of bits result, providing the analogy to a computer. However, it is possible that there are the intermediate states (due to the rate control), which will result in tremendous advantages for devising novel computationally efficient and robust processing paradigms. The organization and topology of adaptive networked information (data) transfer and processing are of great interest with regard to the examination of information propagation (transfer) and parallel processing (computing).

6.4.2 Topology Aggregation

There is a critical need to study the topology aggregation in the hierarchically distributed nanocomputer and nervous systems that compress, route, and network the information in the optimal manner. Utilizing the neuroscience concept, one can examine a hierarchical network with multiple paths and routing domain functions. Figure 6.10 shows the principle of topology aggregation of neurons in the possible

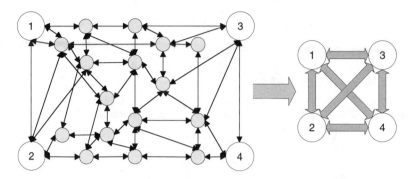

FIGURE 6.10   Network topology aggregation.

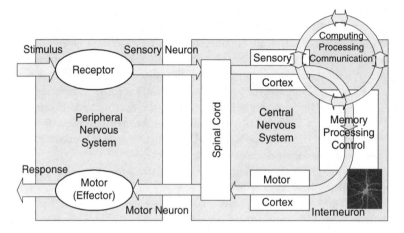

FIGURE 6.11   The vertebrate nervous system.

nanocomputer architecture. For example, the documented network consists of four routers (buffers) numbered 1, 2, 3, and 4, and interior (core) routers are illustrated. After the aggregation, a meshed network with four nodes represents the same topology. The difficulty in topology aggregation is to calculate the link metrics to attain the accuracy and to synthesize optimized network topologies. The aggregated link metrics can be computed using the methods used for multilayer artificial networks. The bandwidth and delay are integrated to optimize the hierarchically distributed networks.

The vertebrate nervous system architecture is documented in Figure 6.11.

We believe that the *distributed memory* central nervous system is adaptively configured and reconfigured based upon *processing*, *communication*, and *control* (instruction) *parallelisms*. This principle can be very effectively used in solving the nanocomputer synthesis problem.

6.5   Nanocomputer Architecture

System-level performance analysis is based on mathematical models used to examine nanocomputers and optimize their architectures, organizations, topologies, and parameters, as well as to perform optimal hardware–software codesign. Computers process inputs, producing outputs accordingly. To examine the performance, one should develop and apply the mathematical model of a nanocomputer that comprises central processing, memory, I/O, control, and other units. However, one cannot develop the mathematical model without explicitly specifying the nanocomputer architecture, identifying the optimal organization, and synthesizing the topologies. There are different levels of abstraction for nanocomputer architectures, organizations, models, etc.

Advanced nanocomputer architectures (beyond Von Neumann computer architecture) can be devised to guarantee superior processing and communication speed. Novel nanodevices that utilize new effects and phenomena to perform computing and communication are sought. For example, through quantum computing, the information can be stored in the phase of wave functions. This concept leads to utilization of massive parallelism of these wave functions to achieve enormous processing speed.

The CPU executes sequences of instructions and operands, which are fetched by the program control unit (PCU), executed by the data processing unit (DPU), and then placed in memory. In particular, caches (high-speed memory, where data is copied when it is retrieved from the random access memory, improving the overall performance by reducing the average memory access time) are used. The instructions and data form instruction and data streams that flow to and from the processor. The CPU may have two or more processors and coprocessors with various execution units, multilevel instructions, and data caches. These processors can share or have their own caches. The CPU *datapath* contains nanoICs to perform arithmetic and logical operations on words such as fixed- or floating-point numbers. The CPU design involves the trade-off analysis among hardware, speed, and affordability. The CPU is usually partitioned on the control and datapath units, and the control unit selects and sequences the data processing operations. The core interface unit is a switch that can be implemented as autonomous cache controllers operating concurrently and feeding the specified number (32, 64, or 128) of bytes of data per cycle. This core interface unit connects all controllers to the data or instruction caches of processors. Additionally, the core interface unit accepts and sequences information from the processors. A control unit is responsible for controlling data flow between controllers that regulate the *in* and *out* information flows. There is an interface to I/O devices. On-chip debug, error detection, sequencing logic, self-test, monitoring, and other units must be integrated to control a pipelined nanocomputer. The computer performance depends on the architecture and hardware components which are discussed in this chapter. Figure 6.12 illustrates the possible nanocomputer organization.

The current CMOS technology has been advanced to the 90- and 130-nm processes, 2.4 GHz with 4 GB/sec, 400 MHz bus, and the Intel Pentium 4 die (two-dimensional) as illustrated in Figure 6.13.

In general, nanodevices ensure high density, superior bandwidth, high switching frequency, low power, etc. It is envisioned that in the near future nanocomputers will allow one to increase the computing speed by a factor of millions compared with the existing CMOS IC-based computers. Three-dimensional, multilayered, high-density nanoIC assemblies, shown in Figure 6.14, are envisioned to be used. Unfortunately, a number of formidable fundamental, applied, experimental, and technological challenges arise: robust operation and characteristics of nanoICs are significantly affected by the *second-order* effects (gate oxide and bandgap tunneling, energy quantization, electron transport, etc., and, furthermore, the operating principles for nanodevices can be based on the quantum effects), noise vulnerability, complexity,

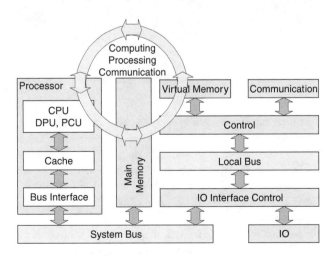

FIGURE 6.12　Nanocomputer organization similar to the human brain.

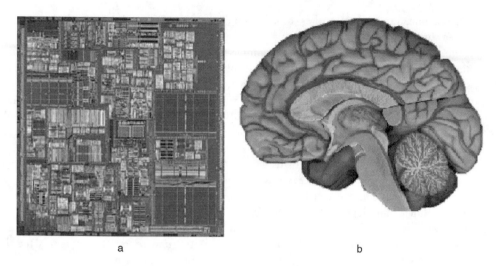

FIGURE 6.13   (a) Two-dimensional 478-pin Pentium 4 processor on 217 mm² die with 42 million transistors and six levels of aluminum interconnect, and (b) three-dimensional human brain.

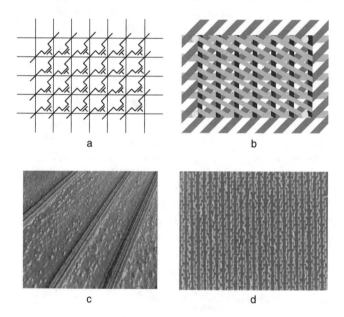

FIGURE 6.14   Three-dimensional, multiple-layered, high-density nanoIC assemblies (crossbar switching, logic or memory arrays), 3 nm-wide parallel (six-atom-wide) erbium disilicide (ErSi₂) nanowires (Hewlett-Packard), and carbon nanotube array.

etc. It is well known that high-fidelity modeling, data-intensive analysis, heterogeneous simulation, and other fundamental issues even for a single nanodevice are not completely solved yet. In addition, the currently existing fabrication processes and technologies do not allow one to fabricate ideal nanodevices and nanostructures. In fact, even molecular wires are not perfect. Different fabrication technologies, processes, and materials have been developed to attain self-assembly and self-ordered features in fabrication of nanoscale devices[11–18] (Figure 6.14). As an example, the self-assembled and aligned carbon nanotube array is illustrated in Figure 6.14.

One of the basic components of the current computers is CMOS transistors fabricated with different technologies and distinct device topologies. However, the CMOS technology and transistors fabricated with even the most advanced CMOS processes reach the physical limits. Therefore, the current research

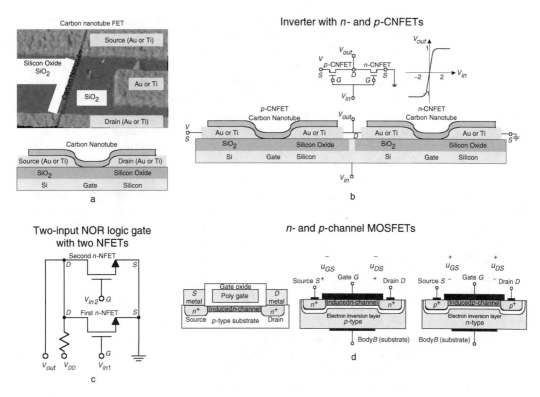

FIGURE 6.15 (a) Carbon nanotube FETs; (b) inverter with CNFETs; (c) NOR gate with NFETs; (d) *n*- and *p*-channel MOSFETs.

developments have been concentrated on the alternative solutions. Leading companies (IBM, Intel, Hewlett-Packard, etc.), academia, and government laboratories are developing novel nanotechnologies. We study computer hardware. Nanodevices (switches, logics, memories, etc.) can be implemented using the three-dimensional nanoelectronics arrays illustrated in Figure 6.14. It must be emphasized that the extremely high-frequency logic gates can be fabricated using carbon nanotubes, which are from 1 to 10 nm in diameter (100,000 times less then the diameter of a human hair). *P*- and *n*-type carbon nanotube field-effect transistors (CNFETs) were fabricated and tested with single- and multi-wall carbon nanotubes as the channel.[19,20] The atomic-force microscope image of a single-wall CNFET (50 nm total length) and CNFET are documented in Figure 6.15a.

The two-dimensional carbon nanotube structure can be utilized to devise and build different transistors with distinct characteristics utilizing different phenomena.[11,19,20] For example, twisted carbon nanotubes can be used. Carbon nanotubes can be grown on the surface using chemical vapor deposition, deposited on the surface from solvent, etc.[19–21] Photolithography can be used to attain the device-level structural and functional integration connecting source, drain, gate, etc. One concludes that different transistor topologies and configurations are available, and these results are reported elsewhere.[11–22] Taking note of this fact, we use NFET to synthesize and analyze the nanoICs. The carbon nanotube inverter, formed using the series combination of *p*- and *n*-CNFETs, is illustrated in Figure 6.15.b. The gates and drains of two CNFETs are connected together to form the input and output. The voltage characteristics can be examined studying the various transistor bias regions. When the inverter input voltage V_{in} is either a logic 0 or a logic 1, the current in the circuit is zero because one of the CNFETs is cut off. When the input voltage varies in the region $V_{threshold} < V_{in} < V - |V_{threshold}|$, both CNFETs are conducting and a current exists in the inverter.

The current-control mechanism of the field-effect transistors is based on an electrostatic field established by the voltage applied to the control terminal. Figure 6.15d shows *n*- and *p*-channel enhancement-type MOSFETs with four terminals (gate, source, drain, and base). The substrate terminals are denoted as G, S, D, and B. Consider the n-channel enhancement-type MOSFETs. A positive voltage u_{GS}, applied to the gate, causes the free positively charged holes in the n-channel region. These holes are pushed downward into the p-base (substrate). The applied voltage u_{GS} attracts electrons from the source and drain regions into the channel region. The voltage is applied between the drain and source, and the current flows through the induced n-channel region. The gate and body form a parallel-plate capacitor with the oxide layer. The positive gate voltage u_{GS} causes the positive charge at the top plate, and the negative charge at the bottom is formed by the electrons in the induced n-channel region. Hence, the electric field is formed in the vertical direction. The current is controlled by the electric field applied perpendicularly to both the semiconductor substrate and the direction of current (the voltage between two terminals controls the current through the third terminal). The basic principle of the MOSFET operation is the metal–oxide–semiconductor capacitor, and high-conductivity polycrystalline silicon is deposited on the silicon oxide. As a positive voltage u_{DS} is applied, the drain current i_D flows in the induced n-channel region from source to drain, and the magnitude of i_D is proportional to the effective voltage $u_{GS} - u_{GSt}$. If u_{GS} is greater than the threshold value u_{GSt}, the induced n-channel is enhanced. If one increases u_{DS}, the n-channel resistance is increased, the drain current i_D saturates, and the saturation region occurs as one increases u_{DS} to the u_{DSsat} value. A sufficient number of mobile electrons must be accumulated in the channel region to form a conducting channel if the u_{GS} is greater than the threshold voltage u_{GSt}, which usually is 1 V — that is, thousands of electrons are needed. It should be emphasized that in the saturation region, the MOSFET is operated as an amplifier, while in the triode region and in the cut-off region, the MOSFET can be used as a switch. The p-channel enhancement-type MOSFETs are fabricated on the n-type substrate (body) with p^+ regions for the drain and source. Here, the voltages applied u_{GS} and u_{DS} and the threshold voltage u_{GSt} are negative.

We use and demonstrate the application of nanoscale field-effect transistors (NFETs) to design the logic nanoICs. The NOR logic can be straightforwardly implemented when the first and second n-NFETs are connected in parallel (Figure 6.15.c). Different flip-flops usually are formed by cross-coupling NOR logic gates. If two input voltages are zero, both n-NFETs are cut off, and the output voltage V_{out} is high. Specifically, $V_{out} = V_{DD}$, and if $V_{DD} = 1$ V, then $V_{out} = 1$ V. If $V_{in1} \neq 0$ (for example, $V_{in1} = 1$ V) and $V_{in2} = 0$, the first n-NFET turns on and the second n-NFET is still cut off. The first n-NFET is biased in the nonsaturation region, and V_{out} reaches the low value ($V_{out}=0$). By changing the input voltages such that $V_{in1} = 0$ and $V_{in2} \neq 0$ ($V_{in2} = 1$ V), the first n-NFET becomes cut off, while the second n-NFET is biased in the nonsaturation region. Hence, V_{out} has the low value, $V_{out} = 0$. If $V_{in1} \neq 0$ and $V_{in2} \neq 0$ ($V_{in1} = 1$ V and $V_{in2} = 1$ V), both n-NFET become biased in the nonsaturation region, and V_{out} is low ($V_{out} = 0$). Table 6.1 summarizes the result and explains how the NFET can be straightforwardly used in nanocomputer units.

The series–parallel combination of the NFETs is used to synthesize complex logic gates. As an example, the resulting carbon nanotube circuitry using n-NFETs to implement the Boolean output function $V_{out} = f(\overline{A \cdot B + C})$ is illustrated in Figure 6.16.

The n-NFETs executive OR logic gate $V_{out} = A \otimes B$ can be made (Figure 6.16). If $A = B = $ logic 1, the path exists from the output to ground through NFET A and NFET B transistors, and the output goes low. If $A = B = $ logic 0 ($\overline{A} = \overline{B} = $ logic 1), the path exists from the output to ground through NFET A1 and NFET B1 transistors, and the output goes low. For all other input logic signal combinations, the output is isolated from ground, and, hence, the output goes high.

TABLE 6.1   Two-Input NOR Logic Circuit with Two NFETs

V_{in1}	V_{in2}	V_{out}
0	0	1
1	0	0
0	1	0
1	1	0

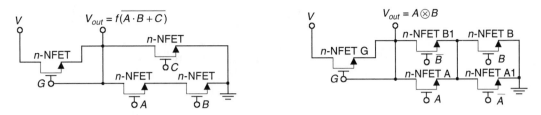

FIGURE 6.16 Static logic gates $V_{out} = \overline{f(A \cdot B + C)}$ and $V_{out} = \mathbf{A} \otimes \mathbf{B}$ synthesized using n- and p-NFETs.

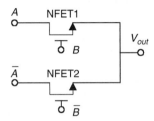

FIGURE 6.17 NanoIC's pass network.

Two logic gates $V_{out} = \overline{f(A \cdot B + C)}$ and $V_{out} = A \otimes B$ synthesized are the static nanoICs (the output voltage is well defined and is never left floating). The static nanoICs can be redesigned adding the clock.

The nanoIC's pass networks can be easily implemented. Consider the nanoICs with two n-NFETs as illustrated in Figure 6.17. The nanoIC's output V_{out} is determined by the conditions listed in Table 6.2. In states 1 and 2, the transmission gate of NFET2 is biased in its conduction state (NFET2 is on), while NFET1 is off. For state 1, $\overline{A}$ = logic 1 is transmitted to the output, and $V_{out} = 1$. For state 2, $\overline{A}$ = logic 0, and, although A = logic 1, because NFET1 is off, $V_{out} = 0$. In states 3 and 4, the transmission gate of NFET1 is biased, and NFET2 is off. For state 3, A=logic 0, we have V_{out}=0 because NFET2 is off ($\overline{A}$ = logic 1). In contrast, in state 4, A=logic 1 is transmitted to the output, and V_{out}=1. For other states, the results are reported in Table 6.2. Thus, the output V_{out} is a function of two variables, i.e., gate control and input (logic) signals.

The clocked nanoICs are dynamic circuits that, in general, precharge the output node to a particular level when the clock is at logic 0. The generalized clocked nanoIC is illustrated in Figure 6.18, where F_N is the NFET's network that performs the particular logic function $F_N(X)$ of i variables; here $X = (x_1, x_2, \ldots, x_{i-1}, x_i)$. The set of X inputs to the logic nanoIC's $F_N(X)$ is derived from the outputs of other static and dynamic nanoICs. When C_{clock} = logic 0, the outputs of all inverters are logic 0 during the precharged cycle, and during the precharged cycle all x variables of $X = (x_1, x_2, \ldots, x_{z-1}, x_z)$ are logic 0. During the precharge phase, all NFETs are cut off, and the transistor outputs are precharged to V. The transitions are possible only during the evaluation phase. The output of the nanoIC's buffer

TABLE 6.2 Two-Input NOR Logic Circuit with Two NFETs

| State | Input gate control | | Input gate control | | NFET1 | NFET2 | V_{out} |
	A	$\overline{A}$	B	$\overline{B}$			
1	0	1	0	1	off	on	1
2	1	0	0	1	off	on	0
3	0	1	1	0	on	off	0
4	1	0	1	0	on	off	1
5	0	1	0	1	off	on	1
6	1	0	0	1	off	on	0
7	0	1	1	0	on	off	0
8	1	0	1	0	on	off	1

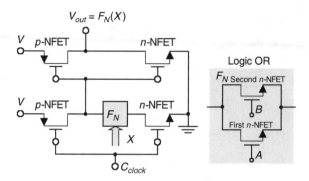

FIGURE 6.18   Dynamic generalized clocked nanoIC with logic function $F_N(X)$.

changes from 0 to 1. Specifically, the logic OR function is illustrated to demonstrate the generalized clocked nanoIC in Figure 6.18.

A combination of logic gates is used to perform logic functions, addition, subtraction, multiplication, division, multiplexing, etc. To store the information, the memory cell nanoICs are used. A systematic arrangement of memory cells and peripheral nanoICs (to address and write the data into the cells as well as to delete data stored in the cells) constitutes the memory. The NFETs can be used to build superior static and dynamic random access memory (RAM is the read-write memory in which each individual cell can be addressed at any time) and programmable and alterable read-only memory (ROM is commonly used to store instructions of a system's operating system). The static RAM (implemented using six NFETs) consists of basic flip-flop nanoICs with two stable states (0 and 1), while dynamic RAM (implemented using one NFET and storage capacitor) stores one bit of information charging the capacitor. As an example, the dynamic RAM (DRAM) cell is documented in Figure 6.19. In particular, the binary information is stored as the charge on the storage capacitor C_s (logic 0 or 1). The DRAM cell is addressed by turning on the pass n-NFET via the word line signal S_{wl}, and charge is transferred into and out of C_s on the data line (C_s is isolated from the rest of the nanoICs when the n-NFET is off, but the leakage current through the n-NFET requires the cell refreshment to restore the original signal).

Dynamic shift registers are implemented using transmission gates, and inverters and flip-flops are synthesized by cross-coupling NOR gates, while delay flip-flops are built using transmission gates and feedback inverters.

For many nanoICs (we do not consider the optical, single-electron, and quantum nanoICs), synthesis and design can be performed using the basic approaches, methods, and computer-aided design for conventional ICs (viable results exist for MOSFETs); but the differences must be outlined. The major problems arise because the developed technologies cannot guarantee the fabrication of high-yield perfect nanoICs and different operating principles. In general, the secondary effects must be integrated (for example, even for NFETs, which are comparable to MOSFETs, the switching and leakage phenomena are different). Correspondingly, the simple scaling of ICs with MOSFETs to nanoICs cannot be performed.

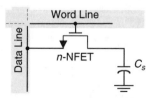

FIGURE 6.19   One n-NFET dynamic RAM with the storage capacitor.

The direct self-assembly, nanopatterning, nanoimprinting, nanoaligning, nanopositioning, overlayering, margining, and annealing have been shown to be quite promising.[11,13–22] However, it is unlikely that near-future nanotechnologies will guarantee the reasonable repeatable characteristics, high quality, satisfactory geometry uniformity, suitable failure tolerance, and other important specifications and requirements imposed on nanodevices and nanostructures.[11–23] Different novel concepts were recently reported in the literature.[24–26] Therefore, syntheses of robust defect-tolerant adaptive (reconfigurable) nanocomputer architectures (hardware) and software to accommodate failures, inconsistencies, variations, nonuniformities, and defects are critical.

6.6 Hierarchical Finite-State Machines and Their Use in Hardware and Software Design

Simple register-level systems perform a single data-processing operation, e.g., summation $X:=x_1+x_2$ or subtraction $X:x_1-x_2$. To do different complex data-processing operations, multifunctional register-level systems should be synthesized. These multifunctional register-level systems are partitioned as a data processing unit (*datapath*) and a controlling unit (control unit). The control unit is responsible for collecting and controlling the data-processing operations (actions) of the *datapath*. To design the register-level systems, one studies a set of operations to be executed and then designs nanoICs using a set of register-level components that implement the desired functions while satisfying the affordability and performance requirements. It is very difficult to impose meaningful mathematical structures on register-level behavior or structure using Boolean algebra and conventional gate-level design. Due to these difficulties, the heuristic synthesis is commonly accomplished as sequential steps listed below:

1. Define the desired behavior as a set of sequences of register-transfer operations (each operation can be implemented using the available components) making up the algorithm to be executed.
2. Examine the algorithm to determine the types of components and their number to attain the required datapath. Design a complete block diagram for the datapath using the components chosen.
3. Examine the algorithm and datapath in order to derive the control signals with the ultimate goal of synthesizing the control unit for the found datapath that meets the algorithm's requirements.
4. Analyze and verify the synthesis performing modeling and detail simulations (VHDL, Verilog, ModelSim, and other environments are commonly used).

Let us perform the synthesis of virtual control units that ensures extensibility, flexibility, adaptability, robustness, and reusability. The synthesis will be performed using the hierarchical graphs (HGs). A most important problem is to develop straightforward algorithms that ensure implementation (nonrecursive and recursive calls) and utilize hierarchical specifications. We will examine the behavior, perform logic synthesis, and implement reusable control units modeled as hierarchical finite-state machines with virtual states. The goal is to attain the top-down, sequential, well-defined decomposition to develop complex robust control algorithms.

We consider datapath and control units. The datapath unit consists of memory and combinational units. A control unit performs a set of instructions by generating the appropriate sequence of micro instructions that depend on intermediate logic conditions or intermediate states of the datapath unit. To describe the evolution of a control unit, behavioral models have been developed.[28,29] We use the direct-connected HGs containing nodes. Each HG has an entry (*Begin*) and an output (*End*). Rectangular nodes contain micro instructions, macro instructions, or both.

A micro instruction U_i includes a subset of micro operations from the set $U = \{u_1, u_2, ..., u_{u-1}, u_u\}$. Micro operations $\{u_1, u_2, ..., u_{u-1}, u_u\}$ force the specific actions in the datapath (Figures 6.20 and 6.21). For example, one can specify that u_1 pushes the data in the local stack, u_2 pushes the data in the output stack, u_3 forms the address, u_4 calculates the address, u_5 pops the data from the local stack, u_6 stores the data from the local stack in the register, u_7 pops the data from the output stack to external output, etc.

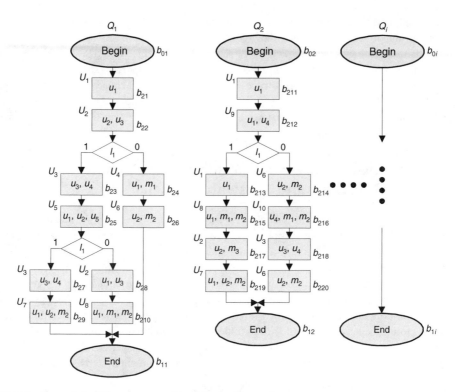

FIGURE 6.20   Control algorithm represented by HGs Q_1, Q_2, ..., Q_i.

A macro operation is the output causing an action in the datapath. Any macro instruction incorporates macro operations from the set $M = \{m_1, m_2, ..., m_{m-1}, m_m\}$. Each macro operation is described by another lower-level HG. Assume that each macro instruction includes one macro operation. Each rhomboidal node contains one element from the set $L \cup G$. Here, $L = \{l_1, l_2, ..., l_{l-1}, l_l\}$ is the set of logic conditions, while $G = \{g_1, g_2, ..., g_{g-1}, g_g\}$ is the set of logic functions. Using logic conditions as inputs, logic functions are calculated examining a predefined set of sequential steps that are described by a lower-level HG. Directed lines connect the inputs and outputs of the nodes.

Consider a set $E = M \cup G$, $E = \{e_1, e_2, ..., e_{e-1}, e_e\}$. All elements $e_i \in E$ have HGs, and e_i has the corresponding HG Q_i, which specifies either an algorithm for performing e_i (if $e_i \in M$) or an algorithm for calculating e_i (if $e_i \in G$). Assume that $M(Q_i)$ is the subset of macro operations and $G(Q_i)$ is the subset of logic functions that belong to the HG Q_i. If $M(Q_i) \cup G(Q_i) = \varnothing$, the well-known scheme results.[28,29] The application of HGs enables one to gradually and sequentially synthesize complex control algorithms, concentrating the efforts at each stage on a specified level of abstraction because specific elements of the set E are used. Each component of the set E is simple and can be checked and debugged independently. Figure 6.20 shows HGs Q_1, Q_2, ..., Q_i which describe the control algorithm.

The execution of HGs is examined by studying complex operations $e_i = m_j \in M$ and $e_i = g_j \in G$. Each complex operation e_i that is described by an HG Q_i must be replaced with a new subsequence of operators that produces the result executing Q_i. In the illustrative example, shown in Figure 6.21, Q_1 is the first HG at the first level Q^1, the second level Q^2 is formed by Q_2, Q_3 and Q_4, etc. We consider the following hierarchical sequence of HGs: Q_1 (level 1) $\Rightarrow Q^2$ (level 2) $\Rightarrow ... \Rightarrow Q^{q-1}$ (level q–1) $\Rightarrow Q^q$ (level q). All Q_i (level i) have the corresponding HGs. For example, Q^2 is a subset of the HGs that are used to describe elements from the set $M(Q_1) \cup G(Q_1) = \varnothing$, while Q^3 is a subset of the HGs that are used to map elements from the sets $\bigcup_{q \in Q^2} M(q)$ and $\bigcup_{q \in Q^2} G(q)$. In Figure 6.21, $Q^1 = \{Q_1\}$, $Q^2 = \{Q_2, Q_3, Q_4\}$, $Q^j = \{Q_2, Q_4, Q_5\}$, etc.

Micro operations u^+ and u^- are used to increment and to decrement the stack pointer (SP). The problem of switching to various levels can be solved using a stack memory (Figure 6.22). Consider an algorithm for $e_i \in M(Q_1) \cup G(Q_1) = \varnothing$. The SP is incremented by the micro operation u^+, and a new register

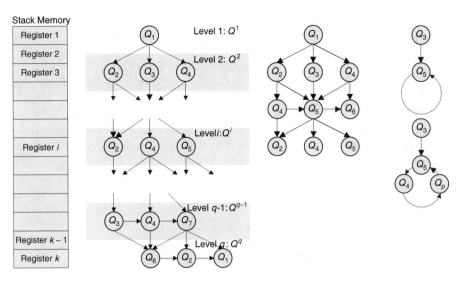

FIGURE 6.21 Stack memory with multilevel sequential HGs with illustration of recursive call.

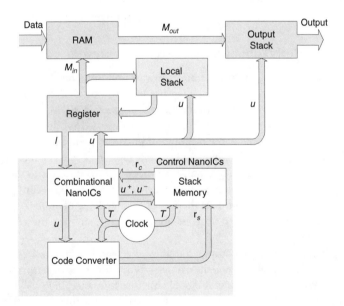

FIGURE 6.22 Hardware implementation.

of the stack memory is set as the current register. The previous register stores the state when it was interrupted. New Q_i becomes responsible for the control until terminated. After termination of Q_i, the micro operation u^- is generated to return to the interrupted state. As a result, control is passed to the state in which Q_f is called.

The synthesis problem can be formulated as: for a given control algorithm A, described by the set of HGs, construct the FSM that implements A. In general, the synthesis includes the following steps:

1. Transforming the HGs to the state transition table
2. State encoding
3. Combinational logic optimization
4. Final structure design

The first step is divided into substeps: (1) mark the HGs with b labels (Figure 6.20); (2) record all transitions between the labels in the extended state transition table; and (3) convert the extended table to ordinary form.

The labels b_{01} and b_{11} are assigned to the nodes *Begin* and *End* of the Q_1. The label $b_{02}, ..., b_{0i}$ and $b_{12}, ..., b_{1i}$ are assigned to nodes *Begin* and *End* of $Q_2, ..., Q_i$, respectively. The labels $b_{21}, b_{22}, ..., b_{2j}$ are assigned to other nodes of HGs, inputs and outputs of nodes with logic conditions, etc. Repeating labels is not allowed. The labels are considered the states. The extended state transition table is designed using the state evolutions due to inputs (logic conditions) and logic functions which cause the transitions from $x(t)$ to $x(t+1)$. All evolutions of the state vector $x(t)$ are recorded, and the state $x_k(t)$ has the label k. It should be emphasized that the table can be converted from the extended to the ordinary form. To program the code converter, one records the transition from the state x_1 assigned to the *Begin* node of the HG Q_1, i.e., $x_{01} \Rightarrow x_{21}(Q_1)$. The transitions between different HGs are recorded as $x_{ij} \Rightarrow x_{nm}(Q_j)$. For all transitions, the data-transfer instructions are synthesized.

The hardware implementation is illustrated in Figure 6.22.

Robust control algorithms are synthesized and implemented with the HGs using the hierarchical behavior specifications and top-down decomposition. The reported method guarantees exceptional adaptation and reusability features through reconfigurable hardware and reprogrammable software.

6.7 Adaptive (Reconfigurable) Defect-Tolerant Nanocomputer Architectures, Redundancy, and Robust Synthesis

6.7.1 Reconfigurable Nanocomputer Architectures

To design nanocomputers, specific hardware and software solutions must be developed and implemented. For example, ICs are designed by making use of hardware description languages, e.g., Very High Speed Integrated Circuit Hardware Description Language (VHDL) and Verilog. Making the parallel to the conventional ICs, the programmable gate arrays (PGAs) developed by Xilinx, Altera, Actel, and other companies can serve as the meaningful inroad in design of reconfigurable nanocomputers. These PGAs lead one to the on-chip reconfigurable logic concept. The reconfigurable logic can be utilized as a functional unit in the datapath of the processor, having access to the processor register file and to on-chip memory ports. Another approach is to integrate the reconfigurable part of the processor as a coprocessor. For this solution, the reconfigurable logic operates concurrently with the processor. Optimal design and memory port assignments can guarantee the coprocessor reconfigurability and concurrency.

In general, the reconfigurable nanocomputer architecture synthesis emphasizes a high-level design, rapid prototyping, and reconfigurability to reduce time and cost and improve performance and reliability. The goal is to devise, design, and fabricate affordable high-performance, high-yield nanoIC arrays and application-specific nanoICs (ASNICs). These ASNICs should be testable to detect defects and faults. The design of ASNICs involves mapping application requirements into specifications implemented by nanoICs. The specifications are represented at every level of abstraction including the system, behavior, structure, physical, and process domains. The designer should be able to differently utilize the existing nanoICs to meet the application requirements. User-specified nanoICs and ASNICs must be developed to attain affordability and superior performance.

The PGAs can be used to implement logic functions utilizing millions of gates. The design starts by converting the application requirements into architectural specifications. As the application requirements are examined, the designer translates the architectural specifications into behavior and structure domains. Behavior representation means the functionality required as well as the ordering of operations and completion of tasks in specified times. A structural description consists of a set of nanodevices and their interconnections. Behavior and structure can be specified and studied using hardware description languages. This nanoIC Hardware Description Language (NHDL) should efficiently manage very complex hierarchies that may include millions of logic gates. Furthermore, NHDLs should be translated into net-lists of library components using synthesis software. The NHDL software, which is needed to describe hardware and must permit concurrent operations, should perform the following major functions:

1. Translate text to a Boolean mathematical representation
2. Optimize the representation-based specified criteria (size, delays, optimality, reliability, testability, etc.)
3. Map the optimized mathematical representation to a technology-specific library of nanodevices

Reconfigurable nanocomputers should use reprogrammable logic units such as PGAs to implement a specialized instruction set and arithmetic units to optimize performance. Ideally, reconfigurable nanocomputers can be reconfigured in real time (runtime), enabling the existing hardware to be reused depending on its interaction with external units, data dependencies, algorithm requirements, faults, etc. The basic PGA architecture is built using the programmable logic blocks (PLBs) and programmable interconnect blocks (PIBs) (Figure 6.23). The PLBs and PIBs will hold the current configuration setting until adaptation is accomplished. The PGA is programmed by downloading the information in the file through a serial or parallel logic connection. The time required to configure a PGA is called the *configuration time* (PGAs could be configured in series or in parallel). Figure 6.23 illustrates the basic architectures from which most multiple PGA architecture can be derived (pipelined architecture with the PGAs interfaced one to another is well suited for functions that have streaming data at specific intervals, while arrayed PGA architecture is appropriate for functions that require a systolic array). A hierarchy of configurability is different for the various PGAs architectures.

The goal is to design reconfigurable computer architectures with corresponding software to cope with less-than-perfect, entirely or partially defective, and faulty nanoscale devices, structures, and connects (i.e., nanoICs) encountered in arithmetic and logic, control, I/O, memory, and other units. To achieve our objectives, the redundant nanoIC units can be used (the redundancy level is determined by the nanoscale IC's quality and software capabilities). Hardware and software evolutionary learning, adaptability, and reconfigurability can be achieved through decision making, diagnostics, health monitoring, analysis and optimization of software, as well as pipelining, rerouting, switching, matching, and controlling of hardware. We concentrate our research on how to devise, design, optimize, build, test, and configure nanocomputers. The overall objective can be achieved guaranteeing the evolution (behavior) matching between the ideal (C_I) and fabricated (C_F) nanocomputers, their units, systems, or components. The nanocompensator (C_{F1}) can be designed for the fabricated C_{F2} such that the response of the C_F will

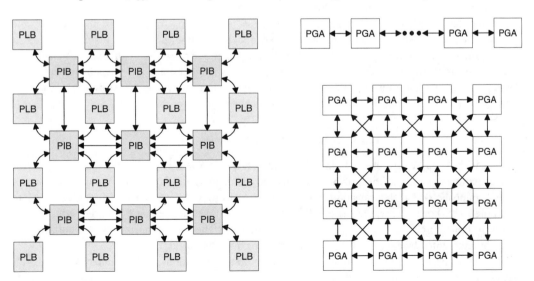

FIGURE 6.23 Programmable gate arrays and multiple PGA architectures.

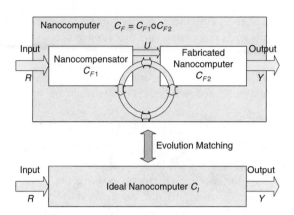

FIGURE 6.24   Nanocomputers and evolution matching.

match the evolution of the C_I (Figure 6.24). The C_I gives the reference ideal (evolving) model, which analytically and/or experimentally maps the ideal (desired) I/O behavior; and the nanocompensator C_{F1} should modify the evolution of C_{F2} such that C_F, described by $C_F = C_{F1} \circ C_{F2}$ (series architecture), matches the C_I behavior. Figure 6.24 illustrates the concept. The necessary and sufficient conditions for strong and weak evolution matching based on C_I and C_{F2} must be derived.

6.7.2   Mathematical Model of Nanocomputers

To address analysis, control, diagnostics, optimization, and design problems, the explicit mathematical models of nanocomputers must be developed and applied. There are different levels of abstraction in nanocomputer modeling, simulation, and analysis. High-level models can accept streams of instruction descriptions and memory references, while the low-level (device-level) logic gates or memory modeling can be performed by making use of streams of input and output signals as well as nonlinear transient behavior of devices. The system- and subsystem-level modeling (medium-level) also can be formulated and performed. It is evident that one subsystem can contain millions of nanodevices. For example, computer systems can be modeled as queuing networks.[3] Different mathematical modeling frameworks exist and have been developed for each level. In this section we concentrate on the high-, medium- and low-level modeling of nanocomputer systems using the finite-state machine concept.

A computer accepts the input information, processes it according to the stored instructions, and produces the output. There are numerous computer models, e.g., Boolean models, polynomial models, and information-based models. However, all mathematical models are the mathematical idealization based upon the abstractions, simplifications, and hypotheses made. In fact, it is virtually impossible to develop and apply the complete mathematical model due to complexity and numerous obstacles.

It is possible to concurrently model nanocomputers by the six-tuple:

$$\mathbf{C} = \{\mathbf{X, E, R, Y, F, X_0}\},$$

where X is the finite set of states with initial and final states $x_0 \in X$ and $x_f \subseteq X$; E is the finite set of events (concatenation of events forms a string of events); R and Y are the finite sets of the input and output symbols (alphabets); F is the transition function mapping from $X \times E \times R \times Y$ to X (denoted as F_X), to E (denoted as F_E) or to Y (denoted as F_Y), $F \subseteq X \times E \times R \times Y$. (We assume that $F = F_X$. For example, the transition function defines a new state to each quadruple of states, events, references, and outputs, and F can be represented by a table listing the transitions or by a state diagram.)

The nanocomputer evolution is due to inputs, events, state evolutions, and parameter variations (as explained in the end of this subsection), etc.

We formulate two useful definitions.

Definition 1

A vocabulary (or an alphabet) A is a finite nonempty set of symbols (elements). A word (or sentence) over A is a string of a finite length of elements of A. The empty (null) string is the string that does not contain symbols. The set of all words over A is denoted as A_w. A language over A is a subset of A_w.

Definition 2

A finite-state machine with output $C_{FS} = \{X, A_R, A_Y, F_R, F_Y, X_0\}$ consists of a finite set of states X, a finite input alphabet A_R, a finite output alphabet A_Y, a transition function F_Y that assigns a new state to each state and input pair, an output function F_Y that assigns an output to each state and input pair, and initial state X_0.

Using the input–output map, the evolution of C can be expressed as:

$$E_C \subseteq R \times Y.$$

That is, if the computer in state $x \in X$ receives an input $r \in R$, it moves to the next state $f(x,r)$ and produces the output $y(x,r)$. Nanocomputers can be represented as the state tables that describe the state and output functions. In addition, the state transition diagram (direct graph whose vertices correspond to the states, edges correspond to the state transitions, and each edge is labeled with the input and output associated with the transition) is frequently used.

The quantum computer is described by the seven-tuple $QC = \{X, E, R, Y, H, U, X_0\}$, where H is the Hilbert space and U is the unitary operator in the Hilbert space that satisfies the specific conditions.

6.7.3 Nanocomputer Modeling with Parameters Set

Nanocomputers can be modeled using the parameters set P. Designing reconfigurable fault-tolerant nanocomputer architectures, sets P and P_0 should be integrated, and we have:

$$C = \{X, E, R, Y, P, F, X_0, P_0\}.$$

It is evident that the nanocomputer evolution depends upon P and P_0. The optimal performance can be achieved through adaptive synthesis, reconfiguration, and diagnostics. For example, one can vary F and variable parameters P_v to attain the best possible performance.

The nanocomputer evolution, considering states, events, outputs, and parameters, can be expressed as:

$$\overset{\text{evolution 1}}{(x_0, e_0, y_0, p_0) \Rightarrow} (x_1, e_1, y_1, p_1) \overset{\text{evolution 2}}{\Rightarrow} \dots \overset{\text{evolution } j-1}{\Rightarrow} (x_{j-1}, e_{j-1}, y_{j-1}, p_{j-1}) \overset{\text{evolution } j}{\Rightarrow} (x_j, e_j, y_j, p_j).$$

The input, states, outputs, events, and parameter sequences are aggregated within the model $C = \{X, E, R, Y, P, F, X_0, P_0\}$. The concept reported allows us to find and apply the minimal (but complete) functional description of nanocomputers. The minimal subset of state, event, output, and parameter evolutions (transitions) can be used. That is, the partial description $C_{partial} \subset C$ results, and every essential nanocomputer quadruple (x_i, e_i, y_i, p_i) can be mapped by $(x_i, e_i, y_i, p_i)_{partial}$. This significantly reduces the complexity of modeling, simulation, analysis, and design problems.

Let the transition function F map from $X \times E \times R \times Y \times P$ to X, i.e., $F : X \times E \times R \times Y \times P \to X$, $F \subseteq X \times E \times R \times Y \times P$. Thus, the transfer function F defines a next state $x(t+1) \in X$ based upon the current state $x(t) \in X$, event $e(t) \in E$, reference $r(t) \in R$, output $y(t) \in Y$, and parameter $p(t) \in P$. Hence,

$$x(t + 1) = F(x(t), e(t), r(t), p(t))$$

for $x_0(t) \in X_0$ and $p_0(t) \in P_0$.

The robust adaptive control algorithms must be developed. The control vector $u(t) \in U$ is integrated into the nanocomputer model. We have $C = \{X, E, R, Y, P, U, F, X_0, P_0\}$.

Different synthesis methods are reported in the literature,[1-10] and some issues in the control algorithms design are reported in the previous and current sections.

6.7.4   Nanocompensator Synthesis

In this section we will design the nanocompensator. Two useful definitions that allow one to precisely formulate and solve the problem are formulated.

Definition 3

The strong evolutionary matching $C_F = C_{F1} \circ C_{F2} = {}_B C_I$ for given C_I and C_F is guaranteed if $E_{C_F} = E_{C_I}$. Here, $C_F = {}_B C_I$ means that the behaviors (evolution) of C_I and C_F are equivalent.

Definition 4

The weak evolutionary matching $C_F = C_{F1} \circ C_{F2} \subseteq {}_B C_I$ for given C_I and C_F is guaranteed if $E_{C_F} \subseteq E_{C_I}$. Here, $C_F \subseteq {}_B C_I$ means that the evolution of C_F is contained in the behavior C_I.

The problem is to derive a nanocompensator $C_{F1} = \{X_{F1}, E_{F1}, R_{F1}, Y_{F1}, F_{F1}, X_{F10}\}$ such that, for given $C_I = \{X_I, E_I, R_I, Y_I, F_I, X_{I0}\}$ and $C_{F2} = \{X_{F2}, E_{F2}, R_{F2}, Y_{F2}, F_{F2}, X_{F20}\}$, the following conditions:

$$C_F = C_{F1} \circ C_{F2} = {}_B C_I \text{ (strong behavior matching)}$$

or

$$C_F = C_{F1} \circ C_{F2} \subseteq {}_B C_I \text{ (weak behavior matching)}$$

are satisfied.

Here we assume that the following conditions are satisfied:

- Output sequences generated by C_I can be generated by C_{F2}
- The C_I inputs match the C_{F1} inputs

It must be emphasized that the output sequences involve the state, event, output, and/or parameters vectors, e.g., x, e, y, p.

> ***Lemma 1.*** If there exists the state modeling representation $\gamma \subseteq X_I \times X_F$ such that $C_I^{-1} \subseteq {}_B^\gamma C_{F2}^{-1}$ (if $C_I^{-1} \subseteq {}_B^\gamma C_{F2}^{-1}$, then $C_I \subseteq {}_B^\gamma C_{F2}$), then the evolution matching problem is solvable. The nanocompensator C_{F1} solves the strong matching problem $C_F = C_{F1} \circ C_{F2} = {}_B C_I$ if there exist the state modeling representations $\beta \subseteq X_I \times X_{F2}$, $(X_{I0}, X_{F20}) \in \beta$, and $\alpha \subseteq X_{F1} \times \beta$, $(X_{F10}, (X_{I0}, X_{F20})) \in \alpha$ such that $C_{F1} \subseteq {}_B^\alpha C_\beta^I$ for $\beta \in \Gamma = \{\gamma | (C_I^{-1} \subseteq {}_B^\gamma C_{-1}^{F2})\}$. Furthermore, the strong matching problem is tractable if there exist C_I^{-1} and C_{F2}^{-1}.

The nanocomputer can be decomposed using algebraic decomposition theory, which is based on the closed partition lattice. For example, consider the fabricated nanocomputer C_{F2} represented as $C_{F2} = \{X_{F2}, E_{F2}, R_{F2}, Y_{F2}, F_{F2}, X_{F20}\}$.

A partition on the state set for C_{F2} is a set $\{C_{F2\,1}, C_{F2\,2}, \ldots, C_{F2\,i}, \ldots, C_{F2\,k-1}, C_{F2\,k}\}$ of disjoint subsets of the state set X_{F2} whose union is X_{F2}, i.e.,

$$\bigcup_{i=1}^{k} C_{F2i} = X_{F2} \text{ and } C_{F2i} \cap C_{F2j} = \varnothing \text{ for } i \neq j.$$

Hence, we can design and implement the nanocompensators (hardware) $C_{F1\,i}$ for given $C_{F2\,i}$.

6.7.5   Information Propagation Model

To model, analyze, and optimize the information propagation (data flow) and other phenomena in nanocomputers, we concentrate on the communication model. A mathematical model should provide the explicit mathematical model for arrival rate, message length, transmission capacity, buffer size, delays, losses, etc. We represent the nanocomputer using connected M nodes with L links (channels). The stochastic difference equation with queueing delays is given as:

$$x_i^d(t+1) = Ax_i^d(t) + Bx_i^d(t)u_{ii}^d(t) + F\sum_{j \in M} x_i^d(t)u_{ij}^d(t - \tau_{ij}^d) + \Xi\xi_i^d(t),$$

where x_i^d is the state vector (messages) at node i with destination node d; u_{ii}^d and u_{ij}^d are the control vectors (message to be sent from node i and messages to be delivered to node i); ξ_i^d is the noise; τ_{ij}^d is the delay; and A, F, and Ξ are the matrices or nonlinear maps.

It is impossible to solve the optimization problem using the classical methods (dynamic programming, Lyapunov theory, etc.), and the multi-input/multi-output nonlinear problem can be solved only using probabilistic methods.

6.8   Information Theory, Entropy Analysis, and Optimization

The basic problem in classical information theory is to measure the information. If X is a random variable that has value x with probability $p(x)$, then the information content of X is:

$$S(X) = S(\{p(x)\}) = -\sum_x p(x)\log p(x),$$

where $S(x)$ is the entropy (function of the probability distribution of values of X, i.e., the information content of X), and $S(x) > 0$ because $p(x) \leq 1$.

The entropy concept is used to express the data compression equation, and for noiseless channel coding using Shannon's theorem we have $n_{bit} = S(X)$.

The probability that $Y = y$, given that $X = x$, is $p(y|x)$; and the conditional entropy $S(Y|X)$ is

$$S(Y|X) = -\sum_x p(x)\sum_y p(y|x)\log p(y|x) = -\sum_x\sum_y p(x,y)\log p(y|x)$$

where $p(x,y)$ is the probability that $X=x$ and $Y=y$, $p(x,y) = p(x)p(y|x)$.

The conditional entropy $S(Y|X)$ measures how much information would remain in Y if we were to learn X and, furthermore, in general $S(Y|X) \leq S(Y)$ and usually $S(Y|X) \neq S(X|Y)$. The conditional entropy is used to define the mutual information in order to measure how much X and Y contain information about each other, i.e.,

$$I(X;Y) = S(X) - S(X|Y) = \sum_x\sum_y p(x,y)\log\frac{p(x,y)}{p(x)p(y)}.$$

If X and Y are independent, then $p(x,y) = p(x)p(y)$, and $I(X;Y) = 0$. The information content $S(X,Y)$ is

$$S(X,Y) = S(X) + S(Y) - I(X;Y).$$

The data processing is studied. If the Markov chain is $X \rightarrow Y \rightarrow Z$ (Z depends on Y, but not directly on X, $p(x, y, z) = p(x)p(y|x)p(z|y)$), the following inequalities are satisfied:

$$S(X) \geq S(X;Y) \geq S(X;Z) \text{ (data processing inequality)},$$

$$S(Z;X) \leq S(Z;Y) \text{ (data pipelining inequality)},$$

$$I(X;Z) \leq I(X;Y),$$

i.e., the device Y can pass on to Z no more information about X than it has received.

The channel capacity $C(p)$ is defined to be the maximum possible mutual information $I(X;Y)$ between the input and output of the channel maximized over all possible sources. That is,

$$C(p) = \max_{\{p(x)\}} I(X;Y)$$

,

and for binary channels,

$$0 \le C(p) \le 1.$$

Complex optimization problems can be solved by applying the results reported elsewhere.[30] Furthermore, for binary symmetric channels, $C(p) = 1 - S(X|Y)$ must lie between zero and one.

6.9   Some Problems in Nanocomputer Hardware–Software Modeling

In this section we will discuss the mathematical model development issues. While the importance of mathematical models in analysis, simulation, design, optimization, and verification is evident, some other meaningful aspects should be emphasized. For example, hardware and software codesign, integration, and verification are important problems to be addressed. However, these problems interact with mathematical modeling and analysis. The synthesis of concurrent nanocomputer architectures (collection of units, systems, subsystems, and components that can be software programmable and adaptively reconfigurable) is among the most important issues. It is evident that software is highly dependent upon hardware (and vice versa), and the nanocomputer architectronics concurrency means hardware and software compliance and matching. Due to inherent difficulties in fabricating high-yield ideal (perfect) nanodevices and nanostructures integrated in subsystems, it is unlikely that the software can be developed for configurations which are not strictly defined and must be adapted, reconfigured, and optimized within the devised hardware architectures. The not-perfect nanodevices lead to other problems such as diagnostics, estimation, and assessment analysis to be implemented through robust adaptive software. The mathematical models of nanocomputers were reported and discussed in this chapter. The systematic synthesis, analysis, optimization, and verification of hardware and software, as illustrated in Figure 6.25, are applied to advance the design methodology and refine mathematical models. In fact, software must also be mapped and modeled, but this problem is not studied in this chapter.

The analysis and modeling can be formulated and examined only if the nanocomputer architecture is devised. The optimal design and redesign cannot be performed without mathematical modeling. Therefore, it is very important to start the design process from a high-level, but explicitly defined, abstraction domain which should:

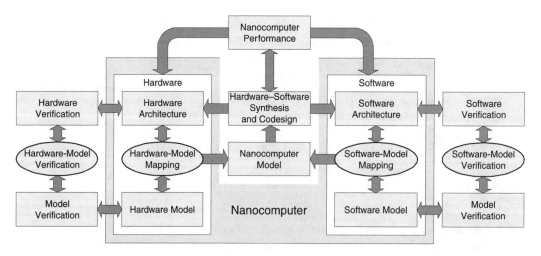

FIGURE 6.25   Nanocomputer mathematical model, performance, and hardware–software codesign.

- Comprehensively capture the functionality and performance
- Allow one to verify the correctness of the functional specifications and behavior of units, systems, and subsystems with respect to desired properties
- Depict the specification in different architectures examining their adaptability, reconfigurability, optimality, etc.

System-level models describe the nanocomputer as a hierarchical collection of units, systems, and subsystems. More specifically, steady-state and dynamic processes in nanocomputer components are studied, examining how these components perform and interact. It was illustrated that states, events, outputs, and parameters evolutions describe behavior and transients. Different discrete events, process networks, Petri nets, and other methods have been applied to model computers. It appears that models based on synchronous and asynchronous finite-state machine paradigms with some refinements ensure meaningful features and map essential behavior for different abstraction domains. Mixed control, data flow, data processing (encryption, filtering, and coding), and computing processes can be modeled. Having synthesized the nanocomputer architectures, the nanocomputer hardware mathematical model is developed as reported in Section 6.7.

Hardware description languages (HDLs) can be used to represent nanoIC diagrams or high-level algorithmic programs that solve particular problems. The structural or behavioral representations are meaningful ways of describing a model. In general, HDL can be used for documentation, verification, synthesis, design, simulation, analysis, and optimization. Structural, data flow, and behavioral domains of hardware description are studied. For conventional ICs, VHDL and Verilog are standard design tools.[31–33]

In VHDL, a design is typically partitioned into blocks. These blocks are then connected together to form a complete design using the schematic capture approach. This is performed using a block diagram editor or hierarchical drawings to represent block diagrams. In VHDL, every portion of a VHDL design is considered as a block. Each block is analogous to an off-the-shelf IC and is called an *entity*. The entity describes the interface to the block, schematics, and operation. The interface description is similar to a pin description and specifies the inputs and outputs to the block. A complete design is a collection of interconnected blocks.

Consider a simple example of an entity declaration in VHDL. The first line indicates a definition of a new entity, e.g., latch. The last line marks the end of the definition. The lines between, called the *port clause*, describe the interface to the design. The port clause provides a list of interface declarations. Each interface declaration defines one or more signals that are inputs or outputs to the design. Each interface declaration contains a list of names, mode, and type. For the example declaration, two input signals are defined as in1 and in2. The list to the left of the colon contains the names of the signals, and to the right of the colon are the mode and type of the signals. The mode specifies whether this is an input (in), output (out), or both (inout). The type specifies what kind of values the signal can have. The signals in1 and in2 are of mode in (inputs) and type bit. Signals ou1 and ou2 are defined to be of the mode out (outputs) and of the type bit (binary). Each interface declaration, except the last one, is followed by a semicolon and the entire port clause has a semicolon at the end.

```
entity latch is
     port (in1,in2: in bit;
           ou1,ou2: out bit);
end latch;
```

Signals and variables are used. In the example, the signals are defined to be of the type bit and, thus, logic signals have two values (0 and 1). As the interface declaration is accomplished, the architecture declaration is studied. The example of an architecture declaration for the latch entity with NOR gate is examined. The first line of the declaration is a definition of a new architecture called dataflow, and it belongs to the entity named *latch*. Hence, this architecture describes the operation of the latch entity. The lines between (begin and end) describe the latch operation.

```
architecture dataflow of latch is
begin
      ou1<=in2 nor ou2;
      ou2<=in1 nor ou1;
end dataflow;
```

As the basic building blocks using entities and their associated architectures are defined, one can combine them together to form other designs. For example, the structural approach is illustrated:

```
architecture structure of latch is
component nor_gate
      port (i1,i2: in bit;
            o1: out bit);
      end component;
begin
      name1: nor_gate
            port map (in2,ou2,ou1);
      name2: nor_gate
            port map (in1,ou1,ou2);
end structure;
```

The lines between the first line and the keyword begin represent a *component declaration*, which describes the interface of the entity nor_gate, which is used as a component. Between the begin and end keywords, the first two lines as well as the second two lines define two *component instances*. The component nor_gate has two inputs (i1 and i2) and one output (o1). There are two instances of the nor_gate component in this architecture. The first line of the component instantiation statement gives this instance a name, name1, and specifies that it is an instance of the component nor_gate. The second line describes how the component is connected to the rest of the design using the port map clause. The port map clause specifies what signals of the design connect to the interface of the component in the same order as they are listed in the component declaration. The interface is specified in order as i1, i2, and then o1. Hence, this instance connects in2 to i1, ou2 to i2, and ou1 to o1. The second instance, named name2, connects in1 to i1, ou1 to i2, and ou2 to o1 of a different instance of the same nor_gate component. We can also specify exact connections in the port map without depending on the order of the signals in the port list. The structural description of a design is a textual description of a schematic. A list of components and their connections is usually called a *netlist*.

In the data flow domain, ICs are described by indicating how the inputs and outputs of built-in primitive components or pure combinational blocks are connected together. Thus, one describes how signals (data) flow through ICs. For example, in:

```
entity latch is
      port (in1,in2: in bit;
            ou1,ou2: out bit);
end latch;
architecture dataflow of latch is
begin
      ou1<=in2 nor ou2;
      ou2<=in1 nor ou1;
end dataflow;
```

There are four signals, in1, in2, ou1, and ou2, that are accessible externally to the design. For the signals, we set the data type as bit (logic values 0 an 1). The architecture part describes the internal operation of the design. In the data flow domain, one specifies how data flows from the inputs to the outputs. In VHDL this is accomplished with the signal assignment statement. The example architecture consists of two signal assignment statements. A signal assignment statement describes how data flows from the signals on the right side of the <= operator to the signal on the left side. The first signal assignment indicates that the data coming from signals in2 and ou2 flow through a NOR gate to determine the value of the signal ou1. The nor represents a built-in component called an operator, because it operates on some data to produce new data. The second signal assignment specifies that the signal ou2 is produced from data (in1 and ou1) flowing through (or processed by) the NOR operator. The right side of the <= operator is called an *expression*. The evaluation of the expression is performed by substituting the values of the signals in the expression and computing the result of each operator in the expression.

The scheme used to model a VHDL design is called *discrete event time simulation*. When the value of a signal changes, this means that an event has occurred on that signal. If data flows from signal x1 to signal x2, and an event has occurred on signal x1 (i.e., x1 changes), then we need to determine a new value of x2. The values of signals are only updated when discrete events occur. Because one event causes another, simulation proceeds in rounds. The simulator maintains a list of events that need to be processed. In each round, all events in a list are processed, and any new events that are produced are placed in a separate list (scheduled) for processing in a later round. Each signal assignment is evaluated once, when simulation begins, to determine the initial value of each signal. For the studied example, data flows from in2 and ou2 to ou1. Thus, ou1 depends on in2 and ou2. In general, given any signal assignment statement, the signal on the left side of the <= operator depends on all the signals appearing on the right side. If a signal depends on another signal on which an event has occurred, then the expression in the signal assignment is reevaluated. If the result of the evaluation is different from the current value of the signal, an event will be scheduled (added to the list of events to be processed) in order to update the signal with the new value. Thus, if an event occurs on in2 or ou2, then the nor operator is evaluated; and if the result is different from the current value of ou1, then an event will be scheduled to update ou1. For the latch example, the signals are in1 = 0, in2 = 0, ou1 = 1 and ou2 = 0. Assume that the value of the signal in2 changes due to some event to the value 1. Because ou1 depends on in2, we must reevaluate the expression in2 nor ou2, which now is 0. Because the value of ou1 must be changed to 0, a new event will be scheduled on the signal ou1. During the next round the event scheduled for ou1 is processed, and the ou1 value is updated to be 0. Because ou2 depends on ou1, the expression in1 nor ou1 must be reevaluated. The result is 1, and an event is scheduled to update the value of ou2. During the next round, when the event on ou2 is processed, the expression for ou1 will be evaluated again because it depends on ou2. However, the result will be 0, and no new event will be scheduled because ou1 is already 0. Because no new events were scheduled, there are no more events that will occur internally to the latch. Let an external event cause in2 to return to the value 0. The ou1 depends on in2 and in2 nor ou2 and is evaluated again. The result is 0, and ou1 is already 0. Hence, events are not scheduled. This correctly models the latch.

The logical operators NOT, AND, OR, NAND, NOR, and XOR can be used with any bit type or bit_vector. When used as operators on bits, they have their conventional meanings. When used with bit_vectors, the bit_vectors must have the same number of elements, and the operation is performed bitwise. For example, 00101001 xor 0110100 results in 01001101. The typical algebraic operators are available for integers. In particular, +, −, *, and /. Although these operations are not built in for bit_vectors, they are provided in libraries. They are used with bit_vectors by interpreting them as a binary representation of integers, which may be added, subtracted, multiplied, or divided. Also predefined are the normal relational operators, e.g., =, /=, <, <=, >, and >=. The result of all these operators is a Boolean value (TRUE or FALSE).

The behavioral domain for modeling hardware components is different from the other two methods because it does not reflect how the design is implemented. It can be viewed as the black-box approach to modeling. It accurately models what happens on the inputs and outputs of the black box, but what is inside the box and how it works are irrelevant. Behavioral descriptions are supported with the process statement. The process statement can appear in the body of an architecture declaration as the signal assignment statement. The contents of the process statement can include sequential statements such as those found in software programming languages. These statements are used to compute the outputs of the process from its inputs. In general, sequential statements are more powerful but may not have the direct correspondence to a hardware implementation. The process statement can contain signal assignments in order to specify the outputs of the process. A group of processes will be executed concurrently because the process is a concurrent statement and usually contains sequential statements. We can also group a block of concurrent statements and specify when they are to be executed. Consider the following process statement:

```
compute_xor: process (i2,o1)
begin
      i1<=i2 xor o1;
end process;
```

This example process contains the signal assignment evaluated only when events occur on the signals regardless of which signals appear on the right side of the <= operators. Thus, it is critical to make sure the proper signals are in the sensitivity list. The statements in the body of the process are performed (executed) in order from first to last. When the last statement has been executed, the process is finished and is said to be *suspended*. When an event occurs on a signal in the sensitivity list, the process is said to be *resumed* and the statements will be executed from top to bottom. Each process is executed once during the beginning of a simulation to determine the initial values of the outputs.

There are two major kinds of objects used to hold data. The signal is used in structural and data flow descriptions and integrates processes. This variable assignment assigns the value of b to a, i.e., making use of a:=b, the b is copied to a.

There are several statements that can be used in the body of a process. These statements are called *sequential statements* because they are executed sequentially. For example, *if* and *for* loops are commonly used. A signal assignment schedules an event to occur on a signal and does not have an immediate effect. When a process is resumed, it executes from top to bottom and no events are processed until the process is complete. Hence, if an event is scheduled on a signal during the execution of a process, that event can be processed after the process has completed.

Using the VHDL and Verilog analogies, the HDLs can be developed for nanoICs to design subsystems, systems, and units of nanocomputers.

References

1. J. Carter, *Microprocessor Architecture and Microprogramming, A State Machine Approach*, Prentice-Hall, Englewood Cliffs, NJ, 1996.
2. V.C. Hamacher, Z.G. Vranesic, and S.G. Zaky, *Computer Organization*, McGraw-Hill, New York, 1996.
3. J.P. Hayes, *Computer Architecture and Organizations*, McGraw-Hill, Boston, 1998.
4. J.L. Hennessey and D.A. Patterson, *Computer Architecture: A Quantitative Approach*, Morgan Kaufman, San Mateo, CA, 1990.
5. K. Hwang, *Computer Arithmetic*, Wiley, New York, 1978.
6. P.M. Kogge, *The Architecture of Pipelined Computers*, McGraw-Hill, New York, 1981.
7. D.A. Patterson and J.L. Hennessey, *Computer Organization and Design — The Hardware/Software Interface*, Morgan Kaufman, San Mateo, CA, 1994.
8. L.H. Pollard, *Computer Design and Architecture*, Prentice-Hall, Englewood Cliffs, NJ, 1990.
9. A.S. Tanenbaum, *Structured Computer Organization*, Prentice-Hall, Englewood Cliffs, NJ, 1990.

10. R.F. Tinder, *Digital Engineering Design: A Modern Approach*, Prentice-Hall, Englewood Cliffs, NJ, 1991.

11. S.E. Lyshevski, *MEMS and NEMS: Systems, Devices, and Structures*, CRC Press, Boca Raton, FL, 2002.

12. E.K. Drexler, *Nanosystems: Molecular Machinery, Manufacturing, and Computations*, Wiley-Interscience, New York, 1992.

13. Y. Chen, D.A.A. Ohlberg, G. Medeiros-Ribeiro, Y.A. Chang, and R.S. Williams, Self-assembled growth of epitaxial erbium disilicide nanowires on silicon (001), *Appl. Phys. Lett.*, 76, 4004–4006, 2000.

14. J.C. Ellenbogen and J.C. Love, Architectures for molecular electronic computers: logic structures and an adder designed from molecular electronic diodes, *Proc. IEEE*, 88, 3, 386–426, 2000.

15. W.L. Henstrom, C.P. Liu, J.M. Gibson, T.I. Kamins, and R.S. Williams, Dome-to-pyramid shape transition in Ge/Si islands due to strain relaxation by interdiffusion, *Appl. Phys. Lett.*, 77, 1623–1625, 2000.

16. T.I. Kamins and D.P. Basile, Interaction of self-assembled Ge islands and adjacent Si layers grown on unpatterned and patterned Si (001) substrates, *J. Electron. Mater.*, 29, 570–575, 2000.

17. T.I. Kamins, R.S. Williams, Y. Chen, Y.L. Chang, and Y.A. Chang, Chemical vapor deposition of Si nanowires nucleated by $TiSi_2$ islands on Si, *Appl. Phys. Lett.*, 76, 562–564, 2000.

18. T.I. Kamins, R.S. Williams, D.P. Basile, T. Hesjedal, and J.S. Harris, Ti-catalyzed Si nanowires by chemical vapor deposition: microscopy and growth mechanism, *J. Appl. Phys.*, 89, 1008–1016, 2001.

19. V. Derycke, R. Martel, J. Appenzeller, and P. Avouris, Carbon nanotube inter- and intramolecular logic gates, *Nano Letters*, 2001.

20. R. Martel, H.S.P. Wong, K. Chan, and P. Avouris, Carbon nanotube field-effect transistors for logic applications, *Proc. Electron Devices Meeting, IEDM Technical Digest*, 7.5.1–7.5.4, 2001.

21. R. Saito, G. Dresselhaus, and M.S. Dresselhaus, *Physical Properties of Carbon Nanotubes*, Imperial College Press, London, 1999.

22. J. Appenzeller, R. Martel, P. Solomon, K. Chan, P. Avouris, J. Knoch, J. Benedict, M. Tanner, S. Thomas, L.L. Wang, and J.A. Del Alamo, A 10 nm MOSFET concept, *Microelectron. Eng.*, 56, 1-2, 213–219, 2001.

23. W.T. Tian, S. Datta, S. Hong, R. Reifenberger, J.I. Henderson, and C.P. Kubiak, Conductance spectra of molecular wires, *Int. J. Chem. Phys.*, 109, 7, 2874–2882, 1998.

24. S.C. Goldstein, Electronic nanotechnology and reconfigurable computing, *Proc. Comp. Soc. Workshop VLSI*, 10–15, 2001.

25. P.L. McEuen, J. Park, A. Bachtold, M. Woodside, M.S. Fuhrer, M. Bockrath, L. Shi, A. Majumdar, and P. Kim, Nanotube nanoelectronics, *Proc. Device Res. Conf.*, 107–110, 2001.

26. K. Tsukagoshia, A. Kanda, N. Yoneya, E. Watanabe, Y. Ootukab, and Y. Aoyagi, Nano-electronics in a multiwall carbon nanotube, *Proc. Microprocesses Nanotechnol. Conf.*, 280–281, 2001.

27. K. Likharev, Riding the crest of a new wave in memory, *IEEE Circuits Devices Mag.*, 16, 4, 16–21, 2000.

28. S. Baranov, *Logic Synthesis for Control Automata*, Kluwer, Norwell, MA, 1994.

29. V. Sklyarov, *Synthesis of Finite State Machines Based on Matrix LSI*, Science, Minsk, Belarus, 1984.

30. S.A. Yuzvinski, Metric properties of the endomorphisms of compact groups, *Acad. Sci. USSR, Math.*, 29, 1295–1328, 1967.

31. S. Yalamanchili, *Introductory VHDL: From Simulation to Synthesis*, Prentice-Hall, Upper Saddle River, NJ, 2001.

32. *IEEE Standard For VITAL ASIC (Application Specific Integrated Circuit) Modeling Specification*, IEEE Std. 1076.4–2000, 2001.

33. *IEEE Standard Verilog Hardware Description Language*, IEEE Std. 1364–2001, 2001.

7

Architectures for Molecular Electronic Computers[1]

CONTENTS

James C. Ellenbogen
The MITRE Corporation

J. Christopher Love
Harvard University

---

[1]This chapter has appeared in the *Proceedings of the IEEE* (J.C. Ellenbogen and J.C. Love, Architectures for molecular electronic computers: 1. Logic structures and an adder designed from molecular electronic diodes, *Proc. IEEE*, 88, 3, 386–426, 2000). It is reprinted here with permission.

Abstract

Recently there have been significant advances in the fabrication and demonstration of individual molecular electronic wires and diode switches. This chapter reviews those developments and shows how demonstrated molecular devices might be combined to design molecular-scale electronic digital computer logic. The design for the demonstrated rectifying molecular diode switches is refined and made more compatible with the demonstrated wires through the introduction of intramolecular dopant groups chemically bonded to modified molecular wires. Quantum mechanical calculations are performed to characterize some of the electrical properties of the proposed molecular diode switches. Explicit structural designs are displayed for AND, OR, and XOR gates that are built from molecular wires and molecular diode switches. The diode-based molecular electronic logic gates are combined to produce a design for a molecular-scale electronic half adder and a molecular-scale electronic full adder. These designs correspond to conductive monomolecular circuit structures that would be one million times smaller in area than the corresponding micron-scale digital logic circuits fabricated on conventional solid-state semiconductor computer chips. It appears likely that these nanometer-scale molecular electronic logic circuits could be fabricated and tested in the foreseeable future. At the very least, such molecular circuit designs constitute an exploration of the ultimate limits of electronic computer circuit miniaturization.

7.1   Introduction

Significant new advances have been made in the fabrication and demonstration of molecular electronic wires[1-4] and of molecular electronic diodes — two-terminal electrical switches made from single molecules.[5-14] There also have been advances in techniques for making reliable electrical contact with such electrically conducting molecules.[3,15] These promising developments and others[16-19] in the field of nanoelectronics suggest that it might be possible to build and to demonstrate somewhat more complex molecular electronic structures that would include two or three molecular electronic diodes and that would perform as digital logic circuits.

It is the purpose of this chapter to provide and to explain novel designs for several such simple molecular electronic digital logic circuits: a complete set of three fundamental logic gates (AND, OR, and XOR gates), plus an adder function built up from the gates via the well-known principles of combinational logic. En route to the design of these molecular electronic logic gates and functions, we also propose and characterize designs for simpler, polyphenylene-based molecular electronic rectifying diode switches that should be easy to integrate in logic circuits with other previously demonstrated[1-3,5-7,10-13] polyphenylene-based molecular electronic wires and devices. The design for rectifying molecular diode switches is refined herein to be more compatible with the demonstrated wires through the introduction of intramolecular dopant groups chemically bonded to modified molecular wires.

A key idea of this work is that the implementation of elementary molecular electronic logic gates and elementary molecular electronic computational functions does not necessarily require the use of three-terminal molecular-scale transistors. Only a few two-terminal molecular diode switches are essential to implement simple molecular electronic computer logic circuits.

Moreover, such molecular diode switches already have been demonstrated.[5-11,13,14] It is only necessary that two or three of them be assembled, either mechanically or chemically, and brought into appropriate electrical contact in order to test and to demonstrate molecular electronic logic.

The molecular electronic logic structures proposed here would occupy an area *one million times* smaller than analogous logic structures that currently are implemented in micron-scale, solid-state semiconductor integrated circuits. Because of their extremely small sizes, at the very least, these molecular logic designs constitute an exploration of the limits of electronic computer circuit miniaturization. Further, as detailed below, all these molecular-scale electronic logic structures are built up simply from a few molecular diode switches similar to those already demonstrated experimentally[2,3,5,7,9–12,15,20] as well as theoretically.[21–28] Thus, it might be possible to fabricate and to test experimentally such ultra-small, ultra-dense, molecular electronic digital logic in the relatively near future. (Note for 2003 CRC edition: see, for example, References 133 and 134.)

To begin to explain why such dramatic developments now are likely, in Section 7.2 of this chapter we review and analyze in detail the recent research results that have begun to introduce and demonstrate functioning molecular-scale electronic switches and wires. In Section 7.3, we propose how to integrate these earlier molecular devices with some new compatible diode switches introduced in Section 7.4. This integration allows the proposal of molecular electronic digital circuit structures for which explicit designs are provided in Section 7.5. Then, Section 7.6 discusses the strengths and weaknesses of these novel molecular circuit designs. They are considered relative to a set of architectural issues that must be addressed in order to design, fabricate, and operate an entire molecular electronic digital computer that might be built from logic structures and functions similar to or evolved from those described in this work.

Finally, Section 7.7 considers possible interface and integration strategies for such molecular circuits as well as their possible applications and technological impacts. Three appendices provide more details about important quantitative aspects of the structure, function, and scaling, respectively, for the molecular diode switches and the molecular logic gates.

As outlined above, the work presented here establishes a qualitative framework for molecular circuit architectures and identifies architectural issues associated with this approach. Therefore, it is an additional purpose of the present work to define an agenda and to provide a direction for future, detailed, systematic, quantitative investigations that the authors and others will perform to explore the validity of this proposed approach to designing molecular electronic computers.

7.2 Background

Recent developments indicate that the physical limits of miniaturization for bulk-effect semiconductor transistors are likely to be encountered sooner and on larger scales than previously had been expected.[29,30] This eventuality makes it more important than ever to explore alternatives, such as molecular electronics, for the continued miniaturization of electronic devices and circuits down to the nanometer scale.

Presently there are two primary types of molecules that have been proposed and demonstrated for use as the potential basis or backbone for current-carrying, molecular-scale electronic devices. These two types of molecular backbones are (1) polyphenylene-based chains and (2) carbon nanotubes.

In addition, there has been much speculation in the literature about still a third possible type of conductive backbone involving biomolecules. To provide a sound basis for explaining the molecular circuit designs proposed here, the principal concepts and experimental data concerning each of these three possible conductive molecular-scale backbones is reviewed and discussed separately below.

7.2.1 Polyphenylene-Based Molecular-Scale Electronic Devices

Polyphenylene-based molecular wires and switches involve chains of organic aromatic benzene rings. Such chains are shown in Figure 7.1(d).

Until recently, it was an open question whether such small molecules, taken individually, had appreciable conductance. However, over the last several years, several different groups have shown experimentally that individual polyphenylene molecules can conduct small electrical currents.[1–3] Motivated by such experimental advances, several groups of theorists, including Datta et al.;[26–28] Ratner and his collaborators;[21–23] and Hush, Reimers, and their collaborators[31,32] have begun to establish a set of computational approaches and a qualitative mechanistic framework[24,25] for the detailed interpretation of electrical conduction in small organic molecules. Calculations using these and related methods have confirmed the experiments by explaining in detail the features of the electrical conduction observed in short polyphenylene wires.[28,33–36]

Further, substituted polyphenylenes and similar small organic molecules have been shown experimentally to be capable of switching small currents.[5–11,13]

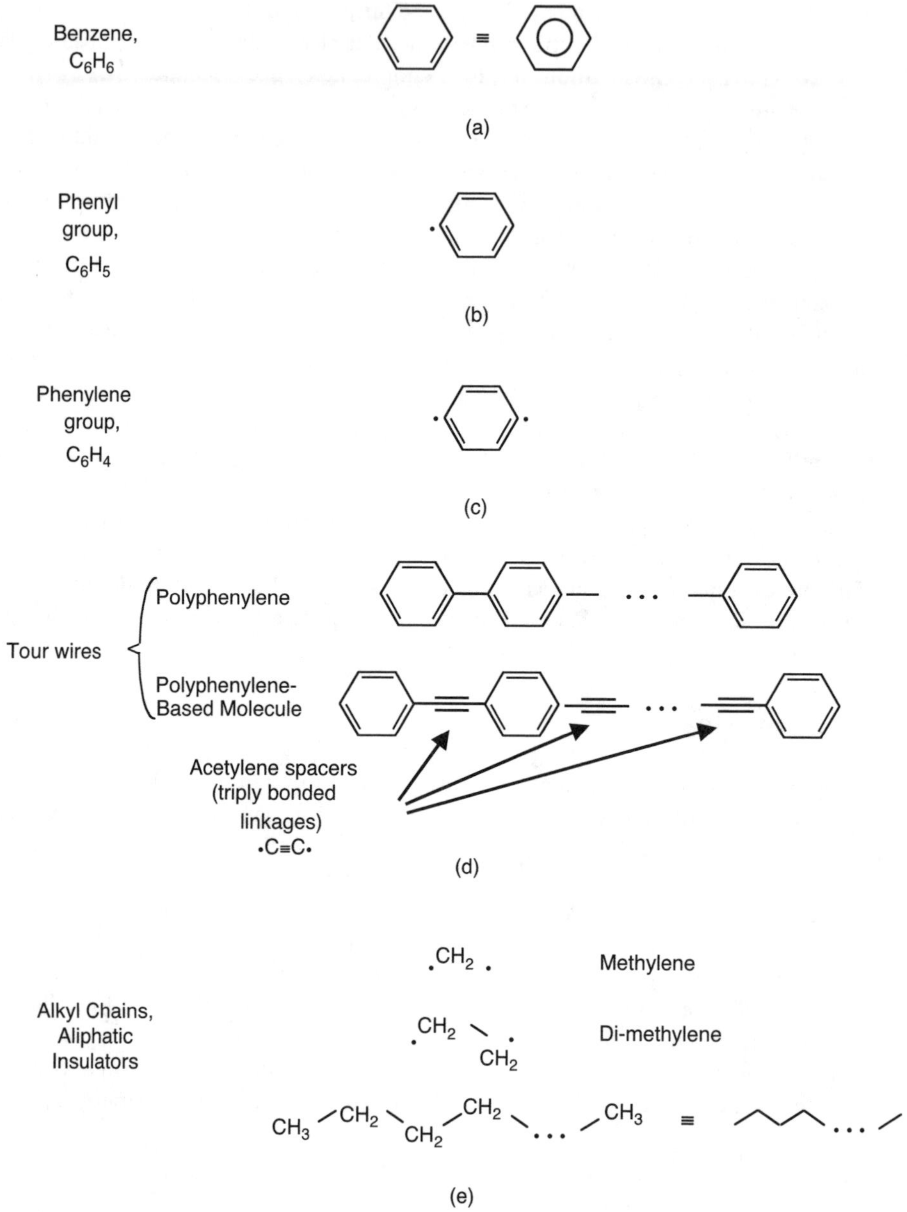

FIGURE 7.1   Aromatic molecular conductors and aliphatic molecular insulators.

7.2.1.1   Conductors or Wires: Conjugated Aromatic Organic Molecules

An individual benzene ring having the chemical formula C_6H_6 is shown in Figure 7.1(a). A benzene ring with one of the hydrogens removed to form C_6H_5 so that it can be bonded as a group to other molecular components is shown in Figure 7.1(b). Such a ring-like substituent group is termed a *phenyl group*. By removing two hydrogens from benzene, one obtains the structure C_6H_4 or *phenylene*, a ring which has two free binding sites as shown in Figure 7.1(c).

By binding phenylenes to each other on both sides and terminating the resulting chain-like structures with phenyl groups, one obtains a type of molecule known as a *polyphenylene*. Polyphenylenes, such as the upper molecular structure in Figure 7.1(d), may be made in a number of different shapes and lengths. In addition, one may insert into a polyphenylene chain other types of molecular groups (e.g., singly bonded aliphatic groups, doubly bonded ethenyl groups, and triply bonded ethynyl or acetylenic groups) to obtain polyphenylene-based molecules with very useful structures and properties. An example of such a polyphenylene-based molecule is the lower structure in Figure 7.1(d). Molecules such as benzene and polyphenylenes, which incorporate benzene-like ring structures as significant components, are termed *aromatic*.

As discussed briefly above, various investigators recently have performed sensitive experiments in which one or a few polyphenylene-based molecules have been shown to conduct electricity. The results of several of those experimental efforts are summarized in Table 7.1. In one example of such an experiment, Reed and his collaborators passed an electrical current through a monolayer of approximately 1000 polyphenylene-based molecular wires which were arrayed in a nanometer-scale pore and adsorbed to metal contacts on either end.[8] The system was carefully prepared such that all of the molecules in the *nanopore* were identical three-benzene-ring polyphenylene-based chain molecules. These chain molecules each had a form similar to the molecules diagrammed in Figure 7.1(d), including triply bonded, conductive, acetylenic "spacers" bonded between each pair of benzene rings. The total current passed through this assembly of molecules measured 30 μA, which converts to roughly 30 nA of electrical current passing through each molecule. This current, listed in the fourth column of Table 7.1, corresponds to approximately 200 billion electrons per second transmitted across each short polyphenylene-based molecular wire.

In tests of molecular electronic devices, such as those cited in Table 7.1, small lengths of molecular wire are sandwiched between narrow Schottky-type potential barriers, where the molecule joins but is imperfectly bound to the closely spaced metal contacts. Thus, while tunneling can occur through such barriers, the current that can pass through the molecule is greatly reduced and the Schottky barriers have a large impact on the entire molecular electronic system.[21,24,37] One can anticipate, however, that further research and development will produce a larger conductive molecule which integrates molecular wires, molecular switches, etc., covalently bonded into a single molecular electronic circuit. In the operation of the larger molecular electronic circuit, any Schottky barriers would be much farther apart (i.e., at least ten times), and the flow of current between devices within the molecule would be relatively less affected by the Schottky barriers. Therefore, the current or signal could pass among all the devices of a molecular electronic circuit with much less impedance than might be inferred from the experimental results displayed in Table 7.1.

For comparison with the data in Table 7.1 in the case of the polyphenylene-based wires and switches, we note that a somewhat larger molecule, a carbon nanotube, or *buckytube*, has been measured transmitting a current between 10 and 100 times greater than that measured for the simple polyphenylene-based chains. Those measurements, tabulated in Table 7.1, correspond to a current of approximately 20 to 100 nanoamps, or approximately 120 billion to 620 billion electrons per second. Strengths and limitations of these highly conductive carbon nanotube molecules are discussed further in Section 7.2.2.

While polyphenylene-based molecular wires like those depicted in Figure 7.1(d) do not carry as much current as carbon nanotubes, polyphenylenes and their derivatives are much smaller molecules. Thus, because of their very small cross-sectional areas, they do have very high current densities. This is seen in Table 7.1, where approximate current densities are calculated for several selected molecular electronic devices as well as for a typical macroscopic copper wire. We observe that the current density for a polyphenylene-based molecular wire is approximately the same as for a carbon nanotube and as much as one million times greater than that for the copper wire.

Also, polyphenylene-based molecules have the significant advantages of a very well-defined chemistry and great synthetic flexibility, based upon more than a century of experience accumulated by organic chemists in studying and manipulating such aromatic compounds. Recently, Tour has refined the synthetic techniques for conductive polyphenylene-based chains to produce enormous numbers of these molecules (approximately 10^{23}), every one of which is of exactly the same structure and length.[38,39] Thus, such polyphenylene-based molecular wires with the chemical structures shown schematically in Figure 7.1(d) have come to be known as *Tour wires*.

TABLE 7.1 Approximate Current Densities in Electrons per Second per Square Nanometer Calculated from Experimental Data for Selected Molecular Electronic and Macroscopic Metal Devices

Quantity	Units	Molecular Electronic Device				Copper Wire
		1,4-Dithiol Benzene	3-Ring Poly-phenylene Wire	Poly-phenylene RTD (5 rings)	Carbon Nanotube	
Applied Voltage	Volts	1	1	1.4 (peak)	1	2×10^{-3} (10 cm wire)
Current Measured in Experiment	Amperes	2×10^{-8}	3.2×10^{-5}	1.4×10^{-11}	1×10^{-7}	1 (approx.)
Current Inferred per Molecule	Amperes	2×10^{-8}	3.2×10^{-8}	1.4×10^{-14}	1×10^{-7}	–
	Electrons per Sec	1.2×10^{11}	2.0×10^{11}	8.7×10^{4}	6.2×10^{11}	–
Estimated Cross-Sectional Area per Molecule	nm^2	~0.05	~0.05	~0.05	~3.1 (Radius $\approx$ 1 nm)	~3.1×10^{12} (Radius $\approx$ 1 mm)
Current Density	Electrons per Sec-nm^2	~2×10^{12}	~4×10^{12}	~2×10^{6}	~2×10^{11}	~2×10^{6}
Reference		(7)	(8)	(5,6)	(4)	

[a] Conversion factor for amperes to electrons per second is 1 Ampere ≡ 1 Coulomb/sec = $(1.6 \times 10^{-19})^{-1}$ electrons/sec = 6.2×10^{18} electrons/sec.

[b] In order to estimate the current densities per molecule from the published data on the room temperature nanopore measurements in References 5, 6, and 8, it was determined that the samples in the monolayer in the nanopore contained on the order of 1000 molecules per monolayer. This estimate is based on an average nanopore diameter of 30 nm and an estimated molecular diameter on the order of approximately 1 nm.

[c] Common copper wire generally is regarded as being highly conductive. Therefore, data for 10 cm of 1mm diameter (18 gauge) copper wire is included only for comparison as a familiar, conductive, macroscopic reference system. A current on the order of 1 ampere is the maximum recommended for such wire to avoid undue heating and danger of fire.

Sources of current measurements: (4) S.J. Tans et al., Individual single-wall carbon nanotubes as quantum wires, *Nature*, 386, 474–477, 1997; (5) M.A. Reed, Electrical Properties of Molecular Devices, presented at 1997 DARPA ULTRA Program Review Conference, Santa Fe, NM, October, 1997; (6) M.A. Reed, Molecular-scale electronics, *Proc. IEEE*, 87, 652–658, 1999; (7) C. Thou, M.R. Deshpande, M.A. Reed, and J.M. Tour, Nanoscale metal/self-assembled monolayer/metal heterostructures, *Appl. Phys. Lett.*, 71, 611–613, 1997; (8) C. Zhou, Atomic and Molecular Wires, Ph.D. dissertation, Yale University, 1999. With permission.

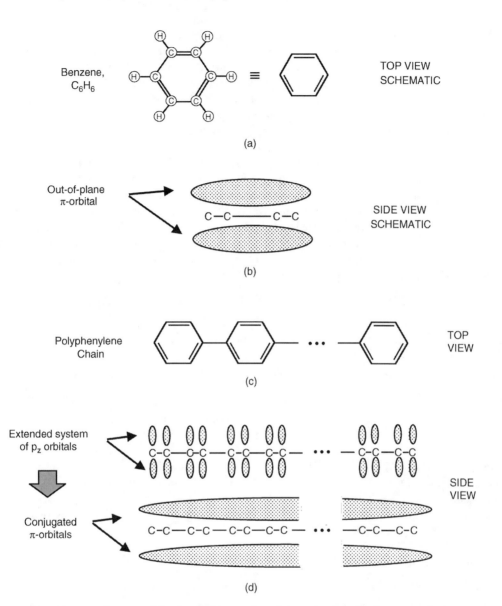

FIGURE 7.2 Schematic diagrams of the chemical structures and the molecular orbital structures for benzene and for polyphenylene molecules.

The source of the conductivity for a polyphenylene-based wire is a set of π-type (i.e., "pi"-type) molecular orbitals that lie above and below the plane of the molecule when it is in a planar or near-planar conformation.[40–42] An example of a π-orbital is illustrated in Figure 7.2. In a planar conformation, the π-orbitals associated with each individual atom overlap or *conjugate* in various combinations to create a set of extended π-orbitals that spans the length of the molecule. This occurs because there is a significant energetic advantage that arises from delocalizing valence (and conduction) electrons in orbitals that span or nearly span the length of the entire molecule. The lowest energy π-orbital available in the molecule is depicted in Figure 7.2(d). Higher energy π-orbitals differ somewhat in that they do have nodal planes (planes where the orbital vanishes) oriented perpendicularly to the axis of a wire-like molecule. The higher the energy of the orbital, the more nodal planes it will contain.

Long π-orbitals are located out of the plane of the nuclei in the molecule and are relatively diffuse compared with the in-plane σ-type ("sigma"-type) molecular orbitals. Thus, one or more such long-unoccupied or only partially occupied π-orbitals can provide *channels* that permit the transport of additional electrons from one end of the molecule to the other when it is under a voltage bias. The mechanisms of such transport have been discussed in detail by Ratner and Jortner,[24] based upon formal considerations explored earlier by Mujica, Ratner, and their collaborators[21–23] and by Datta.[27] In order to align our discussion and terminology more closely with solid-state electronics, we will refer to the collection or manifold of available molecular channels that is widely spaced in energy — i.e., sparse unoccupied π-orbitals — as a *conduction band*, although the term *band* usually is reserved in physics and electrical engineering for an interval of energy where there is a nearly continuous set of allowed quantized energy states.

A delocalized π-orbital usually extends across one or more of the neighboring aromatic rings in the molecule, as well as across other intervening multiply bonded groups. The component π-orbitals from the several substituent aromatic rings and multiply bonded groups can sum or merge to form a number of larger, molecule-spanning π-orbitals, each having a different nodal structure and energy. Several such separate, unoccupied π-orbitals that are low-lying in energy are thought to be primarily responsible for the conduction through polyphenylene-based molecular wires. When such a molecule is placed under an externally applied voltage bias, an unoccupied π-orbital that is lower in energy is likely to be a more effective pathway for electron transport because:

1. It can be brought into energetic coincidence more easily with still lower energy occupied orbitals.
2. A lower energy unoccupied orbital yields a more stable pathway for a conduction electron.

Generally speaking, it can be expected that conjugated aromatic molecules such as polyphenylenes will conduct currents. This also is true of polyphenylene-based molecules with other multiply bonded groups (such as ethenyl, –HC=CH–, or ethynyl, –C≡C–) inserted between the aromatic rings, so long as conjugation among the π-bonded components is maintained throughout. Thus, in practice, as shown in Figure 7.1(d), triply bonded ethynyl or acetylenic linkages often are inserted as spacers between the phenyl rings in a Tour wire.[5,6,43–45] These spacers eliminate the steric interference between hydrogen atoms bonded to adjacent rings. Otherwise this steric interference would force the component rings in the Tour wire to rotate into a non-planar conformation. That would reduce the extent of π-orbital overlap between adjacent rings, break up the electron channels, and decrease the conductivity of the molecular wire. The acetylenic linkages themselves permit the conductivity to be maintained throughout the length of the molecule because of their own out-of-plane π-electron density. This is true despite the fact that acetylenic groups seem to be terminated by single bonds. When the triply bonded structure is inserted into the aromatic chain, the out-of-plane π-orbitals in the acetylenic triple bond delocalize and overlap sufficiently with those on neighboring aromatic rings so that the acetylenic linkage maintains and becomes a part of a molecule-spanning electron channel.

7.2.1.2   Insulators: Aliphatic Organic Molecules

On the other hand, *aliphatic* organic molecules serve as insulators. Examples are sketched in Figure 7.1(e). These are singly bonded molecules that contain only sigma bonds that lie along the axes of the atoms that are joined to form the backbone of the molecule. Such bonds do not form an uninterrupted channel outside the plane of the nuclei in the molecule. The positively charged atomic nuclei are obstacles to negatively charged electrons traveling along the axis or plane of the molecule. Therefore, such singly bonded structures cannot easily transport an unimpeded electron current when they are placed under a voltage bias. For this reason, aliphatic molecules or groups act as insulators.

It follows that, when a small aliphatic group is inserted into the middle of a conductive polyphenylene chain, it breaks up the conductive channel and forms a *barrier* to electron transport. This fact plays a significant role in later considerations, as does the notion that such barriers act somewhat like electrical resistors in a molecular circuit.

The data in Table 7.1 can be used to derive an approximate indication of the effectiveness of aliphatic insulating groups. Note in Columns 3 and 4 of the table that a single-ring and a three-ring length of polyphenylene-based wire are measured to conduct approximately the same current. However, in Column 5, the data show that the polyphenylene-based RTD, which is simply a five-ring polyphenylene-based molecular wire with two separate one-carbon aliphatic methylene groups inserted into it, conducts approximately 250,000 times less current than the somewhat shorter three-ring wire with no such insulating aliphatic groups inserted into it. (The structure and electrical behavior of the polyphenylene-based RTD is discussed in detail below in Section 7.2.1.4.) Assuming, then, that each methylene group reduces the current in the wire by the same factor, this data suggests that each one-carbon insulating group reduces the electron current by a factor of as much as the square root of 250,000 — i.e., by a factor of between 100 and 1000.[46]

7.2.1.3 Diode Switches: Substituted Aromatic Molecules

A diode is simply a two-terminal switch. It can turn a current on or off as it attempts to pass through the diode from the *in* to the *out* terminal. These terminals usually are termed, respectively, the *source* and the *drain* for the current. Unlike more familiar three-terminal switching devices (e.g., triodes and transistors), diodes lack a third or *gate* terminal via which a small current or voltage can be used to control a larger current or voltage passing from the source to the drain. Thus, unlike a transistor, a single diode switch, by itself, cannot produce current amplification or *power gain*. (See further discussion of this point in Section 7.6.1.5, however.)

Two types of molecular-scale electronic diodes have been demonstrated recently and are discussed here: (1) rectifying diodes and (2) resonant tunneling diodes. Each of these types of molecular diodes is modeled after more familiar solid-state analogs.

7.2.1.3.1 Molecular Rectifying Diodes ("Molecular Rectifiers")

A rectifying diode incorporates structures that make it more difficult to induce electric current to pass through it in one direction, usually termed the *reverse* direction from terminal B to A, than in the opposite *forward* direction from A to B. This *one-way* behavior is suggested by the standard schematic symbol for the rectifying diode, shown in Figure 7.3(a), and by the current vs. voltage plots for an "ideal" diode and a non-ideal Zener diode shown in Figures 7.3(b) and 7.3(c), respectively. (Note: the reader should be aware that, in accordance with a long-standing convention in electronics, the direction of *current* flow in a diode or any other circuit element is taken to be the direction of flow of the *positive* charge, which is the opposite of the direction of flow for the *electrons*. That is the convention used here.)

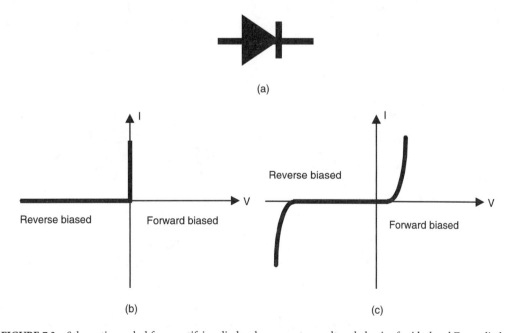

FIGURE 7.3 Schematic symbol for a rectifying diode, plus current vs. voltage behavior for ideal and Zener diodes.

7.2.1.3.1.1 Solid-State Rectifying Diodes — Diode switches implemented using vacuum tubes or bulk-effect solid-state devices (e.g., p-n junction diodes) have been elements of analog and digital circuits since the beginning of the electronics revolution very early in the twentieth century. The behavior of junction diodes was a very important influence in the development of the transistor.[47] A detailed discussion of the function of such solid-state devices may be found elsewhere.[48]

7.2.1.3.1.2 Aviram and Ratner Rectifying Diodes and the Origin of Molecular Electronics — Diode switches likewise have had a seminal role in the formulation and testing of strategies for miniaturizing electronics down to the molecular scale. Rectifying diode switches were the topic of the first scientific paper about molecular electronics, Aviram and Ratner's influential 1974 work, "Molecular Rectifiers."[49] Aviram and Ratner based their suggestion for the structure of a molecular diode switch on the operational principles for the solid-state, bulk-effect p-n junction diode.

Aviram and Ratner's proposal and theoretical specification for the structure and function of a molecular rectifying diode was made a quarter century ago, and there have been a number of experimental efforts to demonstrate a molecule with such electrical properties in the intervening years. Building on an earlier experimental demonstration by Martin et al. in 1993,[50] two separate groups in 1997, one led by Metzger at the University of Alabama[9] and another led by Reed at Yale University[7] demonstrated molecular rectifiers in independent experiments.

The structures of the molecular rectifiers they demonstrated are shown in Figures 7.4 and 7.5, respectively. Figure 7.4(a) shows the molecular structure for Metzger's molecular diode, Figure 7.4(b) shows the experimental setup, and Figure 7.4(c) shows the measured electrical properties of the molecular rectifying diode. In Figure 7.5(a) are shown the molecules and the experimental setup for Reed's molec-

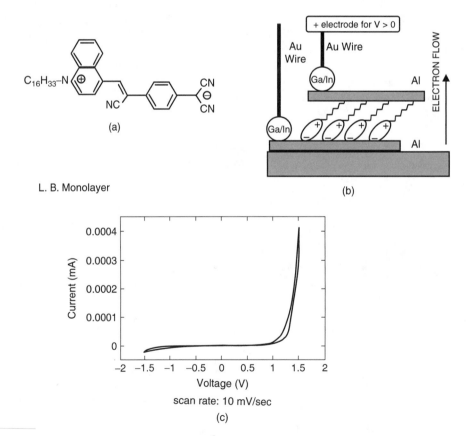

FIGURE 7.4   Molecular structure, schematic of apparatus, and electrical behavior in experiment by Metzger et al. demonstrating molecular rectification in a Langmuir–Blodgett (LB) film.[9]

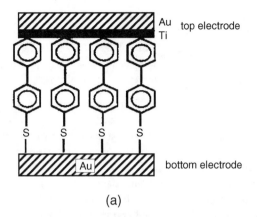

(a)

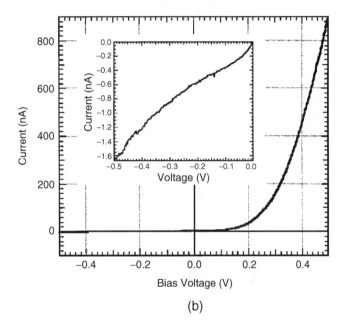

(b)

FIGURE 7.5 Experiment by Reed and his collaborators demonstrating molecular rectification (i.e., a molecular rectifying diode) in a monolayer of Tour wires between electrodes of two different metals (see Reference 7). (a) Schematic showing Tour-type molecular wires between Ti/Au top electrode and Au bottom electrode; (b) I vs. V plot showing characteristic rectifying behavior for molecular diode.

ular rectification experiment, while in Figure 7.5(b) the measured electrical properties of the rectifier are shown. In comparing Figures 7.4(c) and 7.5(b) to Figure 7.3, the reader should remain aware that, as noted above, the direction of *current* flow represented in Figure 7.3 is the direction of flow of positive charge or *holes*, which is opposite to the direction of electron flow.

Impressive as are the demonstrations of Metzger's and Reed's molecular rectifying diodes, the molecular rectifiers they demonstrated cannot be readily integrated with Tour's polyphenylene-based wires to produce a compact, purely molecular circuit. Thus, in Section 7.4 of this chapter, the authors propose, characterize, and explain the operation of new polyphenylene-based molecular rectifiers. These would operate according to the same principles expounded by Aviram and Ratner, and as were demonstrated by Metzger et al. and by Reed et al. However, the molecular rectifying diodes proposed in this work should be more appropriate in structure for building molecular-scale digital circuitry.

7.2.1.3.2 Molecular Resonant Tunneling Diodes

A resonant tunneling diode (RTD) or negative differential resistance (NDR) switch takes advantage of energy quantization to permit the amount of voltage bias across the source and drain contacts of the diode to switch on and off an electric current traveling from the source to the drain. Unlike the rectifying diode, current passes equally well in both directions through the RTD.

7.2.1.3.2.1 Structure of the Molecular RTD — Depicted in Figure 7.6(a) is a molecular resonant tunneling diode that has only very recently been synthesized by Tour[45,51] and demonstrated by Reed.[5,6] Structurally and functionally, the device is a molecular analog of much larger solid-state RTDs that for the past decade commonly have been fabricated in III-V semiconductors and employed in solid-state, quantum-effect circuitry.[17] However, the Tour-Reed molecular RTD, or NDR switch, is the first experimental demonstration of a working molecular electronic device of this type.

Based upon a Tour-wire backbone, as shown in Figure 7.6(a), Reed's and Tour's polyphenylene-based molecular RTD is made by inserting two aliphatic methylene groups into the wire on either side of a single aliphatic ring. Because of the insulating properties of the aliphatic groups, as discussed above, they act as potential energy barriers to electron flow. They establish the aromatic ring between them as a

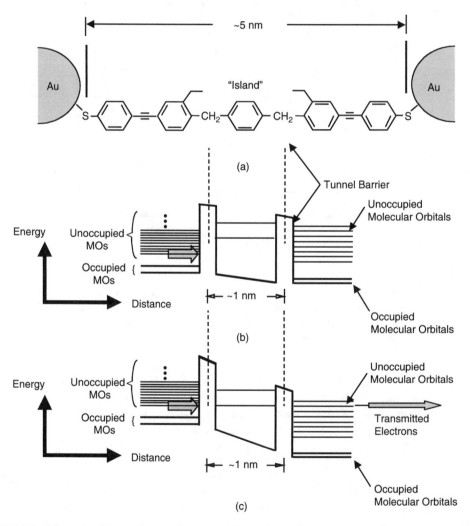

FIGURE 7.6 Schematic of the structure and operation for the molecular resonant tunneling diode demonstrated by Reed, Tour, and their collaborators.[5,6,45,51]

narrow (approximately 0.5 nm) "island" of lower potential energy through which electrons must pass in order to traverse the length of the molecular wire. (Note that solid-state RTDs have islands approximately 10 nm wide or approximately 20 times as wide as that in the molecular variant.)

The molecular RTD and the other molecular devices illustrated in this work often are attached at their ends to gold electrodes by thiol (–SH) groups that adsorb (i.e., form tight bonds) to the gold lattice. This is seen, for example, in Figure 7.6. Groups of this type and others that bind molecules tightly to metals have been termed *molecular alligator clips* by Tour.[52] Ideally, these molecular alligator clips would not only bind the molecules tightly to the metal, as does the thiol, but also promote conduction between metal and molecules.

Thiols on gold are the most common "clips" for attaching molecules to metal substrates, but they may not form the optimal connections for molecular electronic devices. The geometry of the orbitals on the sulfur does not permit the conjugated π-orbitals from the organic chain to interact strongly with the conduction orbitals of the gold. The mismatch of orbitals creates a potential energy barrier at each end of the wire, raising the resistance of the wire. It has been suggested that other clip combinations might make a better electrical contact between the metal and the organic molecule. However, alternative metal contacts may present other experimental difficulties such as surface oxidation and side reactions with the organic molecules.

7.2.1.3.2.2 Operation of the Molecular RTD — Whenever electrons are confined between two such closely spaced barriers, quantum mechanics restricts their energies to one of a finite number of discrete "quantized" levels.[40,41,53] This energy quantization is the basis for the operation of the resonant-tunneling diode.[17,53]

The smaller the region in which the electrons are confined, the farther apart in energy are the allowed quantized energy levels. Thus, on the island between the barriers of the RTD illustrated in Figure 7.6, the allowed unoccupied π-type energy levels are relatively far apart. Also shown in the illustration, the unoccupied π-type energy levels are more densely spaced in energy in the less confining low-potential-energy regions of the molecule to the left and right of the barriers surrounding the island. Electrons are injected under a voltage bias into the lowest unoccupied energy levels on the left-hand side of the molecule.

The only way for electrons with these moderate kinetic energies to pass through the device is to "tunnel," quantum mechanically, through the two barriers surrounding the island. The probability that the electrons can tunnel from the left-hand side onto the island is dependent on the energies of the incoming electrons compared with the widely spaced unoccupied energy levels on the island.

As illustrated in Figure 7.6(b), if the bias across the molecule produces incoming electrons with kinetic energies that differ from the unoccupied energy levels available inside the potential well on the island, then current does not flow. The RTD is switched off.

However, as illustrated in Figure 7.6(c), if the bias voltage is adjusted so that the energy of the incoming electrons aligns with one of the island's energy levels, the energy of the electrons outside the well is *in resonance* with the allowed energy inside the well. In that case an electron can tunnel from the left-hand region onto the island. If this can occur, and, simultaneously, if the bias is such that the energy level inside the well also is in resonance with one of the many unoccupied energy levels in the region to the right of it, then electrons flow through the device from left to right. That is, the RTD is switched on.

7.2.1.3.2.3 Current vs. Voltage Behavior of the Molecular RTD — To understand the utility of such RTDs in logic circuits, it essential to be familiar with the characteristic current vs. voltage plot that is produced by the operational mechanism described above.

Figure 7.7(a) shows the dependence of the transmitted current on the voltage bias in the measurements by Reed and his collaborators for the polyphenylene-based RTD with the structure given in Figure 7.6(a).[5,6] It shows the transmitted current rising to a characteristic *peak* at the resonance voltage, then falling off into a *valley* as the bias sweeps past the resonance voltage. The effectiveness of the operation of a particular RTD often is characterized by how well defined are the peak and valley in the current vs. voltage plot. This is measured by the peak-to-valley current ratio. In Figure 7.7(a), the measured peak-to-valley ratio is approximately 1.3:1.

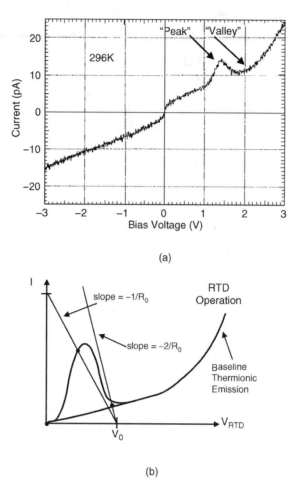

(a)

(b)

FIGURE 7.7   Behavior of the molecular resonant tunneling diode depicted in Figure 7.6 as (a) measured by Reed,[5,6] and (b) as abstracted by the present authors for the analysis of the logic circuit depicted in Figure 7.14.

In very recent work, however, Tour and Reed have modified their molecular RTD using an intramolecular doping strategy based upon the one outlined below in this chapter for rectifying diodes. Experimentally, they found that the dopants seemed to enhance greatly the peak-to valley-ratio (to approximately 1000:1), much improving the switching of the molecular RTD.[12,13]

The asymmetry of the current vs. voltage plot in Figure 7.7(a) about the vertical, zero-bias-voltage axis merits some comment. Generally speaking, one would expect that such a plot for an RTD would be symmetric about the zero voltage axis. However, in the difficult experimental demonstration of this molecular RTD, a number of them (approximately 1000) were aligned vertically in a nanopore, adsorbed in a monolayer between two metal electrodes. To permit the nanofabrication of this system, the electrodes on the top and on the bottom necessarily were of different compositions.[5–7] The bottom electrode was gold, while the top electrode consisted of layers of titanium and gold (as is also the case in the earlier experiment illustrated in Figure 7.5). It is believed, based upon similar asymmetric performance curves observed in experiments upon both micron-scale and nanometer-scale devices, that the asymmetry in the composition of the nanopore's electrical contacts induced a degree of rectification in the experimental system.[7] If so, this would produce the observed asymmetry in the molecular RTD's performance plot. Under a negative bias, the rectifier in which the molecular RTDs were embedded would suppress the negative peak current in the region of the negative resonance voltage. However, under a forward bias, the rectifier would not similarly suppress the peak current that does manifest itself in the positive voltage direction in Figure 7.7(a).[43]

For some RTDs (e.g., in III-V semiconductors) the first current peak observed at either positive or negative voltage bias may even be followed by subsequent current peaks observed at higher voltages corresponding to resonances with higher energy unoccupied energy levels that are defined by the potential well on the island. For the purpose of developing the logic circuits described in following sections, however, only one resonance peak is essential. With this in mind, we abstract the current vs. voltage plot of the molecular RTD as shown in Figure 7.7(b). This current vs. voltage behavior is a key contributor to the operation of the molecular XOR gate for which a design is proposed in Section 7.5.1.2.

7.2.2 Carbon-Nanotube-Based Molecular-Scale Electronic Devices

The second type of demonstrated molecular electronic backbone structure is the carbon nanotube, also known as a fullerene tube or "buckytube."[54] This type of structure can make an extremely conductive wire.[4,55–59] In addition, carbon nanotubes of different diameters and chiralities (character-istic "twists" in their structures) have been predicted and demonstrated to exhibit different semicon-ducting electrical properties ranging from those of excellent conductors (i.e., semi-metals) to those of near insulators.[56,57,60–62] Taking advantage of these properties, carbon nanotubes that incorporate a junction between two chiral structures of different types have been shown to exhibit electrical recti-fication, behaving as a diode switch.[14,63] Also, carbon nanotube wires have been shown to exhibit transistor-like electrical switching properties when employed on a micropatterned semiconductor surface that can be charged to act as a gate.[64]

On the other hand, because fullerene structures only have been discovered and characterized within the past two decades, the chemistry for producing and manipulating them is not as well defined or as flexible as might be desired. Remarkable progress has been made in developing the chemistry of fullerene nanotubes, especially very recently. Advances are being made toward establishing rules and methods for substitution of other chemical groups[65–70] and also toward extending the range of available structures.[71] Nonetheless, these rules are not well established yet, nor is it known yet how to make with specificity and selectivity a single-walled carbon nanotube structure of a particular desired geometry and chirality, to the exclusion of others.

Further, carbon nanotubes tend to be very stable structures when formed, such that their chemical formation and manipulation in bulk chemical processes occurs only under relatively extreme conditions. Such reactions have tended not to be very selective or precise in the range of carbon nanotube structures they produce at one time. Many carbon nanotube molecules with a range of different structures usually are produced in a synthesis. However, small electronic circuits require components with precisely spec-ified, uniform structures. At this stage, to achieve the requisite precision and uniformity still requires the use of physical inspection and manipulation of the molecules one-by-one in order to segregate and select carbon nanotubes for use in electronic research devices. Bulk chemical methods have not been developed or refined yet for this purpose, although some advances in this direction are being made using finely patterned catalytic surfaces.[14,72–74]

One might be able to speculate with some accuracy on the prospective electrical properties of a complex carbon nanotube structure or substructure, such as a chemically bonded "T-junction" or "Y-junction" among two or three carbon nanotube wires[75–78] that one might need in order to build carbon nanotube-based molecular electronic circuits.[79] Some such carbon nanotube junction structures have been made (Y-junctions), but only very recently.[71] Others (e.g., T-junctions) still have not been reported. Also, the junction structures that have been made by design are composed of multi-walled carbon nanotubes, not the single-walled junction structures of specific chirality that are likely to be necessary to have sufficiently sensitive control over small electrical currents. Research in the designed synthesis of junction structures is just beginning to exhibit useful results, and it is not yet certain which complex carbon nanotube structures can be made chemically, let alone how to make them easily and with specificity.

Thus, the design and development of carbon nanotube logic architectures is complicated and hampered presently by the fact that carbon nanotubes are relatively inert chemically and also that the chemistry of carbon nanotubes is relatively immature. This places the molecular electronics community in a situation where the molecular backbone that can be used reliably to make suitably branched structures, polyphenylene, is not the most conductive, while the most conductive molecular backbone, the carbon nanotube, does not yet have a reaction chemistry flexible enough for making the desired branched, chiral structures. Despite some promising new work in this direction,[65–71,80–83] there does not appear to be a method yet by which one can make reliably and with specificity all the structures that are likely to be needed for purely carbon nanotube-based electronic logic structures. From the point of view of the molecular circuit designer, in particular, the range of accessible carbon nanotube structures has not been clear, although some such carbon-nanotube-based molecular circuit design has proceeded anyway.[79]

The rapid advances in carbon nanotube chemistry and in the demonstration of carbon-nanotube-based electrical devices, especially very recently, are a cause for optimism, though. With further advances in the chemistry of carbon nanotubes, in the foreseeable future it might be possible to fabricate and to demonstrate purely molecular carbon nanotube logic structures similar to those proposed below for polyphenylene-based molecular circuits. Even sooner, it might be possible to build one-of-a-kind prototype carbon-nanotube-based logic circuits using mechanical nanoprobes to pick and place onto microfabricated contacts the several separate carbon nanotube molecular switches and wires that would be necessary. (Note for 2003 CRC edition: this feat was actually accomplished by Dekker et al. in 2001, 2 years after this text was originally written. See References 133 and 135.)

7.2.3   Biomolecules as Possible Molecular Electronic Devices

For completeness, we also review here the status of investigations toward using biomolecules as the conductive backbone for electronic wires, switches, and circuits. This third possibility for a conductive monomolecular backbone has much appeal because biomolecules are ubiquitous and come in a wide range of structures, and advances in biotechnology over the past several decades have provided sensitive methods for their precise chemical manipulation and assembly.

DNA molecules, for example, with their long, wire-like shape, have been suggested as a possible backbone structure. One very early effort considered in some detail a design that used DNA as the basis for an electronic memory cell.[84]

Encouragement for the possibility of using DNA in molecular electronics has come from experimental work on the bulk kinetics of electron transfer processes through DNA molecules in solution. Results from such experiments carried out in the middle 1990s suggested that DNA molecules might be conductive enough for use in wires and other electrical devices. This has been a matter of some debate and confusion, though, as is well summarized elsewhere.[85] New experimental and theoretical work seems to be on the verge of resolving this debate in the case of the kinetic experiments, though, by describing how the measured flow of charge results from a combination of mechanisms that produces different results depending on the length, composition, and environment of the DNA.[86–89] Disappointingly, from the point of view of using pure DNA as a wire in a molecular circuit, this theoretical work also suggests that DNA's conductance should fall off very rapidly with increasing length of the molecule, unless there is very significant thermal dissipation.

Of great importance, also, Dekker and his collaborators very recently carried out a remarkable experiment in which they measured directly the conductance of one DNA molecule (or, at most, a few molecules) suspended between two metal electrodes. They found that a DNA molecule can transport charge at high bias voltages, but that it shows negligible conductance at low bias voltages up to a few volts.[90]

Despite these discouraging results for DNA, other biomolecules have shown some promise recently for use in molecular-scale electronic systems. Porphyrins and metallo-porphyrins, in particular, are another type of biomolecule[91] that is being considered for use as the conductive backbone of molecular electronic circuits.[92] This is not especially surprising, as well-known porphyrins, such as hemoglobin, chlorophyll, and the cytochromes, perform their biological roles in processes that chiefly involve them in storing and transferring electrons. Among their useful electrical properties, it has been calculated that porphyrinic molecular substituents will exhibit relatively high capacitances when considered in comparison to polyphenylene-based molecular components in the same circuit.[93]

Nonetheless, experimental progress on using biomolecules as components in molecular electronic devices and circuits lags well behind that on using polyphenylenes and carbon nanotubes.

7.2.4 An Advantage of Polyphenylene-Based Structures for Logic Design

Given the primitive state of molecular electronics experiments involving biomolecules and the evolving state of carbon nanotube chemistry, as outlined above, there are some very significant advantages in considering logic structures that would be built from polyphenylene-based molecules in this first effort to design realizable conductive molecular electronic logic circuits. Using polyphenylenes, it is much easier to propose the more complex molecular electronic structures required for digital logic and to know in advance with some certainty that they can be synthesized. There exists a rational approach to planning the total synthesis of such discrete structures using organic chemistry.

Also, as shown in Table 7.1, polyphenylene-based molecular electronic devices can be expected to conduct a very appreciable current of electrons for such small structures. The absolute magnitude of the current they conduct may not be as large as that for a carbon nanotube. However, the current density of electrons transported through a polyphenylene-based molecular wire is of the same order of magnitude as for a carbon nanotube, because the polyphenylene-based device is so much smaller and has a smaller cross-sectional area.

The fact that these polyphenylene-based molecules are much smaller than carbon nanotubes provides yet a further reason for basing molecular circuit designs upon Tour wires in this initial effort to design monomolecular electronic digital logic. That is, if electronic logic structures and functions based upon Tour wires can be synthesized chemically (or assembled mechanically) and made to operate, they represent the ultimate in miniaturization for digital electronic logic. Virtually any other structure will be as large or larger. If we or others who follow demonstrate that these structures will not operate for any reason, the present effort at least will have established a lower limit on the size and an upper limit on the density for electronic digital logic.

7.3 Approach and Objectives

As has been summarized above, in Sections 7.1 and 7.2, previous theoretical and experimental efforts have provided us with a small repertoire of experimentally demonstrated polyphenylene-based molecular electronic devices upon which to build. These devices, which can transport a large current density, are relatively straightforward analogs of well-understood existing solid-state microelectronic and nanoelectronic devices. Moreover, as is discussed in detail above, the design of molecular-scale electronic devices and logic based upon chemically modified or "functionalized" polyphenylene-based Tour wires follows a set of simple structural principles and heuristics.

For these reasons, in the subsequent sections of this chapter we shall focus our attention on designing polyphenylene-based structures for building molecular electronic digital logic circuits. In another paper, building upon this preliminary effort, we propose alternative structures for carbon nanotube-based molecular electronic digital logic.[79]

With such objectives in mind, we take as our working hypothesis the premise that molecular electronic digital logic circuits can be designed in analogy to microelectronic circuitry using Tour wires as a conductive backbone. Associated with this hypothesis, classically observed and derived principles such as Kirchoff's circuit laws and the independent function of discrete circuit elements are assumed to apply on the molecular scale.

In taking that approach to this first effort at molecular circuit design, we have thereby attempted to restrict the consideration of quantum effects to the device level and neglected such effects at the circuit level. At the next higher level of complexity, therefore, one might lift all such simplifying classical assumptions, applying quantum principles at both the circuit and the device level for the molecular electronic logic structures proposed below.[94,95] For example, since the completion of this research, other investigators have published results that consider some of the likely impacts of quantum effects at the circuit level.[96–100,133,134] In order to make progress step by step, though, for the time being we have adopted a semiclassical approach that restricts the consideration of quantum effects to the switching devices.

Thus, we proceed from the hypothesis above to build in a simple logical manner upon the work of those who have used a polyphenylene-based backbone to make and to demonstrate individual molecular wires and individual molecular-scale electrical switching devices (primarily diode switches). Our approach also attempts to extend to the molecular domain circuit designs and principles of combinational logic well established in the microelectronic domain.

First, in Section 7.4, we shall propose and explore novel theoretical designs for a set of polyphenylene-based rectifying diodes. These new diode switching devices use the same operational principles as the Aviram and Ratner (A&R)-type molecular rectifiers just recently demonstrated,[7,9] but the ones proposed here should be easier to combine chemically with single molecular polyphenylene-based wires — Tour wires — in molecular-scale electrical circuits.

Second, in Section 7.5, we posit plausible designs for molecular structures that take the next step: they connect via Tour wires first two, then three molecular diode switches to make the three elementary logic gates, AND, OR, and XOR (or NOT). Further, we shall propose a design that combines two of these logic gates to make a molecular electronic adder — i.e., a single molecule that will add two binary numbers.

Third, in Section 7.6, we consider the strengths and limitations of such diode-based molecular electronic digital circuit structures in the context of a number of architectural issues that must be considered and addressed in order to build a working molecular electronic computer using these or related designs. Finally, in Section 7.7, we conclude by considering how such small, diode-based molecular logic circuits and functions might be employed in extended circuitry integrated on the nanometer scale.

7.4 Polyphenylene-Based Molecular Rectifying Diode Switches: Design and Theoretical Characterization

In this section, designs are proposed for a set of polyphenylene-based rectifying diodes that use the same operational principles as the A&R-type molecular rectifiers recently demonstrated,[7,9] but with structures that should be easier to combine chemically with single molecular polyphenylene-based wires to make molecular-scale electrical circuits. Specifically, rather than connect disaggregate switching molecules together with polyphenylene-based wires, we propose to make the molecular wires themselves into molecular electronic switches. This solves an important design problem, because it eliminates the need for incorporating other, much less conductive switching molecules into molecular circuits.

7.4.1 Proposed Polyphenylene-Based Designs for A&R-Type Rectifying Diodes

7.4.1.1 Basic Structure

Figure 7.8 schematically depicts the molecular structure and the energy level structure of the authors' proposed variant of Aviram and Ratner's basic concept for a molecular rectifying diode. The unique feature of this variant is that it uses *chemically doped* polyphenylene-based molecular wires as the conductive backbone. In this chapter, electron-donating and electron-withdrawing substituent groups chemically bound to a single molecule shall be termed *intramolecular dopants*. This terminology serves to emphasize the similarity in influence and function that these substituent groups have to dopant atoms which commonly are introduced as randomly distributed impurities in solid-state semiconductors in order to control their electrical properties. However, the modifier *intramolecular*, which is applied in a manner consistent with its usage in other chemical contexts, also serves to distinguish the dopant substituents used here, which are precisely placed and bound in position on an individual molecule, from their analogs in solids.

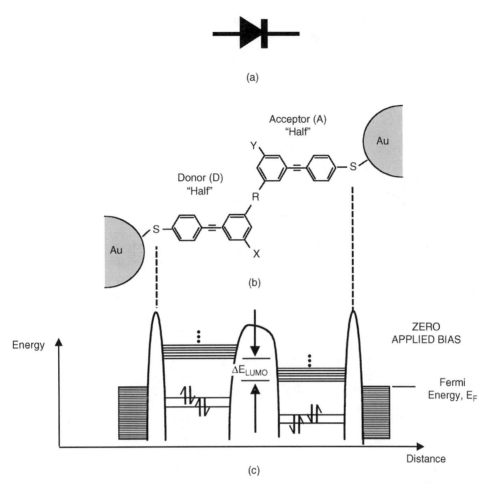

FIGURE 7.8. Molecular structure and schematic of electron orbital energy levels for a proposed polyphenylene-based molecular rectifying diode switch.

Applying this terminology, the structure shown in Figure 7.8 has two intramolecular dopant groups: the electron-donating substituent group, X, and the electron-withdrawing substituent group, Y. The first of these is an intramolecular analog of an n-type dopant in solid-state semiconductors and the second is an intramolecular analog of a p-type dopant. Additionally, the terminology *intramolecular* distinguishes the dopants employed in this chapter from *extramolecular* dopants such as those applied in metal contacts to which otherwise undoped molecules are adsorbed. For example, in the work of Zhou et al.,[7] a different type of molecular rectifier is produced via such extramolecular dopants.

In contrast, the characteristic general structure of an A&R-type monomolecular rectifying diode features an electron donor subcomplex at one end (i.e., the donor "half" of the molecule) and an electron acceptor subcomplex at the other end (the acceptor "half").[49] The donor subcomplex consists of an electrically conductive molecular backbone with one or more electron-donating intramolecular dopant substituents covalently bound to it. The acceptor subcomplex consists of a similar backbone with one or more electron-withdrawing intramolecular dopant substituents covalently bound to it. These donor and acceptor subcomplexes are separated within the diode structure by a semi-insulating bridging group, to which they both are chemically bonded. Usually, this three-part *donor–acceptor complex* is envisioned or implemented as being in electrical contact with metal terminals at both ends.

Consistent with this concept, in Figure 7.8(b) the polyphenylene-based A&R diode is shown to consist of an electron donor subcomplex (the Tour wire on the left contains the electron-donating intramolecular dopant substituent X) that is separated by a semi-insulating group R from an electron acceptor subcomplex (the Tour wire on the right contains the electron-withdrawing dopant substituent Y). Figure 7.8(c) shows a schematic of the molecular orbital energy diagram associated with the polyphenylene-based A&R diode.

Observe that the insulating group R in the middle of the molecule is associated with a potential energy barrier. This group serves as an insulating bridge or barrier between the donor and acceptor halves of the diode. The barrier is intended to preserve the potential drop induced by the donor and acceptor substituents, X and Y, respectively, which serve in the role of chemically bonded intramolecular dopants. The barrier group prevents the differing electron densities in the substituted complexes on either side from coming to equilibrium, while it still permits added electrons under a voltage bias to tunnel through.

There are also barriers between the molecule and the gold (Au) contacts at either end due to the thiol linkages, as discussed above. These barriers serve to maintain a degree of electrical isolation between the different parts of the structure, sufficient to prevent the equilibration of the electron densities (and the associated one-electron energy levels) of the parts on either side. However, none of these barriers is so wide or high as to completely prevent electrons under a suitable voltage bias from tunneling through them.

The most likely candidates for R are aliphatic groups such as sigma-bonded methylene groups (R = $-CH_2-$) or dimethylene groups (R = $-CH_2CH_2-$). Their insulating properties are discussed above and have been demonstrated by Reed and Tour. For the polyphenylene-based rectifying diode designs used here, the aliphatic dimethylene group is selected as the central bridging group because it is the smallest nonconducting, aliphatic group that can serve as a narrow insulating tunnel barrier R but still permit the aromatic rings on either side to be aligned easily in the same plane. Coplanarity of the donor-substituted and acceptor-substituted aromatic rings is desirable in order to enhance the extent of π-orbital conjugation between the rings and thereby increase the conductivity of the diodes in the preferred direction. The dimethylene bridging barrier group also permits internal rotations wherein the donor- and acceptor-substituted benzene rings are moved to different planes but still remain parallel. This provides useful flexibility in circuit designs where it may be necessary for one molecular wire to cross over another without intersecting it. Further, as is shown below in Section 7.4.2, rotations of this sort do not necessarily change the energies of the key orbitals upon which the operation of a polyphenylene-based rectifying diode depends.

7.4.1.2 Operational Principles for Polyphenylene-Based and Other A&R-Type Rectifying Diodes

The principles of operation for these proposed polyphenylene-based, molecular rectifying diodes are similar to those of other A&R-type molecular rectifiers.[49] To understand, qualitatively, these principles of operation and how they dictate such a molecular structure, we must examine in detail the dependence of the molecule's energy levels on its structure and how the energy levels change when a voltage bias is applied across the molecule.

Figure 7.8 shows the correspondence of the molecular structure to the molecular energy levels when there is no externally applied voltage bias. By contrast, Figure 7.9 schematically depicts the operational principles of the polyphenylene-based molecular rectifier by showing how the energy levels change under an externally applied bias.

In Figure 7.9 and in similar following figures, it is assumed and depicted implicitly that the voltage drop across a molecular diode is divided equally between its two halves. This may not be exactly true, but it is a useful simple approximation.

7.4.1.2.1 *Energy Structure of the Donor Half of the Polyphenylene-Based Molecular Rectifier*

Observe in Figure 7.8(c) that, to the left of the central barrier, on the side or half of the molecule associated with the electron-donating dopant substituent group X, the valence energy levels are elevated in energy. This energy elevation affects all the molecular orbitals that are localized on the left-hand side of the molecule. This includes the highest occupied molecular orbital (HOMO), the lowest unoccupied molecular orbital (LUMO), and the associated low-energy unoccupied pi orbitals (collectively, the "LUMOs") on the left-hand, donor side of the molecule.

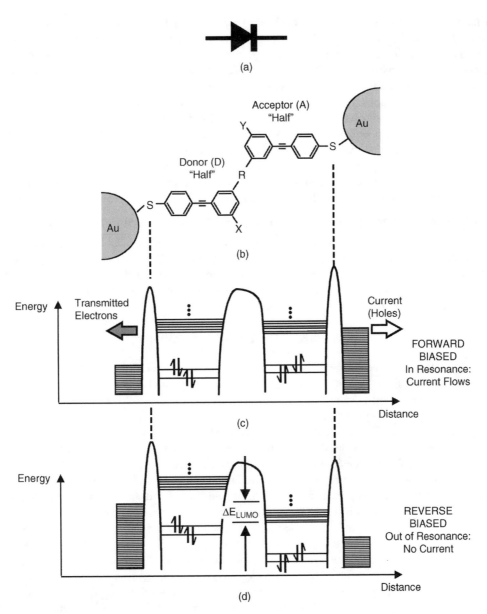

FIGURE 7.9 Schematic describing operation of a proposed polyphenylene-based molecular rectifying diode switch under two opposite externally applied voltages.

These effects are due to the influence of the electron-donating substituent group X. Substituent groups with this characteristic behavior have been known for decades to organic chemists due to their effect on the stability of aromatic reactive intermediates,[42] independent of any considerations having to do directly with molecular conductivity. The common electron donating substituents X are $-NH_2$, $-OH$, $-CH_3$, $-CH_2CH_3$, etc.

An electron-donating group bonded to an aromatic ring tends to place more electron density upon the ring (or upon a group of neighboring conjugated aromatic rings). This increases the mutual repulsion among the electrons in the molecular orbitals associated with the (conjugated) ring structure. In the case of the conjugated ring structure to the left of the central barrier shown in Figure 7.8(b), these additional repulsive interactions raise the total energy, as well as the component orbital energies, as suggested in Figure 7.8(c).

7.4.1.2.2   Energy Structure of the Acceptor Half of the Polyphenylene-Based Molecular Rectifier
Conversely, to the right of the central barrier, all the valence energy levels are lowered by the presence
of an electron-withdrawing dopant substituent group Y. This energy lowering affects all the molecular
orbitals localized primarily on the right-hand side of the molecule. This includes the HOMO and the
LUMOs that are localized on the right-hand side of the molecule.

As is the case for the donating group X discussed above, substituent groups with electron-withdrawing
characteristic behavior also have been well known for decades to organic chemists.[42] The common
electron-withdrawing substituents Y are $-NO_2$, $-CN$, $-CHO$, $-COR'$, etc., where R' is an aliphatic chain.

An electron-withdrawing group bonded to a ring (or to several neighboring conjugated aromatic rings)
tends to remove electron density from the ring, thereby reducing electron repulsion within the (conjugated) ring structure. These diminished repulsive interactions lower the total energy of the structure to
the right of the central barrier, as well as its component orbital energies, as is shown in Figure 7.8(c).

7.4.1.3   Forward-Bias Operation of the Polyphenylene-Based Molecular Rectifier

As shown in Figure 7.8(c) and described above, even with no external applied bias, there is a dopant-
induced difference in the relative energetic positions of the π orbitals in the donor and acceptor halves
of the molecule in Figure 7.8(b). This energy-level difference is analyzed quantitatively in detail in
Appendix A. Here, we observe simply that the dopant-induced energy-level difference in the rectifying
diode may be assessed in terms of the quantity:

$$\Delta E_{LUMO} \equiv E_{LUMO}(Donor) - E_{LUMO}(Acceptor), \tag{7.1}$$

where $E_{LUMO}(Donor)$ is the energy of the lowest unoccupied orbital localized on the donor side of the
central barrier and $E_{LUMO}(Acceptor)$ is the energy of the lowest unoccupied orbital localized on the
acceptor side. Under most circumstances, the latter should be expected to be the LUMO for the entire
molecule, as well. The energy difference ΔE_{LUMO} is depicted in Figure 7.8(c).

This dopant-induced difference in the energy levels localized respectively on the donor and the acceptor
portions of the molecule provides the foundation for the operation of the polyphenylene-based A&R-
type molecular rectifying diode. This operation is illustrated schematically in Figure 7.9.

Imagine first that a *forward* voltage bias is placed upon the system in Figure 7.9(b), as depicted in
Figure 7.9(c), with the higher voltage on the left-hand gold contact and the lower voltage on the right-
hand contact. As shown, the process of applying the field shifts the electrons on the right-hand contact
to higher energies and the electrons on the left-hand contact to lower energies, because the energy change
for the electrons is of the *opposite sign* from the applied voltage. The energy differential induces the
electrons in the occupied quantum levels of the high-energy right-hand contact to attempt to flow from
right to left through the molecule to reach the lower energy left-hand contact.

Note that the very densely spaced *occupied* quantum levels in the valence band of each contact are
represented by the closely spaced horizontal lines at the far left and far right of Figure 7.8(c) and
Figure 7.9(c). The highest of these occupied levels, the Fermi level, in the metal contact has an energy
known as the *Fermi energy* (E_F). Above this energy is also a very dense band of unoccupied energy levels
that are not depicted in the figure. Applying a bias voltage tends to raise the Fermi level in the low-voltage
contact and lower the Fermi level in the other one.

For this right-to-left flow of electrons to occur under forward bias, the voltage bias must be sufficient
to raise the Fermi energy of the electrons in the occupied levels of the external gold contact on the right
at least as high as the energy of the LUMO π orbital in the right-hand acceptor portion of the molecule.
Then the electrons can tunnel from the right contact into the empty LUMOs for the acceptor.

The electrons that have migrated from the right-hand contact into the acceptor LUMOs can tunnel
once again to the left through the central insulating barrier to the unoccupied manifold of molecular
orbitals in the donor half of the complex (which above a certain threshold applied voltage are sufficiently
lowered in energy so that one or more of them match an orbital in the unoccupied manifold in the
acceptor half to the left). From this point, resonant transmission into the left-hand contact is assured,

because the unoccupied manifold of the metal is very dense. This situation is depicted in Figure 7.9(c), where the transmitted electrons are shown flowing from the right to the left and the positively charged current is shown flowing from the left to the right.

To summarize the forward-bias mechanism, due to the application of a high voltage applied on the left contact and low voltage on the right contact: (1) the Fermi energy in the right contact exceeds the LUMO energy in the right-hand acceptor half of the molecule and (2) the LUMOs on the left-hand side of the molecule align with those on the right-hand side. This permits electrons to pass all the way through the molecule via resonant transmission.

Fortunately, in the forward bias case, only a relatively small voltage bias is required to raise the Fermi energy of the right-hand contact sufficiently to exceed the LUMO energy of the acceptor. This is because all the energy levels of the acceptor have been lowered beforehand by the presence of the intramolecular dopant group Y, as shown in Figure 7.8(c). Also, in the forward bias case, it is very important that the applied potential tends to pull the higher energy unoccupied orbitals on the donor half of the diode *down* in energy toward the energies of the acceptor LUMOs, as illustrated in Figure 7.9(c). This tendency of the forward bias to bring the donor LUMOs in coincidence with the acceptor LUMOs enhances electron tunneling from the acceptor to the donor through the central barrier of the molecule.

7.4.1.4 Reverse-Bias Operation of the Polyphenylene-Based Molecular Rectifier

In contrast, imagine now that a *reverse* voltage bias has been placed upon the system in Figure 7.9(b), with the higher voltage on the right-hand gold contact and the lower voltage on the left-hand contact, driving up the Fermi energy on the left and depressing it on the right.

Analogous to the forward bias case described above, for the electrons in the external left-hand contact to begin to flow from left to right through the molecule, the reverse voltage bias must be sufficient to raise the Fermi energy of the gold contact on the left so that it is at least as high as the energy of the LUMO π orbitals in the left-hand, donor portion of the molecule.

However, in the reverse bias case, the amount of voltage that must be applied is considerably greater than in the forward bias case in order to raise the Fermi energy of the contact sufficiently to exceed the LUMO energy of the adjoining portion of the molecule. This is because, as shown in Figure 7.8(b), all the energy levels of the donor half have been raised by the presence of the substituent group X. Figure 7.8(d) confirms that simply applying the same amount of voltage in the reverse direction as is used to induce a current in the forward direction is insufficient to allow electrons to tunnel from the left contact into the LUMO energy levels of the molecule.

Thus, more voltage must be applied in the reverse (right-to-left) direction than in the forward (left-to-right) direction in order to get electrons to flow through the molecule. This is the classic behavior of a rectifying diode, as represented by the schematic symbol shown in Figure 7.9(a). Thus, this behavior and this symbol may be associated with the molecule.

Furthermore, the reverse bias tends to drive up the energy of the LUMOs on the donor portion of the molecule relative to the LUMOs on the acceptor to the right. As depicted in Figure 7.9(d), this increases the separation of the lowest lying unoccupied orbitals on the two sides of the molecule to a value greater than the unbiased value ΔE_{LUMO}, rather than decreasing their energy separation. This makes it difficult at moderate bias voltages to bring the energy levels of the donor-half LUMOs in coincidence with the low-energy unoccupied manifold that is localized on the acceptor half to its right. Consequently, tunneling through the central barrier from left to right, from donor to acceptor, is impeded rather than enhanced by a reverse bias.

The discussion above suggests why it is highly likely that a doped polyphenylene-based wire of the general molecular structure shown in Figures 7.8(b) and 7.9(b) should behave as an A&R-type molecular rectifying diode switch.

7.4.1.5 Additional Considerations: Nonresonant Electron Transport under Reverse Bias

To ensure the effective performance of the doped polyphenylene molecular rectifiers discussed above, it is essential that charge transport under a forward bias be made as large as possible, while charge transport is kept as small as possible under reverse bias. Toward that latter goal, it is desirable to understand how current might flow in the reverse direction, in order to see how to prevent this. By considering analogies to the better-known mechanisms that can permit charge to flow in larger, conventional solid-state diodes under a reverse bias, some qualitative insight may be gained into the possible ways that charge might flow under reverse bias in a molecular rectifying diode.

When solid-state p-n junction rectifying diodes are placed under a reverse bias, there is a small current — corresponding to a nearly constant baseline conductance — that occurs at all reverse voltages, even small ones. This reverse flow usually is several orders of magnitude smaller than the large currents that flow when the rectifier is under a forward bias. The small reverse-bias *leakage current* arises due to a *drift* mechanism that is different from the diffusion of electrons that dominates the forward-bias electron flow.[48]

Similarly, in molecular rectifiers, one can expect that there will be reverse leakage largely due to electron transport via *nonresonant* mechanisms. These are mechanisms different from the resonant transport that one desires to dominate electron flow in the forward direction — and to shut it off in the reverse direction. Nonresonant mechanisms can result, for example, from the coupling of the vibrational and electronic modes of the molecules (*vibronic* coupling). Vibronic coupling seems analogous in certain ways to the drift mechanism of solids. Other nonresonant transport mechanisms, such as electron hopping, also could contribute.[20,24,25]

At sufficiently high reverse-bias voltages, above the *breakdown voltage*, very large currents will flow through solid-state p-n junction rectifiers. This breakdown corresponds to voltages so high that large numbers of electrons in the localized orbitals in the valence band of the n-doped solid on one side of the rectifier are given additional energy greater than the central barrier (or *depletion region*) in the diode. In that case, electrons are ripped out of valence orbitals of atoms in the n-doped solid and injected in quantity into the delocalized orbitals of the conduction band on the p-doped side of the rectifier. This results in large currents in the reverse direction.[48]

Likewise, in molecules, as is suggested by Figure 7.8, a sufficiently large applied voltage bias in the reverse direction could raise the energies of electrons from the localized valence orbitals on the donor half of a rectifier so that they become greater than the maximum energy of the central barrier in the molecule. (This is the energy above which orbitals do not show characteristics that tend to localize the associated electron density primarily on one side of the barrier.) When valence π-electrons from the donor half of the molecule are promoted above this barrier energy, they may be injected into the unoccupied, delocalized conduction orbitals that lie above the energy barrier on the acceptor half of the molecular rectifier. Large currents should result. This would be the equivalent of breakdown for molecular rectifiers, and these devices must be operated at voltages below this threshold.

In view of the discussion above, qualitative design steps that seem likely to enhance the function of polyphenylene molecular rectifiers are those that would disfavor nonresonant transport in the reverse direction and/or raise the voltage at which reverse breakdown occurs. Disfavoring nonresonant transport requires, for example, ensuring that the structure of the diode is such that the vibrational modes of motion in the molecule are not excited by moderate voltages applied across the molecule in the reverse direction. This would reduce reverse transport by vibronic mechanisms.

The reverse breakdown voltage is raised by using central aliphatic insulating bridge structures and intramolecular dopants that tend to enhance localization of the π-type charge density on the opposite sides of this central insulator. This tends to keep high the central energy barrier shown in Figure 7.8. One wishes to keep this energy barrier relatively high for valence electrons on the donor half of the molecule without severely impairing the flow of current under a forward bias. (Resonant tunneling transport through an energy barrier in the forward direction also is disfavored by a higher central barrier.)

This tendency toward charge localization on opposite sides of the central insulator in the molecular diode also should be assisted by using intramolecular dopant substituents that couple or bond particularly strongly with the π-orbitals of the aromatic rings and thereby make the value of ΔE_{LUMO} large. Requirements for this condition are explored in detail in the next section.

7.4.2 Results of Quantum Calculations for the Selection of Particular Polyphenylene-Based Rectifier Molecules

Detailed quantum mechanical calculations were performed[101] to characterize quantitatively the electrical properties of several molecular electronic rectifying diodes with a structure of the type shown in Figures 7.8 and 7.9. In combination with the experimental conductivity demonstrations for polyphenylene-based Tour wires, these calculations indicate that our proposed polyphenylene-based A&R diodes are likely to have the properties required to produce molecular rectification. The results of these calculations were used to select a particular polyphenylene-based molecule most likely to have strong rectifying properties.

First, in performing the calculations, the barrier group R in the middle of the structure was chosen to be a dimethylene group $-CH_2CH_2-$. As noted above, it is the shortest aliphatic insulating chain that would still permit the aromatic components on either side to occupy the same plane and, thereby, be able to maintain a relatively high conductivity through the molecule. One might expect, though, that the dimethylene group would reduce the conductivity of the wire by approximately the same factor of 10^5 as do the two methylene groups in the molecular RTD. (See Table 7.1 and Section 7.2.1.2 for details.)

Second, the quantum calculations were used to determine the donor substituent group(s) X and the acceptor substituent group(s) Y that would yield a molecule with a relatively large intrinsic energy drop ΔE_{LUMO} across the barrier. Candidate molecules were considered that were doubly substituted in X and in Y, as well as molecules that were only singly substituted in those groups, as shown in Figures 7.8 and 7.9.

The geometry of each such molecular diode structure was optimized, subject to certain constraints, via an energy minimization procedure prior to the use of the structure for the quantum calculation of the properties pertinent to its function as a diode switch. Except for the aforementioned constraints, the optimal geometries of the diode molecules usually would have been nonplanar. The geometry constraints we used forced the planes of the two aromatic benzene rings in the diode to be at least parallel, and in some cases forced the two rings into the same plane. This was done because, as will be seen below, it is envisioned that these diode rectifier molecules will be embedded in a larger molecular circuit structure, and that larger structure or its supporting medium would enforce or nearly enforce geometry constraints of the type just described.

Other details of the calculations are presented in Appendix A at the end of this chapter. The primary conclusion of the calculations is the choice of the dimethoxy-dicyano substituted polyphenylene-based rectifying diode, the structure for which is shown in Figure 7.10. As indicated by the computational results for this structure, which appear in Table 7.2, the molecule has an intrinsic voltage drop from donor to acceptor of $\Delta E_{LUMO} = 1.98$ eV. This was nearly the largest value of ΔE_{LUMO} calculated for the various diode structures considered here, and, as indicated in Table 7.2, this value was particularly stable with respect to out-of-plane deformations of the molecule.

For example, the results in the second column of Table 7.2 correspond to an out-of-plane deformation in the molecule in Figure 7.10 that places the dimethylene bridge that connects the aromatic rings at a

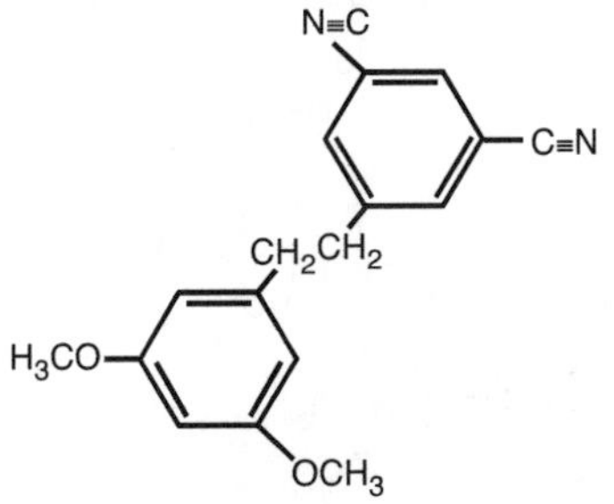

FIGURE 7.10 Dimethylene bridge connecting aromatic rings at a 90° angle.

TABLE 7.2   Results of HF STO 3-21G Molecular Orbital Calculations to Determine ΔE_{LUMO} between Donor and Acceptor Halves of the Dimethoxy–Dicyano Polyphenylene-Based Molecular Rectifying Diode (Shown in Figure 7.10)

Molecular Orbital	Two Di-Substituted Benzene Rings are *Co-Planar*			Two Di-Substituted Benzene Rings are *Non-Planar*, but Parallel		
	Orbital Energy	Localization D	A	Orbital Energy	Localization D	A
HOMO	−9.23 eV	X		−9.24 eV	X	
LUMO	1.52		X	1.50		X
LUMO+1	2.17		X	2.12		X
LUMO+2	3.49	X		3.49	X	
LUMO+3	3.69	X		3.67	X	
ΔE_{LUMO}	1.97 eV			1.99 eV		
$\Delta E_{LUMO}(R = \infty)$	2.28 eV			2.28 eV		

TABLE 7.3   Results of HF STO 3-21G Molecular Orbital Calculations to Determine ΔE_{LUMO} between Donor and Acceptor Halves of Dimethyl–Dicyano Polyphenylene-Based Molecular Rectifying Diode (Shown in Figure 7.11)

Molecular Orbital	Two Di-Substituted Benzene Rings are *Co-Planar*			Two Di-Substituted Benzene Rings are *Non-Planar*, but Parallel		
	Orbital Energy	Localization D	A	Orbital Energy	Localization D	A
HOMO	−9.11 eV	X		−8.99 eV	X	
LUMO	1.74		X	1.59		X
LUMO+1	2.36		X	2.22		X
LUMO+2	3.79	X		3.74	X	
LUMO+3	3.945	X		3.80	X	
ΔE_{LUMO}	2.05 eV			2.15 eV		
$\Delta E_{LUMO}(R = \infty)$	2.59 eV			2.59 eV		

full 90° angle out of the parallel planes of the two aromatic rings. This deformation results in an interplane separation of fully 1.46 Angstroms (or 0.146 nm). Nonetheless, as shown in Table 7.2, the key orbital energies associated with this nonplanar molecular conformation are very close to those tabulated in the first column of the table for the planar conformation.

Additional computational results, displayed in Table 7.3, showed that a d*imethyl*-dicyano substituted polyphenylene-based rectifying diode, the structure shown in Figure 7.11, would be likely to have rectification properties comparable to that for the dimethoxy-dicyano molecule in Figure 7.10. The calculations indicate that either molecule might serve as an effective rectifier, but it is the diode molecule in Figure 7.10 that will be used to build the molecular circuits depicted in subsequent sections of this work.

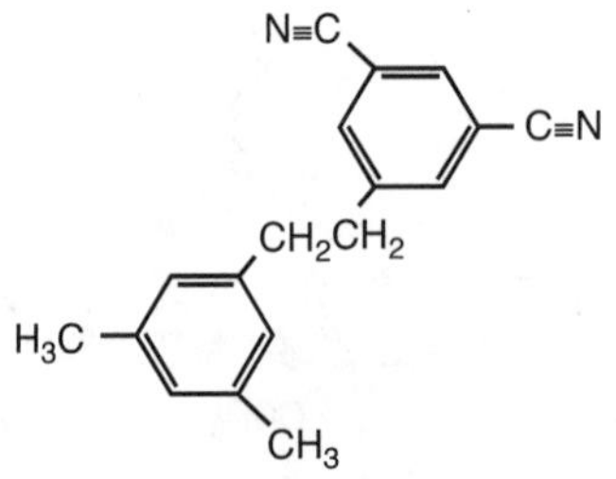

FIGURE 7.11   Structure of proposed dimethyl–dicyano polyphenylene-based molecular rectifying diode for use in the design of molecular electronic diode–diode logic circuits.

7.5 Novel Designs for Diode-Based Molecular Electronic Digital Circuits

7.5.1 Novel Diode-Based Molecular Electronic Logic Gates

Based upon the foregoing development, in this section the authors propose what they believe to be among the first designs for nanometer-scale logic gates and among the first designs for nanometer-scale arithmetic functions based upon the conduction of electrical current through a *single molecule*.

Tour, Kozaki, and Seminario (TKS)[51] recently proposed molecular-scale logic gates based upon a mechanism that involves perturbing the equilibrium electric charge distribution to change the electrostatic potential of a molecule. However, that work specifically and deliberately avoids employing nonequilibrium currents moving through a molecule under the influence of a strong externally imposed field or voltage. Therefore, the Tour group proposes logic structures significantly different from those proposed and discussed here.

Because of its reliance on the equilibrium (or near-equilibrium) charge distribution for computation rather than nonequilibrium electrical currents, the molecular-scale logic in the TKS proposal is more closely analogous to the micron-scale implementations[102–104] of the quantum cellular automata approach to computing suggested by Lent and Porod[18,105] than it is to micron-scale conductive logic circuitry. The present work, on the other hand, is and is intended to be the molecular-scale analog of micron-scale conductive circuitry, with the exception that the circuits proposed in this chapter explicitly take advantage of quantum effects, such as tunneling, that would impair the operation of micron-scale circuits.

As explained in the preceding sections, various conducting molecular-scale circuit components — wires and diode switches — have been demonstrated and/or simulated to function in a manner somewhat analogous to well-known solid-state electrical components. This suggests that assembling the molecular components in a circuit according to a schematic that normally is applied to a solid-state circuit would yield a molecular circuit that performs in a manner somewhat analogous to the corresponding solid-state, micron-scale digital logic circuits.

Thus, molecular diodes and molecular wires are used below in this way to design a functionally complete set of logic devices. This includes AND gates, OR gates, and XOR gates (or NOT gates) from which any more complex binary digital function may be designed and constructed, at least in principle. (To understand why an XOR gate is useful in making a complete set of logic gates, note that an XOR gate is readily adapted to produce a NOT gate, as is explained below.)

7.5.1.1 Molecular AND and OR Gates Using Diode–Diode Logic

Circuits for AND and OR digital logic gates based upon diodes, so-called *diode–diode* logic structures, have been known for decades. Schematic diagrams for these two circuits are shown, respectively, in Figures 7.12(a) and 7.13(a).[106,107] The operation of these logic circuits is explained in Appendix B, Sections B.1 and B.2.

Part (b) of each of these figures shows a structure for a corresponding novel *molecular* implementation of each of these logic gates. It is noteworthy that these molecular logic gates each would measure only about 3 nm × 4 nm, which is approximately one million times smaller in area than the corresponding logic element fabricated on a semiconductor chip using transistor-based circuits. (See Appendix C for a detailed size comparison.)

In the molecular implementations of both logic gates, the doped polyphenylene-based rectifying diode structures proposed above and depicted schematically in Figure 7.10 are connected together via conductive polyphenylene-based Tour wires, respecting the geometric constraints imposed by the chemical bonding behavior of organic molecules. Insulating aliphatic chains, as described in connection with Figure 7.1(e), are used for the purely resistive elements of the circuit. In particular, the three-methylene chains at the lower right of the molecular structure diagrams for both the AND gate and the OR gate correspond to the large resistances R specified in the schematics immediately above them in Figures 7.12 and 7.13, respectively. As is explained further in Appendix B, these large resistances serve to reduce power dissipation and to maintain a distinct output voltage signal at C when the inputs at A and B cause the diodes in either logic gate to be forward biased so that a current flows through them.

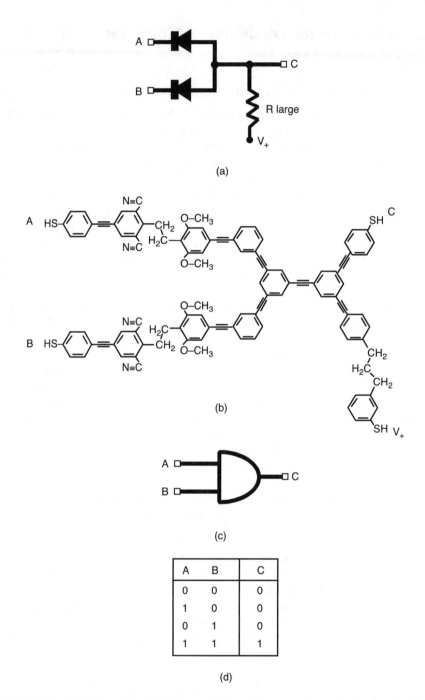

FIGURE 7.12 Diode–diode-type molecular electronic AND gate incorporating polyphenylene-based molecular rectifying diodes embedded in Tour wires.

The primary structural difference between the AND gate in Figure 7.12(b) and the OR gate in Figure 7.13(b) is that the orientation of the molecular diodes is reversed. Figures 7.12(d) and 7.13(d) show the truth tables of the inputs and outputs for each gate.

Similar diode–diode logic gates might be built from molecules like that shown in Figure 7.4(a), which recently were demonstrated by Metzger as having the properties of rectifying diodes. Metzger's experiment was conducted using a Langmuir-Blodgett film of such molecules, as sketched in part (b) of the same figure.

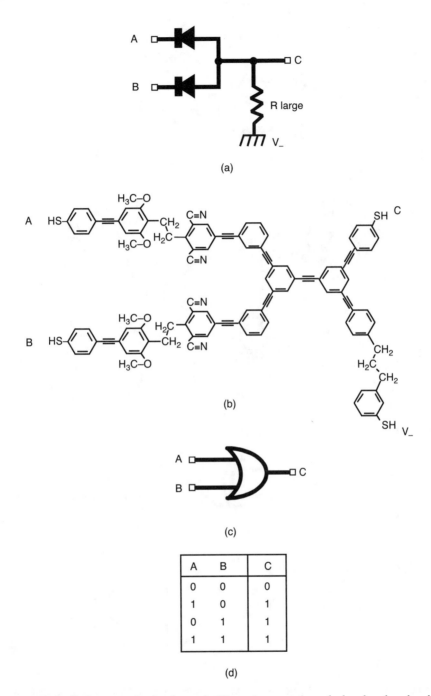

FIGURE 7.13 Diode–diode-type molecular electronic OR gate incorporating polyphenylene-based molecular rectifying diodes embedded in Tour wires.

Metzger et al. are attempting to demonstrate an AND gate and an OR gate based upon the present authors' suggestion of diode–diode logic,[108] as shown schematically in Figures 7.12(a) and 7.13(a).[106,107]

7.5.1.2 Molecular XOR Gate Using Molecular RTDs and Molecular Rectifying Diodes

AND gates and OR gates alone are not sufficient to permit the design of more complex digital functions. A NOT gate is required to complete the diode-based family of logic gates. However,

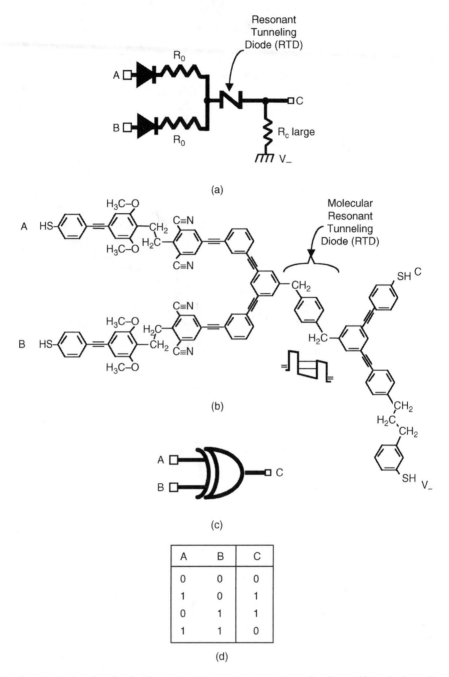

FIGURE 7.14 Diode-based molecular electronic XOR gate incorporating molecular rectifying diodes and molecular resonant tunneling diodes.

rectifying diodes alone are not sufficient to build a NOT gate.[107] To make a NOT gate with diodes, and thereby make it possible to build any higher function, we also must employ *resonant tunneling* diodes in the logic circuits.

7.5.1.2.1 Structure for the Molecular XOR Gate
A schematic diagram for a diode-based XOR gate is shown in Figure 7.14(a). This circuit schematic is due to Mathews and Sollner of the MIT Lincoln Laboratory, who employed it for the purpose of

developing diode-based solid-state logic circuits.[109] In the circuit schematic, the device symbol that resembles an "N" or a "Z" lying on its side represents an RTD.

Our proposed molecular implementation of this diode-based XOR circuit is shown in Figure 7.14(b). The structure, which would have dimensions of approximately 5 nm x 5 nm, is built using one of the Reed–Tour molecular RTDs and two of the polyphenylene-based molecular rectifying diodes proposed above. All three of these switching devices are built utilizing a backbone of polyphenylene-based Tour wires. The molecular circuit for the XOR gate is similar to that for the OR gate shown in Figure 7.13(b), except for the insertion of the molecular RTD.

This extra inserted element makes the output from the XOR gate different from the OR gate in the case that both inputs are binary 1s. The truth table for the XOR circuit is shown in Figure 7.14(d).

7.5.1.2.2 Overview of the Operation of the Molecular XOR Gate

Stated simply, the diode-based XOR gate operates like the diode-based OR gate except in the case that both of the XOR gate's inputs A and B are "1" — i.e., high voltages at both inputs. Those inputs put the operating point of the RTD into the valley region of the I-V curve shown schematically in Figure 7.7(b). This shuts off the current flowing through the RTD and makes the voltage low or "0" at the output C of the XOR gate. The detailed principles of operation for the XOR gate are explained in Appendix B of this chapter, Section B.3.

The addition of the molecular XOR gate to the set including the molecular AND and OR gates produces a functionally complete set, which can be made logically equivalent to the complete set AND, OR, and NOT. This is because an XOR gate can be adapted readily to produce a NOT gate simply by fixing one of the input terminals of the XOR gate to binary 1, as may be seen from the truth table in Figure 7.14(c). However, an XOR gate is better than just a simple NOT gate, because the XOR makes it possible to construct some higher functions in a particularly simple manner, as is shown below.

7.5.2 Molecular Electronic Half Adder

Given the complete set of molecular logic gates introduced above and displayed in Figures 7.12 through 7.14, the well-established principles of combinational logic[110] suggest designs that bond together several molecular logic gates to make larger molecular structures that implement still higher binary digital functions. For example, by combining the structures for the molecular AND gate and the molecular XOR gate given in Figures 7.12 and 7.14, respectively, one may build a structure for a molecular electronic half adder, as shown in Figure 7.15.

The well-known combinational logic circuit for a binary half adder is shown in Figure 7.15(a).[110] The novel design for the molecular structure corresponding to that combinational logic diagram is displayed in Figure 7.15(c). It arises from the substitution of the molecular structures proposed above for the component AND and XOR gates, the use of a framework of Tour wires, and an accounting for the geometry and steric constraints imposed by the bonding and shapes of the organic molecules. The resulting structure should have dimensions of about 10 nm x 10 nm, and it should behave in accordance with the truth tables shown in Figure 7.15(b). That is, the structure in Figure 7.15(c) is a molecule that should add two binary numbers electronically when the currents and voltages representing the addends are passed through it.

In Figure 7.15(c), A and B represent the one-bit binary inputs to the adder, while S and C represent the one-bit outputs, the sum and the carry bits, respectively. Currents introduced into the half adder structure via either of the input leads A or B on the left are divided to pass into both of the function's component molecular logic gates.

As is suggested by part (b) of Figure 7.15, the logic operation performed by the XOR gate forms the sum of two bits and outputs it at lead S, with the "excluded" XOR operation on a 1 input to the adder at A and a 1 input at B, representing the arithmetic binary sum 1+1 = 0. Simultaneously, the AND gate component of the half adder can form the carry bit from the same two input 1 bits, and it provides this result as an output at C.

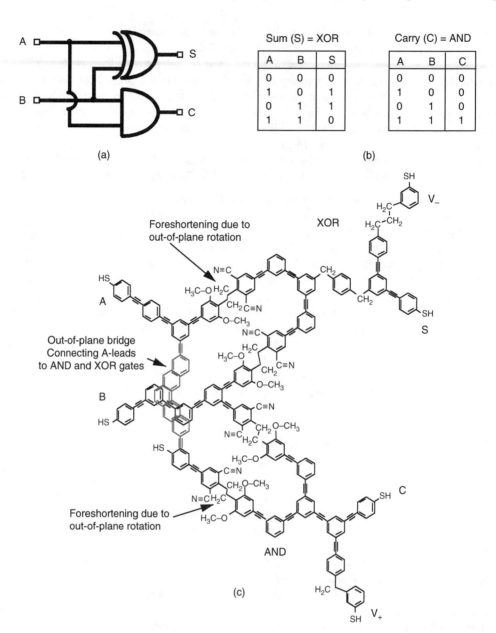

Sum (S) = XOR		
A	B	S
0	0	0
1	0	1
0	1	1
1	1	0

Carry (C) = AND		
A	B	C
0	0	0
1	0	0
0	1	0
1	1	1

(a)   (b)

(c)

FIGURE 7.15   Design for a molecular electronic half adder built from two diode-based molecular logic gates.

The Tour-type molecular wire that constitutes the adder's input lead B branches in the plane of the molecule immediately to the right of B to connect to the lower input lead of the XOR gate and the upper input lead of the AND gate. The input signal current through lead B would be split accordingly between the two component gates, just as is suggested by the more schematic combinational logic diagram in Figure 7.15(a).

The other input signal, which passes through the molecular half adder's lead A, similarly would be split between both the AND gate and the XOR gate. The Tour wire that begins on the left at A in Figure 7.15(c) connects directly to the upper lead of the XOR gate in the plane of the molecule. However, the out-of-plane, linked-ring, arc-like aromatic molecular structure is then necessary to pass over the in-plane molecular wire for lead B and then connect lead A to the lower lead of the AND gate. This out-of-plane molecular connector corresponds to the arc in the schematic in Figure 7.15(a), which also is used to pass over input lead B.

Thus, the very small input through each molecular input lead is immediately split and therefore halved. This should not be a major problem, though, for small diode-based molecular circuits such as the half-adder. For one thing, the half-adder molecule also recombines the split signals at the right of the structure, so signal loss should not be overwhelming. Some amount of signal strength almost certainly would be dissipated, though, into the ground structure labeled V– at the top right of Figure 7.15. The three-methylene aliphatic chain shown embedded in the output lead to the ground is intended to form a resistor that will make this dissipation small.

Still, in the absence of three-terminal molecular electronic devices (i.e., molecular transistors[119]) to effect signal restoration, branched diode-based molecular electronic circuits probably will have to remain small in order to function with a satisfactory signal-to-noise ratio. This point is suggested by Figure 7.16 and in the next section, where a larger molecular circuit structure, a molecular electronic full adder, is shown and discussed.

7.5.3 Molecular Electronic Full Adder

Extending the combinational design process begun above yields designs for molecular electronic structures that perform still higher functions. Combinational logic incorporating two half adders like that shown in the preceding section, plus an OR gate, is sufficient to produce the molecular electronic full adder shown in Figure 7.16. Figure 7.16(a) shows the well-known combinational logic schematic for a full adder based upon two half adders,[110] while Figure 7.16(b) shows the novel design proposed here for the corresponding molecule based upon molecular electronic diodes. This full adder molecule would have dimensions of approximately 25 nm ×25 nm.

The full adder circuit adds the carry bit C_{IN} from a previous addition to the one-bit sum S_1 of the addition of two new bits A and B. The three inputs C_{IN}, A, and B are shown at the left of Figures 7.16(a) and 7.16(b). The results shown at the right of both figures are the one-bit sum S, a carry bit C_2 associated with that sum, as well as a third bit C_{OUT} that is the result of the logical OR operation used to combine carry bits C_1 and C_2 from each of the two half-additions.

To interpret in detail the function of the two diode-based molecular half adder circuits and the diode-based molecular OR gate when they are combined to make the full adder circuit, one needs to be conscious of the relatively high resistance of molecular diodes, even when they are forward biased or "on." Otherwise, the explanation of the function of the full adder in Figure 7.16(b) follows directly from the discussion in the preceding section of its major subcomponent, the half adder, although some patience is required to absorb all the details.

Even a glance, however, at the molecular structure proposed in Figure 7.16(b) to perform this full addition reveals the rapid growth in the size and complexity of the molecular structure that is required merely to perform relatively simple arithmetic on numbers that contain only a very few bits. The implications of this observation are discussed further below in Section 7.6.1.5.

7.6 Discussion

The approach to conductive molecular electronic digital circuit design in this work makes use of and builds upon designs developed for and widely employed in familiar macroscopic and micron-scale electronic digital systems. Above, it also is illustrated how familiar, well-tested combinational principles might be employed to build up specific larger digital circuits from the smaller logic modules. Thus, an approach to larger scale computer architectures that employs such combinational principles is implicit in the explicit designs and molecular structures displayed and explained above in the small modular circuits for logic gates and functions.

It follows that this work raises a number of issues at both (1) the design level and (2) the architectural level. Issues at both levels are considered separately below.

7.6.1 Further Design Challenges for Molecular Electronic Circuits

Several issues that complicate the design of monomolecular conductive logic circuitry are

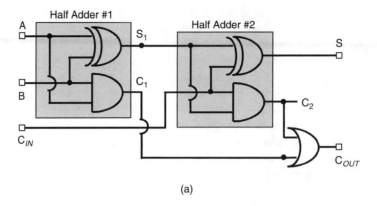

(a)

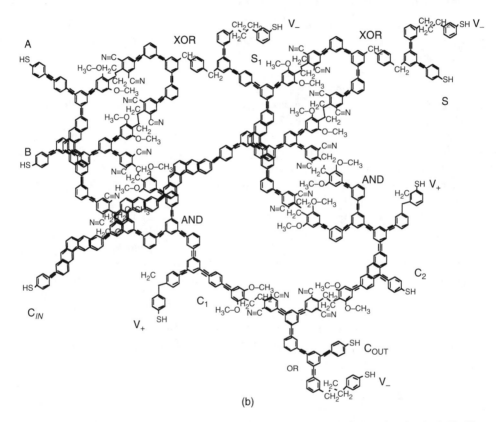

(b)

FIGURE 7.16 Design for a molecular electronic full adder built from two diode-based molecular half adders and a diode-based molecular OR gate.

1. The problem of combining devices on the molecular scale without altering their individual electrical responses as a result of quantum mechanical effects and other close-range interactions
2. The several different mechanisms that can produce conductance through the molecule
3. The nonlinear I–V behavior of individual molecular devices
4. The dissipation of electron energy into vibrational and other modes of motion in the molecule
5. The difficulty of achieving gain in extended circuits
6. The limitations on the operating speed of a molecular electronic computer

The discussion above of proposed logic circuits neglected these details in order to focus on the primary point of this chapter: building elementary molecular electronic logic circuits may require simply assembling a few previously demonstrated individual devices. However, to implement the resultant designs proposed here may require special attention to the specific issues enumerated above.

7.6.1.1 Combining Individual Devices

To simplify the initial design of molecular-scale logic functions, we have proposed that existing individual elementary devices — molecular switches and wires — might be bonded together to create circuits with more complex functionality, much in the same way as is done in conventional solid-state circuits. Solid-state circuit design exploits the fact that the electrical properties of individual devices do not change significantly when they are linked together in a complex circuit. Thus, the behavior of the whole macroscopic or micron-scale circuit is defined simply by the combination of the electrical properties of the isolated individual components used in the circuit.

Unlike bulk solid-state electronics, however, when small sections of molecular-scale wires and switches are combined into larger molecular circuits, these molecular components probably will not behave in circuits in the same way they do in isolation. Instead, strong Coulombic effects at short range and quantum wave interference among the electrons in all the components will alter the characteristic properties of each device. The potential impact of such quantum effects was suggested above in Section 7.3. Quantum wave interference manifests itself most obviously in the fact that the number of one-particle quantum levels in the combined molecular circuit is the sum of the number of levels from the two or more separate molecular devices. This change is likely to increase the relatively sparse density of states around the molecular Fermi level — especially in the low-lying conduction manifold of molecular orbitals. There is a similar summing of one-particle states when two solid-state circuit elements are combined, but the density of states is already so large relative to the number of electrons that the impact is not likely to be as noticeable as one might expect in a conductive molecular assembly.

However, the simple summation of the number of levels contributed by each device does not fully characterize the combined circuit either. When the energy levels from each device mix, the magnitude of the coupling between the levels will determine whether the electron density associated with particular orbitals remains localized primarily in one region of the entire molecule. Also, redistribution of the electron densities in the molecule can alter the Fermi level relative to the center of the HOMO–LUMO bandgap. This is somewhat analogous to the redistribution of electrons in p-n doped semiconductor junctions.

Recently published theoretical work by Magoga and Joachim[96,97] and by Yaliraki, Ratner, et al.[98,99] emphasizes the possibility of strong influences upon the overall function of a molecular electronic circuit due to the mutual quantum interference of nearby devices. This is being explored elsewhere analytically by one of the present authors.[94] In addition, modeling software is now being developed to take account of quantum mechanical issues in order to predict the responses of entire molecular electronic circuits.[95]

7.6.1.2 Mechanisms of Conductance

Electrons can flow through molecules by several possible mechanisms, as discussed in great detail by Ratner and others.[21–27] Different mechanisms with different current vs. voltage responses may dominate in larger molecular structures more than in smaller ones. More than one type of mechanism may be manifested in extended structures.

In the simplest monomolecular electronic systems, the electron transport mechanisms of interest are those governing the flow of electrons through a molecular wire attached at both ends to a metal electrode. Experiments to date suggest that transmission of electrons in polyphenylene-based molecular wires can occur via a resonance pathway — i.e., coherent transport though a small number of one-particle states aligned in energy.[5,6,12] For very short molecules between electrodes, the transport mechanism more readily observed is electron tunneling from one metal electrode to another modulated by the molecular orbitals available on the molecule.

This nonresonant *superexchange* mechanism has the advantage that dissipation is low, but the rate of transport falls off exponentially with distance as in standard tunneling.[111] Hence, the molecule does not behave like the familiar ohmic wire (i.e., one governed by Ohm's Law). This is because the transported electrons reside primarily in the continuum orbitals of the two metal contacts. In superexchange, these orbitals do not mix directly (in first order of perturbation theory) with the orbitals of the molecule. The molecule only influences the electron transport indirectly (in second order). For conductance to occur by this mechanism, the Fermi level of the metal contact must fall between the energies of the HOMO and LUMO of the molecule.[24,25]

In the absence of resonant transport, conductance in extended molecular systems must be based on incoherent mechanisms, with the electrons hopping across the molecule along a chain of localized orbitals. This incoherent electron transport, which has been both predicted and observed, much improves the distance dependence of the conductance.[20] The resistance in an ohmic-type wire with incoherent transport increases only linearly as the wire gets longer, so that electron transport falls off only linearly with the inverse of the length. Further, in long, nonrigid molecules, inelastic electron scattering off the nuclear framework of the molecule can inhibit coherent transport. Such scattering could be a factor in long polyphenylene-based systems with inserted, flexible alkyl chain potential barriers. That would make it desirable to find ways to enhance the incoherent transport mechanisms.

In large polyphenylene-based molecular circuits, both mechanisms of conductance discussed above may be important. In the active regions of the circuit such as the diodes, experiments indicate that it should be possible to take advantage of coherent resonant transport. However, between switching devices, molecular structures may have to be optimized to ensure that the electrons can move via incoherent transport.

7.6.1.3   Nonlinear I–V Behavior

In bulk solid-state electronics, the current–voltage (I–V) behavior of wires and resistors varies linearly according to Ohm's law, $V = IR$. As an initial approximation, this rule has been used in the design of the proposed circuits presented here. Actually, though, the I–V behavior of a molecular wire is observed and calculated to be nonlinear.[27,28] Still, for small variations in the voltage, Ohm's law is likely to be a good approximation. The I–V behavior for molecular electronic circuits operating by incoherent transport ought to be nearly linear (ohmic). However, novel devices may also take advantage of the nonlinearity inherent to small molecules, allowing new architectures for computation that have no meaningful analogs in bulk-effect solid-state electronics.

7.6.1.4   Energy Dissipation

Energy dissipation will diminish the effectiveness of electron transport and signal transmission through a molecular structure. Also, dissipation results in heating that could have a particularly negative cumulative impact in an ultra-dense molecular electronic system with many closely spaced wires, switches, gates, and functions. As electrons move through a molecule, some of their energy can be transferred or dissipated to the motions of other electrons and to the motions of the nuclei in the molecule, such as the internal molecular vibrations and rotations. The amount of energy transferred is dependent on how strongly the electronic energy levels of the molecule couple (or interact) with the vibrational modes of the molecule. The molecular motion can, in turn, relax by dissipating energy or heat to the substrate or other surroundings.

The amount of energy dissipation in the structure is closely tied to the mechanism of conductance. For example, in the superexchange mechanism discussed above, where electrons are not directly coupled to the wire, the dissipation within the molecule should be small because the electrons are not thought to interact strongly enough with the molecule to transfer much energy. In larger molecules, where incoherent transport mechanisms dominate and the electrons are localized in orbitals on the molecule itself, dissipation of energy to the molecular framework should be significantly larger.

The loss of energy to the lattice vibrations and the surroundings decreases the signal strength and, in the extreme, could break bonds in the structure, destroying the device. Because it is likely that extended molecular circuits will utilize at least some incoherent transport mechanisms, it will be necessary to implement circuit designs with robust structures or a built-in means of transferring the molecular motion out of the circuit molecule itself.

7.6.1.5 Necessity for Gain in Molecular Electronic Circuits

Generation of the design for the molecular electronic full adder shown in Figure 7.16(b) is a useful exercise revealing the size of molecular structures that will be required to perform useful computation. Such large molecular structures made from narrow wires and devices with relatively high resistances will likely require some form of power gain to achieve signal restoration and to compensate for the dissipative losses in the signal.

In solid-state logic, it is possible to achieve power gain using only diodes. Rectifying diodes cannot achieve this. However, RTDs organized in a circuit known as a Goto pair may be used for this purpose.[112–114] This also might be an option for molecular-scale electronic circuits. If so, molecular diode switches like those described and employed in this work might be sufficient to build complex molecular electronic computer logic, without the absolute requirement for molecular transistors to permit operation of the extended circuitry. In the near term, this might have some particular practical advantages. Presently it is much easier to make electrical contact with two-terminal molecular switches than with molecular switches having three or more terminals.

In conventional, solid-state digital electronics, it is more common, of course, to achieve power gain and signal restoration by employing three-terminal devices — i.e., transistors. This is because it is much easier to accomplish these goals with transistors. Transistorized solid-state circuits tend to be less sensitive and contain fewer switching devices than diode–diode circuits. This seems likely to be the case for molecular-scale electronic circuits, too.

Thus, it will be important to develop three-terminal molecular devices with power gain — i.e., a molecular electronic transistor. It will be important, also, to refine methods for making electrical contact with large numbers of densely spaced three-terminal molecular devices. Such advances would make it possible to achieve signal isolation, maintain a large signal-to-noise ratio, and achieve fan-out in molecular circuits. In addition, solid-state logic gates based upon three-terminal devices usually can achieve much more reliable latching than can diode-based logic.[115,116]

Several groups of investigators have suggested how such molecular electronic transistors can be made with hybrid structures in which molecules are adsorbed to solids.[64,117,118] However, few, if any, of the molecular-scale switches proposed elsewhere involve purely molecular structures.

Building upon the principles and structures introduced in this chapter, subsequent work will show molecular structures for a three-terminal molecular electronic amplifying switch, as well as designs for logic gates and functions based upon this molecular-scale three-terminal device.[119] Nonetheless, the authors anticipate that molecular diode switches — molecular two-terminal devices — and diode-based logic are likely to be important components of future densely integrated molecular electronic digital circuits, demanding many fewer interconnects than three-terminal devices.

7.6.1.6 Potentially Slow Speeds

The speed at which operations can be performed by a molecular electronic circuit is closely related to the issue of energy dissipation in the system. Strong dissipative couplings could decrease the signal-to-noise ratio dramatically. Such reduction in signal strength would require a greater total charge flow to ensure the appropriate reading of a bit, thereby requiring more time. Moreover, an examination of Table 7.1 shows that the currents measured in electrons per second that pass through each molecule, while amazingly large, nonetheless limit the speed of signal transmission to a range between approximately 10 kHz and 1 GHz. These upper limits assume that only one electron per bit is required to transmit a signal reliably. Obviously, if more than one electron were required per bit of signal, as would certainly be the case in the presence of dissipation and noise, the number of electrons per bit necessarily would rise, and the speed of the computer would decrease. If as many as 10 or 100 electrons were required to transmit a single bit, as is possible, molecular computer circuits like those described in this chapter could be no faster than conventional microelectronic computers that presently operate in the several hundred MHz range. From the current-based arguments alone, molecular electronic computers might even be much slower.

Also, it is likely that extended molecular circuits, including internal molecular interconnections and external metal contacts, will have a significant capacitance. The corresponding RC time constants for the components of the molecular circuit could be large, and that would also slow the speed at which the system could be clocked.

On the other hand, the synthesis of a very small (micron-scale to millimeter-scale) and very dense (~10,000 devices per sq. micron) molecular electronic computer ought to be possible, in principle. Such machines might even be made three-dimensional to increase the density and to shorten interdevice communication delay. At that scale and density, such molecular electronic computers may not need to be very fast to be extremely useful.

The human brain is a massively parallel computer capable of performing 100 million MIPS (million instructions per second).[120,121] For comparison, a Pentium chip only performs at approximately 750 MIPS. Thus, it could require more than 100,000 Pentium chips wired in parallel in order to achieve the number of instructions per second carried out by the human brain. Many molecular electronic processors wired in parallel, each with only the computational power of a Pentium, would offer greater computational power per unit volume than is possible with the present state-of-the-art parallel processing. The area of a Pentium processor is approximately 1 cm^2. Using the same scaling ratio estimated for the molecular half adder in Appendix C (1 million times smaller in area or 1000 times smaller linearly), a molecular electronic equivalent of a Pentium would measure approximately 10 μm × 10 μm. Although each micron-scale molecular electronic processor might operate more slowly, in only a two-dimensional layout, 100,000 molecular electronic Pentium-equivalent circuits would require only one tenth of the space required by an existing Pentium. Very small and relatively slow, but powerful, distributed computing could be realized using molecular circuitry.

The design challenges discussed above will require serious consideration in implementing the logic structures proposed here or variants of them. Overcoming these challenges is likely to require further innovations.

7.6.2 Architectural Approaches and Architectural Issues for Molecular Electronics

Present architectural approaches for using small molecules to make electronic digital logic structures fall into two broad categories:

- Those that rely on small electrical currents to transfer and to process information
- Those that rely on the deformation of the molecular electronic charge density to transfer and process information

Examples of the latter category include both the TKS approach[51] discussed in Section 7.5.1 as well as other possible molecular implementations of quantum cellular automata.[18,105] The present work obviously falls in the former category that relies on electrical currents.

In either architectural approach, however, one must overcome a formidable array of architectural issues. Among the most important of these are

1. Designing logic gates, functions, and extended circuitry using molecules
2. Making reliable, uniform electrical contacts with molecules
3. For current-based architectures: minimizing resistance in narrow molecular wires; for charge-based architectures: avoiding trapping in metastable states
4. Interconnect issues, geometric and dynamic: as discussed above, the dynamic interconnect problem involves the slowdown of circuits due to high capacitances in tiny junctions, while the geometric problem involves the problem of laying out the many, many closely spaced junctions required for dense logic in a manner such that they do not conflict
5. Assembly strategies for extended systems of smaller molecular logic units
6. Fault tolerance: evolving strategies for *mitigating* effects of errors and structural imperfections
7. Dissipation: reducing heating in extended ultra-dense circuitry or charge receptacles and also the cooling of such tiny systems

In the preceding sections of this chapter, only the first architectural issue has been addressed in detail. However, the approach proposed here for designing diode-based electronic logic gates, functions, and extended circuitry using molecules also may assist in reducing the number of interconnects required, one of the problems specified in the fourth issue listed above.

Elements of an approach to addressing the fifth issue are outlined below in Sections 7.7.2 and 7.7.3. Investigators elsewhere also are working toward addressing at least Issues 2, 5, and 6 for molecular-scale computer circuitry.[43,44,122,123] In subsequent papers, the present authors and their collaborators will discuss strategies for addressing some of the other architectural issues listed above, systematically working toward the design and implementation of electronic digital computers integrated on the molecular scale.[133,134]

7.7 Summary and Conclusions

7.7.1 The Next Logical Step in Molecular Electronics

This chapter outlines an architectural approach for building molecular electronic digital logic structures. Designs for a complete set of molecular electronic logic gates and molecular adders are given. These molecular electronic digital circuit designs require only simple arrangements of a few molecular electronic devices — switches and wires — variants of which already have been demonstrated individually by other investigators. Thus, it is likely that some of these molecular circuits can be fabricated and tested experimentally soon as the next logical step in the development of molecular electronics.

7.7.2 Environment and Interface for Molecular Electronic Circuits

In order to test and to operate such molecular circuits, it will be necessary to provide an electrically benign environment for them, to support them structurally, and to make contact with them electrically. However, judging from the molecular electronics experiments to date, circuits based upon conductive polyphenylene and carbon nanotube molecules should not need to be held in a vacuum or cooled to cryogenic temperatures in order to operate. They should operate in open air at room temperature.

Generally speaking, though, it would be desirable to support the molecular circuits above a relatively nonconductive dielectric layer. This layer would be penetrated only by the conductive structures that will serve as their contacts, or else their conductive contacts also would lie upon and be supported by this layer.

Supporting molecular electronic systems on very thin, nonconductive layers would offer the obvious advantage of likely reducing the occurrence of electrical noise and errors in the system. In addition, this method of support might make it possible to stack the very thin layers of molecular circuitry to produce an ordered three-dimensional electronic computer processor or memory.

Several approaches to making contact with molecular circuits supported above a dielectric layer are suggested by previous experimental work with molecular-scale electronic devices, as discussed below.

7.7.2.1 Grid of Metal Nanowires

Williams, Kuekes, Heath, and their collaborators at Hewlett-Packard Corporation in Palo Alto and at the University of California at Los Angeles have demonstrated a method for assembling closely spaced parallel lines of conductive metal wires, each of which is only approximately one nanometer wide.[124] They envision that these can be used to make a grid of addressable contacts for molecular devices by pressing together two layers of perpendicular nanowires, with an insulating molecular monolayer or pad at each of the points of intersection.[125,134]

The different terminal ends of a molecular device or molecular circuit would be adsorbed via their molecular alligator clips (e.g., the –SH group) to different nanowires that bound each square cell in the grid. Because four nanowires bound each cell, which could be as small as only 5 nm across, a number of molecular devices and circuits with up to four terminal ends might be supported, contacted electrically, and addressed in such an extended, grid-like, conductive nanostructure.

7.7.2.2   Molecular Electronic *Breadboard*

In this approach a regularly spaced array of nanofabricated or self-assembled, conductive metal *posts* would simultaneously make electrical contact with the molecular circuits and provide support for them. If the posts were made of gold, then the thiol molecular alligator clips (–SH groups) on the terminal ends of the molecular circuit structures would bind to the posts. The posts might be self-assembled quantum dots or metal nanocrystallites. An interesting approach for using gold nanocrystallites to make an array of conductive posts has been pioneered by an interdisciplinary group at Purdue University.[3]

In the Purdue work, the posts were tightly spaced over a conductive substrate, but refinements of the Purdue work might allow a less dense regular array of such structures to be self-assembled over the conductive nanopatterned substrate. The pattern on the substrate would allow the posts to be selectively addressed — turned on and off. A nonconductive molecular monolayer might be self-assembled over the exposed regions of the substrate between the posts. Finally, the molecular circuits could be arranged on top of the dielectric layer and connected to the conductive posts where they penetrated above the dielectric layer.

7.7.2.3   Carbon Nanotubes

While carbon nanotubes are not very reactive on their sidewalls, it is relatively easy to substitute other organic molecules on their open or closed ends. Thus, it might be possible to use these very conductive, nanometer-scale structures as wires and contacts to address the terminals of molecular electronic devices that are made out of smaller molecules. For example, one might use several carbon nanotubes arranged on an insulating substrate to address and to support a polyphenylene-based molecular electronic adder, of the type described above in Section 7.5.2. A possible nanostructure of this type is sketched in Figure 7.17.

Such a molecular circuit structure would have the advantage that it uses each type of molecule in a role for which it is most suited. Small, finely articulated structures are made out of aromatic organic molecules, which are more reactive than carbon nanotubes. Current is carried over long distances by carbon nanotubes, which are more conductive than small, polyphenylene-based molecules.

In a similar spirit, if controlled sidewall substitution of nanotubes were to become easier and more routine, it might even be possible to affix small conductive molecules to a grid-like carbon nanotube contact structure analogous to the grid of metal nanowires envisioned above.

7.7.2.4   Nanopores

Nanopores, such as have been fabricated by Reed and his collaborators,[7] also are an option for supporting and making contact with molecular electronic devices and circuits. In such a nanopore, the conductive contacts are at the top and the bottom of the pore, which means that the molecules are arranged vertically rather than tiled horizontally on a substrate. This increases the potential packing density of molecular devices in a horizontal layer, but only two-terminal devices can be attached to and addressed by the two contacts of an individual nanopore.

It might be possible, though, to effect hybrid molecular-solid diode–diode logic circuits by building electrically connected ultra-dense circuits and systems of nanopores that contain molecular diode switches. Such circuitry might even be embedded on top of microelectronic devices in order to build, for example, a dense, low-power preprocessor or cache memory right on top of a microelectronic processor.

7.7.3   Circuits One Million Times Denser Than Microelectronics

As discussed in Section 7.7.6 and immediately above, much remains to be learned about the mechanisms of conductance for small molecular wires, as well as about the means for manipulating, bonding, and ordering them in extended circuit-like structures.[37] However, when such molecular electronic digital logic circuits are fabricated and tested, they are likely to be on the order of one million times smaller in area than the corresponding conventional silicon semiconductor logic circuits. For example,

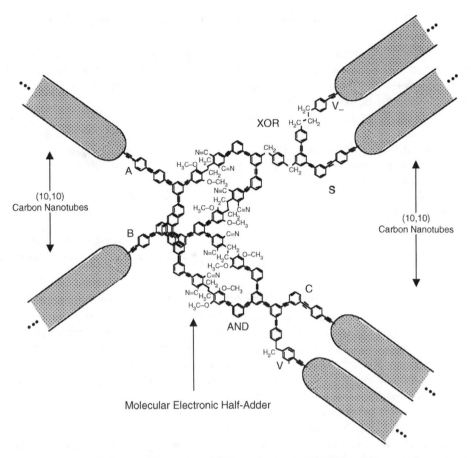

FIGURE 7.17 Conceptual diagram of a molecular electronic circuit in which carbon nanotube molecules are employed to make electrical contact with and support a polyphenylene-based molecular electronic half adder.

the OR gate displayed in Figure 7.13 is only about 3 nm by 4 nm or, conservatively stated, it is on the order of 10 nm on a side. As detailed in Appendix C, this may be compared with the same type of logic gate made from micron-scale solid-state silicon semiconductor transistorized logic. A microelectronic OR gate occupies (approximately) a square on the order of 10 micrometers or 10,000 nm on a side. The size advantages of the molecular electronic logic circuits laid out in the preceding section therefore are on the order of a factor of 1000 in linear dimensions and a factor of 1 million in area. On that small scale, they constitute a lower limit on the miniaturization of conductive electronic logic. For this reason, at least, intensive study and experiments upon such logic designs are essential.

7.7.3.1 Gate and Function-Level Integration to Produce Molecular-Scale Electronic Computers

Using one of the interface strategies enumerated and discussed in Section 7.7.2, small monomolecular logic gates and functions like those described in Section 7.5, Figures 7.12 through 7.15, might someday be linked together to build complex extended molecular-scale electronic computer circuitry. This could be described as molecular gate-level and molecular function-level integration. Even working in only two dimensions, given the small, approximately 10 nm ×10 nm size of those molecular building blocks, it can be estimated that as many as 10,000 logic gates would fit in an area of only one square micron. Thus, molecular integration at this level might fit the circuitry for an entire computer in the area that now is occupied on a microchip by only one micron-scale, solid-state transistor.

7.7.3.2 Device-Level Integration for More Rapid Prototyping of Molecular-Scale Electronic Logic

Alternatively, with only a little sacrifice in size and density, one might use interface strategies like those discussed in Section 7.7.2 to connect together several separate molecular diode switches in accordance with the diode–diode logic schematics shown above. This molecular device-level integration might afford a strategy for the more rapid fabrication and testing of prototype molecular diode–diode logic gates, without having to take on immediately the challenge of synthesizing them in their most compact and most densely integrated, monomolecular structure. Only two or three molecular diode switches need be connected together in this way to demonstrate a molecular-scale electronic logic gate. Thus, molecular electronic integration of this sort is just a small step further along a path that already is being traveled successfully by some experimentalists.[14]

7.7.4   Future Prospects

The physical characteristics of the molecular electronic devices and logic structures considered in detail in this work suggest future applications and trends in advanced computer electronics as it evolves toward integration on the molecular scale.

7.7.4.1 Key Advantages and Key Applications for Proposed Molecular Electronic Logic

Because of their extremely small sizes, molecular electronic digital logic structures such as those suggested here might permit a great increase in the density of digital electronics, well beyond the limits of conventional silicon-based circuitry. In that case, such ultra-high-density molecular circuitry could be useful in increasing the power of conventionally sized computer chips, a few centimeters on a side. This could make available a whole new class of desktop supercomputers and open new vistas for computationally intensive applications.

However, ultra-high-density molecular circuitry might have an even greater impact by making it possible to shrink dramatically the area and volume occupied by a computer with moderate computational power comparable to a conventional Intel Pentium chip or an even older, simpler microprocessor chip. A whole new domain of applications might be foreseen for such a Pentium on a pinhead. Such computers necessarily would draw less electrical power than conventional computers of comparable computational capability, and they would open the possibility of much smaller, lighter, and smarter portable devices. They might even be used as controllers — "brains" — for micro-sensors and the millimeter-scale robots that are just now coming into development.[126]

7.7.4.2 Matter As Software

The small molecular logic gates for which designs are proposed in this work, and other similar structures, embody the potential for a kind of molecular "macro" language. Each molecular logic gate might be bonded to a slightly larger, nonconductive, supportive molecular "handle" to permit the molecules to be manipulated and assembled quickly and easily, as needed, in any desired extended circuit structure or program. Entire complex programs still could be embodied in molecular-scale structures much smaller and just as flexible as any material structure in which conventional microcomputers store a single bit of information. Thus, matter — modular molecular logic structures — might become the basic nanoscopic elements of future software, eroding once and for all the technological and the economic distinctions between computer hardware and computer software.[127]

Acknowledgments

For their invaluable material assistance to this work in so generously sharing their ideas and their unpublished research results with us, the authors wish to express particular thanks to Prof. M. Reed of Yale University and Prof. J. Tour of Rice University, Dr. G. Sollner and Dr. R. H. Mathews of MIT-Lincoln Lab, as well as to Dr. A. Seabaugh of Notre Dame University and J. Soreff of the IBM Corporation, East Fishkill, NY. Several of our colleagues reviewed an earlier, September 1998 draft version of this chapter and made very helpful comments and suggestions. These have been incorporated into this final version and have improved it greatly. We are grateful to those reviewers, who are as follows: D. Allara, Penn State University; D. Goldhaber–Gordon, Harvard University; R. Lytel and D. Naegle, Sun Microsystems, Inc.; P. MacDougall, Middle Tennessee State University; R. Merkle, Xerox Palo Alto Research Center; M. Ratner, Northwestern University; J. Seminario, University of South Carolina; W. Tolles, Miniaturization Science and Technology; J. Tour, Rice University; G. Whitesides, Harvard University; and T. Bollinger, G. Tseng, and A. Wissner–Gross, The MITRE Corporation. Additional thanks in this regard are due to the referees of this final version of the chapter. D. Naegle of Sun Microsystems, Inc., graciously contributed the analysis and estimate of the size of a silicon CMOS half adder, which appears in Appendix 7.C. The authors wish to express their appreciation for his generous contribution of that effort. Thanks are due, as well, to many of our other colleagues and collaborators in the nanoelectronics community for their positive comments and their encouragement of the investigations upon which this chapter is based. Dr. G. Pomrenke, Dr. B. Gnade, and Dr. W. Warren of the Defense Advanced Research Projects Agency have been especially generous in this way, and we are grateful. In addition, we are grateful to the other members of the MITRE Nanosystems Group, especially G. Tseng, D. Moore, K. Wegener, and K.H. Hanson Wong, for their collaboration in the nanoelectronics research investigations which supported or led up to the one upon which this chapter is based. We also are indebted to many other colleagues at MITRE for their long-standing support of this research, notably D. Lehman, S. Huffman, E. Palo, C. Cook, W. Hutzler, and K. Pullen.

References

1. M.A. Reed, C. Zhou, C.J. Muller, T.P. Burgin, and J.M. Tour, Conductance of a molecular junction, *Science*, 278, 252–254, 1997.
2. L.A. Bumm et al., Are single molecular wires conducting? *Science*, 271, 1705–1707, 1996.
3. R.P. Andres et al., Coulomb staircase at room temperature in a self-assembled molecular nano-structure, *Science*, 272, 1323–1325, 1996.
4. S.J. Tans et al., Individual single-wall carbon nanotubes as quantum wires, *Nature*, 386, 474–477, 1997.
5. M.A. Reed, Electrical Properties of Molecular Devices, presented at 1997 DARPA ULTRA Program Review Conference, Santa Fe, NM, October, 1997.
6. M.A. Reed, Molecular-scale electronics, *Proc. IEEE*, 87, 652–658, 1999.
7. C. Zhou, M.R. Deshpande, M.A. Reed, and J.M. Tour, Nanoscale metal/self-assembled monolayer/metal heterostructures, *Appl. Phys. Lett.*, 71, 611–613, 1997.
8. C. Zhou, Atomic and Molecular Wires, Ph.D. dissertation, Yale University, 1999. [9]
9. R.M. Metzger et al., Unimolecular electrical rectification in hexadecylquinolinium tricyanoquin-odimethanide, *J. Am. Chem. Soc.*, 119, 10455–10466, 1997.
10. Dhirani, R. Zehner, P.-H. Lin, L R. Sita, and P. Guyot-Sionnest, Self-assembled molecular rectifiers, *J. Chem. Phys.*, 106, 5249–5253, 1997.
11. L. Ottaviano, et al., Rectifying behavior of silicon-phthalocyanine junctions investigated with scanning tunneling microscopy/spectroscopy, *J. Vac. Sci. Technol. A*, 15, 1014–1019, 1997.
12. M.A. Reed, Progress in Molecular-Scale Devices and Circuits, presented at 57th Annual IEEE Device Research Conference, Santa Barbara, CA, 28–30 June 1999.
13. J. Chen, M.A. Reed, A.M. Rawlett, and J.M. Tour, Large on–off ratios and negative differential resistance in a molecular electronic device, *Science*, 286, 1550–1552, 1999.
14. Z. Yao, H.W.C. Postma, L. Balents, and C. Dekker, Carbon nanotube intramolecular junctions, *Nature*, 402, 273–276, 1999.
15. C. Zhou, C.J. Muller, M.A. Reed, T.P. Burgin, and J.M. Tour, Mesoscopic phenomena studied with mechanically controllable break junctions at room temperature, in J. Jortner and M. Ratner (Eds.), *Molecular Electronics*, Blackwell Science, London, 1997.

16. J.C. Ellenbogen, A Brief Overview of Nanoelectronic Devices, presented at Proceedings of the 1998 Government Microelectronics Conference (GOMAC98), Arlington, VA, 13–16 March 1998. Available online at http://www.mitre.org/technology/nanotech/GOMAC98_article.html

17. D. Goldhaber–Gordon, M.S. Montemerlo, J.C. Love, G.J. Opiteck, and J.C. Ellenbogen, Overview of nanoelectronic devices, *Proc. IEEE*, 85, 521–540, 1997. Available online at http://www.mitre.org/technology/nanotech/IEEE_article.html

18. M. Montemerlo, J.C. Love, G.J. Opiteck, D.J. Goldhaber–Gordon, and J.C. Ellenbogen, *Technologies and Designs for Electronic Nanocomputers*, MITRE Technical Report No. 96W0000044, The MITRE Corporation, McLean, VA, 1996. Available online at http://www.mitre.org/technology/nanotech/review_article.html

19. The Nanoelectronics and Nanocomputing Home Page, The MITRE Corporation, McLean, VA 1996–1998. Available online at http://www.mitre.org/technology/nanotech. (This site provides References 13–15 and 82 as downloadable documents.)

20. W.B. Davis, W.A. Svec, M.A. Ratner, and M.R. Wasielewski, Molecular-wire behaviour in p-phenylenevinylene oligomers, *Nature*, 396, 60–63, 1998.

21. V. Mujica, M. Kemp, and M.A. Ratner, Electron conduction in molecular wires. I. A scattering formalism, *J. Chem. Phys.*, 101, 6849–6855, 1994.

22. V. Mujica, M. Kemp, and M.A. Ratner, Electron conduction in molecular wires. II. Application to scanning tunneling microscopy, *J. Chem. Phys.*, 101, 6856–6864, 1994.

23. V. Mujica, M. Kemp, A. Roitberg, and M. Ratner, Current–voltage characteristics of molecular wires: eigenvalue staircase, Coulomb blockade, and rectification, *J. Chem. Phys.*, 104, 7296–7305, 1996.

24. M.A. Ratner and J. Jortner, Molecular electronics: some directions, in J. Jortner and M. Ratner (Eds.), *Molecular Electronics*, Blackwell Science, London 1997, pp. 5–72.

25. M.A. Ratner et al., Molecular wires: charge transport, mechanisms, and control, in A. Aviram and M. Ratner (Eds.), *Molecular Electronics: Science and Technology*, a special issue of *Ann. N.Y. Acad. Sci.*, 852, 22–37, 1998.

26. M.P. Samanta, W. Tian, S. Datta, J.I. Henderson, and C.P. Kubiak, Electronic conduction through organic molecules, *Phys. Rev. B*, 53, 7626–7629, 1996.

27. S. Datta, *Electron Transport in Mesoscopic Systems*, Cambridge University Press, Cambridge, U.K, 1995.

28. S. Datta et al., Current–voltage characteristics of self-assembled monolayers by scanning-tunneling microscopy, *Phys. Rev. Lett.*, 79, 2530–2533, 1997.

29. D.A. Muller et al., The electronic structure at the atomic scale of ultrathin gate oxides, *Nature*, 399, 758–761, 1999.

30. P.A. Packan, Pushing the limits, *Science*, 285, 2079–2081, 1999.

31. J.R. Reimers and N.S. Hush, Electron transfer and energy transfer through bridged systems, III, *J. Photochem. Photobiol. A*, 82, 31–46, 1994.

32. L.E. Hall, J.R. Reimers, N.S. Hush, and K. Silverbrook, Formalism, analytical model, and *a priori* Green's function-based calculations of the current–voltage characteristics of molecular wires, *J. Chem. Phys.*, 112, 1510–1521, 2000.

33. E.G. Emberly and G. Kirczenow, Electrical conduction through a molecule, in A. Aviram and M. Ratner (Eds.), *Molecular Electronics: Science and Technology*, a special issue of *Ann. N.Y. Acad. Sci.*, 852, 54–67, 1998.

34. E.G. Emberly and G. Kirczenow, Theoretical study of electrical conduction through a molecule connected to metallic nanocontacts, *Phys. Rev. B.*, 58, 10911–10920, 1998.

35. E.G. Emberly and G. Kirczenow, Electrical conductance of molecular wires, *Nanotechnology*, 10, 285–291, 1999

36. S.T. Panteleides, M. DiVentra, and N.D. Lang, *Ab Initio* Simulation of Molecular Devices, presented at 1999 DARPA Molecular Electronics (Moletronics) Program Review Conference, Ashburn, VA, 8–9 July 1999.

37. G.M. Whitesides, private communications, October 1998–January 1999.
38. J.S. Schumm, D.L. Pearson, and J.M. Tour, Iterative divergent/convergent approach to linear conjugated oligomers by successive doubling of the molecular length: a rapid route to a 128 Å-long potential molecular wire, *Angew. Chem. Int. Ed. Engl.*, 33, 1360–1363, 1994.
39. J.M. Tour, R. Wu, and J.S. Schumm, Extended orthogonally fused conducting oligomers for molecular electronic devices, *J. Am. Chem. Soc.*, 113, 7064–7066, 1991.
40. P.W. Atkins, *Quanta: A Handbook of Concepts*, 2nd ed., Oxford University Press, Oxford, U.K., 1992.
41. P.W. Atkins, *Molecular Quantum Mechanics*, 3rd ed., Oxford University Press, Oxford, U.K., 1997.
42. R.T. Morrison and R.N. Boyd, *Organic Chemistry*, 2nd ed., Allyn and Bacon, Boston, 1966. Also see R.T. Morrison and R.N. Boyd, *Organic Chemistry*, 6th ed., Prentice-Hall, New York, 1992.
43. M.A. Reed, private communication, October, 1997.
44. J.M. Tour, private communication, December, 1998.
45. J.M. Tour, Chemical Synthesis of Molecular Electronic Devices, presented at 1997 DARPA ULTRA Review Conference, Santa Fe, NM, 26–31 October, 1997.
46. P.S. Weiss et al., Probing electronic properties of conjugated and saturated molecules in self-assembled monolayers, in A. Aviram and M. Ratner (Eds.), *Molecular Electronics: Science and Technology*, a special issue of *Ann. N.Y. Acad. Sci.*, 852, 145–168, 1998.
47. M. Riordan and L. Hoddeson, *Crystal Fire: The Birth of the Information Age*, Norton, New York, 1997.
48. L. Edwards–Shea, *The Essence of Solid-State Electronics*, Prentice-Hall-Europe, Hertfordshire, U.K., 1996.
49. I. Aviram and M.A. Ratner, Molecular rectifiers, *Chem. Phys. Lett.*, 29, 277–283, 1974.
50. A.S. Martin, J.R. Sambles, and G.J. Ashwell, Molecular rectifier, *Phys. Rev. Lett.*, 70, 218–221, 1993.
51. J.M. Tour, M. Kozaki, and J.M. Seminario, Molecular scale electronics: a synthetic/computational approach to digital computing, *J. Am. Chem. Soc.*, 120, 8486–8493, 1998.
52. J.M. Seminario, A.G. Zacharias, and J.M. Tour, Molecular alligator clips for single molecule electronics. Studies of group 16 and isonitriles interfaced with Au contacts, *J. Am. Chem. Soc.*, 121, 411–416, 1998.
53. D.K. Ferry, *Quantum Mechanics: An Introduction for Device Physicists and Electrical Engineers*, IOP Publishing Ltd., London, U.K., 1995.
54. B.I. Yakobson and R.E. Smalley, Fullerene nanotubes: C(1,000,000) and beyond, *Am. Sci.*, 85, 324–337, 1997.
55. S.J. Tans, Ph.D. Dissertation, Delft Technical University, 1998.
56. J.W.G. Wildöer, L.C. Venema, A.G. Rinzler, R.E. Smalley, and C. Dekker, Electronic structure of atomically resolved carbon nanotubes, *Nature*, 391, 59–62, 1998.
57. T.W. Odom, J.-L. Huang, P. Kim, and C.M. Lieber, Atomic structure and electronic properties of single-walled carbon nanotubes, *Nature*, 391, 62–64, 1998.
58. M. Bockrath et al., Single-electron transport in ropes of carbon nanotubes, *Science*, 275, 1922–1925, 1997.
59. H. Dai, E.W. Wong, and C.M. Lieber, Probing electrical transport in nanomaterials: conductivity of individual carbon nanotubes, *Science*, 272, 523–526, 1996.
60. H. Yorikawa and S. Muramatsu, Electronic properties of semiconducting graphitic microtubules, *Phys. Rev. B*, 50, 12203–12206, 1994.
61. C.T. White, D.H. Robertson, and J.W. Mintmire, Helical and rotational symmetries of nanoscale graphitic tubules, *Phys. Rev. B*, 47, 5485–5488, 1993.
62. R. Saito, G. Fujita, G. Dresselhaus, and M.S. Dresselhaus, Electronic structure of chiral graphene tubules, *Appl. Phys. Lett.*, 60, 2204–2206, 1992.
63. P.G. Collins et al, Nanotube nanodevice, *Science*, 278, 100–104, 1997.
64. S.J. Tans, A.R.M. Verschueren, and C. Dekker, Single nanotube-molecule transistor at room temperature, *Nature*, 393, 49–51, 1998.
65. J. Chen et al., Solution properties of single-walled carbon nanotubes, *Science*, 282, 95–98, 1998.

66. M.A. Hamon et al., Dissolution of single-walled carbon nanotubes, *Adv. Mater.*, 11, 834–840, 1999.

67. E.T. Mickelson et al., Fluorination of single-wall carbon nanotubes, *Chem. Phys. Lett.*, 296, 188–194, 1998.

68. E.T. Mickelson et al., Solvation of fluorinated single-wall carbon nanotubes in alcohol solvents, *J. Phys. Chem. B*, 103, 4318–4322, 1999.

69. P. Boul et al., Reversible sidewall functionalization of buckytubes, *Chem. Phys. Lett.*, 310, 367–372, 1999.

70. K.S. Kelly et al., Insight into the mechanism of sidewall functionalization of single-walled nanotubes: an STM study, *Chem. Phys. Lett.*, 313, 445–450, 1999.

71. J. Li, C. Papadopoulos, and J. Xu, Growing Y-junction carbon nanotubes, *Nature*, 402, 253–254, 1999.

72. Z.P. Huang et al., Growth of highly oriented carbon nanotubes by plasma-enhanced hot filament chemical vapor deposition, *Appl. Phys. Lett.*, 73, 3845–3847, 1998.

73. Z.F. Ren et al., Synthesis of large arrays of well-aligned carbon nanotubes on glass, *Science*, 282, 1105–1107, 1998.

74. Z.F. Ren et al., Growth of a single freestanding multiwall carbon nanotube on each nanonickel dot, *Appl. Phys. Lett*, 75, 1086–1088, 1999.

75. G.E. Scuseria, Negative curvature and hyperfullerenes, *Chem. Phys. Lett.*, 195, 534–536, 1992.

76. L. Chico, V.H. Crespi, L.X. Benedict, S.G. Louie, and M.L. Cohen, Pure carbon nanoscale devices: nanotube heterojunctions, *Phys. Rev. Lett.*, 76, 971–974, 1996.

77. M. Menon and D. Srivastava, Carbon nanotube "T junctions": nanoscale metal-semiconductor-metal contact devices, *Phys. Rev. Lett.*, 79, 4453–4456, 1997.

78. M. Menon and D. Srivastava, Carbon nanotube based molecular electronic devices, *J. Mater. Res.*, 13, 2357–2361, 1998.

79. M.S. Smith, M.H. Schleier–Smith, G.Y. Tseng, and J.C. Ellenbogen, *Architectures for Molecular Electronic Computers. 4. Designs for Digital Logic Structures Constructed from Carbon Nanotubes*, No. MP 99W0000165, The MITRE Corporation, McLean, VA, August 1999, in review.

80. J. Kong, A. Cassell, and H. Dai, Chemical vapor deposition of methane for single-walled carbon nanotubes, *Chem. Phys. Lett.*, 292, 567, 1998.

81. J. Kong, H.T. Soh, A. Cassell, C.F. Quate, and H. Dai, Synthesis of single single-walled carbon nanotubes on patterned silicon wafers, *Nature*, 395, 878–881, 1998.

82. H. Dai, N. Franklin, and J. Han, Exploiting the properties of carbon nanotube for nanolithography, *Appl. Phys. Lett.*, 73, 1508–1510, 1998.

83. S. Fan et al., Self-oriented regular arrays of carbon nanotubes and their functional devices, *Science*, 283, 512–514, 1999.

84. B.H. Robinson and N. Seeman, The design of a biochip: a self-assembling molecular-scale memory device, *Protein Eng.*, 1, 295–300, 1987.

85. E.K. Wilson, DNA conductance convergence?, *Chem. Eng. News*, 43–48, 1999,

86. E. Meggers, M.E. Michel–Beyerle, and B. Giese, Sequence dependent long-range hole transport in DNA, *J. Am. Chem. Soc.*, 120, 12950–12955, 1998.

87. P.T. Henderson, D. Jones, G. Hampikian, Y. Kan, and G.B. Schuster, Long-distance charge transport in duplex DNA: the phonon-assisted polaron-like hopping mechanism, *Proc. Natl. Acad. Sci. USA*, 96, 8353–8358, 1999.

88. J. Jortner, M. Bixon, T. Langenbacher, and M.E. Michele–Beyerle, Charge transfer and transport in DNA, *Proc. Natl. Acad. Sci. USA*, 95, 12759–12765, 1998.

89. Y.A. Berlin, A.L. Burin, and M.A. Ratner, On the long-range charge transfer in DNA, *J. Phys. Chem. B*, in press.

90. D. Porath, A. Bezryadin, S. de Vries, and C. Dekker, Direct measurement of electrical transport through DNA molecules, *Nature*, 403, 635–638, 2000.

91. L.R. Milgrom, *The Colours of Life: An Introduction to the Chemistry of Porphyrins and Related Compounds*. Oxford University Press, Oxford, U.K., 1997.

92. D.F. Bocian, W.G. Kuhr, and J.S. Lindsey, manuscript in preparation.

93. G.Y. Tseng and J.C. Ellenbogen, *Architectures for Molecular Electronic Computers. 3. Design for a Memory Cell Built from Molecular Electronic Devices*, No. MP 99W0000138, The MITRE Corporation, McLean, VA, October 1999. In review.

94. M.S. Ullagaddi, K. Wegener, and J.C. Ellenbogen, Molecular Electronic Circuit Analysis. 1. Analysis of a Molecular XOR Gate, manuscript in preparation.

95. MolSPICE (tentative name), a computer software program being developed at The MITRE Corporation in McLean, VA.

96. M. Magoga and C. Joachim, Minimal attenuation for tunneling through a molecular wire, *Phys. Rev. B*, 57, 1820–1821, 1998.

97. M. Magoga and C. Joachim, Conductance of molecular wires connected or bonded in parallel, *Phys. Rev. B*, 59, 16011–16021, 1999.

98. S.N. Yaliraki and M.A. Ratner, Molecule-interface coupling effects on electronic transport in molecular wires, *J. Chem. Phys*, 109, 5036–5043, 1998.

99. S.N. Yaliraki, A.E. Roitberg, C. Gonzalez, V. Mujica, and M.A. Ratner, The injecting energy at molecule/metal interfaces: implications for conductance of molecular junctions from an ab initio molecular description, *J. Chem. Phys*, 111, 6997–6702, 1999.

100. The authors thank the referees for their helpful comments in regard to this point in the discussion and for pointing out additional relevant references in the literature.

101. D. C. Moore, of the MITRE Nanosystems Group, assisted in refining and extending early quantitative estimates of the properties of polyphenylene-based molecular rectifying diodes.

102. A.O. Orlov, I. Amlani, G.H. Bernstein, C.S. Lent, and G.L. Snider, Realization of a functional cell for quantum-dot cellular automata, *Science*, 277, 928–930, 1997.

103. I. Amlani, A.O. Orlov, G.L. Snider, C.S. Lent, and G.H. Bernstein, Demonstration of a six-dot quantum cellular automata system, *Appl. Phys. Lett.*, 72, 2179–2181, 1998.

104. I. Amlani et al., Digital logic gate using quantum-dot cellular automata, *Science*, 284, 289–291, 1999.

105. C.S. Lent, P.D. Tougaw, W. Porod, and G.H. Bernstein, Quantum cellular automata, *Nanotechnology*, 4, 49–57, 1993.

106. G. Epstein, *Multiple-Valued Logic Design: An Introduction*, Institute of Physics Publishing, Bristol, U.K., 1993.

107. R.C. Jaeger, *Microelectronic Circuit Design*, McGraw-Hill, New York, 1997.

108. R.M. Metzger, private communication, March 1998.

109. R.H. Mathews and T.C.L.G. Sollner, MIT Lincoln Laboratory, unpublished data. The authors thank Dr. Gerry Sollner for providing them in October 1997 with a copy of their unpublished circuit schematic, as shown in Figure 7.14a. The schematic describes an XOR gate that incorporates only rectifying diodes and resonant tunneling diodes, but no transistors. The MIT Lincoln Laboratory team led by Dr. Sollner had been using this circuit as part of a strategy to implement diode-based logic in solid semiconductors. The authors of the present chapter are responsible for the novel use of this schematic circuit to generate a molecular electronic XOR gate.

110. A.E.A. Almaini, *Electronic Logic Systems*, 2nd ed., Prentice-Hall International, Hertfordshire, U.K., 1989. See especially Chapter 6 on "Arithmetic Logic Circuits" for a description of the principles of combinational logic that are required to generate a half adder and a full adder using AND, OR, and XOR gates.

111. M.A. Ratner, private communication, June 1999.

112. J. Soreff, private communication, November 1999.

113. W.F. Chow, Tunnel diode logic circuits, in J.M. Carroll (Ed.), *Tunnel Diode and Semiconductor Circuits*, McGraw-Hill, New York, 1963, pp. 101–105.

114. H.C. Liu and T.C.L.G. Sollner, High-frequency resonant-tunneling devices, in R.A. Kiehl and T.C.L.G. Sollner (Eds.), *Semiconductors and Semimetals*, Academic Press, Boston, 1994, pp. 359–418.

115. A.C. Seabaugh, private communications, October–November, 1997.

116. T.C.L.G. Sollner, private communication, March 1998.
117. M.A. Reed, Conductance of Molecular Junctions, presented at 1998 DARPA Molecular Electronics Workshop, Reston, VA, 2–3 February 1998. This presentation proposed a novel hybrid molecular-solid state transistor.
118. M.A. Reed, Sub-Nanoscale Electronic Systems and Devices, U.S. Patent No. 5,475,341 (12 December 1995).
119. J.C. Ellenbogen, Architectures for Molecular Electronic Computers: 2. Logic Structures Using Molecular Electronic FETs, in preparation. See also J.C. Ellenbogen, Monomolecular Electronic Device, U.S. Patent No. 6,339,227, issued to the MITRE Corporation, January 15, 2002.
120. H. Moravec, When will computer hardware match the human brain? *J. Transhumanism*, Vol. 1, 1998. Available online at http://www.transhumanist.com/volume1/moravec.htm.
121. R.C. Merkle, Energy Limits to the Computational Power of the Human Brain, Xerox Palo Alto Research Center, Palo Alto, CA, 1989. Available online at http://www.merkle.com/brainlimits.html.
122. D. Naegle, private communication, July 1998.
123. J.R. Heath, P.J. Kuekes, G.S. Snider, and R.S. Williams, A defect-tolerant computer architecture: opportunities for nanotechnology, *Science*, 280, 1716–1721, 1998.
124. J. Markoff, Computer scientists are poised for revolution on a tiny scale, in *New York Times*, 1999, p. C1. Available online at http://www.nytimes.com/library/tech/99/11/biztech/articles/01nano.html
125. R.S. Williams and P.J. Kuekes, private communication, October 1999.
126. D.A. Routenberg, and J.C. Ellenbogen, Design for a Millimeter-Scale Walking Robot, MITRE Report No. MP0W00000010, The MITRE Corporation, McLean, VA, 2000.
127. J.C. Ellenbogen, Matter As Software, presented at the Software Engineering and Economics Conference, McLean, VA, 2–3 April 1997. Also published as MITRE Report No. MP 98W00000084 and is available online at http://www.mitre.org/technology/nanotech/SWEE97_article.html
128. MacSpartan Plus1.1, Quantum chemistry molecular modeling software developed and distributed by Wavefunction, Inc., 18401 Von Karman Avenue, Suite 370, Irvine, CA 92612.
129. *NIST Chemistry Webbook*, National Institute of Standards and Technology, Bethesda, MD. Available online at http://webbook.nist.gov/chemistry/.
130. D.V. Bugg, *Circuits, Amplifiers, and Gates*, Institute of Physics Publishing, Bristol, U.K., 1991.
131. *IBM ASIC-12 Databook*, International Business Machines Corp., 1998.
132. J.C. Ellenbogen, Advances Toward Molecular-Scale Electronic Digital Logic Circuits: A Review and Prospectus, presented at Ninth Great Lakes Symposium on VLSI, Ypsilanti, MI, 4–6 March 1999. To appear in proceedings that will be published by the IEEE Computer Society.
133. G.Y. Tseng and J.C. Ellenbogen, Toward nanocomputers, *Science*, 294, 1293–1294, 2001. See also references cited therein.
134. K. Kwok and J.C. Ellenbogen, Moletronics: future electronics, *Materials Today*, 5(2): 28–37, February 2002. See also references cited therein.
135. A. Bachtold, P. Hadley, T. Nakanishi, and C. Dekker, Logic circuits with carbon nanotube transistors, *Science*, 294, 1317–1320, November 9, 2001.

Appendix 7.A
Ab Initio Quantum Calculations of Rectifying Properties for Polyphenylene-Based Molecular Rectifiers

To characterize the likely electrical properties of the proposed polyphenylene-based molecular electronic rectifying diodes, we have conducted *ab initio* quantum mechanical molecular orbital calculations.[101] As described below, these calculations were integral to the selection of the most favorable polyphenylene-based structure to employ as a rectifying diode in molecular electronic digital circuits.

Rectifying diode behavior has not yet been demonstrated experimentally for molecules of the structure shown in Figure 7.8(b). However, the likelihood that such a structure will function as a molecular rectifier is suggested by the rectifying diode behavior that was demonstrated using a monolayer of many Tour-type molecular wires adsorbed at either end to two different thin metal layers in the experiment conducted by Reed's group at Yale University.[7] In addition, we have conducted molecular orbital calculations and assembled related experimental data for molecular structures of the type shown in Figure 7.8(b). These calculations, as detailed below, suggest that these *single molecules* should have the electronic structure and the behavior characteristic of an A&R-type rectifying diode, as shown in Figures 7.8(c), 7.9(c), and 7.9(d). In combination with the experimental conductivity demonstrations for polyphenylene-based Tour wires, these calculations indicate that our proposed polyphenylene-based A&R diodes are likely to have the properties required to produce molecular rectification.

Displayed in Tables 7.A.1 through 7.A.3 of this appendix and in Tables 7.2 and 7.3 of the main text are the results of our theoretical calculations to estimate the electrical properties of the proposed polyphenylene-based A&R-type rectifying diodes. The calculations are premised on the notion that the energy gap ΔE_{LUMO} determines the static, asymmetric energy step along the length of an A&R-type rectifying diode switch of the general structure shown in Figure 7.8(b). Consequently, the energy gap ΔE_{LUMO} measures the general effectiveness of the molecule as an A&R diode. As depicted in Figure 7.8, the quantity ΔE_{LUMO} is the difference in energy between the LUMOs localized separately on the aromatic donor and acceptor halves of the molecule. In Figure 7.8(b), the aromatic donor and acceptor halves are located, respectively, on the left and right sides of the central aliphatic insulating barrier group R. (Such aliphatic barrier groups are discussed in Section 7.1.1.2.)

The calculations applied the *ab initio* Hartree–Fock self-consistent field (SCF) molecular orbital method[40,41] as implemented in the MacSpartan Plus 1.1 commercial quantum chemistry simulation program.[128] In performing these calculations, we had assistance from our collaborator in the MITRE Nanosystems Group, David Moore. The calculations proceeded in several stages, as described below.

TABLE 7.A.1 HOMO and LUMO Energies Determined from Molecular Orbital Theory Calculations for Benzene and Benzene Rings Monosubstituted with Donor and Acceptor Substituent Groups

| Mono-substituted Benzene | Structure | Experimental Ionization Potential (IP)[a] | Results of SCF Molec Orb. Calc'ns[b] | | | |
| | | | STO 3-21 G Basis | | STO 6-31 G Basis[c] | |
			E_{HOMO}	E_{LUMO}	E_{HOMO}	E_{LUMO}
Methoxybenzene $C_6H_5\text{-}OCH_3$	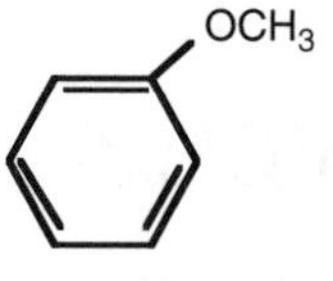	8.20 eV	−8.93 eV	3.86 eV	−8.75 eV	3.85 eV
Methlybenzene $C_6H_5\text{-}CH_3$	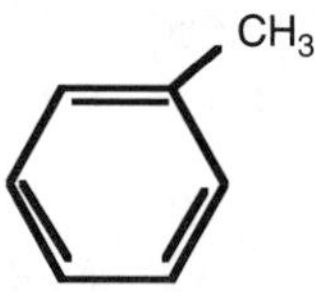	8.83 eV	−8.88 eV	4.15 eV	−8.69 eV	4.09 eV
Benzene C_6H_6	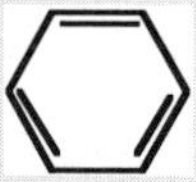	9.24 eV	−9.20 eV	4.02 eV	−8.98 eV	4.00 eV
Trifluoromethyl-Benzene $C_6H_5\text{-}CF_3$	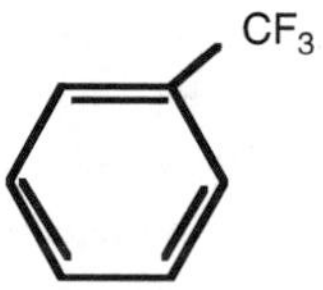	9.69 eV	−9.98 eV	2.73 eV	−9.69 eV	2.87 eV
Benzonitrile $C_6H_5\text{-}CN$	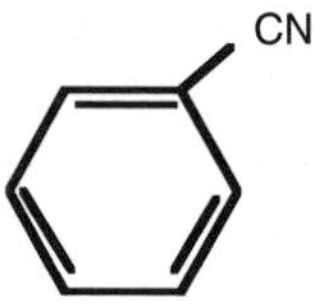	9.73 eV	−9.71 eV	2.33 eV	−9.58 eV	2.27 eV

Donor Substituents (X)

Acceptor Substituents (Y)

Increasing IP

[a] Source of experimental ionization potentials: NIST Chemistry Webbook. National Institute of Standards and Technology, Bethesda, MD. Available online at http://webbook.nist.gov/chemistry/.

[b] SCF molecular orbital calculations performed using the commercial MacSpartan Plus, version 1.1, quantum chemistry computer simulation program. See References 101 and 128.

[c] For most properties, calculations using larger STO 6-31G Gaussian basis set would be anticipated to be more accurate than calculations using smaller STO 3-21G basis set.

7.A.1 Calculations Upon Isolated Benzene Rings Monosubstituted with Donor and Acceptor Substituents

First, as displayed in Table 7.A.1, we determined computationally the one-electron energies of the HOMO and the LUMO molecular orbitals for benzene and for benzene rings that had been singly substituted with the one of the two donor substituents X = −OCH$_3$, −CH$_3$ or with one of the two acceptor substituents Y = −CF$_3$, −CN. The molecular orbital energies were determined using the commonly applied STO 3–21G Gaussian atomic basis set and also in more time-consuming calculations with the larger STO 6–31G Gaussian atomic basis set. The second set of calculations with the larger basis set was used to assess

TABLE 7.A.2 Orbital Energies and LUMO Energy Differences Calculated via 3-21G SCF Molecular Orbital Theory[a] for Archetypal Polyphenylene-Based Diodes Assembled from Singly Substituted Benzene Components Separated by an Insulating Dimethylene Group

Acceptor Substituents (Y)

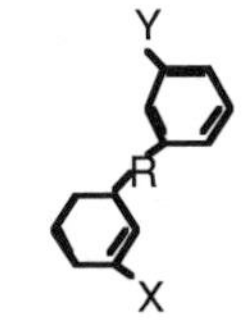

	Trifluoromethyl-Benzene C_6H_5-CF_3 E_{LUMO} = 7.15 eV	Benzonitrile C_6H_5-CN E_{LUMO} = 6.40 eV
Methoxybenzene C_6H_5-OCH_3 E_{LUMO} = 7.90 eV	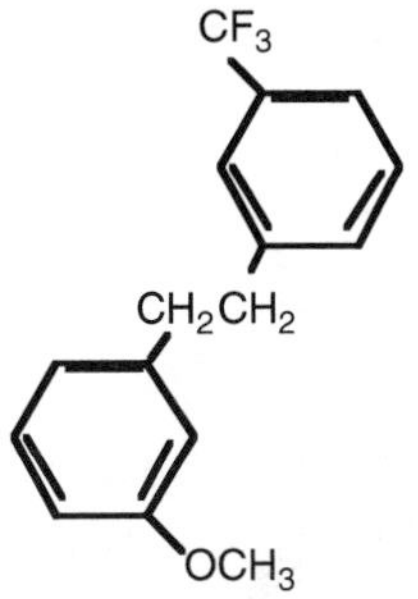	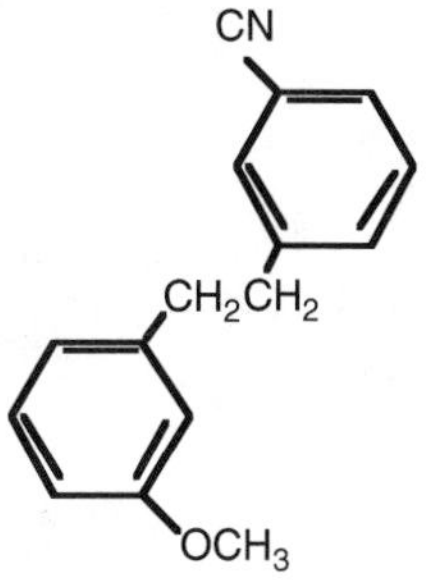

Donor Substituents (X)

Molecular Orbital	Orbital Energy		Localization D	Localization A	Orbital Energy		Localization D	Localization A
HOMO	−8.59 eV		X		−8.98 eV		X	
LUMO	3.03	⎤		X	2.38	⎤		X
LUMO+1	3.42	ΔE_{LUMO}		X	3.24	ΔE_{LUMO}		X
LUMO+2	3.83	⎦	X		3.71	⎦	X	
LUMO+3	4.21		X		3.91		X	
ΔE_{LUMO}	0.80 eV				1.33 eV			
ΔE_{LUMO} (R = ∞)	0.75 eV				1.50 eV			

Methylbenzene C_6H_5-CH_3 E_{LUMO} = 8.73 eV	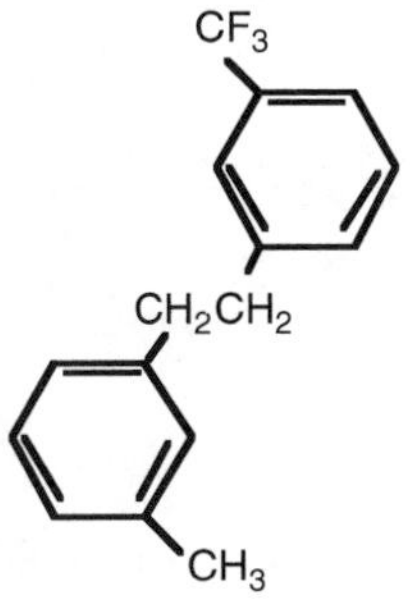	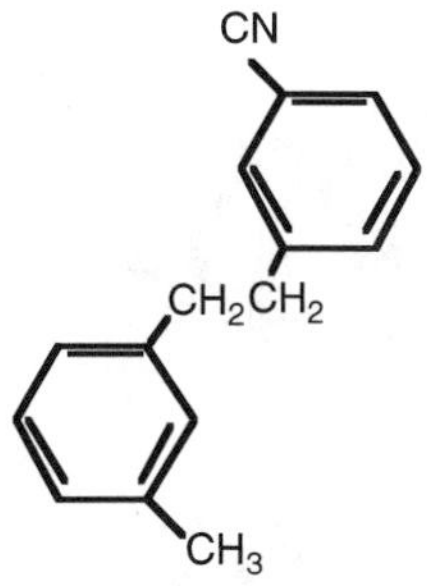

Molecular Orbital	Orbital Energy		Localization D	Localization A	Orbital Energy		Localization D	Localization A
HOMO	−8.82 eV		X		−8.87 eV		X	
LUMO	3.02	⎤		X	2.42	⎤		X
LUMO+1	3.42	ΔE_{LUMO}		X	3.31	ΔE_{LUMO}		X
LUMO+2	3.83	⎦	X		3.79	⎦	X	
ΔE_{LUMO}	0.81 eV				1.37 eV			
ΔE_{LUMO} (R = ∞)	1.58 eV				2.33 eV			

[a] See References 101 and 128.

TABLE 7.A.3 HOMO and LUMO Energies Determined from Molecular Orbital Theory Calculations for Benzene and Benzene Rings *Disubstituted* with Donor and Acceptor Substituent Groups

Di-substituted Benzene	Structure	Experimental Ionization Potential (IP)[a]	SCF MO Calc'ns[b] STO 3-21 G Basis E_{HOMO}	E_{LUMO}
1,3-Dimethoxybenzene C_6H_4-$(OCH_3)_2$	OCH₃ / OCH₃	8.16 eV	−8.93 eV	3.79 eV
1,3-Dimethylbenzene C_6H_4-$(CH_3)_2$	CH₃ / CH₃	8.55 eV	−8.66 eV	4.10 eV
Benzene C_6H_6	(benzene ring)	9.24 eV	−9.20 eV	4.02 eV
1,3-Dicyanobenzene C_6H_4-$(CN)_2$	CN / CN	10.40 eV	−10.32 eV	1.51 eV

Left margin labels: Increasing IP (arrow downward); Donor Substituents (X); Acceptor Substituents (Y)

* Source of experimental ionization potentials: NIST Chemistry Webbook. National Institute of Standards and Technology, Bethesda, MD. Available online at http://webbook.nist.gov/chemistry/.

** SCF molecular orbital calculations performed using the commercial MacSpartan Plus, version 1.1, quantum chemistry computer simulation program. See References 101 and 128.

whether the smaller basis set would be sufficient for an estimate of the LUMO energies, which would be relatively insensitive to the choice of basis set. In this connection, we observed from the values listed in Table 7.A.1 that the calculations with the larger basis set yielded LUMO energies that were very nearly the same as those with the smaller STO 3–21G basis set.

In general, though, Hartree–Fock calculations, such as we have performed here, do not give an accurate estimate of the value of the LUMO energy for a single atom or molecule. Such calculations usually give values that are too high for the difference between the energies of the HOMO and LUMO orbitals. This is because Hartree–Fock calculations do not accurately reflect the relaxation energy associated with the other electrons due to an electron that is added to the neutral species in the LUMO orbital. However, we apply the Hartree–Fock method here as a simple way to determine the trend in the value of ΔE_{LUMO}, assuming a constant error in the relaxation energy across a sequence of calculations.

As also may be observed from Table 7.A.1, in applying Koopmans' Theorem,[40,41]

$$E_{HOMO} = -I, \qquad (7.A.1)$$

the smaller STO 3–21G basis set provided HOMO energies E_{HOMO} that generally agreed well with the experimental[129] molecular ionization potentials. Thus, we determined that we could rely upon the smaller STO 3–21G basis set for the estimation of the HOMO and LUMO energies. We employ it for all subsequent calculations.

The calculated LUMO energies given in Table 7.A.1 for the electron-acceptor-substituted aromatic rings are lower than the LUMO energies for the donor-substituted rings. This is expected from the notion that the electron-acceptor substituents (Y = –CF3, –CN) should reduce the electron density on the ring slightly and thereby lower the positive energy contribution due to the repulsive interaction between electrons on the ring. Contrariwise, electron-donor substituents (X = –OCH3, –CH3) should tend to raise the LUMO energies, as is observed for the methyl-substituted benzene in Table 7.A.1, by enhancing the electron density on the ring slightly and thereby increasing the positive repulsive energies among the electrons on the ring.

To conclude the first part of the theoretical calculations, the STO 3–21G Hartree–Fock orbital energy estimates for E_{LUMO} in Table 7.A.1 were used to approximate crudely the relative magnitudes of the anticipated potential drops ΔE_{LUMO} across the polyphenylene-based diodes. This was done by taking the difference:

$$\Delta E_{LUMO}(R = \infty) = E_{LUMO}(C_6H_5\text{-}X) - E_{LUMO}(C_6H_5\text{-}Y) \qquad (7.A.2)$$

between $E_{LUMO}(C_6H_5\text{-}X)$ determined for an *isolated* benzene monosubstituted with donor substituent X and the LUMO energy $E_{LUMO}(C_6H_5\text{-}Y)$ for an *isolated* benzene monosubstituted with acceptor substituent Y.

These approximations $\Delta E_{LUMO}(R = \infty)$ are displayed in Table 7.A.2 for the four possible donor and acceptor pairs. The values of $\Delta E_{LUMO}(R = \infty)$ range from 0.75 to 2.33 electron volts (eV). Of these, the pairings of the methyl-substituted benzene donor molecules with acceptor molecules yielded the largest two values of $\Delta E_{LUMO}(R = \infty)$. Further, the cyano-substituted acceptor molecule (benzonitrile) produced a larger approximate potential drop than the trifluoromethyl-substituted acceptor molecule when each was paired with the methyl-substituted benzene donor.

In actuality, we would expect these LUMO energy differences or potential drops across the molecule all to be somewhat smaller than is estimated by the values of $\Delta E_{LUMO}(R = \infty)$, which yield an estimate based upon the isolated donor and acceptor components of the diode structure. Some equilibration of the electron density across the central insulating barrier, with a corresponding reduction in the chemical potential difference, should be expected when the aromatic donor and acceptor halves of the molecule each are bound chemically to the same central insulating group R. Also, if these diodes were embedded in molecular wires, the other large undoped molecular components would reduce ΔE_{LUMO} still further.

7.A.2 Calculations upon Bonded Polyphenylene-Based Donor-Acceptor Complexes Built from *Mono*substituted Aromatic Rings

To estimate more accurately the trend in the potential drops across the several proposed molecular rectifying diodes, in the second part of the theoretical analysis, STO 3–21G *ab initio* Hartree–Fock self-consistent field molecular orbital calculations were performed upon several chemically bonded polyphenylene-based donor–acceptor complexes. These four complexes, illustrated in Table 7.A.2, were built from pairs of the monosubstituted benzene rings shown in Table 7.A.1. A donor-benzene ring and an acceptor-benzene ring are chemically bonded to the opposite ends of an insulating central dimethylene barrier group (i.e., R = –CH2CH2–).

The results of the *ab initio* calculations upon the complete donor–acceptor complexes bonded to either side of the dimethylene bridge are displayed in Table 7.A.2. For each of the four molecules, values are given for the uppermost molecular orbital energy in the occupied manifold (i.e., E_{HOMO}) and for the lower several orbitals in the unoccupied orbital manifold (E_{LUMO}, E_{LUMO+1}, E_{LUMO+2}, etc.). Here, we use the notation E_{LUMO+k} to specify the energy of the k-th unoccupied orbital above the LUMO, taking the orbitals in order of increasing orbital energy.

Observe in Table 7.A.2 that, for each of the orbitals listed for each molecule, we specify with an "X" in the appropriate column of Table 7.A.2 whether the orbital is calculated to be localized primarily on the donor (D) side of the molecule or on the acceptor (A) side of the molecule. Note that the LUMO orbital always is localized on the acceptor side of the molecule.

It follows, consistent with the definition in Equation 7.1 in the body of this chapter, that the potential drop ΔE_{LUMO} across each polyphenylene-based diode structure is determined in Table 7.A.2 by subtracting E_{LUMO} from the lowest value E_{LUMO+k} for an unoccupied orbital localized on the opposite (donor) side of the molecule from the LUMO — i.e.,

$$\Delta E_{LUMO} \equiv E_{LUMO}(\text{Donor}) - E_{LUMO}(\text{Acceptor})$$

$$= E_{LUMO+k}(\text{Donor LUMO}) - E_{LUMO}. \tag{7.A.3}$$

In this way, it is determined that the largest potential drop $\Delta E_{LUMO} = 1.37$ eV is produced by the pairing of the methyl-substituted benzene donor molecule with cyano-substituted benzene acceptor molecule. However, contrary to our expectations from the tabulated values of $\Delta E_{LUMO}(R = \infty)$, there is nearly as large a potential drop $\Delta E_{LUMO} = 1.33$ eV for the diode produced by the pairing of the meth*oxy*-substituted benzene donor molecule with cyano-substituted benzene acceptor molecule.

Moreover, in view of its large value for ΔE_{LUMO}, closer examination of the results calculated for methoxybenzene-cyanobenzene complex reveals that it is likely to be preferable to the methylbenzene-cyanobenzene complex for use as a rectifying diode switch. This is because the localization of the unoccupied orbitals is more pronounced in the methoxybenzene-cyanobenzene complex. Specifically, it is observed that the methoxy-substituted complexes described in Table 7.A.2 have *four* unoccupied orbitals (LUMO, LUMO+1, LUMO+2, and LUMO+3) for which strong localization on one or the other side of the molecule is seen. For the methyl-substituted complexes, however, there are only three strongly localized orbitals in the unoccupied manifold — LUMO+3 and higher orbitals are calculated to be delocalized across the entire molecule.[101] This is shown in Table 7.A.2.

A larger number of localized unoccupied orbitals in the "conduction band" of the diode is a significant advantage because it provides additional channels for efficient resonant transport through the central barrier of the switching structure. A structure with fewer such unoccupied orbitals operates efficiently over a narrower range of voltages. (See Reference 24 and Section 7.6.1 above for a more detailed discussion of the mechanisms of electron transport through molecular wires and switches.)

7.A.3 Calculations upon Bonded Polyphenylene-Based Donor-Acceptor Complexes Built from Disubstituted Aromatic Rings

The methoxybenzene-cyanobenzene donor–acceptor complex and the methylbenzene-cyanobenzene complexes built from *mono*substituted aromatic rings do exhibit significant intrinsic internal potential drops, as was anticipated. However, in each case the magnitude of the drop is less than 1.5 eV, as indicated by the calculated values of ΔE_{LUMO}. Larger permanent potential drops might be desirable to ensure robust operation under a range of voltages in molecular electronic circuits.

To attempt to increase the intrinsic internal potential drop across the molecular diodes (i.e., to restore the amount of the value of ΔE_{LUMO} "lost" due to equilibration of electron densities across the central barrier upon binding to it), one might use *two* donor substituents and *two* acceptor substituents on the respective aromatic component halves of the diode molecule, instead of only one each. This is the strategy adopted in this chapter.

Thus, as shown in Table 7.A.3, along with Tables 7.2 and 7.3, in the third and final phase of the computational analysis of the polyphenylene-based rectifying diodes, we repeat for disubstituted donor–acceptor complexes calculations analogous to those that are reported in Tables 7.A.1 and 7.A.2 for complexes built from only monosubstituted benzenes. For the central insulator, we continue to employ a dimethylene bridging group.

Specifically, these further calculations analyze a dimethoxybenzene–dicyanobenzene donor–acceptor complex and a dimethylbenzene–dicyanobenzene complex. The structure for these molecules is shown in Figures 7.10 and 7.11 in the main body of this chapter. As shown in Tables 7.2 and 7.3, both of these doubly disubstituted molecular structures are calculated to have similar large values of ΔE_{LUMO}. This is a favorable indication of the likelihood that each doubly disubstituted molecule is a good molecular rectifier. The computed values of ΔE_{LUMO} for the doubly disubstituted rectifier molecules shown in Tables 7.2 and 7.3 exhibit significant increases of approximately 50% (or 0.7 eV) relative to the proposed monosubstituted diodes depicted in Table 7.A.2.

Additionally, both of the doubly disubstituted diodes described by the data in Tables 7.2 and 7.3 have four unoccupied orbitals localized on either the donor or the acceptor half of the molecule. Thus, each of these diode molecules has channels for robust resonant electron transport across it.

One difference between the dimethoxy and the dimethyl diodes, however, is that the dimethoxy diode may be slightly less sensitive to rotations out of the plane. This can be inferred from the data near the bottom of Tables 7.2 and 7.3, where we note that ΔE_{LUMO} changed by only 0.02 eV for an out-of-plane rotation of the dimethoxy substituted diode, but it changed by 0.1 eV for the out-of-plane rotation of the dimethyl-substituted diode.

Low sensitivity to such out-of-plane rotations is desirable, as they prove to be necessary in some of the designs proposed below for molecular electronic digital circuits. Thus, in the designs below we choose to use the dimethoxy-dicyano-substituted diode structure shown in Figure 7.10 in preference to the dimethyl–dicyano-substituted diode shown in Figure 7.11.

Nonetheless, the calculations here indicate that both of the doubly disubstituted molecular structures depicted in Figures 7.10 and 7.11 should perform well as molecular rectifying diodes in such circuits. Both of these proposed diode structures have a large intrinsic internal potential drop of approximately 2.0 eV, and both have an unoccupied manifold with two unoccupied orbitals localized on either side of the central barrier to ensure robust efficient electron transport in the preferred direction under a forward bias.

Appendix 7.B
Explanation of
Diode–Diode Logic

The development of molecular electronic logic circuits described in the body of this work depends heavily upon the schematics and operational principles for diode-based circuits developed originally for implementation on much larger scales. Thus, for completeness, in this appendix we explain in detail the operation of the diode–diode AND, OR, and XOR gates. In formulating these explanations, we had assistance from our collaborator in the MITRE Nanosystems Group, Kevin Wegener.

7.B.1 Operation of the Diode-Based AND Gate

A diode-based molecular AND gate is described above in Figure 7.12 and Section 7.5.1.1. Diode-based logic also is described in detail elsewhere,[106,107] but a brief explanation of the operational principles for the diode-based AND gate is presented here.

A rectifying diode is a two-terminal switch that can be "on" or "off" depending on the applied voltage. A diode is said to be *forward biased* when the voltage difference across the diode is greater than or equal to zero. As shown in Figure 7.3(b), an ideal diode is "on" when it is forward biased — i.e., current flows. The diode is switched "off" when it is reverse biased — i.e., no current flows. Thus, the ideal rectifying diode is a simple switch that is either open or closed depending on the bias applied. (Note that the circuits described in this appendix behave in the same manner when real diodes are used, but for a simple qualitative explanation, the rectifying diodes are assumed here to be ideal.)

Figure 7.B.1 shows how a diode can be utilized in simple circuits to give a particular output voltage, V_C. The direction the diode is positioned relative to the input is the same as would be the case for the AND gate. The voltage across the diode is the difference between the voltages at nodes A and C with respect to ground (taken to be zero volts). There are two relevant examples to examine how a rectifying diode behaves under a voltage bias.

In Example 1 shown in Figure 7.B.1(a), the potential at A with respect to ground is V_0. The potential at C is V_0 less the voltage drop across the resistor, V_R. Thus, the voltage across the diode is:

$$V_{Diode} = V_C - V_A = (V_0 - V_R) - V_0 = -V_R \qquad (7.B.1)$$

Assuming $V_R > 0$, the voltage across the diode is less than zero. If the bias is less than zero, the diode must be reversed biased or switched "open." There is no current flow in the system. The potential at C corresponds to the measured output, V_C:

$$V_C = V_0 - V_R = V_0 - IR = V_0 \qquad (7.B.2)$$

That is, because the diode is reverse biased or "open," it prevents the current I from flowing, and the potential or output at C is V_0, which corresponds to a "high" voltage or a binary bit "1."

In the circuit Example 2, shown in Figure 7.B.1(b), the potential at A with respect to ground is set at 0 volts. Now, the voltage across the diode is:

$$V_{Diode} = V_C - V_A = (V_0 - V_R) - 0 = V_0 - V_R \qquad (7.B.3)$$

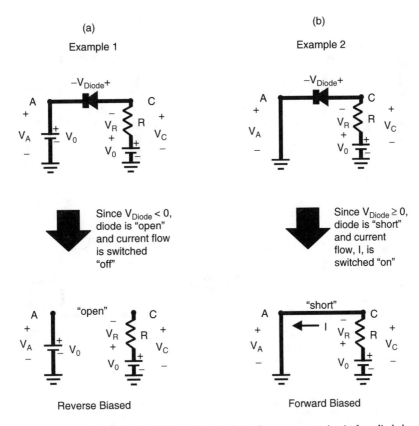

FIGURE 7.B.1 Two modes of operation for ideal rectifying diodes in the component circuits for a diode-based AND gate.

This voltage is greater than or equal to zero, implying that there is a forward bias (assuming $V_0 = V_R$). A forward-biased ideal diode behaves like a closed switch, allowing current to flow in the circuit. The direction of the current is illustrated in the lower portion of Figure 7.B.1(b). The potential at C with respect to ground is $V_C = V_A = 0$. In other words, the potential at C is zero volts because it is connected directly to ground through the forward biased diode. Thus, the output is a "low" voltage or a binary "0."

The more complex circuit of an AND logic gate is implemented with diode logic simply by connecting in parallel two of the elementary diode circuits described above. This parallel circuit layout is depicted in Figure 7.B.2. Note that V_A and V_B can be set independently to "on" (i.e., "1") or "off" (i.e., "0") via two-way switches. This allows the different combinations of binary inputs to be represented.

There are three possible pairs of binary inputs to the circuit as shown in Figure 7.B.2. In Case 1, both inputs are switched to low voltage or "off." Then, both inputs represent the binary number "0." Following the same operational principles described above for the simple circuit in Example 2, Figure 7.B.1(b), both V_A and V_B are less than V_C, so both diodes A and B are forward biased. Thus, current flows and the potential at C drops to 0 as expected for the output of the entire AND gate circuit — i.e., 0 AND 0 = 0.

In Case 2, Figure 7.B.2(b), either input A or input B is set to "off" and the other to "on." The potential applied in the "on" setting corresponds to a binary "1" input. Because one of the two inputs is set to "off," there is a forward biased diode in the circuit allowing current to flow. Therefore, the potential at C is zero volts, corresponding to an output of "0" for the entire circuit — e.g., 0 AND 1 = 0, as expected.

In Case 3, Figure 7.B.2(c), where both A and B are switched to "on," the inputs are both "1." As described above for Example 1, Figure 7.B.1(a), when V_A is greater than V_C, the diode is reverse biased. The diode behaves like an open switch that prevents current from flowing. When both diodes are reversed biased, there is no pathway for current to flow — i.e., I = 0. The potential at C is V_0 or a binary "1."

Case 1: Both A and B are "off"—i.e., A = 0 and B = 0

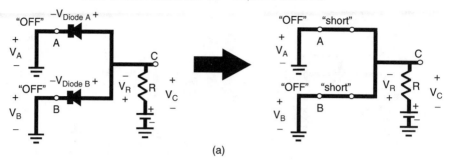

(a)

Case 2: Either A or B is "on"—e.g., A = 0 and B = 1

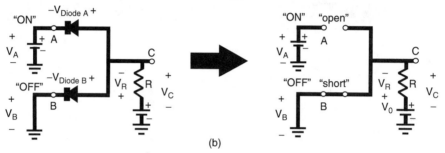

(b)

Case 3: Both A and B are "on"—i.e., A = 1 and B = 1

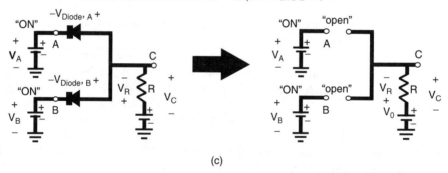

(c)

FIGURE 7.B.2   Three cases for operation of a diode-based AND gate.

Thus, of the three possible sets of inputs, only this combination of inputs yields an output of a binary "1," matching the correct output for an AND gate, namely 1 AND 1 = 1.

Using considerations related to those above, it is also possible to determine that the resistance R should be large in the AND gate when it is implemented using non-ideal diodes. For example, when either or both diodes are forward biased and the output voltage of the AND gate is supposed to be approximately zero or LOW, representing a binary "0," a large value of R ensures that the voltage from the power source (not the input signal) is dropped across this resistor rather than across the diode. Therefore, the voltage at the output is kept relatively small, following the analysis above. Then, if more AND gates were attached to the output, the equivalent resistance of those circuits and the diode is smaller, and the majority of the voltage still is dropped across the resistor R. This serves to increase fan-out, which is the number of circuits that can be connected to the output. A value of R at least one order of magnitude greater than the nonzero resistance of the non-ideal diodes in the circuit is essential for this purpose.

The appropriate magnitude of resistance R for the resistor in the AND gate also can be determined by examining the expected power dissipation from the diode-based AND gate. Power is defined as:

$$P = IV = V^2/R = I^2R \qquad (7.B.4)$$

When either A or B or both are "off" and I > 0, the power dissipated is V_0^2/R watts. For the case where both A and B are "on" and no current flows, then the power dissipated is 0 watts. A large resistance R makes P small. That is, when R is large, V_0 will supply less current because I is inversely proportional to R by Ohm's law for a given V_0 — e.g., $I = V_0/R$.

7.B.2 Operation of the Diode-Based OR Gate

A diode-based molecular OR gate is described above in Figure 7.13 and Section 7.5.1.1. Diode-based logic has been discussed previously elsewhere,[106,107] but a brief explanation of the operational principles for the diode-based OR gate is presented here.

The OR gate consists of two simple diode circuits wired in parallel much like the AND gate described above. For the OR gate, the simple circuit differs from the one used to make the AND gate in the direction the diode is positioned relative to the input signals, as seen in Figure 7.B.3. Thus, in order for the diode to be forward biased in the example circuits, the difference between the potentials at nodes A and C with respect to ground must be greater than or equal to zero. (Note that this is reversed from the required bias direction for the simple circuits that compose the AND gate described above.)

In Example 1 of Figure 7.B.3(a), the potential at A is V_0 while the potential at C is V_R. If the initial assumption is that the diode is forward biased, then current would be flowing from A to C. Because the diode is ideal, there is no voltage drop across the diode itself; and the potential measured at C with

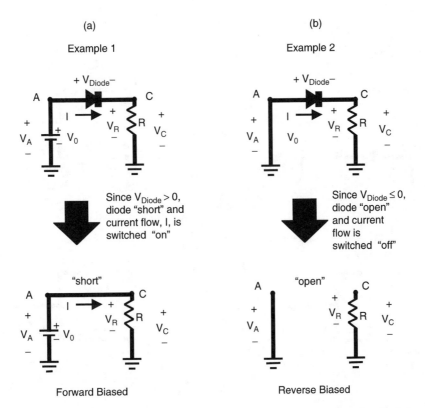

FIGURE 7.B.3 Two modes of operation for ideal rectifying diodes in the component circuits for a diode-based OR gate.

respect to ground is equal to the potential measured at A, so $V_C = V_A = V_0$. In other words, the assumption of a forward bias on the diode was correct and the output at C would be "on" or a binary "1."

In Example 2 of Figure 7.B.3(b), the potential at A with respect to ground is 0 volts. If the initial assumption is that there is current flowing, then the voltage drop across the resistor by Ohm's law would be $V_R = IR$. This implies that the potential difference between A and C across the diode would be $-V_R$ (i.e., a negative voltage). This is a reverse biased condition for the diode, so the initial assumption that current was flowing was incorrect. The reverse biased diode acts as an open circuit. The potential measured at C with respect to ground is $V_C = IR$, but because $I = 0$, $V_C = 0$. Therefore, the output is "off" or a binary "0."

The OR logic gate is implemented with diode logic simply by connecting in parallel two of the elementary diode circuits described in Figure 7.B.3. This parallel circuit layout is depicted in Figure 7.B.4 with the three possible input cases. Note that V_A and V_B can be set independently to "on" (i.e., "1") or "off" (i.e., "0") via two-way switches. This allows the different combinations of inputs to be represented.

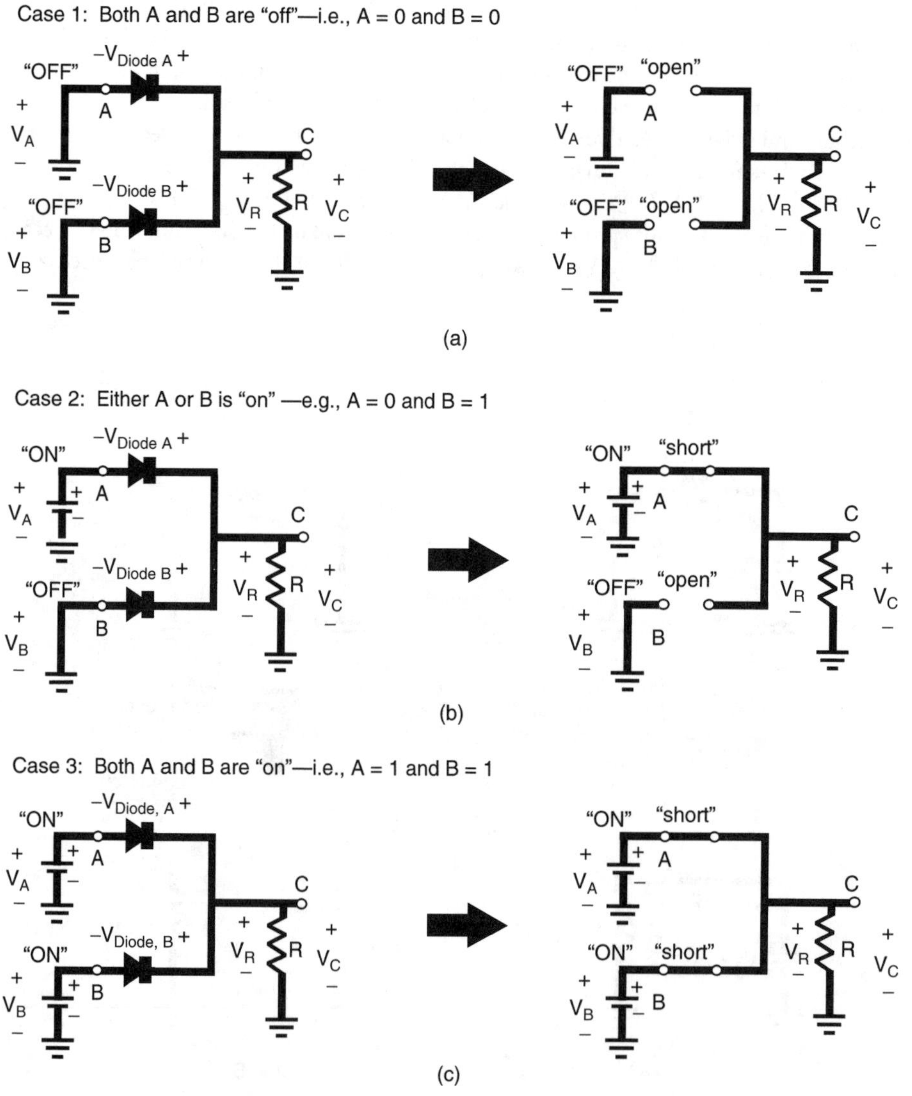

(a)

(b)

(c)

FIGURE 7.B.4   Three cases for operation of a diode-based OR gate.

There are three possible pairs of binary inputs to the circuit as shown in Figure 7.B.4. In Case 1, both inputs A and B are switched to a "low" voltage, or "off." These inputs both represent the binary number "0." Following the same operational principles described above for the simple circuit in Example 2, Figure 7.B.3(b), both diodes A and B are reverse biased. Because both diodes act like "open" switches and no current flows, the potential at C is "0" as expected for the output of the entire OR gate circuit — i.e., 0 OR 0 = 0.

In Case 2, Figure 7.B.4(b), either input A or input B is set to "off" and the other to "on." The potential applied in the "on" setting corresponds to a binary number "1" input. Because one of the two inputs is set to "on," there is a forward biased diode in the circuit allowing current to flow as described above in Example 1 in Figure 7.B.3(a). As there is no voltage drop across the ideal diode, the potential at C is $V_C = V_A = V_0$, corresponding to an output of a binary "1" as expected for the OR function — e.g., in Figure 7.B.4(b), 1 OR 0 = 0.

In Case 3, Figure 7.B.4(c), where both A and B are switched to "on," the inputs are both "1." As described above for Example 1, Figure 7.B.3(a), when V_A is greater than V_C, the diode is forward biased. The diode behaves like a wire that allows current to flow. When both diodes are forward biased, there is a pathway for current to flow. Because there is no voltage drop across the ideal diode, the potential at C is $V_C = V_A = V_0$ or a binary "1." The output of the circuit is "on" when either A or B or both are "on" as expected for the entire circuit — e.g., in Figure 7.B.4(c), 1 OR 1 = 1.

Using considerations related to those above, it is also possible to determine that the resistance R should be large in the OR gate when it is implemented using non-ideal diodes. For example, when either or both diodes are forward biased and the output voltage of the OR gate is supposed to be approximately V_0 or HIGH, representing a binary "1," a large value of R ensures that the voltage from the input signal (not the power source) is dropped across this resistor rather than across the diode. Therefore, the voltage at the output is kept relatively large, following the analysis above. Then, if more OR gates are attached to the output, the equivalent resistance of those circuits and the diode is smaller, and the majority of the voltage still is dropped across the resistor R. This serves to increase fan-out, which is the number of circuits that can be connected to the output. A value of R at least one order of magnitude greater than the nonzero resistance of the non-ideal diodes in the circuit is essential for this purpose.

The appropriate magnitude of resistance R for the resistor in the OR gate can be determined by examining the expected power dissipation from the diode-based OR gate. Power is defined as:

$$P = IV = V^2/R = I^2 R \qquad (7.B.5)$$

For the case where both A and B are "off" and no current flows, the power dissipated is 0 watts. When either A or B or both are "on," the power dissipated is V_0^2/R watts. A large resistance R means the input signal V_0 will supply less current because I and R are inversely proportional by Ohm's law for a given V_0 — i.e., $I = V_0/R$. Therefore, to minimize power loss, R should be relatively large.

7.B.3 Operation of the Diode-Based XOR Gate

A diode-based molecular XOR gate is described above in Figure 7.14 and in Section 7.5.1.2. This molecular electronic logic gate is based upon a novel schematic credited to Mathews and Sollner.[109] The principles of operation for this novel circuit deserve some explanation.[116,130]

The first key idea is that the known total voltage drop V_T that is applied across the XOR gate is the sum of the effective voltage drop V_{IN} across the input resistances that are associated with the rectifying diodes at the left in Figures 7.14(a) and 7.14(b), plus the voltage drop V_{RTD} across the RTD to the right in those figures; i.e.,

$$V_T = V_{IN} + V_{RTD}. \qquad (7.B.6)$$

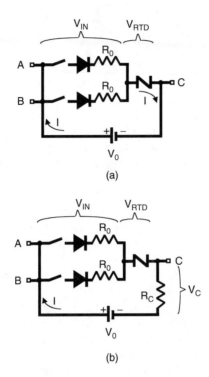

FIGURE 7.B.5 Circuit schematics explaining the operation of the molecular electronic XOR gate.

This relation is depicted in the circuit schematic in Figure 7.B.5(a), which shows the XOR gate as it might have current and voltage provided to it in a larger circuit. From Equation 7.B.6, the voltage drop across the RTD may be written

$$V_{RTD} = V_T - V_{IN}. \tag{7.B.7}$$

A second key idea is that the effective resistance of the part of the logic gate containing the rectifying diodes differs depending upon whether one or both of the two identical parallel inputs is "on." If *only one* of the inputs is "on" (e.g., A is 1 and B is 0), the effective resistance of the left-hand, input portion of the XOR gate shown in Figure 7.14 or in Figure 7.B.5(a) is R_0. However, if *both* of the inputs are "on" (i.e., A is 1 and B is 1), the aggregate effective resistance of the left-hand, input portion of the gate is only $R_0/2$.

Thus, if a (variable) current I is passed through only one input to the gate, then the voltage drop across the input resistance is $V_{IN} = IR_0$. However, if the total current I is split evenly between the two inputs A and B, then the voltage drop across the input resistances is $V_{IN} = IR_0/2$. It follows from these ideas and from Equation 7.B.7 that there are two possible operating equations for the gate. Depending upon the inputs, these two equations are approximately:

$$V_{RTD} = V_T - IR_0, \text{ if only one input is "1"} \tag{7.B.8a}$$

$$V_{RTD} = V_T - IR_0/2, \text{ if both inputs are "1"} \tag{7.B.8b}$$

By solving these two equations for the variable operating current through the XOR circuit, we derive approximate equations for the two possible operating lines for the circuit, which often are termed the *load lines* for the RTD in the XOR gate:

$$I = (V_{RTD} - V_T)(-1/R_0), \text{ if one input is "1"} \tag{7.B.9a}$$

$$I = (V_{RTD} - V_T)(-2/R_0), \text{ if both inputs are "1"} \qquad (7.B.9b)$$

These two straight lines, which have different slopes but the same intercept on the V-axis, are depicted in the I vs. V plot for the RTD that appears in Figure 7.7(b). In Figure 7.7(b), it may be noted that these two approximate operating lines for the circuit have two different points of intersection with the curved operating line for the RTD.

Which of these two common operating points applies plainly depends upon the inputs to the gate. In the case that only one input is "1," Equation 7.B.8a yields the less steep load line and the operating point on the *peak* of the RTD operating curve, where the RTD is "on" and the output voltage of the gate is therefore a "1."

However, in the case that both inputs to the XOR gate are "1," Equation 7.B.8b and the steeper load line applies. The corresponding operating point is in the *valley* of the RTD operating curve, where the RTD is "off." *Thus, the output of the XOR gate is likewise off or "0" for two "1s" on input.* This result, which is opposite to that produced by the same inputs in the ordinary or *inclusive* OR gate, is a consequence of the valley current shut-off that is manifested by the RTD past its first peak on the I-V curve.

Thus far, the analysis has ignored other resistances in the circuit that also would affect its operation — especially the resistance R_C at the lower right of the circuit associated with the measured output voltage. However, by adjusting the value of R_0, other resistances such as R_C may be inserted into the circuit, as illustrated in Figure 7.B.5(b), without affecting the essential current response of the circuit, which is, in summary:

- Having only the A input set to "1" or only the B input set to "1," but not both, yields V_{RTD} as a low voltage, while it produces a relatively high value for the current I.
- Having both A and B set to "1" yields a relatively high value of V_{RTD}, but a low output current.

Thus, the low current in the latter case ensures that the measured output voltage drop $V_C = IR_C$ in the latter case with both inputs as "1" will be lower than in the former case with only one input as "1." Of course, this is precisely the output voltage response desired for an XOR gate.

A final observation about the analysis above in Appendix 7.A concerns our implicit and explicit use of Ohm's law, $V = IR$, thereby assuming both constant resistances R for the molecular circuit elements and a linear relationship between the voltage V and the current I. In fact, neither of these assumptions is exactly true for molecular wires.[1,28] However, over narrow ranges of values for the current, these assumptions are likely to be nearly true. The current through a molecular wire still rises monotonically and nearly linearly as the voltage is swept through a narrow range. Thus, the sense of the preceding analysis is likely to be true, although the details may require adjustment when such a circuit actually is constructed and its current–voltage behavior is measured.

Appendix 7.C
Size Estimate for CMOS Half Adder and Scaling Estimate for a Molecular Adder

7.C.1 Motivation

It is of interest to have a quantitative estimate of the degree of scaling that might be provided by molecular electronic circuits such as the approximately 3 nm × 4 nm molecular logic gates and the approximately 10 nm × 10 nm adder function for which designs are given in the body of this work. Ideally, we would compare these dimensions with those for the analogous functions on commercial CMOS logic circuits. Data on the actual measurements of structures on commercial computer chips often is regarded as proprietary information, though. Thus, such data have proven to be difficult to obtain.

However, David Naegle of Sun Microsystems, Inc., in Palo Alto, CA, has very graciously performed for us an approximate calculation of the sizes for the microelectronic implementations of the analogous logic functions when they are fabricated using conventional CMOS technology, assuming 250 nm or 0.25 μm lithographic linewidths. His estimate, including his detailed reasoning, appears immediately below.

7.C.2 Naegle's Estimate of the Sizes for CMOS Logic Structures

The 1998 edition of the *IBM ASIC SA-12 Databook*[131] can be used to compute cell sizes of two-input AND, OR, and XOR gates for their smallest implementation (lowest drive) by an IBM 0.25 μm SA-12 process.

The areas of the individual logic gates are not available directly in the format adopted by the databook. Instead of quoting an area for each of the active devices by itself, the number of "cells" that the collection of subcomponent devices occupies is quoted. This appears to account for some amount of the interconnect structure that is needed to wire together devices within the gate. For example, quoting from p. 20 of the databook:

"Cell Units = Area of [gate name], in number of cell units. A cell unit is 2 Metal-2 ×16 Metal-1 wiring channels for non-IO cells.... Each wiring channel is 1.26 μm wide."

On that basis, assuming that both the metal 1 and the metal 2 wiring channels have a width of 1.26 μm, a cell unit is:

$$2 \times 16 \times (1.26 \ \mu m^2) = 50.8032 \ \mu m^2$$

This evaluation of the area of a cell permits the tabulation of the areas in μm^2 of the CMOS AND, OR, and XOR gates from the data given for the areas of these gates in cells. The tabulation is as follows:

2-input AND (p. 274):	2 cells	= 101.6064 μm^2
2-input OR (p. 291):	2 cells	= 101.6064 μm^2
2-input XOR (p. 296):	4 cells	= 203.2128 μm^2

For completeness, I also tabulate the estimated areas for NAND, NOR, and Invert logic functions, as follows:

2-input NAND (p. 281):	2 cells = 101.6064 μm^2	
2-input NOR (p. 286):	2 cells = 101.6064 μm^2	
Inverter (p. 279):	1 cell = 50.8032 μm^2	

The pages quoted above are from the *Databook*.[131]

The lack of a difference in area for AND and NAND gates in the results above is somewhat troubling if one hopes to understand in detail the areas involved in implementing these cells. The unit cell described above clearly is too coarse a measure of area to get a very accurate idea of the actual sizes of the gates. Nevertheless, this is the best area a designer can achieve in the SA-12 standard cell library. It may be expected that these numbers are not off by more than a factor of two or three. That is, a custom layout by hand of the equivalent standard cells probably would not be smaller than one half or one third the quoted size, especially if you take into account the size of the p-well for the n-channel transistors.

7.C.3 Conclusions

From David Naegle's calculation, the approximate 100 μm^2 area of the silicon CMOS microelectronic logic gates would be about 6 to 8 million times greater in area than the approximate 12 nm^2 molecular AND, OR, and XOR gates for which designs are proposed in the body of this chapter. Allowing for the approximate nature of the calculation above, this implies a scale-down factor of approximately 10^6 or 10^7 for the molecules.

A circuit for a half adder function, be it solid state or molecular, consists of at least two logic gates of the types listed above. Thus, on the basis of Naegle's results, it would appear likely that the 100 nm^2 molecular electronic half adder would be at least one million to 10 million times smaller in area than a conventional silicon CMOS half adder.

8

Spintronics — Spin-Based Electronics

CONTENTS

S.A. Wolf
*Defense Advanced Research Projects
Agency and Naval Research
Laboratory*

A.Y. Chtchelkanova
Strategic Analysis, Inc.

D.M. Treger
Strategic Analysis, Inc.

Abstract

Spintronics is an acronym for a *spin transport electronics* and was originally used as a name for the Defense Advanced Research Projects Agency (DARPA) program. In 1994, DARPA started to develop magnetic field sensors and memory based on the giant magnetoresistance (GMR) effect and spin-dependent tunneling. The overall goal for the Spintronics program was *to create a new generation of electronic devices where the spin of the carriers would play a crucial role in addition to or in place of the charge.* Spintronic devices can be used as magnetic memories, magnetic field sensors, spin-based switches, modulators, isolators, transistors, diodes, and perhaps some novel devices without conventional analogues that can perform logic functions in new ways. Spintronics brings together specialists in semiconductors, magnetism, and optical electronics studying spin dynamics and transport in semiconductors, metals, superconductors, and heterostructures. As always, the key question is whether any potential benefit of such technology will be worth the production costs.

This paper will provide an overview of the field and the reference material to the original papers for the future in-depth reading. In Section 8.1 we briefly describe spin transport electronics in metallic systems and commercially available devices utilizing magnetization and spin transport properties in metals.[1,2,3] Section 8.2 addresses the issues in semiconductor spin electronics,[4] which have to be resolved to create successful devices — efficient spin injection into semiconductors and heterostructures and the search for new spin-polarized materials. It also mentions effects potentially important for spintronics devices including optical and electrical manipulation of ferromagnetism, current-induced magnetization switching and precessing, long decoherence time for optically excited spins in semiconductors, etc. Section 8.3 briefly describes proposed devices utilizing spins, and Section 8.4 covers quantum computing schemes relying on spins.

Progress in spintronics would not be possible without the maturity of electron-beam and ion-beam fabrication, molecular beam epitaxy (MBE), and the ability to manufacture devices at the nanoscale with nanoimprint lithography. Advances in magnetic microscopy for direct, real-space imaging demonstrated for the past decade[5] have also played a crucial role in new materials and device characterization.

8.1 Spin Transport Electronics in Metallic Systems

Conventional electronic devices are based on charge transport, and their performance is limited by the speed and dissipation of the energy of the carriers (electrons). Prospective spintronic devices utilize the direction and coupling of the spin of the electron in addition to or in place of the charge. Spin orientation along the quantization axis (magnetic field) is dubbed as *up-spin* and in the opposite direction as *down-spin*. Electron spin is a major source for magnetic fields in solids.

8.1.1 High-Density Data Storage and High-Sensitivity Read Heads

The spin has been an important part of magnetic high-density data storage technology for many years. Hard disk drives (HDD) store information as tiny magnetized regions along concentric tracks. Magnetization pointing in one direction denotes a zero bit, and in the opposite direction a one bit. Areal density, expressed as billions of bits per square inch of disk surface area (Gbits/in^2), is the product of linear density (bits of information per inch of track) multiplied by track density (tracks per inch) and varies with disk radius. The recent progress in the increasing storage areal density is due to high-sensitivity read heads made possible after the discovery of the GMR effect. The first commercially available disk drive using a GMR sensor head was the 1998 IBM Deskstar 16GP disk drive that had 2.69 Gbits/in^2 areal density. The current commercially available density is up to 40 Gbits/in^2 and 110 Gbits/in^2 is under investigation.

8.1.2 The Giant Magnetoresistance Effect

The GMR effect was discovered in the late 1980s when two research groups performed magnetoresistivity studies[6,7] of heterostructures comprised of alternating thin (10–100 Å) metallic layers of magnetic and nonmagnetic metals in the presence of high magnetic fields (2T) at low temperatures (4.2 K). They saw resistance changes up to 50% between the resistivities at zero field and in the saturated state. At room temperature the GMR effect was smaller but still significant. The resistance of the GMR structure is lower if the directions of the magnetization of the ferromagnetic (FM) layers are aligned than when they are anti-aligned. At zero field with the thickness of nonmagnetic metal in the range of 8–18 Å, indirect electron exchange provides a mechanism to induce an anti-parallel alignment of the magnetic layers.[8] Application of an external magnetic field (100–1000 Oe) overcomes the anti-parallel coupling between the magnetic layers and, as a result, the resistance changes in the range of 10–60%. The GMR effect is measured as the change in resistivity divided by the resistivity at large fields, typically termed $\Delta R/R$. GMR is observed when the current is in the plane of the layers (CIP) and perpendicular to the plane of the layers (CPP). GMR is attributed to spin-dependent scattering at the interfaces and spin-dependent conductivity.

It has long been known that electrons in metals have two spin states and that, when electric field is applied to a metal, two approximately independent currents flow. In nonmagnetic metals these two spin channels are equivalent because they have the same Fermi energy density of states (DOS) and the same electron velocities, but in ferromagnetic transition metals they are quite different. *Spin polarization* is defined as the ratio of the difference between the up- and down-spin channel population to the total number of carriers:

$$P = (n\uparrow - n\downarrow)/(n\uparrow + n\downarrow)$$

For most electronics applications, only the total conductivity resulting from these two parallel currents or spin channels is important. Recent technologically important effects such as GMR take advantage of the differences in electron transport between the two spin-channels. The physics of spin-dependent transport in magnetic multilayers is explained in Reference 9.

8.1.3 Spin Valves

A GMR structure widely used in HDD read heads is a spin valve (SV), originally proposed by IBM in 1994.[10] An SV has two ferromagnetic layers (alloys of nickel, iron, and cobalt) sandwiching a thin nonmagnetic metal (usually copper) with one of the two magnetic layers being "pinned," i.e., the magnetization in that layer is relatively insensitive to moderate magnetic fields.[11] The other magnetic layer is called the *free* layer, and its magnetization can be changed by application of a relatively small magnetic field. As the alignment of the magnetizations in the two layers changes from parallel to anti-parallel, the resistance of the spin valve typically rises from 5% to 10%. Pinning is usually accomplished by using an antiferromagnetic layer that is in close contact with the pinned magnetic layer. In SV read heads for high-density recording, the magnetic moment of the pinned layer is fixed along the transverse direction by exchange coupling with an antiferromagnetic layer (FeMn), while the magnetic moment of the free layer rotates in response to signal fields. The resultant spin-valve response is given by $\Delta R \sim cos(\theta_1 - \theta_2)$, where θ_1 and θ_2 are the angles between the magnetization direction of the free layer and pinned layer and the direction parallel to the plane of the media magnetization, as seen in Figure 8.1. When a weak magnetic field, such as from a bit on a hard disk, passes beneath the read head, the magnetic orientation of the free layer changes its direction relative to the pinned layer, generating a change in electrical resistance due to the GMR effect. Because an SV is so important for industrial applications, there have been many improvements in recent years. The simple pinned layer is replaced with a synthetic antiferromagnet — two magnetic layers separated by a very thin (~10Å) nonmagnetic conductor, usually ruthenium.[12] The magnetizations in the two magnetic layers are strongly anti-parallel coupled and are thus effectively immune to outside magnetic fields. The second innovation is the nano-oxide layer or NOL, which is formed at the outside surface of the soft magnetic film. This layer reduces resistance due to surface scattering,[13] thus reducing background resistance and thereby increasing the percentage change in magnetoresistance of the structure. The magnetoresistance of spin valves has increased dramatically from about 5% in early heads to about 15 to 20% today, using synthetic antiferromagnets and NOLs.

8.1.4 Magnetic Tunnel Junctions

A magnetic tunnel junction (MTJ) is a device in which a pinned layer and a free layer are separated by a very thin insulating layer, commonly aluminum oxide. The tunneling resistance is modulated by the magnetic field in the same way as in a spin valve, and recently it has been shown to exhibit more than a 50% change[14] in the magnetoresistance while requiring a saturating magnetic field equal to or somewhat less than that required for spin valves. Spin-dependent tunneling (STD) between FM materials separated by an insulator (I) was first studied in 1975.[15] In 1995, FM-I-FM tunneling was measured in $Fe/Al_2O_3/Fe$,[16] $Fe/Al_2O_3/Ni_{1-x}Fe_x$,[17] $CoFe/Al_2O_3/Co$, and NiFe junctions.[18] It was proposed that these junctions have potential use as low-power field sensors and memory elements. Because the

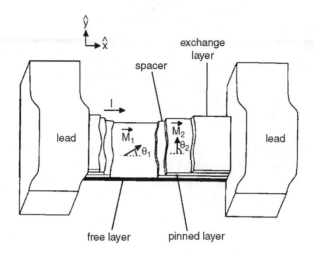

FIGURE 8.1   Schematic of an IBM GMR spin valve sensor used in read head. I = the direction of the current, M = magnetization, θ_1 and θ_2 = angles from the longitudinal direction. The read head is moving perpendicular to the XY plane. (Adapted from C. Tsang et al., Design, fabrication and testing of spin-valve read heads for high density recording. *IEEE Trans. Magn.* 30, 3801, 1994. With permission.)

tunneling current density is usually small, MTJ devices tend to have high resistances. The resistance of an MTJ exponentially depends on the thickness of the insulating layer. Uniformity over the MTJ insulating layer is crucial to device operation. Also, reproducibility from MTJ to MTJ is important for the proper working of arrays of such devices. MTJs currently have been made as small as 0.3×0.8 μm^2, and the thickness of the insulating layer is typically 1 nm. Another potential issue is the reproducibility of the magnetic states of the ferromagnetic layers separated by 1 nm. Surprisingly, this issue has been resolved by careful device processing.

8.1.5   Device Applications for Spin Valves and Magnetic Tunnel Junctions

Spin valve and MTJ device applications are expanding and already include magnetic field sensors, read heads for hard drives, galvanic isolators, and magnetoresistive random access memory (MRAM). General-purpose GMR sensors have been introduced in the past 7 years[19] (Figure 8.2), and several companies are producing GMR sensors for internal consumption. A new magnetic field sensor utilizing CPP geometry, a spin-valve transistor, was demonstrated in 1995,[20] but it was not commercially produced. No commercial sensors using MTJ structures are available, but one is under development.[21]

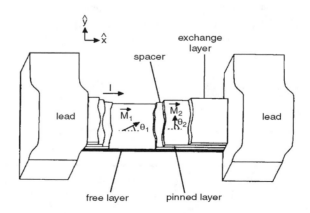

FIGURE 8.2   GMR sensor. (Courtesy of NVE Corporation, Eden Prairie, MN.)

FIGURE 8.3   256 Kb Motorola MRAM. (From S.A. Wolf et al., Spintronics: a spin-based electronics vision for the future. *Science.* 294, 1488, 2001. With permission.)

The GMR-based galvanic isolator is a combination of an integrated coil and a GMR sensor on an integrated circuit chip. GMR isolators introduced in 2000 eliminate ground noise in communications between electronics blocks, thus performing a function similar to that of opto-isolators — providing electrical isolation of grounds between electronic circuits. The GMR isolator is ideally suited for integration with other communications circuits and the packaging of a large number of isolation channels on a single chip. Complex multichannel, bidirectional isolators are currently in production. The speed of the GMR isolator is currently 10 times faster than today's opto-isolators and can eventually be 100 times faster, with the principal speed limitations identified as the switching speed of the magnetic materials and the speed of the associated electronics.

8.1.6   Magnetoresistive Random Access Memory

MRAM (Figure 8.3) uses magnetic hysteresis to store data and magnetoresistance to read data. GMR-based MTJ[14,22] or pseudo-spin valve[23] memory cells are integrated on a semiconductor chip and function like a static semiconductor RAM chip with the added feature that the data is retained with power off, i.e., nonvolatility. Potential advantages of the MRAM compared with silicon Electrically Erasable Programmable Read-Only Memory (EEPROM) and flash memory are:

- One thousand times faster write times
- No wearout with write cycling (EEPROM and flash memories wear out with about one million write cycles)
- Lower energy for writing

MRAM data access times are about 1/10,000 that of hard disk drives. MRAM is not yet available commercially, but production of at least 4 MB MRAM is anticipated in 2–3 years. See Table 8.1 for a description of how MRAM tracks with the Semiconductor Industries Association's roadmap for memory technologies. Excellent reviews of the current status of MRAM with detailed descriptions of the working principles of the devices and progress in incorporating them into existing semiconductor technology are given in References 24, 25, and 26. In just a few years we have seen devices develop very rapidly. Theory suggests that, with new materials or structures and with switching controlled by magnetism or current, further improvements in the magnetoresistance effect[27–29] can be achieved from the 15% to 40% available today in GMR and MTJ structures to hundreds of percentage point changes at room temperature.

TABLE 8.1 Comparison of Memory Technologies for the Year 2011

Technology	CMOS			
	DRAM	Flash	SRAM	MRAM
Reference	SIA 1999	SIA 1999	SIA 1999	
Generation at Introduction	64 GB	64 GB	180 MB/cm²	64 GB
Circuit Speed	150 MHz	150 MHz	913 MHz	>500 MHz
Feature Size	50 nm	50 nm	35 nm	<50 nm
Access Time	10ns	10 ns	1.1 ns	<2 ns
Write Time	10 ns	10 μs	1.1 ns	<10 ns
Erase Time	<1 ns	10 μs	1.1 ns	N/A
Retention Time	2–4 s	10 years	N/A	Infinite
Endurance Cycles	Infinite	10^5	Infinite	Infinite
Operating Voltage (V)	0.5–0.6 V	5 V	0.5–0.6 V	<1 V
Voltage to Switch State	0.2 V	5 V	0.5–0.6 V	<50 mV
Cell Size	2.5 F²*/bit	2F²/bit	12F²/bit	2F²/bit
	0.0005 μm²			

* F = minimal lithographic feature size.

8.2 Issues in Spin Electronics

8.2.1 Spin Injection

To be commercially useful, a spintronic device has to work at room temperature and be compatible with existing semiconductor-based electronics. Almost every imaginable spintronics device should have means of spin injection, manipulation, and detection. Well-known sources of spin-polarized electrons are ferromagnetic materials (FM). The magnetic field of the FM interacts with the spins of electrons; as a result, the majority of electrons are in the states such that their spins are aligned with the local magnetization. Spin-polarized current injection was achieved from FM into superconductors,[30] from FM into normal metals,[31] between two FM separated by a thin insulating film,[15] from a normal metal using magnetic semiconductors as an aligner into nonmagnetic semiconductors[32–34] and hole injection from p-type FM into nonmagnetic semiconductor.[35]

8.2.1.1 Electrical Injection

The best electrical injection (forming an ohmic contact) from FM into semiconductors reported to date[36–38] reached 4.5% at T < 10 K. To date the best room temperature electric injection from an FM into a semiconductor was reported by Hammar et al.,[39–42] but it is not clear if experimental data establishes conclusively electrical injection into the semiconductor.[43–45]

The main obstacle to effective electrical injection is the conductivity mismatch between the FM electrode and the semiconductor.[46] The effectiveness of the spin injection depends on the ratio of the (spin-dependent) conductivities of the FM and nonferromagnet (NFM) electrodes, σ_F and σ_N, respectively. When $\sigma_F = \sigma_N$ as in the case of a typical metal, then efficient and substantial spin injection can occur; but when the NFM electrode is a semiconductor, $\sigma_F \gg \sigma_N$ and the spin-injection efficiency will be very low. Only for a ferromagnet where the conduction electrons are 100% spin-polarized can efficient spin injection be expected in diffusive transport.

8.2.1.2 Tunnel Injection

Insertion of a tunnel contact (T) at a FM-to-normal conductor interface can be a solution of the conductivity mismatch problem.[47] Two possible configurations considered were FM-T-semiconductor and Shottky barrier diode. A 2% room temperature tunneling spin injection was achieved from Fe into GaAs in the Shottky diode configuration.[48] Measurements of spin polarization of the current transmission across an FM-insulator-2DEG junction yield 40% with little dependence over the range 4 K to 295 K.[49] A 30% injection efficiency was achieved from an Fe contact into a semiconductor light-emitting diode structure (T = 4.5 K) and persisted to almost room temperature (4% at T = 240 K).[50]

The Shottky barrier formed at the Fe/GaAs interface provides a natural tunnel barrier for injecting spin-polarized electrons under reverse bias.

Tunnel injection of "hot" (energy much greater than E_F) electrons into a ferromagnet can be used to create spin-polarized currents. Because inelastic mean free paths for majority and minority electrons differ significantly, hot electron passage through a 3-nm Co layer can result in 90% spin-polarized currents.[20,51,52] The highly polarized current then can be used for further injection into the semiconductor. The disadvantage of hot electron injection is that the overall efficiency is low.

8.2.2 Research in New Materials for Spintronics

Progress in new materials engineering and research is very important because, as was mentioned above, one can expect efficient spin injection if the source of spin-polarized electrons is 100% spin polarized. Measurements in a variety of metals, half-metals, metallic binary oxides, Heusler alloys, and other compounds have shown a 35 to 90% range of spin polarization.[53–56] The measurement methods employed were spin-polarized photoemission, spin-polarized tunneling, and superconducting point of contact (Andreev reflection). This section describes materials as sources of spin-polarized carriers.

8.2.2.1 Magnetic Semiconductors

Materials combining ferromagnetic and semiconducting properties, magnetic semiconductors, can be a very attractive option as a source of spin-polarized carriers because there are no interface problems or conductivity mismatch. To achieve large spin polarization in semiconductors, the Zeeman splitting of the conduction (valence) band must be greater than the Fermi energy (E_F) of the electrons (holes). In concentrated materials, this occurs easily because the net magnetization upon ordering is proportional to the concentration of magnetic species. If the concentration of magnetic species is low, ~ 5% and below, large externally applied magnetic fields and low temperatures are required to produce large polarization.

8.2.2.1.1 Mixed-Valence Manganese Perovskites with the General Formula

$A_{1-x}B_xMnO_3$ (where A = La, Nd, or Pr, and B = Ca, Ba, or Sr) have been extensively studied[57] for their magnetic and transport properties resulting in *colossal magnetoresistance* (CMR) observed at the vicinity of the Curie Temperature, T_C. Spin ordering in these materials can be obtained by applying external magnetic fields and/or lowering the temperature. Magnetoresistance values of more than 400% at low temperatures were measured in all-oxide spin valves in which $La_{0.7}Sr_{0.3}MnO_3$ electrodes were separated by a thin insulating layer, and spin polarization was estimated to be at least 83% at Fermi level.[58] For many of the mixed-valence manganese perovskites the T_C is above room temperature, but photoemission data show that the spin polarization decreases to almost zero at room temperature.

8.2.2.1.2 Europium Chalcogenides

Europium chalcogenides with the general formula EuX (X = O, S, Se, Te), in which the magnetic ion Eu^{2+} resides on every lattice site, have low T_C (~80 K) with little hope of improvement.[59]

8.2.2.1.3 Diluted Magnetic Semiconductors (DMS)

DMS are alloys in which atoms of a group-II element of a II-VI compound (CdTe or ZnS) semiconductor are randomly replaced by magnetic atoms (e.g., Mn). However, II-VI-based DMS have been very difficult to dope to create *p*- and *n*-type semiconductors used in electronic applications. At high magnetic atom concentration and low temperature, antiferromagnetic interaction between magnetic ions creates a magnetically ordered phase. Below the T_C, DMS exhibit ferromagnetic behavior. So far for II-VI semiconductors (e.g., CdMnTe), the ferromagnetic phase has been observed below 2 K.[60]

The equilibrium solubility of transitional atoms in III-V compound semiconductors is low, and only sophisticated nonequilibrium growth conditions of low-temperature molecular beam epitaxy allowed the successful preparation of (In,Mn)As[61] for which the hole-induced ferromagnetic ordering was detected below 35 K.[62] A GaAs-based DMS was fabricated in 1996,[63] and T_C around 110 K was reported for some samples in 1998.[64] For a detailed review of the properties of ferromagnetic III-V semiconductors, see Ohno.[65] GaMnAs heterostructures can be epitaxially grown with abrupt interfaces and with atomically controlled layered thickness.[64–66] Tanaka and Higo reported large tunneling magnetoresistance in GaMnAs/AlAs/GaMnAs ferromagnetic semiconductor tunneling junctions.[67] Although the total change in the magnetoresistance was 44% at 4.2 K, the TMR ratio due to SV effect was estimated to be only 15 to 19%.[68] From a technological point of view, the use of semiconducting magnetic elements may reduce the large currents required in all-metal spin valves.

So far Mn is the only successfully used ion for doping DMS. Some groups have started to use Cr ions to dope GaAs and grow new DMS $Ga_{1-x}Cr_xAs$ by MBE.[69]

A thin-film magnetic system consisting of nanoscale $Mn_{11}Ge_8$ ferromagnetic clusters (T_C ~ room temperature) embedded into a Mn_xGe_{1-x} DMS semiconductor matrix exhibits magnetoresistance at low fields (2 kOe) and at low temperatures (22 K).[70] A group IV DMS Mn_xGe_{1-x} was successfully grown using MBE in which the Curie temperature was found linearly dependent on the Mn concentration from 25 to 116 K.[71] Theoretical calculations predict even higher T_C than was measured.

T_C calculations[72] predicting above room temperature ferromagnetism prompted active interest in DMS GaMnN.[73] Ferromagnetic behavior was recently confirmed at 300 K,[74,75] and T_C is estimated to be 940 K.[76]

8.2.2.2 Half-Metallic Ferromagnets

Half-metallic ferromagnets (HMF) are another class of potential sources of spin-polarized electrons. The Fermi level of HMF intersects the majority spin electron band while the minority band has an energy gap near the Fermi level. The HMFs have simultaneously both metallic and semiconducting characteristics, and theory predicts that the conduction electrons are 100% spin polarized.[77] As a result the magnetoresistance in magnetic multilayers or tunneling junctions is expected to be significantly higher than with conventional ferromagnetic materials.

8.2.2.2.1 Heusler Alloys

Spin-polarized tunneling from NiMnSb was studied in a trilayer $NiMnSb/Al_2O_3/Al$ TMJ.[78] A 58% spin-polarization for Ni_2MnGa was measured via Andreev reflection[82] and TC ~340 K was measured by Dong et al.[79] There are T_C measurements available for other HMF[79] such as Ni2MnGe (~320 K) and Ni2MnIn (~290 K), but no spin-polarization data is mentioned.

8.2.2.2.2 Half-Metallic Oxides

Half-metallic oxides were predicted[80] to be potential sources of fully spin-polarized electrons. Spin-resolved photoemission from polycrystalline CrO_2 films have shown[81] a spin polarization of nearly +100% with confirmation by point contact Andreev reflection measurements.[82,83] Large TMR effects were observed for magnetite Fe_3O_4[84] (T_C ~860 K) at helium temperatures, but at room temperature the effect is less than 1%.

8.2.2.2.3 Transition Metal Pnictides

Transition metal pnictides MnAs, MnSb, CrAs, CrSb can be easily incorporated into existing semiconductor technology if proven to be useful as spin-polarized sources. For magneto-optical device applications, such as an optical isolator, high optical transmission and large magneto-optical effects must be realized at room temperature. The GaAs:MnAs nanocluster system was fabricated and characterized by Shimizu and Tanaka.[85] Zinc blende structure CrAs thin films were synthesized on GaAs(001) substrates by MBE and show a ferromagnetic behavior at room temperature. Calculations predict a highly spin-polarized electronic band structure.[86]

Thin films of CrSb grown by MBE on GaAs, (Al,Ga)Sb, and GaSb are found to have a zinc blende structure and $T_C > 400$.[87]

A thousand-fold magnetoresistance effect was discovered in granular MnSb films[88] at room temperature with low magnetic fields (less than 0.5 T). A 20% positive, photoinduced magnetoresistance effect was observed in GaAs with inclusions of MnSb nanomagnets when irradiated with photons with energies above the band gap of GaAs. This effect is presumably due to an enhancement of the tunneling probability between MnSb islands by photogenerated carriers in the GaAs matrix.[89]

8.2.2.2.4 *Diluted Magnetic Semiconductor Oxide*

Epitaxial thin films of diluted magnetic semiconductor oxide, Mn-doped ZnO, fabricated by pulse-laser deposition, showed considerable magnetoresistance at low temperature.[90] Following the prediction that ZnO has a T_C above room temperature[72] and would become ferromagnetic by doping with $3d$ transition elements, intensive combinatorial work began in Japan on dilute magnetic oxides. Room-temperature ferromagnetism was reported in transparent Co-doped (up to 8%) Ti_2O atanase films.[91] This material is transparent to visible light and might be of great importance to optoelectronics.

8.2.2.3 Borides

Electronic structure calculations for La-doped CaB_6[92] ($T_C \sim 900$ K) showed that it has a semiconductor band structure and can be considered as a new semiconducting material for spin electronics.

8.2.3 Optically and Electrically Controlled Ferromagnetism

Optically and electrically controlled ferromagnetism is now a low-temperature effect, but extension to higher temperatures may have important implications in areas ranging from optical storage to photonically and electrically driven micromechanical elements.

Ferromagnetism was induced by photogenerated carriers in an MBE-grown p-(In,Mn)As/GaSb film at temperatures below 35 K.[93] The order was preserved even after the light was switched off and recovered to the original paramagnetic condition above 35 K. The results were explained in terms of hole transfer from GaSb to InMaAs, which then enhanced the ferromagnetic spin exchange between Mn ions in the heterostructures.

Electric-field control of hole-induced ferromagnetism was demonstrated in (In,Mn)As using an insulating-gate field-effect transistor (FET) at temperatures below 20 K[94] in 2000. Manganese substitutes indium and provides a localized magnetic moment and a hole. These holes mediate magnetic interactions resulting in ferromagnetism. Changing the hole concentration by applying a gate voltage modifies the ferromagnetic properties in the DMS below the transition temperature. It was also found that the new group IV DMS, Mn_xGe_{1-x}, allows control over ferromagnetic order by applying a 0.5 V gate voltage.[71]

8.2.4 Current-Induced Magnetization Switching and Spin Wave Generation

Current-induced magnetization switching and spin wave generation are recently discovered effects allowing manipulation of magnetization of the FM layers in heterostructures. The change in scattering of the electrons traversing alternating ferromagnetic and nonmagnetic multilayers depends on the relative orientation of the magnetization. The scattering of the electrons within the alternating layers of FM and regular metals can affect the moments in the magnets. Theoretical calculations indicate[95–98] that spin-polarized currents perpendicular to the layers can transfer angular momentum between layers, causing torque on the magnetic moments and, as a result, causing rotation and possibly even precession of the layer magnetization and high-frequency switching. A few independent experiments have demonstrated current-induced magnetization rotation.[99–105] Current-induced magnetization switching has potential applications to high-speed, high-density GMR-based MRAM as a convenient writing process.[106] Other possible applications might include spin-filter devices and spinwave-emitting diodes.[107] In the presence of a large magnetic field, the spin rotation can become a high-frequency coherent precession of the moments. This effect, labeled *spin amplification by simulated emission of radiation* (SASER), was predicted by Berger,[95] and the feasibility of it was studied by Tsoi et al.[105]

8.2.5 Optically Excited Spin States in Semiconductors

Optically excited spin states in semiconductors have been studied by optoelectronic manipulation of spins allowing spatial selectivity and temporal resolution. This is a way to study spin transport in bulk semiconductors and across junctions in heterostructures. For a very detailed description of electron spin and optical coherence in semiconductors see Awschalom and Kikkawa in Reference 108.

8.2.5.1   The Measurement of Long-Spin Lifetimes

Time-resolved optical experiments[109–112] have revealed a remarkable persistence of coherent electron spin states. These insensitivities to environmental sources of decoherence in a wide variety of direct bandgap semiconductors and long-spin lifetimes in bulk semiconductors have been shown to exceed a hundred nanoseconds. Optical pulses are used both to create a superposition of the basis spin states defined by an applied magnetic field and to follow the phase, amplitude, and location of the resulting electronic spin precession (coherence) in bulk semiconductors, heterostructures, and quantum dots. In heterostructures and quantum dots, nanosecond dynamics persist to room temperature, providing pathways toward practical coherent quantum electronics.

8.2.5.2   Transfer of Spin Coherence across Heterojunctions

Data show that spin coherence across heterojunctions[113] can be preserved when a "pool" of coherent spins is crossing the interface between two semiconductors. A phase shift of spins on opposite sides of the interface can be set by the difference in electron g-factors between the two materials and can be controlled by utilizing epitaxial growth techniques. More recent measurements have established an increase in spin injection efficiency with bias-driven transport: relative increases of up to 500% in electrically biased structures, and 4000% in *p-n* junctions with intrinsic bias have been observed[114] relative to the unbiased interfaces.

8.2.5.3   Defects and Spin Coherence

Another important aspect for the development of spin-based electronics is the effect of defects on spin coherence. In this context the III-V semiconductor GaN is intriguing because it combines a high density of charged threading dislocations with high optical quality, allowing optical investigation of the effects of momentum scattering on coherent electronic spin states. Despite an increase of eight orders of magnitude in the density of charged threading dislocations, studies reveal electron spin coherence times in GaN epilayers reach ~20 ns at T = 5 K, with observable coherent precession at room temperature.[115] Detailed investigations reveal a dependence on both magnetic field and temperature qualitatively similar to previous studies in GaAs, suggesting a common origin for spin relaxation in these systems.

8.3   Potential Spintronics Devices

The devices described below are highly experimental and can be considered as potential proof of concept that integrated spintronic devices can be built.

8.3.1   Light-Emitting Diode

A spin-polarized light-emitting diode (spin-LED), schematically depicted in Figure 8.4, is a very important device used primarily to measure the effectiveness of the injection of spin-polarized currents into semiconductor heterostructures. In a conventional LED, electrons and holes recombine in the vicinity of a *p-n* junction under forward bias, producing light. The resulting light is unpolarized because all spin-state carriers are equally populated and allow transitions to occur with equal probability. In a spin-LED, recombination of spin-polarized carriers results in the emission of right (σ^-) or left (σ^+) circularly polarized light in the direction normal to the surface according to selection rules.[116] Polarization analysis of the resulting electroluminescence (EL) provides quantitative measurement of the injection efficiency. A spin-LED was used to measure the electrical spin injection of electrons into a GaAs quantum well at low temperatures (4.2K) and with an external magnetic field up to 8 T. It was demonstrated using a diluted magnetic semiconductor spin-aligner concept.[32] Unpolarized electrons are injected from a metal into a DMS incorporating Mn magnetic ions. Even a small external magnetic field produces Zeeman splitting of the conduction band states, and the injected electrons became highly polarized because of the *sp-d* exchange interaction with the Mn ions. The resulting spin-polarized electrons diffuse into the nonmagnetic semiconductor. Two groups[33,34] used Mn-doped ZnSe quaternary DMS alloys as spin

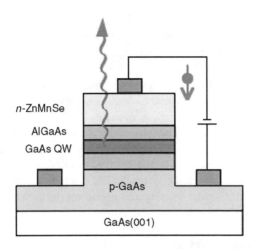

FIGURE 8.4 Schematic cross-sections of the samples in spin LED. (Adapted from B.T. Jonker et al., Robust electrical spin injection into a semiconductor heterostructure. *Phys. Rev. B* 62, 8180, 2000. With permission.)

aligners and achieved 90% and 50% injection efficiency, respectively. Another group electrically injected holes[35] into III-V heterostructures based on GaAs at T < 52 K. Injection into nonmagnetic semiconductors is achieved at zero field using a *p*-type ferromagnetic semiconductor (Ga,Mn)As as the spin polarizer. It was measured that hole spins can be injected and transported across the interfaces over 200 nm, which confirms the idea that spin-polarized transport can survive the length of the device. At this time, however, there are no plans for mass production of a spin-LED, but they are very useful in the measurement of spin injection despite low temperatures and high magnetic field requirements. If progress in new materials development brings to fruition an *n*-doped high Curie temperature FS, then the role of a mass-produced spin-LED can be reconsidered.

8.3.2 Field-Effect Transistor

The spin-polarized field-effect transistor (spin-FET), schematically depicted in Figure 8.5, was proposed by Datta and Das in 1990.[117] The device was proposed as an electronic analog of the electro-optic modulator. We will briefly describe how the electro-optic modulator works. As light passes through the electro-optic material, the two components of polarization (z and y) undergo a phase shift because dielectric constants, ε_{zz} and ε_{yy}, differ. The gate voltage can control the phase shift between polarization components by changing the dielectric constants. As a result, the output power of the light coming out of the analyzer can be modulated by changing the gate voltage. In a proposed spin-FET, the spin-polarized electrons are injected and collected by the FM electrodes. Two physical phenomena are important in understanding the principle of the spin-FET. First, at the Fermi level in FM materials, the density of states for electrons with one spin exceeds that for the other; and as a result, the electrode preferentially injects and detects electrons with a particular spin. Second, in a 2DEG in narrow gap semiconductors, there is an energy splitting between electrons with up-spin and down-spin even in the absence of an

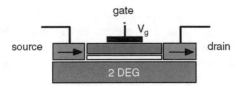

FIGURE 8.5 Schematic of spin-FET. The current modulation is controlled by a gate voltage which affects the spin precession due to spin orbit coupling in 2DEG semiconductor. FM source and drain are used to inject and detect specific spin orientation. The magnetization direction can be considered as *4th terminal*.

external magnetic field. The main effect responsible for *zero-field splitting* is the Rashba term in the effective mass Hamiltonian.[118] The electric field perpendicular to the 2DEG at the heterojunction interface yields an effective magnetic field for moving electrons and lifts the spin degeneracy. The spin–orbit interaction is proportional to the value of the electric field at the interface and can be controlled by the applied gate voltage. A spin–orbit interaction causes the spins of the carriers to precess. The modulation of the current can be controlled by changing the alignment of the carrier spins with the magnetization vector of the collector electrode. By changing the interface electric field, the gate electrode on the top of the device can be used to control spin–orbit interaction. In 1997, gate control of the spin–orbit interaction in an inverted $In_{0.53}Ga_{0.47}As/In_{0.52}Al_{0.48}As$ heterostructures was reported.[119] Another spintronics semiconductor field-effect transistor operating at low temperatures was designed in 1999.[37] In this device, resistance modulation was achieved through the spin-valve effect; but it was done by having ferromagnetic contacts with different coercivities and varying an applied magnetic field. At this time both devices operate at low temperatures and achieve very small changes in resistance.

8.3.3 Resonant Tunneling Diode

The resonant tunneling diode, RTD, is normally a vertical tunneling diode with the vertical dimensions produced by growth and the lateral dimensions produced by lithography. If a quantum well is placed between two thin barriers, the tunneling probability is greatly enhanced when the energy level in the quantum well coincides with the Fermi energy (resonant tunneling). The typical dimensions in the tunneling direction are a few atomic layers thick and determine the current and power dissipation. To produce lower power devices, smaller dimensions are required, and issues of control of the uniformity of the tunneling layer become very important because the tunneling current depends exponentially on the thickness of the tunnel barrier.

Introduction of ferromagnetic materials into RTDs can greatly enhance their functionality. Magnetization-controlled resonance tunneling in GaAs/ErAs[120] RTDs shows splitting and enhancement of the resonant channels, which depend on the orientation of the external magnetic field with respect to the interface. Theoretical calculations predict that polarization of the transmitted beam can achieve 50%[121] or even higher.[122] Spin RTDs can be used both as spin filters and energy filters. Current–voltage characteristics of AlAs/GaAs/AlAs double-barrier resonant tunneling diodes with ferromagnetic *p*-type (Ga, Mn)As on one side and *p*-type GaAs on the other, have been studied.[64,123] A series of resonant peaks have been observed in both polarities, i.e., injecting holes from *p*-type GaAs and from (Ga, Mn)As. When holes are injected from the (Ga, Mn)As side, spontaneous resonant peak splitting has been observed below the ferromagnetic transition temperature of (Ga, Mn)As without magnetic field. The temperature dependence of the splitting is explained by the spontaneous spin splitting in the valence band of ferromagnetic (Ga, Mn)As.

Introducing a ferromagnetic quantum well in a ferromagnetic junction is shown to greatly enhance the tunneling magnetoresistance effect[124] due to spin filtering as well as energy filtering.

8.3.4 New Spintronic Device Proposals

New spintronic devices are highly speculative and have not been implemented yet. One proposal[125] is considering the possibility of constructing *unipolar* electronic devices by utilizing ferromagnetic semiconductor materials with variable magnetization directions. Such devices should behave very similarly to *p-n diodes* and *bipolar transistors* and could be applicable for magnetic sensing, memory, and logic.

Another theoretical device model[126] is a *spin-polarized p-n junction* with spin polarization induced either optically (in which case it is a solar cell) or electronically to majority or minority carriers. It is demonstrated that spin polarization can be injected through the depletion layer by both minority and majority carriers, making semiconductor devices such as spin-polarized solar cells and bipolar transistors feasible. Spin-polarized *p-n* junctions allow for spin-polarized current generation, spin amplification, voltage control of spin polarization, and a significant extension of spin diffusion range.

Spin filters can possibly be constructed by carefully engineering[127] ordered interfaces between FM and S, and a ballistic spin-filter transistor[128] can be added to the growing list of possible spintronic devices.

8.4 Quantum Computation and Spintronics

The introduction of coherent spins into ferromagnetic structures could lead to a new class of *quantum spintronics*. As an example, some proposed quantum computation schemes rely on the controllable interaction of coherent spins with ferromagnetic materials to produce quantum logic operations.[129] There is also experimental evidence that ferromagnetic materials can be used to imprint nuclear spins in semiconductors, offering a way of manipulating and storing information at the atomic scale.[130]

Semiconductor-based quantum spin electronics is focused on developing solid-state quantum information processing devices. Nuclear spins have been proposed as candidates for storing both classical and quantum information due to spin lifetimes that exceed those of electrons by at least several orders of magnitude.

8.4.1 Nuclear Spin Quantum Computer

In a nuclear spin quantum computer proposal, individual phosphorous nuclei ^{31}P embedded in silicon are treated as quantum bits (qubits),[131] as shown in Figure 8.6. By placing a gate electrode (A) over a qubit and applying a bias, one can control the overlap of the bound electron with the nucleus and thus the hyperfine interaction between nuclear spin and electron spin (controlled one-bit rotations). Another gate (J) controls the potential barrier between neighboring nuclear spins, allowing them to interact via electron-spin exchange (entanglement).

8.4.2 Spin-Resonance Transistor

Spin-resonance transistor (SRT) from Si-Ge compounds,[132] seen in Figure 8.7, was proposed to sense and control a single donor (^{31}P) electron spin. The choice of group-IV semiconductors has the advantage

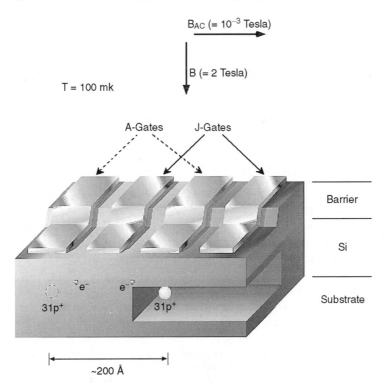

FIGURE 8.6 Schematic representation of two cells in one-dimensional array containing ^{31}P donors and electrons in Si host separated by metal gates. A-gates control the resonant frequency of the nuclear spin qubits, and J-gates control the electron-mediated coupling between adjacent nuclear spins. The ledge over which the gates cross localizes the gate electric field in the vicinity of the donors. (Adapted from B. Kane, *Nature*, 393, 133, 1998. With permission.)

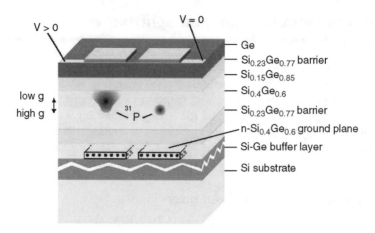

FIGURE 8.7 Spin resonance transistor. The left transistor gate is biased $V > 0$, producing single-qubit unitary transformations in the left SRT. The right gate is unbiased. The n-$Si_{0.4}Ge_{0.6}$ ground plane is a counterelectrode to the gate. It is used for sensing the spin. (Adapted from R. Vrijen et al., Electron spin-resonance transistors for quantum computing in silicon-germanium heterostructures. *Phys. Rev. A* 62, 012306, 2000. With permission.)

of reduced spin–orbit coupling that could lead to longer spin coherence times. One- and two-qubit operations are performed by applying a gate bias. The electric field bias pulls the electron wave function away from the dopant ion into layers of different alloy compositions. Because different layers have different g-factors, this displacement changes the spin Zeeman energy, allowing single-bit operations. By displacing the electron even further, the overlap with neighboring qubits is achieved, allowing two-qubit operations.

8.4.3   NMR Quantum Computer

An NMR quantum computer was proposed to be constructed from isolated chains of ^{29}Si (spin 1/2) embedded in ^{28}Si or ^{30}Si (spin 0) and attached to a thin bridge structure. The nuclei within each chain are distinguished by a large magnetic field gradient created by a micromagnet. Magnetic moment of the ensemble of nuclei causes mechanical force on the flexible bridge, which can be detected by Magnetic Resonance Force Microscope. Issues of initialization, manipulation, computation, and readout are addressed in Reference 133.

8.4.4   Quantum Dots as Quantum Bits

It has been proposed that the spin of an electron confined to quantum dots is a promising candidate for quantum bits and that arrays of quantum dots can be used in principle to implement a large-scale quantum computer.[134,135] Quantum operations in these proposals are provided by the coupling of electron spins in neighboring quantum dots by an exchange interaction between them. This interaction can be switched by applying controlled gate voltage pulses, thus allowing realization of fundamental quantum gates such as the exclusive OR. The read-out of such a spin-qubit can be performed efficiently as a spin-polarized electric current passing through the dot[136] or optically through integration in solid-state microcavities.[137] Alternatively, qubit rotations can be implemented by local electrostatic shifting of the electron into a region with a different effective magnetic field, such as that which occurs at heterointerfaces and in magnetic semiconductor structures.

Direct optical manipulation of charge-based coherent wave packets has been achieved in individual quantum dots.[138] Many proposals exist for a hybrid technique of spin-to-charge conversion that may be desirable for combining the longer spin-coherent lifetimes with the sensitivity of charge detection. Recent experiments have revealed that the transverse and longitudinal relaxation times for electron spins in insulating quantum dots are in the nanosecond regime and offer promise for their utilization as computing elements in quantum electronics.[111,139] The challenge of performing a suitably large number of qubit rotations within the spin-coherence time has been addressed by a new technique developed in quantum wells that produces rotations of electron spins on 100 femtosecond time scales.[140] In these experiments, intense laser pulses energetically tuned below the semiconductor bandgap generate a light-induced effective magnetic field via the optical Stark effect and successfully operate on quantum-confined electron spin states.

All proposals briefly described above are very far from being implemented, and it is very early to predict which are going to be the winners.

8.5   Conclusion

Spintronics is a rapidly evolving field that will have increasing impact as it matures in the decades to come. GMR devices have revolutionized magnetic disk storage, and spin-dependent tunneling devices will soon have a similar impact on random access memory. Using the spin degree of freedom in semi-conductor heterostructures, we are reaching the stage of development in which devices are appearing, and many more are on the horizon. The real dream of spintronics is the use of spin-phase coherence to develop totally new devices and methods for computation and communication because the coupling of spin coherence and coherent light is remarkable. This concept, taken to the extreme of coherence between single spins, may lead to the next revolution in computing and communication — namely, quantum information science and the quantum internet.

Acknowledgments

This paper was inspired and based on the review article.[1] We would like to express our gratitude to David Awschalom, Robert Buhrman, Jim Daughton, and Stephan von Molnár for their contributions to the original paper. We would like also to thank Berend Jonker for providing help with references and figures. We are grateful to Jim Daughton, Saied Tehrani, Bruce Kane, Eli Yablonovitch, and Ching Tsang for providing original figures.

References

1. S.Wolf et al., Spintronics: a spin-based electronics vision for the future. *Science.* 294, 1488 (2001).
2. G. Prinz, Magnetoelectronics. *Science.* 282, 1660 (1998).
3. G. Prinz, Magnetoelectronics applications. *J. Magn. Magn. Mater.* 200, 57 (1999).
4. H. Ohno, F. Matsukura, and Y. Ohno, Semiconductor spin electronics. *JSAP Intl.* 5, 4 (2002).
5. M. Freeman and B.C. Choi, Advances in magnetic microscopy. *Science.* 294, 1484 (2001).
6. M. Baibich et al., Giant magnetoresistance of (001)Fe/(001)Cr magnetic superlattices. *Phys. Rev. Lett.* 61, 2472 (1988).
7. J. Barnas, A. Fuss, R. Camley. P. Gruenberg, and W. Zinn, Novel magnetoresistance effect in layered magnetic structures: theory and experiment. *Phys. Rev. B.* 42, 8110 (1990).
8. S.S.P. Parkin, N. More, and K. Roche, *Phys. Rev. Lett.* 64, 2304 (1991).
9. W.H. Butler and X.-G. Zhang, Spin-dependent transport in magnetic multilayers, *Magnetic Interactions and Spin Transport*, Stuart Wolf and Yves Idzerda (Eds.), *Spintronics*, Kluwer, Dordrecht (2002).
10. C. Tsang et al., Design, fabrication and testing of spin-valve read heads for high density recording. *IEEE Trans. Magn.* 30, 3801 (1994).
11. B. Dieny et al., Magnetotransport properties of magnetically soft spin-valve structure. *J. Appl. Phys.* 69, 4774 (1991).
12. S.S.P. Parkin and D. Mauri, Spin engineering: direct determination of the Ruderman–Kittel–Kasuya–Yosida far-field range function in ruthenium. *Phys. Rev. B.* 44, 7131 (1991).
13. W.F. Egelhoff, Jr. et al., Specular electron scattering in metallic thin films. *J. Vac. Sci. Technol. B.* 17, 1702 (1999).

14. P. Naji, M. Durlam, S. Tehrani, J. Calder, and M. DeHerrera, A 256kb 3.0V MTJ nonvolatile magnetoresistive RAM. *ISSCC Visuals Supplement and Dig. Tech. Papers*, pp. 122–123, February 2001.

15. M. Jullière, Tunneling between ferromagnetic films. *Phys. Lett. A.* 54, 225 (1975).

16. T. Miyazaki and N. Tezuka, Giant magnetic tunneling effect in $Fe/Al_2O_3/Fe$ junction. *J. Magn. Magn. Mater.* 151, 403 (1995).

17. N. Tezuka and T. Miyazaki, Magnetic tunneling effect in $Fe/Al_2O_3/Ni_{1-x}Fe_x$ junctions. *J. Appl. Phys.* 79, 6262 (1996).

18. J.S. Moodera, L.R. Kinder, T.M. Wong, and R. Meservey, Large magnetoresistance at room temperature in ferromagnetic thin film tunnel junctions. *Phys. Rev. Lett.* 74, 3273 (1995).

19. J. Daughton, J. Brown, E. Chen, R. Beech, A. Pohm, and W. Kude, Magnetic field sensors using GMR multiplayer. *IEEE Trans. Magn.* 30, 4608 (1994).

20. D.J. Monsma, J.C. Lodder, Th.J.A. Popma, and B. Dieny, Perpendicular hot electron spin-valve effect in a new magnetic field sensor: the spin-valve transistor. *Phys. Rev. Lett.* 74, 5260 (1995).

21. M. Tondra, J. Daughton, D. Wang, and A. Fink, Picotesla field sensor design using spin-dependent tunneling devices. *J. Appl. Phys.* 83, 6688 (1998).

22. R.E. Scheuerlein et al., 2000 IEEE ISSCC Dig. Tech. Papers, Cat. No.00CH37056, 128 (2000).

23. R. Katti et al., Paper FD-04 presented at the 8th Joint MMM-Intermag Conference, San Antonio, Texas, January 7–11, 2001.

24. S. Tehrani, E. Chen, D. Durlam, M. DeHerrera, J.M. Slaughter, J. Shi, and G. Kerszykowski, High density submicron magnetoresistive random access memory (invited). *J. Appl. Phys.* 85, 5822 (1999).

25. S.S.P. Parkin et al., Exchange-biased magnetic tunnel junctions and application to nonvolatile magnetic random access memory (invited). *J. Appl. Phys.* 85, 5828 (1999).

26. M. Johnson, Magnetoelectronic memories last and last. *IEEE Spectrum.* 37, 33, (2000).

27. J.J. Verslujs and J.M.D. Coey, Magnetotransport properties of Fe_3O_4 nanocontacts. *J. Magn. Magn. Mater.* 226, 688 (2001).

28. H. Akinaga, M. Mizuguchi, K. Ono, and M. Oshima, Room-temperature thousandfold magnetoresistance change in MnSb granular films: magnetoresistive switch effect. *Appl. Phys. Lett.* 76, 357 (2000).

29. M.-H. Jo, N.D. Mathur, N.K. Todd, and M.G. Blamire, Very large magnetoresistance and coherent switching in half-metallic manganite tunnel junctions. *Phys. Rev. B.* 61, R14905 (2000).

30. R. Meservey, P.M. Tedrow, and P. Fulde, Magnetic field splitting of the quasiparticle states in superconducting aluminium films. *Phys. Rev. Lett.* 25, 1270 (1970).

31. M. Johnson and R. Silsbee, Interfacial charge-spin coupling: injection and detection of spin magnetization in metals. *Phys. Rev. Lett.* 55, 1790 (1985).

32. M. Oestreich et al., Spin injection into semiconductors. *Appl. Phys. Lett.* 74, 1251 (1999).

33. R. Fiederling et al., Injection and detection of a spin-polarized current in a light-emitting diode. *Nature.* 402, 787 (1999).

34. B.T. Jonker et al., Robust electrical spin injection into a semiconductor heterostructures. *Phys. Rev. B.* 62, 8180 (2000).

35. Y. Ohno et al., Electrical spin injection in a ferromagnetic semiconductor heterostructure. *Nature.* 402, 790–792 (1999).

36. F.G. Monzon and M. Roukes, Spin injection and the local Hall effect in InAs quantum wells. *J. Magn. Magn. Mater.* 198–199, 632 (1999).

37. S. Gardelis et al., Spin-valve effects in a semiconductor field-effect transistor: a spintronics device. *Phys. Rev. B.* 60, 7764 (1999).

38. C.-M. Hu et al., Spin-polarized transport in a two-dimensional electron gas with interdigital-ferromagnetic contacts. *Phys. Rev. B.* 63, 125333 (2001).

39. P.R. Hammar, B.R. Bennett, M.J. Yang, and M. Johnson, Observation of spin injection at a ferromagnet-semiconductor interface. *Phys. Rev. Lett.* 83, 203 (1999).

40. P.R. Hammar and M. Johnson, Potentiometric measurements of the spin-split subbands in a two-dimensional electron gas. *Phys. Rev. B.* 61, 7202 (2000).

41. P.R. Hammar et al., Observation of spin-polarized transport across a ferromagnet-two-dimensional electron gas interface (invited). *J. Appl. Phys.* 87, 4665 (2000).

42. P.R. Hammar and M. Johnson, Detection of spin-polarized electrons injected into a two-dimensional electron gas. *Phys. Rev. Lett.* 88, 066806 (2002).

43. F.G. Monzon et al., Magnetoelectronic phenomena at a ferromagnet-semiconductor interface. *Phys. Rev. Lett.* 84, 5022 (2000).

44. B.J. Wees, Comment on observation of spin injection at a ferromagnet-semiconductor interface. *Phys. Rev. Lett.* 84, 5023 (2000).

45. P.R. Hammar et al., Reply. *Phys. Rev. Lett.* 84, 5024 (2000).

46. G. Schmidt et al., Fundamental obstacle for electrical spin injection from a ferromagnetic metal into a diffusive semiconductor. *Phys. Rev. B.* 62 R4790 (2000).

47. E.I. Rashba, Theory of electrical spin injection: tunnel contacts as a solution of the conductivity mismatch problem. *Phys. Rev. B.* 62, R16267 (2000).

48. H.J. Zhu et al., Room-temperature spin injection from Fe in to GaAs. *Phys. Rev. Lett.* 87, 016601 (2001).

49. P.R. Hammar and M. Johnson, Spin-dependent current transmission across a ferromagnet-insulator-two-dimensional gas junction. *Appl. Phys. Lett.* 79, 2591 (2001).

50. A. Hanbiki et al., Electrical spin injection from a magnetic metal/tunnel barrier contact into a semiconductor. *Appl. Phys. Lett.* 80, 83 (2002).

51. W.H. Rippard and R.A. Buhrman, Spin-dependent hot electron transport in Co/Cu thin films. *Phys. Rev. Lett.* 84, 971 (2000).

52. R. Jansen et al., The spin-valve transistor: fabrication, characterization, and physics. *J. Appl. Phys.* 89, 7431 (2001).

53. J.W. Dong et al., Spin-polarized quasiparticle injection devices using $Au/Yba_2Cu_3O_7/LaAlO_3/Nd_{0.7}Sr_{0.3}MnO_3$ heterostructures. *Appl. Phys. Lett.* 71, 1718 (1997).

54. R.J. Soulen et al., Measuring the spin polarization of a metal with a superconducting point contact. *Science.* 282, 85 (1998).

55. D.C. Worledge and T.H. Geballe, Spin-polarized tunneling in $La_{0.67}Sr_{0.33}MnO_3$. *Appl. Phys. Lett.* 76, 900 (2000).

56. D.J. Monsma and S.S.P. Parkin, Spin polarization of tunneling current from ferromagnet/Al_2O_3 interfaces using copper-doped aluminum superconducting films. *Appl. Phys. Lett.* 77, 720 (2000).

57. J.M. Coey, M. Viret and S. von Molnár, Mixed-valence manganites. *Advances Phys.* 48, 167 (1999).

58. M. Viret et al., Spin polarized tunneling as a probe of half metallic ferromagnetism in mixed-valence manganites. *J. Magn. Magn. Mater.* 198–199, 1 (1999).

59. F. Holtzberg, T. McGuire, S. Methfessel, and J. Suits, Effect of electron concentration on magnet exchange interactions in rare earth chalcogenides. *Phys. Rev. Lett.* 13, 18 (1964).

60. A. Haury et al., Observation of a ferromagnetic transition induced by two-dimensional hole gas in modulation-doped CdMnTe quantum wells. *Phys. Rev. Lett.* 79, 511 (1997).

61. H. Munekata, H. Ohno, S. von Molnár, A. Segmüller, L. Chang, and L. Esaki, Diluted magnetic III-V semiconductors. *Phys. Rev. Lett.* 63, 1849 (1989)

62. H. Ohno, H. Munekata, S. von Molnár, and L. Chang, Magnetotransport properties of *p*-type (In,Mn)As diluted magnetic semiconductors. *Phys. Rev. Lett.* 68, 2664 (1992).

63. H. Ohno et al., (Ga,Mn)As: a new diluted magnetic semiconductor based on GaAs. *Appl. Phys. Lett.* 69, 363–365 (1996).

64. H. Ohno, Making nonmagnetic semiconductor ferromagnetic. *Science.* 281, 951 (1998).

65. H. Ohno, Properties of ferromagnetic III-V semiconductors. *J. Magn. Magn. Mater.* 200, 100 (1999).

66. M. Tanaka, Epitaxial growth and properties of III-V magnetic semiconductor (GaMn) As and its heterostructures. *J. Vac. Sci. Technol. B.* 16, 2267 (1998).

67. M. Tanaka and Y. Higo, Large tunneling magnetoresistance in GaMnAs/AlAs/GaMnAs ferromagnetic semiconductor tunneling junctions. *Phys. Rev. Lett.* 87, 26602 (2001).

68. T. Hayashi et al., Tunneling spectroscopy and tunneling magnetoresistance in (Ga,Mn)As ultrathin heterostructures. *J. Cryst. Growth.* 201/202, 689 (1999).

69. H. Saito et al, Transport properties of a III-V $Ga_{1-x}Cr_xAs$ diluted magnetic semiconductor, presented at Spintech1 International Conference, Maui, Hawaii, May 2001.

70. Y. D. Park et al., Magnetoresistance of Mn:Ge ferromagnetic nanoclusters in a diluted magnetic semiconductor matrix. *Appl. Phys. Lett.* 78, 2739 (2001).

71. Y. D. Park et al., A group IV ferromagnetic semiconductor Mn_xGe_{1-x}. *Science.* 295, 561 (2002).

72. T. Dietl, F. Matsukura, J. Cibert, and D. Ferrand, Zener model description of ferromagnetism in zinc-blende magnetic semiconductors. *Science.* 287, 1019 (2000).

73. M. Overberg et al., Epitaxial growth of dilute magnetic semiconductors: GaMnN and GaMnP. *Mat. Res. Soc. Symp. Proc.* 674, T6.5.1 (2001).

74. M.L. Reed et al., Room temperature magnetic (Ga,Mn)N: a new material for spin electronic devices. *Mater. Lett.* 51, 500 (2001).

75. M.L. Reed et al., Room temperature ferromagnetic properties of (Ga,Mn)N. *Appl. Phys. Lett.* 79, 3473 (2001).

76. S. Sonoda et al., Molecular beam epitaxy of wurtzite (Ga,Mn)N films on sapphire(0001) showing the ferromagnetic behaviour at room temperature. *J. Cryst. Growth.* 237–239, 1358 (2002).

77. R. A. de Groot et al., New class of materials: half-metallic ferromagnets. *Phys. Rev. Lett.* 50, 2024 (1983).

78. C.T. Tanaka and J.S. Moodera, Spin-polarized tunneling in half-metallic ferromagnets. *J. Appl. Phys.* 79, 6265 (1996).

79. J.W. Dong et al., MBE growth of ferromagnetic single crystal Heusler alloys on (001) (001) $Ga_{1-x}In_xAs$. *Physica E.* 10, 428 (2001).

80. K. Schwartz, CrO_2 predicted as a half-metallic ferromagnet. *J. Phys. F.* 16, L211 (1986).

81. K.P. Kaemper et al., CrO_2 – a new half-metallic ferromagnet. *Phys. Rev. Lett.* 59, 2788 (1988).

82. R.J. Soulen et al., Measuring the spin polarization of a metal with a superconducting point contact. *Science.* 282, 85 (1998).

83. Y. Ji et al., Determination of the spin polarization of half-metallic CrO_2 by point contact Andreev reflection. *Phys. Rev. Lett.* 86, 5585 (2001).

84. J.J. Versluijs, M.A. Bari, and J.M.D. Coey, Magnetoresistance of half-metallic oxide nanocontacts. *Phys. Rev. Lett.* 87, 026601 (2001).

85. H. Shimizu and M. Tanaka, Magneto-optical properties of semiconductor-based superlattices having GaAs with MnAs nanoclusters. *J. Appl. Phys.* 89, 7281 (2001).

86. H. Akinaga, T. Manago, and M. Shirai, Material design of half-metallic zinc-blende CrAs and the synthesis by molecular beam epitaxy. *Jap. J. Appl. Phys.* 39, L1120 (2000).

87. J. Zhao et al., Room-temperature ferromagnetism in zinc blende CrSb grown by molecular beam epitaxy. *Appl. Phys. Lett.* 79, 2776 (2001).

88. H. Akinaga, M. Mizuguchi, K. Ono, and M. Oshima, Room-temperature thousand-fold magnetoresistance change in MnSb granular films: magnetoresistive switch effect. *Appl. Phys. Lett.* 76, 357 (2000).

89. H. Akinaga, M. Mizuguchi, K. Ono, and M. Oshima, Room-temperature photoinduced magnetoresistance effect in GaAs including MnSb nanomagnets. *Appl. Phys. Lett.* 76, 2600 (2000).

90. T. Fukumura Z. Jin, A. Ohtomo, H. Koinuma, and M. Kawasaki, An oxide-diluted magnetic semiconductor Mn-doped ZnO. *Appl. Phys. Lett.* 75, 3366 (2000).

91. Y. Matsumoto et al., Room-temperature ferromagnetism in transparent transition metal-doped titanium dioxide. *Science.* 291, 854 (2001).

92. H.J. Tromp et al., CaB6: a new semiconducting material for spin electronics. *Phys. Rev. Lett.* 87, 016401 (2001).

93. S. Koshihara et al., Ferromagnetic order induced by photogenerated carriers in magnetic III-V semiconductor heterostructures of (In,Mn)As/GaSb. *Phys. Rev. Lett.* 78, 4617 (1997).

94. H. Ohno et al., Electric-field control of ferromagnetism. *Nature.* 408, 944 (2000).

95. L. Berger, Emission of spin waves by a magnetic multiplayer traversed by a current. *Phys. Rev. B.* 54, 9353 (1996).

96. J.C. Slonczewski, Current-driven excitation of magnetic multilayers. *J. Magn. Magn. Mater.* 159, L1 (1996).

97. J.C. Slonczewski, Excitation of spin waves by an electric current. *J. Magn. Magn. Mater.* 195, L261 (1999).

98. Ya.B. Bazaliy, B. Jones, and S.-C. Zhang, Modification of the Landau–Lifshitz equation in the presence of a spin-polarized current in colossal- and giant-magnetoresistive materials. *Phys. Rev. B.* 57, R3213 (1998).

99. M. Tsoi et al., Excitation of a magnetic multiplayer by an electric current. *Phys. Rev. Lett.* 80, 4281 (1998); 81, 493 (1998) (E).

100. E.B. Myers, D.C. Ralph, J.A. Katine, R.N. Louie, and R.A. Buhrman, Current-induced switching of domains in magnetic multiplayer devices. *Science.* 285, 867 (1999).

101. J.E. Wegrowe et al., Current-induced magnetization reversal in magnetic nanowires. *Europhys. Lett.* 45, 626 (1999).

102. J.Z. Sun, Current-driven magnetic switching in manganite trilayer junctions. *J. Magn. Magn. Mater.* 202, 157 (1999).

103. S.M. Rezende et al., Magnon excitation by spin injection in thin Fe/Cr/Fe films. *Phys. Rev. Lett.* 84, 4212 (2000).

104. J.A. Katine et al., Current-driven magnetization reversal and spin-wave excitation in Co/Cu/Co pillars. *Phys. Rev. Lett.* 84, 3149 (2000).

105. M. Tsoi, A.G.M. Jansen, J. Bass, W.-C. Chiang, V. Tsoi, and P. Wyder, Generation and detection of phase-coherent current-driven magnons in magnetic multilayers. *Nature.* 406, 46-48 (2000).

106. J. Slonszewski, U.S. patent 5,695,864. December, 9, 1997.

107. L. Berger, Multilayer as a spin-wave emitting diode. *J. Appl. Phys.* 81, 4880 (1997).

108. D.D. Awschalom and J.M. Kikkawa, Electron spin and optical coherence in semiconductors. *Phys. Today* 52, 33 (1999).

109. J.M. Kikkawa, D.D. Awschalom, I.P. Smorchkova, and N. Samarth, Room-temperature spin memory in two-dimensional electron gases. *Science.* 277, 1284 (1997).

110. J.M. Kikkawa and D.D. Awschalom, Resonant spin amplification in *n*-type GaAs. *Phys. Rev. Lett.* 80, 4313 (1998).

111. J.M. Kikkawa and D.D. Awschalom, Lateral drag of spin coherence in gallium arsenide. *Nature.* 397, 139 (1999).

112. J.A. Gupta, D.D. Awschalom, A.P. Alivisatos, and X. Peng, Spin coherence in semiconductor quantum dots. *Phys. Rev. B.* 59, R10421 (1999).

113. I. Malajovich et al., Coherent transfer of spin through a semiconductor heterointerface. *Phys. Rev. Lett.* 84, 1015 (2000).

114. I. Malajovich et al., Persistent sourcing of coherent spins for multifunctional semiconductor spintronics. *Nature.* 411, 770 (2001).

115. B. Beschoten et al., Spin coherence and dephasing in GaN. *Phys. Rev. B.* 63, R121202 (2001).

116. B.T. Jonker, U.S. patent #5,874,749.

117. S. Datta and B. Das, Electronic analog of the electro-optic modulator. *Appl. Phys. Lett.* 56, 665 (1990).

118. Y.A. Bychkov and E.I. Rashba, Oscillatory effects and the magnetic susceptibility of carriers in inversion layers. *Pis'ma Zh. Eksp. Teor. Fiz.* 39, 66 (1984); *JETP Lett.* 39, 78 (1984).

119. J. Nitta, T. Akazaki, H. Takayangi, and T. Enoki, Gate control of spin-orbit interaction in an inverted $In_{0.53}Ga_{0.47}As/In_{0.52}Al_{0.48}As$ heterostructure. *Phys. Rev. Lett.* 78, 1335 (1997).

120. D.E.Brenner et al., Resonant tunneling through ErAs semimetal quantum wells. *Appl. Phys. Lett.* 67, 1268 (1995).

121. E.A. de Andrada e Silva and G. C. La Rocca, Electron spin polarization by resonant tunneling. *Phys. Rev. B.* 59, R15583 (1999).

122. A.G. Petukhov, D.O. Demchenko, and A.N. Chantis, Spin-dependent resonant tunneling in double-barrier magnetic heterostructures. *J. Vac Sci. Technol. B.* 18, 2109 (2000).

123. H. Ohno, N. Akiba, F. Matsukura, A. Shen, K. Ohtani, and Y. Ohno, Spontaneous splitting of ferromagnetic (Ga,Mn)As valence band observed by resonant tunneling spectroscopy. *Appl. Phys. Lett.* 73, 363 (1998).

124. T. Hayashi, M. Tanaka, and A. Asamitsu, Tunneling magnetoresistance of a GaMnAs-based double barrier ferromagnetic junction. *J. Appl. Phys.* 87, 4673 (2000).

125. M.E. Flatte and G. Vignale, Unipolar spin diodes and transistors. *Appl. Phys. Lett.* 78, 1273–1275 (2001).

126. I. Žutić et al., Spin injection through the depletion layer: a theory of spin-polarized *p-n* junctions and solar cells. *Phys. Rev. B.* 64, 121201R (2001).

127. G. Kirczenow, Ideal spin filters: a theoretical study of electron transmission through ordered and disordered interfaces between ferromagnetic metals and semiconductors. *Phys. Rev. B.* 63, 054422 (2001).

128. D. Grundler, Ballistic spin-filter transistor. *Phys. Rev. B.* 63, R161307 (2001).

129. D.P. DiVincenzo et al., Universal quantum computation with the exchange interaction. *Nature.* 408, 339 (2000).

130. R.K. Kawakami et al., Ferromagnetic imprinting of nuclear spins in semiconductors. *Science.* 294, 131 (2001).

131. B. Kane, Silicon-based nuclear spin quantum computer. *Nature.* 393, 133 (1998).

132. R. Vrijen et al., Electron spin-resonance transistors for quantum computing in silicon-germanium heterostructures. *Phys. Rev. A.* 62, 012306, (2000).

133. T. Ladd, J. Goldman, F. Yamaguchi, Y. Yamamoto, E. Abe, and K. Itoh, An all silicon quantum computer, quant-ph/0109039.

134. D. Loss and D.P. DiVincenzo, Quantum computation with quantum dots. *Phys. Rev. A.* 57, 120 (1998).

135. G. Burkard, H. Engel, and D. Loss, Spintronics and quantum dots for quantum computing and quantum communication. *Fortschr. Phys.* 48, 965 (2000).

136. H.-A. Engel and D. Loss, Detection of single spin decoherence in a quantum dot via charge currents. *Phys. Rev. Lett.* 86, 4648 (2001).

137. A. Imamoglu et al., Quantum information processing using quantum dot spins and cavity QED. *Phys. Rev. Lett.* 83, 4204, (1999).

138. N.H. Bonadeo et al., Coherent optical control of the quantum state of a single quantum dot. *Science.* 282, 1473 (1998).

139. N. Paillard et al., Spin relaxation quenching in semiconductor quantum dots. *Phys. Rev. Lett.* 86, 1634 (2001).

140. J.A. Gupta et al., Ultrafast manipulation of electron spin. *Science.* 292, 2458 (2001).

9

QWIP: A Quantum Device Success

Kwong-Kit Choi
U.S. Army Research Laboratory

Abstract

Quantum well infrared photodetector (QWIP) technology has become increasingly mature as useful products are emerging around the world. QWIP focal plane array (FPA) cameras are used in a wide variety of settings to acquire high-resolution infrared imageries. Remarkably, the operation of the detector is entirely based on the nanometer material thickness. The small thickness creates one-dimensional particle-in-a-box quantization, enabling large infrared intersubband absorption in this material. The intersubband absorption is absent in the original host materials due to the conservation of momentum. In this chapter, we will discuss the current status of the QWIP technology and its associated nanoscience.

9.1   Introduction

Since the initial demonstration of quantum well infrared photodetectors (QWIPs) in the late 1980s,[1,2] there has been a worldwide sustained effort to develop the detector into a mainstream infrared technology. At present, the technology is on the verge of acceptance as the commercial standard for high-performance and large-format long wavelength infrared detection. Remarkably, the biggest advantage of the technology relies on its simple detector operation: the one-dimensional particle-in-a-box quantization, the most elementary quantum phenomenon. This well-known physical mechanism allows a simple, reliable, and robust detector design and a straightforward interpretation of the experiment observation. It is, therefore, explicable that its development cycle was short, and volume production began just a few years after its invention.

The basic QWIP active material consists of a number of alternate material layers stacked together, each with a different band gap. The material layers are grown sequentially by molecular beam epitaxy (MBE) in mono-atomic layer accuracy. Due to band gap misalignment, a potential well appears in the conduction band of the smaller band gap material. Because each potential well is only a few nanometers thick, the quantization energy along this dimension is substantial; and the layer becomes a quantum well (QW). The energy spacing in a QW can be adjusted freely by changing the material structural parameters and hence can be designed to fall into the infrared regime for radiation detection.

Although the energy levels in a QW are discrete along the growth direction, the energy along the plane of the layer is still continuous. Each energy level is therefore also a subband, and the total energy of an electron in a subband is continuous. However, it turns out that the in-plane momentum is conserved during an optical intersubband transition, which makes the associated in-plane energy a constant in the process. Electrons with different in-plane momentum will therefore have the same transition energy equal to the quantized level spacing, making the existence of the extra dimensions irrelevant to the detector optical behavior. The material unit can thus be conveniently treated as a one-dimensional (1D) quantum well, typically described as the first example of quantum mechanics. With the MBE technology, one can then treat QWs as 1D artificial atoms in constructing new "materials," such as artificial molecules and superlattices to formulate the desirable optoelectronic properties of the detector for practical applications. At the same time, these nanostructures can also serve as the physical realization of 1D quantum wells heuristically described in the usual texts. The main focus of this article is concentrated on the physics of one-dimensional semiconductor nanostructures as illustrated by the examples of QWIPs. The described properties will be precursors to the expected behaviors of three-dimensionally confined structures. We will also discuss the utilization of the discovered physics for device applications.

9.2 QWIP Focal Plane Array Technology

Before discussing the nanoscience of QWIPs, I would like to mention the current status of the focal plane array (FPA) development. Some of these products are commercially available, and they illustrate one of the quantum device successes. For example, Gunapala et al.[3] demonstrated a 320 × 256 palm-sized, hand-held QWIP camera for long-wavelength infrared (LWIR) detection. It operates at 70 K with the noise equivalent temperature difference NEΔT of 33 mK. This thermal resolution is approximately the same as the state-of-the-art LWIR HgCdTe FPAs. The cutoff wavelength λ_c is 8.8 μm. The group also constructed a 640 × 480 interlaced dual-band camera. In this camera, the odd rows are for one color and the even rows are for another color. Operated at 40 K, the NEΔT is 29 mK for the 9.1 μm cutoff band and 74 mK for the 15 μm cutoff band. The FPAs are marketed through QWIP Technology LLC.

Schneider et al.[4] used a photovoltaic low-noise QWIP structure to achieve a NEΔT of 7.4 mK with 20 ms integration time and 5.2 mK with 40 ms. This thermal sensitivity is the highest so far for all the LW infrared technologies. In addition, they demonstrated a standard 256 × 256 LWIR FPA with NEΔT < 10 mK and a 640 × 512 LWIR FPA with NEΔT < 20 mK. They also reported a 3–5 μm mid-wavelength infrared (MWIR) FPA with NEΔT of 14 mK with 20 ms integration time under $f/1.5$ optics and operated at 88 K.[5] The FPA technology is marketed by AEG Infraot-Module GmbH.

Bois et al.[6] adopted an approach of using two stacks of identical QWIPs to subtract out the detector thermal current. With this detector structure, the FPA is expected to be operable at much higher temperatures and at longer integration times. They expected NEΔT to be 10 mK at 85 K for λ_c = 9.3 μm and $f/1$ optics. In addition, the more conventional QWIP FPAs with 384 × 288 and 640 × 480 formats are also planned by THOMSON-CSF/LCR and SOFRADIR.

Sundaram et al.[7] reported the fabrication and characterization of two-color FPAs with different MW and LW combinations. The two-color QWIPs are vertically stacked with 3 indium bumps per pixel, allowing for pixel-registered and simultaneously integrated detection. For the 8.7/11.4 μm cutoff array, the typical NEΔT measured at 40 K is 23/43 mK. The integration time is 4–10 ms under *f*/3 optics. For the 5.4/9.4 μm cutoff array, NEΔT is less than 35 mK for both colors operated at 65K and *f*/2 optics. For the 4.2/5.1 μm array, NEΔT is 41/32 mK operated at 90 K with *f*/3 optics, and integration time is 8–10 ms.

Hirschauer et al.[8] described a European consortium in the research and development of QWIP FPAs and the associated read-out circuits. They emphasized the high quantum efficiency of 50% in their grating design. The 320 × 240 FPAs are marketed through FLIR systems as the ThermaCAM™ SC3000 model.

The above FPA developments used diffractive gratings for light coupling.[9] Choi et al.[10] used triangular surface corrugations to couple light and obtained 22 mK for a 11.2 μm cutoff array and 73 mK for a 16.4 μm cutoff array. The FPA format is 256 × 256.

For QWIP FPA applications, Goldberg et al.[11] used a color fusion scheme on the 5.4/9.4 μm cutoff two-color array made by Sundaram et al. and demonstrated improved scene visibility based on the difference of the emissivity between the two colors. The color fusion scheme also enables a cooler object such as a vehicle represented by a different color than a hotter object such as the plume. The camera also shows different scene signatures in the two infrared bands during a rocket launch.

Medical infrared imaging is potentially an important application for QWIPs. Fauci et al.[12] of Omni-Corder Technologies, Inc. pointed out that there is a convergence of three technological developments that may open a new era for infrared imaging in medical screening and diagnostics. They are the advent of the QWIP technology, the availability of high-speed and large-capacity personal computers, and the shift of the clinical modality from anatomical imaging to functional imaging. The advantages of QWIPs in this area include high LW sensitivity, high uniformity, small number of dead pixels, high thermal and spatial resolution, short acquisition time, compatibility with large focal depth optics, and affordable price.

Besides medical imaging, other explored applications are fire fighting, volcano monitoring, infrared astronomy, national and tactical missile defense, geological heat emission monitoring, hyperspectral imaging, LWIR/near IR imaging, and preventive maintenance. From laboratory demonstrations to field tests, QWIPs have been proven to be sensitive, versatile, and reliable. Because the detector spectral range can span from the near infrared to the far infrared, it greatly expands the human perception. QWIPs will continue to find applications in other aspects of human life. Figures 9.1 and 9.2 are the infrared imageries

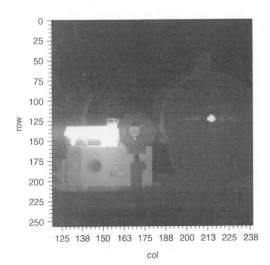

FIGURE 9.1 Infrared images are taken by a 256 × 256 corrugated QWIP focal plane array. The operating temperature is 65 K and the cutoff wavelength is at 8.2 μm. The polarizer in front of the hot plate shows the polarization sensitivity of the FPA.

FIGURE 9.2   The infrared image taken by a 256×256 corrugated QWIP FPA at 38 K, which has a cutoff at 16.4 μm.

taken by the corrugated QWIP FPAs with 8.2 and 16.4 μm cutoff wavelengths, respectively.[10] In the remaining article, we will turn to the nanoscience of QWIPs.

9.3   Optical Properties of Semiconductor Nanostructures

In this section, we will describe the use of a transfer-matrix method to calculate the quantized levels in one-dimensional quantum wells. It allows one to accurately predict the detection wavelengths of QWIPs that are made of $In_yGa_{1-y}As/Al_xGa_{1-x}As$, the most commonly used detector material in both MW and LW detection. We will also present superlattice structures to tailor the absorption width of the MW detectors.

9.3.1   Transfer-Matrix Method

Among different theoretical approaches, the transfer-matrix method is the most efficient method in solving multiple-layer structures. In the following, we will outline this approach specific to the present material system. Within this approach, the electron envelope wave function Ψ^n in the nth material layer can be written as:

$$\Psi^n = A_n e^{ik_n z} + B_n e^{-ik_n z} \tag{9.1}$$

where A_n and B_n are two constants to be determined, z is spatial coordinate along the growth direction, k_n is the wave vector given by:

$$k_n(E) = \frac{\sqrt{2m_n^*(E)}}{\hbar} \sqrt{E - \Delta E_n} \tag{9.2}$$

$m_n^*(E)$ is the energy-dependent effective mass, and E and ΔE_n are the electron energy and the conduction band offset measured from the GaAs conduction band edge, respectively. The constants A_n and B_n relate to A_{n+1} and B_{n+1} of the adjacent $(n+1)$th layer through the boundary conditions by:[13]

$$\begin{bmatrix} A_n \\ B_n \end{bmatrix} = \frac{1}{2} \begin{bmatrix} (1+\gamma_{n,n+1})e^{i(k_{n+1}-k_n)d_{n,n+1}} & (1-\gamma_{n,n+1})e^{-i(k_{n+1}+k_n)d_{n,n+1}} \\ (1-\gamma_{n,n+1})e^{i(k_{n+1}+k_n)d_{n,n+1}} & (1+\gamma_{n,n+1})e^{-i(k_{n+1}-k_n)d_{n,n+1}} \end{bmatrix}$$

$$= \frac{1}{2} M_{n,n+1} \begin{bmatrix} A_{n+1} \\ B_{n+1} \end{bmatrix} \tag{9.3}$$

where $\gamma_{n,n+1} = (m_n^* k_{n+1})/(m_{n+1}^* k_n)$ and $d_{n,n+1}$ is the spatial coordinate of the interface. The energy-level structure of a multi-layer material system can be examined by studying the global transmission coefficient T_G of a plane wave through the material. When the incident energy of the wave coincides with one of the energy level E_s of the quantum well structure, strong resonant transmission occurs and T_G attains a local maximum. Assuming the plane wave is incident from the left, the coefficients A_1 and B_1 for the first material layer on the left and the coefficients A_p and B_p of the last pth layer are connected by a series of matrix multiplications, and are related by:

$$\begin{bmatrix} A_1 \\ B_1 \end{bmatrix} = \frac{1}{2^{p-1}} M_{1,p} \begin{bmatrix} A_p \\ B_p \end{bmatrix} \tag{9.4}$$

where $M_{1,p} = M_{1,2} M_{2,3} \ldots M_{p-1,p}$, and $M_{n,n+1}$ is defined in Equation (9.3). Assuming that there is no wave traveling to the left in the last layer, one can set $A_p = 1$ and $B_p = 0$, with which all the coefficients in the layers are determined. The value of T_G at a given E is then equal to:

$$T_G(E) = \frac{1}{|A_1(E)|^2} \frac{v_p(E)}{v_1(E)} = \frac{2^{2p-2}}{|a_{11}(E)|^2} \frac{m_1^*(E) k_p(E)}{m_p^*(E) k_1(E)} \tag{9.5}$$

where v is the electron group velocity and a_{11} is the first diagonal element of $M_{1,p}$. The local maxima in $T_G(E)$ determine the locations of E_s. In general, these local maxima are less than unity unless the material structure is symmetric. Calculating $A_n(E)$ and $B_n(E)$ at $E = E_s$ will yield the wave function Ψ_s of that energy level.

After obtaining the wave functions and energies of the quantum well states, the dipole oscillator strength $f(\hbar\omega)$, which is directly proportional to the quantum efficiency η, is generally given by:

$$f(\hbar\omega) = \frac{2\hbar}{m^*\omega} \left| \left\langle \Psi_f \left| \frac{\partial}{\partial z} \right| \Psi_i \right\rangle \right|^2 \tag{9.6}$$

where $\hbar\omega$ is the incoming photon energy, m^* is the effective mass of the ground state, Ψ_i is the ground state wave function subjected to the normalization condition $\langle \Psi_i | \Psi_i \rangle = 1$ within the quantum well structure, and Ψ_f is the normalized excited state wave function having an energy $\hbar\omega$ above the ground state. In this chapter, we are interested only in the shape of the spectrum but not the absolute magnitude of the absorption. It is then sufficient to evaluate f for all the available excited states, with which the absorption spectral lineshape can be obtained. Its peak location determines the absorption peak wavelength λ_{ap}. Furthermore, the responsivity R is given by:

$$R = \frac{\eta}{\hbar\omega} eg\gamma \tag{9.7}$$

where g is the photoconductive gain and γ is the photoelectron tunneling probability. Assuming the $g\gamma$ product to be constant in $\hbar\omega$, which is generally true for small bias,[14] the spectral lineshape of $R(\hbar\omega)$ will be determined by the functional form of $f(\hbar\omega)/\hbar\omega$. Its maximum determines the responsivity peak wavelength λ_{rp}.

The calculation can be handled efficiently by a personal computer. The material parameters required for the calculation were collected and summarized by Shi et. al.[15] With these material parameters, the quantities $m_n^*(E)$ and $k_n(E)$ in Equations (9.4) and (9.5) are known; and solving these equations will give the energy levels E_s and the state wave functions Ψ_s in terms of $A_n(E_s)$ and $B_n(E_s)$. Subsequently, $f(\hbar\omega)/\hbar\omega$ can be computed.

9.3.2 Multiple Quantum Wells

We apply the calculation to several common detector structures.[16] Figure 9.3 shows the band diagrams and the relevant energy values of three detectors. Because barriers are thick, the ground state wave functions are localized. The structures are referred to as *multiple quantum well* (MQW) structures. Detectors A, B, and C are respectively bound-to-bound, bound-to-continuum, and bound-to-quasi-bound detectors.

In the actual experiment, each detector consists of 20 QW periods. But we apply the theoretical model to a single quantum well (SQW) unit and a four quantum well (4QW) unit. The 4QW unit is shown in Figure 9.3. The SQW unit can be considered as the *atomic* species of the material, and the 4QW unit represents the constituted *molecule*. Figure 9.4 shows $T_G(E)$ of the SQW and the 4QW of detector A above the barriers. The maxima indicate the locations of the atomic or molecular levels in the respective case. In addition, there are also bound states below the barriers. Figure 9.5 shows the *atomic orbitals* of the lowest 10 energy levels. The wave functions can be clearly separated into even and odd parities. Similarly, Figure 9.6 shows the *molecular orbitals* of the lowest three molecular or *mini* bands.

From the shapes of these wave functions, the quantity $f(\hbar\omega)/\hbar\omega$ of detector A can be computed. Interestingly, $f(\hbar\omega)/\hbar\omega$ is indistinguishable between the two structures in the presence of line broadening σ in the transitions. The calculation with more QWs also yields the same result. This situation indicates that the properties of the entire MQW are determined by its single-well constituent. Hence, only $f(\hbar\omega)/\hbar\omega$ for the 4QW structure is plotted in Figure 9.7 as the dash–dotted curve. In this plot, we have assumed a σ of 10.5 meV to fit the experimental data, which are shown as circles. A finite σ is due to solid-state interactions and material nonuniformity. The theoretical curve can be brought to overlap with experiment if the wavelength is shifted by 0.165 μm. This small wavelength discrepancy can be caused by the uncertainties in the theoretical material parameters or the actual material calibrations. Nevertheless, this example illustrates the fact that the infrared absorption of a QWIP can be explained by 1D quantization.

In a 4QW structure, there are actually many hybridized excited levels as indicated in Figure 9.6. The calculated lineshape is a superposition of all the transitions to these individual levels. In the plot for detector B in Figure 9.7, we show the theoretical responsivity spectra for two values of σ, the dash–dotted curve for 10.5 meV, and the fast-changing curve for 0.5 meV. The large σ curve is to fit the data.

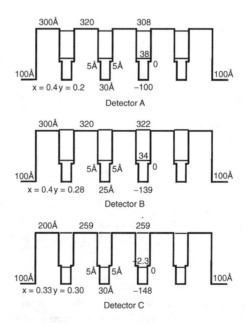

FIGURE 9.3 The band structures of the detectors A, B, and C. Each represents a four-quantum well structure truncated from the actual detector. The numerical valves without units are the energies in the unit of meV.

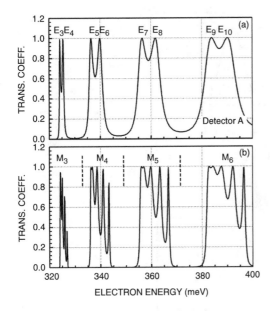

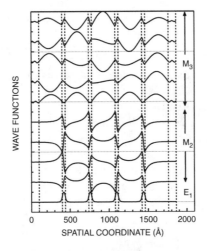

FIGURE 9.4 The transmission coefficients of (a) an SQW and (b) a 4QW for detector A above the barrier height.

FIGURE 9.5 The wave functions of the SQW of detector A. The vertical lines show the locations of the quantum well region and the contact regions.

FIGURE 9.6 The wave functions of the 4QW of detector A for the ground state E_1 and the minibands M_2 and M_3. The vertical dashed lines show the locations of the quantum well regions and the contact regions.

The small σ curve is to reveal each transition location and its absorption strength. These sharp molecular absorption lines are to be expected in the observation for the 4QW if there were no solid-state broadenings, as in the case of natural atomic systems. In the presence of the line broadening, however, the difference among the different number of QWs cannot be distinguished. For detector B, one can observe that the agreement in the lineshapes is excellent after a wavelength shift, proving the origin of the absorption.

The agreement is further tested in detector C, which has thinner barriers. For this detector, the inter-QW coupling is stronger, which leads to more widely distributed excited minibands and produces subsidiary peaks. The agreement in this case is still maintained, apart from a small wavelength difference. In general, the predicted wavelengths in Figure 9.7 agree with the measurements to within ± 0.2 μm.

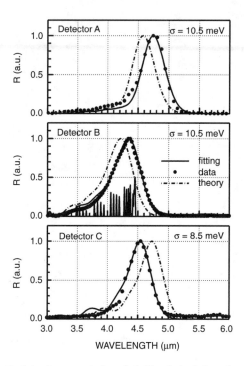

FIGURE 9.7 The responsivity R of the detectors A, B, and C. The dashed–dotted curve is the calculated spectrum for the 4QW structure with the indicated line broadening σ. The solid curve is the calculated curve with a finite wavelength shift in order to fit the experimental data that are represented by circles. The rapidly changing curve in the plot for Detector B assumes a smaller σ of 0.5 meV to reveal the individual transition locations.

Because either positive or negative shift is needed in the curve fitting, the discrepancy is more likely due to the uncertainty in the material growth. In this plot, there is another fitting parameter σ. But in considering that the spectrum is basically determined by the individual transitions and the purpose of σ is only to smooth out the individual peaks, the major absorption characteristics can actually be predicted without free adjusting parameters.

With the above agreement between theory and experiment, we plot the calculated absorption peak λ_{ap} as a function of x and y in Figure 9.8. The structure of the material in a QW unit is assumed to be 5Å GaAs/25Å In$_y$Ga$_{1-y}$As/5Å GaAs/200Å Al$_x$Ga$_{1-x}$As. Figure 9.8 is a useful guide to the detector design in the MW regime. Similar plots for the LW regime were given in Reference 17.

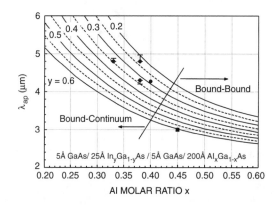

FIGURE 9.8 The calculated absorption peak wavelength of a 200Å Al$_x$Ga$_{1-x}$As/5Å GaAs/25Å In$_y$Ga$_{1-y}$As/5Å GaAs/ 200Å Al$_x$Ga$_{1-x}$As structure. The dashed curves are for $y = 0.25$, 0.35, 0.45, and 0.55, respectively.

9.3.3 Superlattices

When the thickness of the barriers is thinner than the decay length of the ground state wave function in the barriers, the ground states of different wells overlap and form a coherent miniband. The modulation of its phase factor changes the absorption characteristics of an MQW structure. In the following, we study several detectors with miniband-to-miniband transitions. The practical reason for this study is an attempt to widen the spectral width Γ and shift λ_{rp} to a shorter wavelength. Note that the detectors commonly employed in MW detection as that shown in Figure 9.3 have a rather small Γ around 0.5 μm and λ_{rp} around 4.5 μm, which may not be adequate for broadband detection.

Figure 9.9 shows the detector structures studied. For superlattice detectors, the ground miniband is also highly conducting. Therefore, thick blocking barriers are needed to separate the superlattice into groups to prevent a large dark current. Each detector in the experiment consists of eight superlattice groups. To carry the atomic analogy further, the ground miniband can be considered as the valence band of a semiconductor and the first excited miniband as the conduction band. Current transport is possible only if electrons are excited from the ground miniband to the excited miniband. The thick blocking barriers can then be thought to be doped impurities, only in this case to impede rather than to enhance conduction. For detectors D and E, each group contains five wells and four barriers as shown in Figure 9.9. They are analogous to *doped elemental semiconductors*. For detector F, the QWs have alternate well widths to create two nearby minibands M_1 and M_2, both serving as ground states for absorption. Likewise, detector F can be regarded as a *doped compound semiconductor*. With the present doping density of 2×10^{18} cm^{-3} in the wells and also in the QW barriers for F, M_2 is, however, only partially filled, which limits its absorption bandwidth. In the calculation, the blocking barriers of F are replaced by thinner QW barriers to facilitate the calculation.

In order to study the intrinsic superlattice transition without the effects of the blocking layers, we first assume all the barriers in detector D have the same 30-Å thickness. The calculated T_G is shown in Figure 9.10(a). The splitting of the ground miniband M_1 states is well resolved. There are five states in each miniband. After calculating the corresponding wave functions, the oscillator strengths among different miniband states can be found. It turns out that there are five dominant transitions out of the 25 possible transitions. They are the $(1, 5)$, $(2, 4)$, $(3, 3)$, $(4, 2)$ and $(5, 1)$ transitions, where (i, j) denotes the transition

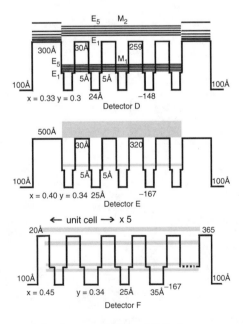

FIGURE 9.9 The band structures of the detectors D, E, and F. Each represents a four-quantum well structure truncated from the actual detector. The numerical valves without units are the energies in the unit of meV.

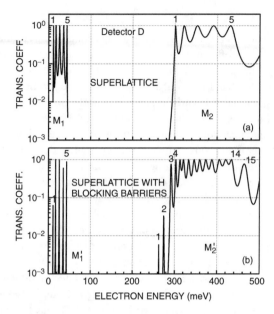

FIGURE 9.10 The transmission coefficients of Detector D for (a) a 5QW superlattice unit and (b) a 5QW superlattice unit with 300Å blocking barriers at both ends.

between the ith state in M_1 and the jth state from M_2. The dominance of the five transitions follows the conservation of miniband wave vector in the transition.

The corresponding calculated R (i.e., $f(\hbar\omega)/\hbar\omega$ in arbitrary unit) with $\sigma = 27.5$ meV and 0.5 meV are plotted in Figure 9.11(a) to fit the data and to reveal the individual transitions, respectively. In this fitting, no wavelength shift is needed. Note that the overall absorption width of this detector is set mainly by the level distribution. The observed absorption width Γ is 1.8 μm, which is much larger than that of the

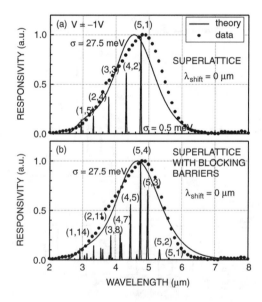

FIGURE 9.11 (a) The responsivity R of detector D for the combined transitions in the superlattice with different σ (solid curves) and the experimental data (circles). (b) The calculated NR for the superlattice with blocking barriers with different σ (solid curves) and the experimental data (circles).

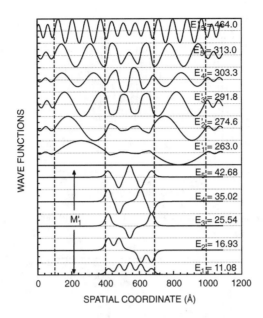

FIGURE 9.12 The wave functions of Detector D with blocking barriers for the first two minibands. Only selected wave functions in M'_2 are shown. The vertical dashed lines show the locations of the blocking layers and the contact regions.

MQW detectors. The reason for a larger Γ for superlattice detectors is the increasing coupling among the well states, both the ground and the excited states, through thinner barriers.

Next, we calculate the structure with the original blocking layers for detector D. The blocking layers introduce their own quantization levels ("impurity levels") above the barrier. The new miniband structure (labeled as M') is shown in Figure 9.10(b). Comparing with that in Figure 9.10(a), the blocking barriers have no significant effects on M_1 but create four new levels (1, 2, 3, and 15) outside the original M_2 miniband. It turns out that T_G of these four states are always less than unity regardless the precision of the incoming energy. For the present choice of the calculated structure, the superlattice acts as a single potential barrier to these states, resulting in finite transmission attenuation. If one chooses a *double-barrier* structure in the calculation, in which the thick blocking layer is surrounded by two superlattices, there will be unity resonant transmission through these states. However, as with the usual bound states in a QW, if the superlattice is thicker than the decay length of the wave functions, the coupling among these states will be small; and they will easily become localized in the presence of inhomogeneity. Therefore, there may exist localized states above the barrier height in this structure.

Figure 9.12 shows the selected miniband wave functions with thick blocking barriers. The wave functions of the first miniband are practically unchanged by the blocking layers. For the M'_2 miniband, the Ψ_1, Ψ_2, Ψ_3, and Ψ_{15} have attenuated wave amplitudes in the superlattice region, indicating that the superlattice is acting as a potential barrier to these states. Ψ_1, Ψ_2, and Ψ_3 nevertheless acquire finite spectral weights from the original M_2E_1 state, thus allowing for finite optical transition from the M'_1E_5 state. The overall expected R for this QW structure, with the same $\sigma = 27.5$ meV, is plotted in Figure 9.11(b). Due to the hybridization of the thick barrier states, there are no clear conservation rules to the transitions, as indicated by the large number of individual transitions. Nevertheless, the two spectral shapes are still very similar. The major difference is created by the new transitions, the (5, 3), (5, 2), and (5, 1) transitions, to the three blocking layer bound states. These new transitions increase the long wavelength response and improve the overall agreement with the experiment. Therefore, the insertion of *impurity* layers into a superlattice can create a finite *extrinsic* optical response.

Detector E is a further attempt to shorten the detection wavelength by increasing the band offset. The T_G curve through the detector E with blocking layers is shown in Figure 9.13. Because the blocking barrier is thicker than that in detector D, the number of hybridized levels in M_2 is greatly increased. In fact, the

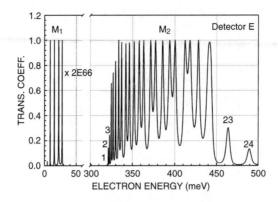

FIGURE 9.13 The transmission coefficient of a 5QW superlattice unit with 500Å blocking barriers at both ends for Detector E.

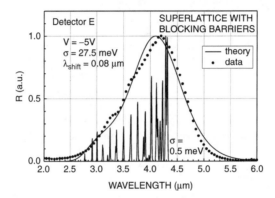

FIGURE 9.14 The responsivity R for the combined transitions in Detector E with blocking barriers and with different σ (solid curves). The circles are the experimental data.

excited states of detector E form a quasi-continuum above the barrier with no clear distinction among different minibands. Each superlattice unit can actually be considered as a single quantum well *molecular species* embedded in the barrier material, analogous to that in Section 9.3.2. With five states in M_1 and 24 states in M_2 between 320 to 500 meV, there are 120 possible transitions. Figure 9.14 shows that, out of all these possible transitions, 22 of them have appreciable oscillator strengths. Because these transitions are derived from the five unperturbed miniband-to-miniband transitions, the oscillator strength follows the similar rising trend with λ as in Figure 9.11(a). The value of Γ with σ = 27.5 meV is 1.15 μm, and the $λ_{rp}$ is at 4.1 μm, the center of the 3 to 5 μm band. The smaller Γ of this detector compared with detector D is due to the fact that $Γ = λ^2 δE/hc$, where $δE$ is the absorption width in energy. For a constant $δE$, Γ decreases as $λ^2$ as the detection is pushed to a shorter wavelength. The expected $Γ_E/Γ_D$ between detectors E and D is therefore equal to $(4.1/4.9)^2 = 0.70$, while the observed ratio is 1.15/1.8 = 0.64, consistent with the same $δE$.

Although $λ_{rp}$ of detector E is located at the center of 3 to 5 μm band, the detection bandwidth falls short in covering the entire MW band. Further reduction in the barrier width will increase Γ, but $λ_{rp}$ will also shift to a longer wavelength due to reduced quantum confinement. In order to substantially increase Γ, detector F is designed to have different well widths in the alternate wells as shown in Figure 9.9. The entire structure consists of eight superlattice units separated by 500 Å $Al_{0.35}Ga_{0.65}As$ blocking barriers. Each superlattice consists of five subunits. Due to the more complex miniband structure of this binary superlattice, the calculation will assume the same barrier thickness as in the superlattice. We showed in Figure 9.11 that such an approximation would not change the absorption spectrum significantly.

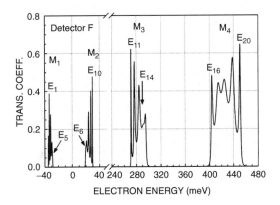

FIGURE 9.15 The transmission coefficient of Detector F.

T_G of Figure 9.15 shows four transmission minibands below 480 meV. Unlike the symmetric QW structures considered before, all the T_G maxima of detector F are less than unity, indicating that even if the incident wave coincides with one of the QW levels, the transmission of the wave is less than unity. In this asymmetrical structure, the reflected wave cannot cancel the incoming wave exactly because its phase is not always 180° out of phase with the incoming wave. Therefore, the less-than-unity transmission does not necessarily indicate a weaker resonant level at that energy. A calculation of the wave functions at the peaks and shoulders, as that shown in Figures 9.16 and 9.17, does reveal large wave amplitudes inside the structure; and thus, these peaks and shoulders of T_G continue to reflect the energy level structure of the system. The curve of T_G can still be useful for level searching in asymmetric structures.

In Figure 9.15, the energy levels are assembled into four minibands. The wave functions in each miniband have very similar unit cell Bloch functions. They are different only in the miniband wave

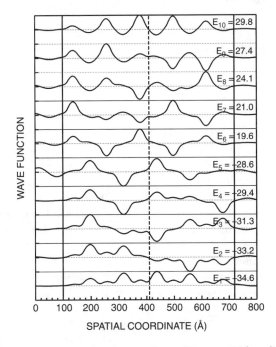

FIGURE 9.16 The wave functions of the minibands M_1 and M_2 of Detector F. The solid vertical lines indicate the contact regions and the dashed line indicates the center of the structure. The wave functions indicate the significant ionic nature of the bonding.

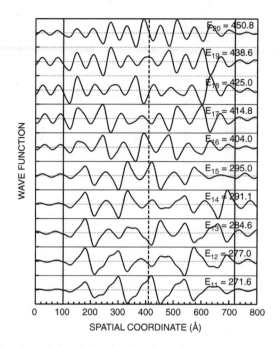

FIGURE 9.17 The wave functions of the minibands M_3 and M_4 of Detector F.

vectors. For the first two (more tightly bounded) minibands, the wave amplitudes are more asymmetrically localized in the wells. Therefore, the chemical bonding between a quantum well pair is partially ionic, and its ionicity depends on the filling of the minibands. This binary superlattice thus resembles a *polar compound semiconductor*. The existence of optical transitions from the two lower minibands to the two upper minibands greatly widens the absorption width of this detector.

However, the wide band coverage only occurs when both M_1 and M_2 are fully populated with electrons. Based on the present doping level in the wells and barriers, the Fermi energy E_F at a low temperature can be estimated from the following expression:

$$\sum_{n=1}^{10} g_{2d}(E_F - E_n) = N_d L_S \tag{9.8}$$

where $g_{2d} = 1.401 \times 10^{36}$ J^{-1}m^{-2} is the two-dimensional density of states of the InGaAs wells, E_n is the state energy, N_d is the doping density, and L_s is the length of the superlattice between the two blocking layers. From Equation (9.8), E_F is 50.0 meV above the GaAs conduction band. The electron density in each miniband state E_n will be directly proportional to $E_F - E_n$.

In Figure 9.18, the value of R for each transition is modified by a factor proportional to $E_F - E_n$, which strengthens the M_1 transitions relative to the M_2 transitions. The calculated λ_{rp} is located at 4.0 μm with Γ of 1.5 μm. Despite the partially filled M_2, this value of Γ is still 30% larger than that of detector E for the same peak wavelength. More significantly, the transitions among different minibands created a non-Lorentzian lineshape, which substantially increases the responsivity near its tails. The theoretical λ_{rp} needs to be shifted by 0.1 μm in this case, and the fitted σ is 33 meV. The predicted lineshape is in good agreement with the experiment except in the long wavelength tail. The discrepancy can be due to the lack of electron occupation of the higher M_2 miniband states because of carrier freeze-out. In the data, there are four indicated peaks that can be attributed to the four main groups of miniband transitions. In addition, there is also an unidentified peak around 4.3 μm. This peak may originate from transitions from M_1 to the localized blocking layer states below M_3, which have been neglected in this calculation. Comparing the experimental spectral responses between detectors E and

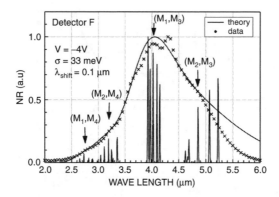

FIGURE 9.18 The normalized responsivity *NR* for the combined transitions in the superlattice with different σ (solid curves) assuming the two lowest minibands to have a common Fermi level. The crosses are the experimental data.

F, one can observe that the binary superlattice design has greatly enhanced the long wavelength response of the detector.

In this section, we have shown that one can use molecular beam epitaxy to create 1D nanostructures. The small layer thickness creates well-defined quantized levels. The optical transitions between two of these levels provide a means to detect radiation. One can use the tailored quantum wells as basic building blocks to construct desirable infrared materials for different applications.

9.4 Transport Properties of Semiconductor Nanostructures

In the last section, we learned that the infrared properties of QWIPs are created purely by quantum quantization in 1D nanostructures. In this section, we will examine their electron transport properties. Because both the dark current and the photocurrent of the detector are conducting perpendicular to the layers, the small dimension of the layers could affect their transport characteristics and the associated noise.

9.4.1 Hot-Electron Spectroscopy

Due to the small dimension (~5–50 nm) of each QW period and the large potential drop (≥ 50 mV/period) in the detector during operation, the conduction electrons may gain sufficient energy from the applied field and attain a nonequilibrium energy distribution. In order to study the electron energy distribution in these nanostructures, a quantum barrier is grown next to the QWIP structure to serve as an electron energy analyzer. The quantum barrier can be a high-pass filter[18] or a band-pass filter.[19] The band structure of an exemplified device with a high-pass filter is shown in Figure 9.19. For high-pass

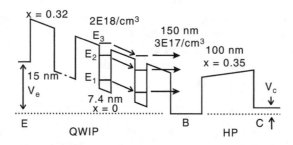

FIGURE 9.19 The band diagram of a QWIP with a high-pass filter placed next to one of the QWIP contact layer, referred as the base B. During the usual operation, the base is grounded, and the emitter voltage V_e is varied while the collector V_c is fixed. The hot-electron distribution injected from the QWIP can be analyzed when the distribution is shifted up by V_e and more electrons transfer into the collector.

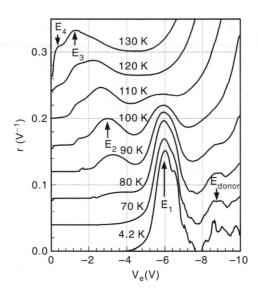

FIGURE 9.20 The hot-electron distribution measured in the dark at different temperatures for the detector structure shown in Figure 9.19. Note the vertical shifts of the curves for clarity.

filters, the hot-electron energy distribution $\rho(E)$ injected from the QWIP and passing through the base is directly proportional to $r \equiv d\alpha/dV_e$ at a constant V_c or $d\alpha/dV_c$ at a constant V_e, where the current transfer ratio α is J_c/J_e, J_c and J_e are the current densities measured at the collector and emitter, respectively, and V_e and V_c are the emitter- and collector-applied voltage, respectively, in the common base configuration. For band-pass filters, which are typically made of double-barrier structures, $\rho(E)$ is directly proportional to α when the filter-pass band is much narrower than the hot-electron distribution width.

Figure 9.20 shows $d\alpha/dV_e$ of the device depicted in Figure 9.19 at different operating temperatures T in the absence of radiation. In this plot, the electrons from the lower energy levels appear at larger V_e because they need a larger electrical potential to lift their energies over the filter barrier. The observed $\rho(E)$ is dependent of the QWIP structural parameters. For the present relatively thin (150 Å) QW barriers, direct tunneling among the ground states dominates the current flow at low T, resulting in a single pronounced peak injected from E_1 at –6 V. Besides this main peak, there is also a much smaller peak from the partially populated donor state at higher V_e. At higher T, conduction along the higher subbands becomes more important. Although the Fermi distribution at these higher energy levels is still very small, the large increase in the tunneling probability of these upper levels overcompensates for the decreasing population and leads to a dominant current flow at elevated temperatures. (Notice that the curves have been shifted vertically in Figure 9.20 for clarity.)

Another conclusion obtained from Figure 9.20 is that the number of observed peaks is equal to the number of thermally occupied subbands. Ballistic transport is therefore not observed in this GaAs/AlGaAs QWIP detector. In other words, thermal equilibrium is maintained in each QW despite the presence of a large applied field. The absence of ballistic transport can be explained by the long tunneling time for the low-energy electrons and the presence of low-lying L-valleys (300 meV above the GaAs band edge) for the high-energy electrons. The electrons from the excited states of the wells can accelerate into these valleys in the downstream, where their energy and momentum can be efficiently randomized. The poor vertical transport thus helps to thermalize the conducting electrons.

On the other hand, the electrons from the last QW may be able to travel ballistically across the base, depending on the base parameters. Near the base region, the electric field is usually too low to transfer electrons into the higher energy satellite valleys. For other material systems, if the satellite valleys are well above the barriers, ballistic peaks from the last few QWs near the base were indeed observed, such as that in InGaAs/AlGaAs systems with low Al contents.[20] If the thickness of the QW barriers is increased,

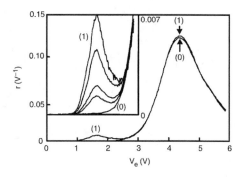

FIGURE 9.21 The hot-electron distribution of a QWIP with (denoted as curve #1) and without (curve #0) mono-chromatic light illumination at low temperature. The insert shows the photocurrent peak at different light intensities.

direct tunneling decreases. When the thickness is larger than 30 nm, tunneling via the impurity states inside the barriers becomes dominant. In this case, the low-energy injection peak decomposes into several sharper peaks, indicative of the energy positions of the impurities inside the barriers.[21]

In the presence of other excitation sources such as light, the electron distribution will be modified. Figure 9.21 shows the hot-electron distribution of another detector structure with and without illumination from a CO_2 laser. The measurement was done at a low T where the thermal peaks are absent. Without the light source, only the tunneling peak at $|V_e| = 4.3$ V is observed. Under illumination, a new peak appears at $|V_e| = 1.65$ V. Its peak position is similar to that of E_2 observed in a thermal excitation experiment, but with a narrower linewidth.[22] In addition, the peak magnitude varies with the laser intensity as shown in the insert of Figure 9.21. This new peak is thus identified as the photocurrent peak. The width of the peak is estimated to be 35 meV in energy. With the present monochromatic light source, the finite energy distribution of the photocurrent is attributed to the inhomogeneous broadening of E_1 and the subsequent elastic scattering in the base.

From these hot-electron spectroscopy studies, although the optical transitions of the electrons are quantum mechanical, the transport of the electrons across the barriers of the QWs is classical in nature. The heavy scattering in the L-valleys and the presence of other scattering mechanisms in the barriers such as the alloy scattering lead to large energy relaxation and thus destroy the coherent current transport. The magnitude of the photocurrent I_p is then well described by the drift-diffusion model, in which:

$$I_p = eAn_p\gamma v_d = e\frac{\eta}{\hbar\omega}\gamma gP \tag{9.9}$$

where A is the detector area, n_p is the photoelectron density, γ is the photoelectron tunneling probability, v_d is the drift velocity, g is the photoconductive gain, and P is the incident optical power. Note that only when the electrons are in thermal and optical equilibrium (except for a small displacement of the Fermi circle in an applied field), the photocurrent is independent of the number of QW periods as observed.[23]

In this section, we have shown that one can perform energy-resolved current transport measurements in nanostructures, which are difficult to achieve in bulk materials.

9.4.2 Infrared Hot-Electron Transistors

In Section 9.4.1, we learned that the photocurrent injection peak is rather sharp even after passing through a 150 nm-wide base. Its position is generally different from that of the dark current peaks. One can then design a narrow band-pass filter to selectively collect the photoelectrons from a QWIP at the collector and reject both the high- and low-energy dark currents. The detector is known as the *infrared hot-electron transistor* (IHET). Such a detector should be able to improve the QWIP sensitivity and allow higher operating temperatures.

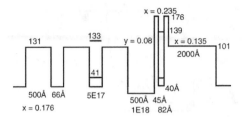

FIGURE 9.22 The band diagram of an IHET with a band-pass filter. The numerals without units are the energies in meV.

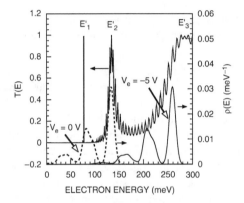

FIGURE 9.23 The transmission coefficient $T(E)$ of the filter shown in Figure 9.22 and the deduced hot-electron distribution at different V_e.

One example of the IHET structures is shown in Figure 9.22.[24] The QWIP consists of 30 QW periods. It has a responsivity peak at 13.2 μm and a cutoff at 14.0 μm. The transmission $T(E)$ of the band-pass filter is plotted in Figure 9.23. For this filter, E_1' is blocked by the thick 200-nm barrier. The filtering function is thus performed by E_2' at 139 meV. Its bandwidth is around 10 meV. At higher energies, the filter transmission shows a bandgap followed by a wider resonance at $E_3' = 280$ meV.

Figure 9.24 (a) and (b) show the emitter and collector current–voltage characteristics at $V_c = 0$ V, respectively. The solid curves represent the dark current J_d at different T, while the dashed curves are the 300 K background window photocurrent J_p (using a 45° edge coupling). Both photocurrents are measured at 10 K with 36° field of view. Notice the occurrence of the plateau regions at $T \sim 50$ K and $V_e \sim -1$ V for the collector dark current J_{cd} and the photocurrent J_{cp}, even though the corresponding emitter currents rise rapidly in this bias regime. These characteristics indicate the filter is blocking electrons with energies higher than its pass-band. The energy of injected electrons can also be raised by the operating temperature rather than by bias. At 77 K, the dark current is composed of mostly high-energy electrons. At this high temperature, J_{cd} is reduced by three orders of magnitude from that of J_{ed} at $V_e = -0.5$ V. Due to the selective filtering, the background limited temperature T_{BLIP}, at which $J_p = J_d$, is able to increase from 52 K to 63 K at the same voltage.

To understand better the filtering characteristics, the current transfer ratios α_d and α_p are plotted in Figure 9.25. The α_d curve is peaked at −1.4 V at 40 K, and this peak is due to the thermally assisted tunneling. It moves to a lower V_e at higher T as the energy of the electrons increases. Above 60 K, the dark current injection peak is above the E_2' pass-band of the filter even without bias; therefore, it is not observable in the E_2' window. Since:

$$\alpha(V_e) = c \int_0^\infty \rho(V_e, E) T(E) dE \qquad (9.10)$$

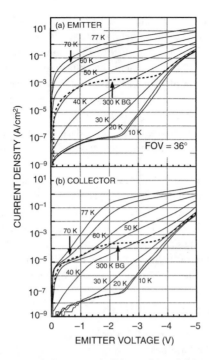

FIGURE 9.24 The measured (a) emitter dark current (solid curves) and (b) collector dark current (solid curves) as a function of emitter voltage at different temperatures. The collector voltage is kept at 0 V. The dashed curves are the 300 K background photocurrent under 36° field of view. The ratio of photocurrent to dark current is seen to be larger at the collector than at the emitter.

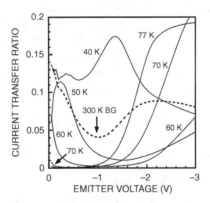

FIGURE 9.25 The dark current transfer ratio (solid curves), defined as the ratio of the collector dark current density to the emitter dark current density, as a function of emitter voltage at different temperatures. The collector voltage is 0 V. The dashed curve is that under 300 K background.

where c is the collection efficiency of the filter accounting for the electrons not traveling in the forward direction. When the width of $T(E)$ (~10meV) is much less than that of the thermal peak (~40–70 meV), $T(E) \approx \delta(E - E_2')$, and α_d is directly proportional to $\rho_d(V_e, E_2')$. The α_d shown in Figure 9.25 thus directly indicates the peaked nature of the dark current injection without performing a derivative. At higher V_e, α_d rises again. The reason for the rise will be discussed in the next section.

For the background photocurrent, the peak of α_p is observed at $V_e \sim -0.3$ V, which indicates the photocurrent peak just below the filter pass-band without a bias. From the movement of the thermal peak, the photocurrent should be degenerate with the thermal current at $T \sim 52$ K. These results indicate that for a 13.2 μm detector, the thermal current may overlap or even be higher in energy than the photocurrent at certain T. By using the present band-pass filter, one can suppress this high-energy current component as well as the low-energy thermally assisted tunneling current, thereby achieving a higher %BLIP ($\equiv \sqrt{[I_p/(I_p+I_d)]}$). For example, at $T = 77$ K and $V_e = -0.5$ V, the %BLIP for the QWIP is 4.5%, whereas that measured at the collector is 45%, a factor of 10 improvement. The filtering does not change the spectral responsivity appreciably due to the interdiffusion of photoelectrons of different energies in the base.[24]

Due to the large dark current suppression at 77 K, the detectivity D^* of the IHET is improved by a factor of 1.75 compared with the associated QWIP and is equal to 7.0×10^9 cm$\sqrt{\text{Hz}}$/W at -0.5 V. Combined with the increased integration time because of the lower dark current, NEΔT can be improved by a factor of 55 at this temperature. When the detector is operated at the T_{BLIP} of 62 K, $D^* = 3.2 \times 10^{10}$ cm$\sqrt{\text{Hz}}$/W at $V_e = -1.4$ V.

In conclusion, we have shown that by aligning the filter pass-band precisely with the photoelectron peak, one can obtain a high degree of dark current rejection while maintaining a sufficient photoelectron collection. D^* and T_{BLIP} are thus improved. Because the filter bandwidth can be made arbitrarily small by structural design, the current level of a detector and its impedance at a particular temperature become dissociated from its cutoff wavelength. A high-temperature operation is thus more feasible, especially for a CMOS read-out circuit because of its limited charge handling capacity. The present example demonstrates that an IHET can be operated at 77 K with a 45% BLIP for a cutoff wavelength of 14.0 μm.

9.4.3 Hot-Electron Energy Relaxation

In order to understand in detail the current transfer characteristics of an IHET, the energy relaxation of the hot-electron injection in the base needs to be taken into account. In the previous IHET, the base doping density of 1×10^{18} cm^{-3} is relatively high; and the base is made of a lower band-gap material, resulting in extra electron accumulation. Heavy electron–electron scattering in this base structure is expected. In the following, we present a calculation on the evolution of $\rho(E)$ as a function of time t and the traveling distance L, and compare with the experimental result in Figure 9.25.[25] We calculate $\rho(E)$ for those electrons that suffered only a few small angle collisions. Those that suffered large angle scattering will lose most of their energy in the direction of injection and will not be collected at the collector. The small angle scattering turns out to be the dominant scattering process for hot-electrons with large initial wave vector k because the scattering angle is proportional to q/k, where q is the wave vector transfer. In the calculation, we did not take into account the effect of elastic scattering from the ionized dopant impurities. While it is significant for low-energy electrons, impurity scattering is less important for hot-electrons because the scattering rate is inversely proportional to k^3. The estimated elastic mean free path for a 250 meV electron is 130 nm with $N_d = 1 \times 10^{18}$ cm^{-3}, and it is negligible in the present base ($L = 50$ nm) structure.

In considering energy relaxation of ballistic electrons, the two most important energy relaxation mechanisms are the optical phonon emission and the plasmon emission. The acoustic phonon scattering is two orders of magnitude lower than the optical phonon scattering and is ignored in the present calculation. The operating temperature of an IHET is generally below 90 K, at which $kT \ll \hbar\omega_{ph}$ and $\hbar\omega_{pl}$, where $\hbar\omega_{ph} = 36.25$ meV is the longitudinal phonon energy in GaAs, $\omega_{pl} = (N_d e^2/m^* \varepsilon_0 \varepsilon_\infty)^{1/2}$ is the plasma frequency, ε_0 is the vacuum permittivity, and ε_∞ is the high-frequency dielectric constant. At these low temperatures, one needs to consider only the spontaneous emission processes. The total emission rate $1/\tau_T$ can then be expressed as:

$$\frac{1}{\tau_T(E_k)} = -\frac{2}{\hbar} \int \frac{d^3q}{(2\pi)^3} \frac{e^2}{\varepsilon_0 q^2} \text{Im} \frac{1}{\varepsilon_T(q, \omega)} \tag{9.11}$$

where $\varepsilon_T(q, \omega) = \varepsilon_\infty + \varepsilon_{ee} + \varepsilon_{ph}$ is the total dielectric function, ε_{ee} is the contribution from electron–electron interaction, and ε_{ph} is the contribution from electron–phonon interaction. Although the contributions to ε_T are combined linearly, the fact that the excitation spectrum depends on $1/\varepsilon_T$ instead of ε_T produces a coupled mode spectrum. A full-scale computation of $1/\varepsilon_T$, however, turns out to be very complex. The evolution of $\rho(E)$ is usually obtained by Monte Carlo simulations. Unfortunately, the functional dependence of individual parameters is not always apparent in these simulations. For this reason, a simplified analytical solution is described here.

When N_d is very small, only phonon emission is possible, in which case:

$$\mathrm{Im}\frac{1}{\varepsilon_T(q,\omega)} = -\left(\frac{1}{\varepsilon_\infty}-\frac{1}{\varepsilon_1}\right)\frac{\pi}{2}\hbar\omega_{\mathrm{ph}}\delta(\hbar\omega-\hbar\omega_{\mathrm{ph}})$$
$$\equiv -\gamma_{\mathrm{ph}}\delta(\hbar\omega-\omega_{\mathrm{ph}}) \tag{9.12}$$

where ε_l is the low-frequency dielectric constant. On the other hand, if only electron–electron interaction is considered, ε_T is given by the Lindhard dielectric function, and $\mathrm{Im}(1/\varepsilon_T)$ consists of contributions from single particle excitations and plasma oscillations. Because the scattering is mostly confined to small q values due to the q^{-2} dependence in the bare potential, single particle excitations are negligible. The contribution from plasmon emission, within plasmon-pole approximation, is given by

$$\mathrm{Im}\frac{1}{\varepsilon_T(q,\omega)} = \frac{\beta}{\varepsilon_\infty}\frac{\pi}{2}\hbar\omega_{\mathrm{pl}}\delta(\hbar\omega-\hbar\omega_{\mathrm{pl}})$$
$$\equiv -\gamma_{\mathrm{pl}}\delta(\hbar\omega-\omega_{\mathrm{pl}}) \tag{9.13}$$

In this approximation, γ_{pl} is q independent, which means equal strength of excitation is given to all q values, an approximation known to overestimate the plasmon emission rate. For this reason, we introduce a fitting parameter β in Equation (9.13). β is expected to be less than unity.

Strictly speaking, because electrons and phonons are coupled, a more accurate description should be that of the coupled mode excitation. However, it is simpler to describe the scattering in terms of their original identities. The combined scattering rate turns out to be very close to the coupled mode picture. Basically, there are two major effects in mode coupling. First, mode coupling modifies the individual excitation energies and, according to Equations (9.12) and (9.13), changes the individual scattering rates. However, the energy shifts of the two excitations are always in opposite directions. When one scattering rate increases, the other decreases, making the total scattering rate approximately unchanged. The second effect of mode coupling is mutual screening. A detailed analysis shows that one excitation is screened out by another only when the former has a lower natural frequency. Again, according to Equations (9.12) and (9.13), only the weaker scattering process will be screened and become even weaker, but this situation does not substantially affect the total scattering rate. As a result, the total scattering rate can be expressed as the sum of the two individual scattering rates:

$$\frac{1}{\tau_T(E_k)} = \frac{m^*e^2}{2\pi^2\hbar^3}\frac{1}{k}\left[\gamma_{\mathrm{ph}}\ln\left(\frac{k+k'_{\mathrm{ph}}}{k-k'_{\mathrm{ph}}}\right)+\gamma_{\mathrm{pl}}\ln\left(\frac{k+k'_{\mathrm{pl}}}{k-k'_{\mathrm{pl}}}\right)\right] \tag{9.14}$$

where k' is the final state wave vector after emitting a respective quantum. Equation (9.14) gives the total scattering rate in all possible directions. In the case of high-energy injected electrons, however, scattering is mostly confined to the forward direction because of the large initial k in the direction of injection and the small q transfer in the scattering process. In this case, the final k' also concentrates in the forward direction, and the electron transport is once again pertaining to a one-dimensional problem. Denoting $f(k)$ as the single spin energy distribution function at time t_n, $f(k, t_{n+1})$ can be expressed as:

$$f(k, t_{n+1}) = f(k)+\frac{f(k''_{\mathrm{ph}})[1-f(k)]}{\tau_{\mathrm{ph}}(k''_{\mathrm{ph}})}\Delta t+\frac{f(k''_{\mathrm{pl}})[1-f(k)]}{\tau_{\mathrm{pl}}(k''_{\mathrm{pl}})}\Delta t$$
$$-\frac{f(k)[1-f(k'_{\mathrm{ph}})]}{\tau_{\mathrm{ph}}(k'_{\mathrm{ph}})}\Delta t-\frac{f(k)[1-f(k'_{\mathrm{pl}})]}{\tau_{\mathrm{pl}}(k'_{\mathrm{pl}})}\Delta t \tag{9.15}$$

where k'' is the state wave vector one quantum above the state E_k and $\Delta t = t_{n+1} - t_n$.

To explain the photocurrent transfer ratio observed in Figure 9.25, we rewrite Equation (9.10) for α_p as:

$$\alpha_p(V_e) = c \int_0^\infty \rho(V_e, E, 500) T(E) dE \tag{9.16}$$

where $\rho(V_e, E, 500)$ is the photoelectron energy distribution at a particular V_e and at a fixed L of 500 Å. The initial photoelectron distribution $\rho(V_e, E, 0)$, defined as $f(E, t = 0)/\int f(E, t = 0)dE$, is assumed to be Gaussian. Its width, determined from the emitter spectral responsivity measurement, is 13 meV. The effect of V_e is to lift the electron distribution to higher energies so that $\rho(V_e, E, 500) = \rho(0, E - e\sigma|V_e|, 500)$, where σ is a proportionality constant between the injection energy shift and the applied eV_e. Because the injection energy depends on the potential drop at the quantum well unit next to the base, σ is expected to be 0.033 if the potential drop across the emitter is linear.

$\rho(0, E, 500)$ at zero bias is first computed from Equation (9.15) with two adjustable parameters, β and N_d, which determine the shape of the hot-electron distribution in the front boundary of the filter. The phonon energy is assumed to be the same as that in GaAs. $\rho(V_e, E, 500)$ at finite bias is then equal to $\rho(0, E - e\sigma|V_e|, 500)$ with a free parameter σ, which controls the rate of the energy lifting relative to the filter. The value of $\alpha_p(V_e)$ in Equation (9.16) can then be computed with one more adjustable parameter c, which controls the absolute magnitude of the observed transfer. The best fit is shown as circles in Figure 9.26. It is derived by setting $\beta = 0.25$ and $N_d = 1.3 \times 10^{18}$ cm^{-3} in the distribution calculation. The fitted value of β is consistent with other experiments.[25] The slightly higher N_d than the nominal doping density is expected from electron accumulation in the InGaAs base layer because of its lower conduction band edge. The electron energy distributions for two V_e (= 0 V and −5 V) are shown in Figure 9.23. From fitting α_p below −2 V, $c = 0.42$, and $\sigma = 0.025$. Because the fitted σ is slightly less than 0.033, the potential drop in the MQW detector is not completely linear even at this low bias. The MQW potential drop near the base is less than the other periods as mentioned previously. At higher V_e, the nonlinearity is seen to increase, delaying the experimental turn-on voltage for electron injection to the higher filter pass-band E_3'. This observation is consistent with the self-consistent potential distribution calculated by Thibaudeau et al. under optical illumination.[26] In Figure 9.26, we also plot the fitted α_p if we consider each individual scattering alone. From the shape of the theoretical curves, it is apparent that neither scattering can explain the data alone, and plasmon emission is more important in this higher doping sample.

In summary, we have established an approach to calculate the hot-electron energy evolution in the nanometer scale. It can be used to describe the electron transfer across the base of an IHET.

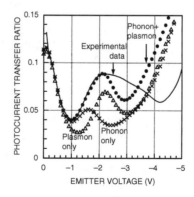

FIGURE 9.26 The measured photocurrent transfer ratio (solid curve) as a function of emitter voltage. The triangles represent the fitted theoretical curve if only plasmon emission is considered. The crosses represent that if only the phonon emission is considered. The circles are the fitted theoretical curve if both plasmon and phonon are included. The discrepancy between theory and experiment beyond −2V is attributed to nonuniform potential drop across the QWIP. The smaller potential drop near the base delays the rise of the transfer ratio from −3 V to −5 V.

From the fitting analysis, we showed that the initial α_p peak appeared at ~ -0.2 V is due to the transfer of the ballistic electrons through the E_2' pass-band. The first rise of α_p between -1 and -2 V is due to the transfer of the electrons whose energies are one quantum below the ballistic peak. The second rise between -4 and -5 V is due to the transfer of ballistic electrons through the E_3' pass-band. We have also determined the relative strength of phonon scattering and plasmon scattering, with which the characteristics of current transfer in nanometer hot-electron devices can be predicted and optimized.

9.5 Noise in Semiconductor Nanostructures

9.5.1 Noise of QWIPs

Although the photocurrent of a QWIP can be well described by the classical drift–diffusion model of Equation (9.9), the value of the photoconductive gain g depends on the detailed electron capture and release processes in the excited QW states. The photoconduction process consists of several steps, in which (1) an electron absorbs a photon and promotes to an excited state, (2) after a time period, the electron either recombines in the same well or scatters into one of the barrier states, (3) the electron propagates across the barrier through drift–diffusion, and (4) the electron scatters into the excited state of the next well; thereafter the electron either recombines into that well or continues to pass onto another barrier. If step (2) is a quantum transition process, then the dwell time in the original well will be the same as that (denoted by t_w) in all the other wells. If the electron drifts out of the well in a continuous motion, then the time spent in the original well will be $t_w/2$ because the photoelectrons are created at the center of the well on the average.

In the following, in addition to the above two cases, which will be referred to as *quantum* and *classical* cases, we also consider a *direct* case in which the electron is directly excited into the barrier so that the initial dwell time is zero. With these three cases, we can introduce an inhomogeneous electron decay function $F(t)$ to account for the discrete structure of the QWIP. $F(t)$ describes the probability of a hot-electron surviving from recombination after its creation. The hot-electron can be either a photoelectron or a thermal electron. Within this model, the electron subjects to recombination only in the quantum well region, with an intrinsic lifetime τ_i; and there is no recombination in the barrier region, resulting in a nonuniform decay process. The $F(t)$ in the insert of Figure 9.27 represents (a) the quantum case, (b) the direct case, and (c) the classical case.

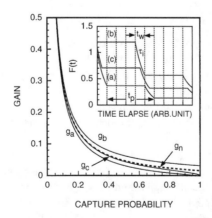

FIGURE 9.27 The theoretical photoconductive gains vs. the capture probability for three different assumed dwell times in the original well where the photo-excitation occurs. In this figure, g_a assumes equal dwell time for the original well and all other wells, g_c assumes half the dwell time for the original well compared with all other wells, and g_b assumes zero dwell time for the original well. g_n is the noise gain, independent of the assumed dwell times. The insert shows the corresponding photoelectron decay functions vs. time.

With these three different $F(t)$, the average lifetime τ_{av} of an electron in each case can be calculated using:

$$\tau_{av} = \int_0^\infty tP(t)dt \tag{9.17}$$

where $P(t) = -dF(t)/dt$ is the probability of an electron's being captured at time t. For example, in the quantum case,

$$P(t) = \frac{1}{\tau_i}e^{-(t - nt_p + nt_w)/\tau_i} \quad \text{for } nt_p < t < nt_p + t_w, \tag{9.18}$$

and 0 otherwise. In Equation (9.18), n is the well index starting from 0, and t_p is the transit time per period. As expected, the calculated τ_{av} is the smaller in the quantum case because the hot-electron spends most time in the original well where recombination can start instantly. The direct case has the longest τ_{av} because recombination will not start until the electron travels to the next well.

With the calculated τ_{av}, g can be obtained using the definition $g \equiv \tau_{av}/(Nt_p)$ and expressed in terms of the capture probability p_c, which is defined as the fraction of electrons recombined in a single well other than the original well, i.e.,

$$p_c = 1 - e^{-t_w/\tau_i} \tag{9.19}$$

For a general structure, g is also an explicit function of t_p; but for detectors with very thick barriers, the expression for g can be simplified. It turns out that g in the quantum case is $g_a = (1-p_c)/(Np_c)$, g in the classical case is $g_c = (1-p_c/2)/(Np_c)$ for $p_c \neq 1$,[27] and g in the direct case is $g_b = 1/(Np_c)$. Therefore, for a given p_c, the value of g is model-dependent with $g_a < g_c < g_b$.

Similarly, one can also calculate the noise gain g_n in the g–r noise for each case using:[28]

$$i_n^2 = 2egIB\int_0^\infty \left(\frac{t}{\tau_{av}}\right)^2 P(t)dt \tag{9.20}$$

$\equiv 4eg_n IB$, where i_n and I are the noise current and the average current, respectively, and B is the noise bandwidth. Strikingly, g_n in the three cases are identical, and $g_n = (1-p_c/2)/(Np_c)$. The identical g_n is not accidental but follows a general principle that, if a random variable g is increased by a constant value c, the average value will be increased from $<g>$ to $<g> + c$, such that $g_b = g_a + 1/N$; but the variance $<[(g+c)-(<g>+c)]^2> = <(g-<g>)^2>$ should remain unchanged. Physically, while the electrons recombining in the original well do not contribute to the average photocurrent, the fluctuation of these electrons away from the mean value does contribute to the noise.

Figure 9.27 shows the calculated g and g_n as a function of p_c for $N = 32$. From Figure 9.27, it is seen that g_n is equal to g_c in a wide range of p_c, i.e., the photoconductive gain and the noise gain will be the same if the photoelectrons conduct through the entire MQW structure including the QW region through drift–diffusion. On the other hand, if the photoelectron transfer out of a well is a quantum real-space transfer, the photoconductive gain represented by g_a will be smaller than g_n. The direct case will yield $g_b > g_n$. Through an analysis of experimental data of Xing et al.,[29] we concluded that $g < g_n$,[27] and thus the electron transfer between a well and the adjacent barrier is a quantum process. As seen in Figure 9.27, when p_c is small, the photoelectron is unlikely captured by the original well. In this case, there will be practically no difference among different models.

9.5.2 Classical Noise in IHETs

While the noise of a QWIP is identical to that of a classical detector, the noise of an IHET is substantially modified by the quantum transport through the filter. In this section, however, we will first discuss the noise if the electron transmission is classical. In a macroscopic device such as an electron vacuum tube, a uniform probability p can be assigned to each electron able to reach the anode, and a probability ($1-p$) to each captured by the grid. Likewise, in an IHET, one can assign a uniform probability α, which is the current transfer ratio, for each injected electron able to reach the collector, and a probability ($1-\alpha$) for each captured by the base. As in a vacuum tube, the statistical nature of this current division process introduces current fluctuation, which is known as *partition noise*. The noise measured at the collector of an IHET, therefore, comprises two independent and uncorrelated noise components. One is the generation–recombination (g—r) noise from the QWIP that is transferred into the collector, and the other is the partition noise generated at the base.

The dark current-induced g–r noise of a QWIP, which forms the emitter part of an IHET, is given by

$$i_{ne} = \sqrt{4egI_eB} \tag{9.21}$$

where g in this section denotes the noise gain, I_e is the emitter dark current, and B is the noise bandwidth. (The distinction between the photoconductive gain and the noise gain discussed in the last section is not relevant in the following discussion.) In deriving Equation (9.21), it is important to realize that it is the number of electrical pulses N_p received at the contact for a given integration time τ_d, each of which contains an electrical charge of ge, but not the number of physical electrons collected ($= gN_p$), that follows Poisson distribution.

To simplify the derivation, we consider a detector structure where the emitter area A_e and the collector area A_c are the same. With this detector geometry, the dark current transfer ratio α_d is I_c/I_e, if the filtering process does not introduce additional noise, the collector noise current i_{tr}, referred to as transfer noise, is simply $\alpha_d i_{ne}$, and the filter in this case acts as a noiseless amplifier. However, the electron filtering is a statistical process subject to fluctuation. The standard deviation of these fluctuations σ obeys binomial statistics and is equal to $\sqrt{[N_p\, \alpha_d\, (1-\alpha_d)]}$. Thus, the partition noise current i_{cp} is equal to

$$i_{cp} = \sqrt{2}\frac{\sigma}{\tau_d}eg = [4eg\alpha_d(1-\alpha_d)I_eB]^{1/2} \tag{9.22}$$

where the extra factor of $\sqrt{2}$ is related to the fact that g is not an absolute constant but a statistical average. Because i_{tr} and i_{cp} are uncorrelated, the total collector noise current is

$$i_{nc} = \sqrt{i_{tr}^2 + i_{cp}^2} = \sqrt{\alpha_d^2 i_{ne}^2 + i_{cp}^2} = \sqrt{4eg\alpha_d I_eB} = \sqrt{4egI_cB} \tag{9.23}$$

Equation (9.23) happens to be the same as the standard expression for g–r noise for an average current I_c, and it can be obtained directly from the facts that the average number of electrical pulses arrived at the collector is $N_p\alpha_d$ and the pulse arrival follows Poisson distribution.

9.5.3 Separation of g–r Noise and Partition Noise

By measuring i_{nc}, the total collector noise caused by the two noise sources can be obtained. Such a measurement, however, cannot distinguish the noise from different origins. Chen et al.[30] made use of the fact that, because the two noise components are not correlated, they are separable by noise correlation measurements. In particular, the transfer noise current will always be in phase with the emitter noise current. Therefore, by measuring the cross-power spectrum $S_{ec} \equiv i_{ne}(i_{nc})_{\text{in-phase}}/B$, the magnitude of the transfer noise can be determined independently.

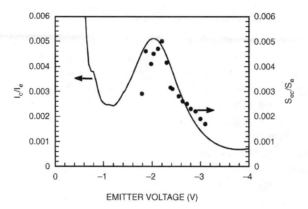

FIGURE 9.28 The measured dark current ratio I_c/I_e and the ratio of S_{ec} and S_e (defined in text) as a function of emitter voltage.

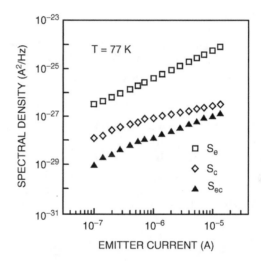

FIGURE 9.29 The noise power spectral densities S_e, S_c, and S_{ec} as a function of the emitter current. The observation that $S_c > S_{ec}$ indicates those electrons accepted to the collector have a higher current gain.

The experimental IHET consists of a band-pass filter, resulting in a pronounced peak in I_c/I_e as shown in Figure 9.28. Because $A_e > A_c$ ($A_e/A_c = 3.52$) in the experimental structure, $I_c/I_e = \alpha_d A_c/A_e$ where α_d follows the standard definition of the current density ratio. At the same time, the spectral power densities S_e ($\equiv i_{ne}^2/B$), S_c ($\equiv i_{nc}^2/B$), and S_{ec} are measured and shown in Figure 9.29. S_{ec} is much smaller than S_e, showing that a very small percentage of emitter g–r noise is transferred into the collector, consistent with the small α_d shown in Figure 9.28. In order to verify the in-phase component originated from the emitter, we write

$$S_{ec} = \frac{1}{B} i_{ne} (i_{nc})_{in-phase} = \frac{1}{B} i_{ne} \left(\frac{A_c \alpha_d}{A_e} \right) i_{ne} = \left(\frac{I_c}{I_e} \right) S_e \qquad (9.24)$$

The ratio of S_{ec}/S_e should be the same as I_c/I_e. In Figure 9.28, the data of S_{ec}/S_e are also shown. The similarity between the two ratios indicates that the in-phase noise component indeed comes from the emitter, and the noise transfer ratio is the same as α_d.

Although the magnitude of S_{ec} relative to S_e is in agreement with the expectation, the relative magnitude between S_c and S_{ec} is unexpected, because:

$$S_c = 4egI_c = 4eg\left(\frac{I_c}{I_e}\right)I_e = \left(\frac{I_c}{I_e}\right)S_e = S_{ec} \tag{9.25}$$

The large difference in the experimental S_c and S_{ec} in Figure 9.28 thus indicates that other factors need to be considered, which will be described in the next section.

9.5.4 Quantum Partition Noise

It turns out that Equation (9.23) is valid only for devices with g and α_d both independent of electron energy E, which is not the case for the present detectors. In an IHET, the thermally activated current has a relatively wide energy distribution of 40 to 70 meV. This energy spread introduces two effects in the noise consideration. One is a lifetime effect: the hot-electrons with higher E are expected to have a longer lifetime and hence a larger gain. The g value in Equation (9.23) is actually an average value across the dark current energy distribution. Because the collector accepts only the higher energy electrons, the gain measured at the collector g_c can be larger than the emitter gain g_e. Another effect is a quantum effect: α_d should also be different for electrons with different energies governed by the transmission coefficient of the filter $T(E)$. According to the theory proposed by Büttiker[31] and Beenakker and van Houten,[32] partition noise should be applied to separate quantum channels, which in the present case are represented by different energy values. Therefore, σ in Equation (9.22) is more accurately expressed as:

$$\sigma^2 = N_p\int_0^\infty \rho(E)T(E)[1 - T(E)]dE \equiv N_pZ \tag{9.26}$$

where $\rho(E)$ is the hot-electron energy distribution at the front boundary of the filter. The factor Z depends on $T(E)$. If the resulting $T(E)$ is slow varying, Z approaches $\alpha_d(1-\alpha_d)$, and Equation (9.22) is recovered. In another extreme, if $T(E)$ assumes only values 0 and 1, i.e., the transmission is deterministic, then $Z = 0$, in which case there is no partition noise although the average transmission may still be the same. Incidentally, i_{nc} and the base noise current, i_{nb}, are totally correlated in the deterministic case with $i_{ne} = i_{nc} + i_{nb}$, instead of the usual expression $i_{ne}^2 = i_{nc}^2 + i_{nb}^2$ for uncorrelated noise. Accounting only for the quantum effect, i_{nc} is given by

$$i_{nc}^2 = \alpha_d^2 4egI_eB + 4egZI_eB \tag{9.27}$$

In order to incorporate the lifetime effect, let us consider only those electrons with E larger than the filter transmission threshold E_{th}; the lower energy electrons are not relevant in the present discussion. We denote the portion of the emitter dark current carried by these electrons to be I_{eh} and the average transfer ratio to be α_h with $I_c = \alpha_h I_{eh}$. Obviously, the average current gain of these electrons is equal to g_c, and the emitter noise current carried by these electrons is $\sqrt{(4egI_{eh}B)}$. Therefore, including the lifetime effect, i_{nc} is now given by

$$\begin{aligned}
i_{nc} &= \sqrt{4eg_c(\alpha_h + Z_h/\alpha_h)\alpha_h I_{eh}B} \\
&= \sqrt{4eg_c(\alpha_d + Z_h/\alpha_d)_h I_cB} \\
&= \sqrt{4eg_c'I_cB}
\end{aligned} \tag{9.28}$$

where g_c' is the measured apparent collector noise gain including both effects discussed above, and $(\alpha_d + Z/\alpha_d)_h$ is evaluated for $E > E_{\text{th}}$. Because the two effects oppose each other, g_c' can be larger or smaller than g_e. With these two additional effects, Equation (9.25) should read

$$S_c = 4eg_c'I_c = 4eg_e\left(\frac{g_c'}{g_e}\right)\left(\frac{I_c}{I_e}\right)I_e = \left(\frac{g_c'}{g_e}\right)S_{ec} \tag{9.29}$$

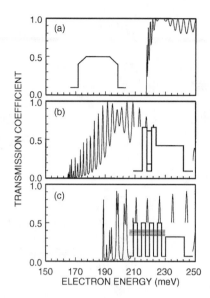

FIGURE 9.30 The transmission coefficients of three filter structures under study. The one that has larger transmission fluctuation between 0 and 1 offers a smaller partition noise.

At $I_e = 10^{-5}$ A (i.e., $V_e = -3$ V), S_c/S_{ec} is measured to be 3.0 from Figure 9.29. On the other hand, g_e and g_c' can be deduced separately from S_e and S_c, respectively. At the same V_e, g_e is measured to be 0.15, whereas g_c' is 0.42, which gives $g_c'/g_e = 2.8$, consistent with the ratio of S_c/S_{ec} and thus confirming Equation (9.29). Note that S_{ec} in Equation (9.24) is independent of the difference between the emitter and the collector noise gains because the cross-spectrum is a coincidence measurement, in which only those fluctuation events occurring simultaneously in the emitter and the collector are measured. The ratio of the fluctuation (S_{ec}/S_e) is thus equal to the ratio of the mean (I_c/I_e).

In order to further investigate the partition noise, we present the noise characteristics of three IHETs with different filter structures but with similar emitter structures.[33] The filter transmission characteristics $T(E)$ of these filters are shown in Figure 9.30. Because of the graded barrier structure of Figure 9.30(a), $T(E)$ is very close to 1 when E is 15 meV above the transmission threshold E_{th}. The quantum partition noise is expected to be small for this filter, and hence it should offer a low noise performance. However, it turns out that due to the high E_{th} of this filter and the small potential drop near the base for this QWIP, the hot-electron distribution barely reaches the top of the filter barrier. Even at a high bias of $V_e = -3$ V, only the upper 1% distribution is accepted into the collector. As a result, only the turn-on characteristics of the filter are relevant in the noise consideration. Because near the transmission threshold, $T(E)$ of this filter is far from either 0 or 1, it will not reduce the partition noise significantly. In comparison, filters in (b) and (c) of Figure 9.30 offer a larger degree of suppression because $T(E)$ is closer to either 0 or 1 near the transmission threshold.

Figure 9.31(a) shows the measured g_e and g_c' for detector in Figure 9.30(a). In this plot, g_e increases linearly with V_e below −2 V and saturates at a value of 0.5. For the collector, the measured g_c' is slightly higher than g_e as expected from the lifetime effect. A computation of the factor $(\alpha_d + Z/\alpha_d)_h$ in Equation (9.28), assuming a constant $\rho(E)$ up to 20 meV above E_{th}, is equal to 0.96, showing the quantum effect indeed negligible for this filter structure.

Figure 9.31(b) shows the noise characteristics for detector in Figure 9.30(b) that fall into an intermediate case, where the lifetime effect is larger than the quantum effect below −1.6 V and vice versa at higher V_e. For this filter, because of the much lower energy threshold E_{th} as shown in Figure 9.30, the hot-electron distribution can be above the threshold at large bias, reducing the lifetime effect yet maintaining a small current transfer because of the small α_d near the threshold. The calculated $(\alpha_d + Z/\alpha_d)_h$ factor is 0.82 for the same energy range above E_{th}, bringing g_c' from an expected value of ≈ 0.5 to the measured value of 0.4 at large V_e.

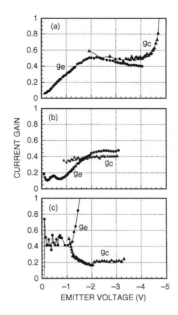

FIGURE 9.31 The current gains measured at the emitter and the collector of the IHETs having the filter structures depicted in Figure 9.30.

Figure 9.31(c) shows the noise characteristics for the detector in Figure 9.30(c) that are mostly dominated by the quantum effect. For this detector, the number of QW period is much less than that of the last two detectors (5 vs. 30). Therefore, a large g_e of 0.45 is observed at a small V_e. For higher V_e, the apparent g_e increases rapidly, which is caused by the large $1/f$ noise at large applied field. If one subtracts out the $1/f$ noise from the total observed noise, g_e is expected to be the same as the low voltage value. On the other hand, the $1/f$ noise was found to be absent at the collector because the filter is able to filter out the low-energy impurity-related tunneling currents. Consequently, the value of g_c' is unaffected at high V_e. For this detector, the lifetime effect is small because most of the injected electrons are above the filter threshold at this high field. As a result, g_c' decreases from 0.5 at low bias to 0.21 at high bias, where only the quantum effect remains. The calculated $(\alpha_d + Z/\alpha_d)_h$ factor for this filter is 0.66, consistent with the observed value of 0.47 based on the ratio of g_c' and g_e.

In summary, the noise properties of current filtering in nanostructures have been elucidated. We found that a filter whose $T(E)$ fluctuates close to 0 and 1 offers a better noise performance. In fact, a noise factor can be assigned to each filter design, analogous to a signal amplifier. The sensitivity of an IHET can be improved using an appropriate filter with a low noise factor.

9.5.5 Energy-Resolved Noise in Nanostructures

In a classical photoconductor operated at temperature T such that kT is much smaller than the bandgap energy, the dark current initiated by thermionic emission (TE) composes a very narrow range of energy due to the rapidly decreasing Fermi distribution tail. In this case, the g–r noise is accurately described by a single noise gain g for all the conducting electrons. In QWIPs, however, the subband spacing is much smaller, resulting in a wider energy spread of the TE current at the typical operating temperature. In addition, the dark current of a QWIP is also contributed by the thermally assisted tunneling through the barriers. Under this transport mechanism, the decreasing Fermi distribution function is partially compensated by the increasing barrier transmission as the electron energy E increases, leading to an even wider energy distribution. With different E, the noise gain $g(E)$ at a specific E is expected to be different as both the carrier lifetime and the transit time are energy-dependent. It is therefore interesting to determine the energy dependence of g in QWIPs.

In the previous sections we described the energy-resolved current measurements and discussed the change of the observed gain due to energy filtering. In this section, we will describe an energy-resolved noise measurement using a square-barrier high-pass filter.[34] Different from the last section, we will be changing V_c instead of V_e to alter the amount of emitter current into the collector for noise analysis. Because the transmission through a square barrier is far from unity, especially under a tilting field, the quantum effect on the noise partition is negligible and the energy dependence of the QWIP noise gain can be examined.

When the current division process is classical, the factor $(\alpha_d + Z/\alpha_d)_h$ is equal to unity and $S_c = 4eg_cI_c$. Under a collector bias, electrons originated from the QWIP with E larger than the threshold energy $\Delta E_c + eV_c$ can be accepted into the collector, where ΔE_c is the filter barrier height and V_c is the magnitude of a negative collector bias. Other electrons with lower energy will be drained through the base. When V_c is increased by δV_c, the injected I_c will drop by δI_c and a concomitant decrease in S_c by δS_c with:

$$\delta S_c = 4eg(E)\delta I_c \tag{9.30}$$

where $E = \Delta E_c + eV_c$. The value of $g(E)$ as a function of E can thus be obtained by measuring the change of I_c and S_c as a function of V_c, while keeping I_e constant.

The dc $I_c - V_c$ characteristics of the device are shown in Figure 9.32 for two different constant I_e levels (80 and 120 μA). The corresponding V_e is −1.46 and −1.58 V, respectively, when $V_c = 0$ V. To keep I_e constant, V_e is approximately changed by 5% at the highest V_c applied due to the finite base resistance. Figure 9.32 shows a decreasing I_c with increasing V_c as expected. Figure 9.33 shows S_e and S_c of the device in the two I_e conditions. Throughout the V_c bias, S_e remains approximately constant, indicating that the transport process within the QWIP is not significantly affected by the collector bias. From the values of

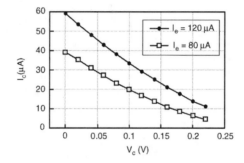

FIGURE 9.32 The collector I-V characteristics of an IHET at different emitter current levels.

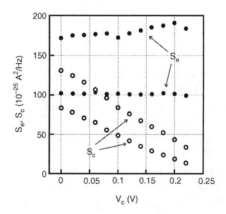

FIGURE 9.33 The change of S_e and S_c as a function of collector voltage. S_c is reduced at a larger V_c because a smaller number of electrons is accepted to the collector.

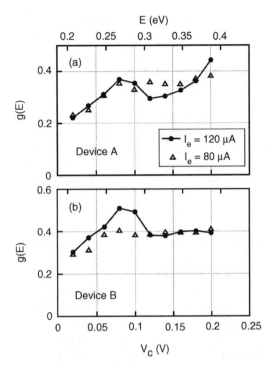

FIGURE 9.34 The current gain as a function of collector voltage, which can be translated into the energy by $E = \Delta E_c + eV_c$. The $g(E)$ minimum indicates the location of the L-valley.

S_e and I_e, the average emitter noise gain g_e, with $S_e \equiv 4eg_eI_e$, can be obtained. g_e is deduced to be 0.22, approximately the same for both I_e levels. For another device from the same wafer, g_e is 0.23. On the other hand, S_c decreases monotonically in Figure 9.33. Using Equation (9.30), $g(E)$ can be deduced, and the results are shown in Figure 9.34 for the two devices from the same wafer material. Both devices show similar noise behavior. In particular, $g(E)$ of both devices increases generally with E except in the energy range between 267 meV and 307 meV, where $g(E)$ decreases. The decrease is more pronounced for the larger I_e (i.e., larger V_e) value.

The characteristics of $g(E)$ [$\equiv \tau(E)/\tau_{tr}(E)$] can be explained by first examining the parameter τ_{tr}. Under an electric field F, the thermal electrons drift across the barriers with a drift velocity $v_d = \mu F$, where μ is the mobility. When the electron is far below the L-valleys, $\mu(E)$ and thus $\tau_{tr}(E)$ are relatively insensitive to E. However, when E approaches the L-valley energy E_L, the electrons can accelerate into the L-valley under a small field and lower its mobility. As a result, $\tau_{tr}(E)$ increases and $g(E)$ decreases as E approaches E_L. The minimum at $V_c = 120$ meV is interpreted as the location of the L-valleys. The corresponding $E_L = \Delta E_c + eV_c = 307$ meV is almost the same as that (310 meV) expected from an $Al_{0.25}Ga_{0.75}As$ quantum well barrier, confirming the significance of intervalley scattering.

For the parameter τ, it is expected to be longer for a larger E because the electron scattering time is proportional to $\sqrt{E}$ for both optical phonon and plasmon emissions as shown in Equation (9.14), and it takes more scattering events to bring a higher energy electron into the well. The increasing τ leads to a generally increasing g as E increases, which explains the generally increasing trend with V_c. On the other hand, τ is insensitive to F because the hot-electron distribution is insensitive to the applied field. The slightly higher g under a larger I_e or V_e is thus attributed to a slightly shorter τ_{tr} at higher F.

In summary, in a semiconductor nanostructure, one can analyze the noise produced by the individual current components according to their energies. We found that the noise gain generally increases with electron energy due to a longer lifetime. The gain starts to decrease as the energy approaches a satellite valley because of a longer transit time. $g(E)$ forms a minimum at the valley minima, and hence the value of E_L can be identified by the noise measurement. The presence of the L-valleys has a larger effect at a larger field because more electrons are accelerated into the valleys. The noise gain of those electrons with $E > \Delta E_c$ is larger than the average noise gain measured at the emitter, implying that the gain with $E < \Delta E_c$ is less than the average.

9.6 Voltage-Tunable QWIPs

One of the important uses of two-color infrared detection is remote temperature sensing.[35] For this application, it is desirable to have a detector with two distinctive detection wavelengths. A two-color QWIP with its detection wavelength tunable by a bias is useful for this purpose. It can be integrated with a time-multiplexed read-out circuit to greatly simplify the focal plane array production. Voltage-tunable two-color QWIPs have been investigated in the past based on two different mechanisms. The first approach is based on Stark shift in asymmetric multiple QW (MQW) structures.[36–38] The second approach relies on electron transfer between coupled QWs under an applied bias.[39,40] In this section, we describe a two-color QWIP based on the latter approach.[41] Its peak detection wavelength can be switched between 7.5 μm and 8.8 μm by reversing its bias V_b on the top contact at 4 V. The responsivities at both wavelengths are found to be very similar, suitable for read-out integration.

The two-color QWIP consists of 36 periods of QW pairs sandwiched between a 0.5 μm-thick n⁺–GaAs top contact layer and a 1.5 μm-thick n⁺–GaAs bottom contact layer. Each period consists of a 44 Å $Al_{0.05}Ga_{0.95}As$ left QW (LQW) coupled to a 44 Å GaAs right QW (RQW) through a 200 Å $Al_{0.3}Ga_{0.7}As$ barrier and separated from the next set of coupled wells by a 350 Å graded $Al_xGa_{1-x}As$ barrier (x graded from 0.3 to 0.25 along the growth direction). The $Al_{0.05}Ga_{0.95}As$ wells are uniformly doped (7×10^{17} cm⁻³ Si doping), and the GaAs wells and the barriers are undoped. The energy band diagram under different top contact bias is shown in Figure 9.35.

In this design, we use a linearly graded barrier between each unit of the MQW to improve the transport of the lower energy photoelectrons. With a graded barrier, the photoelectrons in the longer wavelength peak do not need tunneling to produce a photocurrent. The use of $Al_{0.05}Ga_{0.95}As$ instead of GaAs in one

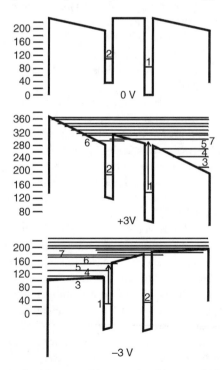

FIGURE 9.35 The band structure of a voltage-tunable two-color detector, which is based on the electron transfer between the two quantum wells. The horizontal lines indicate the extent of the wave functions.

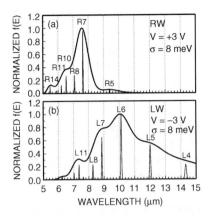

FIGURE 9.36 The predicted intersubband absorption spectra of the two-color detector when (a) the electrons are in the right QW under positive bias and (b) the electrons are in the left QW under negative bias.

of the coupled wells allows the longer wavelength detection with a thicker well width to improve the absorption oscillator strength. The selective doping of this well further ensures the occupation of this well under appropriate bias.

The energy scale in Figure 9.35 is referenced to the right edge of the RQW, and the origin is set at the GaAs conduction band edge. The lowest 14 energy levels are indicated with lines. The length of these lines represents the extent of the wave functions. For positive bias and small negative bias, the lowest energy state E_1 is in the RQW. For $V_b < -3$ V, E_1 moves to the LQW. At $V_b = -3$ V, E_1 is 4 meV below E_2 and, therefore, there are electrons in both wells. For $V_b < -3.6$ V, E_1 is at least 11 meV lower than E_2, and only the LQW is occupied. We calculate the oscillator strength of optical transitions from the computed wave functions. At $V_b = 3$ V, the largest calculated oscillator strength f is at 7.6 μm as shown in Figure 9.36(a), which corresponds to the E_1 to E_7 transition (we call this transition R7 because E_1 is in the RQW). The next significant transition is the R10 transition at 6.5 μm with $f = 0.36 f_{R7}$. Similarly, at $V_b = -3$ V, the E_1 to E_6 (L6) transition at 10.1 μm in Figure 9.36(b) has the largest oscillator strength. Other significant transitions are L7 ($\lambda = 8.9$ μm), L8 ($\lambda = 8.3$ μm), and L11 ($\lambda = 7.4$ μm) with $f = 65$, 20, and 20% of f_{L6}, respectively. Figure 9.35 also shows that all of these transitions are bound-to-extended transitions and, hence, should have comparable photoconductive gain.

The measured ac spectral photoresponse at 10 K is shown in Figure 9.37. For positive bias, the peak detection wavelength λ_p is 7.5 μm, close to the designed value of R7 at 7.6 μm. The linewidth Γ is 0.6 μm. The cut-off wavelength λ_c is 7.8 μm. There is also a small peak at 6.5 μm, which can be assigned to the R10 transition. Under negative bias, there is a narrow responsivity peak at 7.5 μm and a broad peak centered around 8.5 μm at $V_b = -1$ V. With increasing $|V_b|$, the size of the long wavelength peak increases

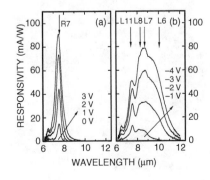

FIGURE 9.37 The observed responsivity spectra of the two-color detector under different bias. The arrows indicate the expected locations of the absorption peaks.

more than that of the short wavelength peak as electrons transfer to the left well. For V_b larger than -2 V, the dominant peak switches to 8.8 μm with a broad linewidth of 3.5 μm. The cut-off wavelength in this case extends to 10.5 μm. The broadband response under this bias is due to the absorption of the nearby transitions L6, L7, L8, and L11 as indicated in Figure 9.36(b), modified by the respective gains. For this detector, switching the bias polarity at $|V_b| \geq 3$ V leads to a 1.3 μm shift in λ_p and a 2.7 μm change in λ_c. This clearly demonstrates the voltage tunability of the two-color detector.

It is interesting to note that this two-color detector has a short-circuit photocurrent at zero bias peaked at 7.5 μm. This photovoltaic behavior of the detector arises from the presence of a built-in electric field in the graded barrier regions at zero bias.[38] Note that because E_2 is 24 meV higher than E_1 at zero bias, their difference is larger than E_F of 11 meV. The left QW is therefore unpopulated, which explains the absence of the photovoltaic effect in the long wavelength.

In this section, we have shown that the voltage dependence of the detector photoresponse can be accurately predicted by 1D quantization in an electric field. We also showed an example of switching the electron location in a nanostructure to accomplish new detector functionality of two-color detection.

9.7   Quantum Grid Infrared Photodetectors

In the previous sections we discussed the QWIP properties pertaining to the nanometer scale of the material thickness. In this section, we describe an attempt to create lateral quantum confinement within the material layers. This approach employs electron-beam lithographic techniques to generate desirable two-dimensional patterns on the QWIP material surface.[42,43] The detector material is then etched down to the bottom contact layer by low-damage reactive ion beam etching (RIBE).[44] Some examples of the patterns are shown in Figure 9.38. Due to the physical confinement by the detector sidewalls as well as the potential confinement perpendicular to the material layers, the electrons in the detector active volume will then be subject to different dimensional confinements depending on the pattern geometry, the electron in-plane locations, and their energies. For GaAs/AlGaAs material system, it is known that there is a 0.8 eV upward band bending along the material–air interface. The band bending creates a depletion layer at the interface, and for a physically exposed line with two sidewalls, a parabolic potential well will form along its width. Therefore, the electrons are also electrically confined in this direction. Because the depletion layer thickness is about a few hundred nanometers, which is a function of the doping density, quantum confinement effects are to be expected when the line width is in the submicron regime.

There are two main classes of patterns. One consists of material posts, and another is a grid structure. The post geometry should produce more different confinement geometries such as dots, conjoined dots, polygons of dots, and rings. However, as a photodetector, the photocurrent produced in individual posts

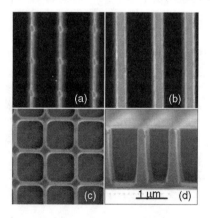

FIGURE 9.38 The SEM pictures of QGIPs with (a) connected dots, (b) lamellar lines, and (c) a crossed grid. (d) shows the cross-section of the lamellar sample.

needs to be collected by a single electrical lead, which is more difficult to achieve than the grid approach. In the grid structure, all the lines are interconnected, and because the top contact layer is thick and highly doped, it can continue to serve as a common bulk contact to the quantum structure underneath. Therefore, our first approach is of the grid patterns, and the detector is called the *quantum grid infrared photodetector* (QGIP). Note that there is no significant difference between the posts and the grids, if the pattern is a lamellar pattern, except that the lines are joined together at both ends in the grids. In a two-dimensional grid pattern, the electrons are usually subject to two-dimensional confinement, which is perpendicular to the layers and the grid lines. However, if the grid lines are much thinner than the depletion width, the lines can be devoid of electrons except those forming the intersections of the grid. At the intersections, because there are no exposed sidewalls, pockets of electrons can still exist; and these electrons are subject to potential confinement in all directions. Therefore, three-dimensional confined quantum dots (QDs) are possible in QGIP structures.

Besides the fact that the QGIPs are the natural extensions of QWIPs in nanoscience research, there are also practical reasons for their studies. It is well known that QWIP material does not absorb normal incident light. The intersubband transition is induced by an electric field perpendicular to the layers. The in-plane quantization in QGIPs will allow intrinsic normal incident absorption. The grid patterns can further be designed into photonic crystals to localize the incoming field for larger absorption. Other than the improved optical properties, the completely discrete energy levels under three-dimensional confinement may also increase the photoelectron lifetime of the detector. As we discussed in Section 9.4.3, the two major electron energy relaxation mechanisms are optical phonon emission and plasmon emission. If the energy spacing of the electron levels in the QDs is different from the optical phonon energy, phonon emission will be suppressed. At the same time, the plasmon mode in a quantum dot will be absent, leading to a much longer electron recombination lifetime. The longer lifetime allows a larger photocurrent relative to the dark current and hence will increase the detector's operating temperature. At present, the QGIPs and other quantum dot-related research is still in its infancy. Many of the advantages have yet to be realized.

In the following, we will discuss the experimental results on QGIPs with lamellar grid patterns. The QWIP material consists of 20 periods of QWs sandwiched between two thick contact layers. To vary depletion layer thickness w_d, two wafers are grown. The doping density in wafer #1 is 5×10^{17} cm^{-3} and that in wafer #2 is 5×10^{18} cm^{-3}. The top contact to the grid is achieved by forming a narrow metal bridge over the edge of the etched region and connected to a 10×15 μm^2 unetched region, which is negligible to the 146×146 μm^2 grid area. The electrical junction is protected from shorting by a 130 nm-thick SiO$_2$ insulating layer. The physical width w_l is varied from 0.1 to 4.0 μm in different samples. The spacing s between lines is kept constant at 1.0 μm. Ni metal is used as RIBE etching mask, which produces almost vertical sidewalls as shown in Figure 9.38(d).

Dark current I_d of the detectors at 77 K is measured. Assuming the dark current density J_d is the same for all the samples, the line width w_e that is actually occupied by electrons can be deduced. J_d can be obtained from a regular QWIP detector without a grid structure. Specifically, $w_e = I_d/J_d/L/N$, where L is the length of the lines and N is the number of lines in the grid. Figure 9.39 shows the deduced w_e vs. w_l. For both wafers, w_e is directly proportional to w_l with a finite offset, indicating a constant depletion width $w_d = (w_l - w_e)/2$. For the lower doping wafer in Figure 9.39(a), $w_d \approx 150$ nm. For the higher doping wafer in Figure 9.39(b), it is about 110 nm. The higher doping wafer maintains a small but finite dark current even when w_l is less than $2w_d$. The depletion hence is incomplete in this very high doping sample. If one assumes the same J_d for those with $w_l < 2w_d$, $w_e \approx 16$ nm for all of these samples.

In order to compare the performance of a QGIP and a regular QWIP under 45° edge light coupling, Figure 9.40 shows the NR ratio of the QGIPs with different physical width w_l. The NR ratio is defined as $R(\text{QGIP})/R(\text{QWIP}) \times J_d(\text{QWIP})/J_d(\text{QGIP})$, where R is the responsivity. The NR ratio is directly proportional to the $\eta\tau$ product of the detector, where η is the quantum efficiency and τ is the lifetime. An NR ratio of unity means $\eta\tau$ of the QGIP is equal to that of a QWIP under 45° edge coupling. The triangular symbols are for wafer #1, and the circles are for wafer #2. For the higher doping sample, the incomplete depletion allows the detector characteristics to be probed with w_l as narrow as 82 nm. For the lower

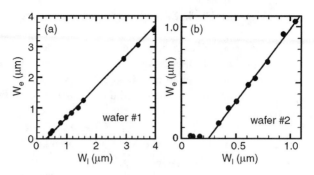

FIGURE 9.39 The conducting width w_e vs. the physical width w_l for a wafer (#1) with doping density of 5×10^{17} cm^{-3} and a wafer (#2) of 5×10^{18} cm^{-3} The x-intercepts equal $2w_d$, where w_d is the depletion width.

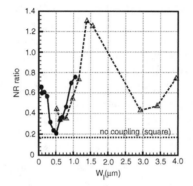

FIGURE 9.40 The measured normalized responsivity ratios for wafer #1 (dashed curve) and wafer #2 (solid curve) as a function of the physical width. The rising of the NR ratio below $w_l = 0.5$ μm indicates possible lateral quantum confinement effects. The dotted line shows the signal from a regular QWIP under normal incidence, which indicates the presence of a small amount of strayed light.

doping sample, the lines are fully depleted if w_l is less than 300 nm. From Figure 9.40, it is clear that the characteristics of the grids can be separated into two distinctive regimes. When the line width is smaller than 0.5 μm (i.e., the period p is less than 1.5 μm), the confinement effect is significant, which leads to an increasing *NR* ratio with decreasing w_l. At the smallest w_l tested, the $\eta\tau$ product is equal to 0.68 times that of a QWIP in the active volume. This result shows that either the normal incident η or τ or both are significantly increased by the lateral confinement. Further experiments are needed to determine their respective contributions and the exact conducting widths of the samples.

The experimental QWIP also has a small but finite normal incident photoresponse as indicated in Figure 9.40, which may indicate the presence of oblique stray light in the present experimental setup. However, subtracting out such a background from both the QWIP and the QGIPs will only revise the performance factor slightly from 0.68 to 0.63 and, hence, it will not affect our conclusion. When w_l is larger than 0.5 μm (i.e., when the grid parameters approach the wavelength of light in the material), optical diffraction and scattering effects are dominant, and *NR* ratio is peaked at $w_l = 1.3$ μm. In this case, the grid acts as an efficient grating for conventional light coupling.

9.8 Conclusion

In this chapter we pointed out that the QWIP represents one of the successes of quantum devices for practical applications. It improves the sensitivity of the existing infrared technology, expands the range of spectral coverage, and offers new detector functionality. Currently it is in mass production for various military, civilian, and scientific applications. Interestingly, the operation of QWIPs is critically dependent on the physics in the nanometer scale. We have described several topics of nanoscience that are related to the QWIP technology. For example, the infrared absorption of the detector is created by the one-dimensional quantization in the quantum wells. Different optical properties can therefore be formulated by synthesizing appropriate combinations of quantum well species. From the hot-electron spectroscopy studies, we learned that the electrons are conducted by tunneling through the ground states at low temperatures. At higher temperatures, the sequential population of the upper states determines the thermal activation characteristics. The ballistic electron transport in the nanometer base of an IHET provides a new means to study electron transport and its noise as a function of electron energy. The selective nature of energy filtering at the collector, on the other hand, can be used to improve the detector sensitivity. The IHET structure can also be used to elucidate the noise associated with quantum partition of electrical currents. The switching of electron locations in alternate QWs can be applied to two-color detection. Finally, we have also proposed the QGIP structure as a means for further three-dimensional quantization studies. The QGIP research is still in its preliminary stage. The change in the detector optoelectronic properties from that of the QWIPs should shed light on the intrinsic properties of the nanostructures as they cross over from 1D quantum confinement to fully three-dimensional quantum confinement. The interesting properties are the dark current, the optical absorption and photocurrent, the noise, and the magnetic properties of the QGIPs. In the last decade, we have witnessed the great opportunities offered by QWIPs in the 1D nanoscience studies and in high-performance long wavelength infrared detection. We hope that the QGIPs and other related quantum nanostructures will offer even greater opportunities in science and technology in years to come.

Acknowledgments

I would like to thank many of my colleagues who have contributed to the original work described in this article. They are C. J. Chen, L. P. Rohkinson, D. C. Tsui, A. C. Goldberg, S. V. Bandara, S. D. Gunapala, W. K. Liu, J. M. Fastenau, L. Fotiadis, P. G. Newman, G. J. Iafrate, C. H. Kuan, J. Yao, C. Y. Lee, R. P. Leavitt, and A. Majumdar.

References

1. B.F. Levine, K.K. Choi, C.G. Bethea, J. Walker, and R.J. Malik, *Appl. Phys. Lett.* 50, 1092, 1987.
2. K.K. Choi, B.F. Levine, C.G. Bethea, J. Walker, and R.J. Malik, *Appl. Phys. Lett.* 50, 1814, 1987.
3. S.D. Gunapala et al., *Inf. Phys. Tech.* 42, 267–282, 2001.
4. H. Schneider, P. Koidl, M. Walther, J. Fleissner, R. Rehm, E. Diwo, K. Schwarz, and G. Weimann, *Inf. Phys. Tech.* 42, 283–290, 2001.
5. H. Schneider et al., *Proc. ITQW'01 Conf.*, Asilomar, CA, September 10-14, 2001.
6. P. Bois, E. Costard, X. Marcadet, and E. Herniou, *Inf. Phys. Tech.* 42, 291–300, 2001.
7. M. Sundaram, S.C. Wang, M.F. Taylor, A. Reisinger, G.L. Milne, K.B. Reiff, R.E. Rose, and R.R. Marin, *Inf. Phys. Tech.* 42, 301–308, 2001.
8. B. Hirschauer et al., *Inf. Phys. Tech.* 42, 329–332, 2001.
9. J.Y. Andersson and L. Lundqvist, *J. Appl. Phys.* 71, 3600, 1992.
10. K.K. Choi, C.J. Chen, L.P. Rohkinson, D.C. Tsui, N.C. Das, M. Jhabvala, M. Jiang, and T. Tamir, in H.-D. Shih (Ed.), *Proceedings of the 1999 Meeting of the MSS Speciality Group on Infrared Materials,* Infrared Information Analysis Center, Lexington, MA, 1999, p. 243.
11. A. Goldberg, T. Fischer, S. Kennerly, W. Beck, V. Ramirez, and K. Garner, *Inf. Phys. Tech.* 42, 309–322, 2001.
12. M.A. Fauci, R. Breiter, W. Cabanski, W. Fick, R. Fick, R. Koch, J. Ziegler, and S.D. Gunapala, *Inf. Phys. Tech.* 42, 337–344, 2001.
13. E. Merzbacher, *Quantum Mechanics*, John Wiley & Sons, New York, 1970, Chapter 7.
14. K.K. Choi, C.J. Chen, and D.C. Tsui, *J. Appl. Phys.* 88, 1612, 2000.
15. J.-J. Shi and E.M. Goldys, *IEEE Trans. Elect. Dev.* 46, 83–88, 1999.

16. K.K. Choi, S.V. Bandara, S.D. Gunapala, W.K. Liu, and J.M. Fastenau, *J. Appl. Phys.* 91, 551, 2002.
17. K.K. Choi, *J. Appl. Phys.* 73, 5230, 1993.
18. K.K. Choi, L. Fotiadis, P.G. Newman, and G.J. Iafrate, *Appl. Phys. Lett.* 57, 76, 1990.
19. K.K. Choi, M. Dutta, R.P. Moekirk, C.H. Kuan, and G.J. Iafrate, *Appl. Phys. Lett.* 58, 1533, 1991.
20. K.K. Choi, S.W. Kennerly, J. Yao, and D.C. Tsui, *Inf. Phys. Tech.* 42, 221, 2001.
21. K.K. Choi, *The Physics of Quantum Well Infrared Photodetectors*, World Scientific, NJ, 1997, p. 249.
22. K.K. Choi, M. Dutta, P.G. Newman, L. Calderon, W. Chang, and G.J. Iafrate, *Phys. Rev. B.* 42, 9166, 1990.
23. A.G. Steele, H.C. Liu, M. Buchanan, and Z.R. Wasilewski, *J. Appl. Phys.* 73, 1062, 1992.
24. C.Y. Lee, K.K. Choi, R.P. Leavitt, and L.F. Eastman, *Appl. Phys. Lett.* 66, 90, 1995.
25. K.K. Choi, M.Z. Tidrow, and W.H. Chang, *Appl. Phys. Lett.* 68, 358, 1996.
26. L. Thibaudeau, P. Bois, and J.Y. Duboz, *J. Appl. Phys.* 79, 446, 1996.
27. K.K. Choi, *J. Appl. Phys.* 80, 1257, 1996.
28. R.H. Kingston and D.L. MacAdam, *Detection of Optical and Infrared Radiation*, Springer, New York, 1978, p. 56.
29. B. Xing, H.C. Liu, P.H. Wilson, M. Buchanan, Z.R. Wasilewski, and J. G. Simmons, *J. Appl. Phys.* 76, 1889, 1994.
30. C.J. Chen, Ç. Kurdak, D.C. Tsui, and K.K. Choi, *Appl. Phys. Lett.* 68, 2535, 1996.
31. M. Büttiker, *Phys. Rev. Lett.*, 65, 2901, 1990.
32. C.W. J. Beenakker and H. van Houten, *Phys. Rev. B.* 43, 12066, 1991.
33. C.H. Kuan, K.K. Choi, W.H. Chang, C.W. Farley, and F. Chang, *Appl. Phys. Lett.* 64, 238, 1994.
34. J. Yao, C.J. Chen, K.K. Choi, W.H. Chang, and D.C. Tsui, *Appl. Phys. Lett.* 72, 453, 1998.
35. C.J. Chen, K.K. Choi, W.H. Chang, and D.C. Tsui, *Appl. Phys. Lett.* 72, 7, 1998.
36. K.K. Choi, B.F. Levine, C.G. Bethea, J. Walker, and R.J. Malik, *Phys. Rev. B.* 39, 8029, 1989.
37. B.F. Levine, C.G. Bethea, V.O. Shen, and R.J. Malik, *Appl. Phys. Lett.* 57, 383, 1990.
38. E. Martinet, F. Luc, E. Rosencher, P. Bois, and S. Delaitre, *Appl. Phys. Lett.* 60, 895, 1992; E. Martinet, E. Rosencher, F. Luc, P. Bois, E. Costard, and S. Delaitre, *Appl. Phys. Lett.* 61, 246, 1992.
39. K.K. Choi, B.F. Levine, C.G. Bethea, J. Walker, and R.J. Malik, *Appl. Phys. Lett.* 52, 1979, 1988.
40. V. Berger, N. Vodjdani, P. Bois, B. Vinter, and S. Delaitre, *Appl. Phys. Lett.* 61, 1898, 1992.
41. A. Majumdar, K.K. Choi, J.L. Reno, L.P. Rokhinson, and D.C. Tsui, *Appl. Phys. Lett.* 80, 707, 2002.
42. K.K. Choi, Quantum Grid Infrared Photodetector, U.S. patent. No. 5485015, January 16, 1996.
43. L. Rohkinson, C.J. Chen, D.C. Tsui, G.A. Vawter, and K.K. Choi, *Appl. Phys. Lett.* 74, 759, 1999.
44. J.R. Wendt, G.A. Vawter, R.E. Smith, and M.E. Warren, *J. Vac. Sci. Technol. B.* 13, 2705, 1995.

Section 3

Molecular Electronics: Fundamental Processes

10

Molecular Conductance Junctions: A Theory and Modeling Progress Report

Vladimiro Mujica
Universidad Central de Venezuela

Mark A. Ratner
Northwestern University

Abstract

Conductance in molecular wire junctions — the passing of current between two electrodes through a molecular bridge — is overviewed. This is the simplest problem in molecular electronics, but it contains much of the richness that molecular circuits and molecular electronics in general imply. The discussion is based on use of a scattering formalism to calculate charge transport between electrodes. Specific topics include quantized conductance, coherent and incoherent limits of charge transport, incoherent hopping, comparison with electron transfer measurements, and more complex junction structures.

10.1 Introduction

The disciplinary area of molecular electronics is concerned with electronic phenomena that are controlled by the molecular organization of matter. The two simplest thematic parts of this research world are molecular wire junctions (in which an electrical current between two electrodes is modulated by a single molecule or a few molecules that form a junction) and molecular optoelectronics (in which a molecule or complex multimolecular structure acts either as a light-emitting diode producing light from electrical input or as a photo-conversion device producing electrical output from incident photons). Several of the fundamental problems and attributes of these systems are the same: they depend on electrical transduction between a continuum electrode and a discrete molecular structure. This fundamental theoretical problem in molecular electronics will be our entire focus in this chapter.

Because the molecular structures are characterized by discrete energy levels and well-defined wave functions, the problem of transport in molecular wire junctions is inherently quantum mechanical. Therefore, in this chapter we will discuss the simplest quantal formulations of conductance properties in molecular wire junctions, how calculations of such conductance might be completed, and how they compare with the available experimental data.

Molecular electronics[1–10] deals with two kinds of materials. In the discrete molecular wires themselves, the characterization is most simply done in terms of a molecular wave function; in independent particle models such as Huckel, Kohn–Sham, or Hartree–Fock, these can be characterized in terms of single delocalized molecular orbitals. The electrodes, on the other hand, are continuum structures. Simple metallic electrodes can be characterized by their band structures, and semiconductor electrodes involve multiple band structures and band bending. All of these phenomena are crucial in determining the conductance, and all must therefore be included in a simple conductance formulation.

To construct a molecular wire junction such as that schematically shown in Figure 10.1, it is necessary to make the molecules, to assemble the junction, and to measure the transport. The issues of assembly and synthesis are crucial to the whole field of molecular electronics, but we will not deal with them in this chapter. Our aim here is a conceptual review of how to calculate and understand molecular conductance in wire junctions and some of the mechanisms that underlie it. As Figure 10.1 makes clear, there are several problems involved in moving charge between the macroscopic anode and macroscopic cathode. These processes include charge injection at the interface, charge transport through the molecule, modulation of charge by external fields, and the nature of the electrostatic potential acting at finite voltages.

Molecular electronics is a very young field, and neither the experimental generalities nor the appropriate computational/theoretical formulation is entirely agreed upon. The viewpoint taken here is that

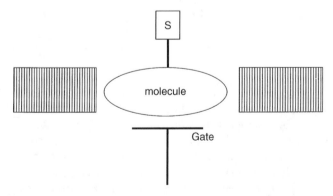

FIGURE 10.1 Schematic of a molecular junction consisting of a single molecule suspended between source and drain macroscopic electrodes. The transport through the molecule can be modulated by tuning the electrostatic potential through the gate or by bringing up another molecule or ionic species, denoted as S, that interacts covalently with the molecule.

the role of theory and modeling is to aid the interpretation and prediction of experiments. Mechanistic insight is crucial and will be our focus. Nevertheless, the development of accurate quantitative methodology to compare with accurate quantitative experimental data must be one of the aims of the entire endeavor of molecular electronics. Only under those conditions can this subject become a well-grounded science as opposed to a series of beautiful episodic measurements and interpretations.

These topics are sufficiently important to have been surveyed several times. References 1–17 are some of the overviews and surveys that have appeared in the recent literature; other articles in this book provide descriptions of the experimental situation, device applications, and preparation methodologies.

The current chapter is organized as follows: Section 10.2 recaps very briefly some of the relevant experiments in simple molecular wire junctions. Section 10.3 is devoted to the simplest behavior — that of coherent conductance as formulated by Landauer — and its generalization to molecular wire junctions. Section 10.4 briefly discusses electronic control and gating processes. Section 10.5 deals with the onset of inelastic behavior and how that modulates the current in tunneling junctions. Section 10.6 is devoted to the relationship between intramolecular nonadiabatic electron transfer and molecular wire junction conductance. The transition from coherent to incoherent behavior and the appearance of hopping mechanisms are discussed in Section 10.7. Section 10.8 lists some issues that have been so far unaddressed in the understanding of molecular wire junctions. Finally, in Section 10.9 we discuss very briefly the possible development of a reliable structure/function interpretation of molecular transport junctions.

10.2 Experimental Techniques for Molecular Junction Transport

Because this entire book is devoted to subjects involving molecular and nanoscale electronics, mostly with an experimental focus, we limit ourselves here to a very brief discussion of the measurements that have been made and are being made, and the challenges that they pose to computation.

Molecular electronics really began with the pioneering work of Kuhn[18] and collaborators, who measured current through molecular adlayer films. This work raised very significant questions concerning transport in molecular junctions, but the methodology both for their preparation and for their measurement was quite crude by contemporary standards.

The two great experimental advances that heralded the growth of molecular electronics as a science were the development of self-assembly methodologies[2,19] for preparation of molecular nanostructures and of scanning probe methodologies both for their preparation and measurement.[20] These are reviewed extensively elsewhere; suffice it here to say that without such techniques, the subject would never have developed. This is not to say that the techniques are ideal: self-assembled monolayers give monolayers on surfaces, not isolated molecules. It is entirely clear (changing work functions of the metal, interaction among wires, mutual polarization of the surface, long-range electrostatic interactions among charges on different wires) that transport through a single molecule within a molecular adlayer structure is not the same as transport through a single isolated molecular junction. This issue has not been very well attacked either experimentally or computationally, and it is one to which the field will have to return if the understanding of junctions is to be complete.

Similarly, the use of scanning probe methodologies for measuring transport is not entirely straightforward. If either a scanning atomic force microscope (AFM) tip or a scanning tunneling microscope (STM) tip directly contacts the film, the junction is very poorly defined geometrically; and such geometric effects may completely overwhelm the intrinsic transport that one wishes to measure. Perturbation in the structure caused by this scanning probe tip can also be important. The issue of geometries in scanning probe junctions is one that has not been effectively addressed and will be dealt with briefly in Sections 10.3 and 10.8.

It is also important to distinguish soft molecular structures from hard molecular structures. In hard molecular transport junctions, particularly those based on nanowires and carbon nanotubes,[21–23] the problems are intrinsically different from those in soft junctions using traditional organic molecules. Table 10.1 sketches some of the differences. Using nanotubes, tremendous progress has been made in addressing important issues such as the limit of quantized conductance, multiple channels, the sensitivity of the measurements to the nature of the contacts, field effects, modulations, and a panoply of other

TABLE 10.1 Junctions In Different Limits

	Soft (Molecular) Junction	Hard (Semiconductor) Junction
Typical species	p-benzene dithiol, alkene thiols, DNA	Carbon nanotubes, InP nanowires
Atomic count	Accurate to one atom	Accurate to 10^2–10^4 atoms
Interfacial binding	Coordinate covalent	Metallic or physisorptive
Orbital structure	Discrete molecular orbitals	Dense orbital structure
Transport mechanism	Nonresonant tunneling, resonant tunneling, hopping	Ballistic transport, quantized conductance

fascinating topics.[21–24] The current chapter will not discuss these structures; its scope will be limited to soft molecular junctions.

The simplest issue, then, is that of transport through a junction containing one molecule, as schematically illustrated in Figures 10.1 and 10.3. Several measurements, notably molecular break junctions,[25,26] may well measure current through single molecules. Such measurements will be easiest to interpret, and indeed some of the early results of Reed's work using molecular break junctions have generated the most computational effort in the single molecule transport situation — the primary focus in this chapter.[27–29]

More interesting and complex behavior certainly will arise in molecular junctions. Otherwise, they would be interesting only as rather poor interconnects that are dominated by their surface properties and, as such, much less interesting from the point of view of applications and technology. Molecular junctions can exhibit fascinating properties. These include applications as diodes (molecular rectifiers)[30–33] and triodes (molecular field-effect transistors).[34–36] There are also significant applications of molecular wires as sensors, as switches, and even as logic structures.[37–38] Molecules have been critical for all development of organic light-emitting diodes, and the general area of optoelectronics and photo conversion is clearly a significant one involving molecular wire junctions.[38–40]

To keep this review at a reasonable length, we have chosen to focus directly on simple single wire interconnect structures. These have been studied using a variety of techniques including STM,[41] AFM,[42] nanopore,[32] break junction,[32] crossed wires,[43] ESDIAD,[44] and direct nanoscale measurements. While all of these measurements raise some issues of interpretation, they are the raw data with which it is best to compare computations of molecular junction transport.

One important remark should be made here and is amplified a bit in Section 10.6: the fundamental process of molecular junction transport consists of electron transfer. It is therefore closely related both to measurements and models of the intramolecular nonadiabatic electron transfer phenomena in molecules and to measurements on microwave conductivity in isolated molecular structures. The huge difference between those measurements, which are made directly on molecules, and measurements on molecules in junctions is that in the junction structures the electrodes are crucial to the structure, the performance, and the conductance spectrum. This is a major focus of this chapter, and the most important lesson from this work is a straightforward one: transport in molecular tunnel junctions is not a measurement of the conductance of a molecular wire but rather a measurement of the conductance of the molecular wire junction that includes the electrodes. This is most obviously clear from the formal analysis involving the overall system, given in the next section.

10.3 Coherent Transport: The Generalized Landauer Formula

The simplest mesoscopic circuit involving a molecular wire consists of a molecule coupled to two or more nanoelectrodes, or contacts, which in turn are connected to an external voltage source. This basic circuit, where charge and energy transport occur, has been the subject of a very intensive research effort in the last 10 years. It can be used as a model for molecular imaging via the STM, for break junctions, and for molecular conductance.[45–52]

A remarkable feature of the research effort in the mesoscopic domain is that it has brought up the existence of a common theoretical framework for transport in very different systems: quantum dots, tunneling junctions, electrostatic pores, STM and AFM imaging, and break junctions. Concepts such as conductance, once ascribed only to bulk matter, can now be extended to a single molecular junction, revealing a deep connection between the models used in intramolecular electron transfer (ET) and those for transport.

10.3.1 Length Scales in Mesoscopic Systems

The word *mesoscopic* was coined by van Kampen [53] to refer to a system whose size is intermediate between microscopic and macroscopic. Microscopic systems are studied using either quantum mechanics or semi-classical approximations to it. Usually a system approaches macroscopic behavior once its size is much larger than all relevant correlation lengths ξ characterizing quantum phenomena.[54] Important systems for electronics are usually *hybrid* — that is, they involve both microscopic and macroscopic parts. This property is, to a large extent, responsible for some of the most interesting and challenging aspects of both fabrication and understanding of such devices.

In its original meaning, mesoscopic systems were studied to understand the macroscopic limit and how it is achieved by building up larger and larger clusters to go from the molecule to the bulk. In current research, many novel phenomena exist that are intrinsic to mesoscopic systems. An example is provided by very small conducting systems that show coherent quantum transport; that is, an electron can propagate across the whole system without inelastic scattering, thereby preserving phase memory.[54]

Roughly speaking, for a system of size L, there are four important lengths, in increasing magnitude, that characterize the system and define the type of transport: the Fermi wavelength λ_F, the electronic mean free path I_e, the coherence length ξ, and the localization length L_ϕ. For L between λ_F and L_ϕ the system is considered to be mesoscopic and the transport mechanism goes from ballistic to diffusive and then localized, depending on whether $\lambda_F \leq L \leq I_e$, $I_e \leq L \leq \xi$, or $\xi \leq L \leq L_\phi$, respectively. In all these regimes, transport must be described quantum mechanically. For $L > L_\phi$, the system is considered to be macroscopic; and a Boltzmann equation is a good approximation to the transport equation.[54]

We will concentrate on electron/hole transport in mesoscopic systems. For hybrid systems, both the contacts and the molecular intervening medium must be taken into account within a single framework. Contacts can be either metallic or semiconductors, and they bring in the continuum of electronic states that is necessary to have conduction. Molecular systems by themselves show gaps distributed in the whole energy spectrum that would, in principle, preclude transport.

From size considerations only, it is clear that transport through a mesoscopic system will involve the three transport mechanisms mentioned above. But a molecule has an electronic structure of its own that must be taken into account for the description of the current. This means that considering electron transport through an individual molecular junction brings in a much richer structure because of the discrete nature of the energy spectrum and the fact that, upon charging, a molecule virtually becomes a new chemical species.[50–51,55–57]

10.3.2 Landauer Transmission Model of Conductance and Its Extensions to Molecular Wires

Landauer[54,58–61] provided the most influential work in the study of charge transport in mesoscopic systems. The basic idea in Landauer's approach is to associate conductance with transmission through the inter-electrode region. This approach is especially suitable for mesoscopic transport and requires the use of the full quantum mechanical machinery, because now the carriers can have a coherent history within the sample. The physics of Landauer's model is essentially a generalization of models for quantum tunneling along several paths. These paths interfere according to the rules of quantum mechanics; and in orbital models of a molecule, they arise from molecular orbitals.

Conductance, g, is defined by the simple relation:

$$g = \frac{j}{V},$$
(10.1)

where j is the net current flow and V is the external voltage. For an ideal conducting channel, with no irregularities or scattering mechanisms along its length, and with the additional assumption that the tube is narrow enough so that only the lowest of the transverse eigenstates in the channel has its energy below the Fermi level of the contacts, the resulting conductance is given by the quantum of conductance:

$$g = \frac{e^2}{\pi\hbar}. \tag{10.2}$$

This is the conductance of an ideal ballistic channel with transmission equal to unity. The potential drop, associated with the resistance $r = g^{-1}$, occurs at the connections to the contacts. Therefore, Equation (10.2) specifies a limiting value of the contact resistance, equal to 12.8 kΩ, that is known as the *Sharvin limit*.[62] This alone is a most remarkable result that corresponds to the impossibility of totally short-circuiting a device operating through quantum channels — an event that for a classical circuit would correspond to zero resistance. Equation (10.2) also expresses the fact that conductance is quantized, a result that has been confirmed experimentally.

If a scattering obstacle, for instance a molecule, is inserted into the channel, the transmission probability T generally becomes smaller than one and the conductance is reduced accordingly to

$$g = \frac{e^2 T}{\pi\hbar}. \tag{10.3}$$

The Landauer formalism, in the simple version presented here, assumes that both the temperature and the voltage are very small and there are no incoherent processes involved in the transport. These constraints can be released in generalized versions of the theory.[60–67]

The resistance associated to Equation (10.3) can be written as:[68]

$$r = \frac{\pi\hbar}{e^2}\left[1 + \frac{1-T}{T}\right] = \frac{\pi\hbar}{e^2}\left[1 + \frac{R}{T}\right], \tag{10.4}$$

where R is the reflection coefficient. Equation (10.4) makes it explicit that, in the simple case considered here, the total resistance associated to a channel consists of a contact and a molecular resistance. Under some conditions the molecular resistance can be brought to zero — a situation that is entirely equivalent to a transparent barrier in tunneling phenomena, leaving only the contact resistance associated to the molecule contact interface.

A very convenient way to adapt Landauer's formalism to molecular circuits is the use of stationary scattering theory to compute the transmission coefficient associated with each channel.[45,46] This provides a description of the stationary current driven by an external voltage V through a molecular circuit. The relevant quantity is the total transition probability per unit time that an electron undergoes a transition from one electrode to the other as time evolves from the remote past to the remote future. This quantity, integrated over the whole energy range accessible for the tunneling electrons and multiplied by the electron charge, is the stationary current in the circuit[33]:

$$I(V) = \frac{2e}{h}\int_{\mu_L}^{\mu_R} d\varepsilon\, T(\varepsilon, \phi(V)). \tag{10.5}$$

where $T(\varepsilon, \phi(V))$ is a dimensionless transmission function; μ_L and μ_R are the chemical potential on the left and right electrodes corresponding to a voltage difference V, i.e., $\mu_L - \mu_R = eV$; and $\phi(V)$ is a function that describes the spatial profile of the voltage drop across the interface, which must be determined self-consistently for the calculation of the current.

The transmission function is given explicitly by[33]

$$\text{Tr}[G(\varepsilon, \phi(V))\Delta_L(\varepsilon, \phi(V))G^T(\varepsilon, \phi(V))\Delta_R(\varepsilon, \phi(V))], \tag{10.6}$$

with $\Delta_{L(R)}$ being the spectral density corresponding to each contact, and $G(\varepsilon, V)$, the Green function matrix, defined by

$$G(z) = (zS - H_M - \Sigma)^{-1} \tag{10.7}$$

where S and Σ are the overlap and self-energy matrices, respectively, and H_M is the molecular Hamiltonian. The self-energy matrix is defined by

$$\Sigma(z) = V_{ME}(z - H_E)^{-1}V_{EM}, \tag{10.8}$$

where V_{ME} is the matrix specifying the coupling between the molecule and the electrodes, whereas H_E is the Hamiltonian matrix or the electrodes. Δ_L is related to the self-energy by

$$\Delta_L = -\frac{1}{\pi}\text{Im}(\Sigma_L), \tag{10.9}$$

and a similar expression holds for the right electrode.

Using this formalism we can derive an expression for the differential conductance, $g = \partial I/\partial V$, that can be used in both the linear and nonlinear voltage regimes and that includes the effect of the reservoirs in a more complete fashion. Equation (10.5) has been used to compute the conductance of real molecules, to study nonlinear effects, and to include the local variation of the external field through the use of a self-consistent procedure to solve Poisson and Schrödinger equations simultaneously[70] or via density functional theory.[29] Green's function techniques of the type associated with Equation (10.5) constitute a powerful tool to describe an open system where the total Hamiltonian has been partitioned so that the reservoirs modify the molecular Hamiltonian through the self-energy term that represents the influence of the contacts. This approach takes its most general form in the Keldysh formalism.[68,71–72]

The use of the scattering description permits a natural interpretation of molecular orbitals, or delocalized molecular states, as the analogue of the transverse channels in electrostatic junctions. The conductance associated with a molecular channel can be much lower than the corresponding one for a ballistic channel because it depends both on the specific nature of the chemical bond between the wire and the metal surface and on the delocalization in the molecular orbital. This corresponds to the nonresonant regime.[46,57] The resonant regime, on the other hand, is entirely analogous to the ballistic behavior, and the wire-electrode conductance becomes equal to the Sharvin conductance multiplied by the number of degenerate molecular channels. These ideas are fundamental to the understanding of the I-V curves involving molecular wires.[46,57,73–84]

Some of the concepts discussed in terms of the Landauer model are illustrated in Figures 10.2, 10.3, 10.4, and 10.5. Figure 10.2 demonstrates the simplest form of the Landauer formula — the only contribution to the transport comes from scattered waves that proceed through the molecular junction constriction, while back-scattered waves do not contribute. This is precisely in accord with the Landauer formulation of Equation (10.3), where the transmission probability T describes the probability for a wave to be scattered forward through the junction.

Figure 10.3 schematically shows transport through a molecular junction. Figure 10.3A interprets the results of Equation (10.6): the spectral density Δ describes the interface between the electrode and the molecule, while the Green's function G describes transport through the molecule between the interfaces. Generally, one cannot limit oneself only to the molecular component — the so-called *extended molecule* (Figure 10.3B) structure[85] keeps a few metallic atoms in its complete orbital representation to describe the binding of the molecule to the metallic interface.

Figure 10.4 shows characteristic conductances as a function of voltage for the parabenzene dithiol bridge between gold electrodes. The two graphs show the measurement reported by Reed's group using molecular break junctions[25] and the current/voltage characteristics calculated using self-consistent den-

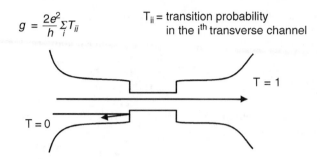

FIGURE 10.2 Illustration of the Landauer coherent conductance picture. Two waves are incident upon a molecular constriction — the wave that passes through has a transition probability of 1, and the wave that is back-scattered has a transition probability of zero. The overall conductance is the atomic unit of conductance times the sum of the transition probabilities.

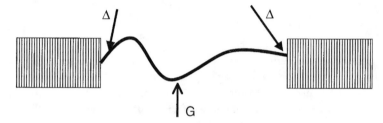

FIGURE 10.3A Physical situation underlying the generalized coherent Landauer transport for a molecular wire junction. The molecule itself, indicated by the solid line, interacts with the electrodes at the left and the right. The spectral densities denoted Δ describe interaction at the interfaces, while the delocalization over the molecule is described by Green's function G.

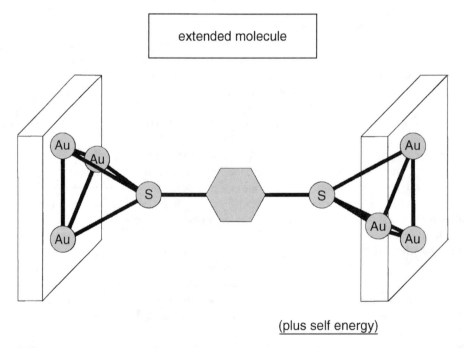

FIGURE 10.3B Schematic of the extended molecule used in the calculation of coherent transport through the benzene dithiol structure. The six gold atoms are treated as valence species along with the benzene dithiol. The bulk electrodes appear as a self-energy term.

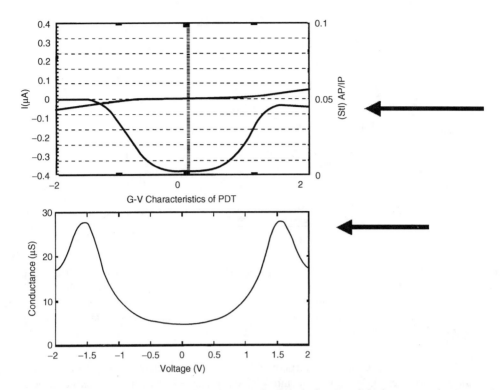

FIGURE 10.4 Comparison of the measured conductance through the benzene dithiol structure of Figure 10.3b as measured by Reed and collaborators (see Figure 10.9 caption) using break junctions (upper curve) with that calculated by Xue (A. Xue and M.A. Ratner, to be published) using the extended molecule scheme in a nonequilibrium Green's function formulation with full density functional treatment. The maxima occur at roughly the same voltage, near 1.5 volts, but the measured reported conductance is nearly a factor of 600 smaller than the calculated one.

sity functional theory by Xue.[86] Note that the voltage corresponding to the peak in the conductance is roughly the same in the calculated and experimental data, but that the conductance as measured is roughly a factor of 600 smaller than the conductance as calculated. We believe that this has to do with the nature of the electrode interfaces in Reed's break junction measurements. Work by Schön and collaborators[87] indeed suggests that with optimal interfacial transport, conductance values quite similar to those calculated here are observed. Note also the symmetric nature of this computed and measured transport: this is because the junction is at least putatively totally symmetric. The first peak corresponds to transport through the occupied (hole-type super exchange) levels in the molecular structure.

In Figure 10.5,[88] results are presented for the calculation of a gold wire consisting of discrete gold atoms between gold interfaces. The two lines show the transport corresponding to gold atoms within a covalent radius of the interface and precisely the same junction with the gold atoms extended away from the interface. With poor interfacial mixing, the conductance is reduced from the quantum conductance — even in a situation where the electrode and metal are made of the same material — so that no Schottky (polarization) barriers exist at the interface.

The calculations and measurements discussed in this section are characteristic for the Landauer (coherent transport) regime in molecular wires.

10.4 Gating and Control of Junctions: Diodes and Triodes

Some of the earliest suggestions of molecular electronic behavior involved molecular rectifiers[30] — that is, the possibility for molecules to act as actual electronic active devices, rather than simply to be used only as wires or interconnects (previously examined by Kuhn[18]). Accordingly, much of the recent activity

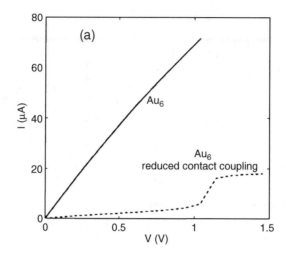

FIGURE 10.5 Calculated current/voltage characteristics for a wire consisting of six gold atoms. The upper curve shows near-perfect conductance quantization, with good contacts (covalent distances) between the gold wire and the bulk gold. The dotted line corresponds to artificially weakened interfacial mixing and shows highly nonlinear current/voltage characteristics due to the interfacial mixing. (From P.S. Damle, A.W. Ghosh and S. Datta, *Phys. Rev. B*, 64, 201403, 2001. With permission.)

has indeed focused on active molecular structures. We have already referred to work on light-emitting diodes and photovoltaics. Other activities include molecular magnetic structures, ferroelectrics, actuators, and other behaviors.[89]

Keeping with our limited focus, it is worth mentioning the modification of current flow by molecular binding or by electrostatic potential interactions that can change the transport. This is schematically indicated in Figure 10.1, where the molecule is shown with two possible functionalization sites. The schematic label S involves the binding of a substrate, while the electrostatic gate is similar to what is done in ordinary field-effect transistors. Both of these schemes have been used for sensing capabilities as well as switches; large gain can be exhibited by either FET or CHEMFET structures, while binding of molecules has also been demonstrated to change the current substantially either because of wave function mixing or (more generally and more easily) by modification of structure.[90–91]

The simplest way to understand this behavior is in terms of Equations (10.5) and (10.6). Writing the Green's function (that corresponds to the molecular component) as:

$$G_{\ell r} = \sum_s \frac{\langle l|s\rangle \langle s|r\rangle}{E_s - \eta - \sum_{(\ell r)}^{(S)}(\eta)} \tag{10.10}$$

we can easily understand how the current is modified. We recall that in Equation (10.10), the state |s> is an eigenstate of an extended molecule, with energy E_s. When the electrostatic potential is changed (FET behavior), then the state |s> will be changed by the electrostatic potential. This in turn will modify both the overlap integrals in the numerator of Equation (10.10) and the energies of the denominator. Figure 10.6 shows some very early calculations by Mao,[92] using an extended Huckel model for the molecule and a simple Newns–Anderson model for the electrode. We see that the current at zero voltage and its shape with changing injection energy are substantially modified by the presence of what is effectively an external potential — in this case resulting from the reduction of a quinone species to its di-anion. Because the charging is local, the effects can in fact be very large.

The general scheme of utilizing nearby molecular structures to change the electrostatic potential indeed underlies both the historical work on CHEMFET structures[34] involving binding at macroscopic surfaces, and more recent work on both molecular and nanowire logic.[37] In all these cases, the simplest theoretical

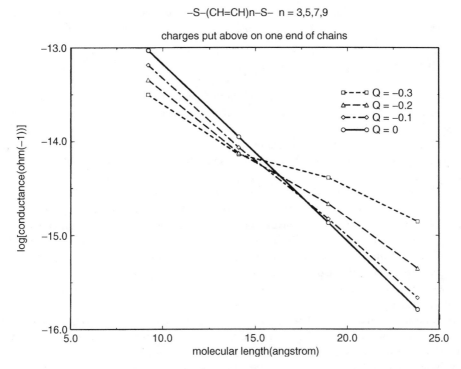

–S–(CH=CH)n–S– n = 3,5,7,9

charges put above on one end of chains

FIGURE 10.6 Results of an early calculation on molecular-type field-effect transistors. A quinone molecule is placed above one end of the conjugated dithiol structure, and the conductance was calculated using a simple Huckel-type representation and the Landauer conductance formula of Section 10.3. The charge Q is simply charge localized on the quinone moiety suspended above one end of the chain. Notice the substantial modification in current due to the external potential arising from charged quinone species. (From Y. Mao, Ph.D. thesis, Northwestern University. 2000.)

analysis is simply based on modification of the wave function in the numerator of Equation (10.10) and the energy in the denominator.

Switching that involves actual geometric change is also of major interest. In particular, work by the UCLA–Hewlett Packard collaboration[93] has been based on substantial geometric change in rotaxane structures. Here the modifications are much larger, because actual chemical entities (in this case cyclic aromatic compounds) have changed position. The switching is relatively slow because it involves major molecular motion, but it also has several advantages. Indeed, the work in California has been very successful in producing real molecular electronic switches and architectures based on this motif. Computationally, what is required to understand the switching is again quite similar to Equation (10.10); but now the states in the numerator and energy denominator will be changed, not because of an external field but because of quite different chemical binding.

In extended molecular wire structures, the coherent formulation will fail, and the analysis in terms of Equation (10.10) is no longer adequate. It is then necessary to employ electron transfer hopping models that describe the motion from one site along the wire to the next.[15] This is almost certainly the appropriate limit to describe experiments on extended conductive polymer molecular structures[94] that can be gated by binding or by electrostatic potentials. In this case, the change in current arises because of change in the local hopping rates. Because these are understood using generalized Marcus/Hush/Jortner theory,[95] the role of the gating will appear in the calculation of the three major factors that enter into that model (reorganization energies, tunneling barriers, and exoergicities). Some initial attempts in this area have been completed.[96]

While there have been few theoretical efforts to describe with quantitative accuracy gating and switching processes, it seems quite clear that the overall formulation of the rate processes for both the coherent and incoherent limits will permit, quite directly, analysis of these phenomena. These should be one of the more exciting and important areas in the molecular electronics of the next 5 years.

10.5 The Onset of Inelasticity

Inelastic transport arises in mesoscopic systems due to a number of processes where the energy of the scattered electron is not preserved. Chief among these are electron–electron and electron–phonon scattering. The former involves many-particle correlation, and the latter is related to the coupling between vibrational and electronic degrees of freedom. They may have important consequences in modifying the transport properties of molecular devices and also in inducing charging, desorption, and chemical reactions.

Inelasticity is also related to the loss of coherence. This is clearly the case for electron–electron interaction, whereas electron–phonon coupling can induce both coherent and incoherent inelastic transport. Here we comment on the description of inelastic processes caused by electron–phonon interaction.

Electron–phonon coupling can be discussed as a pure quantum problem or in the context of semi-classical models. In either description, the most relevant physical parameter in assessing the influence of electron–phonon coupling in the problem of molecular conductance is the Buttiker–Landauer tunneling time. As will be discussed in Section 10.7, this time is longer in the resonant regime and scales with the length of the bridge.

Electron–phonon coupling is responsible for a number of collective excitations that influence the transport properties of many materials.[97] In metals, it accounts for the resistivity behavior and metal–insulator transitions; in semiconductors and other polarizable media, it is related to the formation of polarons; in conducting polymers, it induces the formation of solitons, polarons, and bipolarons. Ness et al.[98] have presented a quantum description of coherent electron–phonon coupling and polaron-like transport in molecular wires. They show that when the electron–lattice interaction is taken into account, transport in the wire is due to polaron-like propagation. The electron transmission can be strongly enhanced in comparison with the case of elastic scattering through the undistorted molecular wire. A static model of this type of coupling had been previously presented in Reference 99.

Burin et al.[100] have developed a semi-classical theory of off-resonance tunneling electrons interacting with a single vibrational mode of the medium. Depending on the difference between the tunneling time and the vibrational period, two different regimes are found: one that corresponds to coherent superexchange and a second involving polaron-like vibronic transport. Both results are qualitatively in agreement with the quantum model despite the fact that the semi-classical treatment is inherently incoherent because the quantum–classical coupling introduces random dephasing of the electronic wave function.

Figure 10.7 shows one of the major results of the semi-classical study. The energy exchange between the tunneling electron and the bridge vibrations is strongest for moderately slow tunneling. In this regime, the electron tunneling process is inelastic, and (in the case shown here) energy is lost to the vibrations of the medium.

Incoherent processes leading to dephasing can be introduced via a density matrix approach to transport. In a series of papers, Nitzan and co-workers[15,101–103] have studied the effect of dephasing on electron transfer and conduction. The main result of these studies is that in addition to the coherent superexchange mechanism, there appears an incoherent path that is closely related to activated hopping.[104–105] This topic is the subject of Section 10.7.

10.6 Molecular Junction Conductance and Nonadiabatic Electron Transfer

Electron transfer is one of the most important areas of chemical kinetics, biochemical function, materials chemistry, and chemical science in general.[106–107] In electron transfer reactions, electronic charge is moved either from one molecular center to another or from one section of a molecule to another section of that

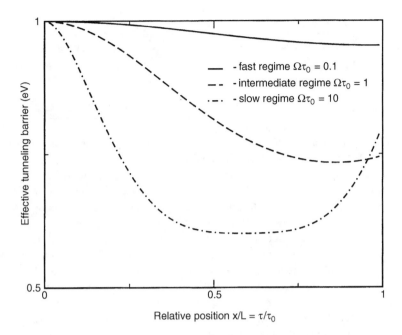

FIGURE 10.7 The effective tunneling barrier for an electron passing through a rectangular barrier in the potential. The modifications in height are due to interactions between the electron and polarizing environment. The tunnel barrier and the applied voltage are each taken to be two volts; the nearly straight line corresponds to fast tunneling, while the two lower curves correspond to slower tunneling in which the electron has more opportunity to polarize the environment. Notice on the far right that the barrier is higher at the end of the reaction than at the beginning; this is because electrons, tunneling through the barrier, have deposited energy in the vibrational medium. (From A.L. Burin, Y.A. Berlin, and M.A. Ratner, *J. Phys. Chem. A*, 105, 2652. 2001. With permission.)

same molecule. The latter processes are called *intramolecular electron transfer* and are advantageous for study because one does not have to worry about the work terms involved in assembling the molecular pair.

Intramolecular electron transfer reactions pervade both natural and synthetic systems. A common structure is the so-called *triad structure*, consisting of a donor, a bridge, and an acceptor. One can then study electron transfer either in the ground state (by first reducing the electron donor, and then watching the kinetics as the electron moves to the acceptor) or photochemically (by photo-exciting the donor to form D*BA, which then decays by intramolecular photo-excited electron transfer to D$^+$BA$^-$, which can then decay by back-transfer to the original reactant, DBA).

In both molecular junctions and electron transfer reactions, the fundamental process is that of an electron moving from one part of space into another. Not surprisingly, then, the two processes can be closely related to one another. In particular, we consider the so-called nonadiabatic electron transfer situation. This is the condition that is obtained when the electronic mixing between donor and acceptor through the bridge is relatively small compared with, for example, typical vibrational frequencies or characteristic temperatures. Under these conditions, perturbation theory shows that the rate for moving from the donor to the acceptor site depends both on the electronic mixing and on the vibronic coupling that permits energy to be exchanged between the electronic and the nuclear motion systems. Under these conditions, the nonadiabatic electron transfer rate, k_{et}, is given by:[106–109]

$$k_{ET} = \frac{2\pi}{\hbar}V_{DA}^2(DWFC) \tag{10.11}$$

This form follows from the golden rule of perturbation theory, expressing the electron transfer (ET) rate in terms of the electronic mixing V_{DA} between the donor and the acceptor sites and a density of states-weighted Franck–Condon factor (DWFC) that describes vibrational overlap between the initial and final

TABLE 10.2

Molecular Junction Transport	Nonadiabatic Electron Transfer
Electron tunneling	Electron tunneling
Electrode continuum	Vibronic continuum
Conductance	Rate constant
Landauer formula	Marcus/Jortner/Hush expression

states, weighted by the number of states. For thermal electron transfer reactions, the Marcus/Hush/Jortner form (Equation 10.11) is broadly applicable and very useful.

Because both nonadiabatic electron transfer and molecular junction tunneling transport from the metal occur by means of electron tunneling, they can be compared in a useful fashion. Table 10.2 shows the similarities and differences. The great similarity is that electron tunneling is the process by which charge is actually transferred. Everything else differs: in particular, the continuum of states (that causes relaxation and chemical kinetics to occur) arises from the multiple electronic states of the electrode in the junction transport problem, but from the vibrations and solvent motions in the nonadiabatic electron transfer situation. The observables also differ: conductance of the junction and rate constants in electron transfer systems. The appropriate analytical expressions are also quite different: the Marcus/Jortner/Hush formula of Equation 10.11 derives from second-order perturbation theory and is valid when the electron mixing is small. The Landauer form of Section 10.3 derives from an identity between scattering and transport and makes no particular assumptions about the smallness of any given term.

Nitzan has developed the similarity substantially farther.[15] Indeed, he has extended it beyond the coherent tunneling regime discussed in Section 10.3 to include the inelasticity discussed in Section 10.5, and even the hopping limits that we will discuss in the next section. Nitzan finds that the electron transfer rate and the transmission can be usefully compared.[15,110] In particular, he finds that:

$$g = \frac{e^2}{\pi\hbar} \frac{k_{ET}}{(DWFC)} \frac{8\hbar}{\pi\Delta_L\Delta_R} \tag{10.12}$$

Here the relationship is between the conductance g and the electron transfer rate constant k_{ET}. The tunneling terms are common to both and therefore disappeared from this expression. What appears instead is the ratio of the bath densities: the spectral density terms $\Delta_L\Delta_R$ describe the mixing of the molecular bridge with the electrodes on the left and right, whereas DWFC describes the mixing of nonadiabatic electron transfer tunneling with the vibrations of the molecular environment.

The assumptions underlying Equation 10.12 are not general, and Nitzan remarks on conditions under which they will and will not hold. Still, it is an extremely useful rule of thumb. Moreover, because semi-classical expressions are available for DWFC, one can choose typical values for the parameters to get a relationship between the conductance and the electron transfer rate. Nitzan derives the form:

$$g(\Omega^{-1}) \sim 10^{-17} k_{ET}(S^{-1}) \tag{10.13}$$

This means that for typical fast nonadiabatic electron transfers of the order of 10^{-12}/s, one would expect a conductance of the order $10^{-5}/\Omega$.

The fundamental observation that electron tunneling is a common process between junction transport and nonadiabatic electron transfer permits not only the semi-quantitative relationships of Equations (10.12) and (10.13) but also comparison of the mechanisms. If the two processes were fundamentally controlled by state densities, dephasing processes, and electron tunneling, then the mechanisms that should inhere in one might also be important in the other. This has led to a very fruitful set of comparisons between the mechanistic behaviors corresponding to electron transfer and molecular junction conductance. Because electron transfer has been studied for so much longer, because the experiments have been so much better developed, and because of the signal importance of electron transfer as a field in chemistry, materials, physics, biology and engineering, it is possible to learn a substantial amount about the mechanisms that one expects in molecular junction conductance from the behaviors involved in electron transfer. This fundamental identification is significant and is the topic of the next section.

10.7 Onset of Incoherence and Hopping Transport

The Landauer approach of Section 10.3 assumes that all scattering is elastic. Section 10.5 dealt with the onset of inelasticity and with simple dynamic polarization of the environment by the tunneling electron. More generally, one expects that sufficiently strong interaction between the tunneling electron and its environment can lead to loss of coherence and to a different mechanism for wire transport.

A significant qualitative insight can be gained by estimating the so-called *contact time*; this is not the inverse of the rate constant for motion but rather an estimate of the actual time for the tunneling process to take place — that is, the time for the tunneling electron to be in contact with the medium with which it will interact. Landauer and Buttiker originally developed this concept[111] and analyzed tunnel processes for heterostructures. In molecular wires, the idea of a tunneling barrier is probably less appropriate than a tight binding picture, effectively because molecules are better described in terms of Huckel-type models than simple barrier structures.

Utilizing the Huckel model for the wire and continuum damping pictures for the electrodes, the general development of Landauer and Buttiker can be modified[112] to give a very simple form for the contact time. This simple form is:

$$\tau_{LB} \approx N\hbar/\Delta E_G \qquad (10.14)$$

Here the Landauer Buttiker time, τ_{LB}, is roughly equal to the uncertainty product times the length of the wire. More precisely, ΔE_G is the gap energy separating the electrostatic potential of the electrode with the appropriate frontier orbital energy on the molecule, and N is the number of subunits within the molecule. A more general form, one that goes smoothly to the classical Landauer Buttiker limit, can be derived by making corrections in the denominator.[112] The important point is that for large gaps and short wires, the Landauer Buttiker time will be substantially shorter than any characteristic vibrational or orientational period of the molecular medium. Under these conditions one expects simple tunneling processes, and the coherent limit of Section 10.3 should be relevant. As the gap becomes smaller and the length longer, the LB time will increase. One then expects that when the LB time becomes comparable to the time scale of some of the motions of the molecular medium in which the electron finds itself, inelastic events should become more important, and the coherent picture needs to be generalized.

The two dominant interaction processes between the molecular bridge and the molecular environment both arise from polarization of the environment by the electronic charge. (A third polarization interaction, image effects, has not been properly dealt with and is mentioned in Section 10.8.) These polarizations include whatever solvent may be present and polarization of chemical bond multipoles by electronic charge. These result in both relaxation processes (energy transfer between the environment and the electronic system) and in dephasing processes (energy fluctuations in the electronic levels). The most straightforward way to treat such structures is through the use of density matrices[113–116] The density matrix describes the evolution of a system that may or may not consist of a pure state (that is, a single wave function). It is ideal for discussion of systems in which environments that are not of primary interest interact strongly with electronic systems that are of primary interest. They are familiar, for example, in the treatment of NMR problems, where the longitudinal and transverse relaxation times (T_1 and T_2) arise because the nuclear spin interacts with its environment.[117] The diagonal elements of the density matrix correspond to populations in electronic levels; and the off-diagonal elements of the density matrix describe so-called coherences, which are effectively bond orders between electronic sites. In the presence of interactions between the system and the environment, it is straightforward to write the time evolution of the density matrix as:

$$\dot{\rho} = \frac{i}{\hbar}[H, \rho] + \dot{\rho})_{\text{diss}} \tag{10.15}$$

Here the first term on the right is the causal one arising from behavior of the system Hamiltonian, while the second term refers to dissipative interactions between the system of interest and its environment. While there are many approaches for analyzing the effects of the dissipative term (and none of them is exact[118]), the simplest and most frequently used approach is that taken by Bloch for magnetic resonance.[117] Here one differentiates the relaxation times for pure dephasing, called T_2^*, and for energy relaxation, called T_1. One then assumes that the populations relax at a rate inversely proportional to T_1, and the coherences dephase at a rate inversely proportion to T_2^* (here δ_{ij} is the Kronecker δ, which vanishes unless i and j are the same, in which case it equals unity):

$$\rho_{ij})_{\text{diss}} = (1 - \delta_{ij})\rho_{ij}/T_2^* + \delta_{ij}\rho_{ii}/T_1 \tag{10.16}$$

In the absence of dissipation terms, the solutions to Equation (10.15) are multiply oscillatory, so equilibrium is never approached. When the dissipative terms are added, the systems indeed approach equilibrium; and a rate is determined both by the frequency values due to the Hamiltonian and by the relaxation processes T_1 and T_2^*.

For an N-level system, the equation system of (10.15) involves N^2 equations and is generally complex to solve. In some limits, however, the solution can be obtained analytically.[15,101] It has been shown that, for a very simple model of bridge-assisted transfer in which only dephasing occurs and in the limit where the dephasing is relatively weak, the overall rate constant for electron transfer reaction between donor and acceptor actually breaks into a sum of two independent contributions:

$$k_{\text{ET}} = k_{\text{coh}} + k_{\text{incoh}} \tag{10.17}$$

Here the coherent term corresponds to electron tunneling through the bridge sites and will generally behave or decay exponentially with distance as expected by the McConnell relationship.[119]

$$k_{\text{coh}} \sim e^{-\beta R} \tag{10.18}$$

Here R is the length between donor and acceptor (or in the molecular junction between the molecule/metal interfaces), and β is a characteristic fall-off parameter. The incoherent term arises from the dephasings and can generally be written as:[15,103]

$$k_{\text{incoh}} \sim \frac{1}{(b + cR + dR^2)} F(T) \tag{10.19}$$

The constants b, c, and d depend on the nature of the experiment and the relative magnitudes of the Hamiltonian parameters. Very commonly, the first term of the denominator will dominate, so that there is effectively no decrease of the incoherent transfer rate with length. The temperature dependence is determined by the function F(T), which in the simple activated case becomes

$$F(T) = \exp(-\Delta G^{\ddagger}/k_B T) \tag{10.20}$$

Equation (10.17) therefore predicts that one should see two different rates, one arising from coherent tunneling and the other from incoherent dephased motion, that can become thermally activated hopping when equilibrium is obtained (that is, when the longitudinal relaxation time T_1 is large).

The expectation of Equations (10.17) to (10.20) is shown in Figure 10.8, for the case where the first term of the denominator of Equation (10.19) indeed dominates. Based on the arguments of Equation (10.14), we would expect the relative magnitude of the incoherent and coherent term to depend on the parameters: for long wires, small gaps, and high temperatures, the incoherent term should be favored.

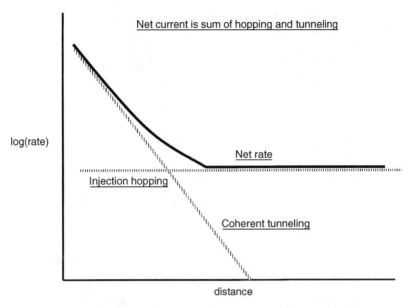

FIGURE 10.8   Schematic of the rate behavior through a molecular wire bridge in the presence of interactions between the system and its environment. The two limiting cases of injection hopping (fully incoherent) and tunneling (fully coherent) are indicated by the dashed lines. The net rate is the sum of those two contributors.

At very low temperatures and with large gaps, one would expect coherent tunneling as is generally seen for donor/bridge/acceptor electron transfer for short bridges.

Because the onset of incoherence really corresponds to vibronic coupling, it is possible to successfully derive more detailed predictions, utilizing an actual coupling model of the spin/boson type to describe the interaction between the tunneling electron and its environment.[104,105] This sort of analysis has been reported by Segal and Nitzan,[103] who also make important remarks on the relationships between subsystems and on the nature of the energy flow that occurs when transport undergoes the mechanism modification from tunneling to hopping.

In polymeric structures, the hopping mechanism has long been known to dominate. *Solitons, polarons,* and *bipolarons* are terms describing particular polarization characteristics, in which charge polarizes the vibrational environment, causing trapping and localization.[94,120] Charge motion then corresponds to the motion of these quasi particles that consist of the electronic carrier plus its accompanying polarization cloud. For molecular transport junctions, this same sort of behavior should be seen for long distances, and the situation has been treated in a simple model. An important set of observations by Reed and his collaborators[32] in extended molecular wires in nanopore junctions indicates that, indeed, one can observe both hopping and tunneling processes, depending on the nature of the energy barriers and the applied potential (Figure 10.9).

The transitions between coherent and incoherent motion have been observed extensively in donor/bridge/acceptor systems, although much less work has to date been reported in tunnel junctions. One particularly marked case is DNA, where Figures 10.10 and 10.11 show clear limiting cases for the dependence of the intramolecular electron transfer rates on distance. Figure 10.10[121] shows photoexcited electron transfer in DNA hairpin structures, where both the charge separation and charge recombination rates are indeed exponential, with characteristic values for the β of Equation (10.18). Figure 10.11[122] shows a comparison with a model very much like that of Equations (10.15) to (10.17) with results for differential cleavage reactions reported by Giese, Michel–Beyerle, and collaborators.[123] The tight binding model parameters corresponding to the energy gap and the local (Huckel) electron tunneling amplitude have been calculated by the Munich and Tel Aviv groups.[124] The fit is quite good, and the two different regimes (tunneling at short distances and hopping at long distances) appear quite straightforward.

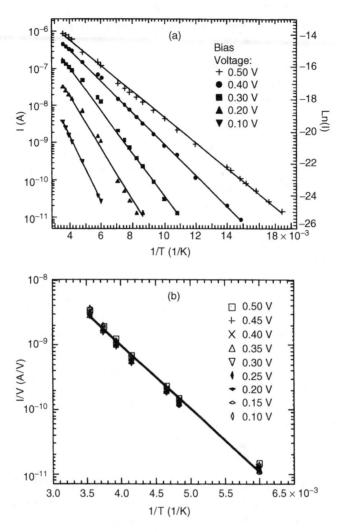

FIGURE 10.9 Temperature dependence of the current at forward (a) and backward (b) biases for a nanopore junction containing the p-biphenylthiol molecule sandwiched between titanium and gold electrodes, with the sulfur next to the gold. In (a), the forward bias is fitted to the thermionic emission through an image potential, while in (b), a simple hopping model is used. (From M.A. Reed et al., *Ann. N.Y. Acad. Sci.*, 852, 133, 1998. With permission.)

10.8 Advanced Theoretical Challenges

While there has been extensive and (in some cases) excellent theory devoted to transport in molecular wire junctions, if molecules simply act as interconnects, their role in molecular electronics is finite. There are far more sophisticated and challenging possibilities that have not yet been extensively addressed in the theoretical community. These processes go beyond conductance in simple wire circuits and still make up major challenges for the community. Some of these issues include:

10.8.1 Electrostatic Potentials and Image Charges

When a molecule is placed between electrodes and voltage is applied, there is clearly a changing chemical potential between the two ends of the molecular junction. A very important issue then becomes exactly how this charge is distributed across the molecular interface. The three simplest approximations are to assume that all the charge drops at the interfaces,[125–126] that the charge drops smoothly across the molecular

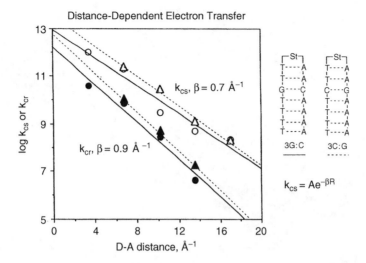

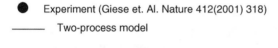

FIGURE 10.10 Rate behavior as a function of distance for hairpin DNA compounds. Initially, the bridging stilbene (noted St) was photoexcited; the results are given for the charge separation (CS) and charge recombination (CR). Both are fully coherent and display an exponential decay with length. (From F.D. Lewis et al., *Science* 227, 673, 1997. With permission.)

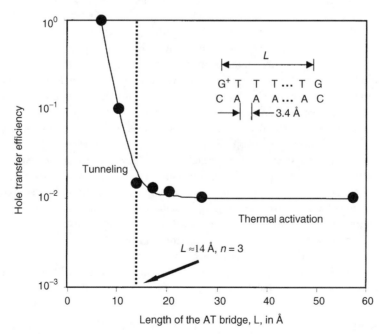

FIGURE 10.11 Comparison of calculated hole transfer efficiency (proportional to rate constant) for extended DNA structures. The tunneling regime at short distances and the thermal hopping at long distances are indicated. The solid dots refer to the experimental work, and the line through them corresponds to theoretical treatment utilizing the two processes of Figure 10.10. All parameters were evaluated theoretically. (Experiment from B. Giese et al., *Nature*, 412, 318, 2001; modeling from Y.A. Berlin et al, *Chem. Phys.*, 275, 61, 2002. With permission.)

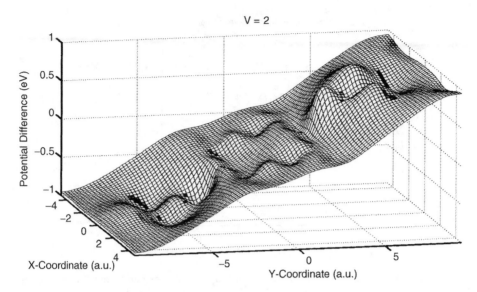

FIGURE 10.12 Calculated electrostatic potential, from a full self-consistent density functional theory calculation on the benzene dithiol bridge between gold electrodes at a voltage of two volts. Notice the substantial flatness near the interface, and the highly structured potential, arising from a combination of electrostatic forces and nuclear trapping. (From A. Xue, M.A. Ratner, and S. Datta, to be published. With permission.)

junction in a linear fashion,[127] and that the electrostatic potential changes across the molecular junction in a fashion determined self-consistently with the total charge density.[28,86,88,128–130] The third of these seems the most elegant and has been used in self-consistent density functional theory calculations, with the usual Green's function approximation used for the electrodes. It is then generally found (Figure 10.12 shows one example) that the electrostatic potential is complex, a substantial drop at the interfaces with substantial structure within the molecule itself. The suggestion that most of the voltage drops at the interfaces arises from the relatively poor contact between molecule and interface, compared with delocalization along well-conjugated molecules. It was used in some of the early investigations[125] in the field and probably represents a reasonable approximation in many situations, especially STM measurement. Assuming that the voltage drops linearly across the molecular junction, which is a solution to the Poisson equation in the absence of all charge, is the simplest way to treat voltage and has been used in the analysis of localization phenomena in charged wires.[127] This seems too bold a simplification for any quantitative work.

Image charges occur, in classical electrodynamics, because of the polarization of the charge within the metal by external charges. Most calculations on molecular wire junctions either ignore the image charge altogether or hope that the image charge is taken into account either by the metallic particles included in the extended molecule treatment or by the self-consistent solution to the density functional. It is not at all clear that this is correct — image charge stabilization is a substantial contribution to the energy, and in areas such as molecular optoelectronics, the image effects can lead to substantial barriers at the interface,[131] which either have to be tunneled through or hopped over. One expects similar behavior in molecular wire junctions,[132] and no explicit treatment of image charge has, to our knowledge, yet been given.

10.8.2 Geometry

Two major geometric questions are involved in trying to understand what happens in molecular wire junctions. For the zero-field conductance, one normally assumes that the molecule in the junction is of the same geometry as the free molecule. This is generally unjustified and is assumed for simplicity. Even if this molecular geometry is unchanged, there is the issue of binding to the electrode interfaces. In thiol/gold (the most investigated situation), it has been normally assumed that the sulfur is either within a three-fold site on the surface or directly over a metal atom. These give very different calculated conductances.[27]

Recent experimental data by Weber and colleagues[26] suggest that, even in the gold/thiol interaction, the geometries are not necessarily regular and may evolve dynamically in time. *Ab-initio* calculations[133] also suggest that the geometry of the gold/thiol interface is highly fluid and not at all regular and that the potential energy surface for this motion is fairly flat. Therefore, the assumptions on the binding that have been used to compute self-energies are not necessarily highly accurate.

Once the current is applied, the situation can be even worse. It is known that with enough current, bonds can break. Molecules in very high electromagnetic fields such as those that occur at the molecular junction interface (exceeding 10^6 volts/centimeter) may well have geometries that are substantially different from those in the unperturbed molecule. Simple calculations of Stark changes in the geometry suggest that this is definitely true for molecules with large dipole moments. Therefore, it is really necessary to understand the geometry of the molecule within the junction before junction conductance can be appropriately calculated.

10.8.3   Reactions

The behavior of molecules with large amounts of current passing through may well be a new kind of spectroscopy, and even a new kind of catalytic chemistry. It is known that reactions very often occur when currents are passing through the molecules: the entire idea of positioning using scanning tunneling microscopes is predicated on the fact that one can break the surface bonding, move the molecule or atom to the tip, and then replace it.[134] This obviously means that either the electrostatic field or the current is causing bond breaking. Such bond-breaking kinetics has not really been discussed in connection with molecular junctions but might be one of the fascinating applications of them.

In nanotubes, Avouris and his colleagues[135] have developed a general cleaning technique in which currents are passed through the nanotubes and particular adlayer structures are excised by bond breaking. Similarly, Wolkow and collaborators[136] have demonstrated that under substantial current conditions, alkane species remain conductive, while unsaturated species can be thermally destroyed. Some excellent theoretical modeling has been done on these problems by Seideman.[137]

Actual chemical reactions in scanning probe environments have been reported by Ho[138] (Figure 10.13) and by other groups. The bond cleavage here is due to the current passing through the molecule, and

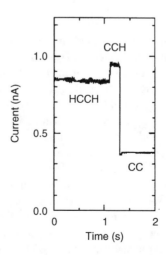

FIGURE 10.13 The current as a function of time for a molecular junction consisting of a single acetylene molecule. After roughly one second, enough energy has been deposited in the molecule to break the bond, giving the CCH radical. After roughly half a second more, the second hydrogen is removed and the bridge consists only of two carbons. (From L.J. Lauhon et al, *Phys. Rev. Lett.*, 84, 1527, 2000. With permission.)

development of appropriate models is still in its infancy. Similarly, Sagiv and Maoz[139] have shown that in the presence of a conductive AFM current, alkane adlayers can be oxidized by oxygen and water to form carboxylic acid structures. This sort of electrochemistry in a tip-field has not been effectively modeled but might represent both a very important application of molecular junction behavior and a challenging theoretical problem.

10.8.4   Chirality Effects

Single electron currents might be very weakly affected by local chirality because of broken symmetry structures. One expects such relativistic interactions to be weak, but Naaman and collaborators[140] have shown substantial effects of chiral adlayers on the transport of polarized spins. This problem is, once again, experimentally challenging and theoretically largely unaddressed.

Closely related to chirality is the issue of magnetic effects on transport. More sophisticated theory[141] and experiments are needed to understand the possible effects of magnetic interactions on currents through molecular junctions.

10.8.5   Photoassisted Transitions

In semiconductor electrode structures, photoassisted chemistry is very important — for example, a so-called Graetzel cell,[142] in which photoexcited dye molecules transfer electrons into the conduction band of nanoscale titania, is of major interest both intrinsically and for applications. The more general problem of photoassisted behavior in molecular junctions is beginning to be addressed[143] theoretically using several techniques; but once again, without comparison with direct experiments, these are incomplete.

It is clear that despite the high interest and beautiful work in the general area of molecular electronics, some of the major theoretical problems have not yet been properly approached.

10.8.6   Dynamical Analysis

Like spectroscopy,[109] current flow can be evaluated dynamically. This has several advantages, including (in particular) increasing understanding of how currents actually flow in molecule-conductor junctions. The first such calculations[144] are starting to appear.

10.9 Remarks

Transport in molecular wire junctions and the fundamental scientific and technological issues that it raises have produced extensive scientific excitement and some very impressive work. In the area of carbon nanotubes, a great deal is now understood: effects of interfacial mixing at the electrode can dominate transport, the behavior can be ballistic in well-formed single-walled structures, both semiconducting and metallic tubes are known and understood, and issues such as defects, adsorbates, branches, and crossings can now be discussed and calculated. Therefore, in the nanotube area, computational capabilities have meshed with excellent experimental work to produce a situation in which we understand quite well what the expected accuracies and errors of any given computational approach might be.

The field of transport through more traditional molecular junctions is much more problematic. In discussing the comparison of theory with experiment for simple break junction measurements on straightforward organic molecules, we saw that disagreements in the conductance by factors of 100 or more between theory and experiment are common and that neither the theoretical approaches nor the experimental ones are necessarily converged. Even the simple suggestion that under favorable conditions, transport through organic molecules can show quantized conductance has never been reported experimentally until 2001,[145] although theory straightforwardly predicts such behaviors in the absence of strong vibronic coupling when contact resistances are sufficiently small.[12,46,146,147]

More sophisticated problems, such as those discussed in the last section, have been approached using model calculations. Significant insight into mechanistic behaviors has been gleaned, but these calculations have not been normed — there is really no independent verification of their validity because the appropriate comparison of quantitative calculations with quantitative measurements has not yet been made.

One major challenge for the community over the next few years, then, will be the development of well-defined model systems in which such comparisons can be made. The notion of defining a model science[148] — by comparison of a well-defined set of accurate experimental data with a series of calculations involving different approximations to the wave function, the mixing, the geometrics, the level of theory, and the method used — is necessary if quantitative estimates and predictions of transport behavior are to be obtained. A number of experimental groups[26,32,42,43] are now pursuing this issue, using reliable geometries and methodologies that can, perhaps, yield accurate and reliable data for a series of molecular wire structures.

In addressing the challenge of understanding molecular wire junctions, then, the community has made great progress. The next step will entail systematic comparison with good data, refinement of methodologies, and extension of the simple schemes described in this chapter toward the understanding of more complex, and more fascinating, molecular junction behavior.

Note Added in Proof

In the months since completion of this overview, several important advances in molecular conductance junctions have been reported. Probably the most significant has been the development,[A1,A2] by groups at Harvard and Cornell, of methods to measure transport in individual molecules. The technology is based on use of a break junction that is prepared by rapid heating of a gold thin wire. The break occurs with characteristic gap sizes of roughly 10 nM; subsequently, individual molecules can be positioned at the break, and their conductance properties measured. These measurements can be controlled by a third, gate electrode that is positioned on a layer below the gold wire.

These measurements allow solving the band lineup problem, because the gate electrode can be used to bring the molecular levels into resonance with the Fermi levels of the source and drain. Quantized conductance has indeed been observed, as discussed above in the theory section. Additionally, Kondo transport (conductance maxima at zero applied source-drain voltage due to pinning of the unpaired occupancy orbitals at the Fermi level) and spin effects on the molecule have been observed.

Work by Seideman, Guo and collaborators[A3] has employed perturbation methods to calculate both the elastic and inelastic responses in junctions containing buckyball and small-molecule junctions. This is significant for the development of approaches to the issues concerning inelastic scattering and chemical reactivity referred to in Section 10.8.

Acknowledgments

We are grateful to the Northwestern Molecular Electronics group, especially Abe Nitzan, Alex Burin, Alessandro Troisi, and Alex Xue for useful discussions. This work was sponsored by the Chemistry Divisions of the NSF and ONR, by the DARPA Moletronics program, and by DoD/MURI collaboration.

References

1. M.C. Petty, M.R. Bryce, and D. Bloor, *Introduction to Molecular Electronics* (Oxford University Press, New York, 1995).
2. C.A. Mirkin and M.A. Ratner, *Ann. Revs. Phys. Chem.* 43, 719 (1992).
3. A. Aviram (Ed.), *Molecular Electronics — Science and Technology* (AIP, New York, 1992).
4. A. Aviram (Ed.), *Molecular Electronics* (Engineering Foundation, New York, 1989).
5. A. Aviram and M.A. Ratner (Eds.), Molecular electronics: science and technology, *Ann. N.Y. Acad. Sci.* 852, (1998).
6. A. Aviram, M.A. Ratner, and V. Mujica (Eds.), Molecular electronics, *Ann. N.Y. Acad. Sci.* 960 (2002).

7. J. Jortner and M.A. Ratner (Eds.), *Molecular Electronics* (Blackwell, Oxford, 1997).
8. S. Datta, *Superlattice Microst.* 28, 253 (2000).
9. M.D. Ward, *J. Chem. Educ.* 78, 1021 (2001).
10. C. Joachim, J.K. Gimzewski, and A. Aviram, *Nature*, 408, 541 (2000).
11. M.A. Ratner and M.A. Reed, *Encyclopedia of Science and Technology*, Third ed. 10, 123 (2002).
12. A. Troisi and M.A. Ratner, to be published.
13. V. Mujica, *Rev. Mex. Phys.* 47, 59 Suppl. (2001).
14. F. Zahid, M. Paulsson, and S. Datta, H. Morkoc (Ed.), in *Advanced Semiconductors and Organic Nano-Techniques*, (Academic Press, New York, in press).
15. A. Nitzan, *Ann. Rev. Phys. Chem.* 52, 681 (2001).
16. *Chemical Physics*, Special Issue, to be published.
17. J.C. Ellenbogen and J.C. Love, *Proc. IEEE* 88, 386 (2000).
18. B. Mann and H. Kuhn, *J. Appl. Phys.* 42, 4398 (1971).
19. Y.N. Xia, J.A. Rogers, K.E. Paul, and G.M. Whitesides, *Chem. Rev.* 99, 1823 (1999).
20. H.-J. Güntherodt and R. Wiesendanger (Eds.), *Scanning Tunneling Microscopy I*, (Springer-Verlag, Berlin, 1992).
21. C. Dekker, *Phys. Today* 52, 22 (1999).
22. M.S. Dresselhaus, G. Dresselhaus, and P. Avouris (Eds.), *Carbon Nanotubes: Synthesis, Structure, Properties, and Applications* (Springer, Berlin, 2001).
23. T.W. Odom, J.H. Hafner, and C.M. Lieber, *Top. Appl. Phys.* 80, 173 (2001).
24. R. Martel, V. Derycke, C. Lavoie, J. Appenzeller, K.K. Chan, J. Tersoff, and P. Avouris, *Phys. Rev. Lett.* 87, 256805 (2001).
25. M.A. Reed, C. Zhou, C.J. Muller, T.P. Burgin, and J.M. Tour, *Science* 278, 252 (1997).
26. J. Reichert, R. Ochs, D. Beckmann, H.B. Weber, M. Mayor, and H. von Lohneysem, *Phys. Rev. Lett.* 88(17): 176804, April 29 (2002).
27. M. Di Ventra, S.T. Pantelides, and N.D. Lang, *Phys. Rev. Lett.* 84, 979 (2000).
28. C.K. Wang, Y. Fu, and Y. Luo, *Phys. Chem. Chem. Phys.* 3, 5017 (2001).
29. Y. Xue and M.A. Ratner, *J. Chem. Phys.*, submitted.
30. A. Aviram and M.A. Ratner, *Chem. Phys. Lett.* 29, 277 (1974).
31. R.M. Metzger et al., *J. Phys. Chem.* B105, 7280 (2001).
32. M.A. Reed, C. Zhou, M.R. Deshpande, and C.J. Muller, in A. Aviram and M.A. Ratner (Eds.), *Molecular Electronics: Science and Technology, Ann. N.Y. Acad. Sci.* 852 (1998).
33. V. Mujica, A. Nitzan, and M.A. Ratner. *Chem. Phys.* in press.
34. D.G. Wu, D. Cahen, P. Graf, R. Naaman, A.Nitzan, and D. Shvarts, *Chem. Eur. J.* 7, 1743 (2001).
35. A. Bachtold, P. Hadley, T. Nakanishi, and C. Dekker, *Science* 294, 1317 (2001); M. Ouyang, J.L. Huang, and C.M. Lieber, *Phys. Rev. Lett.* 88, 066804 (2002).
36. J.H. Schon and Z. Bao, *Appl. Phys. Lett.* 80, 332 (2002).
37. G.Y. Tseng and J.C. Ellenbogen, *Science* 294, 1293 (2001).
38. M.A. Ratner, *Mater. Today* 5, 20 (2002).
39. U. Mitschke and P. Bauerle, *J. Mater. Chem.* 10, 1471 (2000).
40. D.B. Mitzi, K. Chondroudis, and C.R. Kagan, *IBM J. Res. Dev.* 45, 29 (2001).
41. M. Dorogi, J. Gomez, R. Osifchin, and R.P.A. Andres, *Phys. Rev. B* 53, 9071 (1995).
42. X.D. Cui, A. Primak, X. Zarate, J. Tomfohr, O.F. Sankey, A.L. Moore, T.A. Moore, D. Gust, G. Harris, and S.M. Lindsay, *Science* 294, 571 (2001).
43. R. Shashidhar et. al., submitted.
44. J.G. Lee, J. Ahner, and J.T. Yates, Jr., *J. Chem. Phys.* 114, 1414 (2001).
45. V.Mujica, M. Kemp, and M.A. Ratner, *J. Chem. Phys.* 101, 6849 (1994).
46. V.Mujica, M. Kemp, and M.A. Ratner, *J. Chem. Phys.* 101, 6856 (1994).
47. M. Sumetskii, *Phys. Rev. B* 48, 4586 (1993).
48. M. Sumetskii, *J. Phys.: Condens. Matter* 3, 2651 (1991).
49. C. Joachim and J.F. Vinuesa, *Europhys. Lett.* 33, 635 (1996).

50. M.P. Samanta, W. Tian, S. Datta, J.I. Henderson, and C.P. Kubiak, *Phys. Rev. B* 53, R7626 (1996).

51. S. Datta, W. Tian, S.Hong, R. Reifenberger, J.J. Henderson, and C.P. Kubiak, *Phys. Rev. Lett.* 79, 2350 (1997).

52. C. Joachim, J.K. Gimzewski, R.R. Schlittler, and C. Chavy, *Phys. Rev. Lett.* 74, 2102 (1995).

53. N.G. van Kampen, *Stochastic Processes in Physics and Chemistry* (Elsevier Science, New York, 1983).

54. Y. Imry, *Introduction to Mesoscopic Physics* (Oxford University Press, New York 1997).

55. V. Mujica, A. Nitzan, Y. Mao, W. Davis, M. Kemp, A. Roitberg, and M.A. Ratner, *Adv. Chem. Phys. Series*, 107, 403 (1999).

56. E.G. Emberly and G. Kirczenow, *Phys. Rev. B* 58, 10911 (1998).

57. C. Kergueris, J.P. Bourgoin, S. Palacin, D. Esteve, C. Urbina, M. Magoga, and C. Joachim, *Phys. Rev. B* 59, 12505 (1999).

58. R. Landauer, *Philos. Mag.* 218, 863 (1970).

59. R. Landauer, *IBM J. Res. Dev.* 32, 306 (1988).

60. R. Landauer, *Physica Scripta* T42, 110 (1992).

61. Y. Imry and R. Landauer, *Rev. Mod. Phys.* 71, S306 (1999).

62. Y.V. Sharvin, *JETP* 48, 984 (1965).

63. J.L. D'Amato and H. Pastawski, *Phys. Rev. B* 41, 7411 (1990).

64. H. Pastawski, *Phys. Rev. B* 44, 6329 (1991).

65. H. Pastawski, *Phys. Rev. B* 46, 4053 (1992).

66. Y. Meir and N.S. Wingreen, *Phys. Rev. Lett.* 68, 2512 (1992).

67. N.S. Wingreen, A.P. Jauho, Y. Meir, *Phys. Rev. B* 48, 8487 (1993).

68. S. Datta, *Electronic Transport in Mesoscopic Systems* (Cambridge University Press, New York, 1995).

69. S.N.Yaliraki, A.E. Roitberg, C. González, V. Mujica, and M.A. Ratner, *J. Chem. Phys.* 111, 6997 (1999).

70. V. Mujica, A. Roitberg, and M.A. Ratner, *J. Chem. Phys.* 112, 6834 (2000).

71. L.V. Keldysh, *Soviet Phys. JETP* 20, 1018 (1965).

72. G.D. Mahan, *Many-Particle Physics*, (Plenum Press, New York 1990).

73. L.E. Hall, J.R. Reimers, N.S. Hush, and K. Silverbrook, *J. Chem. Phys.* 112, 1510 (2000).

74. W. Tian, S. Datta, S. Hong, R. Reifenberger, J.I. Henderson, and C.P. Kubiak, *J. Chem. Phys.* 109, 2874 (1998).

75. W. Häusler, B. Kramer, and J. Masek, *Z. Phys. B* 85, 435 (1991).

76. N.D. Lang, *Phys. Rev. B* 55, 4113 (1997).

77. A. Nakamura, M. Brandbyge, L.B. Hansen, and K.W. Jacobsen, *Phys. Rev. Lett.* 82, 1538 (1999).

78. H. Nakatsuji and K. Yasuda, *Phys. Rev. Lett.* 76, 1039 (1996).

79. P.A. Serena and N. García (Eds.), *Nanowires*, (Kluwer Academic Publishers, Dordrecht, 1997).

80. G. Treboux, P. Lapstun, and K. Silverbrook, *J. Phys. Chem. B* 102, 8978 (1998).

81. G. Treboux, P. Lapstun, Z. Wu, and K. Silverbrook, *Chem. Phys. Lett.* 301, 493 (1999).

82. M. Dorogi, J. Gomez, R.P. Andres, and R. Reifenberger, *Phys. Rev. B* 52, 9071 (1995).

83. P. Andres, S. Datta, D.B. Janes, C.P. Kubiak, and R. Reifenberger, in H.S. Nalwa (Ed.), *The Handbook of Nanostructured Materials and Technology* (Academic Press, New York, 1998).

84. C. Joachim and S. Roth (Eds.), *Atomic and Molecular Wires* (Kluwer Academic, Dordrecht, 1997).

85. M. Magoga and C. Joachim, *Phys. Rev. B* 56, 4722 (1997).

86. Y. Xue and M.A. Ratner, *Chem. Phys.* in press.

87. J.H. Schon, H. Meng, and Z. Bao, *Nature* 413, 713 (2001).

88. P.S. Damle, A.W. Ghosh, and S. Datta, *Phys. Rev. B* 64, 1403 (2001).

89. A.P. Alivisatos et al,. *Adv. Mat.* 10, 1297 (1998).

90. C.P. Kubiak et al., *J. Phys. Chem. B.* in press.

91. V. Mujica, S. Datta, A. Nitzan, C.P. Kubiak, and M.A. Ratner, *J. Phys. Chem.*, submitted.

92. Y. Mao, Ph.D. Thesis, Northwestern University (2000).

93. C.P. Collier, J.O. Jeppesen, Y. Luo et al., *J. Am. Chem. Soc.* 123, 12632 (2001); C.P. Collier, E.W. Wong et al., *Science* 285, 391 (1999).

94. H.S. Nalwa (Ed.), *Handbook of Organic Conductive Molecules and Polymers* (Wiley, New York, 1997).
95. M. Bixon, and J. Jortner, *Adv. Chem. Phys.* 106, 35 (1999).
96. G. Hutchinson, Y. Berlin, M.A. Ratner, and J. Michl, *J. Phys. Chem.*, submitted.
97. O. Madelung, *Introduction to Solid-State Theory* (Springer, Berlin, 1978).
98. H. Ness, S.A. Shevlin, and A.J. Fisher, *Phys. Rev. B* 63, 125422 (2001).
99. M. Olson, Y. Mao, T. Windus, et al., *J. Phys. Chem. B* 102, 941 (1998).
100. A.L. Burin, Y.A. Berlin, and M.A. Ratner, *J. Phys. Chem. A* 105, 2652 (2001).
101. D. Segal, A. Nitzan, W.B. Davis, M.R. Wasielewski, and M.A. Ratner, *J. Phys. Chem. B* 104, 3817 (2000).
102. D. Segal, A. Nitzan, M. Ratner, and W.B. Davis, *J. Phys. Chem. B* 104, 2709 (2000).
103. D. Segal, A. Nitzan, W.B. Davis et al., *J. Phys. Chem. B* 104, 3817 (2000).
104. S.S. Skourtis and S. Mukamel, *Chem. Phys.* 197, 367 (1995).
105. A.K. Felts et al., *J. Phys. Chem.* 99, 2929 (1995).
106. M. Bixon, and J. Jortner (Eds.), Special number on electron transfer, *Adv. Chem. Phys.* 106 (1999).
107. P.F. Barbara, T.J. Meyer, and M.A. Ratner, *J. Phys. Chem.* 100, 13148 (1996).
108. J. Jortner, *J. Chem. Phys.* 64, 4860 (1976).
109. G.C. Schatz and M. A Ratner, *Quantum Mechanics in Chemistry* (Dover, New York, 2002), Chapter 10.
110. A. Nitzan, *J. Phys. Chem. A* 105, 2677 (2001).
111. M. Buttiker, and R. Landauer, *Phys. Scripta* 32, 429 (1985).
112. A. Nitzan, J. Jortner, J. Wilkie, A.L. Burin, and M.A. Ratner, *J. Phys. Chem. B* 104, 3817 (2000).
113. K. Blum, *Density Matrix Theory and Applications*, (Plenum Press, New York, 1981).
114. G.C. Schatz and M.A Ratner, *Quantum Mechanics in Chemistry* (Dover, New York, 2002), Chapter 11.
115. S. Mukamel, *Principles of Nonlinear Optical Spectroscopy* (Oxford University Press, New York, 1995).
116. M.D. Fayer, *Elements of Quantum Mechanics* (Oxford University Press, New York, 2001).
117. C.P. Slichter, *Principles of Magnetic Resonance* (Springer-Verlag, New York, 1990).
118. D. Kohen, C.C. Marston, and D.J. Tannor, *J. Chem. Phys.* 107, 5236 (1997).
119. H.M. McConnell, *J. Chem. Phys.* 35, 508 (1961).
120. A.J. Heeger, *Rev. Mod. Phys.* 73, 681 (2001).
121. F.D. Lewis et al., *Science* 227, 673 (1997).
122. Y.A. Berlin, A.L. Burin, and M.A. Ratner, *Chem. Phys.* 275, 61 (2002).
123. B. Giese, *Accounts Chem. Res.* 33, 631 (2000); F.D. Lewis, *Accounts Chem. Res.* 34, 159 (2001); G.B. Schuster, *Accounts Chem. Res.* 33, 253 (2000).
124. A.A. Voityuk et al., *J. Phys. Chem. B* 104, 9740 (2000).
125. S. Datta, *Superlattice Microst.* 28, 253 (2000).
126. R.A. Marcus, *J. Chem. Soc. Faraday Trans.* 92, 3905 (1996).
127. V. Mujica, M. Kemp, A. Roitberg, and M.A. Ratner, *J. Chem. Phys.* 104, 7296 (1996).
128. C. Gonzalez, V. Mujica, and M.A. Ratner, *Ann. N.Y. Acad. Sci.* 960, 163–176 (2002).
129. S.T. Pantelides, M. Di Ventra, and N.D. Lang, *Physica B* 296, 72 (2001).
130. T.W. Kelley and C.D. Frisbie, *J. Vac. Sci. Technol. B* 18, 632 (2000).
131. G. Ingold et al., *J. Chem. Phys.*, submitted.
132. Y.Q. Xue, S. Datta, and M.A. Ratner, *J. Chem. Phys.* 115, 4292 (2001).
133. H. Basch and M.A. Ratner, *J. Chem. Phys.*, submitted.
134. M.F. Crommie et al., *Physica D* 83, 98 (1995).
135. P.C. Collins, M.S. Arnold, and P. Avouris, *Science* 292, 706 (2001).
136. R.A. Wolkow, *Annu. Rev. Phys. Chem.* 50, 413 (1999).
137. S. Alavi and T. Seideman, *J. Chem. Phys.* 115, 1882 (2001).
138. L.J. Lauhon and W. Ho, *J. Phys. Chem. B* 105, 3987 (2001).
139. R. Maoz and J. Sagiv, *Langmuir* 3, 1034 (1987).
140. R. Naaman, private communication.

141. E.G. Petrov, I.S. Tolokh, and V. May, *J. Chem. Phys.* 108, 4386 (1998).

142. M. Gratzel, *Nature* 414, 338 (2001).

143. J.T. York, R.D. Coalson, and Y. Dahnovsky, *Phys. Rev. B* 65(23): 235321, June 15, 2002; A. Tikhonov, R.D. Coalson, and Y. Dahnovsky, *J. Chem. Phys.* 117(2): 567–580, July 8 (2002).

144. R. Baer and R. Gould, *J. Chem. Phys.* 114, 3385 (2001).

145. J.H. Schon and Z. Bao, *Appl. Phys. Lett.* 80, 847 (2002).

146. Y.V. Sharvin, *Zh. Eksp. Teor. Fiz.* 48, 984 (1965).

147. L.E. Hall et al., *J. Chem. Phys.* 112, 1510 (2000).

148. J.A. Pople, *Rev. Mod. Phys.* 71, 1267 (1999).

A1. W.J. Liang, M.P. Shores, M. Bockrath, J.R. Long, and H. Park, *Nature* 417(6890): 725–729, June 13 (2002).

A2. J. Park, A.N. Pasupathy, J.L. Goldsmith, C. Chang, Y. Yaish, J.R. Petta, M. Rinkoski, J.P. Sethna, H.D. Abruna, P.L. McEuen, and D.C. Ralph, *Nature* 417(6890): 722–725, June 13 (2002).

A3. S. Alavi, R. Rousseau, G.P. Lopinski, R.A. Wolkow, and T. Seideman, *Faraday Discussions* 117: 213–229 (2000); S. Alavi, B. Larade, J. Taylor, H. Guo, and T. Seideman, *Chem. Phys.* in press.

11

Modeling Electronics at the Nanoscale

Narayan R. Aluru
University of Illinois

Jean-Pierre Leburton
University of Illinois

William McMahon,
University of Illinois

Umberto Ravaioli
University of Illinois

Slava V. Rotkin
University of Illinois

Martin Staedele
Infineon Technologies

Trudy van der Straaten
University of Illinois

Blair R. Tuttle
Penn State University

Karl Hess
University of Illinois

CONTENTS

11.1 Introduction

Nanostructure research is defined by a scale — the nanometer length scale. Simulation of nanostructures, however, must be multiscale in its very nature. It is not the nanostructures themselves that open the horizon to new opportunities and applications in all walks of life; it is the integration of nanostructures into large systems that offers the possibility to perform complex electrical, mechanical, optical, and chemical tasks.

Conventional electronics approaches to nanometer dimensions and simulation techniques must increasingly use atomistic methods to compute, for example, tunneling and size quantization effects as well as the features of the electronic structure of the solids that define the nanometer-sized device. The atomistic properties need then to be linked to macroscopic electromagnetic fields and to the equations of Maxwell and, ultimately, to systems performance and reliability. The transition from the quantum and atomistic scale to the classical macroscopic scale is of great importance for the accuracy of the simulation. It can be described by the Landauer–Buettiker formalism, by Bardeen's transfer Hamiltonian method, or by more demanding methods such as the Schroedinger Equation Monte Carlo approach.[1] To encompass all of these scales and transitions, a hierarchy of methods (sets of equations) that supply each other with parameters is needed even for conventional silicon technology. Similar hierarchical

0-8493-1200-0/03/$0.00+$1.50

approaches will be needed for future devices and their integration in electronics as well as electrome-chanics. One can already anticipate the demand for simulation methods that merge electronics, mechanics, and optics as well as the highly developed methods of chemistry.

A theoretical tool of ever-increasing use and usefulness is density functional theory (DFT). DFT describes, for example, the electrical and optical properties of a quantum dot (the prototype for future electron devices) and is also widely used in chemistry. The simulation methods become altogether more fundamental and powerful as simulation of nanostructure technology, both present and more futuristic, progresses. For example, the same simulation methods that have been developed in the last decade for silicon technology can also be applied to some biological systems, the carbon-based devices of nature, and the newly emerging field of carbon nanotubes. In turn, the methods developed in biochemistry become increasingly useful to answer questions in electronics and electromechanics at the nanoscale.

It is currently not possible to give an overview of all these opportunities in the limited space of this chapter. We present therefore only four vignettes that demonstrate the wide range of knowledge that is needed in nanostructure simulations and what can be anticipated in the future for simulations ranging from silicon-based electronics and nanoelectromechanics to biological systems such as protein-based ion channels.

11.2   Nanostructure Studies of the Si-SiO$_2$ Interface

In this section we discuss modeling of the Si-SiO$_2$ interface, mostly in the context of Metal-Oxide-Semiconductor Field-Effect Transistors (MOSFETs). New insights are gained by explicitly calculating material properties using nanostructure and atomic-level techniques. This section thus offers an example of how nanostructure simulation is already necessary for conventional silicon technology as encountered in the highly integrated chips of today.

11.2.1   Si-H Bonds at the Si-SiO$_2$ Interface

Hydrogen has long been used in the processing of MOSFETs in order to passivate electrically active defects that occur, for instance, at the Si-SiO$_2$ interface. The Si-H binding energy was commonly assumed to be the threshold energy for H-related degradation in MOSFETs.[2] We have used density functional calculations to investigate the energetics of the hydrogen dissociation process itself.[3–8] These calculations show that there are several mechanisms by which hydrogen can desorb through processes that involve much lower energies than the Si-H binding energy of ~3.6 eV. These results explain continued hot-electron degradation in MOSFETs even as operating voltages have been scaled to below 3.6 eV.[5] Moreover, a distribution of dissociation energies due to disorder at the interface is expected. Such a distribution indicates that the probability of degradation will increase dramatically as MOSFETs are scaled to sub-100 nanometer channel lengths.[9,10]

11.2.1.1   Density Functional Calculations

Density functional theory (DFT) has become the leading theoretical tool for understanding nanoscale phenomena in physics and chemistry. This is because DFT allows an accurate determination of electronic structure and also efficiently scales with the number of atoms in a calculation. We have performed a comprehensive DFT study of the mechanisms of Si-H bond breaking at the Si-SiO$_2$ interface. We have used several atomic models of the interface including the cluster model shown in Figure 11.1. These studies demonstrate how DFT can be used to model electronics on the nanoscale.

Our main results for Si-H at the Si-SiO$_2$ interface are as follows. The energy needed to dissociate an isolated silicon–hydrogen bond (placing the hydrogen in a vacuum state at infinity) is found to be ~3.6 eV. For an Si-H bond at the Si-SiO$_2$ interface, if the dissociated hydrogen atom enters bulk SiO$_2$, then the dissociation or dissociation energy is also 3.6 eV because atomic hydrogen interacts only weakly with the rather open, insulating oxide. However, the Si-H dissociation energy can be significantly reduced for Si-H bonds at the Si-SiO$_2$ interface because hydrogen can desorb by first entering bulk silicon. The energy

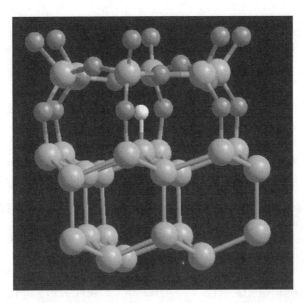

FIGURE 11.1 Atomic ball-and-stick model of an isolated Si-H bond at the Si-SiO$_2$ interface. Smaller balls represent oxygen atoms, and larger balls represent silicon.

needed to place a neutral hydrogen atom, arising from the silicon dangling bond site, into bulk silicon far from any defects is ~2.5 eV. As hydrogen diffuses to a surface or interface, it can passivate other defects or combine with another hydrogen atom to form H$_2$. At a surface or an open interface such as the Si-SiO$_2$ interface, H$_2$ molecules can easily diffuse away, leaving behind the silicon dangling bonds. Experimentally, the thermally activated dissociation of hydrogen from the (111)Si-SiO$_2$ interface is measured at 2.56 eV. This is consistent with our calculated mechanism with H entering bulk silicon before leaving the system as H$_2$.

In addition to the above considerations, the threshold energy for hot-electron degradation can be greatly reduced if dissociation occurs by multiple vibrational excitations. For low voltages, Si-H dissociation involving multiple vibrational excitations by the transport electrons becomes relatively more likely. Because hydrogen is very light, the hydrogen in an Si-H bond is a quantum oscillator. Hot-electrons can excite the hydrogen quantum oscillator from the ground state into an excited state. Because the Si-H

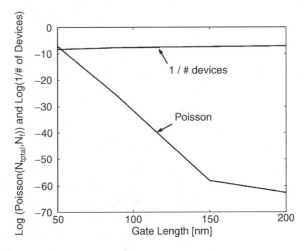

FIGURE 11.2 Example of failure function for interface trap generation.

vibrational modes are well above the silicon phonon modes, the excited state will be long-lived, allowing for multiple vibrational excitation. In this case, the Si-H dissociation can take place at channel electron energies lower than 2.5 eV and perhaps as low as 0.1 eV, the vibrational energy of the Si-H bending mode.[11]

11.2.2 Reliability Considerations at the Nanoscale

11.2.2.1 Increasing Effect of Defect Precursor Distribution at the Nanoscale

For micron-sized devices, many Si-H bonds at the Si-SiO$_2$ interface must be broken before the device has significantly degraded. For nanoscale devices, a much smaller number of defects (possibly on the order of 10s or lower) could cause a device to fail. Because of the smaller number of defects required, there is an increasing probability with decreasing device size of having a significant percentage of defect precursors with lifetimes in the short-lifetime tail of the Si-H dissociation energy distribution mentioned above. This results in an increasing number of short-time failures for smaller devices. In order to quantify this result, the shape of the distribution of dissociation energies must be known. Fortunately, this shape can be determined from the time dependence of trap generation under hot-electron stress, and from this the effect of deviations from this distribution on the reliability can be calculated. To understand how the reliability can be understood, we specifically look at the example of interface trap generation in nMOSFETs.

11.2.2.2 Hot-Electron Interface Trap Generation for Submicron nMOSFETs

A sublinear power law of defect generation with time is observed for the generation of interface traps at the Si-SiO$_2$ interface. Because the hydrogen is relatively diffusely spread throughout the interface (with only around one silicon–hydrogen bond for every hundred lattice spacings), it is clear that any process which breaks these bonds will be first order in the number of Si-H bonds. That is, the rate equation for this process can be written

$$\frac{dN(E_b)}{dt} = -\frac{N(E_b)}{\tau(E_b)}$$

where $N(E_b)$ is the number of silicon–hydrogen bonds (which must be a function of E_b, the bond energy), and τ is some lifetime that, for a hot-electron-driven process, would involve an integration of the electron distribution with the cross-section for defect creation, also a function of the bond energy. In order to get the true number of defects as a function of time for an average device, the solution to this rate equation must be integrated over the distribution of bond energies. This gives

$$N_{tot}(t) = \int_{0}^{\infty} f(E_b) N_0 \exp\left(-\frac{t}{\tau(E_b)}\right) dE_b$$

This integral is what produces the sublinearity of the time dependence of the generation of interface traps. The importance of this integral lies in the fact that it relates the sublinearity of the time dependence of the generation of interface traps with the average distribution of defect energies, which can be related to the distribution of defect generation lifetimes. This distribution can be used to determine the failure function for the failure mode involving this type of defect. One can extract the defect activation energy distribution from this integral once one knows the sublinear power law for defect generation with time.[12]

11.2.2.3 Reliability from Defect Precursor Distribution

Again utilizing the assumption of independent defects, the failure function of a device (defined as the probability of having a sufficient number of defects that will fail before some time t) will be a binomial or, approximately, a Poisson distribution. One of the characteristics of this failure function is an exponential increase in the probability of failure as the number of defects required to cause failure gets small. This is demonstrated in Figure 11.2, where we compare the reciprocal of the number of devices on a chip (which gives an idea of how much the reliability of a single device on that chip must increase) with the variation in the Poisson distribution with the number of interface traps required for the failure of a device. This is done for four gate lengths, with the gate lengths and number of devices at a given gate length taken from the semiconductor roadmap. The number of defects required for device failure comes from ISE-TCAD simulations. Notice the number of failures increases exponentially as the gate length is reduced below 100 nanometers.

Using knowledge of the type of defect involved in a degradation process, one can analytically derive the expected failure function for that type of degradation. This is not restricted to interface trap generation by Si-H dissociation, as the assumptions that went into the model are very few: first-order kinetics and a distribution of dissociation energies.

11.2.3   Tunneling in Ultra Thin Oxides

The thickness of gate oxides in MOSFETs is approaching 1–2 nm, i.e., only a few Si-O bond lengths. Consequently, gate leakage currents have become a major design consideration. For such ultra thin oxides, it is increasingly important to understand the *influence* of microscopic structure and composition of the oxide and its interface with silicon *on* the magnitude of oxide transmission probabilities and tunneling currents.

To fully explore the microscopic nature of gate leakage currents, an atomic-orbital formalism for calculating the transmission probabilities for electrons incident on microscopic models of Si-SiO$_2$-Si heterojunction barriers was implemented.[13–15] Subsequently the magnitude of leakage currents in *real* MOSFETs was calculated by incorporating the incident electron density from device simulations.[16,17] Such an approach allows one to examine the influence of atomic structure on tunneling. Significant results include assessing the validity of the bulk band structure picture of tunneling, determining the energy dependence of tunneling effective mass, and quantifying the nature of resonant tunneling through defects. Below, we will briefly discuss the most important details and results.

The microscopic supercell models of Si[100]-SiO$_2$-Si[100] heterojunctions that have been used were constructed by sandwiching unit cells of (initially) tridymite or *beta*-quartz polytype of SiO$_2$ between two Si[100] surfaces. The models are periodic in the plane perpendicular to the interface with periodic lengths of 0.5–1.5 nm. As more detail is desired, e.g., to examine the effects of interfacial morphology, the lateral periodic length scale can be increased with added computational costs. As an example, Figure 11.3 shows a ball-and-stick skeleton of a tridymite-based cell.

Reflection and transmission coefficients of the supercells described above were calculated using a transfer-matrix-type scheme embedded in a tight-binding framework. We solve the Schrodinger equation with open boundary conditions for the whole junction at a fixed energy E (measured relative to the silicon conduction band minimum on the channel side of the oxide) and in-plane momentum $k_\parallel$ (that is a good quantum number due to the lateral periodicity) in a *layer-orbital basis*. An empirical sp^3 tight-binding basis with second-nearest neighbor interactions for both silicon and the oxide were used. The tight-binding parameters were chosen to yield experimental bulk band gaps and to reproduce density functional calculations of the effective masses of the lowest conduction bands. An electron state propagating toward the oxide from the channel side of the junction, characterized by E, $k_\parallel$ and its wavevector component normal to the interface ($k_{perp,in}$), is scattered into sets of reflected and transmitted states (characterized by wavevector components $k_{perp,out}$). From the scattering wavefunctions, transmission amplitudes and dimensionless transmission coefficients are obtained.

The present microscopic models allow one to predict the intrinsic decay properties of the wavefunctions into the gate oxide. Because of the local nature of bonding in the oxide, a bulk picture of tunneling persists qualitatively even for the thinnest oxide barriers. We have analyzed the complex bands of the present bulk oxide models and find that (1) only one single complex band is relevant for electron tunneling; (2) several different bands are involved in hole tunneling; and (3) all complex oxide bands are highly nonparabolic. Because of the mismatch in the Brillouin zones for the oxide on top of the silicon, the bulk silicon $k_\parallel$ is not conserved and different states have differing decay constants. The energy dependence of the integrated transmission is shown for oxide thicknesses between 0.7 and 4.6 nm in Figure 11.4, which also includes effective–mass-based results with a constant (EM) and the

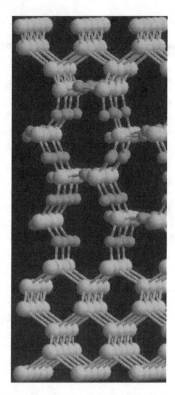

FIGURE 11.3 Ball-and-stick model of an Si[100]-SiO$_2$-Si[100] model heterojunction based on the 1.3 nm thin gate oxide based on the tridymite polytype of SiO$_2$. (Dark = oxygen and light = silicon.)

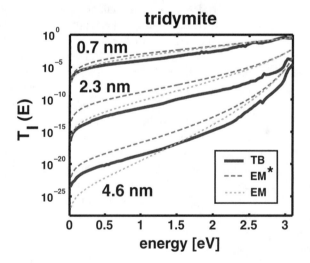

FIGURE 11.4 Integrated transmission (T$_I$) vs. the energy of the incident electron for tridymite-based oxides with thicknesses of 0.7, 2.3, and 4.6 nm. Results are for calculations with our atomic-level tight-binding method (TB, solid line), effective mass approaches with constant (EM, dotted line), and energy dependent (EM*, dashed line) effective masses.

energy-dependent (EM*) electron mass, which was fitted to our tight-binding complex band structures. The parabolic effective mass approximation overestimates the transmission for oxides thinner than ~1 nm. As oxide thicknesses increase, the tight-binding transmission is underestimated at low energies and overestimated at higher energies. The higher slope of the transmission obtained in the parabolic effective mass approximation is consistent with the findings for the tunneling masses and explains previous errors in oxide thicknesses derived from tunneling experiments and a constant parabolic effective mass model.[13] Using the correct tight-binding dispersion of the imaginary bands in an effective mass calculation (i.e., the EM* results in Figure 11.4) leads to qualitatively correct slopes for transmission; however, the absolute values are typically overestimated by one to two orders of magnitude. A possible reason for much of this discrepancy may be that the effective-mass-based transmission calculation underestimates the full band structure mismatch of silicon and its oxide.

The transmission coefficients were combined with electron densities and the corresponding distribution functions at the Si-SiO$_2$ interface of prototypical MOSFETs with channels of 50 nm and 90 nm. These quantities were obtained from full-band Monte Carlo simulations and were used to calculate the absolute magnitudes for gate leakage currents which, for oxide thicknesses smaller than ~4 nm, are dominated by tunneling of cold electrons in the source and drain contacts for defect-free oxides. As a consequence, the tunneling current densities (integrated over the entire gate length) decrease upon applying a drain-source voltage. The elastic gate leakage currents were recalculated including oxygen vacancies for a given energy E_{vac} in the oxide band gap from 0 to 3 eV above the silicon conduction band edge. The leakage currents at an arbitrary vacancy density were calculated using an interpolation formula.[16,17] Interestingly, we find that for all possible combinations of vacancy energy and density, the gate currents are still dominated by cold electrons originating in the contact regions. We have calculated the direct gate current densities from the source contact for the 50 nm transistor with a 1.3 nm oxide and the 90 nm transistor with a 2.9 nm oxide for defect densities in the range of 10^{10}–10^{13} cm^{-2} and a homogeneous as well as various Gaussian distributions of E_{vac} in energy space. The magnitude of the defect-induced current increase is very sensitive to the density and the energy distribution of the defects. For defect densities greater than 10^{12} cm^{-2}, the enhancement can be as high as 2–3 orders of magnitude. Also, the resonant effects are somewhat less pronounced for the thinner oxide.

We regard this work as the first steps toward the full understanding of oxide tunneling from a microscopic point of view. The theoretical approach presented here[13,16,17] could certainly be applied to other systems; and there are other methods to calculate electron transport at the atomic scale, which are of general interest for those interested in modeling nanoelectronic devices.[18–20]

11.3   Modeling of Quantum Dots and Artificial Atoms

The quantum dot is, in a way, the prototype of any future device that is designed to occupy a minimum of space. It is important in this context that quantum dots can be arranged and interconnected in three dimensions, at least in principle. In the last 10 years, the physics of quantum dots has experienced considerable development because of the manifestation of the discreteness of the electron charge in single-electron charging devices, as well as the analogy between three-dimensionally quantum confined systems and atoms.[21] Early studies were motivated by the observation of single-electron charging in granular metallic islands containing a "small" number of conduction electrons (N~100–1000) surrounded by an insulator characterized by a small capacitance C.[22,23] In metallic dots however, quantum confinement is relatively weak; and the large effective mass of conduction electrons makes the energy spectrum a quasi-continuum with negligible separation between electron states even at low temperature, $\Delta E << k_B T$. Hence, the addition of an electron to the island requires the charging energy $e^2/2C$ from a supply voltage source to overcome the electrostatic repulsion or Coulomb blockade from the electrons present in the dot, with negligible influence of the energy quantization in the system.[24]

Advances in patterning and nanofabrication techniques have made possible the realization of semiconductor quantum dots with precise geometries and characteristic sizes comparable to the de Broglie wavelength of charge carriers.[25] These quantum dots are realized in various configurations by combining heterostructures and electrostatic confinement resulting from biased metal electrodes patterned on the semiconductor surfaces. In three-dimensional confined III-V compound semiconductors, the small effective mass of conduction electrons results in an energy spectrum of discrete bound states with energy

(a)

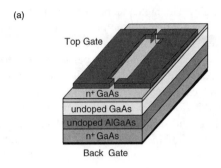

(b)

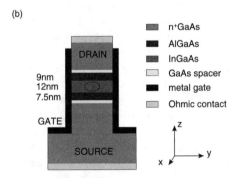

FIGURE 11.5   (a) Schematic representation of a planar quantum dot structure with layered materials; the dark areas represent the confining metallic gates at the surface. (b) Vertical quantum dot structure with different constituting materials; the vertical dark areas on the side represent the controlling metallic gate.

separation comparable to, or even larger than, the charging energy $e^2/2C$. The ability to vary the electrostatic potential over large voltage ranges allows for fine-tuning of the quantum dot charge of just a few electrons ($N\sim1$–10).[26] Early experiments on single-electron charging were made with layered AlGaAs/GaAs structures by patterning several Schottky metal gates on top of a two-dimensional electron gas to achieve lateral confinement. A back gate controls the number of electrons in the two-dimensional gas and the dot (Figure 11.5a).[25] In these planar structures, the current flows parallel to the layers, and the tunneling barriers between the dot and the two-dimensional gas are electrostatically modulated by the top gates. In vertical quantum dots, the electrons are sandwiched vertically between two tunneling heterobarriers, while the lateral confinement results from a vertical Schottky achieved by deep mesa etching of the multilayer structure (Figure 11.5b).[27,28] In this case current flows perpendicularly to the two-dimensional gas between the two hetero-barriers, which are usually high and thin because they are made of different semiconductor materials, e.g., InGaAs and AlGaAs. In general, planar dots have a poor control of the exact number of electrons, while vertical dots lack the barrier tenability of lateral structures.

In semiconductor quantum dots, discrete energy levels with Coulomb interaction among electrons for achieving the lowest many-body state of the system is reminiscent of atomic structures. In cylindrical quantum dots, shell structures in the energy spectrum and Hund's rule for spin alignment with partial shell filling of electrons have recently been observed.[28] One of the peculiarities of these nanostructures is the ability to control not only the shape of the dot but also the number of electrons through gate electrodes.[29] Hence, *artificial atoms* can be designed to depart strongly from the three-dimensional spherical symmetry of the central Coulomb potential and its nucleus charge. In this context, the physics of a few electrons in quantum dots offers new opportunities to investigate fundamental concepts such as the interaction between charge carriers in arbitrary three-dimensional confining potentials and their elementary excitations. Moreover, because Hund's rule is the manifestation of spin effects with shell filling in quantum dots, the electron spin can, in principle, also be controlled by the electric field of a transistor gate.[30] The idea of controlling spin polarization, independent of the number of electrons in quantum dots, has practical consequences because it provides the physical ingredients for processing quantum information and making quantum computation possible.[31] In addition, spin degrees of freedom can be utilized for storing information in new forms of memory devices. Aside from the investigation of basic quantum phenomena, artificial atoms are also promising for applications in high-functionality nanoscale electronic and photonic devices such as ultra-small memories or high-performance lasers.[22,32,33]

11.3.1 The Many-Body Hamiltonian of Artificial Atoms

The electronic spectrum of N-electron quantum dots are computed by considering the many-body Hamiltonian:

$$\hat{H} = \sum_i \hat{H}_{0i} + \sum_{i,j} \hat{H}_{ij} \tag{11.1}$$

where H_{0i} is the single-particle Hamiltonian of the ith electron and:

$$H_{ij} = \frac{e^2}{\varepsilon |\vec{r}_i - \vec{r}_j|} \tag{11.2}$$

is the interaction Hamiltonian describing the Coulomb interaction between carriers. Here ε is the dielectric constant of the material. In the second term of Equation (11.1), the sum is carried out for i/j, avoiding the interaction of carriers with themselves. Quite generally, the Hamiltonian Equation (11.1) is used for solving the Schroedinger equation for the many-particle energies and wavefunctions,

$$E = E_N(1, 2, 3, ..., N)$$

$$\Psi = \Psi_N(\vec{r}_1, \vec{r}_2, \vec{r}_3, ..., \vec{r}_N)$$

which, given the two-body interaction Equation (11.2), can only be solved exactly for $N = 2$. In this section we will describe a natural approach toward the solution of this problem for a general number N of electrons by considering successive approximations.

11.3.1.1 Single-Particle Hamiltonian and Shell Structures

We start by considering a system of independent and three-dimensional confined electrons in the conduction band. By neglecting the interaction H_{ij}, the Hamiltonian Equation (11.1) is reduced to a sum of single particle Hamiltonians, each of the same form:

$$\hat{H}_{0i} = \hat{H}_i = \frac{\hat{p}_{xi}^2 + p_{yi}^2 + p_{zi}^2}{2m^*} + \hat{V}(\vec{r}_i)$$

Here we assume the electrons can be described with an effective mass m^*; p_{xi}, p_{yi}, and p_{zi} are the components of the ith electron momentum, and $V(r_i)$ is the external potential that contains several contributions according to the confinement achieved in the quantum dot. We will assume that the quantum dot is realized by confinement of the electrons in a heterostructure quantum well along the y-direction and electrostatic confinement in the x–z plane (Figure 11.6a). The latter confinement results usually from dopant atoms in neighboring semiconductor layers and from the fringing field of the metal electrodes on the semiconductor surface. This configuration is most commonly achieved in planar quantum dots and vertical quantum dots and results in a first approximation in a two-dimensional parabolic potential in the x–z plane (Figure 11.6b). Confinement at the heterostructure along the y-direction is generally strong (~10 nm) with energy separation of the order of 50–100 meV, while the x–z

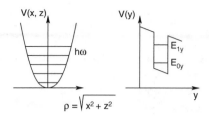

FIGURE 11.6 Schematic representation of (a) a two-dimensional parabolic potential with cylindrical symmetry in the x–z plane showing equally spaced energy levels, and (b) the square potential with the first two quantized levels in the y direction with $E_{2y}-E_{1y} \gg \hbar\omega$.

planar confinement is much weaker with energy separation of the order of 1 meV over larger distance ($> \sim 100$ nm). In that case, the external potential is separable in a first approximation,

$$\hat{V}(\vec{r}) = \hat{V}_1(x, z) + \hat{V}_2(y),$$

which results in the energy spectrum $E_{\nu, nx, nz} = E_\nu + E_{n_x, n_z}$ with corresponding wavefunctions $\Psi_\nu(y) \Psi_{n_x, n_z}(x, z)$, where $E_\nu(E_{n_x, n_z})$ is the spectrum resulting from the y-potential (x–z potential). Hence, each value of the ν-quantum number gives a series of x–z energy levels. At low temperature, given the large separation between the E_ν energy states, only the first levels of the lowest series $\nu = 0$ are occupied by electrons. If one further assumes that the $\hat{V}_1(x, z)$ potential is cylindrically symmetric, the $\nu = 0$ energy spectrum is written as:[34]

$$E_{0, n_x, n_z} = E_{0, m, l} = E_0 + m\hbar\omega$$

where ω is frequency of the cylindrical parabolic potential. Here each m-level is $2m$-times degenerate, with the factor 2 accounting for the spin degeneracy. The number m ($= 1,2,3...$) is the radial quantum number, and the number l ($= 0,\pm1,\pm2,...$) is the angular momentum quantum number. Hence the two-dimensional cylindrical parabolic potential results in two-dimensional s,p,d,f,...-like orbitals supporting 2, 4, 6, 8... electrons, which give rise to shell structures filled with 2, 6, 12, 20, ... particles, thereby creating a sequence of numbers that can be regarded as the two-dimensional analogues of *magic numbers* in atomic physics.[28,35]

In the absence of cylindrical or square symmetry, the parabolic potential is characterized by two different frequencies, ω_x and ω_z, which lift the azimuthal degeneracy on the l-number of the two-dimensional artificial atoms. Therefore, electronic states are spin-degenerate only and determine a sequence of shell filling numbers 2, 4, 6, 8, ... of period or increment 2. Only when the ratio ω_x/ω_z is commensurable does the sequence of filling numbers deviate from the period 2 and provide a new sequence of numbers for particular combinations of the n_x and n_z quantum numbers in the case of accidental degeneracy.[36]

Another important class of three-dimensional confined systems includes quantum dots obtained by self-assembled or self-organized Stranski–Krastanov (SK) epitaxial growth of lattice-mismatched semiconductors, which results in the formation of strained-induced nanoscale islands of materials. InAs and InGaAs islands on GaAs have been obtained with this technique in well-controlled size and density.[33,37,38] For these materials, shapes vary between semispherical and pyramidal form, and the size is so small that these quantum dots only contain one three-dimensional fully quantized level for conduction electrons.

11.3.1.2 Hartree–Fock Approximation and Hund's Rules

The natural extension of the atomic model for independent three-dimensional confined electrons is the consideration of the Coulomb interaction between particles in the Hartree–Fock (HF) approximation. The HF scheme has the advantage of conserving the single-particle picture for the many-body state of the system by representing the total wave function as a product of single-particle wavefunctions in a

Slater determinant that obeys Fermi statistics. The main consequence of the HF approximation for the Coulomb interaction among particles is a correction of two terms to the single-particle energies derived from the H_o Hamiltonian[39]

$$E_i = E_{v, n_x, n_z} + \frac{e^2}{\varepsilon} \sum_{j \neq i} \int d\vec{r}_i d\vec{r}_j \frac{\left| \Psi_i(\vec{r}_i) \right|^2 \left| \Psi_j(\vec{r}_j) \right|^2}{\left| \vec{r}_i - \vec{r}_j \right|}$$

$$- \frac{e^2}{\varepsilon} \sum_{j \neq i} \int d\vec{r}_i d\vec{r}_j \frac{\Psi_i^*(\vec{r}_i) \Psi_j(\vec{r}_i) \Psi_i(\vec{r}_j) \Psi_j^*(\vec{r}_j)}{\left| \vec{r}_i - \vec{r}_j \right|}$$

where the first sum is the Hartree energy carried on all occupied *j*-states different from the *i*-state, irrespective of their spins, and accounts for the classical repulsion between electrons. The second term is the attractive exchange interaction that occurs among carriers with parallel spins. In this scheme, the wavefunctions $\Psi_i(\vec{r}_i)$ satisfy the HF integro-differential equation where the Coulomb interaction term depends upon all the other single-particle wavefunctions of the occupied states. The HF equation is therefore nonlinear and must be solved self-consistently for all wavefunctions of occupied states.

One of the important consequences of the HF approximation for interelectron interaction Equation (11.2) is the prediction of spin effects in the shell filling of artificial atoms similar to Hund's rules in atomic physics.[28] These effects are illustrated in the charging energy of a few electron quantum dots with a cylindrical parabolic potential as achieved in planar or vertical quantum structures.[26] In Figure 11.7a, we show schematically the Coulomb staircase resulting from charging a quantum dot with a few electrons as a function of the charging energy or voltage between the metal electrode or gate and the semiconductor substrate. The relative step sizes of the staircase represent the amount of energy needed to put an additional electron in the dot. The arrows on each step represent the spin of each individual electron on the successive orbitals during the charging process. The filling of the first shell (s-orbital with 2 electrons) consists of one electron with spin-up followed by an electron with spin-down. The step size of the spin-up electron measures the charging energy needed to overcome the Coulomb repulsion against the spin-down electron, which is only the Hartree energy between the two particles. The larger step size of the second (spin-down) electron is due to the fact that the charging of the third electron requires the charging energy augmented by the energy to access the next quantized level, which is the first p-orbital. The latter

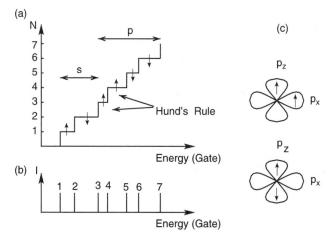

FIGURE 11.7 (a) Coulomb staircase as a function of the charging energy with the spin states of each electron. N is the number of electrons, and the horizontal two-head arrows indicate the occupation of the s- and p-orbitals in the dot. (b) Electron current through the dot vs. the charging energy. (c) Two-dimensional p-orbitals illustrating the two possible occupations of two electrons with parallel (top diagram) and anti-parallel spins (bottom diagram).

process starts the second shell filling with the third electron on either one of the degenerate $l = \pm 1$ orbitals of either spin (here we choose the $l = -1$ and the spin-up). At this stage, the configuration with the fourth electron on the $l = 1$ orbital with a parallel spin becomes more favorable because it minimizes the Hartree energy between orbitals of different quantum numbers and results in an attractive exchange energy between the two electrons (Figure 11.7c). This is the reason the third step is smaller than the first and the second steps, requiring less energy and demonstrating Hund's rule in the electron filling of the two-dimensional artificial atom. The fourth step is long because the addition of the fifth electron on either of the p-orbitals with $l = \pm 1$ must correspond to a spin-down electron that undergoes a repulsion from the two other p-electrons without benefiting from the exchange because its spin is anti-parallel. Figure 11.7b shows the current peaks resulting from the single-electron charging of the quantum dot, which is obtained by differentiating the Coulomb staircase. Current characteristics with similar structure have recently been observed in gated double-barrier GaAs/AlGaAs/InGaAs/AlGaAs/GaAs vertical quantum dot tunnel devices, which revealed the shell structure for a cylindrical parabolic potential as well as spin effects obeying Hund's rule in the charging of the dot.[28]

Hence the HF approximation provides a reasonable picture of the contribution of electron–electron interaction and spin effects in the spectrum of quantum dots. However, it is well known from atomic physics and theoretical condensed matter physics that this approximation suffers from two important drawbacks: neglect of electron correlation and overestimation of the exchange energy.[40] Moreover, it leads to tedious solution of the self-consistent problem when involving a large number of electrons.

11.3.1.3 Full-Scale Simulation of Quantum Dot Devices

Advances in computer simulation combine the sophistication of realistic device modeling with the accuracy of computational physics of materials based on the density functional theory (DFT).[41–45] These powerful methods provide theoretical tools for analyzing fine details of many-body interactions in nanostructures in a three-dimensional environment made of heterostructures and doping, with realistic boundary conditions. *Microscopic* changes in the quantum states are described in terms of the variation of *macroscopic* parameters such as voltages, structure size, and physical shape of the dots without *a priori* assumption on the confinement profile. Consequently, engineering the exchange interaction among electrons for achieving controllable spin effects in quantum devices becomes possible.

The implementation of a spin-dependent scheme for the electronic structure of artificial molecules involves the solution of the Kohn–Sham equation for each of the spins, i.e., up ($\uparrow$) and down ($\downarrow$). Under the local spin density approximation within the DFT, the Hamiltonian $H^{\uparrow(\downarrow)}$ for the spin $\uparrow$ ($\downarrow$) electrons reads[40,46]

$$\hat{H}^{\uparrow(\downarrow)} = -\frac{\hbar^2}{2}\nabla\left[\frac{1}{m^*(\vec{r})}\nabla\right] + E_c(\vec{r}) + \mu_{xc}^{\uparrow(\downarrow)}[n]$$

where $m^*(r)$ is the position-dependent effective mass of the electron in the different materials, $E_c(r) = e\phi(r) + \Delta E_{os}$ is the effective conduction band edge, $\phi(r)$ is the electrostatic potential which contains the Coulomb interaction between electrons, and ΔE_{os} is the conduction band offset between GaAs and AlGaAs. The respective Hamiltonians are identical in all respects, except for the exchange-correlation potential, which is given by

$$\mu_{xc}^{\uparrow(\downarrow)} = \frac{d(n\varepsilon_{xc}[n])}{dn^{\uparrow(\downarrow)}}$$

where ε_{xc} is the exchange-correlation energy as a function of the total electron density $n(r) = n^{\uparrow}(r) + n^{\downarrow}(r)$ and the fractional spin polarization $\xi = (n^{\uparrow} - n^{\downarrow})/n$, as parameterized by Ceperley and Alder.[47] While it is known that the DFT underestimates the exchange interaction between electrons, which leads to incorrect energy gaps in semiconductors, it provides a realistic description of spin–spin interactions in quantum nanostructures, as shown in the prediction of the addition energy of vertical quantum dots (see Figure 11.8).

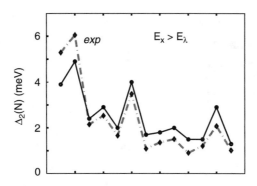

FIGURE 11.8 Addition energy of a vertical single quantum dot. (Data from Tarucha, S., Austing, D.G., Honda, T., van der Hage, R.J., and Kouwenhoven, L.P., *Phys. Rev. Lett.* 77, 3613, 1996; Nagaraja, S., Leburton, J.P., and Martin, R.M., *Phys. Rev. B* 60, 8759, 1999.)

The three-dimensional Poisson equation for the electrostatic potential $\phi(r)$ reads

$$\vec{\nabla}[\varepsilon(\vec{r})\vec{\nabla}\phi(\vec{r})] = -\rho(\vec{r})$$

Here $\varepsilon(r)$ is the permittivity of the material, and the charge density ρ is comprised of the electron and hole concentrations as well as the ionized donor and acceptor concentrations present in the respective regions of the device. The dot region itself is undoped or very slightly p-doped. At equilibrium, the electron concentrations for each spin in the dots are computed from the wavefunctions obtained from the respective Kohn–Sham equations, i.e., $\rho(r) = en(r)$ with $n^{\uparrow(\downarrow)} = \Sigma_i |\psi_i^{\uparrow(\downarrow)}(r)|^2$. In the region outside the dots, a Thomas–Fermi distribution is used so that the electron density outside the dot is a simple local function of the position of the conduction band edge with respect to the Fermi level, ε_F. The various gate voltages —V_{back}, V_t, and those on the metallic pads and stubs — determine the boundary conditions on the potential $\phi(r)$ in the Poisson equation. For the lateral surfaces in the x–y plane on Figure 11.5, vanishing electric fields are assumed.

Self-consistent solution of the Kohn–Sham and Poisson equations proceeds by solving the former for both spins, calculating the respective electron densities and exchange correlation potentials, solving the Poisson equation to determine the potential $\phi(r)$, and repeating the sequence until the convergence criterion is satisfied.[44] Typically, this criterion is such that variations in the energy levels and electrostatic potential between successive solutions are below 10^{-6} eV and 10^{-6} V, respectively.

The determination of N_{eq}, the number of electrons in the dots at equilibrium for each value of the gate and tuning voltages, is achieved by using Slater's transition rule:[48]

$$E_T(N+1) - E_T(N) = \int_0^1 \varepsilon_{LOA}(n)\,dn \simeq \varepsilon_{LOA}\left(\frac{1}{2}\right) - \varepsilon_F$$

where $E_T(N)$ is the total energy of the dot for N electrons and $\varepsilon_{LOA}(1/2)$ is the eigenvalue of the lowest-available orbital when it is occupied by 0.5 electron. From the latter equation, it is seen that if the right-hand side is positive, $N_{eq} = N$; otherwise, $N_{eq} = N+1$. Thus the $N{\to}N+1$ transition points are obtained by populating the system with $N+0.5$ electrons and varying V_{back} until $\varepsilon_{LOA}(1/2) - \varepsilon_F$ becomes negative. It should be noted that the approximation made in the latter equation is valid only if ε_{LOA} varies linearly with N. This approach has been very successful in the analysis of the electronic spectra and charging characteristics of vertically confined quantum dots.[49] Figure 11.8 shows the addition energy spectrum of a single vertical quantum dot as a function of the number N of electrons in the dot. The addition energy measures the energy required to add a new electron in the dot given the presence of other electrons already in the dot and the restriction imposed by the Pauli principle on the electron energy spectrum.

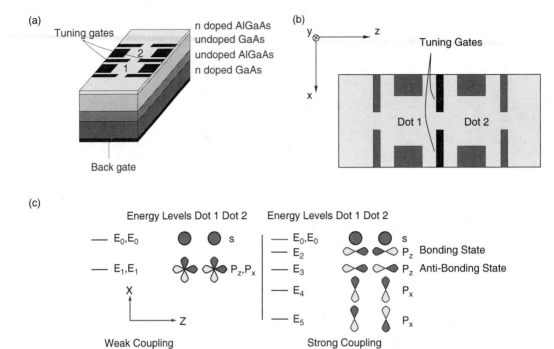

FIGURE 11.9 Schematic representation of the planar coupled quantum dot device. (a) Layer structure with top and back gates; (b) top view of the metal gate arrangement with sizes and orientations; (c) schematic representation of the six lowest orbitals in the weak (left-hand side) and strong (right-hand side) coupling regimes. In both cases, the s-states are strongly localized in their respective dot. Left: p_x- and p_z-like orbitals are degenerate within each dot and decoupled from the corresponding state in the other dot. Right: Increasing coupling lifts the p-orbitals degeneracy with a reordering of the states. The dark and light orbitals indicate positive and negative parts of the wave functions, respectively.

The peaks at $N = 2$, 6, and 12 are the signature of the existence of a two-dimensional shell structure in the dot, while the secondary peaks at $N = 4$ and 9 reflect the existence of Hund's rule at half-filled shells. The agreement between theory and experimental data is excellent for the position of the peaks as well as for their magnitude.

The technique is useful in designing double quantum dots with variable interdot barrier for controlling electron–electron interactions.[50] The devices have a planar geometry made of GaAs/AlGaAs hetero-structure that contains a two-dimensional electron gas (2DEG). The dots are defined by a system of gate pads and stubs that are negatively biased to deplete the 2DEG, leaving two pools of electrons that form two quantum dots connected in series (Figure 11.9).[46] In these *artificial diatomic molecules*, electron states can couple to form covalent states that are delocalized over the two dots, with electrons tunneling between them without being localized to either.[51] These *bonding states* have lower energy than the constituent dot states by an amount that is equivalent to the binding energy of the molecule. In our case, the dimensions are such that the electron–electron interaction energy is comparable to the single-particle energy level spacing. The number of electrons in the dot, N, is restricted to low values in a situation comparable to a light diatomic molecule such as H-H or B-B. The coupling between dots can be adjusted by varying the voltage on the tuning gates V_t to change the height of the barrier between the two dots. The number of electrons N in the double dot is varied as the 2DEG density, with the back gate voltage V_{back} for a fixed bias on the top gates. Hence, controllable exchange interaction that gives rise to spin polarization can be engineered with this configuration by varying N and the system spin, independently.*

---

*This is not the case in shell filling of single quantum dots because the total spin of the electronic system is directly related to the number N of electrons in the dot.

11.3.2 Quantum Modeling of Artificial Molecules and Exchange Engineering

In order to simulate these effects, we consider a structure that consists of a 22.5 nm layer of undoped $Al_{0.3}Ga_{0.7}$ as followed by a 125 nm layer of undoped GaAs, and finally an 18 nm GaAs cap layer (Figure 11.9a). The latter is uniformly doped to $5.10^{18}cm^{-3}$ so that the conduction band is immediately above the Fermi level at the boundary between the GaAs cap layer and the undoped GaAs. The inverted heterostructure is grown on the GaAs substrate. The lateral dimensions of the gates and spacing are shown in Figure 11.9b. Figure 11.9c shows the schematic of the lowest four states with their wavefunctions in the double dot for two different tuning voltages.[46] For both values of V_t, the ground state in the individual dots is s-like and forms a degenerate pair. Here, we borrow the terminology of atomic physics to label the quantum dot states. The first excited states, which are px- and pz-like, are degenerate for weak interdot coupling, whereas for strong coupling, the pz-like states mix to form symmetric (*bonding*) and antisymmetric (*anti-bonding*) states that are lower in energy than the px-like states as seen in Figure 11.9c. This reordering of the states has an important bearing on the spin polarization of the double-dot system, as shall be fully explained below.

In the present double-dot structure, we focus on the spin states of the electron system and allow N to vary from zero to eight for two values of V_t: $V_t = -0.67V$, defined as the weak coupling regime, and $V_t = -0.60V$, defined as the strong coupling regime. Electron spin states that are relevant in this analysis are designated by $s_1^{\uparrow(\downarrow)}$, $s_2^{\uparrow(\downarrow)}$ (lower energy s-states in dots 1 and 2) and $p_{x1}^{\uparrow(\downarrow)}$, $p_{x2}^{\uparrow(\downarrow)}$, $p_{z1}^{\uparrow(\downarrow)}$, and $p_{z2}^{\uparrow(\downarrow)}$- (higher energy p-states where the x- and z-indices indicate the orientation of the wavefunctions). For $N = 0$, in the weak coupling limit, the computer model shows that s-states in dots 1 and 2 have negligible overlap because of the relatively high and wide barrier. Indeed, the bonding–antibonding energy separation resulting from the coupling between these states is orders of magnitude smaller than the Coulomb charging energy so that s-electrons are practically localized in each dot. A similar situation arises for the p_x- and p_z-states, which, although experiencing slight overlap because of higher energy, they see a lower and thinner barrier and are quasi-degenerate within each dot. In fact, in the weak coupling limit, p_{z1}- and p_{z2}-states that are oriented along the coupling direction between dots experience a bigger overlap than the corresponding p_x-states and consequently lie slightly lower in energy than the latter. Hence, as far as the lower s- and p-states are considered, the double-dot system behaves as two quasi-independent dots (Figure 11.9c, left). In addition, because of the large distance separating the two lower s-states for $V_t = -0.67V$, Coulomb interaction between electrons in dots 1 and 2 is negligibly small, and both dots can be charged simultaneously through double charging[46] to completely fill the s1 and s2 states. Therefore, for $N = 4$, there is no net spin polarization in the double dot, because both contain equal numbers of spin $\uparrow$ and spin $\downarrow$ electrons.

When the double dot is charged with a fifth electron, the latter occupies either the p_{z1}- or p_{z2}-state (e.g., $\uparrow$ spin i.e., $p_{z1}^{\uparrow}$ or $p_{z2}^{\uparrow}$) that has the lowest available energy. The sixth electron takes advantage of the nonzero p-state overlap and occupies the other p_z-state with a parallel $\uparrow$ spin. The seventh and eighth electrons find it energetically favorable to occupy successively $p_{x1}^{\uparrow}$ and $p_{x2}^{\uparrow}$, but not any of the spin $\downarrow$ states, because of the attractive nature of the exchange-correlation energy among the spin $\uparrow$ electrons that results from the nonzero p-state overlap, lowering the energy of the double dot. This particular high-spin configuration among p-orbitals in the "artificial" diatomic molecule deserves special attention because it appears to violate one of the Zener principles on the onset of magnetism in transition elements; this principle forbids spin alignment for electrons on similar orbitals in adjacent atoms.[52] Therefore, it could be argued that the high-spin configuration obtained in the calculation is the consequence of a DFT artifact. Recently, however, Wensauer et al. confirmed the DFT results based on a Heitler–London approach.[53] Similar conclusions have also been obtained by an "exact" diagonalization technique on vertically coupled quantum dots for $N = 6$ electrons.[54,55] Let us point out that Zener's principle is purely empirical, as it is based on the observation of the magnetic properties of natural elements that lacks the tunability of artificial systems. Therefore, the total spin of the double dot can possibly steadily increase by $1/2\hbar$ for each electron added after the fourth electron to $2\hbar$ for $N = 8$, and there is no contradiction with Zener's principle applied to

TABLE 11.1 Spin of the Double Dot for Various Occupation Numbers in the Two Coupling Regimes

V_t $\diagdown$ N	1	2	3	4	5	6	7	8	9	10	11	12
Spin −0.67 V	1/2	?	1/2	0	1/2	1	3/2	2	3/2	1	1/2	0
($\hbar$) −0.67 V	1/2	0	1/2	0	1/2	0	1/2	0	1/2	0	1/2	0

The question mark at N = 2 in the weak coupling regime indicates that the spins are uncorrelated.

natural elements. After all, high-spin configurations have been shown to compete for the ground state of light diatomic molecules such as B_2.[56]

The variation of the total spin S in the double dot with N is shown in Table 11.1. It is also seen that as N increases above eight electrons, the spin ↓ states start to be occupied, thereby decreasing S by $1/2\hbar$ for each additional electron, forming anti-parallel pairs to complete the shell until $N = 12$ when S is reduced to zero. The sequence of level filling with the occupation of degenerate states by electrons of parallel spin is observed in atoms and is governed by Hund's rules; it is therefore impressive that similar rules successfully govern level filling in the double dot in the weak coupling regime. Let us point out that even though $S = 2\hbar$ is the most favored state of the double dot, energetically it is not significantly lower than other competing states for $N = 8$. For instance, the excited states with $S < 2\hbar$ for $N = 8$ are only about 0.1 meV higher in energy. Consequently, for this particular double-dot structure, any attempt to observe experimentally the parallel alignment of the spins of unpaired electrons is restricted to low temperatures for which $k_B T << 0.1$ meV, or any kinds of electrostatic fluctuations smaller than this value. However, it must be noted that the structure is not optimized and that the evidence of spin polarization among p-states in the double-dot system may be achieved in smaller dots with stronger exchange interaction. The key issue here is the fact that the quantum mechanical coupling between the two dots in this bias regime is not strong enough to lift the spatial *quasi-degeneracy* among p_x- and p_z-states which, for our particular configuration, were separated by no more than a few microelectron volts. Stronger coupling between the quantum dots eliminates this effect. Accordingly, if V_t increases to −0.60 V, also referred to as the strong coupling regime, the p-state spatial degeneracy is completely lifted, while deeper s-states also couple, although to a slighter extent to lead to the spectrum of Figure 11.9c, right. Therefore, the spin sequence as a function of N is alternatively $S = 1/2\hbar$ for odd N when the last occupying electron is unpaired and $S = 0$ for even N when it pairs up with an electron of the opposite spin (Table 11.1).

The variation of inter-dot coupling by varying V_t provides a control of direct exchange interaction between p-like electrons in the two dots that may be more robust than for s-electrons. Hence, a lowering of the inter-dot barrier results in a reordering of the single-particle levels, thereby transforming the double-dot (for $N = 8$) from a spin-polarized $S = 2\hbar$ to an unpolarized state $S = 0$. An important result from Table 11.1 is that the Loss–Di Vincenzo scheme for quantum computing with double dots could also be achieved for $N = 6$ electrons, where the control of qubit entanglement for a quantum control-not (XOR) gate operation would be realized with the $S = 1/2\hbar$ spin states of two p-electrons instead of two single electrons ($N = 2$) in the original scenario.[57]

The electrostatic nature of the confinement potential, specifically the coupling barrier, is central to the occurrence of the effects mentioned above. Indeed, the barrier is not uniform but is wider (and higher) for the lower quantum dot s-states than for the higher p-states (Figure 11.9c). This situation is similar to the electronic properties of natural diatomic molecules, where the strongly localized s-states correspond to atomic core states and the delocalized p-states to covalent bonding states. It is therefore possible to engineer exchange interaction in the *artificial molecule* by suitably tailoring the coupling barrier between quantum dots. This is achievable by proper device design, i.e., by adjusting gate size and shape, the doping profiles, and the distance between the GaAs/AlGaAs heterojunction and the control gates, and possibly by choosing other III-V semiconductor systems to optimize the energy separation between singlet and multiplet states.

11.4   Carbon Nanotubes and Nanotechnology

Device aspects of carbon nanotubes represent an interesting new area of nanoscience and nanotechnology. Simulation in this area requires special methods that lie in between the methods used for the periodic solid and the methods applied to quantum dots. Various (carbon, nitride, and chalcogenide) nanotubes are promising for applications because of their unusual mechanical and electronic properties, stability, and functionality.

The lattice structure of single-wall carbon nanotubes follows the lattice structure of graphene (monolayer of graphite): a hexagonal pattern is repeated with translational symmetry along the tube axis and with axial (chiral) symmetry along the tube circumference. Nanotubes are labeled using two numbers [n,m]. These are components of the vector that generates the tube circumference after scrolling, in terms of basic vectors of the graphene lattice (see Figure 11.10A). It is easy to find that only two types of single-wall nanotubes (SWNTs) have a pure axial symmetry: so-called armchair (A) and zigzag (Z) nanotubes. The graphene rectangle shown in Figure 11.10A, gives an armchair (A) nanotube when wrapped from top to bottom (Figure 11.10B) and a zigzag (Z) nanotube when wrapped from left to right. Any other type of nanotube is chiral, which means that it belongs to a screw-axis symmetry group.

Graphite-like systems and materials, such as fullerenes, nanotubes, nanographites, and organic macromolecules, are well known to have valence/conduction band systems generated by pi and sigma valence electrons.[58] The latter ones are localized and, normally, contribute only to the mechanical properties of the graphitic material. In contrast, pi-electrons are mobile and highly polarizable and define transport, electrical, and electromechanical properties. The pi-electronic structure of a monolayer of graphite (graphene) is shown in Figure 11.11. It has a six Fermi points that separate an empty conduction band from an occupied (symmetrical) valence band. A simple but correct picture of the electronic structure of a SWNT follows from a band-folding argument: an additional space quantization for the pi-electrons appears due to confinement in the circumferential direction. It can be thought of as a mere cross sectioning of the electronic structure of graphene along the nanotube symmetry direction. Depending on the lattice symmetry of the tube, three different situations can be realized:

1. The armchair SWNT has a cross-section passing through the Fermi point (Figure 11.12, left). In this case the SWNT is metallic and the conduction band merges with the valence band (Figure 11.13, left).

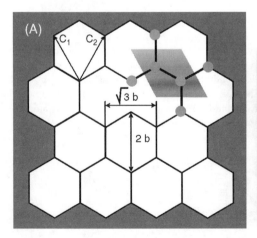
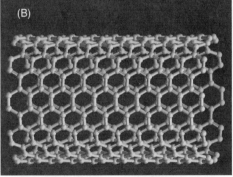

FIGURE 11.10   (A) Honeycomb lattice structure of graphene has a rhombic unit cell with two carbon atoms. Translated along basal vectors, c_1 and c_2, it forms two interconnected sublattices. The carbon–carbon bond length, b, is ~0.14 nm. The edge direction, in basal vectors, is denoted by two integers (shown fragment has left/right edge of type [2,2]). (B) Lattice structure of [10,10] armchair SWNT. Wrapping honeycomb lattice along some chosen axis will form a nanotube.

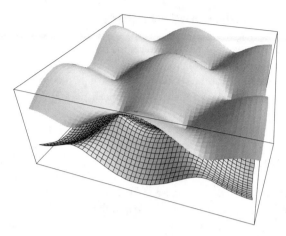

FIGURE 11.11   TBA electronic structure of valence bands of a monolayer of graphite. Threefold symmetry of the lattice results in six Fermi points where the conduction band meets the valence band.

FIGURE 11.12   Lowest conduction sub-band and highest valence sub-band of a metallic tube (left) compared with semiconductor nanotube (right). A nanotube quantization condition makes cut from the cone-shaped bands of graphite. In the case of metallic nanotube, this cross-section passes through the Fermi point and no gap develops between sub-bands. The electron dispersion is linear in longitudinal wave vector. In the case of semiconductor nanotube, the cross-section is shifted away from Fermi point. The carrier dispersion is a hyperbola.

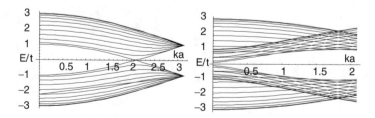

FIGURE 11.13   Electronic structure of a metallic armchair [10,10] nanotube (left) and a semiconductor zigzag [17,0] nanotube (right). The pi-electron energy is plotted in units of hopping integral, t~2.7 eV vs. dimensionless product of the longitudinal wave vector, k, and the bond length, a~0.14 nm (half of the Brilloine zone is shown).

2. The zigzag/chiral nanotube cross-section is distant from the Fermi point (Figure 11.12, right). This tube has a nonzero gap as shown in Figure 11.13, right, and it is a semiconductor tube.
3. One third of zigzag and chiral nanotubes have a very small gap, which follows from arguments other than simple band-folding. In our simplified picture these SWNTs will have a zero band gap and be metals.

The band gap of a semiconductor nanotube depends solely on the tube radius, R — a simple rule follows from the band-folding scheme: $E_g = t \, b/R$, here $t{\sim}2.7$ eV is the hopping integral for pi-electrons. This gap dependence was experimentally measured by Scanning Tunneling Spectroscopy[59] and resonance Raman spectroscopy.[60]

Nanotubes with conductivity ranging from metallic to semiconductor were indeed synthesized. The temperature dependence of the conductivity of the nanotube indicates reliably the metallic or semiconductor character. The field effect is also very useful to distinguish between two types. To measure this effect, the nanotube is placed between two electrodes on top of a back-gate contact that is covered with an insulating layer. After synthesis and purification, a SWNT is normally p-type, i.e., the majority of carries are holes. A typical density of ${\sim}10^7$ cm^{-1} holes defines the conductivity in the ON state of a SWNT when operated as a field-effect transistor at zero gate voltage. External positive voltage, applied to the back gate, can deplete the holes and switch the SWNT–FET into the OFF state. Experiments have shown a drop of 5 orders of magnitude of the source-drain current when the gate voltage was changed by 3 volts for very thin insulator layers (thickness < 2 nm).[61] In the case of metallic nanotube bridging two electrodes, only a weak dependence, if any, of the conductivity on the gate voltage is seen.

The electronic structure of SWNT is highly sensitive to external fields, and lattice distortions cause changes in the electronic structure. A lattice distortion moves the Fermi point of graphite and results in the closing/opening of an energy gap, a change in the electron density, and charging of the tube. This opens many possibilities for application of nanotubes to use as nano-biosensors, mesoscopic devices, and nanoelectromechanical systems.

In this section we focus on a particular application of carbon nanotubes — nanoelectromechanical (NEM) switches (Figure 11.14) and nanotweezers (Figure 11.15). The three basic energy domains that

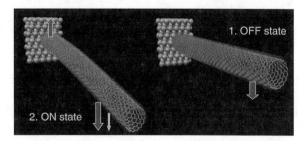

FIGURE 11.14 ON and OFF states of a nanotube electromechanical switch. Arrows show applied forces: electrostatic, van der Waals, and elastic forces.

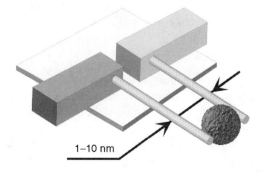

FIGURE 11.15 Nanotube nanotweezers device.

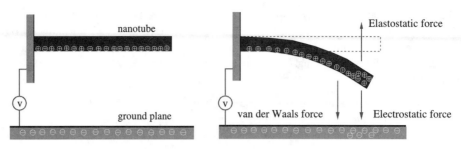

FIGURE 11.16 Force balance for a nanotube over a ground plane: (left) Position of the tube when V = 0; (right) deformed position of the tube when V/0.

describe the physical behavior of NEM switches — mechanics, electrostatics, and van der Waals — are described below.

11.4.1 Operation of Nanoelectromechanical Switches

Shown in Figure 11.16 is the nanoelectromechanical operation of a carbon nanotube-based cantilever switch. The key components are a moveable structure, which can be a single wall or a multiwall carbon nanotube, and a fixed ground plane, which is modeled by a graphite bulk. When a potential difference is created between the moveable structure and the ground plane, electrostatic charges are induced on both the movable structure and the ground plane. The electrostatic charges give rise to electrostatic forces, which deflect the movable tube. In addition to electrostatic forces, depending on the gap between the moveable tube and the ground plane, the van der Waals forces also act on the tube and deflect it. The directions of the electrostatic and van der Waals forces are shown in Figure 11.16. Counteracting the electrostatic and van der Waals forces are elastic forces, which try to restore the tube to its original straight position. For an applied voltage, an equilibrium position of the tube is defined by the balance of the elastic, electrostatic, and van der Waals forces. As the tube deflects, all forces are subject to change, and a self-consistent analysis is necessary to compute the equilibrium position of the tube.

When the potential difference between the tube and the ground plane exceeds a certain critical value, the deflection of the tube becomes unstable and the tube collapses onto the ground plane. The *potential*, which causes the tube to collapse, is defined as the *pull-in voltage* or the *collapse voltage*. When the pull-in voltage is applied, the tube comes in contact with the ground plane, and the device is said to be in the ON state (Figure 11.14.2). When the potential is released and the tube and the ground plane are separated, the device is said to be in the OFF state (Figure 11.14.1).

When compared with microelectromechanical switches, the operation of nanoelectromechanical switches is different because of the importance of the van der Waals forces, which can be neglected at the micrometer scale. The sticking of NEM devices becomes an increasing problem at the nanoscale and can limit the range of operability of NEMS. If the gap between the cantilever tube and the ground plane is very small, even without an applied voltage, the tube can collapse onto the ground plane because of the van der Waals forces. In addition, the separation of the tube from the ground plane after the contact becomes an issue as the van der Waals forces will tend to keep the tube and the ground plane together.

11.4.2 Nanotube Mechanics

Mechanical and structural properties of nanoscale systems have been studied both theoretically and experimentally over the last decade.[62] The strong correlation between the structure and electronic properties of a nanosystem requires a proper understanding of the nanomechanical and nanoelectromechanical behavior of nanotubes. Such studies can lead to new design tools for microscopy and characterization studies as well as development of highly sensitive detectors. The mechanical behavior of a small structure differs from that of a bulk structure. New phenomena such as super-low friction,[63] super-high stiffness,[64] and high cohesion at small distances[65] are encountered.

The mechanical behavior of nanotubes can be modeled either by simple continuum approaches or by more complex atomistic approaches based on molecular dynamics simulations. The elastic properties of pure single-wall and multiwall nanotubes were studied by, for example, Sanchez–Portal et al.[66] and Yakobson and Avouris.[67] Atomistic approaches have the advantage of capturing the mechanical behavior accurately; however, they require large computational resources. Continuum theories, when properly parameterized and calibrated, can be more efficient to understand the mechanical behavior of nanotubes. A simple continuum approach to model the mechanical behavior of nanoelectromechanical switches is based on the beam theory. The beam equation is given by:

$$EI\frac{\partial^4 r}{\partial x^4} = q$$

where r is the gap between the conductor and the ground plane, x is the position along the tube, q is the force per unit length acting normal to the beam, E is the Young's modulus, and I is the moment of inertia and, for nanotubes, can be estimated as:

$$I = \frac{\pi}{4}(R_{ext}^4 - R_{int}^4)$$

where R_{int} is the interior radius and R_{ext} is the exterior radius of the nanotube.

The beam theory can, however, suffer from several limitations. For very large loads, the stress concentration at the edges of the nanotubes may cause the tube to buckle and form kinks. In such cases, the deflection deviates from the beam theory locally. The buckling happens at a certain strain depending on the device geometry, the nanotube symmetry, and the load. If buckling is to be simulated, one can try advanced continuum theories such as a shell theory or a full elasticity theory.[68]

Many-body corrections to van der Waals interactions from semiclassical Casimir forces were calculated and applied in the continuum modeling of nanotube mechanics.[69] The basic analysis of a role of van der Waals terms in electromechanical systems has demonstrated its significance at the sub-nanometer scale.[70] A recent theory[71] of van der Waals interaction for shells of pure carbon is based on universal principles formulated in 1930s.[72] The new approach is based on the quantum electrodynamical description of the van der Waals/Casimir forces. A simple and effective model has been developed to estimate the many-body contribution due to collective modes (plasmons).[69] This contribution is believed to be a major portion of the total van der Waals energy because of the high oscillator strength of the plasmons. The theory reveals many-body terms that are specific for various low-dimensional graphite nanostructures and are not taken into account by standard one-body calculations within the dispersionless model by Lennard–Johns.[72] We have demonstrated the use of the model for several systems (shown in Figure 11.17): a double-wall nanotube (A), a nanotube on the surface (B), and a pair of single-wall tubes (C). A significant difference has been shown for the dependence of the van der Waals energy on distance, which is a consequence of our quantum correction.[69]

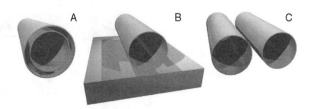

FIGURE 11.17 Geometry of nanotube systems for which a quantum correction to van der Waals forces has been calculated: (A) Double-wall nanotube; (B) single-wall nanotube on a surface; and (C) two single-wall nanotubes.

11.4.3   Electrostatics

The three-dimensional character of the electromagnetic eigenmodes and one-dimensional charge density distribution of a SWNT system result in a weak screening of the Coulomb interaction and the external field. We present a quantum mechanical calculation of the polarizability of the metallic [10,10] tube. The nanotube polarizability is not defined solely by the intrinsic properties of the tube.[73] It depends also on the geometry of the nanotube and closest gates/contacts. Hence, the charge distribution has to be treated self-consistently. Local perturbations of the electronic density will influence the entire system unlike in common semiconductor structures. For example, a point charge placed near the tube surface will generate an induced-charge density along the tube length, which decays very slowly with the distance from the external charge.

Figure 11.18 (after[73]) is a sketch of the depolarization of the tube potential (induced-charge density) by the side electrode and the back-gate (the right part of Figure 11.18 shows the geometry of the device simulated). The continuous line is the statistical approximation (Boltzmann–Poisson equations) which coincides well with the quantum mechanical result (dotted line) except for the quantum beating oscillations at the tube end. The depolarization manifests itself as a significant nonuniformity of the charge along the tube length. This effect is described by the self-consistent compact modeling, which is outlined below.

The potential ϕ^{act} that is induced by a charge density, ρ^{ind}, in one-dimensional systems is proportional to the charge density. Thus, for a degenerate electronic structure of a metallic nanotube in the low-temperature limit, the Poisson equation is effectively reduced to[73,74]

$$\rho^{ind}(z) = -e^2 v_M \phi^{act}(z).$$

Here v_M stands for the nanotube density of states, which is constant in a studied voltage range. We have demonstrated that $e^2 v_M$ acts as an atomistic capacitance of an SWNT:

$$C_A^{-1} = \frac{1}{e^2 v_M}$$

(a similar quantity for a two-dimensional electron gas system has been introduced by Luryi[75]) and the geometric capacitance:

$$C_m^{-1} = 2\log\left(\frac{2h}{R}\right)$$

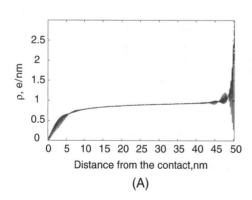
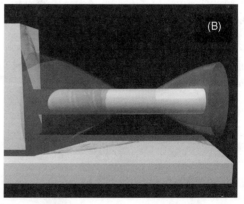

(A)

(B)

FIGURE 11.18 (A) Self-consistent charge density of a [10,10] armchair nanotube at 5 V voltage applied between side and back-gate contacts. (B) Sketch of the simulated device geometry. The distance between the tube center and the back gate is 5 nm, and the tube radius and length are 0.6 nm and 60 nm.

is a function of distance to the back-gate and SWNT radius. In case of the straight SWNT (as in Figure 11.18) the geometric capacitance is a logarithmic function of the distance between the tube and the gate. In equilibrium, we have the following relation between the equilibrium charge density and external potential (gate voltage), which comprises both the atomistic and geometric capacitances:

$$\rho_\infty = \left(-\frac{\varphi^{xt}}{C_m^{-1} + C_A^{-1}}\right) \simeq -\varphi^{xt} C_m\left(1 - \frac{C_m}{C_A}\right)$$

This equation is still valid for a nanotube of arbitrary shape although no simple expression for the geometric capacitance can be written.

11.4.4 Analytical Consideration for the Pull-In

We finish this section with an analytical model that can be used for a quick estimation of pull-in voltages of the nanotube system within continuum modeling. Assuming that the elastic energy of the NEMS device is given by

$$T = k(h - x)^2/2$$

and the external (electrostatic) force is the gradient of the energy component given by

$$V = C\varphi^2/2$$

we can calculate elastic and electrostatic forces. Then we include the van der Waals energy term:

$$W \simeq \varepsilon x^{-\alpha}$$

and write analytically the pull-in voltage and pull-in gap as functions of the device stiffness, k, the device capacitance, C, and van der Waals energy, W:

$$x_o = hA_1\frac{1}{2}\left(1 + \sqrt{1 + A_2\frac{W(x_o)}{kh^2}}\right)$$

$$V_o = B_1\frac{\sqrt{2kh}}{\sqrt{C(x_o)}}\sqrt{\frac{1}{2} - B_2\frac{W(x_o)}{kh^2} + \frac{1}{2}\sqrt{1 + A_2\frac{W(x_o)}{kh^2}}}$$

Here four constants A_1, B_1, A_2, and B_2 are describing the specific dependence of C and W on x, the dynamic gap or the internal coordinate of the NEMS device. In case of a planar switch and the Lennard–Jones potential, these constants are 3/2, $\sqrt{2/3}$, 36, and 36, respectively.

As a result of the van der Waals attraction to the gate, the NEMS device cannot operate at very small gaps, h. The critical gap, h_c, (at which $x_o = 0$) is about 2 nm for the switch with $k \sim W/1$ nm^2, and C $\sim$2 $k^{1/2}/(3$ V/nm). Next, Figures 11.19 and 11.20 show that, by neglecting the van der Waals correction to the pull-in gap, x_o, one underestimates the critical pull-in voltage by 15%.

The self-consistent solution for the pull-in gap is plotted in Figure 11.20. Again, neglecting the van der Waals terms results in an un-physical divergence of the pull-in gap when approaching the critical distance h_c.

11.4.5 Outlook

Development of fast and precise approaches for three-dimensional device modeling of nanotube systems becomes clearly important after recent successes of the IBM and Delft groups in creating prototypes for nanotube electronics.[61,76] The physics of carbon nanotube devices is rather distinct from the physics of

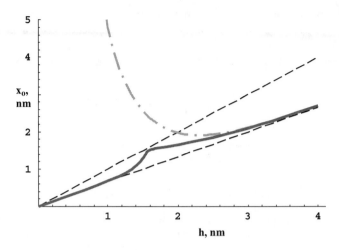

FIGURE 11.19 The pull-in gap as a function of the initial device gap. Solid curve represents the self-consistent analytical result. Dash-dotted curve shows the dependence in neglecting the van der Waals correction.

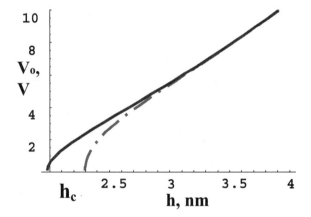

FIGURE 11.20 The pull-in voltage as a function of the gap. Solid curve represents the analytical result explained in the text. Dash-dotted curve shows the dependence in neglecting the van der Waals correction for the pull-in gap.

standard semiconductor devices, and it is unlikely that semiconductor device modeling tools can be simply transferred to nanotube device modeling.

Development of device modeling tools for nanotubes can be very complicated because of the breakdown of continuum theories. Molecular mechanics (MM) and molecular dynamics (MD) can be used reliably when continuum theories break down. However, both MM and MD can be very computer time-consuming. A good compromise is to develop a multiscale approach where continuum theories are combined with atomistic approaches. Multiscale methods can be accurate and more efficient compared with atomistic approaches. The highest level in the multiscale hierarchy is represented by the quantum mechanical result for the single-tube polarizability, which is the atomistic analog of the bulk dielectric function. It contains the complete information for the electronic structure and charge distribution and gives the means for calculating the screened Coulomb and the van der Waals/Casimir forces.[69] The main difficulty here is the requirement to solve the problem for device structures. The electronic structure and the polarizability change during device operation, and this requires a self-consistent treatment. At the intermediate level, classical molecular dynamics provides a detailed knowledge for geometry and material parameters of the system. This is a prerequisite for calculating the mechanical response of the system. It also supplies proper boundary conditions for electrostatic calculations through the actual device geometry. At the lowest level of the simulation hierarchy, the only level that can be used to simulate and understand larger systems of devices, continuum theories must and can be applied. The parameters of the continuum models will, of course, need to be derived from the higher level simulations.

Modeling and simulation of large-scale nanoscale circuits where carbon nanotubes are interconnected with other nanoelectronic, nanomechanical, chemical, and biological molecules are beyond the capability of currently existing supercomputers. Development of compact models for nanotubes and other nanodevices can enable the design of large-scale nanocircuits for breakthrough engineering applications.

11.5 Simulation of Ionic Channels

Nature has created many forms of nanostructures. Ion channels are of particular importance and have become accessible to the simulation methods that are widely used in computational electronics and for the nanostructures that have been described above. We therefore add this section to emphasize the importance of merging the understanding of biological (i.e., carbon-based) and silicon-based nanostructures.

Found in all life forms, ion channels are in a class of proteins that forms nanoscopic aqueous tunnels in the otherwise almost impermeable membranes of biological cells. An example of an ion channel, *ompF* porin, which resides in the outer membrane of the *E. coli* bacterium, is illustrated in Figure 11.21. Every ion channel consists of a chain of amino acids carrying a strong and rapidly varying permanent electric charge. By regulating the passive transport of ions across the cell membrane, ion channels maintain the correct internal ion composition that is crucial to cell survival and function. Ion channels directly control electrical signaling in the nervous system, muscle contraction, and the delivery of many clinical drugs.[77] Most channels have the ability to selectively transmit or block a particular ion species, and many exhibit switching properties similar to electronic devices. From a device point of view, ion channels can be viewed as transistors with unusual properties: exquisite sensitivity to specific environment factors, ability to self-assemble, and desirable properties for large-scale integration such as the infinite ON/OFF current ratio. By replacing or deleting one or more of the amino acids, many channels can be mutated, altering the charge distribution along the channel.[78] Engineering channels with specific conductances and selectivities are thus conceivable, as well as incorporating ion channels in the design of novel bio-devices.

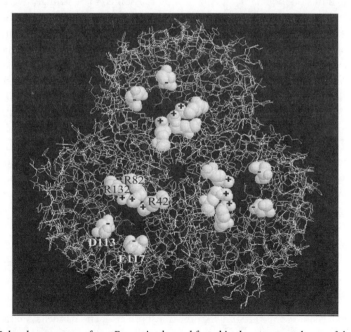

FIGURE 11.21 Molecular structure of *ompF*, a porin channel found in the outer membrane of the *E. coli* bacterium. This projection along the length of the channel shows the threefold symmetry of the trimer. Several ionized amino acids in the constriction region of each pore are highlighted.

Experimentally, the electrical and physiological properties of ion channels can be measured by inserting the channel into a lipid bilayer (membrane) and solvating the channel/membrane in an electrolyte solution. An electrochemical gradient is established across the membrane by immersing electrodes and using different concentrations of salt in the baths on either side of the membrane.

11.5.1 Hierarchical Approach to Modeling Ion Channels

Detailed simulation of ion transport in protein channels is very challenging because of the disparate spatial and temporal scales involved. A suitable model hierarchy is desirable to address different simulation needs. Continuum models, based on the drift-diffusion equations for charge flow, are the fastest approach; but they require large grids and extensive memory to resolve the three-dimensional channel geometry. Ion traversal of the channel is a very rare event on the usual time scale of devices, and the flow is actually a granular process. Continuum models, therefore, are useful mainly to probe the steady-state of the system. We suppose the system to be ergodic. At a given point in the simulation domain, the steady-state ion concentration represents the probability of ion occupation at that position, averaged over very long times or, equivalently, averaged over many identical channels at any given instant. Despite some limitations, continuum models can be parameterized to match current-voltage characteristics by specifying a suitable space and/or energy-dependent diffusion coefficient, which accounts for the ions' interactions with the local environment.

A step above in the hierarchy we find particle models, where the trajectories of individual ions are computed. The simpler model is based on a Brownian Dynamics description of ion flow, in which ion trajectories evolve according to the Langevin Equation. Ions move in the local electric field, calculated from all the charges in the system as well as any externally applied fields. The energy dissipated via ion-water scattering is modeled by including a simple frictional term in the equation of motion, while the randomizing effect of the scattering is accounted for by including a zero-mean Gaussian noise term.[79] Ionic core repulsion can also be included by adding a suitable repulsive term (e.g., Lennard–Jones) to the total force acting on the ion. When the latter is neglected, the simulation is equivalent to a discrete version of the drift-diffusion model.[80] If the ion motion is assumed to be strongly overdamped, relatively long time steps can be used (e.g., picoseconds), making this a very practical approach.

At the next level in the hierarchy are particle models, where the ion flow is resolved with a self-consistent transient, following Monte Carlo or MD approaches, as they are known in semiconductor device simulation. MD simulations resolve the motion and forces among all particles, both free (ions and water molecules) and bound (e.g., protein atoms) in the system. Bound particles are modeled as charged balls connected by springs (chemical bonds). The entire system is brought to a simulated experimental temperature and then equilibrated by allowing the system to evolve according to Newtonian mechanics.[81] While this methodology is the most complete, due to the extreme computational costs involved, it can only be applied today to very small systems on very short time scales of simulation. Monte Carlo methods, originally developed for semiconductor device simulation, provide a more practical compromise. Water and protein are treated as a background dielectric medium, as is done with Brownian Dynamics, and only the individual ion trajectories are resolved. The key difference between Brownian Dynamics and Monte Carlo techniques lies in the way the ion dynamics are handled. In Monte Carlo models the ion trajectories evolve according to Newtonian mechanics; but individual ion-water collision interactions are replaced with an appropriate scattering model, which is resolved on the natural time scales of the problem.[82] In the limit of high friction, both approaches should give the same result.

11.5.2 Drift-Diffusion Models

Drift-diffusion models are useful for studying ion transport in open-channel systems over time scales that cannot be resolved practically by detailed particle models. Water, protein, and membrane are treated as uniform background media with specific dielectric constants; and the macroscopic ion current in the water is resolved by assigning an appropriate space or energy-dependent mobility and diffusion coefficient

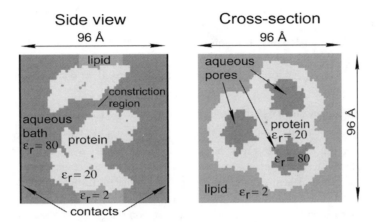

FIGURE 11.22 Mesh representation of the *ompF* trimer *in situ* in a membrane, immersed in a solution of potassium chloride — longitudinal and cross-sectional slices through the three-dimensional computational domain generated on a uniform rectilinear grid (1.5Å spacing). Electrodes immersed in the baths maintain a fixed bias across the channel/membrane system.

to each ionic species. The solution of Poisson's equation over the entire domain provides a simple way to include external boundary conditions and image force effects at dielectric discontinuities. Complete three-dimensional models of flow in ionic channels can be implemented with the established tools of semiconductor device simulation.

In order to define the various regions of the computational domain, the molecular structure of the protein must be mapped onto a grid. Protein structures are known with atomic resolution for a number of important channels, but considerable processing is still necessary to determine the charge and the dielectric permittivity distribution corresponding to the individual molecular components. With this information, one can assemble a grid defining the boundaries between water and protein, as illustrated in Figure 11.22. The current density $j_\pm$ arising from the flow of ions down the electrochemical gradient in the aqueous region of the domain is given by the drift-diffusion equation:

$$\vec{j}_\pm = -(\mu_\pm \rho_\pm \vec{\nabla}\varphi - D_\pm \vec{\nabla}\rho_\pm)$$

where $\rho_\pm$ are the ionic charge densities and $\mu_\pm$ and $D_\pm$ are, respectively, the mobilities and diffusion coefficients of each ionic species. For the purposes of this discussion, we restrict ourselves to systems with only two ionic species of opposite charge, but the same treatment can be extended to allow for multiple ionic species by including a drift-diffusion equation for each additional species. Conservation of charge is enforced by a continuity equation for each species, given by

$$\vec{\nabla} \cdot \vec{j}_\pm + \frac{\partial \rho_\pm}{\partial t} = S_\pm$$

The term $S_\pm$ is set to zero for simple transport simulation, but it can be set to any functional form to describe higher order effects, such as the details of ion binding and other chemical phenomena that populate or deplete the ion densities. The electrostatic potential φ is described by Poisson's Equation:

$$\vec{\nabla} \cdot (\varepsilon \vec{\nabla}\varphi) = -(\rho_{fixed} + \rho_+ + \rho_-)$$

where ρ_{fixed} represents the density of fixed charge residing within and on the surface of the protein. When solved simultaneously, this system of coupled equations provides a self-consistent description of ion flow in the channel. The equations are discretized on the grid and solved iteratively for steady-state conditions, subject to specific boundary conditions for applied potential and for ionic solution concentrations in the baths at the ends of the channel. In a typical semiconductor device, the mobile charge in the contacts is originated by fixed ionized dopants. In an ion-channel system, the salt concentration in the electrolyte far from the protein determines the density of mobile ionic charges, which, from an electrical point of view, behave similarly to the intrinsic electron/hole concentrations in an undoped semiconductor at a given temperature.

11.5.2.1 Application of the Drift-Diffusion Model to Real Ion Channels

Complete three-dimensional drift-diffusion models have been implemented using the computational platform PROPHET[83] and used to study transport in ion channels like gramicidin and porin, for which detailed structure and conductivity measurements are available. Porin in particular presents a very challenging problem because, as shown in Figure 11.21, the channel is a *trimer* consisting of three identical parallel channels, connected through a common anti-chamber region. Memory requirements for continuum simulations of porin are currently at the limit of available workstation resources; however, simulations are now performed routinely on distributed shared memory machines. Figure 11.23 compares the current–voltage curves computed with a three-dimensional drift-diffusion simulation with those measured experimentally.[84] These results were generated in approximately 8 hours on an SGI origin2000.

11.5.3 Monte Carlo Simulations

The Monte Carlo simulation technique, as it is known in the tradition of semiconductor device simulation, can be coupled with a particle-mesh model to provide a self-consistent, time-resolved picture of ion dynamics in a channel system.[82,85] The starting point is the grid, which defines the regions accessible to ions as well as the dielectric topography of the system. In reality the boundaries between aqueous, protein, and membrane regions are not static but move over atomic length scales due to the thermal fluctuations of the atoms of the protein. Such fluctuations, which are resolved in MD simulations, are ignored in Monte Carlo simulations (although in principle they could be included).

11.5.3.1 Resolving Single-Ion Dynamics

Ions are distributed throughout the aqueous region according to a given initial concentration profile. The charge of each mobile ion, and of each static charge within the protein, is interpolated to the grid using a prescribed weighting scheme to construct a charge density at the discrete grid points. The

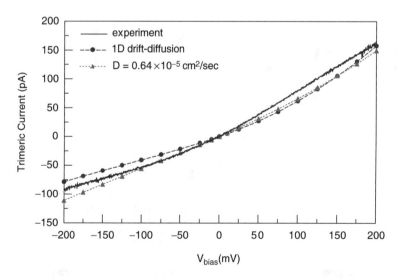

FIGURE 11.23 Comparison of measured and computed current–voltage curves for *ompF* in 100 mM potassium chloride, assuming a spatially dependent diffusion coefficient and a spatially uniform diffusion coefficient.

electrostatic field due to the charge density distribution, as well as any externally applied field, is found by solving Poisson's equation on the grid. The field at the grid points is interpolated back to the ion positions and used to move the ions forward in time by integrating Newton's Second Law over small time steps. At the end of the time step, the new ion positions are used to recalculate a new charge density distribution and hence a new field to advance the ions over the next timestep. This cycle is iterated either until a steady-state is reached or until quantities of interest (e.g., diffusion coefficient) have been calculated. The effects of ion volume can also be incorporated by including an ionic core repulsive term in the force, acting on each ion as is done in Brownian Dynamics.

11.5.3.2 Modeling Ion-Water Interactions

Ion motion is treated as a sequence of free flights interrupted by collisions with water molecules, which are modeled by assuming a particular ion-water scattering rate $v(E_{ion}(t))$, generally a function of ion energy. The scattering rate represents the average number of collisions per unit time that an ion would experience if it maintained a constant energy. The probability for an ion to travel for a time t without scattering is given by

$$P(t) = \exp\left(-\int_0^t v(E_{ion}(t'))dt'\right)$$

The probability density function (probability per unit time) for a flight to have duration t is given by

$$p(t) = v(E_{ion}(t))P(t)$$

Ion flight times can be randomly selected from the probability density function by integrating the latter over the (unknown) flight time T_f and equating the integral to a uniformly distributed random number r on the unit interval. Thus,

$$-\log(r) = \int_0^{T_f} v(E_{ion}(t))dt$$

The integral on the right-hand side is trivial only for constant scattering rates, but in general it cannot be performed analytically. A number of methods have been introduced to solve the integral; an extended discussion is given at the Internet location given in Reference 82.

11.6 Conclusions

The combination of a three-dimensional drift-diffusion and three-dimensional Monte Carlo approach provides the essential hierarchy for looking at biological systems from the point of view of device-like applications. There are, however, significant differences between solid-state devices and biological systems, which require different choices in the definition of a Monte Carlo simulation strategy. In a typical device, the ensemble must include many thousands of particles; but a reasonable steady-state is reached after several picoseconds of simulation (on the order of ten to twenty thousand time steps). In a practical simulation domain for a biological channel, only a very small number of ions is present in the system; but because the ion traversal of the channel is a rare event, measurable current levels can only be established by extending the simulation to the millisecond range. Because the number of time steps required to resolve ion dynamics is typically on the order of tens of femtoseconds, this would require a number of iteration steps on the order of 10^{12}, which is still extremely expensive. For a fully self-consistent simulation, the solution of Poisson's Equation in three-dimensions presents the real bottleneck, while the computational cost of resolving the few particle trajectories is minimal. Alternative schemes for evaluating the electrostatic potential self-consistently include precalculating the potential for various ion

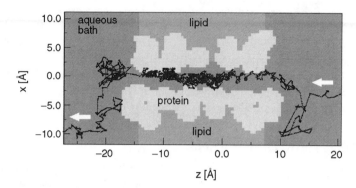

FIGURE 11.24 Geometric representation of the gramicidin channel as used in three-dimensional Monte Carlo transport model. The successful trajectory of a single sodium ion traversing the channel is shown.

pair configurations, storing the results in look-up tables, and employing the superposition principle to reconstruct the potential at the desired point by interpolating between table entries.[86] A prototype Monte Carlo simulation of sodium chloride transport in the gramicidin channel has been successfully implemented,[87] adapting the grid developed for the continuum simulations, as shown in Figure 11.24. For this simulation Poisson's Equation is solved approximately every 10 time steps using an accurate conjugate gradient method.

Acknowledgments

N. R. Aluru is thankful to M. Dequesnes for technical assistance and allowing the use of some of his results. J.-P. Leburton is indebted to G. Austing, R. M. Martin, S. Nagaraja, and S. Tarucha for fruitful discussions, and to P. Matagne for technical assistance. U. Ravaioli and T. van der Straaten are indebted to Bob Eisenberg (Rush Medical College) and Eric Jakobsson (University of Illinois) for introducing them to the subject of ion channels and for many useful discussions; to R. W. Dutton, Z. Yu, and D. Yergeau (Stanford University) for assistance with PROPHET; and T. Schirmer (University of Basel) for providing molecular structures for the porin channel. S. V. Rotkin is indebted to L. Rotkina and I. Zharov for fruitful discussions and to K. A. Bulashevich for technical support. N. R. Aluru, J.-P. Leburton, and S. V. Rotkin acknowledge the support of the CRI grant of UIUC. The work of J.-P. Leburton was supported by NSF grants DESCARTES ECS-98-02730, the Materials Computational Center (DMR-99–76550) and the DARPA-QUIST program (DAAD19–01–1–0659). The work of U. Ravaioli and T. van der Straaten was funded in part by NSF Distributed Center for Advanced Electronics Simulation (DesCArtES) grant ECS 98–02730, DARPA contract F30 602-012-0513 (B.E.), and NSF KDI grant to the University of Illinois. S. V. Rotkin acknowledges support of DoE grant DE-FG02-01ER45932 and the Beckman Fellowship from the Arnold and Mabel Beckman Foundation. For his calculations, B. R. Tuttle primarily utilized the SGI-ORIGIN2000 machines at the National Center for Supercomputing Applications in Urbana, IL. K. Hess acknowledges the Army Research Office (DAAG55-98-1-03306) and the Office of Naval Research (NO0014-98-1-0604) and ONR–MURI.

References

1. Hess, K., *Advanced Theory of Semiconductor Devices*, IEEE Press, 2000.
2. Hu, C., Tam, S.C., Hsu, H., Ko, P., Chan, T., and Terrill, K.W., *IEEE Trans. Electron. Devices* ED-32, 375, 1985.
3. Tuttle, B.R., Hydrogen and PB defects at the Si(111)-SiO2 interface: an *ab initio* cluster study, *Phys. Rev. B* 60, 2631, 1999.
4. Tuttle, B.R. and Van de Walle, C., Structure, energetics and vibrational properties of Si-H bond dissociation in silicon, *Phys. Rev. B* 59, 12884, 1999.

5. Hess, K., Tuttle, B.R., Register, L.F., and Ferry, D., Magnitude of the threshold energy for hot-electron damage in metal oxide semiconductor field-effect transistors by hydrogen desorption, *Appl. Phys. Lett.* 75, 3147, 1999.

6. Tuttle, B.R., Register, L.F., and Hess, K., Hydrogen related defect creation at the Si-SiO2-Si interface of metal-oxide-semiconductor field-effect transistors during hot electron stress, *Superlattices Microstruct.* 27, 441, 2000.

7. Tuttle, B.R., McMahon, W., and Hess, K., Hydrogen and hot electron defect creation at the Si(100)-SiO2 interface of metal-oxide-semiconductor field effect transistors, *Superlattices Microstruct.* 27, 229, 2000.

8. Tuttle, B.R., Energetics and diffusion of hydrogen in SiO2, *Phys. Rev. B* 61, 4417, 2000.

9. Hess, K., Register, L.F., McMahon, W., Tuttle, B.R., Ajtas, O., Ravaioli, U., Lyding, J., and Kizilyalli, I.C., Channel hot carrier degradation in MOSFETs, *Physica B* 272, 527, 1999.

10. Hess, K., Haggag, A., McMahon, W., Cheng, K., Lee, J., and Lyding, J., The physics of determining chip reliability, *IEEE Circuits Device* 17 (3), 33–38, 2001.

11. Van de Walle, C. and Tuttle, B.R., Microscopic theory of hydrogen in silicon devices, *IEEE Trans. Electron. Devices* 47, 1779, 2000.

12. Haggag, A., McMahon, W., Hess, K., Cheng, K., Lee, J., and Lyding, J., *IEEE Intl. Rel. Phys. Symp. Proc.*, 271, 2001.

13. Staedele, M., Tuttle, B.R., and Hess, K., Tunneling through ultrathin SiO$_2$ gate oxide from microscopic models, *J. Appl. Phys.* 89, 348, 2002.

14. Staedele, M., Fischer, B., Tuttle, B.R., and Hess, K., Influence of defects on elastic gate tunneling currents through ultrathin SiO2 gate oxides: predictions from microscopic models, *Superlattices Microstruct.* 28, 517, 2000.

15. Staedele, M., Tuttle, B.R., and Hess, K., Tight-binding investigation of tunneling in thin oxides, *Superlattices Microstruct.* 27, 405, 2000.

16. Staedele, M., Tuttle, B.R., Fischer, B., and Hess, K., Tunneling through ultrathin thin oxides — new insights from microscopic calculations, *Intl. J. Comp. Electr.*, 2002.

17. Staedele, M., Fischer, B., Tuttle, B.R., and Hess, K., Resonant electron tunneling through defects in ultrathin SiO2 gate oxides in MOSFETs, *Solid State Electronics*, 46, 1027–1032 (2001).

18. Klimeck, G., Bowen, R.C., and Boykin, T., Off-zone-center or indirect band-gap-like hole transport in heterostructures, *Phys. Rev. B* 63, 195310, 2001.

19. DiVentra, M. and Pantelides, S., Hellmann–Feynman theorem and the definition of forces in quantum time-dependent and transport problems, *Phys. Rev. B* 61, 16207, 2000.

20. Damle, P.S., Ghosh, A.W., and Datta, S., Unified description of molecular conduction: from molecules to metallic wires, *Phys. Rev. B* 64, 16207, 2001.

21. Kasner, M.A., *Phys. Today* 46, 325, 1993.

22. Averin, D.V. and Likharev, K.K., in Lee, P.A., and Webb, R.A. (Eds.), *Mesoscopic Phenomena in Solids*, Elsevier, Amsterdam, 1991, pp. 173–271.

23. Devoret, M.H. and Grabert, H., in Devoret, M.H. and Grabert, H. (Eds.), *Single Charge Tunneling: Coulomb Blockade Phenomena in Nanostructures*, Plenum Press, 1991, pp. 1–19.

24. Likharev, K.K., *Proc. IEEE* 87, 606, 1999.

25. Meirav, V. and Foxman, E.B., *Semicond. Sci. Technol.* 10, 255, 1995.

26. Ashoori, R.C., Stoermer, H.L., Weiner, J.S., Pfeiffer, L.N., Pearton, S.J., Baldwin, K.W., and West, K.W., *Phys. Rev. Lett.* 68, 3088, 1992.

27. Ashoori, R., *Nature* 379, 413, 1996.

28. Tarucha, S., Austing, D.G., Honda, T., van der Hage, R.J., and Kouwenhoven, L.P., *Phys. Rev. Lett.* 77, 3613, 1996.

29. Austing, D.G., Honda, T., and Tarucha, S., *Jpn. J. Appl. Phys.* 36, 4151, 1997.

30. Sze, S.M., *Physics of Semiconductor Devices*, 2nd ed., John Wiley & Sons, New York, 1981.

31. Vincenzo, D.P.D., *Nature* 393, 113, 1998.

32. Arakawa, Y. and Yariv, A., *IEEE J. Quant. Electron.* 22, 1887, 1986.

33. Bimberg, D., Grundmann, M., and Ledentsov, N.N., *Quantum Dot Heterostructures*, Wiley, London, 1998.

34. Johnson, N.F., *J. Phys. Condens. Matter* 7, 965, 1995.

35. Maccucci, M., Hess, K., and Iafrate, G.J., *J. Appl. Phys.* 77, 3267, 1995.

36. Nagaraja, S., Matagne, P., Thean, V.Y., Leburton, J.-P., Kim, Y.-H., and Martin, R.M., *Phys. Rev. B* 56, 15752, 1997.

37. Leonard, D., Pond, K., and Petroff, P.M., *Phys. Rev. B* 50, 11687, 1994.

38. Miller, M.S., Malm, J.O., Pistol, M.E., Jeppesen, S., Kowalski, B., Georgsson, K., and Samuelson, L., *Appl. Phys. Lett.* 80, 3360, 1996.

39. Madelung, O., *Introduction to Solid State Physics*, Springer Verlag, Berlin, 1978.

40. Jones, R.O. and Gunnarson, O., *Rev. Mod. Phys.* 61, 689, 1989.

41. Kumar, A., Laux, S.E., and Stern, F., *Phys. Rev. B* 42, 5166, 1990.

42. Stopa, M., *Phys. Rev. B* 54, 13767, 1996.

43. Koskinen, M., Manninen, M., and Rieman, S.M., *Phys. Rev. Lett.* 79, 1389, 1997.

44. Jovanovic, D. and Leburton, J.-P., *Phys. Rev. B* 49, 7474, 1994.

45. Lee, I.H., Rao, V., Martin, R.M., and Leburton, J.-P., *Phys. Rev. B* 57, 9035, 1998.

46. Nagaraja, S., Leburton, J.-P., and Martin, R.M., *Phys. Rev. B* 60, 8759, 1999.

47. Perdew, J.P. and Zunger, A., *Phys. Rev. B* 23, 5048, 1981.

48. Slater, J.C., *Quantum Theory of Molecules and Solids*, McGraw-Hill, New York, 1963.

49. Matagne, P., Leburton, J.-P., Austing, D.G., and Tarucha, S., to be published.

50. Blick, R.H., Haug, R.J., Weis, J., Pfannkuche, D., Klitzing, K., and Eberl, K., *Phys. Rev. B* 53, 7899, 1996.

51. Kouwenhoven, L.P., *Science* 268, 1440, 1995.

52. Zener, C., *Phys. Rev.* 81, 440, 1951.

53. Wensauer, A., Steffens, O., Suhrke, M., and Roessler, U., *Phys. Rev. B* 62, 2605, 2000.

54. Imamaura, H., Maksym, P.A., and Aoki, H., *Phys. Rev. B* 59, 5817, 1999.

55. Rotani, M., Rossi, F., Manghi, F., and Molinari, E., *Solid State Commun.* 112, 151, 1999.

56. Bender, C.F. and Davidson, E.R., *J. Chem. Phys.* 46, 3313, 1967.

57. Loss, D. and Di Vincenzo, D.P., *Phys. Rev. A* 57, 1998.

58. Dresselhaus, M.S., Dresselhaus, G., and Eklund, P.C., *Science of Fullerenes and Carbon Nanotubes*, Academic Press, 1996.

59. Wildoer, J.W.G., Venema, L.C., Rinzler, A.G., Smalley, R.E., and Dekker, C., Electronic structure of atomically resolved carbon nanotubes, *Nature* 391 (6662), 59–62, 1998.

60. Kuzmany, H., Plank, W., Hulman, M., Kramberger, C., Gruneis, A., Pichler, T., Peterlik, H., Kataura, H., and Achiba, Y., Determination of SWCNT diameters from the Raman response of the radial breathing mode, *Eur. Phys. J. B* 22 (3), 307–320, 2001.

61. Bachtold, A., Hadley, P., Nakanishi, T., and Dekker, C., Logic circuits with carbon nanotube transistors, *Science* 294 (5545), 1317–1320, 2001.

62. Ebbesen, T.W., Potential applications of nanotubes, in *Carbon Nanotubes*, CRC Press, Boca Raton, FL, 1997, pp. 296.

63. Falvo, M.R., Steele, J., Taylor, R.M., and Superfine, R., Gearlike rolling motion mediated by commensurate contact: carbon nanotubes on HOPG, *Phys. Rev. B* 62 (16), R10665–R10667, 2000.

64. Yu, M.F., Files, B.S., Arepalli, S., and Ruoff, R.S., Tensile loading of ropes of single wall carbon nanotubes and their mechanical properties, *Phy. Rev. Lett.* 84 (24), 5552–5555, 2000.

65. Hertel, T., Walkup, R.E., and Avouris, P., Deformation of carbon nanotubes by surface van der Waals forces, *Phys. Rev. B-Condensed Matter* 58 (20), 13870–13873, 1998.

66. Sanchez–Portal, D., Artacho, E., Solar, J. M., Rubio, A., and Ordejon, P., *Ab initio* structural, elastic, and vibrational properties of carbon nanotubes, *Phys. Rev. B-Condensed Matter* 59 (19), 12678–12688, 1999.

67. Yakobson, B.I. and Avouris, P., Mechanical properties of carbon nanotubes, *Carbon Nanotubes: Synthesis, Structure, Properties, Applications* 80, 287–327, 2001.

68. Yakobson, B.I., Brabec, C.J., and Bernholc, J., Nanomechanics of carbon tubes — instabilities beyond linear response, *Phys. Rev. Lett.* 76 (14), 2511–2514, 1996.

69. Rotkin, S.V. and Hess, K., Many-body terms in van der Waals cohesion energy of nanotubules, *J. Comp. Electr.*, 1, 294–297 (2002).

70. Dequesnes, M., Rotkin, S.V., and Aluru, N.R., Calculation of pull-in voltages for carbon nanotube-based nanoelectromechanical switches, *Nanotechnology* 13, 120–131, 2002.

71. Girifalco, L.A., Hodak, M., and Lee, R.S., Carbon nanotubes, buckyballs, ropes, and a universal graphitic potential, *Phys. Rev. B-Condensed Matter* 62 (19), 13104–10, 2000.

72. Lennard–Jones, J.E., Perturbation problems in quantum mechanics, *Proc. R. Soc London, Ser. A* 129, 598–615, 1930.

73. Rotkin, S.V., Bulashevich, K.A., and Aluru, N.R., Atomistic models for nanotube device electrostatics, in *ECS Centennial Meet*, ECS, Philadelphia, 2002, p. V5–1164.

74. Odintsov, A.A. and Tokura, Y., Contact phenomena and Mott transition in carbon nanotubes, *J. Low Temp. Phys.* 118 (5–6), 509–518, 2000.

75. Luryi, S., Quantum capacitance devices, *Appl. Phys. Lett.* 52 (6), 501–503, 1988.

76. Derycke, V., Martel, R., Appenzeller, J., and Avouris, P., Carbon nanotube inter- and intramolecular logic gates, *Nano Lett.* 1 (9), 453–456, 2001.

77. Hille, B., *Ionic Channels of Excitable Membranes*, Sinauer Associates, Massachusetts, 1992.

78. Phale, P.S., Philippsen, A., Widmer, C., Phale, V.P., Rosenbusch, J.P., and Schirmer, T., Role of charged residues at the ompF porin channel constriction probed by mutagenesis and simulation, *Biochemistry* 40, 6319–6325, 2001.

79. Reif, F., *Fundamentals of Statistical and Thermal Physics*, McGraw-Hill, Singapore, 1987.

80. Schuss, Z., Nadler, B., and Eisenberg, R.S., Derivation of Poisson and Nernst-Planck equations in a bath and channel from a molecular model, *Phys. Rev. E* 64, 036116-(1–14), 2001.

81. Allen, M.P. and Tildesley, D.J., *Computer Simulation of Liquids*, Clarendon, Oxford, 1987.

82. http://www.ncce.ce.g.uiuc.edu/ncce.htm

83. http://www.tcad.stanford.edu/

84. van der Straaten, T., Varma, S., Chiu, S.-W., Tang, J., Aluru, N., Eisenberg, R., Ravaioli, U., and Jakobsson, E., Combining computational chemistry and computational electronics to understand protein ion channels, in *The 2002 Intl. Conf. Computational Nanosci. Nanotechnol.*, 2002.

85. Hockney, R.W. and Eastwood, J.W., *Computer Simulation Using Particles*, McGraw-Hill, 1981.

86. Chung, S.-H., Allen, T.W., Hoyles, M., and Kuyucak, S., Permeation of ions across the potassium channel: Brownian dynamics studies, *Biophys. J.* 77, 2517–2533, 1999.

87. van der Straaten, T. and Ravaioli, U., Self-consistent Monte-Carlo/P3M simulation of ion transport in the gramicidin ion channel, unpublished.

12

Resistance of a Molecule

CONTENTS

Magnus Paulsson
Purdue University

Ferdows Zahid
Purdue University

Supriyo Datta
Purdue University

12.1 Introduction

In recent years, several experimental groups have reported measurements of the current–voltage (I-V) characteristics of individual or small numbers of molecules. Even three-terminal measurements showing evidence of transistor action have been reported using carbon nanotubes[1,2] as well as self-assembled monolayers of conjugated polymers.[3] These developments have attracted much attention from the semiconductor industry, and there is great interest from an applied point of view to model and understand the capabilities of molecular conductors. At the same time, this is also a topic of great interest from the point of view of basic physics. A molecule represents a quantum dot, at least an order of magnitude smaller than semiconductor quantum dots, which allows us to study many of the same mesoscopic and/ or many-body effects at far higher temperatures.

So what is the resistance of a molecule? More specifically, what do we see when we connect a short molecule between two metallic contacts as shown in Figure 12.1 and measure the current (I) as a function of the voltage (V)? Most commonly we get I-V characteristics of the type sketched in Figure 12.2. This has been observed using many different approaches including breakjunctions,[4–8] scanning probes,[9–12] nanopores,[13] and a host of other methods (see, for example, Reference 14). A number of theoretical models have been developed for calculating the I-V characteristics of molecular wires using semi-empirical[12,15–18] as well as first-principles[19–25] theory.

Our purpose in this chapter is to provide an intuitive explanation for the observed I-V characteristics using simple models to illustrate the basic physics. However, it should be noted that molecular electronics is a rapidly developing field, and much of the excitement arises from the possibility of discovering novel physics beyond the paradigms discussed here. To cite a simple example, very few experiments to date[3] incorporate the gate electrode shown in Figure 12.1, and we will largely ignore the gate in this chapter.

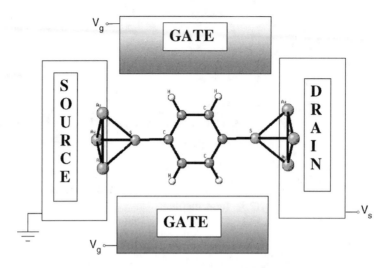

FIGURE 12.1 Conceptual picture of a "molecular transistor" showing a short molecule (Phenyl dithiol, PDT) sandwiched between source and drain contacts. Most experiments so far lack good contacts and do not incorporate the gate electrodes.

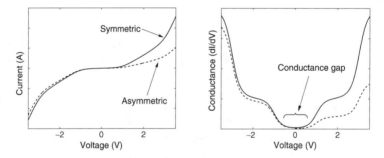

FIGURE 12.2 Schematic picture, showing general properties of measured current–voltage (I-V) and conductance (G-V) characteristics for molecular wires. Solid line, symmetrical I-V. Dashed line, asymmetrical I-V.

However, the gate electrode can play a significant role in shaping the I-V characteristics and deserves more attention. This is easily appreciated by looking at the applied potential profile U_{app} generated by the electrodes in the absence of the molecule. This potential profile satisfies the Laplace equation without any net charge anywhere and is obtained by solving:

$$\nabla \cdot (\varepsilon \nabla U_{app}) = 0 \tag{12.1}$$

subject to the appropriate boundary values on the electrodes (Figure 12.3). It is apparent that the electrode geometry has a significant influence on the potential profile that it imposes on the molecular species, and this could obviously affect the I-V characteristics in a significant way. After all, it is well known that a three-terminal metal/oxide/semiconductor field-effect transistor (MOSFET) with a gate electrode has a very different I-V characteristic compared with a two terminal n-i-n diode. The current in a MOSFET saturates under increasing bias, but the current in an n-i-n diode keeps increasing indefinitely. In contrast to the MOSFET, whose I-V is largely dominated by classical electrostatics, the I-V characteristics of molecules is determined by a more interesting interplay between nineteenth-century physics (electrostatics) and twentieth-century physics (quantum transport); and it is important to do justice to both aspects.

We will start in Section 12.2 with a qualitative discussion of the main factors affecting the I-V characteristics of molecular conductors, using a simple toy model to illustrate their role. However, this toy model misses two important factors: (1) shift in the energy level due to *charging effects* as the molecule

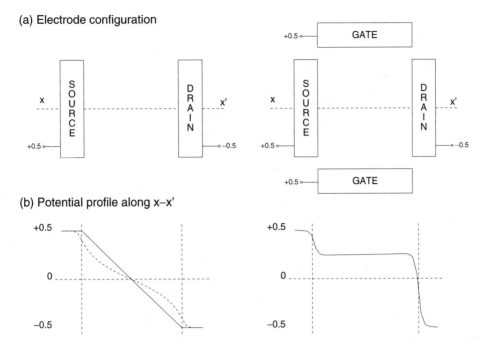

FIGURE 12.3 Schematic picture, showing potential profile for two geometries, without gate (left) and with gate (right).

loses or gains electrons and (2) *broadening* of the energy levels due to their finite lifetime arising from the coupling (Γ_1 and Γ_2) to the two contacts. Once we incorporate these effects (Section 12.3) we obtain more realistic I-V plots, even though the toy model assumes that conduction takes place independently through individual molecular levels. In general, however, multiple energy levels are simultaneously involved in the conduction process. In Section 12.4 we will describe the nonequilibrium Green's function (NEGF) formalism, which can be viewed as a generalized version of the one-level model to include multiple levels or conduction channels. This formalism provides a convenient framework for describing quantum transport[26] and can be used in conjunction with *ab initio* or semi-empirical Hamiltonians as described in a set of related articles.[27,28] Then in Section 12.5 we will illustrate the NEGF formalism with a simple semi-empirical model for a gold wire, *n* atoms long and one atom in cross-section. We could call this a Au$_n$ molecule, though that is not how one normally thinks of a gold wire. However, this example is particularly instructive because it shows the lowest possible *resistance of a molecule* per channel, which is $\pi\hbar/e^2 = 12.9$ kΩ.[29]

12.2   Qualitative Discussion

12.2.1   Where Is the Fermi Energy?

Energy-Level Diagram: The first step in understanding the current (I) vs. voltage (V) curve for a molecular conductor is to draw an energy-level diagram and locate the Fermi energy. Consider first a molecule sandwiched between two metallic contacts but with very weak electronic coupling. We could then line up the energy levels as shown in Figure 12.4 using the metallic work function (WF), the electronic affinity (*EA*), and ionization potential (*IP*) of the molecule. For example, a (111) gold surface has a work function of ~ 5.3 eV, while the electron affinity and ionization potential, EA_0 and IP_0, for isolated phenyl dithiol (Figure 12.1) in the gas phase have been reported to be ~ 2.4 eV and 8.3 eV, respectively.[30] These values are associated with electron emission and injection to and from a vacuum and may need some modification to account for the metallic contacts. For example, the actual *EA, IP* will possibly be modified from EA_0, IP_0 due to the image potential W_{im} associated with the metallic contacts:[31]

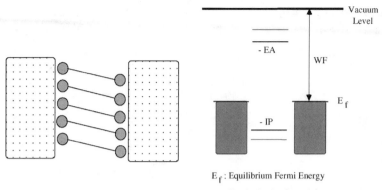

E_f : Equilibrium Fermi Energy

IP : Ionization Potential

EA : Electron Affinity

FIGURE 12.4 Equilibrium energy level diagram for a metal–molecule–metal sandwich for a weakly coupled molecule.

$$EA = EA_0 + W_{im} \tag{12.2}$$

$$IP = IP_0 - W_{im} \tag{12.3}$$

The probability of the molecule losing an electron to form a positive ion is equal to $e^{(WF-IP)/k_BT}$, while the probability of the molecule gaining an electron to form a negative ion is equal to $e^{(EA-WF)/k_BT}$. We thus expect the molecule to remain neutral as long as both $(IP - WF)$ and $(WF - EA)$ are much larger than k_BT, a condition that is usually satisfied for most metal–molecule combinations. Because it costs too much energy to transfer one electron into or out of the molecule, it prefers to remain neutral in equilibrium.

The picture changes qualitatively if the molecule is chemisorbed directly on the metallic contact (Figure 12.5). The molecular energy levels are now broadened significantly by the strong hybridization with the delocalized metallic wave functions, making it possible to transfer fractional amounts of charge to or from the molecule. Indeed there is a charge transfer, which causes a change in the electrostatic potential inside the molecule; and the energy levels of the molecule are shifted by a contact potential (CP), as shown.

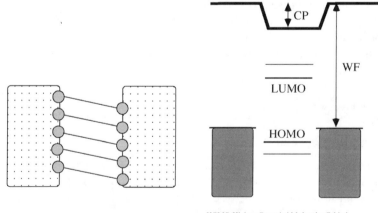

HOMO:Highest Occupied Molecular Orbital

LUMO:Lowest Unoccupied Molecular Orbital

WF : Metal Work Function

CP : Contact Potential

FIGURE 12.5 Equilibrium energy level diagram for a metal–molecule–metal sandwich for a molecule strongly coupled to the contacts.

It is now more appropriate to describe transport in terms of the HOMO–LUMO levels associated with incremental charge transfer[32] rather than the affinity and ionization levels associated with integer charge transfer. Whether the molecule–metal coupling is strong enough for this to occur depends on the relative magnitudes of the single electron charging energy (U) and energy level broadening (Γ). As a rule of thumb, if $U \gg \Gamma$, we can expect the structure to be in the Coulomb Blockade (CB) regime characterized by integer charge transfer; otherwise it is in the self-consistent field (SCF) regime characterized by fractional charge transfer. This is basically the same criterion that one uses for the Mott transition in periodic structures, with Γ playing the role of the hopping matrix element. It is important to note that, for a structure to be in the CB regime, both contacts must be weakly coupled, as the total broadening Γ is the sum of the individual broadening due to the two contacts. Even if only one of the contacts is coupled strongly, we can expect $\Gamma \sim U$, thus putting the structure in the SCF regime. Figure 12.14 in Section 12.3.2 illustrates the I-V characteristics in the CB regime using a toy model. However, a moderate amount of broadening destroys this effect (see Figure 12.17), and in this chapter we will generally assume that the conduction is in the SCF regime.

Location of the Fermi energy: The location of the Fermi energy relative to the HOMO and LUMO levels is probably the most important factor in determining the current (I) vs. voltage (V) characteristics of molecular conductors. Usually it lies somewhere inside the HOMO–LUMO gap. To see this, we first note that E_f is located by the requirement that the number of states below the Fermi energy must be equal to the number of electrons in the molecule. But this number need not be equal to the integer number we expect for a neutral molecule. A molecule does not remain exactly neutral when connected to the contacts. It can and does pick up a fractional charge depending on the work function of the metal. However, the charge transferred (δn) for most metal–molecule combinations is usually much less than one. If δn were equal to $+1$, the Fermi energy would lie on the LUMO; while if δn were -1, it would lie on the HOMO. Clearly for values in between, it should lie somewhere in the HOMO–LUMO gap.

A number of authors have performed detailed calculations to locate the Fermi energy with respect to the molecular levels for a phenyl dithiol molecule sandwiched between gold contacts, but there is considerable disagreement. Different theoretical groups have placed it close to the LUMO[16,21] or to the HOMO.[12,19] The density of states inside the HOMO–LUMO gap is quite small, making the precise location of the Fermi energy very sensitive to small amounts of electron transfer — a fact that could have a significant effect on both theory and experiment. As such it seems justifiable to treat E_f as a fitting parameter within reasonable limits when trying to explain experimental I-V curves.

Broadening by the contacts: Common sense suggests that the strength of coupling of the molecule to the contacts is important in determining the current flow — the stronger the coupling, the larger the current. A useful quantitative measure of the coupling is the resulting broadening Γ of the molecular energy levels, see Figure 12.6. This broadening Γ can also be related to the time τ it takes for an electron placed in that level to escape into the contact: $\Gamma = \hbar/\tau$. In general, the broadening Γ could be different for different energy levels. Also it is convenient to define two quantities Γ_1 and Γ_2, one for each contact, with the total broadening $\Gamma = \Gamma_1 + \Gamma_2$.

One subtle point: Suppose an energy level is located well below the Fermi energy in the contact, so that the electrons are prevented from escaping by the exclusion principle. Would Γ be zero? No, the broadening would still be Γ, independent of the degree of filling of the contact as discussed in Reference 26. This observation is implicit in the NEGF formalism, though we do not invoke it explicitly.

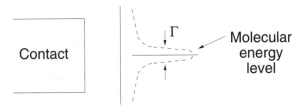

FIGURE 12.6 Energy level broadening.

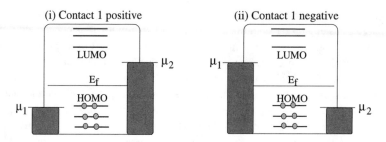

FIGURE 12.7 Schematic energy level diagram of metal–molecule–metal structure when contact 1 is (i) positively biased and when contact 1 is (ii) negatively biased with respect to contact 2.

12.2.2 Current Flow as a Balancing Act

Once we have drawn an equilibrium energy-level diagram, we can understand the process of current flow, which involves a nonequilibrium situation where the different reservoirs (e.g., the source and the drain) have different electrochemical potentials μ (Figure 12.7). For example, if a positive voltage V is applied externally to the drain with respect to the source, then the drain has an electrochemical potential lower than that of the source by eV : $\mu_2 = \mu_1 - eV$. The source and drain contacts thus have different Fermi functions, and each seeks to bring the active device into equilibrium with itself. The source keeps pumping electrons into it, hoping to establish equilibrium. But equilibrium is never achieved as the drain keeps pulling electrons out in its bid to establish equilibrium with itself. The device is thus forced into a balancing act between two reservoirs with different agendas that send it into a nonequilibrium state intermediate between what the source and drain would like to see. To describe this balancing process we need a kinetic equation that keeps track of the in- and out-flow of electrons from each of the reservoirs.

Kinetic equation: This balancing act is easy to see if we consider a simple one-level system, biased such that the energy ε lies in between the electrochemical potentials of the two contacts (Figure 12.8). An electron in this level can escape into contacts 1 and 2 at a rate of $\Gamma_1/\hbar$ and $\Gamma_2/\hbar$, respectively. If the level were in equilibrium with contact 1, then the number of electrons occupying the level would be given by

$$N_1 = 2(\text{for spin}) \, f(\varepsilon, \mu_1) \tag{12.4}$$

where

$$f(\varepsilon, \mu) = \frac{1}{1 + e^{\frac{\varepsilon - \mu}{k_B T}}} \tag{12.5}$$

is the Fermi function. Similarly, if the level were in equilibrium with contact 2, the number would be:

$$N_2 = 2(\text{for spin}) \, f(\varepsilon, \mu_2) \tag{12.6}$$

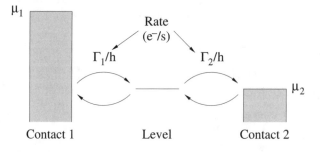

FIGURE 12.8 Illustration of the kinetic equation.

Under nonequilibrium conditions, the number of electrons N will be somewhere in between N_1 and N_2. To determine this number we write a steady-state *kinetic equation* that equates the net current at the left junction:

$$I_L = \frac{e\Gamma_1}{\hbar}(N_1 - N) \tag{12.7}$$

to the net current at the right junction:

$$I_R = \frac{e\Gamma_2}{\hbar}(N - N_2) \tag{12.8}$$

Steady state requires $I_L = I_R$, from which we obtain

$$N = 2\frac{\Gamma_1 f(\varepsilon, \mu_1) + \Gamma_2 f(\varepsilon, \mu_2)}{\Gamma_1 + \Gamma_2} \tag{12.9}$$

so that from Equation (12.7) or Equation (12.8) we obtain the current:

$$I = \frac{2e}{\hbar}\frac{\Gamma_1\Gamma_2}{\Gamma_1 + \Gamma_2}(f(\varepsilon, \mu_1) - f(\varepsilon, \mu_2)) \tag{12.10}$$

Equation (12.10) follows very simply from an elementary model, but it serves to illustrate a basic fact about the process of current flow. No current will flow if $f(\varepsilon, \mu_1) = f(\varepsilon, \mu_2)$. A level that is way below both electrochemical potentials μ_1 and μ_2 will have $f(\varepsilon, \mu_1) = f(\varepsilon, \mu_2) = 1$ and will not contribute to the current, just like a level that is way above both potentials μ_1 and μ_2 and has $f(\varepsilon, \mu_1) = f(\varepsilon, \mu_2) = 0$. It is only when the level lies between μ_1 and μ_2 (or within a few k_BT of μ_1 and μ_2) that we have $f(\varepsilon, \mu_1) \neq f(\varepsilon, \mu_2)$, and a current flows. Current flow is thus the result of the *difference in opinion* between the contacts. One contact would like to see more electrons (than N) occupy the level and keeps pumping them in, while the other would like to see fewer than N electrons and keeps pulling them out. The net effect is a continuous transfer of electrons from one contact to another.

Figure 12.9 shows a typical I vs. V calculated from Equation (12.10), using the parameters indicated in the caption. At first the current is zero because both μ_1 and μ_2 are above the energy level.

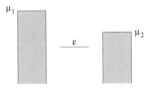

Once μ_2 drops below the energy level, the current increases to I_{max}, which is the maximum current that can flow through one level and is obtained from Equation (12.10) by setting $f(\varepsilon, \mu_1) = 1$ and $f(\varepsilon, \mu_2) = 0$:

$$I_{max} = \frac{2e}{\hbar}\Gamma_{eff} = \frac{2e}{\hbar}\frac{\Gamma_1\Gamma_2}{\Gamma_1 + \Gamma_2} \tag{12.11}$$

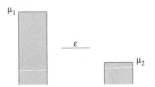

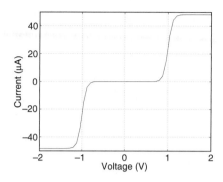

FIGURE 12.9 The current–voltage (I-V) characteristics for our toy model with $\mu_1 = E_f - eV/2$, $\mu_2 = E_f + eV/2$, $E_f = -5.0$ eV, $\varepsilon_0 = -5.5$ eV and $\Gamma_1 = \Gamma_2 = 0.2$ eV. MATLAB code in Section 12.A.1 ($U = 0$).

Note that in Figure 12.9 we have set $\mu_1 = E_f - eV/2$ and $\mu_2 = E_f + eV/2$. We could, of course, just as well have set $\mu_1 = E_f - eV$ and $\mu_1 = E_f$. But the average potential in the molecule would be $-V/2$ and we would need to shift ε appropriately. It is more convenient to choose our reference such that the average molecular potential is zero, and there is no need to shift ε.

Note that the current is proportional to Γ_{eff}, which is the parallel combination of Γ_1 and Γ_2. This seems quite reasonable if we recognize that Γ_1 and Γ_2 represent the strength of the coupling to the two contacts and as such are like two *conductances in series*. For long conductors we would expect a third conductance in series representing the actual conductor. This is what we usually have in mind when we speak of conductance. But short conductors have virtually zero resistance, and what we measure is essentially the contact or interface resistance.[*] This is an important conceptual issue that caused much argument and controversy in the 1980s. It was finally resolved when experimentalists measured the conductance of very short conductors and found it approximately equal to $2e^2/h$, which is a fundamental constant equal to 77.8 μA/V. The inverse of this conductance $h/2e^2 = 12.9$ kΩ is now believed to represent the minimum contact resistance that can be achieved for a one-channel conductor. Even a copper wire with a one-atom cross-section will have a resistance at least this large. Our simple one-level model (Figure 12.9) does not predict this result because we have treated the level as discrete, but the more complete treatment in later sections will show it.

12.3 Coulomb Blockade?

As we mentioned in Section 12.2, a basic question we need to answer is whether the process of conduction through the molecule belongs to the Coulomb Blockade (CB) or the Self-Consistent Field (SCF) regime. In this section, we will first discuss a simple model for charging effects (Section 12.3.1) and then look at the distinction between the simple SCF regime and the CB regime (Section 12.3.2). Finally, in Section 12.3.3 we show how moderate amounts of level broadening often destroy CB effects, making a simple SCF treatment quite accurate.

12.3.1 Charging Effects

Given the level (ε), broadening (Γ_1, Γ_2), and the electrochemical potentials μ_1 and μ_2 of the two contacts, we can solve Equation (12.10) for the current I. But we want to include charging effects in the calculations. Therefore, we add a potential U_{SCF} due to the change in the number of electrons from the equilibrium value ($f_0 = f(\varepsilon_0, E_f)$):

---

[*] Four-terminal measurements have been used to separate the contact from the device resistance (see, for example, Reference 33).

$$U_{SCF} = U(N - 2f_0) \tag{12.12}$$

similar to a Hubbard model. We then let the level ε float up or down by this potential:

$$\varepsilon = \varepsilon_0 + U_{SCF} \tag{12.13}$$

Because the potential depends on the number of electrons, we need to calculate the potential using the self-consistent procedure shown in Figure 12.10.

Once the converged solution is obtained, the current is calculated from Equation (12.10). This very simple model captures much of the observed physics of molecular conduction. For example, the results obtained by setting $E_f = -5.0$ eV, $\varepsilon_0 = -5.5$ eV, $\Gamma_1 = 0.2$ eV, $\Gamma_2 = 0.2$ eV are shown in Figure 12.11 with ($U = 1.0$ eV) and without ($U = 0$ eV) charging effects. The finite width of the conductance peak (with $U = 0$) is due to the temperature used in the calculations ($k_B T = 0.025$ eV). Note how the inclusion of charging tends to broaden the sharp peaks in conductance, even though we have not included any extra level broadening in this calculation. The size of the conductance gap is directly related to the energy difference between the molecular energy level and the Fermi energy. The current starts to increase when the voltage reaches 1 V, which is exactly $2|E_f - \varepsilon_0|$, as would be expected even from a theory with no charging. Charging enters the picture only at higher voltages, when a chemical potential tries to cross the level. The energy level shifts in energy (Equation (12.13)) if the charging energy is nonzero. Thus, for a small charging energy, the chemical potential easily crosses the level, giving a sharp increase of the current. If the charging energy is large, the current increases gradually because the energy level follows the chemical potential due to the charging.

What determines the conductance gap? The above discussion shows that the conductance gap is equal to $4(|E_f - \varepsilon_0| - \Delta)$, where Δ is equal to $\sim 4k_B T$ (plus $\Gamma_1 + \Gamma_2$ if broadening is included, see Section 12.3.3); and ε_0 is the HOMO or LUMO level, whichever is closest to the Fermi energy, as pointed out in Reference 34.

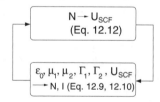

FIGURE 12.10 Illustration of the SCF problem.

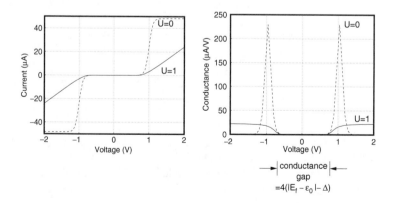

FIGURE 12.11 The current–voltage (I-V) characteristics (left) and conductance–voltage (G-V) (right) for our toy model with $E_f = -5.0$ eV, $\varepsilon_0 = -5.5$ eV and $\Gamma_1 = \Gamma_2 = 0.2$ eV. Solid lines, charging effects included ($U = 1.0$ eV). Dashed line, no charging ($U = 0$). MATLAB code in section 12.A.1.

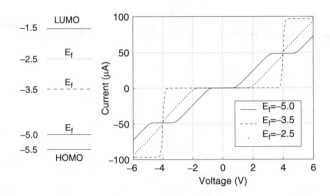

FIGURE 12.12 Right, the current–voltage (I-V) characteristics for the two level toy model for three different values of the Fermi energy (E_f). Left, the two energy levels (LUMO = –1.5 eV; HOMO = –5.5 eV) and the three different Fermi energies (–2.5, –3.5, –5.0) used in the calculations. (Other parameters used $U = 1.0$ eV; $\Gamma_1 = \Gamma_2 = 0.2$ eV) MATLAB code in Section 12.A.1.

This is unappreciated by many who associate the conductance gap with the HOMO–LUMO gap. However, we believe that what conductance measurements show is the gap between the Fermi energy and the nearest molecular level.* Figure 12.12 shows the I-V characteristics calculated using a two-level model (obtained by a straightforward extension of the one-level model) with the Fermi energy located differently within the HOMO–LUMO gap giving different conductance gaps corresponding to the different values of $|E_f - \varepsilon_0|$. Note that with the Fermi energy located halfway in between, the conductance gap is twice the HOMO–LUMO gap; and the I-V shows no evidence of charging effects because the depletion of the HOMO is neutralized by the charging of the LUMO. This perfect compensation is unlikely in practice, because the two levels will not couple identically to the contacts as assumed in the model.

A very interesting effect that can be observed is the asymmetry of the I-V characteristics of $\Gamma_1 \neq \Gamma_2$ as shown in Figure 12.13. This may explain several experimental results which show asymmetric I-V,[6,12] as discussed by Ghosh et al.[35] Assuming that the current is conducted through the HOMO level ($E_f > \varepsilon_0$), the current is less when a positive voltage is applied to the strongly coupled contact (Figure 12.13a). This is due to the effects of charging as has been discussed in more detail in Reference 35. Ghosh et al. also show that this result will reverse if the conduction is through the LUMO level. We can simulate this situation by setting ε_0 equal to –4.5 eV, 0.5 eV above the equilibrium Fermi energy E_f. The sense of

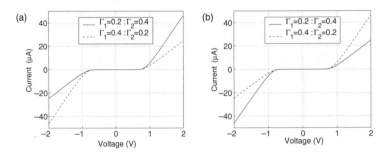

FIGURE 12.13 The current–voltage (I-V) characteristics for our toy model ($E_f = –5.0$ eV and $U = 1.0$ eV). (a) Conduction through LUMO ($E_f < \varepsilon_0 = –4.5$ eV). Solid lines, $\Gamma_1 = 0.2$ eV $< \Gamma_2 = 0.4$ eV. Dashed lines, $\Gamma_1 = 0.4$ eV $> \Gamma_2 = 0.2$ eV. MATLAB code in Section 12.A.1. Here positive voltage is defined as a voltage that lowers the chemical potential of contact 1.

---

*With very asymmetric contacts, the conductance gap could be equal to the HOMO–LUMO gap as commonly assumed in interpreting STM spectra. However, we believe that the picture presented here is more accurate unless the contact is so strongly coupled that there is a significant density of metal-induced gap states (MIGS).[28]

asymmetry is now reversed as shown in Figure 12.13b. The current is larger when a positive voltage is applied to the strongly coupled contact. Comparing with STM measurements seems to favor the first case, i.e., conduction through the HOMO.[35]

12.3.2 Unrestricted Model

In the previous examples (Figures 12.11 and 12.13) we have used values of $\Gamma_{1,2}$ that are smaller than the charging energy U. However, under these conditions one can expect CB effects, which are not captured by a *restricted solution*, which assumes that both spin orbitals see the same self-consistent field. However, an unrestricted solution, which allows the spin degeneracy to be lifted, will show these effects.[*] For example, if we replace Equation (12.13) with ($f_0 = f(\varepsilon_0, E_f)$):

$$\varepsilon_\uparrow = \varepsilon_0 + U(N_\downarrow - f_0) \tag{12.14}$$

$$\varepsilon_\downarrow = \varepsilon_0 + U(N_\uparrow - f_0) \tag{12.15}$$

where the up-spin level feels a potential due to the down-spin electrons and vice versa, then we obtain I-V curves as shown in Figure 12.14.

If the SCF iteration is started with a spin-degenerate solution, the same restricted solution as before is obtained. However, if the iteration is started with a spin-nondegenerate solution, a different I-V is obtained. The electrons only interact with the electron of the opposite spin. Therefore, the chemical potential of one contact can cross one energy level of the molecule because the charging of that level only affects the opposite spin level. Thus, the I-V contains two separate steps separated by U instead of a single step broadened by U.

For a molecule chemically bonded to a metallic surface, e.g., a PDT molecule bonded by a thiol group to a gold surface, the broadening Γ is expected to be of the same magnitude or larger than U. This washes out CB effects as shown in Figure 12.17. Therefore, the CB is not expected in this case. However, if the coupling to both contacts is weak, we should keep the possibility of CB and the importance of unrestricted solutions in mind.

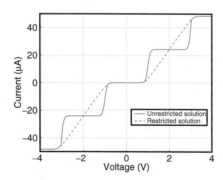

FIGURE 12.14 The current–voltage (I-V) characteristics for restricted (dashed line) and unrestricted solutions (solid line). $E_f = -5.0$ eV, $\Gamma_1 = \Gamma_2 = 0.2$ eV and $U = 1.0$ eV. MATLAB code in Sections 12.A.1 and 12.A.4.

---

[*] The unrestricted one-particle picture discussed here provides at least a reasonable qualitative picture of CB effects, though a complete description requires a more advanced many particle picture.[36] The one-particle picture leads to one of many possible states of the device depending on our initial guess, while a full many-particle picture would include all states.

12.3.3 Broadening

So far we have treated the level ε as discrete, ignoring the broadening $\Gamma = \Gamma_1 + \Gamma_2$ that accompanies the coupling to the contacts. To take this into account we need to replace the discrete level with a Lorentzian density of states $D(E)$:

$$D(E) = \frac{1}{2\pi}\frac{\Gamma}{(E-\varepsilon)^2 + (\Gamma/2)^2} \tag{12.16}$$

As we will see later, Γ is in general *energy-dependent* so that $D(E)$ can deviate significantly from a Lorentzian shape. We modify Equations (12.9) and (12.10) for N and I to include an integration over energy:

$$N = 2\int_{-\infty}^{\infty} dE\,D(E)\frac{\Gamma_1 f(E,\mu_1) + \Gamma_2 f(E,\mu_2)}{\Gamma_1 + \Gamma_2} \tag{12.17}$$

$$I = \frac{2e}{\hbar}\int_{-\infty}^{\infty} dE\,D(E)\frac{\Gamma_1\Gamma_2}{\Gamma_1 + \Gamma_2}(f(E,\mu_1) - f(E,\mu_2)) \tag{12.18}$$

The charging effect is included as before by letting the center ε, of the molecular density of states, float up or down according to Equations (12.12) and (12.13) for the restricted model or Equations (12.14) and (12.15) for the unrestricted model.

For the restricted model, the only effect of broadening is to smear out the I-V characteristics as evident from Figure 12.15. The same is true for the unrestricted model as long as the broadening is much smaller than the charging energy (Figure 12.16). But moderate amounts of broadening can destroy the Coulomb Blockade effects completely and make the I-V characteristics look identical to the restricted model (Figure 12.17). With this in mind, we will use the restricted model in the remainder of this chapter.

12.4 Nonequilibrium Green's Function (NEGF) Formalism

The one-level toy model described in the last section includes the three basic factors that influence molecular conduction, namely, $E_f - \varepsilon_0$, $\Gamma_{1,2}$, and U. However, real molecules typically have multiple levels that often broaden and overlap in energy. Note that the two-level model (Figure 12.12) in the last section treated the two levels as *independent*, and such models can be used only if the levels do not overlap. In general, we need a formalism that can do justice to multiple levels with arbitrary broadening and overlap. The nonequilibrium Green's function (NEGF) formalism described in this section does just that.

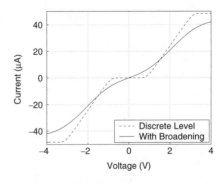

FIGURE 12.15 The current–voltage (I-V) characteristics: Solid line, include broadening of the level by the contacts. Dashed line, no broadening, same as solid line in Figure 12.11. MATLAB code in Sections 12.A.3 and 12.A.1 ($E_f = -5.0$ eV, $\varepsilon_0 = -5.5$, $U = 1$, and $\Gamma_1 = \Gamma_2 = 0.2$ eV).

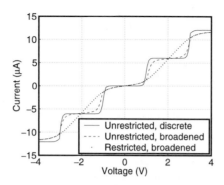

FIGURE 12.16 Current–voltage (I-V) characteristics showing the Coulomb blockade: discrete unrestricted model (solid line, MATLAB code in Section 12.A.4) and the broadened unrestricted mode (dashed line, 12.A.5). The dotted line shows the broadened restricted model without Coulomb blockade (12.A.3). For all curves the following parameters were used: $E_f = -5.0$, $\varepsilon_0 = -5.5$, $U = 1$, and $\Gamma_1 = \Gamma_2 = 0.05$ eV.

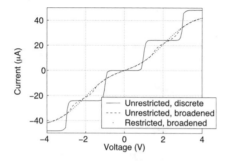

FIGURE 12.17 Current–voltage (I-V) characteristics showing the suppression of the Coulomb blockade by broadening: discrete unrestricted model (solid line) and the broadened unrestricted model (dashed line). The dotted line shows the broadened restricted model, $\Gamma_1 = \Gamma_2 = 0.2$ eV.

In the last section we obtained equations for the number of electrons N and the current I for a one-level model with broadening. It is useful to rewrite these equations in terms of the Green's function $G(E)$, defined as follows:

$$G(E) = \left(E - \varepsilon + i\frac{\Gamma_1 + \Gamma_2}{2}\right)^{-1} \tag{12.19}$$

The density of states $D(E)$ is proportional to the spectral function $A(E)$ defined as:

$$A(E) = -2\text{Im}\{G(E)\} \tag{12.20}$$

$$D(E) = \frac{A(E)}{2\pi} \tag{12.21}$$

while the number of electrons N and the current I can be written as:

$$N = \frac{2}{2\pi} \int_{-\infty}^{\infty} dE(|G(E)|^2 \Gamma_1 f(E, \mu_1) + |G(E)|^2 \Gamma_2 f(E, \mu_2)) \tag{12.22}$$

$$I = \frac{2e}{h} \int_{-\infty}^{\infty} dE \Gamma_1 \Gamma_2 |G(E)|^2 (f(E, \mu_1) - f(E, \mu_2)) \tag{12.23}$$

In the NEGF formalism the single energy level ε is replaced by a Hamiltonian matrix $[H]$, while the broadening $\Gamma_{1,2}$ is replaced by a complex energy-dependent self-energy matrix $[\Sigma_{1,2}(E)]$ so that the Green's function becomes a matrix given by

$$G(E) = (ES - H - \Sigma_1 - \Sigma_2)^{-1} \tag{12.24}$$

where S is the identity matrix of the same size as the other matrices, and the broadening matrices $\Gamma_{1,2}$ are defined as the imaginary (more correctly as the anti-Hermitian) parts of $\Sigma_{1,2}$:

$$\Gamma_{1,2} = i(\Sigma_{1,2} - \Sigma_{1,2}^\dagger) \tag{12.25}$$

The spectral function is the anti-Hermitian part of the Green's function:

$$A(E) = i(G(E) - G^\dagger(E)) \tag{12.26}$$

from which the density of states $D(E)$ can be calculated by taking the trace:

$$D(E) = \frac{\mathrm{Tr}(AS)}{2\pi} \tag{12.27}$$

The density matrix $[\rho]$ is given by, c.f., Equation (12.22):

$$\rho = \frac{1}{2\pi} \int_{-\infty}^{\infty} [f(E, \mu_1)G\Gamma_1 G^\dagger + f(E, \mu_2)G\Gamma_2 G^\dagger] dE \tag{12.28}$$

from which the total number of electrons N can be calculated by taking a trace:

$$N = \mathrm{Tr}(\rho S) \tag{12.29}$$

The current is given by (c.f., Equation (12.23)):

$$I = \frac{2e}{h} \int_{-\infty}^{\infty} [\mathrm{Tr}(\Gamma_1 G\Gamma_2 G^\dagger)(f(E, \mu_1) - f(E, \mu_2))] dE \tag{12.30}$$

Equations (12.24) through (12.30) constitute the basic equations of the NEGF formalism, which have to be solved self-consistently with a suitable scheme to calculate the self-consistent potential matrix $[U_{SCF}]$ (c.f., Equation (12.13)):

$$H = H_0 + U_{SCF} \tag{12.31}$$

where H_0 is the bare Hamiltonian (like ε_0 in the toy model) and U_{SCF} is an appropriate functional of the density matrix ρ:

$$U_{SCF} = F(\rho) \tag{12.32}$$

This self-consistent procedure is essentially the same as in Figure 12.10 for the one-level toy model, except that scalar quantities have been replaced by matrices:

$$\varepsilon_0 \rightarrow [H_0] \tag{12.33}$$

$$\Gamma \rightarrow [\Gamma], [\Sigma] \tag{12.34}$$

$$N \rightarrow [\rho] \tag{12.35}$$

$$U_{SCF} \rightarrow [U_{SCF}] \tag{12.36}$$

The sizes of all these matrices is $(n \times n)$, n being the number of basis functions used to describe the *molecule*. Even the self-energy matrices $\Sigma_{1,2}$ are of this size although they represent the effect of infinitely large contacts. In the remainder of this section and the next section, we will describe the procedure used to evaluate the Hamiltonian matrix H, the self-energy matrices $\Sigma_{1,2}$, and the functional F used to evaluate the self-consistent potential U_{SCF} (see Equation (12.32)). But the point to note is that once we know how to evaluate these matrices, Equations (12.24) through (12.32) can be used straightforwardly to calculate the current.

Nonorthogonal basis: The matrices appearing above depend on the basis functions that we use. Many of the formulations in quantum chemistry use nonorthogonal basis functions; and the matrix Equations (12.24) through (12.32) are still valid as is, except that the elements of the matrix $[S]$ in Equation (12.24) represent the overlap of the basis function $\phi_m(\vec{r})$:

$$S_{mn} = \int d^3 r \phi_m^*(\vec{r}) \phi_n(\vec{r}) \tag{12.37}$$

For orthogonal bases, $S_{mn} = \delta_{mn}$ so that S is the identity matrix as stated earlier. The fact that the matrix Equations (12.24) through (12.32) are valid even in a nonorthogonal representation is not self-evident and is discussed in Reference 28.

Incoherent Scattering: One last comment about the general formalism: the formalism as described above neglects all incoherent scattering processes inside the molecule. In this form it is essentially equivalent to the Landauer formalism.[37] Indeed, our expression for the current (Equation (12.30)) is exactly the same as in the transmission formalism, with the transmission T given by $\mathrm{Tr}(\Gamma_1 G \Gamma_2 G^\dagger)$. But it should be noted that the real power of the NEGF formalism lies in its ability to provide a first-principles description of incoherent scattering processes — something we do not address in this chapter and leave for future work.

A practical consideration: Both Equations (12.28) and (12.30) require an integral over all energy. This is not a problem in Equation (12.30) because the integrand is nonzero only over a limited range, where $f(E, \mu_1)$ differs significantly from $f(E, \mu_2)$. But in Equation (12.28), the integrand is nonzero over a large energy range and often has sharp structures, making it numerically challenging to evaluate the integral. One way to address this problem is to write

$$\rho = \rho_{eq} + \Delta\rho \tag{12.38}$$

where ρ_{eq} is the equilibrium density matrix given by

$$\rho_{eq} = \frac{1}{2\pi} \int\limits_{-\infty}^{\infty} f(E, \mu)[G\Gamma_1 G^\dagger + G\Gamma_2 G^\dagger]dE \tag{12.39}$$

and $\Delta\rho$ is the change in the density matrix under bias:

$$\Delta\rho = \frac{1}{2\pi} \int\limits_{-\infty}^{\infty} G\Gamma_1 G^\dagger [f(E, \mu_1) - f(E, \mu)] + G\Gamma_2 G^\dagger [f(E, \mu_2) - f(E, \mu)]dE \tag{12.40}$$

The integrand in Equation (12.40) for $\Delta\rho$ is nonzero only over a limited range (like Equation (12.30) for I) and is evaluated relatively easily. The evaluation of ρ_{eq} (Equation (12.39)), however, still has the same problem; but this integral (unlike the original Equation (12.28)) can be tackled by taking advantage of the method of contour integration as described in References 38 and 39.

12.5 An Example: Quantum Point Contact (QPC)

Consider, for example, a gold wire stretched between two gold surfaces as shown in Figure 12.18. One of the seminal results of mesoscopic physics is that such a wire has a quantized conductance equal to $e^2/$

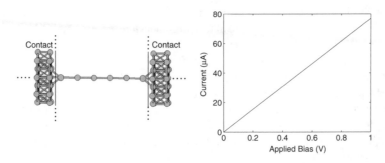

FIGURE 12.18 Left, wire consisting of six gold atoms forming a Quantum Point Contact (QPC). Right, quantized conductance $(I = (e^2/\pi\hbar)V)$.

$\pi\hbar \sim 77.5\ \mu\text{A/V} \sim (12.9\ \text{k}\Omega)^{-1}$. This was first established using semiconductor structures[29,40,41] at 4 K, but recent experiments on gold contacts have demonstrated it at room temperature.[42] How can a wire have a resistance that is independent of its length? The answer is that this resistance is really associated with the interfaces between the narrow wire and the wide contacts. If there were scattering inside the wire, it would give rise to an additional resistance in series with this fundamental interface resistance. The fact that a short wire has a resistance of 12.9 kΩ is a nonobvious result that was not known before 1988. This is a problem for which we do not really need a quantum transport formalism; a semiclassical treatment would suffice. The results we obtain here are not new or surprising. What is new is that we treat the gold wire as an Au_6 molecule and obtain well-known results commonly obtained from a continuum treatment.

In order to apply the NEGF formalism from the last section to this problem, we need the Hamiltonian matrix $[H]$, the self-energy matrices $\Sigma_{1,2}$, and the self-consistent field $U_{SCF} = F([\rho])$. Let us look at these one by one.

Hamiltonian: We will use a simple semi-empirical Hamiltonian which uses one s-orbital centered at each gold atom as the basis functions, with the elements of the Hamiltonian matrix given by

$$
\begin{aligned}
H_{ij} &= \varepsilon_0 && \text{if } i = j \\
&= -t && \text{if } i, j \text{ are nearest neighbors}
\end{aligned}
\tag{12.41}
$$

where $\varepsilon_0 = -10.92$ eV and $t = 2.653$ eV. The orbitals are assumed to be orthogonal, so that the overlap matrix S is the identity matrix.

Self Energy: Once we have a Hamiltonian for the entire molecule-contact system, the next step is to partition the device from the contacts and obtain the self-energy matrices $\Sigma_{1,2}$ describing the effects of the contacts on the device. The contact will be assumed to be essentially unperturbed relative to the surface of a bulk metal so that the full Green's function (G_T) can be written as (the energy E is assumed to have an infinitesimal imaginary part $i0^+$):

$$
G_T = \begin{pmatrix} ES - H & ES_{dc} - H_{dc} \\ ES_{cd} - H_{cd} & ES_c - H_c \end{pmatrix}^{-1} = \begin{pmatrix} G & G_{dc} \\ G_{cd} & G_c \end{pmatrix}
\tag{12.42}
$$

where c denotes one of the contacts (the effect of the other contact can be obtained separately). We can use straightforward matrix algebra to show that:

$$
G = (ES - H - \Sigma)^{-1}
\tag{12.43}
$$

$$
\Sigma = (ES_{dc} - H_{dc})(ES_c - H_c)^{-1}(ES_{cd} - H_{cd})
\tag{12.44}
$$

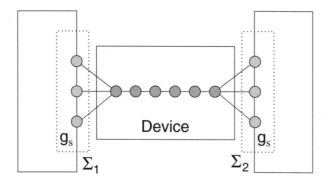

FIGURE 12.19 Device, surface Green's function (g_s) and self-energies (Σ).

The matrices S_{dc}, S_c, H_{dc}, and H_c are all infinitely large because the contact is infinite. But the element of S_{dc}, H_{dc} is nonzero only for a small number of contact atoms whose wave functions significantly overlap the device. Thus, we can write:

$$\Sigma = \tau g_s \tau^\dagger \tag{12.45}$$

where τ is the nonzero part of $ES_{dc} - H_{dc}$, having dimensions $(d \times s)$ where d is the size of the device matrix, and s is the number of surface atoms of the contact having a nonzero overlap with the device. g_s is a matrix of size $s \times s$ which is a subset of the full infinite-sized contact Green's function $(ES_c - H_c)^{-1}$. This surface Green's function can be computed exactly by making use of the periodicity of the semi-infinite contact, using techniques that are standard in surface physics.[31] For a one-dimensional lead, with a Hamiltonian given by Equation (12.41), the result is easily derived:[29]

$$g_s(E) = -\frac{e^{ika}}{t} \tag{12.46}$$

where ka is related to the energy through the dispersion relation:

$$E = \varepsilon_0 - 2t \cos(ka) \tag{12.47}$$

The results presented below were obtained using the more complicated surface Green's function for an FCC (111) gold surface as described in Reference 28. However, using the surface Green's function in Equation (12.46) gives almost identical results.

Electrostatic potential: Finally we need to identify the electrostatic potential across the device (Figure 12.20) by solving the Poisson equation:

$$-\nabla^2 U_{tot} = \frac{e^2 n}{\varepsilon_0} \tag{12.48}$$

with the boundary conditions given by the potential difference V_{app} between the metallic contacts (here ε_0 is the dielectric constant). To simplify the calculations we divide the solution into an applied and self-consistent potential ($U_{tot} = U_{app} + U_{SCF}$), where U_{app} solves the Laplace equation with the known potential difference between the metallic contacts:

$$\nabla^2 U_{app} = 0 \quad U_{app} = -eV_n \text{ on electrode n} \tag{12.49}$$

Thus, U_{tot} solves Equation (12.48) if U_{SCF} solves Equation (12.48) with zero potential at the boundary.

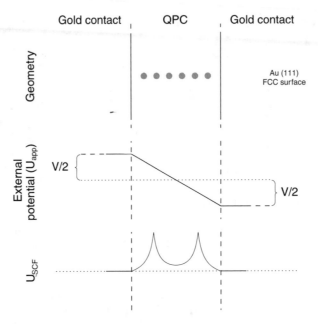

FIGURE 12.20 The electrostatic potentials divided into the applied (U_{app}) and the self-consistent field (U_{SCF}) potentials. The boundary conditions can clearly be seen in the figure; U_{SCF} is zero at the boundary and $U_{app} = \pm V_{app}/2$.

$$-\nabla^2 U_{SCF} = \frac{e^2 n}{\varepsilon_0} \quad U_{SCF} = 0 \text{ on all electrodes} \tag{12.50}$$

In the treatment of the electrostatic we assume the two contacts to be semi-infinite classical metals separated by a distance (W). This gives simple solutions to both U_{app} and U_{SCF}. The applied potential is given by (capacitor)

$$U_{app} = \frac{V}{W}x \tag{12.51}$$

where x is the position relative to the midpoint between the contacts. The self-consistent potential is easily calculated with the method of images where the potential is given by a sum over the point charges and all their images. However, to avoid the infinities associated with point charges, we adopt the Pariser–Parr–Pople (PPP) method[43,17] in the Hartree approximation. The PPP functional describing the electron–electron interactions is

$$H_{ij}^{e-e} = \delta_{ij} \sum_k (\rho_{kk} - \rho_{kk}^{eq})\gamma_{ik} \tag{12.52}$$

where ρ is the charge density matrix, ρ^{eq} the equilibrium charge density (in this case ρ_{ii}^{eq} as we are modeling the s-electrons of gold), and the one-center two-electron integral γ_{ij}. The diagonal elements γ_{ii} are obtained from experimental data, and the off-diagonal elements (γ_{ij}) are parameterized to describe a potential that decreases as the inverse of the distance ($1/R_{ij}$):

$$\gamma_{ij} = \frac{e^2}{4\pi\varepsilon_0 R_{ij} + \dfrac{2e^2}{\gamma_{ii} + \gamma_{jj}}} \tag{12.53}$$

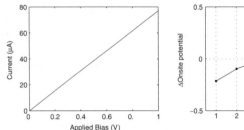
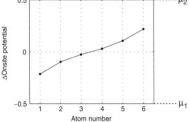

FIGURE 12.21 I-V (left) and potential drop for an applied voltage of 1 V (right) for a six-atom OPC connected to two contacts. The potential plotted is the difference in onsite potential from the equilibrium case.

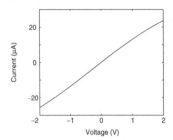
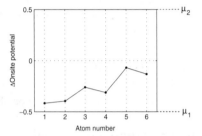

FIGURE 12.22 I-V and potential drop for an applied voltage of 1 V for the QPC asymmetrically connected the gold contacts. The coupling to the right contact used is $-0.2t$.

Calculations on the QPC: The results for the I-V and potential for a QPC are shown in Figure 12.21. The geometry used was a linear chain of six gold atoms connected to the FCC (111) surface of the contacts in the center of a surface triangle. The Fermi energy of the isolated contacts was calculated to be $E_f = -8.67$ eV, by requiring that there is one electron per unit cell.

As evident from the figure, the I-V characteristics are linear and the slope gives a conductance of 77.3 μA/V close to the quantized value of $e^2/\pi\hbar \sim 77.5$ μA/V as previously mentioned. What makes the QPC distinct from typical molecules is the strong coupling to the contacts, which broadens all levels into a continuous density of states; and any evidence of a conductance gap (Figure 12.2) is completely lost. Examining the potential drop over the QPC shows a linear drop over the center of the QPC with a slightly larger drop at the end atoms. This may seem surprising because transport is assumed to be *ballistic*, and one expects no voltage drop across the chain of gold atoms. This can be shown to arise because the chain is very narrow (one atom in cross-section) compared with the screening length.[19]

We can easily imagine an experimental situation where the device (QPC or molecule) is attached asymmetrically to the two contacts with one strong and one weak side. To model this situation we artificially decreased the interaction between the right contact and the QPC by a factor of 0.2. The results of this calculation are shown in Figure 12.22. A weaker coupling gives a smaller conductance, as compared with the previous figure. More interesting is that the potential drop over the QPC is asymmetric. Also in line with our classical intuition, the largest part of the voltage drop occurs at the weakly coupled contact with smaller drops over the QPC and at the strongly coupled contact. The consequences of asymmetric voltage drop over molecules has been discussed by Ghosh et al.[35]

12.6 Concluding Remarks

In this chapter we have presented an intuitive description of the current–voltage (I-V) characteristics of molecules using simple toy models to illustrate the basic physics (Sections 12.1–12.3). These toy models were also used to motivate the rigorous Nonequilibrium Green's Function (NEGF) theory (Section 12.4). A simple example was then used in Section 12.5 to illustrate the application of the NEGF formalism.

The same basic approach can be used in conjunction with a more elaborate Hückel Hamiltonian or even an *ab initio* Hamiltonian. But for these advanced treatments we refer the reader to References 27 and 28.

Some of these models are publicly available through the Purdue Simulation Hub (www.nanohub.pur-due.edu) and can be run without any need for installation. In addition to the models discussed here, there is a Hückel model which is an improved version of the earlier model made available in 1999. Further improvements may be needed to take into account the role of inelastic scattering or polaronic effects, especially in longer molecules such as DNA chains.

Acknowledgments

Sections 12.2 and 12.3 are based on material from a forthcoming book by Supriyo Datta entitled *Quantum Phenomena: From Atoms to Transistors*. It is a pleasure to acknowledge helpful feedback from Mark Ratner, Phil Bagwell, and Mark Lundstrom. The authors are grateful to Prashant Damle and Avik Ghosh for helpful discussions regarding the *ab initio* models. This work was supported by the NSF under Grant No. 0085516–EEC.

12.7.A MATLAB® Codes

The MATLAB codes for the toy models can also be obtained at www.nanohub.purdue.edu.

12.7.A.1 Discrete One-Level Model

```
% Toy model, one level
% Inputs (all in eV)
E0 = -5.5;Ef = -5;gam1 = 0.2;gam2 = 0.2;U = 1;
%Constants (all MKS, except energy which is in eV)
hbar = 1.06e-34;q = 1.6e-19;IE = (2*q*q)/hbar;kT = .025;
% Bias (calculate 101 voltage points in [-4 4] range)
nV = 101;VV = linspace(-4,4,nV);dV = VV(2)-VV(1);
N0 = 2/(1+exp((E0-Ef)/kT));
for iV = 1:nV% Voltage loop
      UU = 0;dU = 1;
      V = VV(iV);mu1 = Ef-(V/2);mu2 = Ef+(V/2);
      while dU>1e-6%SCF
            E = E0+UU;
            f1 = 1/(1+exp((E-mu1)/kT));f2 = 1/(1+exp((E-mu2)/kT));
            NN = 2*((gam1*f1)+(gam2*f2))/(gam1+gam2);% Charge
            Uold = UU;UU = Uold+(.05*((U*(NN-N0))-Uold));
            dU = abs(UU-Uold);[V UU dU];
      end
      curr = IE*gam1*gam2*(f2-f1)/(gam1+gam2);
      II(iV) = curr;N(iV) = NN;[V NN];
end
G = diff(II)/dV;GG = [G(1) G];% Conductance
h = plot(VV,II*10^6,'k');% Plot I-V
```

12.7.A.2 Discrete Two-Level Model

```
% Toy model, two levels
% Inputs (all in eV)
Ef = -5;E0 = [-5.5 -1.5];gam1 = [.2 .2];gam2 = [.2 .2];U = 1*[1 1;1 1];
```

```
% Constants (all MKS, except energy which is in eV)
hbar = 1.06e-34;q = 1.6e-19;IE = (2*q*q)/hbar;kT = .025;
n0 = 2./(1+exp((E0-Ef)./kT));
nV = 101;VV = linspace(-6,6,nV);dV = VV(2)-VV(1);Usc = 0;
for iV = 1:nV
     dU = 1;
     V = VV(iV);mu1 = Ef+(V/2);mu2 = Ef-(V/2);
     while dU>1e-6
          E = E0+Usc;
          f1 = 1./(1+exp((E-mu1)./kT));f2 = 1./(1+exp((E-mu2)./kT));
          n = 2*(((gam1.*f1)+(gam2.*f2))./(gam1+gam2));
          curr = IE*gam1.*gam2.*(f1-f2)./(gam1+gam2);
          Uold = Usc;Usc = Uold+(.1*(((n-n0)*U')-Uold));
          dU = abs(Usc-Uold);[V Usc dU];
     end
     II(iV) = sum(curr);N(iV,:) = n;
end
G = diff(II)/dV;GG = [G(1) G];
h = plot(VV,II);% Plot I-V
```

12.7.A.3 Broadened One-Level Model

```
% Toy model, restricted solution with broadening
% Inputs (all in eV)
E0 = -5.5;Ef = -5;gam1 = 0.2;gam2 = 0.2;U = 1.0;
% Constants (all MKS, except energy which is in eV)
hbar = 1.06e-34;q = 1.6e-19;IE = (2*q*q)/hbar;kT = .025;
% Bias (calculate 101 voltage points in [-4 4] range)
nV = 101;VV = linspace(-4,4,nV);dV = VV(2)-VV(1);
N0 = 2/(1+exp((E0-Ef)/kT));
for iV = 1:nV% Voltage loop
     UU = 0;dU = 1;
     V = VV(iV);mu1 = Ef-(V/2);mu2 = Ef+(V/2);
     nE = 400;% Numerical integration over 200 points
     id = diag(eye(nE))';
     EE = linspace(-10,0,nE);dE = EE(2)-EE(1);
     f1 = 1./(1+exp((EE-id*mu1)/kT));
     f2 = 1./(1+exp((EE-id*mu2)/kT));
     while dU>1e-4% SCF
          E = E0+UU;
          g = 1./(EE-id*(E+i/2*(gam1+gam2)));
          NN = 2*sum(g.*conj(g).*(gam1*f1+gam2*f2))/(2*pi)*dE;
          Uold = UU;UU = Uold+(.2*((U*(NN-N0))-Uold));
          dU = abs(UU-Uold);[V UU dU];
     end
     curr = IE*gam1*gam2*sum((f2-f1).*g.*conj(g))/(2*pi)*dE;
     II(iV) = real(curr);N(iV) = NN;[V NN curr E mu1 mu2];
end
G = diff(II)/dV;GG = [G(1) G];% Conductance
h = plot(VV,II,'.'); Plot I-V
```

12.7.A.4 Unrestricted Discrete One-Level Model

```
% Toy model unrestricted solution
% Inputs (all in eV)
E0 = -5.5;Ef = -5;gam1 = 0.2;gam2 = 0.2;U = 1;
% Constants (all MKS, except energy which is in eV)
hbar = 1.06e-34;q = 1.6e-19;IE = (q*q)/hbar;kT = .025;
% Bias (calculate 101 voltage points in [-4 4] range)
nV = 101;VV = linspace(-4,4,nV);dV = VV(2)-VV(1);
N0 = 1/(1+exp((E0-Ef)/kT));
for iV = 1:nV% Voltage loop
     U1 = 0;U2 = 1e-5;dU1 = 1;dU2 = 1;
     V = VV(iV);mu1 = Ef-(V/2);mu2 = Ef+(V/2);
     while (dU1+dU2)>1e-6% SCF
          E1 = E0+U1;E2 = E0+U2;
          f11 = 1/(1+exp((E1-mu1)/kT));f21 = 1/(1+exp((E1-mu2)/kT));
          f12 = 1/(1+exp((E2-mu1)/kT));f22 = 1/(1+exp((E2-mu2)/kT));
          NN1 = ((gam1*f12)+(gam2*f22))/(gam1+gam2);
          NN2 = ((gam1*f11)+(gam2*f21))/(gam1+gam2);
          Uold1 = U1;Uold2 = U2;
          U1 = Uold1+(.05*((2*U*(NN1-N0))-Uold1));
          U2 = Uold2+(.05*((2*U*(NN2-N0))-Uold2));
          dU1 = abs(U1-Uold1);dU2 = abs(U2-Uold2);
     end
     curr1 = IE*gam1*gam2*(f21-f11)/(gam1+gam2);
     curr2 = IE*gam1*gam2*(f22-f12)/(gam1+gam2);
     (iV) = curr1;I2(iV) = curr2;
     N1(iV) = NN1;N2(iV) = NN2;[V NN1 NN2];
end
G = diff(I1+I2)/dV;GG = [G(1) G];% Conductance
h = plot(VV,I1+I2,'-'); Plot I-V
```

12.7.A.5 Unrestricted Broadened One-Level Model

```
% Toy model, unrestricted solution with broadening
% Inputs (all in eV)
E0 = -5.5;Ef = -5;gam1 = 0.2;gam2 = 0.2;U = 1;
% Constants (all MKS, except energy which is in eV)
hbar = 1.06e-34;q = 1.6e-19;IE = (q*q)/hbar;kT = .025;
% Bias (calculate 101 voltage points in [-4 4] range)
nV = 101;VV = linspace(-4,4,nV);dV = VV(2)-VV(1);
N0 = 1/(1+exp((E0-Ef)/kT));
nE = 200;% Numerical integration over 200 points
id = diag(eye(nE))';
EE = linspace(-9,-1,nE);dE = EE(2)-EE(1);
for iV = 1:nV% Voltage loop
     U1 = 0;U2 = 1;dU1 = 1;dU2 = 1;
     V = VV(iV);mu1 = Ef-(V/2);mu2 = Ef+(V/2);
     f1 = 1./(1+exp((EE-id*mu1)/kT));
     f2 = 1./(1+exp((EE-id*mu2)/kT));
     while (dU1+dU2)>1e-3% SCF
          E1 = E0+U1;E2 = E0+U2;
```

```
            g1 = 1./(EE-id*(E1+i/2*(gam1+gam2)));
            g2 = 1./(EE-id*(E2+i/2*(gam1+gam2)));
            NN1 = sum(g1.*conj(g1).*(gam1*f1+gam2*f2))/(2*pi)*dE;
            NN2 = sum(g2.*conj(g2).*(gam1*f1+gam2*f2))/(2*pi)*dE;
            Uold1 = U1;Uold2 = U2;
            U1 = Uold1+(.2*((2*U*(NN2-N0))-Uold1));
            U2 = Uold2+(.2*((2*U*(NN1-N0))-Uold2));
            dU1 = abs(U1-2*U*(NN2-N0));dU2 = abs(U2-2*U*(NN1-N0));
        end
        curr = IE*gam1*gam2*sum((f2-f1).*(g1.*conj(g1)+...
        g2.*conj(g2)))/(2*pi)*dE;
        II(iV) = real(curr);N(iV) = NN1+NN2;
        [V NN1 NN2 curr*1e6 E1 E2 mu1 mu2];
end
G = diff(II)/dV;GG = [G(1) G];% Conductance
h = plot(VV,II,' - ');% Plot I-V
```

References

1. P. Avouris, P.G. Collins, and M.S. Arnold. Engineering carbon nanotubes and nanotube circuits using electrical breakdown. *Science*, 2001.

2. A. Bachtold, P. Hadley, T. Nakanishi, and C. Dekker. Logic circuits with carbon nanotube transistors. *Science*, 294:1317, 2001.

3. H. Schön, H. Meng, and Z. Bao. Self-assembled monolayer organic field-effect transistors. *Nature*, 413:713, 2001.

4. R.P. Andres, T. Bein, M. Dorogi, S. Feng, J.I. Henderson, C.P. Kubiak, W. Mahoney, R.G. Osifchin, and R. Reifenberger. "Coulomb staircase" at room temperature in a self-assembled molecular nanostructure. *Science*, 272:1323, 1996.

5. M.A. Reed, C. Zhou, C.J. Muller, T.P. Burgin, and J.M. Tour. Conductance of a molecular junction. *Science*, 278:252, 1997.

6. C. Kergueris, J.-P. Bourgoin, D. Esteve, C. Urbina, M. Magoga, and C. Joachim. Electron transport through a metal–molecule–metal junction. *Phys. Rev. B*, 59(19):12505, 1999.

7. C. Kergueris, J.P. Bourgoin, and S. Palacin. Experimental investigations of the electrical transport properties of dodecanethiol and α, ω bisthiolterthiophene molecules embedded in metal–molecule–metal junctions. *Nanotechnology*, 10:8, 1999.

8. J. Reichert, R. Ochs, H.B. Weber, M. Mayor, and H.V. Löhneysen. Driving current through single organic molecules. *cond-mat/0106219*, June 2001.

9. S. Hong, R. Reifenberger, W. Tian, S. Datta, J. Henderson, and C.P. Kubiak. Molecular conductance spectroscopy of conjugated, phenyl-based molecules on au(111): the effect of end groups on molecular conduction. *Superlattices Microstruct.*, 28:289, 2000.

10. J.J.W.M. Rosink, M.A. Blauw, L.J. Geerligs, E. van der Drift, and S. Radelaar. Tunneling spectroscopy study and modeling of electron transport in small conjugated azomethine molecules. *Phys. Rev. B*, 62(15):10459, 2000.

11. C. Joachim and J.K. Gimzewski. An electromechanical amplifier using a single molecule. *Chem. Phys. Lett.*, 265:353, 1997.

12. W. Tian, S. Datta, S. Hong, R. Reifenberger, J.I. Henderson, and P. Kubiak. Conductance spectra of molecular wires. *J. Chem. Phys.*, 109(7):2874, 1998.

13. J. Chen, W. Wang, M.A. Reed, A.M. Rawlett, D.W. Price, and J.M. Tour. Room-temperature negative differential resistance in nanoscale molecular junctions. *Appl. Phys. Lett.*, 77(8):1224, 2000.

14. D. Porath, A. Bezryadin, S. de Vries, and C. Dekker. Direct measurement of electrical transport through DNA molecules. *Nature*, 403:635, 2000.

15. M. Magoga and C. Joachim. Conductance of molecular wires connected or bonded in parallel. *Phys. Rev. B*, 59(24):16011, 1999.

16. L.E. Hall, J.R. Reimers, N.S. Hush, and K. Silverbrook. Formalism, analytical model, and *a priori* Green's function-based calculations of the current–voltage characteristics of molecular wires. *J. Chem. Phys.*, 112:1510, 2000.

17. M. Paulsson and S. Stafström. Self-consistent field study of conduction through conjugated molecules. *Phys. Rev. B*, 64:035416, 2001.

18. E.G. Emberly and G. Kirczenow. Multiterminal molecular wire systems: a self-consistent theory and computer simulations of charging and transport. *Phys. Rev. B*, 62(15):10451, 2000.

19. P.S. Damle, A.W. Ghosh, and S. Datta. Unified description of molecular conduction: from molecules to metallic wires. *Phys. Rev. B*, 64:201403(r), 2001.

20. J. Taylor, H. Gou, and J. Wang. *Ab initio* modeling of quantum transport properties of molecular electronic devices. *Phys. Rev. B*, 63:245407, 2001.

21. M. Di Ventra, S.T. Pantelides, and N.D. Lang. First-principles calculation of transport properties of a molecular device. *Phys. Rev. Lett.*, 84:979, 2000.

22. P. Damle, A.W. Ghosh, and S. Datta. First-principles analysis of molecular conduction using quantum chemistry software. *Chem. Phys.*, 281, 171, 2001.

23. Y.Q. Xue, S. Datta, and M.A. Ratner. Charge transfer and "band lineup" in molecular electronic devices: a chemical and numerical interpretation. *J. Chem. Phys.*, 115:4292, 2001.

24. J.J. Palacios, A.J. Pérez–Jiménez, E. Louis, and J.A. Vergés. Fullerene-based molecular nanobridges: a first-principles study. *Phys. Rev. B*, 64(11):115411, 2001.

25. J.M. Seminario, A.G. Zacarias, and J.M. Tour. Molecular current–voltage characteristics. *J. Phys. Chem. A*, 1999.

26. S. Datta. Nanoscale device modeling: the Green's function method. *Superlattices Microstruct.*, 28:253, 2000.

27. P.S. Damle, A.W. Ghosh, and S. Datta. Molecular nanoelectronics, in M. Reed (Ed.), *Theory of Nanoscale Device Modeling*. (To be published in 2002; for a preprint, e-mail: datta@purdue.edu.)

28. F. Zahid, M. Paulsson, and S. Datta. Advanced semiconductors and organic nano-techniques, in H. Markoc (Ed.), *Electrical Conduction through Molecules*. Academic Press. (To be published in 2002; for a preprint, e-mail: datta@purdue.edu.)

29. S. Datta. *Electronic Transport in Mesoscopic Systems*. Cambridge University Press, Cambridge, UK, 1997.

30. S.G. Lias et al. Gas-phase ion and neutral thermochemistry, *J. Phys. Chem.* Reference Data. American Chemical Society and Americal Institute of Physics, 1988.

31. C. Desjoqueres and D. Spanjaard. *Concepts in Surface Physics*. 2nd ed. Springer-Verlag, Berlin, 1996.

32. R.G. Parr and W. Yang. *Density Functional Theory of Atoms and Molecules*, Oxford University Press, 1989, p.99

33. de Picciotto, H.L. Stormer, L.N. Pfeiffer, K.W. Baldwin, and K.W. West. Four-terminal resistance of a ballistic quantum wire. *Nature*, 411:51, 2001.

34. S. Datta, W. Tian, S. Hong, R. Reifenberger, J.I. Henderson, and C.P. Kubiak. Current–voltage characteristics of self-assembled monolayers by scanning tunneling microscopy. *Phys. Rev. Lett.*, 79:2530, 1997.

35. A.W. Ghosh, F. Zahid, P.S. Damle, and S. Datta, Insights from I-V asymmetry in molecular conductors. Preprint. *cond-mat/0202519*.

36. Special issue on single-charge tunneling, *Z. Phys.B.*, 85, 1991.

37. P.F. Bagwell and T.P. Orlando. Landauer's conductance formula and its generalization to finite voltages. *Phys. Rev. B*, 40(3):1456, 1989.

38. M. Brandbyge, J. Taylor, K. Stokbro, J.-L. Mozos, and P. Ordejon. Density functional method for nonequilibrium electron transport. *Phys. Rev. B*, 65(16):165401, 2002.

39. R. Zeller, J. Deutz, and P. Dederichs. *Solid State Commun.*, 44:993, 1982.

40. Y. Imry. *Introduction to Mesoscopic Physics*. Oxford University Press, Oxford, 1997.

41. D.K. Ferry and S.M. Goodnick. *Transport in Nanostructures*. Cambridge University Press, London, 1997.

42. K. Hansen, E. Laegsgaard, I. Stensgaard, and F. Besenbacher. Quantized conductance in relays. *Phys. Rev. B*, 56:1022, 1997.

43. J.N. Murrell and A.J. Harget. *Semi-Empirical SCF MO Theory of Molecules*. John Wiley & Sons, London, 1972.

Section 4

Manipulation and Assembly

13

Nanomanipulation: Buckling, Transport, and Rolling at the Nanoscale

Richard Superfine
University of North Carolina

Michael Falvo
University of North Carolina

Russell M. Taylor, II
University of North Carolina

Sean Washburn
University of North Carolina

CONTENTS

13.1 Introduction

The study of novel materials produces many challenges in the areas of synthesis, modeling, and characterization. For the latter, one would like to be able to determine mechanical, electrical, and dynamical properties and correlate them with structure. Correspondingly, a new perspective is emerging in biology —the significance of studying individual proteins and macromolecular structures is now appreciated as a way to understand the details of molecular binding, transport, and kinetic pathways. Enabling this revolution in nanoscale science has been the development of microscopy techniques that measure structure and many material properties. The atomic force microscope (AFM) provides a wide range of characterization capabilities (electrical, mechanical, chemical, etc.) on the nanometer scale, while correlating these with structure in the form of detailed topography.[1] Impressive as these capabilities are, the AFM can also be used as a nanometer-scale manipulation tool. The ability to manipulate objects efficiently on surfaces makes available a wide variety of experiments on the interactions between the sample and substrate,[2–4] on the physical properties of individual objects, and on the creation of unusual devices incorporating the nanometer objects.[5,6]

In the laboratory the scientist must use an instrument to probe the nanoscale world. The challenge for the instrument developer is to enable ever-increasing resolution of material structure and properties. The advances in instrumentation are often characterized as ever-increasing spatial resolution, smaller force detection, and smaller electronic current measurement. Beyond these obvious considerations, two other challenges emerge as equally important: the interface between the scientist and the instrument, and the combination of characterization techniques.[7] The ultimate goal of an interface is to make the instrument transparent for the scientist. Actions to be performed within the nanoscale world should feel as natural as performing them on a tabletop object. Also, the visualization of data should be as natural as looking at the object under study as if it were held in hand. As we continue to probe molecular systems, we need to correlate a wide range of physical and chemical properties with structure, all at the same time. This necessitates the ability to measure mechanical, electrical, dynamical, and structural properties all within the same instrument.

In the following chapter, we describe work performed at the University of North Carolina–Chapel Hill (UNC) in the development of microscopy instrument systems, including a natural interface for scanned probe microscopy we call the *nanoManipulator*. We describe the principle design features of the instrument system including the visual display of data, the haptic (force-feedback) control, and display capabilities. Second, we describe the combination of microscopy and manipulation in a joint scanning electron microscopy/scanning probe microscopy system. These systems have been used for studies of nanotube mechanical, dynamical, and electrical properties[8] and for the study of biological macromolecular structures such as viruses, fibers (pili, fibrin, microtubules, etc.), and molecules (DNA).[9] We describe examples of these studies drawn from our work on nanotubes and viruses.

13.2 Instrumentation Systems: The NanoManipulator and Combined Microscopy Tools

13.2.1 The NanoManipulator: The Scientist as an Actor in Nanoscale Science

The goal of our interface development has been to allow the scientist to be an actor in the nanoscale world. We have combined microscopy with a virtual reality interface to provide the intuitive display of instrument data and natural control of the instrument functions. The significance of the virtual reality interface to the SPM is that it gives the scientist simulated *presence* on the sample surface. The benefits of this are improved perception of three-dimensional structures, more effective exploration of the sample, the ability to observe dynamic processes in near-real time, and the ability to interactively modify the surface. To put it in plain language: when you are present somewhere, you can look around, you can look at things from different angles, you can feel interesting things at arm's length, you can watch the behavior of things that move or change, you can pick up things and rearrange them, and you can tweak things to see how they respond. The ideal human interface for a scanning probe microscope (SPM) might present its user with a scaled-up three-dimensional representation of the surface that can be probed and modified with a physical hand-held tool. The control system would translate tool motion into motion of the SPM tip and translate measured surface parameters into force pushing back on the tool, as well as visual and auditory representations of surface data. When using such a system, a scientist would seem to be interacting directly with the surface itself. Natural motions of head and hand would be used to investigate and sculpt the surface as if it were physically present at the scale of the scientist. This would allow the scientist to concentrate fully on investigating the surface and its features rather than on programming the interface. A complete description of virtual reality interfaces for microscopy is available elsewhere.[7] Here we provide a brief description of the representative features of this system, including the display of information through visualization and haptics and the control of the instrument. The nanoManipulator, as diagrammed in Figure 13.1, consists of a scanning probe microscope with its controller, a PHANToM force-feedback device with its controller, and a PC with graphics card. These three computers communicate across an IP-based network.

13.2.1.1 Display of Data: Visualization and Haptics

The SPM is a highly refined instrument for exploration and manipulation at the nanometer scale. Many experiments have been done using preprogrammed or open-loop control of the SPM tip and feedback parameters during modification, and using pseudo-color or line-drawn images to display the collected

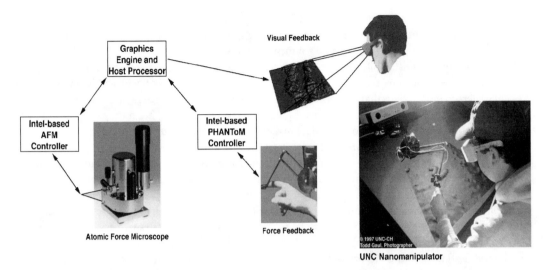

FIGURE 13.1   NanoManipulator system diagram with the system in use. The nanoManipulator comprises a scanning probe microscope (SPM) with its controller, a PHANToM force-feedback device with its controller, and a PC with graphics card. These three computers communicate across an IP-based network. This system combines head-tracked stereo–video with a force-feedback input device to allow the user to see and feel a three-dimensional representation of the surface under study with an SPM. The user is in direct control of the tip's lateral motion and force, allowing the user to manipulate the surface. This picture shows physics graduate student Scott Paulson using the system to investigate virus particles.

data.[10] In a later section we describe the various modes of control of the microscope. Here we focus on the display of data. In addition to surface topography data, SPMs can also acquire many other data sets. These include conductance and current/distance measures (scanning tunneling microscopy), lateral force and adhesion (AFM), laser transmission at various wavelengths and polarizations (near field optical microscopy), magnetic properties (magnetic force microscopy), temperature, and other parameters.[11] Few if any of these have natural mappings to visual channels such as color or height, but nonetheless the scientist may want to view the correlations between these data sets and topography. In these cases, it is useful to display the parameter visually in a nonrealistic but still useful manner. For scientific visualization, the usefulness of a technique is more important than its realism. We have studied and developed many approaches for the visualization of single and multiple data sets.[12,13] Here we review the simple issue of the three-dimensional rendering of a surface as an example of the considerations involved.

In the simplest case, the sample data is the surface topography. This is most naturally displayed as a three-dimensional data set: a directionally illuminated surface. Using off-the-shelf PCs, it is now possible to have real-time rendering of surfaces as the data is acquired from the microscope. We provide real-time interaction with viewing parameters such as illumination direction, viewing angle, and scaling through the force-feedback pen. The three-dimensional view is not always better: for flat surfaces with small features, a pseudo-color view is superior to the shaded three-dimensional view. This is true for two reasons: the pseudo-coloring devotes the entire intensity range to depth, and it also obscures small fluctuations caused by noise in the sampled image. For features that are significant only for their height difference from the rest of the image, the pseudo-color image devotes its entire range of intensities to showing this difference. This is equivalent to drawing the surface from above using only ambient lighting with intensity based on height. Any three-dimensional shading of the image uses some of this range to accommodate the diffuse and specular components at the expense of the ambient component. This shift from ambient to diffuse/specular is a shift from displaying height to displaying slope (because angle is what determines the brightness of the diffuse and specular components). This works well for a noise-free image, but real-world images contain noise. For SPM images, a flat surface has the highest percentage of noise as a fraction of feature size. The noise causes fluctuations in illumination of the same magnitude as those caused by features, sometimes completely obscuring them. However, small features on surfaces with other height variations are better brought out using specular highlighting of a three-dimensional surface (and are often imperceptible in a two-dimensional display).[7]

Furthermore, the scientist's ability to recognize specific molecular structures within the noisy, sampled data is improved by using stereoscopic, shaded, three-dimensional color graphics with specular highlights. Providing stereoscopic rather than monoscopic viewing is useful to the scientist because the stereo provides a direct perception of depth for nearby virtual objects. Allowing accurate perception of the three-dimensional spatial structure of STM data makes it possible for scientists to use their own specialized knowledge to recognize structures and features of interest in the data.

It is intriguing to think of giving the scientist the ability not only to see the surface under the SPM but actually to reach out and touch it. Conceptually, this is equivalent to a telerobotic system that operates across a great difference in scale rather than over a great distance. We employ a force-feedback pen that senses the three-dimensional position of the pen tip (with additional capability to sense a total of 6° of freedom) while also applying force to the user's hand. The position of the pen in space can control various visualization features such as virtual "grabbing" of the surface to change viewing angle. However, it is during modification that force-feedback has proved itself most useful by providing intuitive control of the probe tip position and sensing of the surface contours. This ability allows finer control and enables entirely new types of experiments. Force-feedback has proved essential to finding the right spot to start a modification, finding the path along which to modify, and providing a finer touch than permitted by the standard scan–modify–scan experiment cycle. Force-feedback allows the user to locate objects and features on the surface by feel while the tip is held still near the starting point for modification. Surface features marking a desired region can be located without relying only on visual feedback from the previous scan. This allowed a collaborator to position the tip directly over an adenovirus particle, then increase the force to cause the particle to dimple directly in the center. It also allowed the tip to be placed between two touching carbon filaments in order to tease them apart.

Finally, the nanoManipulator system records all data taken, including topography and external channels such as conductance, as a time-sequenced data set we call a *streamfile*. This file can be replayed at a later time to review the entire experiment, including the complete correlation of property measurements with updating images of the sample structure. The user is not left with a record of the data as determined by decisions made at the time of the experiment. Often the important data is not appreciated until after the experiment is over. This has been widely appreciated in our group, where many of our most exciting insights have been discovered in post-experiment reviews of streamfiles.

13.2.1.2 Instrument Control: Haptics, Virtual Tips, and On-the-Fly Mode Switching

A manipulation can be performed most intuitively by allowing the user's natural hand motions to control the trajectory of the probe tip. We have implemented several strategies for this control. The simplest technique is to allow the computer to control the force applied by the tip on the sample through a feedback loop, while the user's hand motion controls the lateral position of the tip. When the user moves the hand-held pen up and down, a "virtual" surface is felt, such that when the hand is above a designated reference plane, no force is felt. As the hand is lowered and the pen crosses this reference plane, a force is displayed to the hand that reflects the local topography data. Complete control over the tip in three dimensions has been implemented whereby the up-and-down motion of the user's hand controls the motion of the probe tip in the z direction, normal to the surface. To prevent the crashing of the tip into the surface, we have implemented a force limit that returns control of the tip to the computer if the user exceeds a designated applied force.[7]

During the course of a manipulation, it becomes clear that the straightforward approach of pushing the object with the AFM tip is not the most efficient. In the control mode, where the user determines the position of the tip through the hand-held stylus, similar to how the computer moves the tip during a raster, the object is observed to move off to the side of the tip. The reasons for this are obvious, in a naïve sense, to anyone who has pushed a tennis ball with a screwdriver. In performing such an operation, a user is likely to reach for another pushing tool, such as a flat-bladed spatula, which would present a broader surface to the object. While fabrication techniques for tips are reaching a high art, it is clearly impractical to have on hand a range of tip shapes for the range of tasks that might have to be accomplished. We have addressed this issue through the implementation of what we call *virtual tips*, whereby the AFM tip is moved by computer control along a trajectory at a speed much faster than the average translation of the tip. The tip then appears to have shape that can be controlled by the computer. Several of these shapes are described below, including sharp, blunt, sweep, and comb modes. In addition, for AFM-based machining of metal and polymer films, we have implemented sewing mode, whereby the tip is lifted and pressed into the surface in order to machine narrow lines.

1. Sharp mode — no sub-trajectory.
2. Blunt mode — tip moves in a triangular pattern with leading point of triangle oriented along the pen path. The size of the diamond can be varied.
3. Sweep mode — tip moves back and forth along a line. The length and the orientation of the line with respect to the average path are dynamically controlled by the use of two of the angular degrees of freedom of the hand-held force device.
4. Sewing mode — tip moves in the z-direction normal to the surface. It is alternately raised to a pullback position and pressed to a set-point force. This can be implemented as either a fixed-frequency oscillation or a fixed-step size between presses.
5. Comb mode — this is a lateral "tapping" mode where the tip is moved along a line as in sweep mode but takes excursions in the perpendicular direction at proscribed distances. This allows the tip to move across an extended object, such as a tubular virus, and nudge it along its length while always releasing in the local normal direction. In this way the tip does not rub along the object, avoiding possible damage from frictional forces.

The above techniques have been tested in a variety of circumstances. Sharp mode is currently used the most for feeling and for manipulation. Blunt mode has been used in some cases to reduce noise during feeling. Sweep mode is widely used for both manipulation and for lithography, especially the manipulation of stiff objects such as nanotubes, where it allows for controlled translation without rotation. There are several other techniques that might play an important roll in the manipulator's toolbox. These modes are corral mode and roll mode. Corral mode has been implemented to address the manipulation challenges of small clusters where the object is observed to roll off to the side of the tip. A possible remedy is to have the tip move along a trajectory where the tip is hitting the object from a range of angles, with the mean direction determined by the trajectory of the stylus motion.[7]

An AFM typically operates in one of two modes, *contact mode* or *vibrating mode*. Contact mode is described in the introduction, where the tip scrapes along the surface. Vibrating mode oscillates the tip at its resonance frequency (around 100 kHz) and pushes toward the surface until the amplitude of oscillation is reduced to a fraction of its value away from the surface. Vibrating mode does not scrape the surface as much as contact mode does, so it provides damage-free scanning on more surfaces. The nanoManipulator interface allows switching between contact and vibrating modes without retracting the tip.[14] This allows a wider range of interaction forces between imaging and modification. Unfortunately, the positioning elements in the microscope undergo a sharp jump during the transition from one mode to the other. This results in a transient offset, where the surface height appears to jump and then to relax to its new height over several seconds. This is compensated for automatically by the nanoManipulator, avoiding what would otherwise be sharp force discontinuities as the user goes from feeling to modifying the surface.

13.2.2 Combined Scanning Electron/Scanning Probe Microscopy

Most recently, we have combined an AFM, a scanning electron microscope (SEM), and the nanoManipulator interface to produce a manipulation system with simultaneous electron microscopy imaging.[15] Manipulation in the AFM has the limitation of the inability to view the full orientation of the manipulated object during the manipulation. This is because there is only a single tip, and it can either image or manipulate at one time. Only after the manipulation is completed is the tip returned to an imaging mode and the user can understand the results of the manipulation. Scanning Electron Microscopy has been used for a wide range of surface characterization and imaging applications. Unfortunately, the electron beam is incapable of maneuvering objects or measuring their mechanical interactions. With a combined system of the SEM and the AFM, manipulations can be viewed in real time to view material deformations and to precisely place nanoscale objects with respect to each other or within devices. In this way, more sophisticated nanostructures can be created, and measurements can be performed by placing nano-objects within other measuring devices.

Several advances have been made to the user interface to take advantage of these new capabilities. First, we have implemented full three-dimensional control of the AFM tip with the haptic interface, as described above. When the user moves the hand-held pen in three dimensions, the tip follows the corresponding trajectory within the SEM chamber. The force applied by the tip to the sample is conveyed to the user, and a force limit is imposed on the interface so that the user cannot arbitrarily crash the tip. Second, the new system presents challenges in combining two data sets from different imaging techniques, each with its own set of artifacts. These include instrument response functions, drift, and skew of the image, to name a few. We are currently developing strategies for accounting for the limitations of the individual imaging techniques through image modeling, and performing a combining procedure that will provide a best-case reconstruction using the information from the two data sets.

13.3 Nanomanipulation for Mechanical Properties

The mechanical properties of nanoscale materials are of great interest due to the potential lack of defects. In macroscopic samples, mechanical properties can be limited by the presence of defects that can promote sliding of shear planes and fracture. In addition, nanoscale structures may have shapes that are not found in typical materials. An example of this is the nanotube with a hollow core, such as has been found in carbon and noncarbon varieties. In these materials, new modes of deformation may arise that enable the material to perform new functions as objects in device settings or in functional materials. Manipu-lation of nanoscale materials can offer the ability to apply large strains with the accompanying observation of the resulting deformation and its possible reversibility.

We have used the nanoManipulator interface for AFM to perform intricate bending of carbon nanotubes (CNT) to large strains. A carbon nanotube consists of a graphite sheet wrapped seamlessly onto itself to form a tube.[16] They exist in single wall form (SWCNT)[17] and multiwall form (MWCNT)[18] where the tube wall is made of many concentric shells. The SWCNT often come in a *rope* or *bundled* form where many individual tubes are close-packed in parallel. CNTs have been shown to exhibit extraordinary electrical and mechanical properties.[19] The tubular structure takes advantage of the high basal-plane elastic modulus of graphite to produce a fiber predicted to have mechanical properties that surpass those of any previously known material.[20,21] We have observed a variety of behaviors, including bending with and without periodic buckling, as well as large bends that are accommodated either by dramatic, irreversible structural changes or by smooth changes without apparent damage. In the first series of images, a CNT, pinned at one end by carbon debris, was bent into many configurations. Figure 13.2 shows a series of 29 images where the CNT was bent repeatedly back and forth upon itself. The nanotube is never observed to fracture, even with a strain as large as 16% as measured along its outside wall in the tightest bend.[22]

Insight into the remarkable behavior of the nanotube is given through an analysis of another nanotube that had been carefully bent at small radii of curvatures. The inset of Figure 13.3A shows this second tube in its original adsorbed position and orientation. The abrupt vertical step on the left side of the tube was used as a reference mark (s = 0) for feature locations. This CNT was taken through a series of 20 distinct manipulations alternately bending and straightening the tube at various points. We present images from this sequence to highlight specific features. Along with AFM topographic data, we also present the tube height along the tube's center as determined by the cores method[23] along with the calculated curvature. We will refer to raised points on the tube as *buckles*, consistent with the increase in

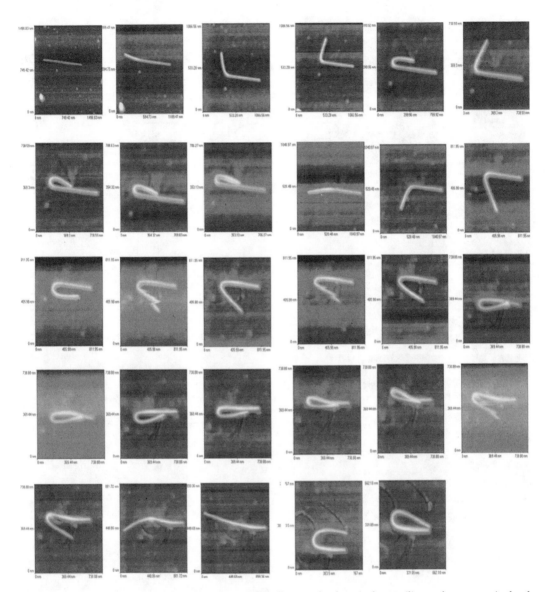

FIGURE 13.2 A sequence of images of a carbon multiwall nanotube deposited on a silicon substrate manipulated by an atomic-force microscope tip, back and forth through strains as high as 16%. No catastrophic fracture of the nanotube is observed.

height expected from the collapse of a shell in response to bending. We observe two behaviors in this sequence: small, regularly distributed buckles at regions of small curvature and large deformation at high curvatures.[22,24,25]

We focus first on the regularly spaced buckles occurring in the more gradually bent region (A, lower right). Figures 13.3B and 13.3C show the tube as it is bent in opposite directions. The location of the buckles has shifted dramatically, with the buckles of Figure 13.3C appearing in regions which had been featureless, and the buckles of Figure 13.3B largely disappearing. The buckles appear with a characteristic interval independent of their absolute positions along the tube. These findings suggest that the buckling is reversible, intrinsic to the CNT, and not mediated by defects. The strong correlation between tube curvature and the location of buckles confirms their role in reducing curvature-induced strain. Buckling in bent shells is well known in continuum mechanics,[26] where two modes of deformation result from

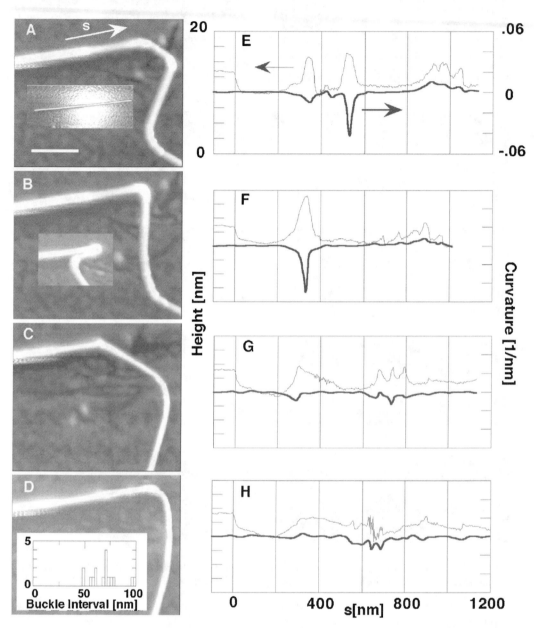

FIGURE 13.3 Curvature and height of buckles along bent carbon nanotube. The white scale bar (A) is 300 nm long and all figures are to the same scale. A 20 nm diameter tube is manipulated with AFM from its straight shape (inset of A) into several bent configurations (A–D). The height and curvature of the bent tube along its centerline (indicated by the arrow in A) is plotted (E–H). The upper trace in each graph depicts the height relative to the substrate and the lower depicts the curvature data. The ripple-like buckles occur between s = 600 nm and 1200 nm and correspond to the tube in the lower right portion of the images in each case. Note that these buckles migrate as the tube is manipulated into different configurations. The appearance and disappearance of the ripple buckles as well as the severe buckle at s~500nm (E–F) suggest elastic reversibility. The distance between ripple buckles was determined for the various bent configurations of the experiment. The average of the buckle interval histogram (D, inset) establishes the dominant interval of 68 nm as the characteristic length of the rippling.

pure bending stresses. The Brazier effect causes the circular cross-section of the tube to become more "ovalized" uniformly over the whole tube length[27] as the bending curvature increases. Bifurcation, on the other hand, leads to a periodic, low-amplitude rippling of the tube wall on the inside of the bend (the portion under compression).

These experiments showed the remarkable strength of the nanotube is accompanied by a resiliency that is promoted by its shape as a hollow cylinder. Upon bending, the nanotube can release the strain on its outer (tensile) and inner (compressive) surfaces by buckling. This indicates that in applications as fillers in composite materials, the nanotube may show elasticity to high strains in both tension and in bending due to the ability of the hollow core to collapse.

13.3.1   Nanoscale Dynamics: Atoms as Gear Teeth

The relative motion of objects in contact is a ubiquitous phenomenon, appearing in the lubricated contact of macroscopic objects and increasingly playing a role in nanometer-scale electromechanical structures.[28,29] Over the past decade, new experimental techniques such as the Surface Force Apparatus (SFA)[30] and atomic force microscopy (AFM) have obtained an atomic-scale view of moving interfaces. Through such studies, the intrinsic dependence of friction on contact area[2,31] and the dependence of friction on crystallographic orientation have been identified. AFM studies have performed friction mapping of surfaces with atomic resolution[32] and the identification of stick-slip motion for nanometer-scale objects.[3] Nanomanipulation provides the opportunity to take an object with atomically smooth and clean surfaces and study its full range of motion in a variety of contexts. Here we describe our experiments with carbon nanotubes which provide a laboratory for atomic lattice interactions.

Friction at the nanoscale differs from macro or micro friction in the degree of order of the contact. As in the case of mechanical properties, in a macroscale or micron-scale contact, the frictional properties can be dominated by defects and represent an average over many crystalline orientations. In a nanometer-scale contact, we have the possibility of having a single crystalline orientation dominate the interfacial dynamics. New behavior emerges due to relative perfection and smoothness of the contacting surfaces[2] and the *ability to tune precisely the relative orientation of contacting atomic lattices.*

In this case the atomic structure matters. The arrangement of the atoms in two interacting surfaces has been shown to play a critical role in the energy loss that occurs when one body slides over a second both in experiment[33,34] and simulation.[29] In particular, in the case of two contacting solid crystalline surfaces, the degree of commensurability has been shown to have a clear effect on friction.[2,35,36] Understanding the effect of these atomic interactions on energy loss will lend insight into the underlying mechanisms of friction. We have established the ability to tune the commensurability of a nanometer contact through AFM manipulation, providing a powerful knob to turn in performing experiments probing the origins of friction. Furthermore, we have established that the electronic transport across the nanometer contact is also modulated smoothly as a function of commensurability.[37] This gives us an independent measure of the structure of the contact and a direct look at the relationship between mechanical and electronic processes in a sliding contact.[29]

When an AFM tip pushes an object, three types of motion can occur: uniform translation, rotation in the plane, and rolling. We have studied all three of these outcomes in the nanotube/flat substrate system. In the simplest case, the nanotube does not roll. Instead, it translates with a rotation in the plane. This is the behavior for multiwall nanotubes on substrates such as mica, oxidized silicon, and MoS_2, and for graphite when the lattices are not interlocked. We have analyzed this motion with a simple analytical model that uses the AFM tip as a positional constraint, with a uniform distributed frictional force applied along the tube by the substrate. When the nanotube is manipulated from the side, the tube undergoes an in-plane rotation about a pivot point that depends on the location of the AFM tip during the push. A unique relation between the point of manipulation and the point of rotation can be derived with no fitting parameters, assuming a uniform friction along the length of the contact.[38] Additional phenomena emerge during end-on pushing of nanotubes. In this case, the lateral force shows a large initial spike before the onset of the uniform translation of the tube — a signature of stick-slip phenomena. These

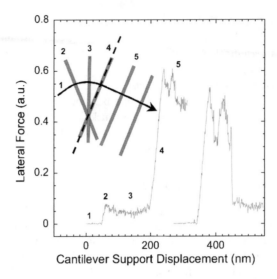

FIGURE 13.4 Lateral force trace as a carbon nanotube (CNT) is rotated into (left trace) and out of (right trace) commensurate contact. The inset shows a top-view schematic of the process for the left trace. (1) The AFM tip is moving along in contact with the graphite substrate. (2) The CNT is contacted and begins rotating in-plane (3). (4) The commensurate state is reached (indicated by the dashed line) and the lateral force rises dramatically before rolling motion begins (5). The right trace begins with the tip on the substrate, the tip then contacts the CNT in the commensurate state, begins rolling, and then pops out of commensurate contact and begins rotating in plane with a corresponding drop in lateral force.

measurements have been shown to be consistent with the shear stress of a graphene-sheet contact, demonstrating that the manipulation of surface-bound objects can provide quantitative measures of the fundamental quantity of tribology and the surface shear stress for any combination of available nanoscale material and substrate while observing complex dynamics.

The manipulation of nanotubes reveals surprises when performed on graphite. If the nanotube starts in an out-of-registry state, it slides smoothly, as described above. However, this in-plane rotational motion is interrupted at discrete in-plane orientations where the nanotube "locks" into a low-energy state, indicated by a ten-fold increase in the force required to move the CNT (Figure 13.4, left). Subsequently, the nanotube rolls with a constant in-plane orientation and characteristic stick-slip modulation in the lateral force.[39,40] This is shown in Figure 13.5, where manipulations from different nanotubes are shown. The characteristic of the rolling motion is (1) no in-plane rotation during the translation of the CNT (as is observed in the nongraphitic substrates above), (2) features in the topography (shape) of the nanotube that reappear in subsequent images, and (3) features in the lateral force that reproduce themselves with a distance that corresponds to the circumference of the nanotube. The lateral force traces all show the expected periodic behavior, while the inset shows a set of images of the changing end cap of the nanotube upon manipulation. This sequence is repeated as the nanotube continues to be rolled across the substrate.

Why does the nanotube roll on graphite and not on other substrates? Most significant, nanotubes can be manipulated to reveal a set of three distinct in-registry orientations separated by ±60°. This behavior is seen in Figure 13.4, where we show lateral force traces as a CNT is rotated into (left trace) and out of (right trace) commensurate contact. The inset shows a top-view schematic of the process for the left trace. (1) The AFM tip is moving along in contact with the graphite substrate. (2) The CNT is contacted and (3) begins rotating in-plane. (4) The commensurate state is reached (indicated by the dashed line), and the lateral force rises dramatically before rolling motion begins (5). The right trace begins with the tip on the substrate, the tip then contacts the CNT in the commensurate state, begins rolling, and then pops out of commensurate contact and begins rotating in plane with a corresponding drop in lateral force. The AFM image sequence of Figure 13.5A–C shows a CNT in three manipulated commensurate

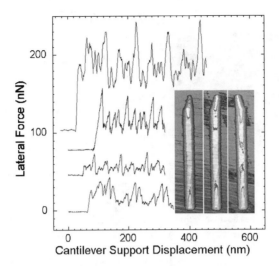

FIGURE 13.5 Periodic lateral force traces indicating rolling motion. The four traces are for four different CNTs. In each case, the periodicity in the traces matches the circumference of the nanotube, indicating rolling without slipping motion. The inset illustrates the topographical evidence for rolling. The top end of the nanotube has an asymmetry that changes in a way that is consistent with rolling motion. The second trace from the top corresponds to the nanotube in the inset sequence.

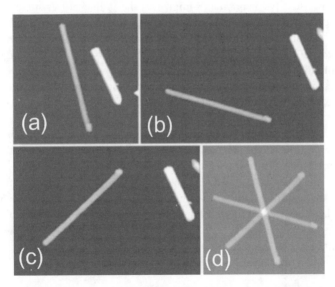

FIGURE 13.6 Commensurate orientations of a CNT on graphite. The nanotube on the left of (A), (B), and (C) is rotated in-plane into three commensurate orientations indicated by pronounced increase in lateral force as shown in Figure 13.4. In (D), the three images were translated and overlaid to emphasize the three-fold symmetry of the commensurate orientations. Similar measurements performed on the shorter CNT demonstrated a registry angle 11° different from the longer tube, most clearly explained as a difference in the chirality of the nanotubes.

orientations. The NT on the left of Figure 13.6(A), (B), and (C) is rotated in-plane into three commensurate orientations indicated by pronounced increase in lateral force as shown in Figure 13.4. In Figure 13.6(D), the three images were overlaid to emphasize the three-fold symmetry of the commensurate orientations. This set of manipulations and quantitative measurements shows that the atomic arrangement of the substrate is responsible for the rolling behavior. Do the atoms of the nanotube play any role?

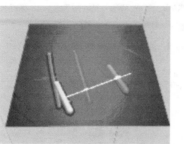 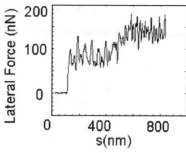

FIGURE 13.7 The right (short) CNT is manipulated by the AFM tip to roll across the graphite substrate. The short CNT collides with the longer CNT, and the two nanotubes continue to roll across the surface. Each maintains its own lock-in angle, with the AFM tip touching only the shorter CNT. The lateral force trace on the right shows a jump at 150 nm when the AFM tip touches the short CNT, and a second jump at 550 nm when the short CNT begins to push (and roll) the longer CNT. The trace to the left of 550 nm shows the lateral force needed to roll both CNTs at the same time.

We have also manipulated two CNTs (CNT 1: 950 nm long, 20 nm diameter; CNT2: 500 nm long, 34 nm diameter) lying in the same immediate area on the graphite substrate. While each nanotube shows the complete set of lock-in behaviors noted above, the two CNT have lock-in orientations that are different from each other by 11°. This difference in lock-in angles implicates the nanotube lattice in nanotube rolling. If the registered orientations are due to atomic registry, the particular set of registered orientations is determined by the nanotube chirality (the wrapping orientation of the outer graphene sheet of the CNT). Taken together, these two experiments imply that the atomic lattices are interlocking to produce rolling. The atoms can act like gear teeth. We have manipulated two CNTs into a collision to demonstrate the robust gear-like motion (Figure 13.7). The orientation of each nanotube is preserved through the collision and subsequent rolling. The lateral force trace after the collision is the sum of the two periodic rolling signals from the individual CNTs (Figure 13.7, right).

Molecular statics and dynamics calculations have shed light on the rolling dynamics.[41,42] They have shown the discrete lock-in orientations, calculated energy barriers for sliding, in-plane rotation, and rolling, all as a function of the nanotube size. Most important is our convergence on an explanation of our rolling data. Our results are consistent with faceted graphene cylinders. First, the lateral force traces are periodic with repeat distance equal to the nanotube circumference. Second, the molecular statics calculations on perfect cylinders indicate that the nanotube should always roll, as rolling has a lower energy barrier than any other motion, whether the tube is in registry or out of registry. The molecular statics calculations that allow relaxation of the nanotube shape show that a sufficiently large multiwall nanotube is faceted. We believe that this faceting plays a key role in our experimental observations. First, the energy cost for rolling now includes a component due to the adhesion of the facet face and the substrate, substantially larger than that for the perfect cylinder. This energy cost for rolling is lower than the energy cost of sliding only when the nanotube is in registry.

13.3.2 Nanoscale Contacts: Momentum Conservation across a Lattice Plane Contact

The lattice plays a critical role in the transport of electrons across an interface. The ability to control the registry angle between the nanotube and the graphite substrate provides us the opportunity to study the electron transport dependence on the orientation of the crystalline axis in the contact. We have measured the resistance of the CNT/graphite contact via a two-probe measurement: the graphite substrate itself serving as one lead, and a conducting AFM tip brought to contact the top of the CNT as the other.[37] Resistance measurements as a function of lattice angle between the nanotube and the graphite substrate are taken in the following manner. The NT is imaged in noncontact (oscillating) mode to locate the tube and measure the angle of the tube. The zero of angle for each tube, designated as the commensurate position, is determined by the lateral force signal that shows the lock-in position. The tip is then engaged

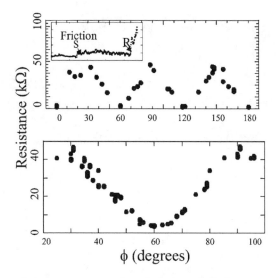

FIGURE 13.8 Measurement of electron transport through the atomic lattice contact of a CNT with a graphite substrate. The resistance of a two-probe measurement from a metallized AFM tip contacting the top of a CNT into the graphite substrate is plotted as a function of the registry angle. The lateral force signature of lattice registry determines the lock-in angles. The resistance is periodic, consistent with a CNT/graphite contact resistance that is 50 times smaller when the lattices are in registry. The lower trace is a higher angle resolution trace showing the detail as the lattices approach registry.

in contact mode on top of the NT with the desired force (~50 nN), and the voltage between the tip and sample is swept from –0.1 to 0.1 V while the current is measured. Data for two different tubes are shown in Figure 13.8. Data from a tube that has been rotated through 180°, with resistance measurements taken approximately every 7–10°, is shown in Figure 13.8 (top). From the data, it is clear the resistance minima occur at the commensurate positions 0, 60, 120, and 180°, and that the data is periodic, repeating every 60°. Figure 13.8 (bottom) shows more dense data, collected approximately every 3° from a different CNT (45 nanometers by 1.6 microns) showing in detail the change in the resistance near the commensurate position.

An analysis of the resistance measurements for the tip against the nanotube and directly applied to the substrate allows us to conclude that the nanotube–graphite contact resistance is changing by a factor of fifty, with the lowest resistance occurring in the in-registry orientation, when the nanotube and substrate have the A–B stacking of crystalline graphite. Why is the resistance lowest in this orientation? Two possible answers are the effect of atomic interlocking and the conservation of momentum. The first has been studied by calculations that show a change in the contact resistance by a factor of two as two single-wall nanotubes are translated past each other.[43] This implies that the conservation of k-vector, the direction of the wave function, is the dominant effect in our measurements. The conduction electrons in graphite are located at the corners of the hexagonal Brilloiun zone, meaning that there are six allowed directions for the electron transport.[44,45] The contact resistance will be lowest when the wave functions in the nanotube and in the substrate have the same value of the k-vector, including direction, and therefore allow transport from one to the other without additional scattering. Therefore, the ability to manipulate the top contact in an atomically clean junction has allowed the striking demonstration of a fundamental concept in transport, *momentum conservation*. This implies that devices based on nanotubes will need to control the relative lattice orientations, or take advantage of this dependence for a new class of sensors and devices.

13.3.3 Nanomanipulation and MEMS Devices

The correlation of structure and properties is the cornerstone of the physical sciences, and the design and control of the nanoworld depends on measurements performed simultaneously with atomic scale

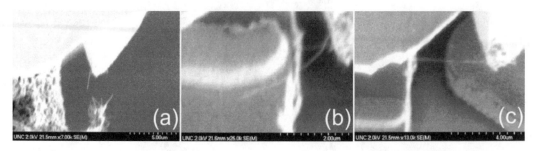

FIGURE 13.9 Manipulation of nanotube onto a MEMS device using a combined SEM/AFM manipulation system for correlation of stress–strain and conductance of individual elements and nanotube/matrix integrity. (a) Nanotube is secured onto AFM tip. (b) Nanotube positioned over MEMS structure, then stretched across open gap (c).

structure determination. The tool of choice for atomic scale structure has been high-resolution transmission electron microscopy (TEM). While the TEM can provide atomic-scale chemical information, the space within the microscope is extremely tight, with sample volumes less than a few millimeters thick. This presents a significant challenge to the design of mechanical stressing stages and force-sensing manipulation stages that can be integrated with the TEM. We have begun to design such stages using Microelectromechanical Systems (MEMS) technology that is widely applied in sensing systems, displays, and optoelectronics.[46,47] This technology can be used to create moving stages with manipulation capability as well as force- and transport-sensing stages for correlated stress/strain/transport/ structure measurements. With such stages in hand, the remaining challenge is to place the nanostructure of interest into the device. We have succeeded in using our SEM/AFM manipulation system to place carbon nanotubes onto MEMS stages, and to use stages designed for stress–strain measurements to grab nanotubes.

The methodology, shown in Figure 13.9, starts with the creation of a nanotube cartridge through electro-deposition of nanotubes onto the edge of a metal foil from a nanotube/dichloromethane suspension.[48] The nanotubes protrude off the edge of the foil and act as a source for manipulation by the AFM tip. The tip is positioned against the nanotubes and it is observed that van der Waals forces are sufficient to adhere them to the silicon AFM tip and pull the tubes away from the foil. The nanotube is clearly observed in the SEM image during the positioning of the tip and the pull-off. The AFM tip is then positioned over the gap in the MEMS device, and the contact of the nanotube is observed as it begins to bend. The tip is then moved laterally across the gap, laying the tube down on the near edge of the device. The SEM is then made to image the location where the CNT lies flat against the surface, depositing a carbon contamination layer that can pin the nanotube and provide electrical contact.[49] The tip is then moved further away until the CNT drops completely off the tip and lies on the far side of the gap. The SEM is then again used to tie the CNT to the surface. Carbon contamination layer produced by the electron bombardment has been shown to be sufficient to secure the nanotube even under stresses that tore the CNT. This methodology of nanosample placement on MEMS testing structures will allow subsequent use of the MEMS test stage in a variety of settings, such as TEM or low-temperature, high-magnetic-field cryostats where *in situ* manipulation is very challenging.

We have also demonstrated that MEMS stages can be used to grab nanosamples during the manipulation of CNT. In this experiment, shown in Figure 13.10, the nanotube, as attached to an AFM tip, was positioned in the gap of a MEMS stress–strain stage. The stage was then closed through the actuation of the comb drive. The tip was then retracted until the nanotube was observed to sever. Transport measurements were taken during the manipulation and the application of stress, and they showed significant changes during the stretching of the CNT. These measurements will be improved by a post-processing step of metalization to provide a metal–nanotube junction within the MEMS device. The ability to manipulate nanoscale structures provides great flexibility in characterization methodologies, including the application of integrated testing stages that can be incorporated into the tight confines of high-resolution transmission electron microscopes and low-temperature cryostats.

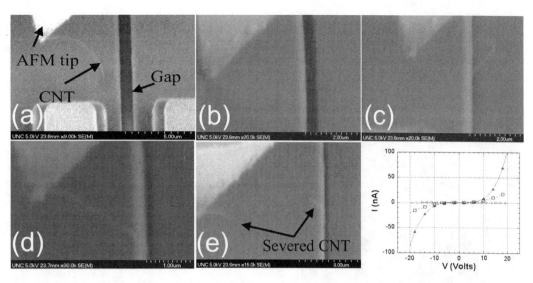

FIGURE 13.10 SEM/nanoManipulator AFM manipulation of carbon nanotube (CNT) into gap in MEMS stress–strain stage. Here the stage is used as a grabber. The tube is maneuvered into the gap (top left, center), and the gap is closed (top right), pinching the CNT. The nanotube is then stretched (lower left) while measuring electron transport until severing occurs (lower center). Lower right shows I-V curves taken before (triangles) and during stretching (open squares). After severing, the I-V showed an open circuit (diamonds).

13.3.4 Nanomanipulation in Biology: Binding and Dynamics

The study of surface interactions and dynamics of biological macromolecules and macromolecular assemblies will provide insight into specific binding that may occur at cell surfaces and the transport of vesicles and viruses. The application of normal forces to measure the energetics of macromolecule extension and antibody–antigen binding has been pursued using AFM.[50,51] Here we discuss the application of lateral forces to measure surface interactions and dynamics of viruses.[52,53] The force necessary to move a virus across the surface can reveal specific and nonspecific interactions and, potentially, through the use of molecularly treated surfaces, probe model cell membrane substrates. In addition, the translation of viruses within the lumen of vessels or on the cell surface is not understood. The measurement of lateral forces during manipulation can shed light on the mode of transport: rolling or sliding.

We have performed preliminary measurements of virus manipulation on solid surfaces under ambient conditions. Our target virus is adenovirus, of interest as a human pathogen and as a vector for gene therapy.[54] Adenoviruses have been deposited onto clean and functionalized silicon substrates, with manipulations performed in liquid and in air under low-humidity conditions. While it is clear that measurements performed in liquid will provide the most direct insight into physiological processes,[55] measurements performed in air are relevant to viral delivery strategies that require the drying of the virion for storage and dosage. Figure 13.11 shows a typical manipulation under ambient conditions, with an initial large manipulation force needed to dislodge the virus before its first translation and a subsequent low force needed to maintain its motion. A second push has a much lower force to initiate motion. We believe that the initial peak is indicative of the strength of surface binding of the virus, with subsequent pushes placing a virion surface in contact with the substrate that is unable to relax to minimize the interfacial energy.

A closer examination of the lateral force traces reveals insight into the mode of virus transport during manipulation. The autocorrelation of the lateral force during manipulation, shown in Figure 13.12, reveals a periodicity in the lateral force of 23.5 nm. The virion, with an icosahedral shape, has a height of 75 nm and a circumference of 235 nm across ten faces.[56] This corresponds to the periodicity in the lateral force trace and reveals a peeling process that is occurring as the virion is rolled from face to face

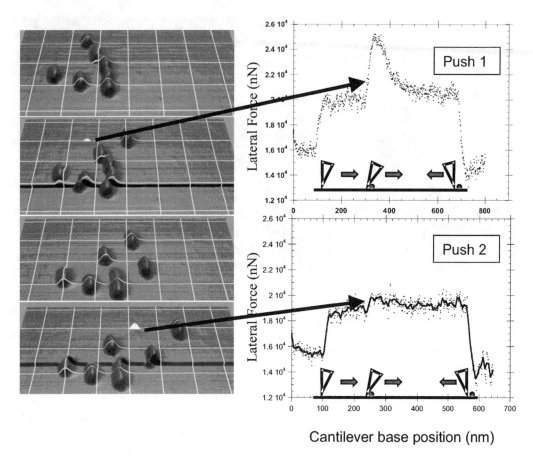

FIGURE 13.11 Adenovirus manipulated in ambient conditions shows lateral force signatures consistent with strong initial binding and subsequent low release force for continued translation. One virus particle is selected and manipulated twice. The top right lateral force trace shows the large release peak at 300 nm where the AFM tip first contacts the virus. The second push of the same virus particle (lower right) shows a significantly lower translation force when the virus is contacted.

as it is manipulated across the surface. Manipulation of control particles — spherical polystyrene bead of similar diameter — showed no signature of periodicity.

The case of the virus bears similarities and differences from the case of carbon nanotubes. Both reveal rolling through a periodicity in the lateral force trace that is consistent with the geometrical features of the object. However, we found that the nanotube involved a balance between sliding and peeling energies that dictated that the nanotube should roll unless the sliding energy was increased through commensurate contact with a substrate. However, the virus was observed to roll on homogeneous substrates where no gear-like interlocking could occur. The molecular origins of the virus/substrate interactions and the competition between sliding and peeling (rolling) await to be uncovered.

13.4   Conclusion

The strongest material buckling like a soda straw, atomic teeth as gears, electron transport facilitated by the interlocking of atomic lattices, rolling viruses — there is much to be learned about the nanoscale world, including how properties such as electrical transport, mechanical properties, and dynamics are affected by the atomic-scale structure of the nano-objects and their interfaces. Nanomanipulation provides exciting insight into these problems by allowing us to probe individual objects with great facility

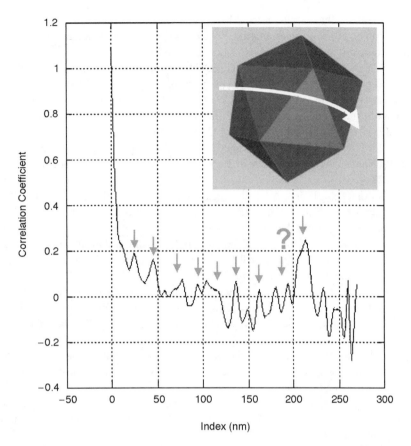

FIGURE 13.12 Evidence for rolling of a virus particle upon translation is revealed through the autocorrelation of the lateral force trace after the initial release of the virus from the surface. The 23.5 nm periodicity indicated in the correlation function corresponds to a pealing of the virus at each facet of the 75 nm high icosahedral capsid.

and to combine property characterization with structural information. Advanced user interfaces will continue to play a critical role in making experiments more transparent to the user and enabling the scientist to be an actor in the nanoscale world.

Acknowledgments

The authors would like to thank the members of the NanoScience Research Group at UNC for their work, and the National Science Foundation, the National Institutes of Health, National Center for Research Resources, the Office of Naval Research, and the Army Research Office for support.

References

1. Guntherodt, H.-J., Anselmetti, D., and Meyer, E., Forces in scanning probe methods, in *NATO ASI Series E, Applied Sciences*, 1st ed., Kluwer Academic Publishers, Dordrecht, The Netherlands, 1995.
2. Sheehan, P.E. and Lieber, C.M., Nanotribology and nanofabrication of MoO_3 structures by atomic force microscopy, *Science* 272 (May 24), 1158–1161, 1996.
3. Luthi, R., Meyer, E., Haefke, H., Howald, L., Gutmannsbauer, W., and Guntherodt, H.-J., Sled-type motion on the nanometer scale: determination of dissipation and cohesive energies of C_{60}, *Science* 266 (December 23), 1979–1981, 1994.

4. Resch, R., Baur, C., Bugacov, A., Koel, B.E., Madhukar, A., Requicha, A.A.G., and Will, P., Building and manipulating three-dimensional and linked two-dimensional structures of nanoparticles using scanning force microscopy, *Langmuir* 14 (23), 6613–6616, 1998.

5. Postma, H.W.C., de Jonge, M., Yao, Z., and Dekker, C., Electrical transport through carbon nanotube junctions created by mechanical manipulation, *Phys. Rev. B* 62 (16), R10653–R10656, 2000.

6. Williams, P.A., Patel, A.M., Papadakis, S.J., Seeger, A., Taylor II, R.M., Helser, A., Sinclair, M., Falvo, M.R., Washburn, S., and Superfine, R., Controlled placement of an individual carbon nanotube onto a MEMS structure, *Appl. Phys. Lett.* 2574–2576, 2002.

7. Taylor II, R.M. and Superfine, R., Advanced interfaces to scanning probe microscopes, in Nalwa, H.S. (Ed.), *Handbook of Nanostructured Materials and Nanotechnology*, Academic Press, New York, 1999, pp. 271–308.

8. Falvo, M.R., Clary, G., Helser, A., Paulson, S., Taylor II, R.M., Chi, V., Brooks Jr., F.P., Washburn, S., and Superfine, R., Nanomanipulation experiments exploring frictional and mechanical properties of carbon nanotubes, *Microsc. Microanal.* 4, 504–512, 1998.

9. Guthold, M., Falvo, M., Matthews, W.G., Paulson, S., Mullin, J., Lord, S., Erie, D., Washburn, S., Superfine, R., Brooks, F.P., and Taylor, R.M., Investigation and modification of molecular structures using the nanoManipulator, *J. Mol. Graph. Model.* 17, 187–197, 1999.

10. Magonov, S.N. and Whangbo, M.-H., *Surface Analysis with STM and AFM: Experimental and Theoretical Aspects of Image Analysis*, VCH Publishers Inc., New York, 1996.

11. Bonnell, D.A., *Scanning Probe Microscopy and Spectroscopy: Theory, Techniques, and Applications*, Wiley, New York, 2000.

12. Weigle, C., Emigh, W.G., Liu, G., Taylor II, R.M., Enns, J.T., and Healey, C.G., Oriented sliver textures: a technique for local value estimation of multiple scalar fields, in *Proc. Graphics Interface 2000*, Montreal, Canada, 2000.

13. Seeger, A., Henderson, A., Pelli, G.L., Hollins, M., and Taylor II, R.M., Haptic display of multiple scalar fields on a surface, in *Proc. Worksh. New Paradigms Inf. Visualization Manipulation, NPIVM 2000*, Washington, D.C., 2000, pp. 33–38.

14. Falvo, M., Finch, M., Chi, V., Washburn, S., Taylor II, R.M., Brooks, Jr., F.P., and Superfine, R., The nanoManipulator: a teleoperator for manipulating materials at the nanometer scale, in *Proc. 5th Intl. Symp. Sci. Eng. Atomically Engineered Mater.*, World Scientific, Richmond, VA, November 5, 1995.

15. Seeger, A., Paulson, S., Falvo, M., Helser, A., Taylor II, R.M., Superfine, R., and Washburn, S., Hands-on tools for nanotechnology, *J. Vac. Sci. Technol. B* 19, 2717–2722, 2001.

16. Iijima, S., Helical microtubules of graphitic carbon, *Nature* 354, 56–58, 1991.

17. Thess, A., Lee, R., and Smalley, R.E., Crystalline ropes of metallic carbon nanotubes, *Science* 273 (July 26), 483, 1996.

18. Ebbesen, T.W. and Ajayan, P.M., Large-scale synthesis of carbon nanotubes, *Nature* 358, 16, 1992.

19. Dresselhaus, M.S., Dresselhaus, G., and Avouris, P., *Carbon Nanotubes: Synthesis, Structure, Properties and Applications*, Springer, New York, 2001.

20. Wong, E.W., Sheehan, P.E., and Lieber, C.M., Nanobeam mechanics: elasticity, strength, and toughness of nanorods and nanotubes, *Science* 277, 1971–1975, 1997.

21. Treacy, M.M.J., Ebbesen, T.W., and Gibson, J.M., Exceptionally high Young's modulus observed for individual carbon nanotubes, *Nature* 381 (June 20), 678–680, 1996.

22. Falvo, M.R., Clary, G.J., Taylor, R.M.I., Chi, V., Brooks, Jr., F.P., Washburn, S., and Superfine, R., Bending and buckling of carbon nanotubes under large strain, *Nature* 389 (October 9), 582–584, 1997.

23. Pizer, S.M., Eberly, D., Morse, B.S., and Fritsch, D.S., Zoom-invariant vision of figural shape: the mathematics of cores, in *Computer Vision and Image Understanding*, 1997.

24. Poncharal, P., Wang, Z.L., Ugarte, D., and de Heer, W.A., Electrostatic deflections and electromechanical resonances of carbon nanotubes, *Science* 283 (March 5), 1513–1516, 1999.

25. Bower, C., Rosen, R., Jin, L., Han, J., and Zhou, O., Deformation of carbon nanotubes in nanotube-polymer composites, *Appl. Phys. Lett.* 74 (22), 3317–3319, 1999.

26. Ju, G.T. and Kyriakides, S., Bifurcation and localization instabilities in cylindrical shells under bending, II. Predictions, *Intl. J. Solids Struct.* 29 (9), 1143–1171, 1992.

27. Axelrad, E.L., On local buckling of thin shells, *Intl. J. Non-Linear Mech.* 20 (4), 249–259, 1985.

28. Bhushan, B., Micro/nanotribology and micro/nanomechanics of MEMS devices, in Bhushan, B. (Ed.), *Micro/Nano Tribology*, CRC Press, Boca Raton, FL, 1998, p. 797.

29. Persson, B.N.J., *Sliding Friction: Physical Principles and Applications*, Springer-Verlag, Berlin, 1998.

30. Israelachvili, J., *Intermolecular and Surface Forces*, 2nd ed., Academic Press, San Diego, 1991.

31. Johnson, K.L., A continuum mechanics model of adhesion and friction in a single asperity contact, in Bhushan, B. (Ed.), *Micro/Nanotribology and its Applications*, Kluwer Academic Publishers, Dordrecht, The Netherlands, 1997, p. 157.

32. Mate, M.C., McClelland, G.M., Erlandsson, R., and Chiang, S., Atomic-scale friction of a tungsten tip on a graphite surface, *Phys. Rev. Lett.* 59 (17), 1942–1945, 1987.

33. Overney, R.M., Takano, H., Fujihira, M., Paulus, W., and Ringsdorf, H., Anisotropy in friction and molecular stick-slip motion, *Phys. Rev. Lett.* 72 (22), 3546–3549, 1994.

34. Liley, M., Gourdon, D., Stamou, D., Meseth, U., Fischer, T.M., Lautz, C., Stahlberg, H., Vogel, H., Burnham, N.A., and Duschl, C., Friction anisotropy and asymmetry of a compliant monolayer induced by a small molecular tilt, *Science* 280, 273–275, 1998.

35. He, G., Muser, M.H., and Robbins, M.O., Adsorbed layers and the origin of static friction, *Science* 284 (4 June), 1650–1652, 1999.

36. Hirano, H., Shinjo, K., Kaneko, R., and Murata, Y., Anisotropy of frictional forces in muscovite mica, *Phys. Rev. Lett.* 67 (19), 2642–2645, 1991.

37. Paulson, S., Falvo, M., Buongiorno–Nardelli, M., Taylor II, R.M., Helser, A., Superfine, R., and Washburn, S., Tunable resistance of a carbon nanotube-graphite interface, *Science* 290, 1742–1744, 2000.

38. Mason, M.T. and Salisbury, J.K.J., *Robot Hands and the Mechanics of Manipulation*, MIT Press, Cambridge, MA, 1985.

39. Falvo, M.R., Taylor II, R.M., Helser, A., Chi, V., Brooks, Jr., F.P., Washburn, S., and Superfine, R., Nanometre-scale rolling and sliding of carbon nanotubes, *Nature* 397 (21 January), 236–238, 1999.

40. Falvo, M., Steele, J., Taylor II, R.M., and Superfine, R., Gearlike rolling motion mediated by commensurate contact: carbon nanotubes on HOPG, *Phys. Rev. B.* 62, R10665–10667, 2000.

41. Schall, J.D. and Brenner, D.W., Molecular dynamics simulations of carbon nanotube rolling and sliding on graphite, *Molecular Simulation* 25 (1–2), 73–79, 2000.

42. Buldum, A. and Lu, J.P., Atomic scale sliding and rolling of carbon nanotubes, *Phys. Rev. Lett.* 83 (24), 5050–5053, 1999.

43. Buldum, A. and Lu, J.P., Contact resistance between carbon nanotubes, *Phys. Rev. B* 63, 161403–161406, 2001.

44. Mintmire, J.W., Dunlap, B.I., and White, C.T., Are fullerenes metallic? *Phys. Rev. Lett.* 68, 631, 1992.

45. Dresselhaus, M.S., Dresselhaus, G., and Eklund, P.C., *Science of Fullerenes and Carbon Nanotubes*, Academic Press, San Diego, 1996.

46. Craighead, H.G., Nanoelectromechanical systems, *Science* 290, 1532–1535, 2000.

47. Bishop, D., Gammel, P., and Randy–Giles, C., The little machines that are making it big, *Phys. Today* 54 (10), 38–44, 2001.

48. Nishijima, H., Kamo, S., Akita, S., Nakayama, Y., Hohmura, K.I., Yoshimura, S.H., and Takeyasu, K., Carbon-nanotube tips for scanning probe microscopy: preparation by a controlled process and observation of deoxyribonucleic acid, *Appl. Phys. Lett.* 74 (26), 4061–4063, 1999.

49. Yu, M.F., Files, B.S., Arepalli, S., and Ruoff, R.S., Tensile loading of ropes of single wall carbon nanotubes and their mechanical properties, *Phys. Rev. Lett.* 84 (24), 5552–5555, 2000.

50. Wang, M.D., Manipulation of single molecules in biology, *Curr. Opinion Biotechnol.* 10 (1), 81–86, 1999.

51. Hansma, H.G. and Hoh, J., Biomolecular imaging with atomic force microscope, *Ann. Rev. Biophys. Biomol. Struct.* 23, 115–139, 1994.

52. Falvo, M.R., Washburn, S., Superfine, R., Finch, M., Brooks, F.P.J., Chi, V., and Taylor, II, R.M., Manipulation of individual viruses: friction and mechanical properties, *Biophys. J.* 72 (March), 1396–1403, 1997.

53. Matthews, G., Guthold, M., Negishi, A., Taylor, R.M., Erie, D., Jr., F.P.B., and Superfine, R., Quantitative manipulation of DNA and viruses with the nanoManipulator scanning force microscope, *Surf. Interface Anal.* 27, 437–43, 1999.

54. Samulski, R.J., *Adeno-Associated Virus Based Vectors for Human Gene Therapy,* World Scientific Publishing Co., Singapore, 1995.

55. Bustamante, C. and Keller, D., Scanning force microscopy in biology, *Phys. Today* 48 (12), 32–38, 1995.

56. Horne, R.W., Brenner, S., Waterson, A.P., and Wildy, P., The icosahedral form of an adenovirus, *J. Mol. Biol.* 1, 84–86, 1959.

14

Nanoparticle Manipulation by Electrostatic Forces

CONTENTS

Michael Pycraft Hughes
University of Surrey

14.1 Introduction

Methods for the manipulation of nanometer-scale objects can be grouped into two philosophies:[1] self-assembly of molecular components, a process widely seen in nature, is described as the *bottom-up approach* (small things make larger things); the opposite is the *top-down approach* where large machines (such as the tools of modern semiconductor manufacture) are used to make much smaller products. While the former is considered to be ultimately preferable, modern technology has only been able to manipulate molecular-scale objects at will using the top-down approach, such as writing using xenon atoms as ink and a scanning tunneling microscope as a pen.[2]

However, with the use of very small electrodes (usually formed by top-down methods), it is possible to generate electric fields with such complex local geometry that the manipulation of single molecules in solution becomes possible — and these fields can be manipulated so that the particles can be steered across the electrodes and used to assemble miniature electric circuits.[3] The means by which this is achieved is through the interaction between time-variant, nonuniform electric fields, which are manifested as a number of sub-phenomena collectively referred to as *electrokinetics*.

Alternating-current (AC) electrokinetic techniques such as dielectrophoresis and electrorotation[4,5] have been used for many years for the manipulation, separation, and analysis of objects with lengths of the order 1 μm–1 mm in solution. The induction of a force (or torque) occurs due to the interaction of an induced dipole with the imposed electric field. This can be used to cause the object to exhibit a variety of motions including attraction, translation, and rotation by changing the nature of the dynamic field. In many ways, these forces may be viewed as an electrostatic equivalent to optical tweezers[6] and optical spanners[7] in that they exert translational and rotational forces on a body due to the interaction between the body and an imposed field gradient.

Recent advances in semiconductor manufacturing technology have enabled researchers to develop electrodes for manipulating proteins,[8,9] to concentrate 14 nm beads from solution,[10] or to trap single viruses and 93 nm-diameter latex spheres in contactless potential energy cages.[11] AC electrokinetic techniques are simple, cheap, and require no moving parts, relying entirely on the electrostatic interactions between the particle and dynamic electric field. In this chapter we will consider the ways in which this form of electrostatic manipulation can benefit nanotechnology and the constraints on the technique due to factors such as Brownian motion, heating of the medium, and electrode dimensions as the electrode array is miniaturized to the nanometer scale.

14.2 Theoretical Aspects of AC Electrokinetics

14.2.1 Dielectrophoresis

If a dielectric particle is suspended in an electric field, it will *polarize* (positive and negative charges will accumulate at the ends of the particle nearest to the electrode of opposite sign); or, to use an alternate phrase, a *dipole* is *induced* in the body. If that electric field is spatially uniform, then the Coulombic attraction induced at either side of the particle is equal and there is no net movement. However, if the electric field strength is higher on one end of the particle than the other, the net force is nonzero and the particle will move toward the region of highest electric field; that is, it will move *up the field gradient*. This is *dielectrophoresis*,[12] shown schematically in Figure 14.1. Note that the force is related to the direction

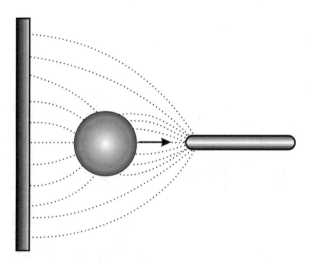

FIGURE 14.1 A schematic of a polarizable particle suspended within a point-plane electrode system. When the particle polarizes, the interaction between the dipolar charges and the local electric field produces a force. Due to the inhomogeneous nature of the electric field, the force is greater in the side facing the point than that on the side facing the plane, and there is net motion toward the point electrode. This effect is called *positive dielectrophoresis*. If the particle is less polarizable than the surrounding medium, the dipole forms counter to the field, and the particle will be repelled from high field regions, called *negative dielectrophoresis*.

of the field gradient and is independent of the direction of the electric field; consequently, we can use AC as well as DC fields.

If the particle is more polarizable than the medium around it (as shown in the figure), the dipole aligns with the field and the force moves up the field gradient toward the region of highest electric field. If the particle is less polarizable than the medium, the dipole aligns against the field and the particle is repelled from regions of high electric field. The principal polarization mechanism is governed by Maxwell–Wagner interfacial polarization; the orientation of the net dipole is governed by the charge on the particle surface and countercharge from the medium. The amounts of these charges, and hence the dipole direction, is frequency dependent. Since the alignment of the field is irrelevant, the use of AC fields has the advantage of reducing any *electrophoretic* force (due to any net particle charge) to zero.

The dielectrophoretic force, $\mathbf{F}_{DEP}$, acting on a spherical body is given by:

$$\mathbf{F}_{DEP} = 2\pi r^3 \varepsilon_m Re[K(\omega)]\nabla E^2 \tag{14.1}$$

where r is the particle radius, ε_m is the permittivity of the suspending medium, ∇ is the Del vector (gradient) operator, E is the *rms* electric field strength, and $Re[K(\omega)]$ is the real part of the Clausius–Mossotti factor, given by:

$$K(\omega) = \frac{\varepsilon_p^* - \varepsilon_m^*}{\varepsilon_p^* + 2\varepsilon_m^*} \tag{14.2}$$

where ε_m^* and ε_p^* are the complex permittivities of the medium and particle, respectively, and $\varepsilon^* = \varepsilon - (j\sigma/\omega)$ with σ the conductivity, ε the permittivity, and ω the angular frequency of the applied electric field. The limiting (DC) case of Equation 14.2 is:

$$k(\omega = 0) = \frac{\sigma_p - \sigma_m}{\sigma_p + 2\sigma_m} \tag{14.3}$$

The frequency-dependence of $Re[K(\omega)]$ indicates that the force acting on the particle varies with frequency. The magnitude of $Re[K(\omega)]$ varies depending on whether the particle is more or less polarizable than the medium. If $Re[K(\omega)]$ is positive, then particles move to regions of highest field strength (positive dielectrophoresis); the converse is negative dielectrophoresis, where particles are repelled from these regions. By careful construction of the electrode geometry that creates the electric field, it is possible to create electric field morphologies so that potential energy minima are bounded by regions of increasing electric field strengths. An example of such an electrode (often called a *quadrupolar* or *polynomial* electrode array) is shown in Figure 14.2. In such electrodes, particles experiencing positive dielectrophoresis are attracted to the regions of highest electric field (typically the electrode edges, particularly where adjacent electrodes are close), while particles experiencing negative dielectrophoresis are trapped in an isolated field minimum at the center of the array.

14.2.2 Dielectrophoretic Behavior of Solid Particles

Dielectrophoretic forces can be induced in a wide range of sub-micrometer particles from molecules to viruses. However, before we study how dielectrophoresis might be applied to manipulating and studying these complex particles, it is wise to consider a simple case to see how the basic principles of dielectrophoresis work at the sub-micrometer range. This is important because, when the diameter becomes significantly smaller than 1 μm, a number of factors that have relatively little effect on the dielectrophoretic response of larger particles (such as cells) increase in importance and begin to dominate the response. In order to understand these effects fully, we shall begin by examining a simple particle, consisting of one material only. By examining this, we can develop a model that can later be adapted for more complex shapes and structures.

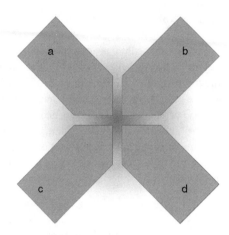

FIGURE 14.2 A schematic of an array of electrodes typically used in nanoscale dielectrophoresis work, known as a *quadrupolar* or *polynomial* electrode array. The gap between opposing electrodes in the center of the array is typically of the order 1–10 µm for nanometer-scale work, but arrays as small as 500 nm have been used. To induce dielectrophoretic motion in particles suspended near the electrode array, electrodes would be energized such that a and c are of the same phase, and b and d are in antiphase to them. Changing the phase relationship allows other effects such as electrorotation to be observed.

One of the most useful tools to understanding the fundamental mechanisms underlying the dielectrophoresis of particles on the nanometer scale is the homogeneous sphere or bead, typically made from polymers such as latex[13] or, occasionally, from metals such as palladium.[3] Most common in dielectrophoresis research are latex spheres. These are (as their name suggests) spherical blobs of latex that have been impregnated with fluorescent molecules, enabling the observation of very small particles (sizes as small as 14 nm diameter are available) with a fluorescence microscope. The primary advantage of using latex spheres is that they are a known quantity. They are solid and homogeneous (that is, they consist of one material and are consistent throughout). The internal conductivity and permittivity are known, as are the surface properties. Because they are spherical, their dielectrophoretic behavior is easy to model. Furthermore, there are straightforward chemical methods for changing those surface properties. Finally, they are readily available in a wide variety of sizes and colors.

Consider a typical experiment in which a sample of latex beads in a solution of known conductivity (such as ultra-pure water with a small quantity of potassium chloride, KCl) is placed on an electrode array written onto a microscope slide and covered with a coverslip. The electrode slide is then placed onto a fluorescence microscope (required in order to see particles this small) and the electrodes are connected, via attached wires, to a power source (typically a benchtop signal generator, providing perhaps 5 V_{pk-pk} at a frequency between 10 kHz and 10 MHz or more). When the voltage is applied, the particles are observed to move quickly to the electrodes. Within a few seconds, collections such as those shown in Figure 14.3 are observed. Whether the particles collect in the interelectrode *arms* (Figure 14.3a) or in the center of the array (Figure 14.3b) depends on the frequency of the applied voltage, occurring at low and high frequencies, respectively. At one specific frequency, the force appears to vanish and the particles float freely. Varying the voltage also changes the force, making the particles travel more quickly or slowly to the trap. If the particles are small enough, then the magnitude of Brownian motion is sufficient to require a large voltage to be applied in order to ensure the particles remain trapped.

The frequency-dependence of ε^*, and hence $Re[K(\omega)]$, implies that the force on the particle also varies with frequency. Under certain conditions, and at a specific value of frequency, the force on the particle goes to zero (i.e., $Re[K(\omega)] = 0$). Above that frequency, a homogeneous particle will experience negative dielectrophoresis; below that, positive dielectrophoresis. The frequency at which the polarizability of the particle (and hence the force exerted on it) goes to zero is commonly referred to as the *crossover frequency*.

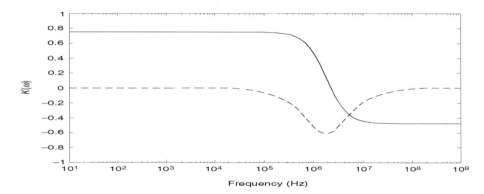

FIGURE 14.3 Fluorescence photograph of 216 nm latex beads collecting in an electrode array, approximately 5 seconds after the application of a $20V_{pk-pk}$ electric field. (a) A 1 MHz signal produces positive dielectrophoresis; (b) a 10 MHz signal gives negative dielectrophoresis. Scale bar in both pictures: 20 μm.

Crossover frequencies are a product of dielectric dispersions which cause the polarizability of the particle to change sign. It is possible to monitor the effects of changing the medium conductivity on the crossover frequency in order to estimate the properties of the particle.

Consider the following example. The polarizability of a homogeneous spherical particle (conductivity 10mSm^{-1}, relative permittivity $2.55\varepsilon0$, radius 216 nm) will exhibit a single dielectric dispersion, such as the one shown in Figure 14.4, if suspended in an aqueous medium of conductivity 1mSm^{-1} as calculated using the Clausius–Mossotti factor (Equation 14.2). This dispersion has the effect of causing the real part of the Clausius–Mossotti factor to change value and the imaginary part to have a peak at the corresponding frequency. As the medium conductivity is increased, the polarizability of the particle compared with the medium drops, resulting in the predispersion (positive) side of the curve having a lower value. Eventually

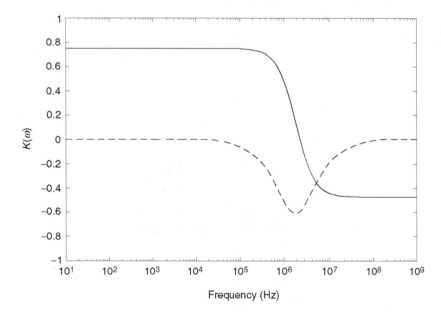

FIGURE 14.4 A plot of the real (solid line) and imaginary (dotted line) parts of the Clausius–Mossotti factor calculated for a 216 nm latex bead in a 1 mSm^{-1} solution, neglecting surface charge effects. The magnitude and signs of the real and imaginary parts govern the magnitude and direction of the dielectrophoretic force and electrorotational torque, respectively.

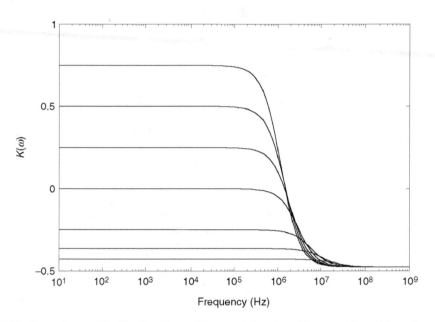

FIGURE 14.5 The real part of the Clausius–Mossotti factor as a function of frequency for a 216 nm diameter latex bead, for different values of suspending medium conductivity. The conductivity varies from 0.1 mSm⁻¹ (top line) to 500 mSm⁻¹ (bottom line). At conductivities above 20 mSm,⁻¹ *Re[K(ω)]* is always negative; that is, the particles always experience negative dielectrophoresis. At lower conductivity, particles cross from positive to negative dielectrophoresis at about 3 MHz.

the low-frequency polarizability becomes so low that it is below zero at all frequencies; that is, the particle always experiences negative dielectrophoresis. This can be seen in Figure 14.5, where the polarizability is plotted for a range of suspending medium conductivities.

If we plot the polarizability as a function of both frequency and conductivity of the suspending medium, we find a plot such as shown in Figure 14.6. Ideally, it would be convenient to directly measure the polarizability as a mechanism for determining the dielectric properties of the particle by curve-fitting data to Equation 14.2, a method often used for the measurements of cells by determining the rate at which particles collect under positive dielectrophoresis for different frequencies. However, this

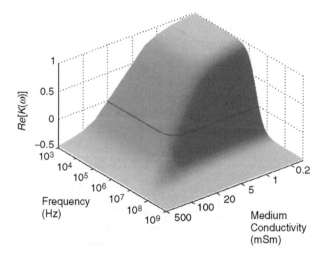

FIGURE 14.6 The data presented in Figure 14.3 plotted with conductivity on a third axis. The combinations of frequency and conductivity where *Re[K(ω)]* = 0 form a distinct shape indicated by the black line.

is not easy in the case of submicrometer particles, where electrohydrodynamic and Brownian motions can easily disrupt the stable collection of particles. While successful attempts have been made to use a modified collection rate technique to study both herpes viruses and latex beads, a far more convenient method of determining dielectric properties is to examine the intercept on the X–Y plane in Figure 14.6 — the plot of frequency where the value of $Re[K(\omega)]$ is zero, against conductivity — and infer the dielectric properties from that graph. This technique has been used to study latex spheres,[13–15] viruses,[16–18] and proteins[19] as well as larger particles such as cells.[20–22] This method is convenient for the measurement of colloids because the zero-force frequency can always be seen quite clearly, even in the presence of disruptive fluid flow or Brownian motion. Data are collected at a range of conductivities (typically at five conductivities per decade), and a best-fit line is used to determine the most likely data set for the experimental data.

For micrometer-scale homogeneous particles in media whose conductivity and permittivity are known, there is a solution for Equation 14.2 that matches the crossover spectrum. However, as the diameter of the particle under study is reduced past 1 μm, this model becomes increasingly inaccurate. The crossover frequency is found to rise with increasing medium conductivity, and above the conductivity threshold, where the crossover drops rapidly and only negative dielectrophoresis should be seen, the particle still exhibits a crossover but at a much lower frequency. The reason for this change in behavior is due to the increasing effect of the surface charge and, more specifically, the electrical double layer.

14.2.3 Double Layer Effects

While simple models of dielectrophoretic behavior using only the Clausius–Mossotti factor will suffice for particles on the micrometer scale and larger, they are relatively poor at predicting the response of smaller particles. This is due to the influence of *surface charge* effects. While these effects can be observed in both micrometer- and nanometer-scale dielectrophoresis, they are much more pronounced in nanoscale particles. This is because the particles are much nearer in size to that of the ion cloud that surrounds them — the *electrical double layer* — and are therefore far more influenced by it.

The electrical double layer is the name given to the cloud of ions that is attracted to any charged surface in solution. The charges on the surface attract ions in the solution that have opposite charge — the *counterions* — while repelling charges of the same sign, or *coions*. At the surface itself, ions and water molecules will be adsorbed onto the surface, forming the *stagnant layer* or *Stern layer*. Between the outside face of the Stern layer (sometimes called the *slip plane*) and the bulk solution (where the ion distribution is unaffected by the presence of the charged surface) lies the *diffuse layer*, where the ion concentrations vary due to the effect of the surface but remain in motion in the solution. The thickness of the diffuse double layer is called the *Debye length*, given the symbol $1/\kappa$. We cannot directly measure the electrostatic potential at the surface, but we can measure the electrostatic potential at the slip plane, which is termed the ζ (zeta) potential.

The actual experimental crossover spectrum for 216 nm latex beads is shown in Figure 14.7. There are a number of significant differences from our original model. The response is not constant over the lower range of conductivities and does not exhibit only negative dielectrophoresis at higher conductivities. The crossover frequency exhibits a rise with increasing conductivity; and when it reaches the threshold and the crossover frequency drops, it only does so by about one order of magnitude.

In order to fit data such as those shown in Figure 14.7, a number of changes must be made to the model. If the only model used were the Clausius–Mossotti factor, then the conductivity of the particle in the model must be much greater than that which we know latex to possess, which is in reality negligibly small. Experiments with latex beads of different sizes show that these effects become increasingly prominent as the size of the particle is decreased. Although there are different causes for these effects, they are all related to the movement of charges in the electrical double layer around the bead — specifically, the movement of charge around the Stern and diffuse layers, and the dielectric dispersion experienced by the charges in the double layer. We will examine these separately to show how they each affect the dielectric response of the particle.

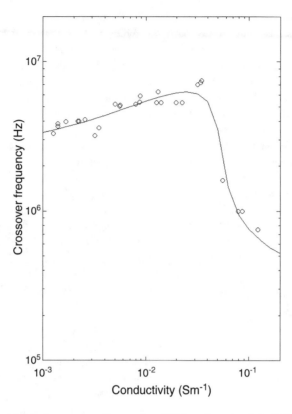

FIGURE 14.7 Experimental data for crossover frequencies of 216 nm latex beads in KCl solutions (circles) with a best-fit line according to the models described in the text.

14.2.3.1 Charge Movement in the Stern Layer

It has been known for some years that the surface charge affects the dielectric response of particles. Studies by Arnold and Zimmermann[23] demonstrated that the electrorotation of latex spheres produced anomalously high values of internal conductivity — in latex spheres this should be near-zero — which was later attributed to the movement of charge *around* the particle. This charge is attracted to the charges on the surface of the particle; when placed in an electric field the charges move in a laminar (i.e., planar) fashion around it. Arnold and Zimmermann determined that the component of aggregate particle conductivity σ_p, which could be attributed to surface charge movement, could be determined using the following equation:

$$\sigma_p = \sigma_{pbulk} + \frac{2K_s}{r} \qquad (14.4)$$

where σ_{pbulk} is the conductivity of the particle interior, K_s is the surface conductance of the particle, and r is the radius of the particle. This formula was used by Gascoyne et al.,[24] for example, in the determination of the surface charge of cells infected by malarial parasites. In our study of nanometer-scale particles, this is very useful as it explains why our models of the behavior of particles indicate a significantly higher conductivity than we know latex to possess. This effect becomes increasingly significant as particle radius is decreased, due to the inverse relationship between radius and the additional conductivity term due to surface conductivity.

For latex spheres, the bulk conductivity is negligible, so that the effective conductivity of the particle is dominated by the surface conductance, K_s, where typical values of K_s are of the order of 1 nS. We can extend this further. According to Lyklema,[25] the surface conductance can be calculated directly from the surface charge density, provided the mobility of the ions in the Stern layer is known and the mobilities of the counterions and coions are approximately equal, using the formula:

$$K_s = u^i \mu^i \tag{14.5}$$

where u^i represents the charge density that exists on the surface of the particle, and μ^i is the mobility of the counterion in the Stern layer, which is usually slightly lower than the value of mobility in the bulk solution. It is possible to measure this by determining the surface charge density by some other means, such as by the use of a Coulter counter[14] and using the value of K_s determined by dielectrophoretic means to establish the Stern layer mobility.

If the ionic mobilities or valances are not equal, then the equation must be adapted. It was first demonstrated by Green and Morgan[14] that latex beads exhibited different behavior in solutions of KCl and KPO_4, despite the fact that the counterion was K^+ in both cases. Further investigation by Hughes and Green[26] suggested that the conductivity of the Stern layer is influenced by the mobility of the counterions *and* coions in the bulk solution, by the equation:

$$K_s^i = \frac{u \mu_s^i \sigma_m}{2z^i F c^i \mu_m^i} \tag{14.6}$$

where μ_s^i and μ_m^i are the mobilities of the ion species in the Stern layer and bulk medium, respectively, c is the electrolyte concentration (mol m^{-3}), z is the valency of the ion, and F is the Faraday constant. If the electrolyte is symmetrical, it is possible to replace the conductivity term and concentration c^i with molar conductivity Λ (S m^{2}mol^{-1}):

$$K_s^i = \frac{u \mu_s^i \Lambda}{2z^i F \mu_m^i} \tag{14.7}$$

Values of Λ are constant for given electrolytes (a table of values is given by Bockris and Reddy[27]). Equation 14.7 reduces to Equation 14.5 only if the values of the mobilities of the coion and counterion are equal, in which case the value of

$$\frac{\Lambda}{2z^i F \mu_m^i}$$

goes to 1, as is the case for solutions such as KCl; in this instance, Equation 14.5 holds. Note that the value for ion mobility in the Stern layer is not equal to that found in the bulk medium, being somewhat lower.[25]

14.2.3.2 Charge Movement in the Diffuse Double Layer

The movement of charge though the Stern layer is an important factor in the contribution of the double layer to the net conductivity of the particle. However, there is a second layer of charge movement in the *diffuse* double layer. This is different and distinct from charge movement in the Stern layer; where the Stern layer charge is bound to the surface of the particle and moves in a laminar manner, charge distributed in the diffuse layer forms an amorphous ionic cloud around the particle. Significantly, the size of this cloud has an inverse relationship with the conductivity of the suspending medium — the greater the ionic strength of the medium, the thinner the diffuse double layer.

Analysis of the surface conductivity of a particle by Hughes et al.[15] led to a model with two separate components, demonstrating that rather than there being a single surface conductance K_s, it contains terms due to both the charge movement in the Stern layer *and* to charge movement in the diffuse part of the double layer.[25] The total surface conductance can then be written as:

$$K_s = K_s^i + K_s^d \tag{14.8}$$

where K_s^i and K_s^d are the Stern layer and the diffuse layer conductances, respectively; this then corresponds to a net particle conductivity given by the expression:

$$\sigma_p = \sigma_{pbulk} + \frac{2K_s^i}{r} + \frac{2K_s^d}{r} \tag{14.9}$$

Unlike the processes within the Stern layer, charge movement in the diffuse part of the double layer is related to *electro-osmotic transport* rather than straightforward conduction. Electro-osmosis is a process of fluid movement due to an applied potential across a nearby charged surface; the countercharge accumulates near the surface and then moves in the electric field due to Coulombic attraction. The presence of the surface creates a viscous drag that impedes the motion of the charges. This effect is widely studied and is a common method for propelling analytes in capillary electrophoresis, where the counterions to the charged glass capillary are attracted to the electrode at the end of the capillary tube, "dragging" the analyte with them by viscous forces.

Lyklema[25] gives the following expression for the effective conductance of the diffuse layer:

$$K_s^d = \frac{(4F^2cz^2D^d(1+3m/z^2))}{RT\kappa}\left(\cosh\left[\frac{zq\zeta}{2kT}\right]-1\right) \tag{14.10}$$

where D^d is the ion diffusion coefficient, z the valence of the counterion, F the Faraday constant, k Boltzmann's constant, R the gas constant, q the charge on the electron, T the temperature, and κ the inverse Debye length given by

$$\kappa = \sqrt{\frac{2czF^2}{\varepsilon RT}} \tag{14.11}$$

c is the electrolyte concentration (mol m⁻³), ζ is the ζ-potential, and the dimensionless parameter m is given by

$$m = \left(\frac{RT}{F}\right)^2 \frac{2\varepsilon_m}{3\eta D^d} \tag{14.12}$$

where η is the viscosity. A key factor in this expression is the relationship between the surface conductance and the concentration of ions in the bulk medium, which appears twice in this expression. There is a c in the expression itself, and a $c^{1/2}$ in the expression for κ. This gives a net contribution of $c^{1/2}$ to the total diffuse layer conductance. Because the concentration defines the medium conductivity, this expression indicates that as the conductivity of the medium is increased, so the conductivity of the particle will increase but by a lesser degree. This is what we see when the crossover frequency of the particle rises as the medium conductivity is increased — the effective conductivity of the particle is *also* increased. The remaining values in the equations are more or less constants; the principal unknown variable is the ζ-potential. This is known to vary slightly as a function of medium ionic strength; but the variation is small, and its mechanism is not fully understood. However, as the concentration of ions is known, determining the diffuse layer conductance allows the direct measurement of ζ-potential.

14.2.3.3 Stern Layer Dispersion

The above formulae describe the way in which the electrical properties of the particle, as represented in the Clausius–Mossotti equation, are augmented by the movement of charge around the particle. However, in order to describe the low-frequency dispersion visible in high-conductivity media (where no positive dielectrophoresis is expected), it is necessary to include an *additional* dispersion. Such additional polarizations follow the Debye model of the form $1/(1+j\omega\tau_e)$, where τ_e is the relaxation frequency of the additional dispersion.[14] One possibility is that this additional term derives from the dielectric dispersion, by surface conduction, of the charge in the Stern layer. Unlike the diffuse layer, the Stern layer is of fixed size and charge, dictated by the surface charge density of the particle. Hence, the frequency of the dielectric dispersion would be expected to be stable over a range of medium conductivities but vary proportionally to the particle radius and the surface conductance, as has been observed.[13-17] A good fit to the published data is given when the dispersion has a relaxation time:

$$\tau_e = \frac{\varepsilon_0 \varepsilon_S a}{K_s^i} \tag{14.13}$$

where ε_s is the relative permittivity of the Stern layer. The best fit is provided when ε_s is approximately $14\varepsilon_0$. Because the Stern layer consists of bound ions and water molecules held in specific orientations by electrostatic interactions with the charged particle surface, this is not unreasonable.

If we consider all the above factors — the Clausius–Mossotti factor, surface conduction in the double layer, and Stern layer relaxation — then we can find best-fit lines which correspond well to our data. For example, we can determine the net effects of all these factors on the 216 nm latex beads shown in Figure 14.7. Superimposed on the data is a best-fit line derived using the above equations and values K_s^i = 0.9nS, ζ = –100mV, ε_S = $14\varepsilon_0$. The relative permittivities of the particle and medium were 2.55 and 78; the internal conductivity of the beads was considered to be negligible. As can be seen, the model accurately predicts the behavior both below and above the decade transition in crossover frequency at 40–50mSm^{-1}.

14.2.4 Modeling of Complex Spheroids: The Multi-Shell Model

Thus far, we have examined the dielectrophoretic response of solid, homogeneous spheres. However, many nanometer-scale particles such as viruses are not solid, homogeneous spheres, consisting of a number of *shells* surrounding a central *core*. In the simplest case, a virus might consist of a protein case enclosing a central space wherein the viral DNA lies; a more complex virus such as herpes simplex encloses that protein shell in a thick protein gel, which is in turn surrounded by a lipid membrane similar to that which encloses a cell. Some viruses are not at all spherical, but are long and cylindrical; we will deal with them later in the chapter.

It is possible to extend the model of dielectric behavior to account for more complex particle structure. Developed by Irimajiri et al.,[28] it works by considering each layer as a homogeneous particle suspended in a medium, where that medium is in fact the layer surrounding it. So, starting from the core we can determine the dispersion at the interface between the core and the layer surrounding it, which we will call shell 1. This combined dielectric response is then treated as a particle suspended in shell 2, and a second dispersion due to that interface is determined; then a third due to the interface between shells 2 and 3, and so on. In this way, the dielectric properties of all the shells combine to give the total dielectric response for the entire particle. This is illustrated schematically in Figure 14.8.

In order to examine this mathematically, let us consider a spherical particle with N shells surrounding a central core. To each layer we assign an outer radius a_i, with a_1 the radius of the core and a_{N+1} the radius of the outer shell (and therefore the radius of the entire particle). Similarly, each layer has its own complex permittivity given by

$$\varepsilon_i^* = \varepsilon_i - j\frac{\sigma_i}{\omega} \tag{14.14}$$

where i has values from 1 to N+1. In order to determine the effective properties of the whole particle, we first replace the core and the first shell surrounding it with a single, homogeneous core. This new core has a radius a_2 and a complex permittivity given by

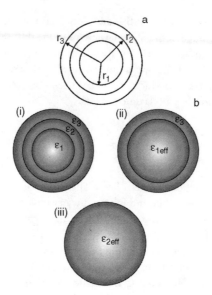

FIGURE 14.8 (a) A sphere comprises a core and inner and outer shells, with radii r1, r2, and r3. (b) These three layers have complex permittivity ε_1^*, ε_2^*, and ε_3^*. We can find the total polarizability of the particle by successively combining the two innermost layers to find the effective combined complex permittivity.

$$\varepsilon_{1eff}^* = \varepsilon_2^* \frac{\left(\dfrac{r_2}{r_1}\right)^3 + 2\dfrac{\varepsilon_1^* - \varepsilon_2^*}{\varepsilon_1^* + 2\varepsilon_2^*}}{\left(\dfrac{r_2}{r_1}\right)^3 - \dfrac{\varepsilon_1^* - \varepsilon_2^*}{\varepsilon_1^* + 2\varepsilon_2^*}} \tag{14.15}$$

We now have a core surrounded by N–1 shells. We then proceed by repeating the above calculation, but combining the new core with the second shell, thus:

$$\varepsilon_{2eff}^* = \varepsilon_3^* \frac{\left(\dfrac{r_3}{r_2}\right)^3 + 2\dfrac{\varepsilon_{1eff}^* - \varepsilon_3^*}{\varepsilon_{1eff}^* + 2\varepsilon_3^*}}{\left(\dfrac{r_3}{r_2}\right)^3 - \dfrac{\varepsilon_{1eff}^* - \varepsilon_3^*}{\varepsilon_{1eff}^* + 2\varepsilon_3^*}} \tag{14.16}$$

If this procedure is repeated a further N–2 times, then the final step will replace the final shell, and the particle will be replaced by a single homogeneous particle with effective complex permittivity ε_{Peff}^* given by

$$\varepsilon_{Peff}^* = \varepsilon_{Neff}^* \frac{\left(\dfrac{r_{N+1}}{r_N}\right)^3 + 2\dfrac{\varepsilon_{(N-1)eff}^* - \varepsilon_{N+1}^*}{\varepsilon_{(N-1)eff}^* + 2\varepsilon_{N+1}^*}}{\left(\dfrac{r_{N+1}}{r_N}\right)^3 - \dfrac{\varepsilon_{(N-1)eff}^* - \varepsilon_{N+1}^*}{\varepsilon_{(N-1)eff}^* + 2\varepsilon_{N+1}^*}} \tag{14.17}$$

This value provides an expression for the combined complex permittivity of the particle at any given frequency ω. It can also be combined with the complex permittivity of the medium to calculate the Clausius–Mossotti factor, as demonstrated by Huang et al.[29] for yeast cells.

14.2.5 Modeling Nonspherical Ellipsoids

A second special case of the formula for dielectrophoretic force which is often required is that of a long cylinder. This is needed in order to model the dielectric response of common nanotechnological components, such as nanotubes and nanowires, as well as complex biological particles such as certain viruses. It is possible in these cases to adapt our model in order to compensate for the change in shape by deriving a general expression for the Clausius–Mossotti factor for elliptical particles, of which the spherical model is a special case; the cylinder can then be approximated to a long, thin ellipsoid, with reasonable accuracy.[18,30,31]

When an elliptical particle polarizes, the magnitude of the dipole moment is different along each axis; for example, a prolate ellipsoid (shaped like a football or rugby ball) will have a dispersion along its long axis of different relaxation frequency to the dispersion across its short (but equal) axis. The dispersion frequency of the dipole formed along the long axis will be of lower frequency than that formed across the shorter axis, but the magnitude of the dipole formed will be greater due to the greater separation between the charges.

Consider an elliptical particle such as that shown in cross-section in Figure 14.9. It consists of two axes in projection, x and y, plus a third axis, z, projecting from the page. The radii of the object along these axes are a, b, and c, respectively. It can be demonstrated[4,31] that the particle will undergo three dispersions at different frequencies according to the thickness of the ellipsoid along each axis. However, in addition to the dielectrophoretic force experienced by the particle, it will also experience a torque acting so as to align the longest nondispersed axis with the field. This phenomenon, often observed in practical dielectrophoresis, is *electro-orientation*.[4] When a nonspherical object is suspended in an electric field (for example, but not solely, when experiencing dielectrophoresis) it rotates such that the dipole along the longest nondispersed axis aligns with the field. Because each axis has a different dispersion, the particle orientation will vary according to the applied frequency. For example, at lower frequencies, a rod-shaped particle experiencing positive dielectrophoresis will align with its longest axis along the direction of electric field; the distribution of charges along this axis has the greatest moment and therefore exerts greatest torque on the particle to force it into alignment with the applied field. As the frequency is increased, the dipole along this axis reaches dispersion, but the dipole formed *across* the rod does not; and the particle will rotate 90° and align perpendicular to the field. This smaller axis has a shorter distance between changes, and hence a smaller capacitance, so the dispersion frequency will be higher; however, the shorter distance means the dipole moment is smaller. This will result in the force experienced by the particle being smaller in this mode of behavior.

When aligned with one axis parallel to the applied field, a prolate ellipsoid experiences a force given by the equation:

$$\mathbf{F}_{DEP} = \frac{2\pi abc}{3}\varepsilon_m Re[X(\omega)]\nabla E^2 \tag{14.18}$$

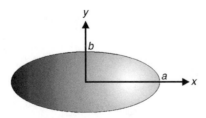

FIGURE 14.9 A schematic diagram of an elliptical particle, showing axes x and y, along which the particle extends by distances a and b. The particle extends along the z-axis (out of the page) by a length c. If $c = b$, the particle is prolate; if $c = a$, the particle is oblate.

where

$$X(\omega) = \frac{\varepsilon_p^* - \varepsilon_m^*}{(\varepsilon_p^* - \varepsilon_m^*)A_\alpha + \varepsilon_m^*} \tag{14.19}$$

where α represents either the x, y, or z axis, and A is the *depolarization factor*. This factor represents the different degrees of polarization along each axis, such that:

$$A_x = \frac{abc}{2}\int_0^\infty \frac{ds}{(s+a^2)\sqrt{(s+a^2)(s+b^2)(s+c^2)}}$$

$$A_y = \frac{abc}{2}\int_0^\infty \frac{ds}{(s+b^2)\sqrt{(s+a^2)(s+b^2)(s+c^2)}} \tag{14.20}$$

$$A_z = \frac{abc}{2}\int_0^\infty \frac{ds}{(s+c^2)\sqrt{(s+a^2)(s+b^2)(s+c^2)}}$$

where s is the variable of integration. The polarization factors are interrelated such that $A_x + A_y + A_z = 1$.[4,32]

The most useful version of these expressions is the simplified one for the case of prolate ellipsoids ($a > b$, $b = c$). This expression is useful because many nanoparticles can be approximated to prolate ellipsoids. In that case, Equation 14.18 may be rewritten as:

$$\mathbf{F}_{DEP} = \frac{2\pi abc}{3}\varepsilon_m Re\left[\frac{\varepsilon_p^* - \varepsilon_m^*}{1 + \left(\frac{\varepsilon_p^* - \varepsilon_m^*}{\varepsilon_m^*}\right)A}\right]\nabla\mathbf{E}^2 \tag{14.21}$$

where A is given by the expansion:

$$A = \frac{1}{3\gamma^2}\left[1 + \frac{3}{5}(1 - \gamma^{-2}) + \frac{3}{7}(1 - \gamma^{-2})^2 + \dots\right] \tag{14.22}$$

and where $\gamma = a/b$. For a spherical particle, $\gamma = 1$ and $A = \frac{1}{3}$, and Equation 14.18 can be rearranged to the expression for the force on a sphere as shown in Equation 14.1.

As with spherical particles, multi-shell prolate ellipsoids may also be modeled provided their dimensions are known. The procedure is exactly as demonstrated earlier for spheroids, but with the expression for the equivalent permittivity of the ellipsoid being replaced by Equation 14.23, thus:

$$\varepsilon_{(i)eff}^* = \varepsilon_i^*\frac{\varepsilon_i^* + (\varepsilon_{(i+1)eff}^* - \varepsilon_i^*)[A_{i\alpha} + v_i(1 - A_{(i-1)\alpha})]}{\varepsilon_i^* + (\varepsilon_{(i+1)eff}^* - \varepsilon_i^*)(A_{i\alpha} + v_i A_{(i-1)\alpha})} \tag{14.23}$$

for i = 1 to N–2 as before, and where:

$$v_i = \frac{a_i b_i c_i}{a_{i-1} a_{i-1} a_{i-1}} \tag{14.24}$$

For a more complete exploration of the mathematics underlying the dielectrophoresis of elliptical particles, readers are referred to the excellent book by Jones.[4]

14.2.6 Phase-Related Effects: Electrorotation and Traveling-Wave Dielectrophoresis

Thus far we have explored only the effects of interactions between an electric field and a dipole that acts in phase with that electric field. However, there is a second class of AC electrokinetic phenomena that depends on the interactions between an *out-of-phase* dipole with a spatially moving electric field. Because an induced dipole may experience a force with both an in-phase and out-of-phase component simultaneously, the induced forces due to these components will be experienced at the same time, with the respective induced forces superimposed.

If a polarizable particle is suspended in a rotating electric field, the induced dipole will form across the particle and should rotate synchronously with the field. However, if the angular velocity of the field is sufficiently large, the time taken for the dipole to form (the *relaxation time* of the dipole) becomes significant and the dipole will lag behind the field. This results in nonzero angle between field and dipole, which induces a torque in the body and causes it to rotate asynchronously with the field; the rotation can be with or against the direction of rotation of the field, depending on whether the lag is less or more than 180°. This phenomenon was called *electrorotation* by Arnold and Zimmermann[33] and is shown schematically in Figure 14.10. The general equation for time-averaged torque Γ experienced by a spherical polarizable particle of radius r suspended in a rotating electric field E is given by

$$\Gamma = -4\pi\varepsilon_m r^3 Im[K(\omega)]E^2 \tag{14.25}$$

where $Im[K(\omega)]$ represents the imaginary component of the Clausius–Mossotti factor shown in Figure 14.4. The minus sign indicates that the dipole moment lags the electric field. When viscous drag is accounted for, the rotation rate $R(\omega)$ of the particle is given by[23]

$$R(\omega) = -\frac{\varepsilon_m Im[K(\omega)]E^2}{2\eta} \tag{14.26}$$

where η is the viscosity of the medium. Note that, unlike the dielectrophoretic force in Equation 14.1, the relationship with the electric field is as a function of the square of the electric field rather than of the *gradient* of the square of the electric field. Furthermore, the torque depends on the *imaginary* rather than the real part of the Clausius–Mossotti factor. A particle may experience both dielectrophoresis and

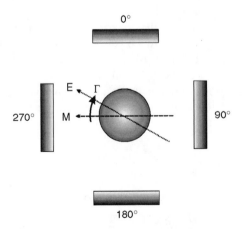

FIGURE 14.10 A schematic of a polarizable particle suspended in a rotating electric field generated by four electrodes with 90° advancing phase. If the electric field E rotates sufficiently quickly, the induced dipole M will lag behind the electric field by an angle related to the time taken for the dipole to form (the relaxation time). The interaction between the electric field and the lagging dipole induces a torque Γ in the particle, causing the particle to rotate. This effect is known as *electrorotation*.

electrorotation simultaneously, and the magnitudes and directions of both are related to the interaction between the dielectric properties of particle and medium; the relative magnitudes of force and torque are proportional to the real and imaginary parts of the Clausius–Mossotti factor.

Electrorotation was first examined scientifically in the early 1960s, for example, by Teixeira-Pinto et al.,[34] but was most fully explored by Arnold and Zimmermann,[23,33] who described the processes causing the observed motion. For cellular-scale objects the principal attraction of electrorotation has been the ability to measure the rotation rate of single cells at different frequencies and thereby gain an accurate value of polarizability not available through crossover measurements where ensembles of particles are more easily observed. However, we may speculate that electrorotation may provide a means for the actuation of nanometer-scale electric motors. Berry and Berg[35] have used electrorotation to drive the molecular motor of *E. coli* bacteria backwards at speeds of up to 2000 Hz. Because the electrodes involved in that work were applying 10 V across interelectrode gaps of 50 μm or more, considerably higher fields (and hence induced torques) could be applied with electrodes with interelectrode gaps of the order of 1 μm. Electric motors using electrorotation have been extensively studied by Hagedorn et al.,[36] who demonstrated that such motors are capable of providing similar power outputs to synchronous dielectric motors for fluid pumping applications. It has been speculated that such nanomotors might be actuated by rotation-mode lasers. There are a number of advantages that favor electrorotation over its optical equivalent, most notably that electrorotation-induced torque is easily controlled by altering the frequency of the electric field and that there does not need to be a direct optical path to the part to be manipulated.

There is a linear analog to electrorotation, which induces a translational force rather than torque in particles, an effect known as *traveling-wave* dielectrophoresis. This shares with dielectrophoresis the phenomenon of translational-induced motion; but rather than acting toward a specific point (that of the highest electric field strength), the force acts to move particles *along* an electrode array in the manner of an electrostatic conveyor belt.

Consider a particle in a sinusoidal electric field which *travels* — that is, rather than merely changing magnitude, the field maxima and minima move through space, like waves on the surface of water.[37] These waves move across a particle, and a dipole is induced by the field. If the speed at which the field crosses the particle is great enough, then there will be a time lag between the induced dipole and the electric field, in much the same way as there is an angular lag in a rotating field which causes electrorotation. This physical lag between dipole and field induces a force on the particle, resulting in induced motion; the degree of lag, related to the velocity (and hence the frequency) of the wave, will dictate the speed and direction of any motion induced in the particle. Such electrostaic waves can be generated using electrodes with phased potentials, such as shown in Figure 14.11. The underlying principle is closely related to electrorotation; it could be argued that the name *traveling-wave dielectrophoresis* is misleading because the origin of the effect is not dielectrophoretic — that is, it does not involve the interaction of dipole and field *gradient*. Instead, the technique is a linear analogue of electrorotation, in a similar manner to the relationship between rotary electric motors and the linear electric motors used to power magnetically levitated trains. As with the rotation of particles, the movement is asynchronous with the moving field, with rates of movement of 100 μm/s reported. The value of the force $\mathbf{F}_{\text{TWD}}$ is given by[29]

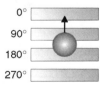

FIGURE 14.11 A schematic showing a polarizable particle suspended in a traveling electric field generated by electrodes on which the applied potential is 90° phase-advanced with respect to the electrode above. If the electric field moves sufficiently quickly, the induced dipole M will lag behind the electric field, inducing a force in the particle. This causes the particle to move along the electrodes, a phenomenon known as *traveling-wave dielectrophoresis*.

$$\mathbf{F}_{\text{TWD}} = \frac{-4\pi\varepsilon_m r^3 Im[K(\omega)]E^2}{\lambda} \tag{14.27}$$

where λ is the wavelength of the traveling wave and is usually equal to the distance between electrodes which have signals of the same phase applied.

Traveling-wave dielectrophoresis was first observed by Batchelder[38] and subsequently by Masuda et al.[39] However, it was not until the work of Fuhr et al.[40] that the phenomenon was fully explored and its origins known. Subsequent work by Huang[37] demonstrated both the equations outlined above and the potential for separation by traveling-wave dielectrophoresis. A large corpus of work now exists, including theoretical studies,[37,41] devices for electrostatic pumping,[42] and large-scale cell separators.[43]

The application of traveling-wave dielectrophoresis is largely used as a means of transporting particles. While the majority of work on traveling-wave dielectrophoresis has concerned micrometer-sized objects such as blood cells,[43] some work has been performed on the concentration of nanoparticles on a surface using so-called *meander* electrodes.[44] These structures use four electrodes in a series of interlocking spirals to generate a traveling wave; at the center of the spiral, the electrodes form a quadrupole-type electrode array. It has been demonstrated that by careful manipulation of the amplitudes and phases of the potentials on these electrode structures, it is possible to "steer" the motion of particles across the array. Tools such as these could be used as the basis for conveyor belts for *factories on a chip*, wherein different chemical processes may be carried out of the same chip, with operations performed by electrostatic or chemical means and the resultant output transferred to a new process by AC electrokinetic means.[45]

14.3   Applications of Dielectrophoresis on the Nanoscale

14.3.1   Particle Separation

Ever since Gascoyne et al.[46] demonstrated the separation of healthy and leukemic mouse blood cells using microelectrodes early in the 1990s, dielectrophoresis has increasingly been used as a tool for the separation of heterogeneous mixtures of particles into homogeneous populations in different parts of a microelectrode array. The method underlying the technique is simple: because polarizable particles demonstrate a crossover frequency that is dependent on those particles' dielectric properties, particles with different properties may under specific conditions exhibit different crossover frequencies. As those particles experience positive dielectrophoresis below the crossover frequency and negative dielectrophoresis above, it follows that at a frequency between the two crossover frequencies of the two particle types, one will experience positive dielectrophoresis while the other experiences negative dielectrophoresis. This will result in one group being attracted to regions of high field strength, with the other group being repelled; hence, the two populations are separated. Such separations are typically carried out using electrode arrays with well-defined regions of high and low electric field strength.

Consider the separation of a mixture of two populations of latex beads, identical except for having different radii. Because the effective conductivity (and hence the polarizability, as expressed in Equation 14.2) of a latex sphere is dependent on the double of surface conductances divided by the particle radius (Equation 14.7), it follows that the value of $Re[K(w)]$ will be strongly affected by particle radius. This is indeed the case, with larger particles exhibiting lower crossover frequencies than smaller (but otherwise identical) particles.

Secondly, particles of identical size and internal composition can be separated according to their surface properties. This was first demonstrated by Green and Morgan in 1997,[47] who reported the separation of 93 nm latex spheres. By using a castellated electrode array with 4 µm feature sizes, the researchers demonstrated that the particles exhibited a narrow range of surface conductances rather than each having an identical value of surface conductance. This meant that the population of particles had crossover frequencies across a narrow frequency window, and by applying a frequency in the middle of that range it was demonstrated that the particles could in fact be separated.

This effect was expanded upon by Hughes et al.[15] by actively modifying the surfaces of latex particles to improve the separation and to identify possible biotechnological applications of the technique. The surfaces of some of the beads were chemically modified using EDAC (1-ethyl-3-(3-dimethylami-nopropyl) carbodiimide), a reagent used for the chemical coupling of protein to the carboxyl surface of the beads. This caused a significant reduction in the crossover frequency, which was found to equate to a similar reduction in surface conductance from 1.1 nS to 0.55 nS. The EDAC-activated beads were then mixed with antibodies, and the crossover behavior was measured again. As the surfaces of the beads were covered by the antibodies, the crossover spectrum exhibited a further drop in frequency equating to a further drop in K_s and ζ potential. However, the crossover frequencies of the IgG-labeled beads varied by up to a factor of 2 between different beads as a result of different amounts of antibody coupling to beads, allowing the separation of beads containing different amounts of protein on their surfaces.

There are a number of potential applications for such a system. First, because the crossover frequency is directly related to the amount of protein attached to the bead surface, it allows the rapid assaying of the amount of protein attached to a sphere, which in turn relates to the amount of protein in the environment, making a single sphere a potential biosensor. In theory, if the system is calibrated such that the crossover frequency in a particular medium that corresponds to a specific protein coverage is known, then observing the frequencies at which a single bead — or an ensemble — changes dielectrophoretic behavior could allow measurement of the protein content in the medium. This could be used for a number of different proteins or other compounds by mixing fluorescent beads of different colors, each with a different surface functionality. By constructing electrodes over a suitable photosensor, systems such as this may form the basis of "lab on a chip" systems.[45]

A second application concerns the fact that very small latex spheres have a large surface-area-to-volume ratio, so that a small volume of beads has a potentially huge surface area. For example, a 1 ml sample containing 1% (by volume) of 200 nm diameter beads (as used by Hughes et al.[15]) has a total surface area of 300 m^2 — which in biosensor terms makes such a sensor exceptionally sensitive. This could, for example, be used to detect very low quantities of target molecules; if a large number of small, activated beads is held near their crossover frequency, then a single molecule attaching to the surface of one bead may change the surface charge of the bead sufficient to cause that bead to pass the crossover frequency and be detected. A similar system, on the micrometer scale, has been developed at the University of Wales at Bangor as a means of detecting waterborne bacteria.[48]

Just as latex beads of different sizes, properties, or surface functionalities can be separated into sub-populations on an electrode array, so we can separate bioparticles with different properties. An example of this is shown in Figure 14.12, which illustrates the separation of herpes simplex virus particles that

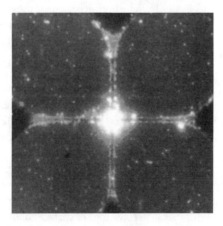

FIGURE 14.12 A separation of herpes simplex virions (in the electrode arms) from herpes simplex capsids (in the ball at the center of the array). The ball is levitated approximately 10 μm above the electrode array. (Courtesy of Drs. M. Hughes and H. Morgan, University of Glasgow.)

have collected in the *arms* of the electrodes and appear pale, from herpes simplex capsids that appear as a bright ball at the center of the array. Similar separations have been performed for herpes simplex virions and tobacco mosaic virions.[49] The procedure for such separations usually follows a categorization of the dielectric response of the two-particle species in order to find the optimum frequency and suspending medium conductivity. Another method of separation was demonstrated by Washizu et al.,[8] who fractioned a mixture of protein and DNA molecules by a combination of field flow and dielectrophoresis. Separation and identification of biological material is perhaps the most important application of dielectrophoresis to nanomedicine — allowing, for example, the point-of-care analysis of blood samples to determine the cause of an infection without the need for lengthy analyses at remote laboratories.

14.3.2 Trapping Single Nanoparticles

One of the many advantages gained by the manipulation of particles on the nanometer scale is the possibility of manipulating single particles. Such a technology could potentially opens up new fields in the study of single-molecule chemistry and molecular biology, and it is presently being pursued by a number of workers (e.g., Hughes and Morgan[11] and Watarai et al.[50]). The majority of this research is performed using optical trapping — so-called *laser tweezers* — in which focused laser beams are used to exert pressure on particles.[51] However, dielectrophoresis offers many advantages over laser tweezers, including the fact that the technique allows trapping from solution followed by contact on a sensor of some sort, allowing an analyte to be studied.

The trapping of single particles is somewhat different and more difficult to achieve than trapping a larger population of particles. In the latter case, particles need only to *tend* to move toward the trapping region; if some particles leave the trap, this is not considered a problem provided greater numbers of particles are moving into the trap. Furthermore, as can be seen in Figure 14.3, dielectrophoretic traps — even those constructed on the micrometer scale — tend to collect large numbers of particles. A spherical trapping volume 1 μm across could contain over 500 particles of diameter 100 nm.

There are two ways to increase the selectivity of the trapping mechanism. One method is to reduce the size of the electrodes to the order of size of the particle to be trapped. The other mechanism is to use a larger trap, but alternate between a regime whereby a single particle is attracted to the electrodes, followed by a second regime to prevent other particles approaching the trap. This can be achieved by either applying negative dielectrophoresis to keep extraneous particles away from an isolated energy well, or removal of the field to prevent further particles being attracted. Both methods have advantages and disadvantages and applications to which they best present themselves.

Different electrode geometries are required for each trapping method. The basis of one electrode design for single-particle applications is the dual need to both *attract* a single particle and to *repel* all others. Unlike bulk nanoparticle trapping, where the aim is merely to attract particles to a region, it is necessary to both attract a particle to a point and trap it while excluding all other particles from that trap. There is a second design strategy that may be used — that the particle experiences no force at all, either by removal of the electric field or by retaining the particle at a point where a field null exists. Either of these two methods may be used to prevent particles nearing the trapping point.

14.3.2.1 Single-Particle Trapping by Positive Dielectrophoresis

This technique was first demonstrated by Bezryadin and co-workers in 1997.[3] The electrode geometry consisted of two needle-type platinum electrodes that faced one another, suspended in free space by etching the silicon substrate beneath the point where the electrodes met. The distance between opposing electrode tips was 4 nm. The potential was applied through a high-value (100 MΩ) resistor; a 4.5 V DC field was used. As we have seen, AC fields are far more common for dielectrophoresis; but this is not a prerequisite.

Colloidal palladium particles with sizes down to 5 nm diameter were introduced in solution. These particles became polarized and were attracted up the field gradient to the electrode tips by positive dielectrophoresis. However, as soon as the first palladium sphere reached the center point between two

FIGURE 14.13 An electron micrograph of a single 17 nm palladium sphere trapped between two electrodes by positive dielectrophoresis. The image measures 200 nm wide and 100 nm high. (Courtesy of Drs. A. Bezryadin, G. Schmid, and C. Dekker, TU Delft, The Netherlands.)

opposing electrodes, a circuit between the electrodes was made. This resulted in current flowing in the circuit, in which the majority of the supply voltage was dropped across the resistor; with virtually no voltage dropped across the electrodes, the magnitude of the electric field generated in the interelectrode gap was diminished, preventing other colloidal particles from reaching the electrode tips. Once in place, the trapped particle was sufficiently attached to the electrodes for the solution to be removed and the assembly to be observed using a scanning electron microscope, as shown in Figure 14.13. Similar principles could be applied to the trapping of single fullerene molecules to form single-molecule transistors.[52]

14.3.2.2 Single-Particle Trapping by Negative Dielectrophoresis

This is more complex than the positive trapping technique described above; however, particles of biological origin such as viruses are nonconducting, and the use of negative dielectrophoresis is therefore more appropriate. Furthermore, because the size of the trapping volume at the center of the arrays is defined not by the electrodes but the geometry of the generated field, the technology required to construct the electrodes is far more readily available. Negative dielectrophoretic trapping of single particles has been achieved by this method using viruses, latex spheres, viral substructures, and macromolecules such as DNA as well as larger structures such as cells.[11, 53,54]

There are a number of drawbacks to this form of trapping; the principal one is the observation of the particle. Unlike positive trapping, where the particle may be detected electrically, particles trapped by negative dielectrophoresis are suspended in the medium at an indeterminate height above the electrode structure. Ultimately, the only means by which such particles can be observed is by fluorescent staining. This is a general problem in the field of single-nanoparticle detection, and other methods such as laser tweezers also require the use of fluorescent staining.

Electrodes used to trap particles are generally quadrupolar arrays such as the design shown in Figure 14.2. This array geometry has the advantage of a well-defined, enclosed field minimum surrounded by regions of high field strength. Ideally, the potential energy minimum would be small enough to contain only one particle. Where this is not the case, if particles are first attracted to the electrode tips by positive dielectrophoresis and the field frequency is then switched to induce negative dielectrophoresis, only those particles on the inward-facing tips of the electrodes will fall into the trap; the others will be repelled into the bulk. It is possible to trap single particles this way, though occasionally two or three particles may fall into the potential energy minimum at the center of the trap. Under these conditions, 93 nm diameter latex beads have been retained in the center of the electrode array while other particles were forced away. Single herpes virus particles could also be held in the same electrode array.

Single particles trapped by this method move within the confines of the electric field cage under the influence of Brownian motion. During trapping of an object with density greater than that of water, such as a single herpes virus, the particle is levitated in a stable vertical position above the electrodes; Brownian motion is balanced against the weight of the particle. However, particles such as the latex spheres (which have a density approximately equal to that of water) are not constrained in this way because Brownian motion causes constant random movement in the z-direction. Such particles may eventually diffuse out of the top of a *funnel*-type or *open* trap, in which the field gradient is generated by one set of planar electrodes "beneath" the trapped particle, though particles have been held for 30 minutes or more. In order to ensure that a particle remains within the trap, a second layer of electrodes may be introduced above the first, so that a closed field *cage* such as those employed by Fuhr and co-workers[53] is created. These have the additional advantage of allowing a degree of three-dimensional positional control by varying the intensity of the field strength at the various positional electrodes, as has been demonstrated using a 1 μm diameter latex bead.[54] Alternatively, a planar (two-dimensional) electrode array can be sufficient, provided that any coverslip used to contain the solution above the electrode is sufficiently close to the electrode plane and that any field trap constraining the particle in the *x-y* plane extends the full height of the solution, creating a force field of cylindrical aspect.

14.3.3 Particle Transport Using Dielectrophoretic Ratchets

There is another separation technique that differs from the above because it is only truly usable for nanometer-scale particles, rather than originally having been developed for cell-sized particles. It operates by exploiting a key property in the definition of a colloid — that Brownian motion plays a significant part in the position of the particle. The concept was originally voiced by Pierre Curie in the late nineteenth century;[55] if particles could be trapped in an asymmetric force field such that they were more likely to move in one direction than another, then a cycle of applying and then removing the field would cause particles to move, then disperse, then move again. Careful field design would allow the development of a *ratchet* mechanism to allow particles to be moved along the field; furthermore, the system could be used to separate particles on either the basis of their response to the field or their rate of diffusion when the field is turned off.

The concept of the thermal ratchet was explored by Ajdari and Prost,[56] who considered the forces acting on a particle in suspension and subject to a Brownian motion. A particle exposed to a potential with *sawtooth* variation in space, when repeatedly applied and removed for finite periods of time, will theoretically show a biased overall motion along the direction in which potential increases for the longest physical distance. This led Chauwin et al. to the assertion that this principle provided "mouvement sans force."[57] Analysis by Magnasco[58] and subsequently by Astumian and Bier[59] illustrated that models of this nature could be devised to explain the motion of proteins along biopolymer chains using thermal noise to advance the smaller molecules though a series of potential ratchets. Other workers have used the principle to explain the mechanism by which molecular motors such as those found in flagellar bacteria might work.

The practical application of this principle, using dielectrophoresis to provide the necessary potential gradient, was first proposed by Ajdari and Prost[56] and subsequently demonstrated experimentally by Rousselet et al.[60] Rousselet and co-workers used latex spheres of varying diameters to attain particle motion of 0.2 μms^{-1} with diffusion rates of particles advancing from one ratchet to the next of 40% per step for significant times of zero applied field. Ajdari and Prost[56] also proposed that this method has applications in the separation of particles according to their relative sizes. The method has since been enhanced to separate particles in a continuous manner according to their relative *dielectric* properties. Furthermore, under the correct conditions it is possible to drive particles of specific dielectric properties *backward* through the ratchet system while other particles are simultaneously being driven *forward* in the manner described previously, allowing full spatial separation.

Consider a quantity of colloidal particles of greater polarizability than the surrounding medium and equally dispersed through the volume under investigation. The volume under study is exposed to an imposed potential energy profile — such as an electric field imparting a positive dielectrophoretic force generated by electrodes which have an asymmetric pattern of tips arranged in a sawtooth pattern, as shown in Figure 14.14. When the field is activated, the particles will move to collect at the highest electric field. Due to the ∇^{E^2} term of the dielectrophoretic force (in Equation 14.1), motion will be directed along a path of increasing local electric field gradient. Due to the asymmetric design of the electrodes, the field gradient is biased such that a greater proportion of the space between successive electrode tips generates dielectrophoretic motion to the right of the diagram rather than to the left. Each electrode tip pair attracts particles from within its own *capture zone*, a volume bounded by the electrode edges on the top and

a

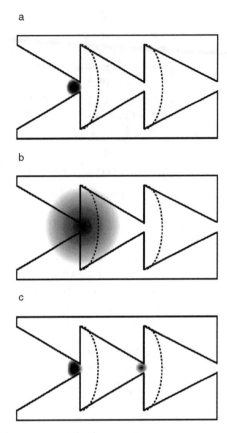

b

c

FIGURE 14.14 Schematic of the electrode geometry used by Rousselet et al. Two sets of electrodes consist of triangular protrusions, which are symmetrical across a line between the electrode tips and thus form large triangular spaces. The electrodes, turned through 90°, resemble Christmas trees and are often referred to as *Christmas tree electrodes*. (a) When an electric field is applied, particles are collected at the electrode tips. (b) When the electric field is released, the particles diffuse outward. Some move over the dotted line into the *capture zone* of the adjacent electrode. (c) When the field is reapplied, some particles are captured by the right-most electrode, giving a net movement of particles to the right.

bottom of the picture, and by the curved lines running approximately vertically. The distance between the electrode tips and the left-most edge of the capture zone is significantly larger than the distance to the right-most edge. If we consider distances along a horizontal line between opposing electrode tips, we can divide the length of a repeating shape d into the length of the regions where motion will be to the left (d_1) and to the right (d_2), respectively. After collection the concentration profile of the particles resembles that shown in Figure 14.14a.

Following the collection of particles over a period τ_{ON}, the potential difference across the electrodes is removed. Under Brownian motion, the particles will then drift from the electrode tips over a period of time. After sufficient time, some particles will drift a distance greater than d_1, which will place those translating to the right within the dielectrophoretic capture zone of the next (right) electrode. At that point, the distribution of particles will resemble Figure 14.14b. After a period of time τ_{OFF}, the electric field is reapplied. Assuming particle dispersion has taken place at an approximately equal rate, all particles except those which have traveled a distance greater than d_1 will be attracted to the same electrode tip. However, those which have moved greater than d_1 to the right will be trapped by the next electrode. Provided no particles have moved a distance greater than d_2 to the left, thereby entering the capture zone of the *previous* electrode, there is a net motion of particles to the right. This is indicated in Figure 14.14c, where some of the particles have now moved to the right-most electrode pair.

Particle diffusion can be calculated using the one-dimensional probability distribution on the Fokker–Planck equations[61] along axis x:

$$\frac{\partial P}{\partial t}(x, t) = -div\, \mathbf{J}(x, t) \tag{14.28}$$

where

$$\mathbf{J}(x, t) = \frac{D}{kT}\, P(x, t)\mathbf{F}(x, t) - D\nabla P(x, t) \tag{14.29}$$

where P indicates the probability density function of the particle location, D is the diffusion coefficient for the particles in solution, and $\mathbf{F}$ is the force due to an imposed field. It has been shown[56] that for optimum transport this diffusion will be bounded by the lower diffusion limiting case of the above expressions, where the diffusion rate is small enough to prevent particles passing beyond a single repeating electrode unit in a single time interval τ_{OFF}. This limiting case is given by the expression:

$$\tau_{OFF} \ll \frac{(d_2 - d_1)^2}{D} \tag{14.30}$$

At this stage, we must consider the electrode geometry required in order to generate our asymmetric field. The most common geometry in use is that of the so-called *Christmas tree* electrodes such as those shown in Figure 14.14. For an ideal case where the condition in Equation 14.30 is met, a two-dimensional isotropic diffusion of particles has a probability of crossing a semicircular boundary of radius d_1 in time τ_{OFF}, and thus the fraction of particles having crossed that boundary at that time, given by the expression:[60]

$$P = \frac{1}{2}\exp\left(\frac{-d_1^2}{4D\tau_{OFF}}\right) \tag{14.31}$$

where the $1/2$-factor indicates isotropic diffusion. However, Rousselet and co-workers[60] determined through experimentation that in practice the boundary between capture zones is only approximately semicircular; but it is not exactly so (as can be seen in Figure 14.14), and they proposed a more accurate empirical model based on their experimental observations:

$$P = 0.9\exp\left(\frac{-d_1'^2}{4D\tau_{OFF}}\right) \tag{14.32}$$

where d_1' is a radius variable chosen to fit the equation from experimental data. The coefficient 0.9 indicates that the diffusion is nonisotropic, with particles in experiments tending not to mount the electrode surfaces and thus more likely to diffuse forward from the collection point.

After the optimum time τ_{OFF}[62] some particles — i.e., those which have traveled a distance d_1 forward (toward the right in Figure 14.14) — will be captured by the next electrode on reapplication of the electric field as shown in Figure 14.14c. If τ_{OFF} is greater than this, some particles will be captured by the previous (left) electrode, and the efficiency of the ratchet will be reduced. Thus, maximum velocity has a defined maximum period time ($\tau_{OFF} + \tau_{ON}$) as shown in the studies of Prost et al.[63] However, the longer this time period lasts, the lower the net velocity $\mathbf{V}$, which cannot exceed a value given by

$$\mathbf{V} = \frac{d_2 + d_1}{\tau_{ON} + \tau_{OFF}} \tag{14.33}$$

Rousselet et al.[60] determined via computer simulation that the boundary between adjacent capture zones is neither circular nor straight, following a more complex pattern. It is therefore difficult to evaluate the efficiency of the geometry in terms of proportion of particles migrating forward within the time τ_{OFF}. Ajdari and Prost[56] proposed a dimensionless factor x as a ratio based on the distance d_1 as a proportion of the total distance d. A similar measure of this ratio Λ can be used for the comparison of different electrode geometries. Λ is expressed as the ratio of the difference between d_1 and d_2 as a proportion of the total distance between tips d. The value Λ may be interpreted as a measure of asymmetry and may take values from 0 (a symmetrical electrode assembly) to 1 (complete asymmetry). In practice $\Lambda = 1$ is unattainable; but by maximizing Λ, ratchet performance may be measured and improved.

There are a number of ways in which this mechanism may be applied to the separation of particles from a heterogeneous population. The first, demonstrated by Faucheux and Libchaber,[64] operated by separating particles according to their size. Because particles of different size will diffuse at different speeds, an appropriately timed ratchet will cause forward motion in one species while the other remains at the same electrode tips. A second method would exploit the fact that particles with differing dielectric properties will respond differently when subjected to electric fields of a given frequency. Nonpolar particles, or those whose crossover frequency coincides with the applied field, will not respond to dielectrophoretic forces; thus, a suspension of polarized and nonpolarized particles will respond within a ratchet assembly by the former being drawn out of the population while the latter remains in place. The final method would use negative dielectrophoresis to extend this. In mixtures of particles having two subpopulations with different crossover frequencies, particles can be simultaneously driven in opposite directions across the array in the window between the two crossover frequencies.[65]

The separation process may be dramatically speeded up by stacking two dielectrophoretic ratchets on top of one another, as described by Chauwin et al. in 1994[66] and demonstrated by Gorre–Talini et al. in 1998.[65] If a second series of ratchet electrodes is placed above the first but displaced by a half-wavelength of the distance between tips, then the points where positive and negative dielectrophoresis occur for one array lie in the correct regions to be further displaced by the second ratchet, and so on. This is demonstrated schematically in Figure 14.15. An array of this type has potential applications similar to methods of continuous dielectrophoretic cell separation proposed elsewhere,[67] but it is appropriate to situations where it is impractical to provide a fluid flow through the particle chamber or where smaller numbers of particles need to be separated.

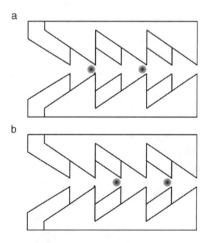

FIGURE 14.15 A continuous separator based on two pairs of stacked ratchet electrodes. When the upper electrode pair is active, particles undergoing positive dielectrophoresis are attracted to the tips of these electrodes. Once collection has taken place, potentials are removed from the lower electrodes and applied to the lower set. The particles at the electrode tips are all held beyond the crossover boundary for these electrodes and are attracted forward. After this collection has taken place, the potentials are returned to their original states. In this manner, 100% of the suspended particles are transported through the electrodes.

14.3.4 Molecular Applications of Dielectrophoresis

It was suggested for many years[12] that due to the action of electrohydrodynamic forces, it would be impossible to successfully manipulate particles on the nanometer scale. It is then ironic that the first research to significantly break the suggested lower particle limit of 1 μm should do so with the manipulation of proteins — macromolecules of the order of a few nanometers across.[8] The manipulation of molecules is somewhat more complex than the manipulation of latex beads, because they contain not only an induced dipole due to the electrical double layer but also a permanent dipole due to the position of fixed charges across the protein molecule. Washizu and co-workers demonstrated not only that dielectrophoresis of proteins was feasible but also that the technique could be used for the separation of proteins of different molecular weight on an electrode array. Another example of manipulation on this scale is the coating of microstructures of micrometer-scale devices using nano-powdered diamond to form cold cathode devices.[68]

Perhaps the greatest macromolecule of interest today is DNA, and many researchers have explored the possible applications of using dielectrophoresis for genomics applications. As with protein studies, the pioneering work was performed by Washizu and co-workers,[69] who demonstrated trapping of DNA molecules in 1990. Subsequent work showed that not only could DNA be trapped, but by careful manipulation of the field, strands of DNA could be stretched across an interelectrode gap and then cut to fragments of specific size using a laser.[70,71] Another demonstration of the application of dielectrophoresis to DNA research was the trapping of single DNA fragments 10 nm long by Bezryadin and co-workers.[72] This group subsequently used the technique to study the electrical properties of single fragments of DNA, which demonstrated interesting electrical properties which may have potential applications in nanometer-scale electronics.

Another application appropriate to nanotechnology is the use of dielectrophoresis to construct nanoscale electronic devices. For example, Velev and Kaler[73] used dielectrophoretic assembly to construct micrometer-sized biosensors using functionalized beads on an electrode array. Another example of nanoconstruction has been the manipulation of large nanowires (conducting or semiconducting threads of material nanometers wide and micrometers long) into functional electronic devices.[74–77] This has been achieved by activating electrodes that pointed toward each other across a central chamber — similar to the quadrupole electrodes array but with more "pointed" electrode structures — and applying a solution containing only p-doped semiconducting nanowires. Only one opposing pair of electrodes was energized in order to trap a single nanowire across the center of the chamber. The solution was removed and a second applied that contained only n-doped nanowires. By then energizing the other electrodes, a second nanowire was attracted, crossing the first at 90°. When the solutions were removed, the device consisting of the crossed nanowires — one p-type, one n-type — remained and was electrically characterized. It demonstrated strong optical characteristics indicating a possible application as a nanoscale light-emitting diode (LED) for nano-optical and nanoelectronic applications. Experiments have also been reported where dielectrophoresis of carbon nanotubes has been observed.[78] We can speculate that this work allows for the possibility of single-molecule electronic devices self-assembled into nanometer-scale computers — a possible future route to the nanotechnological future presented in science fiction novels, with nanotubes connected in three dimensions like plumbing around a ship's engine room.

14.3.5 Dielectrophoresis and Laser Trapping

Dielectrophoresis is not the only method available for the manipulation of nanometer-scale particles in solution. The principal alternative technique is optical trapping by the use of so-called laser tweezers. As we have now amassed sufficient knowledge of how dielectrophoresis can be used for trapping and separating particles, we are in a position to compare the two methods and their relative merits and demerits.

Laser trapping makes use of so-called *optical pressure* to induce force in an optical gradient, as reported by Ashkin in 1986.[6] If a transparent particle is exposed to a focused beam of light such that the particle experiences a gradient in the intensity of light, then the particle experiences a net force toward the direction of increasing gradient. This comprises two components, a *scattering* force and a *gradient* force. The magnitude of the gradient force on a spherical particle is given by the equation:[79]

$$\mathbf{F}_{optical} = \frac{n_m^2 r^3}{2} \left(\frac{n_p^2 - n_m^2}{n_p^2 + 2n_m^2} \right) \nabla E^2 \tag{14.34}$$

where n_p is the refractive index of the medium and n_m is the refractive index of the particle. Comparisons can be made between this expression and Equation 14.1, with the light beam providing the appropriate electric field. Also note the presence of the r^3 term in both expressions and the similarity of the bracketed term to the Clausius–Mossotti factor. However, against this we must set the additional radiation (or scatter) force terms, which do not have a significant dielectrophoretic analog.

In order to generate field gradients of sufficient magnitude, it is necessary to use a focused laser beam, with the particles attracted to the focal point. This technique has been used to manipulate a range of biological nanoparticles, from cells to proteins. There even exists an equivalent of electrorotation, wherein rotating modes in the laser are used to induce a rotational torque in the particle. These *laser spanners*[7] have been suggested as a possible mechanism for driving nanoscopic cams and gears fabricated from carbon nanotubes.[80]

Given that these techniques exist and have gained a widespread reputation, what place is there for dielectrophoresis? Its advantage over laser tweezing lies in the simplicity of operation. Laser tweezers require the use of one or more powerful and expensive lasers and complex optics in order to trap particles of the dimensions we refer to in this book. While it is straightforward to trap particles in solution by either method, it is important in optical trapping that there be a clear line of sight between the laser and the particle, for obvious reasons. On the other hand, with dielectrophoresis, the trapping is performed by small and relatively inexpensive electrode structures, generating electric fields using equipment that can usually be found in any radio repair shop. Provided the electrodes have power lines to the outside world, there is no need for a line of sight to the electrode chamber — which means that fulfilling the aforementioned application of rotating nanoscale gears would be considerably easier by this method. Dielectrophoresis is much more efficient than laser trapping, as both the generation of the laser beam and the dispersal of laser energy in the medium are wasteful of energy. The separation of heterogeneous mixtures into two trapped populations is far more difficult to realize with laser trapping because of the difficulties of creating a closed trap in which particles are repelled; the two-dimensional nature of dielectrophoresis electrodes, and the broad range of geometries that can be fabricated according to the intended application, gives the technique the edge. Finally, unlike tweezers, objects manipulated by dielectrophoresis need not be optically transparent.

The above having been said, it is important to consider that there are other advantages intrinsic to laser trapping. Greatest of these is that, because the particles are attracted to the focal point of the beam, the location of which is dictated by the geometry and position of the focusing objective lens, it is possible to achieve full three-dimensional directed motion of a trapped particle by moving the objective lens. In this respect, dielectrophoresis fares poorly as it is only capable of providing limited control in the vertical plane when particles are trapped by negative dielectrophoresis unless complex field cages are used. Furthermore, laser tweezers offer the ability to be switched to laser *scissors,* where a high-intensity beam is used to burn the contents of the focal point, making a useful tool for nanoconstruction. Ultimately, both techniques have strengths and weaknesses and are, as such, complementary; indeed, both techniques have been used simultaneously. For example, a laser tweezer system has been used to hold a single bacterium in place, while negative dielectrophoresis was used to repel other cells from the area. More commonly, laser tweezers have been used to maintain a steady position of cells within a rotating electric field during electrorotation experiments in order to ensure that the cells do not move within the chamber (thereby experiencing different electric field strengths and, hence, torques).[81]

14.4 Limitations of Nanoscale Dielectrophoresis

14.4.1 Limitations on Minimum Particle Trapping Size

In order to stably trap submicrometer particles, the dielectrophoretic force acting to move the particle into the center of the trap must exceed the action of Brownian motion on the particle, which, if large enough, will cause a particle to escape. In the case of positive dielectrophoresis, the applied force acts toward a single point (that of greatest field strength); and the force attracting it to that point must exceed the action of Brownian motion. For particles trapped in planar electrode arrays by negative dielectrophoresis, the particle is held in a dielectrophoretic forcefield "funnel."

The trapping of particles using positive dielectrophoresis is the simplest case to analyze: particles are attracted to the point of highest electric field strength, rising up the field gradient. Once at that point, the particle remains unless Brownian motion displaces it a sufficient distance that the field is unable to bring it back. The nature of positive dielectrophoresis is such that, given a long enough period of time (whether seconds or years), any particle polarized such that it experiences positive dielectrophoresis will *ultimately* fall into the trap.[82] As the electric field gradient extends to infinity, there will be an underlying average motion which will, over time, cause a displacement toward the high field trap.

The concentration of particles from solution by negative dielectrophoresis is slightly different in concept from the above case. Particles are trapped in regions where the electric field strength is very low, and they are prevented from escaping by a surrounding force-field *wall* that encloses the particles (although it may be in the from of an open-topped funnel). However, because in both the above cases the particle must achieve the same effect — the overcoming of a dielectrophoretic energy barrier which forces the particle into the trap — the mathematical treatment of both cases is similar.

Given this approach to the trapping of particles as overcoming the action of Brownian motion through the application of a quantifiable force, it is possible to determine what the relationship is between the magnitude of the electric field applied by the electrodes and the smallest particle that may be trapped by such electrodes. This approach was pioneered by Smith et al.[83] for determining the smallest particle that may be trapped by laser tweezers — a similar technique to positive dielectrophoresis. However, it is equally applied to both positive and negative dielectrophoresis.

Consider the force on a particle of radius r suspended in a nonuniform electric field, experiencing a trapping dielectrophoretic force $\mathbf{F}_{DEP}$. From Stokes' law, the particle's terminal velocity v is given by

$$v = \frac{\mathbf{F}_{DEP}}{6\pi\eta r} \tag{14.35}$$

Considering a small region of thickness Δd over which the force is constant, then the time t_{DEP} taken for the particle to traverse this region is given by

$$t_{DEP} = \frac{6\pi\eta r\Delta d}{\mathbf{F}_{DEP}} \tag{14.36}$$

Brownian motion acts to displace the particle from its position. From Einstein's equation,[84] the mean time $\langle t_B \rangle$ taken for a particle to move a distance Δd in one dimension is given by

$$\langle t_B \rangle = \frac{3\pi\eta r(\Delta d)^2}{kT} \tag{14.37}$$

For stable trapping to occur, the time for the particle to move along the field gradient by dielectrophoresis should be significantly less than the time taken for the particle to move out from it by Brownian motion, so that any displacement from the trap is immediately countered by dielectrophoresis. A factor of x10 was suggested by Smith et al.,[83] though this is arbitrary. From Equations 14.36 and 14.37, the conditions are

$$\frac{6\pi\eta r\Delta d}{\mathbf{F}_{\mathrm{DEP}}} < \frac{1}{10}\left(\frac{3\pi\eta r(\Delta d)^2}{kT}\right)$$
(14.38)

and:

$$\Delta d > \frac{20kT}{\mathbf{F}_{\mathrm{DEP}}}$$
(14.39)

The smallest particle radius which meets this criterion is given by

$$r > \sqrt[3]{\frac{10kT}{\pi\varepsilon_m\Delta dRe[K(\omega)]\nabla E^2}}$$
(14.40)

The variation r can be calculated from Equation 14.40 as a function of field gradient ∇E^2 and trap width Δd. At a temperature of 300 K, and with $\varepsilon_m = 78\varepsilon_0$ and $Re[K(\omega)] = 1$, this variation is shown in Figure 14.16. In cases where ∇E^2 varies as a function of distance, the trapping efficiency is given as the maximum value of the function $\Delta d\nabla E^2$ for the particular trap.

To determine what this might mean in terms of a given electrode geometry, it is necessary to simulate the electric field gradient around that geometry. The most accurate method of deriving the trapping force is to integrate the force across all given paths from the center of the trap to infinity (or at least to the edges of the trap), thereby determining the value of $\Delta d\nabla E^2$ for all possible escape paths. To do this, numerical models must be employed to determine the nature of the electric field around the electrodes, which is rarely determinable by analytical methods. For example, a numerical model based on the Moments method[85] was used to calculate ∇E^2 around the polynomial electrode array. Figure 14.17 shows a three-dimensional plot of both the *rms* electric field strength and the magnitude of ∇E^2 across the center of the electrode at a height of 7 μm above the electrode array shown in Figure 14.2, where particles are observed to be trapped by negative dielectrophoresis. The simulation was performed with an applied voltage of 5 $V_{\mathrm{pk\text{-}pk}}$. The trap efficiency is governed by the smallest distance that a trapped particle has to travel in order to escape from the trap. In the case of a particle trapped by positive dielectrophoresis, it is principally governed by the magnitude of the electric field as the particle is trapped at the point of highest field, which diminishes with increasing distance from that point. For particles trapped by negative

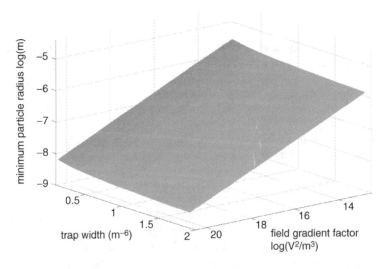

FIGURE 14.16 A graph showing the variation in the minimum radius of particles which could be trapped according to the expression in Equation 14.27.

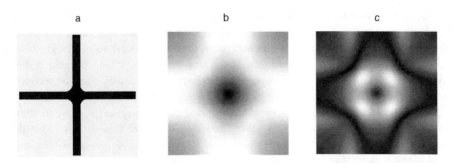

FIGURE 14.17 A simulation of the electric fields and forces generated by quadrupolar electrodes such as those found in Figure 14.2. (a) A schematic of the electrodes covered by the simulation, and the spatial variations of (b). (b) The electric field magnitude. (c) Dielectrophoretic force magnitude across the electrode array. The electrode array was modeled as having gaps of 2 μm between adjacent electrodes and 5 μm across the center of the array, and the simulation shows the field 7 μm above the electrodes. As can be seen, the field strength is greatest above the interelectrode arms of the array; there is a field null at the center of the array, enclosed on all sides by a region of high electric field.

dielectrophoresis, Δd is governed by the magnitude of the field barrier which *endorses* the particle. For example, as can be seen in Figure 14.17, the distribution of the electric field strength (as one might expect) has high field regions in the interelectrode gaps and a field null at the center. However, the force pattern is more complex, with a force barrier surrounding the field null. Repulsion by this barrier prevents escape by particles trapped by negative dielectrophoresis. In order to trap large numbers of particles, the dielectrophoretic force must overcome diffusion, but not necessarily by a factor as great as 10. If the condition is merely $\mathbf{F}_{DEP} > \mathbf{F}_{BROWNIAN}$, then the dielectrophoretic force on particles will be *on average* greater than Brownian motion and there will be a *net* force on the particle mass toward the trap, even if individual particles occasionally escape from the trap.

14.4.2 Brownian Motion, Conduction, and Convection

When attempting to exert a force on nanoparticles in suspension, in order to direct them with great precision in solution, there are a number of factors related to the presence of the suspending medium that act to disrupt the controlled movement. While many of these effects also affect micrometer-scale particles, their influence is far more significant when dealing with nanoparticles. There are two reasons for this: the particles are nearer the size of the molecules of the suspending medium, and the electric field gradient (and therefore the electric field strength) is considerable in order to impart significant forces in particles with small volumes.

The first of these — the relative size of molecule and object — increases the effect of Brownian motion, the movement of colloidal particles due to the impacts of moving water molecules colliding with the surface of the particle. Brownian motion has been widely studied,[84] and within the field of dielectrophoresis it has been examined both as a problem to be overcome[8,11,12] and as a means to provide propulsion.[56–60,62–66] In some cases, statistical analysis is required in order to discriminate between dielectrophoretic and Brownian motion; any force applied to a particle will ultimately result in what Ramos et al.[82] describes as *observable deterministic motion*. But if the force is small, the time taken to observe it may be far longer than the duration of the experiment. However, the forces used in the experiments described in the literature are sufficiently large for the applied force to observably overcome Brownian motion to be of the order of a second or (often significantly) less.

The second major effect that disrupts dielectrophoretic measurement on the nanoscale is the motion of the suspending medium due to the interaction between the water molecules and the electric field — a topic known as *electrohydrodynamics* (EHD). This large and complex subject is beyond the scope of this chapter, and the interested reader is pointed toward excellent in-depth reviews by Ramos et al.[82] and Green et al.[86] The two major EHD forces that are significant in driving fluid flow around electrodes are *electrothermal* and *electro-osmotic* in origin. The former force is due to localized heating of the medium, causing discontinuities of medium conductivity and permittivity, which was found to be insignificant due to the small volumes these discontinuities occupy (as the electrodes are so small). The latter and far more significant force is due to the interaction of the tangential electric field with the diffuse double layer above the electrode surfaces. This creates a fluid pumping action, which can be several orders of magnitude larger than the dielectrophoretic force — suspending medium above quadrupole electrodes, for example, is pumped into the center of the chamber along the electrode surfaces, whereupon it forms a spout that forces the particles up. Under other conditions this fluid flow can be reversed, drawing material to the electrodes. Careful electrode design can allow the EHD forces to be used as an aid to particle trapping and separation, as discussed by Green and Morgan.[87]

14.5 Conclusion

The techniques of AC electrokinetics, and particularly that of dielectrophoresis, have much to offer the expanding science of nanotechnology. Whereas dielectrophoretic methods have been applied to the manipulation of objects in the microscale in the past, there can be no doubt of its tremendous potential for nanoparticle manipulation. This opens a wide range of potential applications for AC electrokinetics to the development of mainstream nanotechnology. The fundamental challenge in the advancement of nanotechnology is the development of precision tools for large-scale self-assembly of nanometer-scale components. The techniques described here go some way in addressing this by providing tools for the trapping, manipulation, and separation of molecules and other nanoparticles using tools (electrodes) on the micrometer scale.

References

1. Drexler, K.E., *Nanosystems: Molecular Machinery, Manufacturing and Computation*, Wiley, New York, 1992.
2. Eigler, D.M. and Schweizer, E.K., Positioning single atoms with a scanning tunnelling microscope, *Nature*, 344, 524, 1990.
3. Bezryadin, A., Dekker, C., and Schmid, G., Electrostatic trapping of single conducting nanoparticles between nanoelectrodes, *Appl. Phys. Lett.*, 71, 1273, 1997.
4. Jones, T.B., *Electromechanics of Particles*, Cambridge University Press, Cambridge, 1995.
5. Zimmermann U. and Neil, G.A., *Electromanipulation of Cells*, CRC Press, Boca Raton, 1996.
6. Ashkin, A., Dziedzic, J.M., Bjorkholm, J.E., and Chu, S., Observation of a single-beam gradient force optical trap for dielectric particles, *Opt. Lett.*, 11, 288, 1986.
7. Simpson, N.B., Dholakia, K., Allen, L., and Padgett, M.J., Mechanical equivalence of spin and orbital angular momentum of light: an optical spanner, *Opt. Lett.*, 22, 52, 1997.
8. Washizu, M., Suzuki, S., Kurosawa, O., Nishizaka, T., and Shinohara, T., Molecular dielectrophoresis of biopolymers, *IEEE Trans. Ind. Appl.*, 30, 835, 1994.
9. Bakewell, D.J.G., Hughes, M.P., Milner, J.J., and Morgan, H., Dielectrophoretic manipulation of Avidin and DNA, in *Proc. 20ᵗʰ Ann. Intl. Conf. IEEE Eng. Med. Biol. Soc.*, 1998.
10. Müller, T., Gerardino, A., Schnelle, T., Shirley, S.G., Bordoni, F., DeGasperis, G., Leoni, R., and Fuhr, G., Trapping of micrometer and sub-micrometer particles by high-frequency electric fields and hydrodynamic forces, *J. Phys. D: Appl. Phys.*, 29, 340, 1996.
11. Hughes, M.P. and Morgan, H., Dielectrophoretic manipulation of single sub-micron scale bioparticles, *J. Phys. D: Appl. Phys.*, 31, 2205, 1998.
12. Pohl, H.A., *Dielectrophoresis*, Cambridge University Press, Cambridge, 1978.
13. Green, N.G. and Morgan, H., Dielectrophoretic investigations of sub-micrometer latex spheres, *J. Phys. D: Appl. Phys.*, 30, 2626, 1997.
14. Green, N.G. and Morgan, H., Dielectrophoresis of submicrometer latex spheres.1. Experimental results, *J. Phys. Chem.*, 103, 41, 1999.

15. Hughes, M.P., Morgan, H., and Flynn, M.F., Surface conductance in the diffuse double-layer observed by dielectrophoresis of latex nanospheres, *J. Coll. Int. Sci.*, 220, 454, 1999.
16. Hughes, M.P., Morgan, H., and Rixon, F.J., Dielectrophoretic manipulation and characterisation of herpes simplex virus-1 capsids, *Eur. Biophys. J.*, 30 268, 2001.
17. Hughes, M.P., Morgan, H., and Rixon, F.J., Measurements of the properties of herpes simplex virus type 1 virions with dielectrophoresis, *Biochim. Biophys. Acta*, 1571, 1, 2002.
18. Morgan, H. and Green, N.G., Dielectrophoretic manipulation of rod-shaped viral particles, *J. Electrostatics*, 42, 279, 1997.
19. Hughes, M.P. and Morgan, H., Dielectrophoretic manipulation of protein molecules in solution, in *Proc. 1st Eur. Worksh. Electrokinetics Electrohydrodynamics Microsyst.*, Glasgow, September 6–7, 2001.
20. Gascoyne, P.R.C., Pethig, R., Burt, J.P.H., and Becker, F.F., Membrane changes accompanying the induced differentiation of Friend murine erythroleukemia cells studied by dielectrophoresis, *Biochim. Biophys. Acta*, 1149, 119, 1993.
21. Huang, Y., Wang, X.-B., Becker, F.F., and Gascoyne, P.R.C., Membrane changes associated with the temperature-sensitive P85$^{gag\text{-}mos}$-dependent transformation of rat kidney cells as determined by dielectrophoresis and electrorotation, *Biochim. Biophys. Acta*, 1282, 76, 1996.
22. Gascoyne, P.R.C., Noshari, J., Becker, F.F., and Pethig, R., Use of dielectrophoretic collection spectra for characterizing differences between normal and cancerous cells, *IEEE Trans. Ind. Appl.*, 30, 829, 1994.
23. Arnold, W.M. and Zimmermann, U., Electro-rotation — development of a technique for dielectric measurements on individual cells and particles, *J. Electrostatics*, 21, 151, 1988.
24. Gascoyne, P.R.C., Pethig, R., Satayavivad, J., Becker, F.F., and Ruchirawat, M., Dielectrophoretic detection of changes in erythrocyte membranes following malarial infection, *Biochim. Biophys. Acta*, 1323, 240, 1997.
25. Lyklema, J., *Fundamentals of Interface and Colloid Science*, Academic Press, London, 1995.
26. Hughes, M.P. and Green, N.G., The influence of Stern layer conductance on the dielectrophoretic behaviour of latex nanospheres, *J. Coll. Int. Sci.*, 250, 266, 2002.
27. Bockris, J.O'M. and Reddy, A.K.N., *Modern Electrochemistry*, Plenum, New York, 1973
28. Irimajiri, A., Hanai, T., and Inouye, V., A dielectric theory of "multi-stratified shell" model with its application to lymphoma cell, *J. Theor. Biol.*, 78, 251, 1979.
29. Huang, Y., Holzel, R., Pethig, R., and Wang X.-B., Differences in the AC electrodynamics of viable and nonviable yeast cells determined through combined dielectrophoresis and electrorotation studies, *Phys. Med. Biol.*, 37, 1499, 1992.
30. Lipowicz, P.J. and Yeh, H.C., Fiber dielectrophoresis, *Aerosol Sci. Technol.*, 11, 206, 1989.
31. Kakutani, T., Shibatani, S., and Sugai, M., Electrorotation of non-spherical cells: theory for ellipsoidal cells with an arbitrary number of shells, *Biochem. Bioenergetics*, 31, 131, 1993.
32. Hasted, J.B., *Aqueous Dielectrics*, Chapman and Hall, London, 1973.
33. Arnold, W.M. and Zimmermann, U., Rotating-field-induced rotation and measurement of the membrane capacitance of single mesophyll cells of Avena sativa, *Z. Naturforsch.*, 37c, 908, 1982.
34. Teixeira–Pinto, A.A., Nejelski, L.L., Cutler, J.L., and Heller, J.H., The behaviour of unicellular organisms in an electromagnetic field, *J. Exp. Cell Res.*, 20, 548, 1960.
35. Berry, R.M. and Berg, H.C., Torque generated by the flagellar motor of escherichia coli while driven backward, *Biophys.J.*, 76, 580, 1999.
36. Hagedorn, R., Fuhr, G., Müller, T., Schnelle, T., Schnakenberg, U., and Wagner, B., Design of asynchronous dielectric micromotors, *J. Electrostatics*, 33, 159, 1994.
37. Huang, Y., Wang, X.-B., Tame, J., and Pethig, R., Electrokinetic behaviour of colloidal particles in traveling electric fields: studies using yeast cells, *J. Phys. D: Appl. Phys.*, 26, 312, 1993.
38. Batchelder, J.S., Dielectrophoretic manipulator, *Rev. Sci. Instrum.*, 54, 300, 1983.
39. Masuda, S., Washizu, M., and Iwadare, M., Separation of small particles suspended in liquid by nonuniform traveling field, *IEEE Trans. Ind. Appl.*, 23, 474, 1987.

40. Fuhr, G., Hagedorn, R., Müller, T., Benecke, W., Wagner, B., and Gimsa, J., Asynchronous traveling-wave induced linear motion of living cells, *Studia Biophysica,* 140, 79, 1991.

41. Hughes, M.P., Pethig, R., and Wang, X.-B., Forces on particles in traveling electric fields: computer-aided simulations, *J. Phys. D: Appl. Phys.,* 29, 474, 1996.

42. Fuhr, G., Schnelle, T., and Wagner, B., Traveling-wave driven microfabricated electrohydrodynamic pumps for liquids, *J. Micromech. Microeng.,* 4, 217, 1994.

43. Morgan, H., Green, N.G., Hughes, M.P., Monaghan, W., and Tan, T.C., Large-area traveling-wave dielectrophoresis particle separator, *J. Micromech. Microeng.,* 7, 65, 1997.

44. Fuhr, G., Fiedler, S., Müller, T., Schnelle, T., Glasser, H., Lisec, T., and Wagner, B., Particle micromanipulator consisting of two orthogonal channels with traveling-wave electrode structures, *Sensors Actuators A,* 41, 230, 1994.

45. Ward, M., Devilish tricks with tiny chips, *New Scientist,* 1st March, 22, 1997.

46. Gascoyne, P.R.C., Huang, Y., Pethig, R., Vykoukal, J., and Becker, F.F., Dielectrophoretic separation of mammalian cells studied by computerized image analysis, *Meas. Sci. Technol.,* 3, 439, 1992.

47. Green, N.G., and Morgan, H., Dielectrophoretic separation of nanoparticles, *J. Phys. D: Appl. Phys.,* 30, L41, 1997.

48. Burt, J.P.H., Pethig, R., and Talary, M.S., Microelectrode devices for manipulating and analysing bioparticles, *Trans. Inst. Meas. Control,* 20, 82, 1998.

49. Morgan, H., Hughes, M.P., and Green, N.G., Separation of submicron bioparticles by dielectrophoresis, *Biophys. J.,* 77, 516, 1999.

50. Watarai, H., Sakamoto, T., and Tsukahara, S., *In situ* measurement of dielectrophoretic mobility of single polystyrene microspheres, *Langmuir,* 13, 2417, 1997.

51. Chiu, D.T. and Zare, R.N., Optical detection and manipulation of single molecules in room-temperature solutions, *Chem. A Eur. J.,* 3, 335, 1997.

52. Park, H., Park, J., Lim, A.K.L., Anderson, E.H., Alivisatos, A.P., and McEuen, P.L., Nanomechanical oscillations in a single-C60 transistor, *Nature,* 407, 57, 2000.

53. Schnelle, T., Hagedorn, R., Fuhr, G., Fiedler, S., and Müller, T., Three-dimensional electric field traps for manipulation of cells — calculation and experimental verification, *Biochim. Biophys. Acta,* 1157, 127, 1993.

54. Schnelle, T., Müller, T., and Fuhr G., Trapping in AC octode field cages, *J. Electrostatics,* 50, 17, 2000.

55. Curie, P., Sur la symétrie dans les phénomènes physiques, *J. Phys.* 383, 111, 1894.

56. Ajdari, A. and Prost, J., Drift induced by a spatially periodic potential of low symmetry: pulsed dielectrophoresis, *C. R. Acad. Sci. Paris,* 315, 1635, 1992.

57. Chauwin, J.-F., Ajdari, A., and Prost, J., Mouvement sans force, in *Proc. SFP 4émes Journées de la Matire Condensèe,* Rennes, 1994.

58. Magnasco, M.O., Forced thermal ratchets, *Phys. Rev. Lett.,* 71, 1477, 1993.

59. Astumian, R.D. and Bier, M., Fluctuation-driven ratchets: molecular motors, *Phys. Rev. Lett.,* 72, 1766, 1994.

60. Rousselet, J., Salome, L., Ajdari, A., and Prost, J., Directional motion of Brownian particles induced by a periodic asymmetric potential, *Nature,* 370, 446 1994.

61. Risken H., *The Fokker–Planck Equation: Methods of Solution and Applications,* Springer-Verlag, Berlin, 1984.

62. Doering, C.R., Horsthemke, W., and Riordan, J., Nonequilibrium fluctuation-induced transport, *Phys. Rev. Lett.,* 72, 2984, 1994.

63. Prost, J., Chauwin, J.-F., Peliti, L., and Ajdari, A., Asymmetric pumping of particles, *Phys. Rev. Lett.,* 72, 2652, 1994.

64. Faucheux, L.P. and Libchaber, A., Selection of Brownian particles, *J. Chem. Soc. Faraday Trans.,* 91, 3163, 1995.

65. Gorre–Talini, L., Spatz, J.P., and Silberzan, P., Dielectrophoretic ratchets, *Chaos,* 8, 650, 1998.

66. Chauwin, J.F., Ajdari, A., and Prost, J., Force-free motion in asymmetric structures — a mechanism without diffusive steps, *Europhys. Lett.,* 27, 421, 1994.

67. Markx, G.H. and Pethig, R., Dielectrophoretic separation of cells: continuous separation, *Biotechnol. Bioeng.*, 45, 337, 1995.
68. Alimova, A.N., Chubun, N.N., Belobrov, P.I., Ya Detkov, P., and Zhirnov, V.V., Electrophoresis of nanodiamond powder for cold cathode fabrication, *J. Vac. Sci. Technol.*, 17, 715, 1999.
69. Washizu, M. and Kurosawa, O., Electrostatic manipulation of DNA in microfabricated structures, *IEEE Trans. Ind. Appl.*, 26, 1165, 1990.
70. Washizu, M., Kurosawa, O., Arai, I., Suzuki, S., and Shimamato, N., Applications of electrostatic stretch and positioning of DNA, *IEEE Trans. Ind. Appl.*, 31, 447, 1995.
71. Yamamoto, T., Kurosawa, O., Kabata, H., Shimamato, N., and Washizu, M., Molecular surgery of DNA based on electrostatic manipulation, *IEEE Trans. Ind. Appl.*, 36, 1010, 2000.
72. Porath, D., Bezryadin, A., de Vries, S., and Dekker, C., Direct measurement of electrical transport through DNA molecules, *Nature*, 403, 635, 2000.
73. Velev, O.D. and Kaler, E.W., *In situ* assembly of colloidal particles into miniaturised biosensors, *Langmuir*, 15, 3693, 1999.
74. Duan, X., Huang, Y., Cui, Y., Wang, J., and Lieber, C.M., Indium phosphide nanowires as building blocks for nanoscale electronic and optoelectronic devices, *Nature*, 409, 66, 2001.
75. Cui, Y. and Leiber, C.M., Functional nanoscale electronic devices assembled using silicon nanowires building blocks, *Science*, 291, 851, 2001.
76. Huang, Y., Duan, X.F., Wie, Q.Q., and Leiber, C.M., Directed assembly of one-dimensional nanostructures into functional networks, *Science*, 291, 630, 2001.
77. Smith, P.A., Nordquist, C.D., Jackson, T.N., Mayer, T.S., Martin, B.R., Mbindyo, J., and Mallouk, T.E., Electric-field assisted assembly and alignment of metallic nanowires, *Appl. Phys. Lett.*, 77, 1399, 2000.
78. Yamamoto, K., Akita, S., and Nakayama, Y., Orientation and purification of carbon nanotubes using AC electrophoresis, *J. Phys. D: Appl. Phys.*, 31, L34, 1998.
79. Smith, P.W., Ashkin, A. and Tomlinson, W.J., Four-wave mixing in an artificial Kerr medium, *Opt. Lett.*, 6, 284, 1981.
80. Srivastava, D., A phenomenological model of the rotation dynamics of carbon nanotube gears with laser electric fields, *Nanotechnology*, 8, 186, 1997.
81. Arai, F., Ogawa, M., and Fukuda, T., Selective manipulation of a microbe in a microchannel using a teleoperated laser scanning manipulator and dielectrophoresis, *Adv. Robotics*, 13, 343, 1999.
82. Ramos, A., Morgan, H., Green, N.G., and Castellanos, A., AC electrokinetics: a review of forces in microelectrode structures, *J. Phys. D: Appl. Phys.*, 31, 2338, 1998.
83. Smith, P.W., Ashkin, A., and Tomlinson, W.J., Four-wave mixing in an artificial Kerr medium, *Opt. Lett.*, 6, 284, 1981.
84. Einstein, A., *Investigations on the Theory of Brownian Movement*, Dover, New York, 1956.
85. Birtles, A.B., Mayo, B.J., and Bennett, A.W., Computer technique for solving three-dimensional electron-optics and capacitance problems, *Proc. IEEE*, 120, 213, 1973.
86. Green, N.G., Ramos, A., Morgan, H., and Castellanos, A., Sub-micrometer AC electrokinetics: particle dynamics under the influence of dielectrophoresis and electrohydrodynamics, *Inst. Phys. Conf. Ser.* 163, 89, 1999.
87. Green, N.G. and Morgan, H., Separation of submicrometer particles using a combination of dielectrophoretic and electrohydrodynamic forces, *J. Phys. D: Apply. Phys.*, 31, L25, 1998.

15

Biologically Mediated Assembly of Artificial Nanostructures and Microstructures

CONTENTS

Rashid Bashir
Purdue University

Abstract

Biologically mediated assembly of nanometer- and micrometer-scale structures can have a profound impact in the fields of nanoelectronics, materials synthesis, and medical diagnostics and therapeutics. Such self-assembled structures can also find applications in microelectromechanical systems and hybrid biosensors, and have the potential to continue the scaling of Moore's law beyond the 50 nm node. While engineers and scientists have been long aspiring to controllably and specifically manipulate structures at the micrometer and nanometer scale, nature has been performing these tasks and assembling complex structures with great accuracy and high efficiency using highly specific biological molecules such as DNA and proteins. The use of such molecules to assemble artificial structures, synthesize new materials, and construct new interfaces is an intense area of research. This chapter describes the motivations and fundamentals behind these assembly concepts, with a focus on biologically mediated assembly, and presents the state of the art in biologically mediated assembly of artificial nanostructures and microstructures.

15.1 Introduction

15.1.1 On Size and Scale

Recent advances in the field of nanotechnology and nanobiotechnology have been fueled by the advancement in fabrication technologies that allow construction of artificial structures that are the same size or smaller than many biological entities. Figure 15.1 shows the size and scale of many biological and artificial structures. It is interesting to note that the minimum feature in modern-day integrated circuits, which was 0.1 μm at the end of 2002, is an order of magnitude smaller than cells and bacteria. The tip of an atomic-force microscope, a key tool in advancing the field of nanotechnology, is smaller than most viruses. The gate insulator thickness of a modern-day metal-oxide semiconductor (MOS) transistor is thinner than one helical turn of a DNA. Thus, it is clear that the top-down fabrication technologies have progressed enough to allow the fabrication of microstructures and nanostructures that can be used to interface, interrogate, and integrate biological structures with artificial structures.

15.1.2 Miniaturization

Since the invention of the junction transistor in 1947 and the subsequent invention of the integrated circuit, the complexity of microelectronic integrated circuits and devices has increased exponentially. Figure 15.2 shows the trends in miniaturization and complexity using silicon CMOS (complementary metal-oxide semiconductor) technology. The minimum feature size has decreased from 2 μm in 1980 to 0.1 μm in 2002 in volume production.[1] In research labs, minimum features sizes, which are a factor of 5–10 smaller, have been demonstrated. The SIA (Semiconductor Industry Association) roadmap projects that these trends will continue for another 15–20 years; but it is becoming increasingly difficult to continue to downscale due to real physical limitations including size of atoms, wavelengths of radiation used for lithography, interconnect schemes, etc. No known solutions currently exist for many of these problems.[2,3]

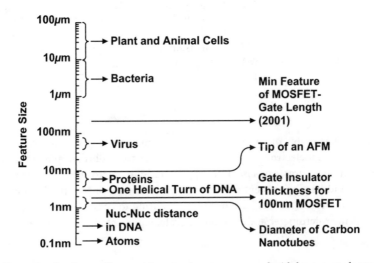

FIGURE 15.1 Size and scale of naturally occurring structures as compared with human-made structures.

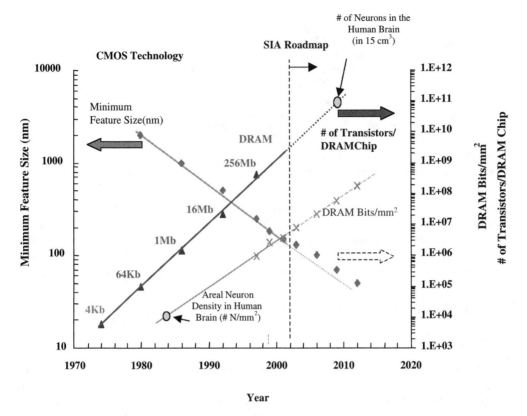

FIGURE 15.2 Trends in miniaturization of integrated circuits in the last 25 years.

As the construction of artificial computational systems, i.e., integrated circuits, continues to become insurmountably difficult, more and more engineers and scientists are turning toward nature for answers and solutions. A variety of extremely sophisticated and complicated molecular systems occur in nature that vary in density, sense, and relay information; perform complex computational tasks; and self-assemble into complex shapes and structures. Two examples can be considered — the human brain and DNA in the nucleus of the cell. There are about 10^{11} neurons in the human brain in a volume of about 15 cm.[3,4] The total number of transistors on a two-dimensional chip will actually reach the number of neurons in the human brain by about year 2010. The area density of the neurons was actually surpassed in the middle 1980s; but as is known, the three-dimensional nature and parallel interconnectivity of the neurons are what make the exquisite functions of the brain possible. So even though humans have achieved or will soon achieve a similar density of basic computational elements to that of the brain, the replication of brain functions is far from reality.

Similarly, the case of DNA is also far-reaching and intriguing. The human DNA is about 6 mm long, has about 2×10^8 nucleotides, and is tightly packed in a volume of 500 μm.[3,4] If a set of three nucleotides can be assumed to be analogous to a byte (a three-codon set from mRNA is used to produce an amino acid), then these numbers represent about 1 Kb/μm (linear density) or about 1.2 Mb/μm³ (volume density). These numbers are not truly quantitative but can give an appreciation of how densely information is stored in the DNA molecules. Certainly, a memory chip based on DNA molecules could have extremely high density.

15.1.3 Engineering and Biology

As the dimensions of artificially machined structures become smaller than biological entities, numerous exciting possibilities arise to solve important and complex problems. One way to categorize the types of

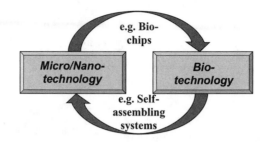

FIGURE 15.3 Microtechnology and nanotechnology can be used to solve problems in life sciences, or life sciences can be used to impact microtechnology and nanotechnology.

possible research is shown in Figure 15.3. Microtechnology and nanotechnology can be used to solve important problems in life sciences. Examples of this category would be biochips, which are now being commercialized. On the other hand, knowledge from life sciences can be used to solve important problems in engineering; examples of this would include bio-inspired self-assembly and biomimitic devices. All biological organisms are essentially self-assembled, beginning from one cell, resulting in entire organisms.

15.2 Bio-Inspired Self-Assembly

15.2.1 Self-Assembly Defined

Self-assembly can be defined as the process of self-organization of one or more entities as the total energy of the system is minimized to result in a more stable state. This process of self-assembly inherently implies: (1) some mechanism where movement of entities takes place using diffusion, electric fields, etc.; (2) the concept of recognition between different elements, or biolinkers, that result in self-assembly; (3) where the recognition results in binding of the elements dictated by forces (electrical, covalent, ionic, hydrogen, van der Waals, etc.), such that the resulting physical placement of the entities results in the state of lowest energy.

The basic idea behind the concept of biologically mediated self-assembly is shown in Figure 15.4. An object is coated with molecules (biolinkers) that recognize and are complementary to other molecules on another object. The two objects are somehow brought in close proximity, and under appropriate conditions, the two linkers would bind together and as a result form an ordered arrangement. This is how cells are captured in an ELISA plate and single strands of DNA hybridize, for example, and how devices can be self-assembled on heterogeneous substrates. The forces that bind the biolinkers could include any of the forces described in more detail in subsequent sections. For example, the molecules could be single strands of DNA that are complementary to each other and help to bind the two objects together. The molecules on one or both objects could also be charged, and electrostatic forces could be used to bring together and bind the two objects.

15.2.2 Self-Assembly Categorized

Significant literature is available on self-assembly. Broadly speaking, the research on self-assembly can be divided into three categories: (1) formation of objects themselves, such as clusters, molecules and macromolecules, quantum dots, nanowires and nanotubes, or crystal growth; (2) formation of two-dimensional/three-dimensional arrays and networks, which could include self-assembled monolayers (SAMs); close-packed arrays of particles, clusters, and shells; and three-dimensional close-packed assemblies, Stranski-Krastanov strain-induced material self-assembly, etc; and (3) directed selective assembly of objects at specific locations due to a variety of biological, chemical, or electrical forces. The discussion in this chapter focuses mostly on the third category, where microscale and nanoscale objects are assembled at specific sites due to biologically inspired forces and phenomenon.

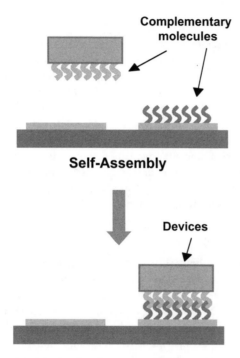

FIGURE 15.4 A basic concept schematic showing the process of self-assembly.

15.2.3 Motivation for Bio-Inspired Self-Assembly

Self-assembly processes are not only interesting from a scientific point of view but can also have a wide variety of applications. These applications can include any case where microscale or nanoscale objects of one type need to be placed or assembled at specific sites on another substrate. As will be evident from Section 15.5 below, the applications can be in the areas of (1) detection and diagnostics, (2) fabrication of novel electronic/optoelectronic systems, and (3) new material synthesis. For example, in the case of detection of DNA, oligonucleotides, avidin-coated gold, or polystyrene beads are assembled onto a biotinylated target DNA to indicate complementary binding. Proteins and DNA, when attached to carbon nanotubes or silicon nanowires or devices, can be used to assemble these devices on substrates for ultra-dense electronics or flexible displays. Heterogeneous integration of materials can be achieved using such a biologically mediated assembly of components. Because these assembly techniques provide microscale and nanoscale placement of objects, the assembly can be repeated multiple times to result in novel three-dimensional material synthesis.

15.3 The Forces and Interactions of Self-Assembly

Molecular and macromolecular recognition processes are key to bio mediated self-assembly, and various forces comprise the molecular recognition processes. We will now review these forces and the resulting chemical bonds briefly.[4] These bonds can be categorized as covalent and noncovalent. Covalent bonds, as is well known, are strong; a single C–C bond has energy of about 90 kcal/mol and is responsible for the bonding in between the subunits of the macromolecules. The much weaker noncovalent bonds in biological molecules are due to (1) van der Waals interactions (~0.1 kcal/mole per atom), (2) hydrogen bonding (~1 kcal/mol), (3) ionic bonding (~3 kcal/mol in water, 80 kcal/mole in vacuum), and (4) hydrophobic interactions. These interactions (or a combination thereof) determine the specific recognition and binding between specific molecules and macromolecules that are used in the self-assembly processes described in this chapter.

15.3.1   van der Waals–London Interactions

The van der Waals–London interactions are generated between two identical inert atoms that are separated from each other by a distance that is large in comparison with the radii of the atom. Because the charge distributions are not rigid, each atom causes the other to slightly polarize and induce a dipole moment. These resulting dipole moments cause an attractive interaction between the two atoms. This attractive interaction varies as the minus sixth power of the separation of the two atoms, $\Delta U = -(A/R^6)$, where A is a constant. The van der Waals–London interaction is a quantum effect and will result between any two charged bodies where the charge distributions are not rigid, so that they can be perturbed in space to inducer a dipole moment. Two atoms will be attracted to each other until the distance between them equals the sum of their van der Waals radii. Brought closer than that, the two atoms will repel each other. Individually, these forces are very weak, but they can play an important role in determining the binding of two macromolecular surfaces.

15.3.2   Hydrogen Bonding

Another important interaction that binds different molecules together is hydrogen bonding. This is how the nucleotides of DNA (described below) complementarily bind to each other. Hydrogen bonding is largely ionic in nature and can result when a hydrogen atom, which is covalently bonded to a small electronegative atom (e.g., F, N, O), develops a positive-induced dipole charge. This positive charge can then interact with the negative end of a neighboring dipole, resulting in a hydrogen bond. This interaction typically varies as $1/R^3$, where R is the distance between the dipoles. Hydrogen bonding is an important part of the interaction between water molecules.

15.3.3   Ionic Bonding

Ionic interaction can also take place in partially charged groups or fully charged groups (ionic bonds). The force of attraction is given by Coulomb's law as $F = K((q^+ q^-)/R^2)$, where R is the distance of separation. When present among water and counterions in biological mediums, ionic bonds become weak because the charges are partially shielded by the presence of counterions. Still, these ionic interactions between groups are very important in determining the recognition between different macromolecules.

15.3.4   Hydrophobic Interactions

Hydrophobic interactions between two hydrophobic groups are produced when these groups are placed in water. The water molecules tend to move these groups close in such a way as to keep these hydrophobic regions close to each other. Hence, this coming together of hydrophobic entities in water can be termed a *hydrophobic bond*. Again, these interactions tend to be weak but are important, for example, when two hydrophobic groups on the surface of two macromolecules come together when placed in water and bind the two surfaces. The surface energy at the interface of a hydrophobic surface and water is high because water tends to be moved away from the surface. Hence, two hydrophobic surfaces will join to minimize the total exposed surface area and hence the total energy. The phenomenon plays a very important part in the protein folding (Section 15.4.2.1) and also to assemble objects in fluids using capillary forces.

15.4   Biological Linkers

Biologically mediated self-assembly utilizes complementary molecules, which can be termed *biolinkers*. These molecules act like a lock and key, binding to each other under appropriate conditions. The binding of these linker molecules (actually macromolecules) is a result of the noncovalent interactions described above. Two possible biolinkers include complementary nucleic acid molecules and protein complexes, and these are described below in more detail.

15.4.1 Deoxyribonucleic Acid (DNA) as Biolinkers

15.4.1.1 DNA Fundamentals

DNA is the basic building block of life. Hereditary information is encoded in the chemical language of DNA and reproduced in all cells of living organisms. The double-stranded helical structure of the DNA is key to its use in self-assembly applications. Each strand of the DNA is about 2 nm wide and composed of a linear chain of four possible bases (adenine, cytosine, guanine, and thymine) on a backbone of alternating sugar molecules and phosphate ions (see Figure 15.5). Each unit of a phosphate, sugar molecule, and base is called a *nucleotide,* and each nucleotide is about 0.34 nm long. The specific binding through hydrogen bonds between adenine, thymine, cytosine, and guanine as shown in Figure 15.5(b) can result in the joining of two complementary single-stranded (ss) DNA to form a double-stranded (ds) DNA. The phosphate ion carries a negative charge in the DNA molecule, a property used in the drift of the molecule under an electric field, e.g., in electrophoresis applications. The negative charges result in electrostatic repulsion of the two strands; hence, to keep the two strands together, positive ions need to be present in the ambient to keep the negative charges neutralized. The joining of two ss DNA through hydrogen bonding (described earlier) to form a ds DNA is called *hybridization.* Two single strands of DNA can be designed to have complementary sequences and made to join under appropriate conditions. If a double-stranded DNA is heated above a certain temperature, called the *melting temperature* (T_m), the two strands will separate into single strands. The melting temperature is a function of temperature, ion concentration of the ambient, and the G–C content in the sequence. When the temperature is reduced, the two strands will eventually come together by diffusion and *rehybridize* or *renature* to form the double-stranded structure as shown in Figure 15.6. These properties of the DNA can be utilized in the ordering and assembly of artificial structures if these structures can be attached to ss DNA. It should also be pointed out that the sequence of the DNA can be chosen and the molecules can now be obtained from a variety of commercial sources.[5]

15.4.1.2 Attachment of DNA to Gold Surfaces

It is also important to review the methodology to attach DNA molecules to surfaces. The most widely used attachment scheme utilizes the covalent bond between sulfur and gold.[6–19] The formation of long-chain, ω-substituted dialkyldisulfide molecules on a gold substrate was first reported in 1983.[6] Films of better quality were formed and reported by the adsorption of alkyl thiols.[7–13] Bain and Whitesides presented a model system consisting of long-chain thiols, $HS(CH_2)_nX$ (where X is the end group), that adsorb from a solution onto gold and form densely packed, oriented monolayers.[7–8] The schematic of the Au–S bond is shown in Figure 15.7.[9] The bonding of the sulfur head group to the gold substrate is in the form of a metal thiolate, which is a very strong bond (~44 kcal/mol), and thus the resulting films are quite stable and suitable for surface attachment of functional groups. For example, the DNA molecule can be functionalized with a thiol (S–H) group at the 3' or 5' end. Upon immersion of clean gold surfaces in solutions of thiol-derivatized oligonucleotides, the sulfur adsorbs on the gold surfaces, forming a single layer of molecules as schematically shown in Figure 15.7, where the hydrocarbon is now replaced with an ss DNA or a ds DNA molecule.[16–19] Chemisorption of thiolated ss DNA leads to surface coverages of about 10^{13} molecules/cm,[2][18] which corresponds to about 1 strand per 10 nm².[18] Hickman et al. also demonstrated the selective and orthogonal self-assembly of disulfide with gold and isocyanide with platinum.[10] This work can be important in the orientation-dependent self-assembly of structures that have both platinum and gold surface exposed for functionalization. Hence, the thiol-based chemistry has served as the fundamental attachment scheme for DNA and oligonucleotides for the self-assembly of artificial nano-structures.

15.4.2 Protein Complexes as Biolinkers

15.4.2.1 Protein Fundamentals

Proteins are made of amino acids. Amino acids are small chemical compounds containing an amino (NH_2) group and a carboxylic group (COOH) connected to a central carbon atom. There are different

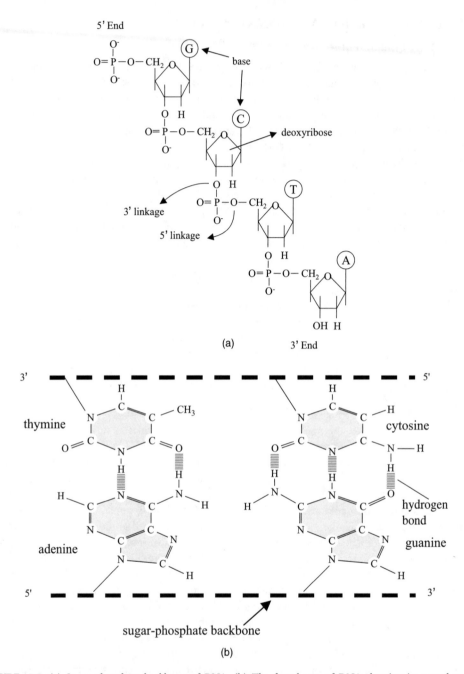

FIGURE 15.5 (a) Sugar-phosphate backbone of DNA. (b) The four bases of DNA showing its complementary binding properties. (Redrawn from Alberts, B., Bray, D., Lewis, J., Raff, M., Roberts, K., and Watson, J.D., *Molecular Biology of the Cell*, 3rd ed., Garland Publishing, New York, 1994.)

side-chains connected to this central carbon atom (called α-carbon), resulting in different amino acids, as shown in Figures 15.8a and 15.8.b. There are many amino acids, but 20 of them are most commonly used in the formation of proteins. Proteins are formed when chains of amino acids are connected by peptide bonds between the amine group of one amino acid and the carboxylic group of another amino acid. Once these chains are placed in water, they fold upon themselves into complex three-dimensional globular structures through covalent and noncovalent interactions (described above in Section 15.3).

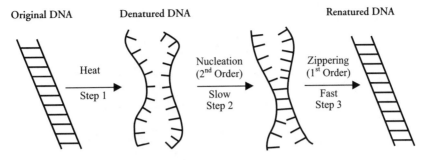

FIGURE 15.6 Schematic showing denaturing and hybridization of DNA.

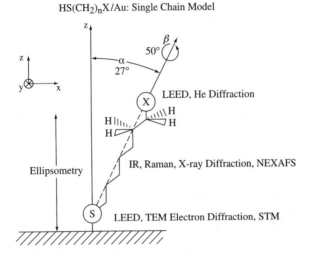

FIGURE 15.7 Schematic of a long-chain thiol molecule on a gold surface. (From Dubois, L. and Nuzzo, R.G., Synthesis, structure, and properties of model organic-surfaces, *Ann. Rev. Phys. Chem.*, 43, 437, 1992. © 1992 Annual Reviews. Used with permission.)

The hydrophobic side-chains of the amino acids tend to cluster in the interior. The overall structure of the proteins is stabilized by noncovalent interactions between the various parts of the amino-acid chain. Antibodies, ligands, etc., are all proteins that have regions on these macromolecules that exhibit specificity to other proteins, receptors, or molecules. The specificity comes about due to a combination of the noncovalent interaction forces described above and can be used for self-assembly. Ligand and receptors can also be used for self-assembly, and the best-known and most widely used ligand/receptor system is that of avidin and biotin. Avidin (a receptor with four binding sites to biotin) is a protein (molecular mass = 68 kDa) found in egg white. Biotin (a vitamin with a molecular mass of 244 D) is the ligand that binds to Avidin with a very high affinity ($K_a = 10^{15}$ M^{-1}).

15.4.2.2   Protein Attachment to Surfaces

The attachment of proteins on micro-fabricated surfaces will be vital to the success of self-assembly that is protein mediated. These attachment strategies are also useful for other applications such as protein chips and attachment and capture of cells on micro-fabricated surfaces. Certainly a lot needs to be learned from the prior work of adsorption of proteins at metal surfaces and electrodes.[20,21] The attachment of proteins on surfaces is a lot more complex when compared with attachment of DNA to surfaces. The proteins have to be attached in such a way that their structure and functionality should be retained. The attachment of antibodies and proteins has been demonstrated on micro-fabricated surfaces using functional groups such as silane,[22,23] amine,[24] carboxyl,[25] and thiols.[26] Attachment of avidin

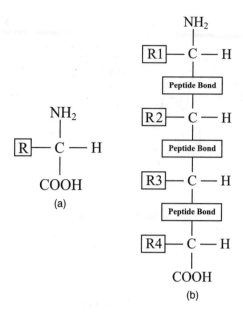

FIGURE 15.8   (a) An amino acid with a side chain R. (b) A chain of amino acids linked by peptide bonds to form a protein.

on micro-fabricated surfaces using bovine serum albumin (BSA) layers has also been demonstrated in such a way that the avidin retains its binding ability to biotin and hence any biotinylated protein.[27] Patterning of ligands has also been demonstrated using alkenethiolate SAMs, which were produced on Au layers; and ligands such as biotin were printed on the SAMs using micro-contact printing.[28] As shown in Figure 15.9, subsequent binding of streptavidin, biotinylated protein G, and fluorescently labeled goat antirabbit IgG protein within 5 μm squares demonstrated the patterning of these proteins. Protein microarrays have been demonstrated where proteins were immobilized by covalently attaching them on glass surfaces that were treated with aldehyde-containing silane reagents.[29] These aldehydes react with the primary amines on the proteins such that the proteins still stay active and interact with other proteins and small molecules. Sixteen hundred spots were produced on a square cm using robotic nanoliter dispensing where each site was about 150–200 μm in diameter. All of the above approaches

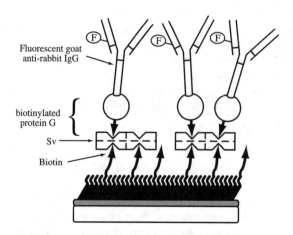

FIGURE 15.9   Schematic of antibody attachment using microcontact printing of a SAM. (From Lahiri, J., Ostuni, E., and Whitesides, G.M., Patterning ligands on reactive SAMs by microcontact printing, *Langmuir*, 15, 2055, 1999. © 1999 American Chemical Society. Used with permission.)

take a protein and devise a technique to attach it to a micro-fabricated surface by functionalizing one end of the protein with chemical groups that have affinity to that particular surface. One of the most exciting techniques, truly bio-inspired, has been demonstrated by Belcher and co-workers, who used natural selection and evolutionary principles with combinatorial phage display libraries to evolve peptides that bind to semiconductor surfaces with high specificity.[30] They screened millions of peptides with unknown binding properties and "evolved" these peptides by re-reacting them with semiconductor surfaces under increasingly stringent conditions. Twelve-mer peptides were developed that showed selective binding affinity to GaAs but not to $Al_{0.98}Ga_{0.02}As$ and Si, and to 100 GaAs but not to 111B GaAs. These proteins can possibly be designed with multiple recognition sites and be used to design nanoparticle heterostructures in two and three dimensions.

15.4.3   Strategies for Assembly with Biolinkers

Different strategies can be employed when using DNA or protein complexes as the biolinker for the self-assembly. These strategies need techniques to attach the molecules to surfaces, as summarized above. Figure 15.10 shows four possible cases of assembling structures using DNA or protein complexes (more strategies are, of course, possible). The first case consists of thiolated DNA, with complementary strands attached to the device and the substrate. Direct hybridization can be used to place and bind the device to the substrate. The second case is indirect hybridization, where a third linking strand, each half being complementary to the two strands, is used to bind the device to the substrate. The third case shows a single-stranded DNA with a thiol at one end, providing the attachment to a gold layer, and a biotin at the other end. The devices or objects to be assembled are coated with avidin and can be captured by the biotin on the DNA strand. The single strand is expected to not be rigid and lie on the Au layer in such a way that the biotin is not physically accessible to avidin for binding. The fourth case shows a single strand attached to the gold layer, which is then hybridized with a complementary strand functionalized with a biotin molecule. The device, which is coated with avidin, can now be captured by the biotin, and the more rigid double strand provides better binding capability between the avidin–biotin complex.

For the first two cases, the DNA attached to the devices can also provide charges (due to the phosphate groups in the backbone) so that the devices can also be brought closer to the binding site using electrostatics. Additional scenarios can be envisioned that utilize long-range electrostatic, magnetic, or capillary forces to bring the structures/devices close to the binding sites and then use the short-range chemical and biological forces to result in intimate binding at the desired site. The use of DNA provides the attractive possibility of making selective addressable sites where different types of devices can be assembled at different sites simultaneously, bringing nanostructure and microstructure "fabrication in a beaker" closer to reality.

15.5   State of the Art in Bio-Inspired Self-Assembly

The work in bio-inspired self-assembly can be grouped into four categories: (1) biological entity mediated, (2) chemically mediated, (3) electrically mediated, and (4) fluidics mediated. Below we will provide a summary of these works reported in literature.

15.5.1   Biological Entity Mediated Self-Assembly

There has been a tremendous interest in recent years to develop concepts and approaches for self-assembled systems for electronic and optical applications. Material self-assembly has been demonstrated in a variety of semiconductor materials (GaAs, InSb, SiGe, etc.) using Stranksi–Krastanov strain-dependent growth of lattice mismatch epitaxial films.[31–34] Periodic structures with useful optical properties have been demonstrated, but no reports of actual electronic functions can be found in literature using this lattice strain-dependent growth and assembly. While significant work continues along that direction, it

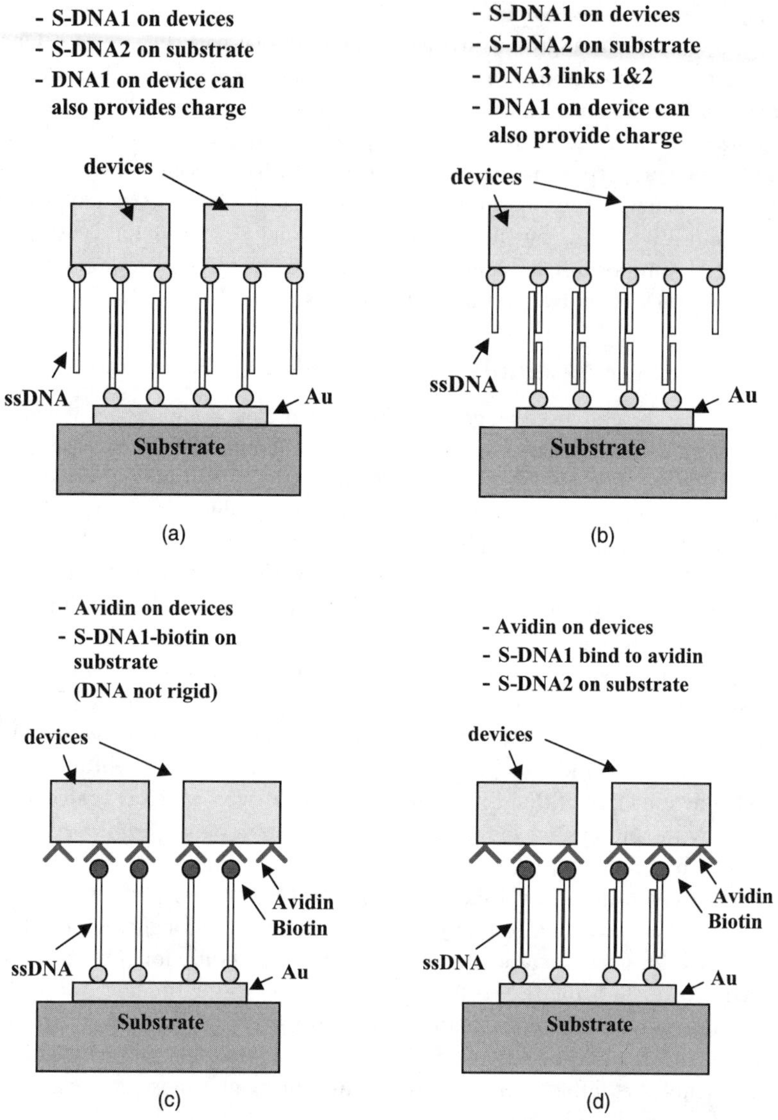
- S-DNA1 on devices
- S-DNA2 on substrate
- DNA1 on device can also provides charge
devices
ssDNA
Au
Substrate
(a)
- S-DNA1 on devices
- S-DNA2 on substrate
- DNA3 links 1&2
- DNA1 on device can also provide charge
devices
ssDNA
Au
Substrate
(b)
- Avidin on devices
- S-DNA1-biotin on substrate
(DNA not rigid)
devices
Avidin
Biotin
ssDNA
Au
Substrate
(c)
- Avidin on devices
- S-DNA1 bind to avidin
- S-DNA2 on substrate
devices
Avidin
Biotin
ssDNA
Au
Substrate
(d)

FIGURE 15.10 Four possible strategies of self-assembly using DNA and ligand/receptor complexes.

has also been recognized by engineers, chemists, and life scientists that the exquisite molecular recognition of various natural biological materials can also be used to form complex networks of potentially useful particles for a variety of optical, electronic, and sensing applications. This approach can be considered a bottom-up approach rather than the top-down approach of conventional scaling, and much work has been reported toward this front.

15.5.1.1 Nanostructures by DNA

Pioneering research extending over 15 years by Seeman has laid a foundation for the construction of structures using DNA as scaffolds, which may ultimately serve as frameworks for the construction of nanoelectronic devices.[35–39] In that work, branched DNA was used to form stick figures by properly choosing the sequence of the complementary strands. Macrocycles, DNA quadrilateral, DNA knots, Holliday junctions, and other structures were designed. Figure 15.11(a) shows a stable branched DNA junction made by DNA molecules. The hydrogen bonding is indicated by dots between the nucleotides. It is also possible to take this structure

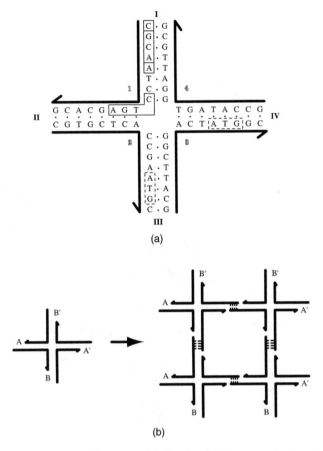

FIGURE 15.11 (a) A four-armed stable branched junction made from DNA molecules. (b) Use of the branched junction to form periodic crystals. (From Seeman, N.C., The use of branched DNA for nanoscale fabrication, *Nanotechnology*, 149, 1991. © 1991 IOP Publishing, Ltd. Used with permission.)

and devise a two-dimensional lattice as shown in Figure 15.11(b) if hybridization regions (*sticky ends*) are provided in region B. It was also pointed out that it was easier to synthesize these structures but more difficult to validate the synthesis. The same group also reported on the design and observation, via AFM, of two-dimensional crystalline forms of DNA double crossover molecules that are programmed to self-assemble by the complementary binding of the sticky ends of the DNA molecules.[39] Single-domain crystal sizes, which were as large as 2 μm $\times$ 8 μm, were shown by AFM images. These lattices can also serve as scaffolding material for other biological materials. It should be noted that in this work, the 2 nm wide stiff DNA molecules themselves are used to form the two- and three-dimensional structures.

15.5.1.2 DNA Mediated Assembly of Nanostructures

Among roles envisioned for nucleic acids in nanoelectronic devices, the self-assembly of DNA conjugated nanoparticles has received the most attention in recent literature. Mirkin et al.[40] and Alivisatos et al.[41] were the first to describe self-assembly of gold nanoclusters into periodic structures using DNA. A method of assembling colloidal gold nanoparticles into macroscopic aggregates using DNA as linking elements has been described.[40] The method involved attaching noncomplementary DNA oligonucleotides to the surfaces of two batches of 13 nm gold particles capped with thiol groups, which bind to gold. When another oligonucleotide duplex — with ends which are complementary to the grafted sequence — is introduced, the nanoparticles self-assemble into aggregates. The process flow is shown in Figure 15.12, and this process could also be reversed when the temperature is increased due to the

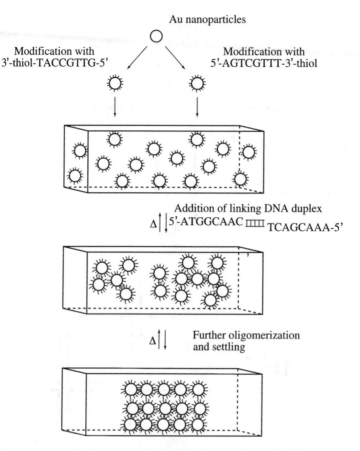

FIGURE 15.12 Fabrication process for the aggregated assembly of DNA conjugated gold nanoparticles. (From Mirkin, C.A., Letsinger, R.L., Mucic, R.C., and Storhoff, J.J., A DNA-based method for rationally assembling nanoparticles into macroscopic materials, *Nature*, 382, 607, 1996. © 1996 Macmillan Magazines, Ltd. Used with permission.)

denaturation of the DNA oligonucleotides. Close-packed assemblies of aggregates with uniform particle separations of about 60Å were demonstrated in this study as shown in Figure 15.13(a).

In the same journal issue, techniques were also reported where discrete numbers of gold nanocrystals are organized into spatially defined structures based on DNA base pair matching.[41] Gold particles, 1.4 nm in size, were attached to either the 3' or 5' of 19 nucleotide-long single-stranded DNA 'codon' molecules through the well-known thiol attachment scheme. Then 37 nucleotide-long, single-stranded DNA 'template' molecules were added to the solution containing the gold nanoparticles functionalized with ss DNA. The authors showed that the nanocrystals could be assembled into dimers (parallel and antiparallel) and trimers upon hybridization of the codon molecules with that of the template molecule. Due to the ability to choose the number of nucleotides, the gold particles can be placed at defined positions from each other as schematically shown in Figure 15.14. TEM results showed that the distance between the parallel and antiparallel dimers were 2.9 nm to 10 nm and 2.0 nm to 6.3 nm, respectively. These structures could potentially be used for applications such as chemical sensors, spectroscopic enhancers, and nanostructure fabrication. These techniques have been used to devise sensitive colorimetric schemes for the detection of polynucleotides based on distance-dependent optical properties of aggregated gold particles in solutions.[42]

Mucic et al. have also described the construction of binary nanoparticle networks composed of 9 nm particles and 31 nm particles, both composed of citrate-stabilized colloidal gold.[43] These 9 (±1) and 31(±3) nm particles were coated with different 12-mer oligonucleotides via a thiol bond. When a third DNA sequence (24-mer) — which was complementary to the oligonucleotides on both particles — was

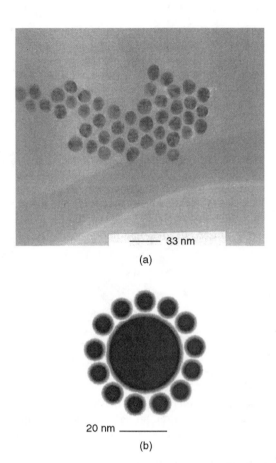

FIGURE 15.13   (a) TEM image of the aggregated DNA/Au colloidal particles. (b) TEM image of a nanoparticle satellite constructed via DNA mediated docking of 9 nm gold particles onto a 31 nm gold particle. (Redrawn from Mucic, R.C., Storhoff, J.J., Mirkin, C.A. and Letsinger, R.L., DNA-directed synthesis of binary nanoparticle network materials, *J. Am. Chem. Soc.* 120, 12674, 1998. © 1998 American Chemical Society. Used with permission.)

added, hybridization led to the association of particles. When the ratio of 9 nm to 31 nm particles was large, the assembly illustrated in Figure 15.13(b) was formed. Loweth et al. have presented further details of the formation of the hetero-dimeric and hetero-trimeric nonperiodic nanocluster molecules based upon earlier work of Alivisatos.[44] The authors showed exquisite control of the placement of 5 nm and 10 nm gold nanoclusters which were derivatized with ss DNA. Various schemes of hetero-dimers and hetero-trimers were designed and demonstrated using TEM images. This nanoparticle DNA mediated hybridization also forms the basis of genomic detection using colorometric analysis.[45] Hybridization of the target with the probes results in the formation of a nanoparticle/DNA aggregate, which causes a red to purple color change in solution due to the red shift in the surface plasmon resonance of the Au nanoparticles. The networks show a very sharp melting transition curve, which allows for single-base mismatches, deletions, or insertions to be detected. The same approach can also be taken for a surface combined with reduction of silver at the site of nanoparticle capture, which allows the use of a conventional flatbed scanner as a reader. Sensitivities that are 100× greater than those of conventional fluorescence-based assays have been described.[46] Reviews on these topics have also been published and can provide more information than what is presented here.[47–49]

Csaki showed the use of gold nanoparticles (with mean diameter 15, 30, and 60 nm) as a means for labeling DNA and characterizing DNA hybridization on a surface.[50,51] Single strands of DNA were attached to unpatterned gold substrates. Colloidal gold particles were labeled with thiolated complementary DNA strands, which were then captured by the strands on the surface. Niemeyer et al.[52] have also shown site-

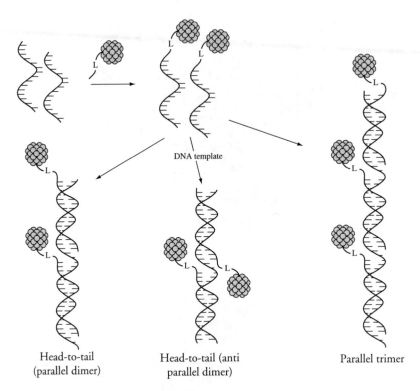

Head-to-tail   Head-to-tail (anti   Parallel trimer
(parallel dimer)   parallel dimer)

FIGURE 15.14 Assembly of nanocrystals to form dimers and trimers based on DNA hybridization. (From Alivisatos, A.P., Johnsson, K.P., Peng, X., Wilson, T.E., Loweth, C.J., Bruchez, M.P. and Schultz, P.G., Organization of "nanocrystal molecules" using DNA, *Nature*, 382, 609, 1996. © 1996 Macmillan Magazines, Ltd. Used with permission.)

specific immobilization of 40 nm gold nanoparticles which were citrate-passivated and then modified with 5'thiol derivatized 24-mer DNA oligomers. The capture DNA was placed at specific sites using nanoliter dispensing of the solution.

Mirkin and co-workers have also demonstrated the formation of supramolecular nanoparticle structures where up to 4 layers of gold nanoparticles have been produced. The scheme is shown in Figure 15.15, where a linking strand can bind the DNA strand, which is immobilized on the surface, and a DNA-derivatized nanoparticle.[53] The linking strand can then be used to bind another layer of nanoparticle, and multilayered network structures can be produced.

15.5.1.3 DNA-Directed Microwires and Nanowires

The concept of DNA mediated self-assembly of gold nanostructures has recently been extended to metallic nanowires/rods.[54–56] The concept, though feasible, has not yet been completely demonstrated. The basic idea behind this work is to fabricate gold and/or platinum metal wires, functionalize these wires with ss DNA, and assemble them on substrates, which have the complementary ss DNA molecules attached at specific sites. Thus, self-assembly of interconnects and wires can be made possible. The metallic wires are formed by electroplating in porous alumina membranes with pores sized about 200 nm.[55] The processes for the formation of alumina films with nanohole arrays have been developed and demonstrated by many authors.[57–58] Metallic rods ranging from 1 to 6 μm in length were produced, depending on the electroplating conditions. The goal of the work would be to form Pt rods with Au at the ends or vice versa. More recently, the same authors showed attachment and quantification of ss DNA on the Au ends of the Au/Pt/Au rods.[59] DNA strands were also attached to gold substrates, and it was shown that the rods attached to the substrates only when the DNA strands were indeed complementary. It should be noted that the attachment of the DNA strands was not patterned, and the complementary binding of the rods was not shown to be site-specific; however, that is the next logical step for this work.

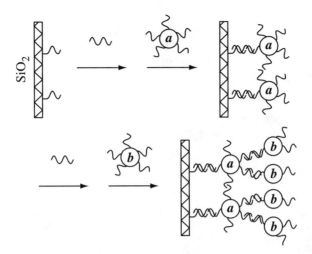

FIGURE 15.15 Synthesis of three-dimensional assembly using DNA and nanoparticles. (From Taton, T.A., Mucic, R.C., Mirkin, C.A., and Letsinger, R.L., The DNA mediated formation of supra-molecular mono- and multilayered nanoparticle structures, *J. Am. Chem. Soc.*, 122, 6305, 2000. © 2000 American Chemical Society. Used with permission.)

The use of DNA as a template for the fabrication of nanowires has been demonstrated through a process described by Braun et al.[60–62] The authors formed a DNA bridge between two gold electrodes, again using a thiol attachment. Once a DNA bridge is formed between the 12–16 μm spacing of the electrodes, a chemical deposition process is used to vectorially deposit silver ions along the DNA through Ag+/Na+ ion exchange and formation of complexes between the gold and the DNA bases (see Figure 15.16). The result is a silver nanowire that is formed using the DNA as a template or skeleton. Current–voltage characteristics were measured to demonstrate the possible use of these nanowires. The authors also reported the formation of luminescent self-assembled poly(p-phenylene vinylene) wires for possible optical applications.[61] The work has a lot of potential, and much room for further research exists to control the wire width, the contact resistances between the gold electrode and the silver wires, and use of other metals and materials. DNA-inspired self-assembly of active devices (complementary strands of DNA on a device and a substrate) has been proposed by Heller and co-workers[63] for assembling optical and optoelectronic components on a host substrate, but the basic concept has not yet been demonstrated.

15.5.1.4 Protein Complex Mediated Assembly of Nanostructures and Microstructures

Semiconductor nanoscale quantum dots have gained a lot of attention in recent years due to the improvement in the synthesis techniques; for example, CdSe or ZnSe quantum dots of predetermined sizes can be synthesized. The optical properties of these structures have been studied in great detail.[64,65] These quantum dots have many advantages over conventional fluorophores such as a narrow, tunable emission spectrum and photochemical stability. The programmed assembly of such quantum dots using DNA has also been reported.[66] Typically these quantum dots are soluble only in nonpolar solvents, which makes it difficult to functionalize them with DNA by a direct reaction. The authors successfully demonstrated the use of 3-mercaptopropionic acid to initially passivate the QD surface and act as a pH trigger for controlling water solubility. The use of DNA-functionalized quantum dots thus allows the synthesis of hybrid assemblies with different types of optical nanoscale building blocks.

The biological tagging of the quantum dots briefly described above is an interesting example of bio-inspired self-assembly. If these optical emitters can be tagged with specific proteins and antibodies, then site-specific markers can be developed. Examples of such applications have been demonstrated in literature where CdSe/CdS core/shell quantum dots with a silica shell were used for biological staining of mouse fibroblast cells.[67] The surfaces of green-colored nanocrystals were modified with trimethoxysilyl-propyl urea and acetate groups so that they were found to selectively go inside the cell nucleus. The authors also incubated fibroblasts with phalloidin–biotin and streptavidin and labeled the actin filaments

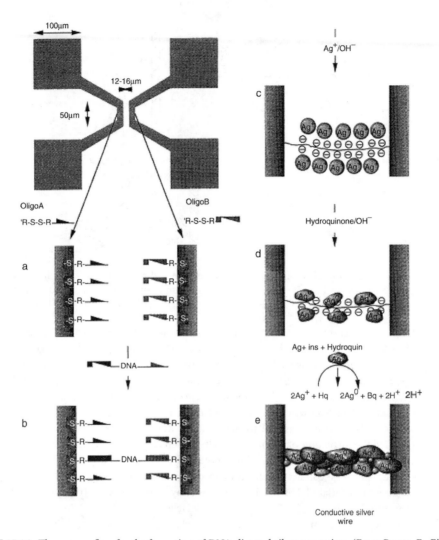

FIGURE 15.16 The process flow for the formation of DNA-directed silver nanowires. (From Braun, E., Eichen, Y., Sivan, U., and Yoseph, G.B., DNA-templated assembly and electrode attachment of a conducting silver wire, *Nature*, 391(19), 775, 1998. ©1998 Macmillan Magazines Ltd. Used with permission.)

of these cells with biotinylated red nanocrystals. Hence, penetration of the green nanoprobes inside the nucleus and the red staining of the actin fibers were simultaneously demonstrated. Covalently coupled protein-labeled nanocrystals for use in biological detection have also been demonstrated.[68] In this case, the ZnS-capped CdSe quantum dot was covalently coupled to a protein by mercaptoacetic acid. The mercapto group binds to the Zn atom. The carboxyl group can be used for covalent coupling of various biological molecules through amine group cross-linking as shown in Figure 15.17. It was postulated that, for steric reasons, perhaps 2–5 protein molecules (100 kD) can be attached to a 5 nm quantum dot. The authors demonstrated receptor-mediated endocytosis and that the quantum dots were transferred inside the cells. These quantum dots can also be made as nanoshells consisting of a dielectric core with a metallic shell of nanometer thickness.[69] By varying the relative dimensions of the core and shell, the optical resonance of these nanoparticles can be varied over hundreds of nanometers in wavelength, across the visible and into the infrared region of the spectrum. When proteins and antibodies are attached to these nanoshells, these can be used as optical markers for biological diagnostic applications.

The assembly of microbeads using DNA and the avidin–biotin complex has also been demonstrated. This scheme is the same as that shown in Figure 15.10(d), where a single strand of capture DNA is first

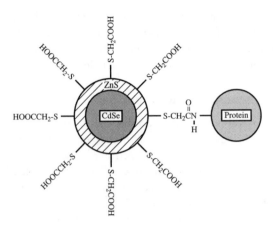

FIGURE 15.17 A CdSe/ZnS core/shell quantum dot with protein attachment scheme. (From Chan, W.C.W. and Nie, S., Quantum dot bioconjugates for ultrasensitive nonisotopic detection, *Science*, 281, 2016, 1998. © 1998 AAAS. Used with permission.)

attached to a substrate. Then a second biotinylated target strand is brought in and, if complementary, will hybridize to the first strand exposing the biotin. Next, beads coated with avidin (or related proteins) are exposed to the surface and, if the target strand has hybridized, then the beads will be captured due to the avidin–biotin interaction. Thus, the presence of beads signals the presence of the complementary strands. Figure 15.18 shows an optical picture of a chip surface where avidin-coated polystyrene beads were captured when the DNA strands were indeed complementary.[70] The same scheme has been used to

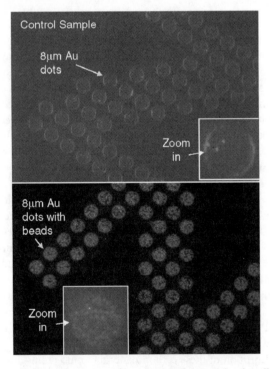

FIGURE 15.18 Optical micrographs of 0.8 μm avidin-coated polystyrene beads collected on biotinylated DNA on a surface. (a) A control sample with no DNA attached, hence beads are not collected. (b) Sample with complementary strands of DNA, hence beads are collected on the pads. (From McNally, H., Pingle, M., Lee, S.W., Guo, D., Bergstrom, D., and Bashir, R., Self-assembly of micro and nano-scale particles using bio-inspired events, submitted to *Applied Surface Science*.)

develop detectors for biological warfare agents by attaching the DNA on thin films exhibiting GMR (giant magneto resistive) effect. Capture of target DNA and the presence of 1 μm sized magnetic beads can be electrically detected using the GMR sensors.[71] The intensity and location of the signal can indicate the concentration and identity of pathogens present in the sample.

15.5.2 Chemically Treated Surfaces and Self-Assembly

The assembly of structures can also be performed using specific surface treatment with chemicals that make two surfaces bind to each other through small molecules. As an example, the assembly of arrays of nanoscale gold clusters and the measurement of electronic properties through the Au/molecule wires have been demonstrated by Andres, Datta, and co-workers.[47,72] They produced 2–5 nm Au clusters, which were initially encapsulated with dodecanethiol. The clusters can be precipitated and cross-linked into an ordered linked cluster network using molecules with a thiol at both ends, which links the adjacent clusters. The molecules used were from the aryl dithiol (ADT) family of molecules. This work demonstrated guided self-assembly, high quality ordering at the nanoscale, and low resistance coupling to a semiconductor surface through the docking molecule.[73]

A powerful approach to chemical self-assembly is the formation of self-assembled monolayers (SAMs) to make surfaces that are hydrophobic and hydrophilic such that these surface forces can be utilized to place structures at specific sites,[74] as shown in Figure 15.19. The authors used microcontact printing to pattern a gold surface into a grid of hydrophobic and hydrophilic SAMs of alkanethiolates. SAMs that were CH_3-terminated were hydrophobic, while those with COOH terminated with hydrophilic. An aqueous solution with $CuSO_4$ or KNO_3 crystals would only attached to the hydrophilic areas, and evaporation of the fluid medium would leave the crystals within those hydrophilic regions. Assembly of crystals and particles with sizes down to 150 nm on a size within a 2×2 μm^2 hydrophilic region was demonstrated.

Whitesides and colleagues have also demonstrated procedures that use capillary forces to form millimeter-scale plastic objects into aggregates and three-dimensional objects.[75–78] Capillary forces scale with the length of a solid–liquid interface, whereas pressure and body forces (gravity) scale with area and volume. Hence, capillary forces become dominant when compared with these other forces. The reported techniques rely on coating selective surfaces of an object with lubricants such as alkanes and photocurable methacrylates to create surfaces of high interfacial energies. These surfaces are then agitated in a second liquid medium in which the lubricant is not soluble. Because the interfacial energy of the liquid medium and the lubricant is high, it is energetically favorable for the surfaces to combine and align to minimize the exposed surface area. Micrometer- and millimeter-scale objects were shown to be assembled using these hydrophobic effects and shape-mediated complementarity. These techniques have also been applied to assemble hydrophobic silicon microstructures on hydrophobic substrates with the use of capillary forces.[79] The patterned substrate was passed through a film of hydrophobic adhesive so that the adhesive coats the hydrophobic binding sites. When the hydrophobic pattern on the micromachined silicon parts ($150 \times 150 \times 15$ μm^3 to $400 \times 400 \times 50$ μm^3) came in contact with the adhesive-coated substrate binding sites, shape-matching self-assembly occurred due to minimization of surface energies. Optical micromirrors for MEMS applications were assembled on silicon and glass substrates, and alignment precisions of less than 0.2 μm and rotational misalignment of 0.3° were demonstrated.

One of the most exciting developments in the field of nanotechnology has been the discovery and development of carbon nanotubes. These tubes are very attractive and promising due to their unique electronic and mechanical properties, and they can possibly be used to make nanoscale transistors and switches.[80] Some work toward site-specific assembly of individual nanotubes has been demonstrated.[81] The authors report the adsorption of these carbon nanotubes onto amino-functionalized surfaces. Trimethylsilyl SAM layers formed on an oxide substrate were patterned using AFM or e-beam lithography, and a second SAM was formed with –NH2 functionality (Figure 15.20). Individual nanotubes were shown to be selectively placed between electrodes on these amino sites. More work needs to be done to control the position and placement of the tubes.

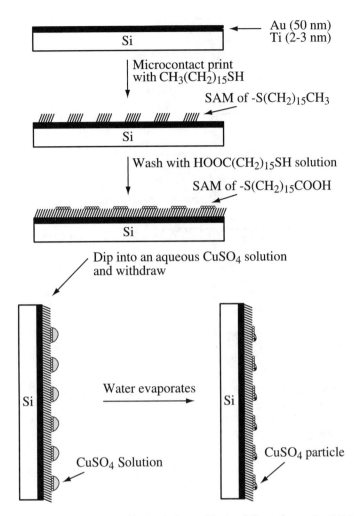

FIGURE 15.19 Process flow for formation of hydrophobic and hydrophilic surfaces using SAMs and microcontact printing. (From Qin, D., Xia, Y., Xu, B., Yang, H., Zhu, C., and Whitesides, G.M., Fabrication of ordered two-dimensional arrays of micro- and nanoparticles using patterned self-assembled monolayers as templates, *Adv. Mater.*, 11(17), 1433, 1999. © 1999 Wiley Interscience. Used with permission.)

15.5.3 Electrically Mediated

A variety of electrically mediated assembly techniques can be found in literature. Basically, these can be divided into assembly due to charged molecules only and assembly due to DC field electrokinetic effects (e.g., electrophoresis and electroosmotic).

Electrostatic forces due to charged molecules can be used to provide site-specific assembly on microfabricated surfaces. The basic idea is simple: the object to be assembled is charged and the site where the device is to be assembled provides the opposite charge. Electrostatic forces, along with external forces such as simple agitation, will then bring the device to the assembly sites. The force acting between the charges is simply given as $F = qE$, where q is the charge and E is the resulting electric field due to the presence of the charges. Figure 15.21 shows the basic concept of the electric field mediated assembly. Tien et al.[82] demonstrated the use of electrostatic interactions to direct the placement and patterning of 10 μm diameter gold disks on substrates functionalized with charged molecules. The disks were produced by electrodeposition of gold in a photoresist pattern and then released using sonication in a charged thiol solution. Meanwhile, substrates with patterned (negative and positive) charged surfaces

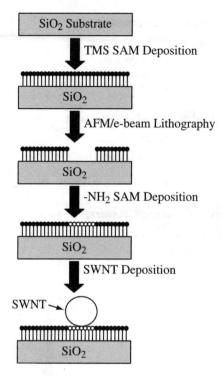

FIGURE 15.20 Chemical self-assembly of a single-walled carbon nanotube. (From Liu, J., Casavant, J., Cox, M., Walters, D.A., Boul, P., Lu, W., Rimberg, A.J., Smith, K.A., Colbert, D.T., and Smalley, R.E., Controlled deposition of individual single walled carbon nanotubes on chemically functionalized templates, *Chem. Phys. Lett.*, 303, 125, 1999. © 1999 Elsevier Science. Used with permission.)

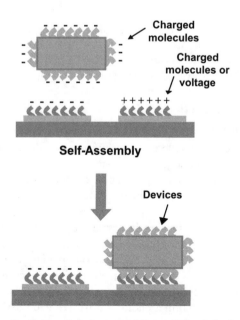

FIGURE 15.21 Concept schematic of charged devices assembling on oppositely charge regions on a substrate.

were produced by microcontact printing and attachment of thiolated molecules. Many solutions were used, such as $HS(CH_2)_{11}NH_3^+Cl^-$, $HS(CH_2)_{11}NMe_2$, $HS(CH_2)_{11}NMe_3^+Br^-$, etc., to provide a positive charge, while $HS(CH_2)_{15}COOH$, $HS(CH_2)_{11}PO_3H_2$ yielded negative charges. Patterned assemblies over 1 cm^2 areas were demonstrated. However, it was noted that in-plane ordering was still lacking, presumably because the objects could not move laterally once they were attached to a site. The required lateral mobility can come from capillary forces at the fluid–fluid interface[82,83] or by simply reducing the size of the charged pattern on the substrate and increasing the concentration of the objects in the solution above the pattern.

The charge on the devices can be provided by molecules, while the charges on the substrate can also be provided by applying a voltage potential, all within a fluid medium. This is the well-known principle of electrophoresis used to separate charged molecules and macromolecules based on the difference in their sizes. The fixed charges on the devices and objects are generally neutralized by the presence of counterions in the fluid; but upon application of a DC voltage, the molecule is polarized and the charged object moves to one electrode, whereas the counterions move to the other electrode. Perhaps the most well-known example is that of Heller and co-workers,[84–87] who have shown the electrophoretic placement of DNA capture strands at specific sites on biochips to realize DNA arrays. Subsequent hybridization of fluorescently labeled target probes at specific sites provides insight into the sequence of the target probes. These active microelectronic array devices allow electrophoretic fields to be used to carry out accelerated DNA hybridization reactions and to improve selectivity for single-nucleotide polymorphism (SNP), short tandem repeat (STR), and point mutation analysis.[88]

The ability to generate electric fields at the microscale allows charged molecules (DNA, RNA, proteins, enzymes, antibodies, nanobeads, and even micron-scale semiconductor devices) to be electrophoretically transported to or from any microscale location on the planar surface of the device.[89] Edman et al.[90] fabricated 20 μm diameter InGaAs LEDs and demonstrated a process for releasing them from the host substrate and an electrophoretic process for assembling them on silicon substrates. No details were given on the charging of devices or the mechanisms behind the device transport. Most likely the mechanism consisted of some variant of electrokinetic transport, where movement of ions in the fluid medium would force the movement of objects along the flow contours. The process will require a higher current density as compared with a purely electrophoretic transport. The devices can generally develop a charge on their surfaces when immersed in a liquid; for example, a clean oxide surface with a native oxide can have Si–OH groups at the surface and develop a negative charge at pH 7 in water.[82] Recently, the use of silicon-on-insulator (SOI) wafers to fabricate trapezoidal-shaped silicon islands — which were 4 μm × 4 μm at the top and about 8 μm × 8 μm at the base and had a thin gold layer on one side — has been demonstrated.[91,92] The Au surface was functionalized with 4-mer DNA or a charged molecule (2-mercaptoethane sulfonic acid sodium salt) to provide negative charges on the islands.[93] The islands were released from the substrate into a fluid medium over an electrode array and then manipulated at the microscale using voltages applied at the electrodes, as schematically shown in Figure 15.22. The molecules provided negative charges on the devices, thus allowing the electrophoretic transport of these devices under electric fields to specific sites.

A third type of effect named *dielectrophoresis* can also be used to manipulate particles at the microscale and nanoscale. Dielectrophoretic forces are developed on a particle of a dielectric constant different from the medium that it is in while it is exposed to a nonuniform AC electric field. Both positive and negative dielectrophoretic forces can be produced and, because the dielectric constant can be a function of the applied signal frequency, the direction of the force on an object can be changed as a function of frequency.[94–96] This effect has been used to assemble and align metallic nanowires within patterns of an interdigitated electrode array, in an effort to develop a molecular-scale interconnect technology.[97]

15.5.4 Fluidics Mediated

Fluidic self-assembly (FSA) is a process for assembling devices and objects using fluidic shear forces along with other forces such as gravity[98–102] or chemical modification.[103,104] Shape-mediated FSA, which uses

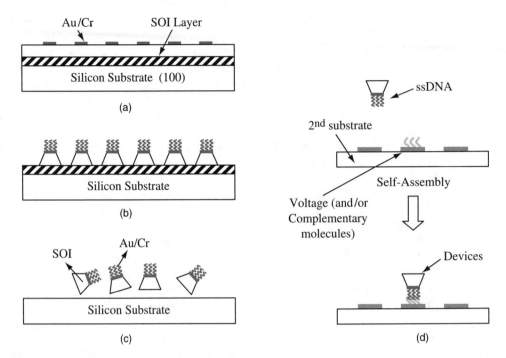

FIGURE 15.22 Process flow for active device fabrication, release. and self-assembly. (a) Define patterns of Au/Cr contacts on an SOI wafer. (b) Etch the silicon down to the buried oxide and attach molecules (DNA, etc.) to the top layer. (c) Release the islands (devices) from the substrate and collect/concentrate. (d) Assembly of the devices on another surface with voltage (or the complementary ss DNA). (Adapted from Bashir, R., DNA mediated artificial nanobiostructures, *Superlattice Microstruct.*, 29, 1, 2001.)

gravity along with fluidics, starts with the fabrication of the devices to be self-assembled and target sites on substrates, which have matching shapes. The technique can also be termed *shape-mediated* assembly. The objects are transported in a fluid and distributed over the target surface, then deposit in the holes in the substrate. Because Brownian motion and diffusion are negligible for large devices (larger than 10 μm or so), fluid motion is used to bring the device close to the assembly site. Gravity and van der Walls forces then hold the device in place. Figure 15.23 shows the basic concept of FSA, which can be applicable to different device types and can also be used for heterogeneous integration of materials.

Smith and co-workers pioneered the FSA method by demonstrating a quasi-monolithic integration of GaAs LEDs on silicon substrates using this fluidic transport and shape mediation for proper orientation and placement.[98–101] The LEDs were designed to be 10 μm × 10 μm at the base and 18 μm × 18 μm at the top and were assembled in trapezoidal wells etched in silicon by anisotropic etching. The authors also extended their work to demonstrate silicon device regions (30 μm to 1000 μm on a side) that were from silicon wafers. These silicon blocks, with prebuilt and tested transistors, were then assembled into wells etched in plastic substrates. The resulting planer surfaces were then metalized to form interconnects at the top side. This technology is now being commercialized for producing flexible displays[102] where the silicon devices are being assembled on plastics substrates within ±1 μm precision to form one side of a liquid crystal display. *Flexible* and *rollable* electronics are also possible using these approaches.

Lieber and co-workers have demonstrated the assembly of one-dimensional nanowires and nano-structures using a combination of fluidic alignment and surface patterning techniques.[103] They fabricated the nanowires of silicon, gallium phosphide, or indium phosphide using laser-assisted catalytic growth. The released nanowires were aligned along the direction of fluid flow as shown in Figure 15.24. It was also shown that the nanowires exhibited more selectivity to the NH_2-terminated monolayers, which have a partial positive charge. The authors went on to show functional electronic elements such as transistors and inverter circuits using these silicon quantum wires.[104] These results are very exciting

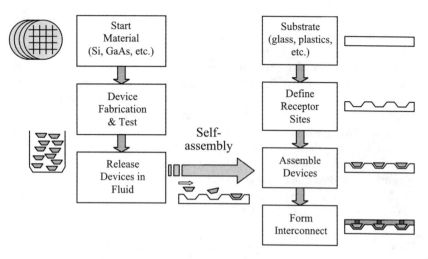

FIGURE 15.23 The process of fluidic self-assembly of silicon regions in plastic substrates. (Adapted from www.alientechnology.com.)

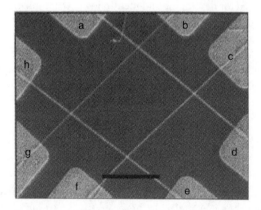

FIGURE 15.24 Assembly of silicon nanowires using fluidics and a surface treatment. (From Huang, Y., Duan, X., Wei, Q., and Lieber, C.M., Directed assembly of one-dimensional nanostructures into functional networks, *Science*, 291, 630, 2001. © 2001 AAAS. Used with permission.)

as they demonstrated the fabrication, assembly, and synthesis of functional circuit blocks using these one-dimensional semiconductor wires.

15.6 Future Directions

15.6.1 Bio-Inspired Active Device Assembly

A lot has been accomplished in the last 10 years toward the biologically mediated assembly of artificial nanostructures and microstructures. Nevertheless, a great deal remains to be done within the next 10 years to bridge the gap between the electronic devices that will be constructed on a 20–50 nm scale and molecules of a few nm or less in size. Much work has been done on assembly of passive electronic components and devices (gold clusters, metal rods, etc.). Work has been done on optical devices (quantum dots, etc.), and recently there are more reports of assembly of carbon nanotubes and quantum wires, which can be used as active devices. The processes for self-assembly of active devices such as semiconductor transistors,[2,3] carbon nanotubes,[80,81] and silicon quantum wires[103,104] need to be pursued more actively. As an example, the entire footprint of current silicon CMOS transistors in production, with W/L = 0.8 μm/0.15 μm, is less than 1.1 μm on a side. Can such active devices of such sizes be assembled in two and three dimensions using bio-inspired assembly with DNA or proteins? Doing so can potentially increase the functionality and three-dimensional integration of microelectronic systems and circuits. The active devices should include scaled-down silicon transistors, silicon nanowires and related materials, and carbon nanotubes, due to their potential promise. Can silicon nanowires or carbon nanotubes be assembled into regular arrays for memory and logic applications? Currently, these one-dimensional devices are either placed one at a time at desired sites by manipulation using AFM tips or grown at specific sites. If millions of these devices are to be used in a circuit, self-assembly using linker molecules could be an attractive way to assemble these devices. It is also important to note that, for use in future scaled nanoelectronic systems, we will need devices as well as interconnects. This direction of research will also result in heterogeneous integration of material at the microscale and nanoscale. For example, can GaAs/InGaAs HBTs be fabricated for RF applications and then assembled on silicon CMOS chips?

15.6.2 New Biolinkers

Two types of biolinkers discussed in this chapter are DNA and protein complexes. There is a need for the development of additional binding strategies for attaching linkers to specific metals (more than just Au and Pt). There is also a need for developing and identifying new complementary biolinkers (molecular glues) that have specific recognition and binding properties that are also electrically conductive and able to carry useful amounts of current densities. Good ohmic contacts to the devices through these linkers are also needed. This way, additional contacts will not be needed, and the linkers themselves could possibly provide good contacts to other devices after self-assembly. It would also be quite interesting and useful if such linkers were reversible to possibly realize *reconfigurable* circuits by causing physical movement of the devices or by reprogramming a link.

15.6.3 New Devices and Switches

In recent years, possible candidates for molecular-scale electronics have been developed such as rotaxane[105] and nitroaromatic[106] molecule-based switches and devices. As the search for molecular devices continues, DNA could also provide possible solutions toward this front. There have been no reports of active device concepts using the DNA molecule; however, the electronic properties of DNA have been under study in recent years. Even though there have been contradictory measurements in literature in regard to DNA conductivity,[107] recent reports have demonstrated that a DNA molecule measured between two electrodes behaves like a large band-gap semiconductor.[108] The resistance of the molecules in the conductive regime was reported to be on the order of $40 \times 10^9/100$Å-long molecule.

The ideas presented below assume that DNA behaves as a conductor in specific voltage–current regimes. The specificity of the Watson–Crick base pairing is a very interesting and useful property that can be exploited to propose novel DNA-based interconnect functionalities and devices. Two complementary strands of DNA can be made to denature by increasing the temperature of the molecule above a characteristic melting temperature (T_m). This concept can be used to make a DNA-based switch as shown in Figure 15.25. Two Au electrodes can be defined, and thiol-derivatized ss DNA can be attached to each electrode such that parts of these strands have a complementary sequence. Once the two molecules have hybridized, current can be passed between the two electrodes. This current will result in joule heating of the molecule, and as the temperature increases beyond T_m, the strands would denature and the current would stop flowing. The temperature increase can be directly from the current flow through the molecule or from an external heater lithographically defined in the region under the two Au electrodes (as shown in the figure). The hybridization can be direct, as shown in Figure 15.25, or indirect where a third DNA strand is used to connect the first two strands. The denaturing phenomenon, which is illustrated in Figure 15.25, also provides the basis to realize DNA-based devices with characteristics that may be suitable as functional elements. If the heating were due to the current

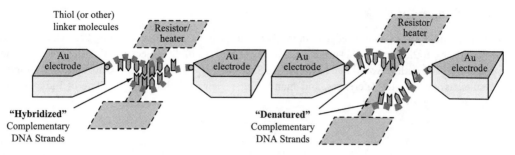

FIGURE 15.25 (a) Single-stranded DNA attached to each electrode using a linker molecule. (b) If $I > I_{critical}$, Joule heating will result in $T > T_m$(DNA), which will denature the DNA and the current will stop flowing. Alternatively, the temperature around the DNA strand can also be increased above T_m by a metal heater/resistor.

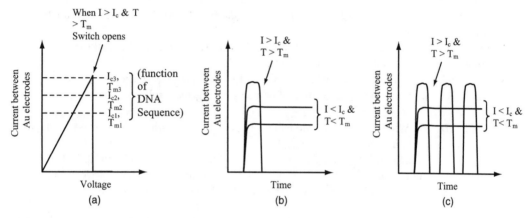

FIGURE 15.26 (a) Expected current vs. voltage. (b) Current vs. voltage. (c) Oscillatory current–time plots for the DNA-based molecular switch device shown in Figure 15.25.

flow through the DNA strands, a negative differential resistance device may be possible. As illustrated in Figure 15.26, such a device could also be used to form an oscillator. Once the two strands have denatured upon internal or external heating of the molecule, they would rejoin due to diffusion in a certain time, given some thermal energy. Thus the strands would denature due to heating, cool due to no current flow, and then hybridize again because they are in close proximity to each other. The rehybridization might be slow at first examination as it is controlled by diffusion. However, because the molecules are in such close proximity to each other, the time for rehybridization might not be as low as expected.

15.7 Conclusions

The field of nanotechnology has emerged as one of the most important areas of research for the future, and biologically and chemically mediated self-assembly has the potential to profoundly impact this field. This chapter presented the motivations and fundamentals behind these assembly concepts with a focus on biologically mediated assembly. Biolinker fundamentals and attachments to surfaces were reviewed. The state of the art in biologically mediated assembly of artificial nanostructures and microstructures was presented. Examples of chemically, electrically, and fluidics-mediated self-assembly were also presented, all with a focus on assembly of nanoscale and microscale objects at specific sites and locations. Work needs to continue in this area to eventually realize bottom-up fabrication to produce fully self-assembling, self-healing, and self-repairing systems that are capable of sensing, computing, and reacting to specific stimuli.

Acknowledgments

The author would like to acknowledge many valuable discussions with Prof. D. Janes, Prof. D. Bergstrom (Co-PI with the author on an NSF-funded active device self-assembly project), Prof. M. Lundstrom, and Prof. S. Datta. The author would like to thank Mr. S. Lee, Dr. H. McNally, Mr. D. Guo, and Dr. M. Pingle for results, assistance with references, and manuscript review, and the NSF, the State of Indiana, and Purdue University for funding.

References

1. Geppert, L., Devices and circuits technology 2000 analysis and forecast, *IEEE Spectrum*, 37(1), 63, 2000.
2. *The National Technology Roadmap for Semiconductors (NTRS)*, SIA (Semiconductor Industry Association), 1997.
3. *The International Technology Roadmap for Semiconductors (ITRS)*, SIA (Semiconductor Industry Association), 1999.
4. Alberts, B., Bray, D., Lewis, J., Raff, M., Roberts, K., and Watson, J.D., *Molecular Biology of the Cell*, 3rd ed., Garland Publishing, New York, 1994.
5. http://alphadna.com/; http://www.biosyn.com/; http://www.genemedsyn.com/
6. Nuzzo, R.G. and Allara, D.L., Adsorption of bifunctional organic disulfides on gold surfaces, *J. Am. Chem. Soc.*, 105, 4481, 1983.
7. Bain, C.D. and Whitesides, G.M., Modeling organic-surfaces with self-assembled monolayers, *Angew. Chem. Int. Ed. Engl.* 28(4), 506, 1989.
8. Bain, C.D., Troughton, E.B., Tao, Y.T., Evall, J., Whitesides, G.M., and Nuzzo, R.G., Combining spontaneous molecular assembly with microfabrication to pattern surfaces — selective binding of isonitriles to platinum microwires and characterization by electrochemistry and surface spectroscopy, *J. Am. Chem. Soc.*, 111, 321, 1989.
9. Dubois, L. and Nuzzo, R.G., Synthesis, structure, and properties of model organic-surfaces, *Ann. Rev. Phys. Chem.*, 43, 437, 1992.
10. Hickman, J.J., Laibinis, P.E., Auerbach, D.I., Zou, C., Gardner, T.J., Whitesides, G.M., and Wrighton, M.S., Toward orthogonal self-assembly of redox active molecules on Pt and Au — selective reaction of disulfide with Au and isocyanide with Pt, *Langmuir*, 8, 357, 1992.
11. Tour, J.M., Jones, L., Pearson, D.L., Lamba, J.J.S., Burgin, T.P., Whitesides, G.M., Allara, D.L., Parikh, A.N., and Atre, S.V., Self-assembled monolayers and multilayers of conjugated thiols, alpha,omega-dithiols, and thioacetyl-containing adsorbates — understanding attachments between potential molecular wires and Au surfaces, *J. Am. Chem. Soc.*, 117, 9529, 1995.
12. Weisbecker, C.S., Merritt, M.V., and Whitesides, G.M., Molecular self-assembly of aliphatic thiols on gold colloids, *Langmuir*, 12, 3763, 1996.
13. Nelles, G., Schonherr, H., Jaschke, M., Wolf, H., Schaur, M., Kuther, J., Tremel, W., Bamberg, E., Ringsdorf, H., and Butt, H.J., Two-dimensional structure of disulfides and thiols on gold(111), *Langmuir*, 14(4), 808, 1998.
14. Jung, C., Dannenberger, O., Xu, Y., Buch, M., and Grunze, M., Self-assembled monolayers from organosulfur compounds: a comparison between sulfides, disulfides, and thiols, *Langmuir*, 14(5), 1103, 1998.
15. Tarlov, J.M. and Newman, J.G., Static secondary ion mass-spectrometry of self-assembled alkane-thiol monolayers on gold, *Langmuir*, 8, 1398, 1992.
16. Peterlintz, K.A., Georgiadis, R.M., Herne, T.M., and Tarlov, M.J., Observation of hybridization and dehybridization of thiol-tethered DNA with two color surface plasmon resonance, *J. Am. Chem. Soc.*, 119, 3401, 1997.
17. Herne, T.M. and Tarlov, M.J., Characterization of DNA probes immobilized on gold surfaces, *J. Am. Chem. Soc.*, 119(38), 8916, 1997.

18. Steel, A.B., Herne, T.M., and Tarlov, M.J., Electrochemical quantitation of DNA immobilized on gold, *Anal. Chem.*, 70, 4670, 1998.

19. Yang, M., Yau, H.C.M, and Chan, H.L., Adsorption kinetics and ligand-binding properties of thiol-modified double-stranded DNA on a gold surface, *Langmuir*, 14, 6121, 1998.

20. Fukuzaki, S., Urano, H., and Nagata, K.N, Adsorption of bovine serum albumin onto metal oxide surfaces, *J. Fermentation Bioeng.*, 81(2), 163, 1996.

21. Roscoe, S.G. and Fuller, K.L., Interfacial behavior of globular proteins at platinum electrode, *J. Colloid Interfacial Sci.*, 152(2), 429, 1992.

22. Britland, S., Arnaud, E.P., Clark, P., McGinn, B., Connolly, P., and Moores, G., Micropatterning proteins and synthetic peptides on solid supports: a novel application for microelectronic fabrication technology, *Biotechnol. Prog.*, 8, 155, 1992.

23. Mooney, J.F., Hunt, A.J., McIntosh, J.R., Liberko, C.A., Walba, D.M., and Rogers, C.T., Patterning of functional antibodies and other proteins by photolithography of silane monolayers, *Proc. Natl. Acad. Sci.*, 93(22), 12287, 1996.

24. Nicolau, D.V., Taguchi, T., Taniguchi, H., and Yoshikawa, S., Micron-sized protein patterning on diazonaphthoquinone/novolak thin polymeric films, *Langmuir*, 14(7), 1927, 1998.

25. Williams, R.A. and Blanch, H.W., Covalent immobilization of protein monolayers for biosensors applications, *Biosensors Bioelectron.*, 9, 159, 1994.

26. Lahiri, J., Isaacs, L., Tien, J., and Whitesides, G.M., A strategy for the generation of surfaces presenting ligands for studies of binding based on an active ester as a common reactive intermediate: a surface plasmon resonance study, *Anal. Chem.*, 71, 777, 1999.

27. Bashir, R., Gomez, R., Sarikaya, A., Ladisch, M., Sturgis, J., and Robinson, J.P., Adsorption of avidin on micro-fabricated surfaces for protein biochip applications, *Biotechnol. Bioeng.*, 73(4), 324, 2001.

28. Lahiri, J., Ostuni, E., and Whitesides, G.M., Patterning ligands on reactive SAMs by microcontact printing, *Langmuir*, 15, 2055, 1999.

29. MacBeath, G. and Schreiber, S.L., Printing proteins as microarrays for high-throughput function determination, *Science*, 289, 1760, 2000.

30. Whaley, S.R., English, D.S., Hu, E.L., Barbara, P.F., and Belcher, A.M., Selection of peptides with semiconductor binding specificity for directed nanocrystal assembly, *Nature*, 405(8), 665, 2000.

31. Madhukar, A., Xie, Q., Cheng, P., and Konkar, Nature of strained InAs 3-dimensional island formation and distribution on GaAs (100), *Appl. Phys. Lett.*, 64(20), 2727, 1994.

32. Moison, J.M., Houzay, F., Barthe, F., and Leprince, L., Self-organized growth of regular nanometer-scale InAs dots on GaAs, *Appl. Phys. Lett.*, 64(2), 197, 1994.

33. Kamins, T.I., Carr, E.C., Williams, R.S., and Rosner, S.J., Deposition of three-dimensional Ge Islands on Si(001) by chemical vapor deposition at atmospheric and reduced pressures, *J. Appl. Phys.*, 81(1), 211, 1997.

34. Bashir, R., Kabir, A.E., and Chao, K., Formation of self-assembled Si1-Xgex islands using reduced pressure chemical vapor deposition and subsequent thermal annealing of thin germanium-rich films, *Appl. Surf. Sci.*, 152, 99, 1999.

35. Seeman, N.C., Nucleic acid junctions and lattices, *J. Theor. Biol.*, 99, 237, 1982.

36. Seeman, N.C., The use of branched DNA for nanoscale fabrication, *Nanotechnology*, 149, 1991.

37. Seeman, N.C., Zhang, Y., and Chen, J., DNA nanoconstructions, *J. Vac. Sci. Technol. A*, 12, 1895, 1994.

38. Winfree, E., Liu, F., Wenzler, L., and Seeman, N.C., Design and self-assembly of two-dimensional DNA crystals, *Nature*, 394, 539, 1998.

39. Seeman, N.C., DNA nanotechnology: novel DNA constructions, *Ann. Rev. Biophys. Biomol. Struct.* 27, 225, 1998.

40. Mirkin, C.A., Letsinger, R.L., Mucic, R.C., and Storhoff, J.J., A DNA-based method for rationally assembling nanoparticles into macroscopic materials, *Nature*, 382, 607, 1996.

41. Alivisatos, A.P., Johnsson, K.P., Peng, X., Wilson, T.E., Loweth, C.J., Bruchez, M.P., and Schultz, P.G., Organization of "nanocrystal molecules" using DNA, *Nature*, 382, 609, 1996.

42. Elghanian, R., Storhoff, J.J., Mucic, R.C., Letsinger, R.L., and Mirkin, C.A., Selective colorimetric detection of polynucleotides based on the distance-dependent optical properties of gold nanoparticles, *Science*, 277, 1078, 1997.

43. Mucic, R.C., Storhoff, J.J., Mirkin, C.A., and Letsinger, R.L., DNA-directed synthesis of binary nanoparticle network materials, *J. Am. Chem. Soc.*, 120, 12674, 1998.

44. Loweth, C.J., Caldwell, W.B., Peng, X., Alivisatos, A.P., and Schultz, P.G., DNA-based assembly of gold nanocrystals, *Agnew. Chem. Int. Ed.*, 38(12), 1808, 1999.

45. Storhoff, J.J., Elghanian, R., Mucic, R.C., Mirkin, C.A., and Letsinger, R.L., One pot colorimetric differentiation of polynucleotides with single base imperfections using gold nanoparticle probes, *J. Am. Chem. Soc.*, 120, 1959, 1998.

46. Taton, T.A., Mirkin, C.A., and Letsinger, R.L., Scanometric DNA array detection with nanoparticle probes, *Science*, 289, 1757, 2000.

47. Andres, R.P., Datta, S., Janes, D.B., Kubiak, C.P., and Reifenberger, R., *The Handbook of Nanostructured Materials and Nanotechnology*, Academic Press, New York, 1998.

48. Niemeyer, C.M., Progress in "engineering up" nanotechnology devices utilizing DNA as a construction material, *Appl. Phys. A.*, 69, 119, 1999.

49. Storhoff, J.J. and Mirkin, C.A., Programmed materials synthesis with DNA, *Chem. Rev.*, 99, 1849, 1999.

50. Csaki, A., Moller, R., Straube, W., Kohler, J.M., and Fritzsche, W., DNA monolayer on gold substrates characterized by nanoparticle labeling and scanning force microscopy, *Nucleic Acid Res.*, 29(16), 81, 2001.

51. Reichert, J., Csaki, A., Kohler, J.M., and Fritzsche, W., Chip-based optical detection of DNA hybridization by means of nanobead labeling, *Anal. Chem.*, 72, 6025, 2000.

52. Niemeyer, C.M., Ceyhan, B., Gao, S., Chi, L., Peschel, S., and Simon, U., Site-selective immobilization of gold nanoparticles functionalized with DNA oligomers, *Colloid Polymer Sci.*, 279, 68, 2001.

53. Taton, T.A., Mucic, R.C., Mirkin, C.A., and Letsinger, R.L., The DNA mediated formation of supramolecular mono- and multilayered nanoparticle structures, *J. Am. Chem. Soc.*, 122, 6305, 2000.

54. Huang, S., Martin, B. Dermody, D., Mallouk, T.E., Jackson, T.N., and Mayer, T.S., *41st Electron. Mater. Conf. Dig.*, 41, 1999.

55. Martin, B.R., Dermody, D.J., Reiss, B.D., Fang, M., Lyon, L.A., Natan, M.J., and Mallouk, T.E., Orthogonal self-assembly on colloidal gold-platinum nanorods, *Adv. Mater.* 11(12), 1021, 1999.

56. Mayer, T.S., Jackson, T.N., Natan, M.J., and Mallouk, T.E., *1999 Mater. Res. Soc. Fall Meet. Dig.*, 157, 1999.

57. Masuda, H. and Fukuda, K., Ordered metal nanohole arrays made by a 2-step replication of honeycomb structures of anodic alumina, *Science*, 268, 1466, 1995.

58. Li, A.P., Muller, F., Birner, A., Nielsch, K., and Gosele, U., Hexagonal pore arrays with a 50–420 nm interpore distance formed by self-organization in anodic alumina, *J. Appl. Phys.*, 84(11), 6023, 1998.

59. Mbindyo, J.K.N., Reiss, B.D., Martin, B.J., Keating, C.D., Natan, M.J., and Mallouk, T.E., DNA-directed assembly of gold nanowires on complementary surfaces, *Adv. Mater.*, 13(4), 249, 2001.

60. Braun, E., Eichen, Y., Sivan, U., and Yoseph, G.B., DNA-templated assembly and electrode attachment of a conducting silver wire, *Nature*, 391(19), 775, 1998.

61. Eichen, Y., Braun, E., Sivan, U., and Yoseph, G.B., Self-assembly of nanoelectronic components and circuits using biological templates, *Acta Polym.*, 49, 663, 1998.

62. Braun, E., Sivan, U., Eichen, Y., and Yoseph G.B., Self-assembly of nanometer scale electronics by biotechnology, *24th Intl. Conf. Phys. Semicond.*, 1998, 269.

63. Ackley, D.E., Heller, M.J., and Edman, C.F., DNA technology for optoelectronics, *Proc. Lasers Electro-Opt. Soc, Annu. Meet. LEOS*, 1, 85, 1998.

64. Alivasatos, A.P., Perspectives on the physical chemistry of semiconductor nanocrystals, *J. Phys. Chem.*, 10, 13226, 1996.

65. Klein, D.L., Roth, R., Kim, A.K.L., Alivisatos, A.P., and McEuen, P.L., A single-electron transistor made from a cadmium selenide nanocrystal, *Nature*, 699, 1997.

66. Mitchell, G.P., Mirkin, C.A., and Letsinger, R.L., Programmed assembly of DNA functionalized quantum dots, *J. Am. Chem. Soc.*, 121, 8122, 1999.

67. Bruchez, M., Moronne, M., Gin, P., Weiss, S., and Alivisatos, A.P., Semiconductor nanocrystals as fluorescent biological labels, *Science*, 281, 2013, 1998.

68. Chan, W.C.W. and Nie, S., Quantum dot bioconjugates for ultrasensitive nonisotopic detection, *Science*, 281, 2016, 1998.

69. Oldenburg, S.J., Averitt R.D., Westcott, S.L., and Halas, N.J., Nanoengineering of optical resonances, *Chem. Phys. Lett.*, 288(2–4), 243, 1998.

70. McNally, H., Pingle, M., Lee, S.W., Guo, D., Bergstrom, D., and Bashir, R., Self-assembly of micro and nano-scale particles using bio-inspired events, submitted to *Applied Surface Science*.

71. Edelstein, R.L., Tamanaha, C.R., Sheehan, P.E., Miller, M.M., Baselt, D.R., Whitman L.J., and Colton, R.J., BARC biosensor applied to the detection of biological warfare agents, *Biosensors Bioelectron.*, 14 (10), 2000.

72. Andreas, R.P., Bielefeld, J.D., Henderson, J.I., Janes, D.B., Kolangunta, V.R., Kubiak, C.P., Mahoney, W.J., and Osifchin, R.G., Self-assembly of a two-dimensional superlattice of molecularly linked metal clusters, *Science*, 273, 1690, 1996.

73. Liu, J., Lee, T., Janes, D.B., Walsh, B.L., Melloch, M.R., Woodall, J.M., Riefenberger, R., and Andres, R.P., Guided self-assembly of Au nanocluster arrays electronically coupled to semiconductor device layers, *Appl. Phys. Lett.*, 77(3), 373, 2000.

74. Qin, D., Xia, Y., Xu, B., Yang, H., Zhu, C., and Whitesides, G.M., Fabrication of ordered two-dimensional arrays of micro- and nanoparticles using patterned self-assembled monolayers as templates, *Adv. Mater.*, 11(17), 1433, 1999.

75. Terfort, A., Bowden, N., and Whitesides, G.M., Three-dimensional self-assembly of millimeter scale objects, *Nature*, 386, 162, 1997.

76. Tien, J., Breen, T.L., and Whitesides, G.M., Crystallization of millimeter scale objects using capillary forces, *J. Am. Chem. Soc.*, 120, 12670, 1998.

77. Breen, T.L., Tien, J., Oliver, S.R.J., Hadzic, T., and Whitesides, G.M., Design and self-assembly of open regular 3-D mesostructures, *Science*, 284, 948, 1999.

78. Terfort, A. and Whitesides, G.M., Self-assembly of an operating electrical circuit based on shape complementarity and the hydrophobic effect, *Adv. Mater.*, 10(6), 470, 1998.

79. Srinivansan, U., Liepmann, D., and Howe, R.T., Microstructure to substrate self-assembly using capillary forces, *J. Microelectromech. Syst.*, 10(1), 17, 2001.

80. Derycke, V., Martel, R., Appenzeller, J., and Avouris, P., Carbon nanotube inter- and intramolecular logic gates, *NanoLetters*, 1(9), 453, 2001.

81. Liu, J., Casavant, J., Cox, M., Walters, D.A., Boul, P., Lu, W., Rimberg, A.J., Smith, K.A., Colbert, D.T., and Smalley, R.E., Controlled deposition of individual single walled carbon nanotubes on chemically functionalized templates, *Chem. Phys. Lett.*, 303, 125, 1999.

82. Tien, J., Terfort, A., and Whitesides, G.M., Microfabrication through electrostatic self-assembly, *Langmuir*, 13, 5349, 1997.

83. Bowden, N., Terfort, A., Carbeck, J., and Whitesides, G.M., Self-assembly of mesoscale objects into ordered two-dimensional arrays, *Science*, 276, 233, 1997.

84. Heller, M.J., An active microelectronics device for multiplex DNA analysis, *IEEE Eng. Med. Biol. Mag.*, 15(2), 100, 1996.

85. Sosnowski, R.G., Tu, E., Butler, W.F., O'Connell, J.P., and Heller, J.J., Rapid determination of single base mismatch mutations in DNA hybrids by direct electric field control, *Proc Natl. Acad. Sci. U.S.A.*, 94(4), 1119, 1997.

86. Huang, Y., Ewalt, K.L., Tirado, M., Haigis, R., Forster, A., Ackley, D., Heller, M.J., O'Connell, J.P., and Krihak, M., Electric manipulation of bioparticles and macromolecules on microfabricated electrodes, *Anal. Chem.*, 73(7), 1549, 2001.

87. www.nanogen.com

88. Heller, M.J., Forster, A.H., and Tu, E., Active microelectronic chip devices which utilize controlled electrophoretic fields for multiplex DNA hybridization and other genomic applications, *Electrophoresis*, 21(1), 157, 2000.

89. Fan, C., Shih, D.W., Hansen, M.W., Hartmann, D., Van Blerkom, D., Esener, S.C., and Heller, M., Heterogeneous integration of optoelectronic components, *Proc. Intl. Soc. Opt. Eng. U.S.A*, 2, 3290, 1997.

90. Edman, C.F., Swint, R.B., Furtner, C., Formosa, R.E., Roh, S.D., Lee, K.E., Swanson, P.D., Ackley, D.E., Coleman, J.J., and Heller, J.J., Electric field directed assembly of an InGaAs LED onto silicon circuitry, *IEEE Photonics Technol. Lett.*, 12(9), 1198, 2000.

91. Bashir, R., Lee, S.W., Guo, D., Pingle, M., Bergstrom, D., McNally, H.A., and Janes, D., *Proc. MRS Fall Meet.*, Boston, MA., 2000.

92. Bashir, R., DNA mediated artificial nanobiostructures, *Superlattice Microstruct.*, 29, 1, 2001.

93. Lee, S., McNally, H., Guo, D., Pingle, M., Bergstrom, D., and Bashir, R., Electric field mediated assembly of silicon islands coated with charged molecules, *Langmuir*, 18, 3383–3386, 2002.

94. Wang, X., Huang, Y., Gascoyne, P.R.C., and Becker F.F., Dielectrophoretic manipulation of particles, *IEEE Trans. Ind. Appl.*, 33(3), 660, 1997.

95. Pethig, R. and Markx, G.H., Application of dielectrophoresis in biotechnology, *TIBTECH*, 15, 426, 1997.

96. Ramos, A., Morgan, H., Green, N.G., and Castellanos, A., AC electrokinetics: a review of forces in microelectrode structures, *J. Phys. D: Appl. Phys.*, 31, 2338, 1998.

97. Smith, P., Nordquist, C.D., Jackson, T.N., Mayer, T.S., Martin, B.R., Mbindyo, J., and Mallouk, T.E., Electric field assisted assembly and alignment of metallic nanowires, *App. Phys. Lett.*, 77(9), 1399, 2000.

98. Yeh, H.J. and Smith, J.S., Fluidic self-assembly for the integration of GaAs light-emitting diodes on Si substrates, *IEEE Photonics Technol. Lett.*, 6(6), 706, 1994.

99. Tu, J.K., Talghader, J.J., Hadley, M.A., and Smith, J.S., Fluidic self-assembly of InGaAs vertical cavity surface emitting lasers onto silicon, *Electron. Lett.*, 31, 1448, 1995.

100. Talghader, J.J., Tu, J.K., and Smith, J.S., Integration of fluidically self-assembled optoelectronic devices using a silicon based process, *IEEE Photonics Technol. Lett.*, 7, 1321, 1995.

101. Smith, J.S., High density, low parasitic direct integration by fluidic self-assembly (FSA), *Proc. 2000 IEDM*, 201, 2000.

102. *Alien Technology*, Morgan Hill, CA., www.alientechnology.com

103. Huang, Y., Duan, X., Wei, Q., and Lieber, C.M., Directed assembly of one-dimensional nanostructures into functional networks, *Science*, 291, 630, 2001.

104. Cui, Y. and Lieber, C.M., Functional nanoscale electronic devices assembled using silicon nanowire building blocks, *Science*, 291, 851, 2001.

105. Collier, C.P., Wong, E.W., Belohradsky, M., Raymo, F.M., Stoddart, J.F., Kuekes, P.J., Williams, R.S., and Heath, J.R., Electronically configurable molecular-based logic, *Science*, 285, 391, 1999.

106. Chen, J., Reed, M.A., Rawlett, A.M., and Tour, J.M., Large on–off ratios and negative differential resistance in a molecular electronic device, *Science*, 286, 1550, 1999.

107. Fink, H-W. and Schonenberger, C., Electrical conduction through DNA molecules, *Nature*, 398, 407, 1999.

108. Porath, D., Bezryadin, A., Vries, S.D., and Dekker, C., Direct measurement of electrical transport through DNA molecules, *Nature*, 403, 635, 2000.

16

Nanostructural Architectures from Molecular Building Blocks

CONTENTS

Damian G. Allis
Syracuse University

James T. Spencer
Syracuse University

16.1 Introduction

The concept of a *molecular building block* (MBB) has been used prominently in describing a particular application of small molecules in the design of macromolecules, such as biomolecules, supramolecular structures, molecular crystal lattices, and some forms of polymeric materials. It is also common to refer to MBBs as "molecular subunits, modular building blocks,[1,2] or synthons, which have been defined as structural units within supermolecules which can be formed and/or assembled by known or conceivable synthetic operations involving intermolecular interactions."[3] MBBs are, therefore, the structural intermediates between atoms, the most basic of all building units, and macromolecules or extended arrays, of which the MBBs are the common structural element. While many MBB approaches are not directed toward the design of nanostructures or nanoscale materials, all share the same design considerations and are consistent with the criteria used to distinguish the MBB approaches considered here from other nanoscale fabrication techniques.

The fabrication of any structure or material from building blocks requires that the design strategy meet specific criteria. First, relying on a building block as the basis of a fabrication process indicates that this starting material is not the smallest possible component from which the manufacturing process can proceed, but it is itself pre-assembled from more fundamental materials for the purpose of simplifying the building process. It is assumed that the subunit, as a prefabricated structure, has been engineered with an important function in the assembly process of a larger, more complex structure. Second, it is assumed that a means to subunit interconnectivity has been considered in the design process. The method of connectivity between subunits may be either intrinsic to the subunit, such as a direct bonding connection between them, or available externally, such as a stabilizing electrostatic force between subunits. Third, it is assumed that the subunit is capable of being positioned correctly and precisely in the fabrication process. Fourth, and perhaps most important from a design perspective, is that the subunit provides an intermediate degree of control in the properties of the larger structure. A fabrication process based upon the manipulation of designed subunits may not provide the ultimate in stability, customizability, or structural detail when compared with the design of a system from the most basic materials, but it certainly offers enough control and flexibility for useful applications.

The defining feature of the MBB approach is the use of a molecular subunit that has incorporated within its covalent framework the means for a directed connectivity between subunits. As the MBB is itself a molecule, its synthesis can be considered among the preparative steps in the overall fabrication process and not necessarily an integral part of the actual supramolecular assembly. If "supermolecules are to molecules and the intermolecular bond what molecules are to atoms and the covalent bond,"[4] then the individual molecule forms the fundamental component in the design of MBB-based nanostructures. The starting point for the final product is the MBB, and the means to assembling the final product is through manipulation of the MBB. The assembly of a nanostructure can be predicted based upon the covalent framework of the MBB and its assembly-forming features. Because this intermolecular connectivity is an integral part of the design process, the means for controlling subunit–subunit interactions needs to be incorporated early in the design of the nanostructure. The self-assembly or self-directing interactions between subunits are based upon the properties of the MBB. The ability to customize the stability and functionality of the resulting materials is, therefore, based upon MBB modification.

The merits and limitations of building block approaches transcend scale. In all cases, the selection of suitable building materials is dictated by their ability to fit together in a precise and controllable manner. Limitations to a particular design or application are imposed partly by the properties of the subunit and partly by the design itself. While all building block designs suffer from one or more limitations, designs can often be successfully employed for a specific application or in a specific environment. For instance, bricks are ideal building materials for the construction of permanent structures, blocks of ice are appropriate for use in below-freezing conditions, and canvas is ideal for structures that require mobility. One would not select ice as a building material in temperate climates, canvas for arctic conditions, or brickwork for temporary residences. Given a set of environmental conditions and the properties of available materials, certain combinations will invariably make more sense than others. In nanostructure design, the important concerns often include solubility, thermal stability, means to assembly, defect tolerance, error correction capabilities, functionality, and chemical reactivity. Chemical environments and ambient conditions limit the feasibility of certain nanostructures just as they limit the choice of molecular subunits. These same issues are key to synthetic chemistry, where factors such as temperature, solvent, reaction duration, and choice of chemical functionalities will always play key roles in the design of chemical pathways and molecular fabrication processes.

All MBB approaches benefit from the ability to accurately predict intermolecular interactions from conceptual and theoretical treatments. Additionally, a vast synthetic background exists from which to make and modify subunits. Experimental precedent for the basic preparative methodology in a number of naturally occurring and man-made systems form a firm foundation for MBB pathways. Not only are the means to nanostructure fabrication facilitated through theoretical investigations, cognizant design strategies, and even Edisonian efforts, but many examples of macromolecular formation exist currently that provide the means for understanding how molecules can be used to construct supermolecular arrays. Concurrent with the design of new nanostructures from MBB approaches is the continued growth of the field of supramolecular chemistry and an enhanced understanding of molecular phenomena "beyond the molecule."[5]

The emphasis on design in molecular nanotechnology from MBBs connotes a certain deliberation in the choice of materials and the means to assembly. It is, therefore, important to stress efforts to engineer macromolecular assemblies from known molecular systems. This chapter begins with a discussion of the chemical and electrostatic interactions important in macromolecular formation. The discussion of these interactions as applied to nanostructure formation begins with two limiting cases in MBB design, covalent and electrostatic connectivity. With the formal groundwork of connectivity and some useful boundaries established to focus the discussion, a few important areas of MBB-based nanostructure formation are presented to demonstrate the application of the approach and related issues. This chapter is not meant to be rigorously complete, but instead provides a broad overview of current techniques involving the use of molecules as building components in larger systems.

16.2 Bonding and Connectivity

A structure is of limited value for an application without a means of maintaining its strength and functional integrity over the duration of its anticipated lifetime. At the macroscale, stabilization may come in the form of interlocking parts, mechanical or adhesive fixtures, fusing or melting at connection points between materials, or, in much larger structures, gravity. At the nanoscale, the role of gravity becomes unimportant in the formation of supramolecular assemblies,[6] and nearly all stability comes from electronic interactions. These interactions take forms ranging from strong covalent bonds to weak intermolecular (noncovalent) interactions. All molecular-based nanostructures incorporate various combinations of these interactions to maintain shape and impart function. It is therefore important to understand the range and form of the stabilization energies associated with these interactions and their relationship to the structures that incorporate them.

16.2.1 Covalent Bonding

Of singular importance in synthetic chemistry is the manipulation of the covalent bond. The design of any nanoscale architecture from simpler molecules must first address the design of the covalent framework of the MBB itself. The means to any macromolecular stabilization is a result of the inclusion of chemical functionalities onto this stable framework. The role of covalent bonds in the MBB approach is then twofold. First, these bonds are required within the subunit to provide the structural integrity necessary for the prediction and synthesis of nanostructures from MBB components. Second, covalent bonds may be employed as one of the means for fastening MBBs together into larger structures.

Covalent bonds are formed by the sharing of pairs of electrons between atoms.[7] The most familiar examples of covalent bonding are the connections between carbons in organic molecules. The importance of organic chemistry as a field underscores our desire to understand and modify the covalent framework of carbon-containing molecules for many important applications. Typical covalent bond energies range from 100 to 500 kJ/mol.[8] In the case of multiple bonds between atoms, the total energy may exceed 1000 kJ/mol. While this is a very large range of energies, even the covalent bonds at the low end of the spectrum are rather strong interactions, especially when compared with the noncovalent energies frequently responsible for macromolecular stabilization (*vide infra*). It is because of these large covalent bond strengths that the subunits involved in MBB approaches provide significant internal structural stability and predictability.

Covalent bonding includes a variety of useful motifs in the structural customization of a subunit. The strong σ-bonds, in which a pair of electrons is shared directly along the interatomic axis of two atoms, provide for low-energy rotation in straight-chain molecules and low-energy twisting in closed-ring systems (Figures 16.1A and 16.1B). In organic molecules, σ-bonding plays the initial role of defining the connectivity and general shape of the structure. The formation of π-bonds in molecules involves electrons in atomic orbitals that are not involved in the σ-bonding framework. In such instances, main group atoms involved in the molecular backbone are either sp²- or sp-hybridized, leaving either one or two p-orbitals through which π-bonding can occur (Figure 16.1C). The π-bonded portion of the molecule is then held planar to maximize p-orbital overlap between atoms. These π-bonds may be delocalized over the entire length of the available π-orbital framework, making them well suited to molecular electronic applications that require both structural stability and electron mobility.[9] Structurally, the π-bonds remove the low-energy rotational freedom from the underlying σ-bond framework. In cases where two π-bonds

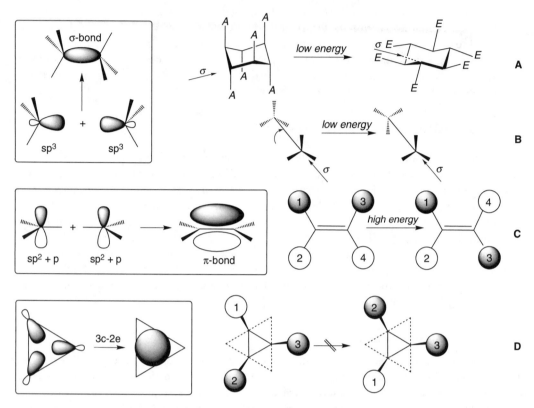

FIGURE 16.1 Rotation in covalent bonds. (A) Ring twisting about a σ-bond (indicated by arrow) with a change in orientation of one set the substituents from axial (A, left) to equatorial (E, right). (B) Free rotation about the σ-bonds in linear chains. (C) High-energy bond breaking is required for rotation about a π-bond. (D) Reorientation of substituents in 3c-2e bonds is highly restricted (extension of cluster framework indicated by dashed lines).

are formed between two adjacent atoms, the resulting π-electron density around the σ-bond is cylindrical, and the molecular fragment behaves as a rigid linear rod.[10] A third motif involves what is often referred to as either *electron deficient* or *three-center-two-electron bonding*. Structural flexibility is fully restricted in three-center-two-electron (3c-2e) bonds. These bonds, observed in main group polyhedra and in many metal clusters, involve three adjacent atoms sharing a single pair of electrons (Figure 16.1D). Molecules employing this mode of bonding are generally three-dimensional, meaning that overall structural flexibility within the molecule is lost due to the cage-like interconnections between atoms. These clusters share some electronic properties with π-bonds, although their delocalized nature is largely limited to their internal skeletal frameworks.[11] As a result, radial bonds from these structures behave very much like typical σ-bonds.

Neglecting the covalent framework of the MBBs and focusing only on the connectivity between subunits, a number of advantages are derived from the application of covalent bonds in nanostructural design. First, intermolecular covalent bonding leads to extremely stable nanostructures. Whereas weaker electrostatic interactions are greatly affected by factors such as temperature and choice of solvent, covalent bonds retain their connectivity until concerted efforts are made to break them. Covalent bonds are, then, structurally dependable, prohibiting the reorganization often observed in the continuous breaking and reforming of the other types of intermolecular interactions. It is this feature that similarly allows the covalent architecture of the subunit to be held constant within the context of the larger nanostructure. Finally, an extensive synthetic precedent also exists for connecting almost any molecular fragment or functional group to another. Where specific types of connections have not been previously addressed, their formation is generally possible by a modification of some other known reaction.

The strength and chemistry of covalent bonding also has some important limitations. First, the strength of these bonds frequently limits the flexibility of larger molecules.[6,10] Weaker electrostatic interactions must be used if motion and structural rearrangement are required. With this greater flexibility in the weaker electrostatic interactions comes a higher degree of error tolerance. An unplanned covalent bond between two subunits in a molecular architecture is difficult to correct, requiring far more intensive efforts than simple thermodynamic manipulation. Structural designs based on covalent bonding must, therefore, be well conceived initially to avoid subsequent problems in the fabrication process. Finally, the use of the covalent bond in nanoscale assembly requires direct chemical manipulation. Consequently, two subunits may be self-directing in the formation of their bond by the choice of functionalities, but they are typically not self-assembling. A chemical workup is generally required to form a covalent bond and, as necessary, isolate a product from a reaction mixture.

16.2.2 Coordination Complexes

Lying between the strong covalent bonds of the smaller main group elements and the variety of noncovalent interactions are the coordination bonds of metal–ligand complexes. The initial descriptions of metal–ligand compounds as *complexes* stems from the ability of metals to coordinate small, electron-donating molecules (ligands) beyond the typical maximum of four-point substitutions possible with many main group elements.[8] Metal complexes are known to exist with the metal coordinated to anywhere from one to 12 ligands, although the vast majority of coordination compounds exist in the four-coordinate to eight-coordinate regime (Figure 16.2). The interest in the properties and applications of metals in discrete molecules has enriched such diverse fields as molecular orbital theory, crystallography, catalysis, molecular electronics, supramolecular chemistry, and medicinal chemistry.[8] The availability of d-orbitals in the transition metals and f-orbitals in the lanthanides and actinides results in an extension of the geometric and structural variety available with main group elements. A well-developed synthetic precedent also provides the means to exploiting this rich structural variety within a single molecule context.[12]

Metal–ligand bonds form either through the covalent association of ligands to pair single electrons in metal orbitals or, most often, through the coordination of paired electrons from ligands to fill the valence shell of the metal. Examples include single-ligand lone-pair/metal bonds (the metal analogue

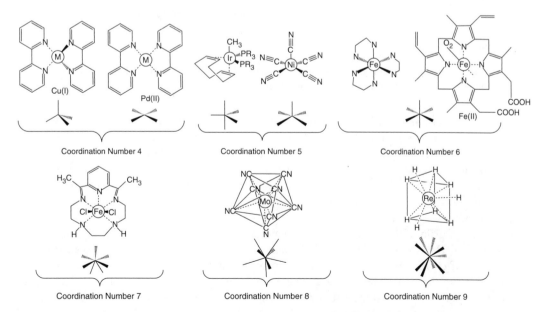

FIGURE 16.2 Examples of coordination geometries among a number of metal complexes.

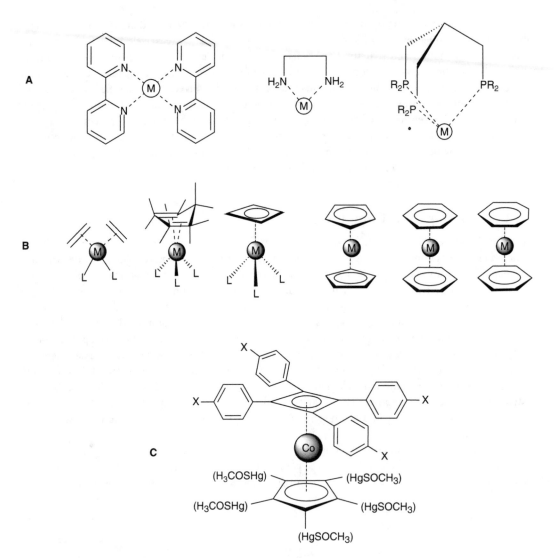

FIGURE 16.3 Metal-ligand bonding. (A) A selection of chelating ligands. (B) Metal–ligand π-interactions including metallocenes. (C) A surface-mounted molecular rotor design.

of a main group σ-bond), chelating ligand bonds (where the ligand is coordinated to the metal by more than one pair of electrons (Figure 16.3A), and metal–ligand π bonds (Figure 16.3B). The low-energy *sharing* of pairs of electrons arises from the coordination sphere of the metal, which can readily accommodate the available electron pairs. Metal–ligand bonds are usually far stronger than other electrostatic interactions because they involve the sharing of pairs of electrons through direct orbital interactions, yet they are generally weaker than the covalent bonds found in organic compounds. To specifically address issues of connectivity, the extensive use of lone-pair coordination to saturate the valence shells of many of the metals adds electron density well in excess of the nuclear charge, pushing the limits of the ability of some metal nuclei to fully accommodate all of the required electrons. Also, the majority of coordinating ligands are stable molecules, and any intermolecular destabilization is typically directed first to the weaker metal–ligand bond. Finally, the molecular volume of the ligand can have a significant effect on the stability of the metal–ligand bond in cases where the metal has a high coordination number, requiring many lone pairs to saturate its valence shell. This last feature of steric saturation is of primary importance in rare earth complexes. In these compounds, orbital interactions between the metal and ligand are significantly attenuated; and stabilization arises primarily from charge balance and steric saturation of the metal center.

A series of metal–ligand coordination complexes are shown in Figure 16.2 to demonstrate some of the structural variety available from metal coordination. Ligand lone-pair coordination is essentially σ-bonding; and the properties of these bonds are consistent with σ-bonding in organic frameworks, including low barriers to rotation and geometric predictability. One special subset of these lone-pair ligands is the chelating ligands, which coordinate to a single metal center through two or more lone-pair donors on the same ligand (Figure 16.3). This class of ligands, driven to higher metal coordination number through entropic effects[8] has a significant role in the design of nanostructures from coordination-based approaches (*vide infra*).

An important case of metal–ligand π-coordination occurs in the metallocenes, where the entire π-system of an organic ring can be coordinated to the metal center[13] (Figure 16.3B). The most familiar of these systems is the neutral ferrocene, which saturates an iron(II) center by the coordination of two five-member aromatic cyclopentadienyl rings ($[C_5H_5]^-$). In the design of some of the smallest functional nanostructures, such π-coordinated molecules have distinct advantages, including (1) high stability, (2) incorporation of organic frameworks with the potential to substitute onto the framework, and (3) very low barriers to rotation about the axis of the metal and ring center. Small, surface-mounted metal-ring compounds have already been demonstrated as potential systems for molecular rotors[14] (Figure 16.3C).

Metal–ligand bonds, as the intermediary between main group covalent bonding and weaker electrostatic interactions, are well suited to the fabrication of many types of macromolecules and nanoscale arrays. First among their advantages are the higher coordination numbers of these atoms. While a single nonmetallic main group atom generally provides the structural flexibility required to link together from one to four substituents, main group molecules are required to achieve higher connectivity. Instead of designing a six-coordinate center from the smallest molecular octahedron, *closo*-$[B_6H_6]^{-2}$, single-metal atoms readily perform the same task (Figure 16.2). A second advantage of metal-based structures is the number of available metals from which to choose, both for structural complexity and functionality. With this large selection of metal atoms also comes an extensive synthetic precedent,[8,12] allowing the selection of a particular coordination geometry for its known structural features, stability, and chemical accessibility. In instances where lone-pair coordination is used to saturate the valence shell of a metal, the required chemical manipulation is typically too mild to affect the covalent structure of the ligand. Furthermore, because ligands coordinate through weaker bonds, they are also often thermally and photochemically labile under moderate conditions. This ability to form stable structures by thermodynamic or photolytic methods, however, also carries with it the disadvantage of having to control the environment carefully in order to maintain the structural integrity of the final products.

16.2.3 Dative Bonds

A dative bond is an intermolecular interaction between a lone pair of electrons on one atom and a vacant, atom-centered orbital on another. These bonds behave as covalent σ-bonds in many respects, making them close analogs to metal–ligand coordinate bonds (the distinction is made here by limiting dative bonding to main group–main group or metal–metal interactions).[15] While a lone pair of electrons and two atom centers are involved, these interactions are relatively weak when compared with the covalent bonding of the main group elements. The molecules involved in these bonds are themselves independently stable species. The strength of the dative bond is determined by several factors, all of which provide their own means to customization depending on the application.

Dative bonds are most common among pairs of molecules incorporating Group III[(13)] and Group V[(15)] atoms.[16] In such cases, the formation of a dative bond requires the presence of the Group III atom, where an empty orbital remains after the σ-bonds are formed from the three available valence electrons (Figure 16.4). Elements including and beyond Group V usually have at least one lone pair available for donation in bond formation. Dative structures are classic examples of Lewis acids and bases,[17] in which the lone-pair donor is the Lewis base and the lone-pair acceptor is the Lewis acid. Among the strongest and most

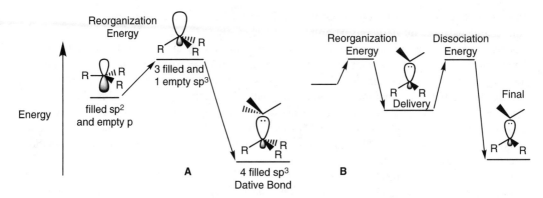

FIGURE 16.4 Dative bonding in main group elements. (A) General pathway for dative bond formation. (B) Energetic considerations of dative bond acceptor "delivery" pathway.

studied dative bonds are those between boron (Group III) and nitrogen (Group V) in cases where the boron is treated as an electron precise (2c-2e) atom.[16] The formation of the three electron precise σ-bonds to boron results in the molecule adopting a trigonal planar conformation, leaving an unoccupied p-orbital to act as an electron pair acceptor (Figure 16.4A). The coupling of an atom with a lone pair of electrons to boron results in a reorganization of the boron center,[18] causing it to change shape and hybridization from trigonal planar (sp^2) to tetrahedral (sp^3). The stability of a dative bond is then dependent upon (1) the choice of lone-pair donor and acceptor, (2) the substituents on the donor and acceptor, and (3) the reorganization energy. These bonds typically range from 50 to 85 kJ/mol, although some have been shown to have bond strengths of 100 kJ/mol.[16] In small systems, such as $H_3B:NH_3$ and $F_3B:NH_3$, the stabilizing energy is large because there is very little steric congestion from the substituents. Among the systems with significant steric congestion, boraadamantane forms uniquely stable dative structures (Figure 16.5).[18] In boraadamantane, the adamantyl framework forces the boron to be sp^3-hybridized regardless of the presence of a lone-pair donor. The reorganization energy is effectively included in the synthesis of the boraadamantane Lewis acid, leaving the entirety of the lone-pair interaction to form a particularly stable dative bond.[18] Dative bonds provide the directionality of covalent bonds with the lower stabilization energy of electrostatic interactions, giving them useful features for nanoscale design. Dative-based molecular assemblies require the selection of building blocks that limit the lone pairs and vacant orbitals to structurally important sites.[19] The design of connectivity is then a matter of limiting dative bonding everywhere else in the subunits. In organic molecules, incorporating the dative components into the "correct" structural sites on the molecule and including only C-H bonds everywhere else effectively accomplishes this. While the chemistry of organoboron compounds might not be as well developed as that of organonitrogen compounds, a considerable synthetic precedent exists for both. The inclusion of active centers for dative design is possible through the addition of many known organic components. The strength of dative bonds can also be tailored by either changing the donor and acceptor substituents or by changing the initial hybridization of the electron pair acceptor.[16]

The limitations of the dative bond approach stem primarily from the lone-pair acceptor. While the lone-pair donor is often unreactive, lone-pair acceptors, such as the many organoboron compounds, are

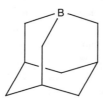

FIGURE 16.5 Boraadamantane.

highly electrophilic and will coordinate with any available electron pairs. Part of the design of these systems must include potential problems with *delivery* to the donor (Figure 16.4B). When a *delivery molecule* is required initially to coordinate to the acceptor prior to assembly, this molecule must be chosen to be more stable than any possible lone-pair donors in solution, yet weakly coordinating enough such that the delivery molecule is easily displaced from the system during assembly. In some instances, the selection of a good delivery molecule can be nontrivial, since an effective choice involves the subtle interplay between the strength of the delivery–acceptor and the final donor–acceptor bond strengths.

16.2.4 π-Interactions

The variety of electrostatic interactions involving the π-systems of aromatic molecules have been shown to play important roles in such diverse areas as the packing of molecules in molecular crystals, the base stacking (as opposed to base pairing) interactions in DNA, polymer chemistry, the structure and reactivity of many organometallic complexes, and the formation, shapes, and function of proteins.[20] The accessible and highly delocalized pool of electrons above and below an aromatic molecular plane is well suited to forming electrostatic interactions with cations, neutral molecular pairs with complementary electron density differences, and other aromatic π-molecular systems.

Because π-systems may be thought of as regions of approachable electron density, noncovalent interactions with aromatic rings occur when a system with a net-positive region is brought within proximity of the aromatic molecule. Three of the most familiar types of π-interactions are (1) aromatic ring/electrophile interactions, where the electrophile is highly positive, (2) phenyl/perfluorophenyl interactions, where the electronegativities of the σ-bond periphery have an overall affect on the charge distribution of the molecule, and (3) lower energy quadrupolar interactions, where weak π-interactions occur based on electron density differences across the molecular plane (Figure 16.6).

Cation-π interactions have long been known to play important roles in molecular recognition, biochemical processes, and catalysis.[21] The most familiar cations used to study these interactions are Group I[(1)] elements and small protonated Lewis bases (such as NR_4^+). The binding energies of these pairs can be quite large, with the strongest interactions approaching the strengths of weak covalent bonds.[21] Important to the nature of these interactions is the ability of the aromatic rings to compete successfully with polar solvents for the cation. Remarkably, the π-system of the nonpolar benzene molecule has been shown to bind K^+ ions more strongly than the oxygen lone pairs in water.[22] The customization of the cation-π binding energy can be controlled by the choice of Group I[(1)] cation or the substitutions on the molecular cations, where bulky substituents tend to lower the stabilization by forcing the cation further from the π-system.

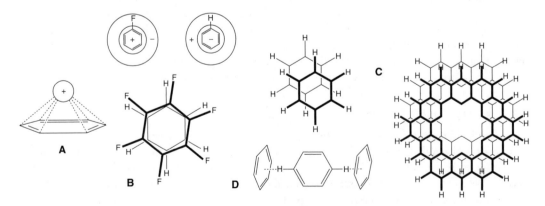

FIGURE 16.6 A selection of π-interactions. (A) π-cation interactions. (B) Ideal π-stacking arrangement of benzene/perfluorobenzene. (C) Staggered π-stacking in benzene (left) and kekulene (right). (D) Preferred herringbone π-stacking configuration of benzene, with hydrogen atoms centered on the π-system of adjacent benzene rings.

Benzene/hexafluorobenzene stacking is a specific example of the general type of π-stabilization that occurs with the pairing of molecules that have large quadrupole moments of opposite sign.[23] In benzene, the regions of highest electron density are the π-system and σ-system of the carbons, leaving the peripheral hydrogen atoms net-positive from inductive effects (Figure 16.6B). In hexafluorobenzene, the charge density is reversed, with the peripheral fluorine atoms containing the highest electron density. Together, the molecular pair is ideally suited for stacking due to the complementary arrangement of their electron densities inside the rings and along the outer periphery of the two molecules. In a classic study of this form of π-stacking, one equivalent of benzene (m.p. 5.5°C) was combined with one equivalent of hexafluorobenzene (m.p. 4°C) to form a mixture with a melting point of 24°C.[24] The actual stacking of these rings was subsequently confirmed by a variety of spectroscopic methods.[25] This same stabilization has been used successfully in the formation of other π-stacking species[26] to align various arenes in molecular crystals, providing a facile means for molecular alignment of thermal and photochemical polymerization reactions.[27]

There are many other important examples and structural motifs in π-stabilizing interactions. These include the stacking of DNA base pairs, the stabilization of tertiary structures in proteins, the aggregation of large porphyrins, and the formation of molecular crystals incorporating aromatic moieties.[20] In benzene and kekulene, for example, stacked structures are most stable when slightly offset, maximizing the overlap of the net-positive periphery and the electron-rich π-system (Figure 16.6C).[20] The offset stacking of the purine and pyrimidine base pairs in DNA plays an important function in stabilizing the double helix. Benzene and many other aromatic systems crystallize as herringbone-shaped structures, with the peripheral hydrogen atom on one ring placed along the central axis of the π-system of a perpendicular ring (Figure 16.6D).[23]

The variety of π-interaction types and structural motifs available among the aromatic rings leads to a number of important features for nanoscale design. First, the π-orbitals exist above and below the molecular plane. An interaction with a π-system is, therefore, often just as likely to occur above the molecular plane as below, allowing these stacking interactions to occur over long distances with many repeating units. For example, crystals of benzene/hexafluorobenzene and the extended π-stacking arrangement in base pairs in a single strand of DNA provide considerable electrostatic stability and alignment. Second, the stability of a π-interaction can be directly controlled by the chemical substituents attached to the ring. The significant change in the properties of benzene/hexafluorobenzene solutions attest to this chemical flexibility.[24] Aromatic heterocycles, such as the purines and pyrimidines in DNA nucleotides, demonstrate the ability to customize these interactions based on directly changing the π-system through hetero-atom substitution. Third, depending upon the surroundings, the π-interactions can be modified by solvent effects. This is demonstrated in the base stacking of DNA, where the stability from heterocycle π-stacking interactions is in addition to the stability gained from minimizing the surface area of the rings exposed to the aqueous environment. Similar arguments have been used to describe the formation of tertiary structure and aggregation of proteins.[20,28] The use of these types of interactions for designing nanostructures is limited, however, by the relatively unpredictable stacking arrangements observed and the sizes of these complex aromatic rings. Stability from aromatic π-stacking requires the use of rings which, when compared to the more direct hydrogen bond or metal–ligand coordination bond, need a larger space and more flexibility to allow for the optimized stacking arrangement to occur.

16.2.5 Hydrogen Bonds

Hydrogen bonding is "the most reliable directional interaction in supramolecular chemistry,"[3] and its role in numerous macromolecular phenomena has been well studied. As a frequently employed electrostatic interaction with vast synthetic and theoretical precedent, a rigorous analysis of this interaction in its many forms is beyond the scope of this discussion on nanoscale design and is left to significantly more detailed treatments in many excellent reviews.[29,30] Important to understanding this type of interaction from a nanoscale design perspective, however, is the nature of the bond, the functional groups responsible for its occurrence, and the relative stabilities that come with different functional groups. Appropriately, these topics are covered here in general with specific examples used to highlight the discussion.

A hydrogen bond is formed when the hydrogen in a polar bond approaches the lone pair of electrons on an ion or atom.[8] A polar bond to hydrogen occurs when the hydrogen is attached to an atom of high electronegativity, such as nitrogen, oxygen, or fluorine. Because hydrogen atoms have no inner core of electrons, the pull of electron density from them exposes a significant positive nuclear charge to interact electrostatically with nearby electron density. This is further strengthened by the very small size of the hydrogen atom. The electronegativity difference between carbon and hydrogen is small enough that a significant dipole is not produced, resulting in very weak hydrogen bonds involving C-H bonds. The strength of the hydrogen bond is determined by the polarity of the bond in which the hydrogen is covalently bound and the electronegativity of the atom to which the hydrogen is electrostatically attracted. Hydrogen bonds can be divided into *strong* (20–40 kJ/mol) and *weak* (2–20 kJ/mol) interactions,[3] each of which is important to certain types of supramolecular assembly.

Hydrogen bonds can be used to stabilize structures ranging from small dimers to extended arrays of massive molecules. The most commonly encountered strong molecular hydrogen bonds tend to favor the use of oxygen or nitrogen, a result of their large electronegativity differences with hydrogen. Also important for MBB assembly is the ability of oxygen and nitrogen to covalently bond to more than one atom, allowing them to be incorporated into larger molecular frameworks. This is in contrast with fluorine, which can only be used to terminate a covalent framework, making its role in typical hydrogen-bonded nanostructures rather limited. There are numerous combinations of hydrogen bonding interactions that can be incorporated into a covalent framework from the available organic precedent for the manipulation of functional groups such as O-H, C = O, N-H, C = N, COOH, NH$_2$, and NOO$^-$ (Figure 16.7A). Weak hydrogen bonds have also been shown to play important roles in the shapes and stabilities

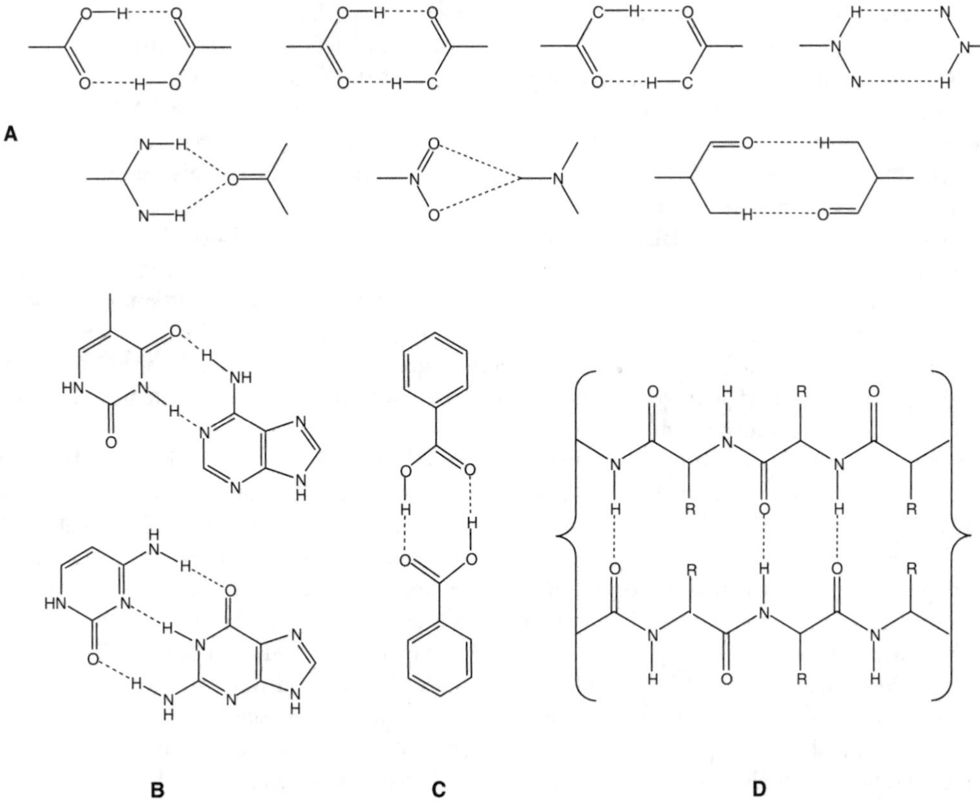

FIGURE 16.7 Hydrogen bonded structures. (A) A selection of hydrogen-bonded structures. (B) Thymine–adenine (top) and cytosine–guanine (bottom) base pairing. (C) Hydrogen-bonded carboxylate dimer. (D) Portion of hydrogen bonding network in peptide β-sheets.

of macromolecular assemblies and crystals that do not include functional groups capable of strong hydrogen bonds.[31,32] These weaker bonds include interactions such as OH···π and NH···π.

A small selection of relevant hydrogen-bonded complexes is provided in Figures 16.7B through 16.7D. The most familiar hydrogen-bonding interaction, outside of ice crystals, occurs in the nucleotide base pairs of DNA, where strongly bonding functional groups are incorporated into small, aromatic heterocycles. The bonds form so as to stabilize particular pairs (thymine/adenine and cytosine/guanine) in the formation of the double-helical structure. Strong hydrogen bonding also occurs between the C=O and N-H groups of amino acids in the formation of the secondary structure of proteins (i.e., α-helices and β-sheets). Artificial superstructures employing hydrogen bonding include simple dimers, linear arrays, two-dimensional networks, and, with the correct covalent framework, three-dimensional structures.

There are many advantages to using hydrogen bonding in the formation of macromolecules and extended arrays. First, these interactions are both self-assembling and self-directing. Stable structures based solely on electrostatic interactions are free to form and dissociate with relatively little energy required. Unlike covalent bonds, which require specific reaction conditions, hydrogen bonds (and other electrostatic interactions) require only the appropriate medium through which to form stable structures. The spontaneity of protein secondary structure formation in aqueous media is, perhaps, the most remarkable example of this phenomenon. Second, there are many functional groups that can act as either hydrogen donors (X-H bond) or acceptors (lone pair). This availability comes from both an extensive synthetic precedent and a large number of different donors and acceptors that can be employed to customize the strengths of hydrogen bonds. Third, hydrogen bonds are typically directed interactions with small steric requirements. Whereas π-stacking requires both a large surface area and very specific electronic distributions in the aromatic rings, hydrogen bonds can form with molecules as small as hydrogen fluoride. Fourth, directional interactions such as hydrogen bonds are relatively easy to incorporate into larger molecules, provided the attached covalent frameworks are shaped correctly to allow the interactions to occur. The pairing of nucleotides in DNA are specific examples of where the selected covalent frameworks determine the optimum orientations of the hydrogen bonding interactions. In crystal engineering, many molecular architectures are based on the inclusion of known pairs of hydrogen-bonding functionalities into organic frameworks.[3] Another advantage that stems from the small size and unidirectional nature of the hydrogen bond is the ability to incorporate multiple interactions within a very small space. Again, base pairing in DNA is an example of where either two (A with T) or three (C with G) hydrogen bonds occur in small heterocytes (Figure 16.7B). The ability to incorporate multiple hydrogen bonds into a single framework also allows orientational specificity to be designed into a structure. Not only do nucleotides pair specifically according to the number of hydrogen bonds (A with T and C with G), but they form stable interactions in only one dimeric conformation.

The greatest limitation in hydrogen bonding comes from the relative stabilities of these bonds and the potential for such bonding throughout an ensemble of molecules. While certain interactions can be predicted to be most stable based on their conformation and functional groups, there are usually many other interactions that form the macromolecular equivalent of metastable structures in solution; and the directing of a single, preferential hydrogen-bonded framework can be difficult to predict or control. In polar solvents, such as water, this predictability becomes even more difficult. The local hydrogen bonds that form with aqueous solvation approach the strengths of many other hydrogen-bonding interactions. Although the formation of the DNA double helix in aqueous media is driven by entropy, the relative stability of nucleotide–water interactions is significant, providing local instabilities in the DNA double helix.[33] This same dynamic equilibrium in DNA between water–nucleotide and nucleotide–nucleotide interactions, however, is also partially responsible for its biological activity, as a DNA helix unable to be destabilized and "unzip" is poorly suited to providing genetic information. As with all of the bonding motifs discussed, the merits and limitations of hydrogen bonding in nanostructural design and formation are sometimes subjective; and the specifics of a system and its surroundings play important roles in determining the best choice of macromolecular stabilization.

16.3 Molecular Building Block Approaches

The overriding goal of the MBB approach is the assembly of nanostructures or nanoscale materials through the manipulation of a subunit by chemical methods or electrostatic interactions. The MBB is selected or designed with this manipulation in mind. The MBB is, ideally, divisible into one or more chemically or electrostatically active regions and a covalent framework, the purpose of which is simply to support the active regions of the subunit. With the division between covalent architectures and lower energy electrostatic systems in mind, the range of MBB designs can be bounded by those systems fabricated through only covalent bonds between subunits and those including only weak interactions between otherwise covalently isolated subunits. Appropriately, these two cases will be considered first. With the definition of the boundaries of what can be done with MBBs in the limit of structural interconnectivity requirements, intermediate systems that balance relative degrees of covalent and electrostatic character, including familiar biological systems, coordination nanostructures, and dendritic systems, are then considered.

16.3.1 Supramolecular Chemistry

Supramolecular chemistry is the science of electrostatic interactions at the molecular level. Direct correlations of structure and function exist between molecular chemistry and supramolecular chemistry, and many parallels can be drawn between the two that highlight the utility and importance of chemical design from noncovalent interactions. The range of covalent bonding and chemical functionalities within a molecular framework gives rise to a range of noncovalent interactions that can be used to form stable structures composed of many molecules. The chemistry of the covalent bond also allows for the engineering of electrostatic interactions. Just as a molecular chemist would employ reaction conditions and various functionalities to direct a particular chemical synthesis, the supramolecular chemist employs the surroundings and the entire molecule to tailor stabilizing interactions into a macromolecular framework. The energies of the interactions between molecules in supramolecular design are far weaker than those interactions within the molecular framework. Consequently, in supramolecular chemistry, the entirety of the covalent framework of the molecular subunit is treated as a whole; and the assembly of the supramolecular array progresses from the MBB just as the synthesis of a molecule is treated as an assembly of discrete atoms.

Supramolecular chemistry is, however, unique in many respects. The formation of new structures in both molecular chemistry and supramolecular chemistry is based upon understanding and predicting chemical interactions. In the case of molecular chemistry, structure formation is based on reaction centers with the covalent framework of the molecule altered to form a new structure. In supramolecular chemistry, superstructure formation is based on interaction centers in which the covalent framework of the molecule as a whole remains unaffected by the stabilizing interactions that occur beyond it. In molecular chemistry, the covalent frameworks of the precursor molecules must be altered through energy-intensive chemical manipulation. Reactions may be self-directing based on the positions of functional groups and reaction conditions; but the actual formation of a molecule requires some form of external manipulation, such as a naturally occurring enzyme or catalyst, or a particular reaction pathway to facilitate the breaking and formation of chemical bonds. A self-assembling molecular reaction is then a fortuitous occurrence of both the correct molecules and the correct chemical environment. In supramolecular chemistry, interactions between molecules are self-directing and spontaneous in solution. Because significant changes to the covalent framework of the subunits are not part of the superstructure formation process, stabilization from noncovalent interactions is based only on localized chemical environments. Provided that the stabilizing interactions between subunits are sufficiently large, molecules will spontaneously form into larger structures. The goal of supramolecular chemistry is the application of this spontaneity in the rational design of larger structures. The total stabilization energy for a supramolecular array from its component molecules is smaller than the total covalent energy between a molecule and its component atoms. Consequently, the formation and degradation of a supramolecular array is far less energy-intensive than the formation and breakage of covalent bonds. In many instances, stabilization in supramolecular designs benefits from the similarities in energy between MBB interactions and the energy of the surroundings, including the stability gained from the interactions between subunits and solvent molecules. The dynamics of proteins in aqueous media are excellent examples of where a macromolecular structure and the environment can be used in concert to create both stability and function in chemically massive molecules.

Supramolecular chemistry broadly encompasses the use of any electrostatic interaction in the formation of larger molecule-based structures. As such, any system that is based on interactions *beyond the molecule* falls under the supramolecular heading. Supramolecular chemistry, as it is then loosely defined, is an outgrowth of many related disciplines which serve to study phenomena beyond the molecular boundary, including biochemistry, crystal engineering, and significant portions of inorganic chemistry. Much of our initial understanding of molecular interactions comes from the study of naturally occurring structures in these well-established fields. To study the secondary structure of proteins and DNA is to study specific examples of the supramolecular aspects of biochemistry. The functions of these macromolecules in the intracellular matrix are based on noncovalent interactions, including the enzymatic activity of proteins on a substrate, the binding of cations to a protein, the dynamics of DNA duplication, and protein folding. The periodic lattices of many molecular crystals provide examples of how electrostatic interactions direct the alignment of molecules in the solid state. For instance, the unique properties of ice crystals relative to liquid water demonstrate how intermolecular interactions can be just as important as intramolecular interactions in defining structure and properties.

As a unique discipline, supramolecular chemistry emphasizes the design of novel molecular architectures based on the rational incorporation of electrostatic interactions into molecular frameworks. The discussion of supramolecular chemistry here will emphasize the design of macromolecules using only electrostatic interactions. Specifically, supramolecular structures formed from hydrogen bonding and π-interactions are detailed. Dative-based designs, while offering a number of attractive properties for noncovalent stabilization, have seen limited application for the design of nanoscale architectures. The division between entirely electrostatic assemblies and mixed covalent/electrostatic assemblies is stressed when possible to examine how specific noncovalent interactions can be used as the primary means to define the shape of supermolecular structures. Specific instances of nanostructure formation employing both covalent and noncovalent bonding are addressed subsequently in two sections, where the importance of both structure and function can be considered in context. The interactions between metal centers and organic ligands for the formation of coordination nanostructures is also treated as separate from this general discussion in order to provide emphasis on this particularly well-defined segment of supramolecular chemistry.

16.3.1.1 Hydrogen Bonding in Supramolecular Design

Hydrogen bonding is used extensively in supramolecular chemistry to provide strength, structural selectivity, and orientational control in the formation of molecular lattices and isolated macromolecules. The advantages inherent to hydrogen bonding interactions are universal among the different areas of supramolecular chemistry, whether the application is in the stabilization of base pairs in DNA or the alignment of synthons in infinite crystal lattices. The functional groups most familiar in hydrogen bonding have significant precedent in organic chemistry and are, therefore, readily incorporated into other molecules through chemical methods.[3] The complementary components of a hydrogen bond can be incorporated into molecules with very different chemical and electronic properties. In benzoic acid, for example, a polar carboxylate group is covalently linked to a nonpolar benzene ring to form a molecule with two distinct electrostatic regions (Figure 16.8A). The formation of benzoic acid dimers in solution is strongly directed by the isolation of polar and nonpolar regions in the individual molecules and the stability that comes with forming hydrogen bonds between the highly directing donor/acceptor groups.[34] The predictability of hydrogen bond formation in solution and the directional control that comes with donor/acceptor pairing allows for MBBs incorporating these functionalities to be divided into distinct structural regions based on their abilities to form strong hydrogen bonding interactions. This simplifies

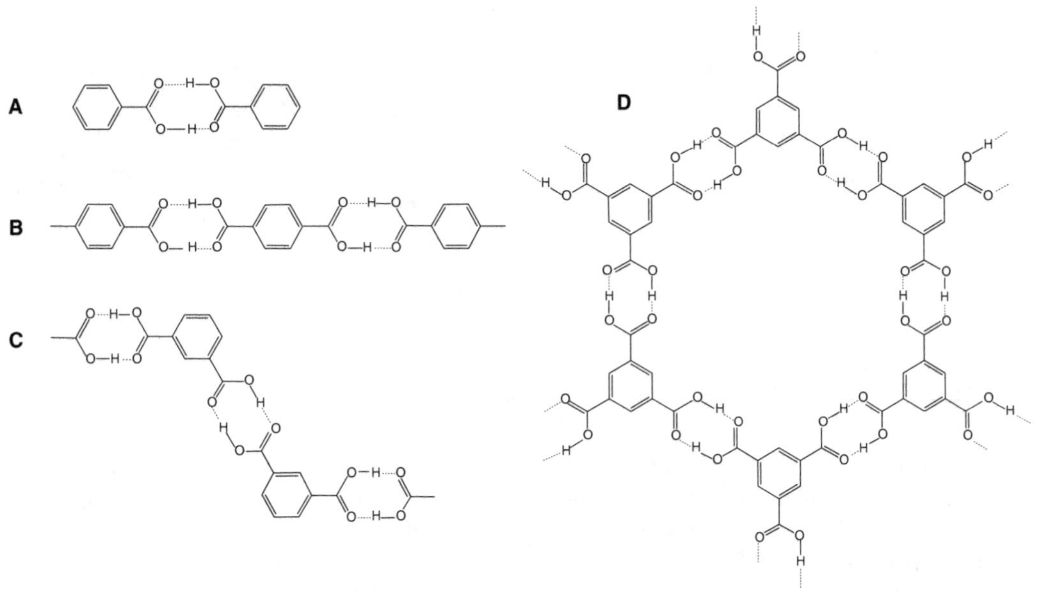

FIGURE 16.8 Hydrogen-bonded aromatic/carboxylic acid assemblies. (A) Benzoic acid dimers. (B) Linear chains of terephthalic acid. (C) Chains of isophthalic acid. (D) Hexagonal arrays of trimesic acid.

the design process in molecules that are tailored to form stable interactions only in specific regions, allowing for the identification of structural patterns in macromolecular formation.[35] The general shape of the nonpolar backbone in benzene, for instance, creates a geometric template from which it becomes possible to predict the shapes of the larger macromolecular structures that result from hydrogen bond formation. To illustrate this template approach with molecular hexagons, a series of examples of both arrays and isolated nanostructures are considered below that use only hydrogen bonding and the shapes of the subunits to direct superstructure formation.

16.3.1.1.1 Crystal Engineering

The hydrogen bond has been used extensively in the design of simple molecular crystals. Crystal engineering has been defined as "the understanding of intermolecular interactions in the context of crystal packing and in the utilization of such understanding in the design of new solids with desirable physical and chemical properties."[36] Many researchers in the field of crystal engineering have been guided by the very predictable and directional interactions that come with hydrogen bonding in its various forms. The cognizant design of extended arrays of hydrogen-bonded structures in molecular crystals is made possible by the broad understanding of these interactions in other systems, especially from the formation of biomolecules and small guest–host complexes. Among those examples that best demonstrate the rational design of molecular crystals from simple subunits and well-understood interactions are the aromatic/carboxylate structures (Figure 16.8). From the very predictable dimerization of benzoic acid in solution comes a number of similar structures whose geometries are singly dependent on the shape of the hexagonal benzene core. Isophthalic acid[37] and terephthalic acid[38] are simple extensions of the benzoic acid motif that form hydrogen-bonded chains (Figure 16.8B, C). The hexagonal trimesic acid structure[39] stems directly from the placement of strong hydrogen bonding groups on the benzene frame, yielding two-dimensional arrays of hexagonal cavities in the solid state (Figure 16.8D). The same chemical design has also been considered with the amide linkages, in which a higher connectivity is possible through four hydrogen bonding positions (Figure 16.9). Linear chains of benzamide[40] form from each amide linkage, forming four strong hydrogen bonds to three adjacent benzamide molecules. The repeating subunit of these chains is a dimer very similar to that of the benzoic acid dimer, with additional hydrogen bonding groups extending perpendicularly from each dimer to facilitate linear connectivity to other pairs (Figure

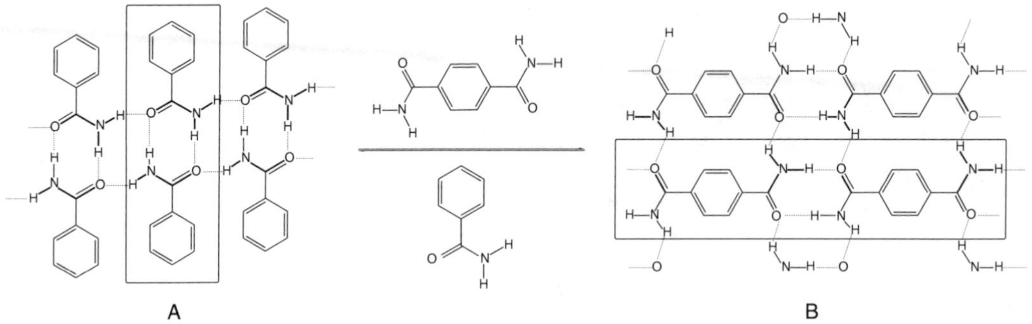

FIGURE 16.9 Hydrogen-bonded structures from amide linkages. (A) Benzamide dimers form linear chains. (B) Terephthalamide forms highly connected sheets. The corresponding aromatic/carboxylate motifs are enclosed in boxes.

16.9A). Planar sheets of terephthalamide[41] form from the same extended linear chain motif found in *para*-substituted terephthalic acid (Figure 16.9B). Again, the perpendicular hydrogen bonding groups direct the connectivity of these linear chains into two-dimensional sheets. The commonality among all of these benzene-based MBBs is the division between the rigid alignment of the functionalities on a covalent framework and the positions of the interaction centers beyond the molecular frame.

16.3.1.1.2 Supramolecular Structures

A number of isolated supramolecular structures are known that use only hydrogen bonding to direct their formation. In some instances, this has been accomplished through modifying the substituents on array-forming MBBs to promote the formation of isolated systems. While unsubstituted isophthalic acid in solution was found to form linear ribbons in the solid state, the addition of bulky substituents at the *meta*-positions of the two carboxylic acid moieties resulted in the formation of isolated molecular hexagons — structures that mimic exactly the hexagonal cavities formed through hydrogen bonding in the trimesic acid arrays[42] (Figure 16.10). For greater control in the formation of complex supermolecules, the engineering of highly directional hydrogen bonding regions is often required. The customization of interactions between MBBs is performed by either attaching more than two hydrogen bonding pairs onto the same framework (to prohibit free rotation when single σ-bonds are used to connect the donor/acceptor assemblies) or by embedding two or more functionalities directly into a covalent framework (Figure 16.11). In both routes, the resulting structures are no longer limited to stable designs based solely on single donor/acceptor pairs or sets of hydrogen bonding fragments isolated to σ-bound molecular fragments.

By fixing the positions of the donor and acceptor groups in a framework, the connectivity of subunits must occur with orientational specificity, creating what are commonly known as *molecular recognition* sites. In hydrogen-bonded systems, each interaction region of the molecular recognition site is clearly identified by the arrangement of the donor/acceptor groups, such as shown in Figure 16.11B. For crystal engineering and nanostructure formation, where stability and the fitting of subunits to one another define the shape of the entire system, both the hydrogen bonding arrangement and the shapes of the molecules are important to the success of a molecular recognition site (Figure 16.11C).

Two specific MBB designs have been extensively used together to illustrate the roles of structure and orientation in the formation of hydrogen-bonded nanostructures. Cyanuric acid and melamine are two highly symmetric molecules with complementary hydrogen bonding regions along each molecular face (Figure 16.12). In solution, 1:1 mixtures of these molecules form insoluble complexes of extended hexagonal cavities[43] (Figure 16.13C). By the removal of a hydrogen bonding interaction from each molecule, two different assemblies have been shown to form. In both instances, cyanuric acid is converted into a barbituric acid-based molecule by the removal of one N-H fragment from the central ring, while the melamine structure is altered by the removal of one nitrogen atom from its central ring (Figures 16.13D and 16.13E). The formation of linear chains has been shown to be favored in the native structures and when the substituents on the MBBs are kept small[44] (Figure 16.13F). The addition

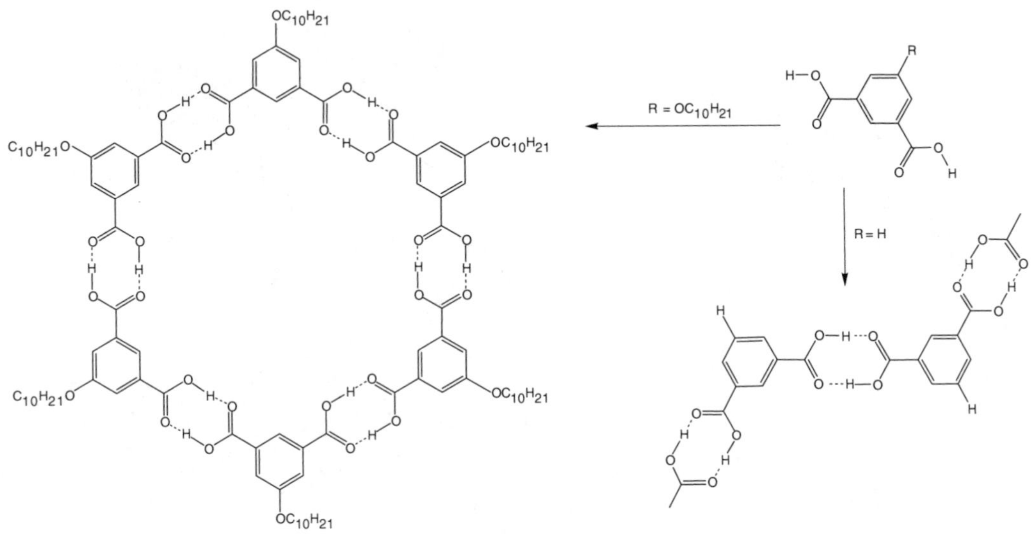

FIGURE 16.10 Isophthalic acid derivatives direct the formation of different hydrogen-bonded networks.

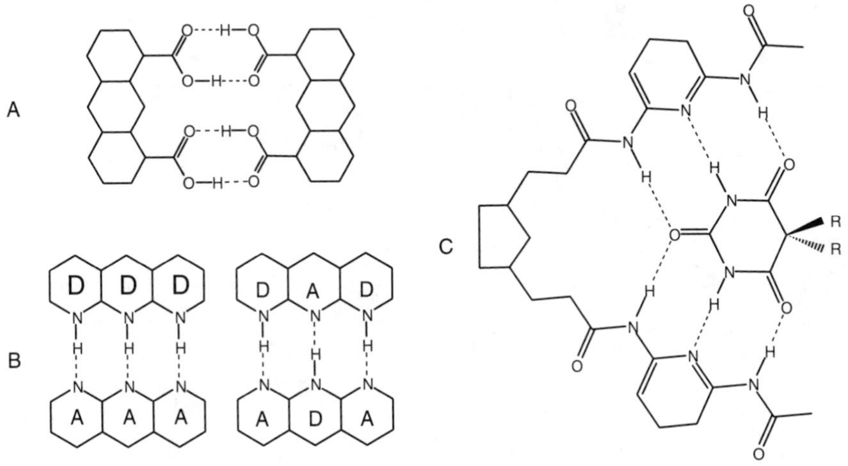

FIGURE 16.11 Engineering orientational specificity into hydrogen-bonded structures. (A) Multiple interaction zones fix the orientation of guest–host complexes by prohibiting rotation. (B) Donor (D) and acceptor (A) interactions between hydrogen-bonded fragments embedded within molecular frameworks. (C) Size and orientation direct the binding of barbituric acid within a molecular recognition zone.

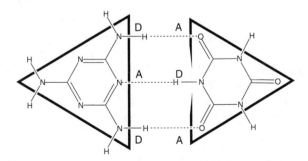

FIGURE 16.12 Complementary DAD:ADA hydrogen bonding in melamine (left) and cyanuric acid (right).

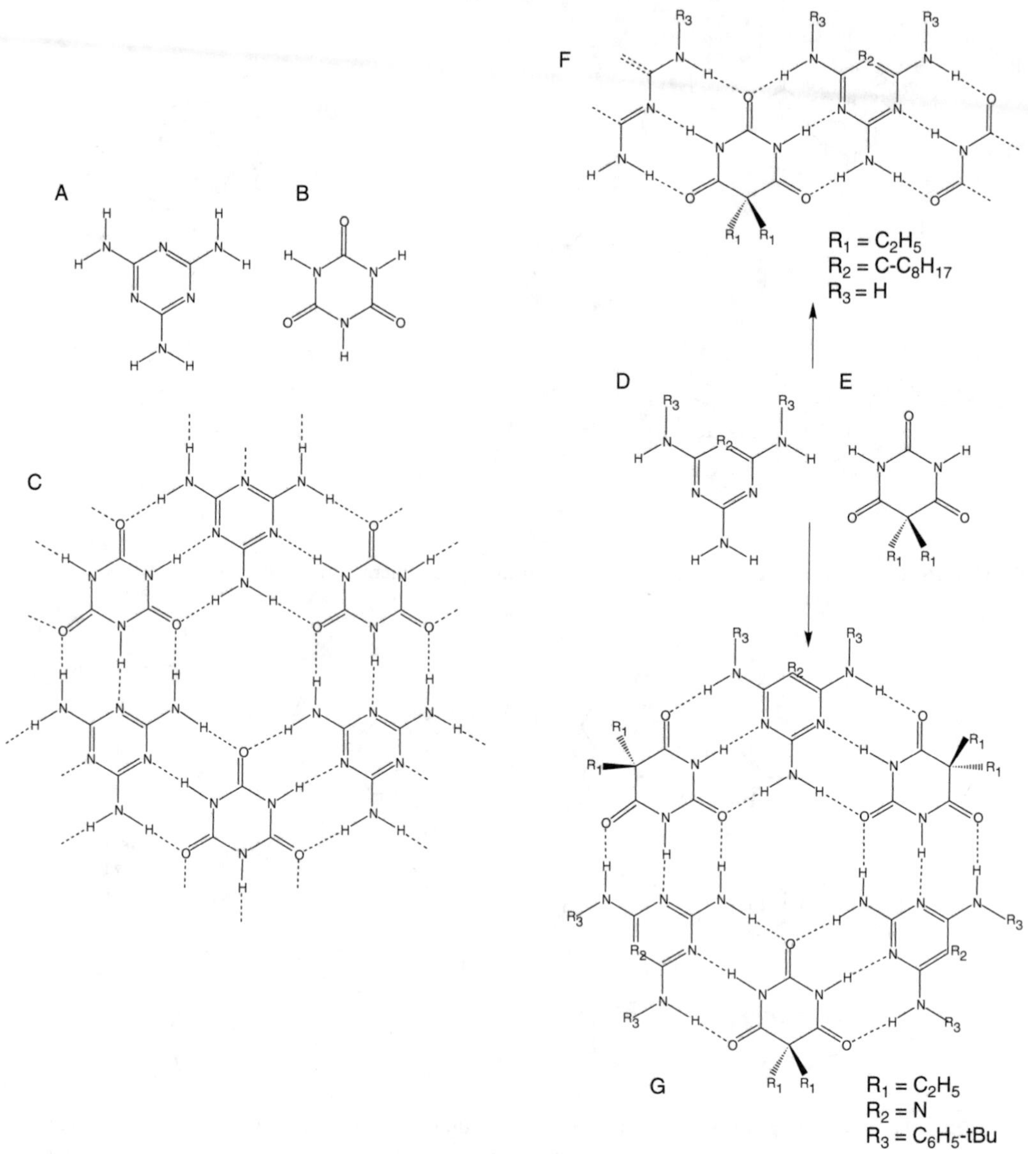

FIGURE 16.13   1:1 mixtures of A and B form extended arrays. Structures utilizing D and E form either linear chains (F) or supramolecular hexagons (G) depending on the choice of R groups.

of bulky substituents to the subunits (similar to the method used to form hexagons of isophthalic acid) directs the hexagonal species shown in Figure 16.13G to self-assemble in solution.[45]

These supramolecular designs are easily rationalized from the shapes of the hexagonal frames to which hydrogen bonding fragments are attached. Hydrogen bonding has been used frequently in the design of smaller guest–host interactions and molecular recognition sites, with much of this work derived from extensive biochemical precedent. Subsequent sections on biomimetic designs and dendrimers illustrate a few of these specific instances of isolated hydrogen bonding interaction in specific MBB designs.

16.3.1.2   π-Interactions

The use of π-interactions has been shown to be important for a number of biological and molecular assembly applications. In biological structures, π-stabilization and the hydrophobicity of the aromatic

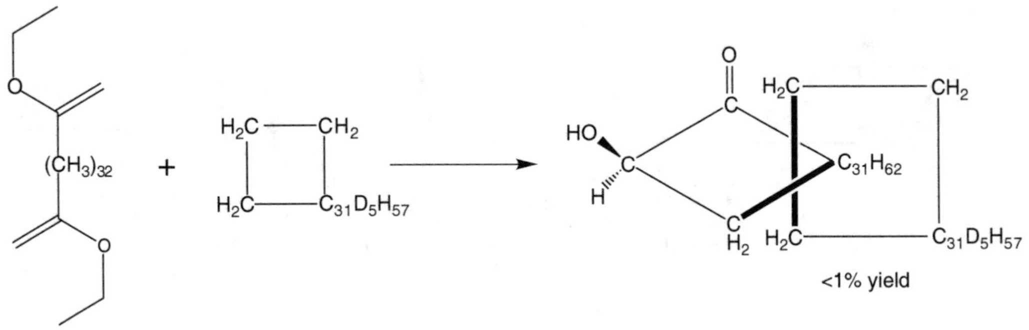

FIGURE 16.14 The first reported catenane.

rings both contribute to the formation of secondary and tertiary (aggregate) structure in DNA and proteins. The stability gained from stacking π-systems with complementary electron densities has been used as a driving force for a number of crystal engineering-based structures. The herringbone stacking pattern of aromatic π-systems with peripheral substituents is a very familiar motif in crystal engineering and has been shown to be responsible for the observed packing of many molecular crystals.[3] For the formation of supramolecular assemblies, however, the role of the π-interaction as a singular driving force is rather limited. Interactions with the π-systems of small aromatic groups are difficult to utilize because the energies of the different orientations can be very similar. Consequently, many supramolecular structures employing π-stacking interactions either use π-stacking in conjunction with other interaction types or use π-π-interactions between highly polarized species to direct the formation of supramolecular structures. Three specific examples are discussed below to illustrate how π-interactions can be employed in the electrostatic-based supramolecular formation of nonbiological structures.

16.3.1.2.1 Catenanes

Catenanes are a unique class of supramolecular structures formed by the interpenetration of two or more macrocycles to form what is often referred to as a *topological bond*. The assembly of interlocked rings has been demonstrated both by statistical and directed techniques. The statistical method used to form the first isolated catenane[46] is shown in Figure 16.14 and gave very poor yields, demonstrating the limitations of self-assembly without direction from electrostatic interactions. The other types of catenanes have been synthesized with far greater success by relying on local stabilization from π-interactions in aromatic rings in conjunction with other electrostatic interaction types.

The formation of two coordination-based catenanes has been proposed to arise from guest–host interactions between π-systems (Figure 16.15). These two structures, identical except for the choice of metal center (either palladium(II) or platinum(II)), form initially as single-ring systems from 1,4-bis(4-pyridylmethyl)benzene. In the palladium(II) complexes,[47] concentration was found to play a key role in determining the relative populations of rings (low concentrations) and catenanes (high concentrations) at ambient temperatures. The equivalent platinum(II) catenane[48] was found to form irreversibly as a function of temperature. Here, raising the temperature of the system to break the strong platinum–nitrogen coordination bond opens the ring systems for monocycle insertion. In both cases, catenane formation is promoted by π-interactions between the ring systems that stabilize the molecular interlocks long enough to allow for the formation of the metal–ligand topological bond.

Perhaps the most familiar catenanes are those composed of paraquat–crown complexes[49] (Figure 16.16). In these systems, the interlocking of a neutral crown ether and a paraquat ring is directed and stabilized by two strong electrostatic interactions. First, strong hydrogen bonding between the crown oxygens and the acidic hydrogens on the aromatic rings of the paraquat serve to fix part of the paraquat within the crown ring. Second, strong π-π-interactions between the crown aromatic rings and the positively charged aromatic rings of the paraquat serve to direct the insertion of the crown ring into the open paraquat assembly prior to its covalent ring closure.

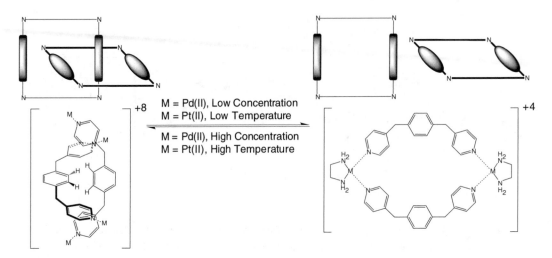

FIGURE 16.15 The effects of metal, concentration, and temperature on the formation of coordination catenanes of palladium(II) and platinum(II).

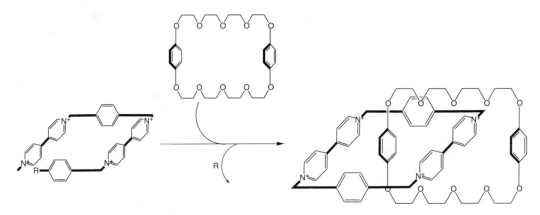

FIGURE 16.16 General crown ether/paraquat catenane and assembly mechanism.

16.3.1.2.2 Molecular Zippers

One of the most interesting pairings of edge-to-face π-π-interactions and hydrogen bonding comes in the form of molecular *zipper* structures formed from amide oligomers[50] (Figure 16.17A). The formation of double strands of the amide oligomers is rationalized based on ¹H NMR titration studies in the nonpolar solvent chloroform and the known structural features of oligomer chain pairs used in the dimerization study. In the general design scheme, the oligomer chains A and B have complementary binding regions capable of forming stable A:A, B:B, or A:B dimers. Based on the chain lengths of the two monomers, however, A:A and B:B dimers are found to not maximize the total possible number of π-π-interactions and amide hydrogen bonds (Figure 16.17B). The A:B dimer maximizes the total number of possible interactions along the entire length of the dimer complex, thereby promoting its formation in solution from equal mixtures of both A and B. Among the number of dimer systems examined, the commonalities to all are the increase in stability with increased oligomer lengths (providing more interactions between dimers) and the decrease in stability in polar solvents, such as methanol, which competitively bind to the polar amide functionalities and weaken the zipper structure.

16.3.1.2.3 Aedemers

The preferential face-to-face stacking of aromatic molecules with complementary ring charge densities has been demonstrated in many instances. The application of this phenomenon to nanoscale design

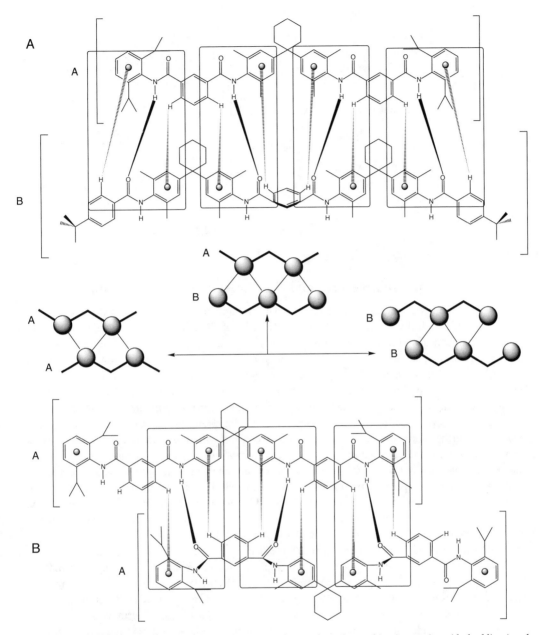

FIGURE 16.17 Molecular zippers from amide oligomers. The number of π-stacking interactions (dashed lines) and hydrogen bonding interactions (bold lines) is maximized with A:B dimers (top).

beyond the alignment of molecules in extended arrays has not, however, been exploited far beyond biological designs. The stabilizing interactions of two complementary π-stacking pairs have been shown to direct the formation of secondary structure in at least one other type of covalently linked macromolecule. The aedemers[51] are synthetic oligomers incorporating π-system donors and acceptors attached by long-chain tethers (Figure 16.18). In aqueous media, the strong π-π-interactions between the donor/acceptor pairs are enhanced by the respective hydrophobicity of the rings and the polar carboxylate groups attached to the tether. In water, the π-systems are found to self-assemble into single stacks of either two or three discrete donor/acceptor pairs.[51]

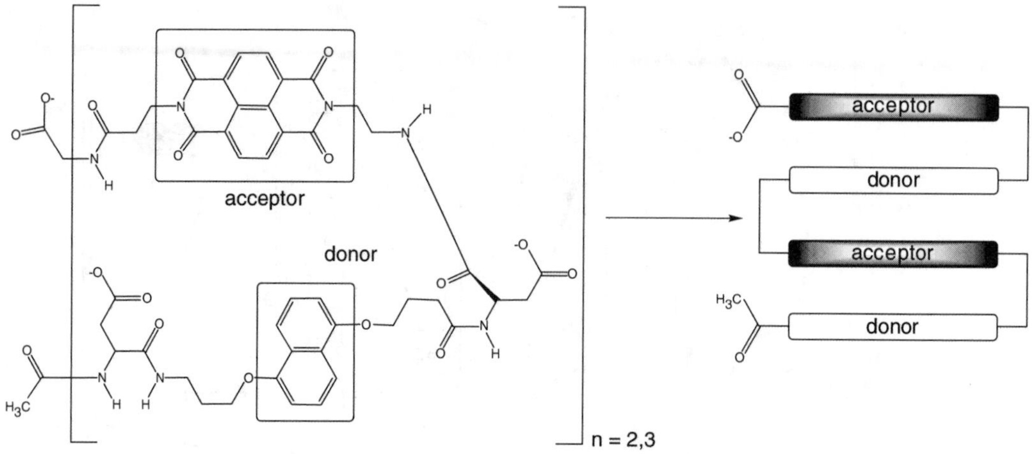

FIGURE 16.18   Aedemer molecules (left) and their directed stacking in water (right).

16.3.2   Covalent Architectures and the Molecular Tinkertoy Approach

The covalent bond is central to all MBB designs. The strength and directionality of these bonds define the shape of the subunits, thereby directing the formation of all larger structures stabilized by covalent bonding or noncovalent interactions. Covalent bonds are typically insensitive to environmental variables, such as the choice of solvent or the ambient temperature. The electrostatic interactions used to stabilize multimolecular structures, in contrast, are often strongly affected by these environmental factors. Covalent bonds offer far greater positional specificity and structural invariance than their noncovalent analogues. The chemical reactions used to form covalent bonds occur preferentially at specific positions on a molecular framework through the placement of suitable functional groups and the control of reaction conditions. Furthermore, covalent bond formation, in contrast with noncovalent interactions, is typically irreversible without concerted efforts to break them. Beyond the formation of the strong connections, the predictability of covalent architectures also allows for control of structure with great accuracy.

The fabrication of larger structures from covalently linked MBBs is based upon the use of individual subunits as rigid building blocks to incrementally build highly stable structures. Covalently linked nanostructures and covalent molecular scaffolding offer the same advantages that stable support structures provide at the macroscale. The shapes of rigidly bound structures are usually reliable over long periods of time. Covalent bond energies for familiar organic structures are an order of magnitude stronger than many of the electrostatic interactions currently employed for the formation of many supramolecular lattices. The continual breaking and reforming of these electrostatic interactions in supramolecular systems, while providing these structures with fault-tolerance and energy-driven self-maintenance, make their interconnectivity very sensitive to their surroundings. Covalently linked structures are themselves structurally stable under similar conditions, and any structural variance comes in the form of deformations instead of bond breaking and reforming. The chemistry involved in forming nanostructures from covalent bonds can be well defined and unidirectional with the correct choices of functional groups and reaction conditions. While the self-assembly methods of supramolecular chemistry provide a means to forming stable structures through the engineering of specific interactions into subunits, covalent connectivity can be directed with great positional control through the rational use of reaction pathways.

The formation of covalently bound nanoscale structures from molecular subunits is common in chemistry and materials science. The most common examples come from polymer chemistry, where small lengths of randomly oriented monomers become long chains of highly interwoven materials as the scale of the system is increased from Angstroms to nanometers and beyond. The formation of highly ordered, covalently bound nanoscale architectures and macromolecules is far less common in chemistry, as the controlled formation of nanoscale structures from covalent bonding is problematic in both of the routes currently proposed. In the engineering-based, top-down approaches, the positional specificity required for fabricating macromolecules from covalent bonds is simply not available, as the MBBs used for their formation are too small to be controlled and placed with any specificity. One might consider the assembly methodology of these approaches to be "too precise" for the selection of MBBs, as the desired level of control places severe restrictions on the design process and the choices of MBBs. In the bottom-up approaches of solvent-based chemistry and atomic manipulation, the reliability of positional accuracy becomes suspect in assemblies formed from rigid, highly stable connections. Errors in the placement of atoms or MBBs within a given framework, because they are irreversible without a level of chemical manipulation that also jeopardizes the structural integrity of the remaining covalent bonds, can potentially render a fabricated assembly useless with a single misplaced bond. Here, the idealized assembly process of solution-based methods may be considered as "too statistical."

To overcome the limits of both approaches, a fabrication process must successfully address positional control, connectivity, and the chemical manipulation of the reaction centers. The basis of supramolecular chemistry is the formation of a macromolecular assembly from weaker, noncovalent interactions; a wealth of examples demonstrates the validity of the approach.[36,52] The means to covalent supramolecular chemistry need not be dissimilar from this already proven approach to macromolecular formation. A covalent-based approach must, however, rigorously control the reaction conditions and the assembly progress of the larger structures to prohibit the unwanted interactions that are, in supramolecular chemistry, easily removed through the control of the ambient conditions. The scope of synthetic chemistry is narrowed considerably when the discussion is limited to the formation of nanoscale architectures from covalent bonding between MBBs instead of only the manipulation of covalent bonds within a single molecule. To illustrate the considerations and limitations of covalent-based nanostructure design from MBBs, one of the most well-developed chemical approaches is detailed below.

The "Molecular Tinkertoy" approach[53] to nanoscale scaffolding is based upon the treatment of molecules as simple, rigid construction components or *modules*. The features of the modules that are considered most important in this approach are those required for the construction of the assembly, such as the module length and the availability of suitable bonding positions on the module for connectivity to other subunits. Within the Tinkertoy paradigm, all of the required components and critical fabrication issues are based upon only covalent bonding. The engineering kit of the modular chemist consists of (1) rigid rod molecules of variable lengths, (2) connectors to act as corners or intersections for the scaffolding, and (3) a chemical means to control the assembly of the rods and connectors[53] (Figure 16.19). Such a kit at the macroscale is already familiar to any student of organic chemistry in the form of molecular

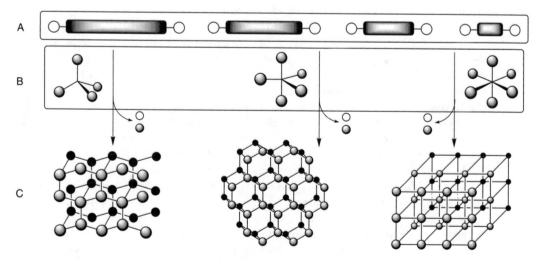

FIGURE 16.19 The engineering kit of the Tinkertoy chemist. (A) Rigid rods of various lengths. (B) Connectors and junctions. (C) A means to covalent assembly to create nanoscale scaffolding.

models, although the construction of scaffolding from the molecular kit is far more challenging than the fitting together of pieces of plastic. Within the context of covalent bonding, each of these three aspects of Tinkertoy design can be treated independently. The fabrication of rigid rods, for instance, can take inspiration from any chemical designs that result in linear structures, regardless of the choice of connectors or the development of the chemical pathways to assemble the nanostructures. The shape of a molecular scaffolding is defined by the connectors; and the engineering of a repeating structural motif, be it a simple cube or a diamondoid-based tetrahedral motif, is accessible based on the choice of the appropriate connector from among the available molecules that allow for the specific connectivity (Figure 16.19). The issue of chemical control becomes the most difficult of the three to handle, as the ordered assembly of extended arrays from simple rods and connectors cannot be controlled from the highly orchestrated procedures used for macroscale scaffolding construction, although the required chemistry is easily applied to the individual connector–junction reactions.

The concepts of the Tinkertoy approach are applicable to all structural features, including the formation of junctions and the assembly of the larger structures in solution. The most exhaustive treatment of the approach thus far has been for the linear, rigid rods used to define the dimensions of the scaffolding. While the number of molecules capable of acting as subunits for linear rods is large, the initial series of proposed subunits has been limited to a select set of twenty-four. The scope of this discussion is limited to the manipulation of these different modules for both the formation of linear rods and the design of molecular junctions. The chemistry of the twenty-four modules has been extensively developed and reviewed in the interest of firmly establishing the precedent for the first components of the engineering kit.[10] These twenty-four linear modules, shown in Figure 16.20, share a number of important characteristics that are briefly described below.

1. Stability

    The most important features to consider with respect to the environment of a nanostructure are the stability and reactivity of its components. Unless chemical functionality is required for an application, the best choices of MBBs are those that will react only during the formation of the covalent architecture. The subunit should, therefore, be inert with respect its chemical environment after assembly.

    The most common structures from among the initial MBBs that provide this level of chemical predictability are the saturated hydrocarbons (Figure 16.20A). These molecules rely exclusively on the use of strong σ-bonding between carbons and hydrogens to form rigid structures and are ideal for rigid rod fabrication. Their interconnected frameworks limit their flexibility while at the same time providing a molecular axis through which linear dimers, trimers, etc., can be formed via single σ-connections. The remaining saturated hydrocarbons (Figure 16.20D) differ from the cage structures by the inclusion of two bonding sites per pair of axial carbons. With these modules, either several σ-bonds can be used to form rigid structures, or both σ- and π-bonding can be used to create single connection points with restricted rotation (Figure 16.20D). The carboranes, a second class of molecules, display extreme stability and unique connectivity within a very small space (Figure 16.20B). The deltahedral framework of the cluster skeleton prohibits appreciable flexibility within the subunit, while the radial bonds of the apical carbons in $C_2B_{10}H_{12}$ and $C_2B_8H_{10}$ provide rigorously linear external linkages. Furthermore, these clusters have been shown in many instances to be remarkably stable compounds under very harsh conditions.[54] The remaining modules contain one or more π-electron systems. While π-systems are more susceptible to chemical reactions than saturated systems, much of this reactivity can be limited through the proper control of the surroundings. The molecules containing π-electrons are the only systems from the original series of subunits that provide a means to form stabilizing electrostatic interactions in solution (e.g., π-interactions, hydrogen bonding, or dative bonds).

2. Size

    Greater control of the size of a nanoscale assembly is possible by using many smaller subunits rather than few larger subunits. The twenty-four initial modules are among the smallest rigid

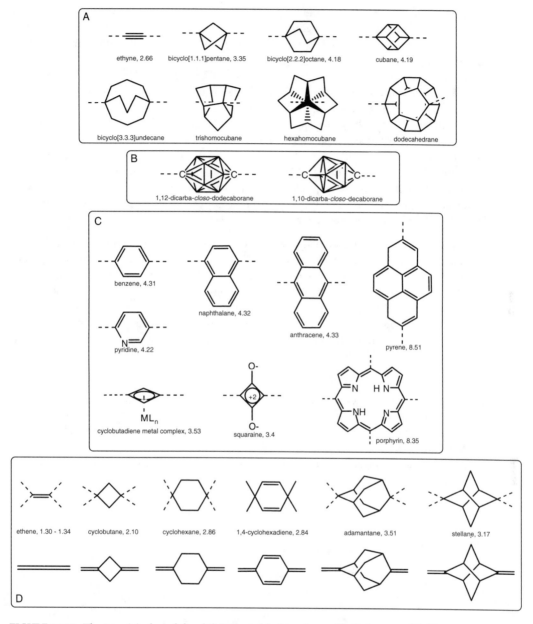

FIGURE 16.20 The 24 original modules. (A) Saturated hydrocarbons. (B) Carboranes. (C) Linear-connecting π-systems. (D) σ/σ and σ/π connectors. As applicable, names and incremental lengths (in Angstroms) are provided.

molecules known, and no single module crosses the nanometer threshold. Ethyne, for instance, is the smallest organic subunit available that provides linear connectivity on both ends through σ-bonding. A linear rod in a nanoscale scaffold can be fabricated from the available modules to "add up" to some required length. The rigid bonding within each module results in the structure having some fixed distance between the axial connection points which, when added to a typical single C-C bond length to account for the extra-module σ-linkage, defines a distance termed an *incremental length* (Figure 16.20). In order to construct a rod of some predetermined length, the only feature that needs to be considered from among the available modules is the incremental length between axial connection points. Having determined which modules are required to fabricate a rod of some predefined length, a chemical pathway can be employed based upon the known reaction chemistry of each subunit. As necessary, the general approach may be applied to any other molecules or combinations of molecular subunits for the fabrication of rods of an absolute length.

3. Chemical Precedent

The design of linear rods from the available modules is both flexible and straightforward. With few exceptions, chemical precedent exists for the syntheses and linking of all twenty-four modules.[10] Furthermore, the chemistry required for linking together different modules has also been demonstrated. Co-oligomers, chains of subunits composed of two or more different modules, are important both for customizing the lengths of the linear rods and for altering the solubility properties of the larger structures. Of particular importance in the linear rod treatment is the ethyne bridge. Ethynyl linkages are ideal for improving the solubility of molecular rods while minimizing the increase in chain length. A great deal of chemical precedent also exists for their inclusion into a number of modular structures.

Many linear molecular rods have been synthesized from the collection of modules. Beyond the formation of the rods is their connection to either two-dimensional junctions to form planar molecular grids or three-dimensional junctions to form molecular scaffolding. While covalent junctions have not been fully addressed, a number of the original twenty-four modules offer both structural flexibility and chemical precedent beyond their useful axial bonding. Specifically, the symmetry and connectivity of certain modules are appropriate for the formation of diamondoid, honeycomb (hexagonal), and cubic molecular lattices through familiar chemical manipulation. These lattices and the modules appropriate for their juncture are discussed below.

16.3.2.1 Diamondoid Scaffolding

Diamondoid structures are networks of tetrahedra in a molecular or macromolecular lattice (Figure 16.21). Within the lattice are two basic structural features. The first and most fundamental feature is the tetrahedral center (Figure 16.21A), to which four adjacent tetrahedra are attached. The smallest tetrahedral-based structural motif in the diamondoid lattice is the adamantanoid framework (Figure 16.21B). In the actual diamond framework, the tetrahedral centers are sp^3-hybridized carbon atoms, and the

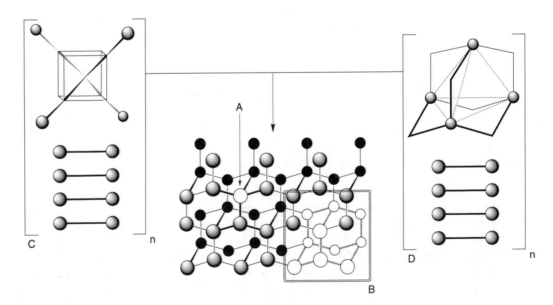

FIGURE 16.21 Diamondoid nanoscaffolding. (A) Tetrahedral module. (B) Adamantyl subunit. (C) Cubane assembly. (D) Adamantane assembly.

repeating motif is the adamantyl frame. The strength of diamond at the macroscale stems from the strength of the carbon–carbon σ-bonds and the extensive connectivity of the carbon atoms within the diamond network. The formation of MBB-based diamondoid frameworks has been explored in a number of coordination and supramolecular designs.[55–57] The noncovalent interactions within these diamondoid lattices offer reasonable strengths, the same high connectivities, and the spontaneous self-assembly of the subunits into rigid lattices. For the fabrication of extended arrays of diamondoid lattices, this self-assembly feature is particularly attractive, because the synthesis of molecular diamond has been limited to small molecules based more on incremental growth of adamantane frames[58] than the actual formation of rigid, covalent arrays.

Covalent diamondoid structures offer structural rigidity and controllable assembly intermediate between molecular diamond and the noncovalent MBB designs. The Tinkertoy approach offers a plausible means to the formation of such covalent diamondoid arrays. To construct these arrays with linear rods, the required molecular junctions must have tetrahedral symmetry elements that can connect through σ-bonds at the tetrahedral centers. Adamantane and cubane provide both the required tetrahedral symmetry elements for the placement of the linear rods and the synthetic precedent for their covalent attachment. Among the modules bicyclo[2.2.2]octane, bicyclo[1.1.1]pentane, bicyclo[3.3.3]undecane, trishomocubane, hexahomocubane, and dodecahedrane, structures with either tetrahedral centers or quasi-tetrahedral bonding positions (threefold rotation axes exist that include the axial connection points for the linear rods), either the chemistry has not been developed for tetrahedral assembly or the structures are too flexible to adequately control the diamondoid assembly. The control of functional group placement at the tetrahedral corners of both adamantane and cubane has been well developed, with many of these same functional groups employed for the syntheses of linear rods from these two modules. The control of tetrahedral adamantane functionalization has already been exploited for the formation of supramolecular building blocks in diamondoid lattice formation. In these MBBs, carboxylate groups are used to form strong hydrogen bonding interactions with neighboring adamantane frames, effectively extending the connectivity of the trimesic acid complex into a third dimension.[55] The covalent attachment of linear rod modules has also been demonstrated by way of a tetraphenyl adamantane derivative (Figure 16.22) that has been used as an MBB for subsequent macromolecular syntheses.[59]

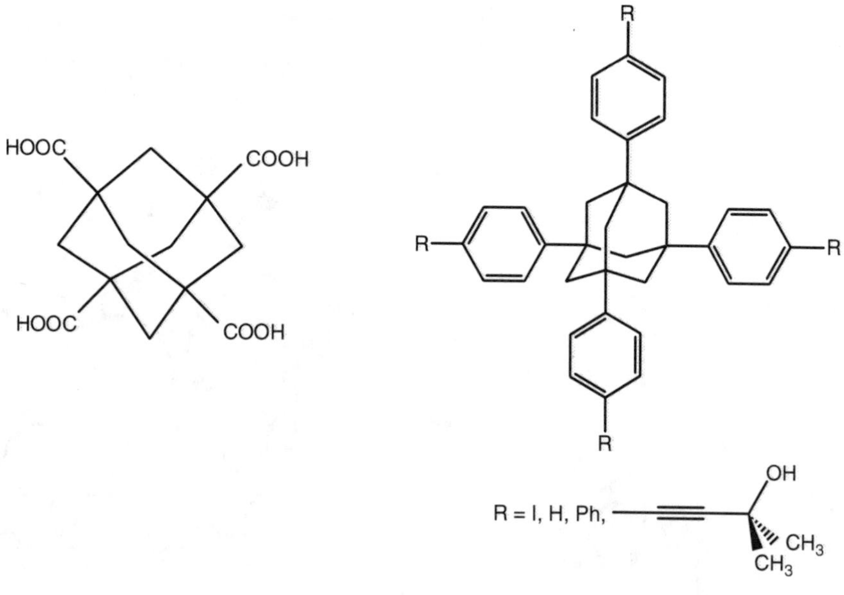

FIGURE 16.22 Adamantane-based MBBs for supramolecular design. (A) Adamantane-1,3,5,7-tetracarboxylic acid for supramolecular designs from hydrogen bonding. (B) Adamantane-based fragment with module linkages and known substituents.

16.3.2.2 Honeycomb Lattices

Macromolecular honeycombs require two different modes of connection (Figure 16.23). The hexagonal planar array is formed by the connection of linear molecules to triangular junctions. With the hexagonal plane formed, the vertical stacking of these structures is performed by attachment of the triangular junctions through chemical bonds perpendicular to the hexagonal plane. The ideal junctions for honeycomb designs are then molecules with trigonal bipyramidal symmetry, providing the ideal connectivity for linear rod structures in all directions. Such junctions are readily available from familiar coordination compounds. These structures, however, do not provide the structural stability of covalently bound junction/rod linkages. Although no single module addresses all of the design issues entirely, three are available that individually account for specific aspects of the honeycomb design.

Planar hexagonal scaffolding has already been addressed in the structure and chemistry of benzene. The placement of functional groups at the 1,3,5-positions of the benzene ring (Figure 16.24) yields the required triangular connectivity for the junctions, while an extensive chemical precedent for benzene functionalization makes the ring ideal for such applications. The propensity of trimesic acid to form

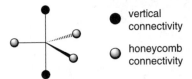

vertical
connectivity

honeycomb
connectivity

FIGURE 16.23 Honeycomb/vertical stacking connectivity in D_3-symmetric modules.

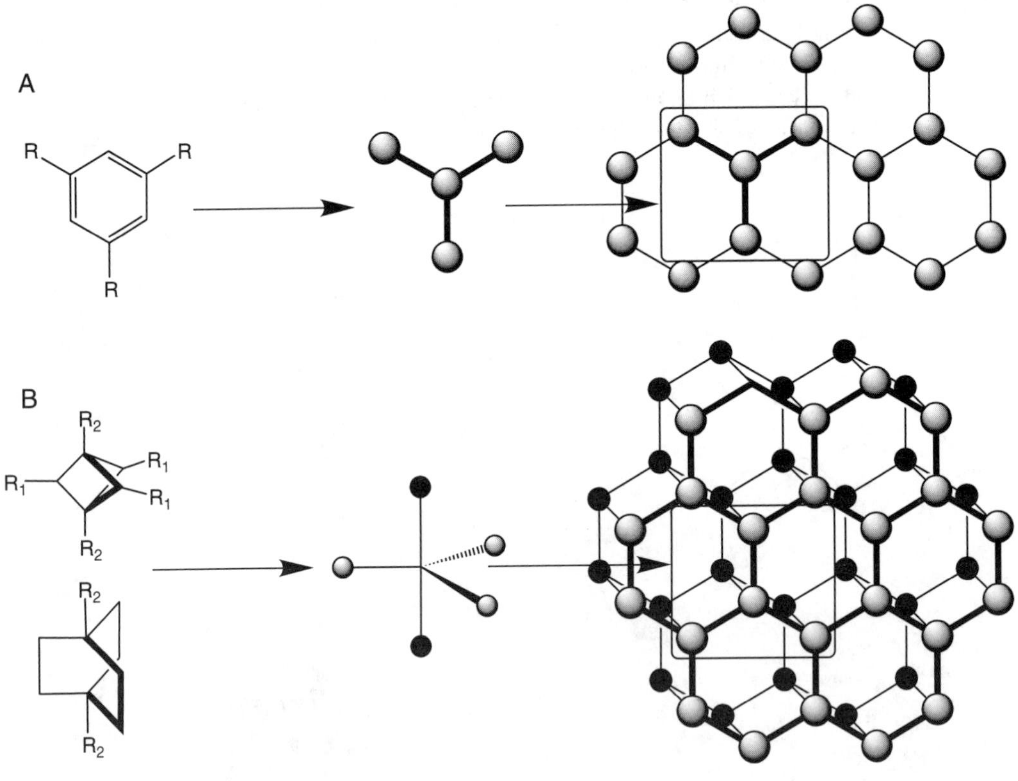

A

B

FIGURE 16.24 Examples of honeycomb scaffolding. (A) Planar structures formed from 1,3,5-substituted benzene rings. (B) bicyclo[1.1.1]pentane and bicyclo[2.2.2]octane modules as potential subunits for three-dimensional honeycomb structures.

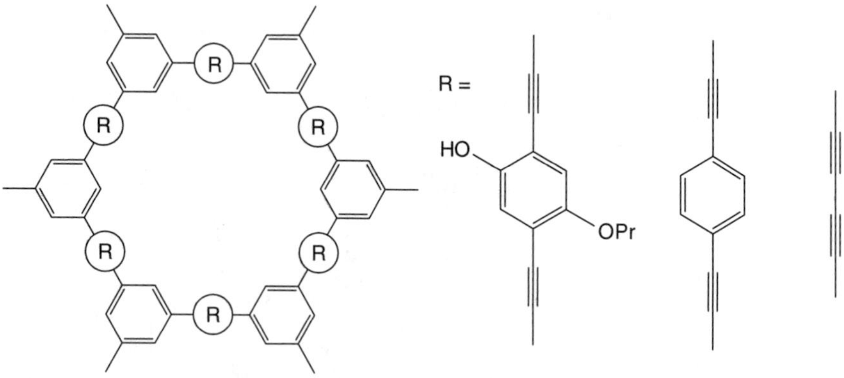

FIGURE 16.25 Isolated hexagonal macromolecules from benzene junctions.

extended arrays of hexagonal cavities from carboxylate-based hydrogen bonding interactions clearly demonstrates the importance of the geometry of the junction in directing the formation of the larger structures in solution (Figure 16.8). This same chemical design can be and has been employed successfully in a number of isolated benzene-based systems employing linear rods (Figure 16.25). Among the many known hexagonal macromolecules employing benzene junctions, many incorporate linear structures similar or identical to rod designs from the selected modules.[60–62]

The limitation of the benzene ring for scaffolding design is its planarity. While π-stacking interactions might be employed to form vertical honeycomb scaffolding, the covalent connection of hexagonal arrays into the third dimension is impossible with the benzene ring alone. From a structural standpoint, however, it is important to note that the only function of the benzene junction is to provide a triangular framework. Any other modules that incorporate equilateral triangles within their covalent frames will perform the same task. From among the remaining modules, the bicyclo[1.1.1]pentane and bicyclo[2.2.2]octane cages provide the correct symmetry and structural elements for the formation of planar arrays and vertical stacking through covalent bonds (Figure 16.25). The bicyclo[1.1.1]pentane is the better choice for designing such systems, as the structure is less flexible than the octane cage, and the carbons used for forming the hexagonal array from the linear rods have their available σ-bonds oriented in the hexagonal plane. The current limitation with the pentane cage for hexagonal designs is the synthetic precedent for the functionalization of the equatorial carbons, although these issues have recently received significant attention.[63]

16.3.2.3 Cubic Scaffolding

Idealized cubic lattices from the molecular Tinkertoy approach share a number of similarities with both the diamondoid and honeycomb designs. Structural connectivity in cubic lattices begins with octahedral junctions (Figure 16.26). Provided the junctions have ideal octahedral symmetry, the cubic lattices appear uniform with respect to all perpendicular sets of axes. The high symmetry of the idealized junction, as was found in diamondoid structures, permits the outward growth of the lattice from a single point by the addition of quantities of junction and linear rod without orientational preference. This simplifies the required control of the growth process relative to honeycomb structures, which have two different types of covalent connectivity that must be considered. Unlike the diamondoid structures, however, lattices formed from octahedral junctions have a very well-defined layering scheme along each axis. Therefore, a plausible growth process for the entire cubic lattice can mimic the same processes used for honeycomb growth, where a single layer is formed from two orthogonal sets of connections, while a third set perpendicular to the growing lattice plane remains unused until vertical connectivity is required. In instances where the growth process is selected to mimic the honeycomb methodology, the idealized octahedral junction can be separated into a square planar component and a perpendicular axial component. The selection of planar or vertical connectivity can be controlled during the growth process by chemical manipulation of the two distinct growth directions (Figure 16.26B).

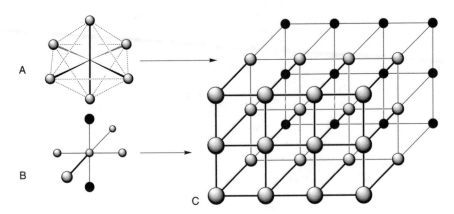

FIGURE 16.26 Cubic scaffolding. (A) Octahedral subunits for uniform structure growth. (B) Square-planar connectivity and vertical stacking connections for deformed cubic lattices. (C) A cubic lattice.

No single module provides the idealized octahedral connectivity required for uniform lattice growth in all directions. The design of two-dimensional square planar lattices can be readily designed from single σ-bond connectivity using porphyrins and cyclobutadiene metal complexes or double σ-bond/mixed σ–π connectivity using cyclobutane rings, stellanes, or adamantanes. Beyond the initial designs, however, the limited chemical precedent of a number of these modules prohibits their current usability. From these initial five modules, the porphyrins have been successfully employed in a number of rectangular and square planar arrays because of their extensive synthetic precedent and the availability of subsequent vertical connectivity through slight structural modification[64–65] (Figure 16.27). A number of linear rods have been used to connect porphyrins together, including ethynyl chains,[66] benzene chains,[67,68] chelating ligands,[69–72] and other porphyrins.[73–75] While the square planar framework has also been demonstrated

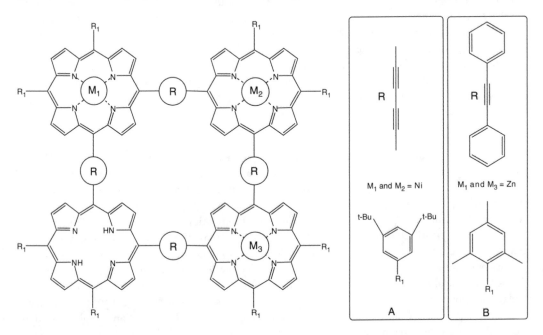

FIGURE 16.27 Porphyrin squares, connecting linear rods, and peripheral substitutions. (Set A from Sugiura, K., Fujimoto, Y., and Sakata, Y., A porphyrin square: synthesis of a square-shaped π-conjugated porphyrin tetramer connected by diacetylene linkages, *J. Chem. Soc., Chem. Commun.*, 1105, 2000; Set B from Wagner, R.W., Seth, J., Yang, S.I., Kim., D., Bocian, D.F., Holten, D., and Lindsey, J.S., Synthesis and excited-state photodynamics of a molecular square containing four mutually coplanar porphyrins, *J. Org. Chem.*, 63, 5042, 1998. With permission.)

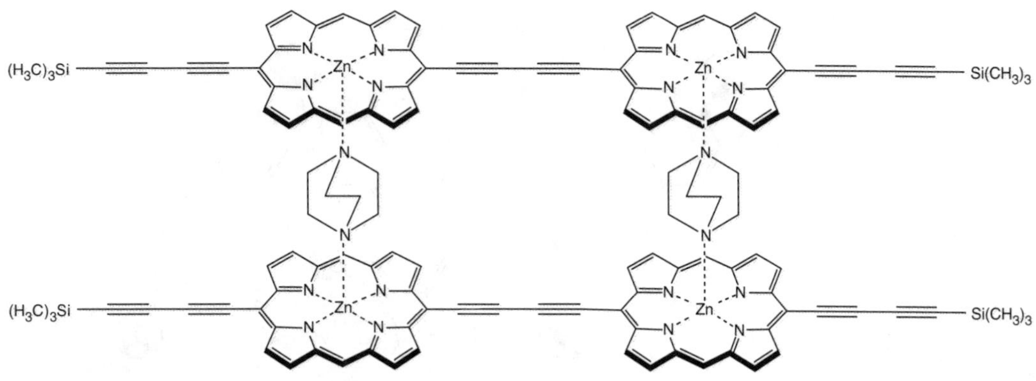

FIGURE 16.28 Metal–ligand coordination stacking design from porphyrins subunits.

with the cyclobutadiene metal complexes, the extension of these arrays into the third dimension is prohibited by the use of metal complexation to stabilize the highly reactive four-member ring.

Because no single module provides a chemically feasible route to vertical stacking after the formation of the square planar array, alternative stacking interactions must be employed for the formation of quasi-octahedral complexes. The porphyrins provide this added functionality by way of metal complexation within the central core. The coordination center within the porphyrin core then requires the use of metal–ligand complexation to form the vertical stacking interactions. The same directionality provided by covalent σ-bonding is still available from metal–ligand coordination, however, and the relative strengths of these stabilizing interactions can be controlled by the choice of metal. While a module-derived dipyridine structure is plausible based on the axial positions of the nitrogen lone pairs, the known vertical stacking motif has been performed with 1,4-diazabicyclo[2.2.2]octane,[76] the axial coordination analogue of bicyclo[2.2.2]octane (Figure 16.28).

The exclusive reliance on covalency for the fabrication of a nanostructure is not without important limitations. One limitation stems from the essentially irreversible formation of covalent bonds. In the formation of larger systems, extreme care must be taken to make chemical reactions as predictable and unidirectional as possible. The thermodynamically driven self-correction mechanisms of biological systems and supramolecular crystals cannot be used to repair an "incorrect" covalent bond without jeopardizing the structural integrity of the remaining structure. When an unwanted covalent bond forms, the means to correcting the error often involves harsh chemical manipulation. Thus, when a chemical route is chosen to correct some structural error, the pathway must be tailored to avoid reacting with any other part of the molecular superstructure. Also, because a chemical reaction is required to form a covalent bond, any structures employing a covalent bond are not strictly self-assembling. In a hydrogen-bonding network, for instance, the lattice forms due to electrostatic interactions between donors and acceptors. The stability that comes with these weak interactions may be small, but the formation of the larger network provides significant stabilization and the structure spontaneously forms. The formation of covalent architectures typically requires control of environmental conditions and subsequent purification of the desired product from the remainder of the reaction mixture.

A variety of chemical considerations associated with the synthesis and characterization of these structures has also been considered within the context of the Tinkertoy approach.[10] First among these considerations is the solubility of the progressively larger structures. The growth of larger structures is often limited by the ability to keep the assembly in solution. The chemical methods most likely to keep a larger structure in solution, such as the addition of side chains or the use of charged species, often have their own drawbacks. For instance, the application of these solvation techniques can affect the function of the nanostructure in unpredictable and undesired ways. With issues of solubility come problems of separation and purification. Such issues are familiar to biochemists, however, and many of the same techniques that have permitted the separation of biomolecules can also be applied to nanostructures.

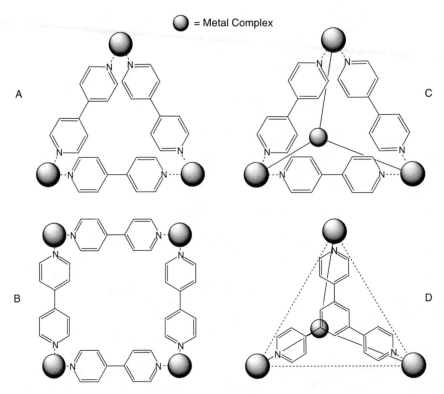

FIGURE 16.29 Metal–ligand structural motifs. A (triangle) and B (square) are two-dimensional structures with the ligands defining the sides. C and D (tetrahedrons) are three-dimensional structures with the ligands defining either the sides (C) or the faces (D).

16.3.3   Transition Metals and Coordination Complexes

One of the great advances in macromolecular design has been in the development of a variety of metal complexation motifs for the formation of two- and three-dimensional nanostructures. The design features here are based on the chemistry of small metal–ligand compounds, where the coordination requirements of the metal direct the attachment and orientation of ligands. The formation of larger geometric structures from metal–ligand compounds typically comes through the use of ligands with two or more separate metal-coordinating regions (Figure 16.29). In two-dimensional designs, the ligands typically constitute the sides of the structure while metal complexes define the corners. In three-dimensional designs, the ligands delineate the faces of the structures with the metals occupying the vertices. The chemistry involved in the formation of these nanostructures is often straightforward. The nanoscale assembly of coordination complexes is typically accomplished by the removal of labile ligands from some coordinately saturated metal complex in solution, a process greatly simplified by the relatively weak strengths of many metal–ligand bonds.[77] Coordination-based methods not only allow for the formation of symmetric molecular nanostructures but also provide for the formation of molecular cavities through ligand encapsulation pathways[78–80] (Figure 16.30).

The vast majority of coordination nanostructures have been based upon the use of chelating organic ligands, with either nitrogen atoms as the lone-pair donors or cyclic ligands with hydroxyl (-OH) groups used to provide metal connectivity through relatively weak covalent metal–oxygen bonds. Nitrogen-based ligands have been used far more often in coordination-based nanostructure design and are preferred among other ligand types for a number of structural reasons. The nitrogen atom is a close structural analogue and is isoelectronic with a covalent C-H fragment, making it quite versatile in the modification of organic ligands for metal complex formation (Figure 16.31). Whether incorporated into a saturated

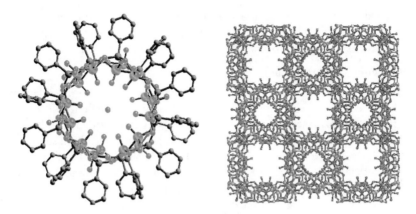

FIGURE 16.30 Structures with metal–ligand coordination cavities and channels.

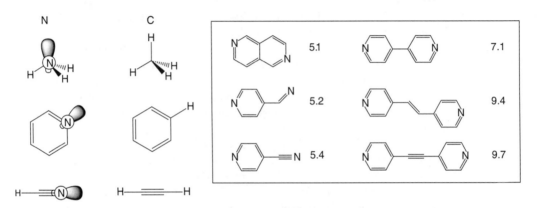

FIGURE 16.31 The inclusion of nitrogen (left) lone-pair donors into simple organic frameworks and a selection of N-N distances (in Angstroms) in common organic ligands.

aliphatic framework or directly into an aromatic ring to form a heterocycle, a nitrogen atom and a C-H fragment are nearly identical in terms of hybridization and geometry except that a hydrogen atom is required to form a complete electronic octet for carbon. The radial orientation of the lone pair from the backbone of many ligands provides accessibility to coordinating metals, while the σ-bond quality of the lone pair provides predictable, unidirectional coordination based on the geometry of the nitrogen atom in the ligand. These organic ligands can be designed such that the nitrogen lone pair is the only site on the ligand available for coordination to the metal center in nanostructure formation. The nitrogen lone pair, in the absence of a Lewis acid, becomes the reaction center for complexation only when the metal center becomes coordinatively unsaturated, typically by chemical methods too gentle to affect the ligand framework. This predictability in nitrogen–metal coordination comes from a vast synthetic precedent, ranging from the simple coordination of NH_3 to the complexation of multidentate ligands that serve to singly saturate the metal coordination sphere. Furthermore, the dissociation of the nitrogen-based ligand from the metal center has little effect on the stability of the ligand itself, providing a thermodynamic means for controlling the formation and self-maintenance of these systems. The chemical modification of these ligands also has significant structural implications for the resulting assemblies. Simple modification to the organic framework of these ligands can target macromolecular structures to within a few Angstroms of some specified size (Figure 16.31). Similar to the molecular Tinkertoy approach, ligands can be modified either step-wise through the addition of linear linkers (such as acetylene) or more subtly using nonlinear linkers, such as either flexible ring systems or saturated organic chains.

Similar to the study of structure and function in biomolecules (*vide infra*), much of the initial work in metal complexation involved the modification of known structures to create new structures. As the field has progressed, the catalogue of structures and reactions has increased to the point where trends and designs have been focused into general strategies for fabricating new structures. The two most actively investigated approaches to designing coordination architectures are discussed below.

16.3.3.1 Molecular Library Model

The molecular library model,[81] also known as the *directional-bond approach*,[77] is the metal–ligand analogue to the molecular Tinkertoy approach. The model addresses the design of nanostructures by using a set of molecular fragments encompassing a wide range of geometric patterns for the fabrication of two- and three-dimensional structures. In this approach, a *geometric fragment* is simply some subunit of a larger structure, such as a corner, a vertex, or a side. To classify a ligand or metal complex into a particular fragment category, a structural analysis is performed to determine the angles among all available coordination sites in the molecular framework. The choice of ligands is typically limited to rigid molecules with monodentate coordination modes (single lone pairs) in order to improve the predictability of the method for nanoscale design.[82] The number of candidate ligands is very large, however, and the restriction to molecules with limited degrees of freedom does not significantly affect the flexibility of the method. The rigidity of both the ligands and the metals is used only to restrict the choices of geometric fragments for particular designs, and a small amount of flexibility in the ligands and metal coordination sphere is expected in the assembly process. An important aspect of this approach is that both ligands and metals can be used as the fragments to form a structural feature. A nonlinear or multi-branched ligand, for instance, can be used as a corner or a vertex just as a metal with axial coordination sites can be used as a side. It is the higher coordination of ligands to a metal center that sets the metal apart from organic systems, however, and the metal is most frequently employed as the more complicated geometric fragment.

The range of available ligands and metal complexes has been divided into two libraries based on the dimensionality of the desired structure[81] (Figure 16.32). For the design of two-dimensional nano-structures, such as regular polygons or polycyclic assemblies, the classification of doubly connecting, or *ditopic*, geometric fragments requires only three points. In the ligand, these points are composed of two lone-pairs and the center of the covalent framework of the ligand (Figure 16.33). In the metal complex, these three points are the two coordination sites for the connected ligands and the metal atom (Figure 16.33). The internal angles of the desired nanostructure then determine which fragments can be used for its fabrication. For the fabrication of cyclic polygons with three to six sides, the internal angles and combinations of geometric fragments required are summarized in Figure 16.32 (A–I). It is important to note that these ditopic classifications define only individual sets of binding angles within a molecule. Within a molecule used as a geometric fragment in a larger structure, it is possible to have independent sets of binding angles. Consequently, structures with multiple planar rings are possible (Figure 16.33).

Three-dimensional nanostructures are fabricated from combinations of tritopic and ditopic geometric fragments. Symmetric three-dimensional structures resulting from various combinations of tritopic and ditopic fragments are shown in Figure 16.32 (J–M). The design strategy for new nanostructures is the same in both two- and three-dimensional systems, except the additional level of complication of three-dimensional structures requires more elaborate geometric fragments for the assembly. In both library sets, the linear linkage serves the important roles of length extender and coupler for identical fragments. It should be noted that length is not a factor considered in the classification process. Modifying the length of a structure is a matter of either modifying the molecular bridge between coordinating regions of a ligand or using linear subunits of the appropriate length with metal complexes at the corners (two dimensions) or vertices (three dimensions).

16.3.3.2 Symmetry Interaction Model

The symmetry interaction model[80,83] is founded in the understanding that many highly symmetric, naturally occurring structures are formed as a consequence of incommensurate lock-and-key interactions between the subunits.[84] The method, as applied to metallocycles, is then retrosynthetic in principle, using

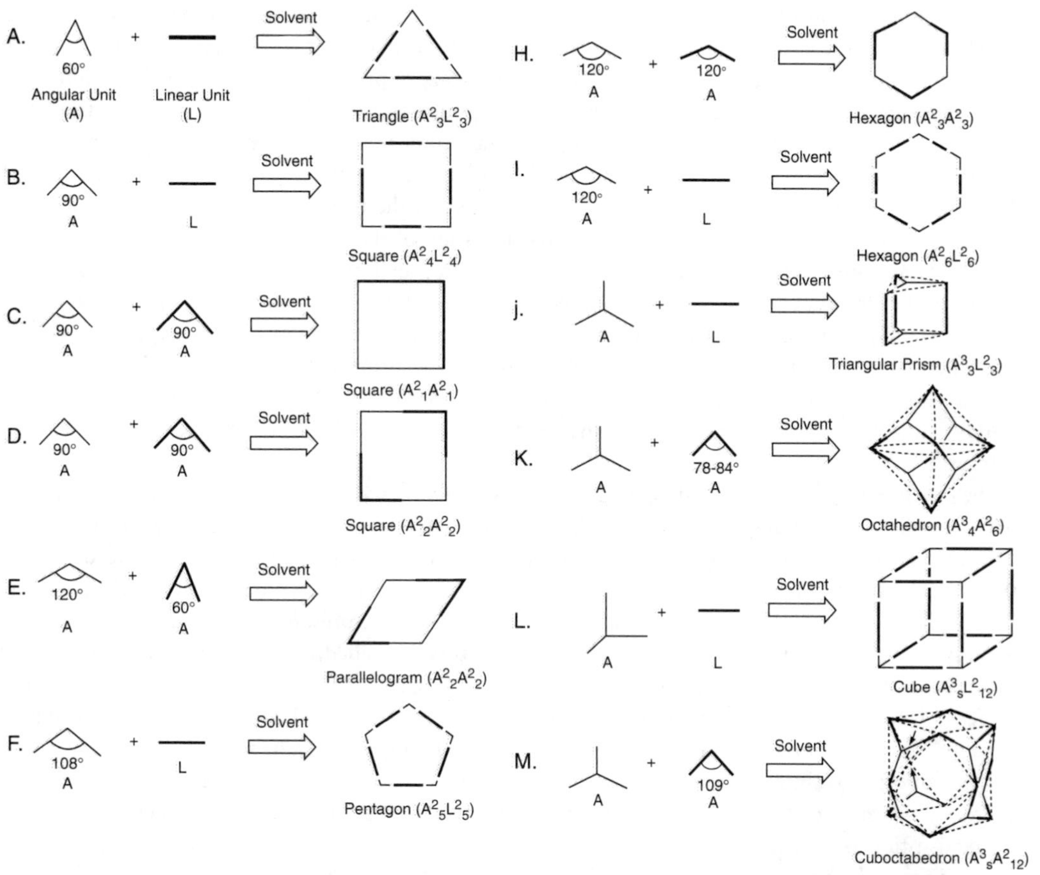

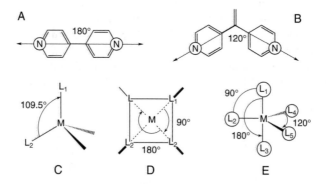

FIGURE 16.32 Metal complex and ligand classifications for two- and three-dimensional coordination nanostructures in the molecular library approach. (From Stang, P.J. and Olenyuk, B., Self-assembly, symmetry, and molecular architecture: coordination as the motif in the rational design of supramolecular metallacyclic polygons and polyhedra, *Acc. Chem. Soc.*, 30, 502, 1997. With permission.)

FIGURE 16.33 Binding angles in a selection of ligands (A, B) and metal complexes (C–E). Note that two unique angles are available in square planar structures (D), and three unique angles are available in trigonal bipyramidal structures (E).

the known geometric features of highly symmetric polyhedra to direct the formation of new metal–ligand assemblies. This model differs from the molecular library model in two important respects. First, there is a definite division between the role of the metal and the role of the ligand in the symmetry interaction model. This is in contrast to the molecular library, where both ligands and metal complexes can be used anywhere within the skeleton of a nanostructure to create corners, vertices, or sides. Second, through the selection of geometric fragments in the molecular library approach, the binding angles within each metal are determined by the orientation of the leaving groups on the metal. The remainder of the metal coordination sphere is saturated with other ligands that retain their coordination positions during the nanostructure fabrication process. The symmetry interaction model relies on the strong binding of chelating ligands to saturate the entire coordination sphere of the metal ion.[82] The coordination sphere of the metal, then, is responsible for defining the orientation of the ligands in the final structure, while the ligands are responsible for forming the sides (between metal–metal pairs) or faces (binding three metals) of the structure (Figure 16.29). The use of chelating ligands in the symmetry interaction model has the benefit of increased stability in the final structures through the formation of multiple coordination bonds per ligand and the inherent kinetic stability that comes from the chelate effect.[8] The important components in the symmetry interaction model are the orientation of the lone pairs of the chelating ligands within the organic framework and the geometry of the coordination sphere of the main group or transition metal atoms.

The development of a rational design strategy for the symmetry interaction model is more complex than for the molecular library model. In the symmetry interaction model, a library of angles and interactions based solely on the choice of metal or ligand is not employed. Instead, the design of nanostructures from this approach requires an understanding of the chelating ligands and the coordination sphere of the metal ion. Among the coordination nanostructures, most of the designs applicable to the symmetry interaction approach are based on the use of tetrahedral (4-coordinate), square planar (4-coordinate), and octahedral (6-coordinate) structures (Figure 16.34). As the ligands themselves are not responsible for imparting dimensionality to these designs, polyhedral coordination nanostructures based on the symmetry interaction approach employ metal ions with octahedral coordination spheres. This limitation does simplify the design process because it is possible to classify the available metal ions according to their coordination numbers. Because this methodology requires that the metal be stripped of ligands prior to nanostructure formation, it is also possible to select metal–ligand starting materials based on the lability of the metal complex ligands under certain reaction conditions.

A means has been developed to understand the spatial relationships between the metal coordination sphere and the attached ligands by defining common geometric features and determining their importance in the fabrication process.[80,84] Because the method relies heavily on the use of symmetry to define the geometric features of both the interactions and the assemblies themselves, highly symmetric structures, such as Platonic solids, can be fully analyzed by considering their vertices. The coordination sphere of the metal, where all of the connectivity and structural determination occurs, is divided into a *Coordinate Vector*, a *Chelate Plane*, and an *Approach Angle* (Figure 16.35). The coordinate vector is defined as the vector between the coordinating atom(s) of the ligand and the metal. In chelating ligands, the bisection point of the lone pairs and the metal atom forms this vector. In monodentate ligands, this vector is simply along the lone pair–metal bond. The coordinate vectors and the rotation axis of the metal that would

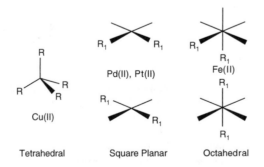

FIGURE 16.34 Coordination geometries for common metal ions.

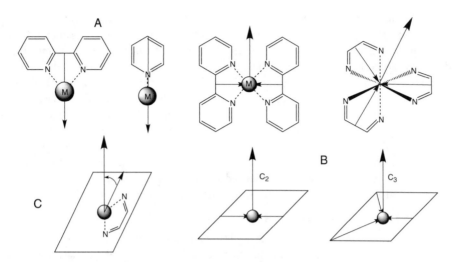

FIGURE 16.35 (A) Coordinate vectors. (B) Chelate plane. (C) Approach angle.

transform one ligand into an adjacent ligand then define the chelate plane. In metals with three ligands, only the three coordinate vectors of the ligands are required to define the chelate plane. In metals with two ligands, defining a third axis perpendicular to the coordinate vectors and the major rotation axis of the metal–ligand complex designates the chelate plane. The approach angle is defined as the angle between the major axis of the metal center and the plane of the chelating ligands in the final structure.

A demonstration of the geometric features in assembled macromolecules is provided here for two structures (Figure 16.36). In a helical, D_3-symmetry structure composed of two metals and three chelating ligands (simplified to M_2L_3), the orientation of the ligands in the coordination sphere of the two metals requires that the two chelate planes be parallel to one another. The C_3-rotation axes and the C_2-rotation axes that bisect each shared ligand of the two metals are then automatically aligned. Within any such helical or rod-like structure of D_{3h} symmetry, the local features of the metal coordination sphere are consistent. The selection of metal–ligand sets can be directed by the known geometric requirements of the structure. The formation of a molecular tetrahedron from chelating ligands and metals can be completed by either an M_4L_6 combination, where six ligands are required to form each of the skeletal components, or an M_4L_4 combination, where each ligand consists of the three chelating sites related by a C_3-rotation axis through the plane of the ligand (Figure 16.36). Each metal is bound to three chelating

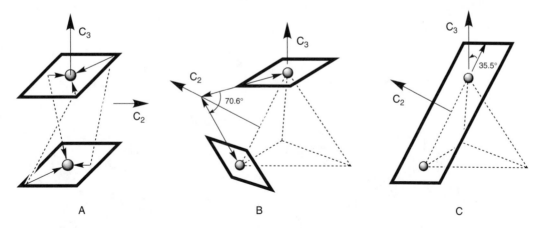

FIGURE 16.36 Structural examples of the symmetry interaction approach. (A) D_3-symmetry helix. (B) Tetrahedron from coordinate vector description. (C) Tetrahedron from approach angle description.

ligands, with the chelate plane defined by the coordinate vectors at each vertex. In the M_4L_6 case, the C_2-rotation axes bisect the ligands along the tetrahedral skeleton. The C_2-rotation axes in the M_4L_4 case are at the same positions in the tetrahedral frame, although the skeletal framework of the tetrahedron is only inferred because all ligands are now facial. The idealized tetrahedron is defined from the coordinate vectors of two coordination centers along a single C_2-axis, making a theoretical angle of 70.6° with respect to the relative orientations of the chelate planes (Figure 16.36). In cases where the chelating ligands are held planar to one another and are orientated antiparallel, the approach angle can be used as the defining feature of the metal–ligand interaction. In a tetrahedron, the use of these ligands requires that the approach angle be 35.3°. Given a coordination nanostructure, all of the isolated metal–ligand complexes can be treated by similar symmetry descriptions. From the geometric relationships of the metal–ligand interactions and the structural features of the isolated metal–ligand complexes, an understanding of many macropolyhedral structures made from metal–ligand interactions becomes possible.

16.3.3.3 Two-Dimensional Structures

Two-dimensional polygons fabricated from metal–ligand interactions have extensive synthetic precedent, and there is virtually no limit to the possible structural combinations that can be made from very simple synthetic modifications.[82] As systematized in the molecular library approach,[81] the specific structural features (side lengths and internal angles) of many regular polygons are chemically accessible by employing bis-monodentate ligands (although bis-bidentate ligands are also appropriate with certain metal centers) and coordinatively unsaturated metal centers. While the control of structural features may not be absolute compared to proposed atomistic[6] or molecular Tinkertoy[53] methods, the approach and the many available structures provide the means for controlling size and shape well below the nanometer threshold in a highly predictable manner. This control over structural features is incorporated both within the extensive synthetic precedent for the organic ligands and the selectivity for coordination number and ligand type among the available metal complexes. With the structural variety of geometric fragments in the molecular library also comes the ability to select for chemical reactivity.

The flexibility of the metal–ligand bond in polygon formation is highlighted here by a selection of palladium- and copper-based systems below (Figures 16.37 and 16.38). The commonly employed palla-

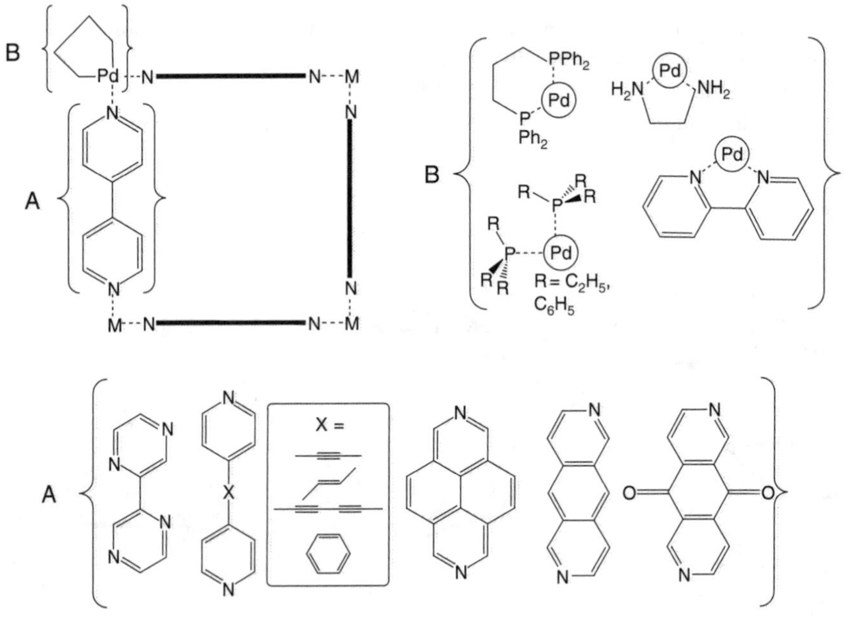

FIGURE 16.37 A selection of known square planar palladium(II) coordination structures. (A) Coordinating ligands. (B) Metal centers with auxiliary ligands.

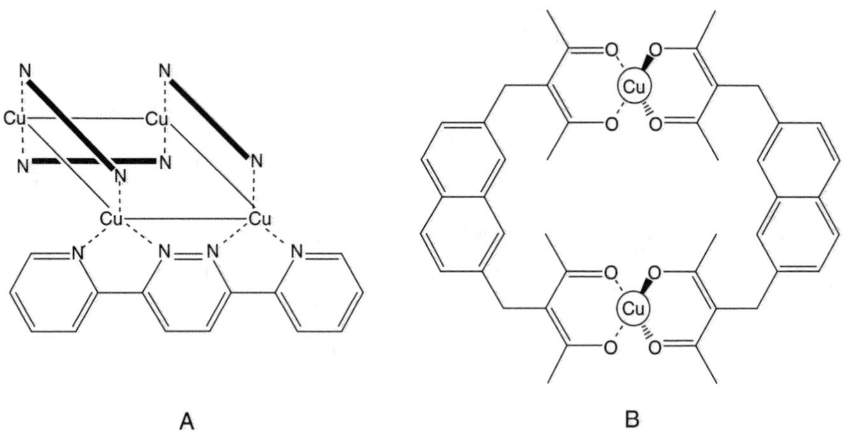

A   B

FIGURE 16.38 Tetrahedral metal–ligand centers for copper(I) complexation structures.

dium(II) ion (as well as platinum(II)) is square planar, providing a perpendicular pair of coordination sites for monodentate ligands. Consequently, molecular squares are a familiar result of their use. The copper(I) ion, as a tetrahedral coordination center, is useful for connecting bidentate ligands perpendicular to one another.[85–89] Besides the obvious differences in ligand placement and orientation come the differences in ligand mobility. In the palladium-based nanostructures, single metal–ligand σ-bonds provide free rotation about the coordination site and greater structural flexibility. With bidentate coordination in the copper complexes, structural flexibility is greatly limited.

A selection of known palladium–nitrogen coordination combinations is shown in Figure 16.37. Although polygons with more sides are possible, structures usually contain from three to six sides.[82] In the structures where the palladium is used as a corner, the metal is delivered to the reaction mixture with one bidentate ligand (typically diamine $[H_2N-R-R-NH_2]$ or diphosphine $[Ph_2P-R-R-PPh_2]$) and two labile ligands. Many nanostructures based on the molecular library model incorporate both strongly binding and weakly binding ligands in the same metal complex to allow for greater control over the ligand coordination position.[82] In tetrahedral metal complexes, where the two labile ligands are always next to one another, the difference in metal–ligand bond strengths serves to control which ligands are removed during nanostructure formation (Figure 16.34). In the square planar and octahedral cases, different isomers place the labile ligands at nonadjacent positions. Both the bond strengths and the ligand positions must be accounted for in the selection of the metal complex. When labile ligands are oriented 180° to one another in the palladium systems, these complexes become linear linkages suitable for use as the sides of polygons.

The formation of the palladium(II) nanostructure begins with the removal of the labile ligands. Two common labile ligands in palladium(II) complexes are triflate (OTf^-) and nitrate (NO_3^-) anions. Their removal leaves both an open coordination site and a positive charge on the metal. The oxidation state of the metal changes as a result of the loss of an unpaired electron to a highly electronegative atom. In OTf^- and NO_3^- the two electrons are lost to form the palladium(II) ion due to the electron-withdrawing oxygens on each ligand. Coordination of bis-monodentate ligands then leads to the formation of the polygon sides. The process is repeated until each metal has lost its labile ligands and coordinated an equal number of nitrogen ligands.

The square planar geometry of the palladium(II) does not limit its applicability to polygons with more or fewer than four sides. The otherwise disfavored formation of strained complexes, such as molecular triangles from square planar palladium(II) cations, can be forced to occur in a system by steric[90] concentration (enthalpy/entropy arguments),[91,92] or guest–complexation effects.[92] For instance, the replacement of the small bidentate ethylenediamine ligand with 2,2′-bipyridine results in the formation of both squares (the preferred structure with the smaller ligand) and triangles in solution from steric effects[92] (Figure 16.39). Both concentration–dependence and guest–complexation were found to play important

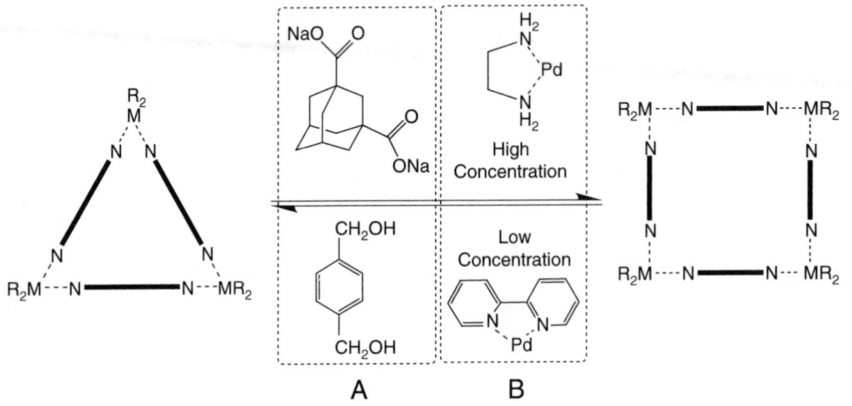

FIGURE 16.39 The direction of triangle or square formation in solution. (A) Steric bulk of polar groups within the nanostructure. (B) Steric bulk of auxiliary metal–ligands or concentration of the coordination nanostructures in solution.

roles in controlling the equilibrium of one palladium(II)-based assembly[47,48] (Figure 16.39). By varying the concentration of palladium(II) complexes (salts of ethylenediaminepalladium with either triflate or nitrate) and bidentate ligands (trans-1,2-bis(4-pyridyl)ethylene), the formation of triangles or squares could be directed.[92] At low concentrations (0.1 mM), triangles were favored due to entropic effects. At higher concentrations (10 mM), the more stable molecular squares were favored. Guest–complexation was found to affect the concentrations of trimer and tetramer in solution by directing either the triangle or the square to form with the addition of p-dimethoxybenzene or a disodium salt of 1,3-adamantanedicarboxylic acid, respectively[92] (Figure 16.39). Alternately, the incorporation of flexible bis-monodentate ligands can be used to form triangles from the square planar palladium(II) ion.[93] In instances where the labile ligands of the palladium(II) complex were oriented 180° from one another, coordination polygons with various numbers of sides were fabricated by altering the binding angle of the ligands. This is the method employed in one instance for forming molecular hexagons and pentagons in solution.[82,94]

Copper(II) has been used as the coupling element for a number of both two- and three-dimensional nanostructures.[85–89] The formation of small molecular squares using four copper(I) ions was made possible by the use of 3,6-bis(2′-pyridyl)pyridazine and copper(I) triflate.[88] In this design, the tetrahedral coordination center of the copper(I) ion fixes two pairs of bidentate ligands perpendicular to one another on opposite sides of the coordination plane of the four metal centers (Figure 16.38A). The characterization of this molecule indicated that the close proximity of the ligand rings to one another allows for a favorable π-stacking interaction, increasing the overall stability of the entire molecule.[88] Two copper(I) ions can also be used with ligands containing flexible bidentate regions to form molecular squares.[89] The free rotation of the bidentate branches in a bis-dione allow for one such dinuclear copper(I) complex[90] (Figure 16.38B). As will be discussed below, the copper(I) ion is very well suited to using the same types of bidentate coordination to form three-dimensional structures.

16.3.3.4 Three-Dimensional Structures

The spontaneous formation of three-dimensional architectures from noncovalent self-assembly is a common occurrence in biological systems. In proteins, the spontaneous formation of structure and function occurs at the most basic level, with hydrogen bonding along the polypeptide chain to form the secondary structure, and also among the largest of the aggregate protein interactions, such as in many viral and bacterial capsids. The tetrahedral bonding in carbon places significant restrictions on the geometric flexibility of the designs. This requires organic systems to use larger molecular subunits, such as amino acids and nucleotides, in order to gain enough structural flexibility to create complex structures. In contrast, the use of metals in the design of nanostructures, especially in smaller nanosystems, offers far greater flexibility for the formation of structurally complex macromolecules from

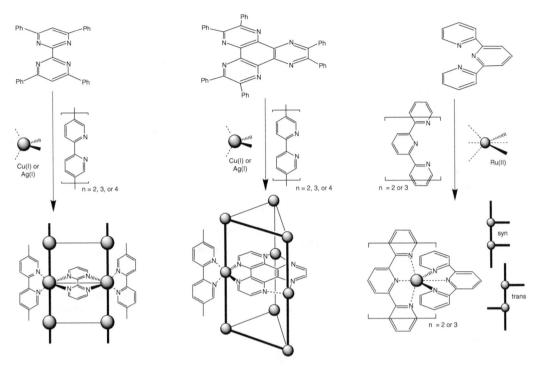

FIGURE 16.40 Ladders (left), rods (middle), and racks (right) formed from metal–ligand coordination.

controllable chemical and electrostatic interactions. The same advantages for creating two-dimensional structures from metal–ligand interactions are also realized in the third dimension, and both mono-dentate and bidentate (chelating) ligands have been used prominently in the formation of three-dimensional nanostructures.

The synthetic precedent for three-dimensional architectures can be divided into two broadly defined categories. The first of these categories includes linear coordination complexes such as ladders, racks, rods, and helices. Such systems are based on the vertical stacking of identical ligand–metal coordination regions and are extendable by modifying the lengths of the subunits that define their walls (Figure 16.40). The second category encompasses the polyhedral macromolecules. These systems confine the coordination regions to vertices instead of linear arrays, resulting in unimolecular architectures with dimensional customizability confined to modification of the ligands that define the sides (bis-chelating ligands) or faces (tri-chelating ligands) (Figure 16.41).

Ladders and rods are fabricated using tetrahedral-coordinating metals that act as *spiro*-centers between two different ligands to lock them in place and perpendicular to one another (Figure 16.40). Both ladders and rods utilize the same coordination center and a repeating sequence of covalently bound bidentate ligands to act as the vertical stabilizers (walls). Two such ladder structures employ a tetraphenyl derivative of the tetradentate bipyrimidine as the horizontal ligands, or rungs, and copper(I) as the metal centers to coordinate the rungs to 2,2′-bipyridine chains[95] (Figure 16.40). Three known rods were fabricated similarly, utilizing tetrahedral metal centers (copper(I) or silver(I)) and the same ligand chains of 2,2′-bipyridine as the vertical supports. These rod structures, however, employ hexaphenyl derivatives of hexaazatriphenylene as the tridentate ligands for the horizontal supports.[86] Molecular racks are based on the same basic ideas. These systems, however, employ tridentate ligands and six-coordinate metal centers to form rigid arrays. Because only one extended vertical chain is required to form these structures, racks can be formed from isomers of the same repeating tridentate motif.[96] A variety of *syn*- and *trans*-isomers of ruthenium(II) racks have been synthesized through thermodynamic self-assembly using vertical chains with both two- and three-tridentate subunits[97,98] (Figure 16.40).

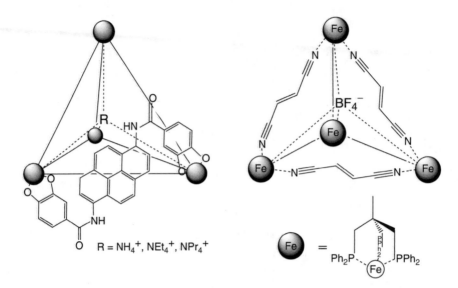

FIGURE 16.41 Coordination tetrahedra and encapsulation of guest molecules. (A) Symmetry interaction-based coordination tetrahedra and a series of encapsulated cations. (B) Molecular library-based tetrahedra and encapsulated boron tetrafluoride ion.

The symmetry interaction approach has been used extensively in the design and study of homodimetallic helicates.[77] The majority of these structures share the same design features, including the utilization of two octahedral coordination centers and three ligands composed of two bidentate regions and various organic bridges that provide unrestricted rotation about the bridge–bidentate bond. The chelate planes of the octahedral metal pairs are held parallel, requiring that each set of three bidentate ligands be provided with enough rotational flexibility to orient themselves along the C_3 rotation axis of the coordination spheres (Figure 16.42). The customization of the helicate shape is limited to modification of the ligand lengths. From among a set of common bidentate motifs, including those shown in Figure 16.42, any of a number of organic structures have been employed as bridges to vary the helicate length.[80]

It is remarkable that often the only requirements for the formation of macromolecular polyhedra are highly coordinating metal centers and multi-branching ligands. Many of the resulting macromolecular polyhedra, because their formation and stability are based only on metal–ligand coordination along the periphery, are skeletal structures with hollow cavities. Because the exteriors of these hollow polyhedra can be often deformed or broken through thermodynamic manipulation, it becomes possible to incor-

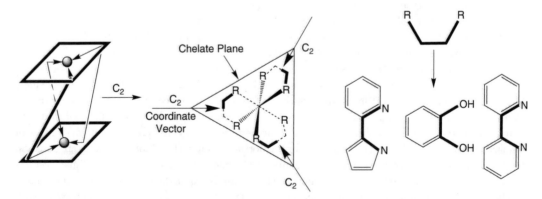

FIGURE 16.42 Helicate formation from the symmetry interaction model. At right is a selection of employed chelating ligand fragments.

porate smaller molecules into the polyhedral cavity in solution. Consequently, these macromolecular coordination polyhedra can act as large host molecules for the incorporation of single-guest molecules or collections of molecules for isolated chemical or structural studies.

Coordination tetrahedra are the smallest of the polyhedral structures to be formed both through the use of monodentate ligands and chelating ligands. The chelate-based systems are ideally suited to formation based on the principles of the symmetry interaction model, and their formation and structural features have been extensively studied.[80] Two basic motifs in coordination tetrahedra exist: those utilizing four metal centers and six ligands to form the edges of the structure (M_4L_6, Figure 16.A1.C) and those using four metal centers and four tridentate ligands to form the faces (M_4L_4, Figure 16.A1.D). It has been shown that small tetrahedral coordination structures can be used as a way of isolating small molecules in solution[99] (Figure 16.41). This work demonstrated two important features of coordination polyhedra. First, it showed that these coordination polyhedra are dynamic, with their metal–ligand bonds continually being broken and reformed in solution in order to establish an equilibrium with the guest molecules. Second, it showed that guest molecules can be preferentially selected and encapsulated within a tetrahedron (in the order $NEt_4^+ > NPr_4^+ > NMe_4^+$). In an example of the molecular library approach to three-dimensional nanostructure formation, the linear bidentate molecule fumaronitrile was used as the linking ligand to form the sides of a tetrahedron employing iron(II) vertices[100] (Figure 16.41). The remainder of the iron(II) coordination sphere was saturated using a tridentate phosphine ligand. Again, the tetrahedron was shown to encapsulate a counterion guest (BF_4^-). In this system, however, the tetrahedral symmetry elements are aligned in conjunction with the symmetry elements of the macrostructure. It is believed that the anion may be acting as a template over which the assembly of the cluster proceeds.[100]

16.3.4 Biomimetic Structures

The most versatile and, arguably, most important use of the MBB approach for the formation of nanostructures occurs in biochemistry, where intricate and highly specialized molecular "machinery" controls the manipulation of simple molecules to create functional structures. Biochemistry, as applied to the synthesis of nanostructures, is a special case of supramolecular chemistry. In these biomolecules, the covalent and electrostatic interactions of individual MBBs are used in concert with their aqueous surroundings to impose a sequential order and preferred orientation in the self-assembly of complex structures. The mechanisms and the raw materials of biochemical nanotechnology are not only self-sustaining, where the means for synthesizing and modifying the subunits are internally available to the system, but also self-regulating, where enzymatic activity controls such features as the availability, degradation, and reconstitution of materials into new macromolecules. The MBB approach, when considered from a biomimetic or biochemically inspired perspective, provides both an extensive background from which to understand design- and preparation-related issues at the nanoscale and a wealth of elegant examples from which to conceive novel structures. By studying the dynamics of biomolecular interactions, the role of the subunit in the formation of larger systems and the effects of environment on the formation and operation of these nanostructures may be better understood within a very important context.

Apart from the structural beauty of biomolecular systems, the greatest advantage of relying on biomimetic approaches to form nanostructures is that entire classes of functional structures already exist for study and modification. Nature has provided both a conceptual scaffolding from which to study structure/property relationships on chemically massive structures and a wealth of example systems that are often easily obtained. The biomimetic approach also has the unusual quality of being based upon a "finished product." The goals of biomimetic design are then achieved through retro-analysis, working from a known model to construct a new system based in biochemical precedent through chemical derivitization of the known subunits, environmental manipulation, or the application of biomimetic principles to other non-biological subunits.

The foundations of biochemical design are well understood from an MBB perspective. A great deal of knowledge of the structure and function of the subunits and a detailed understanding of the electrostatic interactions responsible for imparting secondary structure is available for these systems. The literature on this subject is vast, and more detailed discussions are presented elsewhere.[28,101] A great amount of biochemical detail has been omitted from this discussion in order to focus on the actual MBB aspects of these structures and how the biomimetic approach can be readily applied to new systems. While the intricacies of protein folding, enzymatic activity, and tertiary structure are all important aspects of biochemistry, the fundamental understanding of molecular interactions at the macromolecular level are available from even small segments of DNA or small peptide chains in a protein.

16.3.4.1 DNA

Each nucleic acid molecule, the MBB of DNA, can be divided into three parts, with each portion of the molecule contributing significantly to the structure and electrostatic properties of the resulting DNA double helix (Figure 16.43). The covalent architecture of each helix, the primary structure responsible for maintaining the order of the nucleic acids, is composed of a phosphodiester and a 2'-deoxyribose residue in each subunit. The primary structure of the helix is formed via a condensation reaction between a phosphodiester and the 2'-hydroxyl group of a deoxyribose sugar, resulting in the elimination of one water molecule for each nucleotide linkage. Attached to each deoxyribose is either a monocyclic pyrimidine or dicyclic purine nitrogen base. Base pairs are then stabilized through hydrogen bonding interactions between a purine (adenine or guanine) and a pyrimidine (cytosine or thymine) on different (complementary) helices (Figure 16.44). For the purposes of encoding genetic information, two different purine/pyrimidine pairs (A with T and C with G) occur naturally. It is, however, sufficient to simply define the pyrimidine/purine pairing sequence in order to form the double helical structure of DNA. The complete secondary structure of DNA is a product of two types of electrostatic interactions. First, the formation of the double helix results from the correct hydrogen-bonded pairing of complementary bases between helices. Second, a π-stacking interaction, largely isolated within each helix between adjacent bases, further stabilizes the structure. This stacking is not completely isolated within a single helix, however, as the twisting of the double helix creates a slight overlap between bases on opposite strands.[102]

16.3.4.2 Proteins

Amino acids, the building blocks of proteins, are also composed of three structurally and electrostatically important parts. This division of structure begins with the covalent framework of the amino acid sequence, which is limited to the repeating peptide linkage (N-C-C) formed through rotationally unrestricted σ-interactions (Figure 16.45A). Directly attached to the N-C-C backbone are alternating donor (N-H) and acceptor (O = C) pairs for the formation of hydrogen bonds (Figure 16.45B). Any of a number of possible pendant (R) groups may be incorporated into the structure (Figure 16.45C). These functional substitutions on each amino acid are responsible for some of the secondary structure stabilization and enzymatic activity of the protein. The naturally occurring side-groups fall into four major categories based on their behavior in aqueous media.[101] These categories are (1) hydrophobic, (2) polar, (3) positively

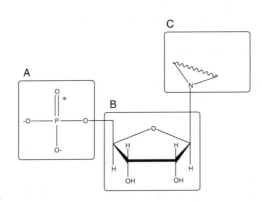

FIGURE 16.43 Structural components of DNA nucleotide bases. (A) Phosphodiester linkage. (B) 2'-ribose sugar. (C) Purine or pyrimidine nitrogen base.

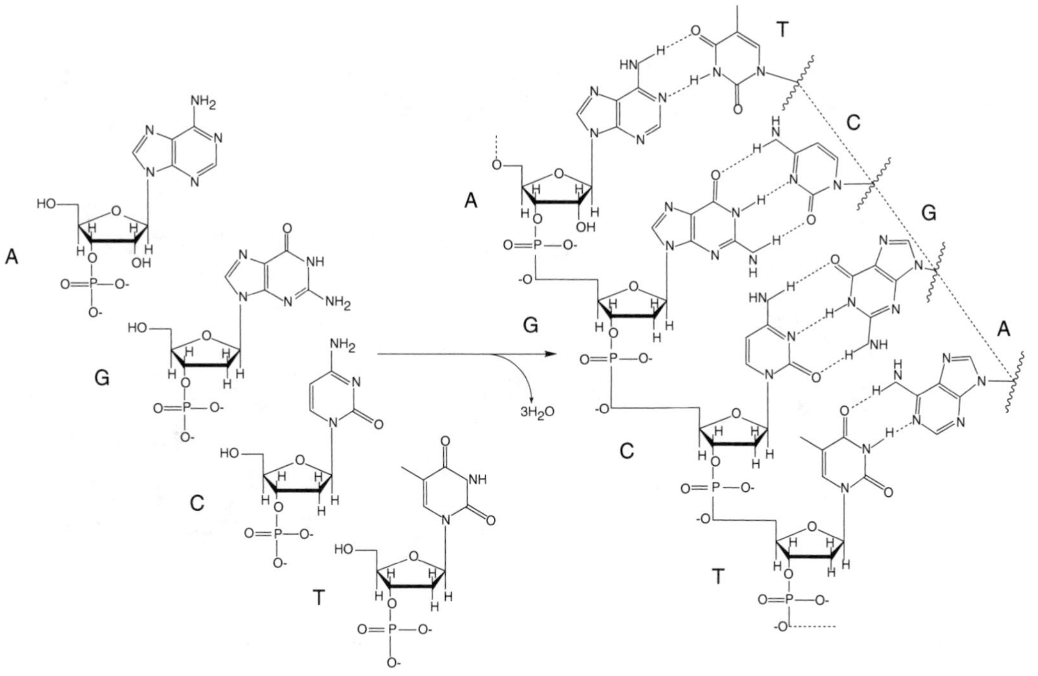

FIGURE 16.44 Connectivity and hydrogen bonding among nucleotide sequences in DNA.

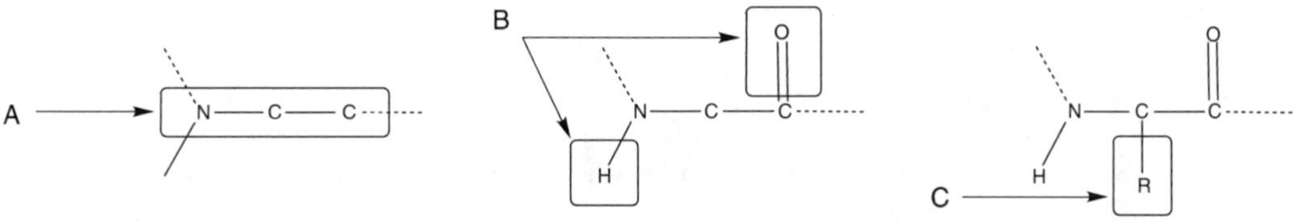

FIGURE 16.45 Structural features of amino acids. (A) N-C-C backbone. (B) Hydrogen bonding regions. (C) Pendant group.

charged, and (4) negatively charged. Since these are all neutral molecules in their isolated forms, their charge is a function of the pH of the intracellular environment.[101] For the general discussion, the R groups can be temporarily neglected, although their importance in imparting function to these structures cannot go unnoticed. As with DNA, the protein backbone is formed through a condensation reaction. The formation of secondary structure then occurs within small sequences of the polypeptide chain through intrachain hydrogen bonds between a N-H hydrogen and a C=O oxygen (α-helices) or between pairs of longer polypeptide chains through intrachain hydrogen bonding between the N-H hydrogens of one chain and the C=O oxygens of another (β-sheets) (Figure 16.46). The twists and bends responsible for the overall three-dimensional structure of proteins are a result of local breaks in the α-helices and β-sheets. The sequences responsible for these local breaks typically extend over many fewer amino acids than do the more regular helices and sheets.[28]

Nucleotides and amino acids share important similarities in subunit design. The formation of polypeptide chains and single helices occur through the removal of water, by far the most prevalent molecule in the intracellular matrix. The availability of subunits for macrostructure formation is regulated by either direct synthesis or modification of externally acquired subunits. In both nucleotides and amino acids, the subunits contain a covalent framework through which to interact with adjacent subunits and a highly directed noncovalent framework capable of stabilizing arrangements of subunits through electrostatic interactions. The majority of all superstructure formation occurs through hydrogen bonding, electrostatic

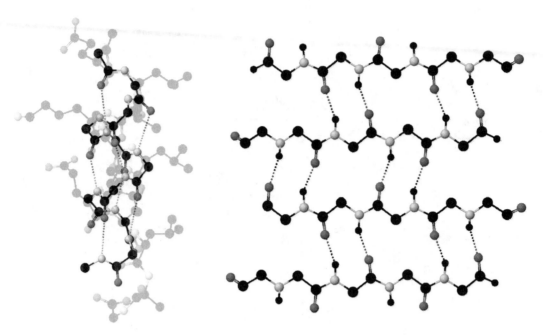

FIGURE 16.46 α-Helix (left) and β-sheet (right) secondary structures from amino acid sequences.

interactions that are easily broken in aqueous solutions. Hydrogen bonding promotes added functionality by allowing for the low-energy error correction of structural mismatches. Both π-π-interactions and the hydrophobic environment promote the π-stacking of the nucleotides, where this stacking serves to minimize the total surface area of the rings in contact with the polar aqueous surroundings. The anionic nature of the phosphodiester backbones at typical *in vivo* pH promotes the solvation of the exterior of the helices and increases stabilization in solution. The aqueous environment also has a destabilizing effect, since broken base pairs can hydrogen-bond to nearby water molecules. The similar strengths of hydrogen bonds between either amino acids or nucleotides and water means that rearrangements of the structures can and do occur dynamically, driving these structures to their energetic minima during their formation and allowing these structures to readily change shape or to be disassembled. While hydrogen bonding with the solvent can occur in the unpaired bases, their correct base pairing pattern provides greater stability, both through entropic affects and the proper alignment of π-stacking pairs. The effect of solvent on structure and biological function is best demonstrated by considering the folding and enzymatic activity of proteins in other solvents. In such nonaqueous instances, significant structural deformations from the aqueous structure and limited function are often observed, which result from improper folding.

This biochemical precision in design and function provides a very complete model from which to design similar macromolecules. Approaches to nanoscale design based on biomimicry begin with the realization that the possible variations of structure and function are enormous. The structures and functions of many of the naturally occurring biomolecules are still being investigated, and much more work still needs to be done to understand how these molecules interact with one another in the intracellular matrix. The extension of the biomimetic approach beyond biochemistry provides researchers with both a synthetic framework and a familiar nanoscale motif from which to design new structures. Two very broad methodologies based on the current understanding of structure and function in biochemistry have emerged in recent years. In one methodology, the known chemistry of nucleotides and amino acids are being exploited to develop novel, nonbiological nanostructures. In this approach, the *biochemical properties* of the subunits are being applied in new ways to form structures based in biochemistry but without any direct biological relevance. In the other methodology, the fundamental principles of biomolecular formation are being applied to new synthetic subunits. The emphasis on biomimetic design leads to the use of molecular subunits that are designed to behave like nucleotides and amino acids based on covalency and electrostatic stabilization features. In these designs, the choice of subunit can include anything from nucleotides that have been slightly altered from their naturally occurring forms to completely novel molecules applied in a biomimetic fashion. It is important to note here that the design approach from the synthetic subunits is directed specifically toward biomolecule mimicry, even when the subunits are ideal for other designs. These two approaches are closely related within the biomimetic context, as both approaches are founded directly from the guiding principles of biochemistry.

16.3.4.3 New Designs from Old Subunits

Nucleotide sequences and polypeptide chains are simply large molecules made up of a series of connected, structurally similar subunits. A particular order of nucleotides in DNA leads to the complementary pairing of bases and the storage of genetic information for the formation of specific polypeptide sequences. A specific order of amino acids in polypeptide chains is responsible for directing the spontaneous formation of a secondary structure by way of hydrogen bond-directed folding. From the final product of this protein formation comes a macromolecule with biological activity. In both DNA and proteins, a limited number of different combinations of nucleotides and amino acids control every biochemical process that occurs in an organism. In all other possible combinations of these MBBs, the potential exists to form a macromolecule with some unique nonbiological structure or function. In instances where a new sequence is nearly identical to a natural sequence, one might expect the structures and functions of both to be very similar. This is often the case, although examples exist where the substitution of one key subunit by another leads to the complete loss of biological activity. As more deviations from a natural sequence are incorporated into a synthetic sequence, the new structure loses these similarities. As a new structure, however, its properties may prove ideal to some other function. As general MBBs, there is essentially no limit on their application to the creation of other macromolecules or nanoscale materials.

Great structural variety and chemical function are available from different combinations of amino acids. The current limits on our ability to understand their interactions, however, prohibit the design of very complex structures. In contrast, the interactions responsible for the formation of DNA double helices are well understood because the separation of covalent backbone and electrostatic moieties is pronounced in the nucleotides. Our understanding of the noncovalent interactions of nucleotide bases with one another are specifically relevant in this respect. Consequently, the cognizant design of new structures from naturally occurring nucleotides has proven to be far more manageable than similar efforts from amino acids. Among the many nanostructural designs employing DNA as a key structural element, the most intriguing of these designs uses DNA as a construction element in the same way that molecular Tinkertoy approaches use linear molecules as components in skeletal frameworks. Many complex supermolecular structures from simple DNA fragments have been synthesized by relying on the strength of the double helix and the very predictable interactions of nucleotide bases. In these approaches, the DNA strands are divisible into rigid sections of stable base pairs and sticky sections of unpaired bases (Figure 16.47). In the fabrication of materials, rigid sections are responsible for defining the sides of structures while the manipulation of the sticky ends are responsible for forming and stabilizing corners.

One of the first structural applications of rigid/sticky nanoscale assembly was in the formation of tetravalent DNA junctions[103] (Figure 16.48). Each junction is composed of three regions which facilitate the self-assembly of two-dimensional lattices in solution. The formation of one base-paired arm exposes

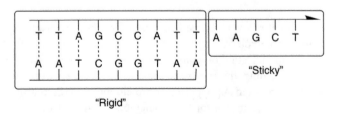

FIGURE 16. 47 *Rigid* (paired) and *sticky* (unpaired) regions of DNA building blocks.

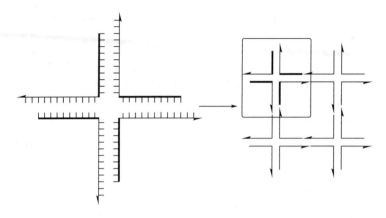

FIGURE 16.48 Two-dimensional DNA junctions from *rigid* and *sticky* engineering.

a sticky region and aligns two single sequences that will become a perpendicular set of arms. The perpendicular arms have identical base sequences and are unable to pair with each other. An arm/unpaired fragment with complementary bases to the unpaired arms of another fragment then pair to form the remainder of the rigid portion of the junction. Self-assembly of these junctions into a two-dimensional lattice occurs with the pairing of the extended sticky ends at each junction corner.

This same DNA design strategy of engineering strongly binding regions within junctions and incorporating unpaired strands to the ends of these junctions has been used for the formation of corners or vertices in a number of complex geometric structures, including isolated polygons and a number of polyhedral nanostructures.[104,105] All cases thus far demonstrate the importance of a rational design approach to the formation of DNA-based structures, because the extension of base pairing beyond two dimensions requires that base-pair complementarity be precisely controlled in order to direct structural formation beyond simple linear sequences. Most recently, a nanomechanical rotary device has been shown to operate by way of conformational changes between the device DNA strands and a second set of strongly binding DNA fragments.[106] The strong noncovalent binding of trigger fragments to the device strands causes conformational changes in regions that find new energetic minima through rotation. In effect, a DNA device has been created which is powered by a very site-specific kind of DNA "fuel."[106] Among other applications of DNA for nanostructural formation are those that rely solely on the complementary binding of strands to direct and stabilize other structures. For instance, complementary binding has been used as the noncovalent stabilizer to direct the formation of simple polygons from oligonucleotide/organic hybrid[105] (Figure 16.49).

16.3.4.4 Old Designs from New Subunits

The reproduction of biomolecular structures by synthetic subunits provides chemists with both an interesting challenge in supramolecular chemistry and a well-established set of guiding principles. The emphasis on designing subunits for the sole purpose of reproducing bioarchitectures is founded in our increased understanding of structural interactions within DNA and proteins. In both DNA and proteins, the vast majority of this structural precedent is based at the subunit level. Much of the work has been based on the use of subunit modification for the purpose of understanding the formation of secondary structures.

At one end of the biomimetic design regime is the use of synthetic nucleotides and amino acids to alter the properties of familiar biomolecules and to make novel structures based on the known interactions of these subunits.[107–111] The modification of amino acids in peptide sequences has been extensively used as a means to study protein folding, the enzymatic processes of these structures, and novel molecular scaffolding designs based on common supramolecular protein motifs, such as artificial α-helices and β-sheets. The design advantages responsible for the proliferation of nanostructures based on nucleotide interactions have also been responsible for the extensive modification of nucleotides as a direct means

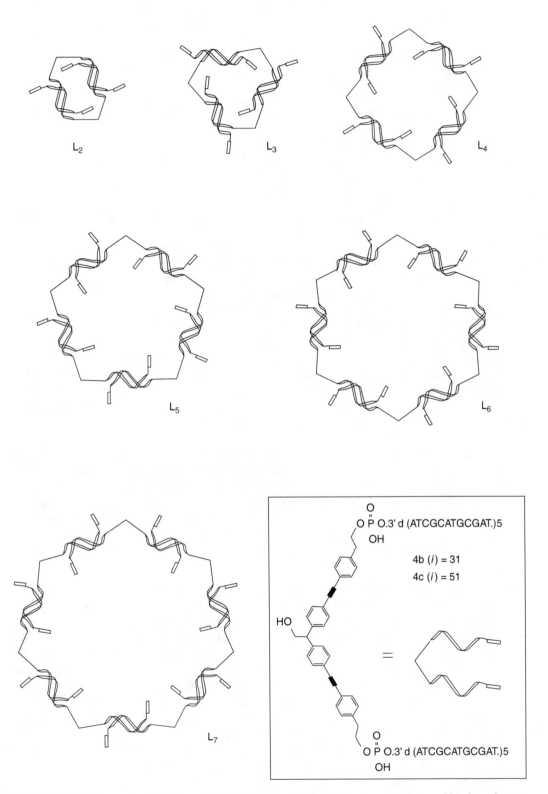

FIGURE 16.49 Polygons from DNA/organic hybrids. (From Shi, J. and Bergstrom, D.E., Assembly of novel DNA cycles with rigid tetrahedral linkers, *Angew. Chem. Intl. Ed. Engl.*, 36, 111, 1997. With permission.)

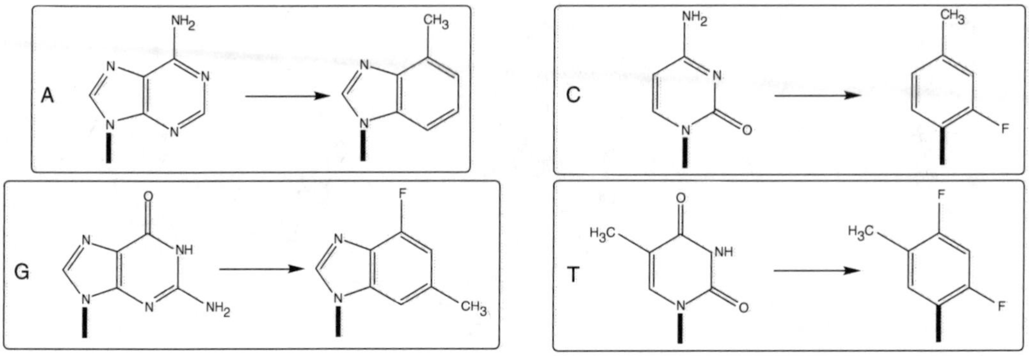

FIGURE 16.50 Nucleotide mimic structures without hydrogen bonding groups (right) from native structures (left).

to study the structure and function of DNA. A selection of synthetic nucleotides and their corresponding natural nucleotides is provided in Figure 16.50. Among these particular designs, the modifications have involved the removal of hydrogen bonding from the nitrogen bases, and they were used specifically to demonstrate the importance of aromatic stacking in the stabilization of the DNA double helix and to provide key insights into the importance of hydrogen bond stacking stabilization in the formation of DNA double helices and the molecular recognition events of DNA replication.[107]

Much of this work, which has emphasized altering the interactions between individual base pairs while causing minimal deformations in the double-helical structure, is also directly applicable to the novel DNA-based design strategies described above, as the modifications are typically rather subtle and the integrity of the nucleotide architecture remains intact. With the structural benefits of artificial MBBs in biomimetic design come many potential biomedical applications, as these synthetic subunits are generally not degraded by enzymatic processes, making them interesting candidates for the synthesis of novel therapeutics and biomaterials.[107]

At the other end of this biomimetic design regime is the reliance on only the biomimetic design strategy for the creation of biomolecular architectures. Such structures follow directly from the implementation of the structure–property relationships found in nucleotides and amino acids as the guide for the synthesis of new MBBs. These new subunits then share many of the same important design features as nucleotides and amino acids but have marginal structural similarities to the native subunits. The design features most important in the biomimetic design of novel subunits include consideration of primary and secondary structural features.

1. Primary Structure

   The covalent backbones of DNA and proteins define the order of the subunits while also providing some degree of structural flexibility to allow the noncovalent assembly of the larger structures. Within each subunit and in the subunit–subunit connections in both structures, this flexibility is incorporated by way of σ-bonding. Positional control is a function of rotation at specific points in the nucleotide/amino acid framework. As one limiting case, the covalent framework of a subunit can be designed to have no structural flexibility except for freedom of rotation at the subunit–subunit connection points and at the point of attachment for the fragment responsible for electrostatic stabilization (Figure 16.51A). This is similar to the freedom of movement in DNA, as the deoxyribose ring does strongly limit the positional freedom of the attached nitrogen base and the phosphodiester linkage. As a second limiting case, only freedom of movement at the subunit–subunit connections is allowed; and the remainder of the structure is rigid with respect to reorientation about the subunit–subunit bond (Figure 16.51B).

2. Secondary Structure

   Secondary structure is determined by the electrostatic stabilization introduced in the subunits and the positional freedom of these subunits as defined by the subunit–subunit connectivity. In DNA, not only are the nitrogen bases connected through a rotationally unrestricted bond to

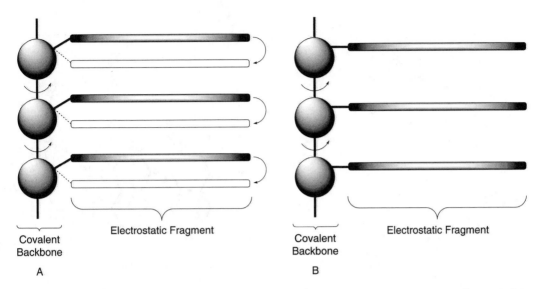

FIGURE 16.51 Limiting cases of MBB flexibility in biomimetic design. (A) Rotational freedom in both the covalent backbone and electrostatic regions. (B) Rotational freedom only in the covalent backbone.

the deoxyribose ring, but a number of pivot points are available for further orientational control. This flexibility is limited, however, by the use of ring structures in both the electrostatic component and the covalent framework. In amino acids, considerable rotational flexibility is available within the covalent backbone, allowing for different structural motifs to form from the subunits (α-helices, β-sheets).

The choice of electrostatic stabilization is also a factor to be considered. In both DNA and proteins, hydrogen bonding predominates. From an engineering perspective, the use of hydrogen bonds is ideal for both the degree of stability required of these structures and the environment in which these structures must function. The interactions of the subunits with the aqueous surroundings are the fundamental means by which all secondary structure formation occurs. Water plays the role of the medium, as it is the solvation of the larger structures that allows for electrostatic interactions to form and reform on the way to a stable minimum. Water, as a small molecule capable of forming stable hydrogen bonds with the noncovalent framework of DNA and proteins, is also responsible for the local destabilization of the larger structures. This local instability is responsible for the structural dynamics of proteins and DNA in solution and can be viewed as an integral part of the function of enzymes in all intracellular processes.

The design of a structural analogue to proteins or DNA from these guidelines must begin with the design of subunits that embody the same fundamental properties as nucleotides or amino acids. While the predictability of protein folding is still difficult, structure and enzymatic activity can be rationalized based on the final structure. DNA, however, has been found to be very amenable to structural manipulation. The predictability of the double helix from naturally occurring sequences has become familiar enough that DNA has been used to build artificial scaffolding and simple devices. Based on the ability to rationally design structures from the familiar structure–property relationships of the DNA nucleotides, alternative helical and double-helical structures based on novel subunits should also offer a certain degree of macromolecular predictability. One example of this approach is provided below.

The design of a new structural subunit employing the limiting cases in bonding and interactions is shown in Figure 16.52. Here, the covalent framework that defines the macromolecular backbone is based on rigid carboranes that are held together through a structurally inflexible five-member ring. By fixing the two inflexible carboranes to one another through the small ring system, structural flexibility within the subunit is greatly diminished. Connectivity between subunits is made by way of either a direct subunit–subunit linkage or through the use of some small, flexible spacer. As a consequence of the design, large-scale flexibility

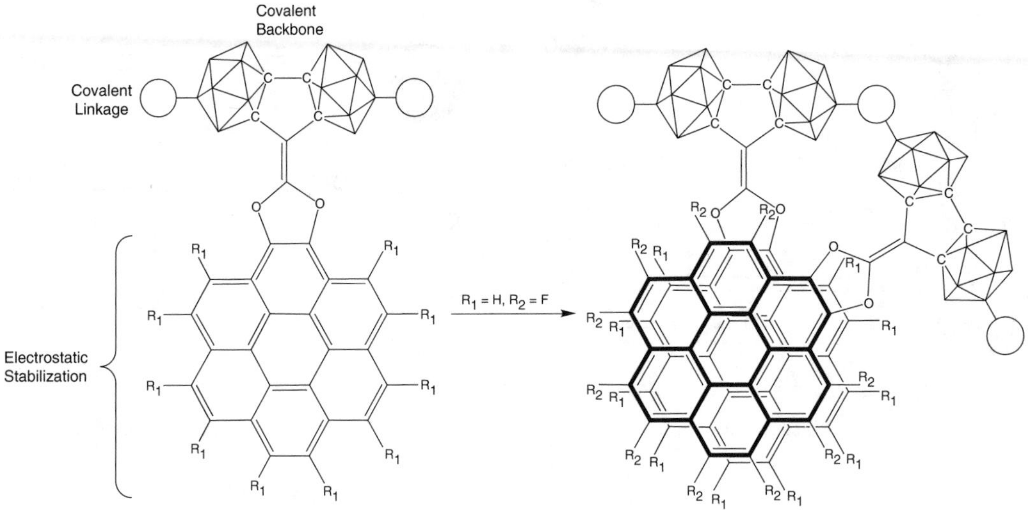

FIGURE 16.52 Synthetic bis-*ortho*-carborane MBB for biomimetic design. Cage boron–hydrogen bonds, oriented in the eclipsed conformation, are shown as unlabeled vertices.

is limited to rotation at a single point in the covalent backbone. All secondary structure formation, therefore, must occur through the rotational reorientation of subunits with respect to one another.

The means to secondary structure stabilization occurs through the interaction of functional groups pendant on the subunit frame. In these structures, the functional groups are placed at the noncarborane-substituted position of the five-member ring. The removal of rotational flexibility in this structure is by way of σ- and π-bonding between the covalent framework and the functional group. With both the interior of the subunit and the functional group held fixed through covalent bonding, interactions between subunits can only occur through rotational interactions.

The reliance on direct interactions, like hydrogen bonding, requires additional degrees of orientational flexibility within the subunit framework in order to form the most stable interactions when the positions of the subunits themselves are not ideally arranged spatially. The reliance on rigorously directional interactions in solution can be removed by the selection of functional groups that do not interact through directed interactions. This route requires the removal of polar interactions as the means to forming stable interactions. The use of π-stacking interactions in the DNA double helix provides both significant stability and direction for the formation of a helical network with unfavorable interactions with the aqueous surroundings. As π-stacking can be engineered to be most favorable with actual stacking of the π-electrons between rings, the use of this type of interaction for the formation of helical structures should be possible. This helical stacking can be accomplished by limiting the positional flexibility of the π-systems to motions that align them in a vertical manner with limited opportunity to form other stable π-stacking arrangements. The rotational limitations of the carborane subunits allow such limited flexibility. Within the subunit formed from the linking of carborane-based MMBs to one another through bonds that only provide rotation and stable π-stacking interactions, the helical structure is both controllable and favored (Figure 16.53A). This preferential formation can be enhanced by the inclusion of polar functionalities on the exterior of the carborane subunits, forcing the π-stacking alignment within the helices by hydrophilic/hydrophobic interactions. Furthermore, the formation of double helices from the same stacking arrangement can be enhanced by the use of π-stacking pairs with alternating ring-periphery electron densities (Figure 16.53B). From the stability shown for benzene–perfluorobenzene pairs, similar MBB designs based on the same covalent subunit framework and π-system containing modified substituents becomes an interesting possibility for directing the formation of such designs. With the exclusive use of π-stacking for the formation of secondary structure, however, a larger space must be employed between stacking moieties.

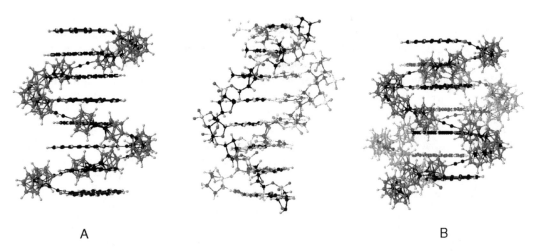

A   B

FIGURE 16.53 Helical (A) and double-helical (B) designs from synthetic MBBs. DNA provided at center (all structures to the same scale).

In carborane-based subunit designs, a double-helical structure can be designed by alternating the covalent backbone of each helix with subunits containing π-stacking functionalities.

16.3.5   Dendrimers

Dendrimers, also commonly referred to as *starburst polymers, cascade polymers*, or *arborols*, compose a special subset of supramolecular chemistry[112–115] that employs an MBB methodology in their formation. Dendrimers can be defined as "highly ordered, regularly branched, globular macromolecules prepared by a stepwise iterative fashion."[116] While the growth process of these structures is based in polymer chemistry, dendrimers offer exceptional control of structural and chemical properties within a predictable, unimolecular architecture. Further, the control of chemical functionality is available both within and along the periphery of dendrimers at any step in the growth process. Consequently, dendrimers can be either synthesized for a specific function or can be designed to behave as a nanoscale chemical environment itself for a number of applications (*vide infra*). With increasing interest in the use of dendrimers in materials science, biomedical applications, and in nanoscale laboratory applications,[114–117] the rapid progress in their development has emphasized both the basic methods for their fabrication and selective methods for the incorporation of function.

A number of structural and synthetic features separate dendritic polymers from the two remaining classes in polymer chemistry: hyperbranched polymers and linear polymers.[116] First, the dimensionality of a dendrimer is controllable from the very beginning of its growth. Linear polymers, while their random assembly in solution is three-dimensional, are formed through one-dimensional bonds. Because their orientation is statistical during this linear assembly process, there is little control over their secondary structure. The dimensionality of a dendrimer is determined from the shape of its structural core, which then directs the polymerization process over a length (one-dimensional), an area (two-dimensional), or a volume (three-dimensional). Because the growth of a dendrimer occurs radially from the inner core, the initial branching of the structure must take on the dimensionality of the inner core. Second, dendrimers are formed through a controllable, iterative process. Both linear and hyperbranched polymers, in contrast, are formed through chaotic, noniterative reactions, limiting both the control of their shapes and the degree of their polymerization. A dendrimer can be grown with no polydispersity, yielding a single, uniform structure of chemically massive unimolecular proportions. The largest of these unimolecular dendrimers have been shown to grow to sizes of up to 100 nm and molecular weights of 10^3 kDa.[117] Third, and perhaps most useful for nanoscale fabrication, is that the growth of a dendrimer can potentially be designed to be self-limiting regardless of the availability of monomer or reaction conditions. This is possible because the exponential addition of monomers to the dendrimer periphery rapidly surpasses the increase in the volume of the final structure, which only increases as the cube of the radius. Consequently, a dendrimer will eventually reach a steric limit past which monomer addition is impossible, a condition known as *De Gennes dense packing*[116] or the more general term *starburst limit*.[118] This steric limitation is based on a theoretical limit, however, and the understanding of dendrimer shape is still an area of significant research interest.

Dendrimers are, perhaps, the most controllable of the covalently bound supermolecular structures because the reactions involved in their formation are both self-directing and statistical in solution. The preparation of dendrimers is based in linear polymer chemistry, where a simple A/B copolymer motif is used to create covalent bonds between complementary reaction pairs. In dendrimers, this reaction pair strategy utilizes both a small molecule from which polymerization begins and an A monomer onto which multiple bonding sites for B monomers are incorporated. The initial A monomer or some other template molecule then becomes the seed, or *focal point*, from which n (typically 2 or 3) branches extend. By defining a dendritic focal point, it is not required that the point from which the growth process occurs be the absolute center of the dendrimer. In fact, dendrimers can be formed with the focal point on almost any type of molecule at almost any position, and a number of structures have been synthesized using aspects of the dendritic growth for purely functional purposes.

The structure of a dendrimer may be divided into a focal point and branched generations (Figure 16.54). A *generation* is simply a shell of B monomers around either the focal point (then referred to as the *inner core*) or a previous growth generation. Uniform dendritic growth then requires the addition of a stoichiometric quantity of B monomers for the number of A regions available along the dendrimer periphery. Uniform dendrimer growth is then most directly limited by the availability of monomer, steric constraints, and solubility.

While few alternative routes are known, the vast majority of all dendrimer syntheses is based on either *divergent* or *convergent* strategies. In the divergent approach,[119–121] the site from which dendritic growth begins becomes the focal point of the entire dendrimer framework (Figure 16.54). Each additional generation of monomer adds such that n of these monomers covalently bond to the dendritic periphery at the tail of each previous generation (which then becomes a local focal point in the growth process). The uniformity of each generation is controllable by the inclusion of chemical functionalities onto the ends of each monomer, rendering the newly added generation incapable of undirected growth. The growth process is then continued by the removal of these chemical functionalities. This control of the periphery during the growth process results in divergent-based dendrimers having limited polydispersity in the final structures. Uniform dendritic growth is halted with the depletion of available monomer or the steric congestion of branches along the periphery. It is important to note that only uniform dendritic growth is stopped due to steric congestion. Irregularities in the peripheral branches result from continued addition of

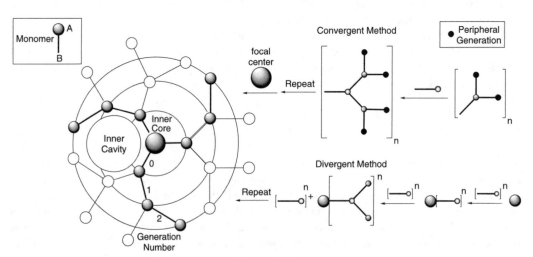

FIGURE 16.54 Dendrimer framework and convergent and divergent synthetic methods.

monomer beyond the starburst limit. Consequently, the control of the absolute size and packing in these structures is difficult to predict with great accuracy beyond a certain generation. It is because of the number of defect structures possible with the radial growth mechanism beyond a certain peripheral steric bulk that the divergent methods have some uncontrollable degree of polydispersity in the final structures.[116]

The convergent approach[116,122,123] to dendrimer formation begins with the peripheral generation and builds inward to a focal point by the coupling of progressively larger branches (Figure 16.54). This reversal from the divergent approach has the effect of switching the important advantages and limitations of the two methods. The fabrication of larger dendrimers is possible by divergent methods, as smaller monomers are added to the periphery of an otherwise sterically congested structure. In convergent methods, large branches are combined with one another, making the proximity of the reaction centers a critical factor in controlling the synthesis of larger structures. Convergent methods lead to greater uniformity of macromolecules, however, as the physical separation of defect structures is a far easier task.[116]

By the divergent method, two dendrimers might have identical molecular weights but great variability in branch lengths due to misdirected polymerization in larger structures. In the convergent methods, large branches either connect together to form a much larger branch or remain unconnected. The resulting increase in mass of bound branches then provides a direct means for separating structures. While the ultimate connection of these branches to the focal center may be a difficult task due to steric crowding, the completed structures are far more massive than any other components left in the reaction mixture and are therefore easier to isolate. The coupling of progressively larger branches, however, does ultimately limit the size of the dendrimers possible by the convergent method.[117]

The design of dendrimers and dendritic structures has begun to move beyond the polymerization chemistry of the branches and into the regime of structure- and application-specific modifications. The design of these functional dendrimers begins with the choice of the inner core. Among the synthesized dendrimers with functionalized cores, some of the most useful interiors for nanoscale applications include those with guest–host binding sites,[124,125] "dendritic probe" potential,[126,127] catalytic activity,[128–130] redox activity,[131–134] and those which employ dendrons, or larger dendritic branches, to act as stoppers for molecular assemblies.[135] A number of these applications are discussed below.

In the design of a dendrimer with an application-specific focal core, the method of dendrimer formation must be chosen carefully. Because dendrimer growth begins at the focal point in divergent methods, the application of a divergent growth scheme requires that the active portion of the core be chemically inert to the polymerization process and that this inertness continue over subsequent polymerization cycles. The convergent method, however, directs the growth of uniform branches until the focal core is ultimately added to the system. As the final formation of the dendrimer in the convergent method requires a chemical step that need not be a polymerization reaction, it is possible to add functional cores with far greater control. Consequently, a number of the discussed functional structures have been synthesized based on convergent approaches.[116]

Beyond the core, the customizability of both the monomers and the periphery has been used to engineer large-scale structural features and functionality into dendrimers. Between the core and the periphery, the inner-branching structure of dendrimers has been found to be highly customizable both for the formation of microenvironments within the cavities formed during the dendrimer growth process and for the inclusion of a number of host–receptors for the selective binding of guest molecules. The chemical modification of the periphery has proven to be a critical feature in the application of dendrimers. The exponential increase in dendrimer growth results in the rapid increase of peripherally bound substituents. As the dendrimer grows, the interactions between the periphery and the environment become the principle features governing dendrimer solubility and morphology. A number of studies have demonstrated that dendrimers incorporating either polar or nonpolar moieties along their periphery have significantly different solubility properties.[136,137] Furthermore, it has been shown that incorporating both highly polar and nonpolar regions into the dendrimer framework gives these structures unique yet controllable molecular encapsulation behavior.[137,138]

Combinatorial strategies have also been used in the dendrimer polymerization process as a means to alter the properties of both their interiors and periphery. A combinatorial approach to dendrimer

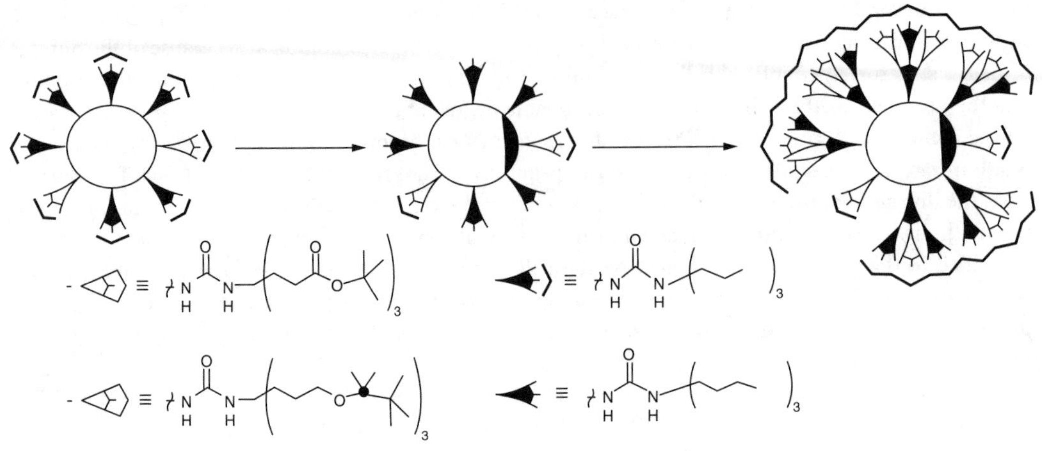

FIGURE 16.55   Dendrimer polycelles from the inclusion of heterogeneous monomers. (From Newkome, G.R., Supra-supermolecular chemistry: the chemistry within the dendrimer, *Pure App. Chem.*, 70, 2337, 1998. With permission.)

synthesis is one in which different monomers are made available during polymerization at various steps in the growth process. In this process, the incorporation of different chemical branches during dendrimer growth can be accomplished either from the very beginning of the dendrimer formation, where the entire dendrimer is then made up of structurally unique branches, or after some number of identical generations have been added. Both approaches result in different local environments within the dendrimer, because the internal cavities typically span multiple generations. These heterogeneous structures, formed by altering the concentrations of different monomers during the growth process, have been termed *polycelles*. The first instances of polycelles employed a selection of isocyanate-based monomers with either reactive or chemically inert ends[139] (Figure 16.55). Not only was it shown that different monomers were readily incorporated into the same dendritic framework, but the combination of reactive and unreactive monomers demonstrated the ability to form dendritic branches with different generation numbers and chemical functionalities.[140] By this method, both the internal cavities and the dendritic periphery form molecular-sized pockets within which encapsulation, trapping, or noncovalent binding can occur.

As a class of supermolecules, dendrimers share similarities in MBB design methodology with both the biomimetic and molecular Tinkertoy approaches. A repeating subunit is connected covalently to other subunits to define a stable, although flexible and highly branched, skeleton. The shape of the final structure is then determined by the interactions of the subunits as constrained by the covalent framework. The reliance on covalency as the principle means of structure formation and the application of covalency within the context of a controlled-growth approach is what gives dendrimers a molecular Tinkertoy quality. Also, the many finger-like projections of the branches that give dendrimers their random, dynamic morphology are still anchored at structurally well-defined focal centers, as in the skeletal framework of rigid architectures. Finally, it is possible to impart structural rigidity to both sets of structures beyond any local stability that comes with noncovalent interactions, although this rigidity in dendrimer design must come at a cost of significant steric congestion, which can make a predictable, uniform growth process difficult.

The MBB similarities between dendritic methods and the biomimetic approach come from the use of subunit properties and interactions to define the secondary structure of the macromolecule, including the customization of both classes of macromolecules to control such features as solubility and aggregate interactions (tertiary structure). Dendrimers, because they are made from simple subunits in solution, can be grown specifically for particular environments. Similar to biomolecules, the electrostatic properties of the monomer can give rise to local environments within the dendrimer itself, as has been demonstrated in many instances by the incorporation of nonpolar/polar functionalities into polar/nonpolar monomers. For example, water-soluble, unimolecular micelles and other large dendritic structures have been synthesized with nonpolar centers by incorporating charged functional groups, such as carboxylate anions, into the periphery.[141] Biomimicry is taken further in dendrimers with the use of redox-active porphyrin focal centers and dendritic outgrowth to model the enzymatic behavior of some proteins.[131,142] By engineering hydrophobic/hydrophilic regions into a macromolecule to direct the formation of secondary structure in solution, this approach is similar to the chemical design of DNA and proteins.

The interactions between subunits that define the final structure in dendrimers are not necessarily based upon the formation of a directed secondary structure (biomimetic approaches) or by fixing the subunits within a larger covalent framework (Tinkertoy approaches). There are no intramolecular features governing the absolute size of the DNA double helix. This holds for proteins to a lesser extent, as it is the intramolecular interactions between the larger subunits (α-helices and β-sheets) in the protein that direct the formation of a localized, three-dimensional structure. Uniform dendrimer growth will, however, eventually succumb to steric crowding along the periphery. Also, the study of dendrimer formation for specific structural applications beyond the radial growth mechanism is still in its infancy. The formation of dendritic superstructures, including monolayer and multilayer formations on surfaces, has been demonstrated as a function of aggregate interactions and general molecular packing. The applicability of these designs, however, are currently limited to "bulk material" uses, such as chemical sensors,[143] catalysis, and chromatographic applications.[144,145]

Dendrimers are not just an interesting class of macromolecular structures. They can be synthesized to include the properties and functions of many customizable monomeric subunits and focal centers. Furthermore, this functionality can be wholly incorporated into a growth generation *via* stoichiometric control of the monomers, introduced statistically by the addition of dissimilar monomers, or performed by post-synthetic modification. Both the *microenvironment* and *functionalization* possibilities of dendrimers have been studied with great success. A brief discussion of two of the applications is provided below.

16.3.5.1 Guest–Host Interactions

One of the functional similarities between dendrimers and nanostructures employing electrostatic interactions, such as molecular crystals and biochemical structures, is the ability to integrate guest–host regions into the covalent skeleton through direct modification of the MBB subunits. A monomer generation can have incorporated into it a region customized to bind a specific molecule or type of chemical functionality. One benefit to introducing chemical functionality by way of monomer-based methods is that the tailoring of noncovalent interactions can be accomplished prior to the incorporation of the monomer into the dendritic framework. Furthermore, as has been demonstrated in the design of dendrimers with polar/nonpolar regions, it is possible to selectively exclude intramolecular or aggregate interactions between the guest–host binding regions from the remainder of the macromolecule simply by the exclusion of certain chemical functionalities from the remaining monomer generations. A dendrimer synthesized with a host interaction designed from strong hydrogen bonds, for instance, can be grown to include large pockets of nonpolar regions (such as long-chain alkanes) in subsequent generations.

The ability to bind molecules in solution by these engineered guest–host interactions depends upon the size of the dendrimer, the amount of branching, and the generation to which the guest–host region is added. In dendrimers with regions of limited steric congestion, it becomes possible to form stable guest–host interactions with many types of molecules. The applications here range from the trapping of molecules in solution by guest–host interactions to the formation of dendrimers themselves by noncovalent means. In both instances, the orientation of the host-binding region with respect to the focal point provides a means for controlling the exact orientation of the bound guest. In the case of dendrimer formation by guest–host interactions, the relative orientation of one branch to another can be controlled. In both cases, however, it is important to note that the size of the dendrimer is a critical feature, as the formation of stable guest–host interactions requires that the guest bind in the absence of steric strain. A number of studies have shown how the steric bulk of the

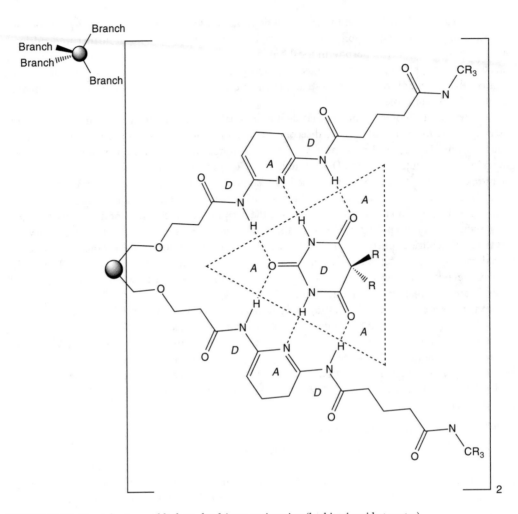

FIGURE 16.56 Guest–host assembly from dendrimer engineering (barbituric acid at center).

dendrimer can affect both the orientational specificity of the guest binding region and the strength of the host–guest interaction.[146–148]

The use of hydrogen bonding within a dendrimer framework for forming stable guest–host inter-actions with small molecules has been demonstrated.[140] In a series of dendritic motifs, a hydrogen-bonding region composed of diacylaminopyridine was introduced early in the growth process (Figure 16.56). The binding pocket of the dendrimer with the inclusion of diacylaminopyridine is then donor–acceptor–donor (DAD) in nature, which can be used to bind selectively to guest molecules with a complementary acceptor–donor–acceptor (ADA) arrangement (Figure 16.56). The molecule selected for studying the guest–host binding interaction in these dendrimers was barbituric acid, which contains two such ADA structures. NMR (^{1}H) titration methods were used to show that pairs of dendrimer arms were able to form stable interactions with a barbituric acid molecule. The assembly of large dendrimers from noncovalent interactions has also been elegantly demonstrated.[149] The focal centers of dendritic branches were engineered with two isophthalic acid fragments incorporated into a small aromatic spacer, providing four hydrogen bonding regions (or eight possible hydrogen bond pairs) per core fragment. Hexameric dendrimers were found to form preferentially in solution by way of strong hydrogen bonding between donor–acceptor pairs at the focal centers of each branch (Figure 16.57).

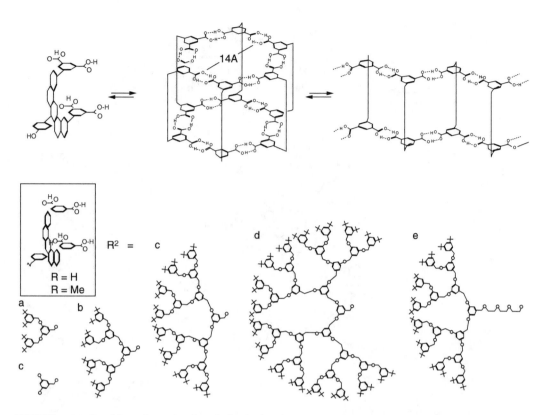

FIGURE 16.57 Dendrimer formation from hydrogen bonding interactions. (From Zeng, F. and Zimmerman, S.C., Dendrimers in supramolecular chemistry: from molecular recognition to self-assembly, *Chem. Rev.*, 97, 1681, 1997. With permission.)

16.3.5.2 Microenvironments

In much the same way that transition metal nanostructures have been shown to encapsulate small molecules, the cavities formed by the overgrowth of generations along the periphery of large dendrimers have been shown to create microenvironments within which molecules can become trapped and bound. The isolation of single molecules or small ensembles of molecules within macromolecular enclosures has obvious utility in nanoscale laboratory applications, a field of chemistry just beginning to develop as an outgrowth of supramolecular chemistry. Molecular cavities within larger dendritic structures benefit from the variety of available monomers, the reproducibility of the cavities using dendritic growth methods, and the wide variety of polar and nonpolar solvents by which to promote solubility and encapsulation. For instance, a macromolecule can be synthesized with multiple regions that behave very differently in different solvents. In dendrimers large enough to encapsulate molecules, the properties of the cavity interior can be very different from the environment at the dendrimer periphery. One notable example of how molecules can be preferentially separated from solution based on polar/nonpolar interactions is provided in the encapsulation of Bengal Rose or 4-nitrobenzoic acid within the nonpolar cavities of a unimolecular micelle composed of long-chain alkane interiors and hydrophilic aliphatic acid exteriors[150,151] (Figure 16.58). Aqueous environments promote the encapsulation of the molecules in the nonpolar interior, while nonpolar solvents, such as toluene, were found to promote their release. Differences in local hydrophilicity/hydrophobicity are easily controlled in dendrimers by either the choice of the initial monomer or the post-synthetic functionalization of the dendrimer periphery. The differences in the spectroscopic properties of many molecules that come with different solvent shells have been the key to studying many encapsulated molecule/dendrimer systems.[117]

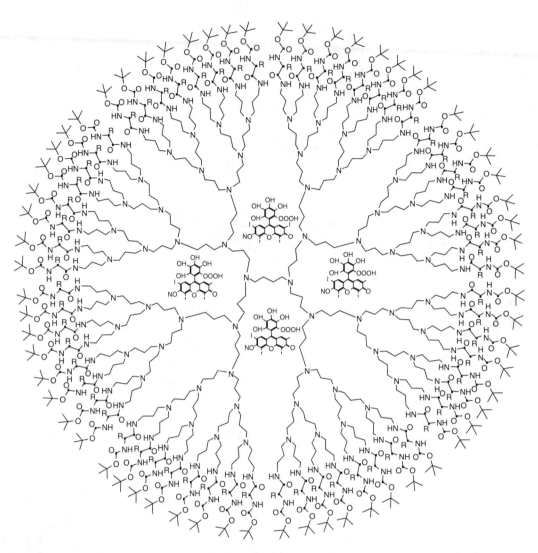

FIGURE 16.58 Bengal Rose encapsulation in dendrimer center. (From Zeng, F. and Zimmerman, S.C., Dendrimers in supramolecular chemistry: from molecular recognition to self-assembly, *Chem. Rev.*, 97, 1681, 1997. With permission.)

References

1. Mullen, K. and Rabe, J.P., Macromolecular and supramolecular architectures for molecular electronics, *Ann. N. Y. Acad. Sci.*, 852, 205, 1998.
2. Lindsey, J.S., Prathapan, S., Johnson, T.E., and Wagner, R.W., Porphyrin building blocks for modular construction of bioorganic model systems, *Tetrahedron*, 50, 8941, 1994.
3. Desiraju, G.R., Supramolecular synthons in crystal engineering – a new organic synthesis, *Angew. Chem. Intl. Ed. Engl.*, 34, 2311, 1995.
4. Lehn, J.-M., Perspectives in supramolecular chemistry: from molecular recognition to molecular information processing and self-organization, *Angew. Chem. Intl. Ed. Eng.*, 29, 1304, 1990.
5. Lehn, J.-M., Perspectives in supramolecular chemistry — from molecular recognition towards self-organization, *Pure Appl. Chem.*, 66, 1961, 1994.
6. Drexler, K.E., *Nanosystems*, Wiley-Interscience, New York, 1992.
7. Ebbing, D.D., *General Chemistry*, 5th ed., Houghton Mifflin, Boston, 1996.

8. Huheey, J.E., Keiter, E.A., and Keiter, R.L., Appendix E of *Inorganic Chemistry*, 4th ed., Harper Collins, New York, 1993.

9. Ellenbogen, J.C. and Love, J.C., Architectures for Molecular Electronic Computers: 1. Logic Structures and an Adder Built from Molecular Electronic Diodes, The Mitre Corp., 1999.

10. Schwab, P.F.H., Levin, M.D., and Michl, J., Molecular rods.1. simple axial rods, *Chem. Rev.*, 99, 1863, 1999.

11. Allis, D.G. and Spencer, J.T., Polyhedral-based nonlinear optical materials. Part 1. Theoretical investigation of some new high nonlinear optical response compounds involving carboranes and charged aromatic donors and acceptors, *J. Organomet. Chem.*, 614–615, 309, 2000.

12. Collman, J.P., Hegedus, L.S., Norton, J.R., and Finke, R.G., *Principles and Applications of Organotransition Metal Chemistry*, University Science Books, Mill Valley, CA, 1987.

13. Long, N.J., *Metallocenes: An Introduction to Sandwich Complexes*, Blackwell Science, Oxford, 1998.

14. Vacek, J. and Michl, J., Molecular dynamics of a grid-mounted molecular dipolar rotor in a rotating electric field, *Proc. Natl. Acad. Sci.*, 98, 5481, 2001.

15. Purcell, K.F. and Kotz, J.C., *Inorganic Chemistry*, W.B. Saunders, London, 1977.

16. Stone, F.G.A., Stability relationships among analogous molecular addition compounds of group III elements, *Chem. Rev.*, 58, 101, 1958.

17. Lewis, G.N., *Valence and the Structure of Atoms and Molecules*, Chemical Catalogue, New York, 1923.

18. Mikhailov, B.M., The chemistry of 1-boraadamantane, *Pure Appl. Chem.*, 55, 1439, 1983.

19. Merkle, R.C., Molecular building blocks and development strategies for molecular nanotechnology, *Nanotechnology*, 11, 89, 2000.

20. Hunter, C.A. and Sanders, J.K.M., The nature of π-π interactions, *J. Am. Chem. Soc.*, 112, 5525, 1990.

21. Ma, J.C. and Dougherty, D.A., The cation-π interaction, *Chem. Rev.*, 97, 1303, 1997.

22. Sunner, J., Nishizawa, K., and Kebarle, P., Ion-solvent molecule interactions in the gas phase. The potassium ion and benzene, *J. Phys. Chem.*, 85, 1814, 1981.

23. Müller-Dethlefs, K. and Hobza, P., Noncovalent interactions: a challenge for experiment and theory, *Chem. Rev.*, 100, 143, 2000.

24. Patrick, C.R. and Prosser, G.S., A molecular complex of benzene and hexafluorobenzene, *Nature*, 187, 1021, 1960.

25. Williams, J.H., Cockcroft, J.K., and Fitch, A.N., Structure of the lowest temperature phase of the solid benzene-hexafluorobenzene adduct, *Angew. Chem. Intl. Ed. Engl.*, 31, 1655, 1992.

26. Collings, J.C., Roscoe, K.P., Thomas, R.L., Batsanov, A.S., Stimson, L.M., Howard, J.A.K., and Marder, T.B., Arene-perfluoroarene interactions in crystal engineering. Part 3. Single-crystal structure of 1:1 complexes of octafluoronaphthalene with fused-ring polyaromatic hydrocarbons, *New J. Chem.*, 25, 1410, 2001.

27. Coates, G.W., Dunn, A.R., Henling, L.M., Dougherty, D.A., and Grubbs, E.H., Phenyl-perfluorophenyl stacking interactions: a new strategy for supermolecule construction, *Angew. Chem. Intl. Ed. Engl.*, 36, 248, 1997.

28. Stryer, L., *Biochemistry*, 4th ed., W.H. Freeman, New York, 1995.

29. Scheiner, S., *Hydrogen Bonding A Theoretical Perspective*, Oxford University Press, New York, 1997.

30. Joesten, M.D., *Hydrogen Bonding*, Marcel Dekker, New York, 1974.

31. Desiraju, G.R., The C-H···O hydrogen bond in crystals: what is it? *Acc. Chem. Res.*, 24, 290, 1991.

32. Steiner, T. and Saenger, W., Geometry of carbon-hydrogen···oxygen hydrogen bonds in carbohydrate crystal structures, analysis of neutron diffraction data, *J. Am. Chem. Soc.*, 114, 10146, 1992.

33. Kool, E.T., Morales, J.C., and Guckian, K.M., Mimicking the structure and function of DNA: insights into DNA stability and replication, *Angew. Chem. Intl. Ed. Engl.*, 39, 990, 2000.

34. Bruno, G. and Randaccio L., A refinement of the benzoic acid structure at room temperature, *Acta Crystallogr.*, B36, 1711.

35. Aakeroy, C.B. and Leinen, D.S., Hydrogen-bond assisted assembly of organic and organic-inorganic solids, in Braga, D., Grepioni, F., and Orpen, A.G. (Eds.), *Crystal Engineering: From Molecules and Crystals to Materials*, NATO Science Series, Kluwer Academic Publishers, London, 1999.

36. Desiraju, G.R., *Crystal Engineering: The Design of Organic Solids*, Elsevier, Amsterdam, 1989.
37. Derissen, J.L., Isophthalic acid, *Acta Crystallogr.*, B30, 2764, 1974.
38. Bailey, M. and Brown, C.J., The crystal structure of terephthalic acid, *Acta. Crystallogr.*, 22, 387, 1967.
39. Duchamp, D.J. and Marsh, R.E., The crystal structure of trimesic acid, *Acta Crystallogr.*, B25, 5, 1969.
40. Penfold, B.R. and White, J.C.B., The crystal and molecular structure of benzamide, *Acta Crystallogr.*, 12, 130, 1959.
41. Cobbledick, R.E. and Small, R.W.H., The crystal structure of terephthalamide, *Acta Crystallogr.*, B28, 2893, 1972.
42. Yang, J., Marendaz, J.-L., Geib, S.J., and Hamilton, A.D., Hydrogen bonding control of self-assembly: simple isophthalic acid derivatives form cyclic hexameric aggregates, *Tetrahedron Lett.*, 35, 3665, 1994.
43. Wang, Y., Wei, B., and Wang, Q., Crystal structure of melamine cyanuric acid complex (1:1) trihydrochloride, MCA.3HCl, *J. Crystallogr. Spectrosc. Res.*, 20, 79, 1990.
44. Lehn, J.-M., Mascal, M., DeCian, A., and Fischer, J., Molecular ribbons from molecular recognition directed self-assembly of self-complementary molecular component, *J. Chem. Soc., Perkin Trans. 2*, 461, 1992.
45. Zerkowski, J.A., Seto, C.T., and Whitesides, G.M., Solid-state structures of "rosette" and "crinkled tape" motifs derived from cyanuric acid–melamine lattice, *J. Am. Chem. Soc.*, 114, 5473, 1992.
46. Wasserman, E., The preparation of interlocking rings: a catenane, *J. Am. Chem. Soc.*, 82, 4433, 1960.
47. Fujita, M., Ibukuro, F., Seki, H., Kamo, O., Imanari, M., and Ogura, K., Catenane formation from two molecular rings through very rapid slippage, a Mobius strip mechanism, *J. Am. Chem. Soc.*, 118, 899, 1996.
48. Fujita, M., Ibukuro, F., Hagihara, H., and Ogura, K., Quantitative self-assembly of a [2]catenane from two preformed molecular rings, *Nature*, 367, 721, 1994.
49. Raymo, F.M. and Stoddart, J.F., Interlocked macromolecules, *Chem. Rev.*, 99, 1643, 1999.
50. Bisson, A.P., Carver, F.J., Eggleston, D.S., Haltiwanger, R.C., Hunter, C.A., Livingstone, D.L., McCabe, J. F., Rotger, C., and Rowan, A.E., Synthesis and recognition properties of aromatic amide oligomers: molecular zippers, *J. Am. Chem. Soc.*, 122, 8856, 2000.
51. Lokey, R.S. and Iverson, B.L., Synthetic molecules that fold into a pleated secondary structure in solution, *Nature*, 375, 303, 1995.
52. Lehn, J.-M., *Supramolecular Chemistry*, VCH, Weinheim, 1995.
53. Michl, J., The "molecular tinkertoy" approach to materials, in Harrod, J.F. and Laine, R.M. (Eds.), *Applications of Organometallic Chemistry in the Preparation and Processing of Advanced Materials*, Kluwer Academic, Netherlands, 1995, p. 243.
54. Grimes, R.N., *Carboranes*, Academic Press, New York, 1970.
55. Ermer, O., Fivefold-diamond structure of adamantane-1,3,5,7-tetracarboxylic acid, *J. Am. Chem. Soc.*, 110, 3747, 1988.
56. Zaworotko, M.J., Crystal engineering of diamondoid networks, *Chem. Soc. Rev.*, 283, 1994.
57. Reddy, D.S., Craig, D.C., and Desiraju, G.R., Supramolecular synthons in crystal engineering. 4. Structure simplification and synthon interchangeability in some organic diamondoid solids, *J. Am. Chem. Soc.*, 118, 4090, 1996.
58. McKervey, M.A., Synthetic approaches to large diamondoid hydrocarbons, *Tetrahedron*, 36, 971, 1980.
59. Mathias, L.J., Reichert, V.R., and Muir, A.V.G., Synthesis of rigid tetrahedral tetrafunctional molecules from 1,3,5,7-tetrakis(4-iodophenyl)adamantane, *Chem. Mater.*, 5, 4, 1993.
60. Tobe, Y., Utsumi, N., Nagano, A., and Naemura, K., Synthesis and association behavior of [4.4.4.4.4.4]metacyclophanedodecayne derivatives with interior binding groups, *Angew. Chem. Intl. Ed. Engl.*, 37, 1285, 1998.

61. Höger, S. and Enkelmann, V., Synthesis and x-ray structure of a shape-persistent macrocyclic amphiphile, *Angew. Chem. Intl. Ed. Engl.*, 34, 2713, 1995.

62. Zhang, J., Pesak, D.J., Ludwick, J.L., and Moore, J.S., Geometrically-controlled and site-specifically functionalized phenylacetylene macrocycles, *J. Am. Chem. Soc.*, 116, 4227, 1994.

63. Levin, M.D., Kaszynski, P., and Michl, J., Bicyclo[1.1.1]pentanes, [n]staffanes, [1.1.1]propellanes, and tricyclo[2.1.0.02,5]pentanes, *Chem. Rev.* 100, 169, 2000.

64. Sugiura, K., Fujimoto, Y., and Sakata, Y., A porphyrin square: synthesis of a square-shaped π-conjugated porphyrin tetramer connected by diacetylene linkages, *J. Chem. Soc., Chem. Commun.*, 1105, 2000.

65. Wagner, R.W., Seth, J., Yang, S.I., Kim., D., Bocian, D.F., Holten, D., and Lindsey, J.S., Synthesis and excited-state photodynamics of a molecular square containing four mutually coplanar porphyrins, *J. Org. Chem.*, 63, 5042, 1998.

66. Arnold, D.P. and James, D.A., Dimers and model monomers of nickel(II) octaethylporphyrin substituted by conjugated groups comprising combinations of triple bonds with double bonds and arenes. 1. Synthesis and electronic spectra., *J. Org. Chem.*, 62, 3460, 1997.

67. Hammel, D., Erk, P., Schuler, B., Heinze, J., and Müllen, K., Synthesis and reduction of 1,4-phenylene-bridged oligoporphyrins, *Adv. Mater.*, 4, 737, 1992.

68. Kawabata, S., Tanabe, N., and Osuka, A., A convenient synthesis of polyyne-bridged porphyrin dimers, *Chem. Lett.*, 1797, 1994.

69. Collin, J.-P., Dalbavie, J.-O., Heitz, V., Sauvage, J.-P., Flamigni, L., Armaroli, N., Balzani, V., Barigelletti, F., and Montanari, I., A transition-metal-assembled dyad containing a porphyrin module and an electro-deficient ruthenium complex, *Bull. Soc. Chim. Fr.*, 133, 749, 1996.

70. Collin, J.-P., Harriman, A., Heitz, V., Obodel, F., and Sauvage, J.-P., Photoinduced electron- and energy-transfer processes occurring within porphyrin-metal-bisterpyridyl conjugates, *J. Am. Chem. Soc.*, 116, 5679, 1994.

71. Odobel, F. and Sauvage, J.-P., A new assembling strategy for constructing porphyrin-based electro- and photoactive multicomponent systems, *New J. Chem.*, 18, 1139, 1994.

72. Harriman, A., Obodel, F., and Sauvage, J.-P., Multistep electron transfer between porphyrin modules assembled around a ruthenium center, *J. Am. Chem. Soc.*, 117, 9461, 1995.

73. Osuka, A. and Shimidzu, H., Meso,meso-linked porphyrin arrays, *Angew. Chem. Intl. Ed. Engl.*, 36, 135, 1997.

74. Yoshida, N., Shimidzu, H., and Osuka, A., Meso-meso linked diporphyrins from 5,10,15-trisubstituted porphyrins, *Chem. Lett.*, 55, 1998.

75. Ogawa, T., Nishimoto, Y., Yoshida, N., Ono, N., and Osuka, A., One-pot electrochemical formation of meso,meso-linked porphyrin arrays, *J. Chem. Soc., Chem. Commun.*, 337, 1998.

76. Anderson, H.L., Conjugated porphyrin ladders, *Inorg. Chem.*, 33, 972, 1994.

77. Holliday, B.J. and Mirkin, C.A., Strategies for the construction of supramolecular compounds through coordination chemistry, *Angew. Chem. Intl. Ed. Engl.*, 40, 2022, 2001.

78. Bonavia, G., Haushalter, R.C., O'Connor, C.J., Sangregorio, C., and Zubieta, J., Hydrothermal synthesis and structural characterization of a tubular oxovanadium organophosphonate, $(H_3O)[V_3O_4)(H_2O)(PhPO_3)_3]\cdot xH_2O(x = 2.33)$, *J. Chem. Soc., Chem. Commun.*, 1998, 2187.

79. Khan, M.I., Meyer, L.M., Haushalter, R.C., Schewitzer, A.L., Zubieta, J., and Dye, J.L., Giant voids in the hydrothermally synthesized microporous square pyramidal-tetrahedral framework vanadium phosphates $[HN(CH_2CH_2)_3NH]K_{1.84}-[V_5O_9(PO_4)_2]\cdot xH_2O$ and $Cs_3[V_5O_9(PO_4)_2]\cdot xH_2O$, *Chem. Mater.*, 8, 43, 1996.

80. Caulder, D.L. and Raymond, K.N., The rational design of high symmetry coordination clusters, *J. Chem. Soc., Dalton Trans.*, 1185, 1999.

81. Stang, P.J. and Olenyuk, B., Self-assembly, symmetry, and molecular architecture: coordination as the motif in the rational design of supramolecular metallacyclic polygons and polyhedra, *Acc. Chem. Soc.*, 30, 502, 1997.

82. Leininger, S., Olenyuk, B., and Stang, P.J., Self-assembly of discrete cyclic nanostructures mediated by transition metals, *Chem. Rev.*, 100, 853, 2000.

83. Albrecht, M., Dicatechol ligands: novel building blocks for metallo-supramolecular chemistry, *Chem. Soc. Rev.*, 27, 281, 1998.

84. Beissel, T., Powers, R.E., and Raymond, K.N., Coordination number incommensurate cluster formation. Part 1. Symmetry-based metal complex cluster formation, *Angew. Chem. Intl. Ed. Engl.*, 35, 1084, 1996.

85. Baxter, P.N.W., Lehn, J.-M., Baum, G., and Fenske, D., The design and generation of inorganic cylindrical cage architectures by metal-ion-directed multicomponent self-assembly, *Chem. Eur. J.*, 5, 102, 1999.

86. Baxter, P.N.W., Lehn, J.-M., Kneisel, B.O., Baum, G., and Fenske, D., The designed self-assembly of multicomponent and multicompartmental cylindrical nanoarchitectures, *Chem. Eur. J.*, 5, 113, 1999.

87. Berl, V., Huc, I., Lehn, J.-M., DeCian, A., and Fischer, J., Induced fit selection of a barbiturate receptor from a dynamic structural and conformational/configurational library, *Eur. J. Org. Chem.*, 11, 3089, 1999.

88. Youinou, M.-T., Rahmouri, N., Fischer, J., and Osborn, J.A., Self-organization of a tetranuclear complex with a planar arrangement of copper(I) ions: synthesis, structure, and electrochemical properties, *Angew. Chem. Intl. Ed. Engl.*, 31, 733, 1992.

89. Maverick, A.W., Ivie, M.L., Waggenspack, J.W., and Fronzek, F.R., Intramolecular binding of nitrogen bases to a cofacial binuclear copper(II) complex, *Inorg. Chem.*, 29, 2403, 1990.

90. Fujita, M., Sasaki, O., Mitsuhashi, T., Fujita, T., Yazaki, J., Yamaguchi, K., and Ogura, K.J., On the structure of transition-metal-linked molecular squares, *J. Chem. Soc., Chem. Commun.*, 1535, 1996.

91. Fujita, M., Supramolecular self-assembly of finite and infinite frameworks through coordination, *Synth. Org. Chem. Jpn.*, 54, 953, 1996.

92. Lee, S.B., Hwang, S.G., Chung, D.S., Yun, H., and Hong, J.-I., Guest-induced reorganization of a self-assembled Pd(II) complex, *Tetrahedron Lett.*, 39, 873, 1998.

93. Schnebeck, R.-D., Randaccio, L., Zangrando, E., and Lippert, B., Molecular triangle from en-Pt(II) and 2,2'-bipyrazine, *Angew. Chem. Intl. Ed. Engl.*, 37, 119, 1998.

94. Stang, P.J., Persky, N., and Manna, J., Molecular architecture via coordination: self-assembly of nanoscale platinum containing molecular hexagons, *J. Am. Chem. Soc.*, 119, 4777, 1997.

95. Baxter, P.N.W., Hanan, G.S., and Lehn, J.-M., Inorganic arrays via multicomponent self-assembly: the spontaneous generation of ladder architectures, *Chem. Commun.*, 2019, 1996.

96. Swiegers, G.F. and Malefetse, T.J., New self-assembled structural motifs in coordination chemistry, *Chem. Rev.*, 100, 3483, 2000.

97. Hanan, G.S., Arana, C.R., Lehn, J.-M., Baum, G., and Fenske, D., Coordination arrays: synthesis and characterization of rack-type dinuclear complexes, *Chem. Eur. J.*, 2, 1292, 1996.

98. Hanan, G.S., Arana, C.R., Lehn, J.-M., Baum, G., and Fenske, D., Synthesis, structure, and properties of dinuclear and trinuclear rack-type Ru(II) complexes, *Angew. Chem. Intl. Ed. Engl.*, 34, 1122, 1995.

99. Johnson, D.W. and Raymond, K.N., The self-assembly of a $[Ga_4L_6]^{+12}$- tetrahedral cluster thermodynamically driven by host-guest interactions, *Inorg. Chem.*, 40, 5157, 2001.

100. Mann, S., Huttner, G., Zsolnia, L., and Heinze, K., Supramolecular host–guest compounds with tripod-metal templates as building blocks at the corners, *Angew. Chem. Intl. Ed. Engl.*, 35, 2808, 1997.

101. Lehninger, A.L., *Biochemistry*, Worth Publishers, New York, 1970.

102. Kool, E.T., Preorganization of DNA: design principles for improving nucleic acid recognition by synthetic oligonucleotides, *Chem. Rev.*, 97, 1473, 1997.

103. Seeman, N.C., Nucleic acid junctions and lattices, *J. Theor. Biol.*, 99, 237, 1982.

104. Seeman, N.C., Nucleic acid nanostructures and topology, *Angew. Chem. Intl. Ed. Engl.*, 37, 3220, 1998.

105. Shi, J. and Bergstrom, D.E., Assembly of novel DNA cycles with rigid tetrahedral linkers, *Angew. Chem. Intl. Ed. Engl.*, 36, 111, 1997.

106. Yan, H., Zhang, X., Shen, Z., and Seeman, N.C., A robust DNA mechanical device controlled by hybridization topology, *Nature*, 415, 62, 2002.

107. Kool, E.T., Morales, J.C., and Guckian, K.M., Mimicking the structure and function of DNA: Insights into DNA stability and replication, *Angew. Chem. Intl. Ed. Engl.*, 39, 990, 2000.

108. Leumann, C.J., Design and evaluation of oligonucleotide analogues, *Chimia*, 55, 295, 2001.

109. Wu, C.W., Sanborn, T.J., Zuckermann, R.N., and Barron, A.E., Peptoid oligomers with α-chiral, aromatic side chains: effects of chain length on secondary structure, *J. Am. Chem. Soc.*, 123, 2958, 2001.

110. Wu, C.W., Sanborn, T.J., Huang, K., Zuckermann, R.N., and Barron, A.E., Peptoid oligomers with α-chiral, aromatic side chains: sequence requirements for the formation of stable peptoid helices, *J. Am. Chem. Soc.*, 123, 6778, 2001.

111. Offord, R.E., *Semisynthetic Proteins*, Wiley Interscience, New York, 1980.

112. Newkome, G.R., Moorefield, C.N., and Vögtle, F., *Dendritic Macromolecules: Concepts, Syntheses, Perspectives*, VCH, Weinheim, Germany, 1996.

113. Matthews, O.A., Shipway, A.N., and Stoddart, J.F., Dendrimers — branching out from curiosities into new technologies, *Prog. Polym. Sci.*, 23, 1, 1998.

114. Voit, B.I., Dendritic polymers — from aesthetic macromolecules to commercially interesting materials, *Acta. Polym.*, 46, 87, 1995.

115. Tomalia, D.A., Naylor, A.M., and Goddard, W.A., III., Starburst dendrimers: control of size, shape, surface chemistry, topology and flexibility in the conversion of atoms to macroscopic materials, *Angew. Chem. Intl. Ed. Engl.*, 29, 138, 1990.

116. Grayson, S.M. and Fréchet, J.M.J., Convergent dendrons and dendrimers: from synthesis to applications, *Chem. Rev.*, 101, 3819, 2001.

117. Zeng, F. and Zimmerman, S.C., Dendrimers in supramolecular chemistry: from molecular recognition to self-assembly, *Chem. Rev.*, 97, 1681, 1997.

118. De Gennes, P.G. and Hervet, H., Statistics of "starburst" polymers, *J. Phys. Lett.*, 44, 351, 1983.

119. Tomalia, D.A., Starburst/cascade dendrimers: fundamental building blocks for a new nanoscopic chemistry set, *Aldrichimica Acta*, 26, 91, 1993.

120. Newkome, G.R., Gupta, V.k., Baker, G.R., and Yao, Z.-Q., Cascade molecules: a new approach to micelles. A [27]-Arborol, *J. Org. Chem.*, 50, 2003, 1985.

121. de Brabander-van de Berg, E.M.M. and Meijer, E.W., Poly-(propylene imine) dendrimers – large-scale synthesis by heterogeneously catalyzed hydrogenations, *Angew. Chem. Intl. Ed. Engl.*, 32, 1308, 1993.

122. Hawker, C.J. and Fréchet, J. M.J., Preparation of polymers with controlled molecular architectures. A new convergent approach to dendritic macromolecules, *J. Am. Chem. Soc.*, 112, 7638, 1990.

123. Xu, Z.F. and Moore, J. S., Stiff dendritic macromolecules. 3. Rapid construction of large-size phenylacetylene dendrimers up to 12.5 nanometers in molecular diameter, *Angew. Chem. Intl. Ed. Engl.*, 32, 1354, 1993.

124. Numata, M., Ikeda, A., Fukuhara, C., and Shinkai, S., Dendrimers can act as a host for [60]fullerene, *Tetrahedron Lett.*, 40, 6945, 1999.

125. Zimmerman, S.C., Wang, Y., Bharathi, P., and Moore, J.S., Analysis of amidinium guest complexation by comparison of two classes of dendrimer hosts containing a hydrogen bonding unit at the core, *J. Am. Chem. Soc.*, 120, 2172, 1998.

126. Hawker, C.J., Wooley, K.L. and Fréchet, J.M.J., Unsymmetrical three-dimensional macromolecules: preparation and characterization of strongly dipolar dendritic macromolecules, *J. Am. Chem. Soc.*, 115, 4375, 1993.

127. Smith, D.K. and Müller, L., Dendritic biomimicry: microenvironmental effects on tryptophan fluorescence, *J. Chem. Soc., Chem. Commun.*, 1915, 1999.

128. Rheiner, P.B. and Seebach, D., Dendritic TADDOLs: synthesis, characterization and use in the catalytic enantioselective addition of Et_2Zn to benzaldehyde, *Chem. Eur. J.*, 5, 3221, 1999.

129. Yamago, S., Furukawa, M., Azumaa, A., and Yoshida, J., Synthesis of optically active dendritic binaphthols and their metal complexes for asymmetric catalysis, *Tetrahedron Lett.*, 39, 3783, 1998.

130. Bhyrappa, P., Young, J. K., Moore, J. S., and Suslick, K.S., Dendrimer-metalloporphyrins: synthesis and catalysis, *J. Am. Chem. Soc.*, 118, 5708, 1996.

131. Dandliker, P.J., Deiderich, F., Gross, M., Knobler, C.B., Louati, A., and Sanford, E.M., Dendritic porphyrins: modulation of the redox potential of the electroactive chromophore by peripheral multifunctionality, *Angew. Chem. Intl. Ed. Engl.*, 33, 1739, 1994.

132. Newkome, G.R., Güther, R., Moorefield, C.N., Cardullo, F., Echegoeyen, L., Pérez-Cordero, E., and Luftmann, H., Chemistry of micelles, routes to dendritic networks: bis-dendrimers by coupling of cascade macromolecules through metal centers, *Ang. Chem. Int. Ed. Engl.*, 34(18): 2023–2026, 1995.

133. Avent, A.G., Birkett, P.R., Paolucci, F., Roffia, S., Taylor, R., and Wachter, N.K., Synthesis and electrochemical behavior of [60]fullerene possessing poly(arylacetylene) dendrimer addends, *J. Chem. Soc., Perkin Trans. 2*, 1409, 2000.

134. Gorman, C.B. and Smith, J.C., Structure–property relationships in dendritic encapsulation, *Acc. Chem. Res.*, 34, 60, 2001.

135. Amabilino, D.B., Ashton, P.R., Balzani, V., Brown, C.L., Credi, A., Frechet, J.M.J., Leon, J.W., Raymo, F.M., Spencer, N., Stoddart, J.F., and Venturi, M., Self-Assembly of [n]rotaxanes bearing dendritic stoppers, *J. Am. Chem. Soc.*, 118, 12012, 1996.

136. Wooley, K.L., Hawker, C.J., and Fréchet, J.M.J., Unsymmetrical three-dimensional macromolecules: preparation and characterization of strongly dipolar dendritic macromolecules, *J. Am. Chem. Soc.*, 115, 11496, 1993.

137. Hawker, C.J., Wooley, K.L., and Fréchet, J.M.J., Unimolecular micelles and globular amphiphiles — dendritic macromolecules as novel recyclable solubilization agents, *J. Chem. Soc., Perkin Trans 1.*, 21, 1287, 1993.

138. Jansen, J.F.G.A., Meijer, E.W., and de Brabander-van den Berg, E.M.M., The dendritic box: shape-selective liberation of encapsulated guests, *J. Am. Chem. Soc.*, 117, 4417, 1995.

139. Newkome, G.R., Suprasupermolecular chemistry: the chemistry within the dendrimer, *Pure App. Chem.*, 70, 2337, 1998.

140. Newkome, G.R., Woosley, B.D., He, E., Moorefield, C.N., Guther, R., Baker, G.R., Escamilla, G.H., Merrill, J., and Luftmann, H., Supramolecular chemistry of flexible, dendritic-based structures employing molecular recognition, *J. Chem. Soc., Chem. Commun.*, 2737, 1996.

141. Stevelmans, S., Van Hest, J.C.M., Jansen, J.F.G.A., Van Boxtel, D.A.F.J., de Brabander-van den Berg, E.M.M., and Meijer, E.W., Synthesis, characterization, and guest–host properties of inverted unimolecular dendritic micelles, *J. Am. Chem. Soc.*, 118, 7398, 1996.

142. Chow, H.-F., Chan, I.Y.-K., Chan, D.T.W., and Kwok, R.W.M., Dendritic models of redox proteins: X-ray photoelectron spectroscopy and cyclic voltammetry studies of dendritic bis(terpyridine) iron(II) complexes, *Chem. Eur. J.*, 2, 1085, 1996.

143. Castagnola, M., Cassiano, L., Lupi, A., Messana, I., Patamia, M., Rabino, R., Rossetti, D.V., and Giardina, B., Ion-exchange electrokinetic capillary chromatography with starburst (PAM–AM) dendrimers — a route towards high-performance electrokinetic capillary chromatography, *J. Chromatogr.*, 694, 463, 1995.

144. Muijselaar, P.G.H.M., Claessens, H.A., Cramers, C.A., Jansen, J.F.G.A., Meijers, E.W., de Brabander-Van den Berg, E.M.M., and Vanderwal, S., Dendrimers as pseudo-stationary phases in electrokinetic chromatography, *HRC J. High. Res. Chromat.*, 18, 121, 1995.

145. Newkome, G.R., Weis, C.D., Moorefield, C.N., Baker, G.R., Childs, B.J. and Epperson, J., Isocyanate-based dendritic building blocks: combinatorial tier construction and macromolecular-property modification, *Angew. Chem. Intl. Ed. Engl.*, 37, 307, 1998.

146. Smith, D.K. and Diederich, F., Dendritic hydrogen bonding receptors: enantiomerically pure dendroclefts for the selective recognition of monosaccharides, *J. Chem Soc., Chem. Commun.*, 22, 2501, 1998.

147. Smith, D.K. Zingg, A., and Diederich, F., Dendroclefts. Optically active dendritic receptors for the selective recognition and chirooptical sensing of monosaccharide guests, *Helv. Chim. Acta.*, 82, 1225, 1999.

148. Cardona, C.M., Alvarez, J., Kaifer, A.E., McCarley, T.D., Pandey, S., Baker, G.A., Bonzagni, N.J., and Bright, F.V. Dendrimers functionalized with a single fluorescent dansyl group attached "off center;" synthesis and photophysical studies, *J. Am. Chem. Soc.*, 122, 6139, 2000.

149. Zimmerman, S.C., Zeng, F.W., Reichert, D.E.C., and Kolotuchin, S.V., Self-assembling dendrimers, *Science*, 271, 1095, 1996.

150. Jansen, J.F.G.A., de Brabander-van der Berg, E.M.M., and Meijer, E.W., Encapsulation of guest molecules into a dendritic box, *Science*, 266, 1226, 1994.

151. Jansen, J.F.G.A., Meijer, E.W., and de Brabander-van der Berg, E.M.M., Bengal rose-at-dendritic box, *Macromol. Symp.*, 102, 27, 1996.

Section 5

Functional Structures
and Mechanics

17

Nanomechanics

CONTENTS

Boris I. Yakobson

Rice University

Abstract

The general aspects and issues of nanomechanics are discussed and illustrated by an overview of the mechanical properties of nanotubes: linear elastic parameters, nonlinear elastic instabilities and buckling, inelastic relaxation, yield strength and fracture mechanisms, and kinetic theory. A discussion of theoretical and computational studies is supplemented by brief summaries of experimental results for the entire range of the deformation amplitudes. Atomistic scenarios of coalescence welding and the role of noncovalent forces (supramolecular interactions) between the nanotubes are also discussed due to their significance in potential applications.

17.1 Introduction

Nanomechanics is an emerging area of research that deals with the mechanical properties and behavior of small materials systems. A span of several nanometers in two dimensions (wires, rods, etc.) or three dimensions (clusters, particles, etc.) is a simplistic criterion for a system to be considered *nano*. This formally may exclude the rich and established science of *surfaces* and interfaces, which are small in only one dimension. One can also argue that a *dislocation* is not a nano-object; while it formally meets the criterion, it is never isolated from a crystal lattice and as such has long been a subject of research in solid-state physics. To distinguish itself from the well-established dynamics of molecules, nanomechanics relies heavily on the heuristics and methods of mechanical engineering and structural mechanics. It mainly deals with objects of distinct geometrical shape and function: rods, beams, shells, plates, membranes, etc. At the same time, due to the small length scale, nanomechanics also relies on physics — specifically, interatomic and molecular forces, methods of quantum chemistry, solid-state physics, and statistical mechanics. With these approaches come a variety of numerical and computational methods (molecular dynamics, Monte Carlo simulations, classical empirical interatomic potentials, tight-binding approximation, density functional theory, etc.). This cross-disciplinary aspect makes this area both complex and exciting for research and education.

Macroscopic mechanics primarily deals with continuum representation of material, neglecting the underlying atomic structure that manifests primarily at a smaller scale. In this context it is interesting to realize that a continuum model of a finite object is not self-contained and inevitably leads to a notion of atom as a discrete building block. Indeed, elastic response of continuum is quantified by its moduli, e.g., Young's modulus, Y (J/m^3). A boundary surface of a material piece of finite size L must be associated with certain extra energy, surface energy γ (J/m^2). A combination γ/Y is dimensional (m) and constitutes a length not contained within such a finite continuum model, which points to some other inherent parameter of the material, a certain size a. The surface energy is the additional work to "overstretch" or tear apart an elastic continuum. Such work equals to the energy of the formed boundary (two boundaries), that is $aY/2 \approx 2\gamma$, and thus $a \approx 4\gamma/Y$. With typical Y = 50 GPa and γ = 1 J/m^2, one gets a = 0.1 nm, a reasonably accurate atomic size.[1,2] The notion of *indivisibles* has been well known since Democritus, and therefore a simple mechanical measurement could yield an estimate of atomic size much earlier than the more sophisticated Brownian motion theory. This discussion also shows that for an object of nanometer scale, its grainy, atomistic structure comes inevitably into the picture of its mechanics and that atomistic or hybrid-multiscale methods are necessary.

Among the numerous subjects of nanomechanics research (tips, contact junctions, pores, whiskers, etc.), carbon nanotubes (CNTs)[3] have earned a special place, receiving much attention. Their molecularly precise structure, elongated and hollow shape, effective absence of a surface (which is no different from the bulk, at least for the single-walled cylinders, or SWNTs), and superlative covalent bond strength are among the traits that put CNTs in the focus of nanomechanics. Discussion of numerous other objects as well as details of the multiscale methods involved in nanomechanics (for example, see recent monograph[4]) is far beyond the scope of this chapter.

It is noteworthy that the term *resilient* was first applied not to nanotubes but to smaller fullerene cages, in the study of high-energy collisions of C_{60}, C_{70}, and C_{84} bouncing from a solid wall of H-terminated diamond. The absence of any fragmentation or other irreversible atomic rearrangement in the rebounding cages was somewhat surprising and indicated the ability of fullerenes to sustain great elastic distortion. The very same property of resilience becomes more significant in the case of carbon nanotubes, because their elongated shape, with the aspect ratio close to a thousand, makes the mechanical properties especially interesting and important due to potential structural applications. An accurate simulation (with realistic interatomic and van der Waals forces) in Figure 17.1[5] vividly illustrates the appeal of CNTs as a nano-mechanical object: well-defined cylindrical shape, compliance to external forces, and an expected type of response qualitatively analogous to a common macroscopic behavior puts these objects among molecular chemical physics, elasticity theory, and mechanical engineering.

The utility of nanotubes as elements in nanoscale devices or composite materials remains a powerful motivation for research in this area. While the feasibility of the practical realization of these applications is currently unknown, another incentive comes from the fundamental materials physics. There is an interesting duality in the nanotubes. CNTs possess, simultaneously, molecular size and morphology as well as sufficient translational symmetry to perform as very small (nano-) crystals with well-defined primitive cells, surfaces, possibility of transport, etc. Moreover, in many respects they can be studied as well defined engineering structures; and many properties can be discussed in traditional terms of moduli, stiffness or compliance, or geometric size and shape. The mesoscopic dimensions (a nanometer diameter), combined with the regular, almost translation-invariant morphology along the micrometer lengths (unlike other polymers, usually coiled), make nanotubes unique and attractive objects of study, including the study of mechanical properties and fracture in particular.

Indeed, fracture of materials is a complex phenomenon whose theory generally requires a multiscale description involving microscopic, mesoscopic, and macroscopic modeling. Numerous traditional approaches are based on a macroscopic continuum picture that provides an appropriate model except at the region of actual failure, where a detailed atomistic description (involving chemical bond breaking) is needed. Nanotubes, due to their relative simplicity and atomically precise morphology, offer the

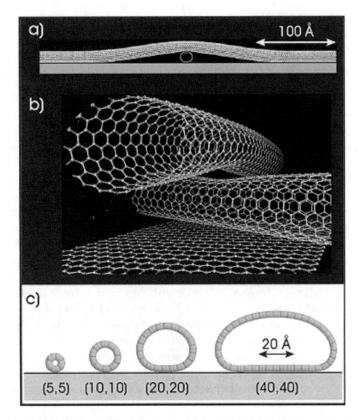

FIGURE 17.1 Molecular mechanics calculations on the axial and radial deformation of single-wall carbon nanotubes. (a) Axial deformation resulting from the crossing of two (10,10) nanotubes. (b) Perspective close up of the same crossing showing that both tubes are deformed near the contact region. (c) Computed radial deformations of single-wall nanotubes adsorbed on graphite. (Adapted from Hertel, T., R.E. Walkup, and P. Avouris, Deformation of carbon nanotubes by surface van der Waals forces. *Phys. Rev. B*, 1998. 58(20): pp. 13870–13873. With permission.)

opportunity of addressing the validity of different macroscopic and microscopic models of fracture and mechanical response. Contrary to crystalline solids, where the structure and evolution of ever-present surfaces, grain-boundaries, and dislocations under applied stress determine the plasticity and fracture of the material, nanotubes possess simpler structure while still able to show rich mechanical behavior within elastic or inelastic brittle or ductile domain. This second, theoretical–heuristic value of nanotube research supplements their import due to anticipated practical applications. A morphological similarity of fullerenes and nanotubes to their macroscopic counterparts, like geodesic domes and towers, makes it compelling to test the laws and intuition of macromechanics in a scale ten orders of magnitude smaller.

Section 17.2 discusses theoretical linear elasticity and results for the elastic moduli, compared wherever possible with the experimental data. The nonlinear elastic behavior, buckling instabilities, and shell model are presented in Section 17.3, with mention of experimental evidence parallel to theoretical results. Yield and failure mechanisms in tensile load are presented in Section 17.4, with emphasis on the combined dislocation theory and computational approach. More recent results of kinetic theory of fracture and strength evaluation in application to CNTs are briefly presented in Section 17.5. Fast molecular tension tests are recalled in the context of kinetic theory. Section 17.6 presents some of the most recent results on CNT "welding," a process essentially the reverse of fracture. In Section 17.7 we also briefly discuss the large-scale mechanical deformation of nanotubes caused by their attraction to each other, and the relation between nanomechanics and statistical persistence length of CNT in thermodynamic suspension. Throughout the discussion we do not attempt to provide a comprehensive review of broad activities in the field. This presentation is mainly based on the author's research started at North Carolina State

University and continued at Rice University. Broader or more comprehensive discussion can be found in other relatively recent reviews by the author[6,7] (see also Chapters 18 and 19 in this book.)

17.2 Linear Elastic Properties

Numerous theoretical calculations are dedicated to linear elastic properties, when displacements (strain) are proportional to forces (stress). We recently revisited[8] this issue in order to compare, with the same method, the elasticity of three different materials: pure carbon (C), boron–nitride (BN), and fluorinated carbon (C_2F). Due to obvious uncertainty in definition of a nanotube cross-section, the results should be represented by the values of in-plane stiffness, C (J/m^2). The values computed with Gaussian-based density functional theory are C = 345 N/m, 271 N/m, and 328 N/m for C, BN, and C_2F, respectively. These values in *ab initio* calculations are almost independent of nanotube diameter and chirality (consistent with the isotropic elasticity of a hexagonal, two-dimensional lattice), somewhat in contrast to previous reports based on tight-binding or classical empirical approximations. Notably, substantial fluorination causes almost no change in the in-plane stiffness, because the bonding involves mainly π-system while the stiffness is largely due to in-plane σ-system. For *material* property assessments, the values of bulk moduli (based on a graphite-type 0.34 nm spacing of layers) yield 1029 GPa, 810 GPa, and 979 GPa — all very high. Knowing the elastic shell parameter C immediately leads to accurate calculation of a nanotube-beam bending stiffness K (proportional to the cube of diameter, ~d^3), as discussed later in Sections 17.3 and 17.7. It also allows us to compute vibration frequencies of the tubules, e.g. symmetric breathing mode frequency, $f \sim 1/d$,[8] detectable in Raman spectroscopy.

An unexpected feature discovered in the course of that study is the localized strain induced by the attachment of fluorine. This shifts the energy minimum of the C_2F shell lattice from an *unstrained* sheet toward the highly curved polygonal cylinders (for C_2F composition of a near square shape, Figure 17.2). Equilibrium free angle is ~72°.

Theoretical values agree reasonably well with experimental values of Young's modulus. It was first estimated[9] by measuring free-standing room-temperature vibrations in a transmission electron microscope (TEM). The motion of a vibrating cantilever is governed by the known fourth-order wave equation,

$$y_{tt} = -(YI/\rho A)y_{xxxx}$$

where A is the cross-sectional area and ρ is the density of the rod material. For a clamped rod the boundary conditions are such that the function and its first derivative are zero at the origin, and

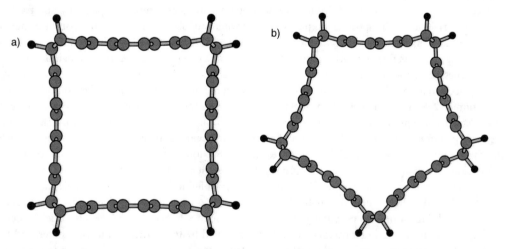

FIGURE 17.2 Geometries of the polygonal fluorinated carbon tubes: (a) square F_4–(10,10), and (b) pentagonal F_5–(10,10). (From Kudin, K.N., G.E. Scuseria, and B.I. Yakobson, C_2F, BN and C nanoshell elasticity by *ab Initio* computations. *Phys. Rev. B*, 2001. 64: p. 235406. With permission.)

the second and third derivatives are zero at the end of the rod. Thermal nanotube vibrations are essentially elastic relaxed phonons in equilibrium with the environment; therefore, the amplitude of vibration changes stochastically with time. The amplitude of those oscillations was defined by means of careful TEM observations of a number of CNTs and yields the values of moduli within a range near 1 Tpa.

Another way to probe the mechanical properties of nanotubes is to use the tip of an AFM (atomic-force microscope) to bend an anchored CNT while simultaneously recording the force exerted by the tube as a function of the displacement from its equilibrium position.[10] Obtained values also vary from sample to sample but generally are close to Y = 1 TPa. Similar values have been obtained with yet another accurate technique[11] based on a resonant electrostatic deflection of a multi-wall carbon nanotube under an external AC field. The detected decrease in stiffness must be related to the emergence of a different bending mode for the nanotube. In fact, this corresponds to a wave-like distortion–buckling of the inner side of the CNT. Nonlinear behavior is discussed in more detail in the next section. Although experimental data on elastic modulus are not very uniform, they correspond to the values of in-plane rigidity C = 340 – 440 N/m, to the values Y = 1.0 – 1.3 GPa for multiwall tubules, and to Y = 4C/d = (1.36 – 1.76) TPa nm/d for SWNTs of diameter d.

17.3   Nonlinear Elasticity and Shell Model

Almost any molecular structure can sustain very large deformations, compared with the range common in macroscopic mechanics. A less obvious property of CNTs is that the specific features of large nonlinear strain can be understood and predicted in terms of continuum theories. One of the outstanding features of nanotubes is their hollow structure, built of atoms densely packed along a closed surface that defines the overall shape. This also manifests itself in dynamic properties of molecules, highly resembling the macroscopic objects of continuum elasticity known as *shells*. Macroscopic shells and rods have long been of interest: the first study dates back to Euler, who discovered the elastic instability. A rod subject to longitudinal compression remains straight but shortens by some fraction ε, proportional to the force, until a critical value (the Euler force) is reached. It then becomes unstable and buckles sideways at $\varepsilon > \varepsilon_{cr}$, while the force almost does not vary. For hollow tubules there is also a possibility of local buckling in addition to buckling as a whole. Therefore, more than one bifurcation can be observed, thus causing an overall nonlinear response to the large deforming forces (note that local mechanics of the constituent shells may well still remain within the elastic domain).

In nanomechanics, the theory of shells was first applied in our early analysis of buckling and since then has served as a useful guide.[12–15] Its relevance for a covalent-bonded system only a few atoms in diameter was far from being obvious. Molecular dynamics (MD) simulations seem better suited for objects that small.

Figure 17.3 shows a simulated nanotube exposed to *axial compression*. The atomic interaction was modeled by the Tersoff–Brenner potential, which reproduces the lattice constants and binding energies of graphite and diamond. The end atoms were shifted along the axis by small steps, and the whole tube was relaxed by a conjugate-gradient method while keeping the ends constrained. At small strains the total energy (Figure 17.3a) grows as $E(\varepsilon) = 1/2\ E''\cdot\varepsilon^2$. The presence of four singularities at higher strains was quite a striking feature and the patterns (b) – (e) illustrate the corresponding morphological changes. The shading indicates strain energy per atom, equally spaced from below 0.5 eV (brightest) to above 1.5 eV (darkest). The sequence of singularities in $E(\varepsilon)$ corresponds to a loss of molecular symmetry from $D_{\infty h}$ to S_4, D_{2h}, C_{2h}, and C_1. This evolution of the molecular structure can be put in the framework of continuum elasticity.

The intrinsic symmetry of a graphite sheet is hexagonal, and the elastic properties of two-dimensional hexagonal structures are isotropic. A curved sheet can also be approximated by a uniform shell with only two elastic parameters — flexural rigidity, D, and its in-plane stiffness, C. The energy of a shell is given by a surface integral of the quadratic form of local deformation:

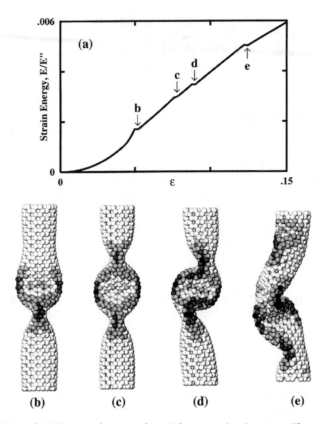

FIGURE 17.3 Simulation of a (7,7) nanotube exposed to axial compression, L = 6 nm. The strain energy (a) displays four singularities corresponding to shape changes. At ε_c = 0.05, the cylinder buckles into the pattern (b), displaying two identical flattenings —"fins" perpendicular to each other. Further increase of ε enhances this pattern gradually until at ε_2 = 0.076 the tube switches to a three-fin pattern (c), which still possesses a straight axis. In a buckling sideways at ε_3 = 0.09, the flattenings serve as hinges, and only a plane of symmetry is preserved (d). At ε_4 = 0.13, an entirely squashed asymmetric configuration forms (e). (From Yakobson, B.I., C.J. Brabec, and J. Bernholc, Nanomechanics of carbon tubes: instabilities beyond the linear response. *Phys. Rev. Lett.*, 1996. 76(14): pp. 2511–2514. With permission.)

$$
\begin{aligned}
E = \frac{1}{2}\iint \{D[(\kappa_x + \kappa_y)^2 - 2(1-v)(\kappa_x\kappa_y - \kappa_{xy}^2)] \\
+ \frac{C}{(1-v^2)}[(\varepsilon_x + \varepsilon_y)^2 - 2(1-v)(\varepsilon_x\varepsilon_y - \varepsilon_{xy}^2)]\}dS
\end{aligned}
\tag{17.1}
$$

where κ is the curvature variation, ε is the in-plane strain, and x and y are local coordinates. In order to adapt this formalism to a graphitic tubule, the values of D and C can be identified by comparison with the detailed *ab initio* and semi-empirical studies of nanotube energetics at small strains. Indeed, the second derivative of total energy with respect to axial strain corresponds to the in-plane rigidity C (Section 17.2). Similarly, the strain energy as a function of tube diameter d corresponds to $2D/d^2$ in Equation (17.1). Using recent *ab initio* calculations, one obtains C = 56 eV/atom = 340 J/m² and D = 1.46 eV. The Poisson ratio v = 0.15 was extracted from a reduction of the diameter of a tube stretched in simulations. A similar value is obtained from experimental elastic constants of single crystal graphite. One can make a further step toward a more tangible picture of a tube as having wall thickness h and Young modulus Y_s. Using the standard relations $D = Y_s h^3/12(1-v^2)$ and $C = Y_s h$, one finds Y_s = 3.9 TPa and h = 0.089 nm. With these parameters, linear stability analysis allows one to assess the nanotube behavior under strain.

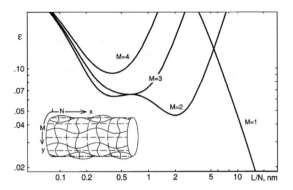

FIGURE 17.4 The critical strain levels for a continuous, 1 nm wide shell tube as a function of its scaled length L/ N. A buckling pattern (M,N) is defined by the number of half-waves 2M and N in y and x directions, respectively; e.g., a (4,4) pattern is shown in the inset. The effective moduli and thickness are fit to graphene. (From Yakobson, B.I., C.J. Brabec, and J. Bernholc, Nanomechanics of carbon tubes: instabilities beyond the linear response. *Phys. Rev. Lett.*, 1996. 76(14): pp. 2511–2514. With permission.)

To illustrate the efficiency of the shell model, consider briefly the case of imposed axial compression. A trial perturbation of a cylinder has a form of Fourier harmonics, with M azimuthal lobes and N halfwaves along the tube (Figure 17.4, inset), i.e., sines and cosines of arguments 2My/d and Nπx/L. At a critical level of the imposed strain, ε_c(M,N), the energy variation (Equation 17.1) vanishes for this shape disturbance. The cylinder becomes unstable and lowers its energy by assuming an (M,N) pattern. For tubes of d = 1 nm with the shell parameters identified above, the critical strain is shown in Figure 17.4. According to these plots, for a tube with L > 10 nm, the bifurcation is first attained for M = 1, N = 1. The tube preserves its circular cross-section and buckles sideways as a whole; the critical strain is close to that for a simple rod

$$\varepsilon_c = 1/2(\pi d/L)^2 \tag{17.2}$$

or four times less for a tube with hinged (unclamped) ends. For a shorter tube the situation is different. The lowest critical strain occurs for M = 2 (and N ≥ 1, see Figure 17.4), with a few separated flattenings in directions perpendicular to each other, while the axis remains straight. For such a local buckling, in contrast to (Equation 17.2), the critical strain depends little on length and estimates to $\varepsilon_c = \sqrt{D/C}\,d^{-1} = (2/\sqrt{3})(1 - v^2)^{-1/2}\,h\cdot d^{-1}$ in the so-called *Lorenz* limit. For a nanotube one finds

$$\varepsilon_c = 0.1\ \text{nm}/d. \tag{17.3}$$

Specifically, for the 1 nm wide tube of length L = 6 nm, the lowest critical strains occur for M = 2 and N = 2 or 3 (Figure 17.4). This is in accord with the two- and three-fin patterns seen in Figure 17.3b,c. Higher singularities cannot be quantified by the linear analysis, but they look like a sideways beam buckling, which at this stage becomes a nonuniform object.

Axially compressed tubes of greater length and/or tubes simulated with hinged ends (equivalent to a doubled length) first buckle sideways as a whole at a strain consistent with (Equation 17.2). After that the compression at the ends results in bending and a local buckling inward. This illustrates the importance of the "beam-bending" mode, the softest for a long molecule and most likely to attain significant amplitudes due to either thermal vibrations or environmental forces. In simulations of *bending*, a torque rather than force is applied at the ends and the bending angle θ increases step-wise. While a notch in the energy plot can be mistaken for numerical noise, its derivative dE/dθ drops significantly. This unambiguously shows an increase in tube compliance — a signature of a buckling event. In bending, only one side of a tube is compressed and thus can buckle. Assuming that it buckles when its local strain, ε = κ·d/2, where κ is the local curvature, is close to that in axial compression, Equation (17.3), we estimate the critical curvature as:

$$\kappa_c = 0.2 \text{ nm/d}^2. \tag{17.4}$$

In simulation of *torsion*, the increase of azimuthal angle ϕ between the tube ends results in abrupt changes of energy and morphology.[12,13,16] In continuum model, the analysis based on Equation (17.1) is similar to that outlined above, except that it involves skew harmonics of arguments such as $N\pi x/L \pm 2My/d$. For overall beam-buckling (M = 1),

$$\phi_c = 2(1+\nu)\pi \tag{17.5}$$

and for the cylinder-helix flattening (M = 2),

$$\phi_c = 0.06 \text{ nm}^{3/2} \text{ L/d}^{5/2}. \tag{17.6}$$

The latter should occur first for $L < 140 \text{ d}^{5/2}$ nm, which is true for all tubes we simulated. However, in simulations it occurs later than predicted by Equation (17.6). The ends, kept circular in simulation, which is physically justifiable, by a presence of rigid caps on normally closed ends of a molecule, deter the through flattening necessary for the helix to form (unlike the local flattening in the case of an axial load).

Experimental evidence provides sufficient support to the notion of high resilience of SWNT. An early observation of noticeable flattening of the walls in a close contact of two multi-walled nanotubes (MWNT) has been attributed to van der Walls forces pressing the cylinders to each other.[17] Collapsed forms of the nanotube (*nanoribbons*), also caused by van der Waals attraction, have been observed in experiment, and their stability can be explained by the competition between the van der Waals and elastic energies.[18] Any additional torsional strain imposed on a tube in an experimental environment also favors flattening[12,13] and facilitates the collapse. Graphically more striking evidence of resilience is provided by bent structures[19] as well as the more detailed observations that actually stimulated our research in nanomechanics.[20] An accurate measurement with the atomic-force microscope (AFM) tip detects the "failure" of a multi-wall tubule in bending, which essentially represents nonlinear buckling on the compressive side of the bent tube. The estimated measured local stress is 15–28 GPa, very close to the calculated value. Buckling and ripple of the outermost layers[21,22] in a dynamic resonant bending have been directly observed and are responsible for the apparent softening of MWNT of larger diameters.[7,23]

17.4 Atomic Relaxation and Failure Mechanisms

The important issue of ultimate tensile strength of CNTs is inherently related to the atomic relaxation in the lattice under high strain. This thermally activated process was first predicted to consist of a sequence of individual bond rotations in the approach based on dislocation theory.[21,24,25] Careful computer simulations demonstrate feasibility of this mechanism and allow us to quantify important energy aspects.[26,27] It has been shown that in a crystal lattice, such as the wall of a CNT, a yield to deformation must begin with a homogeneous nucleation of a slip by the shear stress present. The nonbasal edge dislocations emerging in such slip have a well-defined core, a pentagon–heptagon pair (5/7). Therefore, the prime dipole is equivalent to the Stone–Wales (SW) defect. The nucleation of this prime dislocation dipole *unlocks* the nanotube for further relaxation — either brittle cleavage or a plastic flow. Remarkably, the latter corresponds to a motion of dislocations along the helical paths (glide *planes*) within the nanotube wall. This causes a step-wise (quantized) necking, when the domains of different chiral symmetry (and therefore different electronic structure) are formed, thus coupling the mechanical and electrical properties.[21,24,25] It has further been shown[21,22,24,27–29] that the energy of such nucleation explicitly depends on CNT helicity (chirality).

Below, starting with dislocation theory, we deduce the atomistics of mechanical relaxation under extreme tension. Locally, the wall of a nanotube differs little from a single graphene sheet, a two-dimensional crystal of carbon. When a uniaxial tension σ (N/m — for the two-dimensional wall, it is convenient to use force per unit length of its circumference) is applied, it can be represented as a sum of expansion (locally isotropic within the wall) and a shear of a magnitude $\sigma/2$ (directed at $\pm 45°$ with respect to tension). Generally, in a macroscopic crystal the shear stress relaxes by a movement of *dislocations*, the edges of the atomic extra-planes. Burger's vector **b** quantifies the mismatch in the lattice due to a dislocation. Its glide requires only local atomic rearrangements and presents the easiest way for strain release, with sufficient thermal agitation. In an initially *perfect* lattice such as the wall of a nanotube, a yield to a great axial tension begins with a homogeneous *nucleation* of a slip, when a dipole of dislocations (a tiny loop in three-dimensional case) first has to form. The formation and further glide are driven by the reduction of the applied-stress energy, as characterized by the elastic Peach–Koehler force on a dislocation. The force component along **b** is proportional to the shear in this direction and thus depends on the angle between the Burger's vector and the circumference of the tube

$$f_b = -1/2\sigma|\mathbf{b}|\sin 2\theta \qquad (17.7)$$

The max $|f_b|$ is attained on two $\pm 45°$ lines, which mark the directions of a slip in an isotropic material under tension.

The graphene wall of the nanotube is not isotropic; its hexagonal symmetry governs the three glide planes — the three lines of closest zigzag atomic packing, oriented at $120°$ to each other (corresponding to the $\{10\bar{1}l\}$ set of planes in three-dimensional graphite). At nonzero shear these directions are prone to slip. The corresponding c-axis edge dislocations involved in such slip are indeed known in graphite. The six possible Burger's vectors $1/3a<2\bar{1}\bar{1}0>$ have a magnitude $b = a = 0.246$ nm (lattice constant), and the dislocation core is identified as a 5/7 pentagon–heptagon pair in the honeycomb lattice of hexagons. Therefore, the primary nucleated dipole must have a 5/7/7/5 configuration (a 5/7 attached to an inverted 7/5 core). This configuration is obtained in the perfect lattice (or a nanotube wall) by a $90°$ rotation of a single C–C bond, well known in fullerene science as a Stone–Wales diatomic interchange. One is led to conclude that the SW transformation is equivalent to the smallest slip in a hexagonal lattice and must play a key role in the nanotube relaxation under external force.

The preferred glide is the closest to the maximum-shear $\pm 45°$ lines and depends on how the graphene strip is rolled up into a cylinder. This depends on nanotube helicity specified by the chiral indices (c_1, c_2) or a chiral angle θ indicating how far the circumference departs from the leading zigzag motif a_1. The max $|f_b|$ is attained for the dislocations with $\mathbf{b} = \pm(0,1)$ and their glide reduces the strain energy

$$E_g = -|f_b a| = -Ca^2/2 \cdot \sin(2\theta + 60°) \cdot \varepsilon \qquad (17.8)$$

per one displacement, a. Here ε is the applied strain, and $C = Yh = 342$ N/m can be derived from the Young modulus of $Y = 1020$ GPa and the interlayer spacing $h = 0.335$ nm in graphite. One then obtains $Ca^2/2 = 64.5$ eV. Equation 17.8 allows one to compare different CNTs (assuming similar amount of preexisting dislocations): the more energetically favorable is the glide in a tube, the earlier it must yield to applied strain.

In a pristine nanotube molecule, the 5/7 dislocations have to first emerge as a dipole by a prime SW transformation. Topologically, the SW defect is equivalent to either one of the two dipoles, each formed by a $\sim a/2$ slip. Applying Equation (17.8) to each of the slips, one finds

$$E_{sw} = E_o - A \cdot \varepsilon - B \cdot \sin(2\theta + 30°) \cdot \varepsilon \qquad (17.9)$$

The first two terms, the zero-strain formation energy and possible isotropic dilation, do not depend on chiral symmetry. The symmetry-dependent third term, which can also be derived as a leading term in the Fourier series, describes the fact that SW rotation gains more energy in an armchair ($\theta = 30°$) CNT, making it thermodynamically the weakest and most inclined to SW nucleation of the dislocations, in contrast to the zigzag ($\theta = 0$), where the nucleation is least favorable.

Consider, for example, a (c, c) armchair CNT as a typical representative (we will also see below that this armchair type can undergo a more general scenario of relaxation). The initial stress-induced SW rotation creates a geometry that can be viewed as either a dislocation dipole or a tiny crack along the equator. Once "unlocked," the SW defect can ease further relaxation. At this stage, either brittle (dislo-

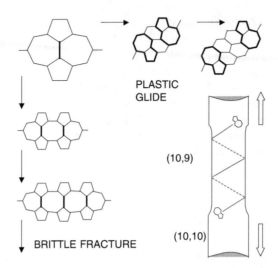

FIGURE 17.5 SW transformations of an equatorially oriented bond into a vertical position create a nucleus of relaxation (top left corner). It evolves further as either a crack — brittle fracture route (left column) — or as a couple of dislocations gliding away along the spiral slip plane (plastic yield, top row). In both cases only SW rotations are required as elementary steps. The step-wise change of the nanotube diameter reflects the change of chirality (bottom right image), causing the corresponding variations of electrical properties (Adapted from Yakobson, B.I., Mechanical relaxation and "intramolecular plasticity" in carbon nanotubes. *Appl. Phys. Lett.*, 1998. 72(8): pp. 918–920. With permission.)

cation pileup and crack extension) or plastic (separation of dislocations and their glide away from each other) routes are possible, the former usually at larger stress and the latter at higher temperatures.

Formally, both routes correspond to a further sequence of SW switches. The 90° rotation of the bonds at the "crack tip" (Figure 17.5, left column) will result in a 7/8/7 flaw and then 7/8/8/7, etc. This further strains the bonds partitions between the larger polygons, leading eventually to their breakage, with the formation of greater openings such as 7/14/7. If the crack, represented by this sequence, surpasses the critical Griffith size, it cleaves the tubule.

In a more interesting distinct alternative, the SW rotation of another bond (Figure 17.5, top row) divides the 5/7 and 7/5, as they become two dislocation cores separated by a single row of hexagons. A next similar SW switch results in a double-row separated pair of the 5/7s, and so on. This corresponds, at very high temperatures, to a plastic flow *inside* the nanotube molecule, when the 5/7 and 7/5 twins glide away from each other driven by the elastic forces, thus reducing the total strain energy (Equation 17.8). One remarkable feature of such glide is due to mere cylindrical geometry: the glide "planes" in the case of nanotubes are actually spirals, and the slow thermally activated Brownian walk of the dislocations proceeds along these well-defined trajectories. Similarly, their extra-planes are only the rows of atoms also curved into the helices.

A nanotube with a 5/7 defect in its wall loses axial symmetry and has a bent equilibrium shape; the calculations show[30] the junction angles <15°. Interestingly, then, an exposure of an even achiral nanotube to the axially symmetric tension generates two 5/7 dislocations; and when the tension is removed, the tube freezes in an asymmetric configuration, S-shaped or C-shaped, depending on the distance of glide, that is, time of exposure. This seemingly "symmetry-violating" mechanical test is a truly nanoscale phenomenon. Of course, the symmetry is conserved statistically, because many different shapes form under identical conditions.

When the dislocations sweep a noticeable distance, they leave behind a tube segment changed strictly following the topological rules of dislocation theory. By considering a planar development of the tube segment containing a 5/7, for the new chirality vector c' one finds

$$(c_1', c_2') = (c_1, c_2) - (b_1, b_2) \qquad (17.10)$$

with the corresponding reduction of diameter, d. While the dislocations of the first dipole glide away, a generation of another dipole results in further narrowing and proportional elongation under stress, thus forming a neck as shown above. The orientation of a generated dislocation dipole is determined every time by the Burger's vector closest to the lines of maximum shear (±45° cross at the endpoint of the current circumference-vector c). The evolution of a (c,c) tube will be: (c,c) → (c,c−1) → (c,c−2) → ... (c,0) → (c−1,1) or (c,−1) → (c−1,0) → (c−2,1) or (c−1,−1) → (c−2,0) → [(c−3,1) or (c−2,−1)] → (c−3,0), etc. It abandons the armchair (c,c) type entirely, but then oscillates in the vicinity of the zigzag kind (c,0), which appears to be a peculiar attractor. Correspondingly, the diameter for a (10,10) tube changes step-wise, d = 1.36, 1.29, 1.22, 1.16 nm, etc.; the local stress grows in proportion, and this quantized necking can be terminated by a cleave at late stages. Interestingly, such plastic flow is accompanied by the change of electronic structure of the emerging domains, governed by the vector (c_1,c_2). The armchair tubes are metallic, and others are semiconducting with the different band gap values. The 5/7 pair separating two domains of different chirality has been discussed as a pure carbon heterojunction, is argued to cause the current rectification detected in a nanotube nanodevice,[31] and can be used to modify, in a controlled way, the electronic structure of the tube. Here we see how this electronic heterogeneity can arise from a mechanical relaxation at high temperature: if the initial tube were armchair-metallic, the plastic dilation would transform it into a semiconducting type irreversibly.[24,25,32]

While the above analysis is based on the atomic picture (structure and interactions), recent developments[33] offer an approach where the fracture nucleation can be described rather elegantly within nonlinear continuum mechanics (based on classical interatomic forces for carbon). Whether this approach can describe change in chirality, temperature dependence, or temporal aspects of relaxation should yet be explored.

The dislocation theory allows one to expand the approach to other materials, and we have recently applied it to boron nitride (BN) nanotubes.[34] While the binding energy in BN is lower than in CNT, the formation of a 5/7/7/5 defect can be more costly due to Coulomb repulsion between emerging BB and NN pairs (Figure 17.6a). (Bonding in BN is partially ionic with a strong preference to chemical neighbor alternation in the lattice.) Another dislocation pair 4/8/8/4 that preserves the alternation must be considered (Figure 17.6b). It turns out that the quantitative results are sensitive to the level of theory accuracy. Tight-binding approximation[35] underestimates the repulsion by almost 3 eV.[34] *Ab initio* DFT calculations show that 5/7/7/5 is metastable lowest energy defect in BN, and its formation energy 5.5 eV is higher than 3.0 eV in carbon,[34] thus suggesting higher thermodynamic stability of BN under tensile load. Relaxation under compression is different as it involves skin-type buckling also investigated recently.[36]

FIGURE 17.6 The geometries of (5,5) BN tubule with (a) 5/7/7/5 defect emerging at high tension and temperature, and (b) 4/8/8/4 dislocation dipole. (From Bettinger, H., T. Dumitrica, G.E. Scuseria, and B.I. Yakobson, Mechanically induced defects and strength of BN nanotubes. *Phys. Rev. B*, 2002. 65 (Rapid Comm.): p. 041406. With permission.)

17.5   Kinetic Theory of Strength

Computer simulations have provided compelling evidence of the mechanisms discussed above. By carefully tuning a *quasi-static* tension in the tubule and gradually elevating its temperature with extensive periods of MD annealing, the first stages of the mechanical yield of CNT have been observed (Figure 17.7).[26,27] In simulation of tensile load, the novel patterns in plasticity and breakage, as described above, clearly emerge. At very *high strain* rate the details of primary defects cannot be seen; and they only begin to emerge at a higher strain level, giving the impression of exceedingly high breaking strain.[16]

Fracture, of course, is a *kinetic* process where time is an important parameter. Even a small tension, as any nonhydrostatic stress, makes material thermodynamically meta-stable and a generation of defects energetically favorable. Thus the important issue of strength remains beyond the defect formation energy and its reduction with the applied tension. Recently we developed kinetic theory in application to CNT.[29,37] In this approach we evaluate conditions (strain ε, temperature T) when the probability P of defect formation becomes significant within laboratory test time Δt:

$$P = \nu \, \Delta t \, N_B/3 \, \Sigma_m \, \exp \, [-E_m(\varepsilon, \chi)/k_b T] \sim 1. \qquad (17.11)$$

Here $\nu = k_b T/h$ is the usual attempted frequency, and N_B is the number of bonds in the sample. Activation barrier $E_m(\varepsilon,\chi)$ must be computed as a function of strain and chirality χ of the tubule, and then the solution of this equation with respect to ε gives the breaking strain values. This approach involved substantial computational work in finding the saddle points and energies (Figure 17.8a) for a variety of conditions and for several transition-state modes (index m in the summation above). Obtained yield strain near 17% (Figure 17.8b)[29,37] is in reasonable agreement with the series of experimental reports.

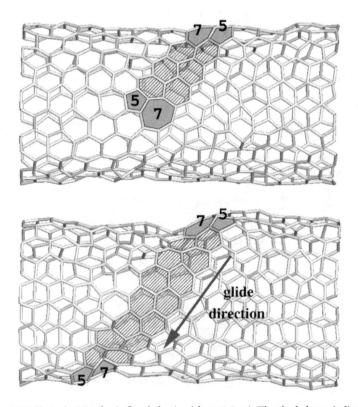

FIGURE 17.7   T = 3000 K, strain 3%, plastic flow behavior (about 2.5 ns). The shaded area indicates the migration path of the 5/7 edge dislocation. (Adapted from Nardelli, M.B., B.I. Yakobson, and J. Bernholc, Brittle and ductile behavior in carbon nanotubes. *Phys. Rev. Lett.*, 1998. 81(21): pp. 4656–4659. With permission.)

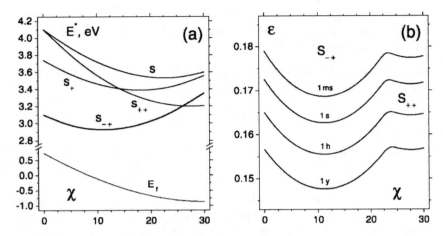

FIGURE 17.8 Activation barrier values (here, computed within classical multibody potential, a) serve as input to the rate Equation (17.11) and the calculation of the yield strain as a function of time (here, from 1 ms to 1 year, b), temperature (here, 300 K) and chiral symmetry (χ). (From Samsonidze, G.G., G.G. Samsonidze, and B.I. Yakobson, Kinetic theory of symmetry-dependent strength in carbon nanotubes. *Phys. Rev. Lett.*, 2002. 88: p. 065501. With permission.)

We currently are implementing a similar approach with an *ab initio* level of saddle point barriers calculations.[38] Preliminary data show higher 8–9 eV barriers, but their reduction with tension is also faster.

Previously performed high strain rate simulations have shown temperature dependence of breaking strain,[13,16] consistent with the kinetic theory.[29] In a constant strain rate arrangement (when the ends of the sample are pulled from the equilibrium with certain rate), the rate equation is slightly modified to its integral form. However, the main contribution comes from the vicinity of the upper limit:

$$P = v\ N_B/3 \int \Sigma_m \exp\{-E_m[\varepsilon(t), \chi]/k_bT\}dt \sim 1 \qquad (17.12)$$

Simple analysis reveals certain invariant $T \times \log(v\Delta t)$ of the time of failure and temperature (provided the constant strain). Detailed simulations could shed additional light on this aspect.[39]

17.6 Coalescence of Nanotubes as a Reversed Failure

Understanding the details of failure mechanism has led us recently[40] to investigate an opposite process, a coalescence of nanoscale clusters analogous to macroscopic sintering or welding. Fusion of smaller components into a larger whole is a ubiquitous process in condensed matter. In molecular scale it corresponds to chemical synthesis, where exact rearrangement of atoms can be recognized. Coalescence or sintering of macroscopic parts is usually driven by well-defined thermodynamic forces (frequently, surface energy reduction), but the atomic evolution path is redundant and its exact identification is irrelevant. Exploring a possibility of the two particles merging with atomic precision becomes compelling in nanometer scale, where one aspires to "arrange the atoms one by one." Are the initial and final states connected by a feasible path of atomic movements, or separated by insurmountable barriers? Direct MD investigation is usually hampered by energy landscape traps and, beyond very few atomic steps, needs to be augmented with additional analysis.

An example of very small particles is offered by fullerene cages and CNTs. Fusion of fullerenes has been previously reported, and the lateral merging (diameter doubling) of CNT has been observed and simulated.[41,42] In contrast, head-to-head coalescence of CNT segments remains unexplored and of particular theoretical and practical interest.

1. Is it permitted by rigorous topology rules to eliminate all the pentagons always present in the CNT ends and thus dissolve the caps completely?

2. Can this occur through a series of well-defined elementary steps, and what is the overall energy change if the system evolves through the intermediate disordered states to the final purely hexagonal lattice of continuous tubule?

If feasible, such "welding" can lead to increase of connectivity in CNT arrays in bundles/ropes and crystals and thus significantly improve the mechanical, thermal, and electrical properties of material. In addition, determining the atomistic steps of small-diameter tubes coalescence (with the end-caps identical to half-buckyball) can shed light on the underlying mechanism of condensed phase conversion or CNT synthesis from C_{60} components.

We have reported[40] for the first time atomically precise routes for complete coalescence of generic fullerene cages: cap-to-cap CNT and C_{60} merging together to form a defectless final structure. The entire process is reduced to a sequence of Stone–Wales bond switches and, therefore, is likely the lowest energy path of transformation. Several other examples of merging follow immediately as special cases: coalescence of buckyballs in peapod, joining of the two (5,5) tubes as in Figure 17.9, welding the (10,10) to (10,10) following Figure 17.10, etc. The approach remains valid for arbitrary tubes with the important constraint of unavoidable grain boundary for the tubes of different chirality. The junction of (n,m) and (n',m') must contain 5/7 dislocations or their equivalent of (n-n',m-m') total Burger's vector.[24] The proposed mechanism[40] has important implications for nanotube material (crystals, ropes) processing and property enhancement, engineering of nanoscale junctions of various types, and possible growth mechanisms with the C_{60} and other nanoparticles as feedstock. In the context of nanomechanics, an interesting feature of the late stages of coalescence is the annealing and annihilation of 5/7 pairs in a process exactly reverse to the formation and glide of these dislocation cores in the course of yield and failure under tension.

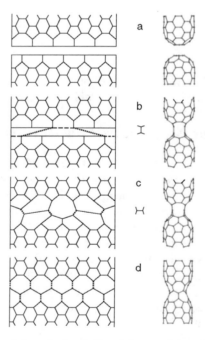

FIGURE 17.9 Two-dimensional geodesic projection (left) and the actual three-dimensional structures (right) show the transformations from a pair of separate (5,5) tubes (a) to a single defect-free nanotube. Primary polymerization links form as two other bonds break (b, dashed lines). The $\pi/2$-rotations of the links (the bonds subject to such SW flip are dotted) and the SW flips of the four other bonds in (c) produce a (5,0) neck (d). It widens by means of another ten SW rotations, forming a perfect single (5,5) tubule (not shown). (From Zhao, Y., B.I. Yakobson, R.E. Smalley, Dynamic topology of fullerene coalescence. *Phys. Rev. Lett.*, 88:185501, 2002. With permission.)

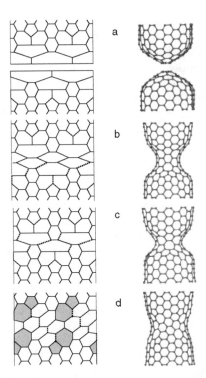

FIGURE 17.10 Two-dimensional projections (left) and the computed three-dimensional intermediate structures (right) in the coalescence of the two (10,10) nanotubes: separate caps (a) in a sequence similar to Figure 17.9 develop a (5,5) junction (b), which then shortens (c) and widens into a (10,5) neck (d). Glide of the shaded 5/7 dislocations completes the annealing into a perfect (10,10) CNT (not shown). Due to the 5-fold symmetry, only two cells are displayed. (From Zhao, Y., B.I. Yakobson, R.E. Smalley, Dynamic topology of fullerene coalescence. *Phys. Rev. Lett.,* 88:185501, 2002. With permission.)

17.7 Persistence Length, Coils, and Random FuzzBalls of CNTS

Van der Waals forces play an important role not only in the interaction of the nanotubes with the substrate but also in their mutual interaction. The different shells of an MWNT interact primarily by van der Waals forces; single-wall tubes form ropes for the same reason. A different manifestation of van der Waals interactions involves the self-interaction between two segments of the same single-wall CNT to produce a closed ring (loop).[43] SWNT rings were first observed in trace amounts in the products of laser ablation of graphite and assigned a toroidal structure. More recently, rings of SWNTs were synthesized with large yields (~50%) from straight nanotube segments. These rings were shown to be loops, not tori.[43] The synthesis involves the irradiation of raw SWNTs in a sulfuric acid–hydrogen peroxide solution by ultrasound. This treatment both etches the CNTs, shortening their length to about 3–4 μm, and induces ring formation.

The formation of coils by CNTs is particularly intriguing. While coils of biomolecules and polymers are well-known structures, they are stabilized by a number of interactions that include hydrogen bonds and ionic interactions. On the other hand, the formation of nanotube coils is surprising given the high flexural rigidity (K = Young's modulus times areal moment of inertia) of CNTs and the fact that CNT coils can only be stabilized by van der Waals forces. However, estimates based on continuum mechanics show that, in fact, it is easy to compensate for the strain energy induced by the coiling process through the strong adhesion between tube segments in the coil. Details of this analysis can be found in the original reports[43] or in our recent review.[7] Here we will outline briefly a different and more common situation where the competition of elastic energy and the intertubular linkage is important. Following our recent work,[44] we will discuss the connection between the nanomechanics of CNTs and their *random* curling in a suspension or a raw synthesized material of buckypaper.

SWNTs are often compared with polymer chains as having very high rigidity and, therefore, large persistence length. In order to quantify this argument, we note that a defectless CNT has almost no static flexing, although 5/7 defects, for example, introduce a small kink-angle of 5–15° and could cause some static curvature; but their energies are high and concentration is usually negligible. Only dynamic elastic flexibility should be considered. If $\mathbf{u}(s)$ is a unit direction vector tangent to the CNT at contour length point s and the bending stiffness is K, then statistical probability of certain shape $\mathbf{u}(s)$ is

$$P[\mathbf{u}(s)] = \exp[-1/2(K/k_bT) \int (\partial \mathbf{u}/\partial s)^2 ds] = \exp[-1/2\ L \int (\partial \mathbf{u}/\partial s)^2 ds] \qquad (17.13)$$

Here persistence length is $L = (K/k_bT)$. For a (10,10) SWNT of radius R = 0.7 nm and the effective wall thickness h = 0.09 nm (see Sections 17.2 and 17.3), the bending stiffness is very close to $K = \pi CR^3$ (C = 345 N/m is the in-plane stiffness, based on *ab initio* calculations). Persistence length at room temperature, therefore, is $L_1[(10,10), 293\ K] \sim 0.1$ mm, in the macroscopic range much greater than for most polymer molecules. The result can be generalized for a single SWNT of radius R as

$$L_1 = (30K/T)(R/0.7nm)^3\ mm \qquad (17.14)$$

or N times more for a bundle of N tubes (assuming additive stiffness for the case of weak lateral cohesion of the constituent SWNTs). For example, for a smallest close-packed bundle of seven (one surrounded by six neighbors), this yields $L_7 = 1$ mm. Such an incoherent bundle and a solid-coherent bundle with perfect lateral adhesion provide the lower and upper limits for the persistence length, $NL_1 < L_N < N^2L_1$. Remarkably, these calculations show that the true thermodynamic persistence length of small CNT bundles or even an individual SWNT is in the macroscopic range from a fraction of a millimeter and greater. This means that highly curved structures often observed in buckypaper mats (Figure 17.11) are attributed not to thermodynamic fluctuations but rather to residual mechanical forces preventing these coils from unfolding. Elastic energy of a typical micron-size (r ~ 1 μm) curl-arc is much greater than

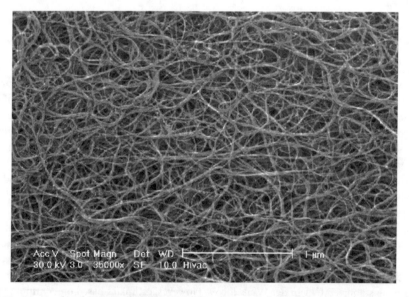

FIGURE 17.11 Raw-produced SWNTs often form ropes/bundles, entangled and bent into a rubbery structure called "buckypaper." The length scale of bends is much smaller than the persistence length for the constituent filaments. Shown here is material produced by HiPco (high pressure CO) synthesis method. (Adapted from O'Connell, M.J. et al., *Chem. Phys. Lett.*, 342, 265, 2001. With permission.)

thermal, $U_{curl} \sim k_b T(L/r)^2 >> k_b T$. At the same time, a force required to maintain this shape is $F_{curl} \sim K/r^2 = N$ pN, several piconewtons, where N is the number of SWNTs in each bundle. This is much less than a force per single chemical bond (~ 1 nN), and therefore any occasional lateral bonding between the tubules can be sufficient to prevent them from disentanglement.

Acknowledgments

The author appreciates the support from NASA Ames, Air Force Research Laboratory, Office of Naval Research DURINT grant, and the Nanoscale Science and Engineering Initiative of the National Science Foundation, award number EEC-0118007.

References

1. Yakobson, B.I., Morphology and rate of fracture in chemical decomposition of solids. *Phys. Rev. Lett.*, 1991. 67(12): pp. 1590–1593.
2. Yakobson, B.I., Stress-promoted interface diffusion as a precursor of fracture. *J. Chem. Phys.*, 1993. 99(9): pp. 6923–6934.
3. Iijima, S., Helical microtubules of graphitic carbon. *Nature*, 1991. 354: pp. 56–58.
4. Phillips, R.B., *Crystals, Defects and Microstructures*. Cambridge: Cambridge University Press, 2001, p. 780.
5. Hertel, T., R.E. Walkup, and P. Avouris, Deformation of carbon nanotubes by surface van der Waals forces. *Phys. Rev. B*, 1998. 58(20): pp. 13870–13873.
6. Yakobson, B.I. and R.E. Smalley, Fullerene nanotubes: $C_{1,000,000}$ and beyond. *Am. Sci.*, 1997. 85(4): pp. 324–337.
7. Yakobson, B.I. and P. Avouris, Mechanical properties of carbon nanotubes, in *Carbon Nanotubes*, M.S. Dresselhaus, G. Dresselhaus, and P. Avouris (Eds.), Springer: Berlin, 2001, pp. 287–327.
8. Kudin, K.N., G.E. Scuseria, and B.I. Yakobson, C_2F, BN and C nanoshell elasticity by *ab Initio* computations. *Phys. Rev. B*, 2001. 64: p. 235406.
9. Treacy, M.M.J., T.W. Ebbesen, and J.M. Gibson, Exceptionally high Young's modulus observed for individual carbon nanotubes. *Nature*, 1996. 381: pp. 678–680.
10. Wong, E.W., P.E. Sheehan, and C.M. Lieber, Nanobeam mechanics: elasticity, strength and toughness of nanorods and nanotubes. *Science*, 1997. 277: pp. 1971–1975.
11. Poncharal, P., Z.L. Wang, D. Ugarte, and W.A. Heer, Electrostatic deflections and electromechanical resonances of carbon nanotubes. *Science*, 1999. 283: pp. 1513–1516.
12. Yakobson, B.I., C.J. Brabec, and J. Bernholc, Nanomechanics of carbon tubes: instabilities beyond the linear response. *Phys. Rev. Lett.*, 1996. 76(14): pp. 2511–2514.
13. Yakobson, B.I., C.J. Brabec, and J. Bernholc, Structural mechanics of carbon nanotubes: from continuum elasticity to atomistic fracture. *J. Comp. Aided Mater. Des.*, 1996. 3: pp. 173–182.
14. Garg, A., J. Han, and S.B. Sinnott, Interactions of carbon-nanotubule proximal probe tips with diamond and graphite. *Phys. Rev. Lett.*, 1998. 81(11): pp. 2260–2263.
15. Srivastava, D., M. Menon, and K. Cho, Nanoplasticity of single-wall carbon nanotubes under uniaxial compression. *Phys. Rev. Lett.*, 1999. 83(15): pp. 2973–2976.
16. Yakobson, B.I., M.P. Campbell, C.J. Brabec, and J. Bernholc, High strain rate fracture and C-chain unraveling in carbon nanotubes. *Comp. Mater. Sci.*, 1997. 8: pp. 341–348.
17. Ruoff, R.S., et al., Radial deformation of carbon nanotubes by van der Waals forces. *Nature*, 1993. 364: pp. 514–516.
18. Chopra, N.G. et al., Fully collapsed carbon nanotubes. *Nature*, 1995. 377: pp. 135–138.
19. Despres, J.F., E. Daguerre, and K. Lafdi, Flexibility of graphene layers in carbon nanotubes. *Carbon*, 1995. 33(1): pp. 87–92.
20. Iijima, S., C.J. Brabec, A. Maiti, and J. Bernholc, Structural flexibility of carbon nanotubes. *J. Chem. Phys.*, 1996. 104(5): pp. 2089–2092.

21. Yakobson, B.I., *Dynamic Topology and Yield Strength of Carbon Nanotubes,* in *Fullerenes*, Electrochemical Society, Pennington, NJ, 1997.

22. Smalley, R.E. and B.I. Yakobson, The future of the fullerenes. *Solid State Commn.*, 1998. 107(11): pp. 597–606.

23. Liu, J.Z., Q. Zheng, and Q. Jiang, Effect of a rippling mode on resonances of carbon nanotubes. *Phy. Rev. Lett.*, 2001. 86: pp. 4843–4846.

24. Yakobson, B.I., Mechanical relaxation and "intramolecular plasticity" in carbon nanotubes. *Appl. Phys. Lett.*, 1998. 72(8): pp. 918–920.

25. Yakobson, B.I., Physical property modification of nanotubes. U. S. Patent 6,280,677 B1, 2001.

26. Nardelli, M.B., B.I. Yakobson, and J. Bernholc, Mechanism of strain release in carbon nanotubes. *Phys. Rev. B*, 1998. 57: p. R4277.

27. Nardelli, M.B., B.I. Yakobson, and J. Bernholc, Brittle and ductile behavior in carbon nanotubes. *Phys. Rev. Lett.*, 1998. 81(21): pp. 4656–4659.

28. Yakobson, B.I., G. Samsonidze, and G.G. Samsonidze, Atomistic theory of mechanical relaxation in fullerene nanotubes. *Carbon*, 2000. 38: p. 1675.

29. Samsonidze, G.G., G.G. Samsonidze, and B.I. Yakobson, Kinetic theory of symmetry-dependent strength in carbon nanotubes. *Phys. Rev. Lett.*, 2002. 88: p. 065501.

30. Chico, L., et al., Pure carbon nanoscale devices: nanotube heterojunctions. *Phys. Rev. Lett.*, 1996. 76(6): pp. 971–974.

31. Collins, P.G., et al., Nanotube nanodevice. *Science*, 1997. 278: pp. 100–103.

32. Tekleab, D., D.L. Carroll, G.G. Samsonidze, and B.I. Yakobson, Strain-induced electronic property heterogeneity of a carbon nanotube. *Phys. Rev. B*, 2001. 64: p. 035419.

33. Zhang, P., Y. Huang, H. Gao, and K.C. Hwang, Fracture nucleation in SWNT under tension: a continuum analysis incorporating interatomic potential. *ASME Trans. J. Appl. Mechanics*, 2002: in press.

34. Bettinger, H., T. Dumitrica, G.E. Scuseria, and B.I. Yakobson, Mechanically induced defects and strength of BN nanotubes. *Phys. Rev. B*, 2002. 65 (Rapid Comm.): p. 041406.

35. Zhang, P. and V.H. Crespi, Plastic deformations of boron-nitride nanotubes: an unexpected weakness. *Phys. Rev. B*, 2000. 62: p. 11050.

36. Srivastava, D., M. Menon, and K.J. Cho, Anisotropic nanomechanics of boron nitride nanotubes: nanostructured "skin" effect. *Phys. Rev. B*, 2001. 63: p. 195413.

37. Samsonidze, G.G., G.G. Samsonidze, and B.I. Yakobson, Energetics of Stone-Wales defects in deformations of monoatomic hexagonal layers. *Comp. Mater. Sci.*, 23:62, 2002.

38. Dumitrica, T., H. Bettinger, and B.I. Yakobson, Stone-Wales barriers and kinetic theory of strength for nanotubes. In progress.

39. Wei, C., K.J. Cho, and D. Srivastava, private communication: xxx.lanl.gov/abs/cond-mat/0202513.

40. Zhao, Y., B.I. Yakobson, and R.E. Smalley, Dynamic topology of fullerene coalescence. *Phys. Rev. Lett.*, 88: 185501, 2002.

41. Nikolaev, P., et al., Diameter doubling of single-wall nanotubes. *Chem. Phys. Lett.*, 1997. 266: p. 422.

42. Terrones, M., et al., Coalescence of single-walled carbon nanotubes. *Science*, 2000. 288: pp. 1226–1229.

43. Martel, R., H.R. Shea, and P. Avouris, Rings of single-wall carbon nanotubes. *Nature*, 1999. 398: p. 582.

44. Yakobson, B.I. and L.S. Couchman. Mechanical properties of carbon nanotubes as random coils: from persistence length to the equation of state. In progress.

18

Carbon Nanotubes

CONTENTS

Meyya Meyyappan
NASA Ames Research Center

Deepak Srivastava
NASA Ames Research Center

18.1 Introduction

Carbon nanotubes (CNT) were discovered by Iijima[1] as elongated fullerenes in 1991. Since then research on growth, characterization, and application development has exploded due to the unique electronic and extraordinary mechanical properties of CNTs. The CNT can be metallic or semiconducting and thus offers possibilities to create semiconductor–semiconductor and semiconductor–metal junctions useful in devices. The high tensile strength, Young's modulus, and other mechanical properties hold promise for high-strength composites for structural applications. Researchers have been exploring the potential of CNTs in a wide range of applications: nanoelectronics, sensors, field emission, displays, hydrogen storage, batteries, polymer matrix composites, body armor, reinforcement material, nanoscale reactors, and electrodes, to name a few. In this chapter, an overview of this rapidly emerging field is provided.

First, the structure of the nanotube and properties are explained. Unlike many other fields in science and engineering, the evolution of carbon nanotubes to its current level owes significantly to the contributions from modeling and simulation. Computational nanotechnology has played an early and major role in predicting as well as explaining the interesting properties of CNTs. Therefore, a section is devoted to modeling and simulation after the description of the properties. Then nanotube growth is covered in detail, followed by material development functions such as purification and characterization. Finally, a review of the current status of various applications is provided.

18.2   Structure and Properties of Carbon Nanotubes

A single-wall carbon nanotube (SWCNT) is best described as a rolled-up tubular shell of graphene sheet (Figure 18.1a) which is made of benzene-type hexagonal rings of carbon atoms.[2] The body of the tubular shell is thus mainly made of hexagonal rings (in a sheet) of carbon atoms, whereas the ends are capped by half-dome-shaped half-fullerene molecules. The natural curvature in the side-walls is due to the rolling of the sheet into the tubular structure, whereas the curvature in the end caps is due to the presence of topological (pentagonal rings) defects in the otherwise hexagonal structure of the underlying lattice. The role of a pentagonal ring defect is to give a positive (convex) curvature to the surface, which helps in closing the tube at the two ends. A multi-wall nanotube (MWCNT) is a rolled-up stack of graphene sheets into concentric SWCNTs, with the ends again either capped by half-fullerenes or kept open. A nomenclature (n,m), used to identify each single-wall nanotube, refers to integer indices of two graphene unit lattice vectors corresponding to the chiral vector of a nanotube.[2] Chiral vectors determine the directions along which the graphene sheets are rolled to form tubular shell structures and axis vectors perpendicular to the tube as explained in Reference 2. The nanotubes of type (n,n), as shown in Figure 18.1b, are commonly called armchair nanotubes because of the _/‾_/ shape, perpendicular to the tube

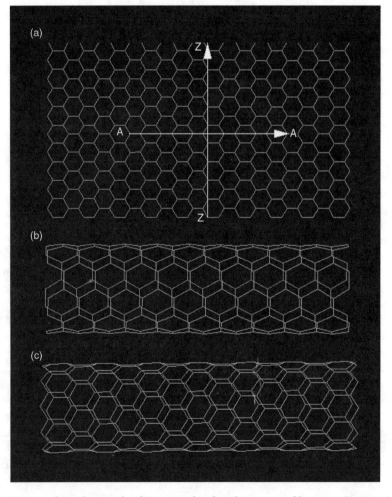

FIGURE 18.1   (a) A graphene sheet made of C atoms placed at the corners of hexagons forming the lattice with arrows AA and ZZ, denoting the rolling direction of the sheet to make (b) a (5,5) armchair nanotube and (c) a (10,0) zigzag nanotube.

(a)  (b)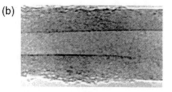

FIGURE 18.2 TEM images of (a) SWCNT and (b) MWCNT.

axis, and have a symmetry along the axis with a short unit cell (0.25 nm) that can be repeated to make the entire section of a long nanotube. Another type of nanotube (n,0) is known as zigzag nanotube (Figure 18.1c) because of the \/\/ shape perpendicular to the axis and as well as the short unit cell (0.43 nm) along the axis.[3] All the remaining nanotubes are known as chiral or helical nanotubes and have longer unit cell sizes along the tube axis. Transmission electron microscopy (TEM) images of an SWCNT and an MWCNT are shown in Figure 18.2. Details of the symmetry properties of the nanotubes of different chiralities are explained in detail in Reference 2.

The single- and multi-wall nanotubes are interesting nanoscale materials for several reasons. A single-wall nanotube can be either metallic or semiconducting, depending on its chiral vector (n,m), where n and m are two integers. A metallic nanotube is obtained when the difference n–m is a multiple of three. If the difference is not a multiple of three, a semiconducting nanotube is obtained. In addition, it is also possible to connect nanotubes with different chiralities creating nanotube heterojunctions, which can form a variety of nanoscale molecular electronic device components.

Single- and multi-wall nanotubes have very good elastomechanical properties because the two-dimensional arrangement of carbon atoms in a graphene sheet allows large out-of-plane distortions, while the strength of carbon–carbon in-plane bonds keeps the graphene sheet exceptionally strong against any in-plane distortion or fracture. These structural and material characteristics of nanotubes point toward their possible use in making the next generation of extremely lightweight but highly elastic and strong composite materials.

Nanotubes are high aspect–ratio structures with good electrical and mechanical properties. Consequently, the applications of nanotubes in field-emission displays, and scanning probe microscopic tips for metrology have begun to materialize in the commercial sector.

Because nanotubes are hollow, tubular, and caged molecules, they have been proposed as lightweight large-surface-area packing material for gas storage and hydrocarbon fuel storage devices, gas or liquid filtration devices, nanoscale containers for molecular drug delivery, and casting structures for making nanowires and nanocapsulates.

Carbon-based materials are ideally suitable as molecular-level building blocks for nanoscale systems design, fabrication, and applications. From a structural or functional materials perspective, carbon is the only element that exists in a variety of shapes and forms with varying physical and chemical properties. For example, diamond and layered graphite forms of carbon are well known, but the same carbon also exists in planar sheet, rolled-up tubular, helical spring, rectangular hollow box, and nanoconical forms. All basic shapes and forms needed to build any complex molecular-scale architectures, thus, are readily available with carbon. Additionally, by coating any carbon-based nanoscale devices with biological lipid layers and/or protein molecules, it may be possible to extend also into the rapidly expanding area of bionanotechnology.

18.3 Computational Modeling and Simulation

The structural, electronic, mechanical, and thermal properties of interacting, bulk, condensed-matter systems were studied in the earlier days with analytical approximation methods for infinite systems. Numerical simulations of the finite sample systems have become more common recently due to the availability of powerful computers. *Molecular dynamics* (MD) refers to an approach where the motion of atoms or molecules is treated in approximate finite difference equations of Newtonian mechanics. The use of classical mechanics is well justified, except when dealing with very light atoms and very low temperatures. The dynamics of complex condensed-phase systems such as metals and semiconductors is described with explicit or implicit many-body force-field functions using Embedded Atom Method (EAM)-type potentials for metals[4] and Stillinger–Weber (SW)[5] and/or Tersoff–Brenner (T–B)[6,7]-type potentials for semiconductors.[8] The T–B-type potentials are parameterized and particularly suited for carbon-based systems (such as carbon nanotubes) and have been used in a wide variety of scenarios yielding results in agreement with experimental observations. However, currently, there is no universal classical force-field function that works for all materials and in all scenarios. Consequently, one needs to be careful where true chemical changes (involving electronic rearrangements) with large atomic displacements are expected to occur.

In its global structure, a general MD code typically implements an algorithm to find a numerical solution of a set of coupled first-order ordinary differential equations given by the Hamiltonian formulation of Newton's Second Law. The equations of motion are numerically integrated forward in finite time-steps using a predictor–corrector method. A major distinguishing feature of the Tersoff–Brenner potential[6,7] is that short-range bonded interactions are reactive so that chemical bonds can form and break during the course of a simulation. Therefore, compared with other molecular dynamics codes, the neighbor list describing the environment of each atom includes only a few atoms and needs to be updated more frequently. The computational cost of the many-body bonded interactions is relatively high compared with the cost of similar methods with nonreactive interactions with simpler functional forms. As a result, the overall computational costs of both short-range interactions and long-range, nonbonding van der Waals interactions are roughly comparable.

For large-scale atomistic modeling (10^5–10^8 atoms), multiple processors are used for MD simulations, and the MD code needs to be parallelized. A route to the parallelization of a standard MD code involves decoupling the neighbor list construction from the computation of the atomic forces and parallelizing each part in the most efficient way possible. Parallelization of the MD code using Tersoff–Brenner potential for carbon atom interactions was attempted and achieved recently. An example of the parallel implementation of this classical MD code is described in detail in Reference 9. This parallelized MD code has been utilized in simulations of mechanical properties of the nanotubes, nanotube–polymer composites, mechanical strain-driven chemistry of carbon nanotubes, and molecular gears and motors powered by laser fields.[3]

In recent years, several more accurate quantum molecular dynamics schemes have been developed in which the forces between atoms are computed at each time-step via quantum mechanical calculations within the Born–Oppenheimer approximation. The dynamic motion for ionic positions is still governed by Newtonian or Hamiltonian mechanics and described by molecular dynamics. The most widely known and accurate scheme is the Car–Parrinello (CP) molecular dynamic method,[10] where the electronic states and atomic forces are described using the *ab-initio* density functional method (usually within the local density approximation, LDA). In the intermediate regimes, the tight-binding molecular dynamics (TBMD)[11] approach for up to a few thousand atoms provides an important bridge between the *ab initio* quantum MD and classical MD methods. The computational efficiency of the tight-binding method derives from the fact that the quantum Hamiltonian of the system can be parameterized. Furthermore, the electronic structure information can be easily extracted from the tight-binding Hamiltonian, which in addition also contains the effects of angular forces in a natural way. In a generalized *nonorthogonal* TBMD scheme, Menon and Subbaswami have used a minimal number of adjustable parameters to develop a transferable scheme applicable to clusters as well as bulk systems containing Si, C, B, N, and H.[12,13] The main advantage of this approach is that it can be used to find an energy-minimized structure of a nanoscale system under consideration without symmetry constraints. The parallelization of the TBMD code involves parallelization of the direct diagonalization part of the electronic Hamiltonian matrix as well as that of the MD part. The parallelization of a sparse symmetric matrix giving many eigenvalues and eigenvectors is a complex step in the simulation of large intermediate-range systems and needs development of new algorithms.

Ab initio or first-principles method is an approach to solve complex quantum many-body Schroedinger equations using numerical algorithms.[14] The *ab initio* method provides a more accurate description of quantum mechanical behavior of materials properties even though the system size is currently limited to only a few hundred atoms. Current *ab initio* simulation methods are based on a rigorous mathematical foundation of the density functional theory (DFT).[15,16] This is derived from the fact that the ground-state total electronic energy is a function of the density of the system. For practical applications, the DFT–LDA method has been implemented with a pseudopotential approximation and a plane-wave (PW) basis expansion of single-electron wave functions.[14] These approximations reduce the electronic structure problem as a self-consistent matrix diagonalization problem. One of the popular DFT simulation programs is the Vienna *Ab initio* Simulation Package (VASP), which is available through a license agreement.[17]

In computational nanotechnology research, these three simulation methods can be used in a complementary manner to improve computational accuracy and efficiency. Based on experimental observations or theoretical dynamic and structure simulations, the atomic structure of a nanosystem can first be investigated. After the nanoscale system configurations are finalized, the electronic behaviors of the system are investigated through static *ab initio* electronic energy minimization schemes[14] or through studies of the quantum conductance[18] behavior of the system. This strategy has been covered in detail in a recent review article focusing on computational nanotechnology.[3]

In the following we describe several representative examples where computational nanotechnology has clearly played an important role in either explaining some recent experimental observations or predicting structures (or properties) that were later fabricated (or measured) in experiments.

18.3.1 Nanomechanics of Nanotubes and Composites

SWCNTs and MWCNTs have been shown to have exceptionally strong and stiff mechanical characteristics along the axis of the tube and very flexible characteristics along the axis of the tube.[19–21] For axial deformations, the Young's modulus of the SWCNTs can reach beyond 1 TPa, and the yield strength can be as large as 120 GPa. The initial investigations, using classical molecular dynamics simulations with Tersoff–Brenner potential, showed that the tubes are extremely stiff under axial compression and that the system remains within elastic limit even for very large deformations (up to 15% strain).[9,19] Nonlinear elastic instabilities, with the appearance of fin-like structures, are observed during these deformations; but the system remains within elastic limits and returns to the original unstrained state as soon as the external constraining forces are removed. As shown in Figure 18.3, when compressed beyond elastic limits, the single- and multi-wall nanotubes undergo sideway bucklings, and plastic deformations occur mainly through extreme bending situations in the sideway-buckled tubes. A significantly different deformation mode, however, was also predicted where the nanotube essentially remains straight, but the structure locally collapses at the location of the deformation[20] (Figure 18.4). This generally occurs for thin tubes and for tube lengths shorter than the Euler buckling limit. The local collapse is driven by graphitic to diamond-like bonding transition at the location of the collapse. Both of the simulated collapsing mechanisms have been observed in experiments.[22] The tensile strain, on the other hand, causes formation of Stone–Wales type topological defects in the tube (Figure 18.5), which leads to thinning and collapse of the tube when stretched further.[23–25] The yielding strains in all scenarios have been simulated in detail and show a strong dependence on the rate at which strain is applied as well as the temperature of the simulation system.[26] In agreement with recent experimental observations,[27] predictions show that at room temperature and at experimentally realizable strain rates, nanotubes typically yield at about 5–10% strain.[26]

Simulations of nanotube–polymer composites are currently in the preliminary stages. A few attempts have been made to characterize mechanical and thermal properties of CNT composite materials with molecular dynamics simulations. It is expected that the superior mechanical, thermal, and electrical characteristics of individual SWCNTs and MWCNTs in a polymer, ceramic, or metal matrix could be imparted to the resulting composite material as well. Several recent experiments on the preparation

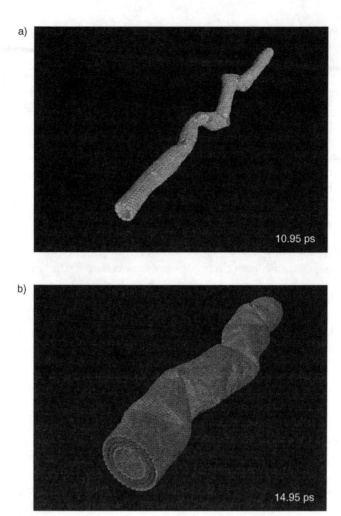

FIGURE 18.3 (a) An axially compressed single-wall nanotube within elastic limit shows sideways buckling and accumulation of strain at the tip of the sideways-buckled structure. (b) Same as in (a) except for a multi-wall nanotube with four walls.

and mechanical characterization of CNT–polymer composites have been reported.[27–31] In preliminary MD investigations of the nanotube–polymer composites above glass transition temperature, the thermal expansion coefficient of the composite matrix and diffusion coefficients of the polymer molecules increase significantly over their bare polymer matrix values. For example, the density vs. temperature profile of the polymer matrix and the composite material in Figure 18.6 shows a greater slope of the changes in density of the composite as compared with that in a bare polymer sample. The thermal expansion coefficient increases by as much as 40% by mixing of 5–10% of nanotubes in the polyethylene polymers.[32] The load transfer for axial straining of the polymer matrix composite is found to be driven by the difference in the Poisson ratio of the constituent materials. An SWCNT is typically a hard material and has a Poisson ratio of about 0.1 to 0.2, whereas polyethylene is a much softer polymeric material with a Poisson ratio of about 0.44. When the composite containing SWCNTs is under tensile strain, there is a resistance from the hard fibers to their surrounding matrix in response to the compression pressure driven by the tensile stress applied to the system. Thus the modulus of a composite can be enhanced with this mechanism even when there is not very good bonding between the SWCNT fibers and the polymer matrix. The Young's modulus is found to increase by as much as 30% for strain below the slippage limit.[32]

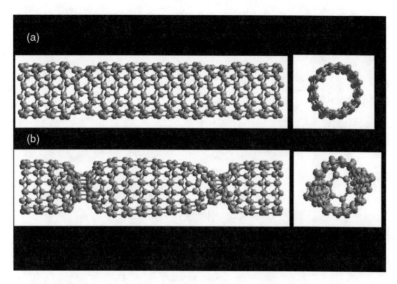

FIGURE 18.4 A 12% axially compressed (8,0) nanotube at (a) the beginning and (b) the end of a spontaneous local plastic collapse of the tube, which is driven by diamond-like bonding transitions at the location of the collapse. (b) Cross-sectional view.

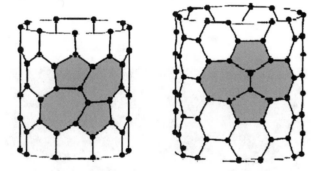

FIGURE 18.5 The Stone–Wales bond rotation on a zigzag and an armchair CNT, resulting in pentagon–heptagon pairs, can lengthen a nanotube, with the greatest lengthening for an armchair tube. (From P. Zhang and V. Crespi, *Phys. Rev. Lett.* 81, 5346, 1998. With permission.)

18.3.2   Molecular Electronics with Nanotube Junctions

The possibility of using carbon in place of silicon in the field of nanoelectronics has generated considerable enthusiasm. The metallic and semiconducting behavior as well as the electronic transport through individual single-wall nanotubes has been extensively investigated. The main thrust has been to see if the individual nanotube (or bundles of nanotubes) could be used as quantum molecular wires for interconnects in future computer systems. The ballistic electron transport through individual nanotubes has been supported by many independent studies, and it is considered to be one of the reasons that nanotubes exhibit high current density as compared with other materials at similar nanoscale.[33]

Inspired by the above, possibilities of connecting nanotubes of different diameter and chirality in nanotube heterojunctions as molecular electronic devices or switching components have also been investigated.[34] The simplest way is to introduce pairs of heptagon and pentagon in an otherwise perfect hexagonal lattice structure of the material. The resulting junction can act like a rectifying diode (Figure 18.7). Two terminal rectifying diodes were first postulated theoretically[34] and recently have been observed in experiments.[35]

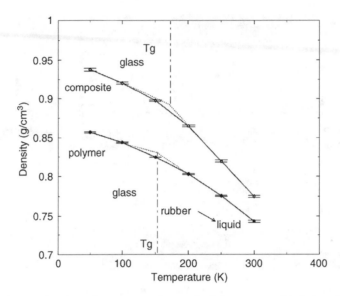

FIGURE 18.6 Density as a function of temperature for a polyethylene system (50 chains with $Np = 10$), and a CNT–polyethylene composite (2nm-long capped (10,0) CNT). The CNT composite has an increase of thermal expansion above T_g. From an MD simulation with van der Waals potential between CNT and matrix. Dihedral angle potential and torsion potential were used for the polyethylene matrix, and Tersoff–Brenner potential was used for carbon atom on the CNT.

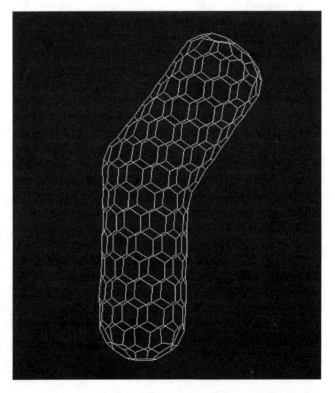

FIGURE 18.7 An example of two-terminal semiconducting (10,0)/(6,6) metallic nanotube junction that shows rectification behavior.

There are two ways to create heterojunctions with more than two terminals: first, connecting different nanotubes through topological defect mediated junctions;[36] second, laying down crossed nanotubes over each other and simply forming physically contacted or touching junctions.[37] The differences in the two approaches are the nature and characteristics of the junctions forming the device. In the first case, the nanotubes are chemically connected through bonding networks forming a stable junction that could possibly give rise to a variety of switching, logic, and transistor applications.[38] In the second case, the junction is merely through a physical contact and will be amenable to changes in the nature of the contact. The main applications in the second category will be in electromechanical bistable switches and sensors.[37] The bistable switches can act as bits in a CNT-based computing architecture.

Novel structures of carbon nanotube T- and Y-junctions have been proposed as models of three-terminal nanoscale monomolecular electronic devices.[36,39] The T-junctions can be considered as a specific case of a family of Y-junctions in which the two connecting nanotubes are perpendicular to each other (Figure 18.8). The pentagon–heptagon defect pair rule was found to be inapplicable in the

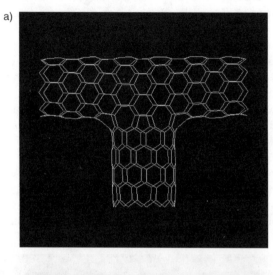

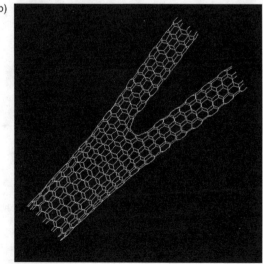

FIGURE 18.8 (a) An example of three-terminal carbon nanotube T-junction connecting metallic (9,0) and semiconducting (10,0) zigzag nanotubes. (b) An example of carbon nanotube Y-junction that is shown to exhibit rectification behavior as a two-terminal device, or a bistable switch and analog OR gate as a three-terminal device.

formation of the Y-junctions.[39] Recently, experimentalists have also succeeded in developing template-based chemical vapor deposition (CVD)[40] and pyrolysis of an organometallic precursor with nickel-ocene and thiophene[41] for the reproducible and high-yield fabrication of multi-wall carbon nanotube Y-junctions. The template-based method reports junctions consisting of large-diameter stems with two smaller branches that have an acute angle between them so they resemble tuning forks. The pyrolysis method reports multiple Y-junctions along a continuous multi-wall carbon nanotube. The electrical conductance measurements on these Y-junctions have been performed and show intrinsic nonlinear and asymmetric I-V behavior with rectification at room temperature. The quantum con-ductivity of a variety of carbon nanotube Y-junctions has also been calculated and shows current rectification under changes in the bias voltage.[42,43] The degree of rectification is found to depend on the type and nature of Y-junctions. Some junctions show good rectification while others show small leakage currents. The presence of rectification indicates, for the first time, the formation of a nanoscale molecular rectifying switch with a robust behavior that is reproducible in a high-yield fabrication method.[40,41] Moreover, simulations also show that the molecular switches thus produced can easily function as three-terminal bistable switches that are controlled by a control or *gate* voltage applied at a branch terminal.[42] Further, under certain biasing conditions, nanotube Y-junctions are shown to work as OR or XOR gates as well. The possible reasons for rectification in single-wall nanotube Y-junctions include constructive or destructive interference of the electronic wave functions through two different channels at the location of the junction; hence, the rectification is strongly influenced by the structural asymmetry across the two branches in a junction.[43]

In considering the architecture for future computing systems, we need not constrain ourselves to the specifications of the silicon-based devices, circuitry, and architecture. For example, a possible alternative architecture could be based on the structure and functioning of dendritic neurons in biological neural logic and computing systems. The tree shown in Figure 18.9 has a four-level branching structure and is made of 14 carbon nanotube Y-junctions. Such a structure is conceptually amenable to fabrication via a template-based CVD method, which is used for growing individual Y-junctions, and provides a first model of a biomimetic neural network made of single- or multi-wall carbon nanotubes.[3] In principle, such a *tree* could be trained to perform complex computing and switching applications in a single pass.

FIGURE 18.9 An illustration of a four-level dendritic neural tree made of 14 symmetric Y-junctions of the type in Figure 18.8.

18.3.3 Endofullerenes as Quantum Bits

The concept for quantum computation is based on quantum bits (qubits) which are the quantum state of a two-level system. A quantum computer with more than 20–30 qubits can outperform conventional classical computers for a certain class of computing tasks and will lead to a revolutionary increase in computing power.[44] Recent experimental demonstration of NMR quantum computation has already proven the feasibility of a device implementation of a real quantum computer.[45] However, these systems are limited in scaling up to a few qubits and would not be suitable as a platform to develop highly scalable quantum computers.

To overcome the scalability problem of NMR quantum computers, recently Kane proposed a solid-state quantum computer based on ^{31}P dopant atoms in bulk crystalline silicon.[46] In this proposal the nuclear spin of a ^{31}P atom is used as a solid-state qubit. The qubit state is controlled by hyperfine coupling to the weakly bound donor electron of the ^{31}P atom in Si lattice, and the quantum computer can be continuously scaled using microfabrication technology. However, the main challenge is that, experimentally, it is not known how to place a single dopant atom at a precise position in a silicon lattice and how to prevent the diffusion of the dopant atoms once placed in a desired position. A possible solution to these positioning and diffusion problems has been proposed using carbon-based nanotechnology.[47] In this new design, carbon diamond lattice is proposed to replace silicon diamond lattice as a host material for ^{31}P dopant atoms. This change of host materials solves both positioning and stability problem as follows.

The basic idea is to fabricate a diamond nanocrystallite with a ^{31}P dopant atom at the center using a bucky onion (multi-shell fullerenes). The sequence of fabrication steps for nanocrystallite is to first encapsulate a ^{31}P atom within a C_{60} fullerene via ion implantation methods to create an endofullerene, $P@C_{60}$, as demonstrated in a recent experiment.[48] Second, use the $P@C_{60}$ as a seed material to grow a bucky onion encapsulating the endofullerene; and third, use an e-beam or ion irradiation on the bucky onion to convert the inner core graphitic layers into a compressed diamond nanocrystallite.[49] Experiment has shown that the third step produces a compressed pristine diamond nanocrystallite with a 2 to 10 nm diameter.[49] Repeating the above fabrication steps with endofullerenes or doped bucky onions will lead to fabrication of doped diamond crystals.

The position control of ^{31}P atom qubits is conceptually feasible by fabricating arrays of 2 to 10 nm sized nanocrystallite (with ^{31}P at the center) qubits in any host dielectric material including diamond lattice. The stability of the dopant atom is ensured by much higher formation energies of vacancy (7 eV) and self-interstitial (~10 eV) defects in diamond than those in silicon lattice. The stability of the P atom is further enhanced by the stability of the P atom at substitutional site relative to interstitial site (by 15 eV). Because of higher formation energies, vacancy and self-interstitial defects are not likely to form during the graphite-to-diamond transformation process at the inner core of bucky onions; so the TED diffusion mechanism of ^{31}P atoms will be suppressed. Figure 18.10 shows the results of *ab initio* simulations of (a) $P@C_{60}$ and (b) a ^{31}P doped at the center of a diamond nanocrystallite. The planar valence electron density (a) and the ^{31}P donor electron density in (111) plane in a diamond lattice (b) are shown in the two cases.[47] The spread of the donor electron density distribution, to define the donor electron mediated interaction between the neighboring qubits, has also been estimated.

18.3.4 Sensors and Actuators

As mentioned earlier, carbon nanotubes have different electronic properties depending on their chiral vectors, ranging from metals to semiconductors (1 eV bandgap). Semiconducting SWCNTs are very promising candidates for novel sensing applications because the surface modifications, due to chemical adsorption or mechanical deformation of the nanotubes, can directly modify the electronic conductance of nanotubes. Recent experimental and theoretical works have proved that SWCNTs are extremely sensitive to gas molecules,[50,51] and both chemical reactivity and electronic properties are strongly dependent on the mechanical deformation of nanotubes.[52,53] These characteristics have given rise to possibilities for use in chemical, vibration, and pressure sensor applications. The role of computational

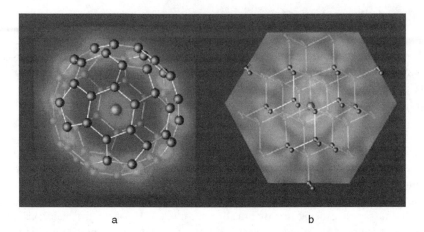

a   b

FIGURE 18.10 (a) The stable position and valance electron charge density of a P atom in a C_{60} fullerene. (b) The position and weakly bound donor electron density of the P atom after it has been encapsulated in a diamond nanocrystallite.

investigations in this case has been to precisely define how chemical reactivity[52] and electronic[53] properties change in a mechanically strained tube, and to determine the effect of gas-phase chemisorption[53] on the electronic characteristics of the system. It has been suggested that the mechanically tunable chemistry of nanotubes[52] will have possible applications in the chemisorption-induced electronics as well.[54] Experiments have shown that the nanotube sensors can detect ppm-level gas molecules at room temperature, and this opens a possibility of developing nanotube biosensors operating at room temperature.

For mechanical and vibration sensor applications, the full range of electronic bandgap changes as functions of axial compression, tensile stretch, torsion, and bending strain have been computed.[53] Additionally, as the cross-section of (8,0) SWCNT is flattened up to 40%, the bandgap of the nanotube decreases from 0.57 eV and disappears at 25% deformation. As the deformation further increases to 40%, the bandgap reopens and reaches 0.45 eV. This strong dependence of SWCNT band structure on the mechanical deformation can be applied to develop nanoscale mechanical sensors.[55]

18.3.5 Molecular Machines

The above examples provide a glimpse of a very broad field of applications and areas that are possible and directly fueled by the efforts and advances in computational nanotechnology. From materials to electronics, to computers and nanomachines, CNTs in fact could provide components of functional molecular-scale machines or robots (nanobots) in the future. Much of the recent progress in this field is in the experimental biomolecular motors arena, where significant understanding has been gained about how the natural biological motor systems work. Also, understanding has advanced in creating interfaces of nanoscale biomolecular motors with synthetic materials in solution-phase environments.[56] Means to power these machines through biomimetic physical and chemical phenomena are also under investigation. Ultimately, one can probably conceptualize nanoscale synthetic machines and motors that could be powered and controlled through external laser, electric, or magnetic fields and could operate in a chemical solution phase or inert gas environment. An example is the proposal of carbon nanotube-based gears[57] and how to power them through external laser fields.[58]

18.4 Nanotube Growth

The earliest approach to produce nanotubes was an arc process[59] as pioneered by Iijima.[1] This was soon followed by a laser ablation technique developed at Rice University.[60] In the last 5 years, chemical vapor deposition (CVD) has become a common technique to grow nanotubes.[61–66] All these processes are described below. The figure of merit for an ideal growth process depends on the application. For development of composites and other structural applications, the expected metric is the ability to produce tons a day. In contrast, the ability to achieve controlled growth (of specified thickness) on patterns is important for applications in nanoelectronics, field emission, displays, and sensors. Even in these cases, such needs for patterned growth — undoubtedly originating from the microelectronics fabrication technology and knowledge base — may become irrelevant if circuits could be assembled from nanotubes in solution, as will be seen later in the section on applications. Regardless of the applications and growth approach, the ability to control the diameter and chirality of the nanotubes is critical to realize the promise of carbon nanotubes.

18.4.1 Arc Process and Laser Ablation

The arc process involves striking a DC arc discharge in an inert gas (such as argon or helium) between a set of graphite electrodes.[1,59] The electric arc vaporizes a hollow graphite anode packed with a mixture of a transition metal (such as Fe, Co, or Ni) and graphite powder. The inert gas flow is maintained at 50–600 Torr. Nominal conditions involve 2000 to 3000°C, 100 amps, and 20 volts. This produces SWCNTs in mixture of MWCNTs and soot. The gas pressure, flow rate, and metal concentration can be varied to change the yield of nanotubes; but these parameters do not seem to change the diameter distribution. Typical diameter distribution of SWCNTs by this process appears to be 0.7 to 2 nm.

In laser ablation, a target consisting of graphite mixed with a small amount of transition metal particles as catalyst is placed at the *end* of a quartz tube enclosed in a furnace.[60] The target is exposed to an argon ion laser beam that vaporizes graphite and nucleates carbon nanotubes in the shockwave just in front of the target. Argon flow through the reactor heated to about 1200°C by the furnace carries the vapor and nucleated nanotubes, which continue to grow. The nanotubes are deposited on the cooler walls of the quartz tube downstream from the furnace. This produces a high percentage of SWCNTs (~70%) with the rest being catalyst particles and soot.

18.4.2 Chemical Vapor Deposition (CVD)

CVD has been widely used in silicon integrated circuit (IC) manufacturing to grow a variety of metallic, semiconducting, and insulating thin films. Typical CVD relies on thermal generation of active radicals from a precursor gas, which leads to the deposition of the desired elemental or compound film on a substrate. Sometimes the same film can be grown at a much lower temperature by dissociating the precursor with the aid of highly energetic electrons in a glow discharge. In either case, catalysts are almost never required. In contrast, a transition metal catalyst is necessary to grow nanotubes from some form of hydrocarbon feedstock (CH_4, C_2H_2, C_2H_4...) or CO; and, in principle, extensive dissociation of the feedstock in the vapor phase is not necessary as the feedstock can dissociate on the catalyst surface. Nevertheless, this approach is still called CVD in the nanotube literature.

A thermal CVD reactor is simple and inexpensive to construct consisting of a quartz tube enclosed in a furnace. Typical laboratory reactors use a 1- or 2-inch quartz tube capable of holding small substrates. The substrate material may be Si, mica, quartz, or alumina. The setup needs a few mass flow controllers to meter the gases and a pressure transducer to measure the pressure. The growth may be carried out at atmospheric pressure or slightly reduced pressures using a hydrocarbon or CO feedstock. The growth temperature is in the range of 700 to 900°C. A theoretical study of CNT formation suggests that a high kinetic energy (and thus a high temperature, ≥ 900°C) and a limited, low supply of carbon are necessary to form SWCNTs.[67] Not surprisingly, CO and CH_4 are the two gases that have been reported to give SWCNTs. MWCNTs are grown using CO, CH_4, and other higher hydrocarbons at lower temperatures of 600 to 750°C. As mentioned earlier, CNT growth requires a transition metal catalyst. The type of catalyst, particle size, and the catalyst preparation techniques dictate the yield and quality of CNTs; this will be covered in more detail shortly.

As in silicon IC manufacturing, the carbon nanotube community has also looked to low-temperature plasma processing to grow nanotubes. The conventional wisdom in choosing plasma processing is that the precursor is dissociated by highly energetic electrons; and as a result, the substrate temperature can be substantially lower than in thermal CVD. However, this does not apply here because the surface-catalyzed nanotube growth needs at least 500°C; and it has not been established if a high degree of dissociation in the gas phase is needed. Nevertheless, several plasma-based growth techniques have been reported;[68–70] and in general, the plasma-grown nanotubes appear to be more vertically oriented than that possible by thermal CVD. This feature is attributed to the electric field in the plasma normal to the substrate. Because the plasma is very efficient in tearing apart the precursors and creating radicals, it is also hard to control and to keep the supply of carbon low to the catalyst particles. Hence, plasma-based growth always results in MWCNTs and filaments.

The plasma reactor consists of a high vacuum chamber to hold the substrate, mass flow controllers, a mechanical (roughing) pump and, if necessary, a turbopump (plasma reactors almost always run at reduced pressure, 0.1 to 50 Torr), pressure gauges, and a discharge source. The latter is the heart of the plasma-processing apparatus; and in CNT growth, microwave sources are widely used, probably because of the popularity of this source in the diamond community. The microwave system consists of a power supply at 2.45 GHz and wave guides coupled to the stainless steel growth chamber. Though the plasma can provide intense heating of the substrate, it is normal to have an independent heater for the substrate holder. Other plasma sources include inductive source and hot filament DC discharge. The latter uses a tungsten, tungsten carbide, or similar filament heated to about 2500 K with the gas flow maintained at 1 to 20 Torr. Simple DC or RF discharges can also be used in nanotube growth. These consist of two parallel-plate electrodes with one grounded and the other connected to either a DC or a 13.56 MHz power supply.

18.4.3 Catalyst Preparation

Several reported catalyst preparation techniques in the literature consist of some form of solution-based catalysts. One recipe for growing MWCNTs is as follows.[63] First, 0.5 g (0.09 mmol) of Pluronic P-123 triblock copolymer is dissolved in 15 cc of a 2:1 mixture of ethanol and methanol. Next, $SiCl_4$ (0.85 cc, 7.5 mmol) is slowly added using a syringe into the triblock copolymer/alcohol solution and stirred for 30 minutes at room temperature. Stock solutions of $AlCl \cdot 6H_2O$, $CoCl_2 \cdot 6H_2O$, and $Fe (NO_3)_3 \cdot 6H_2O$ are prepared at the concentration of the structure-directing agent (SDA) and inorganic salts. The catalyst solutions are filtered through 0.45 µm polytetrafluoroethylene membranes before being applied to the substrate. The substrate with the catalyst formulation is loaded into a furnace and heated at 700°C for 4 hours in air to render the catalyst active by the decomposition of the inorganic salts and removal of the SDA. Figure 18.11 shows SEM images of MWCNTs grown using this formulation in a thermal CVD reactor. A nanotube tower with millions of multi-walled tubes supporting each other by van der Waals force is seen. If the catalyst solution forms a ring during annealing, then a hollow tower results.

Several variations of solution-based techniques have been reported in the literature. All of them have done remarkably well in growing carbon nanotubes. A common problem with the above approaches is that it is difficult to confine the catalyst from solutions within small patterns. Another problem is the excessive time required to prepare the catalyst. A typical solution-based technique for catalyst preparation involves several steps lasting hours. In contrast, physical processes such as sputtering and e-beam deposition not only can deal with very small patterns but are also quick and simple in practice.[66,68,69] Delzeit et al. reported catalyst preparation using ion beam sputtering wherein an underlayer of Al (~10 nm) is deposited first, followed by 1 nm of Fe active catalyst layer.[66] Figure 18.12 shows SWCNTs grown by thermal CVD on a 400-mesh TEM grid used to pattern the substrate. Methane feedstock at 900°C was used to produce these nanotubes. The same catalyst formulation at 750°C with ethylene as the source gas yielded MWCNT towers.

One of the most successful approaches to obtain oriented arrays of nanotubes uses a nanochannel alumina template for catalyst patterning.[71] First, aluminum is anodized on a substrate such as Si or quartz, which provides ordered, vertical pores. Anodizing conditions are varied to tailor the pore diameter, height, and spacing between pores. This is followed by electrochemical deposition of a cobalt catalyst at the bottom of the pores. The catalyst is activated by reduction at 600°C for 4 to 5 hours. Figure 18.13 shows

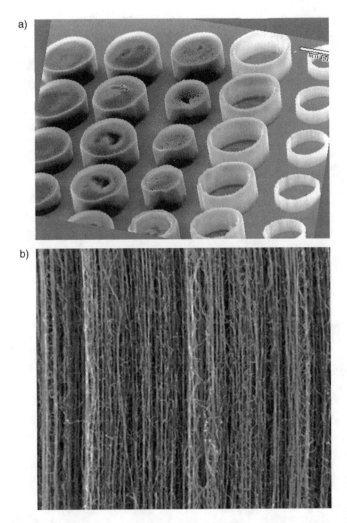

FIGURE 18.11 Multi-wall nanotubes grown by thermal CVD. (a) Different catalyst solution concentrations result in towers and ring-like structures. (b) Close-up view of one of these structures showing a forest of nanotubes supporting each other by van der Waals force, thus resulting in a vertical structure.

an ordered array of MWCNTs (mean diameter 47 nm) grown by CVD from 10% acetylene in nitrogen. The use of a template provides uniformity and vertically oriented nanotubes.

18.4.4 Continuous, High-Throughput Processes

The nanotube growth on substrates is driven by the desire to develop devices. Instead of supported catalysts, nanotubes have also been produced using catalysts floating in the gas phase. This approach is designed for production of large quantities of nanotubes in a continuous process. The earliest report[72] of such a process involved pyrolysis of a mixture containing benzene and a metallocene (such as ferrocene, cobaltocene, or nickelocene). In the absence of metallocene, only nanospheres of carbon were seen, but a small amount of ferrocene yielded large quantities of nanotubes. The growth system uses a two-stage furnace wherein a carrier gas picks up the metallocene vapor at around 200°C in the first stage, and the decomposition of the metallocene as well as pyrolysis of the hydrocarbon and catalytic reactions occur in the second stage at elevated temperatures (> 900°C). Acetylene with ferrocene or iron pentacarbonyl at 1100°C has been shown to yield SWCNTs in a continuous process in the same two-stage system. Andrews et al. also reported a continuous process for multiwall tubes using a ferrocyne–xylene mixture at temperatures as low as 650°C.[73]

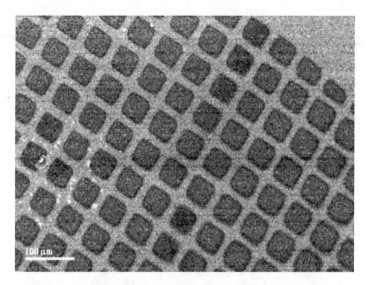

FIGURE 18.12   SWCNTs grown by thermal CVD on a 400-mesh TEM grid to pattern a silicon substrate.

FIGURE 18.13   An ordered array of MWCNTs grown using an alumina template. (From J. Li, C. Papadopoulos, J.M. Xu, and M. Moskovits, *Appl. Phys. Lett.* 75, 367, 1999. With permission.)

Recently, a high-pressure process using CO (HiPCO) has been reported.[74] In this process, the catalyst is generated *in situ* by thermal decomposition of iron pentacarbonyl. The products of this decomposition include iron clusters in the gas phase that act as nuclei for the growth of SWCNTs using CO. The process is operated at pressures of 10 to 50 atm. and temperatures of 800 to 1200°C. SWCNTs as small as 0.7 nm in diameter are produced by this process. The current production rate is about 0.5 g/h, and the process seems to be amenable for scale-up for commercial production.

18.5   Material Development

18.5.1   Purification

The as-grown material typically contains a mixture of SWCNTs, MWCNTs, amorphous carbon, and catalyst metal particles; and the ratio of the constituents varies from process to process and depends on growth conditions for a given process. Purification processes have been developed to remove all the unwanted constituents.[75] The purpose is to remove all the unwanted material and obtain the highest yield of nanotubes with no damage to the tubes. A typical process used in Reference 76 to purify Hipco-derived SWCNTs[74] is as follows. First, a sample of 50 mg is transferred to a 50 ml flask with 25 ml of concentrated HCl and 10 ml of concentrated HNO_3. The solution is heated for 3 hours and constantly stirred with a magnetic stirrer in a reflux apparatus equipped with a water-cooled condenser. This is done to remove iron and graphite nanocrystallites. The resulting suspension is transferred into centrifuge tubes and spun at 3220 g for 30 minutes. After pouring off the supernatant, the solid is resuspended and spun (30 minutes) three times in deionized water. Next, the solid is treated with NaOH (0.01 M) and centrifuged for 30 minutes. This process yields nanotube bundles with tube ends capped by half fullerenes. Finally, the sample is dried overnight in a vacuum oven at 60°C.

18.5.2 Characterization

High-resolution TEM is used (as in Figure 18.2) to obtain valuable information about the nanotube structure such as diameter, open vs. closed ends, presence of amorphous material, defects, and nanotube quality. In MWCNTs, the spacing between layers is 0.34 nm and, therefore, a count of the number of walls readily provides the diameter of the structure. Because the nature of catalyst and particle size distribution appear to be critical in determining the nanotube growth characteristics, researchers have used TEM, EDX, and AFM to characterize the catalyst surface prior to loading in the growth reactor. These techniques together provide information on the particle size and chemical composition of the surface.[66]

Besides TEM, Raman scattering is perhaps the most widely used characterization technique to study nanotubes.[77–80] Raman spectroscopy of the SWCNTs is a resonant process associated with optical transition between spikes in the 1-d electronic density of states. Both the diameter and the metallic vs. semiconducting nature of the SWCNTs dictate the energy of the allowed optical transitions and, hence, this characteristic is used to determine the diameter and nature of the nanotubes. An example from Reference 66 is presented in Figure 18.14. The spectra was obtained in the back-scattering configuration using a 2 mW laser power (633 nm excitation) on the sample with a 1 μm focus spot. The nanotube G band zone is formed through graphite Brilliouin zone folding. The spectrum in Figure 18.14 shows the characteristic narrow G band at 1590 cm^{-1} and signature band at 1730 cm^{-1} for SWCNTs. A strong enhancement in the low frequency is also observed in the low-frequency region for the radial breathing mode (RBM). As shown in Reference 77, 1-d density of state of metallic nanotubes near the Fermi level has the first singularily bandgap between 1.7 and 2.2 eV, which will be resonant with 633 nm (1.96 eV) excitation line to the RBM. Using $\alpha = 248$ cm^{-1} in $\Omega_{RBM} = \alpha/d$, a diameter distribution of 1.14 to 2 nm

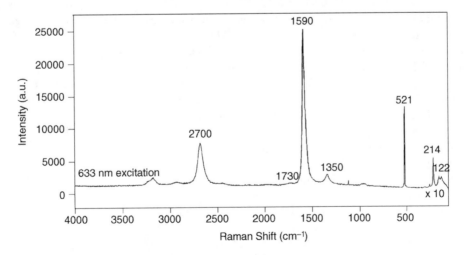

FIGURE 18.14 Raman spectra of CVD-grown SWCNT sample.

is computed for the sample in Figure 18.14, with the dominant distribution around 1.16 nm. The strong enhancement in 633 nm excitation[78] indicates that this sample largely consists of metallic nanotubes.

18.5.3   Functionalization

Functionalization of nanotubes with other chemical groups on the side-wall is attempted to modify the properties required for an application in hand. For example, chemical modification of the side-walls may improve the adhesion characteristics of nanotubes in a host polymer matrix to make functional composites. Functionalization of the nanotube ends can lead to useful chemical and biosensors.

Chen et al.[81] reported that reaction of soluble SWCNTs with dichlorocarbene led to the functionalization of nanotubes with Cl on the side-walls. The saturation of 2% of the carbon atoms in SWCNTs with C–Cl is sufficient to result in dramatic changes in the electronic band structure. Michelson et al.[82] reported fluorination of SWCNTs with F_2 gas flow at temperatures of 250 to 600°C for 5 hours. The fluorine is shown to attach covalently to the side-wall of the nanotubes. Two-point probe measurements showed that the resistance of the fluorinated sample increased to ~20 M Ω from ~15 Ω for the untreated material. While most functionalization experiments use a wet chemical or exposure to high-temperature vapors or gases, Khare et al.[76] reported a low-temperature, cold-plasma-based approach to attach functional groups to the nanotubes. They have been able to show functionalization with atomic hydrogen from an H_2 discharge as evidenced by the C–H stretching modes observed in the FTIR analysis of the samples. They also used a deuterium discharge to functionalize SWCNTs with D and showed C–D stretching modes in the region of 1940 to 2450 cm^{-1}.

18.6   Application Development

18.6.1   CNTs in Microscopy

Atomic-force microscopes (AFM) are widely used now by the research community to image and characterize various surfaces and also by the IC industry as a metrology tool. A typical AFM probe consists of a silicon or silicon–nitride cantilever with a pyramidal-shaped tip. This tip now can be made as small as 10 to 20 nm, offering reasonable resolution. However, the large cone angle of this tip (30 to 35°) makes it difficult for probing narrow and deep features such as trenches in IC manufacturing. Another serious drawback is that the tip is brittle, thus limiting its use in applications; either the tip breaks after only a limited use or it becomes blunt. For these reasons, a CNT probe has become an attractive alternative. Though early CNT probes[83,84] were manually attached to the tip of the AFM cantilever, currently it is possible to directly grow an SWCNT or MWCNT probe by CVD.[85,86] The CNT probe not only offers extraordinary nanometer-scale resolution but is also robust, due to its high strength and the ability to retain structural integrity even after deformation within elastic limit. Figure 18.15 shows an SWCNT probe prepared by CVD with diameter of ~2 nm. In a well-characterized CVD process, it may be possible to control the length of the probe by selecting the growth time. In the absence of knowledge of growth vs. time, it is possible to shorten the tip to the desired length by the application of an electric field to etch away the nanotube. Figure 18.16 shows a line/shape pattern of a polymeric photoresist on a silicon substrate. The image, obtained using a conventional silicon pyramidal tip (not shown here), shows sloping sides for the photoresist, which is an artifact due to the pyramidal shape of the tip. The image in Figure 18.16 was obtained using a MWCNT probe, which shows no such artifact but the vertical walls for the photoresist lines as confirmed by SEM.

In addition to profilometry, CNT tips are also useful in imaging thin films in semiconductor metrology. For example, Figure 18.17 shows an AFM image of a 2 nm film of Ir on mica surface collected with an SWCNT probe (~2 nm tip diameter). This affords a very high lateral resolution with grain sizes as small as 4 to 7 nm. Interestingly, even after continuous scanning for hours, the tip exhibits no detectable degradation in the lateral resolution of the grains. In contrast, the resolution capability becomes worse with time in the case of the silicon probe. As characterization tools in the IC industry, CNT probes would

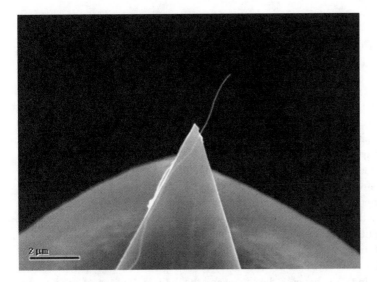

FIGURE 18.15 SWCMT tip grown by CVD at the end of an AFM cantilever.

FIGURE 18.16 AFM profile of a 280 nm line/space photoresist pattern.

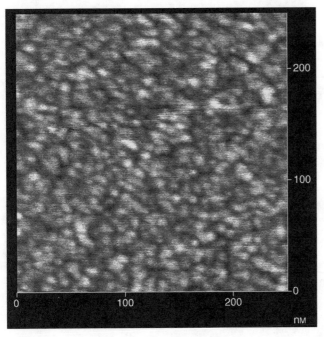

FIGURE 18.17 AFM image of a 2 nm Ir surface collected using an SWCNT tip.

require less frequent changing of probes, thus offering higher throughput. Researchers have also successfully used CNT probes to image biological samples.[83,85]

18.6.2 CNT-Based Nanoelectronics

The unique electronic properties of CNTs and possible devices from theoretical perspectives were discussed in detail in Sections 18.2 and 18.3. Room temperature demonstration of conventional switching mechanisms such as in field-effect transistors first appeared in 1998.[87,88] As shown in Figure 18.18, an SWCNT is placed to bridge a pair of metal electrodes serving as source and drain. The electrodes are defined using lithography on a layer of SiO_2 in a silicon wafer, which acts as the back gate. The variation of drain current with gate voltage at various source–drain biases for a 1.6 nm nanotube[87,88] clearly demonstrate that the gate can strongly control the current flow through the nanotube. In this device, the holes are the majority carriers of current as evidenced by an increase in current at negative gate voltages. Reference 89 provides a theoretical analysis of how the nanotube FETs work and shows that the electrode–nanotube contact influences the subthreshold channel conductance vs. gate voltage.

Since the pioneering demonstration of nanotube FETs, CNT logic circuits have been successfully fabricated. Figure 18.19 shows a CNT-based field-effect inverter consisting of n-type and p-type transistors.[90] The nanotube is grown using CVD between the source and the drain and on top of an SiO_2 layer with a silicon back gate. The as-grown nanotube exhibits p-type doping. Masking a portion of the nanotube and doping the remaining part with potassium vapor creates an n-type transistor. Combining the p- and the n-type transistors, complementary inverters are created. Applying a 2.9 V bias to the V_{DD}

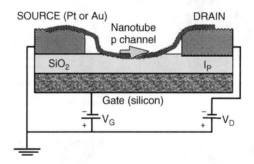

FIGURE 18.18   A carbon nanotube field-effect transistor.

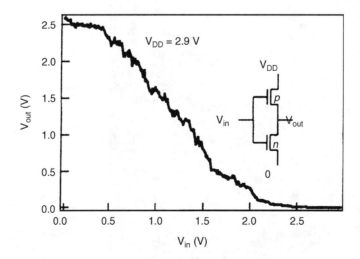

FIGURE 18.19   A carbon nanotube-based inverter.

terminal in the circuit and sweeping the gate voltage from 0 to 2.5 V yields an output voltage going from 2.5 V to 0. The inverter in Figure 18.19 shows a gain of 1.7 but does not exhibit an ideal behavior because, in this early demonstration, the PMOS is suspected to be leaky. Bachtold et al.[91] reported logic circuits with CNT transistors which show high gain (> 10), a large on–off ratio (> 10^5), and room temperature operation. Their demonstration included operations such as an inverter, a logic NOR, a static random-access memory cell, and an AC ring oscillator.

Fabrication of nanotube-based three-terminal devices involves horizontally placing nanotubes between the metal electrodes. While earlier demonstrations carefully transplanted a simple nanotube from bulk material, recent works have used CVD to bridge the electrodes with a nanotube.[90,92] This is indeed anything but routine as growth naturally occurs vertically from a surface. In this regard, successful fabrication of CNT Y-junctions (discussed in detail in Section 18.3) has been reported.[40,41] The I-V measurements on these CVD-grown multi-wall Y-tubes show reproducible rectification at room temperature. Alternatively, a concept proposed by Rueckes et al.[37] plans to use directed assembly of nanotubes from solution. This approach involves a suspended, crossed nanotube geometry that leads to bistable, electrostatically switchable ON/OFF states. Construction of nonvolatile random-access memory and logic functions using this approach is expected to reach a level of 10^{12} elements/cm^2.

18.6.3 Sensors

Significant research is in progress to develop CNT-based chemical, biological, and physical sensors. These efforts can be broadly classified into two categories: one that utilizes certain properties of the nanotube (such as a change in conductivity with gas adsorption) and the second that relies on the ability to functionalize the nanotube (tip and/or side-wall) with molecular groups that serve as sensing elements.

Gas sensors dedicated to sensing NO_2, NH_3, etc. use semiconducting metal oxides (SnO_2 for example) and conducting polymers. Most oxide-based sensors work at temperatures above 200°C but exhibit good sensitivity. Kong et al.[50] reported that the room temperature conductivity of CNTs changes significantly with exposure to NH_3 and NO_2, a property useful for developing sensors. Current–voltage characteristics of a 1.8 nm SWCNT placed between a pair of Ni/Au electrodes were studied before and after exposure to various doses of NH_3 and NO_2. The SWCNT samples used in the experiments were hole-doped semiconductors. Ammonia shifts the valence band of the CNT from the Fermi level and thus reduces the conductance due to hole depletion. The conductance was observed to decrease by a factor of 100 upon exposure to 1% NH_3. The response time, which may be defined as the time needed to see one order of magnitude change in conductance, was 1 to 2 minutes. The sensitivity, defined as the ratio of resistances before and after gas exposure, was 10 to 100. These results at room temperature compare favorably with metal oxides operating at 300 to 500°C.

To study the effect of NO_2, the device was operated with a gate voltage of 4 V at the back gate in a configuration discussed in Figure 18.18 under nanoelectronics (Section 18.6.2). This initially depletes the nanotubes, and exposure to NO_2 shifts the valence band closer to the Fermi level, thus increasing the holes in the nanotubes and conductance. The results for NO_2 are also encouraging. The conductance increase was about 1000 in response to 200 ppm NO_2, and the response time was 2 to 10 seconds. The sensitivity at room temperature was found to be 100 to 1000. One current drawback for deploying CNTs in sensors is the slow recovery of the CNTs to the initial state. In the above experiments, it took up to 12 hours for the conductivity of the CNTs to return to the original value after the source gas was withdrawn. Heating the sensor to 200°C reduced this period to about an hour. Similar large conductance changes upon exposure to oxygen have also been reported.[93] Work on functional sensors using nanotubes is rather limited. Xu[94] proposes IR detection using an ordered CNT array in a broad wavelength range of 1 to 15 μm.

18.6.4 Field Emission

The geometric properties of nanotubes, such as the high aspect ratio and small tip radius of curvature, coupled with the extraordinary mechanical strength and chemical stability, make them an ideal candidate for electron field emitters. CNT field emitters have several industrial and research applications: flat panel displays, outdoor displays, traffic signals, and electron microscopy. De Heer et al.[95] demonstrated the earliest high-intensity electron gun based on field emission from a film of nanotubes. A current density of 0.1 mA/cm^2 was observed for voltages as low as 200 V. For comparison, most conventional field emitter displays operate at 300 to 5000 V, whereas cathode ray tubes use 30000 V. Since this early result, several groups have studied the emission characteristics of SWCNTs and MWCNTs.[96–100]

A typical field emission test apparatus consists of a cathode and anode enclosed in an evacuated cell at a vacuum of 10^{-9} to 10^{-8} Torr. The cathode consists of a glass or PTFE (polytetrafluoroethylene) substrate with metal-patterned lines, where a film of nanotubes can be transplanted after the arc-grown or laser oven material is purified. Instead nanotubes can be directly grown on the cathode in a CVD or plasma CVD chamber described earlier. The anode operating at positive potentials is placed at a distance of 20 to 500 μm from the cathode. The turn-on field, arbitrarily defined as the electric field required for generating 1 nA, can be as small as 1.5 V/μm. The threshold field — the electric field needed to yield a current density of 10 mA/cm^2 — is in the range of 5 to 8 V/μm. At low emission levels, the emission behavior follows the Fowler–Nordheim relation, i.e., the plot of ln (I/V^2) vs. ln (1/V) is linear. The emission current significantly deviates from the F–N behavior in the high field region and, indeed, emission current typically saturates. While most works report current densities of 0.1 to 100 mA/cm^2, very high current densities of up to 4 A/cm^2 have been reported by Zhu et al.[99]

Working full-color flat panel displays (FPD) and CRT lighting elements have been demonstrated by groups from Japan and Korea.[97,100] In the case of FPD, the anode structure consists of a glass substrate with phosphor-coated indium tin–oxide stripes. The anode and cathode are positioned perpendicular to each other to form pixels at their intersections. Appropriate phosphors such as Y$_2$O$_2$S: Eu, ZnS: Cu, Al, and ZnS: Ag, Cl are used at the anode for red, green, and blue colors, respectively. A 4.5 inch display showed a uniform and stable image over the entire 4.5 inch panel.[100] For lighting elements, phosphor screen is printed on the inner surface of the glass and backed by a thin Al film (~100 nm) to give electrical conductivity. A lifetime test of the lighting element suggests a lifetime of over 10000 hours.[97]

18.6.5 Nanotube–Polymer Composites

Using nanotubes as reinforcing fibers in composite materials is still a developing field from both theoretical and experimental perspectives. Several experiments regarding the mechanical properties of nanotube–polymer composite materials with MWCNTs have been reported recently.[29–31,101] Wagner et al.[30] experimentally studied the fragmentation of MWNTs within thin polymeric films (urethane/ diacrylate oligomer EBECRYL 4858) under compressive and tensile strain. They found that the nanotube–polymer interfacial shear stress τ is of the order of 500 MPa, which is much larger than that of conventional fibers with polymer matrix. This has suggested the possibility of chemical bonding between the nanotubes and the polymer in their composites, but the nature of the bonding is not clearly known. Lourie et al.[22] have studied the fragmentation of single-walled CNT within the epoxy resin under tensile stress. Their experiment also suggests a good bonding between the nanotube and the polymer in the sample. Schadler et al.[29] have studied the mechanical properties of 5 wt.% MWCNTs within epoxy matrix by measuring the Raman peak shift when the composites are under compression or under tension. The tensile modulus of the composites, in this experiment, is found to enhance much less than the enhancement of the same composite under compression. This difference has been attributed to the sliding of inner shells of the MWNTs when a tensile stress was applied. In cases of SWNT polymer composites, the possible sliding of individual tubes in the SWCNT rope may also reduce the efficiency of load transfer. It is suggested that for the SWNT rope case, interlocking using polymer molecules might bond SWCNT rope more strongly. Andrews et al.[101] have also studied the composites of 5 wt.% of SWNT embedded in petroleum pitch matrix, and their measurements show an enhancement of the Young's modulus of the composite under tensile stress. Measurements by Qian et al.[31] of a 1 wt.% MWNT–polystyrene composite under tensile stress also show a 36% increase of Young's modulus compared with the pure polymer system. The possible sliding of inner shells in

MWCT and of individual tubes in an SWNT rope is not discussed in the above two studies. There are currently no clean experiments available on SWCNT–polymer composites, perhaps because SWCNTs are not available in large quantities for experimentation on bulk composite materials.

18.6.6 Other Applications

In addition to the fields discussed above, CNTs are being investigated for several other applications. The possibility to store hydrogen for fuel cell development has received much attention.[102] An H_2 storage capability of 8% by weight appears to be the target for use in automobiles. Though interesting basic science of hydrogen storage has been emerging, no technological breakthrough has been reported yet. Filling nanotubes with a variety of metals has also been attempted.[103–106] Most noteworthy of these attempts involves lithium storage for battery applications.[105,106] The thermal conductivity of SWCNTs is reported[107] to be about 3000 W/mK in the axial direction, which is second only to epitaxial diamond. This property can be exploited for cooling semiconductor chips and heat pipes. SWCNTs and MWCNTs are useful as high surface area and high-conductivity electrodes in a variety of applications. CNTs are being studied for gas absorption, for separation, and as support for catalysts. Other interesting applications include nanoscale reactors, ion channels, and drug delivery systems.

18.7 Concluding Remarks

Carbon nanotubes show remarkable potential in a wide variety of applications such as future nanoelectronics devices, field emission devices, high-strength composites, sensors, and in many related fields. This potential has propelled concentrated research activities across the world in growth, characterization, modeling, and application development. Though significant progress has been made in the last 5 years, numerous challenges remain; and there is still a great deal of work to be done before CNT-based products become ubiquitous. The biggest challenge now is to have control over the chirality and diameter during the growth process; in other words, the ability to specify *a priori* and obtain the desired chiral nanotubes is important to develop computing and related applications. Issues related to contacts, novel architectures, and development of inexpensive manufacturing processes are additional areas warranting serious consideration in electronics applications. The major roadblock now to developing structural applications is the lack of availability of raw materials in large quantities. With current bulk production rate of SWCNTs hovering around a few grams per hour, large-scale composite development efforts are nonexistent at present. This scenario will change with breakthroughs in large-scale production of CNTs, bringing the cost per pound to reasonable levels for structural applications. Sensor development efforts are in the early stages; numerous areas, such as functionalization, signal processing and integrity, and system integration, require significant further developments. In all of the above areas, modeling and simulation is expected to continue to play a critical role as it has been since the discovery of carbon nanotubes.

Acknowledgments

The authors thank the members of the experimental group at NASA Ames Research Center for Nanotechnology for their contributions to the material presented in this chapter, and we thank Harry Partridge for his critical review of the manuscript. Deepak Srivastava acknowledges support of his work by NASA contract 704–40–32 to CSC.

References

1. S. Iijima, *Nature*, 354, 56 (1991).
2. M.S. Dresselhouse, G. Dresselhouse, and P.C. Eklund, *Science of Fullerenes and Carbon Nanotubes*, Academic Press, New York, 1996.
3. D. Srivastava, M. Menon, and K. Cho, *Comp. Eng. Sci.*, 42 (2001).

4. M.S. Daw and M.I. Baskes, *Phys. Rev. Lett.* 50, 1285 (1983); S.M. Foiles, M.I. Baskes, and M.S. Daw, *Phys. Rev. B* 33, 7983 (1986).

5. F.H. Stillinger and T.A. Weber, *Phys. Rev. B* 31, 5262 (1985).

6. J. Tersoff, *Phys. Rev. B* 38, 9902 (1988).

7. D.W. Brenner, *Phys. Rev. B* 42, 9458 (1990).

8. B.J. Garrison and D. Srivastava, *Ann. Rev. Phys. Chem.* 46, 373 (1995).

9. D. Srivastava and S. Barnard, in *Proc. IEEE Supercomp. '97* (1997).

10. R. Car and M. Parrinello, *Phys. Rev. Lett.* 55, 2471 (1985).

11. W.A. Harrison, *Electronic Structure and the Properties of Solids*, Freeman, San Francisco, 1980.

12. M. Menon and K.R. Subbaswamy, *Phys. Rev. B* 55, 9231 (1997).

13. M. Menon, *J. Chem. Phys.*, 114, 7731 (2000).

14. Payne, M.C. et al., *Rev. Mod. Phys.* 68, 1045 (1992).

15. P. Hohenberg and W. Kohn, *Phys. Rev.* 136, 864B (1964).

16. W. Kohn and L.J. Sham, *Phys. Rev.* 140, 1133A (1965).

17. Details available at: http://cms.mpi.univie.ac.at/vasp/

18. See for example methodology and references in S.Datta, *Electronic Transport in Mesoscopic Systems*, Cambridge University Press, Cambridge, 1995.

19. B.I. Yakobson, C.J. Brabec, and J. Bernholc, *Phys. Rev. Lett.* 76, 2511 (1996).

20. D. Srivastava, M. Menon, and K. Cho, *Phys. Rev. Lett.* 83, 2973 (1999).

21. B.I.Yakobson and P. Avouris, *Mechanical Properties of Carbon Nanotubes*, M.S. Dresselhaus and P. Avouris (Eds.), Springer-Verlag, Berlin, 2001, p. 293.

22. O. Lourie, D.M. Cox, and H.D. Wagner, *Phys. Rev. Lett.* 81, 1638 (1998).

23. M.B. Nardelli, B.I. Yakobson, and J. Bernholc, *Phys. Rev. Lett.* 81, 4656 (1998).

24. P. Zhang and V. Crespi, *Phys. Rev. Lett.* 81, 5346 (1998).

25. C. Wei, D. Srivastava, and K. Cho, *Comp. Mod. Eng. Sci.*, 3, 255 (2002).

26. C. Wei, K. Cho, and D. Srivastava, *Phys. Rev. Lett.* submitted (2002).

27. M. Yu, O. Lourie, M. Dyer, K. Moloni, T. Kelly, and R. Ruoff, *Science* 287, 637 (2000).

28. B. Vigolo et al., *Science*, 290, 1331 (2000).

29. L.S. Schadler, S.C. Giannaris, and P.M. Ajayan, *Appl. Phys. Lett.* 73, 3842(1998).

30. H.D. Wagner, O. Lourie, Y. Feldman, and R. Tenne, *Appl. Phys. Lett.* 72, 188 (1998).

31. Q. Qian, E.C. Dickey, R. Andrews, and T. Rantell, *Appl. Phys. Lett.* 76, 2868 (2000).

32. C. Wei, D. Srivastava, and K. Cho, *Nano. Lett.*, 2, 647 (2002).

33. P. Collins and P. Avouris, *Sci. Am.* (Dec. 2000), p. 62.

34. L. Chico, V.H. Crespi, L.X. Benedict, S.G. Louie, and M.L. Cohen, *Phys. Rev. Lett.* 76, 971 (1996).

35. Z. Yao, H.W.C. Postma, L. Balants, and C. Dekker, *Nature* 402, 273 (1999).

36. M. Menon and D. Srivastava, *Phys. Rev. Lett.* 79, 4453 (1997).

37. T. Rueckes, K. Kim, E. Joselevich, G.Y. Tseng, C.-L. Cheung, and C.M. Lieber, *Science* 289, 94 (2000).

38. C. Joachim, J.K. Gimzewski, and A. Aviram, *Nature* 408, 541 (2000).

39. M. Menon and D. Srivastava, *J. Mater. Res.* 13, 2357 (1998).

40. C. Papadopoulos, A. Rakitin, J. Li, A.S. Vedeneev, and J.M. Xu, *Phys. Rev. Lett.* 85, 3476 (2000).

41. C. Satishkumar, P.J. Thomas, A. Govindraj, and C.N.R. Rao, *Appl. Phys. Lett.* 77, 2530 (2000).

42. A. Antonis, M. Menon, D. Srivastava, G. Froudakis. and L.A. Chernozatonskii, *Phys. Rev. Lett.* 87, 66802 (2001).

43. A. Antonis, M. Menon, D. Srivastava, and L.A. Chernozatonskii, *Appl. Phys. Lett.* 79, 266 (2001).

44. S.K. Moore, *IEEE Spectrum*, (Oct. 2000), pp. 18–19.

45. I.L. Chuang, L.M.K. Vandersypen, X.L. Zhou, D.W. Leung, and S. Lloyd, *Nature* 393, 143 (1998).

46. B.E. Kane, *Nature* 393, 133 (1998).

47. S. Park, D. Srivastava, and K. Cho, *J. Nanosci. Nanotechnol.*, 1, 75 (2001).

48. C. Knapp et. al., *Molec. Phys.* 95, 999 (1998).

49. F.J. Banhart and P.M. Ajayan, *Nature* 382, 433 (1996).

50. J. Kong, N.R. Franklin, C. Zhou, M.G. Chapline, S. Peng, K.J. Cho, and H. Dai, *Science* 287, 622 (2000).
51. S. Peng and K.J. Cho, *Nanotechnology* 11, 57 (2000).
52. D. Srivastava et al., *J. Phys. Chem. B* 103, 4330 (1999).
53. L. Yang and J. Han, *Phys. Rev. Lett.* 85, 154 (2000).
54. O. Gulseran, T. Yildirium, and S. Ciraci, *Phys. Rev. Lett.* 87, 116802 (2001).
55. S. Peng and K. Cho, *J. Appl. Mech.* in press (2002).
56. C. Montemagno and G. Bachand, *Nanotechnology* 10, 225 (1999).
57. J. Han, A. Globus, R. Jaffe, and G. Deardorff, *Nanotechnology* 8, 95 (1997).
58. D. Srivastava, *Nanotechnology* 8, 186 (1997).
59. C.H. Kiang, W.A. Goddard, R. Beyers, and D.S. Bethune, *Carbon*, 33, 903 (1995).
60. T. Guo, P. Nikolaev, A. Thess, D.T. Colbert, and R.E. Smalley, *Chem. Phys. Lett.* 243, 49 (1995).
61. H. Dai, A.G. Rinzler, P. Nikolaev, A. Thess, D.T. Colbert, and R.E. Smalley, *Chem. Phys. Lett.* 260, 471 (1996).
62. J. Kong, H.T. Soh, A.M. Cassell, C.F. Quate, and H. Dai, *Nature* 395, 878 (1998).
63. A.M. Cassell, S. Verma, J. Han, and M. Meyyappan, *Langmuir* 17, 260 (2001).
64. H. Kind, J.M. Bonard, L. Forro, K. Kern, K. Hernadi, L. Nilsson, and L. Schlapbach, *Langmuir* 16, 6877 (2000).
65. M. Su, B. Zheng, and J. Liu, *Chem. Phys. Lett.* 322, 321 (2000).
66. L. Delzeit, B. Chen, A. Cassell, R. Stevens, C. Nguyen, and M. Meyyappan, *Chem. Phys. Lett.* 348, 368 (2001).
67. H. Karzow and A. Ding, *Phys. Rev. B* 60, 11180 (1999).
68. Y.Y. Wei, G. Eres, V.I. Merkulov, and D.H. Lowndes, *Appl. Phys. Lett.* 78, 1394 (2001).
69. Y.C. Choi, Y.M. Shin, S.C. Lim, D.J. Bae, Y.H. Lee, B.S. Lee, and D.C. Chung, *J. Appl. Phys.* 88, 4898 (2000).
70. C. Bower, W. Zhu, S. Jin, and O. Zhou, *Appl. Phys. Lett.* 77, 830 (2000).
71. J. Li, C. Papadopoulos, J.M. Xu, and M. Moskovits, *Appl. Phys. Lett.* 75, 367 (1999).
72. B.C. Satishkumar, A. Govindraj, R. Sen, and C.N.R. Rao, *Chem. Phys. Lett.* 293, 47 (1998).
73. R. Andrews, D. Jacques, A.M. Rao, F. Derbyshire, D. Qian, X. Fan, E.C. Dickey, and J. Chen, *Chem. Phys. Lett.* 303, 467 (1999).
74. P. Nikolaev, M.J. Bronikowski, R.K. Bradley, F. Rohmund, D.T. Colbert, K.A. Smith, and R.E. Smalley, *Chem. Phys. Lett.* 313, 91 (1999).
75. J. Liu et al., *Science* 280, 1253 (1998).
76. B. Khare, M. Meyyappan, A.M. Cassell, C.V. Nguyen, and J. Han, *Nano Lett.* 2, 73 (2002).
77. A.M. Rao et al., *Science* 275, 187 (1997).
78. M.A. Pimenta et al., *Phys. Rev. B* 58, 16016 (1998).
79. A. Jorio et al., *Phys. Rev. Lett.* 86, 1118 (2001).
80. A.M. Rao et al., *Phys. Rev. Lett.* 84, 1820 (2000).
81. J. Chen et al., *Science* 282, 95 (1998).
82. E.T. Michelson et al., *Chem. Phys. Lett.* 296, 188 (1998).
83. S.S. Wang, J.D. Harper, P.T. Lansbury, and C.M. Lieber, *J. Am. Chem. Soc.* 120, 603 (1998).
84. H. Dai, N. Franklin, and J. Han, *Appl. Phys. Lett.* 73, 1508 (1998).
85. C.L. Cheung, J.H. Hafner, and C.M. Lieber, *Proc. Nat. Acad. Sci.* 97, 3809 (2000).
86. C.V. Nguyen, K.J. Cho, R.M.D. Stevens, L. Delzeit, A. Cassell, J. Han, and M. Meyyappan, *Nanotechnology* 12, 363 (2001).
87. S.J. Tans, A.R.M. Verschueren, and C. Dekker, *Nature* 393, 49 (1998).
88. R. Martel, T. Schmidt, H.R. Shea, T. Hertel, and P. Avouris, *Appl. Phys. Lett.* 2 73, 2447 (1998).
89. T. Yamada, *Appl. Phys. Lett.* 76, 628 (2000).
90. X. Liu, C. Lee, C. Zhou, and J. Han, *Appl. Phys. Lett.* 79, 3329 (2001).
91. A. Bachtold, P. Hadley, T. Nakanishi, and C. Dekker, *Science* 294, 1317 (2001).
92. H. Dai et al., *J. Phys. Chem. B* 103, 11246 (1999).

93. P.G. Collins, K. Bradley, M. Ishigami, and A. Zettl, *Science* 287, 1801 (2000).
94. J.M. Xu, *Infrared Phys. Tech.* 42, 485 (2001).
95. W.A. de Heer, A. Chatelain, and D. Ugarte, *Science* 270, 1179 (1995).
96. P.G. Collins and A. Zettl, *Appl. Phys. Lett.* 69, 1969 (1996).
97. Y. Saito, S. Uemura, and K. Hamaguchi, *Jpn. J. Appl. Phys. Part 2* 37, L 346 (1998).
98. X. Xu and G.R. Brandes, *Appl. Phys. Lett.* 74, 2549 (1999).
99. W. Zhu, C. Bower, O. Zhou, G. Kochanski, and S. Jin, *Appl. Phys. Lett.* 75, 873 (1999).
100. W.B. Choi, Y.H. Lee, D.S. Chung, N.S. Lee, and J.M. Kim, in D. Tomanek and R. Enbody (Eds.), *Science and Application of Nanotubes*, Kluwer Academic/Plenum Publishers, New York, 2000, p. 355.
101. R. Andrews et. al., *Appl. Phys. Lett.* 75, 1329 (1999).
102. Y. Ye et al., *Appl. Phys. Lett.* 304, 207 (1999).
103. M. Terroves et al., *MRS Bull.* p. 43, August 1999.
104. C. Kiang et al., *J. Phys. Chem. B* 103, 7449 (1999).
105. I. Mukhopadhyay et al., *J. Electrochem. Soc.* 149, A39 (2002).
106. J.S. Sakamoto and B. Dunn, *J. Electrochem. Soc.* 149, A26 (2002).
107. M.A. Osman and D. Srivastava, *Nanotechnology* 12, 21 (2001).

19

Mechanics of Carbon Nanotubes[1]

CONTENTS

Dong Qian
Northwestern University

Gregory J. Wagner
Northwestern University

Wing Kam Liu
Northwestern University

Min-Feng Yu
University of Illinois

Rodney S. Ruoff
Northwestern University

Abstract

Soon after the discovery of carbon nanotubes, it was realized that the theoretically predicted mechanical properties of these interesting structures — including high strength, high stiffness, low density, and structural perfection — could make them ideal for a wealth of technological applications. The experimental verification, and in some cases refutation, of these predictions, along with a number of computer simulation methods applied to their modeling, has led over the past decade to an improved but by no means complete understanding of the mechanics of carbon nanotubes. We review the theoretical predictions and discuss the experimental techniques that are most often used for the challenging tasks of visualizing and manipulating these tiny structures. We also outline the computational approaches that have been taken, including *ab initio* quantum mechanical simulations, classical molecular dynamics, and continuum models. The development of multiscale and multiphysics models and simulation tools naturally arises as a result of the link between basic scientific research and engineering application; while this issue is still under intensive study, we present here some of the approaches to this topic. Our

---

[1] This chapter appeared previously in *Applied Mechanics Reviews*, 56, 6, 2002. It is reprinted here with permission of ASME International.

concentration throughout is on the exploration of mechanical properties such as Young's modulus, bending stiffness, buckling criteria, and tensile and compressive strengths. Finally, we discuss several examples of exciting applications that take advantage of these properties, including nanoropes, filled nanotubes, nanoelectromechanical systems, nanosensors, and nanotube-reinforced polymers.

19.1   Introduction

The discovery of multi-walled carbon nanotubes (MWCNTs) in 1991[1] has stimulated ever-broader research activities in science and engineering devoted entirely to carbon nanostructures and their applications. This is due in large part to the combination of their expected structural perfection, small size, low density, high stiffness, high strength (the tensile strength of the outermost shell of an MWCNT is approximately 100 times greater than that of aluminum), and excellent electronic properties. As a result, carbon nanotubes (CNTs) may find use in a wide range of applications in material reinforcement, field-emission panel display, chemical sensing, drug delivery, and nanoelectronics.

Indeed, NASA is developing materials using nanotubes for space applications, where weight-driven cost is the major concern, by taking advantage of their tremendous stiffness and strength. Composites based on nanotubes could offer strength-to-weight ratios beyond any materials currently available. Companies such as Samsung and NEC have invested tremendously and demonstrated product quality devices utilizing carbon nanotubes for field-emission display.[2,3] Such devices have shown superior qualities such as low turn-on electric field, high-emission current density, and high stability. With the advance of materials synthesis and device processing capabilities, the importance of developing and understanding nanoscale engineering devices has dramatically increased over the past decade. Compared with other nanoscale materials, single-walled carbon nanotubes (SWCNTs) possess particularly outstanding physical and chemical properties. SWCNTs are remarkably stiff and strong, conduct electricity, and are projected to conduct heat even better than diamond, which suggests their eventual use in nanoelectronics. Steady progress has been made recently in developing SWCNT nanodevices and nanocircuits,[4,5] showing remarkable logic and amplification functions. SWCNTs are also under intensive study as efficient storage devices, both for alkali ions for nanoscale power sources and for hydrogen for fuel cell applications.

On other fronts, CNTs also show great potential for biomedical applications due to their biocompatibility and high strength. The current generation of composites used for replacement of bone and teeth are crude admixtures of filler particles (often glass) that have highly inadequate mechanical properties compared with skeletal tissue. Fiber-based composites have been investigated in the past but have not worked well because of eventual degradation of the filler/matrix interface, usually due to attack by water. Thus, a major issue is interfacial stability under physiological conditions. Graphitic materials are known to resist degradation in the types of chemical environments present in the body. Recent demonstration of CNT artificial muscle[6] is one step along this direction and promises a dramatic increase in work density output and force generation over known technologies, along with the ability to operate at low voltage. CNTs are also being considered for drug delivery: they could be implanted without trauma at the sites where a drug is needed, slowly releasing a drug over time. They are also of considerable promise in cellular experiments, where they can be used as nanopipettes for the distribution of extremely small volumes of liquid or gas into living cells or onto surfaces. It is also conceivable that they could serve as a medium for implantation of diagnostic devices.

The rapid pace of research development as well as industry application has made it necessary to summarize the current status about what we have known and what we do not know about this particularly interesting nanostructure. A number of excellent reviews on the general properties of CNTs can be found in References 7–9. In this paper we have made this effort with emphasis on the mechanical aspects.

The paper is organized as follows: Section 19.2 focuses on the mechanical properties of CNTs. This includes the basic molecular structure, elastic properties, strength, and crystal elasticity treatment. Section 19.3 summarizes the current experimental techniques that were used in the measurement of the mechanical properties of CNTs. A brief discussion on the challenges in the experiment is also presented. Various simulation methods are discussed in Section 19.4. The length scale spans from quantum level to continuum level, which highlights the multiscale and multiphysics features of the nanoscale problem. Section 19.5 considers a few important mechanical applications of CNTs. Finally, conclusions are made in Section 19.6.

19.2 Mechanical Properties of Nanotubes

19.2.1 Molecular Structure of CNTs

19.2.1.1 Bonding Mechanisms

The mechanical properties of CNTs are closely related to the nature of the bonds between carbon atoms. The bonding mechanism in a carbon nanotube system is similar to that of graphite, as a CNT can be thought of as a rolled-up graphene sheet. The atomic number for carbon is 6, and the atom electronic structure is $1s^2 2s^2 2p^2$ in atomic physics notation. For a detailed description of the notation and the structure, readers may refer to basic textbooks on general chemistry or physics.[10] When carbon atoms combine to form graphite, sp^2 hybridization[10] occurs. In this process, one *s*-orbital and two *p*-orbitals combine to form three hybrid sp^2-orbitals at 120° to each other within a plane (shown in Figure 19.1). This in-plane bond is referred to as a σ-bond (*sigma*-bond). This is a strong covalent bond that binds the atoms in the plane and results in the high stiffness and high strength of a CNT. The remaining *p*-orbital is perpendicular to the plane of the σ-bonds. It contributes mainly to the interlayer interaction and is called the π-bond (*pi*-bond). These out-of-plane, delocalized π-bonds interact with the π-bonds on the neighboring layer. This interlayer interaction of atom pairs on neighboring layers is much weaker than a σ-bond. For instance, in the experimental study of *shell-sliding*,[11] it was found that the shear strength between the outermost shell and the neighboring inner shell was 0.08 MPa and 0.3 MPa according to two separate measurements on two different MWCNTs. The bond structure of a graphene sheet is shown in Figure 19.1.

19.2.1.2 From Graphene Sheet to Single-Walled Nanotube

There are various ways of defining a unique structure for each carbon nanotube. One way is to think of each CNT as a result of rolling a graphene sheet, by specifying the direction of rolling and the circumference of the cross-section. Shown in Figure 19.2 is a graphene sheet with defined roll-up vector *r*. After rolling to form an NT, the two end nodes coincide. The notation we use here is adapted from References 8, 12, and 13. Note that *r* (bold solid line in Figure 19.2) can be expressed as a linear combination of base vectors *a* and *b* (dashed line in Figure 19.2) of the hexagon, i.e.,

$$r = na + mb$$

with *n* and *m* being integers. Different types of NT are thus uniquely defined by the values of *n* and *m*, and the ends are closed with caps for certain types of fullerenes (Figure 19.3).

Three major categories of NT can also be defined based on the chiral angle θ (Figure 19.2) as follows:

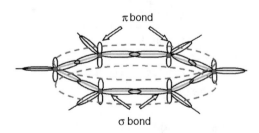

FIGURE 19.1 Basic hexagonal bonding structure for one graphite layer (the *graphene sheet*). Carbon nuclei shown as filled circle, out-of-plane π-bonds represented as delocalized (dotted line), and σ-bonds connect the C nuclei in-plane.

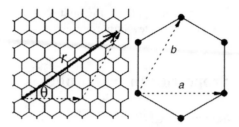

FIGURE 19.2 Definition of roll-up vector as linear combinations of base vectors *a* and *b*.

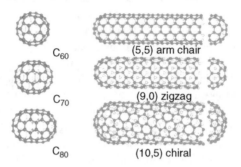

FIGURE 19.3. Examples of zigzag, chiral, and armchair nanotubes and their caps corresponding to different types of fullerenes. (From Dresselhaus MS, Dresselhaus G, and Saito R, (1995), Physics of carbon nanotubes, *Carbon.* 33(7): 883–891. With permission from Elsevier Science.)

$$\theta = 0 \qquad\qquad \textit{Zigzag}$$

$$0 < \theta < 30 \qquad\qquad \textit{Chiral}$$

$$\theta = 30 \qquad\qquad \textit{Arm Chair}$$

Based on simple geometry, the diameter *d* and the chiral angle θ of the NT can be given as:

$$d = 0.783\sqrt{n^2 + nm + m^2}\,\text{Å}$$

$$\theta = \sin^{-1}\left[\frac{\sqrt{3}m}{2(n^2 + nm + m^2)}\right]$$

Most CNTs to date have been synthesized with closed ends. Fujita et al.[14] and Dresselhaus et al.[8,15] have shown that NTs which are larger than (5,5) and (9,0) tubes can be capped. Based on Euler's theorem of polyhedra,[16] which relates the numbers of the edges, faces, and vertices, along with additional knowledge of the minimum energy structure of fullerenes, they conclude that any cap must contain six pentagons that are isolated from each other. For NTs with large radii, there are different possibilities of forming caps that satisfy this requirement. The experimental results of Iijima et al.[17] and Dravid[18] indicate a number of ways that regular-shaped caps can be formed for large-diameter tubes. *Bill*-like[19] and semi-toroidal[17] types of termination have also been reported. Experimental observation of CNTs with open ends can be found in Reference 17.

19.2.1.3 Multi-Walled Carbon Nanotubes and Scroll-Like Structures

The first carbon nanotubes discovered[1] were multi-walled carbon nanotubes (shown in Figures 19.4 and 19.5). Transmission electron microscopy studies on MWCNTs suggest a Russian doll-like structure (nested shells) and give interlayer spacing of approximately ~0.34 nm,[20,21] close to the interlayer separation

FIGURE 19.4 Top left: High-resolution transmission electronic microscopy (HRTEM) image of an individual MWCNT. The parallel fringes have ~0.34 nm separation between them and correspond to individual layers of the coaxial cylindrical geometry. Bottom left: HRTEM image showing isolated SWCNT as well as bundles of such tubes covered with amorphous carbon. The isolated tubes shown are approximately 1.2 nm in diameter. Top right: HRTEM image showing the tip structure of a closed MWCNT. The fringe (layer) separation is again 0.34 nm. Bottom right: The image of an MWCNT showing the geometric changes due to the presence of five and seven membered rings (position indicated in the image by P for pentagon and H for heptagon) in the lattice. Note that the defects in all the neighboring shells are conformal. (From Ajayan PM and Ebbesen TW, (1997), Nanometer-size tubes of carbon, *Rep. Prog. Phys.* 60(10): 1025–1062. With permission.)

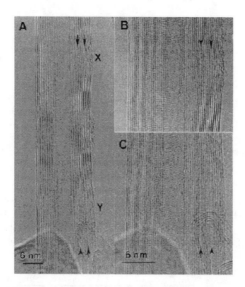

FIGURE 19.5 HRTEM image of an MWCNT. Note the presence of anomalously large interfringe spacings indicated by arrows. (From Amelinckx S, Lucas A, and Lambin P, (1999), Electron diffraction and microscopy of nanotubes, *Rep. Prog. Phys.* 62(11): 1471–1524. With permission.)

of graphite, 0.335 nm. However, Kiang et al.[22] have shown that the interlayer spacing for MWCNTs can range from 0.342 to 0.375 nm, depending on the diameter and number of nested shells in the MWCNT. The increase in intershell spacing with decreased nanotube diameter is attributed to the increased repulsive force as a result of the high curvature. The experiments by Zhou et al.,[21] Amelinckx et al.,[23] and Lavin et al.[24] suggested an alternative *scroll* structure for some MWCNTs, like a cinnamon roll. In fact, both forms might be present along a given MWCNT and separated by certain types of defects. The energetics analysis by Lavin et al.[24] suggests the formation of a scroll, which may then convert into a stable multi-wall structure composed of nested cylinders.

19.2.2 Modeling of Nanotubes as Elastic Materials

19.2.2.1 Elastic Properties: Young's Modulus, Elastic Constants, and Strain Energy

Experimental fitting to mechanical measurements of the Young's modulus and elastic constants of nanotubes have been mostly made by assuming the CNTs to be elastic beams. An extensive summary is given in Section 19.2.2.2. Aside from the use of the beam assumption, there are also experimental measurements that were made by monitoring the response of a CNT under axial load. Lourie and Wagner[27] used micro-Raman spectroscopy to measure the compressive deformation of a nanotube embedded in an epoxy matrix. For SWCNT, they obtained a Young's modulus of 2.8–3.6 TPa, while for MWCNT, they measured 1.7–2.4 TPa.

Yu et al.[28] presented results of 15 SWCNT bundles under tensile load and found Young's modulus values in the range from 320 to 1470 GPa (mean: 1002 GPa). The stress vs. strain curves are shown in Figure 19.6. Another experiment performed by Yu et al.[29] on the tensile loading of MWCNTs yielded a Young's modulus from 270 to 950 GPa (Figure 19.7). It should be pointed out that concepts such as Young's modulus and elastic constants belong to the framework of continuum elasticity; an estimate of these material parameters for nanotubes implies the continuum assumption. Because each individual SWCNT involves only a single layer of rolled graphene sheet, the thickness t will not make any sense until it is given based on the continuum assumption. In the above-mentioned experiments, it is assumed that the thickness of the nanotube is close to the interlayer distance in graphite, i.e., 0.34 nm.

In the following, we focus on the theoretical prediction of these parameters. Unless explicitly given, a thickness of 0.34 nm for the nanotube is assumed. The earliest attempt to predict Young's modulus

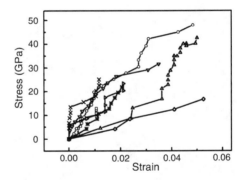

FIGURE 19.6 Eight stress vs. strain curves obtained from the tensile-loading experiments on individual SWCNT bundles. The values of the nominal stress are calculated using the cross-sectional area of the perimeter SWCNTs assuming a thickness of 0.34 nm. (From Yu MF, Files BS, Arepalli S, and Ruoff RS, (2000), Tensile loading of ropes of single-wall carbon nanotubes and their mechanical properties, *Phys. Rev. Lett.* 84(24): 5552–5555. With permission.)

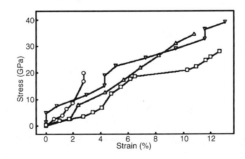

FIGURE 19.7 Plot of stress vs. strain curves for five individual MWCNTs. (Reprinted from Yu MF, Lourie O, Dyer MJ, Moloni K, Kelly TF, and Ruoff RS, (2000), Strength and breaking mechanism of multi-walled carbon nanotubes under tensile load, *Science.* 287(5453): 637–640. © 2000 American Association for the Advancement of Science. With permission.)

theoretically seems to have been made by Overney et al.[30] Using an empirical Keating Hamiltonian with parameters determined from first principles, the structural rigidity of (5,5) nanotubes consisting of 100, 200, and 400 atoms was studied. Although the values are not explicitly given, it was later pointed out by Treacy et al.[31] that the results from Overney et al. implied a Young's modulus in the range of 1.5 to 5.0 TPa. The earliest energetics analysis of CNTs was presented by Tibbetts[32] using elastic theory. He pointed out that the strain energy of the tube is proportional to $1/R^2$ (where R is the radius of the CNT). Ruoff and Lorents[33] suggested the use of elastic moduli of graphite by neglecting the change in the atomic structure when a piece of graphene sheet is rolled into a nanotube. Because the mechanical behavior of single-crystal graphite is well understood, it would be a good approximation to use the in-plane modulus (1.06 TPa of graphite). However, whether such an approximation is good for SWCNTs with small radius was not known.

Robertson et al.[34] examined the energetics and elastic properties of SWCNTs with radii less than 0.9 nm using both Brenner's potential[35] and first-principles total energy methods. Their results showed a consistent linear proportionality to $1/R^2$ of the strain energy, which implies that small deformation beam theory is still valid even for the small radius limit. An elastic constant (C_{11}) close to that of graphite was predicted using the second set of parameters from Brenner's potential;[35] however, it was also shown that the first set of parameters from the same potential results in excess stiffness. Gao et al.[36] carried out a similar study on SWCNTs of larger radius (up to 17 nm) with a potential that is derived from quantum mechanics. A similar linear relation of the strain energy to $1/R^2$ was found. By computing the second derivative of the potential energy, values of Young's modulus from 640.30 GPa to 673.49 GPa were obtained from the MD simulation for closest-packed SWCNTs.

Yakobson et al.[37] compared particular molecular dynamics (MD) simulation results to the continuum shell model[38] and thereby fitted both a value for Young's modulus (~5.5 TPa) and for the effective thickness of the CNTs ($t = 0.066$ nm). Lu[39] derived elastic properties of SWCNTs and MWCNTs using an empirical model in his MD simulation. A Young's modulus of ~1 TPa and a shear modulus of ~0.5 TPa were reported based on a simulated tensile test. Lu also found from his simulation that factors such as chirality, radius, and the number of walls have little effect on the value of Young's modulus. Yao et al.[40] used a similar approach but with a different potential model and obtained a Young's modulus of approximately 1 TPa. In addition, they treated the dependence of Young's modulus on both the radius and chirality of the tube. They employed an MD model that included bending, stretching, and torsion terms, and their results showed that the strain in the tube was dominated by the torsional terms in their model. An alternate method is to derive Young's modulus based on the energy-per-surface area rather than per-volume. This was used in the study by Hernandez et al.[41] Using a nonorthogonal tight-binding scheme, they reported a *surface* Young's modulus of 0.42 TPa-nm, which, when converted to Young's modulus assuming the thickness of 0.34 nm, resulted in a value of 1.2 TPa. This value is slightly higher than that obtained by Lu.[39]

Zhou et al.[42] estimated strain energy and Young's modulus based on electronic band theory. They computed the total energy by taking account of all occupied band electrons. The total energy was then decomposed into the rolling energy, the compressing or stretching energy, and the bending energy. By fitting these three values with estimates based on the continuum elasticity theory, they obtained Young's modulus of 5.1 TPa for SWCNT having an effective wall thickness of 0.71 Å. Note that this is close to the estimate by Yakobson et al.[37] because the rolling energy and stretching energy terms were also included in the shell theory that Yakobson et al. used. In addition, the accuracy of the continuum estimate was validated by the derivation of a similar linear relation of the strain energy to $1/R^2$.

Although no agreement has been reached among these publications regarding the value of the Young's modulus at this moment, it should be pointed out that a single value of Young's modulus cannot be uniquely used to describe both tension/compression *and* bending behavior. The reason is that tension and compression are mainly governed by the in-plane σ-bond, while pure bending is affected mainly by the out-of-plane π-bond. It may be expected that different values of elastic modulus should be obtained from these two different cases unless different definitions of the thickness are adopted, and that is one reason that accounts for the discrepancies described above. To overcome this difficulty, a consistent continuum treatment in the framework of crystal elasticity is needed; and we present this in Section 19.2.4.

19.2.2.2 Elastic Models: Beams and Shells

Although CNTs can have diameters only several times larger than the length of a bond between carbon atoms, continuum models have been found to describe their mechanical behavior very well under many circumstances.[43] Indeed, their small size and presumed small number of defects make CNTs ideal systems for the study of the links between atomic motion and continuum mechanical properties such as Young's modulus and yield and fracture strengths. Simplified continuum models of CNTs have taken one of two forms: simple beam theory for small deformation and shell theory for larger and more complicated distortions.

Assuming small deformations, the equation of motion for a beam is:

$$\rho A \frac{\partial^2 u}{\partial t^2} + EI \frac{\partial^4 u}{\partial x^4} = q(x) \tag{19.1}$$

where u is the displacement, ρ is the density, A the cross-sectional area, E Young's modulus, I the moment of inertia, and $q(x)$ a distributed applied load. This equation is derived assuming that displacements are small and that sections of the beam normal to the central axis in the unloaded state remain normal during bending; these assumptions are usually valid for small deformations of long, thin beams, although deviations from this linear theory are probable for many applications of CNTs.[44,45] The natural frequency of the ith mode of vibration is then given by:

$$\omega_i = \frac{\beta_i^2}{L^2} \sqrt{\frac{EI}{\rho A}} \tag{19.2}$$

where β_i is the root of an equation that is dictated by the boundary conditions. For a beam clamped at one end (zero displacement and zero slope) and free at the other (zero reaction forces), this equation is:

$$\cos \beta_i \cosh \beta_i + 1 = 0 \tag{19.3}$$

Thus, the frequencies of the first three modes of vibration of a clamped-free beam can be computed from Equation (19.3) with $\beta_1 \approx 1.875$, $\beta_2 \approx 4.694$, and $\beta_3 \approx 7.855$.

Measurements on vibrating CNTs can therefore be used to estimate Young's modulus. The first experimental measurements of Young's modulus in MWCNTs were made by Treacy et al.,[31] who used a vibrating beam model of a MWCNT to estimate a modulus of about 1.8 TPa. The authors observed TEM images of MWCNTs that appeared to be undergoing thermal vibration, with a mean-square vibration amplitude that was found to be proportional to temperature. Assuming equipartition of the thermal energy among vibrational modes and a hollow cylinder geometry of the tube, this allowed Young's modulus to be estimated based on the measured amplitude of vibration at the tip of the tube. The spread in the experimental data is quite large, with modulus values for 11 tubes tested ranging from 0.40 to 4.15 TPa with a mean value of 1.8 TPa; the uncertainty was ±1.4 TPa. A similar study by Krishnan et al.[46] of SWCNTs found an average modulus of about 1.3–0.4/+0.6 TPa for 27 SWCNTs.

Rather than relying on estimates of thermal vibrations, Poncharal et al.[44] used electromechanical excitation as method to probe the resonant frequencies of MWCNTs. For tubes of small diameter (less than about 12 nm), they found frequencies consistent with a Young's modulus in the range of 1 TPa. However, for larger diameters, the bending stiffness was found to decrease by up to an order of magnitude, prompting the authors to distinguish their measured *bending modulus* (in the range 0.1 to 1 TPa) from the true Young's modulus. One hypothesis put forth for the decrease in effective modulus is the appearance of a mode of deformation, in which a wavelike distortion or *ripple* forms on the inner arc of the bent MWCNT. This mode of deformation is not accounted for by simple beam theory. Liu et al.[45] used a combination of finite element analysis (using the elastic constants of graphite) and nonlinear vibration analysis to show that nonlinearity can cause a large reduction in the effective bending modulus. However, Poncharal et al. reported that there was no evidence of nonlinearity (such as a shift of frequency with

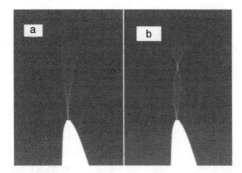

FIGURE 19.8 Scanning electron microscope (SEM) images of electric field-induced resonance of an individual MWCNT at its fundamental resonance frequency (a) and at its second-order harmonic (b).

varying applied force) in their experimental results, which causes some doubt as to whether a large nonlinear effect such as the rippling mode can explain the decrease in bending modulus with increasing tube diameter. Similar resonance excitation of MWCNTs has recently been realized by Yu et al.[47] inside a scanning electron microscope (SEM) (Figure 19.8).

Static models of beam bending can also be used to measure mechanical properties. Wong et al.[48] measured the bending force of MWCNTs using atomic force microscopy (AFM). Assuming the end displacement of an end-loaded cantilevered beam is given by PL3/3EI, where P is the applied force, they fit a Young's modulus of 1.28 ± 0.59 TPa. Salvetat et al.[49,50] measured the vertical deflection vs. the applied force dependence of MWCNT and SWCNT ropes spanning one of the pores in a well-polished alumina nanopore membrane using AFM. They fit values of about 1 TPa for MWCNTs grown by arc discharge, whereas those grown by the catalytic decomposition of hydrocarbons had a modulus 1–2 orders of magnitude smaller.

Govinjee and Sackman[51] studied the validity of modeling MWCNTs with Euler beam theory. They showed the size dependency of the material properties at the nanoscale, which does not appear in the classical continuum mechanics. The beam assumption was further explored by Harik.[52,53] From scaling analysis, he proposed three nondimensional parameters to check the validity of the continuum assumption. The relation between these parameters and MD simulation was discussed. Ru[54–59] has used a shell model to examine the effects of interlayer forces on the buckling and bending of CNTs. It is found that, for MWCNTs, the critical axial strain is decreased from that of a SWCNT of the same outside diameter,[54–56] in essence because the van der Waals forces between layers always cause an inward force on some of the tubes. Note that although the critical axial strain is reduced, the critical axial force may be increased due to the increased cross-sectional area. The phenomenon is also seen when the CNT is embedded in an elastic matrix.[57,58] Ru uses a similar analysis to treat the buckling of columns of SWCNTs arranged in a honeycomb pattern.[59]

19.2.2.3 Elastic Buckling and Local Deformation of NT

Experiments have shown a few cases of exceptional tensile strength of carbon nanotubes (see Section 19.2.3 Strength of Nanotubes). In addition, experiment and theory have addressed structural instability for tubes under compression, bending, or torsion. Buckling can occur in both the axial and transversal direction. Buckling can also occur in the whole structure or locally. Analyses based on continuum theory and the roles of interlayer potential are discussed in Section 19.4.2.2.

Yakobson et al.[37,43] modeled buckling of CNTs under axial compression and used the Brenner potential in their MD simulation. Their simulation also showed four *snap-throughs* during the load process, resulting from instability. The first buckling pattern starts at a nominal compressive strain of 0.05. Buckling due to bending and torsion was demonstrated in References 37, 60, 61, and 62. In the case of bending, the pattern is characterized by the collapse of the cross-section in the middle of the tube, which confirms the experimental observations by Iijima et al.[63] and Ruoff et al.[64,65] using HRTEM

(a)   (b)

FIGURE 19.9   (a) Buckling of SWCNT under bending load. (b) Buckling of SWCNT under torsional load.

and by Wong et al.[48] using AFM. When the tube is under torsion, flattening of the tube, or equivalently a collapse of the cross-section, can occur due to the torsional load. Shown in Figure 19.9 are the simulation results of buckling patterns using a molecular dynamics simulation code developed by the authors. Falvo et al.[66] bent MWCNTs by using the tip of an AFM. They showed that MWCNTs could be bent repeatedly through large angles without causing any apparent fracture in the tube. Similar methods were used by Hertel et al.[67] to buckle MWCNTs due to large bending. Lourie et al.[68] captured the buckling of SWCNTs under compression and bending by embedding them into a polymeric film. Unlike the single kink seen by Iijima et al.,[63] the buckling pattern under bending was characterized by a set of local rippling modes.

The radial deformability for tubes has also been studied. Ruoff et al.[69] first studied radial deformation between adjacent nanotubes (Figure 19.10). Partial flattening due to van der Waals forces was observed in TEM images of two adjacent and aligned MWCNTs along the contact region. This was the first observation that CNTs are not necessarily perfectly cylindrical. Indeed, in an anisotropic physical environment, all CNTs are likely to be, at least to some degree, not perfectly cylindrical due to mechanical deformation. Tersoff and Ruoff[70] studied the deformation pattern of SWCNTs in a closest-packed crystal and concluded that rigid tubes with diameters smaller than 1 nm are less affected by the van der Waals attraction and hardly deform. But for diameters over 2.5 nm, the tubes flatten against each other and form a honeycomb-like structure. This flattening of larger diameter SWCNTs could have a profound effect on factors such as storage of molecular hydrogen in SWCNT crystals composed of such larger diameter tubes, if they can be made, because the interstitial void space is dramatically altered. Lopez et al.[70a] have reported the observation, with HRTEM, of polygonized SWCNTs in contact.

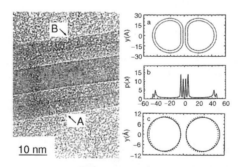

FIGURE 19.10   Left: HRTEM image of two adjacent MWCNTs *a* and *b*. Nanotube *a* has 10 fringes, and nanotube *b* has 22. The average interlayer spacings for inner layers and outer layers belonging to MWCNT *a* are 0.338 nm and 0.345 nm, respectively. For MWCNT *b*, these are 0.343 nm and 0.351 nm, respectively. The 0.07 and 0.08 nm differences, respectively, are due to the compressive force acting in the contact region, and the deformation from perfectly cylindrical shells occurring in both inner and outer portions of each MWCNT. Right from top to bottom: (a) Calculated deformation resulting from van der Waals forces between two double-layered nanotubes. (b) Projected atom density from (a). The projected atom density is clearly higher in the contact region, in agreement with the experimental observation of the much darker fringes in the contact region as compared to the outer portions of MWCNTs *a* and *b*. (c) Calculated deformation for adjacent single-layer nanotubes. (From Ruoff RS, Tersoff J, Lorents DC, Subramoney S, and Chan B, (1993), Radial deformation of carbon nanotubes by van der Waals forces, *Nature*, 364(6437): 514–516. With permission.)

Chopra et al.[71] observed fully collapsed MWCNTs with TEM and showed that the collapsed state can be energetically favorable for certain types of CNTs having a certain critical radius and overall wall thickness. Benedicts et al.[72] proposed the use of the ratio of mean curvature modulus to the interwall attraction of graphite to predict whether the cross-section will collapse and applied their model for a collapsed MWCNT observed in TEM by the same authors. In addition, their experiment also provided the first microscopic measurement of the intensity of the inter-shell attraction.

Hertel et al.[73] and Avouris et al.[74] studied the van der Waals interaction between MWCNTs and a substrate by both experiment (using AFM) and simulation. They found that radial and axial deformations of the tube may lead to a high energy binding state, depending on the types of CNTs present. Fully or partially collapsed MWCNTs on surfaces have also been reported by Yu et al.,[75,76] and their work included a careful analysis of the mechanics of tubes when in contact with surfaces. Lordi and Yao[77] simulated the radial deformation of CNTs due to a local contact and compared the model structures with experimental images of CNTs in contact with nanoparticles. With simulation of both SWCNT and MWCNT up to the radius of a (20, 20) nanotube (13.6 Å), no collapse due to local contact was reported. Instead, the radial deformation was found to be reversible and elastic. Based on these results, they suggested that mechanically cutting a nanotube should be rather difficult, if not impossible.

Gao et al.[36] carried out an energetics analysis on a wide range of nanotubes. They showed the dependence of geometry on the radius of the tube. To be more specific, a circular cross-section is the stable configuration for a radius below 1 nm, in line with the prior results obtained by Tersoff and Ruoff[70] mentioned above. If the radius is between 1 and 2 nm, both circular and collapsed forms are possible. Beyond 3 nm the SWCNT tends to take the collapsed form. However, we should note that the treatment by Gao et al. is of isolated SWCNTs, and their results would be modified by contact with a surface (which might accelerate collapse. Yu et al.,[75,76] on the other hand, as already demonstrated by Tersoff and Ruoff,[70] showed that completely surrounding a SWCNT by similar sized SWCNTs might stabilize against complete collapse.

It is interesting to speculate whether larger diameter SWCNTs would *ever* remain completely cylindrical in any environment. A nearly isotropic environment would be a liquid comprised of small molecules, or perhaps a homogeneous polymer comprised of relatively small monomers. However, because the CNTs have such small diameters, fluctuations are present in the environment on this length scale, locally destroying the time-averaged isotropicity present at longer length scales, may trigger collapse. Alternatively, if full collapse does not occur, the time-averaged state of such a CNT in a (time-averaged) isotropic environment might be perfectly cylindrical. Perhaps for this reason, CNTs might be capable of acting as probes of minute fluctuations in their surrounding (molecular) environment.

Shen et al.[78] conducted a radial indentation test of ~10 nm diameter MWCNTs with scanning probe microscopy. They observed deformability (up to 46%) of the tube and resilience to a significant compressive load (20 μN). The radial compressive elastic modulus was found to be a function of compressive ratios and ranged from 9.0 to 80.0 GPa.

Yu et al.[79] performed a nanoindentation study by applying compressive force on individual MWCNTs with the tip of an AFM cantilever in tapping-mode (Figure 19.11) and demonstrated a deformability similar to that observed by Shen et al. They estimated the effective elastic modulus of a range of indented MWCNTs to be from 0.3 GPa to 4 GPa by using the Hertzian contact model. The reader should note that the difference between this effective elastic modulus and the elastic modulus discussed in Section 19.2.2.1 is that the effective modulus refers to the elastic response to deformation of an anisotropic indentation load applied in the radial direction. We thus also distinguish this type of load from the isotropic load that could, for example, be applied by high pressure for CNTs suspended in a liquid pressure medium, which is more appropriately referred to as *isotropic radial compressive loading*. Yu et al.[76] provided further energetics analysis on MWCNTs that are in configurations of partial or full collapse, or collapsed combined with a twist. They showed that interlayer van der Waals interactions play an important role in maintaining the collapsed configuration.

Chesnokov et al.[80] observed remarkable reversible volume compression of SWCNTs under quasi-hydrostatic pressure up to 3 GPa and obtained a volume compressibility of 0.0277 GPa^{-1}, which suggests the use of CNT as energy-absorbing materials. The volume compressibility of SWCNTs having a diameter

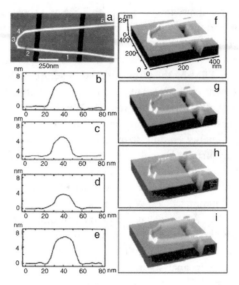

FIGURE 19.11 Deformability of an MWCNT deposited on a patterned silicon wafer as visualized with tapping-mode AFM operated far below mechanical resonance of a cantilever at different set points. The height in this and all subsequent images was coded in grayscale, with darker tones corresponding to lower features. (a) Large-area view of an MWCNT bent upon deposition into a hairpin shape. (b)–(e) Height profiles taken along the thin marked line in (a) from images acquired at different set-point (S/S$_0$) values: (b) 1.0; (c) 0.7; (d) 0.5; (e) 1.0. (f)–(i) Three-dimensional images of the curved region of the MWCNT acquired at the corresponding set-point values as in (b)–(e). (From Yu MF, Kowalewski T, and Ruoff RS, (2000), Investigation of the radial deformability of individual carbon nanotubes under controlled indentation force, *Phys. Rev. Lett.* 85(7): 1456–1459. With permission.)

of 1.4 nm under hydrostatic pressure was also studied by Tang et al.[81] by *in situ* synchrotron X-ray diffraction. The studied SWCNT sample, which consisted primarily of SWCNT bundles and thus not individual or separated SWCNTs, showed linear elasticity under hydrostatic pressure up to 1.5 GPa at room temperature with a compressibility value of 0.024GPa^{-1}, which is smaller than that of graphite (0.028GPa^{-1}). However, the lattice structure of the SWCNT bundles became unstable for pressure beyond ~4 GPa and was destroyed upon further increasing the pressure to 5 GPa.

A very subtle point that comes into the picture is the effect of interlayer interactions when CNTs collapse. This effect can produce some results that cannot be described by traditional mechanics. For instance, Yu et al.[82] observed fully collapsed MWCNT ribbons in the twisted configuration with TEM. One such cantilevered MWCNT ribbon had a twist present in the freestanding segment (Figure 19.12a). Such a configuration cannot be accounted for by elastic theory because no external load (torque) is present to hold the MWCNT ribbon in place and in the twisted form. However, it is known that the difference of approximately 0.012 eV/atom in the interlayer binding energy of the *AA* and the *AB* stacking configurations (Figure 19.12d, c) exists for two rigid graphitic layers spaced 0.344 nm apart. The analysis by Yu et al.[82] suggests that the elastic energy cost for the twist formation of this particular CNT ribbon can be partially compensated by achieving more favorable atomic registry. The observation of the existence of this freestanding twist in the ribbon thus suggests that an energy barrier exists to keep the twisted ribbon from untwisting. The mechanics analysis performed suggests that this twisted structure is metastable. More details on the modeling of interlayer interaction with the account of this registry effect are presented in Section 19.4.2.2.

As a brief summary, compared with their high rigidity in the axial direction, CNTs are observed to be much more compliant in the radial direction. Thus a CNT readily takes the form of a partially or fully collapsed nanoribbon when the radius and wall thickness are in particular ranges, when either isolated in free space, or in contact with a surface. A CNT may also locally flatten when surrounded, as occurs, for example, in the SWCNT bundles.

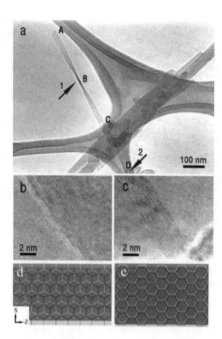

FIGURE 19.12 A free-standing twisted MWCNT ribbon. (a) A TEM image of this ribbon anchored on one end by a carbon support film on a lacy carbon grid. Arrows point to the twists in the ribbon. (b) and (c) Eight resolved fringes along both edges of the ribbon imaged near the anchor point. (d) A schematic depicting the AB stacking between armchair CNT shells (the two layers are the layer having brighter background and black lattices vs. the layer having darker background and white lattices). The AB stacking can be achieved by shifting the layer positions along the x direction that is perpendicular to the long axis of the MWCNT. (e) A schematic depicting the lattice alignment between the zigzag CNT shells by allowing the relative shifting of the layers along the x direction. The AB stacking is not possible; only AA stacking or other stacking (as shown in the schematic) is possible. (From Yu MF, Dyer MJ, Chen J, Qian D, Liu WK, and Ruoff RS, (2001), Locked twist in multi-walled carbon nanotube ribbons, *Phys. Rev. B, Rapid Commn.* 64: 241403R. With permission.)

19.2.3 Strength of Nanotubes

The strength of a CNT will likely depend largely on the distribution of defects and geometric factors. In the case of geometric factors, buckling due to compression, bending, and torsion has been discussed in Section 19.2.2.3. Note that even in these loading cases, it is still possible for plastic yielding to take place due to highly concentrated compressive force. Another geometric factor is the interlayer interaction in the case of MWCNTs and bundles of CNTs.

19.2.3.1 Strength Due to Bond Breaking or Plastic Yielding

Unlike bulk materials, the density of the defects in nanotubes is presumably less; and therefore the strength is presumably significantly higher at the nanoscale. The strength of the CNT could approach the theoretical limit depending on the synthesis process. There are several major categories of synthesis method: carbon evaporation by arc current discharge,[1,83,84] laser ablation,[85,86] or chemical vapor deposition (CVD).[87–89] MWCNTs produced by the carbon plasma vapor processes typically possess higher quality in terms of defects than those produced by the shorter time, lower temperature CVD processes. However, it is not known at this time whether the SWCNTs produced by the laser ablation method, for example, are better than those produced by CVD growth from preformed metal catalyst particles present on surfaces. It is thus worth mentioning the essential role of various catalysts in the production of SWCNTs.[90]

There have been few experimental reports on testing the tensile strength of nanotubes. The idea is simple but seems rather difficult to implement at this stage. Yu et al.[28] were able to apply a tensile load on 15 separate SWCNT bundles and measure the mechanical response (Figure 19.13). The maximum tensile strain they obtained was 5.3%, which is close to the theoretical prediction made by Nardelli et al.[91]

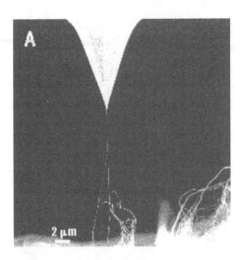

FIGURE 19.13   SEM image of a tensile-loaded SWCNT bundle between an AFM tip and a SWCNT buckytube paper sample. (From Yu MF, Files BS, Arepalli S, and Ruoff RS, (2000), Tensile loading of ropes of single-wall carbon nanotubes and their mechanical properties, *Phys. Rev. Lett.* 84(24): 5552–5555. With permission.)

The *average* SWCNT tensile strength for each bundle ranged from 13 to 52 GPa, calculated by assuming the load was applied primarily on the SWCNTs present at the perimeter of each bundle.

Tensile loading of 19 individual MWCNTs was reported by Yu et al.,[29] (Figure 19.14) and it was found that the MWCNTs broke in the outermost layer by a "sword-in-sheath" breaking mechanism with tensile strain at break of up to 12%. The tensile strength of this outermost layer (equivalent to a large-diameter SWCNT) ranged from 11 to 63 GPa.

By laterally stretching suspended SWCNT bundles fixed at both ends using an AFM operated in lateral force mode, Walters et al.[92] were able to determine the maximum elongation of the bundle and thus the breaking strain. The tensile strength was thus estimated to be 45±7 GPa by assuming a Young's modulus of 1.25TPa for SWCNT. The tensile strength of very long (~2mm) ropes of CVD-grown aligned MWCNTs was measured to be 1.72±0.64 GPa by Pan et al.[93] using a modified tensile testing apparatus, by constantly monitoring the resistance change of the ropes while applying the tensile load. The much lower value they obtained is perhaps to be expected for CVD-grown MWCNTs, according to the authors. But another factor may be the much longer lengths tested, in that a much longer CNT may be more likely to have a critical concentration of defects that could lead to failure present somewhere along their length. The dependence of fracture strength on length has not yet been experimentally addressed.

Another way of applying tensile load is to use the load transferred by embedding the CNTs in a matrix material. Wagner et al.[94] observed fragmentation in SWCNTs using this method and reported a tensile

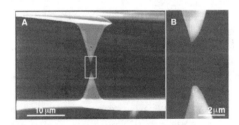

FIGURE 19.14   Tensile loading of individual MWCNTs. (A) An SEM image of an MWCNT attached between two AFM tips. (B) Large magnification image of the indicated region in (A) showing the MWCNT between the AFM tips. (Reprinted from Yu MF, Lourie O, Dyer MJ, Moloni K, Kelly TF, and Ruoff RS, (2000), Strength and breaking mechanism of multi-walled carbon nanotubes under tensile load, *Science.* 287(5453): 637–640. ©2000 American Association for the Advancement of Science. With permission.)

strength of 55 GPa. Tensile strength measurements on a resin-based SWCNT composite were performed by Li et al.[95] Through a treatment that includes modeling the interfacial load transfer in the SWCNT–polymer composite, the tensile strength of the SWCNTs was fit to an average value of ~22 GPa. A compression test of the CNT has been performed by Lourie et al.[68] in which, in addition to the buckling mode that is mentioned in Section 19.2.2.3, they also observed plastic collapse and fracture of thin MWCNTs. The compressive strength and strain corresponding to these cases were estimated to be approximately 100–150 GPa and > 5%, respectively.

Theoretical prediction of CNT strength has emphasized the roles of defects, loading rate, and temperature using MD simulation. Yakobson et al.[37,43,61,96] performed a set of MD simulations on the tensile loading of nanotubes. Even with very high strain rate, nanotubes did not break completely in half; and the two separated parts were instead connected by a chain of atoms. The strain and strength from these simulations are reported to be 30% and 150 GPa, respectively. The fracture behavior of CNT has also been studied by Belytschko et al.[97] using MD simulations. Their results show moderate dependence of fracture strength on chirality (ranges from 93.5 GPa to 112 GPa), and fracture strain between 15.8% and 18.7% were reported. In these simulations, the fracture behavior is found to be almost independent of the separation energy and to depend primarily on the inflection point in the interatomic potential. The values of fracture strains compare well with experimental results by Yu et al.[29]

A central theme that has been uncovered by quantum molecular dynamics simulations of plastic yielding in CNTs is the effect of pentagon/heptagon (or 5/7) defects (Figure 19.15, where the bond rotation leading to the 5–7–7–5 defect initially formed. This is referred to as the *Stone–Wales bond-rotation*[8]) which, for certain types of SWCNTs and at sufficiently high temperature, can lead to plastic yielding.[96] Nardelli et al.[99] studied the mechanism of strain release under tensile loading using both classical and quantum molecular dynamics (MD and QMD). They found that in the case of tension, topological defects such as the 5–7–7–5 defect tend to form when strain is greater than 5% in order to achieve the relaxation of the structure. At high temperature (which for these carbon systems means temperatures around 2000 K) the 5–7–7–5 defect can, for a subset of the SWCNT types, separate into two 5,7 pairs that can glide with respect to each other. Nardelli et al.[91] observed the ductile–to–brittle failure transition as a function of both temperature and strain in their MD simulation. Generally, high strain (15%) and low temperature (1300K) lead to brittle behavior (crack extension or separation), while low strain (3%) and high temperature (3000K) make nanotubes more ductile (dislocation motion without cracking). The reader is reminded of the short time scale of these simulations.

Srivastava et al.[100] demonstrated that a local reconstruction leading to formation of sp^3 bonds can occur under compressive load, and obtained a compressive strength of approximately 153 GPa. Wei

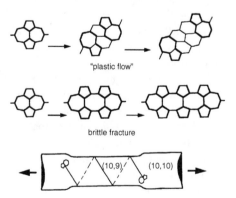

FIGURE 19.15 The 5–7–7–5 dislocation evolves either as a crack (brittle cleavage) or as a couple of dislocations gliding away along the spiral slip plane (plastic yield). In the latter case, the change of the nanotube chirality is reflected by a step-wise change of diameter and by corresponding variations of electrical properties. (From Yakobson BI, 1997, in *Recent Advances in the Chemistry and Physics of Fullerenes and Related Materials*, Kadish KM (Ed.), Electrochemical Society, Inc., Pennington, NJ, 549. With permission of Elsevier Science.)

et al.[101] and Srivastava et al.[102] demonstrated that finite temperature (starting from 300 K) can help the nanotube to overcome a certain energy barrier to achieve plastic deformation. They conducted a similar study of the temperature effect using MD with compressive loading at 12% strain and observed both the formation of sp^3 bonds (at 300K) and 5/7 defects (1600K) for an (8,0) tube. In addition, it was reported[102] that a slower strain rate tends to trigger plastic yield. The conclusion is very similar to the one obtained by Nardelli et al.,[99] but a subtle question is whether the ductile behavior is due to strain or strain rate.

A detailed analysis on the mechanism of defects and dislocation is presented by Yakobson.[61,103] It was shown that the glide direction of the 5/7 defects is dependent on the chirality of the nanotube and, consequently, an irreversible change in the electronic structure also takes place in the vicinity of the chirality change (an armchair tube changes from metallic to semiconducting). Similar conclusions about the effect of strain release on the electronic structure of the tube have been given by Zhang et al.[104] Zhang et al. also observed the dependence of the elastic limit (the onset of plastic yielding) on the chirality of the tube: for the same radius, an $(n,0)$, thus zigzag, tube has nearly twice the limit of an (n,n), thus armchair, tube. This is due mostly to the different alignment of the defects with the principle shear direction. The compressive strength they obtained ranged from 100 to 170 GPa. Another mechanism of defect nucleation is described by Zhang and Crespi,[105] where plastic flow can also occur due to the spontaneous opening of double-layered graphitic patches.

As a brief summary, experimental and simulation efforts have recently been undertaken to assess the strength of CNT. A major limitation for simulation currently is the small time scale that current MD methods can address, much shorter than are actually implemented in experiment. Implicit methods and bridging scale methods show promise in alleviating this difficulty.

19.2.3.2 Strength Due to Interlayer Sliding

Most synthesized nanotubes are either randomly agglomerated MWCNTs or bundles of closest-packed SWCNTs.[85,90] The SWCNTs have a narrow diameter distribution as synthesized and are consequently typically tightly and efficiently packed in bundles.[86,106] The tensile response of SWCNT nanoropes has been measured by Yu et al.;[28] tensile loading of individual MWCNTs was presented by Yu et al.[29] They demonstrated that MWCNTs that are mounted by attachment to the outermost shell and then loaded in tension break in the outermost shell, thus indicating negligible load transfer to the inner shells. This limits the potential of MWCNTs for structural applications — in terms of exhibiting high stiffness and strength in tension, they are a victim of their own high perfection of bonding, as there is evidently no covalent bonding between the nested shells and thus little or no load transfer from the outer shell to the inner shells.

The weak intershell interaction has been measured by Yu et al.[11] (Figure 19.16) and estimated by Cumings and Zettl.[107] In the study of *shell-sliding* of two MWCNTs by Yu et al.,[11] direct measurement of the dependence of the pulling force on the contact length between the shells was performed, and the shear strength between the outermost shell and the neighboring inner shell for two MWCNTs was found to be 0.08 and 0.3 Mpa, respectively. These values are on the low end of the experimental values of the shear strength in graphite samples,[108] thus emphasizing the weak intershell interactions in high-quality, highly crystalline MWCNTs, similar to the type originally discovered by Iijima.[1] Cumings et al.[107] demonstrated inside a TEM that a *telescope process* can be repeated on the MWCNTs that they tested up to 20 times without causing any apparent damage. (The reader should note that assessment of damage at the nanoscale is quite challenging. Here, the authors simply note that there was no apparent change as imaged by high-resolution TEM.) These two studies motivate contemplation of the possibility of a nanoscale-bearing system of exceptional quality. The shear strengths corresponding to static friction and dynamic friction were estimated to be 0.66 and 0.43 MPa, respectively.[107] These values are on the same order as what Yu et al. obtained by experimental measurement.[11]

For SWCNT bundles, there have been theoretical estimates that, in order to achieve load transfer so that the full bundle cross-section would be participating in bearing load up to the intrinsic SWCNT breaking strength (when loading has, for example, been applied only to those SWCNTs on the perimeter), the SWCNT contact length must be on the order of 10 to 120 microns.[109,110] This is much longer than the typical length of individual SWCNTs in such bundles,[111,112] where mean length values on the order

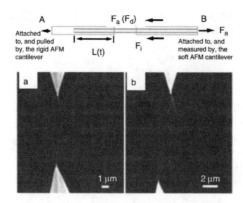

FIGURE 19.16 The forces involved in the shell-sliding experiment can be described by $F_a = F_s + F_i = \pi d\tau L(t) + F_i$, where F_a is the applied pulling force as a function of time, τ is the shear strength, L is the contact length, d the shell diameter, and F_i is a diameter-dependent force originating from both surface tension and edge effects. SEM images showing the sword-in-sheath breaking mechanism of MWCNTs. (a) An MWCNT attached between AFM tips under no tensile load. (b) The same MWCNT after being tensile loaded to break. Notice the apparent overall length change of the MWCNT fragments after break compared to the initial length and the curling of the top MWCNT fragment in (b). (From Yu MF, Yakobson BI, and Ruoff RS, (2000), Controlled sliding and pull-out of nested shells in individual multi-walled carbon nanotubes, *J. Phys. Chem. B.* 104(37): 8764–8767. With permission.)

of a few hundred nanometers have been obtained. This means that load transfer in a parallel bundle containing such relatively short SWCNTs is very likely to be very small and thus ineffective. Quantitative analysis of actual dynamics by modeling using MD requires an interlayer potential that precisely accounts for the interlayer interaction as a function of relative slide position as well as correct incorporation of thermal effects. A general discussion of interlayer potentials is presented in Section 19.3.2.2.

19.2.4 Elastic Properties Based on Crystal Elasticity

It now can be seen that all the theoretical studies presented in Section 19.2.2.1 have either failed to distinguish between the case of infinitesimal strain and finite strain or have directly applied the result at infinitesimal strain to the case of finite strain. In this section, the elastic properties of CNTs at finite strain will be discussed. The significance of this study is obvious from the various deformation patterns due to the load. Of particular interest is the derivation of the elastic constants based on a known atomic model, from either an empirical expression or *ab initio* calculation.

Early attempts to derive elastic constants based on the potential energy of a crystal system were made by Born and Huang.[113] In their treatment, the potential energy was expressed as a function of the displacements of atoms. Because the formulation does not generally guarantee the rotational invariance, the so-called Born–Huang conditions were proposed. Keating[114–117] showed the inconsistencies in the Born–Huang conditions and stated that the potential energy of the crystal system can be alternatively expressed in terms of the variables that intrinsically preserve the invariance property. Based on Brugger's thermodynamic definition[118] of elastic constants, Martin[119–121] derived elastic constants for a crystal system in which the energy density is a sum of the contributions from many-body interactions. Martin's approach is in fact a hyperelastic approach because the energy density was considered as a function of the Lagrangian strain. Based on the embedded-atom method (EAM),[122,123] Tadmor et al.[124,125] derived the corresponding elasticity tensor at finite deformation in their quasicontinuum analysis. Friesecke and James[126] proposed another approach of bridging between continuum and atomic structure, with emphasis on nanostructures in which the size of one dimension is much larger than the other.

The essence of hyperelasticity has been discussed in detail in References 127 through 129. The advantage of the hyperelastic formulation is that it is inherently material frame-invariant; therefore, no special treatment is needed in the large deformation computation.

As a brief summary, if the energy density W of the material is known, the relation between the nominal stress P and the deformation gradient F is given as:[127]

$$P = \frac{\partial W}{\partial F^T} \quad \text{or} \quad P_{ij} = \frac{\partial W}{\partial F_{ji}} \tag{19.4}$$

In Equation (19.4), F_{ij} $\partial x_i/\partial X_j$, in which x and X are, respectively, the spatial and material coordinates. The lower-case subscript denotes the dimension. An equivalent form is to express in terms of the second Piola–Kirchhoff stress S (referred to as 2nd PK stress) and Lagrangian strain E, i.e.,

$$S = \frac{\partial W}{\partial E} \quad \text{or} \quad S_{ij} = \frac{\partial W}{\partial E_{ij}} \tag{19.5}$$

Correspondingly, there are two sets of elasticity tensors:

$$C^{PF} = \frac{\partial^2 W}{\partial F^T \partial F^T} \quad \text{or} \quad C_{ijkl}^{PF} = \frac{\partial^2 W}{\partial F_{ji} \partial F_{lk}} \tag{19.6}$$

and

$$C^{SE} = \frac{\partial^2 W}{\partial E \partial E} \quad \text{or} \quad C_{ijkl}^{SE} = \frac{\partial^2 W}{\partial E_{ij} \partial E_{kl}} \tag{19.7}$$

C^{PF} and C^{SE} are generally referred to as the *first elasticity tensor* and the *second elasticity tensor*, respectively. Their difference is mainly governed by different stress and strain measures that are involved in their definition. It can be shown that the two elasticity tensors are related by

$$C_{ijkl}^{PF} = C_{imnk}^{SE} F_{jm} F_{ln} + S_{ik}\delta_{lj} \tag{19.8}$$

where δ_{ij} is the Kronecker delta.

In addition, there are also the *third elasticity tensor* $C_{ijkl}^{(3)}$ and *fourth elasticity tensor* $C_{ijkl}^{(4)}$. The *third elasticity tensor* relates the velocity gradient $L_{ij} = \partial v_i/\partial x_j$ to the push-forward of the rate of nominal stress P, i.e.,

$$F_{ir}\dot{P}_{rj} = C_{ijkl}^{(3)} L_{kl}^T \tag{19.9}$$

The fourth elasticity tensor is the spatial form of the second elasticity tensor and is defined by

$$C_{ijkl}^{(4)} = F_{im} F_{jn} F_{kp} F_{lq} C_{mnpq}^{SE} \tag{19.10}$$

It can be shown that the fourth elasticity tensor is essentially the tangent moduli in the spatial form and it relates the convective rate of the Kirchhoff stress to the rate of deformation:

$$D_{ij} = \frac{1}{2}(L_{ij} + L_{ji}) \tag{19.11}$$

The convected rate of Kirchhoff stress corresponds to the mathematical concept of Lie derivatives, which consistently define the time derivatives of tensors. Therefore, the fourth elasticity tensor plays an important role in maintaining objectivity during stress update and material stability analysis. More detailed descriptions can be found in Chapters 5 and 6 in the book by Belytschko, Liu, and Moran.[127]

To apply hyperelasticity to the crystal system, the Cauchy–Born rule[124,130,131] must be imposed. This rule assumes that the local crystal structure deforms homogeneously and that the mapping is characterized by the deformation gradient *F*. With this assumption, we can apply the Equations (19.4) to (19.7) to specific atomic models of the nanotube system. We consider the undeformed state of the CNT to be the same as that of a graphene sheet and adopt the classical Tersoff–Brenner model.[35,132–135] The empirical equation is given as follows:

$$\Phi_{ij} = \Phi_R(r_{ij}) - \bar{B}_{ij}\Phi_A(r_{ij}) \tag{19.12}$$

in which *i*, *j* are the indices for carbon atoms. Φ_R and Φ_A represent the repulsive and attractive parts of the potential, respectively. The effect of bonding angle is considered in the term $\bar{B}_{ij}$. The detailed expression for each individual term looks a little tedious due to the consideration of many-body effects. Readers may consult the original paper[35] for parameters and functional forms.

The basic element of the graphite structure is a single hexagon, with each side of the hexagon a result of the covalent bond due to sp^2 hybridization described in Section 19.2.1.1. Because each bond is also shared by the other neighboring hexagon, the total bonding energy Φ for one hexagon can be given as half of the summation of the bonding energy from all six covalent bonds. This summation can be further reduced if the symmetry is considered. The energy density, due to the fact that the single layer of graphite is only a result of repeating the hexagon structure, is given as:

$$W = \frac{\Phi}{A_0} \tag{19.13}$$

with $\Phi = \sum_{l=1}^{3} \Phi_l$. In Equation (19.13), A_0 is the area of the undeformed hexagon and Φ_l is the bond energy for the *I*th bond. According to Brenner's potential:

$$\Phi_l = \Phi_l(r^1, r^{l1}, r^{l2}, \cos\theta^{l1}, \cos\theta^{l2}) \tag{19.14}$$

in which *I*1 and *I*2 refer to the neighboring bonds of bond *I* and Φ_l is a function of bond lengths and bonding angles. Note that *W* is in fact the surface energy density of the system, and the thickness term *t* is not needed because we are dealing with a single layer of graphite. As a result of this, the units for the stress and elasticity tensors are different from those used in the conventional procedures. According to Section 19.2.1.2, the deformation gradient is determined by the roll-up vector, which is composed of the direction and length of rolling operation and the subsequent mechanical relaxation after rolling. These effects are embedded in the deformation gradient F_{ij}. The relaxed structure can be obtained from MD simulation for a specific type of CNT. We have evaluated elastic constants and Young's modulus for various types of nanotubes. For the purpose of comparison, we have converted the elastic constants and Young's modulus by assuming an artificial thickness *t* = 0.34 nm. With this assumption, the second elasticity tensor in the hyper-elastic formulation is given as:

$$C_{ijkl}^{SE} = \frac{\partial S_{ij}}{\partial E_{kl}} \tag{19.15a}$$

with

$$S_{ij} = \frac{1}{A_0 t} \sum_{I=1}^{3} \left(\sum_{N=1, I1, I2} \frac{\partial\Phi}{\partial r^N} \frac{\partial r^N}{\partial E_{ij}} + \sum_{N=1,2} \frac{\partial\Phi_I}{\partial\cos\theta_{IN}} \frac{\partial\cos\theta_{IN}}{\partial E_{ij}} \right) \tag{19.15b}$$

We view the concept of Young's modulus as a tangent modulus that is defined in the deformed configuration (the CNT as a result of rolling the graphene sheet.). Therefore it can be calculated from both the above equation and the expression for the fourth elasticity tensor (Equation (19.10)). The standard definition of Young's modulus is the ratio of the uniaxial stress exerted on a thin rod (a nanotube, in our case) to the resulting normal strain in the same direction. If we define 1 to be the axial direction and 2 and 3 to be the other two orthogonal directions, then it can be shown that the relation between Young's modulus Y and the elastic constants is given as:

$$Y = \frac{C_{11}C_{23}^2 + C_{12}C_{13}^2 + C_{33}C_{12}^2 - C_{11}C_{22}C_{33} - 2C_{12}C_{13}C_{23}}{C_{23}^2 - C_{22}C_{33}} \tag{19.16a}$$

$$C_{ijkl} = F_{im}F_{jn}F_{kp}F_{lq}\left(\frac{\partial S_{mn}}{\partial E_{pq}}\right) \tag{19.16b}$$

Note that the Voigt notation has been used in Equation (19.16a), and elastic constants correspond to the components of the fourth elasticity tensor are defined in Equation (19.16b). By plugging the model parameters and the deformation gradient, the Young's modulus for a (10,10) and (100,100) nanotube are obtained as 0.7 and 1TPa, respectively. It is observed that the Young's modulus from the case of (10,10) is quite different from that of graphite due to the effect of rolling. Such effect becomes small as the radius of CNT increases. Note that this value of Young's modulus is determined *consistently* based on crystal elasticity *combined* with the use of MD simulation. In contrast with some of the approaches described in Section 19.2.2.1, which use purely molecular dynamics to determine the Young's modulus, the approach is semi-analytical and serves as a link between the continuum and atomistic scale. In addition, as we extend to the finite deformation case, the issue of anisotropy naturally arises as a result of this formulation, which can only be qualitatively reproduced by MD. It is emphasized that the thickness assumption is only used in the comparison with the experiments or theoretical predictions that have taken the interlayer separation as the thickness. Clearly this thickness t is not needed in our formulation.

The procedure described above reveals certain limitations of the standard hyperelastic approach and Cauchy–Born rule, which can be described as follows:

- The deformation is in fact not homogenous as the graphene sheet is rolled into CNT. Correspondingly, the energy of the CNT not only depends on the deformation gradient but also on higher order derivatives of F. In such case, a set of high-order elastic constants that belongs to the framework of *multipolar* theory[136] needs to be determined.
- Another aspect that has been missing in the hyperelastic theory is the dependence of the energy on the so-called inner displacement of the lattice, which can be defined as the relative displacement between two overlapping Bravais lattices. Note that the inner displacement can occur without violating the Cauchy–Born rule. According to Cousins,[137] the consideration of these variables results in the so-called *inner elastic constants*.

In general, the factors mentioned above are difficult to evaluate through purely analytical method. A continuum treatment, which accounts for the effect of inner displacement, has been proposed recently in References 138 through 140. For computational implementations, see Section 19.4.3.

19.3   Experimental Techniques

The extremely small dimensions of CNTs — diameters of a few tens of nanometers for MWCNTs and about 1 nm for SWCNTs, and length of a few microns — impose a tremendous challenge for experimental study of mechanical properties. The general requirements for such study include (1) the challenge of CNT placement in an appropriate testing configuration, such as of picking and placing, and in certain cases the fabrication of clamps; (2) the achievement of desired loading; and (3) characterizing and measuring the mechanical deformation at the nanometer length scale. Various types of high-resolution microscopes are indispensable instruments for the characterization of nanomaterials, and recent innovative developments in the new area of *nanomanipulation* based on inserting new tools into such instruments has enhanced our ability to probe nanoscale objects. We discuss several of these below.

19.3.1 Instruments for the Mechanical Study of Carbon Nanotubes

Electron microscopy (EM, SEM, and TEM) and scanning probe microscopy (SPM) have been the most widely used methods for resolving and characterizing nanoscale objects. Electron microscopy uses high-energy electron beams (several keV up to several hundred keV) for scattering and diffraction, which allows the achievement of high resolving power, including down to subnanometer resolution because of the extremely short wavelength (a fraction of an Angstrom) of electrons at high kinetic energy.

We and others have primarily used transmission electron microscopy (TEM) and scanning electron microscopy (SEM) to study nanotube mechanics. In TEM, an accelerated electron beam from a thermal or a field emitter is used. The beam transmits through the sample and passes several stages of electromagnetic lenses, projecting the image of the studied sample region to a phosphor screen or other image recording media. In SEM, a focused electron beam (nanometers in spot size) is rastered across the sample surface; and the amplified image of the sample surface is formed by recording the secondary electron signal or the back scattering signal generated from the sample.

Sample requirements typically differ between TEM and SEM. In TEM, a thin (normally several tens of nanometers or less in thickness) and small (no more than 50mm^2) sample is a requirement as a small sample chamber is available; and a dedicated holder is typically used in these expensive instruments (which are thus typically time-shared among many users) for sample transfer. In SEM there is no strict limit on sample size in principle, and normally a large sample chamber is available so that samples can be surveyed over large areas. As to the difference in the ultimate resolution, TEM is limited by such factors as the spread in energy of the electron beam and the quality of the ion optics; and SEM is limited by the scattering volume of the electrons interacting with the sample material. TEM normally has a resolution on the order of 0.2 nm, while SEM is capable of achieving a resolution up to 1 nm.

An exciting new development in electron microscopy is the addition of aberration correction, also referred to as *corrective ion optics*. We provide no extensive review of this topic here, but note for the reader that a coming revolution in electron microscopy will allow for image resolution of approximately 0.04 nm with TEM and of about 0.1 nm for SEM, and perhaps better. There are two types of corrective ion optics: one corrects for spherical aberration (to correct for aberration in the lenses) and one for achromatic aberration (meant to correct for the spread in the wavelengths of the electrons emanating from the emitter). As newer instruments are installed in the next few years, such as at national laboratories, improvement in the image resolution for mechanics studies of nanosized specimens is an enticing goal (see, for example, http://www.ornl.gov/reporter/no22/dec00.htm and http://www.nion.com/).

The scanning or atomic force microscope (SFM, also referred to as AFM) has also been particularly useful for mechanical studies of CNTs. Since the invention of the STM in 1986, it has been quickly accepted as a standard tool for many applications related to surface characterization. High-resolution (nanometer up to atomic resolution) mapping of surface morphology on almost any type of either conductive or nonconductive material can be achieved with an SFM.

The principle of the microscope is relatively simple. A probe, having a force-sensitive cantilever with a sharp tip, is used as a sensor to physically scan, in close proximity to, the sample surface. The probe is driven by a piezoelectric tube capable of nanometer-resolution translations in the x, y and z directions; and the tip normally has a radius of curvature on the order of 10 nm. The force interaction between tip and sample results in deflection of the cantilever. While scanning the sample surface in the x and y directions, the deflection of the cantilever is constantly monitored by a simple optical method or other approaches. A feedback electronic circuit that reads the deflection signal and controls the piezoelectric tube is responsible for keeping a constant force between the tip and the sample surface, and a surface profile of the sample can thus be obtained. Depending on the type of interaction force involved for sensing, an SPM instrument can include a host of methods, such as AFM, friction-force microscopy (FFM), magnetic-force microscopy (MFM), electric-force microscopy (EFM), and so on. Depending on the mechanism used for measuring the force interaction, scanning probe microscopy also includes many modes of operation, such as contact mode, tapping mode, and force modulation mode. For more information on scanning force microscopy, see, for example, References 141 and 142 as well as http://www.thermomicro.com/spmguide/contents.htm.

19.3.2 Methods and Tools for Mechanical Measurement

Following is a brief summary of the methods used for measuring the mechanical properties of CNTs, and especially isolated individual CNTs; and new tools specifically designed for such tasks will be introduced.

19.3.2.1 Mechanical Resonance Method

Treacy and co-workers[31] deduced values for Young's modulus for a set of MWCNTs by measuring the amplitude from recorded TEM images of the thermal vibration of the free ends of each when naturally cantilevered. Krishnan et al.[46] succeeded in measuring the Young's modulus of SWCNTs (Figure 19.17) using a similar method. The amplitude was measured from the blurred spread of tip positions of the free end of the cantilevered CNT compared with the clamped end. The amplitude can be modeled by considering the excitation of mechanical resonance of a cantilever:

$$\sigma^2 = \frac{16L^3kT}{\pi Y(a^4 - b^4)}\sum_n \beta_n^{-4} \approx 0.4243\frac{L^3kT}{Y(a^4 - b^4)} \tag{19.17}$$

where σ is the amplitude at the free end, L is the length of the cantilevered beam, k is the Boltzmann constant, T is the temperature, Y is the Young's modulus, a is the outer radius and b is the inner radius of CNT, and β_n is a constant for free vibration mode n. The tip blurring originates from thermal activation of vibrations (the CNT behaves classically because of the low frequency modes that are populated by the expected kT of thermal vibrational energy), and the amplitude can be modeled by considering the excitation of mechanical resonance of a cantilever. The image is blurred simply because the frequency is

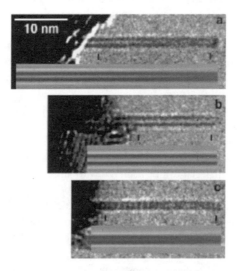

FIGURE 19.17 TEM images of vibrating single-walled nanotubes. Inserted with each micrograph is the simulated image corresponding to the best least-square fit for the adjusted length L and tip vibration amplitude σ. The tick marks in each micrograph indicate the section of the nanotube shank that was fitted. The nanotube length, diameter W, tip amplitude, and the estimated Young's modulus E are (a) $L = 36.8$ nm, $\sigma = 0.33$ nm, $W = 1.50$ nm, $E = 1.33 \pm 0.2$ TPa; (b) $L = 24.3$ nm, $\sigma = 0.18$ nm, $W = 1.52$ nm, $E = 1.20 \pm 0.2$ TPa; and (c) $L = 23.4$ nm, $\sigma = 50.30$ nm, $W = 1.12$ nm, $E = 1.02 \pm 0.3$ TPa. (From Krishnan A, Dujardin E, Ebbesen TW, Yianilos PN, and Treacy MMJ, (1998), Young's modulus of single-walled nanotubes, *Phys. Rev. B.* 58(20): 14013–14019. With permission.)

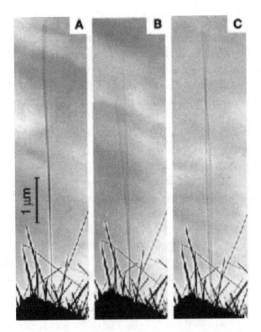

FIGURE 19.18   Electric field-driven resonance of MWCNT. (A) In the absence of a potential, the nanotube tip ($L =$ 6.25 mm, $D = 14.5$ nm) vibrated slightly because of thermal effects. (B) Resonant excitation of the fundamental mode of vibration ($f_1 = 530$ kHz. (C) Resonant excitation of the second harmonic ($f_2 = 3.01$ MHz). For this nanotube, a value of $E_b = 0.21$ TPa was fit to the standard continuum beam mechanics formula. (Reprinted from Poncharal P, Wang ZL, Ugarte D, and de Heer WA, (1999), Electrostatic deflections and electromechanical resonances of carbon nanotubes, *Science.* 283(5407): 1513–1516. ©1999 American Association for the Advancement of Science. With permission.)

high relative to the several-second integration time needed for generating the TEM image. Because the resonance is a function of the cantilever stiffness, and the geometry is directly determined by the TEM imaging, the Young's modulus values could be fit. The advantage of this method is that it is simple to implement without the need for additional instrument modification or development — only a variable-temperature TEM holder is needed. The principle can also be applied for the study of other nanowire-type materials as long as the tip blurring effect is obvious. The drawback is that a model fit is needed to determine the real cantilever length, and human error is inevitable in determining the exact amplitude of the blurred tip. As pointed out by the authors, the error for the Young's modulus estimation using such method is around ±60%.

Poncharal et al.[44] introduced an electric field excitation method (Figure 19.18) for the study of mechanical resonance of cantilevered MWCNTs and measured the bending modulus. In the experiment, a specially designed TEM holder was developed that incorporated a piezo-driven translation stage and a mechanical-driven translation stage. The translation stages allowed the accurate positioning of the MWCNT material inside the TEM, in this case relative to a counter electrode. Electrical connections were made to the counter electrode and the electrode attaching the MWCNT materials, so that DC bias as well as AC sinusoidal voltage could be applied between the counter electrode and the MWCNT. The generated AC electric field interacts with induced charges on the MWCNT, which produces a periodic driving force. When the frequency of the input AC signal matched the mechanical resonance frequency of the MWCNT, obvious oscillation corresponding to the resonance mode of the cantilever MWCNT was observed (Figure 19.18) and the resonance frequency of the MWCNT thereby determined. Using continuum beam mechanics, the bending modulus of the MWCNT was calculated from the measured resonance frequency and CNT geometry according to Equation (19.2). The benefits of such an approach are the efficient method for driving mechanical resonances and, because the whole experiment is done inside a TEM, the ability to analyze the high-aspect ratio nanostructures in detail.

19.3.2.2   Scanning-Force Microscopy Method

The atomic force microscope operated in lateral-force mode, contact mode, or tapping mode, has been the main tool in studying the mechanical response of individual CNTs under static load and when in contact with surfaces. Falvo et al.[66] used a nanomanipulator and contact mode AFM to manually manipulate and bend MWCNTs deposited on a substrate surface. The strong surface force between the MWCNTs and the substrate allowed such an operation to be performed. By intentionally creating large curvature bends in MWCNTs, buckles and periodic ripples were observed. These authors estimated that, based on the local curvature of the bend found, some MWCNTs could sustain up to 16% strain without obvious structural or mechanical failure (Figure 19.19)

Wong et al.[48] measured the bending modulus of individual MWCNTs using an AFM operated in lateral-force mode (Figure 19.20). The MWCNTs, deposited on a low-friction MoS2 surface, were pinned down at one end by overlaying SiO pads using lithography. AFM was then used to locate and measure the dimension of the MWCNT, and lateral force was applied at the different contact points along the length of the MWCNT (Figure 19.20). By laterally pushing the MWCNT, lateral force vs. deflection data were recorded. The data were then analyzed using a beam mechanics model that accounted for the friction force, the concentrated lateral force, and the rigidity of the beam. The bending modulus value for the MWCNT was obtained by fitting the measured force vs. deflection curve. Such a method also allowed the bending strength to be determined by deflecting the beam past the critical buckling point.

Salvetat et al. used another approach to deflect under load MWCNT[49] and SWCNT ropes[50] by depositing them onto a membrane having 200 nm pores (Figure 19.21). By positioning the AFM tip directly on the midpoint of the CNT spanning the pore and applying an indentation force (Figure 19.21), force vs. deflection curves were obtained and compared with theoretical modeling based on beam mechanics. Elastic moduli

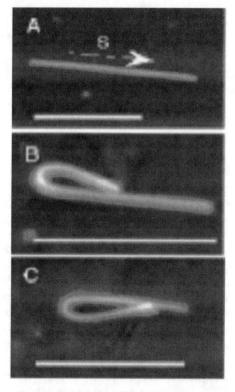

FIGURE 19.19   Bending and buckling of MWCNTs. (a) An original straight MWCNT. (b) The MWCNT is bent upwards all the way back onto itself. (c) The same MWCNT is bent all the way back onto itself in the other direction (Reprinted from Falvo MR, Clary GJ, Taylor RM, Chi V, Brooks FP, Washburn S, and Superfine R, (1997), Bending and buckling of carbon nanotubes under large strain, *Nature.* 389(6651): 582–584. ©1997 Macmillan Publishers. With permission.)

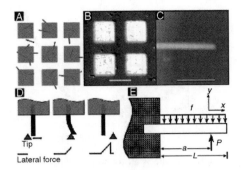

FIGURE 19.20 Overview of one approach used to probe mechanical properties of nanorods and nanotubes. (A) SiC nanorods or carbon nanotubes were deposited on a cleaved MoS_2 substrate, and then pinned by deposition of a grid of square SiO pads. (B) Optical micrograph of a sample showing the SiO pads and the MoS_2 substrate. The scale bar is 8 mm. (C) An AFM image of a 35.3 nm diameter SiC nanorod protruding from an SiO pad. The scale bar is 500 nm. (D) Schematic of beam bending with an AFM tip. The tip (triangle) moves in the direction of the arrow, and the lateral force is indicated by the trace at the bottom. (E) Schematic of a pinned beam with a free end. The beam of length L is subjected to a point load P at $x = a$ and to a distributed friction force f. (Reprinted from Wong EW, Sheehan PE, and Lieber CM, (1997), Nanobeam mechanics: elasticity, strength, and toughness of nanorods and nanotubes, *Science.* 277(5334): 1971–1975. ©1997 American Association for the Advancement of Science. With permission.)

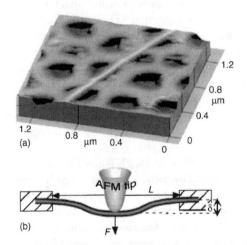

FIGURE 19.21 (a) AFM image of an SWCNT bundle adhered to the polished alumina ultrafiltration membrane, with a portion bridging a pore of the membrane. (b) Schematic of the measurement: the AFM is used to apply a load to the nanobeam and to determine directly the resulting deflection. A closed-loop feedback ensured an accurate scanner positioning. Si_3N_4 cantilevers with force constants of 0.05 and 0.1 N/m were used as tips in the contact mode. (From Salvetat JP, Briggs GAD, Bonard JM, Bacsa RR, Kulik AJ, Stockli T, Burnham NA, and Forro L, (1999), Elastic and shear moduli of single-walled carbon nanotube ropes, *Phys. Rev. Lett.* 82(5): 944–947. With permission.)

for individual MWCNTs and separately for SWCNT bundles were deduced. In principle, such a measurement requires a well-controlled and stable environment to eliminate the errors induced by unexpected tip–surface interactions and instrument instability as well as a very sharp AFM tip for the experiment.

The radial deformability of individual MWCNTs was studied by Shen et al.[78] using an indentation method and Yu et al.[79] using a tapping mode method in AFM. Load was applied along the radial direction (perpendicular to the axial direction that is defined as along the long-axis of the CNT) of MWCNTs, and the applied force vs. indentation depth curve was measured. Using the classic Hertz theory, the deformability of the MWCNT perpendicular to the long axis direction was obtained. The reader should note that this is not the radial compressibility, because the force is not symmetrically applied in the radial direction (not isotropic). Thus, one might think of this as *squashing* the MWCNT locally by indentation.

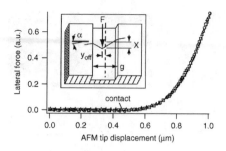

FIGURE 19.22 Lateral force on SWCNT bundle as a function of AFM tip position. The four symbols represent data from four consecutive lateral force curves on the same rope, showing that this rope is straining elastically with no plastic deformation. Inset: The AFM tip moves along the trench, in the plane of the surface, and displaces the rope as shown. (From Walters DA, Ericson LM, Casavant MJ, Liu J, Colbert DT, Smith KA, and Smalley RE, (1999), Elastic strain of freely suspended single-wall carbon nanotube ropes, *Appl. Phys. Lett.* 74(25): 3803–3805. With permission.)

These two approaches are technically different. In the indentation method, the image scan is stopped and the AFM tip is held steady to apply a vertical force on a single point on the MWCNT through the extension and retraction of the piezoelectric tube along the z direction. A force vs. indentation depth curve is obtained by monitoring the AFM cantilever deflection under the extension or retraction. In the tapping mode method, an off-resonance tapping technique is used so that the tapping force can be quantitatively controlled by adjusting the free cantilever amplitude and the set point. The set point is a control parameter in tapping mode AFM for keeping a constant cantilever amplitude (thus a constant distance between the AFM tip and the sample surface) in imaging scan mode. AFM images of each MWCNT are acquired using different set points, and force vs. indentation depth curves are obtained by plotting the curve of the set point vs. MWCNT height. The advantage of using the tapping mode method is that the squashing deformability of the MWCNT along its whole length can be obtained through several image acquisitions, though care must be taken to choose the appropriate tapping mode imaging parameters for such an experiment.

Walters et al.[92] studied the elastic strain of SWCNT nanobundles by creating a suspended SWCNT bundle that was clamped at both ends by metal pads over a trench created with standard lithographic methods. Using an AFM operated in lateral-force mode, they were able to repeatedly stretch and relax the nanoropes elastically as shown in Figure 19.22, including finally stretching to the breaking point to determine the maximum strain. The absolute force used to stretch the SWCNT bundle was not measured, and the breaking strength was estimated by assuming the theoretical value of ~1 TPa for Young's modulus for the SWCNTs in the rope.

19.3.2.3 Measurement Based on Nanomanipulation

The response to axial tensile loading of individual MWCNTs was realized by Yu et al.[29] using a new testing stage based on a nanomanipulation tool operating inside an SEM. The nanomanipulation stage allowed for the three-dimensional manipulation — picking, positioning, and clamping — of individual MWCNTs. The individual MWCNTs were attached to AFM probes having sharp tips by a localized electron beam induced deposition (EBID) of carbonaceous material inside the SEM. An MWCNT so clamped between two AFM probes was then tensile loaded by displacement of the rigid AFM probe (Figure 19.23), and the applied force was measured at the other end by the cantilever deflection of the other, compliant AFM probe. The measured force vs. elongation data were converted, by SEM measurement of the MWCNT geometry, to a stress vs. strain curve; and the breaking strength of each MWCNT was obtained by measuring the maximum tensile loading force at break.

Yu et al.[28] applied a similar approach for the tensile strength measurement of small bundles of SWCNTs. The entangled and web-like agglomeration of SWCNTs in raw samples made it difficult to find an individual SWCNT and resolve it by SEM or to pick out individual SWCNT nanobundles; so a modified

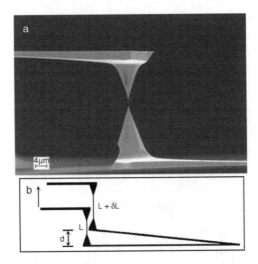

FIGURE 19.23 (a) Individual MWCNT is clamped in place and stretched by two opposing AFM tips. (b) Schematic of the tensile loading experiment. (Reprinted from Yu MF, Lourie O, Dyer MJ, Moloni K, Kelly TF, and Ruoff RS, (2000), Strength and breaking mechanism of multi-walled carbon nanotubes under tensile load, *Science*. 287(5453): 637–640. ©2000 American Association for the Advancement of Science. With permission.)

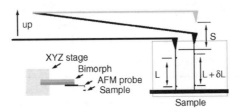

FIGURE 19.24 Schematic showing the principle of the experiment for the measurement of the tensile strength of SWCNT bundles. The gray cantilever indicates where the cantilever would be if no rope were attached on the AFM tip after its displacement upward to achieve tensile loading.

approach was used for the experiment. SWCNT bundles having a strong attachment at one end to the sample surface were selected as candidates for the measurement. The free end of the SWCNT bundle was then approached and attached to an AFM tip by the same EBID method outlined above. The AFM tip was used to stretch the SWCNT bundle to the breaking point, and the same AFM tip also served as the force sensor to measure the applied force (Figure 19.24). Stress vs. strain curves for SWCNT bundles were obtained as well as the breaking strength; these stress vs. strain curves were generated by assuming a model in which only the perimeter SWCNTs in a bundle actually carried the load. The reader is referred to Reference 26 for the full explanation of this model.

The shear strength between the shells of an MWCNT is also an interesting subject for experimental study. Yu et al.[11] were able to directly measure the friction force between the neighboring layers while pulling the inner shells out of the outer shells of an MWCNT, using the same apparatus for measuring the tensile strength of individual MWCNTs. The possibility of such measurement was based on the discovery that tensile-loaded MWCNTs normally broke with a sword-in-sheath breaking mechanism.[29] The separated outer shell can still be in contact with the underlying inner shell in certain cases (in other cases, the *snap back* of the loading and force-sensing cantilevers leads to two separated fragments). The consecutive measurement of force and contact length (the overlap length between the outer shell and its neighbor) provided the necessary data for obtaining the dynamic and static shear strength between the shells.

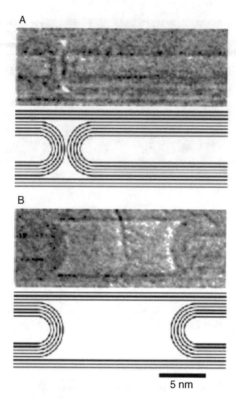

FIGURE 19.25   (A) An as-grown bamboo section. (B) The same area after the core tubes on the right have been telescoped outward. The line drawings beneath the images are schematic representations to guide the eye. (Reprinted from Cumings J and Zettl A, (2000), Low-friction nanoscale linear bearing realized from multi-wall carbon nanotubes, *Science.* 289(5479): 602–604. ©2000 American Association for the Advancement of Science. With permission.)

A similar experiment done in a TEM rather than an SEM was that of Cummings et al.,[107] who used a TEM holder having a piezoelectric-driven translation stage for approaching and opening the end of a MWCNT. The MWCNT cap was opened by *eroding* it away, using an electric discharge method inside the TEM.[143] The end of the exposed core part was then spot-welded to the moving probe using a short electrical pulse, and the MWCNT was telescoped by drawing out the core part from the outer shell housing. It was then possible to disengage the core part from the welding spot and observe the retraction of the core part back into its housing by the surface-driven forces (Figure 19.25). By analyzing the surface force and the friction forces involved in such a retraction using published parameters and modeling, the upper limit values for the dynamic and static friction force between the shells were estimated but not experimentally measured

19.3.3   Challenges and New Directions

The new developments in the area of nanoscale manipulation and measurement as reflected in the studies presented in the last section have certainly helped our understanding of the mechanics of CNTs. Because CNTs possess unique one-dimensional structures that maintain their conformations while being manipulated (in contrast, e.g., to biomolecules or certain polymer systems), they represent a "nanotinkertoy" for manipulation on the nanoscale. Therefore, such types of approaches also provide a window on current capabilities for exploring and exploiting the nanoworld and an avenue for future advancement in methods and tools useful in nanotechnology. In general, mechanical characterization takes a totally different approach than does electrical characterization. Mechanical characterization normally requires dynamic physical interaction with the object. — for example, in the case for CNT, stretching, bending, compressing, and twisting with nanometer positioning accuracy, while performing measurement with nano-Newton force resolution and nanometer dimensional change resolution. Thus, in order for a successful and reliable mechanical study on nanoscale structures, extreme care must be taken to evaluate the three-dimensional stability and accuracy of the instrument and the effect of any significant external factors such as surface contact, mechanical attachment, stray electric charge, and so on. The experiments described in the last section represent the current state of the art in the manipulation and mechanical characterization of CNTs and should provide useful references for further instrument development, which can then be applied to many other low-dimensional nanostructures for mechanical studies.

But what has the community not yet achieved? We have not yet measured the tensile-loading response of an individual SWCNT, nor have we applied a known torque and controllably introduced a twist or series of twists along a CNT. The challenges here include: (1) the visualization, manipulation, precise placement, and fixation of a *flexible* one-dimensional nanostructure, the SWCNT, onto a device having displacement and force-sensing capabilities for the tensile measurement; and (2) a technical breakthrough in generating repeatable coaxial rotation with subnanometer runoff and with sufficient force output for applying torque to CNT. The influence of environment on NT mechanics — such as effects of temperature, chemical environment, or loading rate — has not yet been explored in any detail; nor do we have a clear and detailed picture of the initial defect distribution, or the nucleation, propagation, and ultimate failure resulting from defects. From the experimental perspective, such advances will come with new approaches and automated tools with subnanometer or atomic scale resolving power and stability generated by innovative thinking. It is clear that to attain further advances in nanoscale mechanics, focused effort is necessary in developing new measurement tools that can be integrated into high spatial resolution imaging instruments and that incorporate micro- or nanoelectromechanical system designs.

19.4   Simulation Methods

There are two major categories of molecular simulation methods for NT systems: classical molecular dynamics (MD) and *ab initio* methods. In general, *ab initio* methods give more accurate results than MD, but they are also much more computationally intensive. A hybrid method, tight-binding molecular dynamics (TBMD), is a blend of certain features from both MD and *ab initio* methods. In addition to these methods, continuum and multiscale approaches have also been proposed.

19.4.1   *Ab Initio* and Tight-Binding Methods

The central theme of *ab initio* methods is to obtain accurate solutions to the Schrödinger equation. A comprehensive description of these methods can be found in the book by Ohno et al.[144] Some notation is also adapted in this section. For general background in quantum mechanics, the reader is referred to any of the standard textbooks, such as References 145 through 149. In general, the state of a particle is defined by a wave function ψ based on the well-known wave–particle duality. The Schrödinger equation is

$$H\psi = E\psi \tag{19.18}$$

where H is the Hamiltonian operator of the quantum mechanical system, and ψ is the energy eigenfunction corresponding to the energy eigenvalue E.

Although the phrase *ab initio* is used, analytical or exact solutions are available only for a very limited class of problems. In general, assumptions and approximations need to be made. One of the most commonly used approximations is the Born–Oppenheimer approximation. In this approximation, it is assumed that the electrons are always in a steady state derived from their averaged motion because their positions change rapidly compared to the nuclear motion. Therefore, the motion of the electrons can be considered separately from the motion of the nuclei — as if the nuclei were stationary.

For an N-electron system, the Hamiltonian operator for each electron can be expressed as:

$$H_i = -\frac{1}{2}\nabla_i^2 + \sum_{j>i}^{N}\frac{1}{|r_i - r_j|} + v(r_i) \tag{19.19}$$

The Hamiltonian operator in the above equation is composed of three parts. The first term in Equation (19.19) gives the kinetic energy when operating on the electron wave function; the second term gives the electron–electron Coulomb interaction; and the last term comes from the Coulomb potential from the nuclei. The total Hamiltonian operator of the N-electron system is then:

$$H = \sum_{i=1}^{N}H_i \tag{19.20}$$

and the electron state is solved from the following eigenvalue equation:

$$H\psi_{\lambda_1, \lambda_2, \dots, \lambda_N} = E_{\lambda_1, \lambda_2, \dots, \lambda_N}\psi_{\lambda_1, \lambda_2, \dots, \lambda_N} \tag{19.21a}$$

in which λ_i denotes an eigenstate that corresponds to the one-electron eigenvalue equation:

$$H_i\psi_{\lambda_i} = E_{\lambda_i}\psi_{\lambda_i} \tag{19.21b}$$

Because obtaining exact solutions to Equation (19.21b) is generally very difficult, approximation methods have been developed. In the following, we will introduce two of the most commonly used approaches.

19.4.1.1 The Hartree–Fock Approximation

In the Hartree–Fock approximation,[150–152] the ground state of the Hamiltonian H is obtained by applying the variational principle with a normalized set of wave functions ψ_i. The methodology is identical to the Ritz method, i.e., to seek the solution by minimizing the expectation value of H with a trial function:

$$\langle\psi|H|\psi\rangle = \sum_{S_1}\sum_{S_2}\sum_{S_3}\int\psi^*H\psi\,dr_1dr_2\dots dr_N \tag{19.22}$$

One possible choice of the trial function is the Slater determinant of the single-particle wave functions, i.e.,

$$\psi = \frac{1}{\sqrt{N!}}\begin{vmatrix} \psi_1(1) & \psi_2(1) & \dots & \psi_N(1) \\ \psi_1(2) & \psi_2(2) & & \psi_N(2) \\ \cdot & & & \cdot \\ \cdot & & & \cdot \\ \psi_1(N) & \psi_2(N) & & \psi_N(N) \end{vmatrix} \tag{19.23}$$

In the above equation, the number in the bracket indicates the particle coordinate, which is composed of the spatial coordinate r and internal spin degree of freedom. The subscript denotes the energy level of the wave function, and $\psi_\lambda(i)$ forms an orthonormal set.

The Hamiltonian, on the other hand, is decomposed into a one-electron contribution H_0 and two-body electron–electron Coulomb interaction U as follows:

$$H = \sum_i H_0(i) + \frac{1}{2}\sum_{i,j}U(i,j) \tag{19.24}$$

The one-electron contribution H_0 consists of the kinetic energy and nuclear Coulomb potential:

$$H_0(i) = -\frac{1}{2}\nabla_i^2 + v(r_i) \tag{19.25}$$

in which

$$v(r_i) = -\sum_j \frac{Z_j}{|r_i - R_j|} \tag{19.26}$$

with Z_j being the nuclear charge of the jth atom. The two-body electron–electron interaction is given as the Coulomb interaction:

$$U(i, j) = \frac{1}{|r_i - r_j|} \tag{19.27}$$

With the trial function and decomposition of H, the expectation value of the Hamiltonian can then be rewritten as:

$$\langle\psi|H|\psi\rangle = \sum_{\lambda=1}^{N}\langle\psi_\lambda|H_0|\psi_\lambda\rangle + \frac{1}{2}\sum_{\lambda,v}\langle\psi_\lambda\psi_v|U|\psi_\lambda\psi_v\rangle - \frac{1}{2}\sum_{\lambda,v}\langle\psi_\lambda\psi_v|U|\psi_v\psi_\lambda\rangle \tag{19.28}$$

Applying the variational principle to Equation (19.28), it can be shown that solving the electron state of the system can now be approximated by solving the following equation for the one-electron wave function:

$$H_0\psi_\lambda(i) + \left[\sum_{v=1}^{N}\sum_{S_j}\int\psi_v^*(j)U(i,j)\psi_v(j)dr_j\right]\psi_\lambda(i)$$

$$-\left[\sum_{v=1}^{N}\sum_{S_j}\int\psi_v^*(j)U(i,j)\psi_\lambda(j)dr_j\right]\psi_\lambda(i) = \varepsilon_\lambda\psi_\lambda(i) \tag{19.29}$$

in which ε_j is the Lagrangian multiplier used to enforce the orthonormal condition for the eigenfunction. The Hartree–Fock approximation has been used in many *ab initio* simulations. A more detailed description and survey of this method can be found in Reference 153.

19.4.1.2 Density Functional Theory

The density functional theory was originally proposed in a paper by Hohenberg and Kohn.[154] In this paper, they showed that the ground-state electronic energy is a unique functional of the electronic density. In most cases, the potential due to the external field comes mainly from the nuclei. The electronic energy can then be expressed as:

$$E = T[\rho(r)] + \int\frac{[\rho(r)\rho(r')]}{(|r-r'|)}drdr' + \int V_N(r)dr + E_{XC}[\rho(r)] \tag{19.30}$$

In Equation (19.30), $T[\rho(r)]$ is the kinetic energy and is a function of the electron density; the second term represents the electrostatic potential; the third term denotes the contribution from the nuclei; and the last term is the exchange–correlation functional. Kohn and Sham[155] presented a procedure to calculate the electronic state corresponding to the ground state using this theory, and their method is generally referred to as the *local-density approximation,* or LDA. LDA is another type of widely used *ab initio* method. The term *local density* comes from the assumption that the exchange–correlation function corresponding to the homogeneous electron gas is used. This assumption is only valid locally when the inhomogeneity due to the presence of the nuclei is small.

The essence of the LDA method is to obtain the ground state by introducing the variational principle to the density functional. This leads to a one-electron Schrödinger equation (also called the Kohn–Sham equation) for the Kohn–Sham wave function ψ_λ

$$\left\{ -\frac{1}{2}\nabla^2 + v(r) + \int \frac{\rho(r')}{|r-r'|}dr' + \mu_{XC}[\rho](r) \right\}\psi_\lambda(r) = \varepsilon_\lambda \psi_\lambda(r) \tag{19.31}$$

Note that the term $\mu_{XC}[\rho](r)$ is the derivative of the exchange–correlation functional with respect to the electron density. Different functional forms for the exchange–correlation energy have been proposed.[156–159] The problem is reduced to obtaining the solutions to systems of one-electron equations. Once ψ_λ and ε_λ are solved, the total energy can be obtained from Equation (19.30). The major advantage of using LDA is that the error in the electron energy is second-order between any given electron density and ground-state density.

The solution procedure requires an iterative diagonalization process, which in general involves $O(N^3)$ order of computation. A single electron wave function with a plane-wave basis and pseudo-potential have been used in the application of the LDA method.[160] Major improvements have been made using the Car–Parrinello MD method[161] and conjugate gradient (CG) method.[160] The Car–Parrinello method reduces the order from $O(N^3)$ to $O(N^2)$. As shown in Reference 160, the CG method can even be more efficient.

19.4.1.3 The Tight-Binding Method

The tight-binding theory was originated by Slater and Koster.[162] The advantage of the tight-binding method is that it can handle a much larger system than the *ab initio* method while maintaining better accuracy than MD simulation. A survey of the method can be found in Reference 163. In the tight-binding method, a linear combination of atomic orbitals (referred to LCAO) is adopted in the wave function. Although the exact forms of the basis are not known, the Hamiltonian matrix can be parameterized; and the total energy and electronic eigenvalues can be deduced from the Hamiltonian matrix. The interatomic forces are evaluated in a straightforward way based on the Hellmann–Feynman theorem, and the rest of the procedure is almost identical to the MD simulation. For this reason sometimes the tight-binding method is also referred to as the *tight-binding MD method* or simply *TBMD*.

As shown by Foulkes and Haydock,[164] the total energy can be expressed as the sum of the eigenvalues of a set of occupied nonself-consistent one-electron molecular eigenfunctions in addition to certain analytical functions. The analytical function is usually assumed to take the form of a pair–additive sum. For example, the total energy can be given as

$$E = \sum_i \sum_{j>i} E_{ij} + \sum_k \varepsilon_k \tag{19.32}$$

The first term on the right side is the interatomic interaction, and the second term is the sum of the energies of occupied orbitals. A simple scheme in constructing the second term is to expand the wave function in a localized orthonormal minimal basis with parameterized two-center Hamiltonian matrix elements. The parameterization process can be performed by fitting to results from the *ab initio* methods[165,166] or computing the matrix exactly based on the localized basis.[167–169] A major problem with the TBMD method is the way that the parameterization of the Hamiltonian limits its applicability, or *transferability* as referred to in the computational physics community. As a simple example, when one switches from diamond structure to graphite, the nature of the nearest neighbor changes. In the early development of TBMD, Harrison[163] attempted to use a set of universal parameters; this approach turns out to be neither transferable nor accurate. The solution is then to add in modifications[170,171] or to use a completely different basis.[172]

19.4.2 Classical Molecular Dynamics

Many reviews are available on the subject of classical molecular dynamics.[173-175] MD is essentially a particle method[176,177] because the objective is to solve the governing equations of particle dynamics based on Newton's second law, i.e.,

$$m_i \frac{d^2 \mathbf{r}_i}{dt^2} = -\nabla V \qquad (19.33)$$

in which m_i and $\mathbf{r}_i$ are the mass and spatial coordinates of the ith atom, respectively; V is the empirical potential for the system; and ∇ denotes the spatial gradient. Due to the small time scale involved, explicit integration algorithms such as the Verlet method[178,179] and other high-order methods are commonly used to ensure high-order accuracy.

An alternate but equivalent approach is to solve the Hamiltonian system of ordinary differential equations:

$$dp_i/dt = -\partial H/\partial q_i \qquad (19.34)$$

$$dq_i/dt = -\partial H/\partial p_i \qquad (19.35)$$

in which (q_i, p_i) are the set of canonically conjugate coordinates and momenta, respectively. H is the Hamiltonian function given as:

$$H = \sum_{i=1}^{N} \frac{p_i^2}{2m_i} + V \qquad (19.36)$$

Symplectic integrators[180] have been developed to solve the above Hamiltonian equations of motion. The major advantages of this class of methods are that certain invariant properties of the Hamiltonian system can be preserved,[180] and it is easy to implement in large-scale computations.

19.4.2.1 Bonding Potentials

The basic formulation of MD requires that the spatial gradient of the potential function V be evaluated. Different empirical potentials for an NT system have been developed to satisfy this requirement. Allinger and co-workers developed[181,182] a molecular mechanics force field #2 (MM2) and an improved version known as the MM3 force field. Their model has been applied in the analysis of a variety of organic and inorganic systems. It should be noted that MM2/MM3 is designed for a broad class of problems. It is expected that the model may not work well under certain conditions. For example, the model is known to yield unrealistic results when interatomic distance is in the region of highly repulsive interactions. Mayo et al.[183] presented a generic force field based on simple hybridization considerations. The proposed form of the potential is a combination of bond length (two-body), angle bend (three-body), and torsion (four-body) terms. This empirical model has been used by Guo et al.[184] in the analysis of crystal structures of C_{60} and C_{70}, and by Tuzun et al.[185,186] in the analysis of carbon NT filled with fluid and inert gas atoms. Like the MM2/MM3 model, this force field covers a wide range of nonmetallic main group elements.

Another class of empirical potentials for CNT is characterized by the quantum-mechanical concept of bond order formalism originally introduced by Abell.[187] An alternate interpretation of this formalism can also be found in Reference 188. Using a Morse-type potential, Abell showed that the degree of bonding universality can be well maintained in molecular modeling for similar elements. Tersoff[132-135] introduced this important concept for the modeling of Group IV elements such as C, Si, and Ge, and reasonably accurate results were reported. In Tersoff–Abell bond order formalism, the energy of the system is a sum of the energy on each bond. The energy of each bond is composed of a repulsive part and attractive part. A bond order function is embedded in the formulation. The bond order depends on the local atomic environment such as angular dependency due to the bond angles. Nordlund et al.[189] modified the Tersoff potential such that the interlayer interaction is also considered. Brenner[35] made further improvements to the Tersoff potential by introducing additional terms into the bond order function. The main purpose of these extra terms is to correct the overbinding of radicals. Compared with the Tersoff potential, Brenner's potential shows robustness in the treatment of conjugacy; and it allows for forming and breaking of the bond with the correct representation of bond order. Brenner's potential has enjoyed success in the analysis of formation of fullerenes and their properties,[34,190–192] surface patterning,[193] indentation and friction at nanoscale,[194–203] and energetics of nanotubes.[34] An improved version of Brenner's potential has recently been proposed.[188,204] Based on the approximation of the many-atom expansion for the bond order within the two-center, orthogonal tight-binding (TB) model. Pettifor and Oleinik[205,206] have derived analytical forms that handle structural differentiation and radical formation. The model can be thought of as semi-empirical as it is partly derived from TB. Application of this model to hydrocarbon systems can be found in Reference 205. Depending on the range of applicability, a careful selection of potential model for a specific problem is needed.

19.4.2.2 Interlayer Potentials

Another important aspect of modeling in the analysis of CNT systems is the interlayer interaction. There are two major functional forms used in the empirical model: the inverse power model and the Morse function model. A very widely used inverse power model, the Lennard–Jones (LJ) potential, was introduced by Lennard–Jones[207,208] for atomic interactions:

$$\mu(R_{ij}) = 4\varepsilon\left[\left(\frac{R}{R_{ij}}\right)^{12} - \left(\frac{R}{R_{ij}}\right)^{6}\right] \tag{19.37}$$

where R_{ij} denotes the interatomic distance between atoms i and j, σ is the collision diameter (the inter-atomic distance at which $\phi(R)$ is zero), and ε is the energy at the minimum in $\phi(R_{ij})$. This relationship is shown in Figure 19.26. In the figure, the energy u, interatomic distance R, and interatomic force F are all normalized as $u^* = u/\varepsilon$, $r^* = R_{ij}/R$, and $F^* = FR/\varepsilon$. The corresponding force between the two atoms as a function of interatomic distance is also shown in Figure 19.26 and can be expressed as:

$$F_{ij} = -\frac{\partial u(R_{ij})}{\partial R_{ij}} = 24\frac{\varepsilon}{R}\left[2\left(\frac{R}{R_{ij}}\right)^{13} - \left(\frac{R}{R_{ij}}\right)^{7}\right] \tag{19.38}$$

For the carbon–carbon system, the LJ potential energy has been treated by Girifalco and Lad[209,210] and is given as:

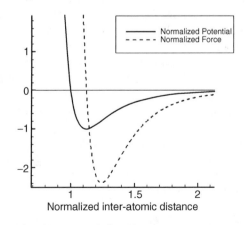

FIGURE 19.26 The pair potential and inter-atomic force in a two-atom system.

$$\phi_i = \frac{A}{\sigma^6}\left[\frac{1}{2}y_0^6\left(\frac{1}{\left(\frac{r_i}{\sigma}\right)^{12}} - \frac{1}{\left(\frac{r_i}{\sigma}\right)^{6}}\right)\right] \tag{19.39}$$

In Equation (19.39), σ is the bond length, y_0 is a dimensionless constant, and r_i is the distance between the ith atom pair. Two sets of parameters have been used: one for a graphite system[209] and the second for an fcc crystal composed of C_{60} molecules.[210] The converted parameters from the original data are given in Table 19.1.

TABLE 19.1 Model Parameters for LJ Potential

Parameter Source	A	Σ	y_0
LJ1 [209]	24.3×10^{-79} J•m^6	1.42 A	2.7
LJ2 [210]	32×10^{-79} J•m^6	1.42 A	2.742

Wang et al.[211] derived the following Morse-type potential for carbon systems based on local density approximations (LDA):

$$U(r) = D_e[(1 - e^{-\beta(r - r_e)})^2 - 1] + E_r e^{-\beta' r} \tag{19.40}$$

where $D_e = 6.50\times10^{-3}$ eV, is the equilibrium binding energy, $E_r = 6.94 \times10^{-3}$ eV, is the hard-core repulsion energy, $r_e = 4.05$ Å, is the equilibrium distance between two carbon atoms, $\beta = 1.00/$Å, and $\beta' = 4.00/$Å. In a comparison study by Qian et al.,[212] it was found that the two LJ potentials yield much higher atomic forces in the repulsive region than the LDA potential; while in the attractive region, the LDA potential gives a much lower value of the binding energy than the two LJ potentials.

Further verification of this LDA potential is obtained by computing the equation of state (EOS) for a graphite system by assuming no relaxation within the graphene plane. The results are compared with published experimental data by Zhao and Spain[213] (referred to as EXP1) and by Hanfland et al. (referred to as EXP2),[214] and with the *ab initio* treatment (which included in-plane relaxation and treated an infinite crystal) by Boetgger.[215] The LDA model fits this experimental data and Boetgger's model reasonably well. The LJ potential, in contrast, deviates strongly from the experimental high-pressure data for graphite (and Boettger's high-level computational treatment) where the relative volume is smaller (Figure 19.27). Based on this comparison, Qian et al.[212] proposed to use LDA for interatomic distances less than 3.3 Å

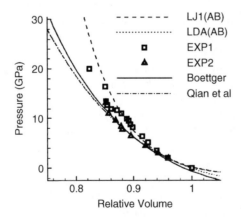

FIGURE 19.27 Comparison of EOS for graphite using different models with experimental data. (From Qian D, Liu WK, and Ruoff RS, (2001), Mechanics of c60 in nanotubes, *J. Phys. Chem. B.* 105: 10753–10758. With permission of Elsevier Science.)

and LJ for interatomic distance greater than 3.4 Å. The transition region is handled by curve fitting to ensure the continuity. A comparison of this model with the rest is also shown in Figure 19.27. Girifalco et al.[216] replaced the discrete sum of the atom pair potentials with a continuous surface integral. In their approach, different model parameters were derived for the cases of two parallel nanotubes and between C_{60} and nanotubes, although the LJ functional form is unchanged.

One of the major disadvantages of the potentials mentioned above is that the difference in interlayer binding energy of the AA and the AB stacking configurations for two rigid graphitic layers spaced at ~0.34 nm is probably not well represented; in short, the *corrugation* in the interlayer energy, due to the *pi* bonds projecting orthogonal to the plane of the layers, is not well captured. In addition, this registration effect also leads to unique nanoscale tribological features. Particularly, the corrugation between neighboring layers will play a central role in the friction present in such a nano-bearing system. The effects of interlayer registration between CNT and the surface it is sliding on have been studied recently in a number of interesting experiments by Falvo et al.[217–220] In these experiments, atomic force microscopy is used to manipulate the MWCNTs on the surface of graphite. A transition of slip-to-roll motion was reported as the MWCNT was moved to certain particular position. This phenomenon uncovers the effects of commensurance at the nanoscale. Both MD[221] and quasi-static simulation[222] have been performed to verify the experiments. Kolmogorov and Crespi[223] developed a new *registry-dependent graphitic potential* which accounts for the exponential atomic-core repulsion and the interlayer delocalization of π orbitals in addition to the normal two-body van der Waals attraction. This model derives an approximate 12 meV/atom difference between the AB stacking and the AA stacking. As discussed in Section 19.2.2.3, Yu et al.[82] applied this model to treat the mechanics of a free-standing twisted MWCNT observed experimentally. In more general settings than, for example, perfectly nested perfect cylinders or two perfectly parallel graphene sheets, the formulation of appropriate models of interlayer interactions remains an important challenge for modeling. Indeed at the nanoscale, the importance of surface interactions cannot be underestimated; and we envision further work by theoreticians and experimentalists to treat problems where some useful discussion between the two camps can be achieved.

19.4.3   Continuum and Multiscale Models

Despite constant increases in available computational power and improvement in numerical algorithms, even classical molecular dynamics computations are still limited to simulating on the order of 10^6–10^8 atoms for a few nanoseconds. The simulation of larger systems or longer times must currently be left to continuum methods. From the crystal elasticity approach in Section 19.2.4, one might immediately see the possibility of applying the finite element method (FEM) to the computational mechanics of nanotubes because the continuum concept of stress can be extracted from a molecular model. However, the fundamental assumption of the continuum approximation — that quantities vary slowly over lengths on the order of the atomic scales — breaks down in many of the most interesting cases of nanomechanics. Thus it would be very useful to have in hand a method that allows the use of a molecular dynamics-like method in localized regions, where quantities vary quickly on the atomic length scale, seamlessly blended with a continuum description of the surrounding material in which, presumably, small scale variations are unimportant or can be treated in an averaged sense.

Several promising methods have been developed toward this goal. The quasicontinuum method, introduced by Tadmor et al.[124,125] and extended and applied to several different problems over the last few years,[224–232] gives a theory for bridging the atomistic and continuum scales in quasistatic problems. In this method, a set of atoms L making up a Bravais lattice has selected from it a subset L_h. A triangulation of this subset allows the introduction of finite element-like shape functions $\varphi_h(l_h)$ at lattice points $l_h \in L_h$, allowing the interpolation of quantities at intermediate points in the lattice. For example, the deformed coordinates q at a lattice point l can be interpolated:

$$q_h(l) = \sum_{l_h \in L_h} \varphi_h(l) q_h(l_h) \qquad (19.41)$$

In this way, the problem of the minimization of energy to find equilibrium configurations can be written in terms of a reduced set of variables. The equilibrium equations then take the form, at each reduced lattice point $\mathbf{l}_h$:

$$\mathbf{f}_h(\mathbf{l}_h) = \sum_{l \in L} \mathbf{f}(\mathbf{l})\varphi_h(\mathbf{l}) = \mathbf{0} \tag{19.42}$$

The method is made practical by approximating summations over all atoms, as implied by the above equation, by using summation rules analogous to numerical quadrature. These rules rely on the smoothness of the quantities over the size of the triangulation to ensure accuracy. The final piece of the method is therefore the prescription of adaptivity rules, allowing the reselection of representative lattice points in order to tailor the computational mesh to the structure of the deformation field. The criteria for adaptivity are designed to allow full atomic resolution in regions of large local strain — for example, very close to a dislocation in the lattice.

The quasicontinuum method has been applied to the simulation of dislocations,[124,125,229,232] grain boundary interactions,[224,227] nanoindentation,[125,227,228] and fracture.[225,226] An extension of the method to finite temperatures has been proposed by Shenoy et al.[233] We have successfully extended the quasicontinuum method to the analysis of the nanotube (CNT) system, although careful treatment is needed. The major challenge in the simulation of CNTs using the quasicontinuum method is the fact that CNT is composed of atomic layers. The thickness of each layer is the size of one carbon atom. In this case, it can be shown that the direct application of the Cauchy–Born rule results in inconsistency in the mapping. Arroyo and Belytshko[234] corrected this inconsistent mapping by introducing the concept of exponential mapping from differential geometry. An alternate approach is to start with the variational principle and develop a method without the Cauchy–Born rule. We have developed the framework of this method with the introduction of a mesh-free approximation.[235–241] For a survey of mesh-free and particle methods and their applications, please see Reference 177; an online version of this paper can be found at http://www.tam.nwu.edu/wkl/liu.html. In addition, two special journal issues[242,243] have been devoted to this topic. Odegard et al.[244] have proposed modeling a CNT as a continuum by equating the potential energy with that of the representative volume element (RVE). This assumption seems to be the same as that in the quasicontinuum method; however, the Born Rule is not used. The method has been recently applied in the constitutive modeling of CNT reinforced polymer composite systems.[245]

The mesh-free method has been directly applied by Qian et al.[212] in the modeling of a CNT interacting with C_{60}. The C_{60} molecule is, on the other hand, modeled with a molecular potential. The interaction between the C_{60} and the continuum is treated based on the conservation of momentum. The interaction forces on the CNT are obtained through the consistent treatment of the weak formulation.

In this approach, a continuous deformation mapping will be constructed through the mesh-free mapping function ϕ. The final form of the discrete equation can be shown to be:

$$f_I^{ext} + f_I^{int} + f_I^{inert} = 0 \tag{19.43}$$

The internal force term that is related to the internal energy is expressed as:

$$f_I^{int} = \sum_{l \in I} w_l f_i(\varphi(r, \theta)) \tag{19.44}$$

Another approach to the coupling of length scales is the FE/MD/TB model of Abraham et al.[246,247] In this method, three simulations are run simultaneously using finite element (FE), molecular dynamics (MD) and semi-empirical tight-binding (TB). Each simulation is performed on a different region of the domain, with a coupling imposed in *handshake* regions where the different simulations overlap. The method is designed for implementation on supercomputers via parallel algorithms, allowing the solution of large problems. One example of such a problem is the propagation of a crack in a brittle material.[246] Here, the TB method is used to simulate bond breaking at the crack tip, MD is used near the crack surface, and the surrounding medium is treated with FE. This method has also been used by Nakano et al. in large-scale simulations of fracture.[248] Rafii-Tabar et al.[249,250] have presented a related method combining FE and MD for the simulation of crack propagation.

A related method, coarse-grained molecular dynamics (CGMD),[251,252] has been introduced as a replacement for finite elements in the FE/MD/TB simulations. In this approach, the continuum-level (or coarse-grained) energy is given by an ensemble average over the atomic motions in which the atomic positions are constrained to give the proper coarse-scale field. In this way, the fine-scale quantities that are not included in the coarse-scale motion are not neglected completely, as their thermodynamic average effect is retained.

We are currently developing a method for the coupling of continuum and MD simulations at finite temperature which, rather than enforcing coupling though boundary conditions in a hand-shake region, allows the continuum and MD representations to coexist in areas of interest in the computational domain. The methodology of the multiscale method can be traced back to the original paper by Liu et al.[253,254] and has been successfully applied in the multiple-scale problems involving strain localization,[255] boundary layers,[256] and coupling of finite elements with mesh-free shape functions.[257] In the current problem setting, this is done by writing a multiple-scale decomposition of the atomic displacements u_α in terms of finite element node displacements d_I and MD displacements d_α; the total scale is given by the usual finite element interpolation, plus the MD displacements, minus the projection of the MD displacements onto the finite element basis. Designating this projection operator as P:

$$u_\alpha = \sum_I N_I(x_\alpha)d_I + d_\alpha - Pd_\alpha \qquad (19.45)$$

where $N_I(x_\alpha)$ is the finite element shape function for node I evaluated at atom α. The key to the method is the subtraction of the projection of d_α, which we term the *bridging scale* and which allows for a unique decomposition into coarse and fine scales. With this decomposition, the Lagrangian (kinetic minus potential energies) can be written

$$L(d_I, \dot{d}_I, d_\alpha, \dot{d}_\alpha) = \sum_{I,J} \frac{1}{2}\dot{d}_I M_{IJ}\dot{d}_J + \sum \frac{1}{2}\dot{d}_\alpha M_{\alpha\beta}\dot{d}_\beta - U(d_I, d_\alpha) \qquad (19.46)$$

where M_{IJ} is the usual finite element mass matrix, $M_{\alpha\beta}$ is a fine-scale mass matrix, and U is the potential energy function. The equations of motion that can be derived from this Lagrangian are the usual finite element equation (plus a contribution to the internal force due to d_α), and the standard MD equations of motion (plus a driving term due to the continuum scale), i.e.,

$$\sum_J M_{IJ}\ddot{d}_J = -\frac{\partial U(d_I, d_\alpha)}{\partial d_I} \qquad (19.47a)$$

$$\sum_\beta M_{\alpha\beta}\ddot{d}_\beta = -\frac{\partial U(d_I, d_\alpha)}{\partial d_\alpha} \qquad (19.47b)$$

These equations can be solved using existing FE and MD codes along with suitable methods for exchanging information about internal forces and boundary conditions. An energy equation can also be derived by considering the continuum-level temperature to be described by the fluctuations in the fine-scale motions. Shown in Figure 19.28 is an example of implementing the multiscale method. It can be seen that the mesh-free discretization and MD coexist and are coupled in the fine-scale region. For the coarse-scale region, a mesh-free approximation is used. For the case shown in Figure 19.28, the error in the bending energy is less than 1% when compared with a purely MD approach. A systematic description of this method is to be described in an upcoming paper.[258]

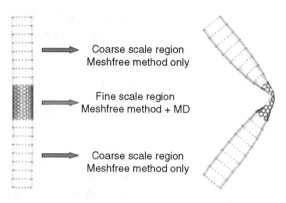

FIGURE 19.28   Multiscale analysis of carbon nanotube.

19.5   Mechanical Applications of Nanotubes

19.5.1   Nanoropes

As discussed in Section 19.2.3.2, the primary product of current methods of SWCNT synthesis contains not individual, separated SWCNTs, but rather bundles of closest-packed SWCNTs.[85,90] Load transfer to the individual SWCNTs in these bundles is of paramount importance for applications involving tensile load bearing. It is estimated that to achieve load transfer so that the full bundle cross-section would be participating in load bearing up to the intrinsic SWCNT breaking strength, the SWCNT contact length must be on the order of 10 to 120 microns. There is strong evidence, however, that the typical length of individual SWCNTs in such bundles is only about 300 nm.[111,112]

From continuum mechanics analysis of other types of wire or fiber forms,[259,260] it is found that twisting the wires or *weaving* the fibers can lead to a cable or rope that has much better load transfer mechanism in tension than a straight bundle would have. Compared with parallel wires, the major difference in terms of the mechanics of a wire rope is that wire ropes have a radial force (in the direction of vector **N** and **N'** as shown in Figure 19.29) that presses the surrounding wires to the core.

Correspondingly, the advantages of having wires in the form of a rope are

- A rope provides better load transfer and structural reliability. For example, when one wire component breaks, the broken sections of that particular wire can still bear load transferred from the other wires through the strong friction force that is a consequence of the radial compression.
- Wire rope has a smaller bending stiffness, therefore it is desirable for applications in which the rope has to be bent frequently. The fatigue life is significantly longer.
- The radial force component gives the rope structure more stability than wires in parallel.

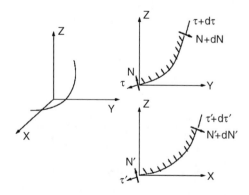

FIGURE 19.29   A three-dimensional section of a wire and force components in the *x-z* and *x-y* planes.

Based on preliminary modeling we have done of a seven-element twisted SWCNT bundle, it is believed that better load transfer in tension can be achieved by making nanoscale ropes and textiles of SWCNTs. We briefly describe some of these new mechanisms and issues below; a detailed analysis is given in Reference 261. We have used MD based on the empirical Tersoff–Brenner potential to analyze a bundle of SWCNTs under twist. A single strand composed of six (10,10) SWCNTs with length of 612 Å surrounding a core (10,10) tube is studied. Both ends of the core tube are fixed, and a twist around the center of the core on the six neighboring SWCNTs is applied at an angular velocity of $20\pi/ns$ (see Figure 19.30 for the cross-section before (a) and after (b) relaxation). The geometric parameters corresponding to the initial configuration are: $r = 6.78$ Å, $d = 3.44$ Å, $R = 2r+d = 17$ Å, $\theta = \pi/3$.

Shown in Figure 19.31 are three snapshots of the deformation of the bundle after twisting is introduced. The corresponding change in the cross-section is shown in Figure 19.32 at the midpoint of the bundle. Clearly, radial deformation strongly depends on the twist angle. Further calculation[262] is performed on a bundle of SWCNTs with the same cross-sectional configuration but with a length of 153.72 Å. The total number of atoms is 17,500. Both ends of the core tube are fixed, and a twist around the center of the core on the six neighboring SWCNTs is applied to achieve a desired angle of twist. After this the whole twisted structure is relaxed to obtain the equilibrium configuration. A constant incremental displacement is then imposed on the core tube while holding the surrounding nanotubes fixed. Plotted in Figure 19.33 is the axial load transferred to the inner tube as a function of the twist angle. We use the transferred axial load as an index for the effectiveness of the load transfer mechanism. As can be seen in Figure 19.33, small twist has very little effect. For the case of 0 degrees (no twist), a force of only ~0.048 eV/Å is transferred to the center tube, and indication of a very smooth inter-tube contact condition; however, for a twist angle of 120 degrees, the transferred load increases to 1.63 eV/Å, about 34 times higher. Our calculation also indicates that too much twist results in unstable structures, i.e., the inner tube is being "squeezed out" when the twisting angle is 180 degrees or higher. This calculation clearly indicates that a great enhancement in the load transfer mechanism can be achieved by making the nanorope. More study on quantitatively determining the effects of various factors is under way.[261] For these significantly collapsed SWCNTs in highly twisted SWCNT bundles, the effective contact area is significantly increased; and this may contribute to better load transfer. In such twisted SWCNT bundles, the load transfer is likely enhanced by an expected increase in shear modulus due to the decrease in the interlayer separation. Recently, Pipes and Hubert[263] applied both textile mechanics and anisotropic elasticity to analyze polymer matrix composites consisting of discontinuous CNTs assembled in helical geometry. The effective elastic properties were predicted, and their study showed the strong dependence of the mechanical behavior on the helical angle of the assembly.

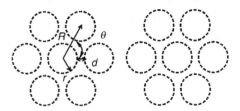

FIGURE 19.30   (Left) geometric parameters used in relaxed nanotube bundles. (Right) Configuration of bundled nanotubes after relaxation.

FIGURE 19.31   Snapshots of twisting of the SWCNT bundle. From a to f, the twisting angles are 30, 60, 90, 120, 150, and 180 degrees, respectively.

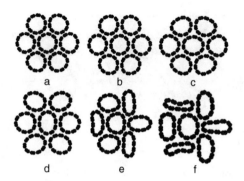

FIGURE 19.32 Change in cross-section at the midpoint of the SWNT bundle as a function of twist angle. From *a* to *f*, the twisting angles are 30, 60, 90, 120, 150, and 180 degrees, respectively.

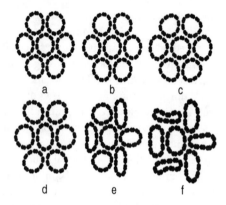

FIGURE 19.33 Transferred load as a function of twisting angle.

In the experiment (Figure 19.34), we attach the bundles using our previously developed nanoclamping methods (Figure 19.35)[11,29,264] and our new method of deposition of low electrical resistance W clamps. One goal involves measuring whether twisting enhances load transfer in terms of increased stiffness and strength. We measure the bundle stiffness without twisting, and then as a function of twisting, in the low-strain regime to assess the effective modulus of the bundle as a function of twist. To study the influence of twisting on strength, we will measure the load at break of similar diameter and length bundles with and without twisting, as a function of diameter, length, and number of twists. The boundary conditions also are assessed. We expect that the carbonaceous deposit made by electron beam-induced decomposition of residual hydrocarbons in the SEM, or the W deposit made, will (largely) be in contact with the outermost (perimeter) SWCNTs in the bundles. In untwisted perfect crystal SWCNT bundles having, for example, little load transfer to core SWCNTs from the perimeter SWCNTs, and with only perimeter SWCNTs clamped (from the method of deposition of our nanoclamp), we should be able to measure if load bearing and breakage occur only at the perimeter SWCNTs. We should be able to differentiate this from the load transfer we hypothesize will take

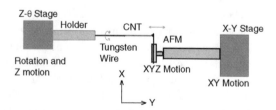

FIGURE 19.34 Proposed experimental stage for twisting the nanoropes.

FIGURE 19.35 Home-built nanomanipulation testing stage, which fits in the palm of the hand and is used in a high-resolution scanning electron microscopy. Further details are available in Yu et al., (2000), *J. Phys. Chem. B.* 104(37): 8764–8767; Yu et al., (2000), *Phys. Rev. Lett.* 84(24): 5552–5555; Yu et al., (2000), *Science.* 287(5453): 637–640.

place upon twisting. One method of monitoring such effects would be to simultaneously monitor the electrical conductivity of the bundle during mechanical loading.

We have recently been developing methods for simultaneous measurement of the electrical conductivity of nanofilaments spanning the opposing AFM tip cantilevers by using conductive AFM tip cantilevers and the W nanoclamp. We propose that such measurements will help to elucidate the dynamics of the nanobundle. The video recording allows for a time resolution of only ~1/30 of a second, the time between video frames. However, electrical conductivity can be measured on a much finer time scale; doing so may allow us to infer when individual SWCNTs have broken prior to the whole bundle breaking, to study interlayer interactions between SWCNTs in the bundle as a function of twisting, and to study changes in conductivity that occur simply due to mechanical deformation of individual tubes in the bundle. Mapping out the electromechanical response of SWCNT bundles as a function of compressive, tensile, and torsional, loading will provide a data base useful for assessing their application as actuators, sensors, and NEMS components. Measuring the electromechanical response in real applications such as cabling in a suspension bridge could also be a useful method of monitoring the reliability of nanorope components in everyday use.

19.5.2 Filling the Nanotubes

The mechanical benefits of filling nanotubes with various types of atoms are the following:

- Filling provides reinforcements for the hollow tube in the radial direction, thus preventing it from buckling, which, from the discussion in Section 19.2.2.3, is known to take place easily.
- Filling provides the smoothest and smallest nano-bearing system.
- Filling also provides an efficient storage system.

Currently there are two major experimental techniques that fill the CNT with foreign materials: arc evaporation, in which foreign materials are put in the anode for their incorporation into the CNTs formed from the carbon plasma, and opening the CNT by chemical agents followed by subsequent filling from either solution-based transport or by vapor transport. Because the chemical and physical environment in the solution-based filling method is less intensive, fragile materials such as biomaterials can also be put into the CNTs using that method.

Early attempts to make filled CNTs using arc evaporation have also resulted in filled carbon nanoparticles rather than filled CNTs. The foreign fillers include most metallic elements,[265–276] magnetic materials,[277–281] and radioactive materials,[282,283] surrounded by a carbon shell. One of the most well-known commercial applications of such filled nanoparticles is the invention of the so-called Technegas,[284,285] which is essentially radioactive material coated with carbon and used as an imaging agent in the detection of lung cancer.

A major breakthrough in filling CNTs was made by Ajayan et al.,[286,287] in which CNTs were opened by an oxidation process and foreign materials (molten lead in their case) were drawn into the CNT due to capillary action. This is the first experiment that showed the possibility of opening the CNT, although

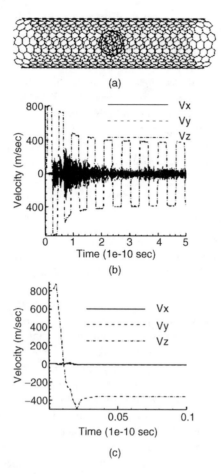

(a)

(b)

(c)

FIGURE 19.36 Computational modeling of C_{60} inside nanotubes. (a) The configuration for the problem. (b) The three velocity components history of the C_{60} as it shuttles through the (10,10) nanotube 20 times (Vz corresponds to the axial direction with an initial value of 0). (c) Same as (b) but for the case of an (8,8) nanotube. (From Wang Y, Tomanek D, and Bertsch GF, (1991), Stiffness of a solid composed of c60 clusters, *Phys. Rev. B.* 44(12): 6562–6565. With permission of Elsevier Science.)

the method used is not universally applicable. Numerous efforts have been undertaken along this direction, including combining oxidation with heating[287–289] and the use of liquid-based approaches, such as nitric acid,[290] hydrochloric acid,[291] or other oxidants.[292,293] Using these processes and the arc evaporation method, different types of materials were also incorporated into CNTs, e.g., compounds of metals or their carbides[268,290,294,295] and biological molecules,[296,297] hydrogen,[298] and argon.[299] Filling of SWCNTs was first reported by Sloan et al.[291,300] Experimental observation of fullerenes inside CNTs has also been reported by Smith et al.[301–303] and Burteaux et al.[304] using pulsed laser evaporation (PLV) of a graphite target containing a catalyst, and by Sloan et al.[305] and Zhang et al.[306] using arc vaporization of carbon with a mixed Ni/Y catalyst.

Filled CNTs have recently been treated by theory and modeling. Pederson et al.[307] discussed the capillary effect at the nanoscale. A major topic is the effect of the change in the dimensionality (from three-dimensional to one-dimensional) and scale (from macroscale to nanoscale) on the resulting physical properties. The effect of the size of the CNT as a geometric constraint on the crystallization process is discussed by Prasad et al.[308] Tuzun et al. studied flow of helium and argon[185] and mixed flow of helium and C_{60}[186] inside CNTs using MD. Berber et al.[309] studied various configurations of putting C_{60} inside a CNT using an *ab initio* method. Stan et al. presented analysis of the hydrogen storage problem.[310] The diffusion properties of molecular flow of methane, ethane, and ethylene inside CNTs were studied using MD by Mao et al.[311]

A general analysis of the statistical properties of the quasi-one-dimensional structure inside a filled CNT is treated by Stan et al.[312] The mechanics of C_{60} inside CNTs was studied by Qian, Liu, and Ruoff et al.[212] using MD simulation. An interesting phenomenon is observed in which C_{60} is sucked into the (10,10) or (9,9) SWCNTs by the sharp surface tension force present in the front of the open end, following which it then oscillates between the two open ends of the nanotube, never escaping. Moreover, the oscillation shows little decay after stabilizing after a few cycles. Both the C_{60} molecule and the nanotube show small deformations as a function of time and position. Furthermore, C_{60}, even when fired on axis with an initial speed up to 1600 m/sec, cannot penetrate into any of the (8,8), (7,7), (6,6) or (5,5) NTs. The simulation results suggest the possibility of using C_{60} in CNTs for making nanodevices such as nanobearings or nanopistons. We continue to study this nanocomposite of fullerenes in CNTs.

19.5.3   Nanoelectromechanical Systems (NEMS)

19.5.3.1   Fabrication of NEMS

Nanoelectromechanical systems (NEMS) are evolving, with new scientific studies and technical applications emerging. NEMS are characterized by small dimensions, where the dimensions are relevant for the function of the devices. Critical feature sizes may be from hundreds to a few nanometers. New physical properties, resulting from the small dimensions, may dominate the operation of the devices. Mechanical devices are shrinking in size to reduce mass, decrease response time, and increase sensitivity.

NEMS systems defined by photolithography processes are approaching the dimensions of carbon nanotubes. NEMS can be fabricated with various materials and integrated with multiphysics systems such as electronic, optical, and biological systems to create devices with new or improved functions. The new class of NEMS devices may provide a revolution in applications such as sensors, medical diagnostics, functional molecules, displays, and data storage. The initial research in science and technology related to nanomechanical systems is taking place now in a growing number of laboratories throughout the world.

The fabrication processes combine various micro/nanomachining techniques including micro/nanostructure fabrication and surface chemistry modification of such fabricated structures to allow site-specific placement of nanofilaments (NF) for actuation and sensing. Different kinds of NFs such as SWCNTs and MWCNTs, nanowhiskers (NWs), and etched nanostructures require different processing steps.

Controlled deposition of individual SWCNT on chemically functionalized templates was first introduced by Liu et al.[313] Reliable deposition of well-separated individual SWCNTs on chemically functionalized nanolithographic patterns was demonstrated. The approach offers promise in making structures and electronic circuits with nanotubes in predesigned patterns.

SWCNTs generally exist as bundles when purchased, so they have to be dispersed into individual SWCNTs in a solution before being used for device experimentation. One controllable procedure is done by diluting the SWCNT suspension into N, N-dimethylformamide (DMF) at a ratio of 1:100. The vial is then sonicated in a small ultrasonic bath for 4–8 hours. Such steps are repeated at least 3 times until the clear suspension contains at least 50% of the SWCNTs completely separated as individual SWCNTs (Figure 19.37).

Recently, Chung, Ruoff, and Lee[314] have developed new techniques for nanoscale gap fabrication (50–500nm) and the consecutive integration of CNTs. These techniques are essential for the batch assembly of CNTs (diameter ~ 10nm) for various applications.[29,264,315–319] Figure 19.38 shows an ideal configuration for chemical sensing by electromechanical transduction involving only a single CNT. So far, they have successfully deposited CNT bundles across the circular gap based on microlithography (Figure 19.39). The fabrication steps do not use serial and time-consuming processes such as e-beam lithography, showing the potential for batch production. It is also noted that these deposition steps were performed at standard environment conditions, thus providing more process freedom. Details of this new fabrication technique can be found in Reference 314.

19.5.3.2   Use of Nanotubes as Sensors in NEMS

The small size and unique properties of carbon nanotubes suggests that they can be used in sensor devices with unprecedented sensitivity. One route to make such sensors is to utilize the electrical properties of

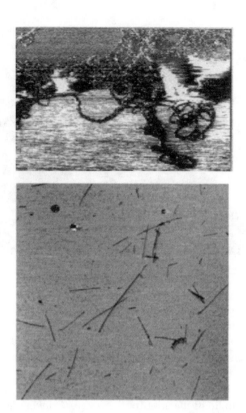

FIGURE 19.37 Before (upper) and after (lower) untangling the nanotubes in the suspension.

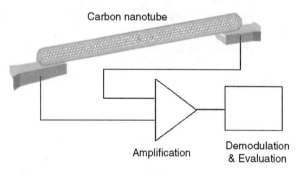

FIGURE 19.38 Carbon nanotube-based sensor.

CNTs. Experiment[320] and computation[321,322] have shown that the conductivity of CNTs can change by several orders of magnitude when deformed by the tip of an atomic force microscope. CNT chemical sensors have also been demonstrated: it has been found experimentally that the electrical resistance of a semiconducting SWCNT changes dramatically upon adsorption of certain gaseous molecules, such as NO_2, NH_3,[318] H_2,[323] and O_2.[317] This phenomenon has been modeled numerically by Peng and Cho[324] using a DFT–LDA technique.

The high sensitivity of the Raman spectra of CNTs to their environment also makes them useful as mechanical sensors. Raman spectra are known from experiment to give shifted peaks when CNTs undergo stress or strain.[325–327] This phenomenon has been used to detect phase transitions and to measure stress fields in polymers with embedded nanotubes.[328,329]

Yet another interesting application of nanotubes as sensors is through the use of the mechanical resonance frequency shift to detect adsorbed molecules or groups of molecules. When the mass of a

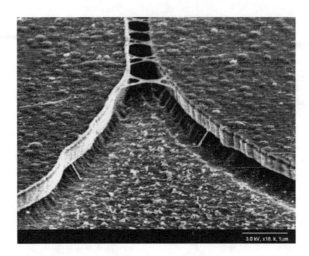

FIGURE 19.39   CNT deposition in a round gap by AC electrophoresis.

vibrating cantilever changes, as by the addition of an adsorbed body, the fundamental frequency of the cantilever as described by Equation (19.2) decreases. Ilic et al.[330] demonstrated micromachined silicon cantilevers with typical dimensions on the order of 20 $\mu m \times$ 320 nm $\times$ 100 μm to detect *E. coli* cells; the device was shown sensitive enough to detect as few as 16 cells under atmospheric conditions. Because the sensitivity of the device is hampered by the mechanical quality factor Q of the oscillating cantilever, even greater sensitivity might be expected in vacuum where the lack of air damping enables higher Q; of course, operation in vacuum limits the utility of the device. Shrinking the size of the design by using a carbon nanotube rather than a micromachined cantilever may give unprecedented sensitivity.

Poncharal et al.[44] were able to use the resonant frequency of a vibrating cantilevered nanotube (Figure 19.18) to estimate the mass of a ~30 fg carbon particle attached to the end of the tube. Our own calculations predict that detection of a 1% change in the resonant frequency of a 100-nm-long (10,10) CNT allows the measurement of an end mass of around 800 amu, approximately the mass of a single C_{60} molecule. The frequency shift of a vibrating cantilevered CNT due to an adsorbed mass at the free end can be estimated using beam theory (Section 19.2.2.2). To the first order approximation, the change in resonance frequency of mode i is

$$\Delta f_i = -\frac{2Mf_i}{\rho AL} \qquad (19.48)$$

where M is the mass of the adsorbed body and $\rho, A,$ and L are, respectively, the density, cross-sectional area, and length of the CNT. Sensitivity may be even further enhanced by taking advantage of critical points in the amplitude–frequency behavior of the vibrating CNT. For example, it is likely that the nonlinear force–deflection curve of the nanotube under certain conditions can result in a bistable response, as has been seen for other vibrating structures at this scale.[331,331a] Recently we have measured four parametric resonances of the fundamental mode of boron nanowires, the first observation of parametric resonance in a nanostructure.[332] It is likely that extraordinarily sensitive molecular sensors can be based on exploiting the parametric resonance of nanostructures (such as nanotubes, nanowires, or nanoplatelets).

19.5.4 Nanotube-Reinforced Polymer

The extremely high modulus and tunable electrical and thermal properties of carbon nanotubes offer an appealing mechanism to dramatically improve both strength and stiffness characteristics, as well as add multifunctionality to polymer-based composite systems. Experimental results to date, however, demonstrate only modest improvements in important material properties amidst sometimes contradictory data.[333–337] Much of the discrepancy in the published results can be attributed to nonuniformity of material samples. In order to obtain optimal property enhancement, key issues to be resolved include improved dispersion of nanotubes, alignment of nanotubes, functionalization of the nanotubes to enhance matrix bonding/load transfer, and efficient use of the different types of nanotube reinforcements (single-wall vs. nanorope vs. multi-wall).

Ongoing investigations are focusing on improved processing and design of nanocomposites with emphasis on controlled nanotube geometric arrangement. Techniques to obtain homogeneous dispersion and significant alignment of the nanotubes include application of electric field during polymerization, extrusion, and deformation methods.[338–340] In addition to dispersion and alignment, recent work[341,342] has demonstrated that the waviness of the nanotubes (see Figure 19.40) in the polymer decreases the potential reinforcing factor by an additional 50% beyond a twofold decrease due to random orientation.

Among the encouraging results is work by Qian[335] for a MWCNT-reinforced polymer, where good dispersion and matrix bonding was achieved. In this case, using only 0.5 vol% nanotube reinforcement with no alignment and moderate NT waviness, elastic stiffness was improved 40% over that of the neat matrix material; and strength values improved nearly 25%. Figure 19.40 shows nanotubes bridging a

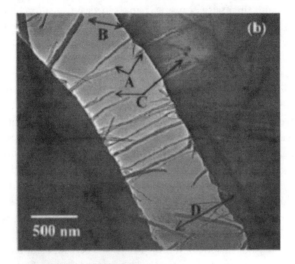

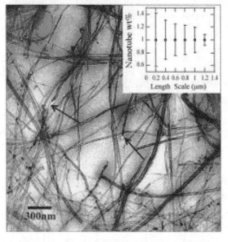

FIGURE 19.40 Upper: *In situ* TEM image of crack in polystyrene film with nanotubes bridging the crack. Lower: arrangement of nanotubes in polymer, good dispersion, random orientation, and moderate waviness. (From Qian D, Dickey EC, Andrews R, and Rantell T, (2000), Load transfer and deformation mechanisms in carbon nanotube–polystyrene composites, *Appl. Phys. Lett.* 76(20): 2868–2870. With permission.)

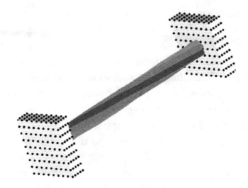

FIGURE 19.41   Multiscale analysis of nanorope-reinforced materials.

matrix crack and demonstrates excellent bonding between nanotubes and matrix material. Viscoelastic properties have also been investigated, with some evidence that well-dispersed nanotubes impact the mobility of the polymer chains themselves, causing changes in glass transition temperature and relaxation characteristics,[343,344] a feature not observed in polymers with a micron-sized reinforcing phase. Limited work on electrical properties shows that percolation can be reached with nanotube volume fractions of less than 1%,[333] leading to dramatic changes in electrical response of the polymer. This enables applications such as polymer coatings with electrostatic discharge capability.[345]

Understanding the mechanisms involved and the degree of property improvement possible for nanotube-reinforced polymers remains a goal of intensive research. Current work emphasizes surface modification of CNTs; control of matrix–NT adhesion; processing methods to control nanotube geometry; hybridization of nanoscale and microscale reinforcements; multifunctional capabilities; and integration of atomistic, micromechanics, and continuum modeling for predictive capability and understanding.

The multiscale method described in Section 19.4.3 shows great potential in the analysis and modeling of nanotube-reinforced material. Shown in Figure 19.41 is an illustration of the approach, combined with the nanorope application discussed in Section 19.5.1. In this method, the domain of the problem is decomposed into three regions: (1) the coarse-scale region, (2) the coupling-scale region, and (3) the fine-scale region. We will solve the coupled equation (Equations (19.47a and 19.47b)) based on the area of interest. A unique advantage of this method is the multiscale decomposition of the displacement field. The governing equations for continuum scale and molecular scale are unified and no special treatment is needed in handling the transition. More details of the modeling approach is included in Reference 346.

19.6   Conclusions

The development of successful methods for the synthesis of high-quality CNTs[85] has led to a worldwide R&D effort in the field of nanotubes. Part of the focus in this paper has been to address the particularly promising mechanical attributes of CNTs as indicated in numerous experiments, with emphasis on the inherent high stiffness and strength of such nanostructures. As stated in Section 19.3.3, much work still remains on the experimental measurement of the mechanical properties of CNTs. Progress made on the theoretical and computational fronts has also been briefly summarized.

It is amazing to see the applicability of continuum elasticity theory even down to the nanometer scale, although this fact should not lead to overconfidence in its use or to any ignorance of the physics that lies beneath. There are also a lot of unresolved issues in theoretical analysis and simulation due to the highly cross-disciplinary nature of the problem. In fact, numerous research efforts are under way in developing multiscale multiphysics simulation schemes. These efforts depend largely on gaining a basic understanding of the phenomena from the quantum level to the continuum level. Therefore, the terms *nanomechanics* or *nanoscale mechanics* indeed go beyond the nanoscale and serve as a manifestation of the link between fundamental science and important engineering applications. This is a major challenge for the engineering community as well as the science community in the sense that the traditional boundary of each field has to be redefined to establish the new forefront of nanotechnology. A number of applications that are described in Section 19.5 serve as the best examples. Yet there are still numerous other likely applications of CNTs that we would like to mention; for instance, there has also been interesting research on the use of CNTs as ferroelectric devices, nanofluidic devices, and energy storage devices; and there exists the possibility of their use in biosurgical instruments. We have not been able to address these topics due to time and space limitations.

As we try to give an up-to-date status of the work that has been done, the field is rapidly evolving, and a tremendous amount of information can be found from various media. It is conceivable that the discovery of carbon nanotubes and other multifunctional nanostructures will parallel the importance of the transistor in technological impact. The size and the mechanical, electrical, thermal, and chemical properties of CNTs — as well as the fact that they are hollow and that matter can thus be located inside them and transported through them — have suggested an astonishingly wide array of potential applications, many of which are under testing now. Statements[347] are being made about factories being set up with ton-level production capabilities. We believe the unique structure and properties of CNTs and related nanomaterials will bring a fundamental change to technology.

Acknowledgments

R.S. Ruoff appreciates support from the NASA Langley Research Center Computational Materials: Nanotechnology Modeling and Simulation Program, and support to G.J. Wagner, R.S. Ruoff, and W.K. Liu from the ONR Miniaturized Intelligent Sensors Program (MIS) and from the grant: Nanorope Mechanics, NSF (Oscar Dillon and Ken Chong, Program Managers). The work of D. Qian and G.J. Wagner has also been supported by the Tull Family Endowment. R.S. Ruoff and W.K. Liu have, for some of the work described here, been supported by several grants from NSF. D. Qian also acknowledges the support of the Dissertation Year Fellowship from Northwestern University. The contributions from Junghoon Lee, Jae Chung, and Lucy Zhang (also supported by ONR MIS) on NEMS and from Cate Brinson (also supported by the NASA Langley Research Center Computational Materials: Nanotechnology Modeling and Simulation Program) on nanotube reinforced polymers are gratefully acknowledged.

References

1. Iijima S, (1991), Helical microtubules of graphitic carbon, *Nature.* 354(6348): 56–58.
2. Normile D, (1999), Technology — nanotubes generate full-color displays, *Science.* 286(5447): 2056–2057.
3. Choi WB, Chung DS, Kang JH, Kim HY, Jin YW, Han IT, Lee YH, Jung JE, Lee NS, Park GS, and Kim JM, (1999), Fully sealed, high-brightness carbon-nanotube field-emission display, *Appl. Phys. Lett.* 75(20): 3129–3131.
4. Bachtold A, Hadley P, Nakanishi T, and Dekker C, (2001), Logic circuits with carbon nanotube transistors, *Science.* 294(5545): 1317–1320.
5. Derycke V, Martel R, Appenzeller J, and Avouris P, (2001), Carbon nanotube inter- and intramolecular logic gates, *Nano Lett.* 10.1021/n1015606f.
6. Baughman RH, Cui CX, Zakhidov AA, Iqbal Z, Barisci JN, Spinks GM, Wallace GG, Mazzoldi A, De Rossi D, Rinzler AG, Jaschinski O, Roth S, and Kertesz M, (1999), Carbon nanotube actuators, *Science.* 284(5418): 1340–1344.
7. Harris PJF, (1999), *Carbon Nanotube and Related Structures: New Materials for the 21st Century.* Cambridge University Press. Cambridge, UK.
8. Dresselhaus MS, Dresselhaus, G, Eklund, PC, (1996), *Science of Fullerenes and Carbon Nanotubes.* Academic Press. San Diego.
9. Dresselhaus MS and Avouris P, (2001), Introduction to carbon materials research, *Carbon Nanotubes*, 1–9.

10. Brown TLL, Bursten BE, and Lemay HE, (1999), *Chemistry: The Central Science.* 8th ed. Prentice Hall. PTR.

11. Yu MF, Yakobson BI, and Ruoff RS, (2000), Controlled sliding and pullout of nested shells in individual multi-walled carbon nanotubes, *J. Phys. Chem. B.* 104(37): 8764–8767.

12. Saito R, Fujita M, Dresselhaus G, and Dresselhaus MS, (1992), Electronic-structure of chiral graphene tubules, *Appl. Phys. Lett.* 60(18): 2204–2206.

13. Dresselhaus MS, Dresselhaus G, and Saito R, (1995), Physics of carbon nanotubes, *Carbon.* 33(7): 883–891.

14. Fujita M, Saito R, Dresselhaus G, and Dresselhaus MS, (1992), Formation of general fullerenes by their projection on a honeycomb lattice, *Phys. Rev. B.* 45(23): 13834–13836.

15. Dresselhaus MS, Dresselhaus, G, Eklund, PC, (1993), Fullerenes, *J. Mater. Res.* 8: 2054.

16. Yuklyosi K, Ed. (1977), *Encyclopedic Dictionary of Mathematics.* The MIT Press. Cambridge.

17. Iijima S, (1993), Growth of carbon nanotubes, *Mater. Sci. Eng. B Solid State Mater. Adv. Technol.* 19(1–2): 172–180.

18. Dravid VP, Lin X, Wang Y, Wang XK, Yee A, Ketterson JB, and Chang RPH, (1993), Buckytubes and derivatives — their growth and implications for buckyball formation, *Science.* 259(5101): 1601–1604.

19. Iijima S, Ichihashi T, and Ando Y, (1992), Pentagons, heptagons and negative curvature in graphite microtubule growth, *Nature.* 356(6372): 776–778.

20. Saito Y, Yoshikawa T, Bandow S, Tomita M, and Hayashi T, (1993), Interlayer spacings in carbon nanotubes, *Phys. Rev. B.* 48(3): 1907–1909.

21. Zhou O, Fleming RM, Murphy DW, Chen CH, Haddon RC, Ramirez AP, and Glarum SH, (1994), Defects in carbon nanostructures, *Science.* 263(5154): 1744–1747.

22. Kiang CH, Endo M, Ajayan PM, Dresselhaus G, and Dresselhaus MS, (1998), Size effects in carbon nanotubes, *Phys. Rev. Lett.* 81(9): 1869–1872.

23. Amelinckx S, Bernaerts D, Zhang XB, Vantendeloo G, and Vanlanduyt J, (1995), A structure model and growth-mechanism for multishell carbon nanotubes, *Science.* 267(5202): 1334–1338.

24. Lavin JG, Subramoney S, Ruoff RS, Berber S, and Tomanek D, (2001), Scrolls and nested tubes in multiwall carbon tubes, unpublished.

25. Ajayan PM and Ebbesen TW, (1997), Nanometre-size tubes of carbon, *Rep. Prog. Phys.* 60(10): 1025–1062.

26. Amelinckx S, Lucas A, and Lambin P, (1999), Electron diffraction and microscopy of nanotubes, *Rep. Prog. Phys.* 62(11): 1471–1524.

27. Lourie O and Wagner HD, (1998), Evaluation of Young's modulus of carbon nanotubes by micro-Raman spectroscopy, *J. Mater. Res.* 13(9): 2418–2422.

28. Yu MF, Files BS, Arepalli S, and Ruoff RS, (2000), Tensile loading of ropes of single wall carbon nanotubes and their mechanical properties, *Phys. Rev. Lett.* 84(24): 5552–5555.

29. Yu MF, Lourie O, Dyer MJ, Moloni K, Kelly TF, and Ruoff RS, (2000), Strength and breaking mechanism of multi-walled carbon nanotubes under tensile load, *Science.* 287(5453): 637–640.

30. Overney G, Zhong W, and Tomanek D, (1993), Structural rigidity and low-frequency vibrational-modes of long carbon tubules, *Zeitschr. Phys. D-Atoms Molecules Clusters.* 27(1): 93–96.

31. Treacy MMJ, Ebbesen TW, and Gibson JM, (1996), Exceptionally high Young's modulus observed for individual carbon nanotubes, *Nature.* 381(6584): 678–680.

32. Tibbetts GG, (1984), Why are carbon filaments tubular? *J. Crystal Growth.* 66(3): 632–638.

33. Ruoff RS and Lorents DC, (1995), Mechanical and thermal properties of carbon nanotubes, *Carbon.* 33(7): 925–930.

34. Robertson DH, Brenner DW, and Mintmire JW, (1992), Energetics of nanoscale graphitic tubules, *Phys. Rev. B.* 45(21): 12592–12595.

35. Brenner DW, (1990), Empirical potential for hydrocarbons for use in simulating the chemical vapor deposition of diamond films, *Phys. Rev. B.* 42(15): 9458–9471.

36. Gao GH, Cagin T, and Goddard WA, (1998), Energetics, structure, mechanical and vibrational properties of single-walled carbon nanotubes, *Nanotechnology.* 9(3): 184–191.

37. Yakobson BI, Brabec CJ, and Bernholc J, (1996), Nanomechanics of carbon tubes: instabilities beyond linear response, *Phys. Rev. Lett.* 76(14): 2511–2514.

38. Timoshenko S and Gere J, (1988), *Theory of Elastic Stability.* McGraw-Hill. New York.

39. Lu JP, (1997), Elastic properties of carbon nanotubes and nanoropes, *Phys. Rev. Lett.* 79(7): 1297–1300.

40. Yao N and Lordi V, (1998), Young's modulus of single-walled carbon nanotubes, *J. Appl. Phys.* 84(4): 1939–1943.

41. Hernandez E, Goze C, Bernier P, and Rubio A, (1998), Elastic properties of c and bxcynz composite nanotubes, *Phys. Rev. Lett.* 80(20): 4502–4505.

42. Zhou X, Zhou JJ, and Ou-Yang ZC, (2000), Strain energy and Young's modulus of single-wall carbon nanotubes calculated from electronic energy-band theory, *Phys. Rev. B.* 62(20): 13692–13696.

43. Yakobson BI and Smalley RE, (1997), Fullerene nanotubes: C-1000000 and beyond, *Am. Sci.* 85(4): 324–337.

44. Poncharal P, Wang ZL, Ugarte D, and de Heer WA, (1999), Electrostatic deflections and electro-mechanical resonances of carbon nanotubes, *Science.* 283(5407): 1513–1516.

45. Liu JZ, Zheng Q, and Jiang Q, (2001), Effect of a rippling mode on resonances of carbon nanotubes, *Phys. Rev. Lett.* 86(21): 4843–4846.

46. Krishnan A, Dujardin E, Ebbesen TW, Yianilos PN, and Treacy MMJ, (1998), Young's modulus of single-walled nanotubes, *Phys. Rev. B.* 58(20): 14013–14019.

47. Yu MF, Dyer MJ, Chen J, and Bray K (2001), Multiprobe nanomanipulation and functional assembly of nanomaterials inside a scanning electron microscope. *Intl. Conf. IEEE-NANO2001,* Maui.

48. Wong EW, Sheehan PE, and Lieber CM, (1997), Nanobeam mechanics: elasticity, strength, and toughness of nanorods and nanotubes, *Science.* 277(5334): 1971–1975.

49. Salvetat JP, Kulik AJ, Bonard JM, Briggs GAD, Stockli T, Metenier K, Bonnamy S, Beguin F, Burnham NA, and Forro L, (1999), Elastic modulus of ordered and disordered multi-walled carbon nanotubes, *Adv. Mater.* 11(2): 161–165.

50. Salvetat JP, Briggs GAD, Bonard JM, Bacsa RR, Kulik AJ, Stockli T, Burnham NA, and Forro L, (1999), Elastic and shear moduli of single-walled carbon nanotube ropes, *Phys. Rev. Lett.* 82(5): 944–947.

51. Govindjee S and Sackman JL, (1999), On the use of continuum mechanics to estimate the properties of nanotubes, *Solid State Comm.* 110(4): 227–230.

52. Harik VM, (2001), Ranges of applicability for the continuum-beam model in the mechanics of carbon-nanotubes and nanorods, *Solid State Comm.* 120(331–335).

53. Harik VM, (2001), Ranges of applicability for the continuum-beam model in the constitutive analysis of carbon-nanotubes: nanotubes or nano-beams? in NASA/CR-2001–211013, also in *Computational Material Science* (Submitted).

54. Ru CQ, (2000), Effect of van der Waals forces on axial buckling of a double-walled carbon nanotube, *J. Appl. Phys.* 87(10): 7227–7231.

55. Ru CQ, (2000), Effective bending stiffness of carbon nanotubes, *Phys. Rev. B.* 62(15): 9973–9976.

56. Ru CQ, (2000), Column buckling of multi-walled carbon nanotubes with interlayer radial displacements, *Phys. Rev. B.* 62(24): 16962–16967.

57. Ru CQ, (2001), Degraded axial buckling strain of multi-walled carbon nanotubes due to interlayer slips, *J. Appl. Phys.* 89(6): 3426–3433.

58. Ru CQ, (2001), Axially compressed buckling of a double-walled carbon nanotube embedded in an elastic medium, *J. Mech. Phys. Solids.* 49(6): 1265–1279.

59. Ru CQ, (2000), Elastic buckling of single-walled carbon nanotube ropes under high pressure, *Phys. Rev. B.* 62(15): 10405–10408.

60. Bernholc J, Brabec C, Nardelli MB, Maiti A, Roland C, and Yakobson BI, (1998), Theory of growth and mechanical properties of nanotubes, *App. Phys. A — Mater. Sci. Proc.* 67(1): 39–46.
61. Yakobson BI and Avouris P, (2001), Mechanical properties of carbon nanotubes, in *Carbon Nanotubes*, 287–327.
62. Qian D, Liu WK, and Ruoff RS, (2002), Bent and kinked multi-shell carbon nanotubes — treating the interlayer potential more realistically. *43rd AIAA/ASME/ASCE/AHS Struct., Struct. Dynamics, Mater. Conf.* Denver, CO.
63. Iijima S, Brabec C, Maiti A, and Bernholc J, (1996), Structural flexibility of carbon nanotubes, *J. Chem. Phys.* 104(5): 2089–2092.
64. Ruoff RS, Lorents DC, Laduca R, Awadalla S, Weathersby S, Parvin K, and Subramoney S, (1995), *Proc. Electrochem. Soc.* 95–10: 557–562.
65. Subramoney S, Ruoff RS, Laduca R, Awadalla S, and Parvin K, (1995), *Proc. Electrochem. Soc.* 95–10: 563–569.
66. Falvo MR, Clary GJ, Taylor RM, Chi V, Brooks FP, Washburn S, and Superfine R, (1997), Bending and buckling of carbon nanotubes under large strain, *Nature.* 389(6651): 582–584.
67. Hertel T, Martel R, and Avouris P, (1998), Manipulation of individual carbon nanotubes and their interaction with surfaces, *J. Phys. Chem. B.* 102(6): 910–915.
68. Lourie O, Cox DM, and Wagner HD, (1998), Buckling and collapse of embedded carbon nanotubes, *Phys. Rev. Lett.* 81(8): 1638–1641.
69. Ruoff RS, Tersoff J, Lorents DC, Subramoney S, and Chan B, (1993), Radial deformation of carbon nanotubes by van der Waals forces, *Nature.* 364(6437): 514–516.
70. Tersoff J and Ruoff RS, (1994), Structural properties of a carbon-nanotube crystal, *Phys. Rev. Lett.* 73(5): 676–679.
70a. Lopez, MJ, Rubio A, Alonso JA, Qin LC, and Iijima S, (2001) Novel polygonized single-wall carbon nanotube bundles, *Phys. Rev. Lett.* 86(14): 3056–3059.
71. Chopra NG, Benedict LX, Crespi VH, Cohen ML, Louie SG, and Zettl A, (1995), Fully collapsed carbon nanotubes, *Nature.* 377(6545): 135–138.
72. Benedict LX, Chopra NG, Cohen ML, Zettl A, Louie SG, and Crespi VH, (1998), Microscopic determination of the interlayer binding energy in graphite, *Chem. Phys. Lett.* 286(5–6): 490–496.
73. Hertel T, Walkup RE, and Avouris P, (1998), Deformation of carbon nanotubes by surface van der Waals forces, *Phys. Rev. B.* 58(20): 13870–13873.
74. Avouris P, Hertel T, Martel R, Schmidt T, Shea HR, and Walkup RE, (1999), Carbon nanotubes: nanomechanics, manipulation, and electronic devices, *Appl. Surf. Sci.* 141(3–4): 201–209.
75. Yu MF, Dyer MJ, and Ruoff RS, (2001), Structure and mechanical flexibility of carbon nanotube ribbons: an atomic-force microscopy study, *J. Appl. Phys.* 89(8): 4554–4557.
76. Yu MF, Kowalewski T, and Ruoff RS, (2001), Structural analysis of collapsed, and twisted and collapsed, multi-walled carbon nanotubes by atomic force microscopy, *Phys. Rev. Lett.* 86(1): 87–90.
77. Lordi V and Yao N, (1998), Radial compression and controlled cutting of carbon nanotubes, *J. Chem. Phys.* 109(6): 2509–2512.
78. Shen WD, Jiang B, Han BS, and Xie SS, (2000), Investigation of the radial compression of carbon nanotubes with a scanning probe microscope, *Phys. Rev. Lett.* 84(16): 3634–3637.
79. Yu MF, Kowalewski T, and Ruoff RS, (2000), Investigation of the radial deformability of individual carbon nanotubes under controlled indentation force, *Phys. Rev. Lett.* 85(7): 1456–1459.
80. Chesnokov SA, Nalimova VA, Rinzler AG, Smalley RE, and Fischer JE, (1999), Mechanical energy storage in carbon nanotube springs, *Phys. Rev. Lett.* 82(2): 343–346.
81. Tang J, Qin LC, Sasaki T, Yudasaka M, Matsushita A, and Iijima S, (2000), Compressibility and polygonization of single-walled carbon nanotubes under hydrostatic pressure, *Phys. Rev. Lett.* 85(9): 1887–1889.
82. Yu MF, Dyer MJ, Chen J, Qian D, Liu WK, and Ruoff RS, (2001), Locked twist in multi-walled carbon nanotube ribbons, *Phys. Rev. B, Rapid Commn.* 64: 241403R.

83. Ebbesen TW and Ajayan PM, (1992), Large-scale synthesis of carbon nanotubes, *Nature*. 358(6383): 220–222.

84. Iijima S, Ajayan PM, and Ichihashi T, (1992), Growth model for carbon nanotubes, *Phys. Rev. Lett.* 69(21): 3100–3103.

85. Thess A, Lee R, Nikolaev P, Dai HJ, Petit P, Robert J, Xu CH, Lee YH, Kim SG, Rinzler AG, Colbert DT, Scuseria GE, Tomanek D, Fischer JE, and Smalley RE, (1996), Crystalline ropes of metallic carbon nanotubes, *Science*. 273(5274): 483–487.

86. Guo T, Nikolaev P, Thess A, Colbert DT, and Smalley RE, (1995), Catalytic growth of single-walled nanotubes by laser vaporization, *Chem. Phys. Lett.* 243(1–2): 49–54.

87. Kong J, Soh HT, Cassell AM, Quate CF, and Dai HJ, (1998), Synthesis of individual single-walled carbon nanotubes on patterned silicon wafers, *Nature*. 395(6705): 878–881.

88. Cassell AM, Raymakers JA, Kong J, and Dai HJ, (1999), Large-scale CVD synthesis of single-walled carbon nanotubes, *J. Phys. Chem. B.* 103(31): 6484–6492.

89. Li WZ, Xie SS, Qian LX, Chang BH, Zou BS, Zhou WY, Zhao RA, and Wang G, (1996), Large-scale synthesis of aligned carbon nanotubes, *Science*. 274(5293): 1701–1703.

90. Dal HJ, Rinzler AG, Nikolaev P, Thess A, Colbert DT, and Smalley RE, (1996), Single-wall nanotubes produced by metal-catalyzed disproportionation of carbon monoxide, *Chem. Phys. Lett.* 260(3–4): 471–475.

91. Nardelli MB, Yakobson BI, and Bernholc J, (1998), Brittle and ductile behavior in carbon nanotubes, *Phys. Rev. Lett.* 81(21): 4656–4659.

92. Walters DA, Ericson LM, Casavant MJ, Liu J, Colbert DT, Smith KA, and Smalley RE, (1999), Elastic strain of freely suspended single-wall carbon nanotube ropes, *Appl. Phys. Lett.* 74(25): 3803–3805.

93. Pan ZW, Xie SS, Lu L, Chang BH, Sun LF, Zhou WY, Wang G, and Zhang DL, (1999), Tensile tests of ropes of very long aligned multi-wall carbon nanotubes, *Appl. Phys. Lett.* 74(21): 3152–3154.

94. Wagner HD, Lourie O, Feldman Y, and Tenne R, (1998), Stress-induced fragmentation of multi-wall carbon nanotubes in a polymer matrix, *Appl. Phys. Lett.* 72(2): 188–190.

95. Li F, Cheng HM, Bai S, Su G, and Dresselhaus MS, (2000), Tensile strength of single-walled carbon nanotubes directly measured from their macroscopic ropes, *Appl. Phys. Lett.* 77(20): 3161–3163.

96. Yakobson BI, Campbell MP, Brabec CJ, and Bernholc J, (1997), High strain rate fracture and c-chain unraveling in carbon nanotubes, *Computational Mater. Sci.* 8(4): 341–348.

97. Belytschko T, Xiao SP, Schatz GC, and Ruoff RS, (2001), Simulation of the fracture of nanotubes, *Phys. Rev. B*, accepted.

98. Yakobson BI, (1997), in *Recent Advances in the Chemistry and Physics of Fullerenes and Related Materials*, Kadish KM (Ed.), Electrochemical Society, Inc., Pennington, NJ, 549.

99. Nardelli MB, Yakobson BI, and Bernholc J, (1998), Mechanism of strain release in carbon nanotubes, *Phys. Rev. B.* 57(8): R4277-R4280.

100. Srivastava D, Menon M, and Cho KJ, (1999), Nanoplasticity of single-wall carbon nanotubes under uniaxial compression, *Phys. Rev. Lett.* 83(15): 2973–2976.

101. Wei CY, Srivastava D, and Cho KJ, (2001), Molecular dynamics study of temperature-dependent plastic collapse of carbon nanotubes under axial compression, *Comp. Modeling Eng. Sci.* 3, 255.

102. Srivastava D, Wei CY, and Cho KJ, (2002) Computational nanomechanics of carbon nanotubes and composites, ASME, *Applied Mechanics Reviews* (special issue on nanotechnology), in press.

103. Yakobson BI, (1998), Mechanical relaxation and intramolecular plasticity in carbon nanotubes, *Appl. Phys. Lett.* 72(8): 918–920.

104. Zhang PH, Lammert PE, and Crespi VH, (1998), Plastic deformations of carbon nanotubes, *Phys. Rev. Lett.* 81(24): 5346–5349.

105. Zhang PH and Crespi VH, (1999), Nucleation of carbon nanotubes without pentagonal rings, *Phys. Rev. Lett.* 83(9): 1791–1794.

106. Bockrath M, Cobden DH, McEuen PL, Chopra NG, Zettl A, Thess A, and Smalley RE, (1997), Single-electron transport in ropes of carbon nanotubes, *Science*. 275(5308): 1922–1925.

107. Cumings J and Zettl A, (2000), Low-friction nanoscale linear bearing realized from multi-wall carbon nanotubes, *Science.* 289(5479): 602–604.

108. Kelly BT, (1981), *Physics of Graphite.* Applied Science. London.

109. Ausman KD and Ruoff RS, (2001), Unpublished.

110. Yakobson BI, (2001), Unpublished.

111. Geohegan DB, Schittenhelm H, Fan X, Pennycook SJ, Puretzky AA, Guillorn MA, Blom DA, and Joy DC, (2001), Condensed phase growth of single-wall carbon nanotubes from laser annealed nanoparticulates, *Appl. Phys. Lett.* 78(21): 3307–3309.

112. Piner RD and Ruoff RS, (2001), Unpublished.

113. Born M and Huang K, (1954), *Dynamical Theory of Crystal Lattices.* Oxford University Press. Oxford.

114. Keating PN, (1966), Theory of third-order elastic constants of diamond-like crystals, *Phys. Rev.* 149(2): 674.

115. Keating PN, (1966), Effect of invariance requirements on elastic strain energy of crystals with application to diamond structure, *Phys. Rev.* 145(2): 637.

116. Keating PN, (1967), On sufficiency of Born–Huang relations, *Phys. Lett. A.* A 25(7): 496.

117. Keating PN, (1968), Relationship between macroscopic and microscopic theory of crystal elasticity. 2. nonprimitive crystals, *Phys. Rev.* 169(3): 758.

118. Brugger K, (1964), Thermodynamic definition of higher order elastic coefficients, *Phys. Rev.* 133(6A): A1611.

119. Martin JW, (1975), Many-body forces in metals and brugger elastic-constants, *J. Phys. C — Solid State Phys.* 8(18): 2837–2857.

120. Martin JW, (1975), Many-body forces in solids and Brugger elastic-constants. 2. INNER elastic-constants, *J. Phys. C — Solid State Phys.* 8(18): 2858–2868.

121. Martin JW, (1975), Many-body forces in solids — elastic-constants of diamond-type crystals, *J. Phys. C — Solid State Phys.* 8(18): 2869–2888.

122. Daw MS and Baskes MI, (1984), Embedded-atom method — derivation and application to impurities, surfaces, and other defects in metals, *Phys. Rev. B.* 29(12): 6443–6453.

123. Daw MS, Foiles SM, and Baskes MI, (1993), The embedded-atom method — a review of theory and applications, *Mater. Sci. Rep.* 9(7–8): 251–310.

124. Tadmor EB, Ortiz M, and Phillips R, (1996), Quasicontinuum analysis of defects in solids, *Philos. Mag. A — Phys. Cond. Matter Struct. Defects Mech. Prop.* 73(6): 1529–1563.

125. Tadmor EB, Phillips R, and Ortiz M, (1996), Mixed atomistic and continuum models of deformation in solids, *Langmuir.* 12(19): 4529–4534.

126. Friesecke G and James RD, (2000), A scheme for the passage from atomic to continuum theory for thin films, nanotubes and nanorods, *J. Mech. Phys. Solids.* 48(6–7): 1519–1540.

127. Belytschko T, Liu WK, and Moran B, (2000), *Nonlinear Finite Elements for Continua and Structures.* John Wiley & Sons, LTD.

128. Marsden JE and Hughes TJR, (1983), *Mathematical Foundations of Elasticity.* Prentice-Hall. Englewood Cliffs, NJ.

129. Malvern LE, (1969), *Introduction to the Mechanics of a Continuous Medium.* Prentice-Hall. Englewood Cliffs, NJ.

130. Milstein F, (1982), Crystal elasticity, in *Mechanics of Solids*, Sewell MJ (Ed.), Pergamon Press, Oxford.

131. Ericksen JL, (1984), in *Phase Transformations and Material Instabilities in Solids*, Gurtin M (Ed.), Academic Press, New York.

132. Tersoff J, (1986), New empirical model for the structural properties of silicon, *Phys. Rev. Lett.* 56(6): 632–635.

133. Tersoff J, (1988), New empirical approach for the structure and energy of covalent systems, *Phys. Rev. B.* 37(12): 6991–7000.

134. Tersoff J, (1988), Empirical interatomic potential for carbon, with applications to amorphous-carbon, *Phys. Rev. Lett.* 61(25): 2879–2882.

135. Tersoff J, (1989), Modeling solid-state chemistry — interatomic potentials for multicomponent systems, *Phys. Rev. B.* 39(8): 5566–5568.

136. Green AE and Rivlin RS, (1964), Multipolar continuum mechanics, *Arch. Rat. Mech. An.* 17: 113–147.

137. Cousins CSG, (1978), Inner elasticity, *J. Phys. C — Solid State Phys.* 11(24): 4867–4879.

138. Zhang P, Huang Y, Geubelle PH, and Hwang KC, (2002), On the continuum modeling of carbon nanotubes. *Acta Mechanica Sinica.* (In press).

139. Zhang P, Huang Y, Geubelle PH, Klein P, and Hwang KC, (2002), The elastic modulus of single-wall carbon nanotubes: a continuum analysis incorporating interatomic potentials, *Intl. J. Solids Struct.* In press.

140. Zhang P, Huang Y, Gao H, and Hwang KC, (2002), Fracture nucleation in single-wall carbon nanotubes under tension: a continuum analysis incorporating interatomic potentials. *J. Appl. Mech.* (In press).

141. Wiesendanger R, (1994), *Scanning Probe Microscopy and Spectroscopy: Methods and Applications.* Cambridge University Press. Oxford.

142. Binnig G and Quate CF, (1986), Atomic force microscope, *Phys. Rev. Lett.* 56: 930–933.

143. Cumings J, Collins PG, and Zettl A, (2000), Materials — peeling and sharpening multi-wall nanotubes, *Nature.* 406(6796): 586–586.

144. Ohno K, Esfarjani K, and Kawazoe Y, (1999), *Computational Material Science: From Ab Initio to Monte Carlo Methods.* Solid State Sciences, Cardona M, et al. (Ed.) Springer. Berlin.

145. Dirac PAM, (1958), *The Principles of Quantum Mechanics.* Oxford University Press. London.

146. Landau LD and Lifshitz EM, (1965), *Quantum Mechanics; Non-Relativistic Theory.* Pergamon. Oxford.

147. Merzbacher E, (1998), *Quantum Mechanics.* Wiley. New York.

148. Messiah A, (1961), *Quantum Mechanics.* North-Holland. Amsterdam.

149. Schiff LI, (1968), *Quantum Mechanics.* McGraw-Hill. New York.

150. Fock V, (1930), Naherungsmethode zur losung des quantenmechanis-chen mehrkorperproblems, *Z. Physik.* 61: 126.

151. Hartree DR, (1928), The wave mechanics of an atom with a non-Coulomb central field, part I, theory and methods, *Proc. Cambridge Phil. Soc.* 24: 89.

152. Hartree DR, (1932–1933), A practical method for the numerical solution of differential equations, *Mem. and Proc. Manchester Lit. Phil. Soc.* 77: 91.

153. Clementi E, (2000), *Ab initio* computations in atoms and molecules (reprinted from *IBM Journal of Research and Development* 9, 1965), *IBM J. Res. Dev.* 44(1–2): 228–245.

154. Hohenberg P and Kohn W, (1964), Inhomogeneous electron gas, *Phys. Rev. B.* 136: 864.

155. Kohn W and Sham LJ, (1965), Self-consistent equations including exchange and correlation effects, *Phys. Rev.* 140(4A): 1133.

156. Perdew JP, McMullen ER, and Zunger A, (1981), Density-functional theory of the correlation-energy in atoms and ions — a simple analytic model and a challenge, *Phys. Rev. A.* 23(6): 2785–2789.

157. Perdew JP and Zunger A, (1981), Self-interaction correction to density-functional approximations for many-electron systems, *Phys. Rev. B.* 23(10): 5048–5079.

158. Slater JC, Wilson TM, and Wood JH, (1969), Comparison of several exchange potentials for electrons in cu+ ion, *Phys. Rev.* 179(1): 28.

159. Moruzzi VJ and Sommers CB, (1995), *Calculated Electronic Properties of Ordered Alloys: A Handbook.* World Scientific. Singapore.

160. Payne MC, Teter MP, Allan DC, Arias TA, and Joannopoulos JD, (1992), Iterative minimization techniques for *ab initio* total-energy calculations — molecular-dynamics and conjugate gradients, *Rev. Mod. Phys.* 64(4): 1045–1097.

161. Car R and Parrinello M, (1985), Unified approach for molecular-dynamics and density-functional theory, *Phys. Rev. Lett.* 55(22): 2471–2474.

162. Slater JC and Koster GF, (1954), Wave functions for impurity levels, *Phys. Rev.* 94(1498).

163. Harrison WA, (1989), *Electronic Structure and the Properties of Solids: The Physics of the Chemical Bond*. Dover. New York.

164. Matthew W, Foulkes C, and Haydock R, (1989), Tight-binding models and density-functional theory, *Phys. Rev. B.* 39(17): 12520–12536.

165. Xu CH, Wang CZ, Chan CT, and Ho KM, (1992), A transferable tight-binding potential for carbon, *J. Phys. Condensed Matter.* 4(28): 6047–6054.

166. Mehl MJ and Papaconstantopoulos DA, (1996), Applications of a tight-binding total-energy method for transition and noble metals: elastic constants, vacancies, and surfaces of monatomic metals, *Phys. Rev. B.* 54(7): 4519–4530.

167. Liu F, (1995), Self-consistent tight-binding method, *Phys. Rev. B.* 52(15): 10677–10680.

168. Porezag D, Frauenheim T, Kohler T, Seifert G, and Kaschner R, (1995), Construction of tight-binding-like potentials on the basis of density-functional theory — application to carbon, *Phys. Rev. B.* 51(19): 12947–12957.

169. Taneda A, Esfarjani K, Li ZQ, and Kawazoe Y, (1998), Tight-binding parameterization of transition metal elements from LCAO *ab initio* Hamiltonians, *Computational Mater. Sci.* 9(3–4): 343–347.

170. Menon M and Subbaswamy KR, (1991), Universal parameter tight-binding molecular-dynamics — application to c-60, *Phys. Rev. Lett.* 67(25): 3487–3490.

171. Sutton AP, Finnis MW, Pettifor DG, and Ohta Y, (1988), The tight-binding bond model, *J. Phys. C — Solid State Phys.* 21(1): 35–66.

172. Menon M and Subbaswamy KR, (1997), Nonorthogonal tight-binding molecular-dynamics scheme for silicon with improved transferability, *Phys. Rev. B.* 55(15): 9231–9234.

173. Haile JM, (1992), *Molecular Dynamics Simulation*. Wiley Interscience, New York.

174. Rapaport DC, (1995), *The Art of Molecular Dynamics Simulation*. Cambridge University Press, London.

175. Frenkel D and Smit, B., (1996), *Understanding Molecular Simulation: From Algorithms to Applications*. Academic Press, New York.

176. Hockney RW and Eastwood, J.W., (1989), *Computer Simulation Using Particles*. IOP Publishing Ltd. New York.

177. Li. SF and Liu WK, (2002), Mesh-free and particle methods, *Appl. Mech. Rev.* 55(1): 1–34.

178. Berendsen HJC and van Gunsteren W.F., (1986), *Dynamics Simulation of Statistical Mechanical Systems*, Ciccotti GPF, Hoover, W.G. (Eds.) Vol. 63. North Holland. Amsterdam. 493.

179. Verlet L, (1967), Computer experiments on classical fluids. I. Thermodynamical properties of Lennard-Jones molecules, *Phys. Rev.* 159(1): 98.

180. Gray SK, Noid DW, and Sumpter BG, (1994), Symplectic integrators for large-scale molecular-dynamics simulations — a comparison of several explicit methods, *J. Chem. Phys.* 101(5): 4062–4072.

181. Allinger NL, (1977), Conformational-analysis.130. MM2 — hydrocarbon force-field utilizing V1 and V2 torsional terms, *J. Am. Chem. Soc.* 99(25): 8127–8134.

182. Allinger NL, Yuh YH, and Lii JH, (1989), Molecular mechanics — the MM3 force-field for hydrocarbons. 1, *J. Am. Chem. Soc.* 111(23): 8551–8566.

183. Mayo SL, Olafson BD, and Goddard WA, (1990), Dreiding — a generic force-field for molecular simulations, *J. Phys. Chem.* 94(26): 8897–8909.

184. Guo YJ, Karasawa N, and Goddard WA, (1991), Prediction of fullerene packing in c60 and c70 crystals, *Nature.* 351(6326): 464–467.

185. Tuzun RE, Noid DW, Sumpter BG, and Merkle RC, (1996), Dynamics of fluid flow inside carbon nanotubes, *Nanotechnology.* 7(3): 241–246.

186. Tuzun RE, Noid DW, Sumpter BG, and Merkle RC, (1997), Dynamics of he/c-60 flow inside carbon nanotubes, *Nanotechnology.* 8(3): 112–118.

187. Abell GC, (1985), Empirical chemical pseudopotential theory of molecular and metallic bonding, *Phys. Rev. B.* 31(10): 6184–6196.

188. Brenner DW, (2000), The art and science of an analytic potential, *Physica Status Solidi B — Basic Res.* 217(1): 23–40.

189. Nordlund K, Keinonen J, and Mattila T, (1996), Formation of ion irradiation induced small-scale defects on graphite surfaces, *Phys. Rev. Lett.* 77(4): 699–702.

190. Brenner DW, Harrison JA, White CT, and Colton RJ, (1991), Molecular-dynamics simulations of the nanometer-scale mechanical properties of compressed buckminsterfullerene, *Thin Solid Films.* 206(1–2): 220–223.

191. Robertson DH, Brenner DW, and White CT, (1992), On the way to fullerenes — molecular-dynamics study of the curling and closure of graphitic ribbons, *J. Phys. Chem.* 96(15): 6133–6135.

192. Robertson DH, Brenner DW, and White CT, (1995), Temperature-dependent fusion of colliding c-60 fullerenes from molecular-dynamics simulations, *J. Phys. Chem.* 99(43): 15721–15724.

193. Sinnott SB, Colton RJ, White CT, and Brenner DW, (1994), Surface patterning by atomically-controlled chemical forces — molecular-dynamics simulations, *Surf. Sci.* 316(1–2): L1055-L1060.

194. Harrison JA, White CT, Colton RJ, and Brenner DW, (1992), Nanoscale investigation of indentation, adhesion and fracture of diamond (111) surfaces, *Surf. Sci.* 271(1–2): 57–67.

195. Harrison JA, White CT, Colton RJ, and Brenner DW, (1992), Molecular-dynamics simulations of atomic-scale friction of diamond surfaces, *Phys. Rev. B.* 46(15): 9700–9708.

196. Harrison JA, Colton RJ, White CT, and Brenner DW, (1993), Effect of atomic-scale surface-roughness on friction — a molecular-dynamics study of diamond surfaces, *Wear.* 168(1–2): 127–133.

197. Harrison JA, White CT, Colton RJ, and Brenner DW, (1993), Effects of chemically-bound, flexible hydrocarbon species on the frictional properties of diamond surfaces, *J. Phys. Chem.* 97(25): 6573–6576.

198. Harrison JA, White CT, Colton RJ, and Brenner DW, (1993), Atomistic simulations of friction at sliding diamond interfaces, *MRS Bull.* 18(5): 50–53.

199. Harrison JA and Brenner DW, (1994), Simulated tribochemistry — an atomic-scale view of the wear of diamond, *J. Am. Chem. Soc.* 116(23): 10399–10402.

200. Harrison JA, White CT, Colton RJ, and Brenner DW, (1995), Investigation of the atomic-scale friction and energy — dissipation in diamond using molecular-dynamics, *Thin Solid Films.* 260(2): 205–211.

201. Tupper KJ and Brenner DW, (1993), Atomistic simulations of frictional wear in self-assembled monolayers, *Abstr. Papers Am. Chem. Soc.* 206: 172.

202. Tupper KJ and Brenner DW, (1993), Molecular-dynamics simulations of interfacial dynamics in self- assembled monolayers, *Abstr. Papers Am. Chem. Soc.* 206: 72.

203. Tupper KJ and Brenner DW, (1994), Molecular-dynamics simulations of friction in self-assembled monolayers, *Thin Solid Films.* 253(1–2): 185–189.

204. Brenner DW, (2001), Unpublished.

205. Pettifor DG and Oleinik II, (1999), Analytic bond-order potentials beyond Tersoff-Brenner. Ii. Application to the hydrocarbons, *Phys. Rev. B.* 59(13): 8500.

206. Pettifor DG and Oleinik II, (2000), Bounded analytic bond-order potentials for sigma and pi bonds, *Phys. Rev. Lett.* 84(18): 4124–4127.

207. Jones JE, (1924), On the determination of molecular fields-I. From the variation of the viscosity of a gas with temperature, *Proc. R. Soc.* 106: 441.

208. Jones JE, (1924), On the determination of molecular fields-II. From the equation of state of a gas, *Proc. R. Soc.* 106: 463.

209. Girifalco LA and Lad RA, (1956), Energy of cohesion, compressibility and the potential energy functions of the graphite system, *J. Chem. Phys.* 25(4): 693–697.

210. Girifalco LA, (1992), Molecular-properties of c-60 in the gas and solid-phases, *J. Phys. Chem.* 96(2): 858–861.

211. Wang Y, Tomanek D, and Bertsch GF, (1991), Stiffness of a solid composed of c60 clusters, *Phys. Rev. B.* 44(12): 6562–6565.

212. Qian D, Liu WK, and Ruoff RS, (2001), Mechanics of c60 in nanotubes, *J. Phys. Chem. B.* 105: 10753–10758.

213. Zhao YX and Spain IL, (1989), X-ray-diffraction data for graphite to 20 gpa, *Phys. Rev. B.* 40(2): 993–997.

214. Hanfland M, Beister H, and Syassen K, (1989), Graphite under pressure — equation of state and first-order Raman modes, *Phys. Rev. B.* 39(17): 12598–12603.

215. Boettger JC, (1997), All-electron full-potential calculation of the electronic band structure, elastic constants, and equation of state for graphite, *Phys. Rev. B.* 55(17): 11202–11211.

216. Girifalco LA, Hodak M, and Lee RS, (2000), Carbon nanotubes, buckyballs, ropes, and a universal graphitic potential, *Phys. Rev. B.* 62(19): 13104–13110.

217. Falvo MR, Clary G, Helser A, Paulson S, Taylor RM, Chi V, Brooks FP, Washburn S, and Superfine R, (1998), Nanomanipulation experiments exploring frictional and mechanical properties of carbon nanotubes, *Microsc. Microanal.* 4(5): 504–512.

218. Falvo MR, Taylor RM, Helser A, Chi V, Brooks FP, Washburn S, and Superfine R, (1999), Nanometre-scale rolling and sliding of carbon nanotubes, *Nature.* 397(6716): 236–238.

219. Falvo MR, Steele J, Taylor RM, and Superfine R, (2000), Evidence of commensurate contact and rolling motion: AFM manipulation studies of carbon nanotubes on hopg, *Tribol, Lett.* 9(1–2): 73–76.

220. Falvo MR, Steele J, Taylor RM, and Superfine R, (2000), Gearlike rolling motion mediated by commensurate contact: carbon nanotubes on hopg, *Phys. Rev. B.* 62(16): R10665-R10667.

221. Schall JD and Brenner DW, (2000), Molecular dynamics simulations of carbon nanotube rolling and sliding on graphite, *Molecular Simulation.* 25(1–2): 73–79.

222. Buldum A and Lu JP, (1999), Atomic scale sliding and rolling of carbon nanotubes, *Phys. Rev. Lett.* 83(24): 5050–5053.

223. Kolmogorov AN and Crespi VH, (2000), Smoothest bearings: Interlayer sliding in multi-walled carbon nanotubes, *Phys. Rev. Lett.* 85(22): 4727–4730.

224. Shenoy VB, Miller R, Tadmor EB, Phillips R, and Ortiz M, (1998), Quasicontinuum models of interfacial structure and deformation, *Phys. Rev. Lett.* 80(4): 742–745.

225. Miller R, Ortiz M, Phillips R, Shenoy V, and Tadmor EB, (1998), Quasicontinuum models of fracture and plasticity, *Eng, Fracture Mech.* 61(3–4): 427–444.

226. Miller R, Tadmor EB, Phillips R, and Ortiz M, (1998), Quasicontinuum simulation of fracture at the atomic scale, *Modelling Sim. Mater. Sci. Eng.* 6(5): 607–638.

227. Shenoy VB, Miller R, Tadmor EB, Rodney D, Phillips R, and Ortiz M, (1999), An adaptive finite element approach to atomic-scale mechanics — the quasicontinuum method, *J. Mech. Phys. Solids.* 47(3): 611–642.

228. Tadmor EB, Miller R, Phillips R, and Ortiz M, (1999), Nanoindentation and incipient plasticity, *J. Mater. Res.* 14(6): 2233–2250.

229. Rodney D and Phillips R, (1999), Structure and strength of dislocation junctions: An atomic level analysis, *Phys. Rev. Lett.* 82(8): 1704–1707.

230. Smith GS, Tadmor EB, and Kaxiras E, (2000), Multiscale simulation of loading and electrical resistance in silicon nanoindentation, *Phys. Rev. Lett.* 84(6): 1260–1263.

231. Knap J and Ortiz M, (2001), An analysis of the quasicontinuum method, *J. Mech Phys. Solids.* 49(9): 1899–1923.

232. Shin CS, Fivel MC, Rodney D, Phillips R, Shenoy VB, and Dupuy L, (2001), Formation and strength of dislocation junctions in fcc metals: a study by dislocation dynamics and atomistic simulations, *J. Physique Iv.* 11(PR5): 19–26.

233. Shenoy V, Shenoy V, and Phillips R, (1999), Finite temperature quasicontinuum methods, in *Multiscale Modelling of Materials*, Ghoniem N (Ed.), Materials Research Society, Warrendale, PA, 465–471.

234. Arroyo M and Belytschko T, (2002), An atomistic-based membrane for crystalline films one atom thick *J. Mech. Phys. Solids.* 50: 1941–1977.

235. Liu WK and Chen YJ, (1995), Wavelet and multiple scale reproducing kernel methods, *Intl. J. Numerical Meth. Fluids.* 21(10): 901–931.

236. Liu WK, Jun S, and Zhang YF, (1995), Reproducing kernel particle methods, *Intl. J. Numerical Meth. Fluids.* 20(8–9): 1081–1106.

237. Liu WK, Jun S, Li SF, Adee J, and Belytschko T, (1995), Reproducing kernel particle methods for structural dynamics, *Intl. J. Numerical Meth. Eng.* 38(10): 1655–1679.

238. Liu WK, Chen YJ, Uras RA, and Chang CT, (1996), Generalized multiple scale reproducing kernel particle methods, *Computer Meth. Appl. Mech. Eng.* 139(1–4): 91–157.

239. Liu WK, Chen Y, Chang CT, and Belytschko T, (1996), Advances in multiple scale kernel particle methods, *Computational Mech.* 18(2): 73–111.

240. Liu WK, Jun S, Sihling DT, Chen YJ, and Hao W, (1997), Multiresolution reproducing kernel particle method for computational fluid dynamics, *Intl. J. Numerical Meth. Fluids.* 24(12): 1391–1415.

241. Liu WK, Li SF, and Belytschko T, (1997), Moving least-square reproducing kernel methods. 1. Methodology and convergence, *Computer Meth. Appl. Mech. Eng.* 143(1–2): 113–154.

242. Liu WK, Belytscho T, and Oden JT. (Eds.), (1996), *Computer Meth. Appl. Mech. Eng.* Vol. 139.

243. Chen JS and Liu WK (Eds.), (2000), *Computational Mech.* Vol. 25.

244. Odegard GM, Gates TS, Nicholson LM, and Wise KE, (2001), *Equivalent-Continuum Modeling of Nano-Structured Materials.* NASA Langley Research Center, NASA-2001-TM 210863.

245. Odegard GM, Harik VM, Wise KE, and Gates TS, (2001), *Constitutive Modeling of Nanotube-Reinforced Polymer Composite Systems.* NASA Langley Research Center, NASA-2001-TM 211044.

246. Abraham FF, Broughton JQ, Bernstein N, and Kaxiras E, (1998), Spanning the continuum to quantum length scales in a dynamic simulation of brittle fracture, *Europhys. Lett.* 44(6): 783–787.

247. Broughton JQ, Abraham FF, Bernstein N, and Kaxiras E, (1999), Concurrent coupling of length scales: methodology and application, *Phys. Rev. B.* 60(4): 2391–2403.

248. Nakano A, Bachlechner ME, Kalia RK, Lidorikis E, Vashishta P, Voyiadjis GZ, Campbell TJ, Ogata S, and Shimojo F, (2001), Multiscale simulation of nanosystems, *Computing Sci. Eng.* 3(4): 56–66.

249. Rafii-Tabar H, Hua L, and Cross M, (1998), Multiscale numerical modelling of crack propagation in two-dimensional metal plate, *Mater. Sci. Tech.* 14(6): 544–548.

250. Rafii-Tabar H, Hua L, and Cross M, (1998), A multi-scale atomistic-continuum modelling of crack propagation in a two-dimensional macroscopic plate, *J. Phys. Condensed Matter.* 10(11): 2375–2387.

251. Rudd RE and Broughton JQ, (1998), Coarse-grained molecular dynamics and the atomic limit of finite elements, *Phys. Rev. B.* 58(10): R5893-R5896.

252. Rudd RE and Broughton JQ, (2000), Concurrent coupling of length scales in solid state systems, *Physica Status Solidi B-Basic Research.* 217(1): 251–291.

253. Liu WK, Zhang Y, and Ramirez MR, (1991), Multiple scale finite-element methods, *Intl. J. Numerical Meth. Eng.* 32(5): 969–990.

254. Liu WK, Uras RA, and Chen Y, (1997), Enrichment of the finite element method with the reproducing kernel particle method, *J. Appl. Mech. Trans. ASME.* 64(4): 861–870.

255. Hao S, Liu WK, and Qian D, (2000), Localization-induced band and cohesive model, *J. Appl. Mech. Trans. ASME.* 67(4): 803–812.

256. Wagner GJ, Moes N, Liu WK, and Belytschko T, (2001), The extended finite element method for rigid particles in stokes flow, *Intl. J. Numerical Meth. Eng.* 51(3): 293–313.

257. Wagner GJ and Liu WK, (2001), Hierarchical enrichment for bridging scales and mesh-free boundary conditions, *Intl. J. Numerical Meth. Eng.* 50(3): 507–524.

258. Wagner GJ and Liu WK, (2002), Coupling of atomistic and continuum simulations (in preparation).

259. Costello GA, (1978), Analytical investigation of wire rope, *Appl. Mech. Rev. ASME.* 31: 897–900.

260. Costello GA, (1997), *Theory of Wire Rope*. 2nd ed. Springer. New York.

261. Qian D, Liu WK, and Ruoff RS, (2002), Load transfer mechanism in nano-ropes, Computational Mechanics Laboratory Research Report (02-03) Dept. of Mech. Eng., Northwestern University.

262. Ruoff RS, Qian D, Liu WK, Ding WQ, Chen XQ, and Dikin D (Eds.), (2002), What kind of carbon nanofiber is ideal for structural applications? *43rd AIAA/ASME/ASCE/AHS Struct. Struct. Dynamics Mater. Conf.* Denver, CO.

263. Pipes BR and Hubert P, (2001), Helical carbon nanotube arrays: mechanical properties (submitted).

264. Yu MF, Dyer MJ, Skidmore GD, Rohrs HW, Lu XK, Ausman KD, Von Ehr JR, and Ruoff RS, (1999), Three-dimensional manipulation of carbon nanotubes under a scanning electron microscope, *Nanotechnology.* 10(3): 244–252.

265. Ruoff RS, Lorents DC, Chan B, Malhotra R, and Subramoney S, (1993), Single-crystal metals encapsulated in carbon nanoparticles, *Science.* 259(5093): 346–348.

266. Tomita M, Saito Y, and Hayashi T, (1993), Lac2 encapsulated in graphite nano-particle, *Jpn. J. App. Phys. Part 2–Lett.* 32(2B): L280-L282.

267. Seraphin S, Zhou D, Jiao J, Withers JC, and Loutfy R, (1993), Selective encapsulation of the carbides of yttrium and titanium into carbon nanoclusters, *Appl. Phys. Lett.* 63(15): 2073–2075.

268. Seraphin S, Zhou D, Jiao J, Withers JC, and Loutfy R, (1993), Yttrium carbide in nanotubes, *Nature.* 362(6420): 503–503.

269. Seraphin S, Zhou D, and Jiao J, (1996), Filling the carbon nanocages, *J. Appl. Phys.* 80(4): 2097–2104.

270. Saito Y, Yoshikawa T, Okuda M, Ohkohchi M, Ando Y, Kasuya A, and Nishina Y, (1993), Synthesis and electron-beam incision of carbon nanocapsules encaging yc2, *Chem. Phys. Lett.* 209(1–2): 72–76.

271. Saito Y, Yoshikawa T, Okuda M, Fujimoto N, Sumiyama K, Suzuki K, Kasuya A, and Nishina Y, (1993), Carbon nanocapsules encaging metals and carbides, *J. Phys. Chem. Solids.* 54(12): 1849–1860.

272. Saito Y and Yoshikawa T, (1993), Bamboo-shaped carbon tube filled partially with nickel, *J. Crystal Growth.* 134(1–2): 154–156.

273. Saito Y, Okuda M, and Koyama T, (1996), Carbon nanocapsules and single-wall nanotubes formed by arc evaporation, *Surf. Rev. Lett.* 3(1): 863–867.

274. Saito Y, Nishikubo K, Kawabata K, and Matsumoto T, (1996), Carbon nanocapsules and single-layered nanotubes produced with platinum-group metals (Ru, Rh, Pd, Os, Ir, Pt) by arc discharge, *J. Appl. Phys.* 80(5): 3062–3067.

275. Saito Y, (1996), Carbon cages with nanospace inside: fullerenes to nanocapsules, *Surf. Rev. Lett.* 3(1): 819–825.

276. Saito Y, (1995), Nanoparticles and filled nanocapsules, *Carbon.* 33(7): 979–988.

277. McHenry ME, Majetich SA, Artman JO, Degraef M, and Staley SW, (1994), Superparamagnetism in carbon-coated Co particles produced by the kratschmer carbon-arc process, *Phys. Rev. B.* 49(16): 11358–11363.

278. Majetich SA, Artman JO, McHenry ME, Nuhfer NT, and Staley SW, (1993), Preparation and properties of carbon-coated magnetic nanocrystallites, *Phys. Rev. B.* 48(22): 16845–16848.

279. Jiao J, Seraphin S, Wang XK, and Withers JC, (1996), Preparation and properties of ferromagnetic carbon-coated Fe, Co, and Ni nanoparticles, *J. Appl. Phys.* 80(1): 103–108.

280. Diggs B, Zhou A, Silva C, Kirkpatrick S, Nuhfer NT, McHenry ME, Petasis D, Majetich SA, Brunett B, Artman JO, and Staley SW, (1994), Magnetic properties of carbon-coated rare-earth carbide nanocrystallites produced by a carbon-arc method, *J. Appl. Phys.* 75(10): 5879–5881.

281. Brunsman EM, Sutton R, Bortz E, Kirkpatrick S, Midelfort K, Williams J, Smith P, McHenry ME, Majetich SA, Artman JO, Degraef M, and Staley SW, (1994), Magnetic-properties of carbon-coated, ferromagnetic nanoparticles produced by a carbon-arc method, *J. Appl. Phys.* 75(10): 5882–5884.

282. Funasaka H, Sugiyama K, Yamamoto K, and Takahashi T, (1995), Synthesis of actinide carbides encapsulated within carbon nanoparticles, *J. Appl. Phys.* 78(9): 5320–5324.

283. Kikuchi K, Kobayashi K, Sueki K, Suzuki S, Nakahara H, Achiba Y, Tomura K, and Katada M, (1994), Encapsulation of radioactive gd-159 and tb-161 atoms in fullerene cages, *J. Am. Chem. Soc.* 116(21): 9775–9776.

284. Burch WM, Sullivan PJ, and McLaren CJ, (1986), Technegas – a new ventilation agent for lung-scanning, *Nucl. Med. Commn.* 7(12): 865.

285. Senden TJ, Moock KH, Gerald JF, Burch WM, Browitt RJ, Ling CD, and Heath GA, (1997), The physical and chemical nature of technegas, *J. Nucl. Med.* 38(8): 1327–1333.

286. Ajayan PM and Iijima S, (1993), Capillarity-induced filling of carbon nanotubes, *Nature.* 361(6410): 333–334.

287. Ajayan PM, Ebbesen TW, Ichihashi T, Iijima S, Tanigaki K, and Hiura H, (1993), Opening carbon nanotubes with oxygen and implications for filling, *Nature.* 362(6420): 522–525.

288. Tsang SC, Harris PJF, and Green MLH, (1993), Thinning and opening of carbon nanotubes by oxidation using carbon-dioxide, *Nature.* 362(6420): 520–522.

289. Xu CG, Sloan J, Brown G, Bailey S, Williams VC, Friedrichs S, Coleman KS, Flahaut E, Hutchison JL, Dunin-Borkowski RE, and Green MLH, (2000), 1d lanthanide halide crystals inserted into single-walled carbon nanotubes, *Chem. Commn.* (24): 2427–2428.

290. Tsang SC, Chen YK, Harris PJF, and Green MLH, (1994), A simple chemical method of opening and filling carbon nanotubes, *Nature.* 372(6502): 159–162.

291. Sloan J, Hammer J, Zwiefka-Sibley M, and Green MLH, (1998), The opening and filling of single walled carbon nanotubes (swts), *Chem. Commn.* (3): 347–348.

292. Hiura H, Ebbesen TW, and Tanigaki K, (1995), Opening and purification of carbon nanotubes in high yields, *Adv. Mater.* 7(3): 275–276.

293. Hwang KC, (1995), Efficient cleavage of carbon graphene layers by oxidants, *J. Chem. Soc. -Chem. Commn.* (2): 173–174.

294. Ajayan PM, Colliex C, Lambert JM, Bernier P, Barbedette L, Tence M, and Stephan O, (1994), Growth of manganese filled carbon nanofibers in the vapor-phase, *Phys. Rev. Lett.* 72(11): 1722–1725.

295. Subramoney S, Ruoff RS, Lorents DC, Chan B, Malhotra R, Dyer MJ, and Parvin K, (1994), Magnetic separation of gdc2 encapsulated in carbon nanoparticles, *Carbon.* 32(3): 507–513.

296. Tsang SC, Davis JJ, Green MLH, Allen H, Hill O, Leung YC, and Sadler PJ, (1995), Immobilization of small proteins in carbon nanotubes – high- resolution transmission electron-microscopy study and catalytic activity, *J. Chem. Soc. Chem. Commn.* (17): 1803–1804.

297. Tsang SC, Guo ZJ, Chen YK, Green MLH, Hill HAO, Hambley TW, and Sadler PJ, (1997), Immobilization of platinated and iodinated oligonucleotides on carbon nanotubes, *Angewandte Chemie.* 36(20): 2198–2200. International Edition in English.

298. Dillon AC, Jones KM, Bekkedahl TA, Kiang CH, Bethune DS, and Heben MJ, (1997), Storage of hydrogen in single-walled carbon nanotubes, *Nature.* 386(6623): 377–379.

299. Gadd GE, Blackford M, Moricca S, Webb N, Evans PJ, Smith AN, Jacobsen G, Leung S, Day A, and Hua Q, (1997), The world's smallest gas cylinders? *Science.* 277(5328): 933–936.

300. Sloan J, Wright DM, Woo HG, Bailey S, Brown G, York APE, Coleman KS, Hutchison JL, and Green MLH, (1999), Capillarity and silver nanowire formation observed in single walled carbon nanotubes, *Chem. Commn.* (8): 699–700.

301. Smith BW and Luzzi DE, (2000), Formation mechanism of fullerene peapods and coaxial tubes: a path to large scale synthesis, *Chem. Phys. Lett.* 321(1–2): 169–174.

302. Smith BW, Monthioux M, and Luzzi DE, (1998), Encapsulated c-60 in carbon nanotubes, *Nature.* 396(6709): 323–324.

303. Smith BW, Monthioux M, and Luzzi DE, (1999), Carbon nanotube encapsulated fullerenes: a unique class of hybrid materials, *Chem. Phys. Lett.* 315(1–2): 31–36.

304. Burteaux B, Claye A, Smith BW, Monthioux M, Luzzi DE, and Fischer JE, (1999), Abundance of encapsulated c-60 in single-wall carbon nanotubes, *Chem. Phys. Lett.* 310(1–2): 21–24.

305. Sloan J, Dunin-Borkowski RE, Hutchison JL, Coleman KS, Williams VC, Claridge JB, York APE, Xu CG, Bailey SR, Brown G, Friedrichs S, and Green MLH, (2000), The size distribution, imaging and obstructing properties of c-60 and higher fullerenes formed within arc-grown single walled carbon nanotubes, *Chem. Phys. Lett.* 316(3–4): 191–198.

306. Zhang Y, Iijima S, Shi Z, and Gu Z, (1999), Defects in arc-discharge-produced single-walled carbon nanotubes, *Philos. Mag. Lett.* 79(7): 473–479.

307. Pederson MR and Broughton JQ, (1992), Nanocapillarity in fullerene tubules, *Phys. Rev. Lett.* 69(18): 2689–2692.

308. Prasad R and Lele S, (1994), Stabilization of the amorphous phase inside carbon nanotubes — solidification in a constrained geometry, *Philos. Mag. Lett.* 70(6): 357–361.

309. Berber S, Kwon YK, and Tomanek D, (2001), Unpublished.

310. Stan G and Cole MW, (1998), Hydrogen adsorption in nanotubes, *J. Low Temp. Phys.* 110(1–2): 539–544.

311. Mao ZG, Garg A, and Sinnott SB, (1999), Molecular dynamics simulations of the filling and decorating of carbon nanotubules, *Nanotechnology.* 10(3): 273–277.

312. Stan G, Gatica SM, Boninsegni M, Curtarolo S, and Cole MW, (1999), Atoms in nanotubes: small dimensions and variable dimensionality, *Am. J. Phys.* 67(12): 1170–1176.

313. Liu J, Casavant MJ, Cox M, Walters DA, Boul P, Lu W, Rimberg AJ, Smith KA, Colbert DT, and Smalley RE, (1999), Controlled deposition of individual single-walled carbon nanotubes on chemically functionalized templates, *Chem. Phys. Lett.* 303(1–2): 125–129.

314. Chung J, Lee JH, Ruoff RS, and Liu WK, (2001), Nanoscale gap fabrication and integration of carbon nanotubes by micromachining (in preparation).

315. Ren Y and Price DL, (2001), Neutron scattering study of h-2 adsorption in single-walled carbon nanotubes, *Appl. Phys. Lett.* 79(22): 3684–3686.

316. Zhang YG, Chang AL, Cao J, Wang Q, Kim W, Li YM, Morris N, Yenilmez E, Kong J, and Dai HJ, (2001), Electric-field-directed growth of aligned single-walled carbon nanotubes, *Appl. Phys. Lett.* 79(19): 3155–3157.

317. Collins PG, Bradley K, Ishigami M, and Zettl A, (2000), Extreme oxygen sensitivity of electronic properties of carbon nanotubes, *Science.* 287(5459): 1801–1804.

318. Kong J, Franklin NR, Zhou CW, Chapline MG, Peng S, Cho KJ, and Dai HJ, (2000), Nanotube molecular wires as chemical sensors, *Science.* 287(5453): 622–625.

319. Yamamoto K, Akita S, and Nakayama Y, (1998), Orientation and purification of carbon nanotubes using ac electrophoresis, *J. Phys. D Appl. Phys.* 31(8): L34-L36.

320. Tombler TW, Zhou CW, Alexseyev L, Kong J, Dai HJ, Lei L, Jayanthi CS, Tang MJ, and Wu SY, (2000), Reversible electromechanical characteristics of carbon nanotubes under local-probe manipulation, *Nature.* 405(6788): 769–772.

321. Maiti A, (2001), Application of carbon nanotubes as electromechanical sensors – results from first-principles simulations, *Physica Status Solidi B-Basic Research.* 226(1): 87–93.

322. Maiti A, Andzelm J, Tanpipat N, and von Allmen P, (2001), Carbon nanotubes as field emission device and electromechanical sensor: results from first-principles dft simulations, *Abstr. Papers Am. Chem. Soc.* 222: 204.

323. Kong J, Chapline MG, and Dai HJ, (2001), Functionalized carbon nanotubes for molecular hydrogen sensors, *Adv. Mater.* 13(18): 1384–1386.

324. Peng S and Cho KJ, (2000), Chemical control of nanotube electronics, *Nanotechnology.* 11(2): 57–60.

325. Wood JR, Frogley MD, Meurs ER, Prins AD, Peijs T, Dunstan DJ, and Wagner HD, (1999), Mechanical response of carbon nanotubes under molecular and macroscopic pressures, *J. Phys. Chem. B.* 103(47): 10388–10392.

326. Wood JR and Wagner HD, (2000), Single-wall carbon nanotubes as molecular pressure sensors, *Appl. Phys. Lett.* 76(20): 2883–2885.

327. Wood JR, Zhao Q, Frogley MD, Meurs ER, Prins AD, Peijs T, Dunstan DJ, and Wagner HD, (2000), Carbon nanotubes: From molecular to macroscopic sensors, *Phys. Rev. B.* 62(11): 7571–7575.

328. Zhao Q, Wood JR, and Wagner HD, (2001), Using carbon nanotubes to detect polymer transitions, *J. Polymer Sci. Part B-Polymer Phys.* 39(13): 1492–1495.

329. Zhao Q, Wood JR, and Wagner HD, (2001), Stress fields around defects and fibers in a polymer using carbon nanotubes as sensors, *Appl. Phys. Lett.* 78(12): 1748–1750.

330. Ilic B, Czaplewski D, Craighead HG, Neuzil P, Campagnolo C, and Batt C, (2000), Mechanical resonant immunospecific biological detector, *Appl. Phys. Lett.* 77(3): 450–452.

331. Carr DW, Evoy S, Sekaric L, Craighead HG, and Parpia JM, (1999), Measurement of mechanical resonance and losses in nanometer scale silicon wires, *App. Phys. Lett.* 75(7): 920–922.

331a. Yu MF, Wagner GJ, Ruoff RS, and Dyer MJ, (2002), Realization of parametric resonances in a nanowire mechanical system with nanomanipulation inside a scanning electron microscope, *Phys. Rev. B*, in press.

332. Turner KL, Miller SA, Hartwell PG, MacDonald NC, Strogatz SH, and Adams SG, (1998), Five parametric resonances in a microelectromechanical system, *Nature.* 396(6707): 149–152.

333. Sandler J, Shaffer MSP, Prasse T, Bauhofer W, Schulte K, and Windle AH, (1999), Development of a dispersion process for carbon nanotubes in an epoxy matrix and the resulting electrical properties, *Polymer.* 40(21): 5967–5971.

334. Schadler LS, Giannaris SC, and Ajayan PM, (1998), Load transfer in carbon nanotube epoxy composites, *Appl. Phys. Lett.* 73(26): 3842–3844.

335. Qian D, Dickey EC, Andrews R, and Rantell T, (2000), Load transfer and deformation mechanisms in carbon nanotube-polystyrene composites, *Appl. Phys. Lett.* 76(20): 2868–2870.

336. Ajayan PM, Schadler LS, Giannaris C, and Rubio A, (2000), Single-walled carbon nanotube-polymer composites: strength and weakness, *Adv. Mater.* 12(10): 750–753.

337. Thostenson ET, Ren ZF, and Chou TW, (2001), Advances in the science and technology of carbon nanotubes and their composites: a review, *Composites Sci. Technol.* 61(13): 1899–1912.

338. Jin L, Bower C, and Zhou O, (1998), Alignment of carbon nanotubes in a polymer matrix by mechanical stretching, *Appl. Phys. Lett.* 73(9): 1197–1199.

339. Haggenmueller R, Gommans HH, Rinzler AG, Fischer JE, and Winey KI, (2000), Aligned single-wall carbon nanotubes in composites by melt processing methods, *Chem. Phys. Lett.* 330(3–4): 219–225.

340. Barrera EV, (2000), Key methods for developing single-wall nanotube composites, *Jom-J. Minerals Metals Mater. Soc.* 52(11): A38-A42.

341. Fischer JE, (2002) Nanomechanics and the viscoelastic behavior of carbon nanotube-reinforced polymers. Ph.D. thesis. Northwestern University, Evanston.

342. Fisher FT, Bradshaw RD, and Brinson LC, (2001), Effects of nanotube waviness on the mechanical properties of nanoreinforced polymers, *Appl. Phys. Lett.* 80(24): 4647–4649.

343. Shaffer MSP and Windle AH, (1999), Fabrication and characterization of carbon nanotube/poly(vinyl alcohol) composites, *Adv. Mater.* 11(11): 937–941.

344. Gong XY, Liu J, Baskaran S, Voise RD, and Young JS, (2000), Surfactant-assisted processing of carbon nanotube/polymer composites, *Chem. Mater.* 12(4): 1049–1052.

345. Lozano K, Bonilla-Rios J, and Barrera EV, (2001), A study on nanofiber-reinforced thermoplastic composites (II): Investigation of the mixing rheology and conduction properties, *J. Appl. Polymer Sci.* 80(8): 1162–1172.

346. Qian D and Liu WK, (2002), In preparation.

347. Mitsubishi chemical to mass-produce nanotech substance, (2001), in *Kyodo News.*

20

Dendrimers — An Enabling Synthetic Science to Controlled Organic Nanostructures

CONTENTS

Donald A. Tomalia
Dendritic Nanotechnologies Limited
Central Michigan University

Karen Mardel
Starpharma Limited

Scott A. Henderson
Starpharma Limited

G. Holan
Starpharma Limited

R. Esfand
Dendritic Nanotechnologies Limited
Central Michigan University

20.1   Introduction

20.1.1   Civilizations, Technology Periods, and Historical Revolutions

Throughout the last 120 centuries of human history, a mere handful of civilizations have emerged to dominate the social order and to define the human condition in the world (Figure 20. 1). These unique cultures generally emerged as a result of certain converging societal parameters which included evolving politics, religion, social order, or major military/scientific paradigm shifts. Historical patterns will show, however, that the ultimate magnitude and duration of the civilizations were inextricably connected to their investment in certain key emerging technologies. These broad technologies not only underpinned these cultures but also defined the cutting edge of human knowledge at that time in history. These advancements are referred to as technology ages or periods (Figure 20.1).

The first 118 centuries (10,000 BC–1800 AD) were distinguished by only three major technology periods — the Stone, Bronze, and Iron Ages. Each period was based on materials (building blocks) derived from natural sources and involved empirical knowledge gained through craftsmanship. Developments in these technology periods provided the resources and intellectual forces for dominance by these earlier civilizations. These successful reigns were followed by regression into the dark ages. Emergence of the European influence concurrent with the Renaissance led to an advancement of the Iron Age, to the initiation of the Chemical Age, and ultimately aligned critical forces leading to the great Industrial Revolution. During the past two centuries, there has been a dramatic proliferation of technologies with at least four major technology ages emerging in the last seven decades; namely, nuclear, plastics, materials, and biotechnology ages. These technical advancements have been aligned primarily with Euro-American societies; however, more recently there has been substantial Asian influence. Such technical advancements are very dependent upon certain critical enabling sciences that include *synthesis, engineering,* and, more recently, *combinatorial/simulation* strategies.

Generally speaking, significant new paradigm shifts have initiated each of these technology periods. Typical of each period has been the systematic characterization and exploitation of novel structural and materials properties. Based on property patterns observed within the complexity boundaries of the specific technology, new scientific rules and principles evolve and become defined. Contemporary society has benefited substantially from both the knowledge and the materials created by these technology periods.

The general trend for succession of these technology periods has involved progression from simpler materials to more complex forms. In this fashion, earlier precursor technologies become the platforms upon which subsequent more complex structures (i.e., symmetries) are based. As described by P.W. Anderson,[1] emerging new properties, principles, and rules are exhibited as a function of *symmetry breaking*. This occurs as one advances from more basic structures and progresses to higher orders of complexity. Such a pattern is noted as one follows the hierarchical/development of complexity that has been observed in biological systems over the past several billion years of evolution. Simply stated — *"the whole becomes not only more than, but different from the sum of its parts."*[1] It is this premise that has driven the human quest for understanding new and higher forms of complexity. It appears that it is from this pattern of inquiry that the industrial revolution was spawned. It may be argued that the convergence of knowledge and new materials created by just five technology areas (Figure 20.1) provided the critical environment and synergy for this historical revolution. Economic benefits and enhancements to the human condition in such areas as transportation, shelter, clothing, energy, and agriculture are now recognized to be immeasurable. It is from this perspective that we now consider the emergence of new technologies such as *genomics, proteomics,* and particularly *nanotechnology* as we examine prospects for the next revolution. The proposed biological revolution, which began with elucidation of the DNA structure by Crick and Watson, is expected to provide viable strategies that sound almost like science fiction at this time. Such predictions include the total elimination of human disease, human life expectancy beyond 100 years, and perhaps extraterrestrial emigration in the next century.

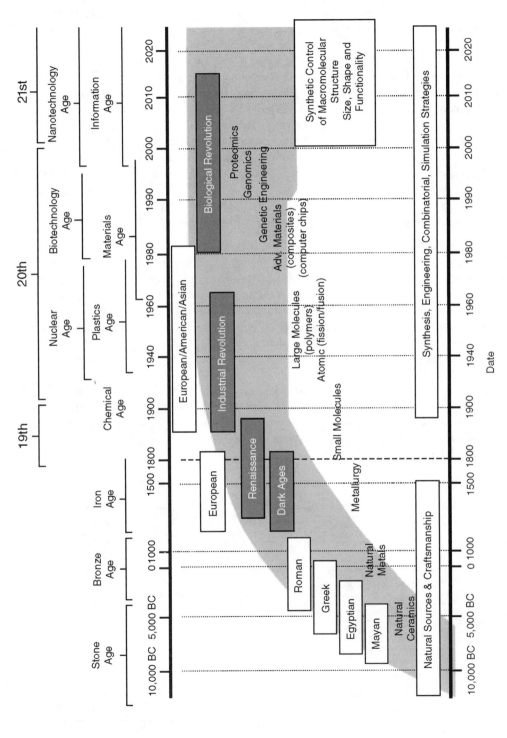

FIGURE 20.1 Civilizations, technology periods (ages), and historical revolutions as a function of time.

20.1.2 Importance of Controlled Organic Nanostructures to Biology

Critical to the successful creation of all biological structures required for life has been the evolutionary development of strategies to produce controlled organic nanostructures. As one reflects on evolutionary progress that defines the dimensional hierarchy of biological matter, it is apparent that it occurred in two significant phases and involved bottom-up synthesis. Clearly, critical parameters such as mass and dimension had to increase in size to define the appropriate building modules. The first phase was abiotic and involved molecular evolution from atoms to small molecules. This began approximately 13 billion years ago and progressed for nearly 8 to 9 billion years. The complexity reached at that point was necessary to set the stage with appropriate building blocks for the subsequent evolution of life, namely macromolecules such as DNA, RNA, and proteins. This latter phase is believed to have begun 3.5 billion years ago. These modules were generally collections of precisely bonded atoms that occupied space with dimensions ranging from 1 to 10^2 nm. Such structures required the controlled assembly of as many as 10^3 to 10^9 atoms and possessed molecular weights ranging from 10^4 to 10^{10} Daltons. These assemblies required the rigorous control of *size, shape, surface chemistries, scaffolding,* and *container properties* as described in Figure 20.2.

20.1.3 The Wet and Dry World of Nanotechnology

According to Nobel laureate R. Smalley,[2] the world of nanotechnology can be subdivided into two major application areas, the *wet* and *dry* sides. The former, of course, includes the biological domain, wherein the water-based science of living entities is dependent upon *hydrophilic* nanostructures and devices that may function within biological cells. Dendritic nanopolymers, especially *dendrimers,* fulfill many applications in the wet world of nanotechnology (Figure 20.3). In contrast, the dry side is expected to include those applications focused on more *hydrophobic* architectures and strategies. Progress in this second area may be expected to enhance the tensile strength of materials, increase the conductivity of electrons, or allow the size reduction of computer chips to levels unattainable with traditional bulk materials.

Although substantial progress has been made concerning the use of fullerenes and carbon nanotubes for dry nanotech applications, their use in biological applications has been hindered by the fact that they are highly hydrophobic and available in only several specific sizes (i.e., usually approximately 1 nm). However, recent advances involving the functionalization of fullerenes may offer future promise for these materials in certain biological applications.[3]

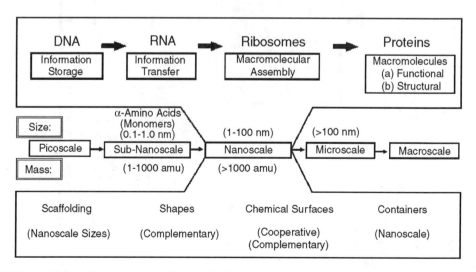

FIGURE 20.2 Biological structure–control strategy leading to nanoscale scaffolding, shapes, chemical surfaces, and containers found in proteins.

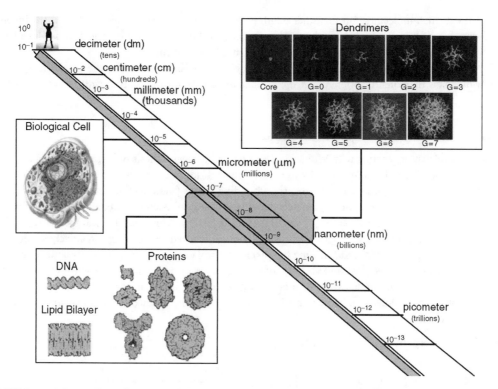

FIGURE 20.3 Comparison of micron-scale biological cells to nanoscale proteins and poly(amidoamine) dendrimers.

20.1.4   Potential Bottom-Up Synthesis Strategies for Organic Nanostructures

At least three major strategies are presently available for covalent synthesis of organic nanostructures: (1) *traditional organic chemistry*, (2) *traditional polymer chemistry*, and, more recently, (3) *dendritic macromolecular chemistry* (Figure 20.4).

Broadly speaking, traditional organic chemistry leads to higher complexity by involving the formation of relatively few covalent bonds between small heterogeneous aggregates of atoms or reagents to give

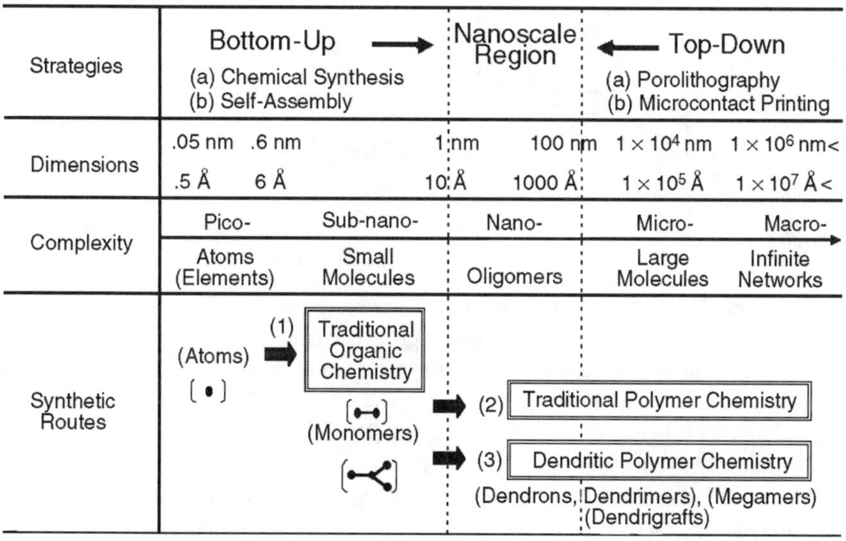

Strategies	Bottom-Up ⟶		Nanoscale Region	⟵ Top-Down	
	(a) Chemical Synthesis (b) Self-Assembly			(a) Porolithography (b) Microcontact Printing	
Dimensions	.05 nm .6 nm		1 nm 100 nm	1 × 10⁴ nm 1 × 10⁶ nm<	
	.5 Å 6 Å		10 Å 1000 Å	1 × 10⁵ Å 1 × 10⁷ Å<	
Complexity	Pico- Sub-nano-		Nano-	Micro- Macro-	
	Atoms (Elements)	Small Molecules	Oligomers	Large Molecules	Infinite Networks
Synthetic Routes	(Atoms) [•]	(1) Traditional Organic Chemistry [↔] (Monomers) (⦓)	(2) Traditional Polymer Chemistry (3) Dendritic Polymer Chemistry (Dendrons, Dendrimers), (Megamers) (Dendrigrafts)		

FIGURE 20.4 Molecular complexity as a function of covalent synthesis strategies and molecular dimensions.

well-defined small molecules. On the other hand, polymerization strategies such as (2) and (3) involve the formation of large multiples of covalent bonds between homogenous monomers to produce large molecules or infinite networks with a broad range of structure control.[4,5]

By comparing these three covalent synthesis strategies, we wish to introduce the third strategy and briefly overview the potential offered by *dendritic macromolecules* for the routine synthesis of controlled organic nanostructures.

20.1.5 Traditional Organic Chemistry

Organic chemistry, originating with Wöhler in 1828, has led to the synthesis of literally millions of small molecules. Organic synthesis involves the use of various hybridization states of carbon combined with specific heteroatoms to produce key hydrocarbon building blocks (modules) or functional groups (connectors). These two construction parameters have been used to assemble literally millions of more complex structures by either *divergent* or *convergent* strategies involving a limited number of step-wise, covalent bond-forming events. Usually product isolation is involved at each stage. Relatively small molecules (i.e., < 1 nm) are produced, allowing the precise control of shape, mass, flexibility, and functional group placement. The divergent and convergent strategies are recognized as the essence of traditional organic synthesis.[6] An example of the divergent strategy may be found in the Merrifield synthesis,[7–10] which involves chronological introduction of precise amino acid sequences to produce structure-controlled, linear architectures.

20.1.5.1 Divergent Multistep Synthesis

$$\boxed{A} \rightarrow \boxed{B} \rightarrow \boxed{C} \rightarrow \boxed{D} \rightarrow \rightarrow \rightarrow \boxed{I}$$

STRUCTURE (20.1)

Many examples of the convergent strategy can be found in contemporary approaches to natural products synthesis. Usually the routes to target molecules (i.e., I) are derived by retro-synthesis from the final product.[6] This involves transformation of the target molecule to lower molecular weight precursors (i.e., A–F).

20.1.5.2 Convergent Multistep Synthesis

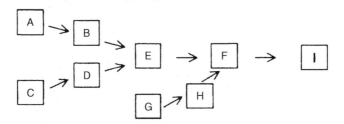

STRUCTURE (20.2)

Mathematically, at least one covalent bond, or in some cases several bonds, may be formed per reaction step (N_i). Assuming high-yield reaction steps and appropriate isolation stages, one can expect to obtain precise monodispersed products. In either case, the total number of covalent bonds formed can be expressed as shown in Scheme 20.1.

Based on the various hybridization states of carbon (Figure 20.5), at least four major carboskeletal architectures are known.[6,11] They are recognized as (I) *linear*, (II) *bridged* (two-dimensional/three-dimensional), (III) *branched*, and (IV) *dendritic*. It should be noted that buckminsterfullerenes, a subset of Class (II), bridged (three-dimensional) structures, and cascade molecules, a low molecular weight subset of Class (IV) dendritic architectures, are relatively recent examples of organic small molecule topologies. The former,

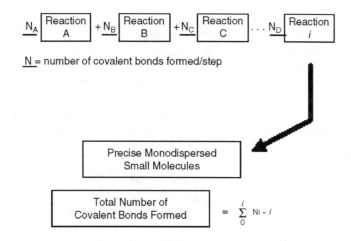

N = number of covalent bonds formed/step

SCHEME 20.1

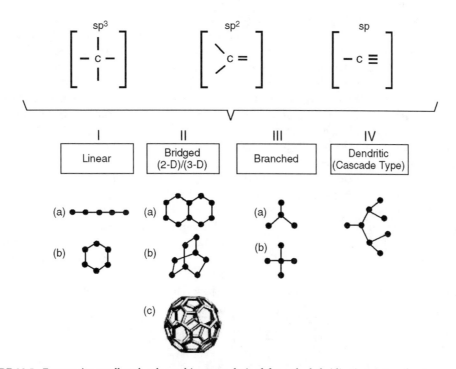

FIGURE 20.5 Four major small molecular architectures derived from the hybridization states of carbon.

recognized as *bucky balls* and *carbon nanotubes*, have enjoyed considerable attention as precise, hydrophobic nanomodules suitable for applications in the dry nanotechnology world. The low molecular weight dendritic architectures have been used as precursors to true dendrimer nanostructures as will be described later.

20.1.6 Traditional Polymer Chemistry

Over the past 70 years, a second covalent synthesis strategy has evolved based on the catenation of reactive small molecular modules or monomers. Broadly speaking, these propagations involve the use of reactive (AB-type monomers) that may be engaged to produce a variety of large, nanoscale molecules with polydispersed masses. Such multiple-bond formation may be driven by a variety of mechanisms including (1) *chain growth*, (2) *ring opening*, (3) *step-growth condensation*, or (4) *enzyme catalyzed processes.* Staudinger first introduced this paradigm in the 1920s[12–14] by demonstrating that reactive monomers could be used to produce a statistical distribution of one-dimensional (linear) molecules with very high molecular weights (i.e., $> 10^6$ Daltons). As many as 10,000 or more covalent bonds may be formed in a single chain reaction of monomers. Although macro/megamolecules with nanoscale dimensions may be attained, structure control of critical macromolecular design parameters (CMDPs) such as *size, molecular shape, positioning of atoms,* or *covalent connectivity* — other than those affording linear or crosslinked topologies — is difficult. However, recent progress has been made using *living polymerization* techniques that afford better control over molecular weight and some structural elements as described elsewhere.[15]

$$n[AB] \text{ (monomers)} \longrightarrow \sim\!\!\{AB\}_n$$

Traditional polymerizations usually involve AB-type monomers based on substituted ethylenes or strained small ring compounds. These monomers may be propagated by using chain reactions initiated by free radical, anionic, or cationic initiators;[16] or, alternatively, AB-type monomers may be used in polycondensation reactions.[5]

Multiple, covalent bonds are formed to produce each macromolecule and, in general, statistical, polydispersed structures are obtained. In the case of controlled vinyl polymerizations, the average length of the macromolecule is determined by monomer-to-initiator ratios. If one views these polymerizations as extraordinarily long sequences of individual reaction steps, the average number of covalent bonds formed may be visualized as shown in Scheme 20.2.

Traditional polymerization strategies generally produce linear architectures; however, branched topologies may be formed either by chain transfer processes or intentionally introduced by grafting techniques. In any case, the linear and branched architectural classes have traditionally defined the broad area of *thermoplastics.* Of equal importance is the major architectural class that is formed by the introduction of covalent bridging bonds between linear or branched polymeric topologies. These cross-linked (bridged) topologies were studied by Flory in the early 1940s and constitute the second major area of traditional polymer chemistry — namely, *thermosets.*

Therefore, approximately 50 years after the introduction of the *macromolecular hypothesis* by Staudinger,[14] the entire field of polymer science could simply be described as consisting of only two major architectural classes: (I) *linear topologies* as found in thermoplastics and (II) *cross-linked architectures* as found in thermosets.[4,16] The major focus of polymer science during the time frame of the 1920s to the 1970s was on the unique architecturally driven properties manifested by either linear or cross-linked topologies. Based on unique properties exhibited by these topologies, many

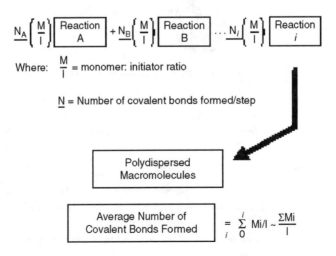

SCHEME 20.2

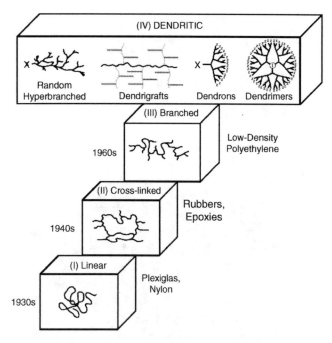

FIGURE 20.6 Three major polymer architectural Classes (I–III) leading to the fourth major class — dendritic (IV) consisting of the subclasses *random hyperbranched, dendrigrafts, dendrons,* and *dendrimers.*

natural polymers critical to success in World War II were replaced with synthetic polymers and were of utmost strategic importance.[4] In the 1960s and 1970s, pioneering investigation into long chain branching (LCB) in polyolefins and other related branching systems began to emerge.[17,18] More recently, intense commercial interest has been focused on new polyolefin architectures based on *random long branched* and *dendritic topologies.*[19] These architectures are produced by *metallocene* and *Brookhart-type* catalysts. In summary, by the end of the 1970s, only three major architectural classes of polymers were recognized and referred to as classical polymers. The fourth major class, dendritic polymers,[166] was only beginning to emerge (Figure 20.6).

20.2 The Dendritic State

Dendritic architecture is one of the most pervasive topologies observed at the micro- and macro-dimensional length scales (i.e., μm to m). At the nanoscale (molecular) level, there are relatively few biological examples of this architecture. Most notable are the glycogen and amylopectin hyperbranched structures that nature uses for energy storage. Presumably, the many chain ends that decorate these macromolecules facilitate enzymatic access to glucose for high-demand bioenergy events.[20] Another nanoscale example of dendritic architecture in biological systems is found in proteoglycans. These macromolecules appear to provide energy-absorbing, cushioning properties and determine the viscoelastic properties of connective tissue (Figure 20.7).

In the past two decades, versatile strategies have been developed that allow the design, synthesis, and functionalization of a wide variety of such dendritic structures.[15]

20.2.1 A Comparison of Organic Chemistry and Polymer Chemistry with Dendritic Macromolecular Chemistry

It is appropriate to compare the well-known concepts of covalent bond formation in traditional organic chemistry with those in classical polymer chemistry and dendritic macromolecular chemistry. This allows

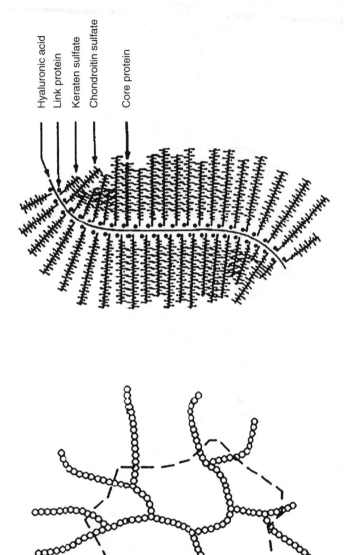

Hyaluronic acid
Link protein
Keraten sulfate
Chondroitin sulfate
Core protein

FIGURE 20.7 Topologies for (a) amylopectin and (b) proteoglycans.

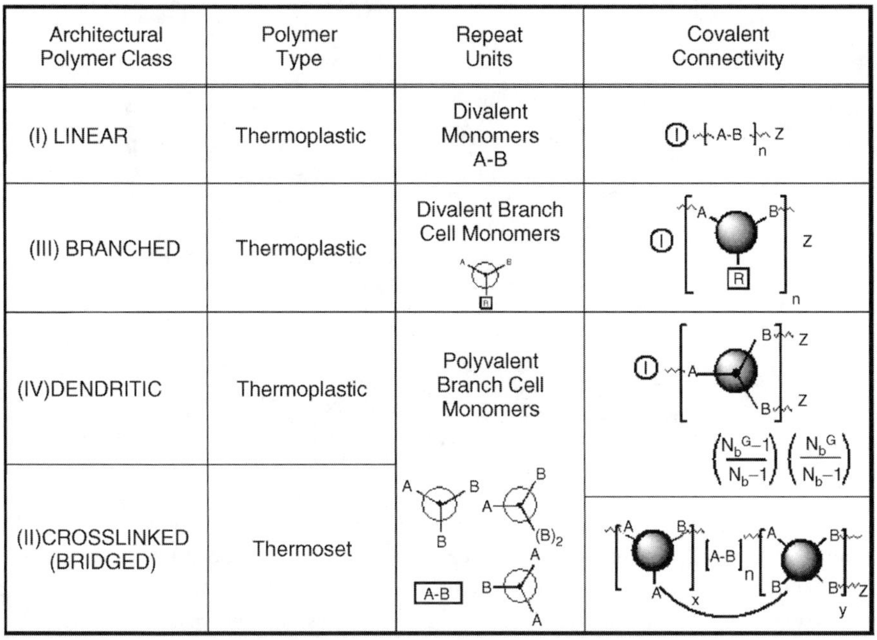

Architectural Polymer Class	Polymer Type	Repeat Units	Covalent Connectivity
(I) LINEAR	Thermoplastic	Divalent Monomers A-B	[diagram: $\text{(I)} \dashv A\text{-}B \dashv_n Z$]
(III) BRANCHED	Thermoplastic	Divalent Branch Cell Monomers	[diagram with R, A, B, Z, subscript n]
(IV) DENDRITIC	Thermoplastic	Polyvalent Branch Cell Monomers	[diagram with A, B, Z] $\left(\dfrac{N_b^{G}-1}{N_b-1}\right)\left(\dfrac{N_b^{G}}{N_b-1}\right)$
(II) CROSSLINKED (BRIDGED)	Thermoset	[diagrams with A, B, $(B)_2$, and boxed $A\text{-}B$]	[diagram with A, B, $A\text{-}B$, subscripts n, x, y, Z]

FIGURE 20.8 Examples of architectural polymer classes (I–IV) polymer type, repeat units, and covalent connectivity associated with architectural classes. (From Esfand, R.; Tomalia, D.A. Laboratory synthesis of poly(amidoamine) (PAMAM) dendrimers. In *Dendrimers and Other Dendritic Polymers*; Fréchet, J.M.J. and Tomalia, D.A., Eds.; John Wiley & Sons: West Sussex, 2001. With permission.)

one to fully appreciate the advantages and shortcomings of these three synthetic strategies in the context of bottom-up routes to organic nanostructures. Covalent synthesis in traditional polymer chemistry has evolved around the use of reactive modules (AB-type monomer) or ABR-type branch reagents that may be engaged in multiple covalent bond formation to produce large one-dimensional molecules of various lengths. Such multiple bond formation may be driven either by chain reactions, ring opening reactions, or polycondensation schemes. These propagation schemes lead to products that are recognized as Class I: *linear* or Class III: *branched* architectures. Alternatively, using combinations and permutations of divalent A-B type monomers and/or A-B$_n$, A$_n$-B polyvalent, branch cell-type monomers produces Class II, *cross-linked (bridged)* architectures.

A comparison of the covalent connectivity associated with each of these architecture classes (Figure 20.8) reveals that the number of covalent bonds formed per step for linear and branched topology is a multiple (n = degree of polymerization) related to the monomer/initiator ratios. In contrast, ideal *dendritic* (Class IV) propagations involve the formation of an exponential number of covalent bonds per reaction step (all termed G = generation), as well as amplification of both mass (i.e., number of branch cells/G) and terminal groups, (Z) per generation (G).

Mathematically, the number of covalent bonds formed per generation (reaction step) varies in an ideal dendron or dendrimer synthesis, according to a power function of the reaction steps, as illustrated in (Scheme 20.3). It is clear that a dramatic amplification of covalent bond formation occurs in all dendritic synthesis strategies. In addition to new architectural consequences, this feature clearly differentiates dendritic growth processes from covalent bond synthesis found in traditional organic and polymer chemistry.[21]

It should be quite apparent that, although all major architectural polymer classes are derived from common or related repeat units, the covalent connectivity is truly discrete and different. Furthermore, mathematical analyses of the respective propagation strategies clearly illustrate the dramatic differences in structure development as a function of covalent bond formation. It should be noted that linear, branched, and dendritic topologies differ substantially in their covalent connectivity and their terminal

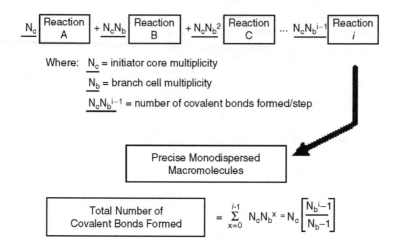

Where: N_c = initiator core multiplicity

N_b = branch cell multiplicity

$N_c N_b^{i-1}$ = number of covalent bonds formed/step

Precise Monodispersed
Macromolecules

$$\text{Total Number of Covalent Bonds Formed} = \sum_{x=0}^{i-1} N_c N_b^x = N_c \left[\frac{N_b^i - 1}{N_b - 1} \right]$$

SCHEME 20.3

group-to-initiator site ratios. In spite of these differences, these open, unlooped macromolecular assemblies clearly manifest thermoplastic polymer-type behavior compared with the looped, bridged connectivity associated with cross-linked thermoset systems. In fact, it is now recognized that these three open assembly-topologies (linear, branched, and dendritic) represent a graduated continuum of architectural intermediacy between thermoplastic and thermoset behavior.[22]

In summary, traditional organic chemistry offers exquisite structure control over a wide variety of compositions up to, but not including, *higher nanoscale structural dimensions* (e.g., buckminsterfullerene diameters $\cong 1$ nm). Furthermore, such all-carbon nanostructures are limited to only one hydrophobic compositional form and are hampered by the difficulties of functionalizing their surfaces to produce hydrophilic structures.[3]

Classical polymer chemistry offers relatively little structural control, but it facilitates access to statistical distributions of polydispersed nanoscale structures. Living polymerizations provide slightly better, but still imperfect, control over product size and mass distribution or polydispersity. In contrast, as will be seen below, dendritic macromolecular chemistry provides essentially all the features required for unparalleled control over *topology, composition, size, mass, shape,* and *functional group placement.*[15] These are features that truly distinguish many successful nanostructures found in nature.[23]

The quest for nanostructures and devices based on the biomimetic premise of architectural and functional precision is intense and remains an ultimate challenge. One must ask: what new options or unique properties does the dendritic state offer to meet the needs of nanoscale science and technology? The rest of this chapter will attempt to overview key features of the dendritic state that address these and other issues.

20.2.2   Dendritic Polymers — A Fourth Major New Architectural Class

Dendritic topologies are recognized as a fourth major class of macromolecular architecture.[24–27,166] The signature for such a distinction is the unique repertoire of new properties exhibited by this class of polymers. Many new synthetic strategies have been reported for the preparation of these materials, thus providing access to a broad range of dendritic structures. Presently, this architectural class consists of three dendritic subclasses as shown in Figure 20.9: (IVa) *random hyperbranched polymers,* (IVb) *dendri-graft polymers,* and (IVc) *dendrimers.* This subset order (IVa–c) reflects the relative degree of structural control present in each of these dendritic architectures.

All dendritic polymers are open, covalent nanoassemblies of branch cells. They may be organized as very symmetrical, monodispersed arrays, as is the case for dendrimers, or as irregular, polydispersed assemblies that typically define random hyperbranched polymers. As such, the respective subclasses and the level of structure control are defined by the propagation methodology as well as by the branch cell (BC) construction

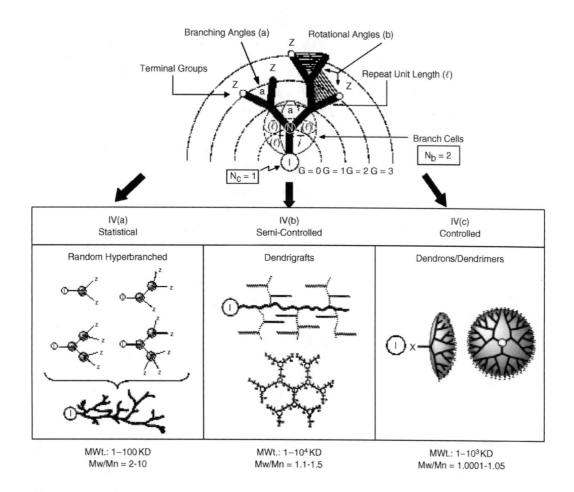

FIGURE 20.9 Branch cell structural parameters: (a) branching angles, (b) rotational angles, (l) repeat units lengths, (Z) terminal groups and dendritic subclasses derived from branches, (IVa) random hyperbranched, (IVb) dendrigrafts, and (IVc) dendrons/dendrimers.

parameters used to produce these assemblies. These BC parameters are determined by the composition of the BC monomers as well as the nature of the *excluded volume* defined by the BC. The excluded volume of the BC is determined by the length of the arms, the symmetry, rigidity/flexibility, the branching, and rotation angles involved within each of the branch cell domains.[28] As shown in Figure 20.9, these dendritic arrays of branch cells usually manifest covalent connectivity relative to some molecular reference marker (I) or core. As such, these branch cell arrays may be very non-ideal/polydispersed (e.g., Mw/Mn $\cong$ 2–10), as observed for random hyperbranched polymers (IVa), or very ideally organized into highly controlled, core–shell-type structures as noted for dendrons/dendrimers (IVc): Mw/Mn $\cong$ 1.01–1.0001. Dendrigraft (arborescent) polymers reside between these two extremes of structure control, frequently manifesting rather narrow polydispersities of Mw/Mn 1.1–1.5 depending on their mode of preparation.

20.2.3 Dendrons and Dendrimers

Dendrons and dendrimers are the most intensely investigated subset of dendritic polymers. In the past decade over 5000 literature references have appeared on this unique class of structure-controlled polymers. The word *dendrimer* is derived from the Greek words *dendri* (branch, tree-like) and *meros* (part of). The term was coined by Tomalia et al. over 15 years ago in the first full paper on poly(amidoamine)

(PAMAM) dendrimers.[29,30,166] Poly(amidoamine) dendrimers constitute the first dendrimer family to be commercialized and undoubtedly represent the most extensively characterized and best understood series at this time. In view of the vast amount of literature in this field, the remaining overview will focus on PAMAM dendrimers and limit the scope to critical property features and unique biomedical applications offered by these fascinating nanostructures.

20.2.3.1 Synthesis — Divergent and Convergent Methods

In contrast to traditional polymers, dendrimers are unique core–shell structures possessing three basic architectural components: (1) *a core*, (2) *an interior of shells (generation)* consisting of repetitive branch cell units, and (3) *terminal functional groups* (i.e., the outer shell or periphery) as illustrated in Figures 20.10 and 20.11.

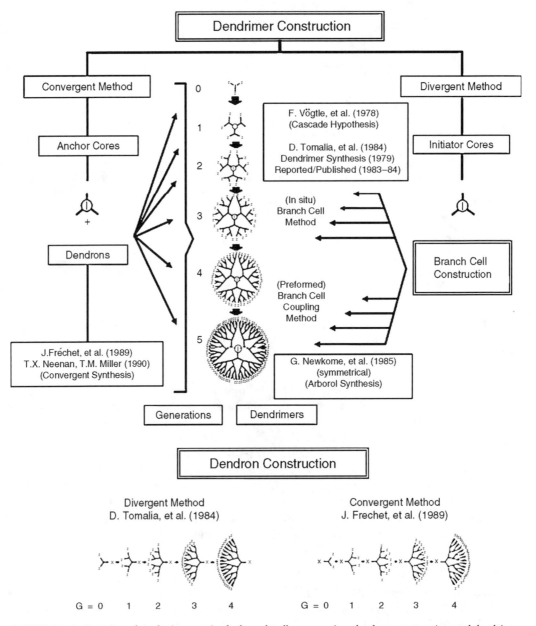

FIGURE 20.10 Overview of synthetic strategies for branch cell construction, dendron construction, and dendrimer construction annotated with discovery scientists.

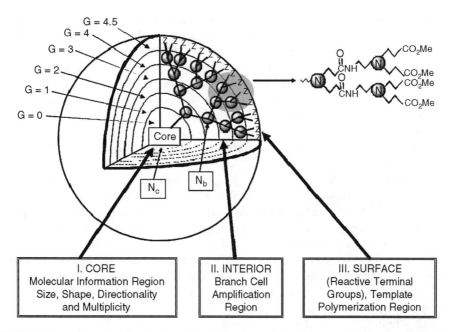

FIGURE 20.11 Three-dimensional projection of dendrimer core-shell architecture for G = 4.5 poly(amidoamine) (PAMAM) dendrimer with principal architectural components (I) core, (II) interior, and (III) surface. (From Esfand, R.; Tomalia, D.A. Laboratory synthesis of poly(amidoamine) (PAMAM) dendrimers. In *Dendrimers and Other Dendritic Polymers*; Fréchet, J.M.J. and Tomalia, D.A., Eds.; John Wiley & Sons: West Sussex, 2001. With permission.)

In general, dendrimer synthesis involves hierarchical assembly strategies that require the construction components shown in Structure 20.3.

STRUCTURE (20.3)

Many methods for assembling these components have been reported; however, they can be broadly categorized as either *divergent* or *convergent* strategies. Within each of these major approaches there may be variations in methodology for branch cell construction (i.e., *in situ* vs. *preformed*) or dendron construction (i.e., *divergent* vs. *convergent*), as overviewed in (Figure 20.10).

Historically, early developments in the field were based on divergent methods. Vögtle et al. (University of Bonn) first reported the synthesis of several low molecular weight (<900 Daltons; G = 0–2) cascade structures[31] using the divergent, *in situ* branch cell method. This synthesis was based on a combination of acrylonitrile and reduction chemistry. As Vögtle reported later,[32] higher generation cascade structures and, indeed, dendrimers could not be obtained by this process due to synthetic and analytical difficulties. Nearly simultaneously, a completely characterized series of high molecular weight (i.e., > 58000 Daltons; G = 0–7), poly(amidoamine) (PAMAM) dendrimers was synthesized by Tomalia et al.[29,30,166] Success with his approach was based on a two-step reaction sequence involving acrylate Michael addition and amidation chemistry.[30,33,34] Historically, this methodology provided the first commercial route to dendrimers as well as the first opportunity to observe unique dendrimer property development that occurs *only* at higher generations (i.e., G = 4 or higher). Many of these observations were described in a publication that appeared in 1985[30] and later reviewed extensively in 1990.[35,36]

The first published use of preformed branch cell methodology was reported in a communication by Newkome et al.[37] This approach involved the coupling of preformed branch cell reagents around a core to produce low molecular weight (i.e., > 2000 Daltons, G = 3) *arborol structures*. This approach has been used to synthesize many other dendrimer families including *dendri-poly(ethers)*,[38] *dendri-*

poly(thioethers),[21] and others.[15,39] Each of these methods involved the systematic divergent growth of branch cells around a core that defined shells within the dendrons. The multiplicity and directionality of the initiator sites (N_c) on the core determine the number of dendrons and the ultimate shape of the dendrimer. In essence, dendrimers propagated by this method constitute groups of molecular trees (i.e., two or more dendrons) that are propagated outwardly from their roots (cores). This occurs in stages (generations), wherein the functional leaves of these trees become reactive precursor templates (scaffolding) upon which to assemble the next generation of branches. This methodology can be used to produce multiples of trees — (dendrimers) — or single trees — (dendrons) as shown in Figure 20.10.

Using a totally novel approach, Hawker and Fréchet[40,41] followed by Neenan and Miller[42] reported the convergent construction of such molecular trees by first starting with the leaves or surface branch cell reagents. By amplifying with these reagents in stages (generations), one produces a dendron possessing a single reactive group at the root or focal point of the structure. Subsequent coupling of these reactive dendrons through their focal points to a common *anchoring core* yields the corresponding dendrimers. Because of the availability of orthogonal functional groups at the focal point and periphery of the dendrons, the convergent synthesis is particularly useful for the preparation of more complex macromolecular architectures[43] such as linear dendritic hybrids, block copolymeric dendrimers, or dendronized polymers. Another significant difference is that the divergent approach requires an exponential increase in the number of coupling steps for generation growth, whereas the convergent method involves only a constant number of reactions (typically two to three) at each stage of the synthesis. Today, several hundred reports utilizing the original poly(ether), Fréchet-type dendron[44] method make this the best understood and structurally most precise family of convergent dendrimers.

Overall, each of these dendrimer construction strategies offers respective advantages and disadvantages. Some of these issues, together with experimental laboratory procedures, are reviewed elsewhere.[15]

20.2.3.2 Dendrimer Features of Interest to Nanoscientists

Dendrimers may be viewed as unique, information processing, nanoscale devices. Each architectural component manifests a specific function while also defining properties for these nanostructures as they are grown generation by generation. For example, the *core* may be thought of as the molecular information center from which *size, shape, directionality,* and *multiplicity* are expressed *via* the covalent connectivity to the outer shells. Within the *interior,* one finds the *branch cell amplification region,* which defines the type and amount of interior void space that may be enclosed by the terminal groups as the dendrimer is grown. Branch cell multiplicity (N_b) determines the density and degree of amplification as an exponential function of generation (G). The interior composition and amount of solvent-filled void space determines the extent and nature of guest–host (endo-receptor) properties that are possible within a particular dendrimer family and generation. Finally, the surface consists of reactive or passive terminal groups that may perform several functions. With appropriate function, they serve as a *template polymerization region* as each generation is amplified and covalently attached to the precursor generation. Secondly, the surface groups may function as passive or reactive gates controlling entry or departure of guest molecules from the dendrimer interior. These three architectural components essentially determine the physical/chemical properties as well as the overall sizes, shapes, and flexibility of dendrimers. It is important to note that dendrimer diameters increase linearly as a function of shells or generations added, whereas the terminal functional groups increase exponentially as a function of generation. This dilemma enhances *tethered congestion* of the anchored dendrons as a function of generation, due to the steric crowding of the end groups. As a consequence, lower generations are generally open, floppy structures — whereas higher generations become robust, less deformable spheroids, ellipsoids, or cylinders, depending on the shape and directionality of the core.

PAMAM dendrimers are synthesized by the divergent approach. This methodology involves *in situ* branch cell construction in step-wise, iterative stages (i.e., generation = 1,2,3…) around a desired core to produce mathematically defined *core-shell* structures. Typically, ethylenediamine (N_c = 4) or ammonia (N_c = 3) are used as cores and allowed to undergo reiterative two-step reaction sequences involving (1) exhaustive alkylation of primary amines (Michael addition) with methyl acrylate and (2) amidation

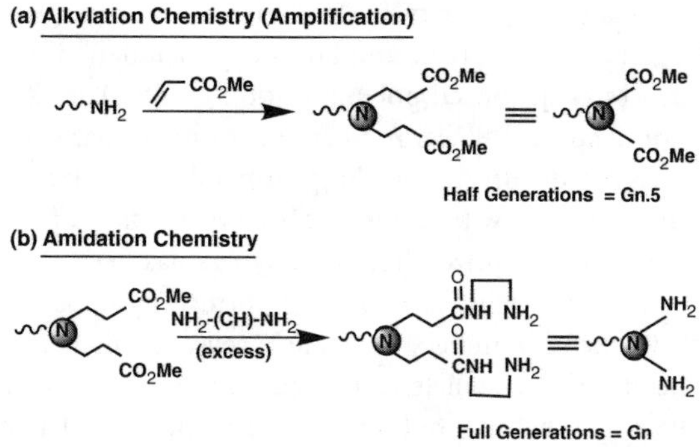

(a) Alkylation Chemistry (Amplification)

Half Generations = Gn.5

(b) Amidation Chemistry

Full Generations = Gn

SCHEME 20.4   (From Esfand, R.; Tomalia, D.A. Laboratory synthesis of poly(amidoamine) (PAMAM) dendrimers. In *Dendrimers and Other Dendritic Polymers*; Fréchet, J.M.J. and Tomalia, D.A., Eds.; John Wiley & Sons: West Sussex, 2001. With permission.)

of amplified ester groups with a large excess of ethylenediamine to produce primary amine terminal groups as illustrated in Scheme 20.4.

This first reaction sequence on the exposed dendron (Figure 20.11) creates G = 0 (i.e., the core branch cell), wherein the number of arms (i.e., dendrons) anchored to the core is determined by N_c. Iteration of the alkylation/amidation sequence produces an amplification of terminal groups from 1 to 2 with the *in situ* creation of a branch cell at the anchoring site of the dendron that constitutes G = 1. Repeating these iterative sequences (Scheme 20.4) produces additional shells (generations) of branch cells that amplify mass and terminal groups according to the mathematical expressions described in Figure 20.12.

Number of Surface Groups	:	$Z = N_c N_b{}^G$	**Polyvalency**

Number of Branch Cells	:	$BC = N_c \left[\dfrac{N_b{}^G - 1}{N_b - 1} \right] =$	Number of Covalent Bonds Formed/Generation

Molecular Weights	:	$MW = M_c + N_c \left[M_{RU} \left(\dfrac{N_b{}^G - 1}{N_b - 1} \right) + M_t N_b{}^G \right]$

Generation	Surface Groups (Z)	Molecular Formula	MW	Diameter (nm)
0	4	$C_{22}H_{48}N_{10}O_4$	517	1.4
1	8	$C_{62}H_{128}N_{26}O_{12}$	1,430	1.9
2	16	$C_{142}H_{288}N_{58}O_{28}$	3,256	2.6
3	32	$C_{302}H_{608}N_{122}O_{60}$	6,909	3.6
4	64	$C_{622}H_{1248}N_{250}O_{124}$	14,215	4.4
5	128	$C_{1262}H_{2528}N_{506}O_{252}$	28,826	5.7
6	256	$C_{2542}H_{5088}N_{1018}O_{508}$	58,048	7.2
7	512	$C_{5102}H_{10208}N_{2042}O_{1020}$	116,493	8.8
8	1,024	$C_{10222}H_{20448}N_{4090}O_{2044}$	233,383	9.8
9	2,048	$C_{20462}H_{40928}N_{8186}O_{4092}$	467,162	11.4
10	4,096	$C_{40942}H_{81888}N_{16378}O_{8188}$	934,720	~13.0

FIGURE 20.12   Dendritic branching mathematics for predicting number of dendrimer surface groups, number of branch cells, and molecular weights. Calculated values for [ethylenediamine core]; *dendri*-poly(amidoamine) series with nanoscale diameters (nm).

It is apparent that both the core multiplicity (N_c) and branch cell multiplicity (N_b) determine the precise number of terminal groups (Z) and mass amplification as a function of generation (G). One may view those generation sequences as quantized polymerization events. The assembly of reactive mono-mers,[33,35] branch cells,[15,35,39] or dendrons[15,40,45] around atomic or molecular cores to produce dendrimers according to divergent/convergent dendritic branching principles, has been well demonstrated. Such systematic filling of space around cores with branch cells, as a function of generational growth stages (branch cell shells), to give discrete, quantized bundles of mass has been shown to be mathematically predictable.[21,46] Predicted molecular weights have been confirmed by mass spectroscopy[47–49] and other analytical methods.[35,40,50–52] Predicted numbers of branch cells, terminal groups (Z), and molecular weights as a function of generation for an ethylenediamine core (N_c = 4) PAMAM dendrimer are shown in Figure 20.12. It should be noted that the molecular weights approximately double as one progresses from one generation to the next. The surface groups (Z) and branch cells (BC) amplify mathematically according to a power function, thus producing discrete, monodispersed structures with precise molecular weights and nanoscale diameter enhancement as described in Figure 20.12. These predicted values are routinely verified by mass spectroscopy for the earlier generations (i.e., G = 4–5); however, with divergent dendrimers, minor mass defects are often observed for higher generations as congestion-induced *de Gennes dense packing* begins to take effect (Figure 20.13).

20.2.3.3. Dendrimer Shape Changes — A Nanoscale Molecular Morphogenesis

As illustrated in Figure 20.13, dendrimers undergo *congestion-induced* molecular shape changes from flat, floppy conformations to robust spheroids as first predicted by Goddard et al.[53] Shape change transitions were subsequently confirmed by extensive photophysical measurements, pioneered by Turro et al.,[54–57] and solvatochromic measurements by Hawker et al.[58] Depending upon the accumulative core and branch cell multiplicities of the dendrimer family under consideration, these transitions were found to occur between G = 3 and G = 5. Ammonia core, PAMAM dendrimers (N_c = 3, N_b = 2) exhibited a molecular morphogenesis break at G = 4.5, whereas the ethylenediamine (EDA) PAMAM dendrimer family (N_c = 4; N_b = 2) manifested a shape change break around G = 3–4,[53] and the Fréchet-type convergent dendrons (N_b = 2) around G = 4.[58] It is readily apparent that increasing the core multiplicity from N_c = 3 to N_c = 4 accelerates congestion and forces a shape change at least one generation earlier. Beyond these genera-tional transitions, one can visualize these dendrimeric shapes as nearly spheroidal or slightly ellipsoidal *core-shell type architectures*.

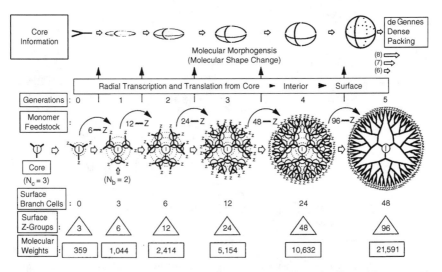

FIGURE 20.13 Comparison of molecular shape change, two-dimensional branch cell amplification, surface branch cells, surface groups (Z), and molecular weights as function of generation: G = 0–5.

20.2.3.4 de Gennes Dense Packing — A Nanoscale Steric Phenomenon

As a consequence of the excluded volume associated with the core, interior, and surface branch cells, steric congestion is expected to occur due to tethered connectivity to the core. Furthermore, the number of dendrimer surface groups, Z, amplifies with each subsequent generation (G). This occurs according to geometric *branching laws*, which are related to core multiplicity, (N_c), and branch cell multiplicity, (N_b).[21,30] These values are defined by the following equation:

$$Z = N_c N_b{}^G$$

Because the radii of the dendrimers increase in a linear manner as a function of G, whereas the surface cells amplify according to $N_c N_b{}^G$, it is implicit from this equation that generational reiteration of branch cells ultimately will lead to a so-called *dense-packed state*.

As early as 1983, de Gennes and Hervet[59,166] proposed a simple equation derived from fundamental principles to predict the dense-packed generation for PAMAM dendrimers. It was predicted that at this generation, ideal branching can no longer occur as available surface space becomes too limited for the mathematically predicted number of surface cells to occupy. This produces a *closed geometric structure*. The surface is *crowded* with exterior groups, which, although potentially chemically reactive, are sterically prohibited from participating in ideal dendrimer growth.

This *critical packing state* does not preclude further dendrimer growth beyond this point in the genealogical history of the dendrimer preparation. On the contrary, although continuation of dendrimer step-growth beyond the dense-packed state cannot yield structurally ideal, next-generation dendrimers, it can nevertheless occur as indicated by further increases in the molecular weight of the resulting products. Predictions by de Gennes[59] suggested that the PAMAM dendrimer series should reach a critical packing state at generations 9–10. Experimentally, we observed a moderate molecular weight deviation from predicted ideal values beginning at generation 4–7 (Figure 20.15). This digression became very significant at generations 7–8 as dendrimer growth was continued to generations 12.[24] The products thus obtained are of imperfect structure because of the inability of all surface groups to undergo further reaction. Presumably a fraction of these surface groups remains trapped or is sterically encumbered under the surface of the newly formed dendrimer shell, yielding a unique architecture possessing two types of terminal groups. This new surface group population will consist of both those groups that are accessible to subsequent reiteration reagents and those that will be sterically screened. The total number of these groups will not, however, correspond to the predictions of the mathematical branching law, but will fall between that value mathematically predicted for the next generations (i.e., G + 1) and that expected for the precursor generation (G). Thus, a mass defective dendrimer generation is formed.

Dendrimer surface congestion can be appraised mathematically as a function of generation, from the following simple relationship:

$$A_z = \frac{A_D}{N_Z} \alpha \frac{r^2}{N_c N_b{}^G}$$

where A_z is the surface area per terminal group Z, A_D the dendrimer surface area, and N_z the number of surface groups Z per generation. This relationship predicts that at higher generations G, the surface area per Z group becomes increasingly smaller and experimentally approaches the cross-sectional area or van der Waals dimension of the surface groups Z. The generation G thus reached is referred to as the *de Gennes dense-packed generation*.[15,21,35] Ideal dendritic growth without branch defects is possible only for those generations preceding this dense-packed state. This critical dendrimer property gives rise to self-limiting dendrimer dimensions, which are a function of the branch cell segment length (*l*), the core multiplicity N_c, the branch cell juncture multiplicity N_b, and the steric dimensions of the terminal group Z (Figure 20.9). Whereas the dendrimer radius r in the above expression is dependent on the branch cell segment lengths *l*, large *l* values delay this congestion. On the other hand, larger N_c, N_b values and larger Z dimensions dramatically hasten it.

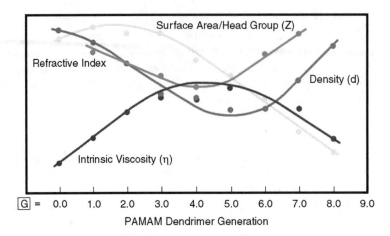

FIGURE 20.14 Comparison of surface area/head group (Z), refractive index, density (d), and intrinsic viscosity (η) as a function of generation: G = 1–9. (From Esfand, R.; Tomalia, D.A. Laboratory synthesis of poly(amidoamine) (PAMAM) dendrimers. In *Dendrimers and Other Dendritic Polymers*; Fréchet, J.M.J. and Tomalia, D.A., Eds.; John Wiley & Sons: West Sussex, 2001. With permission.)

Additional physical evidence supporting the anticipated development of congestion as a function of generation is shown in the composite comparison of property changes in Figure 20.14. Plots of intrinsic viscosity [η],[35,60] density z, surface area per Z group (A_z), and refractive index (η) as a function of generation clearly show maxima or minima at generations = 3–5, paralleling computer-assisted molecular-simulation predictions[53] as well as extensive photochemical probe experiments reported by Turro et al.[54–57]

The intrinsic viscosities [η] increase in a very classical fashion as a function of molar mass (generation) but decline beyond a certain generation because of change from an extended to a globular shape.[35,53] In effect, once this critical generation is reached, the dendrimer begins to act more like an Einstein spheroid. The intrinsic viscosity is a physical property that is expressed in dL/g (i.e., the ratio of volume to mass). As the generation number increases and transition to a spherical shape takes place, the volume of the spherical dendrimer roughly increases in cubic fashion while its mass increases exponentially; hence, the value of [η] must decrease once a certain generation is reached. This prediction has now been confirmed experimentally.[35,60]

The dendrimer density (d) (atomic mass units per unit volume) clearly minimizes between generations 4 and 5. It then begins to increase as a function of generation due to the increasingly larger, exponential accumulation of surface groups. Because refractive indices are directly related to density parameters, their values minimize and parallel the above density relationship.

Clearly, this de Gennes dense-packed congestion would be expected to contribute to sterically inhibited reaction rates and sterically induced stoichiometry.[35] Each of these effects was observed experimentally at higher generations.[34] The latter would be expected to induce dendrimer mass defects at higher generations which we have used as a diagnostic signature for appraising the de Gennes dense-packing effect.

Theoretical dendrimer mass values were compared with experimental values by performing electrospray and MALDI-TOF mass spectral analysis on the respective PAMAM families (i.e., N_c = 3 and 4).[48] Note there is essentially complete shell filling for the first five generations of the (NH_3) core (N_c = 3; N_b = 2) poly(amidoamine) (PAMAM) series (Figure 20.15b). A gradual digression from theoretical masses occurs for G = 5–8, followed by a substantial break (i.e., Δ = 23%) between G = 8 and 9. *This discontinuity in shell saturation is interpreted as a signature for de Gennes dense packing*. It should be noted that shell saturation values continue to decline monotonically beyond this breakpoint to a value of 35.7% of theoretical at G = 12. A similar trend is noted for the EDA core, PAMAM series (N_c = 4; N_b = 2); however, the shell saturation inflection point occurs at least one generation earlier (i.e., G = 4–7, see Figure 20.15a). This suggests that the onset of de Gennes dense packing may be occurring between G = 7 and 8.

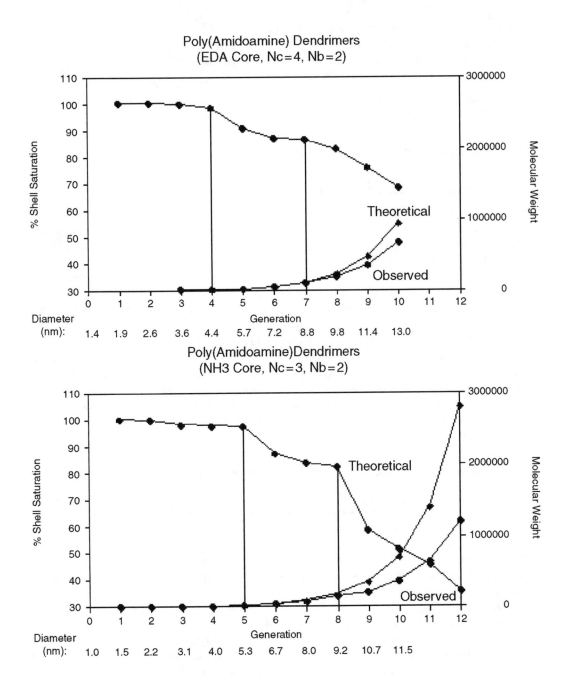

FIGURE 20.15 Top: Comparison of theoretical/observed molecular weights and percentage shell filling for EDA core PAMAM dendrimers as a function of generation: G = 1–10. Bottom: Comparison of theoretical/observed molecular weights and percentage shell filling for NH₃ core PAMAM dendrimers as a function of generation: G = 1–12. (From Esfand, R.; Tomalia, D.A. Laboratory synthesis of poly(amidoamine) (PAMAM) dendrimers. In *Dendrimers and Other Dendritic Polymers*; Fréchet, J.M.J. and Tomalia, D.A., Eds.; John Wiley & Sons: West Sussex, 2001. With permission.)

Unique features offered by the dendritic state that have no equivalency in classical polymer topologies are found almost exclusively in the dendron/dendrimer subset or, to a slightly lesser degree, in the dendrigrafts. They include:

1. Nearly complete nanoscale monodispersity
2. The ability to control nanoscale container/scaffolding properties
3. Exponential amplification and functionalization of dendrimer surfaces
4. Nanoscale dimension and shape mimicry of proteins

These features are captured to some degree with *dendrigraft* polymers, but are either absent or present to a vanishing small extent for random *hyperbranched* polymers.

20.3   Unique Dendrimer Properties

20.3.1   Nanoscale Monodispersity

The monodispersed nature of dendrimers has been verified extensively by mass spectroscopy,[155] size exclusion chromatography, gel electrophoresis,[52] electron microscopy (TEM),[15,167] and atomic force microscopy (AFM).[152] As is always the case, the level of monodispersity is determined by the skill of the synthetic chemist as well as the isolation/purification methods utilized.

In general, convergent methods produce the most nearly isomolecular dendrimers. This is because the convergent growth process allows purification at each step of the synthesis and eliminates cumulative effects due to failed couplings.[40,43] Appropriately purified, convergent dendrimers are probably the most precise synthetic macromolecules that exist today.

As discussed earlier, mass spectroscopy has shown that PAMAM dendrimers (Figure 20.15) produced by the divergent method are very monodisperse and have masses consistent with predicted values for the earlier generations (G = 0–5). Even at higher generations, as one enters the de Gennes dense-packed region, the molecular weight distributions remain very narrow (i.e., 1.05) and consistent, in spite of the fact that experimental masses deviate substantially from predicted theoretical values. Presumably, de Gennes dense packing produces a very regular and dependable effect that is manifested in the narrow molecular weight distribution.

20.3.2   Nanoscale Container/Scaffolding Properties

Unimolecular container/scaffolding behavior appears to be a periodic property that is specific to each dendrimer family or series. These properties are determined by the size, shape, and multiplicity of the construction components that are used for the core, interior, and surface of the dendrimer (Figure 20.9). Higher multiplicity components and those that contribute to tethered congestion will hasten the development of container properties or rigid surface scaffolding as a function of generation. Within the PAMAM dendrimer family, these periodic properties are generally manifested in three phases as shown in Figure 20.16.

The earlier generations (i.e., G = 0–3) exhibit no well-defined interior characteristics, whereas interior development related to geometric closure is observed for the intermediate generations (i.e., G = 4–6/7). Accessibility and departure from the interior is determined by the size and gating properties of the surface groups. At higher generations (i.e., G ≥ 7), where de Gennes dense packing is severe, rigid scaffolding properties are observed, allowing relatively little access to the interior except for very small guest molecules. The site isolation and encapsulation properties of dendrimers have been reviewed recently by Tomalia et al.,[61] Hecht and Fréchet,[62] and Meijer et al.[63]

20.3.3   Amplification and Functionalization of Dendrimer Surface Groups

Dendrimers within a generational series can be expected to present their terminal groups in at least three different modes: *flexible, semi-flexible,* or *rigid functionalized* scaffolding. Based on mathematically defined dendritic branching rules (i.e., $Z = N_c N_b{}^G$), the various surface presentations become more congested and rigid as a function of increasing generation levels. It is implicit that this surface amplification may be designed to control gating properties associated with unimolecular container development.

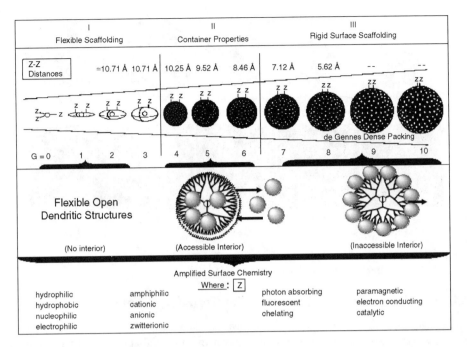

FIGURE 20.16 Periodic properties for PAMAM dendrimers as a function of generation G = 0–10: (I) flexible scaffolding (G = 0–3); (II) container properties (G = 4–6); and (III) rigid surface scaffolding (G = 7–10). Various chemo/physical dendrimer surfaces amplified according to: Z – $N_c N_b{}^G$ where: N_c = core multiplicity, N_b = branch cell multiplicity, G = generation. (From Esfand, R.; Tomalia, D.A. Laboratory synthesis of poly(amidoamine) (PAMAM) dendrimers. In *Dendrimers and Other Dendritic Polymers*; Fréchet, J.M.J. and Tomalia, D.A., Eds.; John Wiley & Sons: West Sussex, 2001. With permission.)

Furthermore, dendrimers may be viewed as versatile, nanosized objects that can be surface functionalized with a vast array of chemical and application features (Figure 20.16). The ability to control and engineer these parameters provides an endless list of possibilities for utilizing dendrimers as modules for nanodevice design.[24,46,64,65] Recent reviews have begun to focus on this area.[35,62,65–67]

20.3.4 Nanoscale Dimensions and Shapes Mimicking Proteins

In view of the extraordinary structure control and nanoscale dimensions observed for dendrimers, it is not surprising to find extensive interest in their use as globular protein mimics.[24] Based on their systematic, dimensional length scaling properties (Figure 20.17) and electrophoretic/hydrodynamic[50,52] behavior, they are referred to as *artificial proteins*.[24,61] Substantial effort has been focused recently on the use of dendrimers for "site isolation" mimicry of proteins[35] and enzyme-like catalysis,[68] as well as for other biomimetic applications,[24,69] drug delivery,[61] surface engineering,[70] and light harvesting.[71,72] These fundamental properties have in fact led to their commercial use as globular protein replacements for gene therapy, immunodiagnostics,[73,74] and a variety of other biological applications described below.

20.4 Dendrimers as Nanopharmaceuticals and Nanomedical Devices

Many promising biomedical applications for these compounds have been discovered and several are currently under development.[69] A current overview of dendrimer applications in the wet nanotechnology area is described below.

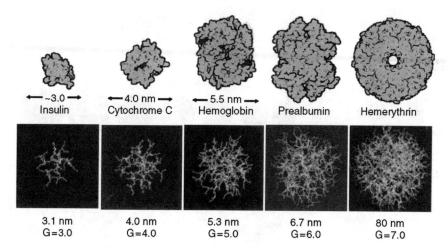

FIGURE 20.17 Comparison of selected proteins showing the close dimensional size/scaling (nm) to respective generations [-ammonia core]; *dendri*-poly(amidoamine) dendrimer.

20.4.1 Dendrimers as Genetic Material Transfer Agents

Antisense oligonucleotides have been shown to down-regulate the expression of specific genes in both *in vivo* and *in vitro* systems.[75,76] Gene regulation has many potential applications in treating disease states such as cancer, tissue graft survival, multiple drug resistance, and other conditions where it would be beneficial to reduce or eliminate messages produced from genes causing the adverse effect. The main drawback with antisense treatment is the delivery of "naked" nucleic material to the cell and, in particular, the cell nucleus. Technologies using viral[77,78] and synthetic cationic lipid preparations[79] have been shown to improve antisense delivery but demonstrate severe limiting side effects *in vivo*. Viral vectors have immunogenicity and cell targeting problems,[80] whereas the cationic lipids cause inflammatory reactions and cytotoxicity.[81,82]

The application of PAMAM dendrimers as transfection agents for antisense oligonucleotides and plasmids was described by Tomalia and co-workers in 1996.[83,84] They found that stable DNA–dendrimer complexes formed and could successfully suppress luciferase expression in an *in vitro* cell culture system.[83] Further work by Kukowska–Latallo et al.,[85] Baker et al.,[86] and DeLong et al.[87] provided additional evidence that the stability of these DNA–dendrimer complexes was dependent on electrostatic charge of the PAMAM dendrimer and that the complexes formed were stable at a range of pH values. These complexes were also stable to restriction endonucleases.

Thus, PAMAM and, to some extent the poly(lysine) dendrimers,[88] overcame some of the problems associated with delivery of nucleic acids intracellularly. In preliminary studies, Roberts et al.[89] showed that several generations of the PAMAM dendrimers elicited no immunogenicity and possessed very low *in vivo* toxicity. The bioavailability results were somewhat unusual, with all methylated dendrimers showing high pancreatic uptake. However, more recent evaluation of the toxicity and bioavailability of PAMAM dendrimers has been reported.[90]

20.4.2 Dendrimer–Carbohydrate Conjugates for Polyvalent Recognition

Carbohydrate–protein interactions play a vital role in many biological functions such as cell adhesion, receptor-mediated events, cellular recognition processes, and microbial–host cell interactions. These individual sugar–receptor complexes are often weak and nonselective.[91] Pioneering work by Roy et al., Stoddart et al., Lindhorst et al., and Okada et al. led to the development of a special class of dendrimers known as carbohydrate dendrimers (see review by Jayaraman et al.[92]). These carbohydrate-linked dendrimers had several sugar groups attached to the outside of the dendrimer shell, which led to a greatly enhanced affinity and selectivity of the sugar–receptor reaction, believed to be mediated via polyvalent (multivalent) interactions (for a review of polyvalent interactions, see Mammen et al.[93]). This enhanced affinity has been demonstrated for carbohydrate-linked poly(lysine),[94] PAMAM,[95] and poly(propylene-imine) (PPI)[96] dendrimers. Many potential biomedical applications exist for the concentration of glyco-dendrimers as anti-viral,[97] bacterial species,[98] and anti-cell adhesion agents in general.[99]

20.4.3 Dendrimers as Targeted Drug Delivery Agents

Dendrimers may be used as hosts or *molecular* or *dendritic boxes* to encapsulate guest molecules and transport them to biological targets.[105] The use of a PPI dendrimer as a host for the dye Bengal Rose was first described by Meijer and co-workers.[100–102] Under certain conditions, the guest molecules can slowly diffuse out of the dendritic host, although large molecules such as Bengal Rose require quite aggressive denaturing of the dendrimer before release occurs. There are many potential applications of dendrimers as drug delivery vehicles, and several of these have been described by Tomalia and co-workers[61,69] and Behr.[103] In a broad patent issued in 1996, Tomalia and colleagues[104] claimed dendrimer polymers as complexing and carrier materials for biological agents, including genes, DNA for transfection, IgG and such agents as the herbicides 2,4-dichlorophenoacetic acid, and abscissic acid.[104]

Additionally, the surface of the dendrimer may possess recognition functionality covalently attached suitable for targeting the dendrimer–guest complexes to a specific biological destination.[105] An example of this application was shown by Moroder and co-workers[106] where antibody-labeled dendrimers containing radioactive boron isotopes were targeted to tumor cells. Barth et al.[107] had previously shown that incurable forms of glioblastoma cancer could be treated with ^{10}B isotope containing dendrimers, which concentrated high levels of neutron-capture agents within the tumor and reduced the tumor size. Duncan and co-workers[108] demonstrated that a PAMAM dendrimer conjugated to cisplatin (producing a dendrimer–platinate) was highly water soluble and released platinum slowly *in vitro*. When tested *in vivo*, the dendrimer–platinate displayed greater anti-tumor activity than naked cisplatin and exhibited lower toxicity. Jansen et al.[109] conjugated a PPI dendrimer with platinum in order to overcome cisplatin resistance in cancer cells. In preliminary *in vitro* tests, this PPI–platinum complex showed low cytotoxicity in a range of human tumor cell lines.

20.4.4 Dendrimers as Magnetic Resonance Imaging Contrast Agents

Certainly dendrimers can play an important role as both tumor diagnosis and therapeutic agents because they are capable of incorporating large amounts of paramagnetic or radioactive metal ions into their interiors. The use of dendrimers to treat tumors as neutron capture agents has been described above; however, a similar principle can be used to image various tumors and cancers.

An example of such a use is that offered by folate-conjugated PAMAM dendrimer chelates containing gadolinium that were used to target tumor cells.[110] These folate-conjugated chelates targeted the complex to folate binding proteins on the tumor cell surfaces. Such high-affinity folate binding sites are up-regulated on the cell surface of many human cancers of epithelial origin.[79]

Margerum et al. have also shown that PAMAM–gadolinium complexes have a reduced liver uptake and increased bioavailability when grafted with polyethylene glycol (PEG) side chains.[111]

20.4.5 Dendrimers as Antiviral Agents

Currently there have been few reports of the use of PAMAM or poly(lysine) dendrimers as potential antiviral or antimicrobial therapeutics. Reuter et al.[97] reported the use of PAMAM dendrimers with sialic acid functional groups as effective inhibitors of influenza virus binding to host cell receptors and noted the potential for these compounds to be used therapeutically. Whitesides and co-workers[112] had previously shown that sialic acid containing polyacrylamide inhibitors prevented the attachment of influenza virus to mammalian erythrocytes. They proposed that the underlying mechanism for this inhibition was due to polyvalent interactions coupled with steric stabilization by the dendrimers to prevent the interaction between the virus and the erythrocyte.

Polyanionic compounds such as dextran sulfate, heparin, and suramin inhibit HIV viral attachment by blocking the binding of the viral envelope glycoprotein gp120 to the cellular CD4 receptor of T lymphocytes. Thus, these polyanionic compounds can inhibit virus binding to cells, virus-induced syncytium formation, and, consequently, viral transmission.[113,114]

Heparin and dextran sulfate, which are both sulfated polysaccharides, have been shown to inhibit the replication of several viruses including HIV, HSV, and cytomegalovirus (CMV), as well as several strains of bacteria.[115–119] The sulfated carbohydrates are believed to block the adhesion to cell surfaces by the pathogens. Several studies suggest that a common cell surface molecule, heparin sulfate (HS), a glycosaminoglycan (GAG), is involved.[119–121]

In recent work a series of PAMAM and poly(lysine) dendrimers were developed and tested as antiviral agents.[122] This dendrimer series contained aryl-sulphonic acid salts on the outer surface and exhibited antiviral/antimicrobial activity to a broad range of viruses and other micro-organisms. It has been postulated that these dendrimers mimic several of the biological activities of heparin[129] but do not possess some of the attributes such as anticoagulant activity, which limits the usefulness of heparin as an anti-infective agent.[130] Wivrouw et al.[122,123] utilized *time of addition* studies to determine the mode of action of these dendrimers. They found that specific PAMAM dendrimers inhibit viral attachment to cells as well as the action of HIV viral reverse transcriptase and virally encoded integrase, which occur intracellularly during the virus replicative cycle. It is interesting to note, however, that some dendrimers showed inhibition of viral attachment only.

In addition to inhibition of HIV viral attachment, the dendrimers also inhibit adsorption of certain enveloped viruses such as herpes simplex virus 2 (HSV)[124] and respiratory syncytial virus (RSV)[125,126] in cell culture at very low concentrations (0.1–1 μg/ml). In recent work, PAMAM and poly(lysine) dendrimers were also shown to inhibit human cytomegalovirus (HCMV), Ebola virus, Hepatitis B virus (HBV), Influenza A and B, Epstein–Barr virus (EbV), and Adeno- and Rhino-viruses.[127]

20.4.6 Dendrimers as Angiogenesis Inhibitors

Angiogenesis, the formation of new blood vessels, is a tightly controlled process that occurs rarely in adult tissue and is limited to wound healing and the female reproductive cycle. Uncontrolled angiogenesis plays a role in the pathogenesis of a number of major diseases including cancer, arthritis, psoriasis, and ocular diseases.

Polyanionic dendrimers are claimed as angiogenesis inhibitors in a patent by Matthews and Holan.[128] It was postulated that these dendrimers mimic heparin or heparin sulfate, sequestering fibroblast growth factor (FGF) and possibly other growth factors, thereby disrupting the formation of new blood vessels.[129]

While having heparin-like properties, polyanionic dendrimers do not possess the anticoagulant activity associated with heparin or other antiangiogenic compounds used as heparin mimetics.[130] These large amino acid-containing structures are nonantigenic when compared with other peptide-based macromolecules.

The results of chorioalantoic membrane (CAM), rat aorta, human umbilical vein endothelial cells (HUVEC), and blood vessel growth inhibition assays demonstrate the potential of these polyanionic dendrimers as angiogenesis inhibitors.[129]

20.4.7 Dendrimers as Antitoxin Agents

Thompson and Schengrund[131,132] reported the use of oligosaccharide-derivatized dendrimers as bacterial enterotoxin inhibitors. The authors have also patented several oligosaccharide-linked PAMAM dendrimers as potential antibacterial and antiviral compounds.[133] They attributed the antitoxin effects of the dendrimers to multivalent (or polyvalent) interactions leading to neutralization of the bacterial toxins. Roy[134] also described the potential use of multivalent glycoconjugates as bacterial toxin inhibitors. Pieters et al.[135] demonstrated that lactose-containing dendrimers bound, in a multivalent fashion, to the cholera toxin B subunit. There have been many recent publications that describe the use of carbohydrate ligands (often toxin receptor analogues) linked to a polymer/dendrimer backbone, which enables a multivalent inhibition of the toxin–cell interaction.[136–139]

In another approach, the use of PAMAM and poly(lysine) dendrimers functionalized with polyanionic groups has been claimed to inhibit a range of viral and bacterial toxins as well as a number of crude snake venoms.[140]

20.4.8   Dendrimers as Antiprion Agents

Recently, Prusiner and co-workers[141] demonstrated the ability of PAMAM and PPI dendrimers to reduce PrP^{Sc}, the protease-resistant isoform of the prion protein, from cultured scrapie-infected neuroblastoma cells to undetectable levels as shown by Western blot analysis.[141] Structure–activity analysis showed that in acidic environments, the dendrimers altered the structure of the PrP^{Sc} and rendered it susceptible to protease degradation.

Further work by Prusiner[142] demonstrated the ability of PAMAM and PPI dendrimers not only to reduce the PrP^{Sc} to undetectable levels but to also eradicate prion infectivity as shown by a bioassay in mice. Additionally, PPI dendrimers denatured the PrP^{Sc} from bovine spongiform encephalopathy-infected brain homogenates; however, PrP^{Sc} from sheep scrapie-infected brain homogenates were resistant. Thus, the dendrimer-induced denaturation of prion protein may be strain-dependent.

This pioneering work by Prusiner and colleagues showed dendrimers to be the first class of compounds that could cure prion infection in living cells. This work has implications not only for the treatment of prion-based diseases but also for important applications in the sterilization of equipment and in diagnostic techniques.

20.4.9   Other Potential Biomedical Applications of Dendrimers

Tam et al.[143,144] demonstrated that peptide sequences that are too light to cause the production of antibodies within a host organism can be functionalized onto a dendrimer backbone to produce a highly antigenic compound called a *multiple antigen peptide* (MAP) dendrimer. This unique use of the dendrimer structure has vast implications for the production of vaccines from important peptide sequences (i.e., haptens from viruses) previously thought incapable of producing immunogenicity.

Dendrimers that act as catalysts, named *dendrizymes*, have been described by Brunner.[145] These dendritic catalysts were made to model enzymes and contained an organometallic active site buried in the branches of the dendritic structures. Initial work carried out by Diedrich and co-workers[146–148] and later by Fréchet and co-workers[149,150] utilized dendrimeric porphyrins as models for enzyme systems such as cytochrome *c*. Jiang and Aida[151] described the use of dendrimeric porphyrins as hemoglobin mimics. Recent advances in the synthesis of peptide dendrimers hold promise that dendrimers could be tailored to act as artificial enzymes in biological systems.

20.5   Dendrimers as Reactive Modules for the Synthesis of More Complex Nanoscale Architectures

20.5.1   Megamers, Saturated Shell, and Partial Shell-Filled Core-Shell Tecto(dendrimers)

In the first full paper published on dendritic polymers,[30] dendrimers were defined as "reactive, structure-controlled macromolecular building blocks." It was proposed that they could be used as repeat units for the construction of a new class of topological macromolecules now referred to as *megamers*.[24,152] Although there is intense activity in the field of dendrimer science, there have been relatively few references focused on this specific concept.[33,46,152–154] The generic term *megamers* has been proposed to describe those new architectures derived from the combination of two or more dendrimer molecules (Figure 20.18).[24,152,153]

Examples of both statistical[152] and structure-controlled[24,153,155] megamer assemblies have been reported and reviewed recently. Covalent oligomeric assemblies of dendrimers (i.e., dimers, trimers, etc.) are well-

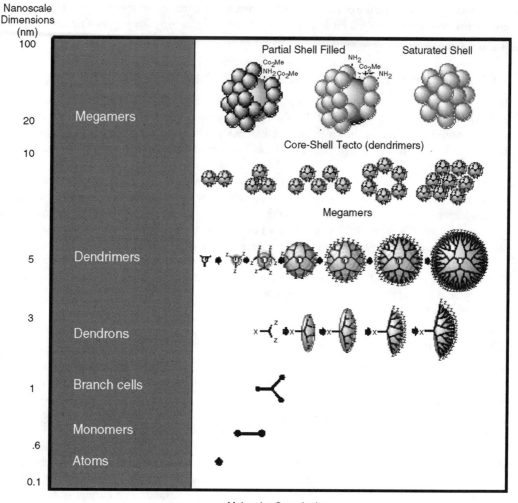

FIGURE 20.18 Approximate nanoscale dimensions as a function of atoms, monomers, branch cells, dendrimers, and megamers.

documented examples of low molecular weight megamers. Statistical megamer assemblies have been reported as both *supramacromolecular*[65,152] and *supermacromolecular*[36] (covalent) topologies.

Recently, mathematically defined, structure-controlled, covalent megamers have been reported. They are a major subclass of megamers referred to as *core–shell tecto(dendrimers)*.[24,155–157] Synthetic methodologies for these new architectures have been reported to produce precise megameric structures that adhere to mathematically defined bonding rules.[24,46,155,158] It appears that *structure-controlled nanoscale complexity beyond dendrimers* is now possible. The demonstrated ability to mathematically predict and synthesize structure-controlled dendrimer assemblies provides a broad concept for the systematic construction of nanostructures with dimensions that could span the entire nanoscale region (Figure 20.18).

20.6   Conclusions

In summary, using strictly abiotic methods, it has been widely demonstrated over the past decade that dendrimers[15] can be routinely constructed with control that rivals the structural regulation found in biological systems. The close scaling of size,[61,159] shape, and quasi-equivalency of surfaces[160–162] observed between nanoscale biostructures and various dendrimer families/generational levels are both striking and provocative.[61,62,159–165] These remarkable similarities suggest a broad strategy based on rational biomimicry as a means for creating a repertoire of structure-controlled, size- and shape-variable dendrimer assemblies. Successful demonstrations of such a biomimetic approach have proven to be a versatile and powerful synthetic strategy for systematically accessing virtually any desired combination of size, shape, and surface in the nanoscale region. Future extensions will involve combinational variation of dendrimer module parameters such as (1) families (interior compositions), (2) surfaces, (3) generational levels, or (4) architectural shapes (i.e., spheroids, rods, etc.).

Acknowledgments

We express deep appreciation to Linda S. Nixon for graphics and manuscript preparation.

References

1. Anderson, P.W. *Science* 1972, *177*, 393–396.
2. Smalley, R. Testimony to U.S. Science and Technology Office, 1999.
3. Richardson, C.F.; Schuster, D.I.; Wilson, S.R. *Org. Lett.* 2000, *2*, 1011–1014.
4. Morawetz, H. *Polymers. The Origin and Growth of a Science*; John Wiley: New York, 1985.
5. Elias, H.-G. *An Introduction to Polymer Science*; VCH: Weinheim, 1997.
6. Corey, E.J.; Cheng, X.-M. *The Logic of Chemical Synthesis*; John Wiley: New York, 1989.
7. Merrifield, R.B.; Barany, G. *The Peptides*; Academic Press: New York, 1980; Vol.2.
8. Merrifield, R.B. *Angew. Chem. Int. Ed. Engl.* 1985, *24*, 799.
9. Bodanszky, M. *Principles of Peptide Synthesis*; Springer-Verlag: Berlin, 1984.
10. Bodanzky, M.; Bodanzky, A.; Springer-Verlag: Berlin, 1984.
11. Berzelius, J. *J. Fortsch. Phys. Wissensch.* 1832, *11*, 44.
12. Staudinger, H. *Schweiz. Chem. Z.* 1919, 105, 28–33, 60–64.
13. Staudinger, H. *Ber.* 1920, *53*, 1073.
14. Staudinger, H. *From Organic Chemistry to Macromolecules, A Scientific Autobiography*; John Wiley & Sons: New York, 1961.
15. Fréchet, J.M.J.; Tomalia, D.A. *Dendrimers and Other Dendritic Polymers*; John Wiley & Sons, Ltd.: West Sussex, 2001.
16. Elias, H.-G. *Mega Molecules*; Springer Verlag: Berlin, 1987.
17. Roovers, J. *Advances in Polymer Science, Branched Polymers I*; Springer-Verlag: Berlin, 1999; Vol. 142.
18. Roovers, J. *Advances in Polymer Science, Branched Polymers II*; Springer-Verlag: Berlin, 2000; Vol. 143.
19. *Metallocene-Based Polyolefins*; John Wiley & Sons, Ltd.: Brisbane, 2000; Vols. 1 and 2.
20. Sunder, A.; Heinemann, J.; Frey, H. *Chem. Eur. J.* 2000, *6*, 2499.
21. Lothian–Tomalia, M.K.; Hedstrand, D.M.; Tomalia, D.A. *Tetrahedron* 1997, *53*, 15495–15513.
22. Dusek, K.; Duskova-Smrckova, M. Formation, structure and properties and the crosslinked state relative to precursor architecture. In *Dendrimers and Dendritic Polymers*; John Wiley & Sons, Ltd.: West Sussex, 2001; pp. 111–145.
23. Goodsell, D.S. *Am. Sci.* 2000, *88*, 230–237.
24. Tomalia, D.A.; Brothers II, H.M.; Piehler, L.T.; Durst, H.D.; Swanson, D.R. *Proc. Natl. Acad. Sci. USA* 2002, 99, 8, 5081–5087.
25. Tomalia, D.A. *Macromol. Symp.* 1996, *101*, 243–255.
26. Tomalia, D.A.; Brothers II, H.M.; Piehler, L.T.; Hsu, Y. *Polym. Mater. Sci. Eng.* 1995, *73*, 75.
27. Naj, A.K. *Persistent Inventor Markets a Molecule*: New York, 1996, *Wall Street Journal*. p. B1.
28. Tomalia, D.A.; Hall, M.; Hedstrand, D.M. *J. Am. Chem. Soc.* 1987, *109*, 1601–1603.
29. Tomalia, D.A.; Dewald, J.R.; Hall, M.J.; Martin, S.J.; Smith, P.B., Kyoto, Japan, Society of Polymer Science (SPSJ), First International Conference Reprints, August 1984, p. 65.

30. Tomalia, D.A.; Baker, H.; Dewald, J.; Hall, M.; Kallos, G.; Martin, S.; Roeck, J.; Ryder, J.; Smith, P. *Polym. J. (Tokyo)* 1985, *17*, 117–132.
31. Buhleier, E.; Wehner, W.; Vögtle, F. *Synthesis* 1978, 155–158.
32. Moors, R.; Vogtle, F. *Chem. Ber.* 1993, *126*, 2133–2135.
33. Tomalia, D.A. *Sci. Am.* 1995, *272*, 62–66.
34. Esfand, R.; Tomalia, D.A. Laboratory synthesis of poly(amidoamine) (PAMAM) dendrimers. In *Dendrimers and Other Dendritic Polymers*; Fréchet, J.M.J. and Tomalia, D.A., Eds.; John Wiley & Sons: West Sussex, 2001; pp. 587–604.
35. Tomalia, D.A.; Naylor, A.M.; Goddard III, W.A. *Angew. Chem. Int. Ed. Engl.* 1990, *29*, 138–175.
36. Tomalia, D.A.; Hedstrand, D.M.; Wilson, L.R. Dendritic polymers. In *Encyclopedia of. Polymer Science and Engineering,* 2nd ed.; John Wiley & Sons:, 1990; Index Volume, pp. 46–92.
37. Newkome, G.R.; Yao, Z.-Q.; Baker, G.R.; Gupta, V.K. *J. Org. Chem.* 1985, *50*, 2003–2004.
38. Padias, A.B.; Hall, Jr., H.K.; Tomalia, D.A. *J. Org. Chem.* 1987, *52*, 5305.
39. Newkome, G.R.; Moorfield, C.N.; Vögtle, F. *Dendritic Molecules*; VCH: Weinheim, 1996.
40. Hawker, C.J.; Fréchet, J.M.J. *J. Am. Chem. Soc.* 1990, *112*, 7638–7647.
41. Hawker, C.J.; Frechet, J.M.J. *J. Chem. Soc. Chem. Commun.* 1990, 1010–1013.
42. Miller, T.M.; Neenan, T.X. *Chem. Mater.* 1990, *2*, 346–349.
43. Fréchet, J.M.J. *Science* 1994, *263*, 1710–1715.
44. Hawker, C.J.; Fréchet, J.M.J. *J. Am. Chem. Soc.* 1990, *112*, 7638.
45. Zeng, F.; Zimmerman, S.C. *Chem. Rev.* 1997, *97*, 1681–1712.
46. Tomalia, D.A. *Adv. Mater.* 1994, *6*, 529–539.
47. Kallos, G.J.; Tomalia, D.A.; Hedstrand, D.M.; Lewis, S.; Zhou, *J. Rapid Commun. Mass Spectrom.* 1991, *5*, 383–386.
48. Dvornic, P.R.; Tomalia, D.A. *Macromol. Symp.* 1995, *98*, 403–428.
49. Hummelen, J.C.; van Dongen, J.L.J.; Meijer, E.W. *Chem. Eur. J.* 1997, *3*, 1489–1493.
50. Brothers II, H.M.; Piehler, L.T.; Tomalia, D.A. *J. Chromatogr. A* 1998, *814*, 233–246.
51. Tomalia, D.A.; Dewald, J. R.: U.S. Patent, 4,587,329 (1986).
52. Zhang, C.; Tomalia, D.A. Gel electrophoresis characterization of dendritic polymers. In *Dendrimers and Other Dendritic Polymers*; Fréchet, J.M.J. and Tomalia, D.A., Eds.; John Wiley & Sons: West Sussex, 2001; pp. 239–252.
53. Naylor, A.M.; Goddard III, W.A.; Keifer, G.E.; Tomalia, D.A. *J. Am. Chem. Soc.* 1989, *111*, 2339–2341.
54. Turro, N.J.; Barton, J.K.; Tomalia, D.A. *Acc. Chem. Res.* 1991, *24 (11)*, 332–340.
55. Gopidas, K.R.; Leheny, A.R.; Caminati, G.; Turro, N.J.; Tomalia, D.A. *J. Am. Chem. Soc.* 1991, *113*, 7335–7342.
56. Ottaviani, M.F.; Turro, N.J.; Jockusch, S.; Tomalia, D.A. *J. Phys. Chem.* 1996, *100*, 13675–13686.
57. Jockusch, J.; Ramirez, J.; Sanghvi, K.; Nociti, R.; Turro, N.J.; Tomalia, D.A. *Macromolecules* 1999, *32*, 4419–4423.
58. Hawker, C.J.; Wooley, K.L.; Frèchet, J.M.J. *J. Am. Chem. Soc.* 1993, *115*, 4375.
59. de Gennes, P.G.; Hervet, H.J. *J. Physique-Lett. (Paris)* 1983, *44*, 351–360.
60. Mourey, T.H.; Turner, S.R.; Rubinstein, M.; Frechet, J.M.J.; Hawer, C.J.; Wooley, K.L. *Macromolecules* 1992, *25*, 2401–2406.
61. Esfand, R.; Tomalia, D.A. *Drug Discovery Today* 2001, *6 (8)*, 427–436.
62. Hecht, S.; Fréchet, J.M.J. *Angew. Chem. Int. Ed.* 2001, *40(1)*, 74–91.
63. Weener, J.-W.; Baars, M.W.P.L.; Meijer, E.W. Some unique features of dendrimers based upon self-assembly and host-guest properties. In *Dendrimers and Other Dendritic Polymers*; Fréchet, J.M.J. and Tomalia, D.A., Eds.; John Wiley & Sons Ltd.: West Sussex, 2001; pp. 387–424.
64. de A.A. Soler-Illia, G.J.; Rozes, L.; Boggiano, M.K.; Sanchez, C.; Turrin, C.-O.; Caminade, A.-M.; Majoral, J.-P. *Angew. Chem. Int. Ed.* 2000, *39*, 4250.
65. Tomalia, D.A.; Majoros, I. Dendrimeric supramolecular and supra*macro*molecular assemblies. In *Supramolecular Polymers*; Ciferri, A., Ed.; Marcel Dekker: New York, 2000; pp. 359–434.

66. Crooks, R.M.; Lemon III, B.; Sun, L.; Yeung, L.K.; Zhao, M. Dendrimer-encapsulated metals and semiconductors: synthesis, characterization, and applications. In *Topics in Current Chemistry, Vol. 212*; Springer-Verlag: Berlin, 2001.

67. Freeman, A.W.; Koene, S.C.; Malenfant, P.R.L.; Thompson, M.E.; Frechet, J.M.J. *J. Am. Chem. Soc.* 2000, *122*, 1285–12386.

68. Piotti, M.E.; Rivera, F.; Bond, R.; Hawker, C.J.; Frèchet, J.M.J. *J. Am. Chem. Soc.* 1999, *121*, 9471.

69. Bieniarz, C. Dendrimers: applications to pharmaceutical and medicinal chemistry. In *Encyclopedia of Pharmaceutical Technology*; Marcel Dekker: 1998; Vol. 18; pp. 55–89.

70. Tulley, D.C.; Fréchet, J.M.J. *Chem. Commun.* 2001, 1229.

71. Androv, A.; Fréchet, J.M.J. *Chem. Commun.* 2000, 1701.

72. Jiang, D.-L.; Aida, T. Dendritic polymers: optical and photochemical properties. In *Dendrimers and Other Dendritic Polymers*; Fréchet, J.M.J. and Tomalia, D.A., Eds.; John Wiley & Sons: West Sussex, 2001; pp. 425–439.

73. Singh, P.; Moll III, F.; Lin, S.H.; Ferzli, C. *Clin. Chem.* 1996, *42 (9)*, 1567–1569.

74. Singh, P. Dendrimer-based biological reagents: preparation and applications in diagnostics. In *Dendrimers and Dendritic Polymers*; Fréchet, J.M.J. and Tomalia, D.A., Eds.; John Wiley & Sons, Ltd.: West Sussex, 2001; pp. 463–484.

75. Goodchild, Y. *Oligonucleotides, Antisense Inhibitors or Gene Expression*; Cohen, J., Ed.; Macmillan: London, 1989; pp. 53–77.

76. Randa, K.; Poteete, A.R. *Genes Dev.* 1993, *7*, 1490–1507.

77. Ch'ng, J.L.; Mulligan, R.C.; Schimmel, P.; Holmes, E.W. *Proc. Natl. Acad. Sci. USA* 1989, *86*, 10006–10010.

78. Roessler, B.Y.; Hartman, J.W.; Vallance, D.K.; Latta, Y.M.; Janich, S.L.; Davidson, B.L. *Hum. Gene Ther.* 1995, *6*, 307–316.

79. Gareis, M.; Harrer, P.; Bertling, W.M. *Cell. Mol. Biol.* 1991, *37*, 191–203.

80. Yang, Y.; Li, Q.; Ertl, H.C.; Wilson, J. M. *J. Virol.* 1995, *69*, 2004.

81. Wagner, R.W.; Matteucci, M.D.; Lewis, J.G.; Gutierrez, A.J.; Moulds, C.; Froehler, B.C. *Science* 1993, *260*, 1510–1513.

82. Logan, J.J.; Bebok, Z.; Walter, L.C.; Peng, S.; Felgner, P.L.; Siegal, G.P.; Frizzell, R.A.; Dong, J.; Howard, M. *Gene Ther.* 1995, *2*, 38.

83. Bielinska, A.; Kukowska–Latallo, J.F.; Johnson, J.; Tomalia, D.A.; Baker, Jr., J.R. *Nucleic Acids Res.* 1996, *24*, 2176–2182.

84. Tomalia, D.A.; Baker, J.R.; Cheng, R.; Bielinska, A.U.; M.J.F.; Hedstrand, D.M.; Johnson, J.J.; Kaplan, D.A.; Klakamp, S.L.; Kruper, Jr., W.J.; Kukowska–Latallo, J.; Maxon, B.D.; Piehler, L.T.; Tomlinson, I.A.; Wilson, L.R.; Yin, R.; Brothers II, H.M. *Bioactive and/or Targeted Dendrimer Conjugates*: U.S. Patent 5,714,166, February 3, 1998.

85. Kukowska–Latallo, J.F.; Bielinska, A.U.; Johnson, J.; Spindler, R.; Tomalia, D.A.; Baker, Jr., J.R. *Proc. Natl. Acad. Sci. USA* 1996, *93*, 4897–4902.

86. Eichman, J.D.; Bielinska, A.U.; Kukowska–Latallo, J.F.; Donovan, B.W.; Baker, Jr., J.R. Bioapplications of PAMAM dendrimers. In *Dendrimers and Other Dendritic Polymers*; Fréchet, J.M.J. and Tomalia, D.A., Eds.; John Wiley & Sons Ltd.: West Sussex, 2001; pp. 441–461.

87. DeLong, R.; Stephenson, K.; Loftus, T.; Fisher, M.; Alahari, S.; Nolting, A.; Juliano, R.L. *J. Pharm. Sci.* 1997, *86*, 762–74.

88. Wu, C.H.; Wilson, J.M.; Wu, G.Y. *J. Biol. Chem.* 1989, *264*, 16985.

89. Roberts, J.C.; Bhalgat, M.K.; Zera, R.T. *J. Biomed. Mater. Res.* 1996, *30*, 53–65.

90. Malik, N. *J. Control. Release* 2000, *65*, 133–148.

91. Kiessling, L.; Pohl, N.L. *Chem. Biol.* 1996, *3*, 71–77.

92. Jayaraman, N.; Nepogodiev, S.A.; Stoddard, J.F. *Chem. – Eur. J.* 1997, *3*, 1193–1199.

93. Mammen, M.; Choi, S.-K.; Whitesides, G.M. *Angew. Chem. Int. Ed.* 1998, *37*, 2754–2794.

94. Roy, R.; Zanini, D.; Meunier, S.J.; Romanowska, A. *J. Chem. Soc. Commun.* 1993, 1869–1872.

95. Zanini, D.; Roy, R. *J. Am. Chem. Soc.* 1997, *119*, 2088–2095.

96. Peerlings, H.W.I.; Nepogodiev, S.A.; Stoddard, J.F.; Meijer, E.W. *Eur. J. Org. Chem.* 1998, 1879–1886.

97. Reuter, J.D.; Myc, A.; Hayes, M.M.; Gan, Z.; Roy, R.; Qin, D.; Yin, R.; Piehler, L.T.; Esfand, R.; Tomalia, D.A.; Baker, Jr., J.R. *Bioconjugate Chem.* 1999, *10*, 271–278.

98. Hansen, H.C.; Haataja, S.; Finne, J.; Magnusson, G. *J. Am. Chem. Soc.* 1997, *119*, 6974–6979.

99. Roy, R. Dendritic and hyperbranched glycoconjugates as biomedical anti-adhesion agents. In *Dendrimer and Dendritic Polymers*; Fréchet, J.M.J. and Tomalia, D.A., Eds.; John Wiley & Sons Ltd.: West Sussex, 2001; pp. 361–385.

100. Jansen, J.F.G.A.; de Brabander–van den Berg, E.M.M.; Meijer, E.W. *Science* 1994, *266*, 1226–1229.

101. Jansen, J.F.G.A.; Meijer, E.W.; de Brabander–van den Berg, E.M.M. *J. Am. Chem. Soc.* 1995, *117*, 4417–4418.

102. Jansen, J.F.G.A.; Janssen, R.A.J.; de Brabender–van den Berg, E.M.M.; Meijer, E.W. *Adv. Mater.* 1995, *7*, 561–564.

103. Behr, J.-P. *Acc. Chem. Res.* 1993, *26*, 274–278.

104. Tomalia, D.A.; Wilson, L.R.; Hedstrand, D.M.; Tomlinson, I.A.; Fazio, M.J.; Kruper, Jr., W.J.; Kaplan, D.A.; Cheng, R.C.; Edwards, D.S.; Jung, C.W. *Dense Star Polymer Conjugates*: U.S. Patent 5,527,524, June 18, 1996.

105. Baker, Jr., J.R.; Quintana, L.; Piehler, L.; Banazak–Holl, M.; Tomalia, D.A.; Raczka, E. *Biomed. Microdevices* 2001, *3*, 61–69.

106. Qualman, B.; Kessels, M.M.; Musiol, H.-J.; Sierralta, W.D.; Jungblut, P.W.; Moroder, L. *Angew. Chem. Intl. Ed. Engl.* 1996, *35*, 909–911.

107. Barth, R.; Soloway, A.; Admas, D.; Alam, F. *Progress in Neutron Capture Therapy for Cancer*; Allen, B.J., Ed.; Plenum Press: New York, 1992; pp. 265–268.

108. Malik, N.; Evagorou, E.G.; Duncan, R. *Anti-Cancer Drugs* 1999, *10*, 767–776.

109. Jansen, B.A.J.; van der Zwan, J.; Reedijk, J.; den Dulk, H.; Brouwer, J. *Eur. J. Inorg. Chem.* 1999, 9, 1429–1433.

110. Wiener, E.C.; Konda, S.; Shadron, A.; Brechbiel, M.; Gansow, O. *Invest. Radiol.* 1997, *32*, 748–754.

111. Magerum, L.D.; Campion, B.K.; Koo, M.; Shargill, N.; Lai, J.-J.; Marumoto, A.; Sontum, P.C. *J. Alloys Compounds* 1997, *249*, 185–190.

112. Mammen, M.; Dahmann, G.; Whitesides, G.M. *J. Med. Chem.* 1995, *38*, 4179–4190.

113. de Clercq, E. Anti-HIV activity of sulfated polysaccharides. In *Carbohydrates and Carbohydrate Polymers, Analysis, Biotechnology, Modification, Antiviral, Medical and Other Applications*; Yalpani, M., Ed.; ATL Press: Mt. Prospect, IL, 1993; pp. 87–100.

114. de Clercq, E. *J. Med. Chem.* 1995, *38*, 2491–2517.

115. Ito, M.; Baba, M.; Pauwels, R.; De Clercq, E.; Shigeta, S. *Antiviral Res.* 1987, *7*, 361–367.

116. Ueno, R.; Kuno, S. *Lancet* 1987, *1*, 1379.

117. Baba, M.; Pauwels, R.; Balzarini, J.; Arnout, J.; Desmyter, J.; De Clerq, E. *Proc. Natl. Acad. Sci. USA* 1988, *85*, 6132–6136.

118. Neyts, J.; Snoeck, R.; Schols, D.; Balzarini, J.; Esko, J.D.; Van Schepdael, A.; de Clercq, E. *Virology* 1992, *189*, 48–58.

119. Herold, B.C.; Siston, A.; Bremer, J.; Kirkpatrick, R.; Wilbanks, G.; Fugedi, P.; Peto, C.; Cooper, M. *Antimicrob. Agents Chemother.* 1997, *41*, 2776–2780.

120. Rostand, K.S.; Esko, J.D. *Infect. Immun.* 1997, *65*, 1–8.

121. Patel, M.; Ynangishita, M.; Rodriquez, G.; Bou Habib, D.C.; Oravecz, T.; Hascall, V.C.; Norcross, M.A. *AIDS Res. Hum. Retroviruses* 1993, *9*, 167–174.

122. Witvrouw, M.; Fikkert, V.; Pluymers, W.; Matthews, B.; Mardel, K.W.; Schols, D.; Raff, J.; Debyser, Z.; DeClercq, E.; Holan, G.; Pannecouque, C. *Molecular Pharmacol.* 2000, *58*, 1100–1108.

123. Witvrouw, M.; Pannecouque, C.; Matthews, B.; Schols, D.; Andrei, G.; Snoeck, R.; Neyts, J.; Leyssen, P.; Desmyter, J.; Raff, J.; De Clercq, E.; Holan, G. *Antiviral Res.* 1999, 41, A41.

124. Bourne, N.; Stanberry, L.R.; Kern, E.R.; Holan, G.; Matthews, B.; Bernstein, D.I. *Antimicrobial Agents Chemother.* 2000, *44*, 2471–2474.

125. Barnard, D.L.; Sidwell, R.W.; Gage, T.L.; Okleberry, K.M.; Matthews, B.; Holan, G. *Antiviral Res.* 1997, 34, A88.

126. Barnard, D.L.; Matheson, J.E.; Morrison, A.; Sidwell, R.W.; Huffman, J.H.; Matthews, B.; Holan, G. *Proc. Intersci. Conf. Antimicrobial Agents Chemother. (ICAAC)* 1998.

127. Holan, G.; Matthews, B.; Korba, B.; DeClercq, E.; Witvrouw, M.; Kern, E.; Sidwell, R.; Barnard, D.; Huffman, J. *Antiviral Res.* 2000, 46, A55.

128. Holan, G.; Matthews, B. *Angiogenic Inhibitory Compounds*: U.S. Patent, 6,426,067 July 30, 2002.

129. Holan, G., Personal Communication.

130. Virgona, C., Personal Communication.

131. Thompson, J.P.; Schengrund, C.-L. *Glycoconjugate J.* 1997, 14, 837–845.

132. Thompson, J.P.; Schengrund, C.-L. *Biochem. Pharmacol.* 1998, 56, 591–597.

133. Schengrund, C.-L.; Thompson, J.P. Branched polymer-linked oligosaccharides and methods for treating and preventing bacterial and viral disease: *PCT Int. Appl.* WO9826662, 1998.

134. Roy, R. *Curr. Op. Struct. Biol.* 1996, 6, 692–702.

135. Vrasidas, I.; de Mol, N.J.; Liskamp, R.M. J.; Pieters, R.J. *Eur. J. Organic Chem.* 2001, 24, 4685–4692.

136. Kitov, P.I.; Sadowska, J.M.; Mulvey, G.; Armstrong, G.D.; Ling, H.; Pannu, N.S.; Read, R.J.; Bundle, D.R. *Nature* 2000, 403, 669–672.

137. Fan, E.; Merritt, E.A.; Verlinde, C.M.J.; Hol, W.G.H. *Curr. Op. Struct. Biol.* 2000, 10, 680–686.

138. Mourez, M.; Kane, R.S.; Mogridge, J.; Metallo, S.; Deschatelets, P.; Sellman, B.R.; Whitesides, G.M.; Collier, R.J. *Nature Biotechnol.* 2001, 19, 958–961.

139. Fan, E.; Zhang, Z.; Minke, W.E.; Hou, Z.; Verlinde, C.L.M.J.; Hol, W.G.J. *J. Am. Chem. Soc.* 2000, 122, 2663–2664.

140. Matthews, B.; Holan, G.; Mardel, K.W. Inhibition of Toxic Materials or Substances: Australian Provisional Patent PP5843/98, 1998.

141. Supattapone, S.; Nguyen, H.O.; Cohen, F.E.; Prusiner, S.B.; Scott, M.R. *Proc. Natl. Acad. Sci. USA* 1999, 96, 14529–14534.

142. Supattapone, W.; Wille, H.; Uyechi, L.; Safar, J.; Tremblay, P.; Szoka, F.C.; Cohen, F.E.; Prusiner, S.B.; Scott, M.R. *J. Virol.* 2001, 75, 3453–3461.

143. Tam, J.P.; Lu, Y.-A. *Proc. Nat. Acad. Sci. USA* 1989, 85, 9084–9088.

144. Tam, J.P. *Immunol. Meth.* 1996, 196, 17–32.

145. Brunner, H. *J. Organomet. Chem.* 1995, 500, 39–46.

146. Dandliker, P.J.; Diederich, F.; Gross, M.; Knobler, A.; Louati, A.; Sanford, E.M. *Angew. Chem. Intl. Ed. Engl.* 1994, 33, 1739–1742.

147. Dandliker, P.J.; Diederich, F.; Gisselbrecht, A.; Louati, A.; Gross, M. *Angew. Chem.* 1995, 107, 2906–2909.

148. Dandliker, P.J.; Diederich, F.; Gisselbrecht, A.; Louati, A.; Gross, M. *Angew. Chem. Intl. Ed. Engl.* 1995, 34, 2725–2728.

149. Pollack, K.W.; Sanford, E.M.; Fréchet, J.M.J. *J. Mater. Chem.* 1998, 8, 519–527.

150. Pollak, K.W.; Leon, J.W.; Fréchet, J.M.J.; Maskus, M.; Abruña, H.D. *Chem. Mater.* 1998, 10, 30–38.

151. Jiang, D.-L.; Aida, T. *J. Macromol. Sci. Pure Appl. Chem.* 1997, A34, 2047–2055.

152. Tomalia, D.A.; Uppuluri, S.; Swanson, D.R.; Li, J. *Pure Appl. Chem.* 2000, 72, 2343–2358.

153. Tomalia, D.A.; Esfand, R.; Piehler, L.T.; Swanson, D.R.; Uppuluri, S. *High Performance Polym.* 2001, 13, S1-S10.

154. Tomalia, D.A.; Durst, H.D. Topics in current chemistry Vol. 165. In *Supramolecular Chemistry I – Directed Synthesis and Molecular Recognition*; Weber, E.W., Ed.; Springer Verlag: Berlin, 1993; pp. 193–313.

155. Uppuluri, S.; Piehler, L.T.; Li, J.; Swanson, D.R.; Hagnauer, G.L.; Tomalia, D.A. *Adv. Mater.* 2000, 12 (11), 796–800.

156. Li, J.; Swanson, D.R.; Qin, D.; Brothers II, H.M.; Piehler, L.T.; Tomalia, D.A.; Meier, D.J. *Langmuir* 1999, 15, 7347–7350.

157. Uppuluri, S.; Swanson, D.R.; Brothers II, H.M.; Piehler, L.T.; Li, J.; Meier, D.J.; Hagnauer, G.L.; Tomalia, D.A. *Polym. Mater. Sci. Eng. (ACS)* 1999, *80*, 55–56.

158. Mansfield, M.L.; Rakesh, L.; Tomalia, D.A. *J. Chem. Phys.* 1996, *105*, 3245–3249.

159. Ottaviani, M.F.; Sacchi, B.; Turro, N.J.; Chen, W.; Jockush, S.; Tomalia, D.A. *Macromolecules* 1999, *32*, 2275–2282.

160. Percec, V.; Johansson, G.; Ungar, G.; Zhou, J.P. *J. Am. Chem. Soc.* 1996, *118*, 9855–9866.

161. Percec, V.; Ahn, C.-H.; Unger, G.; Yeardly, D.J.P.; Moller, M. *Nature* 1998, *391*, 161–164.

162. Hudson, S.D.; Jung, H.-T.; Percec, V.; Cho, W.-D.; Johansson, G.; Ungar, G.; Balagurusamy, V.S.K. *Science* 1997, *278*, 449–452.

163. Goodson III, T. Optical effects manifested by PAMAM dendrimer metal nano-composites. In *Dendrimers and Other Dendritic Polymers*; Fréchet, J.M.J. and Tomalia, D.A., Eds.; John Wiley & Sons: West Sussex, 2001; pp. 515–541.

164. Bosman, A.W.; Janssen, H.M.; Meijer, E.W. *Chem. Rev.* 1999, *99*, 1665–1688.

165. Sayed–Sweet, Y.; Hedstrand, D.M.; Spindler, R.; Tomalia, D.A. *J. Mater. Chem.* 1997, *7 (7)*, 1199–1205.

166. Tomalia, D.A.; Fréchet, J.M.J. *J. Poly. Sci. Part A: Polym. Chem.*, 2002, 40, 2719–2728.

167. Bauer, D.J.; Amis, A.J. Characterization of dendritically branced polymers by small angle neutron scattering (SANS), small angle x-ray scattering (SAXS), and transmission electron microscopy (TEM). In *Dendrimers and Other Dendritic Polymers*; Fréchet, J.M.J. and Tomalia, D.A., Eds.; John Wiley & Sons: West Sussex, 2001.

21

Design and Applications
of Photonic Crystals

CONTENTS

Dennis W. Prather
University of Delaware

Ahmed S. Sharkawy
University of Delaware

Shouyuan Shi
University of Delaware

21.1 Introduction

A major breakthrough in the twentieth century was the ability to engineer the electrical properties of semiconductor materials in such a way as to make them perform in a prescribed way. Based on this, semiconductor devices such as the transistor have revolutionized the field of electronics.

With the end of the twentieth century, similar breakthroughs are needed to control the optical properties of materials. Such breakthroughs should prohibit the propagation of light; in other words, redirect the propagation of light in directions for certain frequencies or localize light to certain areas. The first advance toward such a goal was fiber optic cables, which guide light through total internal reflection (TIR). Fiber optic cables have indeed revolutionized the telecommunications industry by boosting the data transmission rates much higher than did their metallic counterparts and with much lower transmission losses.

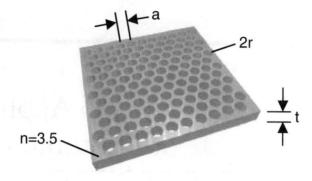

FIGURE 21.1   Periodic structure of air holes in a high-index finite-height substrate. Such a structure exhibits a property of a bandgap for a certain band of frequency called a photonic bandgap.

More recently (1987) E. Yablonovitch[1] and S. John[2,3] proposed the idea that a periodic arrangement of metallic or dielectric objects can possess the property of a bandgap for certain regions in the frequency spectrum, depending on the material from which they are constructed— whether it is metallic for the microwave regime or dielectric for the optical regime. The structure was called a *photonic crystal,* and it can prohibit the propagation of light, over a certain band of wavelengths, while allowing other bands to propagate. Such behavior gives rise to a *photonic bandgap* (PBG), analogous to the electronic bandgap in semiconductor materials. An example of a periodic structure that may exhibit the property of a bandgap is shown in Figure 21.1.

Two-dimensional photonic crystals can be realized using a periodic array of dielectric rods of any shape and/or geometry or by using a perforated dielectric slab of air holes. Such structures can be further optimized to achieve either a wider or a narrower bandgap based on the desired application. Two-dimensional photonic crystals impose periodicity in two dimensions, while the third dimension either is infinitely long (photonic crystal fiber) or has a finite height (photonic crystal slabs). Three-dimensional photonic crystals impose periodicity in all three dimensions. An example of a three-dimensional photonic crystal is shown in Figure 21.2.

Because they are easier to fabricate and analyze, two-dimensional photonic crystals have attracted the attention of a large number of researchers and engineers. An example of a fabricated two-dimensional photonic crystal is shown in Figure 21.3. Planar photonic crystal circuits such as splitters,[4–8] high Q-microcavities,[9–15] and channel drop/add filters[16–19] have been investigated both theoretically[20–22] and experimentally.[23–41]

FIGURE 21.2   A three-dimensional photonic crystal structure (woodpile).

FIGURE 21.3 SEM picture of a periodic array of air holes (photonic crystal).

Once a photonic crystal has been designed, its properties can be engineered in a manner similar to that which is done to an electronic crystal, through the process of doping. In a photonic crystal doping is achieved by either adding or removing dielectric material to a certain area. The dielectric material then acts as a defect region that can be used to localize an electromagnetic wave. Doping a photonic crystal opens a broad range of possibilities for optical device development through the localization of light.

21.2 Photonic Crystals — How They Work

The physical phenomena that clearly describes the operation of a photonic crystal is the localization of light, which is achieved from the scattering and interference produced by a coherent wave in a periodic structure. Upon an incident radiation, the periodic scatterers constructing a photonic crystal could reflect an incident radiation at the same frequency in all directions. Then, wherever in space the radiation interferes constructively, sharp peaks would be observed. This portion of the radiation spectrum is then forbidden to propagate through the periodic structure, and this band of frequencies is what was called a *stop band* or a *photonic bandgap*. On the other hand, wherever in space an incident radiation destructively interferes with the periodic scatterers in a certain directions, this part of the radiation spectrum will propagate through the periodic structure with minimal attenuation; and this band of frequencies is called a *pass band*.

For an electromagnetic wave propagating within dielectric material, scattering takes place on a scale much larger than the wavelength of light. The localization of light occurs when the scale of the coherent multiple scatterers is reduced to the wavelength itself. In this case a photon located in a lossless dielectric media provides an ideal realization of a single excited state in a static medium at room temperature. Unlike electron localization, which requires an electron–electron interaction and electron–phonon interactions, photon localization offers a unique possibility of studying the angular, spatial, and temporal dependence of wave field intensities near localized transitions.

Light localization has fundamental consequences at the quantum level. This can be seen for a periodic array of high dielectrics that have dimensions comparable to the wavelength of light by exhibiting a complete photonic bandgap in certain range, analogous to the electronic energy bandgap in semiconductor material. In a photonic crystal there are no allowed electromagnetic states in the forbidden frequency range.

21.3 Analogy between Photonic and Semiconductor Crystals

In a semiconductor crystal, electron localization can be described using a Schrodinger equation for an electron with an effective mass m^*:

$$\left[\frac{-h^2}{8\pi^2 m^*}\nabla^2 + V(x)\right]\varphi(x) = E\varphi(x) \tag{21.1}$$

where h is Planck's constant, m^* is the effective mass of electron, $V(x)$ is the potential function, $\varphi(x)$ is the wave function, and E is the total energy. The probability of finding an electron at x is given by $|\varphi(x)^2|$. The electron can be trapped by a random potential $V(x)$ in deep local potential fluctuations if the energy E is sufficiently negative. As the energy increases, the probability for the trapped electron to tunnel to a nearby potential fluctuation also increases.

In the case of monochromatic electromagnetic waves of frequency ω propagating in an inhomogeneous but nondissipative dielectric medium, Maxwell's equations are used to describe the wave propagation through space.

Starting with four macroscopic Maxwell equations:

$$\nabla \cdot \mathbf{B} = 0 \tag{21.2}$$

$$\nabla \cdot \mathbf{D} = \rho \tag{21.3}$$

$$\nabla \times \mathbf{E} + \frac{\partial \mathbf{B}}{\partial t} = 0 \tag{21.4}$$

$$\nabla \times \mathbf{H} + \frac{\partial \mathbf{D}}{\partial t} = \mathbf{J}, \tag{21.5}$$

and using the two constitutive equations:

$$\mathbf{D}(\mathbf{r}) = \varepsilon(\mathbf{r})\mathbf{E}(\mathbf{r}) \tag{21.6}$$

$$\mathbf{B}(\mathbf{r}) = \mu(\mathbf{r})\mathbf{H}(\mathbf{r}), \tag{21.7}$$

while keeping in mind that for a dielectric material:

$$\mu(\mathbf{r}) = 1.0 \tag{21.8}$$

we can substitute Equations (21.6), (21.7), and (21.8) into Equations (21.4) and (21.5) and write them in frequency (steady-state) domain form. Doing so results in the following equations:

$$\nabla \times \mathbf{E} + j\omega\mu(\mathbf{r})\mathbf{H}(\mathbf{r}) = 0 \tag{21.9}$$

$$\nabla \times \mathbf{H} - j\omega\varepsilon(\mathbf{r})E(\mathbf{r}) = 0. \tag{21.10}$$

In deriving Equation (21.10) we have assumed that there are no sources of current ($\mathbf{J} = 0$).

Taking the curl of Equation (21.9) and using Equation (21.10) to eliminate $H(r)$ we get

$$\nabla \times [\nabla \times \mathbf{E}(\mathbf{r})] = \omega^2\mu(\mathbf{r})\varepsilon(\mathbf{r})\mathbf{E}(\mathbf{r}) \tag{21.11}$$

$$\nabla \times [\nabla \times \mathbf{H}(\mathbf{r})] = -\omega^2\mu(\mathbf{r})\varepsilon(\mathbf{r})\mathbf{H}(\mathbf{r}). \tag{21.12}$$

The right-hand side of Equation (21.11) can be further expanded using vector identities:

$$-\nabla^2\mathbf{E} + \nabla(\nabla \cdot \mathbf{E}) = \omega^2\mu(\mathbf{r})\varepsilon(\mathbf{r})\mathbf{E}(\mathbf{r}). \tag{21.13}$$

The total dielectric constant $\varepsilon(r)$ can be separated into two parts as:

$$\varepsilon(\mathbf{r}) = \varepsilon_{av} + \varepsilon_{spatial}(\mathbf{r}), \tag{21.14}$$

where ε_{av} is the average value of the dielectric function and $\varepsilon_{spatial}(r)$ is the spatial component of the dielectric function, which is analogous to the potential $V(x)$ in Schrodinger's Equation (21.12), and can be then written as:

$$-\nabla^2 \mathbf{E} + \nabla(\nabla \cdot \mathbf{E}) = \omega^2[\varepsilon_{av} + \varepsilon_{spatial}(\mathbf{r})]\mu(\mathbf{r})\mathbf{E}(\mathbf{r}).$$

The quantity $\varepsilon_{av}\omega^2$ is similar to the total energy E in Schrodinger's equation.

For an electronic system, lowering the electron energy usually enhances the electron localization. For a photonic crystal, lowering the photon energy leads to a complete disappearance of the scattering mechanism itself, where, at a high photon energy, geometric and ray optic theory become more valid and interference corrections to optical transport become less and less effective.

Equation (21.12) can be formulated in the form:

$$-\nabla^2 \mathbf{E} + \nabla(\nabla \cdot \mathbf{E}) = \omega^2[\varepsilon(\mathbf{r}) - 1]\mathbf{E}(\mathbf{r}) = \omega^2\mathbf{E}(\mathbf{r}), \tag{21.15}$$

which is another form of the Schrodinger equation. By comparison it can be seen that positive dielectric scatterers are analogous to regions of negative potential energy in a quantum system.

We can also see from Equation (21.15) that because the increase in dielectric strength is analogous to an increase in the potential well depth of a quantum mechanical system, the overall effect is to lower the frequency of all modes of the system; hence, the band edges will move downward in frequency with a general frequency dependence of $1/\sqrt{\varepsilon_r}$.

21.4 Analyzing Photonic Bandgap Structures

The method initially used for the theoretical analyses of PBG structures is the plane-wave expansion method (PWM), which makes use of an important principle: that normal modes in periodic structures can be expressed as a superposition of a set of plane waves, which is also known as Floquet's theorem.[42] The photon dispersion relations inside a photonic crystal (PhC) have been calculated using the plane-wave expansion method,[43–46] where Equation (21.11) is solved as an eigenvalue problem with $E(r)$ as its eigenfunctions and ω^2 as its eigenvalues. Solution over an irreducible Brillouin zone is plotted in a form of a dispersion diagram, as shown in Figure 21.4. A dispersion diagram is a two-dimensional plot of different eigenmodes for different

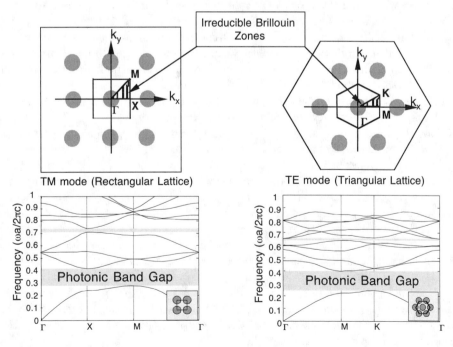

FIGURE 21.4 Plane-wave method (PWM) used to analyze different two-dimensional PBG structures of either rectangular or triangular lattice over an irreducible Brillouin zone. The result is a dispersion diagram, showing the possible eigenmodes for different wave vectors within the PBG lattice.

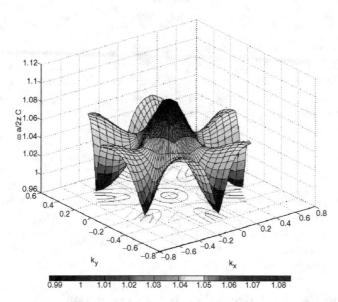

FIGURE 21.5 Dispersion surface. The horizontal plane gives both Bloch wave vector k_x and k_y. The vertical axis gives the normalized frequency $\omega a/2\pi c = a/\lambda$.

wave vectors, or propagation angles, within a photonic crystal lattice. While a two-dimensional dispersion diagram is sufficient to show whether a bandgap may exist for a certain PBG structure, it may not be sufficient for applications where nonlinear behavior of photonic crystals is analyzed. An example is the negative refractive index phenomenon and its applications to the super prismatic effect[47–50] in photonic crystals. For these applications a three-dimensional dispersion diagram, or a dispersion surface, will provide a more detailed view on a photonic crystal spatial response for various bands of frequencies both inside and outside the bandgap. Shown in Figure 21.5 is a plot of a dispersion surface of the first subband. Even though the PWM produces an accurate solution for the dispersion properties of a PhC structure, it is still limited due to the fact that transmission spectra, field distribution, and back reflections cannot be easily extracted.

An alternative approach that has been widely adopted in calculating both the transmission spectra and the field distribution is based on a numerical solution of Maxwell equations using the finite-difference time-domain (FDTD) method. In particular, the FDTD method has been used to analyze multichannel drop/add filters,[16] calculate the transmission through sharp bends,[51] and study a waveguiding mechanism through localized coupled cavities in three-dimensional photonic crystals.[52,53]

Equation (21.11) defines the main design parameters associated with a photonic crystal, such as the fill factor (defined as the ratio of the area of the lattice filled by dielectric to the total area of the whole crystal), the refractive index contrast between the dielectric material and the host material, the ratio of the lattice constant to the radius of the cylinders (for the case of cylindrical rods), and the wavelength-to-lattice constant ratio. These parameters define the location and size of the bandgap, and whether a bandgap may or may not exist for a specific polarization — either transverse electric (TE) field (electric field parallel to the plane) or transverse magnetic (TM) field (magnetic field parallel to the plane). As an example we present the analysis of a two-dimensional photonic crystal structure built on a rectangular lattice or silicon rods ($e_r = 11.56$) on air background ($e_b = 1$) and for a dielectric rod-to-lattice constant ratio of ($r/a = 0.2$), using the FDTD method. Figure 21.6 shows the numerical calculation for transmission coefficient using FDTD, with PML absorbing boundary on two sides of the computational region and block periodic boundary conditions on the other two boundaries (to simulate an infinite crystal). As stated before, within the bandgap transmission is prohibited (highly attenuated) for which a certain PBG structure resembles an optical mirror with a high reflectivity as shown in Figure 21.7.

The horizontal axis represents the normalized frequency, and the vertical axis represents the amplitude transmission coefficient, in units of dB for six layers of dielectric rods built on a rectangular lattice. We can

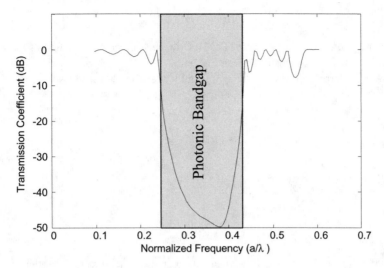

FIGURE 21.6 Attenuation diagram obtained using FDTD to analyze PBG structures.

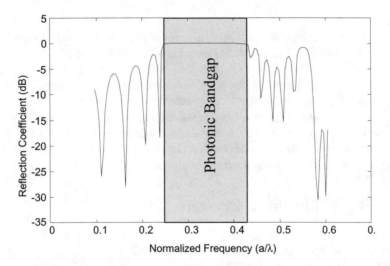

FIGURE 21.7 Plot of reflection coefficient for a certain PBG structure showing that the structure has a high reflectivity within the bandgap.

also see that for the periodic arrangement of the structure mentioned above, a bandgap opens between $a/\lambda_{high} = 0.2452$ and $a/\lambda_{low} = 0.4329$ with a center frequency at $a/\lambda_{center} = 0.339$ and a bandgap size $\Delta f/f_{center}$ equal to 55.4%. This structure can operate for any range of frequencies ranging from microwave frequencies to optical frequencies as all calculations are normalized to the operating wavelength. For example, in the optical frequency regime, where $\lambda = 1550$ nm, the structure will have a lattice constant equal to $a = 525.5$ nm and the silicon rods will have radius $r = 105$ nm. In the following section we discuss electromagnetic localization in photonic crystals through analogy to electron localization in semiconductor crystals.

21.5 Electromagnetic Localization in Photonic Crystals

The analogy between electron and photon localization has led to the scaling theory of localization, where the localization critical point is defined by a condition described by

$$\rho(\omega)D(\omega)1 \cong 1, \tag{21.16}$$

where $\rho(\omega)$ is the photon density of states (DOS), $D(\omega)$ the diffusion coefficient of light in a multiple scattering medium, and 1 the transport mean free path.

If the scattering microstructures do not significantly alter the photon density of states from its free space value:

$$\rho_{vac}(\omega) = \frac{1}{c}\left(\frac{\omega}{c}\right)^2 \tag{21.17}$$

then the localization criterion becomes

$$\left(\frac{\omega}{c}1\right)^2 \approx 1 \tag{21.18}$$

Assuming that $D(\omega) \approx c1$. In general we can interpret the factor $4\pi(\omega/c)^2$ as the total phase space available for propagation of a photon of frequency ω. Thus the localization criterion is given by

$$(phase\ space) \times 1^2 = 4\pi. \tag{21.19}$$

The concepts of density of states, diffusion coefficient, mean free path, phase space, and their interrelationships provide the basis for further understanding photonic bandgaps (PBGs).

The spatial localization of light in a photonic crystal is achieved by introducing defects, which can take the form of a line defect, in which the photonic crystal resembles a waveguide. Photons that lie within the bandgap, which are not allowed to propagate through the crystal, are confined to the defect region as defined by the photonic crystal walls. Another kind of defect is that of a point defect, in which the photonic crystal creates a cavity that confines a single or a multiple of closely separated modes to the spatial location of the point defect, centered within the cavity. In the following section we discuss in more detail the introduction of a point defect into a photonic crystal.

21.6 Doping of Photonic Crystals

In a photonic crystal there exist a dielectric band and an air band, analogous to the valance and the conduction bands, respectively, in semiconductor material. Between the dielectric and air bands is the photonic bandgap, within which no energy state exists and, as a result, propagation is prohibited.

Doping a semiconductor material can be achieved by adding either a donor or an acceptor atom. Both result in a change in the electrical properties of an atomic crystal by either having a p-type or an n-type material. In a similar fashion, the optical properties of a photonic crystal can be changed by introducing point defects, i.e., either adding or removing a certain amount of dielectric material. When dielectric material is added to a unit cell, it behaves like a donor atom in an atomic crystal, which corresponds to a donor mode and has its origin at the bottom of the air band of the photonic crystal. Alternatively removing dielectric material from a unit cell causes it to behave like an acceptor atom in an atomic crystal, which corresponds to an acceptor mode and has its origin at the top of the dielectric band of the photonic crystal, as shown in Figure 21.8. Thus, acceptor modes are preferable for making single-mode laser microcavities, because they allow a single localized mode to oscillate in the cavity. Adding or removing a certain amount of dielectric material to the photonic crystal disrupts the symmetry of the photonic lattice, thereby allowing either a single state or a multiple of closely separated states to exist within the bandgap. This phenomenon of localizing states, by introducing point defects, can be useful in designing high-Q value microcavities in photonic crystals.

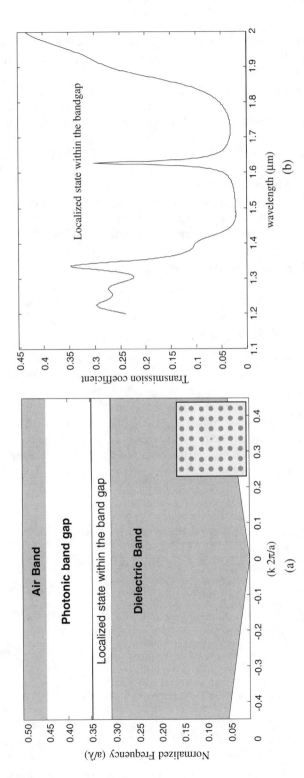

FIGURE 21.8 A photonic crystal with a point defect will allow a single or multiple localized modes to exist within the bandgap as shown in (a) band diagram and (b) attenuation diagram.

21.7   Microcavities in Photonic Crystals

Introducing a point defect in a photonic crystal can make a microcavity. As such, the defect can have any shape, size, or dielectric constant. By varying any one of these parameters, the number of modes and the center frequency of the localized mode or modes inside the cavity can be changed. If we consider the case of a square lattice of cylindrical rods with a difference in dielectric constant much greater than 2 between the host material and the lattice material, we can introduce a point defect by simply changing one of the parameters of a given rod within the crystal. For example, a point defect consisting of a rod with a radius smaller than those surrounding it will guarantee a single mode to be localized at the point defect. Alternatively, as we increase the radius of the defect, to be equal to or greater than those surrounding it, we will introduce a multiple of closely separated modes localized within the cavity.

The quality factor of the microcavity plays a major role in designing a high-density wavelength division multiplexing (WDM) system. The quality factor depends mainly on the size of the crystal. For high-Q values the size of the crystal surrounding the cavity needs to be large. It was also shown that the spectral widths of the defect modes decrease rapidly with an increasing number of lattice layers, which is more favorable in WDM because it maximizes the selectivity of the available bandwidth.

21.8   Photonic Crystal Applications

In this section we review a few applications that have been introduced in the literature using photonic crystals.

21.8.1   Low-Loss Optical Waveguides

When line defects are introduced into a photonic crystal lattice, an electromagnetic wave having a frequency within the bandgap of the structure can be guided through the crystal. In this case the line defect resembles a waveguide as shown in Figure 21.9. In this way line defects can be formed by either adding or removing dielectric material to a certain row or column along one of the directions of the photonic crystal.

To this end, photonic crystal waveguides can be used as an optical wire to guide an optical signal between different points, or devices, within an optical integrated circuit or an optical network. To create such a channel, we can decrease the radius of a certain row to the point that it no longer exists. By doing so we have created a waveguide that has a width of

$$W_{g_{rect}} = (\Omega + 1)a - 2r, \tag{21.20}$$

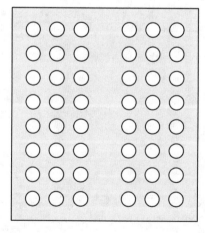

FIGURE 21.9 Waveguide created in a photonic crystal by introducing a line defect by either removing or adding dielectric to a certain row or column.

where Ω is the number of rows or columns where the line defect will be created, and r is the radius of the rods from which the photonic crystal was created. The width of the line defect/waveguide is proportional to the number of guided modes for a certain wave vector.[44,54] Field patterns for every eigenmode as well as energy flow can also be calculated using the FDTD method on the two-dimensional structure presented above.[6]

Waveguides can be also created on a two-dimensional triangular lattice photonic crystal of air holes in silicon (Si) background, for which TE modes can be guided through a line defect. In such a structure the line defect is created in the crystal by increasing the dielectric constant of the line defect, as opposed to decreasing the dielectric constant of the line defect for the two-dimensional rectangular lattice case, which was used to guide TM waves. For the triangular lattice the waveguide width can be calculated using:

$$W_{g_{tri}} = (\Omega + 1)\frac{\sqrt{3}}{2}a - 2r. \tag{21.21}$$

Note that for a perforated dielectric slab, elimination of a single row or column will not be sufficient to have a single mode of propagation through the line defect, and further design considerations must be taken to achieve that goal.[54,55]

By removing a column or a row, we can confine the optical beam to the waveguide in a very similar fashion to the total internal reflection (TIR) concept, which is used to confine the optical signal in optical fibers. However, in photonic crystals the mechanism of in-plane optical confinement for a wave propagating through the defect is through multiple or distributed Bragg reflections (DBR). For finite height photonic crystal structures (slabs), vertical confinement is achieved through TIR at the interface between the PhC slab and lower dielectric constant material, e.g., air.[56,57]

The main idea of operation for this kind of waveguide is that an incident beam with a frequency within the bandgap of the structure will not propagate through the structure, but it will propagate through the waveguide with minimal field leakage. Using this approach, a throughput efficiency as high as 100% can be achieved through the waveguide.[7,20,24,51,58–63] A snapshot of an FDTD simulation of an optical pulse propagating through the above structure is shown in Figure 21.10.

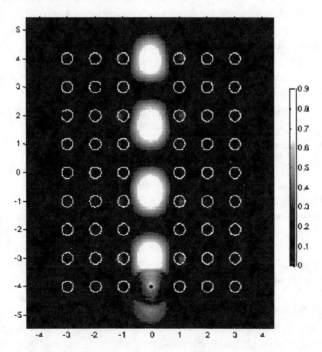

FIGURE 21.10 Steady-state field solution of a TM pulse propagating through a PBG waveguide; simulations were performed using FDTD with PML absorbing boundary conditions.

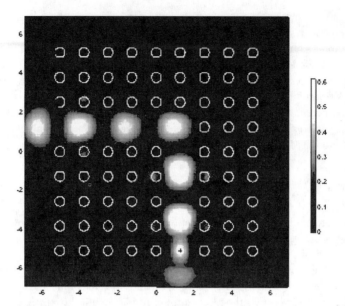

FIGURE 21.11   A sharp waveguide bend built on a PBG.

From Figure 21.10 we can see that, by using only three layers on each side of the PBG channel, we were able to achieve high lateral confinement.

21.8.2   Waveguide Bends

A low-loss waveguide that includes a sharp 90° bend in the two-dimensional photonic crystals has been reported.[6,8,17,50,51,64] Theoretically, it was shown by a simple scattering theory that 100% transmission is possible. Experimentally, over 80% transmission was demonstrated at a frequency of 100 GHz by using a square array of circular alumina rods having a dielectric constant of 8.9 and a radius of $0.2a$, where a is the lattice constant of the array. For $a = 1.27$ mm, the crystal had a large bandgap extending from 76 GHz to 105 GHz. A line defect was created inside the crystal by removing a row of rods. The optical guided mode produced by the defect had a large bandwidth, extending over the entire bandgap. A snapshot of FDTD simulation of the structure described above is shown in Figure 21.11.

21.8.3   High-Q Microcavities

A microcavity can be created in a photonic crystal through doping or the introduction of point defects to a unit lattice as explained previously. If two-dimensional periodicity is broken by a local defect, local defect modes can occur within the forbidden bandgap. The local defect can then be introduced by either adding or removing a certain amount of dielectric material. If extra dielectric material is added to one of the unit cells, the defect will behave like a donor atom in a semiconductor and give rise to donor modes. On the other hand, removing some dielectric materials will introduce defects similar to acceptor atoms in semiconductors that produce acceptor modes. It was found that acceptor modes are particularly well suited for making laser micro-resonator cavities.[10,15,23,54,65]

High-Q microcavities could be constructed with two-dimensional photonic crystals if light scattering in the vertical direction, due to the finite depth of the crystal, is minimized. This was demonstrated by the FDTD calculations of the Q-factor for an optical microcavity defined by a three-layer slab waveguide and a two-dimensional photonic crystal and mirrors. Studying the effect of the finite depth of the crystal on the cavity modes and the loss mechanisms within the cavity optimized the performance of photonic crystal mirror. It was shown that the Q-factor of the cavity mode depends

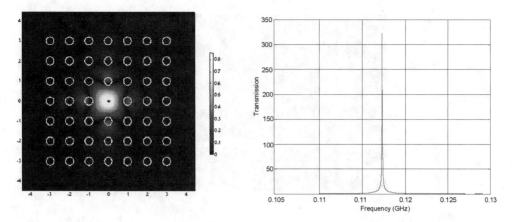

FIGURE 21.12 A microcavity built in a PBG by removing a single rod from the center of a rectangular unit lattice. FDTD simulation is on the left, and the spectral result for the field inside the cavity is on the right.

strongly on the depth of the holes defining the photonic crystal and the refractive index of the material surrounding the waveguide core.[12–14,66–68]

A snapshot of the FDTD simulation, where a point defect was introduced to a photonic crystal for which it was used as a microcavity, is shown in Figure 21.12.

21.8.4 High-Quality Filter

High-quality filters were realized by using a perturbed photonic crystal.[69] The perturbed crystal was constructed by randomly repeated stacking of a number of identical unit cells. Each unit cell was an alternating array of M identical high-dielectric components and M identical low-dielectric components. The unit cells were separated by N spacers made of the same material as the low-dielectric components. It was shown that although most states are localized due to randomness, there existed states of certain wavelengths. This led to high-quality resonant tunneling with a transmission peak much more narrow than that of the tunneling through a perfect crystal.

21.8.5 Channel Drop Filters in Photonic Crystals

A channel drop filter was realized in photonic crystals using two waveguides and an optical resonator system.[18,19,70–72] Maximum transfer between the two waveguides occurs by creating resonant states of different symmetry and by forcing an accidental degeneracy between them. For this case the optical resonator system consists of two high-Q microcavities. A snapshot of the FDTD simulation for the structure explained above is shown in Figure 21.13, and a plot for the frequency response of the time varying field stored at the detector on the dropped channel is shown in Figure 21.14.

Examining the frequency response of the field in the channel, we can see that, even though the above system achieves high selectivity (3 nm line width), there is a considerable amount of interference with the dropped frequency in the channel. This interference can be minimized by narrowing the incident pulse to the extent that the noise is outside the transmission band of the structure; but that will cause a reduction of the number of allowed channels.

21.8.6 Optical Limiter

An effective two-dimensional optical limiter operating at 514.5 nm for pulse duration of 0.1 to 4 ms was investigated.[73] The photonic crystal consisted of 180–230 nm spatial-period nanochannel glass (NCG) containing a thermal nonlinear ethanol–toluene liquid. A dynamic range in excess of 130 was achieved in a single-element device with a threshold current of 200 mJ/cm^2. It was also shown that

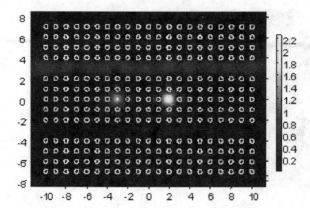

FIGURE 21.13 A channel drop filter realized using a channel, a bus, and two microcavities built on a rectangular PBG lattice.

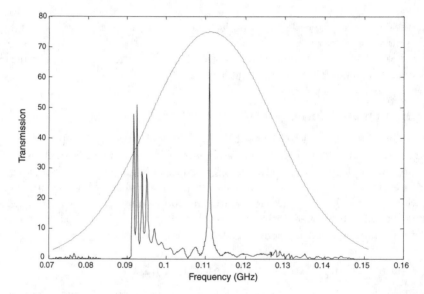

FIGURE 21.14 Frequency spectrum of the time-varying electric field in the channel normalized to the incident field. The incident field is shown by the dashed line, while dropped frequency is shown by the solid line.

the spectral width could be broadened to 100 nm by increasing the index differential between the organic liquid and the glass matrix.

21.8.7 Beam Splitter

Waveguide branching[6,7] and beam splitters[5] have been demonstrated in photonic crystals. Efficient waveguide branching was achieved by satisfying the rate-matching condition:

$$\frac{1}{t_1} = \sum_{n=2}^{N} \frac{1}{t_n}, \, n = 2, 3, \dots N \qquad (21.22)$$

where $1/t_n$ is the decay rate in waveguide number n in an n waveguide branches, and ideal splitting will occur when all $1/t_n$ are equal. This is achieved by placing extra point defects between the input and the output waveguides.

To split electromagnetic waves in three-dimensional photonic crystals, a coupled-cavity photonic crystal waveguide was used.[5] Square-shaped alumina rods with $e_r = 10.24$ were used. Sample dimensions were 0.32 cm, 0.32 cm, 15.25 cm at the microwave frequencies. A center-to-center separation between the rods was 1.12 cm. The structure had a three-dimensional full photonic bandgap extending from 10.6 to 12.8 GHz.

21.8.8 Bragg Reflector

A circular Bragg reflector was made with disk-shaped microcavities of approximately 10 μm^2 in area in a GaAs/AlGaAs waveguide structure by etching deep vertical concentric trenches.[74] The trenches formed the Bragg reflector confining light in two lateral dimensions. From photoluminescence excited in the waveguide, the confinement of discrete disk modes was demonstrated. The wave vector of these discrete disk modes was mainly in the radial direction, in contrast with whispering gallery modes in the tangential direction. The high-Q factor of up to 650 indicated that in-plane reflectivities could approach 90%.

21.8.9 Superprism Phenomenon

The superprism phenomenon was demonstrated at optical frequencies with a three-dimensional periodic structure fabricated on a Si substrate.[47–49,75–81] The extraordinary angular sensitivity of light propagation showed that, with a ±12° change in the incident angle, the direction of the transmitted beam varied from −90° to + 90°. This effect, together with wavelength sensitivity, is two orders of magnitude stronger than that of a conventional prism or grating. It was also shown by photonic band calculations that the angular dependence of negative refraction and multiple-beam branching was due to highly anisotropic dispersion surfaces. The application of this superprism phenomenon enables the fabrication of photonic integrated circuits for wavelength division multiplexing and add–drop filtering.

21.8.10 Spot Size Converter

A spot size converter (SSC) was demonstrated in photonic crystals with a conversion ratio of 10:1 for a 1.0 μm wavelength.[82] The photonic crystal was fabricated by depositing alternate layers of amorphous Si and SiO_2 on a patterned Si substrate having a hexagonal array of holes with a lattice constant of 0.3μm. The replication of the surface holes causes the structure to be self-organized. A polarized light was incident onto the crystal from the edge at an angle of 15° from the Γ–M direction, using a VCSEL with $\lambda = 0.956$ μm. For a 40 μm wide incident beam, the propagating beam inside the photonic crystal reduced to 4 μm.

21.8.11 Zero Cross-Talk

Numerical simulations showed that cross-talk reduction up to 8 orders of magnitude could be achieved with a two-dimensional photonic crystal,[83] as compared with a conventional high-index-contrast waveguide crossing. It was proposed that the design principles could also apply to three-dimensional systems and are not restricted to photonic crystals. Tuning the intersection to reduce radiation losses or using a system of resonators to flatten the resonant peak could improve the performance of these devices.

21.8.12 Waveguiding through Localized Coupled Cavities

Coupled photonic crystal cavities were used as a waveguiding mechanism in three-dimensional photonic bandgap structures.[53,84–86] Here it was shown that photons propagate through strongly localized cavities due to coupling between adjacent cavity modes. Transmission values as high as 100% were observed for various waveguide structures, even for cavities placed along arbitrarily shaped paths.

21.8.13 Self-Collimation

Self-collimation of an electromagnetic wave was achieved with a photonic crystal fabricated on silicon.[49] It was shown that the divergence of the collimated beam is insensitive to that of the incident beam and much smaller than the divergence of conventional Gaussian collimators. The phenomenon was interpreted in terms of highly modulated dispersion surfaces with inflection points, where the curvature changes from downward to upward corresponding to, respectively, a concave/convex lens case.

21.8.14 Two-Dimensional Distributed Feedback Laser Generator

A new type of laser resonator with two-dimensional distributed feedback (DFB) from a photonic crystal was reported.[87] The gain medium consisted of a thin film of organic material doped with Coumarin 490 and DCM. The thickness of the film was 150 nm and the period of the pattern was 400 nm. The planar waveguide consisted of the organic core layer, the air, and Si cladding layers supported by only the lowest-order transverse electric and transverse magnetic modes. With a 337 nm pulsed nitrogen laser as the pump source, laser action was observed in the wavelength range of 580–600 nm. The threshold pump power was ~50 kW/cm^2. Numerical calculations predicted two peaks of different polarization at the wavelengths in close agreement with the experimental results. It was also predicted that it would be possible to achieve laser action from two-dimensional photonic crystals with a complete bandgap by using organic media in conjunction with advanced Si micro fabrication technology.

21.8.15 Photonic Crystal Fiber

Photonic crystal (PhC) fiber was first reported in 1996.[55] The PhC fiber consisted of pure silica core with a higher refractive index surrounded by silica/air PhC material with hexagonal symmetry. It was shown that the fiber supported only a single low-loss guided mode over a very broad spectral range of at least 458–1550 nm. Further experimental investigation and theoretical analysis using the effective index model revealed that such a PhC fiber can be single mode for any wavelength, and its useful single-mode range within the transparency window of silica is bounded by a bending loss edge at both the short and long wavelengths.[88,89] The critical parameter is the ratio d/Γ, where d is the air hold diameter and Γ the spacing between adjacent holes.

21.8.16 Second Harmonic Generation

Second harmonic generation was observed experimentally in a centro-symmetric crystalline lattice of dielectric spheres.[90] The photonic crystal was composed of polystyrene spherical particles of optical dimensions. The inversion symmetry of the centro-symmetric face centered cubic (FCC) lattice is broken locally at the surface of each sphere in such a way that the scattered second harmonic light interferes constructively, leading to a nonvanishing macroscopic field. To enhance the second order nonlinear interaction, a layer of strongly nonlinear molecules is adsorbed on the surface of each sphere of 115 nm diameter. It was also observed that phase matching of the fundamental and second harmonic waves is due to the long-range periodic distribution of dielectric material, which provides the bending of the photon dispersion curve at the edge of the Bragg reflection band of given lattice planes. It was also pointed out that the flexibility in selecting the nonlinear molecules makes the nonlinear photonic crystal very attractive for the study of surface chemical processes and for the improvement of nonlinear optoelectronic devices.

21.8.17 Air-Bridge Microcavity

A new type of high-Q microcavity consisting of a channel waveguide and a one-dimensional photonic crystal was investigated.[91–93] A bandgap for the guided modes is opened, and a sharp resonant state is created by adding a single defect in the periodic system. Numerical analysis of the eigenstates shows that strong field confinement of the defect state can be achieved with a modal volume less than half a cubic half-wavelength. In the structures proposed, coplanar microcavities use index guiding to confine light along two dimensions and a one-dimensional photonic crystal to confine light along the third. The microcavities are made of high-index channel waveguides in which a strong periodic variation of the refractive index is introduced by vertically etching a series of holes through the guide. The guided modes undergo multiple scattering by the periodic array of holes, causing a gap to open between the first and the second guided mode bands. The size of the gap is determined by the dielectric constant of the waveguide and by the size of the holes relative to their central distances. By introducing a defect in the periodic array of holes, a sharp resonant mode can be introduced within the gap. Good confinement of the radiation in the microcavity was achieved by having a large contrast between the waveguide and the substrate, which also kept the mode from extending significantly into the substrate. It was also found that maximum confinement could be reached by completely surrounding the cavity with air.

21.8.18 Control of Spontaneous Emission

An enhancement and suppression of spontaneous emission in thin-film InGaAs/InP photonic crystal have been studied.[1,94] Angular resolved photoluminescence (PL) measurements were used to experimentally determine the band structure of such a photonic crystal and the overall enhancement of spontaneous emission. It was shown that emission into leaky conduction bands of the crystal has the same effect as cavity-enhanced spontaneous emission, provided these bands are flat enough relative to the emission band of the material. A MOCVD-grown $In_{0.47}Ga_{0.53}As/InP$ single quantum well double-hetero structure was used for these experiments.

21.8.19 Enhancing Patch Antenna Performance Using Photonic Crystals

Photonic crystals were used to enhance the performance of a patch antenna.[95] Traditionally, patch antennae have some limitations such as restricted bandwidth of operation, low gain, and potential decrease in radiation efficiency due to surface–wave losses. Using a photonic bandgap substrate for a patch antenna minimized the surface–wave effects compared with conventional patch antennae, thus improving the gain and far-field radiation pattern.

21.8.20 Surface-Emitting Laser Diode Using Photonic Bandgap Crystal Cavity

A surface-emitting laser diode consisting of a three-dimensional photonic bandgap crystal cavity was presented.[28] Spontaneous emission was controlled to radiate in the lasing direction with a narrow radiation angle by introducing a plane phase-shift region into the cavity. The radiation pattern of the localized phase-shift mode in the photonic crystal is analyzed with a plane-wave method and using a two-dimensional model. It was also shown that the radiation angle of the spontaneous emission in the photonic crystal cavity is as narrow as that of the stimulated emission of the conventional surface-emitting laser. The photonic bandgap crystal cavity laser operated as a light source without threshold and spatial emission noise; therefore, this approach is very attractive for use as a light source in spatially integrated optical circuits.

21.8.21 Optical Spectrometer

Both a single-channel and a multiple-channel optical spectrometer were reported.[16] For the theoretical analysis of such devices, the computational region consisted of a two-dimensional photonic crystal having a square lattice with a lattice constant a = 350 nm. The lattice is made of dielectric rods with a dielectric constant of 11.56 (corresponding to Si) and radius r = 70 nm, on an air background. The transmission spectra for the above structure can be obtained using either finite-difference frequency-domain or finite-difference time-domain with periodic boundary conditions. The above structure has a bandgap located between $\lambda = 0.833$ µm and $\lambda = 1.25$ µm.

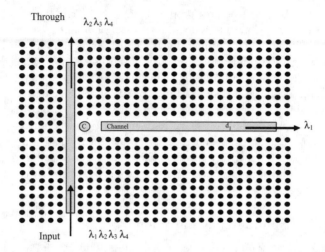

FIGURE 21.15 A single-channel optical spectrometer with a point defect(C) of diameter d = 52.5 nm and dielectric constant $\varepsilon_r = 7$. A broadband incident pulse evanescently coupled to the microcavity will drop a narrow wavelength through a channel toward a detector placed at d_1.

21.8.21.1 Single-Channel Optical Spectrometer

A single-channel optical spectrometer was implemented by combining a photonic crystal waveguide evanescently with a photonic crystal microcavity. For an incident broadband pulse with a frequency content within the bandgap of the structure, a single frequency will be efficiently selected through the microcavity and further directed to its destination through another waveguide as shown in Figure 21.15.

To simulate the structure shown in Figure 21.15, the FDTD method with PML absorbing boundary conditions was used. The computational space had a sampling rate of ($\lambda/40$), where λ is the wavelength of light in vacuum. To this end, a pulse of center wavelength ($\lambda = 1$ μm) and of width ($\Delta\lambda = 0.6$ μm) was transmitted through the waveguide, which excited a single mode of oscillation inside the cavity. The field in the cavity was then coupled to the channel through an evanescent field. A detector was placed inside the channel to obtain the wavelength spectrum of the field in the channel, which is shown in Figure 21.16. The spectrum was obtained by taking the Fourier transform of the time-dependent field at the detector. From Figure 21.16 we can see that the quality factor of the cavity is about 2000; and the point defect of r = 52.5 nm corresponded to a center wavelength of ($\lambda = 1.025$ μm) and had a spectral line width of ($\Delta\lambda = 2$ nm). This means that an incident pulse width of ($\Delta\lambda = 0.6$ μm) can achieve nearly 300 different channels by fine tuning the defect size of the center rod in the cavity while maintaining its dielectric constant at $\varepsilon_r = 7.0$.

21.8.21.2 Multiple-Channel Optical Spectrometer

A multiple-channel optical spectrometer can be achieved by cascading a number of single-channel spectrometers that are branched from a main waveguide channel. In this case six cavities were included, each having a different defect size and its own guiding channel. Each channel was branched from the main waveguide as shown in Figure 21.17. Such a topology allows for better utilization of the structure by maximizing the density of the channels within the computational region. In addition to the previous cavity, which had a point defect size of (r = 52.5 nm), five more cavities were added with different point defects: (r = 8.75 nm), (r = 17.5 nm), (r = 26.25 nm), (r = 35 nm), and (r = 43.75 nm), while maintaining the dielectric constant of all point defects at $\varepsilon_r = 7.0$. A separate analysis for each point defect was performed prior to this case, which corresponded to central wavelengths of ($\lambda_2 = 0.875$ μm), ($\lambda_3 = 0.895$ μm), ($\lambda_4 = 0.925$ μm), ($\lambda_5 = 0.94$ μm), and ($\lambda_6 = 0.96$ μm), respectively. Spectral analysis of the above structure is shown in Figure 21.18. The spectrum was obtained by taking the Fourier transform of the time-dependent field at each detector. They were determined to have Lorentzian line shapes as shown in Figure 21.18. Also shown in Figure 21.18 is that different point defect sizes corresponded to different localized modes or

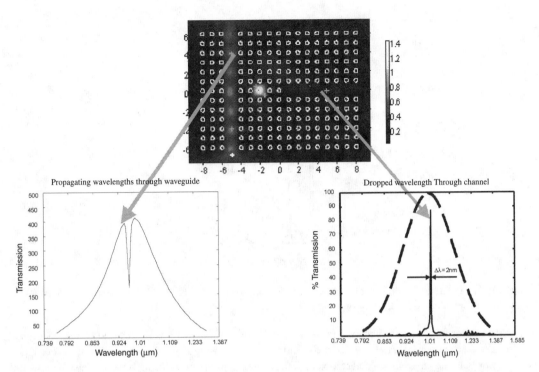

Propagating wavelengths through waveguide

Dropped wavelength Through channel

FIGURE 21.16 FDTD simulation results for the structure shown in Figure 21.15. Shown in the right corner is a selected wavelength from a propagating pulse (dotted), and the left corner shows the propagating pulse after a single wavelength has been detected (dropped) through the cavity.

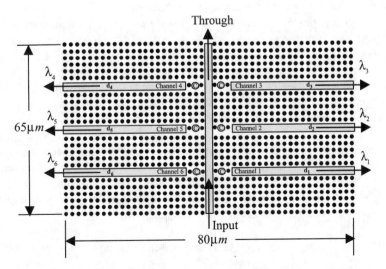

FIGURE 21.17 An optical spectrometer created in a PBG or rectangular lattice. Six channels are shown for simplicity, but the device can be expanded to include N number of channels.

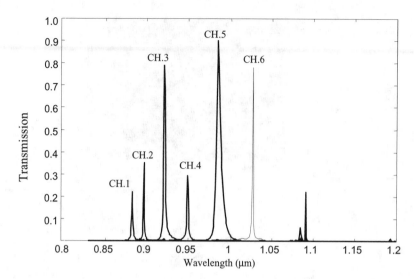

FIGURE 21.18 Spectral results for the optical spectrometer shown in Figure 21.17. Six spectrally separated wavelengths corresponding to six different cavities were detected.

different center frequencies. They were found to match previous calculations for the case of a single cavity. Also note that the central wavelength of each channel is directly proportional to the radius of the defect; in other words, as the radius of the defect increases, the channels are spanned through the available bandwidth of the incident pulse. There is a limit as to how large one can increase the radius of the defect while maintaining a single mode (acceptor mode) inside the cavity. Once the size of the defect starts getting close to that of the lattice rods, in which case more dielectric material is added than removed, multiple modes begin to exist in the cavity. Lastly, the difference in the spectral line widths between different channels is due to the difference in Q-values of different cavities, which can be optimized for equally high Q-values.

This topology offers a more flexible design freedom and fewer constraints in contrast to its prior counterparts, in which achieving a high fan-out will require a much larger area.

21.8.22   Hybrid Photonic Crystal Structures

When a photonic crystal of single crystalline structure is formed, there is a natural matching of the crystal lattice, and a high-quality single crystal layer results. On the other hand, if a photonic crystal is formed from multiple crystalline structures (e.g., rectangular, triangular), the newly created structure is no longer a single crystalline structure but a hybrid structure, which in turn contains the optical characteristics of both structures. Such a structure is called a heterostructure photonic crystal.[96]

Heterostructure photonic crystals were used to further optimize the bandgap size obtained from a single photonic bandgap structure. A bandgap size of 94% was obtained from a heterostructure of a rectangular photonic crystal lattice (bandgap size of 55%) and a triangular photonic crystal lattice (bandgap size of 39% and 19%), as shown in Figure 21.19. Wide bandgap PhCs are advantageous for applications such as wideband optical mirrors, wideband optical matching elements, and wideband optical couplers.

When photonic crystals of different lattice structures are brought together, discontinuities in their respective energy bands are expected, as both structures will have different bandgaps as shown in Figure 21.20. The discontinuities in the dielectric bands ΔE_{diel} and the air band ΔE_{air} accommodate the difference in bandgap between the two photonic crystals ΔE_g, and the barriers:

$$\Delta E_g = \Delta E_{g1} - \Delta E_{g2} \tag{21.23}$$

on either side of the wide bandgap lattice form what is in principle similar to an electronic quantum well in semiconductors, which in this case can be called a *photonic quantum well*. The bands above and

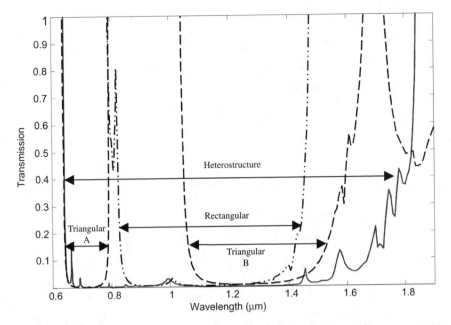

FIGURE 21.19 Band diagram of a rectangular PC lattice (-.-.) with a bandgap size 54.87%. Band diagram of a triangular lattice (--) with a short wavelength bandgap size 19.44% and a long wavelength bandgap size 39.39%. Band diagram of a heterostructure PC lattice (solid line) with a bandgap size 94.02%.

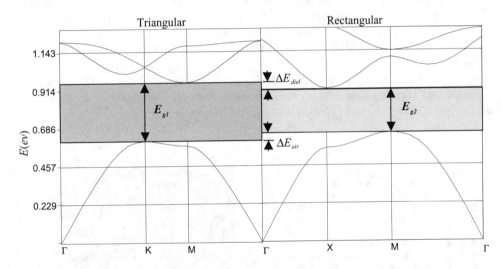

FIGURE 21.20 Band diagram of a triangular PC lattice with E_{g1} and a rectangular PC lattice with E_{g2}. As both lattices are brought together to form heterostructure PC band edge discontinuities, ΔE_{diel} and ΔE_{air} start to appear at both the dielectric and air bands, respectively.

below the bandgap, are generally unoccupied and a transition from one Bloch state to another state is only possible if the phase-matching condition is satisfied.[97]

By careful design of both rectangular and triangular lattices, the discontinuities in the dielectric band as well as the air band shown in Figure 21.20 can be minimized if not completely eliminated. This basically means that bandgap matching between two different lattices can be achieved, and heterostructure PhC can be used as a matching element. Shown in Figure 21.21 is an attenuation diagram for a triangular lattice matched with an attenuation diagram for a rectangular lattice.

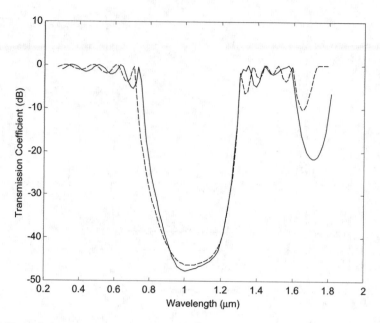

FIGURE 21.21 Transmission spectra for a rectangular lattice (solid line) with a ratio of r/a = 0.4 vs. transmission spectra for a triangular lattice (dashed line) with a ratio of r/a = 0.43.

Another application in which heterostructure PhCs were used to improve the performance of a device was with an optical beam splitter. Shown in Figure 21.22 is an optical beam splitter implemented in a single photonic crystal structure (rectangular). Such a device achieved an overall throughput efficiency of 50%, with most of the loss in transmission contributed by the back reflections at the splitting section.

Using a heterostructure photonic crystal, the device performance increased from 50% to 90% as shown in Figure 21.23. Other examples utilizing heterostructure photonic crystals are a 1-to-4-beam splitter shown in Figure 21.24 and a beam splitter/combiner shown in Figure 21.25.

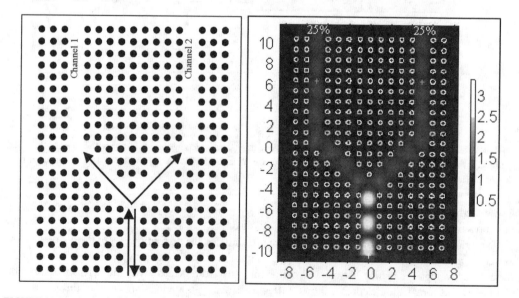

FIGURE 21.22 An optical beam splitter to demonstrate the idea of an optical matching element.

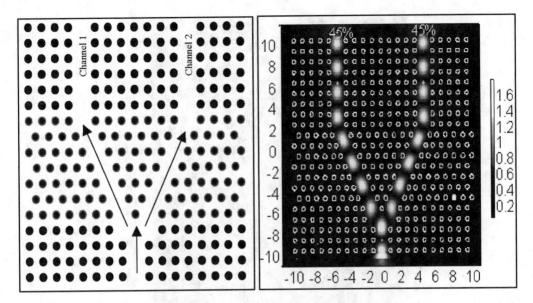

FIGURE 21.23 A schematic diagram of an optical beam splitter using a heterostructure lattice.

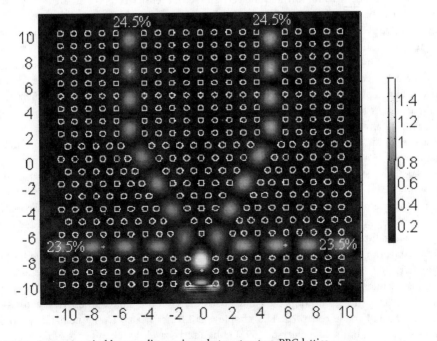

FIGURE 21.24 A 1-to-4 optical beam splitter using a heterostructure PBG lattice.

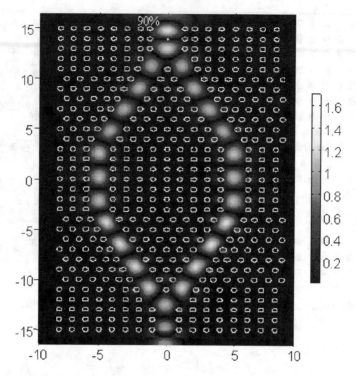

FIGURE 21.25 An optical beam splitter and combiner using a heterostructure PBG lattice.

21.8.23 Tunable Photonic Crystals

Photonic crystals are built using low-loss dielectric material (either rods on an air background or air holes on dielectric background). Once they are built with certain geometry and dimensions, they will behave in a way consistent with those properties. Any desired alteration to their performance requires a new design. However, if one can dynamically vary some of the properties of an existing photonic crystal, one can modify or control its performance. Such a change can be done by changing the index of refraction, which can be achieved by applying an electric field, called the *electro-optic effect*. In this case, applying an electric field to a photonic crystal will change the dielectric constant of the crystal, which in turn will change the transmission properties of the photonic crystal. This way one can tune a photonic crystal without changing any of the geometrical properties.

Tuning the band structures of photonic crystals was first proposed using semiconductor-based photonic crystals.[98] A second method that was later proposed was the use of liquid crystal infiltration.[99,100] Semiconductor-based photonic crystals can be made tunable if the free-carrier density is sufficiently high, and the photonic band structure is strongly dependent on the temperature T and on the impurity concentration N. This was shown for two-dimensional intrinsic InSb or extrinsic Ge photonic crystals. The disadvantage of this technique is that absorption cannot be avoided but can be minimized by careful selection of the materials.

Liquid crystals can also be used to tune a two-dimensional photonic crystal using the temperature-dependent refractive index of a liquid crystal. Liquid crystal E7 was infiltrated into the air pores of a macroporous silicon photonic crystal with a triangular lattice pitch of 1.58 μm and a bandgap wavelength range of 3.3–5.7 μm. After infiltration, the bandgap for the TE mode shifted dramatically to 4.4–6.0 μm, while that of the TM mode collapsed. The sample was further heated to the nematic-isotropic phase transition temperature of the liquid crystal (59°C); the short-wavelength band edge of the TE gap shifted by 70 nm, while the long wavelength edge was constant.

21.8.24 Optical Switch Using PBG

Ultrafast all-optical switching in a silicon-based photonic crystal has been investigated[101] wherein the effect of two-photon absorption with Kerr nonlinearity on the optical properties of photonic crystals made with amorphous silicon and SiO_2 was studied. A stop band appearing near 1.5 μm is monitored with a peak probe beam and modulated by changes in the refractive index caused by a pump pulse at 1.71 μm with 18 GW/cm^2 peak intensity. Nonlinear optical characterization of the sample using Z- scan points to two-photon absorption as the main contributor to free carrier excitation in silicon at that power level. Modulation in the transmittance near the band edge is found to be dominated by the optical Kerr effect within the pulse overlap (~400 fs), whereas free carrier index changes are observed for 12 ps.

Other switching techniques in photonic crystals have also been presented both theoretically as well as experimentally.[101–107]

21.8.25 Photonic Crystal Optical Networks

High-density optical interconnects using PhCs was recently proposed as a technique for three-dimensional optical signal distribution and routing through multiple planes and in different directions.[108] These optical networks offer the ability to guide light analogously to electrical printed circuit boards (PCBs), which transport electrons through electrical networks. Due to their unique ability to confine and control light on the sub-wavelength scale, PhCs have led to a challenging prospect of miniaturization and large-scale integration of high-density optical interconnects.

A schematic diagram showing a cross-sectional view (*xz* plane and *xy* plane) of a two-layer optical network is shown in Figures 21.26 and 21.27, respectively. Here the bottom layer consists of a photonic crystal slab, which is used to guide, split, and process an optical wave between different points within that layer. On the top layer, another photonic crystal slab of either the same dimension (diameter, pitch, and /or thickness) or structure as the one on the bottom layer is again used for different optical guiding

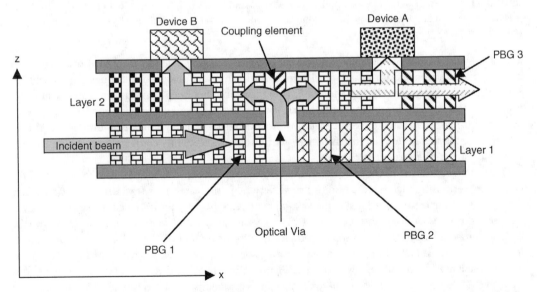

FIGURE 21.26 A cross-sectional view (xz plane) of an on-chip optical network composed of two finite-height planar photonic crystal layers. In layer 1 there are two PBGs predesigned to specific frequency bands, such that an optical beam coming from the right-hand side will propagate through PBG1 (through a waveguide) and will be prohibited from propagation through PBG2. Similar to PBG1 and 2, in layer 2 there exists PBG1, 2, and/or 3, which may or may not be tuned to the same frequency. To efficiently couple between layers 1 and 3, we have used an optical via combined with a coupling element to either enhance or diminish coupling. Lateral confinement is achieved via multiple Bragg reflections through PBGs, while vertical confinement is achieved through total internal reflections (TIR).

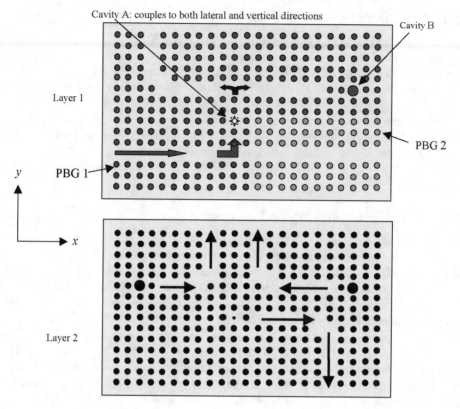

FIGURE 21.27 A cross-sectional view (xy plane) for layers 1 and 2 shown in Figure 21.26. In layer 1, cavity A was used to couple both in the lateral dimension within layer 1 and in the vertical direction to layer 2. Cavity B was used only for vertical coupling between layers 1 and layer 3.

and processing. Finally, optical signals of specific wavelengths are guided to their final destinations at either a specific photodetector or an optoelectronic device, or even guided back through an optical fiber to another chip. In between the top and bottom layers, a confining layer is used to simulate the effect of total internal reflection. For the case of dielectric rods on air background, the confining layer was a PEC layer, and for the perforated dielectric slab, we used a thin layer of SiO_2 as a confining layer. Also in between the top and bottom layers, an optical via was used to efficiently couple a propagating electromagnetic wave from one layer to the next.

One technique that can be used for broadband vertical coupling between adjacent and nonadjacent PBG layered networks is through the use of dielectric deflectors. In this case we use two deflectors of the same dimensions to couple from one layer to the next higher layer. One deflector is located in the original layer, and the other one is located in the next layer, such that a wave propagating in direction perpendicular to the deflector surface in the original layer will be completely reflected vertically to the deflector in the destination layer. This in turn will reflect the wave toward a waveguide in the top layer. A schematic diagram showing an implementation deflector coupling is shown in Figure 21.28, where two deflectors of the same dimensions were used. Note that a PEC cavity is still needed to provide a point-to-point communication between the two deflectors. Deflector coupling can be used for either static coupling, in which the angle is fixed, or dynamic coupling, in which the angle can be varied. Hence, coupling between different layers can be either enhanced or removed based on the system requirements. Dynamic coupling can be achieved by either tuning the angle of the deflector to a certain direction, which can be achieved using MEMS technology or, again, by optically tuning the absorption properties of the deflector to adaptively control coupling to certain layers. Fabricating the deflector from two dielectric materials can provide another way for dynamically controlling the coupling mechanism between two different layers in our optically interconnected network.

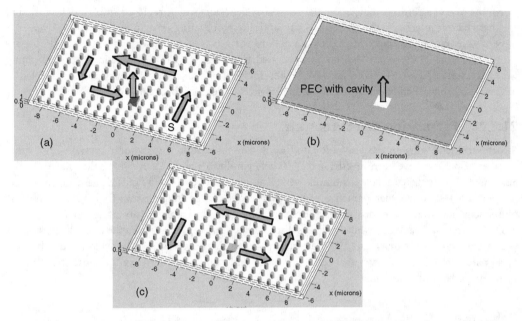

FIGURE 21.28 Dielectric deflector can be used as an optical via to couple light between either adjacent or nonadjacent layers. (a) Layer 1 contains a waveguide and a dielectric deflector. (b) The confining layer contains a cavity for vertical coupling. (c) Layer 2 contains another deflector that will in turn refract an optical beam toward the waveguide in layer 2.

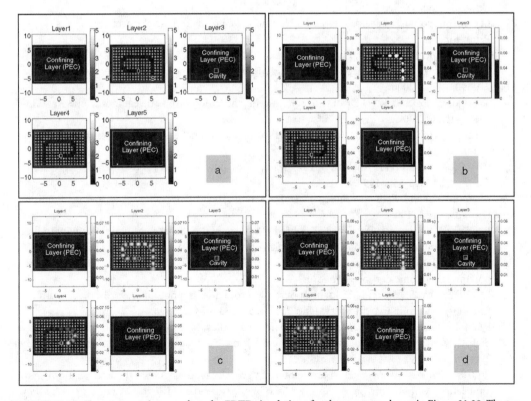

FIGURE 21.29 Four consecutive snapshots for FDTD simulations for the structure shown in Figure 21.28. The top, middle, and bottom layers (1, 3, and 5) are perfectly electric conductor (PEC) for vertical confinement. Layers 2 and 4 are a finite-height photonic crystal waveguide. Source (S) is located at the bottom-right corner.

For the case of a perforated dielectric slab, dielectric deflectors will be replaced with air deflectors; and the dielectric air interface will again redirect an optically propagating wave to its destination layer.

Four consecutive snapshots of FDTD simulations for a two-layer network utilizing a deflector are shown in Figure 21.29, where two layers were used to confine an electromagnetic wave propagating through layer 2 (Figure 21.29a). The wave is then incident on the deflector, which is then coupled to layer 4, also sandwiched between two confining layers (Figures 21.29b, c, and d).

21.8.26 Fabrication Error Analysis

Structural fluctuations on the photonic bandgap during fabrication for photonic crystal with a finite number of periods have been investigated.[109–111] The emphasis was on determining the effects of misalignment of basic structural elements and overall surface roughness because of their general fabrication relevance. It was found that refractive index disorder affects the longer wavelength part of the first photonic band. Interestingly, positional disorder in the cylinders' centers mainly affects the first gap and has little effect on the second, while thickness disorder has a smoothing effect on the band diagram. Even with a misalignment as much as 18%, the bandgap remained as large as 10%. Considering the disorder of all parameters at once shows that refractive index and radius disorder have the most effect on the bandgap for both polarizations

References

1. E. Yablonovitch, Inhibited spontaneous emission in solid-state physics and electronics, *Phys. Rev. Lett.*, 58, 2059–2062, 1987.
2. S. John, Strong localization of photons in certain disordered dielectric superlattices, *Phys. Rev. Lett.*, 58, 2486, 1987.
3. J.D. Joannopoulos, R.D. Meade, and J.N. Winn, *Photonic Crystals*, Princeton University Press, Princeton, NJ, 1995.
4. G. Parker and M. Charlton, Photonic crystals, *Phys, World*, 13, 2000.
5. M. Bayindir, B. Temelkuran, and E. Ozbay, Photonic-crystal-based beam splitters, *Appl. Phys. Lett.*, 77, 3902–3904, 2000.
6. T. Sondergaard and K.H. Dridi, Energy flow in photonic crystal waveguides, *Phys. Rev. B*, 61, 15688–15696, 2000.
7. S. Fan, S.G. Johnson, and J.D. Joannopoulos, Waveguide branches in photonic crystals, *J. Opt. Soc. Am. B*, 18, 162–165, 2001.
8. R.W. Ziolkowski, FDTD Analysis of PBG waveguides, power splitters and switches, *Opt. Quantum Electron.*, 31, 843-855, 1999.
9. *Microcavities and Photonic Bandgaps: Physics and Applications*, Kluwer Academic Publishers, Elounda, Crete, 1995.
10. B. D'Urso, O. Painter, J. O'Brien, T. Tombrello, A. Yariv, and A. Scherer, Modal reflectivity in finite-depth two-dimensional photonic-crystal microcavities, *J. Opt. Soc. Am. B*, 15, 1155-1159, 1998.
11. H.-B. Sun, V. Mizeikis, Y. Xu, S. Juodkazis, J.-Y. Ye, S. Matsuo, and H. Misawa, Microcavities in polymeric photonic crystals, *Appl. Phys. Lett.*, 79, 1–3, 2001.
12. P.R. Villeneuve, S. Fan, and J.D. Joannopoulos, Microcavities in photonic crystals: mode symmetry, tunability, and coupling efficiency, *Phys. Rev. B*, 54, 7837–7842, 1996.
13. S.G. Johnson, F. Shanhui, A. Mekis, and J.D. Joannopoulos, Multipole-cancellation mechanism for high-Q cavities in the absence of a complete photonic band gap, *Appl. Phys. Lett.*, 78, 3388–3390, 2001.
14. R.D. Meade, A. Devenyi, J.D. Joannopoulos, O.L. Alerhand, D.A. Smith, and K. Kash, Novel applications of photonic band gap materials: low-loss bends and high-Q cavities, *J. Appl. Phys.*, 75, 4753–4755, 1994.

15. A. Scherer, O. Painter, B. D'Urso, R. Lee, and A. Yariv, InGaAsP photonic band gap crystal membrane microresonators, *J. Vac. Sci. Tech. B*, 16, 3906–3910, 1998.

16. A. Sharkawy, S. Shi, and D.W. Prather, Multichannel wavelength division multiplexing using photonic crystals, *Appl. Opt.*, 40, 2247–2252, 2001.

17. R. Stoffer, H.J.W.M. Hoekstra, R.M.D. Ridder, E.V. Groesen, and F.P.H.V. Beckum, Numerical studies of 2D photonic crystals: waveguides, coupling between waveguides and filters, *Opt. Quantum Electron.*, 32, 947–961, 2000.

18. R. Fan, R. Villeneuve, J.D. Joannopoulos, and H.A. Haus, Channel drop tunneling through localized states, *Phys. Rev. Lett.*, 80, 960–963, 1998.

19. H.A. Haus, S. Fan, P.R. Villeneuve, and J.D. Joannopoulos, Channel drop filters in photonic crystals, *Opt. Express*, 3, 4–11, 1998.

20. S.G. Johnson, P.R. Villeneuve, S. Fan, and J.D. Joannopoulos, Linear waveguides in photonic-crystal slabs, *Phys. Rev. B*, 62, 8212–8222, 2000.

21. A. Adibi, Y. Xu, R.K. Lee, A. Yariv, and A. Scherer, Properties of the slab modes in photonic crystal optical waveguides, *J. Lightwave Technol.*, 18, 1554–1564, 2000.

22. S.G. Johnson, S. Fan, P.R. Villeneuve, and J.D. Joannopoulos, Guided modes in photonic crystal slabs, *Phys. Rev. B*, 60, 5751–5758, 1999.

23. O. Painter, J. Vuckovic, and A. Scherer, Defect modes of a two-dimensional photonic crystal in an optically think dielectric slab, *J. Opt. Soc. Am. B*, 16, 275–285, 1999.

24. M. Charlton, M.E. Zoorob, G. Parker, M.C. Netti, J.J. Baumberg, J.A. Cox, and H. Kemhadjian, Experimental investigation of photonic crystal waveguide devices and line-defect waveguide bends, *Mater. Sci. Eng.*, B74, 17–24, 2000.

25. M. Gander, R. McBride, J. Jones, D. Mogilevtsev, T. Birks, J. Knight, and P. Russell, Experimental measurement of group velocity dispersion in photonic crystal fibre, *Electron. Lett.*, 35, 63–64, 1999.

26. C.J.M. Smith, R.M.D.L. Rue, M. Rattier, S. Olivier, H. Benisty, C. Weisbuch, P.R. Krauss, R. Houdre, and U. Oesterle, Coupled guide and cavity in a two-dimensional photonic crystal, *Appl. Phys. Lett.*, 78, 1487–1489, 2001.

27. S.-Y. Lin and J.G. Fleming, A three-dimensional optical photonic crystal, *J. Lightwave Technol.*, 17, 1944–1947, 1999.

28. H. Hirayama, T. Hamano, and Y. Aoyagi, Novel surface emitting laser diode using photonic band-gap crystal cavity, *Appl. Phys. Lett.*, 69, 791–793, 1996.

29. Y. Xia, B. Gates, and S.H. Park, Fabrication of three-dimensional photonic crystals for use in the spectral region from ultraviolet to near-infrared, *J. Lightwave Technol.*, 17, 1956–1962, 1999.

30. S. Rowson, A. Chelnokov, C. Cuisin, and J.-M. Lourtioz, Two-dimensional photonic bandgap reflectors for free-propagating beams in the mid-infrared, *J. Opt. A — Pure Appl. Opt.*, 1, 483–489, 1999.

31. S. Rowson, A. Chelnokov, and J.-M. Lourtioz, Two-dimensional photonic crystals in macroporous silicon: from mid-infrared (10um) to telecommunication wavelengths (1.3–1.5um), *J. Lightwave Technol.*, 19, 1989–1995, 1999.

32. S. Noda, N. Yamamoto, and M. Imada, Alignment and stacking of semiconductor photonic bandgaps by wafer-fusion, *J. Lightwave Technol.*, 17, 1948–1955, 1999.

33. J. Moosburger, H. Th., and A. Forchel, Nanofabrication techniques for lasers with two-dimensional photonic crystal mirrors, *J. Vac. Sci. Tech. B*, 18, 3501–3504, 2000.

34. P. Sabouroux, G. Tayeb, and D. Maystre, Experimental and theoretical study of resonant microcavities in two-dimensional photonic crystals, *Opt. Commun.*, 160, 33–36, 1999.

35. C.C. Cheng and A. Scherer, Fabrication of photonic band-gap crystals, *J. Vac. Sci. Tech. B.*, 13, 2696–2700, 1995.

36. C.C. Cheng, A. Scherer, R.-C. Tyan, Y. Fainman, G. Witzgall, and E. Yablonovitch, New fabrication techniques for high quality photonic crystals, *J. Vac. Sci. Technol. B*, 15, 2764–2767, 1997.

37. A.C. Edrington, A.M. Urbas, P. DeRege, C.X. Chen, T.M. Swager, N. Hadjichristidis, M. Xenidou, L.J. Fetters, J.D. Joannopoulos, Y. Fink, and E.L. Thomas, Polymer-based photonic crystals, *Adv. Mater.*, 13, 421–425, 2001.

38. D.J. Norris and Y. Vlasov, Chemical approaches to three-dimensional semiconductor photonic crystals, *Adv. Mater.*, 13, 371–376, 2001.

39. P. Jiang, G.N. Ostojic, R. Narat, D.M. Mittleman, and V.L. Colvin, The fabrication and bandgap engineering of photonic multilayers, *Adv. Mater.*, 13, 389–393, 2001.

40. Y. Xia, B. Gates, and Z.-Y. Li, Self-assembly approaches to three-dimensional photonic crystals, *Adv. Mater.*, 13, 409–413, 2001.

41. G. Feiertag, W. Ehrfeld, H. Freimuth, H. Kolle, H. Lehr, M. Schmidt, M.M. Sigalas, C.M. Soukoulis, G. Kiriakidis, T. Pedersen, J. Kuhl, and W. Koenig, Fabrication of photonic crystals by deep x-ray lithography, *Appl. Phys. Lett.*, 71, 1441–1443, 1997.

42. N.W. Ashcroft and N.D. Mermin, *Solid State Physics*, Holt, Rinehart and Winston, New York, 1976.

43. K.M. Leung and Y. F. Liu, Photon band structures: the plane-wave method, *Phys. Rev. B*, 41, 10188–10190, 1990.

44. L. Liu and J.T. Liu, Photonic band structure in the nearly plane wave approximation, *Eur. Phys. J. B*, 9, 381–388, 1999.

45. Z. Zhang and S. Satpathy, Electromagnetic wave propagation in periodic structures: Bloch wave solution of Maxwell's equations, *Phys. Rev. Lett.*, 65, 2650–2653, 1990.

46. K.M. Ho, C.T. Chan, and C.M. Soukoulis, Existence of a photonic gap in periodic dielectric structures, *Phys. Rev. Lett.*, 65, 3152–3155, 1990.

47. H. Kosaka, T. Kawashima, A. Tomita, M. Notomi, T. Tamamura, T. Sato, and S. Kawakami, Superprism phenomena in photonic crystals, *Phys. Rev. B*, 58, R10096–R10099, 1998.

48. H. Kosaka, T. Kawashima, A. Tomita, M. Notomi, T. Tamamura, T. Sato, and S. Kawakami, Photonic crystals for micro lightwave circuits using wavelength-dependent angular beam steering, *Appl. Phys. Lett.*, 74, 1370–1372, 1999.

49. H. Kosaka, T. Kawashima, A. Tomita, M. Notomi, T. Tamamura, T. Sato, and S. Kawakami, Self-collimating phenomena in photonic crystals, *Appl. Phys. Lett.*, 74, 1212–1214, 1999.

50. M. Tokushima, H. Kosaka, A. Tomita, and H. Yamada, Lightwave propagation through a 120 sharply bent single-line-defect photonic crystal waveguide, *Appl. Phys. Lett.*, 76, 952–954, 2000.

51. A. Mekis, J.C. Chen, I. Kurland, S. Fan, P.R. Villeneuve, and J.D. Joannopoulos, High transmission through sharp bends in photonic crystal waveguides, *Phys. Rev. Lett.*, 77, 3787–3790, 1996.

52. A.R. McGurn, Photonic crystal circuits: a theory for two- and three-dimensional networks, *Phys. Rev. B*, 61, 13235–13249, 2000.

53. M. Bayindir, B. Temmelkuran, and E. Ozbay, Propagation of photons by hopping: a waveguiding mechanism through localized coupled cavities in three-dimensional photonic crystals, *Phys. Rev. B*, 61, R11855–R11858, 2000.

54. E. Yablonovitch and T.J. Gmitter, Donor and acceptor modes in photonic band structures, *Phys. Rev. Lett.*, 67, 3380–3383, 1991.

55. J. Knight, T. Birks, P. Russell, and D. Atkin, All-silica single-mode optical fiber with photonic crystal cladding, *Opt. Lett.*, 21, 1547–1549, 1996.

56. T. Sondergaard, J. Arentoft, A. Bjarklev, M. Kristensen, J. Erland, J. Broeng, and S.E. Barkou, Designing finite-height photonic crystal waveguides: confinement of light and dispersion relations, *Opt. Commun.*, 194, 341–351, 2001.

57. T. Sondergaard, A. Bjarklev, M. Kristensen, J. Erland, and J. Broeng, Designing finite-height two-dimensional photonic crystal waveguides, *Appl. Phys. Lett.*, 77, 785–787, 2000.

58. A. Mekis and J.D. Joannopoulos, Tapered couplers for efficient interfacing between dielectric and photonic crystal waveguides, *J. Lightwave Technol.*, 19, 861–865, 2001.

59. E. Chow, S.Y. Lin, J.R. Wendt, S.G. Johnson, and J.D. Joannopoulos, Quantitative analysis of bending efficiency in photonic-crystal waveguide bends at 1.55 μm wavelengths, *Opt. Lett.*, 26, 286–288, 2001.

60. I. EL-Kady, M.M. Sigalas, R. Biswas, and K.M. Ho, Dielectric waveguides in two-dimensional photonic bandgap materials, *J. Lightwave Technol.*, 17, 2042–2049, 1999.
61. M. Loncar, T. Doll, J. Vuckovic, and A. Scherer, Design and fabrication of silicon photonic crystal optical waveguides, *J. Lightwave Technol.*, 18, 1402–1411, 2000.
62. A.R. McGurn, Intrinsic localized modes in nonlinear photonic crystal waveguides: dispersive modes, *Physica Lett., A*, 260, 314–321, 1999.
63. C.J.M. Smith, H. Benisty, S. Olivier, M. Rattier, C. Weisbuch, T.F. Krauss, R.M.D.L. Rue, R. Houdre, and U. Oesterle, Low-loss channel waveguides with two-dimensional photonic crystal boundaries, *Appl. Phys. Lett.*, 77, 2813–2815, 2000.
64. C. Weisbuch, H. Benisty, S. Olivier, M. Rattier, C.J.M. Smith, and T.F. Krauss, 3D control of light in waveguide-based two-dimensional photonic crystals, *IEICE Trans. Electron.*, E84–C, 660–668, 2001.
65. P. Pottier, C. Seassal, X. Letartre, J.L. Leclercq, P. Viktrorovitch, D. Cassagne, and C. Jouanin, Triangular and hexagonal high Q-factor 2D photonic bandgap cavities on II-V suspended membranes, *J. Lightwave Technol.*, 17, 2058–2062, 1999.
66. R.D. Meade, K.D. Brommer, A.M. Rappe, and J D. Joannopoulos, Photonic bound states in periodic dielectric materials, *Phys. Rev. B*, 44, 13772–13774, 1991.
67. R.D. Meade, A.M. Rappe, K.M. Brommer, J.D. Joannopoulos, and O.L. Alerhand, Accurate theoretical analysis of photonic band-gap materials, *Phys. Rev. B*, 48, 8434–8437, 1993.
68. O. Painter, A. Husain, A. Scherer, J. O'Brien, I. Kim, and P.D. Dapkus, Room temperature photonic crystal defect lasers at near-infrared wavelengths in InGaAsP, *J. Lightwave Technol.*, 17, 2082–2088, 1999.
69. X.-Y. Lei, H. Li, F. Ding, W. Zhang, and N.-B. Ming, Novel application of perturbed photonic crystal: high-quality filter, *Appl. Phys. Lett.*, 71, 2889–2891, 1997.
70. S. Fan, P.R. Villeneuve, and J.D. Joannopoulos, Theoretical analysis of channel drop tunneling process, *Phys. Rev. B*, 59, 15882–15892, 1999.
71. F. Jian and H. Sai-Ling, Analysis of higher order channel dropping tunneling processes in photonic crystals, *Chin. Phys. Lett.*, 17, 737–739, 2000.
72. E. Centeno, B. Guizal, and D. Felbacq, Multiplexing and demultiplexing with photonic crystals, *J. Opt. A: Pure Appl. Opt.*, 1, L10–L13, 1999.
73. H.-B. Lin, R.J. Tonucci, and A.J. Campillo, Two-dimensional photonic bandgap optical limiter in the visible, *Opt. Lett.*, 23, 94–96, 1998.
74. D. Labilloy, H. Benisty, C. Weisbuch, T.F. Krauss, C.J M. Smith, R. Houdre, and U. Oesterle, High-finesse dish microcavity based on a circular Bragg reflector, *Appl. Phys. Lett.*, 73, 1314–1316, 1998.
75. R.A. Shelby, D.R. Smith, and S. Schultz, Experimental verification of a negative index of refraction, *Science*, 292, 77–79, 2001.
76. M. Notomi, Theory of light propagation in strongly modulated photonic crystals: refractionlike behavior in the vicinity of the photonic band gap, *Phys. Rev. B*, 62, 10696–10705, 2000.
77. M. Notomi and T. Tamamura, Direct visualization of photonic band structure for three-dimensional photonic crystals, *Phys. Rev. B*, 61, 7165–7168, 2000.
78. H. Kosaka, T. Kawashima, A. Tomita, M. Notomi, T. Tamamura, T. Sato, and S. Kawakami, Superprism phenomena in photonic crystals: toward microscale lightwave circuits, *J. Lightwave Technol.*, 17, 2032–2034, 1999.
79. H. Kosaka, A. Tomita, T. Kawashima, T. Sato, and S. Kawakami, Splitting triply degenerate refractive indices by photonic crystals, *Phys. Rev. B*, 62, 1477–1480, 2000.
80. S.-Y. Lin, V.M. Hietala, L. Wang, and E.D. Jones, Highly dispersive photonic band-gap prism, *Opt. Lett.*, 21, 1771–1773, 1996.
81. P. Halevi, Photonic crystal optics and homogenization of 2D periodic composites, *Phys. Rev. Lett.*, 82, 719–722, 1999.
82. H. Kosaka, T. Kawashima, A. Tomita, T. Sato, and S. Kawakami, Photonic-crystal spot-size converter, *Appl. Phys. Lett.*, 76, 268–270, 2000.

83. S.G. Johnson, C. Manolatou, S. Fan, P.R. Villeneuve, and J.D. Joannopoulos, Elimination of cross-talk in waveguide intersections, *Opt. Lett.*, 23, 1855–1857, 1998.

84. A.M. Zheltikov, S.A. Magnitskil, and A.V. Tarasishin, Localization and channeling of light in defect modes of a two-dimensional photonic crystals, *JETP Lett.*, 70, 323–328, 1999.

85. M. Bayindir and E. Ozbay, Heavy photons at coupled-cavity waveguide band edges in a three-dimensional photonic crystal, *Phys. Rev. B*, 62, R2247–R2250, 2000.

86. M. Bayindir, B. Temelkuran, and E. Ozbay, Tight-binding description of the coupled defect modes in three-dimensional photonic crystals, *Phys. Rev. Lett.*, 84, 2140–2143, 2000.

87. M. Meier, A. Mekis, A. Dodabalapur, A. Timko, R.E. Slusher, J.D. Joannopoulos, and O. Nalamasu, Laser action from two-dimensional distributed feedback in photonic crystals, *Appl. Phys. Lett.*, 74, 7–9, 1999.

88. T. Birks, J. Knight, and P. Russell, Endlessly single-mode photonic crystal fiber, *Opt. Lett.*, 22, 961–963, 1997.

89. J. Knight, T. Birks, P. Russell, and J.P.D. Sandro, Properties of photonic crystal fiber and the effective index model, *J. Opt. Soc. Am. A*, 15, 748–752, 1998.

90. J. Marorell, R. Vilaseca, and R. Corbalan, Second harmonic generation in photonic crystal, *Appl. Phys. Lett.*, 70, 702–704, 1997.

91. P.R. Villeneuve, S. Fan, and J.D. Joannopoulos, Air-bridge microcavities, *Appl. Phys. Lett.*, 67, 167–169, 1995.

92. A. Taflove and S.C. Hagness, *Computational Electrodynamics: The Finite-Difference Time-Domain Method*, 2nd ed., Artech House, Boston, 2000.

93. D.J. Ripin, K.-Y. Lim, G.S. Petrich, P.R. Villeneuve, S. Fan, E.R. Thoen, J.D. Joannopoulos, E.P. Ippen, and L.A. Kolodziejski, One-dimensional photonic bandgap microcavities for strong optical confinement in GaAs and GaAs/AlxOy semiconductor waveguides, *J. Lightwave Technol.*, 17, 2152–2160, 1999.

94. M. Boroditsky, R. Vrijen, P.R. Krauss, R. Coccioli, R. Bhat, and E. Yablonovitch, Spontaneous emission extraction and purecell enhancement from thin-film 2D photonic crystals, *J. Lightwave Technol.*, 17, 2096–2112, 1999.

95. R. Gonzalo, P.D. Maagt, and M. Sorolla, Enhanced patch-antenna performance by suppressing surface waves using photonic-bandgap substrates, *IEEE Trans. Microwave Theory Tech.*, 47, 2131–2138, 1999.

96. A. Sharkawy, S. Shouyuan, and D.W. Prather, Heterostructure photonic crystals, *J. Lightwave Technol.*, in press, 2001.

97. J.N. Winn, S. Fan, and J.D. Joannopoulos, Intraband transitions in photonic crystals, *Cond. Matter Mater. Phys.*, 59, 1551–1554, 1999.

98. P. Halevi and F.R. Mendieta, Tunable photonic crystals with semiconducting constituents, *Phys. Rev. B*, 85, 1875–1878, 2000.

99. S.W. Leonard, J.P. Mondia, H.M.V. Driel, O. Toader, and S. John, Tunable two-dimensional photonic crystals using liquid-crystal infiltration, *Cond. Matter Mater. Phys.*, 61, R2389–R2392, 2000.

100. D. Kang, J.E. Maclennan, N.A. Clark, A.A. Zakhidov, and R.H. Baughman, Electro-optic behaviours of liquid crystal filled silica opal photonic crystals: effect of liquid crystal alignment, *Phys. Rev. B*, 86, 4052–4055, 2001.

101. A. Hache and M. Bourgeois, Ultrafast all-optical switching in a silicon-based photonic crystal, *Appl. Phys. Lett.*, 77, 4089–4091, 2000.

102. S. Lan, S. Nishikawa, and O. Wada, Leveraging deep photonic band gaps in photonic crystal impurity bands, *Appl. Phys. Lett.*, 78, 2101–2103, 2001.

103. M. Florescu and S. John, Single-atom switching in photonic crystals, *Phys. Rev. A*, 64, 338011–3380121, 2001.

104. P. Tran, Optical switching with nonlinear photonic crystal: a numerical study, *Opt. Lett.*, 21, 1138–1140, 1996.

105. D. Petrosyan and G. Kurizki, Photon–photon correlations and entanglement in doped photonic crystals, *Phys. Rev. A*, 64, 238101–23806, 2001.

106. A.D. Lustrac, F. Gadot, S. Cabaret, J.-M. Lourtioz, T. Brillat, A. Priou, and E. Akmansoy, Experimental demonstration of electrically controllable photonic crystals at centimeter wavelengths, *Appl. Phys. Lett.*, 75, 1625–1627, 1999.

107. V. Lousses and J.P. Vigeneron, Self-consistent photonic band structure of dielectric superlattices containing nonlinear optical materials, *Phys. Rev. E*, 63, 2001.

108. A. Sharkawy, S. Shouyuan, and D.W. Prather, Optical Networks on a Chip Using Photonic Bandgap Materials, presented at SPIE 46th Annual Meeting, San Deigo, CA, USA, 2001.

109. S. Fan, P.R. Villeneuve, and J.D. Joannopoulos, Theoretical investigation of fabrication-related disorder on the properties of photonic crystals, *J. Appl. Phy.*, 78, 1415–1418, 1995.

110. A. Chutinan and S. Noda, Effects of structural fluctuations on the photonic bandgap during fabrication of a photonic crystal: a study of a photonic crystal with a finite number of periods, *J. Opt. Soc. Am. B*, 16, 1398–1402, 1999.

111. A.A. Asatryan, P.A. Robinson, L.C. Botten, R.C. McPhedran, and N.-A.P. Nicorovici, Effects of geometric and refractive index disorder on wave propagation in two-dimensional photonic crystal, *Phys. Rev. E*, 62, 5711–5720, 2000.

22

Nanostructured Materials

CONTENTS

Airat A. Nazarov
Russian Academy of Science

Radik R. Mulyukov
Russian Academy of Science

22.1 Introduction

Nanostructured materials (NSMs), nanocrystals, or nanophase materials are polycrystals with an ultra-fine grain size in the range of 3 to 100 nm. First results on the preparation of NSMs by a compaction of nanometer-sized metal powders and their properties were reported by Russian investigators.[1] However, the idea of a significant modification of the properties of materials by the formation of a nanocrystalline structure was first formulated by Gleiter and co-workers.[2–4] After these publications nanocrystals have gained much interest of materials scientists, because they exhibit a number of improved properties as compared with otherwise the same polycrystals with conventional coarse-grain sizes. The spectrum of properties affected by the nanocrystalline structure ranges from the electronic ones to the macroscopic mechanical behavior. A number of physical and mechanical properties of NSMs have been found to be superior to the properties of conventionally coarse-grained polycrystals and are attractive for engineering applications. The ability of nanocrystalline alloys to deform superplastically at lower temperatures and higher strain rates is expected to significantly lower the costs of metal-forming technologies. High strength, ductility, and wear resistance can make the use of nanocrystalline refractory metals and ceramics as parts of high-temperature devices such as engines. Enhanced soft magnetic properties of nanocrystalline alloys, such as the low coercivity and high magnetic permeability, are already in use in transformers. The NSMs have many orders of magnitude higher diffusion coefficient that can find applications, for example, in hydrogen storage technologies. The nanostructured coatings significantly increase the wear resistance of otherwise soft materials. Modified electronic and electric properties of nanostructured metals and covalent ceramics may find wide applications in electronics. The importance of NSMs for the technology and life of the new century has been recognized by the fact that a review on these advanced materials has been included into the millennium issue of *Acta Materialia*.[5]

Two main factors are considered to determine the properties of NSMs. First, when the grain diameter approaches a few nanometers, the structure modulation length becomes comparable to the characteristic distances of physical processes in solids and the size effects are pronounced. The second important factor is that a significant volume fraction of NSMs is occupied by the atoms at defect sites, mainly in interfaces. Thus, the atomic structure of interfaces is crucial for the properties of these materials.

The methods of synthesis, structural characteristics, properties, and applications of NSMs have been reviewed in a number of papers.[4–16] The present chapter does not aim at a full description of all these issues, as the problem of NSMs has become too wide to elucidate in one book chapter. Instead of this, there will be an attempt to arrive at some general conclusions on the structure of these materials, which is a key issue for an understanding of their properties. Because the structure of nanocrystals depends on the preparation conditions, the most widely used technological synthesis routes of about a dozen existing methods will be considered at the beginning. The body of the chapter will be devoted to an analysis of the structure of the crystallites and interfaces in NSMs prepared by these main methods. In the light of the structural features considered, the diffusion and mechanical properties will be analyzed.

Siegel's paper,[16] which was published after 7 years of intensive studies of the NSMs, was a milestone that closed the period of preliminary studies of the nanocrystals and opened more careful investigations. In the present chapter, attention will be focused on selecting the firmly established facts in this area and comparing what has changed in the last 7 years. Throughout the chapter the authors try to hold neither of the two existing opposite viewpoints on the structure of grain boundaries in NSMs,[4,16] because they both have a significant background and may have a relation to the real nanostructures. The authors realize, however, that this will not always be possible due to the influence of their own preferences.

22.2   Preparation of Nanostructured Materials

Methods of the preparation of NSMs can be divided into the following four main groups: powder technology, severe (or heavy) plastic deformation methods, crystallization from the amorphous state, and deposition methods.[12]

22.2.1   Powder Compaction Methods

Nanometer-sized particles can be obtained by a condensation of metals from a vapor in a helium atmosphere as originally used by Gleiter and co-workers.[2–4] A metal is evaporated inside a chamber filled with a low-pressure helium, where the atoms collide with the gas atoms and condense into particles, the diameter of which depends on the evaporation rate and the pressure of helium. The particles are collected on a cold finger in the chamber. Then they are scraped and compacted in ultrahigh vacuum conditions in order to avoid contamination. This method enables producing NSMs with the grain size down to about 3 nm. The distribution of the grain size is fairly narrow and is well fitted by the log-normal distribution function.[16] Provided that the nanometer-sized powders are produced by gas condensation or any other method, the powder compaction technology allows the production of nanocrystals from any metal, alloy, and ceramic in sufficiently large quantities. The method allows also for an easy control of the mean grain size. However, it suffers from significant drawbacks. In NSMs produced in this way, there is always a residual porosity. The density of the samples is usually in the range of 70 to 97% of the perfect crystal density.[16] Recently developed warm compaction methods have reduced the porosity greatly.[17] Also, in spite of all precautions, inert gas atoms are preserved in compacted metals that can influence the properties of the final product.[18] Finally, the method provides small samples (10 mm in diameter and about 0.1 mm thick) suitable mainly for scientific investigations.

22.2.2 Microstructure Refinement by Severe Plastic Deformation

The nanocrystalline structure can be formed also in a bulk material by refining its microstructure using severe plastic deformation. Four methods of NSM preparation use this as a basic process: ball milling, torsion straining under quasi-hydrostatic pressure, equal-channel angular pressing, and multiple forging. Despite this common basis, ball milling has been considered as a separate method,[9] while the term *severe plastic deformation* has been related to the latter three techniques.[7,8,15]

Ball milling, or mechanical attrition, is a versatile process to make NSMs in large quantities.[9,14,19] In this method, powders of metals with sizes from a few to about 200 micrometers are placed together in a container with the hardened steel or tungsten carbide-coated balls with a diameter about 10 mm and milled for many hours. The optimal ball-to-sample mass ratio is 10:1 to 5:1. During milling the powder particles collide with the balls and are subjected to severe plastic deformation. The particles are repeatedly deformed, fractured, and cold-welded. Heavy deformation introduces a high density of defects into the solid: vacancies, dislocations, stacking faults, and, particularly, interfaces. Structural studies allowed one to establish the following mechanisms of the structure refinement during mechanical attrition.[19] Initially, the deformation is localized in shear bands having the thickness 0.5–1 μm and a high density of dislocations. At higher strains these bands disintegrate to subgrains. These subgrains have sizes in the nanometer range (20 to 30 nm). During further attrition the number of bands increases, they coalesce, the misorientations of the subgrains increase, and the grain size decreases steadily. A balance among the plastic deformation and the recovery and recrystallization gives the lower bound for the grain size, which varies inversely with the melting point and directly with the stacking fault energy. Grain size down to 8 nm can be obtained by this technique. The quantity of powder milled at a time depends on the mill type used: 10–20 g for the SPEX shaker mill, a few hundred grams for planetary ball mills, and 0.5–40 kg for attritor mills.[20] When a mixture of two or more elemental powders is milled, the process is referred to as *mechanical alloying.*[20] The problem when using ball milling is that the vial (milling container), the balls, and the milling atmosphere can contaminate the powder. To avoid this, balls and vial are covered with the metal to mill, and an inert gas atmosphere is used. Some problems with the compaction of the micrometer-sized powders to obtain bulk nanocrystals remain. The advantage of the mechanical attrition/alloying is the very large spectrum of materials treated that makes the method attractive for technological applications.

Torsion straining under quasi-hydrostatic pressure is based on the use of a Bridgeman anvil-type device.[21] The possibility of the formation of a nanocrystalline structure by this method was first demonstrated by Smirnova et al.[22] A disk-shaped sample is put between two anvils and subjected to a pressure of 2–5 GPa (Figure 22.1). Rotation of one of the anvils forces the samples to deform by torsion. Five rotations of the anvil are usually enough to form a homogeneous microstructure with the grain size less than a few hundred nanometers, that is, in the submicrometer range.[7,8] Although the grain size is larger than the one obtainable by powder compaction or ball milling techniques, submicrocrystalline (SMC) materials obtained by this method exhibit practically the same set of properties as the *true* NSMs.[7,8] Moreover, in some metals and alloys with high melting temperature, a grain size as small as 50 nm can be obtained.

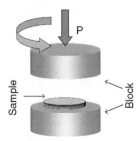

FIGURE 22.1 A schematic rendering of the preparation of nanostructured materials by torsion straining under quasi-hydrostatic pressure.

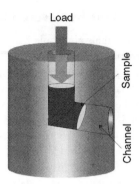

FIGURE 22.2 A schematic representation of equal channel angular pressing as a tool for the preparation of submicrocrystalline materials.

The equal-channel angular pressing was invented as a method for the severe plastic deformation of bulk materials[23] and later was shown to be useful for the formation of an ultrafine-grained structure in metals and alloys.[24] This method allows the deformation of cylinder-shaped samples with a diameter about 10 mm and a length 100 mm. The sample is repeatedly pressed through two channels of the same diameter such that the dimensions of the samples do not change after each pass (Figure 22.2). At the intersection of the channels, the sample undergoes a very large shear strain.[23–25] After each pass the sample can be rotated around its axis to an angle 90° or 180°. If necessary, the treatment can begin at a high temperature and finish at the ambient one. For a number of pure metals and alloys, this method has been demonstrated to yield an SMC structure with the grain size in the range 100...300 nm.[15]

The essence of the multiple forging is in the multiple repeats of simple, free forging operations with a change of the loading axis.[26,27] This method allows the preparation of bulk samples of relatively brittle materials, such as titanium alloys, because the forging starts at elevated temperatures. The microstructure refinement by multiple forging is usually accompanied by dynamic recrystallization.

The advantage of severe plastic deformation methods is the ability of the method to provide pore- and contamination-free bulk nanocrystals; and these methods are considered as promising preparation methods for technological applications.[28]

22.2.3 Crystallization from the Amorphous State

If amorphous alloys are treated in the conditions providing a large number of crystallization centers and low growth rate, a complete crystallization into ultrafine-grained polycrystalline materials occurs.[10] This method was first used to obtain soft magnetic nanocrystals of the alloys Fe-Cu-Nb-Si-B, called FINEMET materials.[29] By controlling the crystallization kinetics through an optimization of the annealing temperature and time, heating rate, etc., the amount of the crystallization, the final grain size, and other structural characteristics can be changed. The method is limited to the elements and alloys in which a glassy structure can be obtained by any method (melt spinning, mechanical alloying, deposition, etc.). As no consolidation process is involved in the preparation process, the nanocrystals obtained by this method are pore-free and contain clean interfaces. The present method is often referred to as a *devitrification of amorphous alloys*.[14]

22.2.4 Electrodeposition

The electrodeposition technique includes conventional DC electroplating, pulsed-current deposition, and co-deposition and is a technologically attractive route to synthesize nanocrystals of pure metals, alloys, and metal matrix composites.[30] It has advantages such as the possibility to produce fully dense samples, few shape and size limitations, high production rates, and a large number of metals, alloys, and composites that can be deposited as nanocrystals. The materials can be deposited as thin films (up to 100 μm) or in the bulk form, as plates of several millimeters' thickness. The grain size can be controlled by changing the electrodeposition variables such as the bath composition, pH, temperature, and current density. The conditions can be chosen that favor the nucleation of new grains rather than the growth of existing ones. One more advantage of the method is an easy transfer from the research laboratory to the technology.

22.3 Structure

22.3.1 Classification of Nanostructured Materials

Gleiter[5] classified the NSMs according to the shapes and chemical composition of crystallites. According to the shapes, three categories of NSMs can be distinguished: layer-shaped, rod-shaped, and equiaxed crystallites. In the classification according to chemical composition, the NSMs of the first family contain crystallites and interfaces of the same chemical composition (e.g., in pure nanostructured metals). The NSMs of the second family contain crystallites of different chemical compositions. NSMs with a compositional variation between the crystallite and interfaces belong to the third family. These are alloys in which the atoms of one type are segregated in grain boundaries. The fourth family of NSMs consists of nanometer-sized crystallites dispersed in a matrix of a differing chemical composition. It is not necessary that all structural elements of NSM are crystalline. The grains or the matrix may be in a glassy phase. Also the NSMs may contain quasicrystalline components. Thus, there are many possibilities for compositional, phase, and microstructural variations of the NSMs that make these materials the best candidates for the satisfaction of the widest aspects of technology needs in the future.

The types of NSMs considered above consist of nanometer-sized building blocks delimited by interfaces. All these materials have a very high free energy stored mainly in interfaces and are far away from the thermodynamic equilibrium. Therefore, they have been classified as nonequilibrium NSMs.[5] Only the nonequilibrium NSMs will be considered in the present chapter.

The preparation methods considered above and other methods are all based on nonequilibrium processes. Changing the method or conditions of preparation also changes the final microstructure. Thus, the structure and properties of NSMs can be manipulated by controlling the preparation routes. This allows the generation of a practically unlimited variety of properties. At the same time, this causes the problems of the controlling, reproducibility, and stability of these properties. That is the main reason for the focusing research efforts on the structure–property relationships for nanocrystals prepared by different techniques.

22.3.2 Interface-to-Bulk Ratio in Nanostructured Materials

From the studies of grain boundaries (GBs) in coarse-grained polycrystals and artificially prepared bicrystals, their widths are known to be on the order of 0.5–1 nm.[31] In the GB region the atomic structure undergoes a transformation from the lattice with one orientation to the lattice with another orientation; thus, interfacial atoms belong neither to one nor to the other lattice assuming some intermediate positions. This makes the atomic structure of interfaces quite different from the crystal lattice. Two opposite viewpoints on this structure have existed in the theory of grain boundaries originating from the models of transition lattice[32] and amorphous cement.[33] Direct high-resolution electron microscopy (HREM) observations and many indirect data show that short-period, special GBs have an ordered atomic structure.[31] For the most general GBs, however, the existence of the crystal-like order is still under debate.[31]

The basic idea of tailoring an NSM put forward by Gleiter[2] was to reduce the grain size to such a scale on which the fraction of atoms residing in interfacial regions was comparable to that of the lattice atoms. Assuming the GB width about $\delta \approx 1$ nm, the volume fraction of interfaces can be estimated as $\Delta \approx 3\delta/d$ where d is the grain size. This gives $\Delta = 30\%$ for $d = 10$ nm. Below 10 nm, nearly half of the atoms are situated in GBs and their junctions.[34] These atoms are in a completely different environment than most atoms in conventional polycrystals. Thus, they are expected to significantly modify the properties of the materials. Also, one can expect that, for the very small size of crystallites, their own structure will be considerably influenced by the interfaces.

22.3.3 Lattice Strains in Nanostructured Materials

Internal elastic strains are one of the key structural features of nanocrystals. These strains have different origins and play an important role in the structural modifications of nanocrystals and in their properties. The most powerful tool for the study of the internal strains is the X-ray diffraction. The X-ray diffraction technique allows the determination of the lattice parameter a_0, root-mean-square (rms) strain $\varepsilon_i = <\varepsilon^2>^{1/2}$, grain size d, and the Debye–Waller parameter B, which depends on the rms displacement of atoms from their equilibrium lattice sites due to defects and thermal oscillations.[35,36]

22.3.3.1 Lattice Parameter

The lattice parameter of isolated nanoparticles is known to be decreased as compared with that of the bulk crystals. For example, the measurements of a_0 for the particles of Ag and Pt with diameters 3.1 and 3.8 nm gave the values, which are 0.7 and 0.5% less than the lattice parameters of bulk Ag and Pt, respectively.[37] This lattice contraction is usually related to the Laplace pressure originating from the surface tension, which can be estimated as $\Delta p = \gamma/d$ where γ is the free energy of the surface.[38] Stresses of the same sign, albeit substantially lower, are expected in compacted NSMs.[39] As a result, lattice contraction would be always observed in nanocrystals. However, the data on the change of lattice parameter in NSMs are controversial. Some experimental data have been analyzed by Lu and Zhao.[40] Out of 12 cases considered in the cited paper, only one[35] supported the decrease of a_0 in NSMs. In all other cases, including NSMs prepared by powder compaction, ball milling, electrodeposition, and devitrification of the amorphous state, an increase of a_0 or, more generally, of the unit cell volume, was observed. Support to the decrease of a_0 comes from the studies of NSMs prepared by severe plastic deformation. In nanostructured Cu prepared by ECAP and torsion straining, the lattice parameter was 0.04 and 0.03% less than in the bulk Cu, respectively.[41] On the other hand, an extended X-ray absorption fine structure spectroscopy (EXAFS) of nanostructured Cu prepared by torsion straining has shown no change of the lattice parameter.[42]

Gorchakov et al.[43] have shown that, in fact, the Laplace pressure cannot result in a contraction of isolated particles. They studied the surface relaxation in small particles and found that this relaxation can either decrease or increase the crystal unit cell, depending on the particle size and material.

The decrease of a_0 in relatively coarse-grained SMC Cu has no satisfactory explanation.[15] The increase of the unit cell volume in NSMs can be explained more easily and seems more natural. In the case of a powder compaction or a crystallization from the melt or amorphous state, the initial volume occupied by atoms is larger than the volume of the fully dense NSM consisting of the same number of atoms. On compacting or crystallization, a part of the excess volume remains in interfaces, making their structure less dense. This can occur, for instance, if a shell of already compacted particles, as has been demonstrated in Figure 22.3a, surrounds a particle. Then, after the hydrostatic compaction or complete crystallization, the inner particle will be a source of dilatation (Figure 22.3b). As known, the dilatation center increases the volume of material.[44] Therefore, a lattice expansion is expected in NSMs prepared by powder compaction and crystallization from the amorphous state as observed. In NSMs prepared by other methods, the lattice expansion may be caused by defects introduced in interfaces. The less dense structure of interfaces can result in some negative pressure on the interfaces as was suggested by Fecht,[45] and this can cause an increase of lattice parameter. Computer simulation studies, which will be considered in more detail in the section devoted to interfaces, have also shown that the lattice parameter of NSMs is approximately 1% larger than that of an ideal crystal.[46]

22.3.3.2 Root Mean Square Elastic Strain

A decrease of the grain size and an increase of the internal elastic strains lead to a broadening of the X-ray diffraction peaks. The diffraction peaks are analyzed to determine d and ε_i by the Sherrer, or Warren–Averbach, or Williamson–Hall methods.[47] These methods are based on the fact that the widths

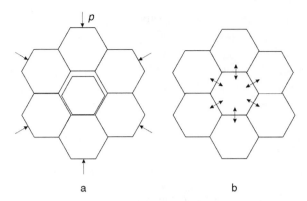

FIGURE 22.3 A group of small particles before (a) and after (b) a compaction. After the compaction and unloading, voids are eliminated and tensile stresses are introduced.

of diffraction lines arising from the small crystallite diameter and from the rms strain vary differently with the scattering angle.

The rms strain in NSMs has been shown to be considerably higher than in conventional cold-worked polycrystals. Internal strains in the range $\varepsilon \approx 0.1$–0.5% were observed in nanocrystalline Pd obtained by the gas-condensation method.[39,48,49] In NC Ru and AlRu prepared by ball milling, the strains reached values as high as 1 and 3%, respectively.[50] A similar value of 1% was observed in nanostructured Ni_3Al prepared by severe plastic deformation.[51] The preparation of nanocrystalline Se by a complete crystallization from the amorphous state also results in internals strains in the range 0.2–0.7% for the grain size in the range 60–10 nm.[52] Contrary to coarse-grained polycrystals, the elastic strains in NSMs are very anisotropic. They have usually a maximum along the <100> direction and a minimum along <111>.[39,52] For the nanocrystals prepared by powder compaction. this anisotropy can be explained by an anisotropy of the elastic moduli and by the presence of a preferred axis for the stresses induced by the compaction.[39] For the same reason, the torsion straining under pressure (Figure 22.1) also introduces an anisotropy in the distribution of internal strains.[51]

Another important feature of the lattice strains in NSMs is their increase with the decrease of the grain size. Analyzing the existing data, Reimann and Würschum[39] showed that ε_i decreased approximately proportionally to d^{-1}. From this analysis it was also concluded that either the strains increased with decreasing d, or size-independent strains were localized in the shells near interfaces.[39]

The origin of high internal strains in NSMs is not fully understood yet. Moreover, they may have been induced by different sources depending on the preparation method. Nevertheless, HREM investigations have provided direct evidence that the strains of nanocrystallites are commonly induced by interfaces.[53,54] Therefore, the possible GB sources of the internal strains will be considered in the next sections.

22.3.4 Interfaces in Nanostructured Materials

As the interfaces occupy a significant fraction of the volume of NSMs, their structure is extremely important for the properties of nanocrystals. Therefore, greatest efforts have been undertaken to understand whether the GBs in NSMs have an atomic structure significantly different from that in coarse-grained materials.

A number of experimental techniques have been applied to study the structure of interfaces in NSMs. Among these, the most widely used are x-ray diffraction (XRD), EXAFS, transmission electron microscopy (TEM), HREM, positron annihilation spectroscopy (PAS), Mössbauer spectroscopy, small-angle neutron scattering (SANS), and Raman spectroscopy. At present the data obtained by nearly all of these methods are available for nanocrystals produced by the techniques of inert gas condensation, severe plastic deformation, ball milling, and crystallization from the amorphous state.

22.3.4.1   Grain Boundaries in Compacted Nanocrystals

Gleiter and co-workers studied the structure of NSMs obtained by the gas condensation methods by the use of XRD, EXAFS, Mössbauer spectroscopy, SANS and PAS.[3,4,55–58]

The XRD study of nanocrystalline Fe has shown a large diffuse scattering background, which is absent in the coarse-grained iron.[3,55] The measured interference function was fitted by the one computed from the following assumptions. The model crystal was a mixture of Fe crystallites with sizes 6 nm and 4 nm and volume fractions 75% and 25%, respectively. In each crystallite the atoms of the outer layer were displaced by 50% of the nearest neighbor distance from their lattice sites in random directions. In order to model the strains at the GB cores, the atoms of the second and third outermost layers were displaced by 25% of the nearest neighbor distance from the lattice sites, also in random directions. The interference function computed on this model reproduced very well the heights and widths of all peaks and the background intensity. On the other hand, computed interference functions assuming a short-range order in the GB structure failed to reproduce all these characteristics correctly.

EXAFS studies yield direct information on the short-range order of a solid in terms of the coordination numbers and nearest neighbor distances.[59] If one assumes, according to the results of XRD investigations, that the structure of interfaces is characterized by a wide variety of interatomic spacings with no short-range order, the oscillations of the EXAFS signal are caused only by the crystalline component.[56] Therefore, the amplitudes of the EXAFS oscillations of a NSM will be less than the amplitudes of a conventional polycrystal of the same mass. EXAFS studies of nanocrystalline Cu have shown that the radii of the coordination shells in conventional and NSM coincide well and, indeed, the amplitudes of the oscillations are significantly reduced.[56] This means that the average first-shell coordination number in NSMs is significantly less than that in the ideal crystal. Therefore, the EXAFS data also seemed to support the conclusion on the absence of the short-range order in the nanograin boundaries.

The Mössbauer spectrum of nanostructured Fe has been interpreted to be made up of two sub-spectra.[57] One subspectrum corresponds to the crystalline Fe, and the other was assigned to the interfacial component. The second subspectrum is characterized by an enhanced hyperfine magnetic field, a larger line width, and an increased isomer shift in comparison with the crystalline component. Because the compression increases the electron density and results in a smaller isomer shift, it has been concluded that the interfacial component has a decreased mass density with a wide distribution of interatomic distances.[57]

From the SANS studies the mass density of interfaces was estimated to be about 60 to 70% of the crystal density.[58]

The above results allowed Gleiter and co-authors to put forward the first model for the atomic structure of interfaces in NSMs. It has been concluded that the GBs in NSMs have a gas-like atomic structure with a considerably reduced mass density, lacking either the long-range or the short-range order.[4]

A strong additional support to the idea on the lower density of interfaces in NSMs comes from the positron annihilation spectroscopy. PAS is particularly well suited for studying the free volumes in crystals.[60] In defect-free crystals the positrons annihilate in a time interval of about 100 ps. Trapping positrons by structural defects, such as vacancies, their agglomerates, and pores, increases the lifetime of the positrons. The larger the defect, the longer is the lifetime of trapped positrons. Positron lifetimes in the ideal lattice, single vacancies, and small vacancy clusters in metals, alloys, and semiconductors are well characterized experimentally and theoretically.[60]

The PAS studies of NSMs prepared by the gas-condensation technique revealed two dominant positron lifetime components τ_1, τ_2 as well as a weak third long-lived component $\tau_3 > 500$ ps.[60–62] The lifetime $\tau_1 = 100$–250 ps is attributed to free volumes of the size of about a lattice vacancy. The lifetime $\tau_2 = 300$–400 ps corresponds to the free volumes with the size of about 10 missing atoms. From the facts that the short lifetime τ_1 is observed even after annealing of the nanostructured Pd, Cu, and Fe up to 600 K, in which lattice vacancies are annealed out in the coarse-grained solids, and that the intensity ratio I_1/I_2 increases with the compacting, it follows that the vacancy-like free volumes are located in interfaces. The larger free volumes with sizes of 0.6–0.7 nm are expected at the junctions of interfaces. The long-lived component τ_3 has been attributed to the ortho-positronium formation in voids of the size of a missing grain.

However, the model of the gas-like GB structure in NSMs was later rejected on the basis of the results from HREM. Direct HREM observations of the structure of nanostructured Pd have shown that the structures of interfaces "are rather similar to those of conventional high-angle grain boundaries."[63,64] From these observations it has been concluded that any significant structural disorder extends no further than 0.2 nm from the interface plane, and any atomic displacement is not more than 12% of the nearest neighbor distance.[63] Although the notion on disordered interfacial structures has been rejected, the HREM studies have indicated a high-energy state of GBs, unlike conventional polycrystals.[64] In most of the cases observed, the lattice planes near the GBs are slightly distorted, indicating local stresses in the GB region. Bending of the grains due to high internal stresses has been observed. The mean dislocation density has been shown to be higher than 10^{15} m^{-2}, rarely reached in plastically deformed metals.

It should be noted that the HREM studies have some limitations and must be interpreted with precautions. The foils for HREM observations are very thin, and their preparation might result in a significant relaxation of the interfacial structure.

Nevertheless, an additional support to the model of GBs not fundamentally different from the GBs in conventional polycrystals has also been obtained by the X-ray diffraction and EXAFS techniques, which initially led to the model of gas-like interfaces. An XRD study of nanostructured Pd has shown no diffuse scattering background and detected only a peak broadening due to the small grain size and high internal strains.[35,65] Therefore, no evidence for the lack of the short- and long-range order in GBs was observed. An EXAFS study of nanostructured Cu, during which special care was given to the preparation of the samples and elimination of experimental artifacts, has shown that the average first-shell coordination number in GBs is not significantly different from the values in the ideal f.c.c. lattice: $z = 11.4 \pm 1.2$.[66] Thus, the GBs in NSMs were stated to be not unusual and to have the short-range order.

As the later studies have indicated on the usual character of GBs in NSMs which are not expected to exhibit extraordinary properties, the significantly modified properties of these materials reported in the first studies[3,4] were associated with extrinsic defects, such as impurities and pores, rather than with the intrinsic behavior of the nanostructures.[6,16]

The controversial results obtained on NSMs prepared by the same technique of gas-condensation have shown that the problem of the interfacial structure in nanocrystals is not simple and requires much more careful analysis of the role of a variety of factors in addition to the sole average grain size: grain size distribution, preparation conditions, age, and annealing states. This controversy stimulated the further very extensive and sophisticated studies of the NSMs involving different preparation methods, new experimental techniques, and development of the existing ones to meet the demands of the new materials' studies and computer simulations. These studies led to the development of both viewpoints. In the next part of this section we will summarize this development and try to extract the most important conclusions.

Löffler and Weissmüller decomposed the radial distribution function (RDF) of atoms in NSMs into intragrain and intergrain parts and proposed a theory of the wide-angle scattering of X-rays in NSMs.[67] The analysis is based on the fact that nonreconstructed GBs, each atom of which occupies a lattice site, and reconstructed ones, in which atoms are displaced from their lattice positions to new, nonlattice equilibrium positions, differently contribute to the X-ray spectrum. They studied nanostructured Pd samples prepared by the gas-condensation method and subjected them to different consolidation, aging, and annealing treatments. It has been found that, in the aged and annealed samples, practically all atoms occupy lattice positions. In the samples investigated within ten days after preparation, approximately 10% of atoms are located on nonlattice sites and have little or no short-range order. Thus, the results suggest that in the aged and annealed nanocrystals, the GBs are ordered and similar to the GBs in coarse-grained polycrystals, while in the as-prepared nanocrystals they are qualitatively different from the latter.

Boscherini, De Panfilis, and Weissmüller[68,69] performed a state-of-the-art *ab initio* analysis of the EXAFS spectra of nanophase materials and characterized the structure of nanostructured Pd prepared and differently treated as for the X-ray diffraction studies.[67] They found no evidence for the large reduction of coordination numbers even in the nanostructured Pd stored at the liquid nitrogen temperature. A small reduction observed can be explained by a size effect due to the high interface-to-bulk ratio. Previous results showing a large reduction of coordination numbers have been explained as an experimental artefact due to the sample thickness inhomogeneities.

The cited authors have pointed out that the terms *ordered* and *disordered* used to describe the GB atomic structure have to be quantified. The experimental distinction between these two extremes is completely based on the analysis of radial distribution functions. In fact, two effects can lead to a conclusion on the disordered atomic structure of GBs: atomic relaxation in the GBs, due to the different local environment than in the bulk, and the different local environment of each GB which, when averaged over all the GBs, can lead to wide peaks in the radial distribution functions.[68] It has been proposed to consider the GBs to have a disordered structure, if there are atoms whose average positions do not belong to the lattice of any of the adjacent crystallites and that have a low average coordination and/or high disorder, leading to significant excess volume.[69] In these terms the EXAFS data mean that the GB atoms in nanostructured Pd all belong to lattice sites, that is, GBs have a nonreconstructed atomic structure.

Such an understanding of the term *disorder* also seems not free of controversies. Indeed, the studies of GBs in bicrystals show that their radial distribution functions exhibit wide peaks.[70] This means the presence of GB atoms at nonlattice sites. It is not clear why such reconstructions should be absent in NSMs. Thus, the term *disorder* needs a further quantification. It is quite possible that the two approaches to the GB structure in NSMs reflect two different aspects of the same more general model. As such, one can propose the classification of the GB structures into the equilibrium and nonequilibrium ones.

As has been noted, XRD studies of nanostructured Pd[67] showed quite different spectra of the as-prepared nanocrystals on the one hand and aged or annealed nanocrystals on the other. Therefore, the GBs in as-prepared NSMs have a nonequilibrium structure, which relaxes toward equilibrium during annealing. This conclusion is confirmed by the data obtained by differential scanning calorimetry (DSC) on NSMs prepared by different techniques. A DSC scan of nanostructured Pt obtained by the gas-condensation technique exhibits two clear peaks of enthalpy release at about 200°C and 500°C.[18] The second peak corresponds to the grain growth, while the first relaxation occurs at a constant grain size. This enthalpy release was attributed to the relaxation in GBs, which in the as-prepared state were in a nonequilibrium state.[18,71] A similar two-stage energy release during the DSC experiments has been observed on nanocrystalline Ru and AlRu prepared by ball milling[50] and Ni_3Al and Cu prepared by severe plastic deformation.[51,72]

The nonequilibrium character of GBs in as-prepared nanocrystals and their relaxation during annealing are also confirmed by the compressibility measurements. The lifetime of positrons in the as-prepared nanocrystalline Pd decreases from 270 ps to 240 ps when increasing the pressure from 0 to 4 Gpa.[60] Annealing at 463 K reduces this effect by half. The Mössbauer spectroscopy of nanocrystalline Fe has shown that the center shift δ_{CS} of the crystalline component decreases with pressure and becomes negative at $p > 100$ Mpa.[73] After annealing at 170°C, δ_{CS} did not change with pressure. This also means that the annealing resulted in an increase in the density of interfaces.

These results indicate that in addition to the average grain size, one of the key structural parameters of the NSMs is the state of GBs. The investigations of the structure and properties of NSMs prepared by different techniques suggested a quite common feature of as-prepared nanocrystals: the GBs in these materials are in a metastable, nonequilibrium state and evolve toward equilibrium during aging or annealing. The concept of the nonequilibrium GB structure has been mostly developed on the basis of the studies of NSMs prepared by severe plastic deformation technique.

22.3.4.2 Grain Boundaries in Nanostructured Solids Prepared by Severe Plastic Deformation

The terms *equilibrium GBs* and *nonequilibrium GBs* come from earlier studies of the GB structure in metals. Experiments show that the GBs in plastically deformed polycrystals trap lattice dislocations that lead to a nonequilibrium GB structure characterized by the presence of long-range stress fields and excess energy.[74,75] It has been demonstrated that the nonequilibrium GB structure plays a critical role in the GB migration and sliding and in the superplastic deformation.[75,76] This concept has been

FIGURE 22.4 Transmission electron micrographs of submicrocrystalline copper prepared by torsion straining under pressure: (eighth as-prepared state; (befitted annealing at 150°C. The micrographs show a significant reduction in the number of extinction contours and the appearance of band contrast on grain boundaries.

naturally accepted for explaining the characteristics of GBs in nano- and (mainly) SMC materials prepared by severe plastic deformation.[7,8,15]

TEM studies of as-prepared SMC materials show a characteristic strain contrast of GBs and extinction contours near the GBs, contrary to the band contrast of GBs in well-annealed polycrystals.[7,8,15] An example is shown in Figure 22.4a. The nonregular extinction contours indicate the presence of high internal stresses. The density of lattice dislocations in SMC materials is usually low; hence, these internal stresses are suggested to arise from the nonequilibrium GBs that had been formed during severe deformation.[7,8,15] Moderate annealings result in the relaxation of the GB structure, which does not involve any significant grain growth (Figure 22.4b). The extinction contours in the grains disappear, and the GBs acquire their usual band contrast.

The fact that the internal stresses observed are induced by GBs is confirmed by direct HREM observations of nanostructured Ni.[54] These studies showed the existence of lattice distortions up to 1–3% in a layer of 6–10 nm thickness near the GBs.

PAS studies of SMC Cu and Ni showed the presence of positron lifetimes τ_1 and τ_2 corresponding to the vacancy-sized and about ten-vacancy-sized free volumes in interfaces and their junctions.[60] The long-lived component τ_3 was not observed, which confirms the absence of porosity in NSMs prepared by severe plastic deformation.

The presence of free volumes is confirmed also by dilatometric studies of an SMC aluminum alloy and Ni.[77,78] During annealing of the as-prepared SMC samples, the relaxation of an excess volume has been observed. The maximum relative volume shrinkage in both materials is approximately 0.1%. Complementary TEM investigations have shown that, at moderate annealing temperatures used in the dilatometric experiments, there was no significant grain growth.[77] Hence, the volume change is mainly due to the relaxation of nonequilibrium GBs.

The Mössbauer spectroscopy of SMC iron has been performed.[79–81] The spectrum consisted of two distinct subspectra that reflect the hyperfine magnetic structure. One of the subspectra corresponded to the bulk crystalline Fe and the second one to the GB atoms, as in nanostructured Fe prepared by the gas-condensation method.[57] As compared with the latter case, however, the second subspectrum was more clear. Its parameters were reproducible from sample to sample. From the relative intensity of the sub-spectra, the volume fraction of the GB atoms has been shown to amount to approximately 10% in the SMC sample with the grain size of 250 nm. Hence, the thickness of the layer responsible for the second spectrum is about 10 nm.

From these results it has been suggested that in SMC materials there is a specific GB phase characterized by a higher dynamic activity of atoms as compared with the atoms in the crystalline phase.[7,79,81] The origin of this phase has been attributed to the elastic distortions due to nonequilibrium GBs. It is assumed that the width of the GB phase depends on the degree of the GB nonequilibrium. The GB phase contributes to the changes of fundamental parameters of ultrafine-grained materials such as the saturation magnetization, Curie temperature. This two-phase model of NSMs prepared by severe plastic deformation has been proved to explain their properties fairly well.[7,8]

This model seems to be supported also by an analysis of the electronic structure of nanostructured Ni and W prepared by torsion straining under pressure. The samples from these metals were studied by the field ion microscopy and field electron spectrometry.[82,83] Two types of the total energy distribution of field-emitted electrons, depending on the selected emission site, were obtained. In the case of areas away from the GB, the distributions are similar to the classical one. Spectra taken from the area containing the GB have an additional peak. Authors explain this behavior by a possible reduction in the electron emission work function in the GBs.

EXAFS studies of nanostructured Cu prepared by severe plastic deformation have shown a significant decrease of the coordination number.[42] The cited authors explained this decrease by a very high concentration (about 1%) of the nonequilibrium vacancies in as-prepared nanocrystals.

The experimental and theoretical data on the structural evolution during large plastic deformation help to elucidate the nature of the GB phase. Plastic deformation leads to the absorption of lattice dislocations by GBs. The significance of this process for the grain refinement during large deformations has been explored.[84–87] Lattice dislocations absorbed by GBs change their misorientations and result in the formation of partial disclinations at the junctions of grains. These defects are the basic elements of the deformed polycrystal, which are responsible for the dividing of the grains. This occurs by the movement of disclinations across the grains that leads to the formation of new boundaries. In fact, the movement of a disclination across a grain is simply the formation of a boundary by lattice dislocations whose glide is caused by the stress field of the disclination. The final stage of the preparation of NSMs by severe plastic deformation is characterized by a stable grain size and balanced density of the GB defects, including the strength of disclinations. These defects are inherited by the as-prepared nanocrystals and are suggested to be the main sources of internal stresses and property modifications of NSMs prepared by severe plastic deformation.[88–92] One can distinguish three basic types of nonequilibrium dislocation arrays which are formed on GBs and their junctions during plastic deformation: disordered networks of extrinsic grain boundary dislocations (EGBDs), arrays of EGBDs with the Burgers vector normal to the GB plane (sessile EGBDs) equivalent to the dipoles of junction disclinations, and arrays of EGBDs with the Burgers vector tangential to the GB plane (gliding EGBDs) as demonstrated in Figure 22.5.

The model of nonequilibrium GBs has been applied to calculate such characteristics of NSMs as the rms strain, excess energy and excess volume.[88–92] For this, a model two-dimensional polycrystal consisting of square-shaped grains has been considered (Figure 22.5). The polycrystal contains the three types of defects mentioned above, which are distributed randomly in GBs of the two orthogonal systems. This distribution is not, however, completely random. Dislocation glide in polycrystals is always confined to the grains. This is illustrated on the upper left part of Figure 22.5. Correspondingly, the three types of defects in the polycrystal with random orientations of grains can be considered to consist of randomly distributed basic elements which come to the GBs and junctions from each grain. These elements are represented in Figure

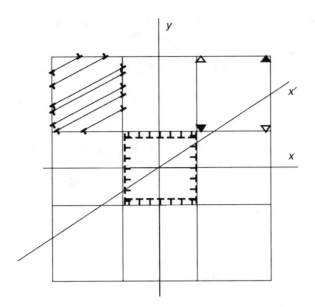

FIGURE 22.5 Components of the nonequilibrium grain boundary structure in a model two-dimensional polycrystal: a random array of EGBDs coupled to dipoles (the upper-left corner), a disclination quadrupole (the upper-right corner), and an array of gliding EGBDs (the central grain).

22.5: dipoles of EGBDs (upper left grain), quadrupoles of junction disclinations (upper right crystallite), and closed arrays of gliding EGBDs with zero net Burgers vector (central grain).

The rms strain ε_i was estimated by calculating the normal strain $\varepsilon_{x'x'}$ along some axis x' (Figure 22.5) and averaging its square over the orientations of the axis x' that effectively corresponds to the averaging over the orientations of the grains in a real polycrystal, over realizations of the structure of disordered EGBD arrays that is done by replacing the square of the stress by its mean square and over the volume of a probe grain, as the stresses and strains depend on the distance from GBs. The rms strains due to the disordered distribution of EGBDs, due to junction disclinations and gliding EGBDs are equal to

$$\varepsilon_i^r \approx 0.23b\sqrt{\frac{\rho_0}{d}\ln\frac{d}{b}}, \quad \varepsilon_i^d \approx 0.1<\Omega^2>^{1/2}, \quad \varepsilon_i^\tau \approx 0.3<\beta^2>^{1/2} \tag{22.1}$$

respectively, where ρ_0 is the mean density of EGBDs (in m^{-1}), $<\Omega^2>^{1/2}$ is the rms strength of junction disclinations, and $<\beta^2>^{1/2}$ is the rms Burgers vector density for gliding EGBDs.

For the random distribution of elements depicted in Figure 22.5, the excess energy due to the GB defects is calculated as a sum of the energies of dislocation dipoles, disclination quadrupoles, and closed cells of gliding EGBDs:

$$\gamma_{ex} = \frac{Gb^2\rho_0}{4\pi(1-\nu)}\ln\frac{d}{b} + \frac{G<\Omega^2>d\ln 2}{1G\pi(1-\nu)} + \frac{G<\beta^2>d(\pi-2\ln 2)}{4\pi(1-\nu)} \tag{22.2}$$

It is known that defects inducing long-range stresses increase the volume of materials.[93] This volume expansion is approximately proportional to the elastic energy of defects and for the case of dislocations was estimated.[93] Using those results, the excess volume of NSMs prepared by severe plastic deformation was calculated as:[92]

$$\xi = \frac{\Delta V}{V} = \Gamma\left[\frac{2.12b^2\rho_0}{d}\ln\frac{d}{b} + 0.37<\Omega^2> + 3.72<(\Delta\beta)^2>\right] \tag{22.3}$$

where Γ is a dimensionless parameter depending on the material and the character of dislocations ($\Gamma \approx$ 0.2 for edge dislocations in Cu and Ni and slightly different for screw dislocations.).[93] The volume expansion by crystal lattice defects is a result of the nonlinearity of the elastic deformation: due to the asymmetry of the interatomic forces with respect to the equilibrium position, internal stresses compensated over the volume of material cause a positive integral strain.

The maximum average density of the disordered EGBDs networks was estimated on the assumption that dense random EGBD arrays can relax at ambient temperature by the climb of dislocations toward equilibrium positions to form an even array.[94] It has been shown that in ultrafine-grained Cu and Ni EGBDs with density of about 10^8 m^{-1} relax in a month or more. The limiting strength of junction disclinations, over which micro-cracks can open at the triple junctions, was estimated to be about 1–3°, that is $\langle\Omega^2\rangle^{1/2} \leq 0.05$.[85] Similar estimates for the gliding components of EGBDs yield the values of the order $\langle\beta^2\rangle^{1/2} \leq 0.02$.[92] For these values of the nonequilibrium GB parameters, the rms total strain can amount to more than 1%, the excess GB energy 1.5 Jm^{-2}, and the excess volume 10^{-3}, in good agreement with the experimental data.

The dislocation and disclination modeling also allowed a study of the annealing kinetics for the NSMs prepared by severe plastic deformation. The excess EGBDs with the Burgers vectors normal and tangential to the GB planes (see Figure 22.5) can leave the boundaries through triple junctions.[95] In the cited paper it has been demonstrated that the relaxation of both these components is controlled by the GB diffusion to the distances of the order of the grain size d and occurs according to an exponential relationship with the characteristic relaxation time:

$$t_0 \approx \frac{d^3 kT}{100 \delta D_b G V_a} \tag{22.4}$$

where δD_b is the GB diffusion width times GB self-diffusion coefficient, and V_a is the atomic volume.

For pure SMC Cu a significant relaxation of the structure and properties occurs during one-hour annealing at a temperature $T = 398$ K.[7,8] A calculation of the characteristic relaxation time according to Equation (22.4) by using the parameter values $\delta D_b = 2.35 \times 10^{-14}$ exp($-107200/RT$) m^3/s,[96] $G = 5 \times 10^4$ MPa, $V_a = 1.18 \times 10^{-29}$ m^3 gives a value $t_0 = 60$ min, which is in excellent agreement with the experimental data.

On the basis of these comparisons one can conclude that the notion of nonequilibrium GBs containing extremely high density of EGBDs inherited from the deformation is a fairly good model for the NSMs prepared by severe plastic deformation.

Nonequilibrium GBs containing extrinsic dislocations and disclinations can exist also in NSMs prepared by the powder compaction technique. The formation of these defects has been considered in detail in the review.[97] The distribution and strength of the defects, however, may differ from those in the SMC materials.

The above described model is essentially mesoscopic, because it does not predict the atomic structure of nonequilibrium GBs. This is not critical for an analysis of the mechanical properties of nanocrystals, but is important for the studies of atomic-level processes such as diffusion. On the basis of HREM studies,[54] it is believed that the GBs in NSMs prepared by severe plastic deformation retain a crystal-like order similarly to the GBs in coarse-grained materials.[15] The shortcomings of an application of the HREM technique to NSMs discussed above should be kept in mind, however, when considering these results. Unfortunately, the results of atomic simulations of the GBs in nanocrystals prepared by deformation are lacking, mainly due to the relatively large grain size of these materials.

22.3.4.3 Nanocrystalline Materials Produced by Crystallization from the Amorphous State

Interfaces in nanocrystals prepared by a crystallization from the amorphous state have been observed by HREM.[10,98] The GBs looked like normal high-angle boundaries; no extended contrast caused by disordered boundaries could be detected.

Quantitative EXAFS studies of nanocrystalline Se with the grain sizes in the range 13 to 60 nm gave no evidence for a significant decrease of coordination numbers.[99]

In nanostructured NiP alloys prepared by the crystallization technique PAS detected three lifetimes as in nanocrystals obtained by the gas-condensation technique: $\tau_1 \approx 150$ ps ($I_1 \approx 0.944$), $\tau_2 \approx 350$ ps ($I_2 \approx 0.045$), and $\tau_3 \approx 1500$ ps ($I_3 \approx 0.011$).[100] The third component was attributed to the ortho-positronium formation on the surfaces of the samples. Similar studies of nanocrystalline alloys $Co_{33}Zr_{67}$, $Fe_{90}Zr_{10}$, and $Fe_{73.5}Cu_1Nb_3Si_{13.5}B_9$ prepared by this technique indicated a single component τ_1.[60] The shortest lifetime component τ_1 corresponds to free volumes slightly smaller than a lattice monovacancy.[60,100] Thus, the results indicate that in NSMs prepared by the crystallization technique, the excess volume is mainly concentrated in GBs as less-than-vacancy-sized free volumes, whereas larger voids corresponding to 10 to 15 vacancies are absent or very rare. This is consistent with this specific preparation method, which allows avoiding the porosity.[10]

It should be noted, however, that amorphous alloys are usually less dense than corresponding crystals. On completing the crystallization process, the final density of an NSM is intermediate between the densities of the amorphous and crystalline alloy. For example, NiP has densities 7.85 g/cm³ and 8.04 g/cm³ in the amorphous and crystalline states, respectively, while the density of the nanocrystalline samples was equal to 8.00 g/cm³.[100] Apparently, the excess volume of the amorphous material is partially locked inside the samples in interfaces, as has been proposed in Figure 22.3. Due to the diffusion-controlled character of carrying the free volumes out of the sample, they will be retained in the as-prepared nanocrystals.

Most probably, this locking of the free volumes in interfaces is responsible for the very specific dependence of the GB energy on the average grain size for NSMs prepared by crystallization. From DSC studies of NiP nanocrystals with $d = 8$–60 nm, it has been concluded that the excess volume and energy of GBs linearly increase with the increasing grain size.[101] The cited authors have suggested that this is an intrinsic behavior of interfaces in NSMs. However, this suggestion contradicts the observations on NSMs prepared by all other techniques, which indicate the opposite behavior — an increase of the GB energy with the decreasing grain size. A plausible interpretation of this point is based again on Figure 22.3. During the crystallization, all volume difference between the amorphous and crystalline phases tends to concentrate on GBs. A part of this volume is retained there after the crystallization. As the grain volume increases proportionally to d^3, while the GB area proportionally to d^2, the larger the grain size of the final polycrystal, the larger is the excess volume per unit area of the interfaces. Interfaces with larger excess volume have a higher energy.[31] Thus, the increase of the interfacial energy with increasing grain size is not an intrinsic property of nanocrystals but is related to the nonequilibrium state of GBs due to the specific preparation conditions. These GBs will relax toward equilibrium-state denser GBs during annealing, when the diffusion of the free volumes out of the sample is allowed.

The free volumes are the most common but not the only defects characterizing the nonequilibrium GB structure in NSMs prepared by crystallization. HREM studies have shown that distorted structures with dislocations are quite common for interfaces in nanocrystalline alloys $Fe_{73.5}Cu_1Nb_3Si_{13.5}B_9$.[102] In fact, due to the complex chemical and phase composition, the nanocrystalline alloys prepared by this method are characterized by a very wide range of nanostructures and interfaces.

22.3.4.4 Nanostructured Materials Prepared by Other Techniques

The above considered types of NSMs have been most widely characterized structurally by the use of available experimental techniques. As far as nanocrystals prepared by other techniques such as ball milling and electrodeposition are considered, their structures are less studied.

The local atomic structure of interfaces in ball-milled nanocrystals has been mainly characterized by Mössbauer spectroscopy. Just as in the case of powder-metallurgy NSMs, two opposite conclusions have been reached on the basis of these studies. Del Bianco et al.[103] resolved two Mössbauer spectra of ball-milled iron with $d = 8$–25 nm, the second one of which was attributed to a GB phase with a low degree of atomic short-range order. These results have been criticized by Balogh et al.,[104] who attributed the

Mössbauer spectra difference between the coarse-grained and nanostructured Fe to Cr contaminations inserted during the ball milling. It has been emphasized that Mössbauer spectroscopy is a very sensitive tool which can detect impurities on the order of 0.1%. Fultz and Frase[105] detected a distinct subspectrum attributable to the GB atoms in ball-milled iron nanocrystals but questioned the existence of disordered regions. Rather, they thought of distorted regions at interfaces. The GB width was estimated to be about 0.5 nm in f.c.c. and 1 nm in b.c.c. alloys — much less than the GB phase width obtained from Mössbauer spectroscopy of SMC metals prepared by severe plastic deformation.[79–81]

The presence of interfacial free volumes seems to be characteristic for ball-milled nanocrystals, too. Comparing the hyperfine magnetic field values from the Mössbauer spectra of ball-milled nanostructured FeAl, Negri et al.[106] found that the density of nanograin boundaries was about 10% less than that of the bulk crystals. Assuming the GB width $\delta \approx 0.5$ nm and the grain size $d \approx 10$ nm, one can estimate the overall excess volume: $(\delta/d) \times 10\% \approx 0.5\%$, which is much less than in powder-compacted nanocrystals but more than in NSMs prepared by severe plastic deformation.

The density of electrodeposited nanostructured Ni determined by the Archimedes densitometry method is 0.6% less than that of the perfect crystal for the grain size 11 nm and 1% less for $d = 18$ nm.[107] The PAS studies of nanocrystalline Pd prepared by this technique have shown that the free volumes are larger than in powder-compacted nanocrystals and correspond either to four missing atoms or to nanovoids of 10–15 missing atoms containing light impurity atoms.[108]

22.3.4.5 Computer Simulation of Nanostructured Materials

With an increase in the capacities of computers, atomic simulations play an increasingly important role in the understanding of the structure of nanocrystals. Several groups have performed sophisticated molecular dynamics (MD) simulation studies of nanostructured metals so far.

Wolf and co-workers simulated the structure of nanocrystals grown *in situ* from the melt.[46,109–112] The cubic simulation box was filled in with a liquid metal or Si. In this liquid differently oriented crystalline seeds were inserted, from which the growth of nanocrystals was simulated. Placing the seeds in the eight octants of the simulation box gave a nanocrystal with eight cube-shaped grains,[46,109] while an f.c.c. arrangement of the seeds provided four dodecahedral grains.[111,112] Orientations of the seeds could be chosen to obtain either special or random GBs. A Lennard–Jones potential has been used to describe atomic interactions in Cu,[46,109] an EAM potential for Pd,[110] and a Stillinger–Weber potential for Si.[111] The growth of a polycrystal was simulated at a temperature well below the corresponding melting temperature. On completing the crystallization, the temperature was slowly decreased to 0 K. Then the samples were annealed for a time interval sufficiently long on the MD scale — about 100 ps. Annealing did not influence the final structures. The authors assert that in this way they obtained the ground state of a nanocrystal with the given grain size, such that the GB structures formed were in a thermodynamic equilibrium.

The results of the simulation studies are summarized as follows. From the studies of bicrystals, it is known that rigid-body translations parallel to the GB plane greatly influence the GB energy.[31] While in individual GBs the rigid-body translation state can be readily optimized, the severe constraints existing in an NSM do not allow for the simultaneous optimization of rigid-body translations for all GBs. Due to this fact, it appears that the GB energies in nanocrystals are significantly higher than in bicrystals.

The GB energy distribution in nanocrystals is much narrower than in coarse-grained polycrystals or bicrystals; that is, all GBs have similar energies. This means that the energies of crystallographically high-energy GBs do not increase much, while the energies of special GBs do when going to the nanocrystalline grain size interval.

Calculated hydrostatic stresses vary significantly along the GBs and across the grains. In some GBs tensile and compressive stresses coexist, while most of grain interiors are slightly compressed. Mean lattice parameter was found to be 0.91%, 0.95%, and 0.65% larger than that of the perfect crystal, along the three axes, such that the mass density of nanocrystalline material is 97.5% of that of the perfect crystal.[46]

The calculation of the system-averaged RDF has shown broadened peaks due to the presence of disorder. In order to catch the similarities of the GB atomic structure with the structure of bulk glass, amorphous Pd was simulated by an extremely rapid quenching. The local RDF has been determined for atoms lying in GBs, triple lines, and point junctions. For the calculation of this RDF, only those atoms of the system, with the highest excess energies have been chosen.[110] This calculated RDF coincided very well with the RDF of bulk Pd glass. Moreover, the study of low-energy special and high-energy general twist GBs in Pd bicrystals has shown that the atomic structure of high-energy GBs is similar to that of bulk glass, too. Quite similar results have been obtained for Si.[111,112]

From these studies it has been concluded that Rosenhain's historic model, considering the GBs as an amorphous cement-like phase with a uniform width and atomic and energy density,[33] describe the interfaces in an NSM very well. The authors state that the presence of such a highly disordered, high-energy, *confined amorphous* GB structure in NSMs is a thermodynamic phenomenon, corresponding to the thermodynamic equilibrium.

Interestingly, the Si melt is 5% denser than the crystal at the same temperature. Due to this, the grain junctions in as-grown nanostructured Si had a density 1% higher than the density of the perfect crystal.[111] It is clear that the opposite phenomenon applies to metal nanocrystals simulated in these studies:[46,109,110] the density of GBs and junctions was less than that of perfect crystal, in accordance with the lower density of the metal melts. This observation seems to confirm the above proposed idea on the interface locking of free volumes in NSM prepared by the crystallization from the amorphous state.

A quite different algorithm has been applied to simulate the NSMs by Schiøtz and co-authors[113] and Van Swygenhoven and co-workers.[114] The simulation box was filled by random points around which Voronoi polyhedra were constructed. This allowed a more realistic simulation of the microstructure of nanocrystals taking into account the distribution of grain sizes. It is worthy to say that this is hardly critical at the present level of our knowledge of nanocrystals; a much stronger effect is expected from other factors than from the distribution of the grain size. Each polyhedron was filled in by a randomly oriented crystallite. When using this generation method, the common problem is that two atoms from neighbor crystallites may get too close to each other. In this case one of the atoms is usually eliminated.[113,114] On the other hand, in other regions of the GBs, the interatomic distances across GBs are larger than in the lattice. Thus, this generation technique introduces a specific RDF artificially cut at the short distances already in the starting configurations and obviously leads to a less dense interfacial structure.

An MD simulation of nanostructured Cu[113] yielded an RDF which resembled that obtained by Phillpot et al.[109] The characteristic feature of this RDF is that it is not equal to zero between the peaks corresponding to the first and second nearest neighbor distances. This makes it difficult to integrate this function in order to calculate the nearest neighbor coordination number. An estimate[113] shows, however, that the average coordination number in simulated nanocrystalline Cu equals approximately 11.9, which is in good agreement with the experimental data obtained.[66]

Two types of nanostructured samples were grown by the Voronoi construction.[114] The samples of one type contained GBs with random misorientations, and those of the second type had only low-angle GBs ($\theta < 17°$). Samples having grain sizes of 5.2 nm and 12 nm and an overall density of 96% and 97.5% of the perfect crystal density, respectively, were generated. The samples with different grain sizes belonging to one type had scalable structures and the same type GBs, because they were obtained by the same grain orientations using only more or less dense distributions of grain centers in the smaller or larger simulation box. This allowed the study of the influence of the grain size on the GB structure.

The structure was characterized by calculating the coordination numbers and investigating the type of each atom using the topological medium-angle order analysis developed by Honneycutt and Andersen.[115] This method allowed the authors to distinguish atoms with f.c.c., h.c.p. order and with non-12 coordinated atoms as well.

A visual analysis of the local structures of the same GBs in nanocrystals with two different grain sizes has shown that:

- In nanocrystals with random misorientations, the GBs exhibit regions of high coherency, which in the case of general GBs are alternated by local regions of high disorder, quite like in the GBs of bicrystals.
- The structures are quite similar for different grain sizes.
- In the samples with low-angle GBs, the boundaries have a dislocation structure as in large-grained polycrystals.

These results allowed the authors to make a conclusion opposite to that of others:[46,109–112] the GBs in NSMs have an ordered structure and can support localized grain boundary dislocations. This discrepancy has been attributed to the fact that originally the local RDF was calculated for GB atoms having only highest energies. The resulting RDF has been shown to depend very much on the criteria of *highest energy*.[114]

The present analysis shows that the controversies existing between the experimental data on the structure of interfaces in NSMs have been transferred to the simulation area as well. There are two reasons for this. First, different structure analysis tools are used to reach conclusions on the atomic structure of simulated nanocrystals, meaning different processing methods for the same experimental data. Second, none of the simulation schemes allows the construction of the ground state of an NSM, i.e., the structure characterized by the lowest free energy, in spite of the statement made by Phillpot et. al.[46] that their scheme does that. Therefore, the obtained GB structures are nonequilibrium ones and will differ from each other.

The latter point is worth analyzing in more detail, because the existence of extremely constrained GBs with non-optimized rigid-body translations having an amorphous atomic structure is considered to be the intrinsic feature of NSMs.[5,112] All simulations included an annealing of the constructed nanocrystalline samples in order to make sure that an equilibrium structure had been obtained. However, the MD annealing time, which is on the order of 100 ps, is many orders of magnitude shorter than the characteristic time for the diffusion-controlled equilibration of the structure of a polycrystal. In fact, the rigid-body translations, which are not optimized in all simulated nanocrystals,[46,109–112] can be optimized in much longer time intervals. For instance, consider two adjacent grains along the boundary between which there is a driving force for the rigid-body translation (Figure 22.6a). Sliding of the upper grain with respect to the lower one decreases the GB energy but induces elastic energy due to the compression in the filled regions and tension in the open ones. These stresses can be released by a diffusion flow of matter from compressed regions to the dilated ones as indicated by arrows in the figure.

The characteristic relaxation time for this process will be on the order of the relaxation time for non-equilibrium EGBD arrays given by Equation (22.4). A calculation for Cu nanocrystals with $d = 10$ nm at room temperature gives a value $\tau \approx 7$ hours. This time is much longer than the time accessible by MD, but quite short for experiments. In fact, Cu nanocrystals studied a day after the preparation may have optimized rigid-body translations and, therefore, an equilibrium GB structure. Therefore, the real nanocrystals may be characterized by relaxed rigid-body translations, contrary to the structures simulated in the cited papers.

Of course, in real polycrystals the rigid-body translations can be optimized not over all GBs and not completely. Another mechanism can be proposed that allows the very fast accommodation of the non-equilibrium translation states. The grains can slide, forming triple-junction dislocations (Figure 22.6b). This will happen when the GB energy decrease due to the translation is larger than the elastic energy of the junction dislocation dipole formed:

$$\frac{d}{2}\Delta\gamma \geq \frac{Gb^2}{4\pi(1-\nu)}\ln\frac{d}{b} \text{ or } d \geq \frac{Gb^2}{2\pi(1-\nu)\Delta\gamma}\ln\frac{d}{b} \tag{22.5}$$

For $\Delta\gamma = 0.5$ Jm^{-2} this gives a critical grain size $d = 10$ nm. Therefore, a significant part of the rigid-body translations of GBs can be accommodated by junction dislocation formation in nanocrystals with the grain size larger than 10 nm.

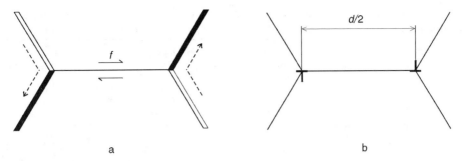

FIGURE 22.6 Optimization of the rigid-body translation of a finite grain boundary bounded by triple junctions by the diffusion flow of atoms (a) and by the formation of junction dislocations (b). The dashed arrows in (a) indicate the directions of vacancy fluxes.

22.4 Properties

22.4.1 Diffusion

Diffusion is one of the key properties of NSMs because it controls the structural stability and many physical properties of these materials. The diffusion coefficient in NSMs is influenced by both the size effects and the specific interfacial structure.

The characteristic diffusion distance for many properties of polycrystals is of the order of the grain size (see, for example, Equation (22.4)). Therefore, a simple decrease of the grain size enhances these properties. A typical example of this is the diffusional creep, which will be considered in the next section.

The effective diffusion coefficient of a polycrystal $<D>$ is determined by averaging the lattice and grain boundary diffusion coefficients D_L and D_b, respectively, over the volume of the polycrystals taking appropriate volume fractions of the components. This coefficient is also much higher in nanocrystals than in coarse-grained polycrystals. Therefore, the NSMs can be applied in many fields of technology where a fast penetration of atoms is required. Most interestingly, however, changes in the GB diffusion coefficient D_b itself are expected on the basis of the structural studies.

Direct measurements of the GB diffusion coefficient in materials are usually performed as follows.[116] A layer of atoms whose diffusion is to be studied is deposited on the surface of material. Then the sample is kept at a sufficiently high temperature referred to as the *diffusion annealing temperature* T_d for a time interval t_d. T_d and t_d are chosen such that the diffusing atoms penetrate into the material to a measurable distance, and at the same time the diffusion length in crystallites is small as compared with the GB width: $\sqrt{D_L t_d} \ll \delta$. In these conditions one observes the C-regime of the GB diffusion, during which the penetration of atoms is completely determined by the GB diffusion.[116] After annealing the concentration of diffusing atoms is determined at different depths either by measuring the radioactivity or by the secondary-ion mass spectrometry. The obtained concentration profiles are fitted to the solution of the diffusion equation for the regime C:

$$c(x, t) = c(0, 0)\operatorname{erfc}\frac{x}{2\sqrt{D_b t_d}} \tag{22.6}$$

Because the diffusion coefficient obeys the Arrenius relation:

$$D_b = D_{b0}\exp\left(-\frac{Q_b}{RT}\right) \tag{22.7}$$

the activation energy Q_b and the pre-exponential factor D_{b0} for the GB diffusion coefficient can be determined from the temperature dependence of D_b expressed in terms of the relation $\ln D_b$ vs. $1/T$.

TABLE 22.1 Characteristics of the Grain Boundary Diffusion in Coarse-Grained (CG) and Nanostructured (NS) Metals Prepared by the Powder Compaction (PC) and Severe Plastic Deformation (SPD) Techniques

Metal	Preparation Method	Mass Density,%	Grain Size, nm	Diffusate	Temperature Interval, K	D_{0b}, m²/s	Q_b, kJ/mole	Reference
CG–Cu	—	100	>1000	^{67}Cu	>0.5T_m	1×10⁻⁵	102	[96]
NS–Cu	PC	91	5–15	^{67}Cu	293–393	3×10⁻⁹	62	[117]
CG–Ni	—	100	>1000	^{63}Ni	>0.5T_m	7×10⁻⁶	115	[96]
NS–Ni	PC	92–93	70	^{63}Ni	293–473	2×10⁻¹²	46	[118]
CG–Ni	—	100	>1000	Cu	773–873	2×10⁻³	128	[119]
NS–Ni	SPD	100	300	Cu	398–448	2×10⁻⁹	43	[119]
CG–Pd*	—	100	>1000	^{59}Fe	—	1×10⁻⁵	143	[120]
NS–Pd*	SPD	100	80–150	^{59}Fe	371–577	1×10⁻¹³	57	[120]
CG–Fe*	—	100	>1000	^{59}Fe	—	1×10⁻²	170	[121]
NS–Fe*	PC	91–96	19–38	^{59}Fe	450–500	3×10⁻⁸	111	[121]

Some experimental data on the parameters D_{b0} and Q_b for nanocrystalline materials are compiled in Table 22.1. For a comparison, the data on the GBs in coarse-grained materials are also presented. For some metals, marked with an asterisk, the values of D_{b0} and Q_b have been recalculated from the graphics presented in corresponding references.

From the table one can see that the activation energy for the GB diffusion coefficient in NSMs is considerably less than that in coarse-grained polycrystals. These values are similar to those for the surface diffusion. For example, the activation energy for the surface diffusion in Cu is Q_s = 66.5 kJ/mole.[117]

The first measurements on the diffusion coefficient in NSMs[117,118] yielded very high values of the effective coefficient <D>, exceeding even D_b for coarse-grained polycrystals. For example, in nanocrystalline Cu with d = 8 nm at T = 393 K <D> = 1.7 × 10⁻¹⁷ m²/s, while D_b = 2.2 × 10⁻¹⁹ m²/s.[4] Such a high value of <D> was later attributed to the presence of a significant porosity in the samples used in the early investigations (the densities of the samples are also indicated in Table 22.1). Bokstein et al.[118] proposed a cluster model according to which the grains in high-porosity nanocrystals are grouped into clusters inside which the GBs are like ordinary ones, while the cluster boundaries have a large excess volume and can carry a surface diffusion. It should be noted that this model has something to do with the situation described by Figure 22.3b.

Schaefer and co-workers have performed the most systematic studies of the GB diffusion in NSMs prepared by powder compaction and severe plastic deformation, its dependence on various factors such as the pre-annealing, diffusion annealing time, etc.[60,120–123] The nanocrystals obtained by powder compaction had a high mass density (about 97% in Pd, 92% in nanocrystalline Fe, and 97% in explosion-densified nanocrystalline Fe). These studies have shown that the diffusivities of GBs in NSMs are similar to or slightly higher than those in ordinary GBs. This has been explained by an equilibrium character of the GBs in the samples investigated due to a relaxation at slightly elevated temperatures.[122] The occurrence of the GB relaxation at the temperatures of diffusion experiments is confirmed by a decrease of the free volume[60] and internal strains.[39]

The role of the GB relaxation processes in the diffusion coefficient is confirmed also by the following observations:

- The measured GB diffusion coefficient depends on the temperature and time of pre-annealing. For example, in nanostructured Pd prepared by severe plastic deformation, the penetration depth of ^{59}Fe atoms at 401 K decreased an order of magnitude after the pre-annealing at 553 K, as compared with the penetration depth after the pre-annealing at 453 K.[120]

- The measured value of D_b depends also on the time of diffusion annealing t_d at a given temperature T_d. For nanocrystalline Fe with d = 19–38 nm at T_d = 473 K, the D_b value calculated from the concentration profile recorded after a 1.5 h annealing is equal to 2 × 10⁻²⁰ m²/s, while after an annealing for 69 h at the same temperature D_b = 3 × 10⁻²¹ m²/s.[121] In SMC Ni the GB diffusivity of Cu was lowered by 3 orders of magnitude due to an annealing at 523 K.[119]

Phenomenologically, the enhancement of the diffusion coefficient in nonequilibrium GBs can be explained on the basis of Borisov's relation:[124]

$$D_b^{ne} = D_b \exp\left(\frac{\alpha \Delta E}{kT}\right) \tag{22.8}$$

where ΔE is the excess energy of the GB per atom, $\alpha \approx 1$. As the number of atoms per 1 m^2 is equal to δ/V_a, Equation (22.8) can be rewritten as:

$$D_b^{ne} = D_b \exp\left(\frac{\Delta \gamma V_a}{kT\delta}\right) \tag{22.9}$$

with $\Delta \gamma$ being the excess GB energy per unit area.

An estimate for $T = 500$ K and $\Delta \gamma = 1$ Jm^{-2} gives $D_b^{ne}/D_b \approx 10^2$. Thus, for the excess GB energies observed in nanocrystalline metals during DSC experiments, one can expect two orders of magnitude increase of the GB diffusion coefficient.

Using Equation (22.4), one can estimate the characteristic time for the relaxation of the GB excess energy and, therefore, of the enhanced diffusion coefficient. Presented in Table 22.2 are the results of such estimates for Ni, Pd, and Fe nanocrystals for the temperatures at which diffusion annealing experiments were performed. From this table one can see that, indeed, there is a correlation between the relaxation of the GB structure and the change of the GB diffusion coefficient. In nanostructured Pd and Fe, the diffusion annealing time is much longer than the GB relaxation time for the diffusion annealing temperatures used in experiments. This can explain the similarity of diffusivities in nanocrystalline and conventional grain boundaries observed.[120–122] In the case of nanostructured Ni, however, t_d is relatively short, and the GBs have a nonequilibrium structure over the whole period of diffusion experiments. Based on the present consideration, one can predict the following exotic behavior of the GB diffusion coefficient in nanocrystals. If one uses comparable diffusion annealing time intervals at different temperatures, it is quite possible that at higher temperatures the GBs have a lower diffusion coefficient. That is, in some temperature intervals (probably narrow) one may observe a negative value of the apparent activation energy for the GB diffusion.

Although the experimental studies have shown a direct relationship between the nonequilibrium GB structure and free volumes, on the one hand, and the GB diffusion coefficient, on the other, the mechanisms of the enhancement of diffusion in nonequilibrium GBs have not yet been well understood.

One of the models explaining the high values of D_b observed in NSMs is based on the consideration of a stress-assisted diffusion.[125] As has been discussed above, NSMs prepared by severe plastic deformation contain GB defects such as triple-line disclinations. These defects induce high internal stresses which can affect the GB diffusion in two ways. First, hydrostatic stresses $\sigma = \sigma_{ii}$ can increase the diffusion coefficient directly according to the relation:[126]

$$D'_b = D_b \exp\left(\frac{\sigma V_{at1}}{kT}\right) \tag{22.10}$$

TABLE 22.2 Values of the Characteristic Time for the Relaxation in Nonequilibrium Grain Boundaries and the Diffusion Annealing Time in Nanostructured Metals

Metal	d, nm	T, K	D_b, m^2/s	t_0, h	t_d, h
Pd	100	430	2×10^{-21}	1	240
		577	1×10^{-18}	0.04	48
Ni	300	398	6×10^{-21}	114	3
		448	3×10^{-19}	3	3
Fe	25	450	1×10^{-22}	4	386
		500	1×10^{-20}	0.04	1.5–69

A calculation of the stresses of junction disclinations shows that only a small region near the junctions is characterized by the stress values higher than 1 GPa. For such stresses Equation (22.10) yields $D'_b/D_b \approx 10$ at $T = 500$ K. Taking into account the small volume fraction of highly stressed regions, the overall enhancement of the GB diffusion coefficient will be very small.

Second, the junction disclinations induce stress gradients which generate additional driving forces for the diffusion of vacancies. In the presence of hydrostatic stress gradients $\partial\sigma/\partial x$, the flow of tracer atoms is equal to $j = -D_b[\partial c/\partial x + (V_a/kT)c(\partial\sigma/\partial x)]$ and the diffusion process will be described by the following equation:

$$\frac{\partial c(x, t)}{\partial x} = D_b \frac{\partial^2 c(x, t)}{\partial x^2} - \frac{D_b V_a}{kT}\frac{\partial\sigma}{\partial x}\frac{\partial c(x, t)}{\partial x} \tag{22.11}$$

For the diffusion from a surface layer the solution to Equation (22.11) is[126]

$$c(y, t) = c(0, 0)\,\text{erfc}\left(\frac{x - Vt}{L_{ef}}\right) \tag{22.12}$$

where $L_{ef} = 2\sqrt{D_b t}$, and $V = (D_b V_a/kT)(\partial\sigma/\partial x)$ is the average drift speed of atoms in the constant stress gradient.

Concentration profiles for this case have been calculated by averaging the drift speed v over many GBs containing disclination dipoles of random strengths[125] and compared with the solutions of the ordinary diffusion equation (Equation (22.6)). If the experimental profiles are processed on the basis of Equations (22.12) and (22.6), the former will give the true diffusion coefficient D_b, while the latter would yield an apparent, or effective, diffusion coefficient D_{ef}. All reported values of the GB diffusion coefficient in nanocrystals are obtained by fitting to Equation (22.6). Therefore, they are all effective diffusion coefficients. In the cited paper it has been demonstrated that in deformation-prepared nanostructured Pd the ratio D_{ef}/D_b can reach the values as high as 100.

Another attempt to explain the high diffusivity of nano-grain boundaries is based on a consideration of the vacancy generation due to a dislocation climb.[127] At slightly elevated temperatures the EGBDs formed in GBs during the preparation of nanocrystals relax toward an equilibrium. The relaxation involves an annihilation of dislocations of opposite signs and an ordering of the rest dislocations. The climb of dislocations decreases the energy of the system that leads to a decrease of the vacancy formation energy near the dislocations. If one considers an annihilation of a dislocation dipole with a separation λ, this decrease is calculated as the change of the dipole energy per one generated vacancy $W_v(\lambda)$.[127] A climb of a dislocation with the Burgers vector b and length d to a distance a comparable to the interatomic distance in the boundary generates d/a vacancies. The energy of the dipole is given by the formula: $W(\lambda) = [Gb^2 L/2\pi(1 - v)][\ln(\lambda/r_0) + Z]$ where $Z \approx 1$ is the term due to the dislocation core-energy. Then, for $\lambda > 2a$, $W_v(\lambda)$ is calculated as:

$$W_v(\lambda) = \frac{a}{d}[W(\lambda) - W(\lambda - a)] = \frac{Gb^2 a}{2\pi(1 - v)}\ln\frac{\lambda}{\lambda - a} \tag{22.13}$$

For $\lambda < 2a$ the authors found $W_v^0(\lambda) \approx Gb^2 a/2\pi(1 - v)$. As one can see from Equation (22.13), $W_v(\lambda)$ decreases with increasing λ. The maximum value $W_v^0(\lambda)$ for $G = 50$ GPa, $a \approx 0.3$ nm, $b \approx a/3$, $v = 1/3$ is equal to $W_v^0(\lambda) = 6 \times 10^{-20}$ J. Averaging $W_v(\lambda)$ over an interval $0 \le \lambda \le 15a$, the authors found that the GB diffusion coefficient near the climbing dislocations can be five orders of magnitude higher than that in equilibrium GBs:

$$D'_b = D_{b0}\exp\left(\frac{\overline{W_v}}{kT}\right) \approx 3 \times 10^5 D_b \tag{22.14}$$

The above calculations are very sensitive to the estimate of $\overline{W}_v$, because the diffusion coefficient depends on it exponentially. On the other hand, the decrease of the energy of a dislocation dipole during its annihilation is spent not only for the generation of vacancies near one of the dislocations, but also for the migration of these vacancies to the other dislocation and for the absorption by this dislocation. It is not clear how the total energy is partitioned between these three consumers. Equation (22.14) seems to give an overestimation of the enhancement of the GB diffusion coefficient.

An important tool to get direct information on the GB diffusion characteristics is the MD simulation. Although direct simulations of the GB diffusion in NSMs have not been performed yet, except for the studies of diffusional creep at very high temperatures,[128] it is clear that much effort will be devoted to this field in the near future. Recent simulations of the diffusion in high-energy GBs of Pd bicrystals[129] seem to be the most closely related to nanocrystals, because, as has been mentioned above, such GBs are expected to exist in NSMs from the structure simulations.[46,109–112] Keblinski et al.[129] simulated two high-energy twist boundaries (110) 50.48° and (113) 50.48° and two relatively high-energy symmetrical tilt GBs $\Sigma = 5(310)$ and $\Sigma = 7(123)$. At the zero temperature the former two boundaries have a confined amorphous-like structure, while the latter two have a crystalline order.[110] The MD simulations demonstrated that above a certain critical temperature T_c, which depends on the geometry and energy of GBs, all investigated boundaries have approximately the same activation energy for diffusion of about 58 kJ/mole. It has been shown that above T_c the high-energy tilt and twist GBs undergo a reversible transition from a low-temperature solid structure to a highly confined liquid GB structure. The liquid GB structure is characterized by a universal value of the activation energy of diffusion given above. Below T_c diffusion occurs in an amorphous or crystalline solid film with a much higher value of the activation energy (e.g., 144.5 kJ/mol for the $\Sigma = 7$ boundary). The tilt boundaries $\Sigma = 5$ and $\Sigma = 7$ exhibited a transition at $T_c = 900$ and 1300 K, respectively. For twist boundaries T_c is below the lowest temperature investigated, 700 K.

Unfortunately, these results cannot be directly compared with the experimental data on nanostructured Pd, as the simulations were performed at higher temperatures than the experiments. Comparing the activation energies with the ones presented in Table 22.1, one can see, however, that the activation energy for $\Sigma = 7$ GB coincides well with that in the coarse-grained Pd, while the activation energy for liquid GBs coincides well with that in nanostructured Pd. If, indeed, the GB structures in nanostructured Pd are not relaxed at the temperatures of diffusion experiments, the results of simulations would seem plausible.

22.4.2 Elasticity

Elastic properties of crystalline solids are usually considered to be structure-insensitive. However, measurements on NSMs showed a significant decrease of the Young's and shear moduli. First measurements for nanocrystalline Pd indicated, for example, that E and G had about 70% of the value for fully dense, coarse-grained Pd (Table 22.3). However, the low moduli measured in NSMs prepared by powder compaction have been attributed to the high level of porosity in these materials. Sanders et al.[132] noted that the data for nanostructured Pd and Cu with different densities collected in Table 22.3 were fitted well by the following relation for the Young's modulus of porous materials:[134]

$$E = E_0 \exp\left(\frac{-\beta \Delta \rho_m}{\rho_m}\right) \tag{22.15}$$

where $\Delta \rho_m$ is the porosity, E and E_0 are the apparent and reference elastic moduli, and $\beta \approx 3–4.5$. The intercept E_0 for Pd (130 GPa) is very near the Young's modulus of coarse-grained Pd (133 GPa), and for Cu (121 GPa) it is about 5% lower than the reference value of 128 Gpa.[130]

In ball-milled nanocrystals the elastic moduli were in the same range as the reference values; but for nanocrystalline Fe with $d < 10$ nm, an approximate 5% decrease of E was observed.[132]

Lebedev et al.[135] extracted the porosity-dependent part from the Young's modulus decrement of nanocrystalline Cu with $d = 80$ nm prepared by powder compaction and found that approximately 2% of the net 8% decrement was intrinsic, due to the small grain size.

TABLE 22.3 Elastic Moduli of Coarse-Grained and Nanostructured Metals Prepared by Different Techniques

Metal	Method of Preparation	d, nm	Mass Density,%	E, GPa	G, GPa	Reference
Pd	—	coarse	100	123	43	[4]
	PC	8	—	88	35	[4]
Cu	—	coarse	100	115–130	—	[130]
	—	coarse	100	128	48	[131]
	PC	10	97.6	108	39.6	[132]
		16	98.6	113	41.7	
		22	98.6	116	42.7	
	BM	26	100	107	—	[133]
	SPD	200	100	115	42	[131]
Ni	—	coarse	100	204–221	—	[130]
	BM	17	100	217	—	[133]

PC = Powder Compaction, BM = Ball Milling, SPD = Severe Plastic Deformation.

Schiøtz et al.[113] calculated the Young's modulus of nanostructured Cu with d = 5.2 nm by MD simulations and found values 90–105 GPa at 0 K, while for the same potential at 0 K, single crystals had E_0 = 150 GPa (the Hill average). Phillpot et al.[46] also found a decrease of E in their simulations.

Thus, there seems to be a decreasing tendency for the elastic moduli of nanocrystals, and the decrease is significant only for very small grain sizes ($d < 10$ nm). This reduction is associated with an increase of the interfacial component, which has lower elastic moduli due to an increased specific volume.

Different changes of E for the same grain size and density can be related to different states of GBs. In nanocrystals with very small grain sizes, for example, these states may differ by the level of rigid-body translations' optimization. Annealing, which eliminates the porosity, also results in more relaxed GBs that contribute to a recovery of the moduli.

The role of the nonequilibrium GB structure in the elastic moduli is demonstrated particularly well on SMC metals prepared by severe plastic deformation.[131] Figure 22.7 represents the dependence of the Young's modulus and grain size of SMC Cu prepared by equal-channel angular pressing on the temperature of annealing for 1 hour. The modulus of the as-prepared SMC Cu ($d \approx 0.2$ μm) is approximately 10% less than that of the coarse-grained Cu (128 GPa); but during annealing at temperatures higher than 200°C, it recovers to the reference value.

The recovery in 99.997% pure nanostructured Cu occurs at temperature 125°C.[131] As has been discussed below of Equation (22.4), the relaxation time calculated for this temperature coincides very well with the experimental annealing time.

The reduced elastic moduli of SMC metals, in which the volume fraction of GBs is much less than in nanocrystals, can be related to the presence of mobile GB dislocations and high internal stresses.[7,8]

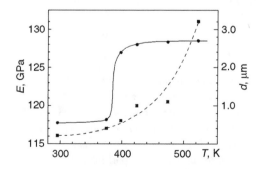

FIGURE 22.7 The dependence of the Young's modulus (filled circles and solid curve) and grain size (filled squares and dashed curve) of 99.98% pure submicrocrystalline copper on the temperature of one hour annealing.

Thus, the studies have shown that the elastic properties are influenced by both effects, the small grain size and the nonequilibrium GB state. The size effects can lead to a significant E reduction for very small grain sizes ($d < 10$ nm). The nonequilibrium state of GBs can reduce E about 10% at even submicron grain sizes.

22.4.3 Hall–Petch Relationship for Nanostructured Materials

The ambient temperature plasticity of polycrystals is usually characterized by the yield stress σ_y, whose dependence of the grain size d obeys the well-known Hall–Petch relationship:[136,137]

$$\sigma_y(d) = \sigma_0 + k_y d^{-1/2} \tag{22.16}$$

where σ_0 is the friction stress and k_y is a positive material's constant. A review of the experimental data supporting this relationship and the corresponding theoretical models can be found in Reference 138.

Data concerning the validity of this relationship for NSMs have largely been obtained by the measurements of the Vickers hardness H_V, which is mainly determined by the yield stress through a relation $H_V \approx 3\sigma_y$.[139] In the large grain size region, the hardness also obeys a Hall–Petch-type relationship:

$$H_V(d) = H_0 + k_H d^{-1/2} \tag{22.17}$$

One of the most important expectations associated with the NSMs is a large increase in σ_y and H_V at very small grain sizes in accordance with Equations (22.16) and (22.17). A decrease of d from 30 μm to 30 nm would result in 33 times increase in the yield stress.

Measurements have shown that the hardness of NSMs indeed exceeds that of coarse-grained polycrystals, but to a much more modest extent — less than 10 times.[6,14] Obviously, the Hall–Petch relationship is violated at grain sizes less than 100 nm. Again, this violation is found to be associated with both the intrinsic size effects and extrinsic defects.

Due to the influence of extrinsic defects such as porosity, dislocations, and other sources of internal stresses, the experimental data on the Hall–Petch relationship for nanocrystals have been controversial. Several authors have reported observations of an inverse Hall–Petch relationship that softening occurred with the decreasing grain size.[140,141] Other studies reported a normal Hall–Petch relationship or the one with a positive but smaller slope k_H.[142–145]

Fougere et al.[146] analyzed the relation between the hardening and softening behaviors and the method used to obtain nanocrystalline samples with different grain sizes. They found that nanocrystalline metals exhibited increased hardness with decreasing grain size when individual samples were compared; that is, no thermal treatment was applied before the hardness measurements. The dependencies of the dimensionless hardness H_V/G on the grain size for such nanocrystals of several materials are presented in Figure 22.8.

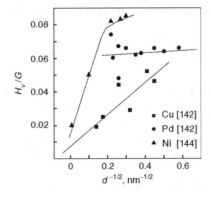

FIGURE 22.8 The grain size dependence of the hardness for nanocrystals which have not been subjected to an annealing after the preparation.

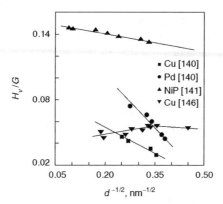

FIGURE 22.9 Negative Hall–Petch relationship for nanocrystals, the grain size of which has been varied by annealing. Annealing of Cu in a wide range of temperatures results in initial strengthening and subsequent softening.

If the grain size is varied by an annealing of the same as-prepared sample, one observes softening with the decreasing grain size. Such cases are collected in Figure 22.9. This figure shows also the behavior of nanostructured Cu, the annealing of which, in a wide range of temperatures, results in initial hardening and subsequent softening, such that a maximum is observed on the $H_V(d^{-1/2})$ dependence.[146] Such a mechanical behavior of nanocrystals prepared by the gas-condensation technique has been attributed to the softening effect of pores, which anneal out when heating the samples to increase the grain size.[147]

A similar effect is observed on NSMs prepared by severe plastic deformation. The hardness of an aluminum alloy and Ni$_3$Al nanocrystals prepared by this method increases during annealing without a significant grain growth.[145,51] These data show that not only the grain size, but also extrinsic defects such as pores and nonequilibrium GBs influence the yield stress and hardness.

The extrinsic factors result in a reversal of the Hall–Petch relationship at relatively large grain sizes (more than 20 nm and even at $d \approx 100$ nm). However, the studies of nanocrystals prepared by electrodeposition[148] and crystallization from the amorphous state[149] have shown that there is a maximum of the hardness at about $d \approx 10$ nm. This behavior is believed to be an intrinsic one and is expected for nanocrystals prepared by any method.

A number of models have been proposed to describe the deviations from the Hall–Petch relationship for NSMs and will be discussed below.

In the first group of models, it is assumed that the deviation from the Hall–Petch relationship is an intrinsic behavior of polycrystals. In one of the models of this kind it is suggested, that for the very small grain size, diffusion creep can occur rapidly at room temperature. The strain rate for Coble creep, which is controlled by the GB diffusion, quickly increases with the decreasing grain size:[150]

$$\dot{\varepsilon} = \frac{148\delta D_b V_a \sigma}{\pi k T d^3} \qquad (22.18)$$

where σ is the applied stress. Chokshi et al.[140] calculated $\dot{\varepsilon}$ for $d = 5$ nm using the GB diffusion data from Reference 117 (see also Table 22.1) and found that for $\sigma = 100$ Mpa, the creep rate could be as high as 6×10^{-3} s^{-1}, that is, of the order of the strain rate during the measurements of the yield stress and hardness.

However, direct measurements of the creep rate of nanostructured metals have yielded much lower values, thus ruling out this explanation. These data will be analyzed in the next section, which is devoted to the creep of nanocrystals.

Scattergood and Koch[151] analyzed the plastic yield of nanocrystals in terms of the dislocation network model, according to which the yield stress is determined by the stress necessary for a glide of dislocations through a dislocation network.[152] The critical stress for the glide at large spacing of the network dislocations is that for the dislocation cut: $\tau = \alpha_{0C} Gb/L$, where $\alpha_{0C} \approx 0.2$–0.4. At small grain sizes the critical stress will be determined by the Orowan bypassing of network dislocations. The stress for Orowan bypassing is proportional to the dislocation line tension.[44] The latter logarithmically depends on the grain size. Thus, for the small grain sizes one has[151]

$$\tau = \left[\frac{1}{2\pi} \ln \frac{d}{r_{eff}}\right] \frac{Gb}{L} \tag{22.19}$$

Here, r_{eff} is an effective cut-off radius.

Thus, there is a critical grain size d_c, at which a transition from the dislocation cut mechanism of deformation to the Orowan bypassing occurs

$$\frac{1}{2\pi} \ln \frac{d_c}{r_{eff}} = \alpha_{0C} \tag{22.20}$$

According to Li,[152] $L \propto \rho^{-1/2} \propto d^{1/2}$. Therefore, for the two grain size regions, above and below the critical size, one will have two different relations:

$$H = H_0 + k_H d^{-1/2} \qquad\qquad (d > d_C)$$
$$H = H_0 + \left(\frac{1}{2\pi\alpha_{0C}} \ln \frac{d}{r_{eff}}\right) k_H d^{-1/2} \quad (d < d_C) \tag{22.21}$$

From these equations it follows that at $d < r_{eff}$ a negative Hall–Petch slope will be observed. For Pd and Cu the experimental data were fitted by Equation (22.21) using $r_{eff} = 5.6$ and 7.7 nm, respectively.

The idea on the grain-size-dependent dislocation line tension was used also by Lian et al.,[153] where it has been suggested that the yield of nanocrystals is controlled by a bow-out of dislocation segments, the length of which scales with the grain size.

The most popular model for the Hall–Petch relationship is the model of dislocation pile-ups.[136,137,154] Pile-ups at GBs induce stress concentrations that can either push the lead dislocation through the boundaries or activate dislocation sources in the adjacent grains, thus resulting in the slip transmission over many grains. According to the classical theory of dislocation pile-ups,[155] the coordinates of dislocations in a pile-up, the leading dislocation of which is locked at $x = 0$ and the others are in equilibrium under the applied shear stress τ, are equal to the zeros of the first derivative of the nth order Laguerre polynomial:

$$L'_n\left(\frac{2\tau x}{Ab}\right) = 0 \tag{22.22}$$

where $A = G/2\pi(1 - \nu)$ for edge dislocations.

For $n \gg 1$ the greatest root of this equation, which is identified with the grain size d, is equal to $d = 4n(Ab/2\tau)$. Because the pile-up of n dislocations multiplies the applied stress τ by n, $\tau = \tau_c/n$, one will have a relation $\tau = (2Ab\tau_c)^{1/2} d^{-1/2}$. Multiplying this by the Taylor factor M ($M = 3.06$ for f.c.c. metals[156]), one obtains the grain-size-dependent part of the yield stress. Hence, $k_y = M(2Ab\tau_c)^{1/2}$.

Pande et al.[157] noted that the exact solution of the pile-up problem deviates from this classical one. Denote d_n the minimum grain size which allows to support a pile-up consisting of n dislocations, and X_n the greatest root of Equation (22.22). Then $d_n = X_n(Ab/2\tau) = nX_n(Ab/2\tau_c) = (nX_n/4)(2MAb/k_y)^2$. The values of X_n are always less than $4n$. In an interval $d_n < d < d_{n+1}$, there is a pile-up containing n dislocations in a grain, and the yield stress is equal to $M\tau_c/n$. Thus, the yield stress will be a staircase function of $d^{-1/2}$, which deviates more and more from the straight line $k_y d^{-1/2}$. Figure 22.10 illustrates this function in terms of the dependence of the normalized yield stress $[2MAb/k_y^2](\sigma_y - \sigma_0) = 1/n$ on $[2MAb/k_y] d_n^{-1/2} = (4/nX_n)^{1/2}$.

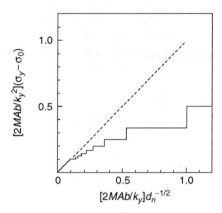

FIGURE 22.10 A deviation from the normal Hall–Petch relationship due to the small numbers of dislocations in pile-ups. The dashed line represents the Hall–Petch relationship extrapolated from the coarse grain size region.

In real NSMs the grain size distribution makes this dependence a smooth curve.[158] A fit by the modified pile-up model accounting for the distribution of grain sizes gives good agreement with the data on electrodeposited nanocrystalline Ni.[144]

In several models an NSM has been considered as a composite consisting of two, three, or even four of these components: crystal interior, GBs, triple junctions, and quadruple nodes.[148,159,160] Each component is characterized by its own strength (σ_c, σ_{gb}, σ_{tj}, σ_{qn}) and volume fraction (f_c, f_{gb}, f_{tj}, f_{qn}) and the strength of the composite is calculated using the rule of mixture:

$$\bar{\sigma} = \sum_{i=1}^{n} f_i \sigma_i \tag{22.23}$$

The models of this type have been very popular, because they allow one to describe easily the softening behavior: when the grain size decreases, the volume fraction of softer regions (grain boundaries, triple junctions, and quadruple junctions) increases and that will result in decreasing $\bar{\sigma}$. However, the physical ground for these models is fairly weak. The strengths of the components other than the grain interior have to be assumed as fitting parameters. It is hardly possible to divide unambiguously the polycrystal into grain and GB (plus other) components and consider separate deformation mechanisms in each of these components. For example, Coble creep and GB sliding cannot be considered as deformation mechanisms related to the GBs alone, as they also involve processes in the grains.

Of particular importance is the role of the GB sliding as a deformation mechanism of ultrafine-grained materials, because there are direct evidences for the occurrence of sliding in nanocrystals coming from experiments[161] and MD simulations.[113,162,163] In SMC Cu prepared by equal-channel angular pressing ($d \approx 200$ nm), a significant amount of GB sliding has been observed during the room temperature deformation.[161]

Schiøtz et al.[113] performed MD simulations of the deformation of nanostructured Cu with mean grain size in the range 3.28 to 13.2 nm and found a clear softening behavior with the decreasing grain size. In this range the main deformation mechanism at 300 K and even at 0 K was the GB sliding, which happens through a large number of small, apparently noncorrelated events. The cited authors observed a weak increase in the dislocation activity when the grain size increased and concluded that the transition from the range of reverse Hall–Petch relationship to the normal one would be beyond the grain size 13.2 nm.

A similar softening behavior has been obtained during the simulations of nanostructured Ni at 300 K.[162,163] Deformation of nanocrystals with $d \leq 10$ nm occurred with no damage accumulation, similarly to the conditions for superplasticity. Accordingly, a viscous deformation behavior has been obtained, with a nonlinear dependence of the strain rate on the stress. This behavior has been interpreted as a result of grain boundary sliding. A transition from the GB sliding-controlled deformation to the dislocation slip-controlled one occurs about a grain size 10 nm for Ni.[163] The cited authors considered also the deformation of a nanocrystal containing only low-angle GBs and found a significant dislocation slip in these samples even at the smallest grain size, 5 nm.

Thus, many experimental and simulation data seem to provide strong evidence for the existence of a critical grain size about $d \approx 10$ nm at which the deformation mechanism of NSMs changes from the dislocation slip at larger sizes to the GB sliding at smaller ones.

In the other, smaller group of the Hall–Petch relationship models for NSMs, it is suggested that the violation of this relationship is caused by extrinsic factors. One of the most common extrinsic factors influencing the yield stress and hardness is the existence of internal elastic strains induced by nonequilibrium GBs. Nazarov[164] analyzed the role of internal stress fields on the stress concentration by dislocation pile-ups and found that these fields can either facilitate or inhibit the slip transfer across the GBs. In a polycrystal containing nonequilibrium GBs there will be a distribution of the GBs by the critical values of the applied stress τ_C, for which they allow the slip propagation; and the distribution function $f(\tau_C)$ will be symmetric with respect to the *transparency* $\tau_C^{(0)}$ of an equilibrium, stress-free GB. The plastic yielding can be considered as a percolation process: macroscopic deformation occurs when the slip propagates through a percolation cluster of GBs transparent for slip.[51,165] The bond percolation in a honeycomb lattice occurs when the concentration of active bonds (transparent GBs) is 35%.[166] For the symmetric distribution function of the bonds, $f(\tau_C)$, this means that the shear yield stress τ_y is less than the yield stress in the absence of internal stresses. The higher the level of internal stresses, the larger is the difference $|\tau_y - \tau_C^{(0)}|$. During the annealing of a polycrystal, the relaxation of internal stresses occurs, the distribution of critical stresses becomes narrower, and the yield stress increases at a constant grain size.

The present model is consistent with the experimentally observed Hall–Petch behavior in SMC alloys.[51,145] It is suggested that this model can be applied also to NSMs prepared by the powder-compaction and ball-milling techniques, as these nanocrystals also contain nonequilibrium GBs with extrinsic dislocation and disclination defects.

22.4.4 Creep

The very small grain size and specific GB structure of NSMs suggest that the high-temperature mechanical properties of these materials will be different from those of coarse-grained polycrystals. The studies of creep can give important information on the deformation mechanisms of nanocrystals at finite temperatures and their relation to the GB diffusion.

As has been mentioned above, for the very small d Coble diffusion creep has been assumed to occur with a considerable rate such that it violates the Hall–Petch relationship. In order to test this hypothesis, Nieman et al.[142] performed room-temperature creep tests of powder-compacted nanocrystalline Pd and Cu with the grain sizes 8 nm and 25 nm, respectively. The creep rates under the stress 150 MPa were about 1.4×10^{-7} s^{-1} for Cu and 0.5×10^{-8} s^{-1} for Pd. Meanwhile, a creep rate as high as 1×10^{-5} s^{-1} was expected on the basis of GB diffusion coefficient measurements.[117] One should point out, however, that Equation (22.18), using the conventional values of D_b obtained from Table 22.1, yields the values $\dot{\varepsilon} = 6 \times 10^{-9}$ s^{-1} for Cu ($d = 25$ nm) and $\dot{\varepsilon} = 2 \times 10^{-9}$ s^{-1} for Pd ($d = 8$ nm), which is in good agreement with the experimental values.

Wang et al.[167] studied the room-temperature creep of nanostructured Ni with three grain sizes: 6, 20, and 40 nm, produced by an electrodeposition method. A nonlinear dependence of the creep rate on the stress has been observed. The stress–strain rate curves could be approximated by $\dot{\varepsilon} \propto \sigma^n$ dependence with the stress exponents $n = 1.2$ for $d = 6$ nm, $n = 2$ for $d = 20$ nm, and $n = 5.3$ for $d = 40$ nm.

The cited authors found that the absolute values of the creep rate at $d = 6$ nm $\dot{\varepsilon} \approx 10^{-7} – 10^{-8}$ s^{-1} were well approximated by the following formula for the strain rate, which had been obtained for deformation by the GB diffusion controlled GB sliding mechanism:[168]

$$\dot{\varepsilon} = 2 \times 10^5 D_b \frac{Gb}{kT} \left(\frac{b}{d}\right)^3 \left(\frac{\sigma}{G}\right)^2 \qquad (22.24)$$

The fact that $n = 1.2 < 2$ was explained by the operation of both mechanisms, Coble creep ($n = 1$) and GB sliding ($n = 2$).

It seems, however, that Equation (22.24) overestimates the creep rate. In good agreement with the experimental data for a wide range of superplastic alloys is the following empirical rate equation:[169,170]

$$\dot{\varepsilon} = 100 \frac{\delta D_b G}{kT} \left(\frac{b}{d}\right)^2 \left(\frac{\sigma}{G}\right)^2 \tag{22.25}$$

This relation yields $\dot{\varepsilon} = 4 \times 10^{-9}$ s^{-1} for the room-temperature creep rate of nanocrystalline Ni with $d = 6$ nm at $\sigma = 1$ GPa, while the Coble creep rate is equal to $\dot{\varepsilon} = 1 \times 10^{-8}$ s^{-1}, that is, of the same order and much less than the observed creep rates.

Thus, it appears that the observed creep rates are about one to two orders of magnitude higher than those predicted by equations. This can be explained by the following two reasons. First, about one to two orders of magnitude error in the estimating of the room-temperature diffusion coefficient is quite possible, as the diffusion data are extrapolated from high temperatures. Second, for the maximum stress applied (1.2 GPa), nanostructured Ni with the grain size 20 nm exhibited creep strain of only 0.4%.[167] For any of the two suggested mechanisms, Coble creep and GB sliding, $\varepsilon \approx x/d$, where x is the displacement normal or parallel to GBs; therefore, one finds that $x \approx \varepsilon d \approx 0.08$ nm. As the GB displacement and shear are less than the GB width $\delta = 0.5$ nm, it is hardly possible to apply the relationships for the steady-state deformation in this case.

Creep test results quite different from those of Nieman et al.[142] have been obtained on nanostructured Cu prepared by an electrodeposition technique.[171,172] The cited authors tested samples with the grain size $d = 30$ nm at temperatures between 20 and 50°C in the stress interval 120–200 MPa. They have found that a threshold stress exists, and the steady-state creep rate is proportional to the effective stress $\sigma_{eff} = \sigma - \sigma_{th}$. The activation energy for creep $Q = 69.4$ kJ/mole is less than that for the GB diffusion and similar to the activation energy for the GB diffusion in nanostructured Cu (Table 22.1). Furthermore, the strain rate calculated from Equation (22.18) with the diffusion data for nanostructured Cu was quite similar to the measured values of 10^{-6} s^{-1}.

Contrary to the data from Nieman et al.,[142] these data seem to provide strong support to the viewpoint that the GB diffusion coefficient in nanocrystals is enhanced. The room-temperature creep rates of nanostructured Cu with nearly the same grain sizes measured elsewhere[142,171] differ by three orders of magnitude. Kong et al.[172] relate this discrepancy to different methods of samples preparation in the two works: the creep rate in fully dense electrodeposited nanocrystals is higher than in powder-compacted nanocrystals. Intuitively, however, one may expect an opposite relation: due to the existence of larger free volumes and voids, the creep rate in powder-compacted nanocrystals would be higher than in the deposited ones.

Therefore, the creep behavior and GB diffusion are not so directly related as assumed in the Coble creep and GB sliding models. Creep in NSMs may have a more complicated mechanism than in ordinary coarse-grained polycrystals. The activation energy for creep need not necessarily coincide with the activation energy for the GB diffusion. For example, diffusion by the interstitial mechanism can make a significant contribution to the creep rate when stress is applied, and at the same time may be less important during the diffusion experiments. In this case one will observe a low value of D_b but high $\dot{\varepsilon}$ and the apparent activation energy for creep will be intermediate between those for the interstitial and vacancy-mediated diffusion mechanisms — that is, less than the activation energy for diffusion in ordinary GBs in which the vacancy mechanism dominates.

Thus, the analysis shows that the studies of the creep of NSMs still remain inconclusive with respect to the operating mechanisms of deformation and relation to the GB diffusion. Further systematic studies are needed, which should, for example, involve structural investigations.

The first MD simulation study has been performed for the very fast diffusional creep ($\dot{\varepsilon} = 10^7 - 10^8$ s^{-1}) of nanostructured Si with the grain sizes $d = 3.8$–7.3 nm at $T = 1200$ K.[128] The stresses necessary for these values of the strain rate were about 1 GPa. It has been shown that Coble's formula generalized for such high stresses as follows:

$$\dot{\varepsilon} \;=\; 94\frac{\delta D_b}{d^3}\sinh\!\left(\frac{\sigma V}{2kT}\right) \tag{22.26}$$

fits the simulation results very well. For the activation volume the value $V \approx 2V_\alpha$ has been found. The activation energy for creep coincides with that for the GB diffusion at high temperatures and is considerably less than the activation energy for the GB diffusion at lower temperatures. Although these results cannot be extended to moderately elevated temperatures, one should expect much progress in the MD simulations of the creep of NSMs.

22.4.5 Superplasticity

Superplasticity is defined as the ability of crystalline materials to deform in tension with elongations as large as several hundreds and thousands percent.[76] Superplasticity occurs usually at temperatures higher than $0.5T_m$ at fairly low strain rates ($10^{-4} - 10^{-3}$ s^{-1}) and requires fine grain sizes (less than 10 µm in alloys and 1 µm in ceramics).

The generalized constitutive equation for superplasticity is given by[76,169]

$$\dot{\varepsilon} \;=\; \alpha\frac{DGb}{kT}\left(\frac{b}{d}\right)^{\!p}\left(\frac{\sigma}{G}\right)^{\!n} \tag{22.27}$$

where D is the lattice or grain boundary diffusion coefficient, b is the lattice Burgers vector, p is the grain size exponent, and n is the stress exponent. In the interval of optimal superplasticity, $n = 2$; correspondingly, the strain rate sensitivity $m = 1/n = 0.5$. Most frequently, $p = 2$. For these values, Equation (22.27) coincides with Equation (22.25), which describes the experimental data on the superplasticity of aluminum alloys, for example, very well, if $D = D_b$.

The main mechanism for superplastic deformation is the GB sliding. The sliding along different GBs can be accommodated by GB diffusion or by slip, through a generation of lattice dislocations at triple junctions.[173]

According to Equation (22.27), one can expect a decrease in the temperature and/or increase in the strain rate at which the alloys can be deformed superplastically, when the grain size decreases to the nanometers range.

The enhanced low-temperature ductility is particularly important for ceramics, in which the superplastic flow occurs at too high temperatures (more than 1500°C), which are difficult to achieve. There are indications that nanocrystalline ceramics can be deformed significantly at temperatures of about $0.5T_m$. For example, nanostructured TiO_2 with the initial grain size 80 nm can be deformed in compression without crack formation up to the total strain 0.6 at 800°C.[4] The strain rate as high as 8×10^{-5} s^{-1} was observed at a stress about 50 MPa. Due to the high temperature of the tests, the grain size increased to 1 µm that led to a reduction in the strain rate.

In general, the grain growth during the deformation is the main concern in the developing of superplastic forming methods for NSMs. To inhibit the grain growth, alloying elements are introduced which can form secondary phase particles or segregate in GBs.

Although very important, the results on the ceramics are not related to the *true* superplasticity, because they have been obtained in compression. The most exciting results on high-strain-rate superplasticity have been obtained on materials of two types.

The first type includes some metallic alloys and their composites.[174] The alloys consist of fine grains with the mean size $d \approx 0.3\text{--}3$ µm, and composites contain second-phase particles with a diameter less than 30 nm. These materials can be deformed superplastically with strain rates $\dot{\varepsilon} = 10^{-2} - 10^{1}$ s^{-1} to total elongations of 500–1500%. The specific origin of the high-strain-rate superplasticity phenomenon in these materials is associated with the formation of a liquid phase at interfaces at the temperatures corresponding to the optimum superplasticity.[174] This liquid forms due to a high concentration of segregants in GBs and helps the accommodation of the GB sliding at triple junctions. Obviously, this type of material cannot exhibit a low-temperature superplasticity.

A combination of both the high-strain-rate and low-temperature superplasticity can be obtained on nanocrystalline and SMC metals prepared by the severe plastic deformation technique.[7,8,15] Summarized in Table 22.4 are the characteristics of the superplastic deformation of NSMs prepared by this method and of their fine-grained counterparts obtained to date. Where available, along with the flow stresses, the values of the strain for which these stresses are measured have been presented in brackets.

From the table one sees that superplastic deformation can be obtained from nanostructured alloys at significantly lower temperatures (by about 300–400°C) than in fine-grained polycrystals traditionally used in the studies of superplasticity. Moreover, at temperatures equal to or slightly lower than the temperature of superplasticity of fine-grained alloys, the nanostructured alloys exhibit a high-strain-rate superplasticity with the strain rates $(1 \times 10^{-2} - 1)s^{-1}$.

Mishra et al.[179,181] noted that the flow stress for optimum superplasticity of nanocrystalline alloys is higher than the stress predicted by the constitutive equation valid for fine-grain sizes (Equation (22.25)). The apparent activation energy for superplastic deformation is significantly higher than the activation energy for the GB diffusion. An important feature of the superplastic deformation of NSMs is also the presence of a significant work-hardening in a wide range of the strain. The higher the strain rate, the larger is the strain-hardening slope $d\sigma/d\varepsilon$. All these features show that the superplastic deformation mechanisms in NSMs may be considerably different from those in fine-grained materials. These mechanisms need further systematic studies.

Interesting results have been obtained on the superplastic deformation of pure nickel. Electrodeposited nanostructured Ni with the initial grain size 30 nm exhibited an 895% elongation at a temperature 350°C equal to $0.36T_m$,[182] which is a large improvement as compared with previously obtained result on nickel (250% elongation at $0.75T_m$).[183]

22.5   Concluding Remarks

The last decade has seen significant progress in the development of the preparation methods, structural characterization, and understanding of the properties of nanostructured materials. It has been well established that the properties of nanocrystals depend not only on the average grain size but also on the preparation route. Differences and common features of nanocrystals prepared by different techniques have been better understood. The nonequilibrium state of grain boundaries, which is characterized by larger free volume and enhanced energy, has proved to be one of the most common structural features of nanocrystals. Nonequilibrium grain boundaries have an excess energy comparable to the equilibrium GB energy, and their role in the properties of NSMs is very important. A combination and exploitation of the size and nonequilibrium structure effects will give an advantage of a very wide variation of the properties of NSMs.

This will require, however, much better understanding of the structure–property relationships for nanocrystals, which are known only in general at present. The present analysis has shown that controversies still exist in the understanding of almost any issue concerning the structure and properties of NSMs. It seems that the scientific research in this area should concentrate on solving these controversies that will require an application of new methods of experimental investigations and a development of simulation methods.

The diffusional and mechanical properties of NSMs are significantly improved as compared with the conventional polycrystalline materials. A high strength and the ability to deform superplastically at lower temperatures and/or higher strain rates make the NSMs engineering materials of the future. Other properties, which have not been considered here due to space limitations, are also considerably modified. To mention only two, the specific heat and electric resistance of nanocrystals is larger than that of the crystal,[4,7] and enhanced magnetic properties are already finding wide applications.[14]

The development of preparation techniques, which would allow the production of bulk NSMs in large quantities, is the most important issue concerning the industrial applicability of these materials. The presently existing techniques mainly produce small samples suitable for laboratory experiments. However, some of the methods, such as ball milling and severe plastic deformation, are becoming very popular and promising methods which seem to be capable of producing technologically applicable NSMs.

TABLE 22.4 Characteristics of Superplastic Deformation of Nanostructured Materials Prepared by Severe Plastic Deformation Methods: Torsion Straining (TS), Equal-Channel Angular Pressing (ECAP) and Multiple Forging (MF)

Alloy	Preparation Method	Initial Grain Size, μm	Final Grain Size, μm	T, K	$\dot{\varepsilon}$, s⁻¹	σ, MPA (ε,%)	$m = \dfrac{\partial \ln \sigma}{\partial \ln \dot{\varepsilon}}$	δ,%	Reference
Al-4%Cu-0.5%Zn	—	8	—	773	3×10^{-4}	13	0.50	800	[175]
	TS	0.3	—	493	3×10^{-4}	23 (20)	0.48	>250	[175]
	ECAP	0.15	—	523	1.4×10^{-4}	—	0.46	850	[176]
Mg-1.5%Mn-0.3%Ce	—	10	—	673	5×10^{-4}	25	0.42	320	[175]
	TS	0.3	0.5	453	5×10^{-4}	33 (20)	0.38	>150	[175]
Ni₃Al	—	6	—	1373	9×10^{-4}	—		641	[177]
	TS	0.05	—	998	1×10^{-3}	750 (200)		560	[178]
Al-1420 (Al-5.5%Mg-2.2%Li-0.12%Zr)	—	6	—	723	4×10^{-4}	5 (50)	0.55	>700	[76]
	TS	0.1	0.3	523	1×10^{-1}	188 (50)	0.28	330	[179]
	TS	0.1	1.0	573	1×10^{-1}	50 (50)	0.38	775	[179]
	ECAP	0.4	1.0	673	1×10^{-1}	30	—	1240	[180]
	ECAP	0.4	—	673	1	—	—	1000	[180]
Ti-6%Al-3.2%Mo	—	5.0	—	1073	5×10^{-4}	80	0.4	600	[27]
	MF	0.06	0.3	823	2×10^{-4}	200	0.33	410	[27]

Figures in brackets in the column for the stress indicate the strain values in percent at which these stresses have been measured.

Acknowledgments

The present work was supported by a grant, Structure and Properties of Nanostructured Materials Prepared by Severe Plastic Deformation, from the Complex Program of the Russian Academy of Science Nanocrystals and Supramolecular Systems. A. Nazarov was supported also by Subcontract No. 1995–0012–02 from the NCSU as a part of the Prime Grant No. N00014–95–1–0270 from the Office of Naval Research.

References

1. Morokhov, I.D., Trusov, L.I., and Chizhik, S.P., *Highly Dispersed Metallic Media*, Atomizdat, Moscow, 1977 (in Russian).
2. Gleiter, H., Materials with ultrafine grain size, in *Proc. 2nd Riso Int. Symp. Metallurgy Mater. Sci.*, Hansen, N., Leffers, T. and Lilholt, H., Eds., Riso, Roskilde, 1981, 15.
3. Birringer, R., Herr, U., and Gleiter, H., Nanocrystalline materials — a first report, *Trans. Jpn. Inst. Metals, Suppl.*, 27, 43, 1986.
4. Gleiter, H., Nanocrystalline materials, *Progr. Mater. Sci.*, 33, 223, 1989.
5. Gleiter, H., Nanostructured materials: basic concepts and microstructure, *Acta Materialia*, 48, 1, 2000.
6. Siegel, R.W., Cluster-assembled nanophase materials, *Annu. Rev. Mater. Sci.*, 21, 559, 1991.
7. Valiev, R.Z., Korznikov, A.V., and Mulyukov, R.R., Structure and properties of metallic materials with a submicrocrystalline structure, *Phys. Metals Metallogr.*, 4, 70, 1992.
8. Valiev, R.Z., Korznikov, A.V., and Mulyukov, R.R., Structure and properties of ultrafine-grained materials produced by severe plastic deformation, *Mater. Sci. Eng. A*, 168, 141, 1993.
9. Koch, C.C., The synthesis and structure of nanocrystalline materials produced by mechanical attrition: a review, *Nanostr. Mater.*, 2, 109, 1993.
10. Lu, K., Nanocrystalline metals crystallized from amorphous solids: nanocrystallization, structure, and properties, *Mater. Sci. Eng. R.*, 16, 161, 1996.
11. Morris, D.G., Mechanical behaviour of nanostructured materials, *Mater. Sci. Found.*, 2, 1, 1998.
12. Andrievskii, R.A. and Glezer, A.M., Size effects in nanocrystalline materials: I. Structure characteristics, thermodynamics, phase equilibria, and transport phenomena, *Phys. Metals Metallogr.*, 88, 45, 1999.
13. Andrievskii, R.A. and Glezer, A.M., Size effects in nanocrystalline materials: II. Mechanical and physical properties, *Phys. Metals Metallogr.*, 89, 83, 1999.
14. Syryanarayana, C. and Koch, C.C., Nanocrystalline materials – current research and future directions, *Hyperfine Interact.*, 130, 5, 2000.
15. Valiev, R.Z., Islamgaliev, R.K., and Alexandrov, I.V., Bulk nanocrystalline materials from severe plastic deformation, *Progr. Mater. Sci.*, 45, 105, 2000.
16. Siegel, R.W., What do we really know about the atomic-scale structures of nanophase materials? *J. Phys. Chem. Solids*, 55, 1097, 1994.
17. Sanders, P.G. et al., Improvements in the synthesis and compaction of nanocrystalline materials, *Nanostruct. Mater.*, 8, 243, 1997.
18. Tschöpe, A. and Birringer, R., Thermodynamics of nanocrystalline platinum, *Acta Metall. Materialia*, 41, 2791, 1993.
19. Fecht, H.J., Nanostructure formation by mechanical attrition, *Nanostr. Mater.*, 6, 33, 1995.
20. Suryanarayana, C., Mechanical alloying and milling, *Progr. Mater. Sci.*, 46, 1, 2001.
21. Bridgeman, P.W., *Studies in Large Plastic Flow and Fracture*, McGraw-Hill, New York, 1952.
22. Smirnova, N.A. et al., Structure evolution of f.c.c. single crystals during large plastic deformations, *Phys. Metals Metallogr.*, 61, 1170, 1986.
23. Segal, V.M. et al., Plastic treatment of metals by simple shear, *Izvestiya AN SSSR. Metally*, 1, 115, 1981 (in Russian).
24. Akhmadeev, N.A. et al., Submicron grained structure formation in copper and nickel by intensive shear deformation, *Izvestiya RAN. Metally*, 5, 96, 1992 (in Russian).

25. Segal, V.M., Equal channel angular extrusion: from macromechanics to structure formation, *Mater. Sci. Eng. A*, 271, 322, 1999.
26. Salishchev, G.A., Valiakhmetov, O.R., and Galeyev, R.M., Formation of submicrocrystalline structure in the titanium alloys VT8 and influence on mechanical properties, *J. Mater. Sci.*, 28, 2898, 1993.
27. Salishchev, G.A. et al., Submicrocrystalline and nanocrystalline structure formation in materials and search for outstanding superplastic properties, *Mater. Sci. Forum*, 170–172, 121, 1994.
28. Horita, Z. et al., Superplastic forming at high strain rates after severe plastic deformation, *Acta Materialia*, 48, 3633, 2000.
29. Yoshizawa, Y., Oguma, S., and Yamaguchi, K., New Fe-based soft magnetic alloys composed of ultrafine grain structure, *J. Appl. Phys.*, 64, 6044, 1988.
30. Erb, U., Electrodeposited nanocrystals: synthesis, properties and industrial applications, *Nanostr. Mater.*, 6, 533, 1995.
31. Sutton, A.P. and Balluffi, R.W., *Interfaces in Crystalline Materials*, Clarendon Press, Oxford, 1995.
32. Hargreaves, F. and Hills, R.J., Work-softening and a theory of intercrystalline cohesion, *J. Inst. Metals*, 41, 257, 1929.
33. Rosenhain, W. and Humphrey, J.C.W., The tenacity, deformation and fracture of soft steel at high temperature, *J. Iron Steel Inst.*, 87, 219, 1913.
34. Palumbo, G., Erb, U., and Aust, K.T., Triple line disclination effects on the mechanical behaviour of materials, *Scripta Metall. Materialia*, 24, 2347, 1990.
35. Eastman, J.A., Fitzsimmons, M.R., and Thompson, L.J., The thermal properties of nanocrystalline Pd from 16 to 300 K, *Philos. Mag. B*, 66, 667, 1992.
36. Aleksandrov, I.V. and Valiev, R.Z., Studies of nanocrystalline materials by x-ray diffraction techniques, *Phys. Metals Metallogr.*, 77, 623, 1994.
37. Wasserman, H.J. and Vermaak, J.S., On the determination of the surface stress of copper and palladium, *Surf. Sci.* 32, 168, 1972.
38. Borel, J.-P. and Châtelain, A., Surface stress and surface tension: equilibrium and pressure in small particles, *Surf. Sci.* 156, 572, 1985.
39. Reimann, K. and Würschum, R., Distribution of internal strains in nanocrystalline Pd studied by x-ray diffraction, *J. Appl. Phys.*, 81, 7186, 1997.
40. Lu, K. and Zhao, Y.H., Experimental evidences of lattice distortion in nanocrystalline materials, *Nanostr. Mater.*, 12, 559, 1999.
41. Zhang, K., Alexandrov, I.V., and Lu, K. The x-ray diffraction study of nanocrystalline Cu obtained by SPD, *Nanostr. Mater.*, 9, 347, 1997.
42. Babanov, Y.A. et al., EXAFS study of short-range order in submicron-grained copper produced by severe plastic deformation, *Phys. Metals Metallogr.*, 86, 559, 1998.
43. Gorchakov, V.I., Nagayev, E.L., and Chizhik, S.P., Does the Laplace pressure compress the solids? *Phys. Solid State*, 30, 1068, 1988.
44. Hirth, J. P. and Lothe, I., *Theory of Dislocations*, Wiley, New York, 1982.
45. Fecht, H.J., Intrinsic instability and entropy stabilization of grain boundaries, *Phys. Rev. Lett.*, 65, 610, 1990.
46. Phillpot, S.R., Wolf, D., and Gleiter, H. Molecular-dynamics study of the synthesis and characterization of a fully dense, three-dimensional nanocrystalline material, *J. Appl. Phys.*, 78, 847, 1995.
47. Klug, H.P. and Alexander, L.E., *X-Ray Diffraction Procedures for Polycrystalline and Amorphous Materials*, Wiley, New York, 1974.
48. Sanders, P.G. et al., Residual stress, strain and faults in nanocrystalline palladium and copper, *Mater. Sci. Eng. A*, 204, 7, 1995.
49. Weissmüller, J., Löffler, J., and Kleber, M., Atomic structure of nanocrystalline metals studied by diffraction techniques and EXAFS, *Nanostr. Mater.*, 6, 105, 1995.
50. Hellstern, E. et al., Structural and thermodynamic properties of heavily mechanically deformed Ru and AlRu, *J. Appl. Phys.*, 65, 305, 1989.

51. Languillaume, J. et al., Microstructures and hardness of ultrafine-grained Ni$_3$Al, *Acta Metall. Materialia*, 41, 2953, 1993.

52. Zhao, Y.H., Zhang, K., and Lu, K., Structure characteristics of nanocrystalline element selenium with different grain sizes, *Phys. Rev. B*, 56, 14322, 1997.

53. Wunderlich, W., Ishida, Y., and Maurer, R. HREM-studies of the microstructure of nanocrystalline palladium, *Scripta Metall. Materialia*, 24, 403, 1990.

54. Valiev, R.Z. and Musalimov, R.S., High-resolution transmission electron microscopy of nanocrystalline materials, *Phys. Metals Metallogr.*, 78, 666, 1994.

55. Zhu, X. et al., X-ray diffraction studies of the structure of nanometer-sized crystalline materials, *Phys. Rev. B*, 35, 9085, 1987.

56. Haubold, T. et al., EXAFS studies of nanocrystalline materials exhibiting a new solid state structure with randomly arranged atoms, *Phys. Lett. A*, 135, 461, 1989.

57. Herr, U. et al., Investigation of nanocrystalline iron materials by Mossbauer spectroscopy, *Appl. Phys. Lett.*, 50, 472, 1987.

58. Jorra, E. et al., Small-angle neutron scattering from nanocrystalline Pd, *Philos. Mag. B*, 60, 159, 1989.

59. Koch, E.E., Ed., *Handbook of Synchrotron Radiation*, North-Holland, New York, 1983, p. 995.

60. Würschum, R. and Schaefer, H.-E., Interfacial free volumes and atomic diffusion in nanostructured solids, in Edelstein, A.S. and Cammarata, R.C., Eds., *Nanomaterials: Synthesis, Properties and Applications*, Inst. Physics Publ., Bristol, 1996, Chapter 11.

61. Schaefer, H.E. et al., Nanometre-sized solids, their structure and properties, *J. Less-Common Metals*, 140, 161, 1988.

62. Würschum, R., Greiner, W., and Schaefer, H.-E., Preparation and positron lifetime spectroscopy of nanocrystalline metals, *Nanostr. Mater.*, 2, 55, 1993.

63. Thomas, G.J., Siegel, R.W., and Eastman, J.A., Grain boundaries in nanophase palladium: high resolution electron microscopy and image simulation, *Scripta Metall. Materialia*, 24, 201, 1990.

64. Wunderlich, W., Ishida, Y., and Maurer, R., HREM studies of the microstructure of nanocrystalline palladium, *Scripta Metall. Materialia*, 24, 403, 1990.

65. Fitzsimmons, M.R. et al., Structural characterization of nanometer-sized crystalline Pd by x-ray-diffraction techniques, *Phys. Rev. B*, 44, 2452, 1991.

66. Stern, E.A. et al., Are nanophase grain boundaries anomalous? *Phys. Rev. Lett.*, 75, 3874, 1995.

67. Löffler, J. and Weissmüller, J., Grain-boundary atomic structure in nanocrystalline palladium from x-ray atomic distribution functions, *Phys. Rev. B*, 52, 7076, 1995.

68. De Panfilis, S. et al., Local structure and size effects in nanophase palladium: an x-ray absorption study, *Phys. Lett. A*, 207, 397, 1995.

69. Boscherini, F., De Panfilis, S., and Weissmüller, J., Determination of local structure in nanophase palladium by x-ray absorption spectroscopy, *Phys. Rev. B*, 57, 3365, 1998.

70. Keblinski, P. et al., Continuous thermodynamic-equilibrium glass transition in high-energy grain boundaries? *Phil. Mag. Lett.* 76, 143, 1997.

71. Tschöpe, A., Birringer, R., and Gleiter, H., Calorimetric measurements of the thermal relaxation in nanocrystalline platinum, *J. Appl. Phys.*, 71, 5391, 1992.

72. Mulyukov, R.R. and Starostenkov, M.D., Structure and physical properties of submicrocrystalline metals prepared by severe plastic deformation, *Acta Materialia Sinica*, 13, 301, 2000.

73. Trapp, S. et al., Enhanced compressibility and pressure-induced structural changes of nanocrystalline iron: *in situ* Mössbauer spectroscopy, *Phys. Rev. Lett.*, 75, 3760, 1995.

74. Grabski, M.W. and Korski, R., Grain boundaries as sinks for dislocations, *Philos. Mag.*, 22, 707, 1970.

75. Valiev, R.Z., Gertsman, V.Y., and Kaibyshev, O.A., Grain boundary structure and properties under external influences, *Phys. Stat. Sol. (a)*, 97, 11, 1986.

76. Kaibyshev, O.A., *Superplasticity of Alloys, Intermetallides and Ceramics*, Springer-Verlag, Berlin, 1992.

77. Musalimov, R.S. and Valiev, R.Z., Dilatometric analysis of aluminium alloy with submicrometre grained structure, *Scripta Metall. Materialia*, 27, 1685, 1992.

78. Mulyukov, Kh. Ya., Khaphizov, S.B., and Valiev, R.Z., Grain boundaries and saturation magnetization in submicron grained nickel, *Phys. Stat. Sol. (a)*, 133, 447, 1992.

79. Valiev, R.Z. et al., Direction of a grain boundary phase in submicrometre grained iron, *Philos. Mag. Lett.*, 62, 253, 1990.

80. Valiev, R.Z. et al., Mössbauer analysis of submicrometer grained iron, *Scripta Metall. Materialia*, 25, 2717, 1991.

81. Shabashov, V.A. et al., Deformation-induced nonequilibrium grain-boundary phase in submicrocrystalline iron, *Nanostr. Mater.*, 11, 1017, 1999.

82. Zubairov, L.R. et al., Effect of submicron crystalline structure on field emission of nickel, *Doklady. Phys.*, 45, 198, 2000.

83. Mulyukov, R.R., Yumaguzin, Yu.M., and Zubairov, L.R., Field emission from submicron-grained tungsten, *JETP Lett.*, 72, 257, 2000.

84. Rybin, V.V., Zisman, A.A., and Zolotarevskii, N.Yu., Junction disclinations in plastically deformed crystals, *Phys. Solid State*, 27, 181, 1985.

85. Rybin, V.V., Zolotarevskii, N.Yu., and Zhukovskii, I.M., Structure evolution and internal stresses in the stage of developed plastic deformation of solids, *Phys. Metals Metallogr.*, 69(1), 5, 1990.

86. Rybin, V.V., Zisman, A.A., and Zolotarevsky, N.Yu., Junction disclinations in plastically deformed crystals, *Acta Metall. Materialia*, 41, 2211, 1993.

87. Romanov, A.E. and Vladimirov, V.I., Disclinations in crystalline solids, in Nabarro, F.R.N., Ed., *Dislocations in Solids*, 9, North-Holland, Amsterdam, 1992, p. 191.

88. Nazarov, A.A., Romanov, A.E., and Valiev, R.Z., On the structure, stress fields and energy of nonequilibrium grain boundaries, *Acta Metall. Materialia*, 41, 1033, 1993.

89. Nazarov, A.A., Romanov, A.E., and Valiev, R.Z., On the nature of high internal stresses in ultrafine-grained materials, *Nanostr. Mater.*, 4, 93, 1994.

90. Nazarov, A.A., Romanov, A.E., and Valiev, R.Z., Models of the defect structure and analysis of the mechanical behavior of nanocrystals, *Nanostr. Mater.*, 6, 775, 1995.

91. Nazarov, A.A., Romanov, A.E., and Valiev, R.Z., Random disclination ensembles in ultrafine-grained materials produced by severe plastic deformation, *Scripta Materialia*, 34, 729, 1996.

92. Nazarov, A.A., Ensembles of gliding grain boundary dislocations in ultrafine grained materials produced by severe plastic deformation, *Scripta Materialia*, 37, 1155, 1997.

93. Seeger, A. and Haasen, P., Density changes of crystals containing dislocations, *Philos. Mag.*, 3, 470, 1958.

94. Nazarov, A.A., Kinetics of relaxation of disordered grain boundary dislocation arrays in ultrafine grained materials, *Annales de Chimie*, 21, 461, 1996.

95. Nazarov, A.A., Kinetics of grain boundary recovery in deformed polycrystals, *Interface Sci.*, 8, 315, 2000.

96. Kaur, I., Gust, W., and Kozma, L., *Handbook of Grain Boundary and Interphase Boundary Diffusion Data*, Ziegler Press, Stuttgart, 1989.

97. Gryaznov, V.G. and Trusov, L.I., Size effects in micromechanics of nanocrystals, *Progr. Mater. Sci.*, 37, 289, 1993.

98. Ping, D.H. et al., High resolution electron microscopy studies of the microstructure in nanocrystalline $(Fe_{0.99}Mo_{0.01})Si_9B_{13}$ alloys, *Mater. Sci. Eng. A*, 194, 211, 1995.

99. Zhao, Y.H., Lu, K., and Liu, T., EXAFS study of structural characteristics of nanocrystalline selenium with different grain sizes, *Phys. Rev. B*, 59, 11117, 1999.

100. Sui, M.L. et al., Positron-lifetime study of polycrystalline Ni-P alloys with ultrafine grains, *Phys. Rev. B*, 44, 6466, 1991.

101. Lu, K., Lück, R., and Predel, B., The interfacial excess energy in nanocrystalline Ni-P materials with different grain sizes, *Scripta Metall. Materialia*, 28, 1387, 1993.

102. Noskova, N.I., Ponomareva, E.G., and Myshlyaev, M.M., Structure of nanophases and interfaces in multiphase nanocrystalline $Fe_{73}Ni_{0.5}Cu_1Nb_3Si_{13.5}B_9$ alloy and nanocrystalline copper, *Phys. Metals Metallogr.*, 83, 511, 1997.

103. Del Bianco, L. et al., Grain-boundary structure and magnetic behavior in nanocrystalline ball-milled iron, *Phys. Rev. B*, 56, 8894, 1997.

104. Balogh, J. et al., Mössbauer study of the interface of iron nanocrystallites, *Phys. Rev. B*, 61, 4109, 2000.

105. Fultz, B. and Frase, H.N., Grain boundaries of nanocrystalline materials – their widths, compositions, and internal structures, *Hyperfine Interactions*, 130, 81, 2000.

106. Negri, D., Yavari, A.R., and Deriu A., Deformation induced transformations and grain boundary thickness in nanocrystalline B2 FeAl, *Acta Materialia.*, 47, 4545, 1999.

107. Haasz, T.R. et al., Intercrystalline density of nanocrystalline nickel, *Scripta Metall. Materialia*, 32, 423, 1995.

108. Würschum, R. et al., Free volumes and thermal stability of electro-deposited nanocrystalline Pd, *Nanostr. Mater.*, 9, 615, 1997.

109. Phillpot, S.R., Wolf, D., and Gleiter, H., A structural model for grain boundaries in nanocrystalline materials, *Scripta Metall. Materialia*, 33, 1245, 1995.

110. Keblinski, P. et al., Structure of grain boundaries in nanocrystalline palladium by molecular dynamics simulation, *Scripta Materialia*, 41, 631, 1999.

111. Keblinski, P. et al., Amorphous structure of grain boundaries and grain junctions in nanocrystalline silicon by molecular-dynamics simulation, *Acta Materialia*, 45, 987, 1997.

112. Keblinski, P. et al., Thermodynamic criterion for the stability of amorphous intergranular films in covalent materials, *Phys. Rev. Lett.*, 77, 2965, 1996.

113. Schiøtz, J. et al., Atomic-scale simulations of nanocrystalline metals, *Phys. Rev. B*, 60, 11971, 1999.

114. Van Swygenhoven, H., Farkas, D., and Caro, A., Grain-boundary structures in polycrystalline metals at the nanoscale, *Phys. Rev. B*, 62, 831, 2000.

115. Honneycutt, J.D. and Andersen, H.C., Molecular dynamics study of melting and freezing of small Lennard–Jones clusters, *J. Phys. Chem.*, 91, 4950, 1987.

116. Kaur, I., Mishin, Yu., and Gust, W., *Fundamentals of Grain and Interphase Boundary Diffusion*, John Wiley, Chichester, 1995.

117. Horvath, J., Birringer, R., and Gleiter, H., Diffusion in nanocrystalline materials, *Solid State Comm.*, 62, 319, 1987.

118. Bokstein, B.S. et al., Diffusion in nanocrystalline nickel, *Nanostr. Mater.*, 6, 873, 1995.

119. Kolobov, Y.R. et al., Grain boundary diffusion characteristics of nanostructured nickel, *Scripta Materialia*, 44, 873, 2001.

120. Würschum, R. et al., Tracer diffusion and crystallite growth in ultra-fine-grained Pd prepared by severe plastic deformation, *Annalles de Chimie*, 21, 471, 1996.

121. Herth, S. et al., Self-diffusion in nanocrystalline Fe and Fe-rich alloys, *Def. Diff. Forum*, 194–199, 1199, 2001.

122. Würschum, R., Brossmann, U., and Schaefer, H.-E., Diffusion in nanocrystalline materials, in *Nanostructured Materials — Processing, Properties and Potential Applications*, Koch, C.C., Ed., William Andrew, New York, 2001, Chapter 7.

123. Tanimoto, H. et al., Self-diffusion and magnetic properties in explosion densified nanocrystalline Fe, *Scripta Materialia*, 42, 961, 2000.

124. Borisov, V.T., Golikov, V.M., and Shcherbedinsky, G.V., On the relation of grain boundary diffusion coefficients to the energy of grain boundaries, *Phys. Metals Mettalogr.*, 17, 881, 1964.

125. Nazarov, A.A., Internal stress effect on the grain boundary diffusion in submicrocrystalline metals, *Philos. Mag. Lett.* 80, 221, 2000.

126. Manning, J.R., *Diffusion Kinetics for Atoms in Crystals*, D. Van Nostrand Co., Toronto, 1968.

127. Ovid'ko, I.A., Reizis, A.B., and Masumura, R.A., Effects of transformations of grain boundary defects on diffusion in nanocrystalline materials, *Mater. Phys. Mech.*, 1, 103, 2000.

128. Keblinski, P., Wolf, D., and Gleiter, H., Molecular-dynamics simulation of grain boundary diffusion creep, *Interface Sci.*, 6, 205, 1998.

129. Keblinski, P. et al., Self-diffusion in high-angle fcc metal grain boundaries by molecular dynamics simulations, *Philos. Mag. A*, 79, 2735, 1999.

130. *Metals Handbook*, 10th ed., V.2, *Properties and Selection-Nonferrous Alloys and Special-Purpose Materials*, ASM International, Materials Park, OH, 1990.

131. Akhmadeev, N.A. et al., The effect of heat treatment on the elastic and dissipative properties of copper with the submicrocrystalline structure, *Acta Metall. Materialia*, 41, 1041, 1993.

132. Sanders, P.G., Eastman, J.A., and Weertman, J.R., Elastic and tensile behavior of nanocrystalline copper and palladium, *Acta Materialia*, 45, 4019, 1997.

133. Shen, T.D. et al., On the elastic moduli of nanocrystalline Fe, Cu, Ni, and Cu-Ni alloys prepared by mechanical milling/alloying, *J. Mater. Res.*, 10, 2892, 1995.

134. Spriggs, R.M., Expression for effect of porosity on elastic modulus of polycrystalline refractory materials, particularly aluminum oxide, *J. Am. Ceramic Soc.*, 44, 628, 1961.

135. Lebedev, A.B. et al., Softening of the elastic modulus in submicrocrystalline copper, *Mater. Sci. Eng. A*, 203, 165, 1995.

136. Hall, E.O., Deformation and aging of mild steel, *Proc. Phys. Soc.*, 64, 747, 1951.

137. Petch, N.J., The cleavage strength of polycrystals, *J. Iron Steel Inst.*, 174, 25, 1953.

138. Lasalmonie, A. and Strudel, J.L., Influence of grain size on the mechanical behaviour of some high strength materials, *J. Mater. Sci.*, 21, 1837, 1986.

139. Tabor, D., *The Hardness of Metals*, Clarendon Press, Oxford, 1951.

140. Chokshi, A.H. et al., On the validity of the Hall–Petch relationship in nanocrystalline materials, *Scripta Metall.*, 23, 1679, 1989.

141. Lu, K., Wei, W.D., and Wang, J.T., Microhardness and fracture properties of nanocrystalline Ni-P alloy, *Scripta Metall. Materialia*, 24, 2319, 1990.

142. Nieman, G.W., Weertman, J.R., and Siegel, R.W., Mechanical behavior of nanocrystalline Cu and Pd, *J. Mater. Res.*, 6, 1012, 1991.

143. Jang, J.S.C. and Koch, C.C., The Hall–Petch relationship in nanocrystalline iron produced by ball milling, *Scripta Metall. Materialia*, 24, 1599, 1990.

144. El-Sherik, A.M. et al., Deviations from Hall–Petch behaviour in as-prepared nanocrystalline nickel, *Scripta Metall. Materialia*, 27, 1185, 1992.

145. Valiev, R.Z. et al., The Hall–Petch relation in submicro-grained Al-1.5%Mg alloy, *Scripta Metall. Materialia*, 27, 855, 1992.

146. Fougere, G.E. et al., Grain-size dependent hardening and softening of nanocrystalline Cu and Pd, *Scripta Metall. Materialia*, 26, 1879, 1992.

147. Weertman, J.R. and Sanders, P.G., Plastic deformation of nanocrystalline metals, *Solid State Phenomena*, 35–36, 249, 1994.

148. Wang, N. et al., Effect of grain size on mechanical properties of nanocrystalline materials, *Acta Metall. Materialia*, 43, 519, 1995.

149. Lu, K. and Sui, M.L., An explanation to the abnormal Hall–Petch relation in nanocrystalline materials, *Scripta Metall. Materialia*, 28, 1465, 1993.

150. Coble, R.L., A model for grain-boundary-diffusion controlled creep in polycrystalline materials, *J. Appl. Phys.*, 34, 1679, 1963.

151. Scattergood, R.O. and Koch, C.C., A modified model for Hall–Petch behavior in nanocrystalline materials, *Scripta Metall. Materialia*, 27, 1195, 1992.

152. Li, J.C.M., Generation of dislocations by grain boundary joints and Hall–Petch relation, *Trans. AIME*, 227, 239, 1961.

153. Lian, J., Baudelet, B., and Nazarov, A.A., Model for the prediction of the mechanical behaviour of nanocrystalline materials, *Mater. Sci. Eng A*, 172, 23, 1993.

154. Li, J.C.M. and Chou, Y.F., The role of dislocations in the flow stress-grain size relationships, *Metall. Trans.*, 1, 1145, 1970.

155. Eshelby, J.D., Frank, F.C., and Nabarro, F.R.N., The equilibrium of linear arrays of dislocations, _Philos. Mag._, 42, 351, 1951.

156. Kocks, U.F., The relation between polycrystal deformation and single-crystal deformation, _Metall. Trans._, 1, 1121, 1970.

157. Pande, C.S., Masumura, R.A., and Armstrong, R.W., Pile-up based Hall–Petch relation for nano-scale materials, _Nanostr. Mater._, 2, 323, 1993.

158. Nazarov, A.A., On the pile-up model of the grain size–yield stress relation for nanocrystals, _Scripta Materialia_, 34, 697, 1996.

159. Gryaznov V.G. et al, On the yield stress of nanocrystals, _J. Mater. Sci._, 28, 4359, 1994.

160. Kim, H.S., A composite model for mechanical properties of nanocrystalline materials, Scripta Materialia, 39, 1057, 1998.

161. Valiev, R.Z. et al., Deformation behavior of ultrafine-grained copper, _Acta Metall. Materialia_, 42, 2467, 1994.

162. Van-Swygenhoven, H. and Caro, A., Plastic behavior of nanophase metals studied by molecular dynamics, _Phys. Rev. B_, 58, 11246, 1998.

163. Van Swygenhoven, H. et al., Competing plastic deformation mechanisms in nanophase metals, _Phys. Rev. B_, 60, 22, 1999.

164. Nazarov, A.A.,. On the role of non-equilibrium grain boundary structure in the yield and flow stress of polycrystals, _Philos. Mag. A_, 69, 327, 1994.

165. Nazarov, A.A. et al., The role of internal stresses in the deformation behavior of nanocrystals, in _Strength of Materials_, JIMIS, Japan, 1994, p. 877.

166. Stauffer, D., _Introduction to Percolation Theory_, Taylor and Francis, London, 1982, p. 17.

167. Wang, N. et al., Room temperature creep behavior of nanocrystalline nickel produced by an electrodeposition technique, _Mater. Sci. Eng. A_, 237, 50, 1997.

168. Lüthy, H., White, R.A., and Sherby, O.D., Grain boundary sliding and deformation mechanism maps, _Mater. Sci. Eng._, 39, 211, 1979.

169. Mukherjee, A.K., Deformation mechanisms in superplasticity, _Ann. Rev. Mater. Sci._, 9, 191, 1979.

170. Perevezentsev, V.N., Rybin, V.V., and Chuvil'deev, V.N., The theory of structural superplasticity – II. Accumulation of defects on the intergranular and interphase boundaries; accommodation of the grain boundary sliding; the upper bound of the superplastic deformation, _Acta Metall. Materialia_, 40, 895, 1992.

171. Cai, B. et al., Interface controlled diffusional creep of nanocrystalline pure copper, _Scripta Materialia_, 41, 755, 1999.

172. Kong, Q.P. et al., The creep of nanocrystalline metals and its connection with grain boundary diffusion, _Defect Diffusion Forum_, 188–190, 45, 2001.

173. Arieli, A. and Mukherjee, A.K., The rate-controlling deformation mechanisms in superplasticity – a critical assessment, _Metall. Trans. A_, 13, 717, 1982.

174. Higashi, K., Recent advances and future directions in superplasticity, _Mater. Sci. Forum_, 357–359, 345, 2001.

175. Valiev, R.Z., Krasilnikov, N.A., and Tsenev, N.K., Plastic deformation of alloys with submicron-grained structure, _Mater. Sci. Eng. A_, 137, 35, 1991.

176. Valiev, R.Z., Superplastic behavior of nanocrystalline metallic materials, _Mater. Sci. Forum_, 243–245, 207, 1997.

177. Mukhopadhyay, J., Kaschner, K., and Mukherjee, A.K., Superplasticity in boron-doped nickel aluminum (Ni_3Al) alloy, _Scripta Metall. Materialia_, 24, 857, 1990.

178. Mishra, R.S. et al., Tensile superplasticity in a nanocrystalline nickel aluminide, _Mater. Sci. Eng. A_, 252, 174, 1998.

179. Mishra, R.S. et al., High-strain-rate superplasticity from nanocrystalline Al alloy 1420 at low temperatures, _Philos. Mag. A_, 81, 37, 2001.

180. Berbon, P.B. et al., Requirements for achieving high-strain-rate superplasticity in cast aluminium alloys, _Philos. Mag. Lett._, 78, 313, 1999.

181. Mishra, R.S., McFadden, S.X., and Mukherjee, A.K., Analysis of tensile superplasticity in nanomaterials, *Mater. Sci. Forum*, 304–306, 31, 1999.
182. McFadden, S.X. et al., Low-temperature superplasticity in nanostructured nickel and metal alloys, *Nature*, 398, 684, 1999.
183. Floreen, S., Superplasticity in pure nickel, *Scripta Metall.*, 1, 19, 1967.

23

Nano- and Micromachines in NEMS and MEMS

CONTENTS

Sergey Edward Lyshevski
Rochester Institute of Technology

Abstract

Nano- and microengineering have experienced phenomenal growth over the past few years. These developments are based on far-reaching theoretical, applied, and experimental advances. The synergy of engineering, science, and technology is essential to design, fabricate, and implement nano- and micro-electromechanical systems (NEMS and MEMS). These are built with devices (components) including nano- and micromachines. These machines are the nano- and microscale motion devices that can perform actuation, sensing, and computing. Recent trends in engineering and industrial demands have increased the emphasis on integrated synthesis, analysis, and design of nano- and micromachines.

Synthesis, design, and optimization processes are evolutionary in nature. They start with biomimicking, prototyping, contemporary analysis, setting possible solutions, requirements, and specifications. High-level physics-based synthesis is performed first in order to devise machines by using a synthesis and classification concept. Then, comprehensive analysis, heterogeneous simulation, and design are performed employing computer-aided design. Each level of the design hierarchy corresponds to a particular abstraction level and has the specified set of evolutionary learning activities, theories, and tools to be developed in order to support the design. The multidisciplinary synthesis and design require the application of a wide spectrum of paradigms, methods, computational environments, and fabrication technologies that are reported in this chapter.

23.1   Introduction to Nano- and Micromachines

Nano- and micromachines, which are the motion nano- and microscale devices, are the important components of NEMS and MEMS used in biological, industrial, automotive, power, manufacturing, and other systems. In fact, actuation, sensing, logics, computing, and other functions can be performed using these machines. *MEMS and NEMS*[1] (Figure 23.1) covers the general issues in the design of nano- and micromachines. Important topics, focused themes, and issues will be outlined and discussed in this chapter.

The following defines the nano- and micromachines under our consideration:

The nano- and micromachines are the integrated electromagnetic-based nano- and microscale motion devices that (1) convert physical stimuli to electrical or mechanical signals and vice versa, and (2) perform actuation and sensing.

It must be emphasized that nano- and microscale features of electromagnetic, electromechanical, electronic, optical, and biological structures as well as operating principles of nano- and micromachines are basic to their operation, design, analysis, and fabrication.

The step-by-step procedure in the design of nano- and microscale machines is:

1. Define application and environmental requirements
2. Specify performance specifications

FIGURE 23.1   *MEMS and NEMS: Systems, Devices, and Structures.*

3. Devise (synthesize) machines, researching operating principles, topologies, configurations, geometry, electromagnetic and electromechanical systems, etc.
4. Perform electromagnetic, mechanical, and sizing–dimension estimates
5. Define technologies, techniques, processes, and materials (permanent magnets, coils, insulators, etc.) to fabricate machines and their nano- and microstructures (components)
6. Develop high-fidelity mathematical models with a minimum level of simplifications and assumptions to examine integrated electromagnetic–mechanical–vibroacoustic phenomena and effects
7. Based upon data-intensive analysis and heterogeneous simulations, perform thorough electromagnetic, mechanical, thermodynamic, and vibroacoustic design with performance analysis and outcome prediction
8. Modify and refine the design, optimizing machine performance
9. Design control laws to control machines and implement these controllers using ICs (this task can be broken down to many subtasks and problems related to control laws design, optimization, analysis, simulation, synthesis of IC topologies, IC fabrication, machine–IC integration, interfacing, communication, etc.).

Before engaging in the fabrication, one must solve synthesis, design, analysis, and optimization problems. In fact, machine performance and its suitability/applicability directly depend upon synthesized topologies, configurations, operating principles, etc. The synthesis with follow-on modeling activities allows the designer to devise and research novel phenomena and discover advanced functionality and new operating concepts. These guarantee synthesis of superior machines with enhanced integrity, functionality, and operationability. Thus, through the synthesis, the designer devises machines that must be modeled, analyzed, simulated, and optimized. Finally, as shown in Figure 23.2, the devised and analyzed nano- and micromachines must be fabricated and tested.

The synthesis and design of nano- and micromachines focuses on multidisciplinary synergy — integrated synthesis, analysis, optimization, biomimicking, prototyping, intelligence, learning, adaptation, decision making, and control. Integrated multidisciplinary features, synergetic paradigms, and heterogeneous computer-aided design advance quickly. The structural complexity and integrated dependencies of machines has increased drastically due to newly discovered topologies, advanced configurations, hardware and software advancements, and stringent *achievable* performance requirements. Answering the demands of the rising complexity, performance specifications, and intelligence, the fundamental theory must be further expanded. In particular, in addition to devising nano- and microscale structures, there are other issues which must be addressed in view of the constantly evolving nature of the nano- and micromachines, e.g., analysis, design, modeling, simulation, optimization, intelligence, decision

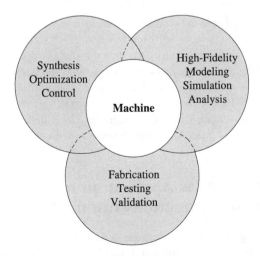

FIGURE 23.2 Synthesis design and fabrication of nano- and microscale machines.

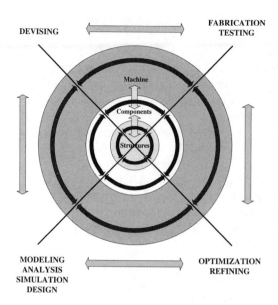

FIGURE 23.3   X-design–flow map with four domains.

making, diagnostics, and fabrication. Competitive *optimum-performance* machines can be designed only by applying advanced computer-aided design using high-performance software that allows the designer to perform high-fidelity modeling and data-intensive analysis with intelligent data mining.

In general, as reported in Reference 1, for nano- and micromachines, the evolutionary developments can be represented as the X-design–flow map documented in Figure 23.3. This map illustrates the synthesis flow from devising (synthesis) to modeling, analysis, simulation, and design; from modeling, analysis, simulation, and design to optimization and refining; and finally from optimization and refining to fabrication and testing. The direct and inverse sequential evolutionary processes are reported. The proposed X map consists of the following four interactive domains:

1. Devising (synthesis)
2. Modeling–analysis–simulation–design
3. Optimization–refining
4. Fabrication–testing

The desired degree of abstraction in the synthesis of new machines requires one to apply this X-design flow map to devise, design, and fabricate novel motion devices which integrate nano- and microscale components and structures. The failure of verify the design for of any machine component in the early phases causes the failure of design for high-performance machines and leads to redesign. The interaction between the four domains as well as integration allows one to guarantee bidirectional *top-down* and *bottom-up* design features applying the low-level component data to high-level design and using the high-level requirements to devise and design low-level components. The X-design flow map ensures hierarchy, modularity, locality, integrity, and other important features allowing one to design high-performance machines.

23.2   Biomimetics and Its Application to Nano- and Micromachines: Directions toward Nanoarchitectronics

One of the most challenging problems in machine design is the topology–configuration synthesis, integration, and optimization, as well as computer-aided design and software developments (intelligent libraries, efficient analytical and numerical methods, robust computation algorithms, tools and environments to perform design, simulation, analysis, visualization, prototyping, evaluation, etc.). The design of state-of-the-art high-performance machines with the ultimate goal to guarantee the integrated synthesis can be pursued through analysis of complex patterns and paradigms of evolutionary developed biological systems. Novel nano- and micromachines can be devised, classified, and designed. It is illustrated in Reference 1 that the biological- and bioengineered-based machines have been examined and made.

Many species of bacterium move around their aqueous environments using flagella, which are the protruding helical filaments driven by the rotating bionanomotor. The *Escherichia coli* (*E. coli*) bacterium was widely studied. The bionanomotors convert chemical energy into electrical energy, and electrical energy into mechanical energy. In most cases, the bionanomotors use the proton or sodium gradient, maintained across the cell's inner membrane as the energy source. In an *E. coli* bionanomotor, the energy conversion, torque production, rotation, and motion are due to the downhill transport of ions. The research in complex chemo-electro-mechanical energy conversion allows one to understand complex torque generation, energy conversion, bearing, and actuation–sensing–feedback–control mechanisms. With the ultimate goal to devise novel organic and inorganic nano- and micromachines through biomimicking and prototyping, one can invent, discover, or prototype unique machine topologies, novel noncontact bearing, new actuation–sensing–feedback–control mechanisms, advanced torque production, energy conversion principles, and novel machine configurations.

Biomimetic machines are the man-made motion devices that are based on biological principles or on biologically inspired building blocks (structures) and components integrated in the devices. These developments benefit greatly from adopting strategies and architectures from biological machines. Based on biological principles, bio-inspired machines can be devised (or prototyped) and designed. This research provides the enabling capabilities to achieve potential breakthroughs that guarantee major broad-based research enterprises.

Nanoarchitectronics concentrates on the development of the NEMS architectures and configurations using nanostructures and nanodevices as the components and subsystems. These components and subsystems must be integrated in the functional NEMS. Through optimization of architecture, synthesis of optimal configurations, design of NEMS components (transducers, radiating energy devices, nanoICs, optoelectronic devices, etc.), biomimicking, and prototyping, novel NEMS as large-scale nanosystems can be discovered. These integrated activities are called *nanoarchitectronics*. By applying the nanoarchitectronic paradigm, one facilitates cost-effective solutions, reducing the design cycle as well as guaranteeing design of high-performance large-scale NEMS. In general, the large-scale NEMS integrate N nodes of nanotransducers (actuators/sensors, smart structures, and other motion nanodevices), radiating energy devices, optoelectronic devices, communication devices, processors and memories, interconnected networks (communication buses), driving/sensing nanoICs, controlling/processing nanoICs, input–output (IO) devices, etc. Different NEMS configurations were synthesized, and diverse architectures are reported in Reference 1.

Let us study a bionanomotor (the component of NEMS or MEMS) in order to devise and design high-performance nano- and micromachines with new topologies, operating principles, enhanced functionality, superior capabilities, and expanded operating envelopes. The *E. coli* bionanomotor and the bionanomotor–flagella complex are shown in Figure 23.4.[1]

The *protonomotive* force in the *E. coli* bionanomotor is axial. However, the *protonomotive* or *magnetomotive* force can be radial as well. Through biomimicking, two machine topologies are defined to be radial and axial. Using the radial topology, the cylindrical machine with permanent magnet poles on the rotor and noncontact electrostatic bearing is shown in Figure 23.5.[1] The electrostatic noncontact bearings allow one to significantly expand the operating envelope, maximizing the angular velocity, efficiency, and reliability; improving ruggedness and robustness; minimizing cost and maintenance; decreasing size and weight; and optimizing packaging and integrity.

The advantage of radial topology is that the net radial force on the rotor is zero. The disadvantages are the difficulties in fabricating and assembling these man-made machines with nano- and microstructures (stator with deposited windings and rotor with deposited magnets). The stationary magnetic field

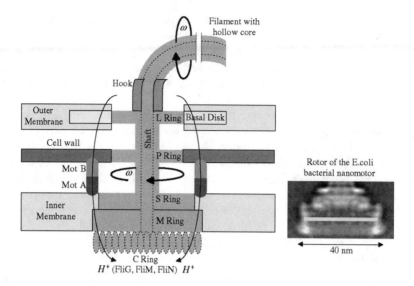

FIGURE 23.4 *E. coli* bacterial bionanomotor–flagella complex and rotor image.

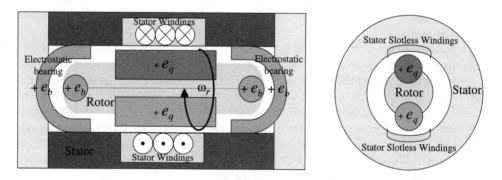

FIGURE 23.5 Radial topology machine with electrostatic noncontact bearings. Poles are $+e_q$ and $-e_q$, and electrostatic bearing is formed by $+e_b$.

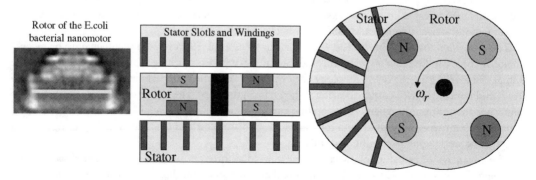

FIGURE 23.6 Axial topology machine with permanent magnets and deposited windings.

is established by the permanent magnets, and microscale stators and rotors can be fabricated using surface micromachining and high-aspect-ratio technologies. Slotless stator windings can be laid out as the deposited microwindings.

Analyzing the *E. coli* bionanomotor, the nano- and microscale machines with axial flux topology are devised. The synthesized axial machine is illustrated in Figure 23.6.[1] The advantages of the axial topology are the simplicity and affordability to manufacture and assemble machines because (1) permanent magnets are flat, (2) there are no strict shape-geometry and sizing requirements imposed on the magnets, (3) there is no rotor back ferromagnetic material required (silicon or silicon–carbide can be applied), and (4) it is easy to deposit the magnets and windings (even molecular wires) on the flat stator. The disadvantages are the slightly lower torque, force, and power densities and decreased winding utilization compared with the radial topology machines. However, nanoscale machines can be feasibly fabricated only as the axial topology motion nanodevices.

23.3   Controlled Nano- and Micromachines

Let us examine the requirements and specifications imposed within the scope of consequent synthesis, design, and fabrication. Different criteria are used to synthesize and design nano- and microscale machines with ICs due to different behaviors, physical properties, operating principles, and performance criteria. The level of hierarchy must be defined, and the design flow is illustrated in Figure 23.7. The similar flow can be applied to the NEMS and MEMS using the *nanoarchitectronic* paradigm.

Automated synthesis can be applied to implement the design flow introduced. The design of machines and systems is a process that starts from the specification of requirements and progressively proceeds to perform a functional design and optimization. The design is gradually refined through a series of sequential synthesis steps. Specifications typically include the performance requirements derived from desired functionality, operating envelope, affordability, reliability, and other requirements. Both *top-down* and *bottom-up* approaches should be combined to design high-performance machines and systems augmenting hierarchy, integrity, regularity, modularity, compliance, and completeness in the synthesis process. The synthesis must guarantee an eventual consensus between behavioral and structural domains as well as ensure descriptive and integrative features in the design.

There is the need to augment interdisciplinary areas as well as to link and place the synergetic perspectives integrating machines with controlling ICs in order to attain control, decision making, signal

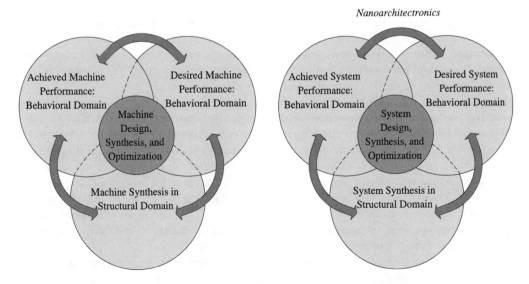

FIGURE 23.7 Design flow in synthesis of nano- and micromachines, NEMS and MEMS.

processing, data acquisition, etc. In fact, nano- and microscale machines must be designed and integrated with the controlling and signal processing ICs, input–output ICs, etc. The principles of matching and compliance are the general design principles which require that the system architectures should be synthesized integrating all components. The matching conditions, functionality, and compliance have to be determined and guaranteed. The system–machines and machine–ICs compliance and operating functionality must be satisfied. In particular, machines devised must be controlled, and motion controllers should be designed. Additional functions can be integrated, e.g., decision making, adaptation, reconfiguration, and diagnostics. These controllers must be implemented using ICs. The design of controlled high-performance nano- and micromachines implies the components' and structures' synthesis, design, and developments. Among a large variety of issues, the following problems must be resolved:

1. Synthesis, characterization, and design of integrated machines and ICs according to their applications and overall systems requirements by means of specific computer-aided-design tools and software
2. Decision making, adaptive control, reconfiguration, and diagnostics of integrated machines and ICs
3. Interfacing and wireless communication
4. Affordable and high-yield fabrication technologies to fabricate integrated machines and ICs

Synthesis, modeling, analysis, and simulation are the sequential activities for integrated machines–ICs, NEMS, and MEMS. The synthesis starts with the discovery of new or prototyping of existing operating principles; advanced architectures and configurations; examining and utilizing novel phenomena and effects, analysis of specifications imposed; study of performance, modeling, and simulation; assessment of the available fundamental, applied, and experimental data; etc. Heterogeneous simulation and data-intensive analysis start with the model developments (based upon machines devised and ICs used to control them). The designer mimics, studies, analyzes, evaluates, and assesses machine–IC behavior using state, performance, control, events, disturbance, decision making, and other variables. Thus, fundamental, applied, and experimental research and engineering developments are used. It should be emphasized that control and optimization of NEMS and MEMS is a much more complex problem.[1]

23.4   Synthesis of Nano- and Micromachines: Synthesis and Classification Solver

The conceptual view of the nano- and micromachines synthesis and design must be introduced in order to set the objectives and goals and to illustrate the need for synergetic multidisciplinary developments. An important problem addressed and studied in this section is the synthesis of nano- and microscale machines (nano- and microscale motion devices). There is a need to develop the paradigm that will allow the designer to devise novel machines and classify existing machines.

In Section 23.2 it was emphasized that the designer synthesizes machines by devising, discovering, mimicking, and/or prototyping new operational principles. To illustrate the procedure, we consider a two-phase permanent magnet synchronous machine as shown in Figure 23.8 (permanent magnet stepper micromotors, fabricated and tested in the middle 1990s, are two-phase synchronous micromachines).[1]

The electromagnetic system is *endless*, and different geometries can be utilized as shown in Figure 23.8. In contrast, the translational (linear) synchronous machines have the *open-ended* electromagnetic system. Thus, machine geometry and electromagnetic systems can be integrated into the synthesis, classification, analysis, design, and optimization patterns. In particular, motion devices can have different geometries (plate, spherical, toroidal, conical, cylindrical, and asymmetrical) and electromagnetic systems. Using these distinct features, we classify nano- and micromachines.

The basic types of electromagnetic-based nano- and micromachines are induction, synchronous, rotational, and translational (linear). That is, machines are classified using a type classifier as given by $Y = \{y : y \in Y\}$.

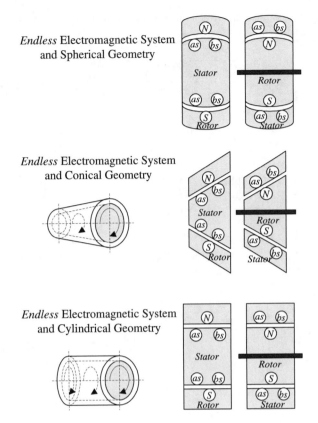

FIGURE 23.8 Permanent magnet synchronous machines with *endless* electromagnetic systems and different geometries.

As illustrated, motion devices are categorized using a geometric classifier (plate P, spherical S, torroidal T, conical N, cylindrical C, or asymmetrical A geometry) and an electromagnetic system classifier (*endless* E, *open-ended* O, or *integrated* I). The machine classifier, as documented in Table 23.1, is partitioned into three horizontal and six vertical strips. It contains 18 sections, each identified by ordered pairs of characters, such as (E, P) or (O, C). In each ordered pair, the first entry is a letter chosen from the bounded electromagnetic system set $M = \{E, O, I\}$. The second entry is a letter chosen from the geometric set $G = \{P, S, T, N, C, A\}$. That is, for electromagnetic machines, the electromagnetic system geometric set is $M \times G = \{(E, F), (E, S), (E, T), \ldots, (I, N), (I, C), (I, A)\}$. In general, we have $M \times G = \{(m, g): m \in M \text{ and } g \in G\}$.

Other categorizations can be applied. For example, multiphase (usually two- and three-phase) machines are classified using a phase classifier $H = \{h: h \in H\}$. Therefore, we have $Y \times M \times G \times H = \{(y, m, g, h): y \in Y, m \in M, g \in G \text{ and } h \in H\}$.

Topology (radial or axial), permanent magnets shaping (strip, arc, disk, rectangular, rhomb, triangular, etc.), thin films, permanent magnet characteristics (BH demagnetization curve, energy product, hysteresis minor loop, etc.), *electromotive force* distribution, cooling, power, torque, size, torque–speed characteristics, bearing, packaging, as well as other distinct features are easily classified. Hence, nano- and micromachines can be devised and classified by an N-tuple as: machine type, electromagnetic system, geometry, topology, phase, winding, sizing, bearing, cooling, fabrication, materials, packaging, etc.

Using the possible geometry and electromagnetic systems (*endless*, *open-ended*, and *integrated*), novel high-performance machines can be synthesized. This idea is very useful in the study of existing machines as well as in the synthesis of an infinite number of innovative motion devices. Using the synthesis and

TABLE 23.1   Classification of Nano- and Micromachines Using the Electromagnetic system – Geometry Using the Synthesis and Classification Solver

		Geometry					
M	**G**	Plate, P	Spherical, S	Torroidal, T	Conical, N	Cylindrical, C	Asymmetrical, A
Electromagnetic System	Endless (Closed), E						Σ
	Open-Ended (Open), O						Σ
	Integrated, I	Σ	Σ	Σ	Σ	Σ	Σ

classification solver, which is documented in Table 23.1, the spherical, conical, and cylindrical geometry of two-phase permanent magnet synchronous machine are illustrated in Figure 23.9.

The cross-section of the slotless radial-topology micromachine, fabricated on the silicon substrate with polysilicon stator (with deposited windings), polysilicon rotor (with deposited permanent magnets), and contact bearing is illustrated in Figure 23.10. The fabrication of this micromotor and the processes were reported in Reference 1.

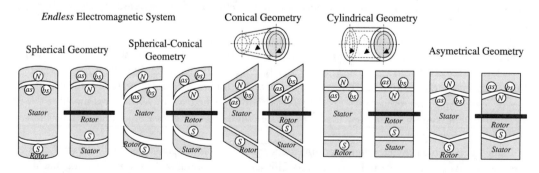

FIGURE 23.9   Two-phase permanent magnet synchronous machine with *endless* system and distinct geometries.

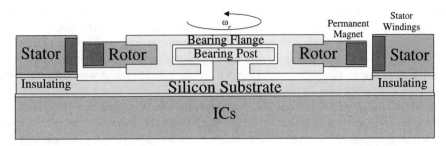

FIGURE 23.10 Cross-section schematics for slotless radial-topology permanent magnet brushless micromachine with controlling ICs.

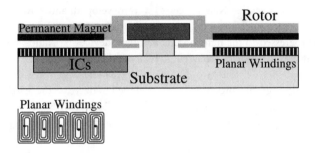

FIGURE 23.11 Axial micromachine (cross-sectional schematics) with controlling ICs.

The major problem is to devise novel motion devices in order to relax fabrication difficulties, guarantee affordability, efficiency, reliability, and controllability of nano- and micromachines. The electrostatic and planar micromotors fabricated and tested to date are found to be inadequate for a wide range of applications due to difficulties associated with low performance and cost.[1–11] Figure 23.11 illustrates the axial topology micromachine that has the *closed-ended* electromagnetic system. The stator is made on the substrate with deposited microwindings (printed copper coils can be made using the fabrication processes described in Reference 1 as well as using double-sided substrate with the one-sided deposited copper thin film made applying conventional photolithography processes). The bearing post is fabricated on the stator substrate, and the bearing holds is a part of the rotor microstructure. The rotor with permanent magnet thin films rotates due to the electromagnetic torque developed. Stator and rotor are made using conventional well-developed processes and materials.

The synthesis and classification solver reported directly leverages high-fidelity modeling, allowing the designer to attain physical and behavioral data-intensive analysis, heterogeneous simulations, optimization, performance assessment, outcome prediction, etc.

23.5 Fabrication Aspects

Nano- and micromachines can be fabricated through deposition of the conductors (coils and windings), ferromagnetic core, magnets, insulating layers, as well as other microstructures (movable and stationary members and their components, including bearing). The subsequent order of the processes, sequential steps, and materials are different depending on the machines devised, designed, analyzed, and optimized. Other sources[1,4,6,12] provide the reader with the basic features and processes involved in the micromachine fabrication. This section outlines the most viable aspects.

Complementary metal–oxide semiconductor (CMOS), high-aspect-ratio (LIGA and LIGA-like), and surface micromachining technologies are key features for fabrication of nano- and micromachines and structures. The LIGA (Lithography–Galvanoforming–Molding or, in German, Lithografie–Galvanik–Abformung) technology allows one to make three-dimensional microstructures with the high aspect ratio (depth vs. lateral dimension is more than 100). The LIGA technology is based on the X-ray lithography, which ensures short wavelength (from few to ten Å) and leads to negligible diffraction effects and larger depth of focus compared with photolithography. The major processes in the machine's microfabrication are diffusion, deposition, patterning, lithography, etching, metallization, planarization, assembling, and packaging. Thin-film fabrication processes were developed and used for polysilicon, silicon dioxide, silicon nitride, and other different materials, e.g., metals, alloys, composites. The basic steps are lithography, deposition of thin films and materials (electroplating, chemical vapor deposition, plasma-enhanced chemical vapor deposition, evaporation, sputtering, spraying, screen printing, etc.), removal of material (patterning) by wet or dry techniques, etching (plasma etching, reactive ion etching, laser etching, etc.), doping, bonding (fusion, anodic, and other), and planarization.[1,4,6,12]

To fabricate motion and radiating energy nano- and microscale structures and devices, different fabrication technologies are used.[1–15] New processes were developed, and novel materials were applied modifying the CMOS, surface micromachining, and LIGA technologies. Currently the state of the art in nanofabrication has progressed to the nanocircuits and nanodevices.[1,16] Nano- and micromachines and their components (stator, rotor, bearing, coils, etc.) are defined photographically, and the high-resolution photolithography is applied to define two-dimensional (planar) and three-dimensional (geometry) shapes. Deep ultraviolet lithography processes were developed to decrease the feature sizes of microstructures to 0.1 μm. Different exposure wavelengths λ (435, 365, 248, 200, 150, or 100 nm) are used. Using the Rayleigh model for image resolution, the expressions for image resolution i_R and the depth of focus d_F are given by:

$$i_R = k_i \frac{\lambda}{N_A}, \, d_F = k_d \frac{\lambda}{N_A^2}$$

where k_i and k_d are the lithographic process constants; λ is the exposure wavelength; N_A is the numerical aperture coefficient (for high-numerical aperture, N_A varies from 0.5 to 0.6).

The g- and i-line IBM lithography processes (with wavelengths of 435 nm and 365 nm, respectively) allow one to attain 0.35 μm features. The deep ultraviolet light sources (mercury source or excimer lasers) with 248 nm wavelength enable one to achieve 0.25 μm resolution. The changes to short exposure wavelength present challenges and new, highly desired possibilities. However, using CMOS technology, 50 nm features were achieved, and the application of X-ray lithography leads to nanometer scale features.[1,6,13,15] Different lithography processes commonly applied are photolithography, screen printing, electron-beam lithography, X-ray lithography (high-aspect ratio technology), and more.

Although machine topologies and configurations vary, magnetic and insulating materials, magnets, and windings are used in all motion devices. Figure 23.12 illustrates the electroplated microstructures (10 μm wide and thick with 10 μm spaced insulated copper microwindings, deposited on ferromagnetic cores), microrotors, and permanent magnets (electroplated NiFe alloy).[1]

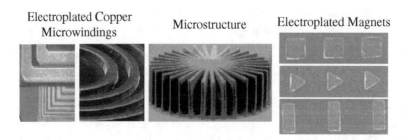

Electroplated Copper Microwindings   Microstructure   Electroplated Magnets

FIGURE 23.12   Deposited copper microwindings, microstructure, and permanent magnets.

23.6 Introduction to Modeling and Computer-Aided Design: Preliminaries

To design nano- and micromachines, fundamental, applied, and experimental research must be performed to further develop the synergetic micro- and nanoelectromechanical theories.[1] Several fundamental electromagnetic and mechanical laws are quantum mechanics, Maxwell's equations, and nonlinear mechanics. The analysis and simulation based upon high-fidelity mathematical models of nano- and micromachines is not a simple task because complex electromagnetic, mechanical, thermodynamic, and vibroacoustic phenomena and effects must be examined in the time domain solving nonlinear partial differential equations. Advanced interactive computer-aided-design tools and software with application-specific toolboxes, robust methods, and novel computational algorithms must be used.

Computer-aided design of nano- and micromachines is valuable due to

1. Calculation and thorough evaluation of a large number of options with data-intensive performance analysis and outcome prediction through heterogeneous simulations in the time domain
2. Knowledge-based intelligent synthesis and evolutionary design that allow one to define optimal solution with minimal effort, time, and cost, as well as with reliability, confidence, and accuracy
3. Concurrent nonlinear quantum, electromagnetic, and mechanical analysis to attain superior performance of motion devices while avoiding costly and time-consuming fabrication, experimentation, and testing
4. Possibility to solve complex nonlinear differential equations in the time domain, integrating systems patterns with nonlinear material characteristics
5. Development of robust, accurate, and efficient rapid design and prototyping environments and tools that have innumerable features to assist the user to set up the problem and to obtain the engineering parameters

The detailed description of different modeling paradigms applied to nano- and microscale systems were reported in Reference 1. In particular, basic cornerstone methods were applied and used, and examples illustrate their application. The following section focuses on the application of quantum and conventional mechanics as well as electromagnetics to nano- and micromachines.

23.7 High-Fidelity Mathematical Modeling of Nano- and Micromachines: Energy-Based Quantum and Classical Mechanics and Electromagnetics

To perform modeling and analysis of nano- and micromachines in the time domain, there is a critical need to develop and apply advanced theories using fundamental laws of classical and quantum mechanics. The quantum mechanics makes use of the Schrödinger equation, while classical mechanics is based upon Newton, Lagrange, and Hamilton laws.[1] Due to the analytic and computational difficulties associated with the application of quantum mechanics, for microsystems, classical mechanics is commonly applied.[1,5] However, the Schrödinger equation can be found using Hamilton's concept, and quantum and classical mechanics are correlated.

Newton's second law $\sum \vec{F}(t, \vec{r}) = m\vec{a}$ in terms of the linear momentum $\vec{p} = m\vec{v}$, is

$$\sum \vec{F} = \frac{d\vec{p}}{dt} = \frac{d(m\vec{v})}{dt} \quad \text{or} \quad \sum \vec{F} = \frac{d\vec{p}}{dt} = m\frac{d\vec{v}}{dt} = m\vec{a}$$

Using the potential energy $\Pi(\vec{r})$, for the conservative mechanical systems we have

$$\sum F(\vec{r}) = -\nabla\Pi(\vec{r})$$

Hence, one obtains

$$m\frac{d^2\vec{r}}{dt^2} + \nabla\Pi(\vec{r}) = 0$$

For the system of N particles, the equations of motion is

$$m_N\frac{d^2\vec{r}_N}{dt^2} + \nabla\Pi(\vec{r}_N) = 0 \text{ or } m_i\frac{d^2(\vec{x}_i, \vec{y}_i, \vec{z}_i)}{dt^2} + \frac{\partial\Pi(\vec{x}_i, \vec{y}_i, \vec{z}_i)}{\partial(\vec{x}_i, \vec{y}_i, \vec{z}_i)} = 0, i = 1, ..., N$$

The total kinetic energy of the particle is $\Gamma = (1/2)mv^2$. For N particles, one has

$$\Gamma\left(\frac{dx_i}{dt}, \frac{dy_i}{dt}, \frac{dz_i}{dt}\right) = \frac{1}{2}\sum_{i=1}^{N} m_i\left(\frac{dx_i}{dt}, \frac{dy_i}{dt}, \frac{dz_i}{dt}\right)$$

Using the generalized coordinates $(q_1, ..., q_n)$ and generalized velocities

$$\left(\frac{dq_1}{dt}, ..., \frac{dq_n}{dt}\right),$$

one finds the total kinetic

$$\Gamma\left(q_1, ..., q_n, \frac{dq_1}{dt}, ..., \frac{dq_n}{dt}\right)$$

and potential $\Pi(q_1, ..., q_n)$ energies.

Thus, for a conservative system, Newton's second law of motion can be given as:

$$\frac{d}{dt}\left(\frac{\partial\Gamma}{\partial\dot{q}_i}\right) + \frac{\partial\Pi}{\partial q_i} = 0, i = 1, ..., n.$$

That is, the generalized coordinates q_i are used to model multibody systems, and:

$$(q_1, ..., q_n) = (\vec{x}_1, \vec{y}_1, \vec{z}_1, ..., \vec{x}_N, \vec{y}_N, \vec{z}_N)$$

The obtained results are connected to the Lagrange equations of motion which are expressed using the total kinetic

$$\Gamma\left(t, q_1, ..., q_n, \frac{dq_1}{dt}, ..., \frac{dq_n}{dt}\right),$$

dissipation

$$D\left(t, q_1, ..., q_n, \frac{dq_1}{dt}, ..., \frac{dq_n}{dt}\right),$$

and potential $\Pi(t, q_1, ..., q_n)$ energies. In particular,

$$\frac{d}{dt}\left(\frac{\partial\Gamma}{\partial\dot{q}_i}\right) - \frac{\partial\Gamma}{\partial q_i} + \frac{\partial D}{\partial q_i} + \frac{\partial\Pi}{\partial q_i} = Q_i$$

Here, q_i and Q_i are the generalized coordinates and the generalized forces (applied forces and disturbances).

The Hamilton concept allows one to model the system dynamics. The equations of motion are found using the generalized momenta p_i,

$$p_i = \frac{\partial L}{\partial \dot{q}_i}$$

The Lagrangian function

$$L\left(t, q_1, \ldots, q_n, \frac{dq_1}{dt}, \ldots, \frac{dq_n}{dt}\right)$$

for conservative systems is the difference between the total kinetic and potential energies. Thus, we have

$$L\left(t, q_1, \ldots, q_n, \frac{dq_1}{dt}, \ldots, \frac{dq_n}{dt}\right) = \Gamma\left(t, q_1, \ldots, q_n, \frac{dq_1}{dt}, \ldots, \frac{dq_n}{dt}\right) - \Pi(t, q_1, \ldots, q_n)$$

Hence,

$$L\left(t, q_1, \ldots, q_n, \frac{dq_1}{dt}, \ldots, \frac{dq_n}{dt}\right)$$

is the function of $2n$ independent variables, and:

$$dL = \sum_{i=1}^{n}\left(\frac{\partial L}{\partial q_i}dq_i + \frac{\partial L}{\partial \dot{q}_i}d\dot{q}_i\right) = \sum_{i=1}^{n}(\dot{p}_i dq_i + p_i d\dot{q}_i)$$

We define the Hamiltonian function as:

$$H(t, q_1, \ldots, q_n, p_1, \ldots, p_n) = -L\left(t, q_1, \ldots, q_n, \frac{dq_1}{dt}, \ldots, \frac{dq_n}{dt}\right) + \sum_{i=1}^{n}p_i \dot{q}_i$$

where

$$\sum_{i=1}^{n}p_i \dot{q}_i = \sum_{i=1}^{n}\frac{\partial L}{\partial \dot{q}_i}\dot{q}_i = \sum_{i=1}^{n}\frac{\partial \Gamma}{\partial \dot{q}_i}\dot{q}_i = 2\Gamma$$

One obtains,

$$dH = \sum_{i=1}^{n}(-\dot{p}_i dq_i + \dot{q}_i dp_i)$$

The significance of the Hamiltonian function is studied analyzing the expression:

$$H\left(t, q_1, \ldots, q_n, \frac{dq_1}{dt}, \ldots, \frac{dq_n}{dt}\right) = \Gamma\left(t, q_1, \ldots, q_n, \frac{dq_1}{dt}, \ldots, \frac{dq_n}{dt}\right) + \Pi(t, q_1, \ldots, q_n)$$

or

$$H(t, q_1, \ldots, q_n, p_1, \ldots, p_n) = \Gamma(t, q_1, \ldots, q_n, p_1, \ldots, p_n) + \Pi(t, q_1, \ldots, q_n)$$

One concludes that the Hamiltonian function, which is equal to the total energy, is expressed as a function of the generalized coordinates and generalized momenta. The equations of motion are governed by the following equations:

$$\dot{p}_i = -\frac{\partial H}{\partial q_i}, \; \dot{q}_i = \frac{\partial H}{\partial p_i}, j = 1, \ldots, n$$

The derived equations are called the Hamiltonian equations of motion.

The Hamiltonian function:

$$H = -\underbrace{\frac{\hbar^2}{2m}\nabla^2}_{\substack{\text{one electron} \\ \text{kinetic energy}}} + \underbrace{\Pi}_{\substack{\text{potential} \\ \text{energy}}}$$

can be used to derive the Schrödinger equation. To describe the behavior of electrons in a media, one uses the N-dimensional Schrödinger equation to obtain the N-electron wave function $\Psi(t, \mathbf{r}_1, \mathbf{r}_2, \ldots, \mathbf{r}_{N-1}, \mathbf{r}_N)$.

The Hamiltonian for an isolated *N*-electron atomic system is

$$H = -\frac{\hbar^2}{2m}\sum_{i=1}^{N}\nabla_i^2 - \frac{\hbar^2}{2M}\nabla^2 - \sum_{i=1}^{N}\frac{1}{4\pi\varepsilon}\frac{e_i q}{|\mathbf{r}_i - \mathbf{r}'_n|} + \sum_{i \neq j}^{N}\frac{1}{4\pi\varepsilon}\frac{e^2}{|\mathbf{r}_i - \mathbf{r}'_j|}$$

where q is the potential due to nucleus; $e = 1.6 \times 10^{-19}$ C (electron charge).

For an isolated *N*-electron, *Z*-nucleus molecular system, the Hamiltonian function is

$$H = -\frac{\hbar^2}{2m}\sum_{i=1}^{N}\nabla_i^2 - \sum_{k=1}^{Z}\frac{\hbar^2}{2m_k}\nabla_k^2 - \sum_{i=1}^{N}\sum_{k=1}^{Z}\frac{1}{4\pi\varepsilon}\frac{e_i q_k}{|\mathbf{r}_i - \mathbf{r}'_k|} + \sum_{i \neq j}^{N}\frac{1}{4\pi\varepsilon}\frac{e^2}{|\mathbf{r}_i - \mathbf{r}'_j|} + \sum_{k \neq m}^{Z}\frac{1}{4\pi\varepsilon}\frac{q_k q_m}{|\mathbf{r}_k - \mathbf{r}'_m|}$$

where q_k are the potentials due to nuclei.

The first and second terms of the Hamiltonian function

$$-\frac{\hbar^2}{2m}\sum_{i=1}^{N}\nabla_i^2 \text{ and } \sum_{k=1}^{Z}\frac{\hbar^2}{2m_k}\nabla_k^2$$

are the multibody kinetic energy operators. The term

$$\sum_{i=1}^{N}\sum_{k=1}^{Z}\frac{1}{4\pi\varepsilon}\frac{e_i q_k}{|\mathbf{r}_i - \mathbf{r}'_k|}$$

maps the interaction of the electrons with the nuclei at **R** (the electron–nucleus attraction energy operator). The fourth term

$$\sum_{i \neq j}^{N}\frac{1}{4\pi\varepsilon}\frac{e^2}{|\mathbf{r}_i - \mathbf{r}'_j|}$$

gives the interactions of electrons with each other (the electron–electron repulsion energy operator). Term

$$\sum_{k \neq m}^{Z}\frac{1}{4\pi\varepsilon}\frac{q_k q_m}{|\mathbf{r}_k - \mathbf{r}'_m|}$$

describes the interaction of the Z nuclei at **R** (the nucleus–nucleus repulsion energy operator).

For an isolated *N*-electron *Z*-nucleus atomic or molecular systems in the Born–Oppenheimer nonrelativistic approximation, we have

$$H\Psi = E\Psi \tag{23.1}$$

The Schrödinger equation is found to be:

$$\left[-\frac{\hbar^2}{2m}\sum_{i=1}^{N}\nabla_i^2 - \sum_{k=1}^{Z}\frac{\hbar^2}{2m_k}\nabla_k^2 - \sum_{i=1}^{N}\sum_{k=1}^{Z}\frac{1}{4\pi\varepsilon}\frac{e_i q_k}{|\mathbf{r}_i - \mathbf{r}'_k|} + \sum_{i \neq j}^{N}\frac{1}{4\pi\varepsilon}\frac{e^2}{|\mathbf{r}_i - \mathbf{r}'_j|} + \sum_{k \neq m}^{Z}\frac{1}{4\pi\varepsilon}\frac{q_k q_m}{|\mathbf{r}_k - \mathbf{r}'_m|} \right]$$

$$\times (\Psi(t, \mathbf{r}_1, \mathbf{r}_2, ..., \mathbf{r}_{N-1}, \mathbf{r}_N) = E(t, \mathbf{r}_1, \mathbf{r}_2, ..., \mathbf{r}_{N-1}, \mathbf{r}_N)\Psi(t, \mathbf{r}_1, \mathbf{r}_2, ..., \mathbf{r}_{N-1}, \mathbf{r}_N))$$

The total energy $E(t, \mathbf{r}_1, \mathbf{r}_2, ..., \mathbf{r}_{N-1}, \mathbf{r}_N)$ must be found using the nucleus–nucleus Coulomb repulsion energy as well as the electron energy.

It is very difficult or impossible to solve analytically or numerically Equation (23.1). Taking into account only the Coulomb force (electrons and nuclei are assumed to interact due to the Coulomb force only), the Hartree approximation is applied expressing the N-electron wave function $\Psi(t, \mathbf{r}_1, \mathbf{r}_2, ..., \mathbf{r}_{N-1}, \mathbf{r}_N)$ as a product of N one-electron wave functions. In particular, $\Psi(t, \mathbf{r}_1, \mathbf{r}_2, ..., \mathbf{r}_{N-1}, \mathbf{r}_N) = \psi_1(t, \mathbf{r}_1)\psi_2(t, \mathbf{r}_2)...\psi_{N-1}(t, \mathbf{r}_{N-1})\psi_N(t, \mathbf{r}_N)$.

The one-electron Schrödinger equation for *j*th electron is

$$\left(-\frac{\hbar^2}{2m}\nabla_j^2 + \Pi(t, \mathbf{r}) \right)\psi_j(t, \mathbf{r}) = E_j(t, \mathbf{r})\psi_j(t, \mathbf{r}) \tag{23.2}$$

In Equation (23.2), the first term $-(\hbar^2/2m)\nabla_j^2$ is the one-electron kinetic energy, and $\Pi(t, \mathbf{r}_j)$ is the total potential energy. The potential energy includes the potential that *j*th electron feels from the nucleus (considering the ion, the repulsive potential in the case of anion, or attractive in the case of cation). It is obvious that the *j*th electron feels the repulsion (repulsive forces) from other electrons.

Assume that the negative electrons' charge density $\rho(\mathbf{r})$ is smoothly distributed in **R**. Hence, the potential energy due interaction (repulsion) of an electron in **R** is

$$\Pi_{Ej}(t, \mathbf{r}) = \int_{R}\frac{e\rho(\mathbf{r}')}{4\pi\varepsilon|\mathbf{r} - \mathbf{r}'|}d\mathbf{r}'$$

Assumptions were made, and the equations derived contradict the Pauli exclusion principle, which requires that the multisystem wave function is antisymmetric under the interchange of electrons. For two electrons, we have:

$$\Psi(t, \mathbf{r}_1, \mathbf{r}_2, ..., \mathbf{r}_j, ..., \mathbf{r}_{j+1}, ..., \mathbf{r}_{N-1}, \mathbf{r}_N) = -\Psi(t, \mathbf{r}_1, \mathbf{r}_2, ..., \mathbf{r}_j, ..., \mathbf{r}_{j+1}, ..., \mathbf{r}_{N-1}, \mathbf{r}_N)$$

To satisfy this principle, the asymmetry phenomenon is integrated using the asymmetric coefficient ±1. The Hartree-Fock equation is

$$\left[-\frac{\hbar^2}{2m}\nabla_j^2 + \Pi(t, \mathbf{r}) \right]\psi_j(t, \mathbf{r}) - \sum_{i}^{N}\int_{R}\frac{\psi_i(t, \mathbf{r})\Psi_j(t, \mathbf{r})\Psi_i(t, \mathbf{r})}{|\mathbf{r} - \mathbf{r}'|}d\mathbf{r}' = E_j(t, \mathbf{r})\Psi_j(t, \mathbf{r}) \tag{23.3}$$

The Hartree–Fock nonlinear partial differential Equation (23.3) is an approximation because the multibody electron interactions should be considered in general. Thus, the explicit equation for the total energy must be used. This phenomenon can be integrated using the charge density function. Furthermore, analytically and computationally tractable concepts are sought because the Hartree–Fock equation is difficult to solve.

23.8   Density Functional Theory

Quantum mechanics and quantum modeling must be applied to understand and analyze nanodevices because usually they operate under the quantum effects. Computationally efficient and accurate procedures are critical to performing high-fidelity modeling of nanomachines. The complexities of the Schrödinger and Hartree equations are enormous even for very simple nanostructures. The difficulties associated with the solution of the Schrödinger equation drastically limit the applicability of the quantum mechanics. The properties, processes, phenomena, and effects in even the simplest nanostructures cannot be studied, examined, and comprehended. The problems can be solved applying the Hohenberg–Kohn density functional theory.[17]

The statistical consideration, proposed by Thomas and Fermi in 1927, gives the viable method to examine the distribution of electrons in atoms. The following assumptions were used: electrons are distributed uniformly, and there is an effective potential field that is determined by the nuclei charge and the distribution of electrons. Considering electrons distributed in a three-dimensional box, the energy analysis can be performed. Summing all energy levels, one finds the energy. Thus, the total kinetic energy and the electron charge density are related. The statistical consideration is used to approximate the distribution of electrons in an atom. The relation between the total kinetic energy of N electrons and the electron density is derived using the local density approximation concept.

The Thomas–Fermi kinetic energy functional is

$$\Gamma_F(\rho_e(\mathbf{r})) = 2.87 \int_R \rho_e^{5/3}(\mathbf{r})d\mathbf{r}$$

and the exchange energy is found to be

$$E_F(\rho_e(\mathbf{r})) = 0.739 \int_R \rho_e^{4/3}(\mathbf{r})d\mathbf{r}$$

Examining electrostatic electron–nucleus attraction and electron–electron repulsion, for homogeneous atomic systems, Thomas and Fermi derived the following energy functional:

$$E_F(\rho_e(\mathbf{r})) = 2.87 \int_R \rho_e^{5/3}(\mathbf{r})d\mathbf{r} - q\int_R \frac{\rho_e(\mathbf{r})}{r}d\mathbf{r} + \iint_{RR} \frac{1}{4\pi\varepsilon} \frac{\rho_e(\mathbf{r})\rho_e(\mathbf{r}')}{|\mathbf{r} - \mathbf{r}'|}$$

applying of the electron charge density $\rho_e(\mathbf{r})$.

Following this idea, instead of the many-electron wave functions, the charge density for N-electron systems can be used. Only the knowledge of the charge density is needed to perform analysis of molecular dynamics in nanostructures. The charge density is the function that describes the number of electrons per unit volume (function of three spatial variables x, y, and z in the Cartesian coordinate system).

The total energy of an N-electron system under the external field is defined in terms of the three-dimensional charge density $\rho(\mathbf{r})$. The complexity is significantly decreased because the problem of modeling of N-electron Z-nucleus systems becomes equivalent to the solution of the equation for one electron. The total energy is given as:

$$E(t, \rho(\mathbf{r})) = \underbrace{\Gamma_1(t, \rho(\mathbf{r})) + \Gamma_2(t, \rho(\mathbf{r}))}_{\text{kinetic energy}} + \underbrace{\int_R \frac{e\rho(\mathbf{r}')}{4\pi\varepsilon|\mathbf{r} - \mathbf{r}'|}d\mathbf{r}'}_{\text{potential energy}} \tag{23.4}$$

where $\Gamma_1(t, \rho(\mathbf{r}))$ and $\Gamma_2(t, \rho(\mathbf{r}))$ are the interacting (exchange) and noninteracting kinetic energies of a single electron in an N-electron Z-nucleus system, and

$$\Gamma_1(t, \rho(\mathbf{r})) = \int_R \gamma(t, \rho(\mathbf{r})) \rho(\mathbf{r}) d\mathbf{r}, \; \Gamma_2(t, \rho(\mathbf{r})) = -\frac{\hbar^2}{2m} \sum_{j=1}^{N} \int_R \psi_j^*(t, \mathbf{r}) \nabla_j^2 \psi_j(t, \mathbf{r})(d\mathbf{r}); \; \gamma(t, \rho(\mathbf{r}))$$

is the parameterization function.

The Kohn–Sham electronic orbitals are subject to the following orthogonal condition:

$$\int_R \psi_i^*(t, \mathbf{r}) \psi_j(t, \mathbf{r}) d\mathbf{r} = \delta_{ij}$$

The state of media depends largely on the balance between the kinetic energies of the particles and the interparticle energies of attraction.

The expression for the total potential energy is easily justified. The term

$$\int_R \frac{e\rho(\mathbf{r}')}{4\pi\varepsilon|\mathbf{r} - \mathbf{r}'|} d\mathbf{r}'$$

represents the Coulomb interaction in **R**, and the total potential energy is a function of the charge density $\rho(\mathbf{r})$.

The total kinetic energy (interactions of electrons and nuclei, and electrons) is integrated into the equation for the total energy. The total energy, as given by Equation (23.4), is stationary with respect to variations in the charge density. The charge density is found taking note of the Schrödinger equation. The first-order Fock–Dirac electron charge density matrix is

$$\rho_e(\mathbf{r}) = \sum_{j=1}^{N} \psi_j^*(t, \mathbf{r}) \psi_j(t, \mathbf{r}) \tag{23.5}$$

The three-dimensional electron charge density is a function in three variables (x, y, and z). Integrating the electron charge density $\rho_e(\mathbf{r})$, one obtains the total (net) electrons charge. Thus,

$$\int_R \rho_e(\mathbf{r}) d\mathbf{r} = Ne$$

Hence, $\rho_e(\mathbf{r})$ satisfies the following properties:

$$\rho_e(\mathbf{r}) > 0, \int_R \rho_e(\mathbf{r}) d\mathbf{r} = Ne, \int_R \left| \sqrt{\nabla \rho_e(\mathbf{r})} \right|^2 d\mathbf{r} < \infty, \int_R \nabla^2 \rho_e(\mathbf{r}) d\mathbf{r} = \infty$$

For the nuclei charge density, we have

$$\rho_n(\mathbf{r}) > 0 \; \text{and} \; \int_R \rho_n(\mathbf{r}) d\mathbf{r} = \sum_{k=1}^{Z} q_k$$

There are an infinite number of antisymmetric wave functions that give the same $\rho(\mathbf{r})$. The minimum-energy concept (energy-functional minimum principle) is applied to find $\rho(\mathbf{r})$. The total energy is a function of $\rho(\mathbf{r})$, and the ground state Ψ must minimize the expectation value $\langle E(\rho) \rangle$. The searching density functional $F(\rho)$, which searches all Ψ in the N-electron Hilbert space to find $\rho(\mathbf{r})$ and guarantee the minimum to the energy expectation value, is expressed as:

$$F(\rho) \leq \min_{\substack{\Psi \to \rho \\ \Psi \in H_\Psi}} \langle \Psi | E(\rho) | \Psi \rangle$$

where H_ψ is any subset of the N-electron Hilbert space.

Using the variational principle, we have

$$\frac{\Delta E(\rho)}{\Delta f(\rho)} = \int_R \frac{\Delta E(\rho)}{\Delta \rho(\mathbf{r'})} \frac{\Delta \rho(\mathbf{r'})}{\Delta f(\mathbf{r})} d\mathbf{r'} = 0 \,,$$

where $f(\rho)$ is the nonnegative function. Thus,

$$\left. \frac{\Delta E(\rho)}{\Delta f(\rho)} \right|_N = \text{constr}$$

The solution to the high-fidelity modeling problem is found using the charge density as given by Equation (23.5). The force and displacement must be found. Substituting the expression for the total kinetic and potential energies in Equation (23.4), where the charge density is given by Equation (23.5), the total energy $E(t, \rho(\mathbf{r}))$ results.

The external energy is supplied to control nano- and micromachines, and one has

$$E_\Sigma(t, \mathbf{r}) = E_{external}(t, \mathbf{r}) + E(t, \rho(\mathbf{r}))$$

Then, the force at position $\mathbf{r}_r$ is

$$\mathbf{F}_r(t, \mathbf{r}) = -\frac{dE_\Sigma(t, \mathbf{r})}{d\mathbf{r}_r} = -\frac{\partial E_\Sigma(t, \mathbf{r})}{\partial \mathbf{r}_r} - \sum_j \frac{\partial E(t, \mathbf{r})}{\partial \psi_j(t, \mathbf{r})} \frac{\partial \psi_j(t, \mathbf{r})}{\partial \mathbf{r}_r} - \sum_j \frac{\partial E(t, \mathbf{r})}{\partial \psi_j^*(t, \mathbf{r})} \frac{\partial \psi_j^*(t, \mathbf{r})}{\partial \mathbf{r}_r} \qquad (23.6)$$

Taking note of

$$\sum_j \frac{\partial E(t, \mathbf{r})}{\partial \psi_j(t, \mathbf{r})} \frac{\partial \psi_j(t, \mathbf{r})}{\partial \mathbf{r}_r} + \sum_j \frac{\partial E(t, \mathbf{r})}{\partial \psi_j^*(t, \mathbf{r})} \frac{\partial \psi_j^*(t, \mathbf{r})}{\partial \mathbf{r}_r} = 0$$

the expression for the force is found from Equation (23.6). In particular, one finds

$$\mathbf{F}_r(t, \mathbf{r}) = -\frac{\partial E_{external}(t, \mathbf{r})}{\partial \mathbf{r}_r} - \int_R \rho(t, \mathbf{r}) \frac{\partial [\Pi_r(t, \mathbf{r}) + \Gamma_r(t, \mathbf{r})]}{\partial \mathbf{r}_r} d\mathbf{r} - \int_R \frac{\partial E_\Sigma(t, \mathbf{r})}{\partial \rho(t, \mathbf{r})} \frac{\partial \rho(t, \mathbf{r})}{\partial \mathbf{r}_r} d\mathbf{r}$$

As the wave functions converge (the conditions of the Hellmann–Feynman theorem are satisfied), we have

$$\int_R \frac{\partial E(t, \mathbf{r})}{\partial \rho(t, \mathbf{r})} \frac{\partial \rho(t, \mathbf{r})}{\partial \mathbf{r}_r} d\mathbf{r} = 0$$

One can deduce the expression for the wave functions, find the charge density, calculate the forces, and study processes and phenomena in nanoscale. The displacement is found using the following equation of motion:

$$m\frac{d^2 \mathbf{r}}{dt^2} = \mathbf{F}_r(t, \mathbf{r})$$

or in the Cartesian coordinate system:

$$m\frac{d^2(\vec{x}, \vec{y}, \vec{z})}{dt^2} = \mathbf{F}_r(\vec{x}, \vec{y}, \vec{z})$$

23.9 Electromagnetics and Quantization

The mathematical models for energy conversion (energy storage, transport, and dissipation) and electromagnetic field (propagation, radiation, and other major characteristics) in micromachines (and many nanomachines) can be found using Maxwell's equations. The vectors of electric field intensity **E**, electric flux density **D**, magnetic field intensity **H**, and magnetic flux density **B** are used as the primary field quantities vectors (variables). It is well known that a set of four nonlinear partial differential Maxwell's equations is given in the time domain. The finite element analysis, which is based upon the steady-state analysis, cannot be viewed as a meaningful concept because it does not describe the important electromagnetic phenomena and effects. The time-independent and frequency-domain Maxwell's and Schrödinger equations also have serious limitations. Therefore, complete mathematical models must be developed without simplifications and assumptions to understand, analyze, and comprehend a wide spectrum of phenomena and effects.[1] This will guarantee high-fidelity modeling features.

Maxwell's equations can be quantized. This concept provides a meaningful means for interpreting, understanding, predicting, and analyzing complex time-dependent behavior at nanoscale without facing difficulties associated with the application of the quantum electrodynamics, electromagnetics, and mechanics.

Let us start with the electromagnetic fundamentals. The Lorentz force on the charge q moving at the velocity v is

$$\mathbf{F}(t, \mathbf{r}) = q\mathbf{E}(t, \mathbf{r}) + q\mathbf{v} \times \mathbf{B}(t, \mathbf{r}).$$

Using the electromagnetic potential A we have the expression for the force as:

$$\mathbf{F} = q\left(-\frac{\partial \mathbf{A}}{\partial \mathbf{t}} - \nabla V + \mathbf{v} \times (\nabla \times \mathbf{A})\right), \mathbf{v} \times (\nabla \times \mathbf{A}) = \nabla(\mathbf{v} \cdot \mathbf{A}) - \frac{d\mathbf{A}}{dt} + \frac{\partial \mathbf{A}}{\partial t}$$

where $V(t, \mathbf{r})$ is the scalar electrostatic potential function (potential difference).

Four Maxwell's equations in the time domain are

$$\nabla \times \mathbf{E}(t, \mathbf{r}) = -\mu \frac{\partial \mathbf{H}(t, \mathbf{r})}{\partial t} - \mu \frac{\partial \mathbf{M}(t, \mathbf{r})}{\partial t}$$

$$\nabla \times \mathbf{H}(t, \mathbf{r}) = \mathbf{J}(t, \mathbf{r}) + \varepsilon \frac{\partial \mathbf{E}(t, \mathbf{r})}{\partial t} + \frac{\partial \mathbf{P}(t, \mathbf{r})}{\partial t}$$

$$\nabla \cdot \mathbf{E}(t, \mathbf{r}) = \frac{\rho_v(t, \mathbf{r})}{\varepsilon} - \frac{\nabla \mathbf{P}(t, \mathbf{r})}{\varepsilon}$$

$$\nabla \cdot \mathbf{H}(t, \mathbf{r}) = 0$$

where J is the current density, and using the conductivity σ, we have $J = \sigma E$; ε is the permittivity; μ is the permeability; ρ_v is the volume charge density.

Using the electric **P** and magnetic **M** polarizations (dipole moment per unit volume) of the medium, one obtains two constitutive equations as:

$$\mathbf{D}(t, \mathbf{r}) = \varepsilon\mathbf{E}(t, \mathbf{r}) + \mathbf{P}(t, \mathbf{r}), \mathbf{B}(t, \mathbf{r}) = \mu\mathbf{H}(t, \mathbf{r}) + \mu\mathbf{M}(t, \mathbf{r})$$

The electromagnetic waves transfer the electromagnetic power. We have:

$$\int_v \nabla \cdot (\mathbf{E} \times \mathbf{H}) dv = \oint_s (\mathbf{E} \times \mathbf{H}) \cdot d\mathbf{s} =$$

total power flowing into volume bounded by s

$$-\int_v \frac{\partial}{\partial t}\left(\frac{1}{2}\varepsilon\mathbf{E} \cdot \mathbf{E} + \frac{1}{2}\mu\mathbf{H} \cdot \mathbf{H}\right) dv - \int_v \mathbf{E} \cdot \mathbf{J} dv - \int_v \mathbf{E} \cdot \frac{\partial \mathbf{P}}{\partial t} dv - \int_v \mu\mathbf{H} \cdot \frac{\partial \mathbf{M}}{\partial t} dv.$$

rate of change of the electromagnetic stored energy in electromagnetic fields   power expended by the field on moving charges   power expended by the field on electric dipoles

The pointing vector $\mathbf{E} \times \mathbf{H}$, which is a power density vector, represents the power flows per unit area. Furthermore, the electromagnetic momentum is found as

$$\mathbf{M} = \frac{1}{c^2} \int_v \mathbf{E} \times \mathbf{H} dv.$$

The electromagnetic field can be examined using the magnetic vector potential $\mathbf{A}$, and:

$$\mathbf{B}(t, \mathbf{r}) = \nabla \times \mathbf{A}(t, \mathbf{r}), \; \mathbf{E}(t, \mathbf{r}) = -\frac{\partial \mathbf{A}(t, \mathbf{r})}{\partial t} - \nabla V(t, \mathbf{r}).$$

Making use of the *Coulomb gauge* equation $\nabla \cdot \mathbf{A}(t, \mathbf{r}) = 0$, for the electromagnetic field in free space (A is determined by the transverse current density), we have

$$\nabla^2 \mathbf{A}(t, \mathbf{r}) - \frac{1}{c^2} \frac{\partial^2 \mathbf{A}(t, \mathbf{r})}{\partial t^2} = 0$$

where c is the speed of light, $c = (1/\sqrt{\mu_0 \varepsilon_0})$, $c = 3 \times 10^8$ m/sec.

The solution of this partial differential equation is

$$\mathbf{A}(t, \mathbf{r}) = \frac{1}{2\sqrt{\varepsilon}} \sum_s a_s(t) \mathbf{A}_s(\mathbf{r})$$

and using the separation of variables technique we have

$$\nabla^2 \mathbf{A}_s(\mathbf{r}) + \frac{\omega_s}{c^2} \mathbf{A}_s(\mathbf{r}) = 0, \; \frac{d^2 a_s(t)}{dt^2} + \omega_s a_s(t) = 0$$

where ω_s is the separation constant which determines the eigenfunctions.

The stored electromagnetic energy

$$\langle W(t) \rangle = -\frac{1}{2v} \int_v (\varepsilon \mathbf{E} \cdot \mathbf{E} + \mu \mathbf{H} \cdot \mathbf{H}) dv$$

is given by

$$\langle W(t) \rangle = \frac{1}{4v} \int_v (\omega_s \omega_s \cdot \mathbf{A}_s \cdot \mathbf{A}_s^* + c^2 \nabla \times \mathbf{A}_s \cdot \nabla \times \mathbf{A}_s^*) a_s(t) a_s^*(t) dv$$

$$= \frac{1}{4v} \sum_{s, s'} (\omega_s \omega_{s'} + \omega_{s'}^2) a_s(t) a_{s'}^*(t) \int \mathbf{A}_s \cdot \mathbf{A}_{s'}^* dv = \frac{1}{2} \sum_s \omega_s^2 a_s(t) a_s^*(t)$$

The Hamiltonian function is

$$H = \frac{1}{2v} \int_v (\varepsilon \mathbf{E} \cdot \mathbf{E} + \mu \mathbf{H} \cdot \mathbf{H}) dv$$

Let us apply the quantum mechanics to examine very important features. The Hamiltonian function is found using the kinetic and potential energies Γ and Π. For a particle of mass m with energy E moving in the Cartesian coordinate system, one has

$$E(x, y, z, t) = \underset{\text{total energy}}{\Gamma(x, y, z, t)} + \underset{\text{kinetic energy}}{\Pi(x, y, z, t)} = \underset{\text{potential energy}}{\frac{p^2(x, y, z, t)}{2m}} + \Pi(x, y, z, t) = \underset{\text{Hamiltonian}}{H(x, y, z, t)}$$

Thus, $p^2(x, y, z, t) = 2m[E(x, y, z, t) - \Pi(x, y, z, t)]$.

Using the formula for the wavelength (Broglie's equation) $\lambda = (h/p) = (h/mv)$, one finds

$$\frac{1}{\lambda^2} = \left(\frac{p}{h}\right)^2 = \frac{2m}{h^2}[E(x, y, z, t) - \Pi(x, y, z, t)].$$

This expression is substituted in the *Helmholtz* equation $\nabla^2\Psi + (4\pi^2/\lambda^2)\Psi = 0$, which gives the evolution of the wave function. We obtain the Schrödinger equation as:

$$E(x, y, z, t)\Psi(x, y, z, t) = -\frac{\hbar^2}{2m}\nabla^2\Psi(x, y, z, t) + \Pi(x, y, z, t)\Psi(x, y, z, t)$$

or

$$E(x, y, z, t)\Psi(x, y, z, t) = \frac{\hbar^2}{2m}\left(\frac{\partial^2\Psi(x, y, z, t)}{\partial x^2} + \frac{\partial^2\Psi(x, y, z, t)}{\partial y^2} + \frac{\partial^2\Psi(x, y, z, t)}{\partial z^2}\right)$$
$$+ \Pi(x, y, z, t)\Psi(x, y, z, t)$$

Here, the modified Plank constant is $\hbar = (h/2\pi) = 1.055 \times 10^{-34}$ J-sec.

The Schrödinger equation:

$$-\frac{\hbar^2}{2m}\nabla^2\Psi + \Pi\Psi = E\Psi$$

is related to the Hamiltonian $H = -(\hbar^2/2m)\nabla + \Pi$, and one has

$$H\Psi = E\Psi$$

The Schrödinger partial differential equation must be solved, and the wave function is normalized using the probability density $\int|\Psi|^2 d\zeta = 1$.

The variables q, p, a, and a^+ are used.[18] In particular, we apply the Hermitian operators q and p, which satisfy the commutative relations $[\mathbf{q}, \mathbf{q}] = 0$, $[\mathbf{p}, \mathbf{p}] = 0$ and $[\mathbf{q}, \mathbf{p}] = i\hbar\delta$.

The Schrödinger representation of the energy eigenvector $\Psi_n(q) = \langle q|E_n\rangle$ satisfies the following equations:

$$\langle q|H|E_n\rangle = E_n\langle q|E_n\rangle, \; \langle q|(1/2m)(\mathbf{p}^2 + m^2\omega^2\mathbf{q}^2)|E_n\rangle = E_n\langle q|E_n\rangle$$

and

$$\left(-\frac{\hbar^2}{2m}\frac{d^2}{dq^2} + \frac{m\omega^2 q^2}{2}\right)\Psi_n(q) = E_n\Psi_n(q)$$

The solution is

$$\Psi_n(q) = \sqrt{\frac{1}{2^n n!}}\sqrt{\frac{\omega}{\pi\hbar}}H_n\left(\sqrt{\frac{m\omega}{\hbar}}q\right)e^{-\frac{m\omega}{2\hbar}q^2}$$

where

$$H_n\left(\sqrt{\frac{m\omega}{\hbar}}q\right)$$

is the Hermite polynomial, and the energy eigenvalues which correspond to the given eigenstates are $E_n = \hbar\omega_n = \hbar\omega(n + 1/2)$.

The eigenfunctions can be generated using the following procedure. Using non-Hermitian operators,

$$\mathbf{a} = \sqrt{\frac{m}{2\hbar\omega}}\left(\omega\mathbf{q} + i\frac{\mathbf{p}}{m}\right) \text{ and } \mathbf{a}^+ = \sqrt{\frac{m}{2\hbar\omega}}\left(\omega\mathbf{q} - i\frac{\mathbf{p}}{m}\right)$$

we have

$$\mathbf{q} = \sqrt{\frac{\hbar}{2m\omega}}(\mathbf{a} + \mathbf{a}^+) \text{ and } \mathbf{p} = -i\sqrt{\frac{m\hbar\omega}{2}}(\mathbf{a} - \mathbf{a}^+)$$

The commutation equation is $[\mathbf{a}, \mathbf{a}^+] = 1$. Hence, one obtains

$$H = \frac{\hbar\omega}{2}(\mathbf{a}\mathbf{a}^+ + \mathbf{a}^+\mathbf{a}) = \hbar\omega\left(\mathbf{a}\mathbf{a}^+ + \frac{1}{2}\right)$$

The Heisenberg equations of motion are

$$\frac{d\mathbf{a}}{dt} = \frac{1}{i\hbar}[\mathbf{a}, H] = -i\omega\mathbf{a} \text{ and } \frac{d\mathbf{a}^+}{dt} = \frac{1}{i\hbar}[\mathbf{a}^+, H] = i\omega\mathbf{a}^+$$

with solutions:

$$\mathbf{a} = \mathbf{a}_s e^{-i\omega t} \text{ and } \mathbf{a}^+ = \mathbf{a}_s^+ e^{i\omega t}$$

and

$$\mathbf{a}|0\rangle = 0, \mathbf{a}|n\rangle = \sqrt{n}|n-1\rangle \text{ and } \mathbf{a}^+|n\rangle = \sqrt{n+1}|n+1\rangle$$

Using the *state vector generating rule*

$$|n\rangle = \frac{\mathbf{a}^{+^n}}{\sqrt{n!}}|0\rangle$$

one has the following eigenfunction generator equation:

$$\langle q|n\rangle = \frac{1}{\sqrt{n!}}\left(\sqrt{\frac{m}{2\hbar\omega}}\right)^n \left(\omega q - \frac{\hbar}{m}\frac{d}{dq}\right)^n \langle q|0\rangle$$

for the equation

$$\Psi_n(q) = \sqrt{\frac{1}{2^n n!}\sqrt{\frac{\omega}{\pi\hbar}}}H_n\left(\sqrt{\frac{m\omega}{\hbar}}q\right)e^{-\frac{m\omega}{2\hbar}q^2}$$

Comparing the equation for the stored electromagnetic energy

$$\langle W(t) \rangle = \frac{1}{2} \sum_s \omega_s^2 a_s(t) a_s^*(t)$$

and the Hamiltonian $H = \omega a^+ a$, we have

$$\sqrt{\frac{\omega_s}{2}} a_s \Rightarrow a = \sqrt{\frac{m}{2\omega}} \left(\omega q + i\frac{p}{m} \right) \text{ and } \sqrt{\frac{\omega_s}{2}} a_s^* \Rightarrow a^+ = \sqrt{\frac{m}{2\omega}} \left(\omega q + i\frac{p}{m} \right)$$

Therefore, to perform the canonical quantization, the electromagnetic field variables are expressed as the field operators using:

$$\sqrt{\frac{\omega_s}{2}} a_s \Rightarrow \sqrt{\frac{m}{2\omega}} \left(\omega \mathbf{q} + i\frac{\mathbf{p}}{m} \right) = \sqrt{\hbar} \mathbf{a}$$

and

$$\sqrt{\frac{\omega_s}{2}} a_s^* \Rightarrow \sqrt{\frac{m}{2\omega}} \left(\omega \mathbf{q} - i\frac{\mathbf{p}}{m} \right) = \sqrt{\hbar} \mathbf{a}^+$$

The following equations finally result:

$$H = \sum_s \frac{\hbar \omega_s}{2} [\mathbf{a}_s(t) \mathbf{a}_s^+(t) + \mathbf{a}_s^+(t) \mathbf{a}_s(t)]$$

$$\mathbf{E}(t, \mathbf{r}) = \sum_s \sqrt{\frac{\hbar \omega_s}{2\varepsilon}} [\mathbf{a}_s(t) \mathbf{A}_s(\mathbf{r}) - \mathbf{a}_s^+(t) \mathbf{A}_{s'}^*(\mathbf{r})]$$

$$\mathbf{H}(t, \mathbf{r}) = c \sum_s \sqrt{\frac{\hbar}{2\mu\omega_s}} [\mathbf{a}_s(t) \nabla \times \mathbf{A}_s(\mathbf{r}) + \mathbf{a}_s^+(t) \nabla \times \mathbf{A}_{s'}^*(\mathbf{r})]$$

$$\mathbf{A}(t, \mathbf{r}) = \sum_s \sqrt{\frac{\hbar}{2\varepsilon\omega_s}} [\mathbf{a}_s(t) \acute{\mathbf{A}}_s(\mathbf{r}) + \mathbf{a}_s^+(t) \acute{\mathbf{A}}_{s'}^*(\mathbf{r})]$$

The derived expressions can be straightforwardly applied. For example, for a single mode field we have

$$\mathbf{E}(t, \mathbf{r}) = i\mathbf{e} \sqrt{\frac{\hbar\omega}{2\varepsilon v}} (\mathbf{a}(t) e^{i\mathbf{k} \cdot \mathbf{r} - i\omega t} - \mathbf{a}^+(t) e^{-i\mathbf{k} \cdot \mathbf{r} - i\omega t})$$

$$\mathbf{H}(t, \mathbf{r}) = i \sqrt{\frac{\hbar\omega}{2\mu v}} \mathbf{k} \times \mathbf{e}(\mathbf{a}(t) e^{i\mathbf{k} \cdot \mathbf{r} - i\omega t} - \mathbf{a}^+(t) e^{-i\mathbf{k} \cdot \mathbf{r} - i\omega t})$$

$$\langle n|\mathbf{E}|n \rangle = 0, \langle n|\mathbf{H}|n \rangle = 0$$

$$\Delta \mathbf{E} = \sqrt{\frac{\hbar\omega}{\varepsilon v} \left(n + \frac{1}{2} \right)}, \Delta \mathbf{H} \sqrt{\frac{\hbar\omega}{\mu v} \left(n + \frac{1}{2} \right)}$$

$$\Delta \mathbf{E} \Delta \mathbf{H} = \sqrt{\frac{1}{\varepsilon\mu} \frac{\hbar\omega}{v} \left(n + \frac{1}{2} \right)}$$

where

$$E_p = \sqrt{\frac{\hbar\omega}{2\varepsilon v}}$$

is the electric field per photon.

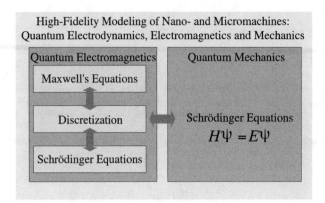

FIGURE 23.13 Quantum electrodynamics, electromagnetics, mechanics, and classical electromagnetics/mechanics for high-fidelity modeling of nano- and micromachines.

The complete Hamiltonian of a coupled system is $H_\Sigma = H + H_{ex}$, where H_{ex} is the interaction Hamiltonian. For example, the photoelectric interaction Hamiltonian is found as:

$$H_{exp} = -e \sum_n \mathbf{r}_n \cdot \mathbf{E}(t, \mathbf{R})$$

where r_n are the relative spatial coordinates of the electrons bound to a nucleus located at R.

The documented paradigm in modeling of nano- and micromachines is illustrated is Figure 23.13.

23.10 Conclusions

In many applications (from medicine and biotechnology to aerospace and power), the use of nano- and micromachines is very important. This chapter focuses on the synthesis, modeling, analysis, design, and fabrication of electromagnetic-based machines (motion nano- and microscale devices). To attain our objectives and goals, the synergy of multidisciplinary engineering, science, and technology must be utilized. In particular, electromagnetic theory and mechanics comprise the fundamentals for analysis, modeling, simulation, design, and optimization, while fabrication is based on the CMOS, micromachining, and high-aspect-ratio technologies. Nano- and microsystems are the important parts of modern confluent engineering, and this chapter provides the means to understand, master, and assess the current trends including the *nanoarchitectronic* paradigm.

References

1. S.E. Lyshevski, *MEMS and NEMS: Systems, Devices, and Structures*, CRC Press, Boca Raton, FL, 2001.
2. C.H. Ahn, Y.J. Kim, and M.G. Allen, A planar variable reluctance magnetic micromotor with fully integrated stator and wrapped coil, *Proc. IEEE Micro Electro Mech. Syst. Worksh.*, Fort Lauderdale, FL, 1993, pp. 1–6.
3. S.F. Bart, M. Mehregany, L.S. Tavrow, J.H. Lang, and S.D. Senturia, Electric micromotor dynamics, *Trans. Electron Devices*, 39, 566–575, 1992.
4. G.T.A. Kovacs, *Micromachined Transducers Sourcebook*, WCB McGraw-Hill, Boston, MA, 1998.
5. S.E. Lyshevski, *Nano- and Micro-Electromechanical Systems: Fundamentals of Micro- and Nano-Engineering*, CRC Press, Boca Raton, FL, 2000.
6. M. Madou, *Fundamentals of Microfabrication*, CRC Press, Boca Raton, FL, 1997.

7. H. Guckel, T.R. Christenson, K.J. Skrobis, J. Klein, and M. Karnowsky, Design and testing of planar magnetic micromotors fabricated by deep x-ray lithography and electroplating, *Tech. Dig. Intl. Conf. Solid-State Sensors Actuators, Transducers 93*, Yokohama, Japan, 1993, pp. 60–64.

8. H. Guckel, K.J. Skrobis, T.R. Christenson, J. Klein, S. Han, B. Choi, E.G. Lovell, and T.W. Chapman, Fabrication and testing of the planar magnetic micromotor, *J. Micromech. Microeng.*, 1, 135–138, 1991.

9. J.W. Judy, R.S. Muller, and H.H. Zappe, Magnetic microactuation of polysilicon flexible structure, *J. Microelectromech. Syst.*, 4(4), 162–169, 1995.

10. M. Mehregany and Y.C. Tai, Surface micromachined mechanisms and micro-motors, *J. Micromech. Microeng.*, 1, 73–85, 1992.

11. M.P. Omar, M. Mehregany, and R.L. Mullen, Modeling of electric and fluid fields in silicon microactuators, *Intl. J. Appl. Electromagn. Mater.*, 3, 249–252, 1993.

12. S.A. Campbell, *The Science and Engineering of Microelectronic Fabrication*, Oxford University Press, New York, 2001.

13. E.W. Becker, W. Ehrfeld, P. Hagmann, A. Maner, and D. Mynchmeyer, Fabrication of microstructures with high aspect ratios and great structural heights by synchrotron radiation lithography, galvanoforming, and plastic moulding (LIGA process), *Microelectron. Eng.*, 4, 35–56, 1986.

14. H. Guckel, Surface micromachined physical sensors, *Sensors Mater.*, 4(5), 251–264, 1993.

15. H. Guckel, K.J. Skrobis, T.R. Christenson, and J. Klein, Micromechanics for actuators via deep x-ray lithography, *Proc. SPIE Symp. Microlithogr.*, San Jose, CA, 1994, pp. 39–47.

16. J.C. Ellenbogen and J.C. Love, *Architectures for Molecular Electronic Computers*, MP 98W0000183, MITRE Corporation, 1999.

17. W. Kohn and R.M. Driezler, Time-dependent density-functional theory: conceptual and practical aspects, *Phys. Rev. Lett.*, 56, 1993 –1995, 1986.

18. A. Yariv, *Quantum Electronics*, John Wiley and Sons, New York, 1989.

24

Contributions of Molecular Modeling to Nanometer-Scale Science and Technology

Donald W. Brenner
North Carolina State University

O.A. Shenderova
North Carolina State University

J.D. Schall
North Carolina State University

D.A. Areshkin
North Carolina State University

S. Adiga
North Carolina State University

J.A. Harrison
United States Naval Academy

S.J. Stuart
Clemson University

CONTENTS

Opening Remarks

Molecular modeling has played a key role in the development of our present understanding of the details of many chemical processes and molecular structures. From atomic-scale modeling, for example, much of the details of drug interactions, energy transfer in chemical dynamics, frictional forces, and crack propagation dynamics are now known, to name just a few examples. Similarly, molecular modeling has played a central role in developing and evaluating new concepts related to nanometer-scale science and technology. Indeed, molecular modeling has a long history in nanotechnology of both leading the way in terms of what in principle is achievable in nanometer-scale devices and nanostructured materials, and in explaining experimental data in terms of fundamental processes and structures.

    There are two goals for this chapter. The first is to educate scientists and engineers who are considering using molecular modeling in their research, especially as applied to nanometer-scale science and technology, beyond the black box approach sometimes facilitated by commercial modeling codes. Particular emphasis is placed on the physical basis of some of the more widely used analytic potential energy expressions. The choice of bonding expression is often the first choice to be made in instituting a molecular model, and a proper choice of expression can be crucial to obtaining meaningful results. The second goal of this chapter is to discuss some examples of systems that have been modeled and the type of information that has been obtained from these simulations. Because of the enormous breadth of molecular modeling studies related to nanometer-scale science and technology, we have not attempted a comprehensive literature survey. Instead, our intent is to stir the imagination of researchers by presenting a variety of examples that illustrate what molecular modeling has to offer.

24.1 Molecular Simulations

The term *molecular modeling* has several meanings, depending on the perspective of the user. In the chemistry community, this term often implies molecular statics or dynamics calculations that use as interatomic forces a valence-force-field expression plus nonbonded interactions (see below). An example of a bonding expression of this type is Allinger's molecular mechanics (Burkert, 1982; Bowen, 1991). The term molecular modeling is also sometimes used as a generic term for a wider array of atomic-level simulation methods (e.g., Monte Carlo modeling in addition to molecular statics/dynamics) and interatomic force expressions. In this chapter, the more generic definition is assumed, although the examples discussed below emphasize molecular dynamics simulations.

In molecular dynamics calculations, atoms are treated as discrete particles whose trajectories are followed by numerically integrating classical equations of motion subject to some given interatomic forces. The numerical integration is carried out in a stepwise fashion, with typical timesteps ranging from 0.1 to about 15 femtoseconds depending on the highest vibrational frequency of the system modeled. Molecular statics calculations are carried out in a similar fashion, except that minimum energy structures are determined using either integration of classical equations of motion with kinetic energies that are damped, by steepest descent methods, or some other equivalent numerical method.

In Monte Carlo modeling, *snapshots* of a molecular system are generated, and these configurations are used to determine the system's properties. Time-independent properties for equilibrium systems are usually generated either by weighting the contribution of a snapshot to a thermodynamic average by an appropriate Boltzmann factor or, more efficiently, by generating molecular configurations with a probability that is proportional to their Boltzmann factor. The Metropolis algorithm, which relies on Markov chain theory, is typically used for the latter. In kinetic Monte Carlo modeling, a list of possible dynamic events and the relative rate for each event is typically generated given a molecular structure. An event is then chosen from the list with a probability that is proportional to the inverse of the rate (i.e., faster rates have higher probabilities). The atomic positions are then appropriately updated according to the chosen event, the possible events and rates are updated, and the process is repeated to generate a time-dependent trajectory.

The interatomic interactions used in molecular modeling studies are calculated either from (1) a sum of nuclear repulsions combined with electronic interactions determined from some first principles or semi-empirical electronic structure technique, or (2) an expression that replaces the quantum mechanical electrons with an energy and interatomic forces that depend only on atomic positions. The approach for calculating atomic interactions in a simulation typically depends on the system size (i.e., the number of electrons and nuclei), available computing resources, the accuracy with which the forces need to be known to obtain useful information, and the availability of an appropriate potential energy function. While the assumption of classical dynamics in molecular dynamics simulations can be severe, especially for systems involving light atoms and other situations in which quantum effects (like tunneling) are important, molecular modeling has proven to be an extremely powerful and versatile computational technique.

For convenience, the development and applications of molecular dynamics simulations can be divided into four branches of science: chemistry, statistical mechanics, materials science, and molecular biology. Illustrated in Figure 24.1 are some of the highlights of molecular dynamics simulation as applied to problems in each of these fields. The first study using classical mechanics to model a chemical reaction was published by Eyring and co-workers in 1936, who used a classical trajectory (calculated by hand) to model the chemical reaction $H+H_2->H_2+H$ (Hirschfelder, 1936). Although the potential energy function was crude by current standards (it produced a stable H_3 molecule), the calculations

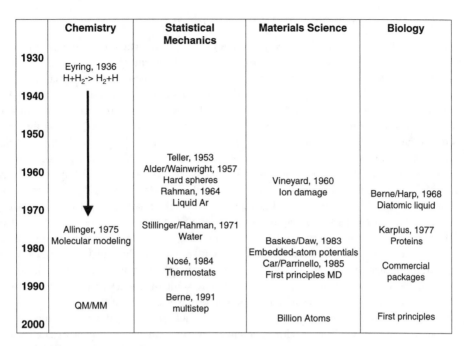

	Chemistry	Statistical Mechanics	Materials Science	Biology
1930				
	Eyring, 1936 $H+H_2 \rightarrow H_2+H$			
1940				
1950				
1960		Teller, 1953 Alder/Wainwright, 1957 Hard spheres Rahman, 1964 Liquid Ar	Vineyard, 1960 Ion damage	Berne/Harp, 1968 Diatomic liquid
1970				
1980	Allinger, 1975 Molecular modeling	Stillinger/Rahman, 1971 Water Nosé, 1984 Thermostats	Baskes/Daw, 1983 Embedded-atom potentials Car/Parrinello, 1985 First principles MD	Karplus, 1977 Proteins Commercial packages
1990				
2000	QM/MM	Berne, 1991 multistep	Billion Atoms	First principles

FIGURE 24.1   Some highlights in the field of molecular dynamics simulations.

themselves set the standard for numerous other applications of classical trajectories to understanding the dynamics of chemical reactions. In the 1950s and early 1960s, the first dynamic simulations of condensed phase dynamics were carried out. These early calculations, which were performed primarily at National Laboratories due to the computational resources available at these facilities, were used mainly to test and develop statistical mechanical descriptions of correlated many-particle motion (Rahman, 1964). A major breakthrough in the application of molecular modeling to statistical mechanics occurred in the 1980s with the derivation of thermostats that not only maintain an average kinetic energy corresponding to a desired temperature but, more importantly, produce kinetic energy *fluctuations* that correctly reproduce those of a desired statistical mechanical ensemble (Nosé, 1984).

The first reported application of molecular modeling to materials science was by Vineyard and co-workers, who in 1960 reported simulations of ions impinging on a solid (Gibson, 1960). The development since then of many-body potential energy functions that capture many of the details of bonding in metals and covalent materials, together with the ability to describe forces using first-principles methods, has allowed molecular modeling to make seminal contributions to our understanding of the mechanical properties of materials.

The application of molecular modeling techniques to biological systems did not begin in earnest until the 1970s, when computers began to get sufficiently powerful to allow simulations of complex hetero-molecules. This research was facilitated by the availability of commercial modeling packages a decade later, which allowed research in areas such as drug design to move from the academic research laboratory to the drug companies. The search for new drugs with specific biological activities continues to be an extremely active area of application for molecular simulation (Balbes, 1994).

24.1.1 Interatomic Potential Energy Functions and Forces

Much of the success of an atomic simulation relies on the use of an appropriate model for the interatomic forces. For many cases relatively simple force expressions are adequate, while for simulations from which quantitative results are desired, very accurate forces may be needed. Because of the importance of the interatomic force model, several of the more widely used potential energy expressions and the derivation of some of these models from quantum mechanics are discussed in detail in the following two subsections.

24.1.2   The Workhorse Potential Energy Functions

There are available in the literature hundreds, if not thousands, of analytic interatomic potential energy functions. Out of this plethora of potential energy functions have emerged about a dozen *workhorse* potentials that are widely used by the modeling community. Most of these potential functions satisfy the so-called *Tersoff test*, which has been quoted by Garrison and Srivastava as being important to identifying particularly effective potential functions (Garrison, 1995). This test is (a) has the person who constructed the potential subsequently refined the potential based on initial simulations? and (b) has the potential been used by other researchers for simulations of phenomena for which the potential was not designed? A corollary to these tests is that either the functional form of the potential should be straightforward to code, or that an implementation of the potential be widely available to the modeling community. An example, which is discussed in more detail below, is the Embedded-Atom Method (EAM) potentials for metals that were originally developed by Baskes, Daw, and Foiles in the early 1980s (Daw, 1983). Although there were closely related potential energy functions developed at about the same time, most of which are still used by modelers, the EAM potentials remain the most widely used functions for metals, due at least in part to the developers' making their computer source code readily available to researchers.

The quantum mechanical basis of most of the potential energy functions discussed in this section is presented in more detail in the next section, and therefore some of the details of the various analytic forms are given in that section. Rather than providing a formal discussion of these potentials, the intent of this section is to introduce these expressions as practical solutions to describing interatomic forces for large-scale atomic simulations. The workhorse potential functions can be conveniently classified according to the types of bonding that they most effectively model, and therefore this section is organized according to bonding type.

24.1.2.1   Metals

A number of closely related potential energy expressions are widely used to model bonding in metals. These expressions are the EAM (Daw, 1983), effective medium theory (Jacobsen, 1987; Jacobsen, 1988), the glue model (Ercolessi 1988), the Finnis–Sinclair (Finnis, 1984), and Sutton–Chen potentials (Sutton, 1990). Although they all have very similar forms, the motivation for each is not the same. The first three models listed are based on the concept of *embedding* atoms within an electron gas. The central idea is that the energy of an atom depends on the density of the electron gas near the embedded atom and the mutually repulsive pairwise interaction between the atomic cores of the embedded atom and the other atoms in the system. In the EAM and glue models, the electron density is taken as a pairwise sum of contributions from surrounding atoms at the site of the atom whose energy is being calculated. There are thus three important components of these potentials: the contribution of electron density from a neighboring atom as a function of distance, the pairwise-additive interatomic repulsive forces, and the embedding function relating the electron density to the energy. Each of these can be considered adjustable functions that can be fit to various properties such as the lattice constant and crystal cohesive energy. In contrast, effective medium theory attempts to use the average of the electron density in the vicinity of the atoms whose energy is being calculated, and a less empirical relationship between this electron density and the energy (Jacobsen, 1988).

Although the functional forms for the Finnis–Sinclair and Sutton–Chen potentials are similar to other metal potentials, the physical motivation is different. In these cases, the functional form is based on the so-called *second moment approximation* that relates the binding energy of an atom to its local coordination through the spread in energies of the local density of electronic states due to chemical binding (see the next section for more details). Ultimately, though, the functional form is very similar to the electron-density-based potentials. The main difference is that the EAM and related potentials were originally developed for face-centered cubic metals, while the Finnis–Sinclair potentials originally focused on body-centered cubic metals.

24.1.2.2   Covalent Bonding

Molecular structure calculations typically used by chemists have relied heavily on expressions that describe intramolecular bonding as an expansion in bond lengths and angles (typically called valence-force fields) in combination with nonbonded interactions. The nonbonded interactions typically use pair-additive functions that mimic van der Waals forces (e.g., Lennard–Jones potentials) and Coulomb forces due to partial charges. Allinger and co-workers developed the most widely recognized version of this type of expression (Burkert, 1982; Bowen, 1991). There are now numerous variations on this approach, most of which are tuned to some particular application such as liquid structure or protein dynamics. While very accurate energies and structures can be obtained with these potentials, a drawback to this approach is that the valence-force expressions typically use a harmonic expansion about the minimum energy configuration that does not go to the noninteraction limit as bond distances get large.

A major advance in modeling interatomic interactions for covalent materials was made with the introduction of a potential energy expression for silicon by Stillinger and Weber (Stillinger, 1985). The basic form is similar to the valence-force expressions in that the potential energy is given as a sum of two-body bond-stretching and three-body angle-bending terms. The Stillinger–Weber potential, however, produces the correct dissociation limits in both the bend and stretch terms and reproduces a wide range of properties of both solid and molten silicon. It opened a wide range of silicon and liquid phenomena to molecular simulation and also demonstrated that a well-parameterized analytic potential can be useful for describing bond-breaking and bond-forming chemistry in the condensed phase. The Stillinger–Weber potential is one of the few breakthrough potentials that does not satisfy the first of the Tersoff rules mentioned above in that a better parameterized form was not introduced by the original developers for silicon. This is a testament to the careful testing of the potential function as it was being developed. It is also one of the few truly successful potentials whose form is not directly derivable from quantum mechanics as described in the next section.

Tersoff introduced another widely used potential for covalent materials (Tersoff, 1986; Tersoff, 1989). Initially introduced for silicon, and subsequently for carbon, germanium, and their alloys, the Tersoff potentials are based on a quantum mechanical analysis of bonding. Two key features of the Tersoff potential function are that the same form is used for structures with high and low atomic coordination numbers, and that the bond angle term comes from a fit to these structures and does not assume a particular orbital hybridization. This is significantly different from both the valence-force type expressions, for which an atomic hybridization must be assumed to define the potential parameters, and the Stillinger–Weber potential, which uses an expansion in the angle bend terms around the tetrahedral angle. Building on the success of the Tersoff form, Brenner introduced a similar expression for hydrogen and carbon that uses a single potential form to model both solid-state and molecular structures (Brenner, 1990). Although not as accurate in its predictions of atomic structures and energies as expressions such as Allinger's molecular mechanics, the ability of this potential form to model bond breaking and forming with appropriate changes in atomic hybridization in response to the local environment has led to a wide range of applications. Addition of nonbonded terms into the potential has widened the applicability of this potential form to situations for which intermolecular interactions are important, such as in simulations of molecular solids and liquids (Stuart, 2000).

To model covalent bonding, Baskes and co-workers extended the EAM formalism to include bond angle terms (Baskes, 1989). While the potential form is compelling, these modified EAM forms have not been as widely used as the other potential forms for covalent bonding described above.

Increases in computing power and the development of increasingly clever algorithms over the last decade have made simulations with forces taken from electronic structure calculations routine. Forces from semi-empirical tight-bonding models, for example, are possible with relatively modest computing resources and therefore are now widely used. Parameterizations by Sutton and co-workers, as well as by Wang, Ho, and their co-workers, have become de facto standards for simulations of covalent materials (Xu, 1992; Kwon, 1994; Sutton, 1996).

24.1.2.3 Ionic Bonding

Materials in which bonding is chiefly ionic have not to date played as large a role in nanometer-scale science and technology as have materials with largely covalent or metallic bonding. Therefore, fewer potentials for ionic materials have been developed and applied to model nanoscale systems. The standard for bulk ionic materials is the shell model, a functional form that includes polarization by allowing ionic charges of fixed magnitude to relax away from the nuclear position in response to a local electric field. This model is quite successful for purely ionic systems, but it does not allow charge transfer. Many ionic systems of interest in nanoscale material science involve at least partial covalent character and, thus, require a potential that models charge transfer in response to environmental conditions. One of the few such potentials was introduced by Streitz and Mintmire for aluminum oxide (Streitz, 1994). In this formalism, charge transfer and electrostatic interactions are calculated using an electronegativity equalization approach (Rappé, 1991) and added to EAM type forces. The development of similar formalisms for a wider range of materials is clearly needed, especially for things like piezoelectric systems that may have unique functionality at the nanometer scale.

24.1.3 Quantum Basis of Some Analytic Potential Functions

The functional form of the majority of the workhorse potential energy expressions used in molecular modeling studies of nanoscale systems can be traced to density functional theory (DFT). Two connections between the density functional equations and analytic potentials as well as specific approaches that have been derived from these analyses are presented in this section. The first connection is the Harris functional, from which tight-binding, Finnis–Sinclair, and empirical bond-order potential functions can be derived. The second connection is effective medium theory that leads to the EAM.

The fundamental principle behind DFT is that all ground-state properties of an interacting system of electrons with a nondegenerate energy in an external potential can be uniquely determined from the electron density. This is a departure from traditional quantum chemical methods that attempt to find a many-body wavefunction from which all properties (including electron density) can be obtained. In DFT the ground-state electronic energy is a unique functional of the electron density, which is minimized by the correct electron density (Hohenberg, 1964). This energy functional consists of contributions from the electronic kinetic, Coulombic, and exchange–correlation energy. Because the form of the exchange–correlation functional is not known, DFT does not lead to a direct solution of the many-body electron problem. From the viewpoint of the development of analytic potential functions for condensed phases, however, DFT is a powerful concept because the electron density is the central quantity rather that a wavefunction, and electron densities are much easier to approximate with an analytic function than are wave functions.

The variational principle of DFT leads to a system of one-electron equations of the form:

$$[T + V_H(\mathbf{r}) + V_N(\mathbf{r}) + V_{xc}(\mathbf{r})]\, \phi_i^{K-S} = \varepsilon_i\, \phi_i^{K-S} \tag{24.1}$$

that can be self-consistently solved (Kohn, 1965). In this expression T is a kinetic energy operator, $V_H(\mathbf{r})$ is the Hartree potential, $V_N(\mathbf{r})$ is the potential due to the nuclei, and ε_i and ϕ_i^{K-S} are the eigenenergies and eigenfunctions of the one-electron Kohn–Sham orbitals, respectively. The exchange–correlation potential, $V_{xc}(\mathbf{r})$, is the functional derivative of the exchange-correlation energy E_{xc}:

$$V_{xc}(\mathbf{r}) = \delta E_{xc}[\rho(\mathbf{r})]/\delta\rho(\mathbf{r}) \tag{24.2}$$

where $\rho(\mathbf{r})$ is the charge density obtained from the Kohn–Sham orbitals. When solved self-consistently, the electron densities obtained from Equation (24.1) can be used to obtain the electronic energy using the expression:

$$E^{KS}[\rho^{sc}(\mathbf{r})] = \Sigma_k \varepsilon_k - \int \rho^{sc}[V_H(\mathbf{r})/2 + V_{xc}(\mathbf{r})]\, d\mathbf{r} + E_{xc}[\rho^{sc}(\mathbf{r})] \tag{24.3}$$

where ρ^{sc} is the self-consistent electron density and ε_k is the eigenvalues of the one-electron orbitals. The integral on the right side of Equation (24.3) corrects for the fact that the eigenvalue sum includes the exchange–correlation energy and double-counts the electron–electron Coulomb interactions. These are sometimes referred to as the double-counting terms.

A strength of DFT is that the error in electronic energy is second-order in the difference between a given electron density and the true ground-state density. Working independently, Harris as well as Foulkes and Haydock showed that the electronic energy calculated from a single iteration of the energy functional,

$$E^{Harris}(\rho^{in}(\mathbf{r})) = \Sigma_k \varepsilon_k^{out} - \int \rho^{in}[V_H^{in}(\mathbf{r})/2 + V_{xc}^{in}(\mathbf{r})] \; d\mathbf{r} + E_{xc}[\rho^{in}(\mathbf{r})] \qquad (24.4)$$

is also second-order in the error in charge density (Harris, 1985; Foulkes, 1989). This differs from the usual density functional equation in that while the Kohn–Sham orbital energies ε_k^{out} are still calculated, the double-counting terms involve only an input charge density and not the density given by these orbitals. Therefore this functional, generally referred to as the *Harris functional*, yields two significant computational benefits over a full density functional calculation — the input electron density can be chosen to simplify the calculation of the double-counting terms, and it does not require self-consistency. This property has made Harris functional calculations useful as a relatively less computationally intensive variation of DFT compared with fully self-consistent calculations (despite the fact that the Harris functional is not variational). As discussed by Foulkes and Haydock, the Harris functional can be used as a basis for deriving tight-binding potential energy expressions (Foulkes, 1989).

The tight-binding method refers to a non-self-consistent semi-empirical molecular orbital approximation that dates back to before the development of density functional theory (Slater, 1954). The tight-binding method can produce trends in (and in some cases quantitative values of) electronic properties; and if appropriately parameterized with auxiliary functions, it can produce binding energies, bond lengths, and vibrational properties that are relatively accurate and that are transferable within a wide range of structures, including bulk solids, surfaces, and clusters.

Typical tight-binding expressions used in molecular modeling studies give the total energy E_{tot} for a system of atoms as a sum of eigenvalues ε of a set of occupied non-self-consistent one-electron molecular orbitals plus some additional analytic function A of relative atomic distances:

$$E_{tot} = A + \Sigma_k \varepsilon_k \qquad (24.5)$$

The idea underlying this expression is that quantum mechanical effects are roughly captured through the eigenvalue calculation, while the analytic function applies corrections to approximations in the electronic energy calculation needed to obtain reasonable total energies. The simplest and most widespread tight-binding expression obtains the eigenvalues of the electronic states from a wavefunction that is expanded in an orthonormal minimal basis of short-range atom-centered orbitals. One-electron molecular orbital coefficients and energies are calculated using the standard secular equation as done for traditional molecular orbital calculations. Rather than calculating multi-center Hamiltonian matrix elements, however, these matrix elements are usually taken as two-center terms that are fit to reproduce electronic properties such as band structures; or they are sometimes adjusted along with the analytic function to enhance the transferability of total energies between different types of structures. For calculations involving disordered structures and defects, the dependence on distance of the two-center terms must also be specified.

A common approximation is to assume a pairwise additive sum over atomic distances for the analytic function:

$$A = \Sigma_i \Sigma_j \theta(r_{ij}) \qquad (24.6)$$

where r_{ij} is the scalar distance between atoms i and j. The function $\theta(r_{ij})$ models Coulomb repulsions between positive nuclei that are screened by core electrons, plus corrections to the approximate quantum mechanics. While a pairwise sum may be justified for the interatomic repulsion between nuclei and core electrons, there is little reason to assume that it can compensate for all of the approximations inherent in tight-binding theory. Nevertheless, the tight-binding approximation appears to work well for a range of covalent materials.

Several widely used potentials have attempted to improve upon the standard tight-binding approach. Rather than use a simple pairwise sum for the analytic potential, for example, Xu et al. assumed for carbon the multi-center expression:

$$A = P[\Sigma_i \Sigma_j \theta(r_{ij})] \qquad (24.7)$$

where P represents a polynomial function and $\theta(r_{ij})$ is an exponential function splined to zero at a distance between the second and third nearest-neighbors in a diamond lattice (Xu, 1992). Xu et al. fit the pairwise terms and the polynomial to different solid-state and molecular structures. The resulting potential produces binding energies that are transferable to a wide range of systems, including small clusters, fullerenes, diamond surfaces, and disordered carbon.

A number of researchers have suggested that for further enhancement of the transferability of tight-binding expressions, the basis functions should not be assumed orthogonal. This requires additional parameters describing the overlap integrals. Several methods have been introduced for determining these. Menon and Subbaswamy, for example, have used proportionality expressions between the overlap and Hamiltonian matrix elements from Hückel theory (Menon, 1993). Frauenheim and co-workers have calculated Hamiltonian and overlap matrix elements directly from density functional calculations within the local density approximation (Frauenheim, 1995). This approach is powerful because complications associated with an empirical fit are eliminated, yet the relative computational simplicity of a tight-binding expression is retained.

Foulkes and Haydock have used the Harris functional to justify the success of the approximations used in tight-binding theory (Foulkes, 1989). First, taking the molecular orbitals used in tight-binding expressions as corresponding to the Kohn–Sham orbitals created from an input charge density justifies the use of non-self-consistent energies in tight-binding theory. Second, if the input electron density is approximated with a sum of overlapping, atom-centered spherical orbitals, then the double-counting terms in the Harris functional are given by

$$\Sigma_a C_a + \frac{1}{2} \Sigma_i \Sigma_j U(r_{ij}) + U_{np} \qquad (24.8)$$

where C_a is a constant intra-atomic energy, $U_{ij}(r_{ij})$ is a short-range pairwise additive energy that depends on the scalar distance between atoms i and j, and U_{np} is a non-pairwise additive contribution that comes from the exchange–correlation functional. Haydock and Foulkes showed that if the regions where overlap of electron densities from three or more atoms are small, the function U_{np} is well approximated by a pairwise sum that can be added to U_{ij}, justifying the assumption of pair-additivity for the function A in Equation (24.6). Finally, the use of spherical atomic orbitals leads to the simple form:

$$V_{xc}(\mathbf{r}) = \Sigma_i V_i(\mathbf{r}) + U(\mathbf{r}) \qquad (24.9)$$

for the one-electron potential used to calculate the orbital energies in the Harris functional. The function $V_i(\mathbf{r})$ is an additive atomic term that includes contributions from core electrons as well as Hartree and exchange–correlation potentials, and $U(\mathbf{r})$ comes from nonlinearities in the exchange–correlation functional. Although not two-centered, the contribution of the latter term is relatively small. Thus, the use of strictly two-center matrix elements in the tight-binding Hamiltonian can also be justified.

Further approximations building on tight-binding theory, namely the second-moment approximation, can be used to arrive at some of the other analytic potential energy functions that are widely used in molecular modeling studies of nanoscale systems (Sutton, 1993). In a simplified quantum mechanical picture, the formation of chemical bonds is due to the splitting of atomic orbital energies as molecular orbitals are formed when atoms are brought together. For condensed phases, the energy and distribution of molecular orbitals among atoms can be conveniently described using a local density of states. The local density of states is defined for a given atom as the number of electronic states in the interval between energy e and e+δe weighted by the "amount" of the orbital on the atom. For molecular orbitals expanded in an orthonormal linear basis of atomic orbitals on each atom, the weight is the sum of the squares of the linear expansion coefficients for the atomic orbitals centered on the atom of interest. The electronic bond energy associated with an individual atom can be defined as twice (two electrons per orbital) the integral over energy of the local density of states multiplied by the orbital energies, where the upper limit of the integral is the Fermi energy. With this definition, the sum of the energies associated with each atom is equal to twice the sum of the energies of the occupied molecular orbitals.

An advantage of using a density of states rather than directly using orbital energies to obtain an electronic energy is that like any distribution, the properties of the local density of states can often be conveniently described using just a few moments of the distribution. For example, as stated above, binding energies are associated with a spread in molecular orbital energies; therefore the second moment of the local density of states can be related to the potential energy of an atom. Similarly, formation of a band gap in a half-filled energy band can be described by the fourth moment (the Kurtosis) of the density of states that quantifies the amount of a distribution in the middle compared with that in the wings of the distribution.

There is a very powerful theorem, called the *moments theorem*, that relates the moments of the local density of states to the bonding topology (Cyrot–Lackmann, 1968; Sutton, 1993). This theorem states that the *nth moment of the local density of states on an atom i is determined by the sum of all paths of n hops between neighboring atoms that start and end at atom i*. Because the paths that determine the second moment involve only two hops, one hop from the central atom to a neighbor atom and one hop back, according to this theorem the second moment of the local densities of states is determined by the number of nearest neighbors. As stated above, the bond energy of an atom can be related to the spread in energy of the molecular orbitals relative to the atomic orbitals out of which molecular orbitals are constructed. The spread in energy is given by the square root of the second moment, and therefore it is reasonable to assume that the electronic contribution to the bond energy E_i of a given atom i is proportional to the square root of the number of neighbors z:

$$E_i \propto z^{1/2} \tag{24.10}$$

This result is called the second-moment approximation.

The definition of neighboring atoms needs to be addressed to develop an analytic potential function from the second-moment approximation. Exponential functions of distance can be conveniently used to count neighbors (these functions mimic the decay of electronic densities with distance). Including a proportionality constant between the electronic bond energy and the square root of the number of neighbors, and adding pairwise repulsive interactions between atoms to balance the electronic energy, yields the Finnis–Sinclair analytic potential energy function for atom i:

$$E_i = \Sigma_j \, 1/2 A e^{-\alpha r_{ij}} - [\Sigma_j B e^{-\beta r_{ij}}]^{1/2} \tag{24.11}$$

where the total potential energy is the sum of the atomic energies (Finnis, 1984). This is a particularly simple expression that captures much of the essence of quantum mechanical bonding.

Another variation on the second-moment approximation is the Tersoff expression for describing covalent bonding (Tersoff, 1986; Tersoff, 1989). Rather than describe the original derivation due to Abell (Abell, 1985), which differs from that outlined above, one can start with the Finnis–Sinclair form for the electronic energy of a given atom i and arrive at a simplified Tersoff expression through the following algebra (Brenner, 1989):

$$E_i^{el} = -B(\Sigma_j \, e^{-\beta \, r_{ij}})^{1/2} \tag{24.12a}$$

$$= -B(\Sigma_j \, e^{-\beta \, r_{ij}})^{1/2} \times [(\Sigma_j \, e^{-\beta \, r_{ij}})^{1/2}/(\Sigma_k \, e^{-\beta \, r_{ik}})^{1/2}] \tag{24.12b}$$

$$= -B(\Sigma_j \, [e^{-\beta \, r_{ij}}]) \times (\Sigma_k \, e^{-\beta \, r_{ik}})^{-1/2} \tag{24.12c}$$

$$= \Sigma_j \; [-Be^{-\beta \; rij} \times (\Sigma_k \; e^{-\beta \; rik})^{-1/2}] \tag{24.12d}$$

$$= \Sigma_j \; [-B \; e^{-\beta \; rij} \times (e^{-\beta \; rij} + \Sigma_{k \neq i,j} \; e^{-\beta \; rik})^{-1/2}] \tag{24.12e}$$

$$= \Sigma_j \; [-B \; e^{-\beta/2 \; rij} \times (1 + \Sigma_{k \neq i,j} \; e^{-\beta \; (rik-rij)})^{-1/2}] \tag{24.12f}$$

This derivation leads to an attractive energy expressed by a two-center pair term of the form:

$$-Be^{-\beta/2 \, rij} \tag{24.13}$$

that is modulated by an analytic bond order function of the form:

$$(1 + \Sigma_{k \neq i,j} \; e^{-\beta \; (rik-rij)})^{-1/2} \tag{24.14}$$

While mathematically the same form as the Finnis–Sinclair potential, the Tersoff form of the second-moment approximation lends itself to a slightly different physical interpretation. The value of the bond order function decreases as the number of neighbors of an atom increases. This in turn decreases the magnitude of the pair term and effectively mimics the limited number of valence electrons available for bonding. At the same time as the number of neighbors increases, the number of bonds modeled through the pair term increases, the structural preference for a given material — e.g., from a molecular solid with a few neighbors to a close-packed solid with up to 12 nearest neighbors — depends on a competition between increasing the number of bonds to neighboring atoms and bond energies that decrease with increasing coordination number.

In the Tersoff bond-order form, an angular function is included in the terms in the sum over k in Equation (24.14), along with a few other more subtle modifications. The net result is a set of robust potential functions for group IV materials and their alloys that can model a wide range of bonding configurations and associated orbital hybridizations relatively efficiently. Brenner and co-workers have extended the bond-order function by adding *ad hoc* terms that model radical energetics, rotation about double bonds, and conjugation in hydrocarbon systems (Brenner, 1990). Further modifications by Stuart, Harrison, and associates have included non-bonded interactions into this formalism (Stuart, 2000).

Considerable additional effort has gone into developing potential energy expressions that include angular interactions and higher moments since the introduction of the Finnis–Sinclair and Tersoff potentials (Carlsson, 1991; Foiles, 1993). Carlsson and co-workers, for example, have introduced a matrix form for the moments of the local density of states from which explicit environment-dependent angular interactions can be obtained (Carlsson, 1993). The role of the fourth moment, in particular, has been stressed for half-filled bands because, as mentioned above, it describes the tendency to introduce an energy gap. Pettifor and co-workers have introduced a particularly powerful formalism that produces analytic functions for the moments of a distribution that has recently been used in atomic simulations (Pettifor, 1999; Pettifor, 2000). While none of these potential energy expressions has achieved the work-horse status of the expressions discussed above, specific applications of these models, particularly the Pettifor model, have been very promising. It is expected that the use of this and related formalisms for modeling nanometer-scale systems will significantly increase in the next few years.

A different route through which analytic potential energy functions used in nanoscale simulations have been developed is effective medium theory (Jacobsen, 1987; Jacobsen, 1988). The basic idea behind this approach is to replace the relatively complex environment of an atom in a solid with that of a simplified host. The electronic energy of the true solid is then constructed from accurate energy calculations on the simplified medium. In a standard implementation of effective medium theory, the simplified host is a homogeneous electron gas with a compensating positive background (the so-called *jellium model*). Because of the change in electrostatic potential, an atom embedded in a jellium alters the initially homogeneous electron density. This difference in electron density with and without the embedded atom can be calculated within the local density approximation of DFT and expressed as a spherical function about the atom embedded in the jellium. This function depends on both the identity of the embedded atom and the initial density of the homogeneous electron gas. The overall electron density of the solid can then be approximated by a superposition of the perturbed electron densities. With this *ansatz*, however, the differences in electron density are not specified until the embedding electron density associated with each atom is known. For solids containing defects Norskov, Jacobsen, and co-workers have suggested using spheres centered at each site with radii chosen so that the electronic charge within each sphere cancels the charge of each atomic nucleus (Jacobsen, 1988). For most metals, Jacobsen has shown that to a good approximation the average embedding density is related exponentially to the sphere radius, and these relationships for 30 elements have been tabulated.

Within this set of assumptions, for an imperfect crystal the average embedding electron density at each atomic site is not defined until the total electron density is constructed; and the total electron density in turn depends on these local densities, leading to a self-consistent calculation. However, the perturbed electron densities need only be calculated once for a given electron density; so this method can be applied to systems much larger than can be treated by full density functional calculations without any input from experiment.

Using the variational principle of density functional theory, it has been shown that the binding energy E_B of a collection of atoms with the assumptions above can be given by the expression:

$$E_B = \Sigma_i E_i (\rho_i^{ave}) + \Sigma_i \Sigma_j E_{ov} + E_{1e} \qquad (24.15)$$

where ρ_i^{ave} is the average electron density for atomic site i, $E_i (\rho_i^{ave})$ is the energy of an atom embedded in jellium with density ρ_i^{ave}, E_{ov} is the electrostatic repulsion between overlapping neutral spheres (which is summed over atoms i and j), and E_{1e} is a one-electron energy not accounted for by the spherical approximations (Jacobsen, 1988). For most close-packed metals, the overlap and one-electron terms are relatively small, and the embedding term dominates the energy. Reasonable estimates for shear constants, however, require that the overlap term not be neglected; and for non-close-packed systems, the one-electron term becomes important. Equation (24.15) also provides a different interpretation of the Xu et al. tight-binding expression Equation (24.7) (Xu, 1992). The analytic term A in Equation (24.7), which is taken as a polynomial function of a pair sum, can be interpreted as corresponding to the energy of embedding an atom into jellium. The one-electron tight-binding orbitals are then the one-electron terms typically ignored in Equation (24.15).

The EAM is essentially an empirical non-self-consistent variation of effective medium theory. The binding energy in the EAM is given by:

$$E^B = \Sigma_i F(\rho_i) + \Sigma_i \Sigma_j U(r_{ij}) \qquad (24.16)$$

where ρ_i is the electron density associated with i, F is called the embedding function, and $U(r_{ij})$ is a pair-additive interaction that depends on the scalar distance r_{ij} between atoms i and j (Daw, 1983). The first term in Equation (24.16) corresponds to the energy of the atoms embedded in jellium, the second term represents overlap of neutral spheres, and the one-electron term of Equation (24.15) is ignored. In the embedded-atom method, however, the average electron density surrounding an atom within a neutral sphere used in effective medium theory is replaced with a sum of electron densities from neighboring atoms at the lattice site i:

$$\rho_i = \Sigma_j \phi(r_{ij}) \qquad (24.17)$$

where $\phi(r_{ij})$ is the contribution to the electron density at site i from atom j. Furthermore, the embedding function and the pair terms are empirically fit to materials properties, and the pairwise–additive electron contributions $\phi^{ij}(r_{ij})$ are either taken from atomic electron densities or fit as empirical functions.

24.2 First-Principles Approaches: Forces on the Fly

The discussion in the previous section has focused on analytic potential energy functions and inter-atomic forces. These types of expressions are typically used in molecular modeling simulations when large systems and/or many timesteps are needed, approximate forces are adequate for the results desired, or computing resources are limited. Another approach is to use forces directly from a total energy calculation that explicitly includes electronic degrees of freedom. The first efficient scheme and still the most widespread technique for calculating *forces on the fly* is the Car–Parrinello method (Car, 1985). Introduced in the middle 1980s, the approach yielded both an efficient computational scheme and a new paradigm in how one calculates forces. The central concepts are to include in the equations of motion both the nuclear and electronic degrees of freedom, and to simultaneously integrate these coupled equations as the simulation progresses. In principle, at each step the energy associated with the electronic degrees of freedom should be *quenched* to the Born–Oppenheimer surface in order to obtain the appro-priate interatomic forces. What the Car–Parrinello method allows one to do, however, is to integrate the equations of motion without having to explicitly quench the electronic degrees of freedom, effectively allowing a trajectory to progress *above*, and hopefully parallel to, the Born–Oppenheimer potential energy surface. Since the introduction of the Car–Parrinello method, there have been other schemes that have built upon this idea of treating electronic and nuclear degrees of freedom on an even footing. Clearly, though, the Car–Parrinello approach opened valuable new avenues for modeling — avenues that will dominate molecular simulation as computing resources continue to expand.

Worth noting is another approach, popular largely in the chemistry community, in which a region of interest, usually where a reaction takes place, is treated quantum mechanically, while the surrounding environment is treated with a classical potential energy expression. Often referred to as the *Quantum Mechanics/Molecular Mechanics* (QM/MM) method, this approach is a computationally efficient com-promise between classical potentials and quantum chemistry calculations. The chief challenge to the QM/MM method appears to be how to adequately treat the boundary between the region whose forces are calculated quantum mechanically and the region where the analytic forces are used.

A similar approach to the QM/MM method has been used to model crack propagation. Broughton and co-workers, for example, have used a model for crack propagation in silicon in which electrons are included in the calculation of interatomic forces at the crack tip, while surrounding atoms are treated with an analytic potential (Selinger, 2000). To treat long-range stresses, the entire atomically resolved region is in turn embedded into a continuum treated with a finite element model. These types of models, in which atomic and continuum regions are coupled in the same simulation, are providing new methods for connecting atomic-scale simulations with macroscale properties.

24.2.1 Other Considerations in Molecular Dynamics Simulations

In addition to the force model, there are other considerations in carrying out a molecular dynamics simulation. For the sake of completeness, two of these are discussed. The first consideration has to do with the choice of an integration scheme for calculation of the dynamics. The equations of motion that govern the atomic trajectories are a set of coupled differential equations that depend on the mutual interaction of atoms. Because these equations cannot in general be solved analytically, numerical inte-gration schemes are used to propagate the system forward in time. The first step in these schemes is to convert the differential equations to difference equations, and then solve these equations step-wise using some finite timestep. In general, the longer the timestep, the fewer steps are needed to reach some target total time and the more efficient the simulations. At the same time, however, the difference equations are better approximations of the differential equations at smaller timesteps. Therefore, long timesteps can introduce significant errors and numerical instabilities. A rule of thumb is that the timestep should be no longer than about 1/20 of the shortest vibrational period in the system. If all degrees of freedom are included, including bond vibrations, timestep sizes are typically on the order of one femtosecond. If vibrational degrees of freedom can be eliminated, typically using rigid bonds, timesteps an order of magnitude larger can often be used. In addition to large integration errors that can occur from step to step, it is also important to consider small errors that are cumulative over many timesteps, as these can adversely affect the results of simulations involving many timesteps. Fortunately, a great deal of effort has gone into evaluating both kinds of integration errors, and integrators have been developed for which small errors tend to cancel one another over many timesteps. In addition, many calculations rely on dynamic simulations purely as a means of sampling configuration space, so that long-time trajectory errors are relatively unimportant as long as the dynamics remain in the desired ensemble.

Another consideration in molecular dynamics simulation is the choice of thermostat. The simplest approach is to simply scale velocities such that an appropriate average kinetic energy is achieved. The drawback to this approach is that fluctuations in energy and temperature associated with a given thermodynamic ensemble are not reproduced. Other methods have historically been used in which frictional forces are added (e.g., Langevin models with random forces and compensating frictional terms), or constraints on the equations of motion are used. The Nosé–Hoover thermostat, which is essentially a hybrid frictional force/constrained dynamics scheme, is capable of producing both a desired average kinetic energy and the correct temperature fluctuations for a system coupled to a thermal bath (Nosé, 1984). Similar schemes exist for maintaining desired pressures or stress states that can be used for simulations of condensed phase systems.

24.3 Applications

Discussed in the following sections are some applications of molecular modeling to the development of our understanding of nanometer-scale chemical and physical phenomena. Rather than attempt to provide a comprehensive literature review, we have focused on a subset of studies that illustrate both how modeling is used to understand experimental results, and how modeling can be used to test the boundaries of what is possible in nanotechnology.

24.3.1 Pumps, Gears, Motors, Valves, and Extraction Tools

Molecular modeling has been used to study a wide range of systems whose functionality comes from atomic motion (as opposed to electronic properties). Some of these studies have been on highly idealized structures, with the intent of exploring the properties of these systems and not necessarily implying that the specific structure modeled will ever be created in the laboratory. Some studies, on the other hand, are intended to model experimentally realizable systems, with the goal of helping to guide the optimization of a given functionality.

Goddard and co-workers have used molecular dynamics simulations to model the performance of idealized planetary gears and neon pumps containing up to about 10,000 atoms (Cagin, 1998). The structures explored were based on models originally developed by Drexler and Merkle (Drexler, 1992), with the goal of the Goddard studies being to enhance the stability of the structures under desired operating conditions. The simulations used primarily generic valence force-field and non-bonded interatomic interactions. The primary issue in optimizing these systems was to produce gears and other mutually moving interfaces that met exacting specifications. At the macroscale these types of tolerances can be met with precision machining. As pointed out by Cagin et al., at the nanometer scale these tolerances are intimately tied to atomic structures; and specific parts must be carefully designed atom-by-atom to produce acceptable tolerances (Cagin, 1998). At the same time, the overall structure must be robust against both shear forces and relatively low-frequency (compared with atomic vibrations) system vibrations caused by sudden accelerations of the moving parts. Illustrated in Figure 24.2 is a planetary gear and a neon pump modeled by the Goddard group. The gear contains 4235 atoms, has a molecular weight of 72491.95 grams/mole, and contains eight moving parts, while the pump contains 6165 atoms. Molecular simulations in which the concentric gears in both systems are rotated have demonstrated that in designing such a system, a careful balance must be maintained between rotating surfaces being too far apart, leading to gear slip, and being too close together, which leads to lock-up of the interface.

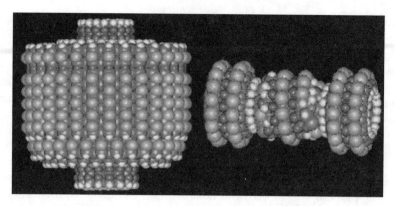

FIGURE 24.2 Illustrations of molecular pump and gear. (Courtesy of T. Cagin and W.A. Goddard, from Cagin, T., Jaramillo–Botero, A., Gao, G., Goddard, W.A. (1998) Molecular mechanics and molecular dynamics analysis of Drexler–Merkle gears and neon pump, *Nanotechnology,* 9, 143–152. With permission.)

Complicating this analysis are vibrations of the system during operation, which can alternately cause slip and lock-up as the gears move. Different modes of driving these systems were modeled that included single impulses as well as time-dependent and constant angular velocities, torques, and accelerations. The simulations predict that significant and rapid heating of these systems can occur depending on how they are driven, and therefore thermal management within these systems is critical to their operation.

Several models for nanometer-scale bearings, gears, and motors based on carbon nanotubes have also been studied using molecular modeling. Sumpter and co-workers at Oak Ridge National Laboratory, for example, have modeled bearings and motors consisting of nested nanotubes (Tuzen, 1995). In their motor simulations, a charge dipole was created at the end of the inside nanotube by assigning positive and negative charges to two of the atoms at the end of the nanotube. The motion of the inside nanotube was then driven using an alternating laser field, while the outer shaft was held fixed. For single-laser operation, the direction of rotation of the shaft was not constant, and beat patterns in angular momentum and total energy occurred whose period and intensity depended on the field strength and placement of the charges. The simulations also predict that using two laser fields could decrease these beat oscillations. However, a thorough analysis of system parameters (field strength and frequency, temperature, sleeve and shaft sizes, and placement of charges) did not identify an ideal set of conditions under which continuous motion of the inner shaft was possible.

Han, Srivastava, and co-workers at NASA Ames have modeled gears created by chemisorbing benzynyl radicals to the outer walls of nanotubes and using the molecules on two nanotubes as interlocking gear teeth (Han, 1997; Srivastava, 1997). These simulations utilized the bond order potentials described above to describe the interatomic forces. Simulations in which one of the nanotubes is rotated such that it drives the other nanotube show some gear slippage due to distortion of the benzynyl radicals, but without bond breakage. In the simulations, the interlocking gears are able to operate at 50–100 GHz. In a simulation similar in spirit to the laser-driven motor gear modeled by Tuzun et al., Srivastava used both a phenomenological model and molecular simulation to characterize the NASA Ames gears with a dipole driven by a laser field (Srivastava, 1997). The simulations, which used a range of laser field strengths and frequencies, demonstrate that a molecular gear can be driven in one direction if the frequency of the laser field matches a natural frequency of the gear. When the gear and laser field frequency were not in resonance, the direction of the gear motion would change as was observed in the Oak Ridge simulations.

Researchers at the U.S. Naval Research Laboratory have also modeled nanotube gears using the same interatomic potentials as were used for the NASA gears (Robertson, 1994). In these structures, however, the gear shape was created by introducing curvature into a fullerene structure via five- and seven-membered rings (see Figure 24.3). The primary intent of these simulations was not necessarily to model a working gear, but rather to demonstrate the complexity in shape that can be introduced into fullerene structures. In the simulations the shafts of the two gears were allowed to rotate while full dynamics of

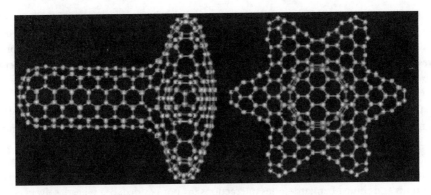

FIGURE 24.3 Illustration of the gears simulated by Robertson and associates.

the cams was modeled. The system was driven by placing the gears such that the gear teeth interlocked, and then rotating the shaft of one of the gears. Different driving conditions were studied, including the rate at which the driving gear was accelerated and its maximum velocity. For example, when the rotational speed of the driving gear was ramped from 0 to 0.1 revolutions per picosecond over 50 picoseconds, the driven gear was able to keep up. However, when the rotation of the driving gear was accelerated from 0 to 0.5 revolutions per picosecond over the same time period, significant slippage and distortion of the gear heads were observed that resulted in severe heating and destruction of the system.

Molecular modeling has recently been used to explore a possible nanoscale valve whose structure and operating concept is in sharp contrast to the machines discussed above (Adiga, 2002). The design is motivated by experiments by Park, Imanishi, and associates (Park, 1998; Ito, 2000). In one of these experiments, polymer brushes composed of polypeptides were self-assembled onto a gold-plated nanoporous membrane. Permeation of water through the membrane was controlled by a helix–coil transformation that was driven by solvent pH. In their coiled states, the polypeptide chains block water from passing through the pores. By changing pH, a folded configuration can be created, effectively opening the pores.

Rather than using polypeptides, the molecular modeling studies used a ball-and-spring model of polymer comb molecules chemisorbed to the inside of a slit pore (Adiga, 2002). Polymer comb molecules, also called molecular bottle brushes, consist of densely grafted side chains that extend away from a polymer backbone (Figure 24.4). In a good solvent (i.e., a solvent in which the side chains can be dissolved), the side chains are extended from the backbone, creating extensive excluded volume interactions. These interactions can create a very stiff structure with a rod configuration. In a poor solvent, the

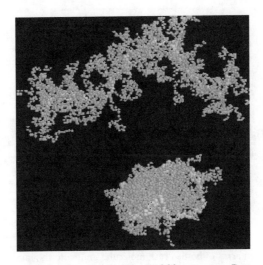

FIGURE 24.4 Illustration of a bottle brush molecule. Top: Rod-like structure. Bottom: Globular structure.

side chains collapse to the backbone, relieving the excluded volume interactions and allowing the total system to adapt a globular configuration. The concept behind the valve is to use this rod–globular transition in comb polymers assembled into the inside of a nanometer-scale pore to open or close the pore in response to solvent quality. The size of the pore can in principle be controlled to selectively pass molecular species according to their size, with the maximum species size that is allowed to pass controlled by solvent quality (with respect to the comb polymers).

Illustrated in Figure 24.5 are snapshots from a molecular modeling simulation of a pore of this type. The system modeled consists of comb polymer molecules 100 units long with sides of 30 units grafted to the backbone every four units. Periodic boundary conditions are applied in the two directions parallel to the grafting surface. The brushes are immersed in monomeric solvent molecules of the same size and mass as those of the beads of the comb molecules. The force-field model uses a harmonic spring between monomer units and shifted Lennard–Jones interactions for nonbonded, bead–bead, solvent–solvent, and bead–solvent forces. Purely repulsive nonbonded interactions are used for the brush–wall and solvent–wall interactions. The wall separation and grafting distances are about 2 times and one-half, respectively, of the end-to-end distance of a single comb molecule in its extended state, and about 8 and 2 times, respectively, of a comb molecule in its globule state. The snapshots in Figure 24.5 depict the pore structure at three different solvent conditions created by altering the ratio of the solvent–solvent to solvent–polymer interaction strength. In the simulations, pore opening due to a change in solvent quality required only about 0.5 nanoseconds. The simulations also showed that oligomer chain molecules with a radius smaller than the pore size can translate freely through the pore, while larger molecules become caught in the comb molecules, with motion likely requiring chain reptation, a fairly slow process compared with free translation. Both the timeframe of pore opening and the size selectivity for molecules passing through the pore indicate an effective nanoscale valve.

A molecular abstraction tool for patterning surfaces, which was first proposed by Drexler, has also been studied using molecular modeling (Sinnott, 1994; Brenner, 1996). The tool is composed of an ethynyl radical chemisorbed to the surface of a scanned-probe microscope tip, which in the case of the modeling studies was composed of diamond. When brought near a second surface, it was proposed that the radical species could abstract a hydrogen atom from the surface with atomic precision. A number of issues related to this tool were characterized using molecular modeling studies with the bond-order potential discussed above. These issues included the time scales needed for reaction and for the reaction energy to flow away from the reaction site, the effect of tip crashes, and the creation of a signal that abstraction had occurred. The simulation indicated that the rates of reaction and energy flow were very fast, effectively creating an irreversible abstraction reaction if the tip is left in the vicinity of the surface

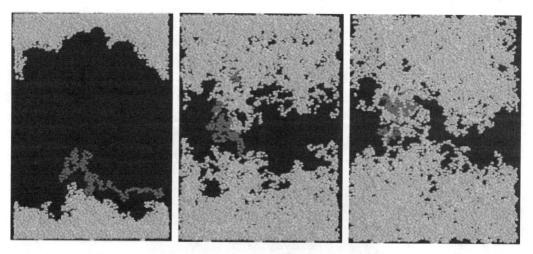

FIGURE 24.5 Illustration of an array of bottle brush molecules grafted to the inside of nanometer-scale slit pores at three different solvent qualities.

from which the hydrogen is abstracted. As expected, however, with the ethynyl radical exposed at the end of the tip, tip crashes effectively destroy the system. To study how this can be avoided, and to create a system from which a signal for abstraction could be detected, a structure was created in the modeling studies in which asperities on the tip surround the ethynyl radical. With this configuration, the simulations predicted that a load on the tip can be detected as the tip comes into contact with a surface, and that abstraction can still occur while the asperities protect the ethynyl from further damage if a larger load is applied to the tip. While this system has not been created experimentally, the modeling studies nonetheless represent a creative study into what is feasible at the nanometer scale.

24.3.2 Nanometer-Scale Properties of Confined Fluids

The study of friction and wear between sliding bodies, now referred to as the field of *tribology*, has a long and important history in the development of new technologies. For example, the ancient Egyptians used water to lubricate the path of sleds that transported heavy objects. The first scientific studies of friction were carried out in the sixteenth century by Leonardo da Vinci, who deduced that the weight, but not the shape, of an object influences the frictional force of sliding. Da Vinci also first introduced the idea of a friction coefficient as the ratio of frictional force to normal load. This and similar observations led, 200 years later, to the development of Amonton's Law, which states that for macroscopic systems the frictional force between two objects is proportional to the normal load and independent of the apparent contact area. In the eighteenth century, Coulomb verified these observations and clarified the difference between static and dynamic friction.

Understanding and ultimately controlling friction, which traditionally has dealt with macroscale properties, is no less important for the development of new nanotechnologies involving moving interfaces. With the emergence of experimental techniques, such as the atomic-force microscope, the surface-force apparatus, and the quartz crystal microbalance, has come the ability to measure surface interactions under smooth or single-asperity contact conditions with nanometer-scale resolution. Interpreting the results of these experiments, however, is often problematic, as new and sometimes unexpected phenomena are discovered. It is in these situations that molecular modeling has often played a crucial role.

Both experiments and subsequent molecular modeling studies have been used to characterize behavior associated with fluid ordering near solid surfaces at the nanometer scale and the influence of this ordering on liquid lubrication at this scale. Experimental measurements have demonstrated that the properties of fluids confined between solid surfaces become drastically altered as the separation between the solid surfaces approaches the atomic scale (Horn, 1981; Chan, 1985; Gee, 1990). At separations of a few molecular diameters, for example, an increase in liquid viscosity by several orders of magnitude has been measured (Israelachvili, 1988; Van Alsten, 1988). While continuum hydrodynamic and elasto-hydrodynamic theories have been successful in describing lubrication by micron-thick films, these approaches start to break down when the liquid thickness approaches a few molecular diameters. Molecular simulations have been used to great advantage at length scales for which continuum approaches begin to fail. Systems that have been simulated include films of spherical molecules, straight-chain alkanes, and branched alkanes confined between solid parallel walls.

Using molecular simulations involving both molecular dynamics and Monte Carlo methods, several research groups have characterized the equilibrium properties of spherical molecules confined between solid walls (Schoen, 1987, 1989; Bitsanis, 1987; Thompson, 1990; Sokol, 1992; Diestler, 1993). These studies have suggested that when placed inside a pore, fluid layers become layered normal to the pore walls, independent of the atomic-scale roughness of the pore walls (Bitsanis, 1990). Simulations have also shown that structure in the walls of the pore can induce transverse order (parallel to the walls) in a confined atomic fluid (Schoen, 1987). For example, detailed analysis of the structure of a fluid within a layer, or epitaxial ordering, as a function of wall density and wall–fluid interaction strengths was undertaken by Thompson and Robbins (Thompson, 1990b). For small values of wall-to-liquid interaction strength, fluid atoms were more likely to sit over gaps in the adjacent solid layer. Self-diffusion within this layer, however, was roughly the same as in the bulk liquid. Increasing the strength of the wall–fluid interactions by a factor of 4.5 resulted in the epitaxial locking of the first liquid layer to the solid. While diffusion in the first layer was too small to measure, diffusion in the second layer was approximately half of its value in the bulk fluid. The second layer of liquid crystallized and became locked to the first *liquid* layer when the strength of the wall–liquid interaction was increased by approximately an order of magnitude over its original value. A third layer never crystallized.

Confinement by solid walls has been shown to have a number of effects on the equilibrium properties of static polymer films (Ribarsky, 1992; Thompson, 1992; Wang, 1993). For example, in simulations of linear chains by Thompson et al., film thickness was found to decrease as the normal pressure on the upper wall increases. At the same time, the simulations predict that the degree of layering and in-plane ordering increases, and the diffusion constant parallel to the walls decreases. In contrast to films of spherical molecules, where there is a sudden drop in the diffusion constant associated with a phase transition to an ordered structure, films of chain molecules are predicted to remain highly disordered, and the diffusion constant drops steadily as the pressure increases. This indicates the onset of a glassy phase at a pressure below the bulk transition pressure. This wall-induced glass phase explains dramatic increases in experimentally measured thin-film relaxation times and viscosities (Van Alsten, 1988; Gee, 1990).

In contrast to the situation for simple fluids and linear chain polymers, experimental studies (Granick, 1995) have not indicated oscillations in surface forces of confined highly-branched hydrocarbons such as squalane. To understand the reason for this experimental observation, Balasubramanian et al. used molecular modeling methods to examine the adsorption of linear and branched alkanes on a flat Au(111) surface (Balasubramanian, 1996). In particular, they examined the adsorption of films of n-hexadecane, three different hexadecane isomers, and squalane. The alkane molecules were modeled using a united atom model with a Lennard–Jones 12–6 potential used to model the interactions between the united atoms. The alkane–surface interactions were modeled using an external Lennard–Jones potential with parameters appropriate for a flat Au(111) substrate. The simulations yield density profiles for n-hexadecane and 6-pentylundecane that are nearly identical to experiment and previous simulations. In contrast, density profiles of the more highly branched alkanes such as heptamethylnonane and 7,8-dimethyltetradecane exhibit an additional peak. These peaks are due to methyl branches that cannot be accommodated in the first liquid layer next to the gold surface. For thicker films, oscillations in the density profiles for heptamethylnonane were out of phase with those for n-hexadecane, in agreement with the experimental observations.

The properties of confined spherical and chain molecular films under shear have been examined using molecular modeling methods. Work by Bitsanis et al., for example, examined the effect of shear on spherical, symmetric molecules, confined between planar, parallel walls that lacked atomic-scale roughness (Bitsanis, 1990). Both Couette (simple shear) and Poiseuille (pressure-driven) flows were examined. The density profiles in the presence of both types of flow were identical to those under equilibrium conditions for all pore widths. Velocity profiles, defined as the velocity of the liquid parallel to the wall as a function of distance from the center of the pore, should be linear and parabolic for Couette and Poiseuille flow, respectively, for a homogeneous liquid. The simulations yielded velocity profiles in the two monolayers nearest the solid surfaces that deviate from the flow shape expected for a homogeneous liquid and that indicate high viscosity. The different flow nature in molecularly thin films was further demonstrated by plotting the effective viscosity vs. pore width. For a bulk material the viscosity is independent of pore size. However, under both types of flow, the viscosity increases slightly as the pore size decreases. For ultrathin films, the simulations predict a dramatic increase in viscosity.

Thompson and Robbins also examined the flow of simple liquids confined between two solid walls (Thompson, 1990). In this case, the walls were composed of (001) planes of a face-center-cubic lattice. A number of wall and fluid properties, such as wall–fluid interaction strength, fluid density, and temperature, were varied. The geometry of the simulations closely resembled the configuration of a surface-force apparatus, where each wall atom was attached to a lattice site with a spring (an Einstein oscillator model) to maintain a well-defined solid structure with a minimum number of solid atoms. The thermal roughness and the response of the wall to the fluid was controlled by the spring constant, which was adjusted so that the atomic mean-square displacement about the lattice sites was less than the Lindemann criterion for melting. The interactions between the fluid atoms and between the wall and fluid atoms were modeled by different Lennard–Jones potentials. Moving the walls at a constant velocity in opposite directions simulated Couette flow, while the heat generated by the shearing of the liquid was dissipated using a Langevin thermostat. In most of the simulations, the fluid density and temperature were indicative of a compressed liquid about 30% above its melting temperature. A number of interesting phenomena were observed in these simulations. First, both normal and parallel ordering in the adjacent liquid was induced by the well-defined lattice structure of the solid walls. The liquid density oscillations also induced oscillations in other microscopic quantities normal to the walls, such as the fluid velocity in the flow direction and the in-plane microscopic stress tensor, that are contrary to the predictions of the continuum Navier–Stokes equations. However, averaging the quantities over length scales that are larger than the molecular lengths produced smoothed quantities that satisfied the Navier–Stokes equations.

Two-dimensional ordering of the liquid parallel to the walls affected the flow even more significantly than the ordering normal to the walls. The velocity profile of the fluid parallel to the wall was examined as a function of distance from the wall for a number of wall–fluid interaction strengths and wall densities. Analysis of velocity profiles demonstrated that flow near solid boundaries is strongly dependent on the strength of the wall–fluid interaction and on wall density. For instance, when the wall and fluid densities are equal and wall–fluid interactions strengths are small, the velocity profile is predicted to be linear with a no-slip boundary condition. As the wall–fluid interaction strength increases, the magnitude of the liquid velocity in the layers nearest the wall increases. Thus the velocity profiles become curved. Increasing the wall–fluid interaction strength further causes the first two liquid layers to become locked to the solid wall. For unequal wall and fluid densities, the flow boundary conditions changed dramatically. At the smallest wall–fluid interaction strengths examined, the velocity profile was linear; however, the no-slip boundary condition was not present. The magnitude of this slip decreases as the strength of the wall–fluid interaction increases. For an intermediate value of wall–fluid interaction strength, the first fluid layer was partially locked to the solid wall. Sufficiently large values of wall–fluid interaction strength led to the locking of the second fluid layer to the wall.

While simulations with spherical molecules are successful in explaining many experimental phenomena, they are unable to reproduce all features of the experimental data. For example, calculated relaxation times and viscosities remain near bulk fluid values until films crystallize. Experimentally, these quantities typically increase many orders of magnitude before a well-defined yield stress is observed (Israelachvili, 1988; Van Alsten, 1988). To characterize this discrepancy, Thompson et al. repeated earlier shearing simulations using freely jointed, linear-chain molecules instead of spherical molecules (Thompson, 1990, 1992). The behavior of the viscosity of the films as a function of shear rate was examined for films of different thickness. The response of films that were six to eight molecular diameters thick was approximately the same as for bulk systems. When the thickness of the film was reduced, the viscosity of the film increased dramatically, particularly at low shear rates, consistent with the experimental observations.

Based on experiments and simulations, it is clear that fluids confined to areas of atomic-scale dimensions do not necessarily behave like liquids on the macroscopic scale. In fact, depending on the conditions, they may often behave more like solids in terms of structure and flow. This presents a unique set of concerns for lubricating moving parts at the nanometer scale. However, with the aid of simulations such as the ones mentioned here, plus experimental studies using techniques such as the surface-force apparatus, general properties of nanometer-scale fluids are being characterized with considerable precision. This, in turn, is allowing scientists and engineers to design new materials and interfaces that have specific interactions with confined lubricants, effectively controlling friction (and wear) at the atomic scale.

24.3.3 Nanometer-Scale Indentation

Indentation is a well-established experimental technique for quantifying the macroscopic hardness of a material. In this technique, an indenter of known shape is loaded against a material and then released, and the resulting permanent impression is measured. The relation between the applied load, the indenter shape, and the profile of the impression is used to establish the hardness of the material on one of several possible engineering scales of hardness. Hardness values can in turn be used to estimate materials properties such as yield strength. Nanoindentation, a method in which both the tip–surface contact and the resulting impression have nanometer-scale dimensions, has become an important method for helping to establish properties of materials at the nanometer scale. The interpretation of nanoindentation data usually involves an analysis of loading and/or unloading force-vs.-displacement curves. Hertzian contact mechanics, which is based on continuum mechanics principles, is often sufficient to obtain properties such as elastic moduli from these curves, as long as the indentation conditions are elastic. In many situations, however, the applicability of Hertzian mechanics is either limited or altogether inappropriate, especially when plastic deformation occurs. It is these situations for which molecular modeling has become an essential tool for understanding nanoindentation data.

Landman was one of the first researchers to use molecular modeling to simulate the indentation of a metallic substrate with a metal tip (Landman, 1989, 1990, 1991, 1992, 1993). In an early simulation, a pyramidal nickel tip was used to indent a gold substrate. The EAM was used to generate the interatomic forces. By gradually lowering the tip into the substrate while simultaneously allowing classical motion of tip and surface atoms, a plot of force vs. tip-sample separation was generated, the features of which could be correlated with detailed atomic dynamics. The shape of this virtual loading/ unloading curve showed a jump to contact, a maximum force before tip retraction, and a large loading–unloading hysteresis. Each of these features match qualitative features of experimental loading curves, albeit on different scales of tip–substrate separation and contact area. The computer-generated loading also exhibited fine detail that was not resolved in the experimental curve. Analysis of the dynamics in the simulated loading curves showed that the jump to contact, which results in a relatively large and abrupt tip–surface attraction, was due to the gold atoms bulging up to meet the tip and subsequently wetting the tip. Advancing the tip caused indentation of the gold substrate with a corresponding increase in force with decreasing tip–substrate separation. Detailed dynamics leading to the shape of the virtual loading curve in this region consisted of the flow of the gold atoms that resulted in the pile-up of gold around the edges of the nickel indenter. As the tip was retracted from the sample, a connective neck of atoms between the tip and the substrate formed that was largely composed of gold atoms. Further retraction of the tip caused adjacent layers of the connective neck to rearrange so that an additional row of atoms formed in the neck. These rearrangement events were the essence of the elongation process, and they were responsible for a fine structure (apparent as a series of maxima) present in the retraction portion of the force curve. These elongation and rearrangement steps were repeated until the connective neck of atoms was severed.

The initial instability in tip–surface contact behavior observed in Landman's simulations was also reported by Pethica and Sutton and by Smith et al (Pethica, 1988; Smith, 1989). In a subsequent simulation by Landman and associates using a gold tip and nickel substrate, the tip deformed toward the substrate during the jump to contact. Hence, the softer material appears to be displaced. The longer-range jump-to-contact typically observed in experiments can be due to longer-ranged surface adhesive forces, such as dispersion and possibly wetting of impurity layers, as well as compliance of the tip holder, that were not included in the initial computer simulations.

Interesting results have been obtained for other metallic tip–substrate systems. For example, Tomagnini et al. used molecular simulation to study the interaction of a pyramidal gold tip with a lead substrate using interatomic forces from the glue model mentioned above (Tomagnini, 1993). When the gold tip was brought into close proximity to the lead substrate at room temperature, a jump to contact was initiated by a few lead atoms wetting the tip. The connective neck of atoms between the tip and the surface was composed almost entirely of lead. The tip became deformed because the inner-tip atoms were pulled more toward the sample surface than toward atoms on the tip surface. Increasing the substrate temperature to 600K caused the formation of a liquid lead layer approximately four layers thick on the surface of the substrate. During indentation, the distance at which the jump to contact occurred increased by approximately 1.5 Å, and the contact area also increased due to the diffusion of the lead. The gold tip eventually dissolved in the liquid lead, resulting in a liquid-like connective neck of atoms that followed the tip upon retraction. As a result, the liquid–solid interface moved farther back into the bulk lead substrate, increasing the length of the connective neck. Similar elongation events have been observed experimentally. For example, scanning tunneling microscopy experiments on the same surface demonstrate that the neck can elongate approximately 2500 Å without breaking.

Nanoindentation of substrates covered by various overlayers have also been simulated via molecular modeling. Landman and associates, for example, simulated indentation of an n-hexadecane-covered gold substrate with a nickel tip (Landman, 1992). The forces governing the metal–metal interactions were derived from the EAM, while a variation on a valence-force potential was used to model the n-hexadecane film. Equilibration of the film on the gold surface resulted in a partially ordered film where molecules in the layer closest to the gold substrate were oriented parallel to the surface plane. When the nickel tip was lowered, the film swelled up to meet and partially wet the tip. Continued lowering of the tip toward the film caused the film to flatten and some of the alkane molecules to wet the sides of the tip. Lowering the tip farther caused drainage of the top layer of alkane molecules from underneath the tip and increased wetting of the sides of the tip, pinning of hexadecane molecules under the tip, and deformation of the gold substrate beneath the tip. Further lowering of the tip resulted in the drainage of the pinned alkane molecules, inward deformation of the substrate, and eventual formation of an intermetallic contact by surface gold atoms moving toward the nickel tip, which was concomitant with the force between the tip and the substrate becoming attractive.

Tupper and Brenner have used atomic simulations to model loading of a self-assembled thiol overlayer on a gold substrate (Tupper, 1994). Simulations of compression with a flat surface predicted a reversible structural transition involving a rearrangement of the sulfur head groups bonded to the gold substrate. Concomitant with the formation of the new overlayer structure was a change in slope of the loading curve that agreed qualitatively with experimental loading data. Simulations of loading using a surface containing a nanometer-scale asperity were also carried out. Penetration of the asperity through the self-assembled overlayer occurred without an appreciable loading force. This result suggests that scanning-tunneling microscope images of self-assembled thiol monolayers may reflect the structure of the head group rather than the end of the chains, even when an appreciable load on the tip is not measured.

Experimental data showing a large change in electrical resistivity during indentation of silicon has led to the suggestion of a load-induced phase transition below the indenter. Clarke et al., for example, report forming an Ohmic contact under load; and using transmission electron microscopy, they have observed an amorphous phase at the point of contact after indentation (Clarke, 1988). Based on this data, the authors suggest that one or more high-pressure electrically conducting phases are produced under the indenter, and that these phases transform to the amorphous structure upon rapid unloading. Further support for this conclusion was given by Pharr et al., although they caution that the large change in electrical resistivity may have other origins and that an abrupt change in force during unloading may be due to sample cracking rather than transformation of a high pressure phase (Pharr, 1992). Using micro-Raman microscopy, Kailer et al. identified a metallic β-Sn phase in silicon near the interface of a diamond indenter during hardness loading (Kailer, 1999). Furthermore, upon rapid unloading they detected amorphous silicon as in the Clarke et al. experiments, while slow unloading resulted in a mixture of high-pressure polymorphs near the indent point.

Using molecular dynamics simulations, Kallman et al. examined the microstructure of amorphous and crystalline silicon before, during, and after simulated indentation (Kallman, 1993). Interatomic forces governing the motion of the silicon atoms were derived from the Stillinger–Weber potential mentioned above. For an initially crystalline silicon substrate close to its melting point, the simulations indicated a tendency to transform to the amorphous phase near the indenter. However, an initially amorphous silicon substrate was not observed to crystallize upon indentation; and no evidence of a transformation to the β-Sn structure was found. In more recent simulations by Cheong and Zhang that used the Tersoff silicon potential, an indentation-induced transition to a body-centered tetragonal phase was observed, followed by transformation to an amorphous structure after unloading (Cheong, 2000). A transition back to the high-pressure phase upon reloading of the amorphous region was observed in the simulations, indicating that the transition between the high-pressure ordered phase and the amorphous structure is reversible.

Smith, Tadmore, and Kaxiras revisited the silicon nanoindentation issue using a quasi-continuum model that couples interatomic forces from the Stillinger–Weber model to a finite element grid (Smith, 2000). This treatment allows much larger systems than would be possible with an all-atom approach. They report good agreement between simulated loading curves and experiment, provided that the curves are scaled by the indenter size. Rather than the β-Sn structure, however, atomic displacements suggest formation of a metallic structure with fivefold coordination below the indenter upon loading and a residual simple cubic phase near the indentation site after the load is released rather than the mix of high-pressure phases characterized experimentally. Smith et al. attribute this discrepancy to shortcomings of the Stillinger–Weber potential in adequately describing high-pressure phases of silicon. They also used a simple model for changes in electrical resistivity with loading involving contributions from both a Schottky barrier and spreading resistance. Simulated resistance-vs.-loading curves agree well with experiment despite possible discrepancies between the high-pressure phases under the indenter, suggesting that the salient features of the experiment are not dependent on the details of the high-pressure phases produced.

Molecular simulations that probe the influence of nanometer-scale surface features on nanoindentation force–displacement curves and plastic deformation have recently been carried out. Simulations of shallow, elastic indentations, for example, have been used to help characterize the conditions under which nanoindentation could be used to map local residual surface stresses (Shenderova, 2000). Using the embedded-atom method to describe interatomic forces, Shenderova and associates performed simulations of shallow, elastic indentation of a gold substrate near surface features that included a trench and a dislocation intersecting the surface. The maximum load for a given indentation depth of less than one nanometer was found to correlate to residual stresses that arise from the surface features. This result points toward the application of nanoindentation for nondestructively characterizing stress distributions due to nanoscale surface features.

Zimmerman et al. have used simulations to characterize plastic deformation due to nanoindentation near a step on a gold substrate (Zimmerman, 2001). The simulations showed that load needed to nucleate dislocations is lower near a step than on a terrace, although the effect is apparently less than that measured experimentally due to different contact areas.

Vashishta and co-workers have used large-scale, multi-million-atom simulations to model nanoindentation of Si_3N_4 films with a rigid indenter (Walsh, 2000). The simulations demonstrated formation of an amorphous region below the indenter that was terminated by pile-up of material around the indenter and crack formation at the indenter corners.

The utility of hemispherically capped single-wall carbon nanotubes for use as scanning probe microscope tips has been investigated using molecular dynamics simulation by Harrison and co-workers and by Sinnott and co-workers (Harrison, 1997; Garg, 1998a). In the work reported by Harrison et al., it was shown that (10,10) armchair nanotubes recover reversibly after interaction with hydrogen-terminated diamond (111) surfaces. The nanotube exhibits two mechanisms for releasing the stresses induced by indentation: a marked inversion of the capped end, from convex to concave, and finning along the tube's axis. The cap was shown to flatten at low loads and then to invert in two discrete steps. Compressive stresses at the vertex of the tip build up prior to the first cap-inversion event. These stresses are relieved by the rapid popping of the three layers of carbon atoms closest to the apex of the tip inside the tube. Continued application of load causes the remaining two rings of carbon atoms in the cap to be pushed inside the tube. Additional stresses on the nanotube caused by its interaction with the hard diamond substrate are relieved via a *finning* mechanism, or flattening, of the nanotube. That is, the nanotube collapses so that opposing walls are close together. These conformational changes in the tube are reversed upon pull-back of the tube from the diamond substrate. The tube recovers its initial shape, demonstrating the potential usefulness of nanotubes as scanning probe microscope tips.

The same capped (10,10) nanotube was also used to indent n-alkane hydrocarbon chains with 8, 13, and 22 carbon atoms chemically bound to diamond (111) surfaces (Tutein, 1999). Both flexible and rigid nanotubes were used to probe the n-alkane monolayers. The majority of the torsional bonds along the

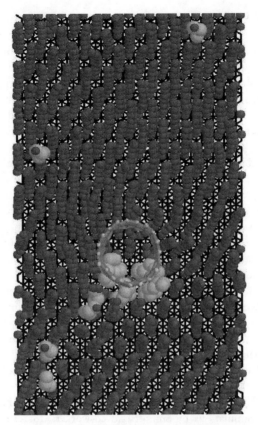

FIGURE 24.6 Illustration of a (10,10) single-wall nanotube that has partially indented a monolayer composed of C_{13} chains on a diamond substrate. Looking down along the tube, it is apparent that gauche defects (light gray, largest spheres) form under and adjacent to the nanotube. Hydrogen atoms on the chains and the tube cap atoms have been omitted from the picture for clarity.

carbon backbone of the chains were in their anti conformation prior to indentation. Regardless of the nanotube used, indentation of the hydrocarbon monolayers caused a disruption in the ordering of the monolayer, pinning of hydrocarbon chains beneath the tube, and formation of gauche defects with the monolayer below and adjacent to the tube (see Figure 24.6). The flexible nanotube is distorted only slightly by its interaction with the softer monolayers because nanotubes are stiff along their axial direction. In contrast, interaction with the diamond substrate causes the tube to fin, as it does in the absence of the monolayer. Severe indents with a rigid nanotube tip result in rupture of chemical bonds with the hydrocarbon monolayer. This was the first reported instance of indentation-induced bond rupture in a monolayer system. Previous simulations by Harrison and co-workers demonstrated that the rupture of chemical bonds (or fracture) is also possible when a hydrogen-terminated diamond asperity is used to indent both hydrogen-terminated and hydrogen-free diamond (111) surfaces (Harrison, 1992).

24.3.4 New Materials and Structures

Molecular modeling has made important contributions to our understanding of the properties, processing, and applications of several classes of new materials and structures. Discussing all of these contributions is beyond the scope of this chapter (a thorough discussion would require several volumes). Instead, the intent of this section is to supplement the content of some of the more detailed chapters in this book by presenting examples that represent the types of systems and processes that can be examined by atomic simulation.

24.3.4.1 Fullerene Nanotubes

Molecular modeling has played a central role in developing our understanding of carbon-based structures, in particular molecular fullerenes and fullerene nanotubes. Early molecular modeling studies focused on structures, energies, formation processes, and simple mechanical properties of different types of fullerenes. As this field has matured, molecular modeling studies have focused on more complicated structures and phenomena such as nonlinear deformations of nanotubes, nanotube functionalization, nanotube filling, and hybrid systems involving nanotubes. Molecular modeling is also being used in conjunction with continuum models of nanotubes to obtain deeper insights into the mechanical properties of these systems.

Several molecular modeling studies of nanotubes with sidewall functionalization have been recently carried out. Using the bond order potential discussed earlier, Sinnott and co-workers modeled CH_3+ incident on bundles of single-walled and multi-walled nanotubes at energies ranging from 10 to 80 eV (Ni, 2001). The simulations showed chemical functionalization and defect formation on nanotube sidewalls, as well as the formation of cross-links connecting either neighboring nanotubes or between the walls of a single nanotube (Figure 24.7). These simulations were carried out in conjunction with experimental studies that provided evidence for sidewall functionalization using CF3+ ions deposited at comparable incident energies onto multi-walled carbon nanotubes.

Molecular modeling has also predicted that kinks formed during large deformations of nanotubes may act as reactive sites for chemically connecting species to nanotubes (Figure 24.8). In simulations by Srivastava et al., it was predicted that binding energies for chemically attaching hydrogen atoms to a nanotube can be enhanced by over 1.5 eV compared with chemical attachment to pristine nanotubes (Srivastava, 1999). This enhancement comes from mechanical deformation of carbon atoms around kinks and ridges that force bond angles toward the tetrahedral angle, leading to radical sites on which species can strongly bond.

Several modeling studies have also been carried out that have examined the mechanical properties of functionalized nanotubes. Simulations by Sinnott and co-workers, for example, have predicted that covalent chemical attachment of $H_2C = C$ species to single-walled nanotubes can decrease the maximum compressive force needed for buckling by about 15%, independent of tubule helical structure or radius (Garg, 1998b). In contrast, similar simulations predict that the tensile modulus of single-walled (10,10)

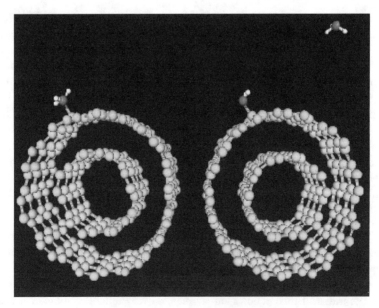

FIGURE 24.7 Illustration of fullerene nanotubes with functionalized sidewalls. (Courtesy of S.B. Sinnott, University of Florida.)

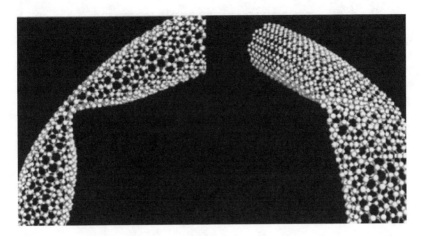

FIGURE 24.8 Illustrations of a kinked fullerene nanotube.

nanotubes is largely unchanged for configurations on which up to 15% of the nanotube carbon atoms (the largest degree modeled) are of the carbon atoms being covalently bonded to CH_3 groups. These simulations also predict a slight decrease in nanotube length due to rehybridization of the nanotube carbon atom valence orbitals from sp^2 to sp^3 (Brenner, 2002).

Several applications of sidewall functionalization via covalent bond formation have been suggested. For example, molecular simulations suggest that the shear needed to start pulling a nanotube on which 1% of carbon atoms are cross-linked to a model polymer matrix is about 15 times that needed to initiate motion of a nanotube that interacts with the matrix strictly via nonbonded forces (Frankland, 2002). This result, together with the prediction that the tensile strength of nanotubes is not compromised by functionalization, suggests that chemical functionalization leading to matrix–nanotube cross-linking may be an effective mode for enhancing load transfer in these systems without sacrificing the elastic moduli of nanotubes (Brenner, 2002). Other applications of functionalized nanotubes include a means for controlling the electronic properties of nanotubes (Brenner, 1998; Siefert, 2000) and a potential route to novel quantum dot structures (Orlikowski, 1999).

The structure and stability of several novel fullerene-based structures have also been calculated using molecular modeling. These structures include nanocones, tapers, and toroids, as well as hybrid diamond cluster-nanotube configurations (Figure 24.9) (Han, 1998; Meunier, 2001; Shenderova, 2001; Brenner, 2002). In many cases, novel electronic properties have also been predicted for these structures.

24.3.4.2 Dendrimers

As discussed in Chapter 20 by Tomalia et al., dendritic polymer structures are starting to play an important role in nanoscale science and technology. While techniques such as nuclear magnetic resonance and infrared spectroscopies have provided important experimental data regarding the structure and relaxation dynamics of dendrimers, similarities between progressive chain generations and complex internal structures make a thorough experimental understanding of their properties difficult. It is in these cases that molecular simulation can provide crucial data that is either difficult to glean from experimental studies or not accessible to experiment.

Molecular modeling has been used by several groups as a tool to understand properties of these species, including their stability, shape, and internal structure as a function of the number of generations and chain stiffness. Using a molecular force field, simulations by Gorman and Smith showed that the equilibrium shape and internal structure of dendrimers varies as the flexibility of the dendrimer repeat unit is changed (Gorman, 2000). They report that dendrimers with flexible repeat units show a somewhat globular shape, while structures formed from stiff chains are more disk-like. These simulations also showed that successive branching generations can fold back, leading to branches from a given generation that can permeate the entire structure.

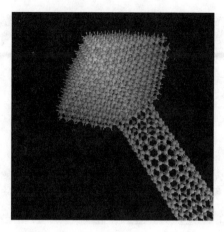

FIGURE 24.9   Illustration of a diamond-nanotube hybrid structure.

In related molecular modeling studies, Karatasos, Adolf, and Davies (Karatasos, 2001) as well as Scherrenberg et al (Scherrenberg, 1998) studied the structure and dynamics of solvated dendrimers, while Zacharopoulos and Economou modeled a melt (Zacharopoulos, 2002). These studies indicate that dendritic structures become more spherical as the number of generations increases, and that the radius of gyration scales as the number of monomer units to the 1/3 power. Significant folding of the chains inside the structures was also observed (as was seen in the Gorman and Smith study).

24.3.4.3   Nanostructured Materials

As discussed in detail in Chapter 22 by Nazarov and Mulyukov, nanostructured materials can have unusual combinations of properties compared with materials with more conventional grain sizes and microstructures. Molecular simulations have contributed to our understanding of the origin of several of these properties, especially how they are related to deformation mechanisms of strained systems.

Using an effective medium potential, Jacobsen and co-workers simulated the deformation of strained nanocrystalline copper with grain sizes that average about 5 nm (Schiotz, 1998). These simulations showed a softening for small grain sizes, in agreement with experimental measurements. The simulations indicate that plastic deformation occurs mainly by grain boundary sliding, with a minimal influence of dislocation motion on the deformation.

Van Swygenhoven and co-workers performed a series of large-scale molecular dynamics simulations of the deformation of nanostructured nickel and copper with grain sizes ranging from 3.5 nm to 12 nm (Van Swygenhoven, 1999). The simulations used a second-moment-based potential as described above, with constant uniaxial stress applied to the systems. The simulations revealed different deformation mechanisms depending on grain size. For samples with grain sizes less than about 10 nm, deformation was found to occur primarily by grain boundary sliding, with the rate of deformation increasing with decreasing grain size. For the larger grain sizes simulated, a change in deformation mechanism was reported in which a combination of dislocation motion and grain boundary sliding occurred. Characteristic of this apparent new deformation regime was that the strain rate was independent of grain size. In subsequent simulations, detailed mechanisms of strain accommodation were characterized that included both single-atom motion and correlated motion of several atoms, as well as stress-assisted free-volume migration (Van Swygenhoven, 2001).

Wolf and co-workers have also carried out detailed studies of the deformation of nanostructured metals. In studies of columnar structures of aluminum, for example, emission of partial dislocations that were formed at grain boundaries and triple junctions was observed during deformation (Yamakov, 2001). The simulations also showed that these structures can be reabsorbed upon removal of the applied stress, which the authors suggest may contribute to the fact that dislocations are not normally observed experimentally in systems of this type after external stresses are released.

Simulations have also been used to characterize the dynamics in nanostructured materials during ion bombardment and to understand the origin of apparently anomalous vibrational modes in nanostructured materials (Derlet, 2001; Samaras, 2002). In studies of the latter, for example, it was shown that enhancements in both the low- and high-vibrational frequencies for nanostructured nickel and copper arise from atoms at the grain boundaries, and that the vibrational frequencies of atoms in the grains are largely unaffected by the grain size.

Wolf and co-workers have recently simulated the dynamics of grain growth in nanocrystalline face-center-cubic metals (Haslam, 2001). Assuming columnar structure and grain sizes of about 15 nm, these simulations indicate that grain rotation can play a role in grain growth that is as equally important as grain boundary migration. The simulations predict that necessary changes in the grain shape during grain rotation in columnar polycrystalline structures can be accommodated by diffusion either through the grain boundaries or through the grain interior (Moldovan, 2001). Based on this result, the authors have suggested that both mechanisms, which can be coupled, should be accounted for in mesoscopic models of grain growth. Moreover, Moldovan et al. have recently reported the existence of a critical length scale in the system that enables the growth process to be characterized by two regimes. If the average grain size is smaller than the critical length, as in a case of nanocrystals, grain growth is dominated by the grain-rotation coalescence mechanism. For average grain sizes exceeding the critical size, the growth mechanism is due to grain boundary migration.

Large-scale simulations of the structure, fracture, and sintering of covalent materials have also been studied using large-scale atomic modeling (Vashishta, 2001). Vashishta and co-workers, for example, have simulated the sintering of nanocluster-assembled silicon nitride and nanophase silicon carbide. The simulations, which used many-body potentials to describe the bonding, revealed a disordered interface between nanograins. This is a common feature of grain boundaries in polycrystalline ceramics, as opposed to more ordered interfaces typical of metals. In the silicon nitride simulations, the amorphous region contained undercoordinated silicon atoms; and because this disordered region is less stiff than the crystalline region, the elastic modulus was observed to decrease in systems with small grain sizes within which more of the sample is disordered. In simulations of crack propagation in this system, the amorphous intergranular regions were found to deflect cracks, resulting in crack branching. This behavior allowed the simulated system to maintain a much higher strain than a fully crystalline system. In the silicon carbide simulations, onset of sintering was observed at 1500 K, in agreement with neutron scattering experiments. This temperature is lower than that for polycrystalline silicon carbide with larger grain sizes, and therefore is apparently due to the nanocrystalline structure of the samples. The simulations also predict that bulk modulus, shear modulus, and Young's modulus all have a power–law dependence on density with similar exponents.

In related studies, Keblinski et al. have used atomic simulation to generate nanocrystalline samples of silicon and carbon (Keblinski, 1997; Keblinski, 1999). For silicon, which used the Stillinger–Weber potential, disordered layers with structures similar to bulk amorphous silicon between grains were predicted by the simulations. This result suggests that this structure is thermodynamically stable for nanocrystalline silicon, and that some mechanical properties can be understood by assuming a two-phase system. In the case of carbon, grains with the diamond cubic structure connected by disordered regions of sp^2-bonded carbon have been revealed by molecular simulation. These disordered regions may be less susceptible to brittle fracture than crystalline diamond.

24.4　Concluding Remarks

By analyzing trends in computing capabilities, Vashishta and co-workers have concluded that the number of atoms that can be simulated with analytic potentials and with first-principle methods is increasing exponentially over time (Nakano, 2001). For analytic potentials, this analysis suggests that the number of atoms that can be simulated has doubled every 19 months since the first liquid simulations using continuous potential by Rahman in the early 1960s (Rahman, 1964). For simulations using first-principles forces, the same analysis suggests that the number of atoms that can be simulated has doubled every 12 months since the Car–Parrinello method was introduced in the middle 1980s (Nakano, 2001). With these extrapolations, modeling a gold interconnect 0.1 µm on a side and 100 µm long is just about feasible now with potentials such as the EAM, while first-principles simulations will have to wait almost two decades (or longer) to attack a problem of this size. However, increases in computing are only one side of a convergence among modeling, experiment, and technology. Over the same period of time it will take for first-principles modeling to rise to the scale of current interest in electronic device technology, there will be a continuing shrinkage of device dimensions. This means that modeling and technological length scales will converge over the next decade. Indeed, a convergence of sorts is already apparent in nanotube electronic properties and in molecular electronics, as is apparent from the chapters on these subjects in this handbook. This is clearly an exciting time, with excellent prospects for modeling and theory in the next few years and beyond.

Acknowledgments

Helpful discussions with Kevin Ausman, Jerzy Bernholc, Rich Colton, Brett Dunlap, Mike Falvo, Dan Feldheim, Alix Gicquel, Al Globus, Chris Gorman, Jan Hoh, Richard Jaffe, Jackie Krim, J.-P. Lu, John Mintmire, Dorel Moldovan, A. Nakano, Airat Nazarov, Boris Ni, John Pazik, Mark Robbins, Daniel Robertson, Chris Roland, Rod Ruoff, Peter Schmidt, Susan Sinnott, Deepak Srivastava, Richard Superfine, Priya Vashishta, Kathy Wahl, Carter White, Sean Washburn, Victor Zhirnov, and Otto Zhou are gratefully acknowledged. The authors wish to acknowledge support for their research efforts from the Air Force Office of Scientific Research, the Army Research Office, the Department of Defense, the Department of Energy, the National Aeronautics and Space Administration, the Petroleum Research Fund, the National Science Foundation, the Office of Naval Research, and the Research Corporation.

References

Abell, G.C. (1985) Empirical chemical pseudopotential theory of molecular and metallic bonding, *Phys. Rev. B* 31, 6184–6196.

Adiga, S.P. and Brenner, D.W. (2002) Virtual molecular design of an environment-responsive nanoporous system, *Nanoletters*, 2, 567–572.

Balasubramanian, S., Klein, M., and Siepmann, J.I. (1996) Simulation studies of ultrathin films of linear and branched alkanes on a metal substrate, *J. Phys. Chem.* 100, 11960.

Balbes, L.M., Mascarella, S.W., and Boyd, D.B. (1994) Perspectives of modern methods in computer-aided drug design, in *Reviews in Computational Chemistry*, K.B. Lipkowitz and D.B. Boyd, Eds., VCH Publishers, New York, 1994, Vol. 5, pp. 337–379.

Baskes, M.I., Nelson, J.S., and Wright, A.F. (1989) Semiempirical modified embedded-atom potentials for silicon and germanium, *Phys. Rev. B* 40, 6085.

Bitsanis, I., Magda, J., Tirrell, M., and Davis, H. (1987) Molecular dynamics of flow in micropores, *J. Chem. Phys.* 87, 1733–1750.

Bitsanis, I., Somers, S.A., Davis, T., and Tirrell, M. (1990) Microscopic dynamics of flow in molecularly narrow pores, *J. Chem. Phys.* 93, 3427–3431.

Bowen, J.P. and Allinger, N.L. (1991) Molecular mechanics: the art and science of parameterization, in *Reviews in Computational Chemistry*, K.B. Lipkowitz and D.B. Boyd, Eds., VCH Publishers, New York, Vol. 2, pp. 81–97.

Brenner, D.W. (1989) Relationship between the embedded-atom method and Tersoff potentials, *Phys. Rev. Lett.* 63, 1022.

Brenner, D.W. (1990) Empirical potential for hydrocarbons for use in simulating the chemical vapor deposition of diamond films, *Phys. Rev. B* 42, 9458

Brenner, D.W., Sinnott, S.B., Harrison, J.A., and Shenderova, O.A. (1996) Simulated engineering of nanostructures, *Nanotechnology*, 7, 161.

Brenner, D.W., Schall, J.D., Mewkill, J.P., Shenderova, O.A., and Sinnott, S.B. (1998) Virtual design and analysis of nanometer-scale sensor and device components," *J. Brit. Interplanetary Soc.*, 51 137 (1998).

Brenner, D.W., Shenderova, O.A., Areshkin, D.A., Schall, J.D., and Frankland, S.-J.V. (2002) Atomic modeling of carbon-based nanostructures as a tool for developing new materials and technologies, in *Computer Modeling in Engineering and Science*, in press.

Burkert, U. and Allinger, N.L. (1982) *Molecular Mechanics*, ACS Monograph 177, American Chemical Society, Washington, D.C., 1982.

Cagin, T., Jaramillo–Botero, A., Gao, G., and Goddard, W.A. (1998) Molecular mechanics and molecular dynamics analysis of Drexler–Merkle gears and neon pump, *Nanotechnology*, 9, 143–152.

Car, R. and Parrinello, M. (1985) Unified approach for molecular dynamics and density-functional theory, *Phys. Rev. Lett.* 55, 2471.

Carlsson, A.E. (1991) Angular forces in group-VI transition metals: application to W(100), *Phys. Rev. B* 44, 6590.

Chan, D.Y.C. and Horn, R.G. (1985) The drainage of thin liquid films between solid surfaces, *J. Chem. Phys.* 83, 5311–5324.

Cheong, W.C.D. and Zhang, L.C. (2000) Molecular dynamics simulation of phase transformations in silicon monocrystals due to nano-indentation, *Nanotechnology* 11, 173.

Clarke, D.R., Kroll, M.C., Kirchner, P.D., Cook, R.F., and Hockey, B.J. (1988) Amorphization and conductivity of silicon and germanium induced by indentation, *Phys. Rev. Lett.* 60, 2156.

Cyrot-Lackmann, F. (1968) Sur le calcul de la cohesion et de la tension superficielle des metaux de transition par une methode de liaisons fortes, *J. Phys. Chem. Solids* 29, 1235.

Daw M.S. and Baskes, M.I. (1983) Semiempirical quantum mechanical calculation of hydrogen embrittlement in metals, *Phys. Rev. Lett.* 50, 1285.

Derlet, P.M., Meyer, R., Lewis, L.J., Stuhr, U., and Van Swygenhoven, H. (2001) Low-frequency vibrational properties of nanocrystalline materials, *Phys. Rev. Lett.* 87, 205501.

Diestler, D.J., Schoen, M., and Cushman, J.H. (1993), On the thermodynamic stability of confined thin films under shear, *Science* 262, 545–547.

Drexler, E. (1992) *Nanosystems: Molecular Machinery, Manufacturing and Computation*, Wiley, New York.

Ercolessi, F., Parrinello, M., and Tossatti, E. (1988) Simulation of gold in the glue model, *Philos. Mag. A* 58, 213.

Finnis M.W. and Sinclair, J.E. (1984) A simple empirical n-body potential for transition metals, *Philos. Mag. A* 50, 45.

Foiles, S.M. (1993) Interatomic interactions for Mo and W based on the low-order moments of the density of states, *Phys. Rev. B* 48, 4287.

Foulkes, W.M.C. and Haydock, R. (1989) Tight-binding models and density-functional theory, *Phys. Rev. B* 39, 12520.

Frankland, S.J.V., Caglar A., Brenner D.W., and Griebel M. (2002) Molecular simulation of the influence of chemical cross-links on the shear strength of carbon nanotube-polymer interfaces, *J. Phys. Chem. B* 106, 3046.

Frauenheim, Th., Weich, F., Kohler, Th., Uhlmann, S., Porezag, D., and Seifert, G. (1995) Density-functional-based construction of transferable nonorthogonal tight-binding potentials for Si and SiH, *Phys. Rev. B* 52, 11492.

Garg, A., Han, J., and Sinnott, S.B. (1998a) Interactions of carbon-nanotubule proximal probe tips with diamond and graphene, *Phys. Rev. Lett.* 81, 2260.

Garg, A. and Sinnott, S.B. (1998b) Effect of chemical functionalization on the mechanical properties of carbon nanotubes, *Chem. Phys. Lett.* 295, 273.

Garrison, B.J. Srivastava, D. (1995) Potential energy surface for chemical reactions at solid surfaces, *Annu. Rev. Phys. Chem.* 46, 373.

Gee, M.L., McGuiggan, P.M., Israelachvili, J.N., and Homola, A.M. (1990) Liquid to solidlike transitions of molecularly thin films under shear, *J. Chem. Phys.* 93, 1895.

Gibson, J.B., Goland, A.N., Milgram, M., and Vineyard, G.H. (1960) Dynamics of radiation damage, *Phys. Rev.* 120, 1229.

Gorman, C.B. and Smith, J.C. (2000) Effect of repeat unit flexibility on dendrimer conformation as studied by atomistic molecular dynamics simulations, *Polymer* 41, 675.

Granick, S., Damirel, A.L., Cai, L.L., and Peanasky, J. (1995) Soft matter in a tight spot: nanorheology of confined liquids and block copolymers, *Isr. J. Chem.* 35, 75–84.

Han, J., Globus, A., Jaffe R., and Deardorff, G. (1997) Molecular dynamics simulations of carbon nanotube-based gears, *Nanotechnology* 8, 95.

Han, J. (1998) Energetics and structures of fullerene crop circles, *Chem. Phys. Lett.* 282, 187.

Harris, J. (1985) Simplified method for calculating the energy of weakly interacting fragments, *Phys. Rev. B* 31, 1770.

Harrison, J.A., Stuart, S.J., Robertson, D.H., and White, C.T. (1997) Properties of capped nanotubes when used as SPM tips, *J. Phys. Chem. B.* 101 9682.

Harrison, J.A., White, C.T., Colton, R.J., and Brenner, D.W. (1992) Nanoscale investigation of indentation, adhesion, and fracture of diamond (111) surfaces, *Surf. Sci.* 271 57.

Haslam, A.J., Phillpot, S.R., Wolf, D., Moldovan, D., and Gleiter, H. (2001) Mechanisms of grain growth in nanocrystalline fcc metals by molecular-dynamics simulation, *Mater. Sci. Eng.* A318, 293.

Hirschfelder, J., Eyring, H., and Topley, B. (1936) Reactions involving hydrogen molecules and atoms, *J. Chem. Phys* 4, 170.

Hohenberg P. and Kohn, W. (1964) Inhomogeneous electron gas, *Phys. Rev.* 136, B864.

Horn, R.G. and Israelachvili, J.N. (1981) Direct measurement of structural forces between two surfaces in a nonpolar liquid, *J. Chem. Phys.* 75, 1400–1411.

Israelachvili, J.N., McGuiggan, P.M., and Homola, A.M. (1988) Dynamic properties of molecularly thin liquid films, *Science* 240, 189–191.

Ito, Y., Park, Y.S., and Imanishi, Y. (2000) Nanometer-sized channel gating by a self-assembled polypeptide brush, *Langmuir* 16, 5376.

Jacobsen, K.W. (1988) Bonding in metallic systems: an effective medium approach, *Comments Cond. Matter Phys.* 14, 129.

Jacobsen, K.W., Norskov, J.K., and Puska, M.J. (1987) Interatomic interactions in the effective-medium theory, *Phys. Rev. B* 35, 7423.

Kailer, A., Nickel, K.G., and Gogotsi, Y.G. (1999) Raman microspectroscopy of nanocrystalline and amorphous phases in hardness indentations, *J. Raman Spectrosc.* 30, 939.

Kallman, J.S., Hoover, W.G., Hoover, C.G., De Groot, A.J., Lee, S.M., and Wooten, F. (1993) Molecular dynamics of silicon indentation, *Phys. Rev. B* 47, 7705.

Karatasos, K., Adolf, D.B., and Davies, G.R. (2001) Statics and dynamics of model dendrimers as studied by molecular dynamics simulations, *J. Chem. Phys.* 115, 5310.

Keblinski, P., Phillpot, S.R., Wolf, D., and Gleiter, H. (1999) On the nature of grain boundaries in nanocrystalline diamond, *Nanostruct. Mater.* 12, 339.

Keblinski, P., Phillpot, S.R., Wolf, D., and Gleiter, H. (1997) On the thermodynamic stability of amorphous intergranular films in covalent materials, *J. Am. Ceramic Soc.* 80, 717.

Kohn, W. and Sham, L.J. (1965) Self-consistent equations including exchange and correlation effects, *Phys. Rev. A* 140, 1133.

Kwon, I., Biswas, R., Wang, C.Z., Ho, K.M., and Soukoulis, C.M. (1994) Transferable tight-binding model for silicon, *Phys. Rev. B* 49, 7242.

Landman, U., Luedtke, W.D., and Ribarsky, M.W. (1989a) Structural and dynamical consequences of interactions in interfacial systems, *J. Vac. Sci. Technol. A* 7, 2829–2839.

Landman, U., Luedtke, W.D., and Ribarsky, M.W. (1989b) Dynamics of tip–substrate interactions in atomic force microscopy, *Surf. Sci. Lett.* 210, L117.

Landman, U., Luedtke, W.D., Burnham, N.A., and Colton, R.J. (1990) Atomistic mechanisms and dynamics of adhesion, nanoindentation, and fracture, *Science* 248, 454.

Landman, U. and Luedtke, W.D. (1991) Nanomechanics and dynamics of tip–substrate interactions, *J. Vac. Sci. Technol. B* 9, 414–423.

Landman, U., Luedtke, W.D., and Ringer, E.M. (1992) Atomistic mechanisms of adhesive contact formation and interfacial processes, *Wear* 153, 3.

Landman, U., Luedtke, W.D., Ouyang, J., and Xia, T.K. (1993) Nanotribology and the stability of nanostructures, *Jpn. J. App. Phys.* 32, 1444.

Menon, M. and Subbaswamy, K.R. (1993) Nonorthogonal tight-binding molecular-dynamics study of silicon clusters, *Phys. Rev. B* 47, 12754.

Meunier, V, Nardelli M.B., Roland, C., and Bernholc, J. (2001) Structural and electronic properties of carbon nanotube tapers, *Phys. Rev. B* 64, 195419.

Moldovan, D., Wolf, D., and Phillpot, S.R. (2001) Theory of diffusion-accommodated grain rotation in columnar polycrystalline microstructures, *Acta Mat.* 49, 3521.

Moldovan, D., Wolf, D., Phillpot, S.R., and Haslam, A.J. (2002) Role of grain rotation in grain growth by mesoscale simulation, *Acta Mat.* in press.

Nakano, A., Bachlechner, M.E., Kalia, R.K., Lidorikis, E., Vashishta, P., Voyiadjis, G.Z., Campbell, T.J., Ogata, S., and Shimojo, F. (2001) Multiscale simulation of nanosystems, *Computing Sci. Eng.,* 3, 56.

Ni, B., Andrews R., Jacques D., Qian D., Wijesundara M.B.J., Choi Y.S., Hanley L., and Sinnott, S.B. (2001) A combined computational and experimental study of ion-beam modification of carbon nanotube bundles, *J. Phys. Chem. B* 105, 12719.

Nosé, S. (1984a) A unified formulation of the constant-temperature molecular dynamics method, *J. Chem. Phys.* 81, 511–519.

Nosé, S. (1984b) A molecular dynamics method for simulations in the canonical ensemble, *Mol. Phys.* 52, 255–268.

Orlikowski, D., Nardelli, M.B., Bernholc, J., and Roland, C. (1999) Ad-dimers on strained carbon nanotubes: A new route for quantum dot formation? *Phys. Rev. Lett.* 83, 4132.

Park, Y.S., Toshihiro, I., and Imanishi, Y. (1998) Photocontrolled gating by polymer brushes grafted on porous glass filter, *Macromolecules* 31, 2606.

Pethica, J.B. and Sutton, A.P. (1988) On the stability of a tip and flat at very small separations, *J. Vac. Sci. Technol. A* 6, 2490.

Pettifor D.G. and Oleinik I.I. (1999) Analytic bond-order potentials beyond Tersoff–Brenner. I. Theory, *Phys. Rev. B* 59, 8487.

Pettifor D.G. and Oleinik I.I. (2000) Bounded analytic bond-order potentials for sigma and pi bonds, *Phys. Rev. Lett.* 84, 4124.

Pharr, G.M., Oliver, W.C., Cook, R.F., Kirchner, P.D., Kroll, M.C., Dinger, T.R., and Clarke, D.R. (1992) Electrical resistance of metallic contacts on silicon and germanium during indentation, *J. Mat. Res.* 7, 961.

Rahman, A. (1964) Correlations in the motion of liquid argon, *Phys. Rev.* 136A, 405.

Rappé, A.K. and Goddard, W.A. (1991) Charge equilibration for molecular dynamics simulations, *J. Phys. Chem.* 95, 3358.

Ribarsky, M.W. and Landman, U. (1992) Structure and dynamics of n-alkanes confined by solid surfaces. I. Stationary crystalline boundaries, *J. Chem. Phys.* 97, 1937–1949.

Robertson, D.H., Brenner, D.W., and White, C.T. (1994) Fullerene/tubule based hollow carbon nanogears, *Mat. Res. Soc. Symp. Proc.* 349 283.

Samaras, M., Derlet, P.M., Van Swygenhoven, H., and Victoria, M. (2002) Computer simulation of displacement cascades in nanocrystalline Ni, *Phys. Rev. Lett.* 88, 125505.

Scherrenberg, R., Coussens, B., van Vliet, P., Edouard, G., Brackman, J., and de Brabander, E. (1998) The molecular characteristics of poly(propyleneimine) dendrimers as studied with small-angle neutron scattering, viscosimetry, and molecular dynamics, *Macromolecules* 31, 456.

Schiøtz, J., Di Tolla, F.D., and Jacobsen, K.W. (1998) Softening of nanocrystalline metals at very small grain sizes, *Nature* 391, 561.

Schoen, M., Rhykerd, C.L., Diestler, D.J., and Cushman, J.H. (1989) Shear forces in molecularly thin films, *Science* 245, 1223.

Schoen, M., Rhykerd, C.L., Diestler, D.J., and Cushman, J.H. (1987) Fluids in micropores. I. Structure of a simple classical fluid in a slit-pore, *J. Chem. Phys.* 87, 5464–5476.

Selinger, R.L.B., Farkas, D., Abraham, F., Beltz, G.E., Bernstein, N., Broughton, J.Q., Cannon, R.M., Corbett, J.M., Falk, M.L., Gumbsch, P., Hess, D., Langer, J.S., and Lipkin, D.M. (2000) Atomistic theory and simulation of fracture, *Mater. Res. Soc. Bull.* 25, 11.

Shenderova O., Mewkill J., and Brenner D.W. (2000) Nanoindentation as a probe of nanoscale residual stresses: atomistic simulation results, *Molecular Simulation* 25, 81.

Shenderova, O.A., Lawson, B.L., Areshkin, D., and Brenner, D.W. (2001) Predicted structure and electronic properties of individual carbon nanocones and nanostructures assembled from nanocones, *Nanotechnology* 12, 291.

Siefert, G., Kohler, T., and Frauenheim, T. (2000) Molecular wires, solenoids, and capacitors and sidewall functionalization of carbon nanotubes, *App. Phys. Lett.* 77, 1313.

Sinnott, S.B., Colton, R.J., White, C.T., and Brenner, D.W. (1994) Surface patterning by atomically controlled chemical forces – molecular dynamics simulations, *Surf. Sci.* 316, L1055.

Slater, J.C. and Koster, G.F. (1954) Simplified LCAO method for the periodic potential problem, *Phys. Rev.* 94, 1498.

Smith, G.S., Tadmor, E.B., and Kaxiras, E. (2000) Multiscale simulation of loading and electrical resistance in silicon nanoindentation, *Phys. Rev. Lett.* 84, 1260.

Smith, J.R., Bozzolo, G., Banerjea, A., and Ferrante, J. (1989) Avalanche in adhesion, *Phys. Rev. Lett.* 63, 1269–1272.

Sokol, P.E., Ma, W.J., Herwig, K.W., Snow, W.M., Wang, Y., Koplik, J., and Banavar, J.R. (1992) Freezing in confined geometries, *Appl. Phys. Lett.* 61, 777–779.

Srivastava, D. (1997) A phenomenological model of the rotation dynamics of carbon nanotube gears with laser electric fields, *Nanotechnology* 8, 186.

Srivastava D., Brenner, D.W., Schall, J.D., Ausman, K.D., Yu, M.F., and Ruoff, R.S. (1999) Predictions of enhanced chemical reactivity at regions of local conformational strain on carbon nanotubes: kinky chemistry, *J. Phys. Chem. B* 103, 4330.

Stillinger, F. and Weber, T.A. (1985) Computer simulation of local order in condensed phases of silicon *Phys. Rev. B* 31, 5262.

Streitz, F.H. and Mintmire, J.W. (1994) Electrostatic potentials for metal-oxide surfaces and interfaces, *Phys. Rev. B* 50, 11996–12003.

Stuart, S.J., Tutein, A.B., and Harrison, J.A. (2000) A reactive potential for hydrocarbons with intermolecular interactions, *J. Chem. Phys.* 112, 6472.

Sutton, A.P. and Chen, J. (1990) Long-range Finnis–Sinclair potentials, *Philos. Mag. Lett.* 61, 139.

Sutton, A.P. (1993) *Electronic Structure of Materials*, Clarendon Press, Oxford.

Sutton, A.P., Goodwin P.D., and Horsfield, A.P. (1996) Tight-binding theory and computational materials synthesis, *Mat. Res. Soc. Bull.* 21, 42.

Tomagnini, O., Ercolessi, F., and Tosatti, E. (1993) Microscopic interaction between a gold tip and a Pb(110) surface, *Surf. Sci.* 287/288, 1041–1045.

Tupper, K.J. and Brenner, D.W. (1994a) Compression-induced structural transition in a self-assembled monolayer, *Langmuir* 10, 2335–2338.

Tupper, K.J., Colton, R.J., and Brenner, D.W. (1994b), Simulations of self-assembled monolayers under compression: effect of surface asperities, *Langmuir* 10, 2041–2043.

Tutein, A.B., Stuart, S.J., and Harrison, J.A. (1999) Indentation analysis of linear-chain hydrocarbon monolayers anchored to diamond, *J. Phys. Chem. B,* 103 11357.

Thompson, P.A. and Robbins, M.O. (1990a) Origin of stick-slip motion in boundary lubrication, *Science* 250, 792–794.

Thompson, P.A. and Robbins, M.O. (1990b) Shear flow near solids: epitaxial order and flow boundary conditions, *Phys. Rev. A* 41, 6830–6837.

Thompson, P.A., Grest, G.S., and Robbins, M.O. (1992) Phase transitions and universal dynamics in confined films, *Phys. Rev. Lett.* 68, 3448–3451.

Tersoff, J. (1986) New empirical model for the structural properties of silicon, *Phys. Rev. Lett.* 56, 632.

Tersoff, J. (1989) Modeling solid-state chemistry: interatomic potentials for multicomponent systems, *Phys. Rev. B* 39, 5566.

Tuzun, R., Noid, D.W., and Sumpter, B.G. (1995a) Dynamics of a laser-driven molecular motor, *Nanotechnology* 6, 52.

Tuzun, R., Noid, D.W., and Sumpter, B.G. (1995b) The dynamics of molecular bearings, *Nanotechnology* 6, 64.

Van Alsten, J. and Granick, S. (1988) Molecular tribometry of ultrathin liquid films, *Phys. Rev. Lett.* 61, 2570.

Van Swygenhoven H. and Derlet, P.M. (2001) Grain-boundary sliding in nanocrystalline fcc metals, *Phys. Rev. B* 64, 224105.

Van Swygenhoven, H., Spaczer, M., Caro, A., and Farkas, D. (1999) Competing plastic deformation mechanisms in nanophase metals, *Phys. Rev. B* 60, 22.

Vashishta, P., Bachlechner, M., Nakano, A, Campbell, T.J., Kalia, R.K., Kodiyalam, J., Ogata, S., Shimojo, F., and Walsh, P. (2001) Multimillion atom simulation of materials on parallel computers —nanopixel, interfacial fracture, nanoindentation and oxidation, *Appl. Surf. Sci.* 182, 258.

Walsh, P., Kalia, R.K., Nakano, A., Vashishta, P., and Saini, S. (2000) Amorphization and anisotropic fracture dynamics during nanoindentation of silicon nitride: a multimillion atom molecular dynamics study, *Appl. Phys. Lett.* 77, 4332.

Wang, Y., Hill, K., and Harris, J.G. (1993a) Thin films of n-octane confined between parallel solid surfaces. Structure and adhesive forces vs. film thickness from molecular dynamics simulations, *J. Phys. Chem.* 97, 9013.

Wang, Y., Hill, K., and Harris, J.G. (1993b) Comparison of branched and linear octanes in the surface force apparatus. A molecular dynamics study, *Langmuir* 9, 1983.

Xu, C.H., Wang, C.Z., Chan, C.T., and Ho, K.M. (1992) A transferable tight-binding potential for carbon, *J. Phys. Condens. Matt.* 4, 6047.

Yamakov, V., Wolf, D., Salazar, M., Phillpot, S.R., and Gleiter, H. (2001) Length-scale effects in the nucleation of extended dislocations in nanocrystalline Al by molecular dynamics simulation, *Acta Mater.* 49 (2001) 2713–2722.

Zacharopoulos, N. and Economou, I.G. (2002) Morphology and organization of poly(propylene imine) dendrimers in the melt from molecular dynamics simulation, *Macromolecules* 35, 1814.

Zimmerman, J.A., Kelchner, C.L., Klein, P.A., Hamilton, J.C., and Foiles, S.M. (2001) Surface step effects on nanoindentation, *Phys. Rev. Lett.* 87, 165507.

Index